Birkhäuser

Operator Theory: Advances and Applications

Volume 259

Founded in 1979 by Israel Gohberg

More information about this series at http://www.springer.com/series/4850

Dario A. Bini • Torsten Ehrhardt
Alexei Yu. Karlovich • Ilya Spitkovsky
Editors

Large Truncated Toeplitz Matrices, Toeplitz Operators, and Related Topics

The Albrecht Böttcher Anniversary Volume

Birkhäuser

Editors
Dario A. Bini
Dipartimento di Matematica
Università di Pisa
Pisa, Italy

Alexei Yu. Karlovich
Departamento de Matemática
Faculdade de Ciências e Tecnologia
Universidade Nova de Lisboa
Caparica, Portugal

Torsten Ehrhardt
Mathematics Department
University of California
Santa Cruz, California, USA

Ilya Spitkovsky
New York University Abu Dhabi
Abu Dhabi, United Arab Emirates

ISSN 0255-0156 ISSN 2296-4878 (electronic)
Operator Theory: Advances and Applications
ISBN 978-3-319-49180-6 ISBN 978-3-319-49182-0 (eBook)
DOI 10.1007/978-3-319-49182-0

Library of Congress Control Number: 2017934911

Mathematics Subject Classification (2010): 15B05, 47B35, 60B20

Printed on acid-free paper

This book is published under the trade name Birkhäuser, www.birkhauser-science.com
The registered company is Springer International Publishing AG
The registered company address is: Gewerbestrasse 11, 6330 Cham, Switzerland

Contents

Preface

This volume is dedicated to Albrecht Böttcher on the occasion of his sixtieth birthday. The first part contains two essays written by Sergei Grudsky and Bernd Silbermann, and several personal recollections written by colleagues, students, and friends of Albrecht. The second part consists of twenty four selected scientific papers devoted to various fields of Analysis and Matrix Theory, including asymptotics of Toeplitz and Hankel matrices, boundary value problems and singular integral operators, numerical ranges and pseudospectra, convolution and pseudodifferential operators. Albrecht Böttcher has been enriching all these areas by fundamental contributions during more than three decades. The volume ends with the contribution by Elias Wegert, who was a classmate of Albrecht in "Spezialklasse für Mathematik und Physik, Hochschule Karl-Marx-Stadt". It tells us about the first steps of Albrecht in Mathematics. We thank all contributors for their enthusiasm when preparing the articles for this volume. We feel happy to honour Albrecht Böttcher by the edition of this volume, and wish him good health and ever greater success in his work.

September 2016

Dario A. Bini,
Torsten Ehrhardt,
Alexei Yu. Karlovich,
Ilya Spitkovsky

Albrecht Böttcher

Operator Theory:
Advances and Applications, Vol. 259, xi–xxvi

Publications of Albrecht Böttcher

Books

1. A. Böttcher and B. Silbermann: *Invertibility and Asymptotics of Toeplitz Matrices*. Mathematical Research, Vol. 17. Akademie-Verlag, Berlin, 1983.
2. A. Böttcher and B. Silbermann: *Analysis of Toeplitz Operators*. Akademie-Verlag, Berlin, 1989 and Springer-Verlag, Berlin, Heidelberg, New York, 1990 (1st edition).
3. A. Böttcher, A. Dijksma, H. Langer, M.A. Dritschel, J. Rovnyak, and M.A. Kaashoek: *Lectures on Operator Theory and Its Applications*. Fields Institute Monographs, Vol. 3. Edited by Peter Lancaster. American Mathematical Society, Providence, RI, 1996.
4. A. Böttcher and Yu.I. Karlovich: *Carleson Curves, Muckenhoupt Weights, and Toeplitz Operators*. Progress in Mathematics, Vol. 154. Birkhäuser Verlag, Basel, 1997.
5. A. Böttcher and B. Silbermann: *Introduction to Large Truncated Toeplitz Matrices*. Universitext, Springer-Verlag, New York, 1999.
6. A. Böttcher and S. Grudsky: *Toeplitz Matrices, Asymptotic Linear Algebra, and Functional Analysis*. Texts and Readings in Mathematics, Vol. 18. Hindustan Book Agency, New Delhi, 2000 and Birkhäuser Verlag, Basel, 2000.
7. A. Böttcher, Yu.I. Karlovich, and I.M. Spitkovsky: *Convolution Operators and Factorization of Almost Periodic Matrix Functions*. Operator Theory: Advances and Applications, Vol. 131. Birkhäuser Verlag, Basel, 2002.
8. A. Böttcher and S.M. Grudsky: *Spectral Properties of Banded Toeplitz Matrices*. SIAM, Philadelphia, 2005.
9. A. Böttcher and B. Silbermann: *Analysis of Toeplitz Operators*. Springer-Verlag, Berlin, Heidelberg, New York, 2006. (2nd edition).

Papers

1. A. Böttcher and B. Silbermann: *Notes on the asymptotic behavior of block Toeplitz matrices and determinants.* Math. Nachr. **98** (1980), 183–210.
2. A. Böttcher and B. Silbermann: *The asymptotic behavior of Toeplitz determinants for generating functions with zeros of integral orders.* Math. Nachr. **102** (1981), 79–105.
3. A. Böttcher and B. Silbermann: *Über das Reduktionsverfahren für diskrete Wiener–Hopf-Gleichungen mit unstetigem Symbol.* Z. Analysis Anw. **1** (1982), 1–5.
4. A. Böttcher: *Toeplitz determinants with piecewise continuous generating function.* Z. Analysis Anw. **1** (1982), 23–39.
5. A. Böttcher and A.E. Pasenchuk: *On the invertibility of Wiener–Hopf operators on the quarter-plane.* In: Differential and Integral Equations, pp. 9–19, Elista, 1982 (in Russian).
6. A. Böttcher: *On two-dimensional Wiener–Hopf integral equations with a degenerate symbol.* Math. Nachr. **109** (1982), 195–213 (in Russian).
7. A. Böttcher: *Toeplitzdeterminanten in der statistischen Physik.* In: Ergebnisse d. Schule Junger Wissenschaftler zur Math. Physik, pp. 19–21, Pruchten, 1982.
8. A. Böttcher and B. Silbermann: *Wiener–Hopf determinants with symbols having zeros of analytic type.* Seminar Analysis 1982/83 (1983), 224–243.
9. A. Böttcher and B. Silbermann: *The finite section method for Toeplitz operators on the quarter-plane with piecewise continuous symbols.* Math. Nachr. **110** (1983), 279–291.
10. A. Böttcher: *Two-dimensional convolutions in angles with kernels supported in a half-plane.* Matem. Zametki **34** (1983), 207–218 (in Russian). Engl. transl. in Math. Notes **34** (1983), 585–591.
11. A. Böttcher: *Fredholm theory and finite section method for two-dimensional Wiener–Hopf operators with piecewise continuous symbols.* Dokl. Akad. Nauk SSSR **273** (1983), 1298–1300 (in Russian). Engl. transl. in Soviet Math. Dokl. **28** (1983), 773–776.
12. A. Böttcher: *Das Reduktionsverfahren für nichtelliptische Wiener–Hopf'sche Integraloperatoren in einer Klasse von topologischen Vektorräumen.* Wiss. Zeitschr. TH Karl-Marx-Stadt **25** (1983), 308–312.
13. A. Böttcher: *Fredholmness and finite section method for Toeplitz operators in* $\ell^p(\mathbb{Z}_+ \times \mathbb{Z}_+)$. Z. Analysis Anw. **3** (1984), 97–110.
14. A. Böttcher: *Fredholmness and finite section method for Toeplitz operators in* $\ell^p(\mathbb{Z}_+ \times \mathbb{Z}_+)$, *II.* Z. Analysis Anw. **3** (1984), 191–202.
15. A. Böttcher: *The finite section method for Wiener–Hopf integral operators with piecewise continuous symbols in the spaces* L^p. Funkts. Anal. Prilozh.

18 (1984), 55–56 (in Russian). Engl. transl. in Funct. Anal. Appl. **18** (1984), 132–133.

16. A. Böttcher: *The finite section method for two-dimensional Wiener–Hopf operators in L^p with piecewise continuous symbols.* Math. Nachr. **116** (1984), 61–73.
17. A. Böttcher and B. Silbermann: *Toeplitz determinants with symbols from the Fisher–Hartwig class.* Dokl. Akad. Nauk SSSR **278** (1984), 13–16 (in Russian). Engl. transl. in Soviet Math. Dokl. **30** (1984), 301–304.
18. A. Böttcher and B. Silbermann: *Toeplitz determinants generated by symbols with one singularity of Fisher–Hartwig type.* Wiss. Zeitschr. TH Karl-Marx-Stadt **26** (1984), 186–188.
19. A. Böttcher and B. Silbermann: *Toeplitz matrices and determinants with Fisher–Hartwig symbols.* J. Funct. Analysis **63** (1985), 178–214.
20. A. Böttcher: *Scalar Toeplitz operators, distance estimates, and localization over subalgebras of $C + H^\infty$.* Seminar Analysis 1985/1986 (1986), 1–17.
21. A. Böttcher: *A remark on the relation between the partial indices of a matrix function and its harmonic extension.* Seminar Analysis 1985/86 (1986), 19–22.
22. A. Böttcher, S. Roch, and B. Silbermann: *Local constructions and Banach algebras associated with Toeplitz operators on H^p.* Seminar Analysis 1985/86 (1986), 23–30.
23. A. Böttcher and B. Silbermann: *Toeplitz operators and determinants generated by symbols with one Fisher–Hartwig singularity.* Math. Nachr. **127** (1986), 95–124.
24. A. Böttcher: *On Toeplitz operators generated by symbols with three essential cluster points.* Preprint P-Math-04/86, 22 pages, Karl-Weierstrass-Institut, Berlin 1986 (never published but often cited).
25. A. Böttcher and B. Silbermann: *Local spectra of approximate identities, cluster sets, and Toeplitz operators.* Wiss. Zeitschr. TH Karl-Marx-Stadt **28** (1986), 175–180.
26. A. Böttcher: *Multidimensional Toeplitz operators with locally sectorial symbols.* Seminar Analysis 1986/87 (1987), 1–16.
27. A. Böttcher and B. Silbermann: *Toeplitz operators with symbols from $C+H^\infty$ in the spaces ℓ^p.* Zap. Nauchn. Sem. LOMI **157** (1987), 124–128 (in Russian). Engl. transl. in J. Soviet Math. **44** (1989), 834–836.
28. A. Böttcher: *Asymptotic formulas for rationally generated block Toeplitz determinants.* Seminar Analysis 1987/88 (1988), 1–13.
29. A. Böttcher and B. Silbermann: *Asymptotics of Toeplitz and Wiener–Hopf operators.* In: Proc. 9th Conf. Probl. and Meth. in Math. Phys., Karl-Marx-Stadt 1988, pp. 27–35, Teubner, 1988.

30. A. Böttcher, N. Krupnik, and B. Silbermann: *A general look at local principles with special emphasis on the norm computation aspect.* Integral Equations and Operator Theory **11** (1988), 455–479.
31. A. Böttcher and S. Dzhirgalova: *On Wiener–Hopf determinants with rational matrix symbols.* Seminar Analysis 1988/89 (1989), 21–39.
32. A. Böttcher: *On Mikaelyan's conjecture in the theory of Wiener–Hopf determinants.* Izv. Armen. Akad. Nauk **24** (1989), 188–192 (in Russian). Engl. transl. in Sov. J. Contemp. Math. Anal., Arm. Acad. Sci. **24** (1989), 85–89.
33. A. Böttcher: *Wiener–Hopf determinants with rational symbols.* Math. Nachr. **144** (1989), 39–64.
34. A. Böttcher: *Status report on rationally generated block Toeplitz and Wiener–Hopf determinants.* Unpublished manuscript (available from the author on request), 1989.
35. A. Böttcher and H. Heidler: *On linear functional equations with two involutions.* Seminar Analysis 1989/90 (1990), 31–43.
36. A. Böttcher: *Truncated Toeplitz operators on the polydisk.* Monatshefte f. Math. **110** (1990), 23–32.
37. A. Böttcher, B. Silbermann, and I. Spitkovsky: *Toeplitz operators with piecewise quasisectorial symbols.* Bull. London Math. Soc. **22** (1990), 281–286.
38. A. Böttcher, S. Roch, B. Silbermann, and I. Spitkovsky: *A Gohberg–Krupnik–Sarason symbol calculus for algebras of Toeplitz, Hankel, Cauchy, and Carleman operators.* Operator Theory: Adv. and Appl. **48** (1990), 189–234.
39. A. Böttcher and I. Spitkovsky: *Toeplitz operators with PQC symbols on weighted Hardy spaces.* J. Funct. Analysis **97** (1991), 194–214.
40. A. Böttcher and H. Wolf: *Finite sections of Segal–Bargmann space Toeplitz operators with polyradially continuous symbols.* Bull. Amer. Math. Soc. **25** (1991), 365–372.
41. A. Böttcher: *Invertible values of elementary operators and projection methods for Toeplitz operators.* In: Proc. Internat. Conf. on Elementary Operators, Blaubeuren 1991, pp. 157–161, World Scientific Publishing Co., Singapore, 1992.
42. A. Böttcher and H. Wolf: *Large sections of Bergman space Toeplitz operators with piecewise continuous symbols.* Math. Nachr. **156** (1992), 129–155.
43. A. Böttcher and H. Wolf: *Collocation methods for Toeplitz operators on Dzhrbashyan spaces.* Bull. Armen. Acad. Sci. **93** (1992), 168–172 (Russian).
44. A. Böttcher and H. Heidler: *Algebraic composition operators.* Integral Equations and Operator Theory **15** (1992), 389–411.
45. A. Böttcher and H. Wolf: *Interpolation in the Bergman space, mean values of harmonic functions, and an inverse problem of potential theory.* Wiss. Z. TU Chemnitz **34** (1992), 31–35.

46. A. Böttcher and I. Spitkovsky: *Wiener–Hopf integral operators with PC symbols on spaces with Muckenhoupt weight.* Revista Matematica Iberoamericana **9** (1993), 257–279.

47. A. Böttcher and H. Wolf: *Galerkin–Petrov methods for Bergman space Toeplitz operators.* SIAM J. Numer. Analysis **30** (1993), 846–863.

48. A. Böttcher and I. Spitkovsky: *On a theorem of Rooney concerning the spectrum of the singular integral operator.* Z. Analysis Anw. **12** (1993), 93–96.

49. A. Böttcher: *Toeplitz operators on the disk with locally sectorial symbols.* Rocky Mountain J. Math. **23** (1993), 803–816.

50. A. Böttcher and H. Heidler: *Classification of finite-dimensional algebras generated by the Calkin image of a composition operator on ℓ^p with weight.* Algebra i Analiz **5** (1993), 69–96 and St. Petersburg Math. J. **5** (1994), 1099–1119.

51. A. Böttcher and B. Silbermann: *Axler–Chang–Sarason–Volberg theorems for harmonic approximation and stable convergence.* In: Linear and Complex Analysis Problem Book 3, Part I (V.P. Havin, N.K. Nikolski, eds.), pp. 340–341, Lecture Notes in Math., Vol. 1573, Springer-Verlag, 1994.

52. A. Böttcher and H. Wolf: *Asymptotic invertibility of Bergman and Bargmann space Toeplitz operators.* Asymptotic Analysis **8** (1994), 15–33.

53. A. Böttcher: *Magnete, Determinanten und Fourier-Summen.* Spektrum der Wissenschaft 3/1994 (1994), 25–27.

54. A. Böttcher, Yu.I. Karlovich, and B. Silbermann: *Singular integral operators with PQC coefficients and freely transformed argument.* Math. Nachr. **166** (1994), 113–133.

55. A. Böttcher, Yu.I. Karlovich, and I. Spitkovsky: *Toeplitz operators with semi-almost periodic symbols on spaces with Muckenhoupt weight.* Integral Equations and Operator Theory **18** (1994), 261–276.

56. A. Böttcher and H. Widom: *Two remarks on spectral approximations for Wiener–Hopf operators.* J. Integral Equations Appl. **6** (1994), 31–36.

57. A. Böttcher and B. Silbermann: *Operator-valued Szegő–Widom limit theorems.* Operator Theory: Adv. and Appl. **71** (1994), 33–53.

58. A. Böttcher, B. Silbermann, and H. Widom: *A continuous analogue of the Fisher–Hartwig formula for piecewise continuous symbols.* J. Funct. Analysis **122** (1994), 222–246.

59. A. Böttcher, B. Silbermann, and H. Widom: *Determinants of truncated Wiener–Hopf operators with Hilbert–Schmidt kernels and piecewise continuous symbols.* Archiv d. Math. **63** (1994), 60–71.

60. A. Böttcher and I. Spitkovsky: *Pseudodifferential operators with heavy spectrum.* Integral Equations and Operator Theory **19** (1994), 251–269.

61. A. Böttcher: *Pseudospectra and singular values of large convolution operators.* J. Integral Equations Appl. **6** (1994), 267–301.

62. A. Böttcher: *The Onsager formula, the Fisher–Hartwig conjecture, and their influence on research into Toeplitz operators.* J. Statistical Physics **78** (Lars Onsager Festschrift) (1995), 575–585.

63. A. Böttcher and H. Wolf: *Analytic element collocation over geodesic circles for Bergman space Toeplitz operators.* In: Proc. Internat. Conf. Differential Geometry, Hamiltonian Systems, and Operator Theory, Kingston, Jamaica, 1994, pp. 69–87, University of the West Indies Press, Mona, 1995.

64. A. Böttcher and H. Heidler: *Algebraic and essentially algebraic composition operators on $C(X)$.* Aequationes Mathematicae **49** (1995), 276–294.

65. A. Böttcher and Yu.I. Karlovich: *Toeplitz and singular integral operators on Carleson curves with logarithmic whirl points.* Integral Equations and Operator Theory **22** (1995), 127–161.

66. A. Böttcher: *Toeplitz operators with piecewise continuous symbols – a neverending story?* Jahresbericht der DMV **97** (1995), 115–129.

67. A. Böttcher and B. Silbermann: *Infinite Toeplitz and Hankel matrices with operator-valued symbols.* SIAM J. Math. Analysis **27** (1996), 805–822.

68. A. Böttcher and H. Wolf: *Polynomial collocation over massive sets for Toeplitz integral equations on the Bergman space.* J. Computational and Applied Math. **66** (1996), 89–96.

69. A. Böttcher and Yu.I. Karlovich: *Submultiplicative functions and spectral theory of Toeplitz operators.* Integral Transforms and Special Functions **4** (1996), 181–202.

70. A. Böttcher: *Mathematischer Beweis versus Computerexperiment.* In: Wiss. Kolloquium "Zum Nutzen von Grundlagenforschung", Villa Hügel, 30. November 1995, pp. 39–42, Krupp-Stiftung, Essen, 1996.

71. A. Böttcher, Yu.I. Karlovich, and V.S. Rabinovich: *Emergence, persistence, and disappearance of logarithmic spirals in the spectra of singular integral operators.* Integral Equations and Operator Theory **25** (1996), 406–444.

72. A. Böttcher, I. Gohberg, Yu.I. Karlovich, N. Krupnik, S. Roch, B. Silbermann, and I. Spitkovsky: *Banach algebras generated by N idempotents and applications.* Operator Theory: Adv. and Appl. **90** (1996), 19–54.

73. A. Böttcher and S. Grudsky: *Toeplitz operators with discontinuous symbols: phenomena beyond piecewise discontinuity.* Operator Theory: Adv. and Appl. **90** (1996), 55–118.

74. A. Böttcher and Yu.I. Karlovich: *Toeplitz and singular integral operators on general Carleson Jordan curves.* Operator Theory: Adv. and Appl. **90** (1996), 119–152.

75. A. Böttcher and H. Heidler: *Characteristic polynomials of composition operators.* In: Proc. European Conf. on Iteration Theory, Batschuns, Austria, 1992, pp. 279–280, World Scientific Publishing Co., Singapore, 1996.

76. A. Böttcher: *Infinite matrices and projection methods.* In: Lectures on Operator Theory and Its Applications (P. Lancaster, ed.), pp. 1–72, Fields Institute Monographs, Vol. 3, Amer. Math. Soc., Providence, RI, 1996.

77. A. Böttcher: *Review of "Spectral Theory of Approximation Methods for Convolution Operators" by R. Hagen, S. Roch, and B. Silbermann.* Bull. Amer. Math. Soc. **33** (1996), 237–243.

78. A. Böttcher and H. Wolf: *Spectral approximation for Segal–Bargmann space Toeplitz operators.* In: Linear Operators, Banach Center Publ., Vol. 38, pp. 25–48, PAN, Warsaw, 1997.

79. A. Böttcher and Yu.I. Karlovich: *The algebra of singular integral operators on the Lebesgue space on a closed Carleson curve.* Dokl. Akad. Nauk **357** (1997), 7–10 (in Russian). Engl. transl. in Doklady Math. **56** (1997), 813–816.

80. A. Böttcher, S. Grudsky, and B. Silbermann: *Norms of inverses, spectra, and pseudospectra of large truncated Wiener–Hopf operators and Toeplitz matrices.* New York J. Math. **3** (1997), 1–31.

81. A. Böttcher: *On the approximation numbers of large Toeplitz matrices.* Documenta Mathematica **2** (1997), 1–29.

82. A. Böttcher and Yu.I. Karlovich: *The algebra of singular integral operators on a Lebesgue space with a Muckenhoupt weight on a closed Carleson curve.* Dokl. Akad. Nauk **359** (1998), 151–154 (in Russian). Engl. transl. in Doklady Math. **57** (1998), 193–196.

83. A. Böttcher, Yu.I. Karlovich, and V.S. Rabinovich: *Mellin pseudodifferential operators with slowly varying symbols and singular integrals on Carleson curves with Muckenhoupt weights.* Manuscripta Mathematica **95** (1998), 363–376.

84. A. Böttcher and S. Grudsky: *On the condition numbers of large semi-definite Toeplitz matrices.* Linear Algebra Appl. **279** (1998), 285–301.

85. A. Böttcher and S. Grudsky: *On the composition of Muckenhoupt weights and inner functions.* J. London Math. Soc. **58** (1998), 172–184.

86. A. Böttcher: *On the corona theorem for almost periodic functions.* Integral Equations and Operator Theory **33** (1999), 253–272.

87. A. Böttcher and Yu.I. Karlovich: *Toeplitz operators with PC symbols on general Carleson Jordan curves with arbitrary Muckenhoupt weights.* Trans. Amer. Math. Soc. **351** (1999), 3143–3196.

88. C.J. Bishop, A. Böttcher, Yu.I. Karlovich, and I. Spitkovsky: *Local spectra and index of singular integral operators with piecewise continuous coefficients on composed curves.* Math. Nachr. **206** (1999), 5–83.

89. A. Böttcher and S. Grudsky: *Toeplitz band matrices with exponentially growing condition numbers.* Electronic Journal of Linear Algebra (ELA) **5** (1999), 104–125.

90. A. Böttcher, S. Grudsky, A. Kozak, and B. Silbermann: *Norms of large Toeplitz band matrices.* SIAM J. Matrix Analysis Appl. **21** (1999), 547–561.

91. A. Böttcher, S. Grudsky, A. Kozak, and B. Silbermann: *Convergence speed estimates for the norms of the inverses of large truncated Toeplitz matrices.* Calcolo **36** (1999), 103–122.

92. A. Böttcher and M. Seybold: *Wackelsatz and Stechkin's inequality for discrete Muckenhoupt weights.* Preprint 99-7, 12 pages, TU Chemnitz, Fakultät für Mathematik, Chemnitz 1999 (will not be submitted for publication; contains full proofs to two basic theorems used in paper No. 98).

93. A. Böttcher and S. Grudsky: *Eighteen old and new asymptotic results on Toeplitz band matrices.* In: Large-Scale Scientific Computations of Engineering and Environmental Problems II (M. Griebel, S. Margenov, P. Yalamov, eds.), pp. 65–71, Vieweg, Braunschweig, 2000.

94. A. Böttcher, Yu.I. Karlovich, and V.S. Rabinovich: *The method of limit operators for one-dimensional singular integrals with slowly oscillating data.* J. Operator Theory **43** (2000), 171–198.

95. A. Böttcher, S. Grudsky, and I. Spitkovsky: *On the Fredholm indices of associated systems of Wiener–Hopf equations.* J. Integral Equations Appl. **12** (2000), 1–29.

96. A. Böttcher, S. Grudsky, and I. Spitkovsky: *Matrix functions with arbitrarily prescribed left and right partial indices.* Integral Equations and Operator Theory **36** (2000), 71–91.

97. A. Böttcher, S. Grudsky, and I. Spitkovsky: *The spectrum is discontinuous on the manifold of Toeplitz operators.* Archiv d. Math. **75** (2000), 46–52.

98. A. Böttcher and M. Seybold: *Discrete Wiener–Hopf operators on spaces with Muckenhoupt weight.* Studia Math. **143** (2000), 121–144.

99. A. Böttcher: *C^*-algebras in numerical analysis.* Irish Math. Soc. Bulletin **45** (2000), 57–133.

100. A. Böttcher and S. Grudsky: *Twenty old and new asymptotic results on Toeplitz band matrices.* In: Structured Matrices: Recent Developments in Theory and Computation (D.A. Bini, E. Tyrtyshnikov, P. Yalamov, eds.), pp. 1–10, Nova Science Publishers, Huntington, NY, 2001.

101. A. Böttcher, Yu.I. Karlovich, and V.S. Rabinovich: *Singular integral operators with complex conjugation from the viewpoint of pseudodifferential operators.* Operator Theory: Adv. and Appl. **121** (2001), 36–59.

102. A. Böttcher, A.V. Chithra, and M.N.N. Namboodiri: *Approximation of approximation numbers by truncation.* Integral Equations and Operator Theory **39** (2001), 387–395.

103. A. Böttcher, Yu.I. Karlovich, and I. Spitkovsky: *Toeplitz operators with semi-almost periodic matrix symbols on Hardy spaces.* Acta Applicandae Mathematicae **65** (2001), 115–136.

104. A. Böttcher: *One more proof of the Borodin–Okounkov formula for Toeplitz determinants.* Integral Equations and Operator Theory **41** (2001), 123–125.

105. A. Böttcher and M. Seybold: *Discrete one-dimensional zero-order pseudodifferential operators on spaces with Muckenhoupt weight.* Algebra i Analiz **13** (2001), 116–129 and St. Petersburg Math. J. **13** (2002), 241–252.

106. A. Böttcher, S. Grudsky, and I. Spitkovsky: *Toeplitz operators with frequency modulated semi-almost periodic symbols.* J. Fourier Analysis Appl. **7** (2001), 523–535.

107. A. Böttcher and S. Grudsky: *Condition numbers of large Toeplitz-like matrices.* Contemp. Math. **280** (2001), 273–299.

108. A. Böttcher and Yu.I. Karlovich: *Cauchy's singular integral operator and its beautiful spectrum.* Operator Theory: Adv. and Appl. **129** (2001), 109–142.

109. A. Böttcher, M. Embree, and V.I. Sokolov: *Infinite Toeplitz and Laurent matrices with localized impurities.* Linear Algebra Appl. **343/344** (2002), 101–118.

110. A. Böttcher, M. Embree, and M. Lindner: *Spectral approximation of banded Laurent matrices with localized random perturbations.* Integral Equations and Operator Theory **42** (2002), 142–165.

111. A. Böttcher and S. Grudsky: *Can spectral value sets of Toeplitz band matrices jump?* Linear Algebra Appl. **351/352** (2002), 99–116.

112. A. Böttcher, S. Grudsky, and A. Kozak: *On the distance of a large Toeplitz band matrix to the nearest singular matrix.* Operator Theory: Adv. and Appl. **135** (2002), 101–106.

113. A. Böttcher: *On the determinant formulas by Borodin, Okounkov, Baik, Deift, and Rains.* Operator Theory: Adv. and Appl. **135** (2002), 91–99.

114. A. Böttcher: *Essay on Bernd Silbermann.* Operator Theory: Adv. and Appl. **135** (2002), 1–12.

115. A. Böttcher, M. Embree, and L.N. Trefethen: *Piecewise continuous Toeplitz matrices and operators: slow approach to infinity.* SIAM J. Matrix Analysis Appl. **24** (2002), 484–489.

116. A. Böttcher, M. Embree, and V.I. Sokolov: *On large Toeplitz band matrices with an uncertain block.* Linear Algebra Appl. **366** (2003), 87–97.

117. D. Bini and A. Böttcher: *Polynomial factorization through Toeplitz matrix computations.* Linear Algebra Appl. **366** (2003), 25–37.

118. A. Böttcher, S. Grudsky, and I. Spitkovsky: *On the essential spectrum of Toeplitz operators with semi-almost periodic symbols.* Operator Theory: Adv. and Appl. **142** (2003), 59–77.

119. A. Böttcher, M. Embree, and V.I. Sokolov: *The spectra of large Toeplitz band matrices with a randomly perturbed entry.* Mathematics of Computation **72** (2003), 1329–1348.

120. A. Böttcher, S. Grudsky, and I. Spitkovsky: *Block Toeplitz operators with frequency-modulated semi-almost periodic symbols.* Int. J. Math. Math. Sci. 2003:**34** (2003), 2157–2176.

121. A. Böttcher and S. Grudsky: *The norm of the product of a large matrix and a random vector.* Electronic Journal of Probability **8** (2003), Paper no. 7, pages 1–29.

122. A. Böttcher and S. Grudsky: *Asymptotic spectra of dense Toeplitz matrices are unstable.* Numerical Algorithms **33** (2003), 105–112.

123. A. Böttcher and S. Grudsky: *Toeplitz matrices with slowly growing pseudospectra.* In: Factorization, Singular Operators and Related Problems in Honour of Georgii Litvinchuk (S. Samko, A. Lebre, A.F. dos Santos, eds.), pp. 43–54, Kluwer Academic Publishers, Dordrecht, 2003.

124. A. Böttcher, Yu.I. Karlovich, and I. Spitkovsky: *The C^*-algebra of singular integral operators with semi-almost periodic coefficients.* J. Funct. Analysis **204** (2003), 445–484.

125. A. Böttcher and S. Grudsky: *Fejér means and norms of large Toeplitz matrices.* Acta Sci. Math. (Szeged) **69** (2003), 889–900.

126. A. Böttcher and B. Silbermann: *Erhard Meister – friend and colleague.* Operator Theory: Adv. and Appl. **147** (2004), p. 67.

127. A. Böttcher and S. Grudsky: *Asymptotically good pseudomodes for Toeplitz matrices and Wiener–Hopf operators.* Operator Theory: Adv. and Appl. **147** (2004), 175–188.

128. A. Böttcher: *Transient behavior of powers and exponentials of large Toeplitz matrices.* Electronic Transactions on Numerical Analysis (ETNA) **18** (2004), 1–41.

129. A. Böttcher, S. Grudsky, and E. Ramírez de Arellano: *Algebras of Toeplitz operators with oscillating symbols.* Revista Matematica Iberoamericana **20** (2004), 647–671.

130. A. Böttcher: *The constants in the asymptotic formulas by Rambour and Seghier for inverses of Toeplitz matrices.* Integral Equations and Operator Theory **50** (2004), 43–55.

131. A. Böttcher, S. Grudsky, and E. Ramírez de Arellano: *Approximating inverses of Toeplitz matrices by circulant matrices.* Methods and Applications of Analysis **11** (2004), 211–220.

132. A. Böttcher and K. Rost: *Topics in the numerical linear algebra of Toeplitz and Hankel matrices.* GAMM-Mitteilungen **27** (2004), 174–188.

133. Z. Hurak and A. Böttcher: *MIMO ℓ^1-optimal control via block Toeplitz operators.* In: Proceedings of 16th International Symposium on Mathematical Theory of Networks and Systems (MTNS'04), Katholieke Universiteit Leuven, 2004.

134. A. Böttcher and D. Wenzel: *How big can the commutator of two matrices be and how big is it typically?* Linear Algebra Appl. **403** (2005), 216–228.

135. A. Böttcher and S. Grudsky: *Structured condition numbers of large Toeplitz matrices are rarely better than usual condition numbers.* Numerical Linear Algebra with Applications **12** (2005), 95–102.

136. A. Böttcher, I. Gohberg, and B. Silbermann: *Georg Heinig (1947–2005) In Memoriam.* Integral Equations and Operator Theory **53** (2005), 297–300.

137. A. Böttcher and P. Otte: *The first Szegő limit theorem for non-selfadjoint operators in the Følner algebra.* Mathematica Scandinavica **97** (2005), 115–126.

138. A. Böttcher and H. Widom: *Two elementary derivations of the pure Fisher–Hartwig determinant.* Integral Equations and Operator Theory **53** (2005), 593–596.

139. A. Böttcher, S. Grudsky, and E. Ramírez de Arellano: *On the asymptotic behavior of the eigenvectors of large banded Toeplitz matrices.* Math. Nachr. **279** (2006), 121–129.

140. A. Böttcher: *On the problem of testing the structure of a matrix by displacement operations.* SIAM J. Numerical Analysis **44** (2006), 41–54.

141. Z. Hurak, A. Böttcher, and M. Sebek: *Minimum distance to the range of a banded lower triangular Toeplitz operator in ℓ^1 and application in ℓ^1-optimal control.* SIAM J. Control Optim. **45** (2006), 107–122.

142. A. Böttcher, B. Hofmann, U. Tautenhahn, and M. Yamamoto: *Convergence rates for Tikhonov regularization from different kinds of smoothness conditions.* Applicable Analysis **85** (2006), 555–578.

143. A. Böttcher: *Review of "Spectra and Pseudospectra: The Behavior of Nonnormal Matrices and Operators" by L.N. Trefethen and M. Embree.* Linear Algebra Appl. **416** (2006), 1098–1101.

144. A. Böttcher and H. Widom: *From Toeplitz eigenvalues through Green's kernels to higher-order Wirtinger–Sobolev inequalities.* Operator Theory: Adv. and Appl. **171** (2006), 73–87.

145. A. Böttcher and H. Widom: *On the eigenvalues of certain canonical higher-order ordinary differential operators.* J. Math. Analysis Appl. **322** (2006), 990–1000.

146. A. Böttcher: *Schatten norms of Toeplitz matrices with Fisher–Hartwig singularities.* Electronic Journal of Linear Algebra (ELA) **15** (2006), 251–259.

147. A. Böttcher and H. Widom: *Szegő via Jacobi.* Linear Algebra Appl. **419** (2006), 656–667.

148. A. Böttcher and D. Wenzel: *On the verification of linear equations and the identification of the Toeplitz-plus-Hankel structure.* Operator Theory: Adv. and Appl. **170** (2007), 43–51.

149. J. Gutiérrez-Gutiérrez, P.M. Crespo, and A. Böttcher: *Functions of banded Hermitian block Toeplitz matrices in signal processing.* Linear Algebra Appl. **422** (2007), 788–807.

150. A. Böttcher, D. Potts, and D. Wenzel: *A probability argument in favor of ignoring small singular values.* Operators and Matrices **1** (2007), 31–43.

151. A. Böttcher, J. Gutiérrez-Gutiérrez, and P.M. Crespo: *Mass concentration in quasicommutators of Toeplitz matrices.* J. Comput. Appl. Math. **205** (2007), 129–148.

152. A. Böttcher and D. Potts: *Probability against condition number and sampling of multivariate trigonometric random polynomials.* Electronic Transactions on Numerical Analysis (ETNA) **26** (2007), 178–189.

153. A. Böttcher and J. Virtanen: *Norms of Toeplitz matrices with Fisher–Hartwig symbols.* SIAM J. Matrix Analysis Appl. **29** (2007), 660–671.

154. A. Böttcher, A. Karlovich, and B. Silbermann: *Generalized Krein algebras and asymptotics of Toeplitz determinants.* Methods of Functional Analysis and Topology **13** (2007), 236–261.

155. A. Böttcher and S. Grudsky: *Uniform boundedness of Toeplitz matrices with variable coefficients.* Integral Equations and Operator Theory **60** (2008), 313–328.

156. A. Böttcher, S. Grudsky, and M. Schwartz: *Some problems concerning the test functions in the Szegő and Avram–Parter theorems.* Operator Theory: Adv. and Appl. **187** (2008), 81–93.

157. A. Böttcher, S. Grudsky, and E.A. Maksimenko: *The Szegő and Avram–Parter theorems for general test functions.* C. R. Math. Acad. Sci. Paris **346** (2008), 749–752.

158. A. Böttcher, S. Grudsky, and E.A. Maksimenko: *Pushing the envelope of the test functions in the Szegő and Avram–Parter theorems.* Linear Algebra Appl. **429** (2008), 346–366.

159. A. Böttcher: *Orthogonal symmetric Toeplitz matrices.* Complex Analysis and Operator Theory **2** (2008), 285–298.

160. A. Böttcher and D. Wenzel: *Rigorous stochastic bounds for the error in large covariance matrices.* Math. Methods Appl. Sci. **31** (2008), 1209–1220.

161. A. Böttcher: *Linear and one-dimensional.* In: Israel Gohberg and Friends (H. Bart, T. Hempfling, M.A. Kaashoek, editors), pp. 291–293, Birkhäuser Verlag, Basel, 2008.

162. A. Böttcher and D. Wenzel: *The Frobenius norm and the commutator.* Linear Algebra Appl. **429** (2008), 1864–1885.

163. A. Böttcher, S. Grudsky, and J. Unterberger: *Asymptotic pseudomodes of Toeplitz matrices.* Operators and Matrices **2** (2008), 525–541.

164. A. Böttcher and M. Lindner: *Pseudospectrum.* Scholarpedia **3**(3):2680 (2008).

165. A. Böttcher and P. Dörfler: *On the best constants in inequalities of the Markov and Wirtinger types for polynomials on the half-line.* Linear Algebra Appl. **430** (2009), 1057–1069.

166. A. Böttcher, S. Grudsky, E.A. Maksimenko, and J. Unterberger: *The first-order asymptotics of the extreme eigenvectors of certain Hermitian Toeplitz matrices.* Integral Equations and Operator Theory **63** (2009), 165–180.

167. A. Böttcher, S. Kunis, and D. Potts: *Probabilistic spherical Marcinkiewicz–Zygmund inequalities.* J. Approx. Theory **157** (2009), 113–126.

168. A. Böttcher and I. Spitkovsky: *Drazin inversion in the von Neumann algebra generated by two orthogonal projections.* J. Math. Analysis Appl. **358** (2009), 403–409.

169. A. Böttcher, S. Grudsky, and E.A. Maksimenko: *On the asymptotics of all eigenvalues of Hermitian Toeplitz band matrices* (Russian). Dokl. Akad. Nauk **428** (2009), 153–156. Engl. transl. in Doklady Math. **80** (2009), 662–664.

170. A. Böttcher and P. Dörfler: *Weighted Markov-type inequalities, norms of Volterra operators, and zeros of Bessel functions.* Math. Nachr. **283** (2010), 40–57.

171. A. Böttcher and I. Spitkovsky: *A gentle guide to the basics of two projections theory.* Linear Algebra Appl. **432** (2010), 1412–1459.

172. A. Böttcher, S. Grudsky, and E.A. Maksimenko: *Inside the eigenvalues of certain Hermitian Toeplitz band matrices.* J. Comput. Appl. Math. **233** (2010), 2245–2264.

173. A. Böttcher and S. Grudsky: *Variable-coefficient Toeplitz matrices with symbols beyond the Wiener algebra.* Operator Theory: Adv. and Appl. **199** (2010), 192–202.

174. A. Böttcher, S. Grudsky, and E.A. Maksimenko: *On the structure of the eigenvectors of large Hermitian Toeplitz band matrices.* Operator Theory: Adv. and Appl. **210** (2010), 15–36.

175. A. Böttcher and P. Dörfler: *On the best constants in Markov-type inequalities involving Laguerre norms with different weights.* Monatshefte f. Math. **161** (2010), 357–367.

176. A. Böttcher, H. Brunner, A. Iserles, and S.P. Nørsett: *On the singular values and eigenvalues of the Fox-Li and related operators.* New York J. Math. **16** (2010), 539–561.

177. M. Bogoya, A. Böttcher, S. Grudsky, and E.A. Maksimenko: *Eigenvalues of Hessenberg Toeplitz matrices generated by symbols with several singularities.* Commun. Math. Analysis, Conf. **03** (2011), 23–41.

178. A. Böttcher and P. Dörfler: *On the best constants in Markov-type inequalities involving Gegenbauer norms with different weights.* Operators and Matrices **5** (2011), 261–272.

179. A. Böttcher, S. Grudsky, and R.M. Porter: *European double-barrier options with a compound Poisson component.* In: Progress in Economics Research, Vol. 18 (A. Tavidze, ed.), pp. 315–331, Nova Science Publishers, Huntington, NY, 2011.

180. A. Böttcher and P. Dörfler: *Inequalities of the Markov type for partial derivatives of polynomials in several variables.* J. Integral Equations Appl. **23** (2011), 1–37.

181. A. Böttcher: *The algebraic Riccati equation with Toeplitz matrices as coefficients.* Electronic Journal of Linear Algebra (ELA) **22** (2011), 348–362.

182. A. Böttcher, S. Grudsky, and A. Iserles: *Spectral theory of large Wiener–Hopf operators with complex-symmetric kernels and rational symbols.* Math. Proc. Cambridge Phil. Soc. **151** (2011), 161–191.

183. A. Böttcher and I. Spitkovsky: *On certain finite-dimensional algebras generated by two idempotents.* Linear Algebra Appl. **435** (2011), 1823–1836.

184. A. Böttcher, S. Grudsky, D. Huybrechs, and A. Iserles: *First-order trace formulae for the iterates of the Fox–Li operator.* Operator Theory: Adv. and Appl. **218** (2012), 207–224.

185. M. Bogoya, A. Böttcher, and S. Grudsky: *Asymptotics of individual eigenvalues of a class of large Hessenberg Toeplitz matrices.* Operator Theory: Adv. and Appl. **220** (2012), 77–95.

186. M. Bogoya, A. Böttcher, S. Grudsky, and E.A. Maksimenko: *Eigenvectors of Hessenberg Toeplitz matrices and a problem by Dai, Geary, and Kadanoff.* Linear Algebra Appl. **436** (2012), 3480–3492.

187. M. Bogoya, A. Böttcher, and S. Grudsky: *Eigenvalues of Hermitian Toeplitz matrices with polynomially increasing entries.* Journal of Spectral Theory **2** (2012), 267–292.

188. A. Böttcher and I. Spitkovsky: *Group inversion in certain finite-dimensional algebras generated by two idempotents.* Indagationes Mathematicae **23** (2012), 715–732.

189. A. Böttcher, S. Grudsky, and A. Iserles: *The Fox–Li operator as a test and a spur for Wiener–Hopf theory.* In: Essays in Mathematics and Its Applications, in Honor of Stephen Smale's 80th Birthday, pp. 37–48, Springer-Verlag, Heidelberg, 2012.

190. A. Böttcher and A. Pietsch: *Orthogonal and skew-symmetric operators in real Hilbert space.* Integral Equations and Operator Theory **74** (2012), 497–511.

191. A. Böttcher and A. Perälä: *The index formula of Douglas for block Toeplitz operators on the Bergman space of the ball.* Operator Theory: Adv. and Appl. **228** (2013), 39–55.

192. A. Böttcher and I. Spitkovsky: *The factorization problem: some known results and open questions.* Operator Theory: Adv. and Appl. **229** (2013), 101–122.

193. A. Böttcher and M. Halwass: *Wiener–Hopf and spectral factorization of real polynomials by Newton's method.* Linear Algebra Appl. **438** (2013), 4760–4805.

194. A. Böttcher and I. Spitkovsky: *Classification of the finite-dimensional algebras generated by two tightly coupled idempotents.* Linear Algebra Appl. **439** (2013), 538–551.

195. A. Böttcher: *An operator theoretic approach to the brickwork Ising model with second-neighbor interactions.* Linear Algebra Appl. **439** (2013), 675–685.

196. A. Böttcher: *Best constants for Markov type inequalities in Hilbert space norms.* In: Recent Trends in Analysis, Proceedings of the Conference in Honor of Nikolai Nikolski, Bordeaux 2011, pp. 73–83, Theta, Bucharest, 2013.

197. A. Böttcher and I. Spitkovsky: *Special types of matrices.* Chapter 22 of the Handbook of Linear Algebra, 2nd edition, edited by Leslie Hogben, Chapman & Hall/CRC, Boca Raton, FL, 2013.

198. A. Böttcher and M. Halwass: *A Newton method for canonical Wiener–Hopf and spectral factorization of matrix polynomials.* Electronic Journal of Linear Algebra (ELA) **26** (2013), 873–897.

199. A. Böttcher: *Borodin–Okounkov and Szegő for Toeplitz operators on model spaces.* Integral Equations and Operator Theory **78** (2014), 407–414.

200. A. Böttcher: *On Hurwitz stable polynomials with integer coefficients.* Computational Methods and Function Theory **14** (2014), 139–156.

201. A. Böttcher and M. Halwass: *Canonical Wiener–Hopf and spectral factorization of large-degree matrix polynomials.* Proceedings in Applied Mathematics and Mechanics (PAMM) **14** (2014), 817–818.

202. M. Bogoya, A. Böttcher, S. Grudsky, and E.A. Maximenko: *Eigenvalues of Hermitian Toeplitz matrices with smooth simple-loop symbols.* J. Math. Analysis Appl. **422** (2015), 1308–1334.

203. A. Böttcher, L. Fukshansky, S.R. Garcia, and H. Maharaj: *Toeplitz determinants with perturbations in the corners.* J. Funct. Analysis **268** (2015), 171–193.

204. A. Böttcher, L. Fukshansky, S.R. Garcia, and H. Maharaj: *On lattices generated by finite Abelian groups.* SIAM J. Discrete Math. **29** (2015), 382–404.

205. M. Bogoya, A. Böttcher, S. Grudsky, and E.A. Maximenko: *Maximum norm versions of the Szegő and Avram–Parter theorems for Toeplitz matrices.* J. Approx. Theory **196** (2015), 79–100.

206. A. Böttcher, D. Kowerko, and R.K.O. Sigel: *Explicit analytic equations for multimolecular thermal melting curves.* Biophysical Chemistry **202** (2015), 32–39.

207. A. Böttcher, L. Fukshansky, S.R. Garcia, and H. Maharaj: *Lattices from Hermitian function fields.* J. Algebra **447** (2016), 560–579.

208. M. Bogoya, A. Böttcher, S. Grudsky, and E.A. Maximenko: *Eigenvectors of Hermitian Toeplitz matrices with smooth simple-loop symbols.* Linear Algebra Appl. **493** (2016), 606–637.
209. A. Böttcher and F. Kittaneh: *The limit of the zero set of polynomials of the Fibonacci type.* J. Number Theory **163** (2016), 89–100.
210. A. Böttcher: *The part of my path I walked together with Sergei Grudsky.* Bol. Soc. Mat. Mex. **22** (2016), 309–327.
211. M. Bogoya, A. Böttcher, and E.A. Maximenko: *From convergence in distribution to uniform convergence.* Bol. Soc. Mat. Mex. **22** (2016), 695–710.
212. A. Böttcher, H. Langenau, and H. Widom: *Schatten class integral operators occurring in Markov-type inequalities.* Operator Theory: Adv. and Appl. **255** (2016), 91–104.
213. A. Böttcher, L. Fukshansky, S.R. Garcia, H. Maharaj, and D. Needell: *Lattices from equiangular tight frames.* Linear Algebra Appl. **510** (2016), 395–420.
214. A. Böttcher and F.-O. Speck: *On the symmetrization of general Wiener–Hopf operators.* J. Operator Theory **76** (2016), 335–349.
215. A. Böttcher, L. Fukshansky, S.R. Garcia, and H. Maharaj: *Lattices from Abelian groups.* Oberwolfach Reports **13** (2016) 121–124.
216. A. Böttcher: *The Duduchava–Roch formula.* Operator Theory: Adv. and Appl. **258** (2017), 1–19.
217. A. Böttcher and C. Rebs: *On the constants in Markov inequalities for the Laplace operator on polynomials with the Laguerre norm.* Asymptotic Analysis, to appear.
218. A. Böttcher, L. Fukshansky, S.R. Garcia, and H. Maharaj: *Lattice theory and Toeplitz determinants.* Operator Theory: Adv. and Appl., to appear.
219. A. Böttcher: *Index formulas for Toeplitz operators, approximate identities, and the Wolf-Havin theorem.* To appear.
220. A. Böttcher, C. Garoni, and S. Serra-Capizzano: *Exploration of Toeplitz-like matrices with unbounded symbols: not a purely academic journey.* To appear.
221. A. Böttcher, J.M. Bogoya, S.M. Grudsky, and E.A. Maximenko: *Asymptotic formulas for the eigenvalues and eigenvectors of Toeplitz matrices.* To appear (in Russian).
222. A. Böttcher, S. Eisenbarth, L. Fukshansky, S.R. Garcia, and H. Maharaj: *Spherical 2-designs and lattices from Abelian groups.* To appear.

Operator Theory:
Advances and Applications, Vol. 259, 1–18

Albrecht Böttcher – 20 Years of Friendship and Joint Work

Sergei Grudsky

To my excellent friend and best co-author Albrecht on his sixtieth birthday

1. Starting joint work

More than twenty years of our friendship with Albrecht Böttcher, as well as joint work on numerous projects (nearly 50 papers and 2 monographs!) have already passed. In the beginning of 1990s, I was already an established scientist, however his influence on me as a mathematician was very substantial.

In fact we met for the first time much earlier, in 1982. At that time, I already had a PhD and worked at the Computer Center of the Rostov-on-Don State University. Albrecht has come from East Germany to study in the PhD program in mathematics under the supervision of Vladimir Borisovich Dybin, who was also my research adviser. We are of the same age with Albrecht, however I and my colleagues felt the differences between the educational systems of the USSR and GDR. Already at that time all of us treated him as a very capable and incredibly active mathematician. Dybin's statement in a conversation with one of his former PhD students, the excellent young mathematician Misha Gorodetsky, is characteristic: "Mikhail Borisovich, your results are just "scratches" on the "building" of results obtained by Albrecht Böttcher". At all extremeness of this statement, which, incidentally, is not shared by Albrecht himself and by me, it shows the attitude of our small community towards Albrecht. Unfortunately, at that time he and I had not become closer friends because my scientific and administrative obligations in hydroacoustics at the laboratory of the Computer Center of the Rostov State University took a lot of time and strength. Later I always regretted that.

In the early 1990s, I spent 3 months in Chemnitz on Albrecht's invitation. At that time he was already full professor of Analysis at the Faculty of Mathematics of the Technical University of Chemnitz. We joined our forces to write a paper about Toeplitz operators with oscillating symbols. Here all remarkable qualities of my friend showed in full. They have formed a basis of further cooperation during many years. First of all, from his point of view, each paper has to be a piece of work with a bright fundamental idea, and not simply a collection of results, even good ones. Sometimes, results that I liked were thrown out ruthlessly as not meeting this criterion. Further, the exposition has to be clear and bright, supported by beautiful pictures and graphs. At the first stage, we did a lot of work searching such an idea, looking at specific interesting problems and their solutions. All this was very intensive, interesting, and emotional. At this time he reminded me of a plane flying at a high speed and browsing the mathematical landscape looking for beautiful problems. But some day during this time of search a miracle happened. After we had been sitting together on Friday, he went home and on Monday morning he brought a completed manuscript 20–30 pages long! He had even taken care of the fine tuning of the text. Such moments have occurred repeatedly since then, and it has always fascinated me when a good mathematical text appeared in 2–3 days from separate thoughts and results and incomplete pieces of proofs. One more feature of the mathematical talent of Albrecht is his ability to find currently interesting problems which, on the one hand, can be solved and, on the other hand, have solutions that lead to other interesting work.

I would like to mention the high mathematical productivity of my friend. He wrote more than two hundred papers and eight monographs. I do not think it is reasonable (and even possible!) to review in one article all papers and research topics in which Albrecht was involved during almost 40 years of his mathematical career. The best way to have an idea of them is to visit his home page designed with extraordinary care and love:

`https://www-user.tu-chemnitz.de/~aboettch/`

In this note I will discuss only some highlights of our joint work which, at least to some extent, illustrate what was said above.

2. Toeplitz operators with oscillating symbols

In an article dedicated to Albrecht it should be said that he is a remarkable lecturer. Each of his lectures devoted to this or that problem gives listeners a great pleasure. His exposition style is clear and always supported by some unexpected graphic illustrations. The story flow is seemingly slow, but suddenly the listener understands that the lecturer taught him a lot.

An indelible impression on me was made by his lecture about our first joint work [10]. Below the reader can see the first four pages of the transparencies prepared for that lecture, which give an idea of “Böttcher’s method”. The lecturer has chosen an image of a pyramid to classify and explain the total body of results of the theory of normal solvability of Toeplitz operators with symbols having oscillating discontinuities and an infinite index.

Toeplitz operators
with discontinuous symbols:
problems, methods, and results
beyond piecewise continuity

A. Böttcher and S. Grudsky
1997

$T(a) = aP + Q$ on $L^2(\mathbb{T})$

$$P = \frac{I+S}{2}, \quad Q = \frac{I-S}{2}, \quad a \in GL^\infty(\mathbb{T})$$

$T(a) = Pa \mid H^2(\mathbb{T})$

$$T(a) = \begin{pmatrix} a_0 & a_{-1} & a_{-2} & \dots \\ a_1 & a_0 & a_{-1} & \dots \\ a_2 & a_1 & a_0 & \dots \\ \dots & \dots & & \end{pmatrix} \quad \text{on } \ell^2(\mathbb{Z}_+)$$

Operators on $L^2(\mathbb{T})$

A normally solvable $\Leftrightarrow$ Im A closed

$\alpha(A) = \dim \operatorname{Ker} A$

$\beta(A) = \dim \operatorname{Coker} A = \dim (\operatorname{Im} A)^{\perp} = \dim \operatorname{Ker} A^{*}$

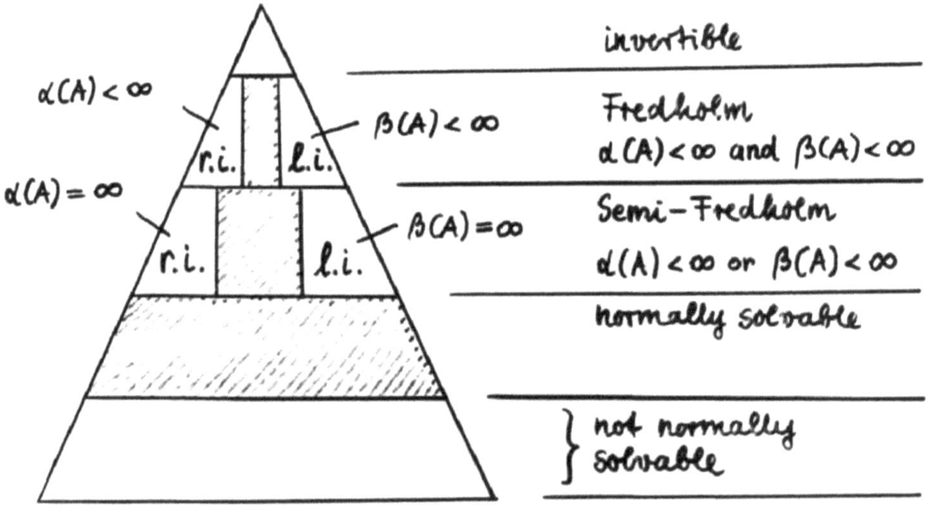

Toeplitz operators on $L^2(\mathbb{T})$

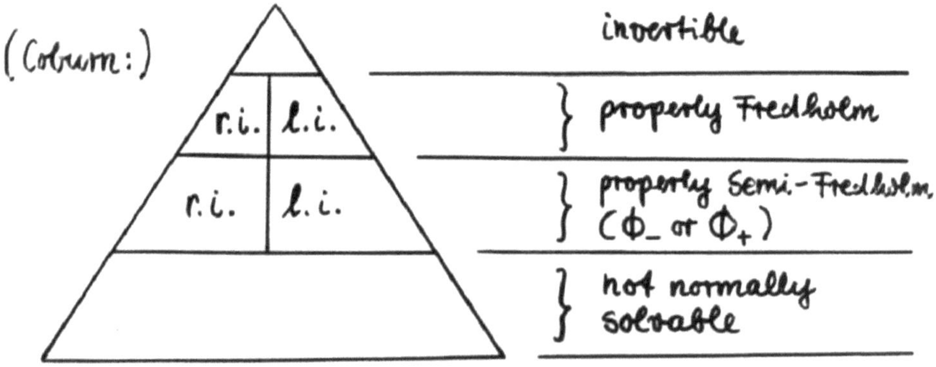

$a \in C(\mathbb{T}) \cap GL^\infty(\mathbb{T})$

(Gohberg 1952)

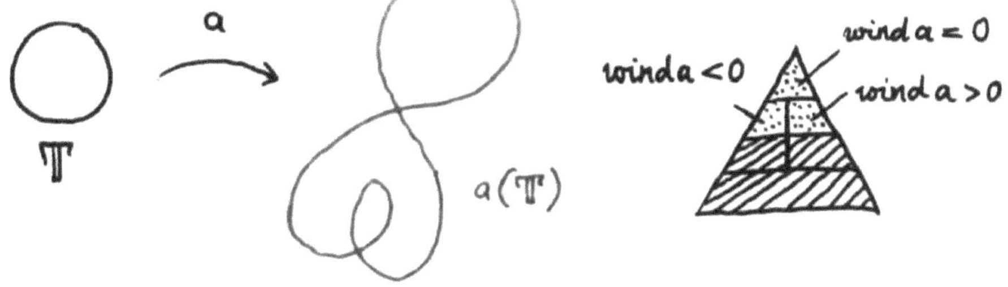

$a \in PC(\mathbb{T}) \cap GL^\infty(\mathbb{T})$ (Widom, Devinatz, Calderón, Spitzer, Simonenko, Gohberg, Krupnik)

$a \in GL^\infty(\mathbb{T}) \setminus PC(\mathbb{T})$?

$a \in \underbrace{GL^\infty(\mathbb{T})} \cap C(\mathbb{T} \setminus \{1\})$

$H^\infty, \overline{H^\infty}$

$C + H^\infty = \mathrm{alg}(C, H^\infty), \; C + \overline{H^\infty} = \mathrm{alg}(C, \overline{H^\infty})$

$QC = (C + H^\infty) \cap (C + \overline{H^\infty}) \supset SO$

$PQC = \mathrm{alg}(PC, QC) \supset PSO$

$AP, \; SAP, \ldots$

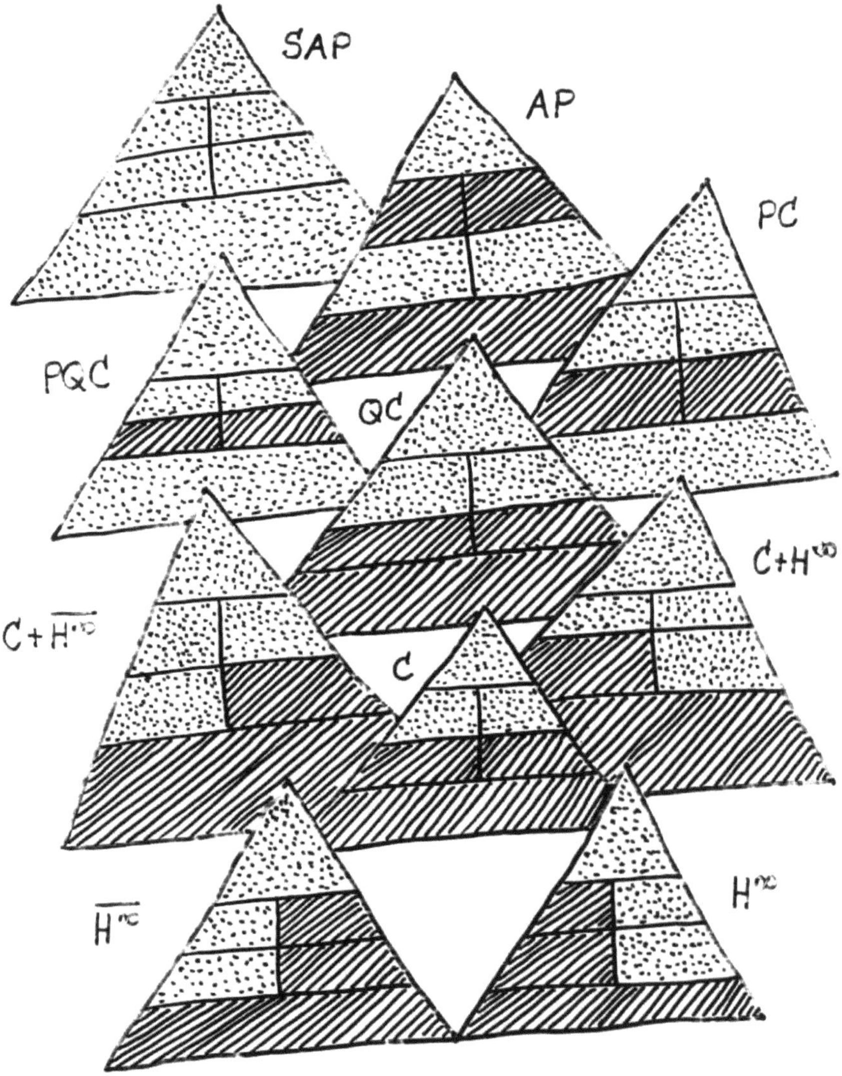

Observations:

(1) $a \notin PC(\mathbb{T}) \Rightarrow \Phi_\pm$ operators appear

(2) $SAP \cap GL^\infty(\mathbb{T})$ is the "simplest" class for which everything is possible.

(3) The existence of operators which are <u>not</u> normally solvable has something to do with "jumps".

3. Compositions of Muckenhoupt weights and inner functions

During my visit to Chemnitz in 1995 there was an especially bright event: we managed to solve a problem which had been open for many years. I learned about it in 1981. The problem is as follows.

Let $\mathbb{T}$ be the unite circle in the complex plane $\mathbb{C}$. For $1 < p < \infty$, let $L^p(\mathbb{T})$ be the usual space of all measurable functions such that

$$\|f\|_p := \left(\int_{\mathbb{T}} |f(t)|^p |dt| \right)^{1/p} < \infty.$$

A measurable function $w : \mathbb{T} \to [0, \infty]$ is called a weight if the set $w^{-1}(\{0, \infty\})$ has measure zero. A weight w is said to belong to the Muckenhoupt class A_p if $w \in L^p(\mathbb{T})$, $1/w \in L^q(\mathbb{T})$, where $1/p + 1/q = 1$, and

$$\sup_I \left(\frac{1}{|I|} \int_I w^p(t)|dt| \right)^{1/p} \left(\frac{1}{|I|} \int_I w^{-q}(t)|dt| \right)^{1/q} < \infty.$$

Here the supremum is taken over all arcs $I \subset \mathbb{T}$ and $|I|$ denotes the length of I. As usual, let $H^\infty(\mathbb{T})$ be the Hardy space of all bounded analytic functions in the unit disk. A nonconstant function $u \in H^\infty(\mathbb{T})$ is called an inner function if $|u(t)| = 1$ a.e. on $\mathbb{T}$.

Problem 3.1. *Let $w \in A_p$ and let u be an inner function. Does the composition $(w \circ u)(t) = w(u(t))$ belong to A_p?*

This problem is interesting by itself. It also has several important reformulations. Let us mention a reformulation related to the theory of Toeplitz operators. For a symbol $a \in L^\infty(\mathbb{T})$, consider the Toeplitz operator

$$T(a) := PaP : H^p(\mathbb{T}) \to H^p(\mathbb{T})$$

acting on the Hardy space

$$H^p(\mathbb{T}) := \{f \in L^p(\mathbb{T}) : \ f_n = 0, \ n < 0\},$$

where f_n is the nth Fourier coefficient of f and

$$P : L^p(\mathbb{T}) \to H^p(\mathbb{T}), \qquad \sum_{n=-\infty}^{\infty} f_n t^n \mapsto \sum_{n=0}^{\infty} f_n t^n$$

is the analytic projection. Problem 3.1 has the following equivalent reformulation.

Problem 3.2. *Let $a \in L^\infty(\mathbb{T})$. Suppose the operator $T(a)$ is invertible on $H^p(\mathbb{T})$ and $u \in H^\infty(\mathbb{T})$ is an inner function. Is the operator $T(a \circ u)$ invertible on $H^p(\mathbb{T})$?*

The conjecture was that the answer to this question is positive. It was proved in many particular cases. For instance, if $p = 2$ and u is an arbitrary inner function, then the answer is "yes". I was sure that the conjecture is true in general and tried to prove it over and over again for 14 years. However, after numerous discussions with Albrecht, we started to understand that there are situations when the conjecture fails. Finally, we constructed a counterexample, which is the heart of our paper [11]. This article has served as an incentive to find sufficient conditions guaranteeing a positive answer to the question posed in Problem 3.2 as well as to study several related questions. As the result of this work, the papers [21–25] have appeared.

4. Stability of spectra of Toeplitz matrices

I want to show the reader the beauty of spectra of finite Toeplitz matrices. The four pictures in Figures 1–4 represent a small piece of huge number of numerical experiments done during our work. Under the influence of Albrecht, we tried to produce an outcome in an attractive shape whenever possible.

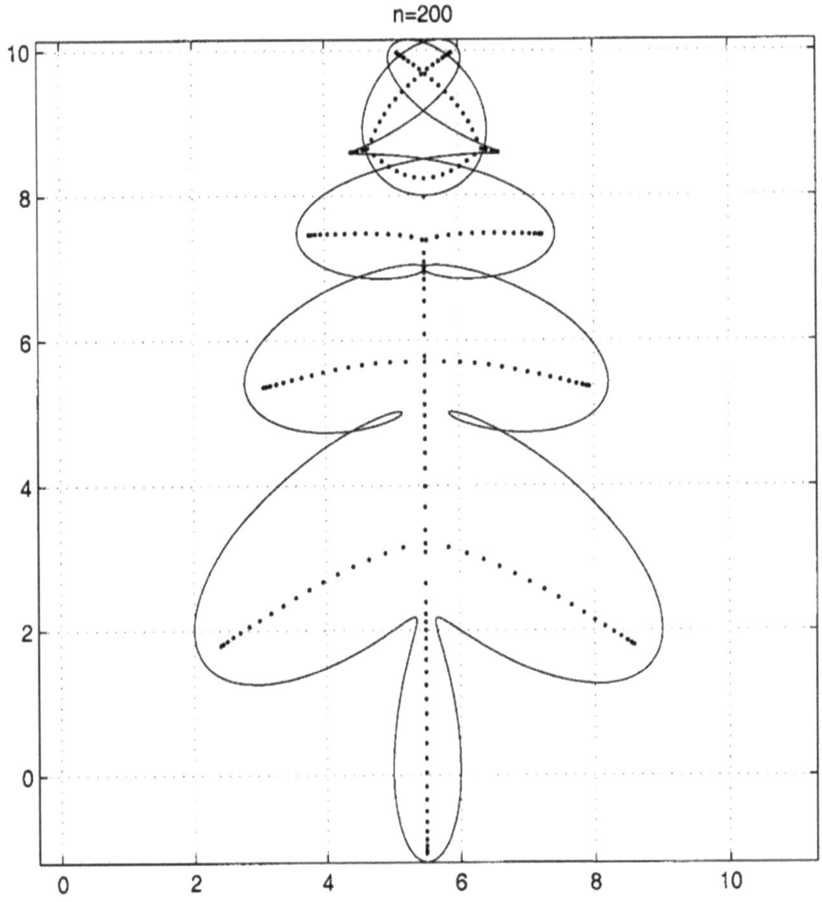

FIGURE 1. Christmas tree.

One of the interesting topics to which we have returned with Albrecht many times is the stability problem for the of spectra of finite Toeplitz matrices $T_n(a)$ when the size n of the matrices goes to infinity.

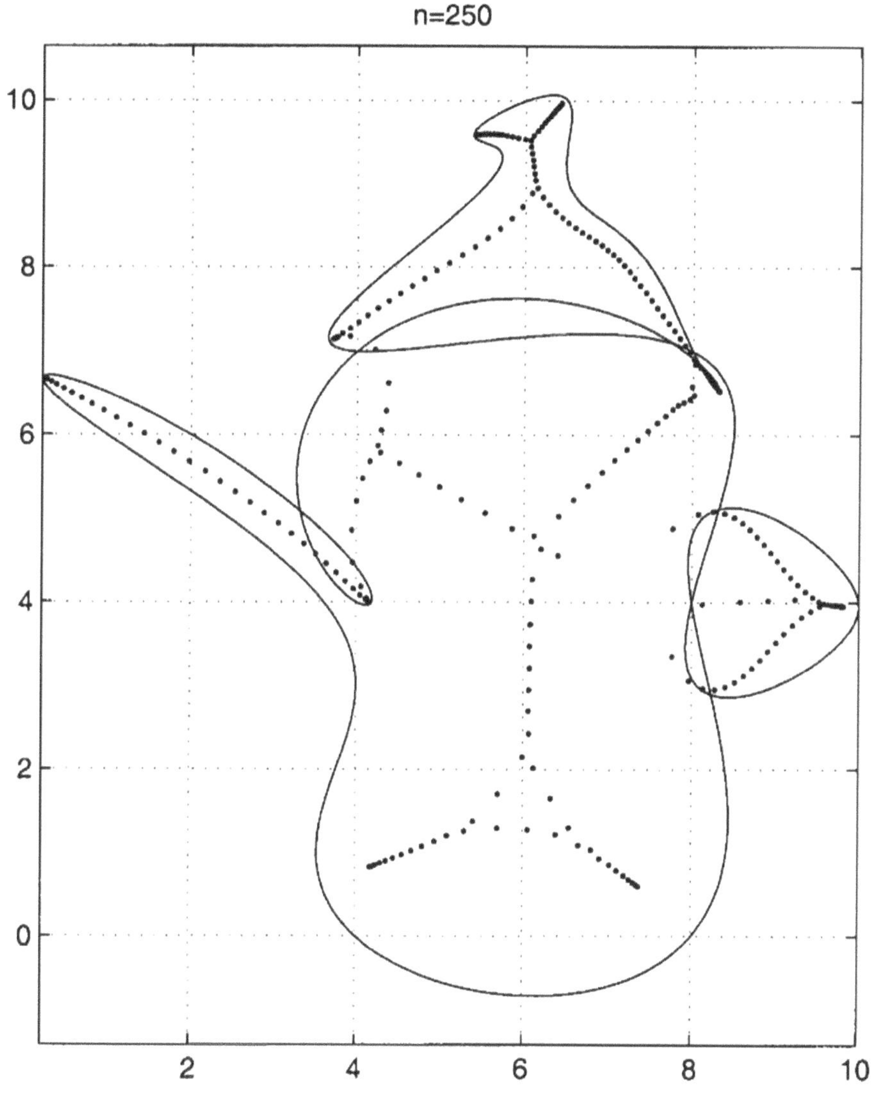

FIGURE 2. Tea pot.

If the finite Toeplitz matrices $T_n(a)$ are generated by a trigonometric polynomial a, then the matrix of the corresponding Toeplitz operator $T(a)$ is banded, that is, it has only a finite number of nonzero diagonals. The description of the limiting set of the spectra of $T_n(a)$ as $n \to \infty$ was given by Schmidt and Spitzer in 1960 in the classical paper [28]. In that paper it was shown that the limiting set is always the union of analytic arcs each pair of which have at most endpoints in common.

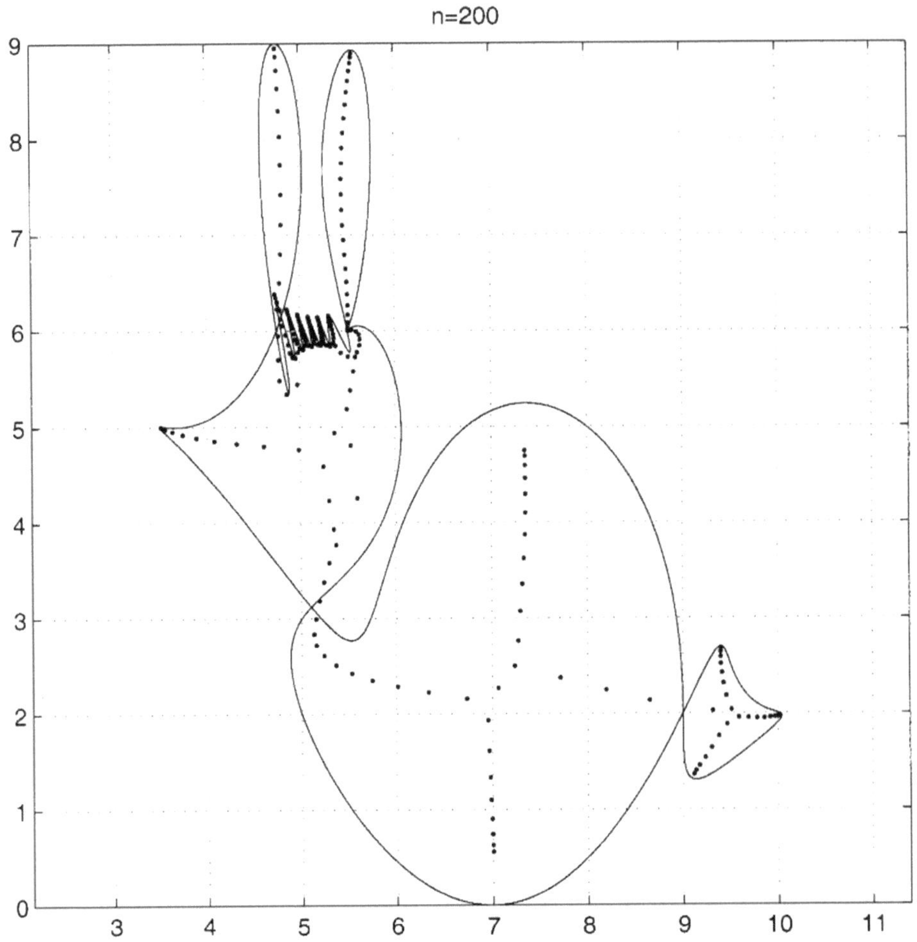

FIGURE 3. Rabbit

The above pictures were all produced by Olga Grudskaya, who in particular invented the symbol functions, performed the calculations, and made all related adjustments. In each picture, we see an outer contour, which is the range $a(\mathbb{T})$ of the corresponding polynomial symbol a, usually of large degree, while the dots represent the eigenvalues of the finite Toeplitz matrix $T_n(a)$ of large size n.

These pictures demonstrate that, as a rule, the limit of the spectra of finite Toeplitz matrices $T_n(a)$ is substantially smaller than the spectrum of the corresponding Toeplitz operator $T(a)$. The latter coincides with the domain bounded by the curve $a(\mathbb{T})$.

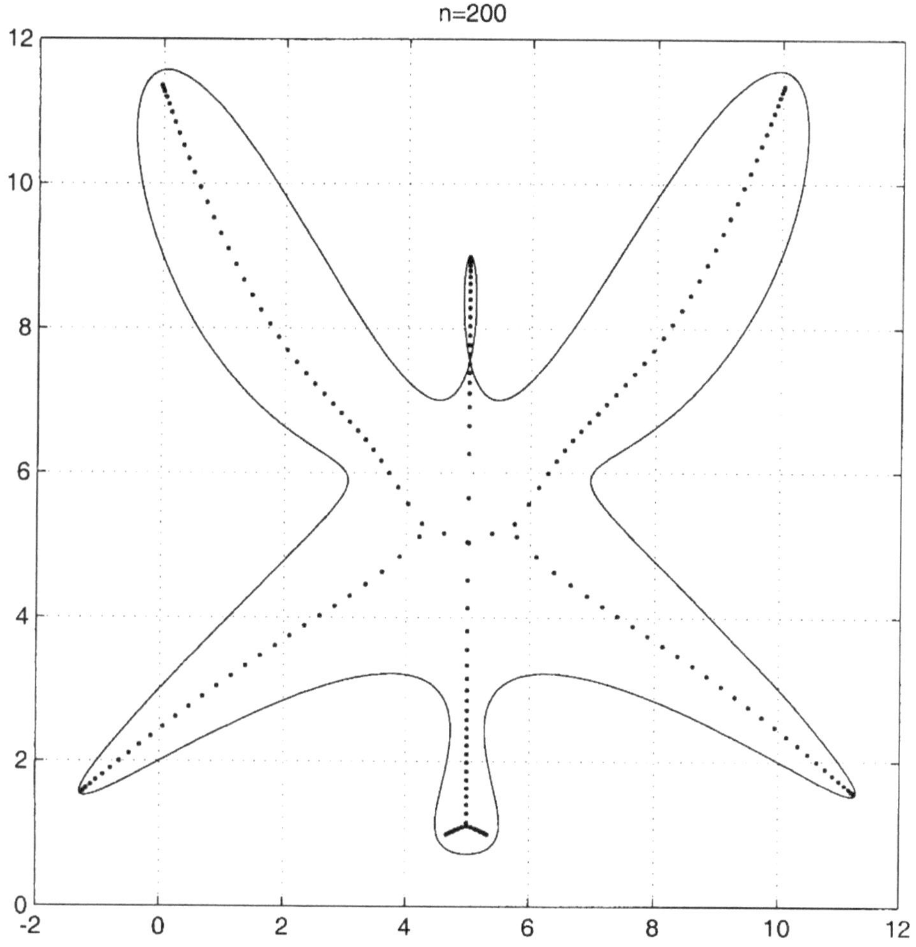

FIGURE 4. Bat

It is well known that the numerical procedure for calculating the eigenvalues of a finite Toeplitz matrix $T_n(a)$ of large size n is unstable. This means that, for large n, small errors in calculations lead to large errors in the values of the eigenvalues. The next picture is taken from the transparencies of Albrecht's lecture "Asymptotic linear algebra illustrated with Toeplitz matrices" delivered by him in 2004.

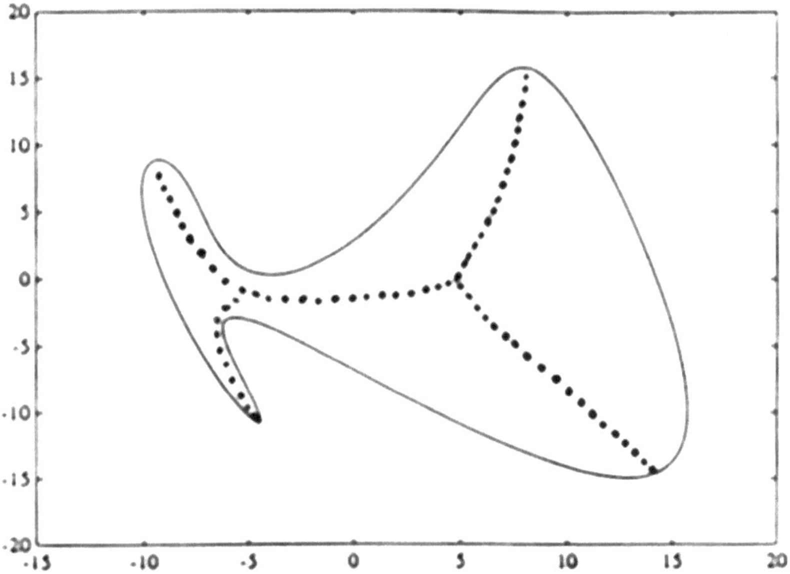

Theorem von Schmidt/Spitzer 1960
(streng bewiesen)

Computerexperiment

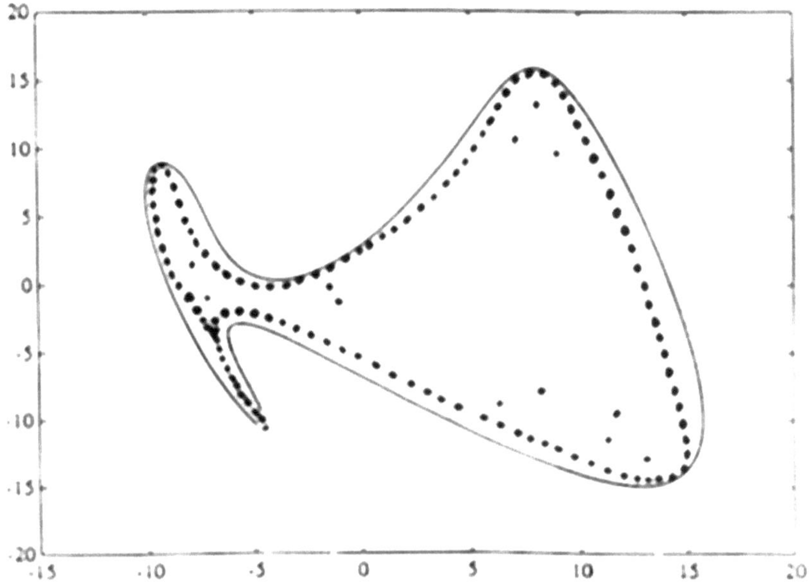

We can see from the above picture how the structure of the spectrum is being destroyed as the dimension of the matrices grows. Note that these calculations were done with "the machine precision". Due to the described effect, we have asked the question whether we can guarantee the stability of the spectra if we restrict ourselves to perturbations within the class of Toeplitz matrices only. A more precise formulation of the problem is as follows.

Let $\operatorname{sp} T_n(a)$ denote the spectrum (= set of the eigenvalues) of $T_n(a)$. The limiting set of the sequence $\{T_n(a)\}_{n=1}^{\infty}$ is the set $\Lambda(a)$ defined by

$$\Lambda(a) = \limsup_{n\to\infty} \operatorname{sp} T_n(a),$$

that is, a point λ belongs to $\Lambda(a)$ if and only if there are $n_1 < n_2 < n_3 < \dots$ and $\lambda_{n_k} \in \operatorname{sp} T_{n_k}(a)$ such that $\lambda_{n_k} \to \lambda$ as $k \to \infty$.

Problem 4.1 (Problem of stability). *Suppose $a \in C(\mathbb{T})$ and $a_n \in C(\mathbb{T})$ is a sequence such that $\|a_n - a\|_\infty \to 0$ as $n \to \infty$. Can one guarantee that*

$$\limsup_{n\to\infty} \Lambda(a_n) = \Lambda(a)\,?$$

If a and all a_n, $n \in \mathbb{N}$, are polynomials, then the answer is "yes" (see [16]). However, in the general case the answer is negative, as was shown in [12]. More precisely, the following result holds.

Theorem 4.2 (Böttcher, Grudsky). *Let*

$$a(t) = t^{-1}(33 - (t + t^2)(1 - t^2)^{3/4}), \quad t \in \mathbb{T},$$

and $a_n = P_n a$ be the nth partial sum of the Fourier series of a given by

$$a_n(t) = t^{-1}\left(33 - (t+t^2)\sum_{j=0}^{n}\binom{3/4}{j}t^{2j}\right), \quad t \in \mathbb{T}.$$

Then $\|a_n - a\|_\infty \to 0$ as $n \to \infty$, but

$$\limsup_{n\to\infty} \Lambda(a_n) \neq \Lambda(a).$$

Note that the function a in the above theorem is piecewise C^∞ but not C^∞. Moreover, $\Lambda(a)$ coincides with the range $a(\mathbb{T})$ of a. Note that $0 \notin \Lambda(a)$. On the other hand, $0 \in \Lambda(a_n)$ for all $n \geq 2$. The pictures below are taken from [12]. The two pictures of Figure 1 show the range of $a(\mathbb{T})$ and the eigenvalues of $T_m(a)$ for $m = 128$ (a) and for $m = 512$ (b). In Figure 2, we plot the eigenvalues of $T_{128}(P_n a)$ for $n = 4, 6, 8, 12$. These eigenvalue distributions mimic the set $\Lambda(P_n a)$ sufficiently well, and it is clearly seen that the limit $\Lambda(P_n a)$ as $n \to \infty$ has nothing in common with $\Lambda(a)$.

The ideas used in [12] became the basis of the cycle of works [1–3] written jointly with Egor Maximenko and my student Manuel Bogoya. These papers are dedicated to the study of the asymptotic behavior of the eigenvalues and the eigenvectors of a Toeplitz matrix in the case where its symbol has a power singularity

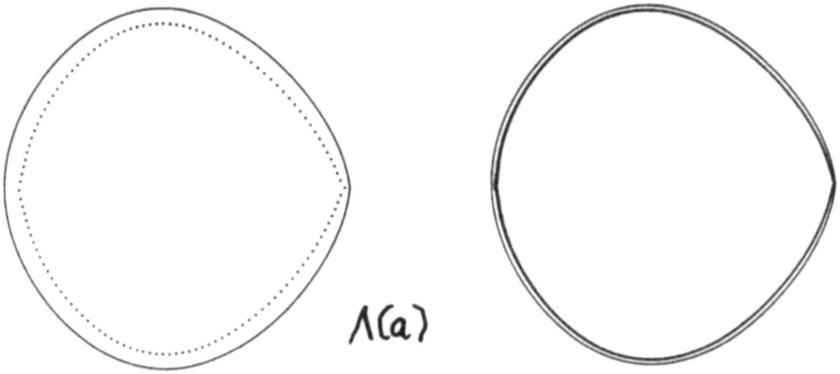

Figure 1. The range $a(\mathbf{T})$ and the eigenvalues of $T_{128}(a)$ (left) and $T_{512}(a)$ (right).

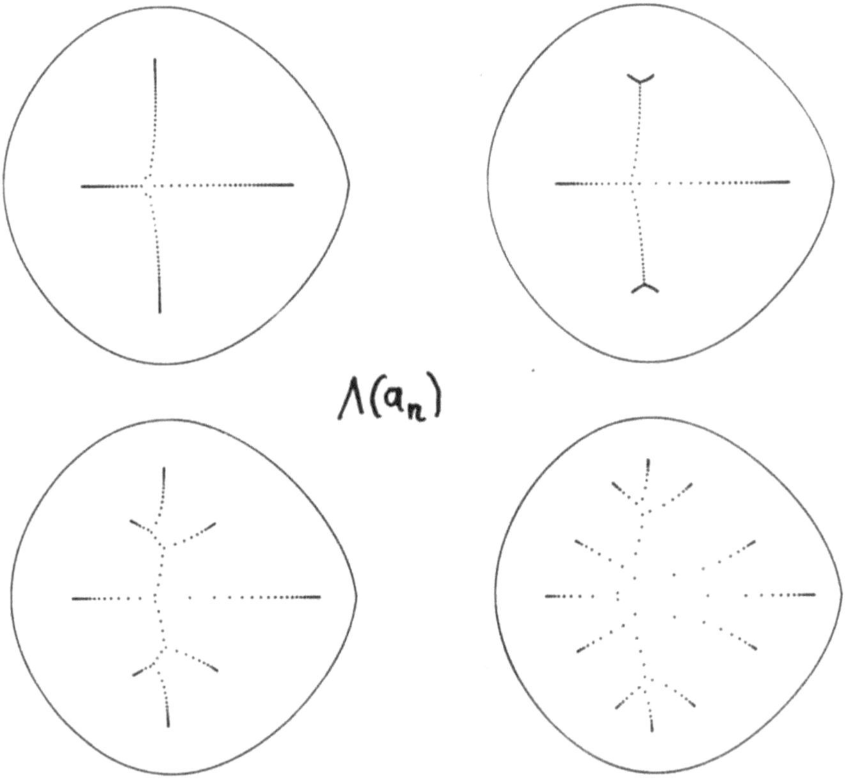

Figure 2. The range $a(\mathbf{T})$ and the eigenvalues of $T_{128}(P_n a)$ for $n = 4, 6, 8, 12$.

or several singularities of power type. Note also that in these works we were able to prove and refine a hypothesis of H. Dai, Z. Geary, and L.P. Kadanoff.

5. Uniform asymptotics of the eigenvalues of finite Toeplitz matrices

In 2008 Albrecht asked me a question about the asymptotics of the extreme eigenvalues of the finite Toeplitz matrices $T_n(a)$ as $n \to \infty$ in the case of some unbounded symbol a important in several applications. I want to remark that this subject is rather developed and well elaborated in the case of bounded symbols. However, the case of unbounded symbols had never been considered before. We could answer the posed question by using more or less known methods. The main credit here goes to our young coauthor Egor Maximenko. The obtained result was published in [20].

During the work on that paper, we realized that despite the abundance of works on the asymptotics of extreme eigenvalues (the smallest and the greatest), there were only very few results on the uniform asymptotics of all the eigenvalues, although the relevance of this problem is beyond any doubt. Reflections on this subject led us to the series of papers [4–6, 17–19], where we succeeded in developing new methods which allowed us to consider the problem for wide classes of symbols. The work in this direction continues, and I am sure that we will publish more than one interesting paper on this subject in the years to come.

6. Asymptotics of the eigenvalues of Wiener–Hopf integral operators on a finite interval

I spent my sabbatical in 2010 at the Chemnitz Technical University. During this visit, Albrecht drew my attention to the problem of the asymptotic behavior of the eigenvalues of Wiener–Hopf integral operators on a finite interval as the length of the interval tends to infinity. This problem arose in the 1960s in applications to the theory of lasers. It was very popular among the numerical analysis community. The very few papers with rigorously proven results on the topic include [9, 26, 27].

Albrecht suggested to employ the ideas and methods used by us in the study of the asymptotic behavior of the eigenvalues of Toeplitz matrices and his old result [7] on the determinants of Wiener–Hopf integral operators on a finite interval with a rational symbol. As the result, our joint work with Arieh Iserles [14] arose. In that paper we obtained and rigorously justified uniform asymptotic formulas for the eigenvalues of Wiener–Hopf operators on a finite interval as the length of the interval goes to infinity in the case of rational symbols.

In my opinion, this was an essential step towards the solution of problems arising in applications. Taking into account our latest progress in the study of asymptotics of eigenvalues of Toeplitz matrices, there is a real hope that we will be able to solve problems important in applications in the near future. Finally, let me mention [13, 15] as two of our papers in this direction.

7. Albrecht the traveler

It is impossible to speak about Albrecht without mentioning his wander-thirst. This passion is hardly inferior to his interest in mathematics. When he visits a new country, he rents a car and travels along a route carefully designed beforehand. Usually he drives 12–14 hours per day making short stops near tourist attractions, which are the purpose of the travel. Albrecht has already made dozens of such trips across Europe and America, Australia and India. Within a week after each trip, he prepares a beautifully decorated album containing many magnificent photos and other interesting details about the taken route. During my visits to Chemnitz, Albrecht always shows me new albums reporting on the tours that occurred after my previous visit. Every time I get a great pleasure!

The greatest impression on me and my wife was made by our joint tour around Mexico in 2003 during Albrecht's visit to CINVESTAV, where I had already worked. As usual, Albrecht rented a car and the four of us (Albrecht and his wife Sylvia, my wife Olya and I) went to a journey. I should say that by now I have lived and worked in Mexico for 14 years, but neither before nor after that trip I have seen so many attractions of this country. During five days we traveled along all South-West of Mexico including the famous resort of Cancun. Still today my Mexican friends think that this is absolutely impossible. While I and Olga were tired by the end of each day, it seemed that Albrecht was driving tirelessly. One late evening, sitting at dinner in a small hotel on the shore of the Gulf of Mexico, our leader was preparing the plan for tomorrow: "If we start at 7 a.m., then we can visit this and that ...". Olya, tired to death, joked: "Should we leave at 6 a.m.?" After few seconds' thought Albrecht told: "Olya, you are absolutely right!". Early next morning, we were already traveling in the predawn darkness at a speed that was possible on the local roads.

8. Epilogue

In this essay I tried to tell about only a few mathematical subjects studied by Albrecht. Certainly, there are many more interesting stories which could be mentioned here. Some of them can be found in Albrecht's paper [8], which I could not resist to mention. Once again, it shows the literary gifts of the hero of the anniversary. In conclusion, I would like to wish my dear friend to keep the undying interest to mathematics and all of the same great energy! Finally, I wish both of us to have many interesting joint projects and to meet more often in spite of the fact that the distance between us is 9724 km!

Aknowledgement

I would like to thank Eugene Shargorodsky and Ilya Spitkovsky for the help in preparation of the English version of the text.

References

[1] J.M. Bogoya, A. Böttcher, and S.M. Grudsky, *Asymptotics of individual eigenvalues of a class of large Hessenberg Toeplitz matrices.* Operator Theory: Advances and Applications **220** (2012), 77–95.

[2] J.M. Bogoya, A. Böttcher, S.M. Grudsky, and E.A. Maksimenko, *Eigenvalues of Hessenberg Toeplitz matrices generated by symbols with several singularities.* Commun. Math. Anal., Conference **03** (2011), 23–41.

[3] J.M. Bogoya, A. Böttcher, and S.M. Grudsky, and E.A. Maksimenko, *Eigenvectors of Hessenberg Toeplitz matrices and a problem by Dai, Geary, and Kadanoff.* Linear Algebra Appl. **436** (2012), 3480–3492.

[4] J.M. Bogoya, A. Böttcher, S.M. Grudsky, and E.A. Maksimenko, *Eigenvalues of Hermitian Toeplitz matrices with smooth simple-loop symbols.* J. Math. Anal. Appl. **422** (2015), 1308–1334.

[5] J.M. Bogoya, A. Böttcher, S.M. Grudsky, and E.A. Maksimenko, *Maximum norm versions of the Szegő and Avram–Parter theorems for Toeplitz matrices.* J. Approx. Theory **196** (2015), 79–100.

[6] J.M. Bogoya, A. Böttcher, S.M. Grudsky, and E.A. Maksimenko, *Eigenvectors of Hermitian Toeplitz matrices with smooth simple-loop symbols.* Linear Algebra Appl. **493** (2016), 606–637.

[7] A. Böttcher, *Wiener–Hopf determinants with rational symbols.* Math. Nachr. **144** (1989), 39–64.

[8] A. Böttcher, *The part of my path I walked together with Sergei Grudsky.* Boletín de la Sociedad Matemática Mexicana **22** (2016), 309–327.

[9] A. Böttcher, H. Brunner, A. Iserles, and S.P. Norsett, *On the singular values and eigenvalues of the Fox–Li and related operators.* New York J. Math. **16** (2010), 539–561.

[10] A. Böttcher and S. Grudsky, *Toeplitz operators with discontinuous symbols: phenomena beyond piecewise discontinuity.* Operator Theory: Advances and Applications **90** (1996), 55–118.

[11] A. Böttcher and S. Grudsky, *On the composition of Muckenhoupt weights and inner functions.* J. London Math. Soc. **58** (1998), 172–184.

[12] A. Böttcher and S. Grudsky, *Asymptotic spectra of dense Toeplitz matrices are unstable.* Numer. Algorithms **33** (2003), 105–112.

[13] A. Böttcher, S. Grudsky, D. Huybrechs, and A. Iserles, *First-order trace formulae for the iterates of the Fox–Li operator.* Operator Theory: Advances and Applications **218** (2012), 207–224.

[14] A. Böttcher, S. Grudsky, and A. Iserles, *Spectral theory of large Wiener–Hopf operators with complex-symmetric kernels and rational symbols.* Math. Proc. Cambridge Philos. Soc. **151** (2011), 161–191.

[15] A. Böttcher, S. Grudsky, and A. Iserles, *The Fox–Li operator as a test and a spur for Wiener–Hopf theory.* In: Essays in Mathematics and Its Applications, in Honor of Stephen Smale's 80th Birthday, Springer, Berlin (2012), 37–48.

[16] A. Böttcher, S. Grudsky and A. Kozak, *On the distance of a large Toeplitz band matrix to the nearest singular matrix.* Operator Theory: Advances and Applications **135** (2002), 101–106.

[17] A. Böttcher, S.M. Grudsky, and E.A. Maksimenko, *On the asymptotics of all eigenvalues of Hermitian Toeplitz band matrices.* Dokl. Math. **80** (2009), 662–664.

[18] A. Böttcher, S.M. Grudsky, and E.A. Maksimenko, *Inside the eigenvalues of certian of Hermitian Toeplitz band matrices.* J. Comput. Appl. Math. **233** (2010), 2245–2264.

[19] A. Böttcher, S.M. Grudsky, and E.A. Maksimenko, *On the structure of the eigenvectors of large Hermitian Toeplitz band matrices.* Operator Theory: Advances and Applications **210** (2010), 15–36.

[20] A. Böttcher, S.M. Grudsky, E.A. Maksimenko, and J. Unterberger, *The first-order asymptotics of the extreme eigenvectors of certain Hermitian Toeplitz matrices.* Integr. Equ. Oper. Theory **63** (2009), 165–180.

[21] A. Böttcher, S.M. Grudsky, and E. Ramírez de Arellano, *Algebras of Toeplitz operators with oscillating symbols.* Rev. Mat. Iberoam. **20** (2004), 647–671.

[22] A. Böttcher, S. Grudsky, and I. Spitkovsky, *Toeplitz operators with frequency modulated semi-almost periodic symbols.* J. Fourier Anal. Appl. **7** (2001), 523–535.

[23] A. Böttcher, S. Grudsky, and I. Spitkovsky, *Block Toeplitz operators with frequency-modulated semi-almost periodic symbols.* Int. J. Math. Math. Sci. **34** (2003), 2157–2176.

[24] S. Grudsky, *Toeplitz operators and the modelling of oscillating discontinuities with the help of Blaschke products.* Operator Theory: Advances and Applications **121** (2001), 162–193.

[25] S. Grudsky and E. Shargorodsky, *Spectra of Toeplitz operators and compositions of Muckenhoupt weights with Blaschke products.* Integr. Equ. Oper. Theory **61** (2008), 63–75.

[26] H.J. Landau, *The notion of approximate eigenvalues applied to an integral equation of laser theory.* Quart. Appl. Math. **35** (1977/78), 165–172.

[27] H.J. Landau and H. Widom, *Eigenvalue distribution of time and frequency limiting.* J. Math. Anal. Appl. **77** (1980), 469–481.

[28] P. Schmidt and F. Spitzer, *The Toeplitz matrices of an arbitrary Laurent polynomial.* Math. Scand. **8** (1960), 15–38.

Sergei Grudsky
CINVESTAV
Departamento de Matemáticas
07360 Ciudad de México, Mexico
e-mail: `grudsky@math.cinvestav.mx`

Operator Theory:
Advances and Applications, Vol. 259, 19–20

Salutatory with Regards from the Mathematics Students of Chemnitz

Jonas Jahns

Dear Professor Böttcher, dear math enthusiasts.

It is a great pleasure for me – and an honor at the same time – that I'm able to express some words of gratitude and acknowledgment from the students' point of view.

When someone decides for mathematics studies, this is usually because one enjoyed the subject in school. Often, this was linked already with certain type of success. Nevertheless, several people soon will discover that the material presented at a university is pretty much different from the things we were taught in school. This may be a reason for the quite high dropout rates in such programs all over the world.

Even though these facts were known to me, I overcame my fears and started studying math together with two dozens kindred spirits here in Chemnitz back in October of 2012. The beginnings were hard, and the result of the first examination was (at least to me) shattering: failed! Should I resign that early and quit like a couple of my fellow students? Nonetheless, for myself, I liked mathematics. It wouldn't let me go anymore.

The main reason for the captivation was our analysis lecturer, Professor Böttcher. His passion for this subject was heartening and catching. Adjusting the talks to the level of knowledge in front of him, he can explain even very complex mathematical issues. Students of any year may confirm this. Moreover, he understands to draw interesting connections to other fields. Also, the given examples and applications were very motivating.

After the first exam (my disastrous one), he encouraged me face-to-face to continue the studies. Professor Böttcher knows his students and is hence able to respond to their needs individually. That's not a matter of course, but exceptionally of help.

After a year, when the analysis lectures had come to an end, Professor Böttcher thanked *us*, the students, for our attention, for our participation, and for our serious efforts in comprehending the definitions and theorems he presented. To him, it was a pleasure to teach us.

Now, the student representatives selected me to tell you that all of us greatly enjoined attending your lectures, too. We wish for you staying so active and dedicated, caring for us and the future students of mathematics in Chemnitz. Your way of teaching is vital for us novices, as it improves the power of endurance that is necessary to successfully get through the challenging program.

The German student bodies in mathematics (Konferenz der deutschsprachigen Mathematikfachschaften) have designed a card game depicting famous mathematicians. For your special day, we created a personalized edition with a very special joker, which we'd like to give you as a present and small piece for remembering. All the best, good health, and still fun with math itself and bringing others close to it. Best wishes for a joyful life from all the students.

Fachschaftsrat Mathematik
Technische Universität Chemnitz
Reichenhainer Str. 41, Zi. 001
D-09126 Chemnitz, Germany
e-mail: `fsrmathe@tu-chemnitz.de`

Operator Theory:
Advances and Applications, Vol. 259, 21–36

Essay on Albrecht Böttcher

Bernd Silbermann

About 40 years ago Albrecht Böttcher stepped into my life as a real person. It was when he began his study in mathematics at the TU Karl-Marx-Stadt (now TU Chemnitz) in the year 1975. My curricular duty was to give the analysis course, and so I met him for the first time at the lectures. But, as a matter of fact, I had already heard about him and his close friend Elias Wegert before. Both finished the so-called "Spezialklasse", which was something like a gymnasium attached to the Department of Mathematics of the university. It was especially meant for gifted young people in order to prepare them for a study in STEM directions (STEM stands here for Science, Technology, Engineering, Mathematics). Albrecht Böttcher, as well as Elias Wegert, turned out to be outstanding pupils. Let me just mention the fact that they both were prize-winners at the International Mathematical Olympiad which took place 1973 in Moscow. The gap between 1973 and 1975 was filled by military service.

Here is the right place to write a few words about Siegfried Prößdorf. He was a PhD student of the famous Soviet mathematician Solomon Grigor'evich Mikhlin, the inventor of the symbol concept for singular operators. By the way, S.G. Mikhlin was awarded Laurea honoris causa by the Karl-Marx-Stadt Politechnic (now TU Chemnitz) in 1968 and was elected into Leopoldina in 1970 and into the Accademia Nazionale dei Lincei in 1981.

Prößdorf came to Chemnitz in 1966. His mathematical interests were mainly aimed at the study of degenerate one-dimensional singular integral operators. At the time, this topic attracted much attention and Solomon Mikhlin suggested it to Siegfried Prößdorf as a possible topic for a PhD thesis. Let me also mention that Mikhlin introduced Prößdorf to Israel Gohberg, another great player in the theory of singular integral operators, especially in the study of Banach algebras generated by such operators. Both Solomon Mikhlin and Israel Gohberg visited the department in Chemnitz several times.

In the simplest case, a degenerate one-dimensional singular integral operator is an operator of the following kind. Let $\mathbb{T}$ denote the unit circle in the complex plane, oriented counterclockwise as usual. The Cauchy singular integral operator is defined on $\mathbb{T}$ by

$$(S\varphi)(t) = \frac{1}{\pi i} \int_{\mathbb{T}} \frac{\varphi(\tau)}{\tau - t} d\tau$$

with the integral taken in the sense of Cauchy's principal value. This operator is a bounded linear operator on the spaces $L^p(\mathbb{T})$, $1 < p < \infty$, and $S^2 = I$. The operator $P = (I + S)/2$ is a projection onto the Hardy space H^p,

$$H^p = \{f \in L^p(\mathbb{T}) \mid \widehat{f}_n = 0,\, n < 0\},$$

where $\{\widehat{f}_n\}_{n\in\mathbb{Z}}$ is the sequence of the Fourier coefficients of f. Now, in the simplest case, a degenerate singular integral operator is an operator of the form

$$A = aI + bS,$$

where a and b are continuous complex-valued functions on $\mathbb{T}$ and the function

$$a^2 - b^2,$$

called the symbol of A, does have zeros on $\mathbb{T}$.

The important point here is that the familiar Fredholm theory is not in force for such operators. With that in mind, the task is to study these operators, which means in particular that one tries to find Banach spaces B of functions such that $B \cap \operatorname{Im} A$ has finite codimension in B. Related problems can be studied also for Toeplitz or Wiener–Hopf operators. Toeplitz operators are close to singular integral operators: for $a \in L^\infty(\mathbb{T})$, the Toeplitz operator $T(a)$ is defined by

$$T(a) : H^p \to H^p, \quad h \mapsto P(ah).$$

The function a is called the generating function, and the matrix representation of $T(a)$ with respect to the basis $\{t^n\}_{n\in\mathbb{Z}_+} \subset H^p$ has the form

$$(\widehat{a}_{i-j})_{i,j=0}^{\infty}, \tag{1}$$

where $\{\widehat{a}_k\}_{k\in\mathbb{Z}}$ is the sequence of the Fourier coefficients of a. The matrix (1) is a Toeplitz matrix: its entries depend only on the difference of the indices.

I joined Siegfried Prößdorf in 1967 and we both worked until 1973/74 on problems of above type. About 1973 Siegfried became interested in numerical analysis. Here the personal influence of Israel Gohberg played a strong role. In this context, it should be mentioned that in 1971 and 1972, I. Gohberg and V.I. Levchenko published two papers concerning the projection method for degenerate Toeplitz operators acting on the space c_0 of all sequences tending to zero. The case of degenerate systems of Toeplitz operators was settled in 1975 by I. Gohberg and S. Prößdorf. So, it is not surprising that about 1973 Siegfried Prößdorf and I started to work on related problems.

But Siegfried was not happy with the situation as it was at the time at the Department of Mathematics in Karl-Marx-Stadt. Continually there were frictions with leading people from the department, the real reasons of which I did not know. In 1974, Siegfried accepted an invitation from the Weierstrass Institute of the Academy of Science to move to Berlin, and he left Karl-Marx-Stadt in 1975. I decided to stay in Karl-Marx-Stadt, and it was completely clear to me that, as a consequence, I had to choose my own way. Quite by chance I came across a paper by Harold Widom entitled

> Toeplitz determinants with singular generating functions. *Amer. J. Math.* **95** (1973), 333–383.

In this paper, Widom proved an important particular case of a conjecture which had been formulated in 1968 by Fisher and Hartwig. Despite the fact that Widom's paper and the research carried out with Prößdorf turned out to be of a very different nature, there was one common point, namely the presence of singular generating functions. Thus I became interested in Toeplitz determinants.

Toeplitz determinants are defined as follows. Let $\{\widehat{a}_n\}_{n\in\mathbb{Z}}$ be the sequence of the Fourier coefficients of an L^1-function a defined on $\mathbb{T}$ and consider the sequence of finite Toeplitz matrices

$$T_n(a) = (\widehat{a}_{j-k})_{j,k=0}^{n}$$

as well as the sequence $D_n(a)$ of the determinants of the matrices $T_n(a)$. About 100 years ago, G. Szegő proved that if $a \in L^1(\mathbb{T})$, $a \geq 0$, and $\log a \in L^1(\mathbb{T})$, then

$$\lim_{n\to\infty} \frac{D_n(a)}{D_{n-1}(a)} = \exp(\widehat{\log a})_0.$$

This result is important in the study of orthogonal polynomials on the unit circle, and it was refined by G. Szegő in 1952. He then showed that, for a positive function $a \in C^{1+\varepsilon}(\mathbb{T})$, where $\varepsilon > 0$, we have

$$D_n(a) \sim G(a)^{n+1} E(a),$$

where $G(a) = \exp(\widehat{\log a})_0$ and $E(a) = \exp \sum_{k=1}^{\infty} k(\widehat{\log a})_k(\widehat{\log a})_{-k}$.

Interestingly enough, this result was initiated by studies concerning the two-dimensional Ising model of ferromagnetism (see, for example, the discussion in Section 5.2 of the book *Introduction to large truncated Toeplitz matrices,* Universitext, Springer-Verlag, New York (1999) by Albrecht and myself). In fact it allowed to simplify the proof of Onsager's formula below the Curie point. The description of the spontaneous magnetization at the Curie point and above the Curie point cannot be carried out with the help of Szegő's theorem. This flaw provided a source of motivation for Fisher and Hartwig to raise the following conjecture (nowadays called the Fisher–Hartwig conjecture). Let a be an $L^1(\mathbb{T})$ function having a representation

$$a(t) = b(t) \prod_{j=1}^{m} |t - t_j|^{2\alpha_j} \varphi_{\beta_j,t_j}(t), \quad t \in \mathbb{T},$$

such that

1. b is a smooth function defined on $\mathbb{T}$ and $b(t) \neq 0$ for all $t \in \mathbb{T}$,
2. $t_1, \ldots, t_m$ are distinct points on $\mathbb{T}$,
3. $\alpha_j \in \mathbb{C}$ and $\operatorname{Re}\alpha_j > -\frac{1}{2}$ for $j = 1, \ldots, m$,

4. $\beta_j \in \mathbb{C}$ for $j = 1, \ldots, m$, and φ_{β_j,t_j} are certain "model" functions defined on $\mathbb{T}$ which are continuous on $\mathbb{T} \setminus \{t_j\}$, have a jump at the point t_j satisfying
$$\frac{\varphi_{\beta_j,t_j}(t_j - 0)}{\varphi_{\beta_j,t_j}(t_j + 0)} = \exp(2\pi i \beta_j), \quad \beta_j \neq 0,$$
and, in addition, do not vanish on $\mathbb{T}$.

If b has winding number zero about the origin and $\alpha_j \pm \beta_j \notin \mathbb{Z}$, the asymptotic behavior of $D_n(a)$ is conjectured to be
$$D_n(a) \sim G(b)^{n+1} n^q \tilde{E}(b; t_1, \ldots, t_m; \alpha_1, \ldots, \alpha_m; \beta_1, \ldots, \beta_m), \tag{2}$$
where $G(b) = \exp(\widehat{\log b})_0$, $\tilde{E}(\ldots)$ is some unknown constant which has to be found, and $q = (\alpha_1^2 - \beta_1^2) + \cdots + (\alpha_m^2 - \beta_m^2)$. The notation $a_n \sim b_n$ means $a_n/b_n \to 1$.

As already mentioned, Albrecht Böttcher started his studies in mathematics in 1975. He turned out to be an excellent student. I remember that I put to him the following question in the analysis exam after the first year: find a function defined on the real line which is continuous at exactly 7 points and differentiable at exactly 3 points. To my surprise, he came up with an example after only a few minutes of thinking. I understood that I had to bring him into research work and so I proposed to work together on a proof for the Fisher–Hartwig conjecture. Albrecht was at that time a student in the second year. Despite the fact that I already had some experience in that field, it was clear that this project was a great venture, not only for Albrecht but also for me. But, it marked the beginning of an unbelievable time of joint research, a time full of emotions and happy moments.

In the early 1980s, we were able to prove the Fisher–Hartwig conjecture under the assumption that
$$|\operatorname{Re}\alpha_j| < \frac{1}{2}, \qquad |\operatorname{Re}\beta_j| < \frac{1}{2}.$$
The asymptotic formula then reads as follows:
$$D_n(a) \sim G(b)^{n+1} n^q E(b) \prod_{j=1}^{m} b_-(t_j)^{-(\alpha_j+\beta_j)} b_+(t_j)^{-(\alpha_j-\beta_j)}$$
$$\times \prod_{j=1}^{m} \frac{\mathsf{G}(1+\alpha_j+\beta_j)\mathsf{G}(1+\alpha_j-\beta_j)}{\mathsf{G}(1+2\alpha_j)} \prod_{1\le i\ne j\le m} \left(1 - \frac{t_i}{t_j}\right)^{-(\alpha_j+\beta_j)(\alpha_i-\beta_i)}.$$
Here G is the so-called Barnes function, and b_-, b_+ come from the Wiener–Hopf factorization $b = G(b)b_-b_+$. This formula reveals that the asymptotics for $D_n(a)$ consists of the product of the asymptotics of the single factors on the one hand and of factors which reflect the interaction of the asymptotics of the single factors on the other. So we recognize some kind of interference. Particular cases were known before. Let me list only a few of them.

1. The case $\alpha_j = \beta_j = 0$ can be traced back to G. Szegő. Much work was done to refine Szegő's original result, that is, to allow also complex-valued generating functions and to relax the smoothness conditions. The impact of Harold Widom's work in that direction should be especially mentioned.

2. $\beta_j = 0$, $j = 1, \ldots, m$: H. Widom (1973). It is the paper I cited above.
3. $\operatorname{Re}\beta_j = 0$, $j = 1, \ldots, m$: E. Basor (1978). It might seem that this result is only a mild generalization of Widom's result. This, however, is by no means the case. If $\beta_j \neq 0$, serious new difficulties arise.
4. $\alpha_j = 0$, $j = 1, \ldots, m$: In this case, the conjecture was proved thrice and independently by E. Basor (1979), A. Böttcher (1982), and P. Blekher (1982).

Without the assumption $|\operatorname{Re}\alpha_j| < \frac{1}{2}$, $|\operatorname{Re}\beta_j| < \frac{1}{2}$, the asymptotics (2) for $D_n(a)$ fails in general. For instance, if $\alpha_j \pm \beta_j \in \mathbb{Z}$, $j = 1, \ldots, m$, then $D_n(a)$ features a completely different asymptotics. The pertinent asymptotic behavior can be found in Section 10.51 of our book *Analysis of Toeplitz operators.* Here is, just to give an impression of what then happens, an extract from that section.

> *Put*
>
> $$\gamma = \sum_{j=1}^m \gamma_j, \quad \delta = \sum_{j=1}^m \delta_j, \quad \mu_j = \gamma_j + \delta_j, \quad \mu = \max\{\mu_1, \ldots, \mu_m\},$$
>
> *and let $b \in W \cap F\ell^p_\lambda$, where $\lambda > \mu$ and p is an arbitrary number satisfying $1 < p < \infty$. Without loss of generality assume $\gamma \leq \delta$ (otherwise consider $D_n(\tilde{a}) = D_n(a)$). Define*
>
> $$\mathfrak{R} = \{s = (s_1, \ldots, s_m) \in \mathbb{Z}^m : 0 \leq s_j \leq \mu_j, s_1 + \ldots + s_m = \gamma\},$$
>
> $$q = \max_{s \in \mathfrak{R}} \sum_{j=1}^m (s_j\mu_j - s_j^2), \quad \mathfrak{R}^* = \left\{ s \in \mathfrak{R} : \sum_{j=1}^m (s_j\mu_j - s_j^2) = q \right\},$$
>
> *and put $b_- = \exp(Q \log b)$, $b_+ = \exp(P \log b)/\exp(\log b)_0$. Then*
>
> $$D_n(a) = G(b)^{n+1} G\left(\frac{b_-}{b_+}\right) E(b) n^q \left\{ 1 + o\left(\frac{1}{n^{\lambda-\mu-\varepsilon}}\right)\right\}$$
> $$\times \left\{ \sum_{s \in \mathfrak{R}^*} A_s \left(\frac{t_1^{s_1} \ldots t_m^{s_m}}{t_1^{\gamma_1} \ldots t_m^{\gamma_m}}\right)^{n+1} + O\left(\frac{1}{n^{\min\{1, \gamma-\mu-\varepsilon\}}}\right)\right\},$$
>
> *where $A_s = A_{s_1,\ldots,s_m}$ is equal to*
>
> $$\prod_{j \neq k} \left(1 - \frac{t_j}{t_k}\right)^{(s_j-\mu_j)s_k} \prod_{j=1}^m \mathrm{G}_{\mu_j - s_j, s_j} \prod_{j=1}^m b_-(t_j)^{-s_j} b_+(t_j)^{s_j-\mu_j},$$
>
> *with*
>
> $$\mathrm{G}_{\mu_j - s_j, s_j} := \frac{\mathrm{G}(1 + \mu_j - s_j)\mathrm{G}(1 + s_j)}{\mathrm{G}(1 + \mu_j)}$$
>
> *and*
>
> $$\mathrm{G}(k) := (k-2)! \ldots 2!1!0!$$
>
> *being the value of the Barnes* G*-function (see* 10.58 *below) at k.*

Example: $D_n\left[(1+t)\left(1 - \frac{1}{t}\right)\right] = \frac{1}{2} + \frac{(-1)^{n+1}}{2}.$

After the middle of the 1980s, the development in the Fisher–Hartwig business paused about 10 years. In 1997, Torsten Ehrhardt proved the Fisher–Hartwig conjecture in all cases where it turns out to be in force. For more on the entire story, from the very beginnings up to the present, I refer to the excellent recent article

P. Deift, A. Its, I. Krasovsky, Toeplitz determinants under the impetus of the Ising model: some history and some recent results. *Comm. Pure Appl. Math.* **66** (2013), no. 9, 1360–1438.

Our proof of the Fisher–Hartwig conjecture was based in particular on the asymptotic invertibility of the finite sections of some operators belonging to the Toeplitz algebra generated by Toeplitz operators with piecewise continuous generating functions (case $p = 2$). This result has a rich and long history in which Banach algebra techniques played a crucial role. It often happens that while solving hard problems, the developed methods open new horizons in other fields. For instance, Banach algebra methods were used by my colleagues Roland Hagen and Steffen Roch to study spline approximation methods for singular integral operators, and by Peter Junghanns to study collocation methods for singular integral operators on intervals. The impact of Siegfried Prößdorf and Andreas Rathsfeld in these developments should be mentioned also.

Let me list a few papers which were written by Albrecht Böttcher and in parts by me in the period of 1979–1985.

1. A. Böttcher, Toeplitzdeterminanten mit singulärer Erzeugerfunktion. *Wiss. Informationen* **13**, TH Karl-Marx-Stadt, 1979.
2. A. Böttcher and B. Silbermann, The asymptotic behavior of Toeplitz determinants for generating functions with zeros of integral orders. *Math. Nachr.* **102** (1981), 79–105.
3. A. Böttcher, Toeplitz determinants with piecewise continuous generating function. *Z. Anal. Anwendungen* **1** (1982), no. 2, 23–39.
4. A. Böttcher and B. Silbermann, *Invertibility and Asymptotics of Toeplitz Matrices.* Mathematical Research, Vol. 17, Akademie-Verlag, Berlin, 1983.
5. A. Böttcher, *The Finite Section Method for the Wiener–Hopf Integral Operator.* Cand. Dissertation, Rostov-on-Don State University, 1984 (Russian).
6. A. Böttcher and B. Silbermann, Toeplitz matrices and determinants with Fisher–Hartwig symbols. *J. Funct. Anal.* **63** (1985), no. 2, 178–214.

Among these publications is Albrecht's PhD thesis, which he defended at the Rostov-on-Don State University in the year 1984. This thesis is written in Russian and the story behind is as follows. Like in West Germany, it was also considered desirable in East Germany that young scientists spend some time abroad. For people from GDR that meant to go to some university in the eastern part of Europe. In that time, that is, about 1980, we had good contacts with Igor Borisovich Simonenko and his group at the Rostov-on-Don University. So it was decided to send Albrecht to I. Simonenko. The actual supervision was in the hands of V. Dybin.

Since Albrecht and his future wife Sylvia were expecting a baby, he stayed in Rostov-on-Don one year but then decided to found a family and to live together with his wife and daughter. Thus, as it has been the case through all history, love

FIGURE 1. Albrecht Böttcher and me, Black Forest (Oberwolfach), 1989

was again a strong reason to change decisions. So he carried out the rest of his PhD work more or less completely in Chemnitz, but he regularly traveled to Rostov-on-Don to work with the mathematicians there, to pass all the PhD exams, and finally to defend his thesis. His thesis is written in Russian, and it is not only a remarkable mathematical work but also brought to light that Albrecht Böttcher possesses a great gift to learn and to use foreign languages.

FIGURE 2. Albrecht Böttcher in Stuttgart (visiting Wolfgang Wendland), 1989/90

Parallel to his thesis, he wrote the book No. 4 of the above list with me (and was engaged in proving Fisher–Hartwig). In the second half of the 1980s, we spent a lot of effort to write our book *Analysis of Toeplitz Operators*, which we finished about 1989 and which appeared 1990 in the Akademie-Verlag, Berlin and, thanks to the support of Israel Gohberg, in that year also in the Springer-Verlag.

The political developments around the former GDR made the year 1989 a special one. I remember with pleasure the 1989 Oberwolfach meeting "Wiener–Hopf-Probleme, Toeplitz-Operatoren und Anwendungen", which was organized by I. Gohberg, E. Meister, and M.A. Kaashoek. This workshop was a special one. Albrecht Böttcher met Israel Gohberg for the first time, and I saw him again the first time after his immigration to Israel. In addition, we met a lot of colleagues who for us on this occasion, finally, got a face. Erhard Meister sent Albrecht and me again an invitation after the opening of the boundary to West Germany. I always remember with gratitude Erhard Meister's help by the new start of our department and his friendship.

FIGURE 3. Albrecht Böttcher and me at birthday celebration of Erhard Meister in Darmstadt, 1989/90

In appreciation of his outstanding mathematical work and his high ability to teach, Albrecht Böttcher became "Hochschuldozent" in 1990, and in 1992 this position was smoothly changed into a C3-professorship. In the same year Albrecht Böttcher was awarded the *"Alfried Krupp-Förderpreis für junge Hochschullehrer"*. This prize is highly endowed and is meant for strong support of young and gifted researchers. The related deed was handed out to Albrecht during a ceremonial act held at Villa Hügel in Essen. The laudation was presented by F. Hirzebruch. Johannes Rau, in that time still the head of Nordrhein-Westfalen's government, also attended this event. I remember Albrecht Böttcher and Johannes Rau meeting for a conversation during a break, of course with cigarettes in their hands.

With the prize at his disposal, Albrecht Böttcher turned towards a new and highly nontrivial issue, namely the study of Banach algebras generated by singular integral operators with piecewise continuous coefficients. If these operators act on spaces $L^p(\Gamma, w)$, $1 < p < \infty$, where Γ is a Lyapunov curve and w a Khvedelidze weight, a complete Fredholm theory was available (H. Widom, I. Gohberg, and N. Krupnik). But for the case when w is a general Muckenhoupt weight and Γ a general Carleson curve, up to 1990 not much else was known than the boundedness of the associated singular integral operator S. In fact, the singular integral operator S was known to be bounded on the space $L^p(\Gamma, w)$ if and only if Γ is a Carleson curve and w is a Muckenhoupt weight (Calderón, David, Hunt, Muckenhoupt, Wheeden).

The prize enabled Albrecht Böttcher to invite Yuri Karlovich to take part in the planned research project. It turned out that in the investigation of the above-mentioned algebras essential spectra of Toeplitz operators serve as a crucial tool, especially in the situation where Γ is a curve in $\mathbb{C}$ which is simultaneously a Carleson and a Jordan curve. If, in addition, w is a Muckenhoupt weight, the singular integral operator on $L^p(\Gamma, w)$ satisfies the equality $S^2 = I$, and therefore the operator $P = (I + S)/2$ is a projection. Its image is denoted by $L^p_+(\Gamma, w)$; for $\Gamma = \mathbb{T}$ and $w = 1$ it is nothing else than the familiar Hardy space H^p. Here always, and as is customary, it is assumed that Γ is oriented counterclockwise.

The Toeplitz operator $T(a)$ with piecewise continuous generating function a, that is $a \in PC(\Gamma)$, is defined by

$$\varphi \mapsto P(a\varphi), \quad \varphi \in L^p_+(\Gamma, w).$$

It is instructive to look at the following examples of essential spectra of Toeplitz operators which were obtained up to 1990.

- $a \in C(\mathbb{T})$, $T(a)$ acting on H^p, $1 < p < \infty$. Then the essential spectrum $\mathrm{sp}_{\mathrm{ess}}\, T(a)$ coincides with the range $\mathcal{R}(a)$ of the function a.
- $a \in PC(\mathbb{T})$, $T(a)$ acting on H^p, $1 < p < \infty$. Then

$$\mathrm{sp}_{\mathrm{ess}}\, T(a) = \mathcal{R}(a) \cup \left(\bigcup_{t \in \mathbb{T}} \mathcal{A}_p(t) \right),$$

where

$$\mathcal{A}_p(t) = \begin{cases} a(t) & \text{if } a \text{ is continuous at } t, \\ \text{is an arc (depending on } p\text{) which joins the point} \\ a(t-0) \text{ with } a(t+0) \text{ if } a \text{ has a jump at } t. \end{cases}$$

If $p = 2$, the arc degenerates to a line segment. Such results are also valid for spaces $L^p(\Gamma, w)$ in case Γ is smooth and w is a Khvedelidze weight, i.e., if w has the form

$$w(t) = \prod_{j=1}^{m} |t - t_j|^{\mu_i}, \quad t, t_j \in \Gamma, \quad -\frac{1}{p} < \mu_j < \frac{1}{q}.$$

Here the pertinent arcs depend, in addition, on w. These results were mainly obtained by H. Widom, I. Gohberg, and N. Krupnik.

- In 1990, Ilya Spitkovsky proved the remarkable result that, in the case of a Muckenhoupt weight being "essentially" different from a Khvedelidze weight, $\mathcal{A}_p(t) = \mathcal{A}_p(t, w)$ has to be replaced by a "horn":

Incidentally, the name "horn" was coined by Albrecht. Naturally, Spitkovsky's result led to the question on what happens if, in addition, the "smooth" curves are replaced by general Carleson curves. Here are two examples of Carleson curves taken from page 12 of the book A. Böttcher, Yu.I. Karlovich, *Carleson Curves, Muckenhoupt Weights, and Toeplitz operators*, Birkhäuser, Basel, 1997:

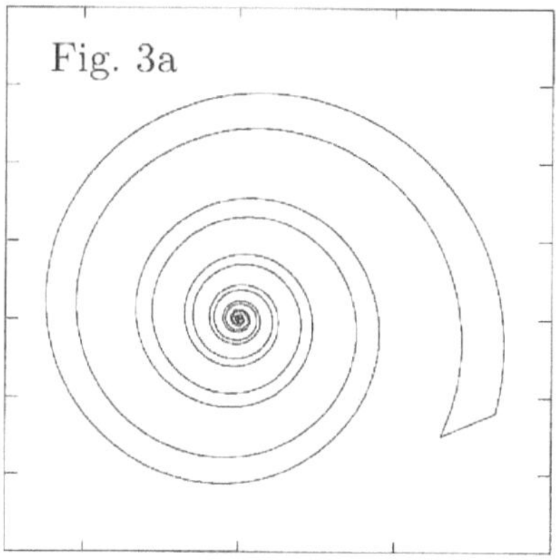

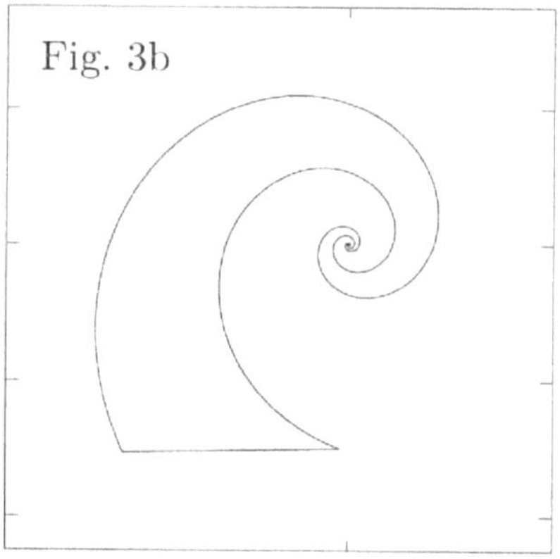

As already mentioned, Albrecht Böttcher and Yuri Karlovich, a well-known mathematician from the school of G. Litvinchuk (Odessa), decided to tackle this problem. An undertaking which was, in my opinion, of what one might call of breakneck type. Nevertheless, they succeeded in completely describing the set

$$\mathcal{A}_p(t) = \mathcal{A}_p(t, \Gamma, w).$$

They showed that it depends not only on the local properties of the curve Γ and the weight w at the point t, but also on the "interference" between Γ and w at t. Here are a few pictures of sets $\mathcal{A}_p(t, \Gamma, w)$ taken from pages 250–251 of the book by Albrecht Böttcher and Yuri Karlovich, *Carleson Curves, Muckenhoupt Weights, and Toeplitz Operators*, Birkhäuser, Basel, 1997.

Fig. 9b

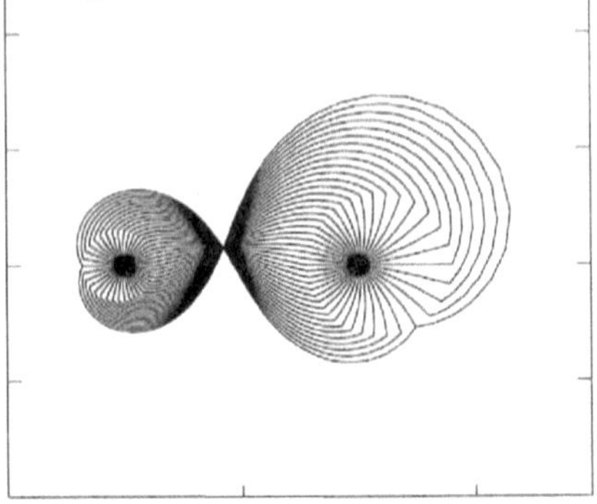

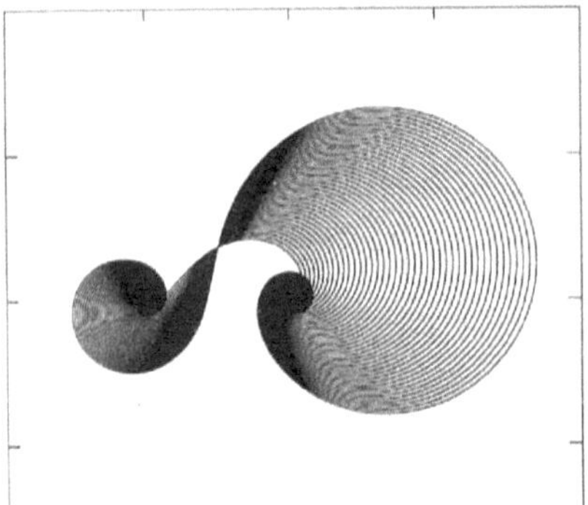

Fig. 10a

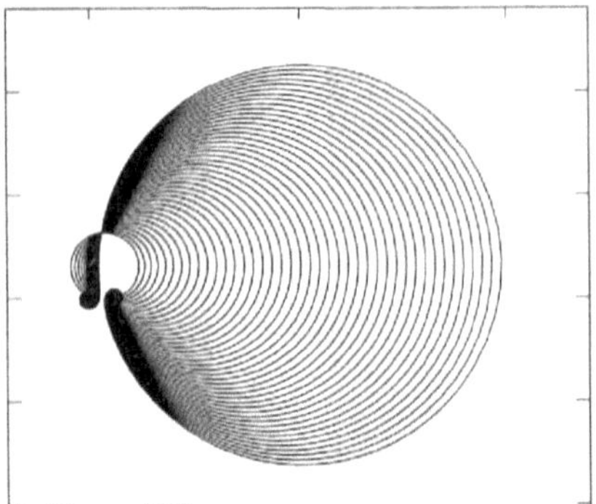

Fig. 10b

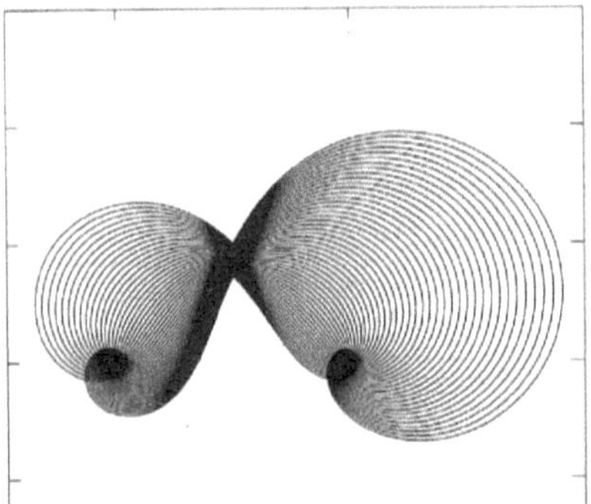

Fig. 11a

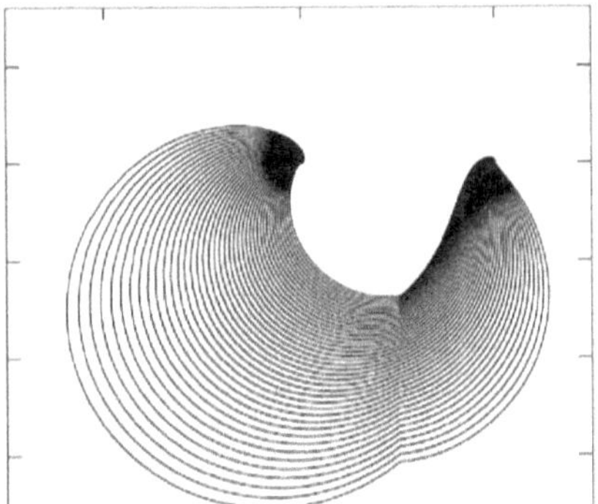

Fig. 11b

The 1997 Böttcher/Karlovich book is one of the most important mathematical works published in the late nineties. So it is well deserved that it was awarded the 1997 *Ferran-Sunyer-i-Balaguer-Prize* of the Institut d'Estudis Catalans.

The theory created by Albrecht Böttcher and Yuri Karlovich is the endpoint of a development which took place over decades. It stands out, not only as far as mathematical content is concerned, but also because of incredible clarity and esthetics.

I have tried, by discussing two topics, to give some insight into the mathematical activities of Albrecht Böttcher. However, his work is by no means confined to those two issues. Up to now, he has about 200 papers, 10 monographs, many by himself alone, many with coauthors. Interestingly enough, the number of coauthors is 70, and they are from 20 countries. He has also been the supervisor of 5 PhD students, and he is on the editorial boards of several renowned journals. So it is not easy to select a further topic from his rich mathematical work. In 2004, on request of the Alfried Krupp Foundation, Albrecht Böttcher indicated that he ranked the following 10 publications (up to 2004) as his most important ones:

1. A. Böttcher and B. Silbermann, The asymptotic behavior of Toeplitz determinants for generating functions with zeros of integral orders. *Math. Nachr.* **102** (1981), 79–105.
2. A. Böttcher and B. Silbermann, The finite section method for Toeplitz operators on the quarter-plane with piecewise continuous symbols. *Math. Nachr.* **110** (1983), 279–291.
3. A. Böttcher and B. Silbermann, Toeplitz matrices and determinants with Fisher–Hartwig symbols. *J. Funct. Anal.* **63** (1985), 178–214.
4. A. Böttcher and I.M. Spitkovsky, Wiener–Hopf integral operators with PC symbols on spaces with Muckenhoupt weight. *Rev. Mat. Iberoamericana* **9** (1993), 257–279.
5. A. Böttcher, B. Silbermann, and H. Widom, Determinants of truncated Wiener–Hopf operators with Hilbert–Schmidt kernels and piecewise continuous symbols. *Arch. Math.* **63** (1994), 60–71.

6. A. Böttcher, Pseudospectra and singular values of large convolution operators. *J. Integral Equations Appl.* **6** (1994), 267–301.
7. A. Böttcher and S.M. Grudsky, On the condition numbers of large semidefinite Toeplitz matrices. *Linear Algebra Appl.* **279** (1998), 285–301.
8. A. Böttcher and Yu.I. Karlovich, Toeplitz operators with PC symbols on general Carleson Jordan curves with arbitrary Muckenhoupt weights. *Trans. Amer. Math. Soc.* **351** (1999), 3143–3196.
9. A. Böttcher, On the determinant formulas by Borodin, Okounkov, Baik, Deift and Rains. *Oper. Theory: Adv. Appl.* **135** (2002), Birkhäuser, Basel, 91–99.
10. A. Böttcher, M. Embree, and V.I. Sokolov, The spectra of large Toeplitz band matrices with a randomly perturbed entry. *Math. Comp.* **72** (2003), 1329–1348.

FIGURE 4. From left to right: A. Böttcher, V. Dybin, I. Gohberg, me, S. Grudsky, Wasserschloß Klaffenbach near Chemnitz, 1999

Maybe, at this point in time, the list would look differently. I am sure, however, that his paper No. 6 about the pseudospectra of large convolution operators would also make it onto a new, updated list. Recall that the ε-pseudospectrum is defined as follows. For $\varepsilon > 0$ and a bounded linear operator A acting on a Banach space, the ε-pseudospectrum is given by

$$\mathrm{sp}_\varepsilon(A) = \left\{ \lambda \in \mathbb{C} \mid \| (A - \lambda I)^{-1} \| \geq \frac{1}{\varepsilon} \right\},$$

where for λ in the spectrum of A, the expression $\|(A - \lambda I)^{-1}\|$ is read as ∞. The famous book by Trefethen and Embree presents the philosophy behind this notion, which, in a nutshell, says that when working with non-normal operators, pseudospectra tell us much more than the familiar spectrum.

Albrecht Böttcher observed that, approximating the ε-pseudospectrum of an operator A by the ε-pseudospectra of operators A_n such that A_n tends in norm or in the sense of strong convergence to A, the following question is crucial. Can the resolvent of a bounded linear operator acting on a Banach space have constant norm on an open subset of $\mathbb{C} \setminus \operatorname{sp} A$? In the Hilbert space setting the answer is "no". This result enabled Albrecht Böttcher to prove that the ε-pseudospectra of the finite sections of a Toeplitz operator $T(a)$, $a \in PC$, acting on H^2, converge in the Hausdorff metric to the ε-pseudospectrum of $T(a)$. Let me mention that the operators $T_n(a)$ converge to $T(a)$ strongly, but not in norm if a is not the zero function. The proof of this result also shows that the following result is true. If a sequence (A_n) of bounded linear operators on a Hilbert space tends in norm to a bounded linear operator A, then for every $\varepsilon > 0$,

$$\operatorname{sp}_\varepsilon(A_n) \to \operatorname{sp}_\varepsilon(A)$$

in the Hausdorff distance. This result is remarkable because it fails for the familiar spectra.

The above question raised by Albrecht Böttcher inspired quite a lot of people to study it in the context of Banach spaces. As we learned later, the question was in fact already posed by J. Globevnik in 1976. However, it was Albrecht who made the question popular. Now, especially due the work of E. Shargorodsky, it is known that the answer heavily depends on the geometry of the Banach space under consideration. That means that the answer can be also "yes". Another interesting question that comes up naturally is: what can be said about the rate of convergence of $\operatorname{sp}_\varepsilon T_n(a)$ to $\operatorname{sp}_\varepsilon T(a)$? It was answered in a paper by A. Böttcher, M. Embree, and L.N. Trefethen for the case when $a \in PC$. They showed that the rate of convergence is spectacularly slow if a has a jump. Interestingly enough, the proof is based on the asymptotic behavior of certain Toeplitz determinants and, once again, the Fisher–Hartwig conjecture enters the scene.

Approaching the end of this essay, let me emphasize that Albrecht Böttcher is not only an excellent mathematician, but also an exceptional writer and lecturer. His papers and talks are not only of high quality, but also pieces of culture and frequently also pieces of art.

I want to summarize that I am extremely thankful to have and to have had Albrecht Böttcher with me, as a good friend and colleague. In December 2014 we celebrated his 60th birthday. This is a good reason to wish Albrecht and his family all the best for the coming years.

FIGURE 5. That's him, from head to toe: independent, confident, mischievous: Albrecht Böttcher shortly after he left a historical mine in Pobershau (Ore Mountains), 2001

Bernd Silbermann
Fakultät für Mathematik
Technische Universität Chemnitz
Germany
e-mail: silbermn.toeplitz@googlemail.com

Operator Theory:
Advances and Applications, Vol. 259, 37–39

Meeting Albrecht the Strong

Frank-Olme Speck

To Albrecht Böttcher on the occasion of his sixtieth birthday

It happened on the 7th February 1989 in Karl-Marx-Stadt (KMS) when we met for the first time, a day which I remember with emotion. Somehow Albrecht reminded me on August den Starken (Augustus II the Strong, Elector of Saxony and King of Poland, 1670–1733) because of his strong personality and his impressive mathematical output (he had 30 publications aged 34 and was about finishing the book on Toeplitz operators with Bernd Silbermann [3]). We walked around the Mathematics Building talking about the quarter-plane problem and further boundary problems. A rare possibility at the time of the iron curtain. Albrecht complained about not being allowed to travel to Western countries. Nevertheless, our mentors brought us together: Erhard Meister, Siegfried Prössdorf and Bernd Silbermann, guided by intuition for good development. Therefore, my emotion resulted not only from very close mathematical interests and the nice review that Albrecht had written about my research notes [1, 7], but also from the political circumstances.

Already in 1983-84 during conferences in Poland, some of us found it fruitful to exchange ideas and to start a German/German cooperation. So I was invited by Siegfried Prössdorf to spend two weeks at the Academy of Sciences in Berlin in February 1989 where we worked busily on a joint paper [6]. Within this period of time, he sent me to his colleagues in KMS to give two seminar talks and to discuss scientific matters with the real specialists, particularly with Bernd Silbermann and Albrecht Böttcher. These meetings influenced my scientific life tremendously.

Next time we met in the Black Forest in December 1989, shortly after the opening of the wall (on 9 November). Albrecht and some of his colleagues were able to attend the conference on Wiener–Hopf-Problems, Toeplitz Operators und Applications at the Mathematische Forschungsinstitut Oberwolfach (3–9 December 1989). The organizers Israel Gohberg, Rien Kaashoek and Erhard Meister managed to find extra places for the unexpected participants. Two German states still existed, but the atmosphere was unique. In the conference report you can read: "For the first time a large group of mathematicians from East Germany and the USSR took part in this Wiener–Hopf conference." The group included

J. Elschner and S. Prössdorf (Berlin), A. Böttcher, G. Heinig, M. Lorenz and B. Silbermann (KMS), L. von Wolfersdorf (Freiberg), R. Duduchava (Tbilissi), I.A. Feldman (Kishinev), G.S. Litvinchuk and N.L. Vasilevski (Odessa). And so it happened that Albrecht and me, we met twice in 1989 which turned out to be a fateful year for both of us.

In summer 1990 a light gray Trabi (small car produced in GDR) with a yellow roof stopped in our street in Darmstadt and Albrecht together with his wife Sylvia visited us on a round trip through the West including a conference in Italy. The time had come to visit many places in the world.

In the autumn of 1990 Germany was unified, shortly after I left the country for a visiting position in Portugal at the Universidade Técnica de Lisboa, together with my beloved wife Gaby and our three children (Ben, Janina and Alexander aged 9-13-16). It was supposed to be a visit of one or two years, however the stay did not find an end until now. Our research groups (around E. Meister, B. Silbermann, A.F. dos Santos) developed several joint German/Portuguese projects, generating plenty of visits, particularly visits of young colleagues and PhD students. Some of the Portuguese finished their PhD with Bernd Silbermann at Chemnitz (after re-naming KMS). Albrecht came to Lisbon several times from 1991 on and was regularly invited to our conferences as main (and wonderful) speaker.

Portugal enjoyed a rapid economic and scientific development in those years, since it joined the European Union in 1986. In 1991, the Technical University of Lisbon offered Albrecht Böttcher a professor position in the Mathematics Department of the Instituto Superior Técnico at Lisbon. We still regret that he did not accept this invitation for the benefit to save a position at Chemnitz.

Conversely I took part in two conferences at the Wasserschloss Klaffenbach near Chemnitz (1999 and 2009) and recently in a summer school at the Technical University of Chemnitz (September 2015). In the mathematical year 2000 Albrecht and me, we had the opportunity to participate together in the organization of the IWOTA 2000 at Faro, publishing the conference proceedings jointly with Rien Kaashoek, Amarino Lebre and António Ferreira dos Santos [4].

In what concerns the scientific work of Albrecht the Strong, which is further surveyed in other contributions of this volume, let me just refer to a sentence from the laudatory speech of Friedrich Hirzebruch on the occasion of the award of the Alfred Krupp Prize to Albrecht in 1993: "Es wird vielfach ... zu wenig beachtet, dass dieser Teil der Operatortheorie für Anwendungen in Mathematischer Physik, Kontrolltheorie, Approximationstheorie usw. von grundlegender Bedeutung ist, sozusagen das Wesen der Dinge beschreibt. (There is often ... too little attention to the fact that this part of the operator theory is of fundamental importance for applications in mathematical physics, control theory, approximation theory, etc., so to speak, describes the essence of things.)" It was a citation of one of the referee's recommendation letters (guess who it was), but standing for the consensus of the community of his colleagues, family and friends, I believe.

Recently I learned what it means to work mathematically with Albrecht: challenge, liveliness and fun. He detected a gap in the proof of the main result of

[8] when writing a review for that article [2]. Hence he contacted me by email to verify the conclusions. After a short discussion, Albrecht proposed to elaborate a new paper related to that gap extending the topic to further interesting results. In a very dynamic exchange of some thirty emails in two months time it was done and resulted in our first joint research paper [5]. During that work he wrote to me (in connection with his opinion about another paper): "The times when it was still interesting to generalize functional analysis and operator theory to more and more abstract settings and to weaken existing assumptions more and more are over. I expect that a present day functional analysis or operator theory paper provides insights into deep mathematical connections or the background or the foundation for the solution of at least one nontrivial concrete problem, preferably with applications in physics, biology, technology, etc." I was very pleased to read those sentences because they reflected exactly the opinion of my PhD supervisor Erhard Meister who tried to guide me along these lines already in the 1970s.

My wishes for Albrecht Böttcher are: To continue strong work for many, many years with good health, pleasure and enthusiasm, frankly speaking: to remain as you are, Albrecht, and to follow your slogan: Let's push things forward!

References

[1] A. Böttcher, *Review on* [7]. Math. Reviews 0790315 (1987).

[2] A. Böttcher, *Review on* [8]. Math. Reviews 3355786 (2016).

[3] A. Böttcher and B. Silbermann, *Analysis of Toeplitz Operators.* 2nd Edition. Springer, Berlin 2006 (first edition 1990).

[4] A. Böttcher, M.A. Kaashoek, A. Lebre, A.F. dos Santos and F.-O. Speck, *Singular Integral Operators, Factorization and Applications.* Proceedings IWOTA-Portugal 2000, Operator Theory: Advances and Applications, Vol. 142, Birkhäuser, Basel 2003.

[5] A. Böttcher and F.-O. Speck, *On the symmetrization of general Wiener–Hopf operators.* J. Operator Theory **76** (2016), 335–349.

[6] F.-O. Speck and S. Prössdorf, *A factorization procedure for two by two matrix functions on the circle with two rationally independent entries.* Proc. Royal Soc. Edinburgh **115 A** (1990), 119–138 (submitted on 7 August 1989).

[7] F.-O. Speck, *General Wiener–Hopf Factorization Methods.* Pitman, London 1985.

[8] F.-O. Speck, *Wiener–Hopf factorization through an intermediate space.* Integr. Equ. Oper. Theory **82** (2014), 395–415.

Frank-Olme Speck
Departamento de Matemática
Instituto Superior Técnico
Universidade de Lisboa
Avenida Rovisco Pais
P-1049-001 Lisboa, Portugal
e-mail: `fspeck@tecnico.pt`

Operator Theory:
Advances and Applications, Vol. 259, 41–46

The Beginning (the Way I Remember it)

Ilya M. Spitkovsky

To Albrecht Böttcher, a dear friend and coauthor, on the occasion of his sixtieth birthday

One Spring day of 1984, Mark Grigor'evich Krein showed me the abstract of a PhD thesis he received in the mail from Rostov-on-Don University. He liked it at first glance, so we set to read it together more carefully, and towards the end of this reading he decided to write a formal (enthusiastically positive) report to the PhD defence committee. Considering the amount of correspondence Professor Krein was receiving daily, this says a lot about the PhD thesis and its author, Albrecht Böttcher. A new player emerged in the field of Wiener–Hopf operators and related topics, and a close attention should be paid to what he had written and would write moving forward.

It is from this abstract that I learned of the book [4], the results of which I then used in [11] when considering the Szegő limit theorem in the case of locally sectorial matrix symbols. And (I am guessing here) in its turn the appearance of [11] prompted an invitation to attend the 1987 "Sommerschule Analysis" of the Mathematics department of Technische Universität Karl-Marx-Stadt in Einsiedel organized by Bernd Silbermann, Albrecht's senior colleague and coauthor. The invitation was received well in advance and, this being already the "perestroika" years, a certain amount of paperwork and several trips to Kiev and Moscow secured (still somewhat unexpectedly) a permission to travel. So, my first trip outside of the Soviet Union transpired in the Fall of 1987. This is where and how I met Albrecht in person.

We did talk about our published and current research; in particular, I remember a rather intense session (joint with Bernd) at which I was asked to explain some skipped steps in the proofs of results in [10] caused by the strict limitation on the number of pages in each journal published then in the USSR. Luckily, I managed... There was no discussion of specific future joint work; however, an invitation to visit again was extended (and gratefully accepted).

My next visit to Karl-Marx-Stadt lasted three weeks and took place about a year later, in the Fall of 1988. I then had the pleasure and privilege to read the manuscript of the forthcoming second book [5] by Albrecht and Bernd. I made some suggestions which the authors took into consideration, but in return learned

FIGURE 1. Albrecht and I in Darmstadt in Summer 1992.

much more. During this visit I also shared with Albrecht my (joint with Tashbaev) results on factorization of piecewise-constant matrix functions, which at the time were submitted for publication only as a short no-proofs version (appeared later as [12]). Albrecht kindly volunteered to translate the full version into English and submit it for publication in Mathematische Nachrichten. He had done it by February 1989, and the paper [14] appeared in due course. More importantly, in these three weeks we also laid the groundwork for our first joint papers [3, 6, 7]. This work was continued via correspondence upon my return to Odessa. For the benefit of younger readers I have to clarify here that there was no such thing as email or internet at the time; Albrecht and I were writing regular letters by hand, keeping carbon copies for ourselves, and it took weeks for a letter to reach the addressee (honestly, I was then rather surprised that the letters reached their destinations at all). The result of [7] was especially evasive, and it took quite a few iterations to finally nail it down. In any case, the first round of our collaboration was completed by July of 1989.

In the same year, important changes were taking place in my life: in February I made the decision to leave the Soviet Union, in November my wife, I, and our two children left Odessa, and in mid March of 1990 we landed in JFK. Four months later, we moved from New York to Williamsburg, Virginia, for what at the time I thought would be a one year visiting position but ended up as a quarter-century long tenure at the College of William and Mary. During the Christmas break of my first academic year there I wrote up the result on the spectra of some operators on spaces with Muckenhoupt weights (which eventually appeared as [13]) and mailed a copy to Albrecht, along with some suggestions on possible further research in this direction. Two more papers [8, 9] followed "by correspondence", still via regular mail.

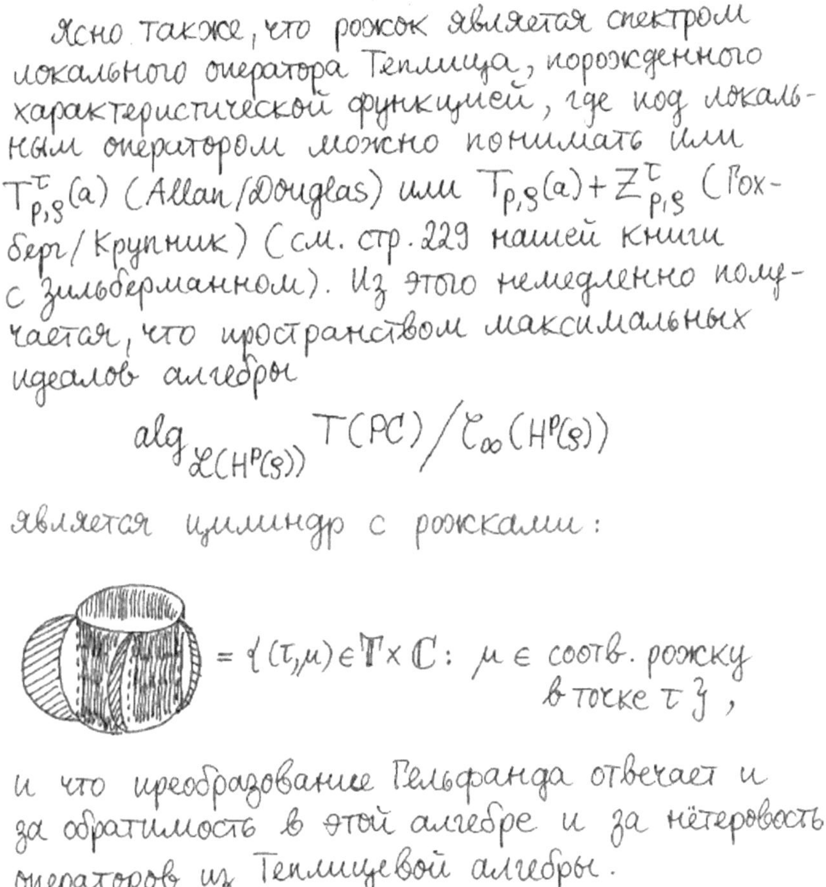

Ясно также, что рожок является спектром локального оператора Теплица, порожденного характеристической функцией, где под локальным оператором можно понимать или $T^{\tau}_{p,\varrho}(a)$ (Allan/Douglas) или $T_{p,\varrho}(a)+Z^{\tau}_{p,\varrho}$ (Гохберг/Крупник) (см. стр. 229 нашей книги с Зильберманном). Из этого немедленно получается, что пространством максимальных идеалов алгебры

$$\mathrm{alg}_{\mathcal{L}(H^p(\varrho))}\, T(PC)/\mathcal{C}_\infty(H^p(\varrho))$$

является цилиндр с рожками:

$= \{(\tau,\mu)\in\mathbb{T}\times\mathbb{C} : \mu \in$ соотв. рожку в точке $\tau\}$,

и что преобразование Гельфанда отвечает и за обратимость в этой алгебре и за нётеровость операторов из Теплицевой алгебры.

FIGURE 2. Fragment of a letter sent to me by Albrecht in August 1991.

In an interesting twist of fate, while I immigrated to the West, Albrecht happened to "move" there without emigration, as a result of the reunification of Germany. In the summer of 1992 I was visiting Professor Ehrhard Meister in Darmstadt and for a couple of days went to Karl-Marx-Stadt (or was it already

Chemnitz then?) from there. So Albrecht and I met again, for the first time in the free world. Moreover, we were joined by Yuri Karlovich, my former colleague and coauthor from Odessa, whom I did not hope to see (or collaborate with) after my emigration. Thankfully, things turned out quite differently: this meeting resulted in the first joint paper by the three of us [1], and eventually we wrote the book [2] together. Later that year, when I was still in Darmstadt, Albrecht and his wife Sylvia were driving to their summer vacations, and since Darmstadt was along the way, they made a stop and visited me briefly for a couple of hours. The photo of Figure 1 is from that short visit.

FIGURE 3. With Albrecht and Yuri at a conference on Madeira Island in January 2002. The book is about to be published.

There were many more meetings with Albrecht: in Chemnitz, Williamsburg, and at various conferences all over the world. More joint projects initiated and (most of the time) successfully executed. So far, we wrote 23 papers together, the (already mentioned) book, and I hope the list is far from being final. It is a great pleasure working with Albrecht, the only drawback of which being an uneasy feeling that he is always contributing much more than his fair share. I wish Albrecht many more years of exciting mathematical adventures, and wish myself to still play some part in them.

FIGURE 4. In Albrecht's apartment in 2002. With Yuri Karlovich and his wife Zhenya, Albrecht and his wife Sylvia, and Sergei Grudsky (clockwise).

References

[1] A. Böttcher, Yu.I. Karlovich, and I.M. Spitkovsky. Toeplitz operators with semi-almost periodic symbols on spaces with Muckenhoupt weight. *Integral Equations and Operator Theory*, 18:261–276, 1994.

[2] A. Böttcher, Yu.I. Karlovich, and I.M. Spitkovsky. *Convolution Operators and Factorization of Almost Periodic Matrix Functions*, volume 131 of *Operator Theory: Advances and Applications*. Birkhäuser Verlag, Basel and Boston, 2002.

[3] A. Böttcher, S. Roch, B. Silbermann, and I. Spitkovsky. A Gohberg–Krupnik–Sarason symbol calculus for algebras of Toeplitz, Hankel, Cauchy, and Carleman operators. *Operator Theory: Advances and Applications*, 48:189–234, 1990.

[4] A. Böttcher and B. Silbermann. *Invertibility and asymptotics of Toeplitz matrices*. Akademie Verlag, Berlin, 1983.

[5] A. Böttcher and B. Silbermann. *Analysis of Toeplitz Operators*. Springer-Verlag, Berlin, Heidelberg, New York, 1990.

[6] A. Böttcher, B. Silbermann, and I.M. Spitkovskiĭ. Toeplitz operators with piecewise quasisectorial symbols. *Bull. London Math. Soc.*, 22(3):281–286, 1990.

[7] A. Böttcher and I. Spitkovsky. Toeplitz operators with PQC symbols on weighted Hardy spaces. *J. Functional Analysis*, 97:194–214, 1991.

[8] A. Böttcher and I. Spitkovsky. Wiener–Hopf integral operators with PC symbols on spaces with Muckenhoupt weight. *Revista Matématica Iberoamericana*, 9(2):257–279, 1993.

[9] A. Böttcher and I. Spitkovsky. Pseudodifferential operators with heavy spectrum. *Integral Equations and Operator Theory*, 19:251–269, 1994.

[10] I.M. Spitkovskiĭ. Factorization of matrix functions from the classes $A_n(p)$ and TL. *Ukrain. Mat. Zh.*, 35(4):455–460, 1983.

[11] I.M. Spitkovskiĭ. Szegő limit theorems in the case of a matrix locally sectorial symbol. *Dokl. Akad. Nauk SSSR*, 284(1):61–65, 1985.

[12] I.M. Spitkovskiĭ and A.M. Tashbaev. Factorization of piecewise constant matrix functions with 3 points of discontinuity in the classes $L_{p,\rho}$ and some of its applications. *Dokl. Akad. Nauk SSSR*, 307(2):291–296, 1989.

[13] I.M. Spitkovsky. Singular integral operators with PC symbols on the spaces with general weights. *J. Functional Analysis*, 105:129–143, 1992.

[14] I.M. Spitkovsky and A.M. Tashbaev. Factorization of certain piecewise constant matrix functions. *Math. Nachr*, 151:241–261, 1991.

Ilya M. Spitkovsky
Division of Science
New York University
Abu Dhabi (NYUAD)
Saadiyat Island
P.O. Box 129188
Abu Dhabi, UAE
e-mail: ims2@nyu.edu
imspitkovsky@gmail.com

Operator Theory:
Advances and Applications, Vol. 259, 47–49

Personal Address on the Occasion of Albrecht Böttcher's 60th Birthday

David Wenzel

Dear Albrecht. All the people holding this volume in their hands do this with the best wishes to your big birthday. And with this event in mind, I want to say some words.

Back in 2000, I started my studies. And because of a very special diploma program with the well-sounding name *Mathematics with extended computer science education*, my choice for the Chemnitz University was obvious. *Obvious* ... that's a word going to cross my ways more often since then. I barely noticed it before, but it turned out to be quite popular in math. However, if one doesn't muse about the contents – convincing oneself that the "obvious" is valid – one easily steps into one trap after another. We soon learned that the simplest looking things, in reality, tend to be pretty complicated ... not seldom even more difficult than one ever would have expected.

Mathematics per se does not belong to these things. This one is clear, accurate, and indeed ... simple. No comparison to what one met in school. And coming right from there, one usually has little to no idea of what to do at a university. But the supposedly strange mother of all sciences will surprise you, if you only give it a try. The same is true for the people taking care of you at the local math department.

A first term student in mathematics normally encounters two professors. In my timetable, behind the linear algebra lecture, was written – as one might have expected – Albrecht Böttcher. Of course, as a beginner, I'd never even heard this name, so far. For completeness, the analysis that time was held by Bernd Silbermann. Later, I learned that Albrecht formerly was a student of him.

New experiences influence us. And you teaching with aplomb, dear Albrecht, have a huge impact. Ninety minutes totally focused ... without any notes, just out of the mind ... week by week, even the vacation won't disturb the flow, and the gap is smoothly bridged as if it wouldn't exist ... always providing comprehensible motivations, illustrative examples, and detailed explanations.

Whereas many rush through umpteen overfull presentation slides, you attach great importance to bringing the content coordinately into the minds. You construct an excellent blackboard image, which most couldn't imitate ... not even with help of a computer. That's the way one understands the new mathematical principles and topics really good. Because you realized: a lecture is no scientific talk.

Several points have to be taken into account for being a successful teacher: knowledge of the contents and especially an insight into the backgrounds, as well as enthusiasm for the subject, and also the wish to pass own experiences to others. Not to forget that one must be aware of the opposites still having to learn the whole stuff – which oneself has no problem with – and they consequently can't know that much by then. Respecting all these aspects is a challenge for every teacher. But you, Albrecht, you are uniting these properties in a rarely seen fashion.

After the basic courses, I could attend some special lectures. This was hard to arrange since, because of his other obligations, Albrecht generally can afford a research course at most once in two years. Nevertheless, I was able to have him somehow around me all the time. Hence, I decided to finish my studies with him and asked for a diploma's subject. Albrecht again found the right mixture of support and demand, so I got quickly to my thesis.

What is even better, thanks to his help, Albrecht and I published a paper before the diploma thesis was actually done. Who else can tell something of such kind? And what might sound a bit crazy, at that time we were not yet able to prove our initial conjecture. But, this is exactly the beauty of mathematics we all know and love. She[1] is free, and that's her strength. One may analyze special cases, make generalizations, or investigate analogues in another world. In principle, one could even prove that something is *not* true. Which other discipline would accept a "negative" result? As a mathematician, you needn't to be good in counting, but you must count on everything.

Luckily, we didn't end up in the worst case. Our intuition was right. But, it should take us more than three years to finally get the solution. Since the problem resisted for so a long time, one can imagine that the original statement was in no way trivial. Undoubtedly, composure is required. Though, stay tuned to it when the first article has started a world-wide run.

So, right after my studies, I started working for a doctoral's degree. Naturally, I switched position and joined Albrecht. On his side, I now could give the exercises I attended as part of the audience shortly before. The new point of view became extremely interesting when the exams conquered the agenda. Over a full week, half an hour each, student by student tells what he believes to have understood. Some proved quite good ... others were less successful. In any case, they should rely on a kind tester, who takes the long, hard procedure as an opportunity for getting to know his students better.

[1] If something is really beautiful, it must be female, no matter what might be common use in the English language.

And all this may develop into a burden, occasionally; it is not easy to keep patient if one is forced to listen to the same story several times. Not everyone is willing to endure the pains – a written test would be a lot easier to correct. But mathematicians like Albrecht do this for the students' good. Still, most of them won't initially believe that orals take a better course. The fear is unjustified ... as long as the professor acts professionally.

I for myself had to operate as an examinee again. About four years ago, my PhD thesis requested for its defence. In order to give it a comprehensive mathematical content, a lot of work was necessary in advance. All along, I could bank on you, Albrecht. And I'm very grateful for your suggestions.

I also appreciate the nature of our collaboration, which you wanted to be not at all one-way, neither for your nor for my sole benefit. Right from the beginning, you established a mutual partnership. Within this real cooperation, we could devise new ideas together and develop them further. Treating young academics as equal cannot be taken as granted, but, to the newcomer, it gives an additional impulse.

That brings us back from our little excursion to the here and now. With this volume, we want to celebrate the sixtieth iteration of your birthday. Some may wonder why this one receives more acclaim than the common, *obvious* half-century. But, the mathematicians know the background of the superiority of 60 over 50. And any non-math readers holding this book in their hands be assured: there is a reason for dividing an hour into sixty minutes since the beginning of time.

The diamond anniversary appears to be connected with the retirement age. However, this is pure coincidence. In fact, none of them necessarily implies the other. I claim to speak for each of your fellows when saying that we wish us you continuing your activities. And anyone willing to study shall feel encouraged to matriculate for mathematics and to jump at the chance of meeting Albrecht Böttcher. Come and listen to his lectures, visit his talks, and enjoy the guidance of my *Doktorvater*.

David Wenzel
Fakultät für Mathematik
Technische Universität Chemnitz
D-09107 Chemnitz, Germany
e-mail: david.wenzel.mathematik@t-online.de

Operator Theory:
Advances and Applications, Vol. 259, 51–77

Asymptotics of Eigenvalues for Pentadiagonal Symmetric Toeplitz Matrices

M. Barrera and S.M. Grudsky

Dedicated to our friend and colleague Albrecht Böttcher on the occasion of his 60th birthday anniversary

Abstract. In this paper, we find uniform asymptotic formulas for all the eigenvalues of certain pentadiagonal symmetric Toeplitz matrices of large dimension. The entries of the matrices are real and we consider the case where the real-valued generating function has a minimum and a maximum such that its fourth derivative at the minimum and its second derivative at the maximum are nonzero. This is not the simple-loop case considered in [1] and [2]. We apply the main result of [7] and obtain nonlinear equations for the eigenvalues. It should be noted that our equations have a more complicated structure than the equations in [1] and [2]. Therefore, we required a more delicate method for its asymptotic analysis.

Mathematics Subject Classification (2010). 15B05, 15A18, 41A60.

Keywords. Toeplitz matrix, eigenvalue, asymptotic expansions.

1. Introduction

The $n \times n$ Toeplitz matrix $T_n(a)$ generated by a function (symbol) a in L^1 on the complex unit circle $\mathbb{T}$ is defined by $T_n(a) = (a_{j-k})_{j,k=1}^{n}$, where a_ℓ is the ℓth Fourier coefficient for a,

$$a_\ell = \frac{1}{2\pi}\int_0^{2\pi} a(e^{i\varphi})e^{-i\ell\varphi}d\varphi \quad (\ell \in \mathbb{Z}).$$

The asymptotic behavior of the eigenvalues of $T_n(a)$ as n goes to infinity has been thoroughly studied by mathematicians and physicists for a long time. Toeplitz matrices arise in particular in many problems of statistical physics and there are questions about the asymptotics of their spectrum, principally their determinants, eigenvalues, and eigenvectors are always at the heart of the matter.

The first author was supported by CONACYT grant 290817.
The second author was partially supported by CONACYT grants 180049 and 238630.

By the first Szegő limit theorem, we can understand the collective behavior of the eigenvalues of Hermitian Toeplitz matrices. This theorem says that, in a sense, the eigenvalues are asymptotically distributed as the values of a. See, for example [8],[18], [20], [23], [24].

The case of non-Hermitian Toeplitz matrices is more complicated; see the papers [17], [22], [15], [19] and the books [4], [5].

In the Hermitian case, there is a lot of known work about the extreme eigenvalues of the matrices $T_n(a)$. In [21], H. Widom considered the class of functions $g(\varphi) := a(e^{i\varphi})$ such that g is real-valued, smooth, even, $M := g(0) = \max g$ and $g''(0) \neq 0$. There, he established an asymptotic formula for the extreme eigenvalues near M. Similarly, in [11], S.V. Parter considered the class of functions $g(\varphi) := a(e^{i\varphi})$ such that g is real-valued, smooth, $m := g(0) = \min g$ and $g^{(2k)}(0) \neq 0$ for $k \in \mathbb{N}$. Once again, there he established an asymptotic formula for the extreme eigenvalues near m. See [3], [9], [10], [12], [13], and [14] for more information about the behavior of extreme eigenvalues.

Results on the individual asymptotic formulas for all (extreme and inner) eigenvalues of Hermitian Toeplitz matrices were obtained only recently in [1], [2], and [6]. In [2] it is assumed that the symbol a is a Laurent polynomial such that $a(e^{i\varphi})$ is strictly increasing from its minimum to its maximum and strictly decreasing from its maximum back to its minimum. Further, $a(e^{i\varphi})$ has nonzero second derivatives at the minimum and the maximum. These requirements were relaxed in [1] where it was considered a class of smooth symbols with 4th derivative in the Wiener algebra. Finally, the recent results of [6] are stated for C^∞ symbols.

In this paper, we present a model case such that the second derivative of the symbol a at the minimum is equal to 0. Therefore, throughout the paper we suppose that the symbol is

$$a(t) = \left(\frac{1}{t} + t - 2\right)^2 \quad (t = e^{i\varphi} \in \mathbb{T}), \tag{1.1}$$

and the generating function

$$g(\varphi) := a(e^{i\varphi}) = 4(\cos\varphi - 1)^2 = M \sin^4\left(\frac{\varphi}{2}\right),$$

where $\varphi \in [-\pi, \pi]$ and $M := 16$. Notice that the Fourier coefficients of the pentadiagonal Toeplitz matrix with symbol a are $a_0 = 6$, $a_{-1} = a_1 = -4$, $a_{-2} = a_2 = 1$, and $a_k = 0$ for $|k| > 2$.

Moreover, pentadiagonal Toeplitz matrices frequently arise from boundary value problems involving fourth-order derivatives, i.e., the Toeplitz matrix with symbol a comes from discretization, by equi-spaced Finite Differences (FD) of precision order two, of the unidimensional fourth derivative with homogeneous boundary conditions both on the function and on the first derivative. See, for example, [16].

The function g has the following properties:

i) The range of g is the segment $[0, M]$ with $M := g(\pi) > 0$.

ii) $g(0) = 0$, $g(-\pi) = g(\pi) = M$, $g'(-\pi) = g'(\pi) = 0$, $g''(-\pi) = g''(\pi) < 0$, $g'(0) = g''(0) = g'''(0) = 0$, and $g^{(4)}(0) > 0$.

The function g is even. Furthermore, $g : [0, \pi] \to [0, M]$ is bijective and increasing. Let $g^{-1} : [0, M] \to [0, \pi]$ be the inverse function of g restricted to $[0, \pi]$. Thus, we get $\lambda = g(\varphi)$ if and only if $\varphi = g^{-1}(\lambda)$.

We will express all the main objects in terms of φ rather than λ itself. Using the main result of [7], we will find formulas for the determinant of $T_n(a - \lambda) = T_n(a) - \lambda I$ that will allow us to obtain asymptotic formulas in terms of φ. Given that $\lambda = g(\varphi)$ if and only if $\varphi = g^{-1}(\lambda)$ we will arrive at the asymptotic formulas for the eigenvalues of $T_n(a)$.

It should be noted that our case is essentially different to the simple-loop case presented in [1], [2], and [6]; because we have a fourth-order zero at the point $\varphi = 0$. Therefore, the asymptotic formulas on a neighborhood of the minimum, have a different form of the formulas presented in [1] and [2]. Furthermore, the nonlinear equations for the eigenvalues of $T_n(a)$ (obtained with the help of the main result of [7]) have a more complicated structure than in [1], [2], and [6]. In particular, since all the terms in these equations depend on n, we require a more delicate method for its asymptotic analysis.

Once again, it is emphasized that we get asymptotic formulas for all eigenvalues of $T_n(a)$. We simplified the asymptotic formulas for inner and extreme eigenvalues and compared our results for extreme eigenvalues with the well-known articles by H. Widom [21] and S.V. Parter [11].

Our model case (1.1) allow us to consider more general smooth symbols satisfying i) and ii). We pretend to embark the topic in a future paper. It should be noted that our results can easily be generalized to symbols

$$A(t) = c_1 a(t) + c_2 \quad (t = e^{i\varphi} \in \mathbb{T}),$$

where $c_1, c_2 \in \mathbb{R}$. In fact, for A we obtain that $T_n(A) = c_1 T_n(a) + c_2$.

The paper is organized as follows. Section 2 contains the main results. In the Section 3, using the main result of [7] for the determinants of finite Toeplitz matrices, we obtain a nonlinear equation which can be employed to derive uniform asymptotic formulas for all eigenvalues. Section 4 contains the proof of our main results. Section 5 contains the comparison between the asymptotic formulas presented in [21] and [11] with our results. Finally, in the Section 6 are presented numerical examples.

2. Main results

In this section, we begin by introducing the functions $\beta : [0, \pi] \to \mathbb{R}$ and $f : [0, \pi] \to \mathbb{R}$, where

$$\beta(\varphi) := 2\log\left(\sin\frac{\varphi}{2} + \sqrt{1 + \sin^2\frac{\varphi}{2}}\right) \quad \text{and} \quad f(\varphi) := \frac{\cos\frac{\varphi}{2}}{\sqrt{1 + \sin^2\frac{\varphi}{2}}}.$$

Notice that $f(\varphi) = \beta'(\varphi)$.

Now we formulate our main results.

Theorem 2.1. *A number λ is an eigenvalue of $T_n(a)$ if and only if there exists a number $\varphi \in (0,\pi)$ such that $\lambda = g(\varphi)$ and*

$$\tan\left(\frac{n+2}{2}\varphi\right) = -f(\varphi)\tanh\left(\frac{n+2}{2}\beta(\varphi)\right), \tag{2.1}$$

or

$$\tan\left(\frac{n+2}{2}\varphi\right) = \frac{1}{f(\varphi)}\tanh\left(\frac{n+2}{2}\beta(\varphi)\right). \tag{2.2}$$

Remark 2.2. In this paper we will express all the main objects in terms of φ rather than λ itself. This approach will simplify the proof of our main results. Since, we have the relation $\lambda = g(\varphi)$ for $\varphi \in [0,\pi]$, then we will obtain the results in terms of λ.

Theorem 2.3. *If n is sufficiently large, then*

i) *the equation*

$$\frac{n+2}{2}\varphi + \arctan\left(f(\varphi)\tanh\left(\frac{n+2}{2}\beta(\varphi)\right)\right) = \pi j \tag{2.3}$$

has exactly one solution $\varphi_{2j-1}^{(n)} \in \left(\frac{(2j-1)\pi}{n+2}, \frac{2\pi j}{n+2}\right)$ for $j = 1,\ldots,[\frac{n+1}{2}]$, with $\lambda_{2j-1}^{(n)} = g(\varphi_{2j-1}^{(n)})$.

ii) *the equation*

$$\frac{n+2}{2}\varphi - \arctan\left(\frac{1}{f(\varphi)}\tanh\left(\frac{n+2}{2}\beta(\varphi)\right)\right) = \pi j \tag{2.4}$$

has exactly one solution $\varphi_{2j}^{(n)} \in \left(\frac{2\pi j}{n+2}, \frac{(2j+1)\pi}{n+2}\right)$ for $j = 1,\ldots,[\frac{n}{2}]$, with $\lambda_{2j}^{(n)} = g(\varphi_{2j}^{(n)})$.

iii) *the eigenvalues of $T_n(a)$ are all distinct and we can write $\lambda_1^{(n)} < \cdots < \lambda_n^{(n)}$.*

Remark 2.4. Theorem 2.3 shows that, for sufficiently large n, the equations (2.3) and (2.4) have a unique solution for each $j \in \{1,\ldots,[\frac{n+1}{2}]\}$ and $j \in \{1,\ldots,[\frac{n}{2}]\}$, respectively. Furthermore, if n is an even number, then the equation (2.3) gives us $\frac{n}{2}$ solutions on $(0,\pi)$ and the equation (2.4) gives us $\frac{n}{2}$ solutions on $(0,\pi)$, as well. Thus, we have exactly n solutions. But, if n is an odd number, then the equations (2.3) and (2.4) give us $\frac{n+1}{2}$ solutions and $[\frac{n}{2}]$ solutions on $(0,\pi)$, respectively. Again, we have exactly n solutions.

Furthermore, we have to notice that the problem of finding the eigenvalues of $T_n(a)$ is reduced to solve the nonlinear equations (2.3) and (2.4) on the intervals $\left(\frac{(2j-1)\pi}{n+2}, \frac{2\pi j}{n+2}\right)$ and $\left(\frac{2\pi j}{n+2}, \frac{(2j+1)\pi}{n+2}\right)$, respectively. Each one of these equations can be solved numerically in many ways.

The following theorems give us the asymptotic approach. For this goal we introduce the parameters $d_- := d_{2j}^{(n)} = \frac{2\pi j}{n+2}$ and $d_+ := d_{2j+1}^{(n)} = \frac{(2j+1)\pi}{n+2}$. We need to consider the equations

$$u = -\arctan(f(d_-)\tanh\Psi_j^-(u)) \qquad u \in \left[-\frac{\pi}{2}, 0\right] \tag{2.5}$$

and

$$w = -\operatorname{arccot}\left(\frac{1}{f(d_+)}\tanh\Psi_j^+(w)\right) \quad w \in \left[-\frac{\pi}{2}, 0\right], \tag{2.6}$$

where $\Psi_j^-(u) := \frac{n+2}{2}\beta(d_-) + \beta'(d_-)u$ and $\Psi_j^+(w) := \frac{n+2}{2}\beta(d_+) + \beta'(d_+)w$.

Theorem 2.5. *If $n \to \infty$, then*

i)
$$\varphi_{2j-1}^{(n)} = d_- + \frac{2u_{1,j}}{n+2} + \frac{4u_{2,j}}{(n+2)^2} + \Delta_1(n,j), \tag{2.7}$$

where $|\Delta_1(n,j)| \le \frac{b_1}{n^3}$ and the constant b_1 does not depend on n and j, for $j = 1, \dots, [\frac{n+1}{2}]$. The coefficient $u_{1,j}$ is the unique solution of (2.5) *in $[-\frac{\pi}{2}, 0]$. The coefficient $u_{2,j}$ is defined in terms of $u_{1,j}$. (See* (4.11).*)*

ii)
$$\varphi_{2j}^{(n)} = d_+ + \frac{2w_{1,j}}{n+2} + \frac{4w_{2,j}}{(n+2)^2} + \Delta_2(n,j), \tag{2.8}$$

where $|\Delta_2(n,j)| \le \frac{b_2}{n^3}$ and the constant b_2 does not depend on n and j, for $j = 1, \dots, [\frac{n}{2}]$. The coefficient $w_{1,j}$ is the unique solution of (2.6) *in $[-\frac{\pi}{2}, 0]$. The coefficient $w_{2,j}$ is defined in terms of $w_{1,j}$. (See* (4.19).*)*

Here is the result for the inner eigenvalues.

Theorem 2.6. *Let $\epsilon > 0$ be a small number such that $\epsilon \ge \gamma_1 \frac{\log n}{n}$ for some positive constant γ_1.*

i) *If $\pi\epsilon < d_- < (1-\epsilon)\pi$, then*

$$\lambda_{2j-1}^{(n)} = g(d_-) + \frac{2u_{1,j}^* g'(d_-)}{n+2} + \frac{2(u_{1,j}^*)^2 g''(d_-) + 4u_{2,j}^* g'(d_-)}{(n+2)^2} + \Delta_3(n,j), \tag{2.9}$$

where $|\Delta_3(n,j)| \le \frac{b_3}{n^3}$ and the constant b_3 does not depend on n and j. Furthermore, the coefficients $u_{1,j}^*$ and $u_{2,j}^*$ are defined by

$$u_{1,j}^* = -\arctan f(d_-) \quad \textit{and} \quad u_{2,j}^* = \frac{f'(d_-)\arctan f(d_-)}{1 + f^2(d_-)}.$$

ii) *If $\pi\epsilon < d_+ < (1-\epsilon)\pi$, then*

$$\lambda_{2j}^{(n)} = g(d_+) + \frac{2w_{1,j}^* g'(d_+)}{n+2} + \frac{2(w_{1,j}^*)^2 g''(d_+) + 4w_{2,j}^* g'(d_+)}{2(n+2)^2} + \Delta_4(n,j), \tag{2.10}$$

where $|\Delta_4(n,j)| \le \frac{b_4}{n^3}$ *and the constant* b_4 *does not depend on* n *and* j. *The coefficients* $w^*_{1,j}$ *and* $w^*_{2,j}$ *are defined by*

$$w^*_{1,j} = -\operatorname{arccot}\frac{1}{f(d_+)} \quad \text{and} \quad w^*_{2,j} = \frac{f'(d_+)\operatorname{arccot}\frac{1}{f(d_+)}}{1+f^2(d_+)}.$$

Now, it is formulated the result for the extreme eigenvalues near zero.

Theorem 2.7.

i) *If* $d_- \to 0$ *as* $n \to \infty$, *then*

$$\lambda^{(n)}_{2j-1} = \frac{(2\pi j + 2u_{1,j})^4}{(n+2)^4} + \frac{16u_{2,j}(2\pi j + 2u_{1,j})^3}{(n+2)^5} + \Delta_5(n,j), \tag{2.11}$$

where $|\Delta_5(n,j)| \le b_5\left(\frac{d_-^3}{n^3} + d_-^8\right)$ *and the constant* b_5 *does not depend on* n *and* j. *Furthermore,* $u_{1,j}$ *is the unique solution on* $[-\frac{\pi}{2}, 0]$ *of the equation* (2.5), *and the coefficient* $u_{2,j}$ *is defined in terms of* $u_{1,j}$. *(See* (4.11).*)*

ii) *If* $d_+ \to 0$ *as* $n \to \infty$, *then*

$$\lambda^{(n)}_{2j} = \frac{((2j+1)\pi + 2w_{1,j})^4}{(n+2)^4} + \frac{16w_{2,j}((2j+1)\pi + 2w_{1,j})^3}{(n+2)^5} + \Delta_6(n,j), \tag{2.12}$$

where $|\Delta_6(n,j)| \le b_6\left(\frac{d_+^3}{n^3} + d_+^8\right)$ *and the constant* b_6 *does not depend on* n *and* j. $w_{1,j}$ *is the unique solution of* (2.6) *in* $[-\frac{\pi}{2}, 0]$ *and the coefficient* $w_{2,j}$ *is defined in terms of* $w_{1,j}$. *(See* (4.19).*)*

Finally, it is presented the result for the extreme eigenvalues near M.

Theorem 2.8.

i) *If* $\pi - d^{(n)}_{2j} \to 0$ *as* $n \to \infty$, *then*

$$\lambda^{(n)}_{2j-1} = M - \frac{M(d^{(n)}_{2j} - \pi)^2}{2} + \frac{2Mf'(\pi)(d^{(n)}_{2j} - \pi)^2}{n+2} + \Delta_7(n,j), \tag{2.13}$$

where $|\Delta_7(n,j)| \le b_7\left((d^{(n)}_{2j} - \pi)^4 + \frac{1}{n^3}\right)$ *and the constant* b_7 *does not depend on* n *and* j.

ii) *If* $\pi - d^{(n)}_{2j+1} \to 0$ *as* $n \to \infty$, *then*

$$\lambda^{(n)}_{2j} = M - \frac{M(d^{(n)}_{2j+1} - \pi)^2}{2} + \frac{2Mf'(\pi)(d^{(n)}_{2j+1} - \pi)^2}{n+2} + \Delta_8(n,j), \tag{2.14}$$

where $|\Delta_8(n,j)| \le b_8\left((d^{(n)}_{2j+1} - \pi)^4 + \frac{1}{n^3}\right)$ *and the constant* b_8 *does not depend on* n *and* j.

Remark 2.9. The formulas in Theorems 2.5–2.8 have 3 asymptotic terms. However, our method allows us to obtain as many terms as we desire.

3. Determinants

The goal of this section is to obtain formulas for the determinant of $T_n(a-\lambda)$ that will allow us to analyze the asymptotic behavior of the eigenvalues. We will derive the equations (2.1) and (2.2) using the main result of [7].

We begin by defining the Chebyshev polynomials $\{\mathcal{T}_n\}$, $\{\mathcal{U}_n\}$, $\{\mathcal{V}_n\}$, and $\{\mathcal{W}_n\}$. They all satisfy the same recurrence relation

$$\mathcal{T}_{n+1}(x) = 2x\mathcal{T}_n(x) - \mathcal{T}_{n-1}(x) \quad \text{for} \quad n = 1, 2, \dots$$

and the different initial conditions are

$$\begin{aligned} \mathcal{T}_0(x) = \mathcal{U}_0(x) = 1, & \qquad 2\mathcal{T}_1(x) = \mathcal{U}_1(x) = 2x, \\ \mathcal{W}_0(x) = \mathcal{V}_0(x) = 1, \quad \text{and} & \quad \mathcal{W}_1(x) = \mathcal{V}_1(x) + 2 = 2x + 1. \end{aligned}$$

Furthermore, we use the following useful properties of them:

$$\begin{aligned} \mathcal{T}_n(\cos\varphi) &= \cos n\varphi, & \mathcal{U}_n(\cos\varphi) &= \frac{\sin(n+1)\varphi}{\sin\varphi}, \\ \mathcal{V}_n(\cos\varphi) &= \frac{\cos(n+\frac{1}{2})\varphi}{\cos\frac{\varphi}{2}}, \quad \text{and} & \mathcal{W}_n(\cos\varphi) &= \frac{\sin(n+\frac{1}{2})\varphi}{\sin\frac{\varphi}{2}}. \end{aligned}$$

Now, we formulate the main results of [7] where the symbol $a(t) = \sum_{k=-r}^{r} a_k t^k$, with $a_r \neq 0$ and $a_k = a_{-k}$, was considered.

Theorem 3.1 ([7] Theorem 1). *Let ξ_j and $\frac{1}{\xi_j}$ $(j = 1, \dots, r)$ be the (distinct) zeros of the polynomial $g_1(t) = t^r a(t)$. Then, for all $p \geq 1$, $\det T_{2p}$ equals*

$$\frac{a_r^{2p}}{2^{r(r-1)}} \times \frac{\begin{vmatrix} \mathcal{V}_p(\alpha_1) & \cdots & \mathcal{V}_p(\alpha_r) \\ \vdots & \ddots & \vdots \\ \mathcal{V}_{p+r-1}(\alpha_1) & \cdots & \mathcal{V}_{p+r-1}(\alpha_r) \end{vmatrix}}{\prod\limits_{1\leq i<j\leq r} (\alpha_j - \alpha_i)} \times \frac{\begin{vmatrix} \mathcal{W}_p(\alpha_1) & \cdots & \mathcal{W}_p(\alpha_r) \\ \vdots & \ddots & \vdots \\ \mathcal{W}_{p+r-1}(\alpha_1) & \cdots & \mathcal{W}_{p+r-1}(\alpha_r) \end{vmatrix}}{\prod\limits_{1\leq i<j\leq r} (\alpha_j - \alpha_i)}$$

and $\det T_{2p+1}$ equals

$$\frac{(-1)^r a_r^{2p+1}}{2^{r(r-2)}} \times \frac{\begin{vmatrix} \mathcal{U}_p(\alpha_1) & \cdots & \mathcal{U}_p(\alpha_r) \\ \vdots & \ddots & \vdots \\ \mathcal{U}_{p+r-1}(\alpha_1) & \cdots & \mathcal{U}_{p+r-1}(\alpha_r) \end{vmatrix}}{\prod\limits_{1\leq i<j\leq r} (\alpha_j - \alpha_i)} \times \frac{\begin{vmatrix} \mathcal{T}_{p+1}(\alpha_1) & \cdots & \mathcal{T}_{p+1}(\alpha_r) \\ \vdots & \ddots & \vdots \\ \mathcal{T}_{p+r}(\alpha_1) & \cdots & \mathcal{T}_{p+r}(\alpha_r) \end{vmatrix}}{\prod\limits_{1\leq i<j\leq r} (\alpha_j - \alpha_i)},$$

where $\alpha_k = \frac{1}{2}\left(\xi_k + \frac{1}{\xi_k}\right)$ $(k = 1, \dots, r)$ are the zeros of the polynomial $h_1(x) = a_0 + 2\sum_{k=1}^{r} a_k \mathcal{T}_k(x)$.

In our case, the coefficients of the polynomial h_1 are $a_0 = 6 - \lambda$, $a_1 = -4$, and $a_2 = 1$. Since $\mathcal{T}_1(x) = x$ and $\mathcal{T}_2(x) = 2x\mathcal{T}_1(x) - \mathcal{T}_0(x) = 2x^2 - 1$, it follows that $h_1(x) = 4x^2 - 8x + (4 - \lambda)$. Notice that h_1 has the zeros $\alpha_1 = 1 - \frac{\sqrt{\lambda}}{2}$ and $\alpha_2 = 1 + \frac{\sqrt{\lambda}}{2}$. Given $t \in \mathbb{T}$ and $\lambda = g(\varphi) = M\sin^4\left(\frac{\varphi}{2}\right)$, we obtain that $\alpha_1 = \cos(\varphi)$ and $\alpha_2 = \cos(\xi)$, where $\xi := \xi(\varphi) = -i\beta(\varphi)$.

Lemma 3.2. *Let a be as in* (1.1), $g(\varphi) = a(e^{i\varphi})$ *and* $\lambda = g(\varphi)$ *for* $\varphi \in (0, \pi)$. *Then, for all* $p \geq 1$, $\det T_{2p}(a - g(\varphi))$ *equals*

$$\frac{\cos((p+\frac{1}{2})\varphi)\cos((p+\frac{3}{2})\xi(\varphi)) - \cos((p+\frac{1}{2})\xi(\varphi))\cos((p+\frac{3}{2})\varphi)}{4\lambda\cos\frac{\varphi}{2}\cos\frac{\xi(\varphi)}{2}}$$
$$\times\frac{\sin((p+\frac{1}{2})\varphi)\sin((p+\frac{3}{2})\xi(\varphi)) - \sin((p+\frac{1}{2})\xi(\varphi))\sin((p+\frac{3}{2})\varphi)}{\sin\frac{\varphi}{2}\sin\frac{\xi(\varphi)}{2}}$$

and $\det T_{2p+1}(a - g(\varphi))$ *equals*

$$(\cos((p+1)\varphi)\cos((p+2)\xi(\varphi)) - \cos((p+1)\xi(\varphi))\cos((p+2)\varphi))$$
$$\times\frac{\sin((p+1)\varphi)\sin((p+2)\xi(\varphi)) - \sin((p+1)\xi(\varphi))\sin((p+2)\varphi)}{\lambda\sin\varphi\sin\xi(\varphi)}.$$

Proof. By Theorem 3.1 $\det T_{2p}$ equals

$$\frac{(\mathcal{V}_p(\alpha_1)\mathcal{V}_{p+1}(\alpha_2) - \mathcal{V}_p(\alpha_2)\mathcal{V}_{p+1}(\alpha_1))(\mathcal{W}_p(\alpha_1)\mathcal{W}_{p+1}(\alpha_2) - \mathcal{W}_p(\alpha_2)\mathcal{W}_{p+1}(\alpha_1))}{4(\alpha_2 - \alpha_1)^2}$$
$$= \frac{\cos((p+\frac{1}{2})\varphi)\cos((p+\frac{3}{2})\xi(\varphi)) - \cos((p+\frac{1}{2})\xi(\varphi))\cos((p+\frac{3}{2})\varphi)}{4\lambda\cos\frac{\varphi}{2}\cos\frac{\xi(\varphi)}{2}}$$
$$\times\frac{\sin((p+\frac{1}{2})\varphi)\sin((p+\frac{3}{2})\xi(\varphi)) - \sin((p+\frac{1}{2})\xi(\varphi))\sin((p+\frac{3}{2})\varphi)}{\sin\frac{\varphi}{2}\sin\frac{\xi(\varphi)}{2}}$$

and $\det T_{2p+1}$ equals

$$\frac{(\mathcal{U}_p(\alpha_1)\mathcal{U}_{p+1}(\alpha_2) - \mathcal{U}_p(\alpha_2)\mathcal{U}_{p+1}(\alpha_1))(\mathcal{T}_{p+1}(\alpha_1)\mathcal{T}_{p+2}(\alpha_2) - \mathcal{T}_{p+1}(\alpha_2)\mathcal{T}_{p+2}(\alpha_1))}{(\alpha_2 - \alpha_1)^2}$$
$$= \frac{\sin((p+1)\varphi)\sin((p+2)\xi(\varphi)) - \sin((p+1)\xi(\varphi))\sin((p+2)\varphi)}{\lambda\sin\varphi\sin\xi(\varphi)}$$
$$\times(\cos((p+1)\varphi)\cos((p+2)\xi(\varphi)) - \cos((p+1)\xi(\varphi))\cos((p+2)\varphi)). \qquad \square$$

Lemma 3.3. *For all* $p \geq 1$ *and* $\varphi \in (0, \pi)$, *we have* $\det T_{2p}(a - g(\varphi)) = 0$ *and* $\det T_{2p+1}(a - g(\varphi)) = 0$ *are equivalent to the equations* (2.1) *and* (2.2).

Proof. From Lemma 3.2 we obtain that $\det T_{2p}(a - g(\varphi)) = 0$ is equivalent to

$$\begin{aligned}&\cos\left(\left(p+\frac{1}{2}\right)\varphi\right)\cos\left(\left(p+\frac{3}{2}\right)\xi(\varphi)\right)\\ &-\cos\left(\left(p+\frac{1}{2}\right)\xi(\varphi)\right)\cos\left(\left(p+\frac{3}{2}\right)\varphi\right) = 0\end{aligned} \tag{3.1}$$

or

$$\begin{aligned} &\sin\left(\left(p+\frac{1}{2}\right)\varphi\right)\sin\left(\left(p+\frac{3}{2}\right)\xi(\varphi)\right) \\ &-\sin\left(\left(p+\frac{1}{2}\right)\xi(\varphi)\right)\sin\left(\left(p+\frac{3}{2}\right)\varphi\right)=0. \end{aligned} \tag{3.2}$$

Since $\cos(-ix)=\cosh(x)$, substituting $\xi(\varphi)=-i\beta(\varphi)$ in (3.1) we arrive at the equation

$$\frac{\cos((p+\frac{3}{2})\varphi)}{\cos((p+\frac{1}{2})\varphi)}=\frac{\cosh((p+\frac{3}{2})\beta(\varphi))}{\cosh((p+\frac{1}{2})\beta(\varphi))}. \tag{3.3}$$

Notice that (3.1) and (3.3) are equivalent. Furthermore, $\cos((p+\frac{3}{2})\varphi)=\cos((p+\frac{1}{2})\varphi)\cos\varphi-\sin((p+\frac{1}{2})\varphi)\sin\varphi$. Therefore, if $\cos((p+\frac{1}{2})\varphi)=0$, then $\varphi=\frac{(2j+1)\pi}{2p+1}$, for some $j\in\mathbb{Z}$. Thus, $\cos((p+\frac{3}{2})\varphi)=(-1)^{j+1}\sin\frac{(2j+1)\pi}{2p+1}$. Since $j\neq p$ it follows that $\cos((p+\frac{3}{2})\varphi)\neq 0$.

The equation (3.3) can be rewritten as follows

$$\frac{\cos((p+1)\varphi+\frac{\varphi}{2})}{\cos((p+1)\varphi-\frac{\varphi}{2})}=\frac{\cosh((p+1)\beta(\varphi)+\frac{\beta(\varphi)}{2})}{\cosh((p+1)\beta(\varphi)-\frac{\beta(\varphi)}{2})}.$$

Taking $p=\frac{n}{2}$ a simple calculation shows that

$$\tan\left(\frac{n+2}{2}\varphi\right)=-\cot\frac{\varphi}{2}\tanh\frac{\beta(\varphi)}{2}\tanh\left(\frac{n+2}{2}\beta(\varphi)\right). \tag{3.4}$$

Given that $\beta(\varphi)=2\log\left(\sin\frac{\varphi}{2}+\sqrt{1+\sin^2\frac{\varphi}{2}}\right)$ and $\tanh x=\frac{e^{2x}-1}{e^{2x}+1}$, we easily see that

$$\cot\frac{\varphi}{2}\tanh\frac{\beta(\varphi)}{2}=\frac{\cos\frac{\varphi}{2}}{\sqrt{1+\sin^2\frac{\varphi}{2}}}=f(\varphi). \tag{3.5}$$

Substituting (3.5) in (3.4) we arrive at the equation (2.1), i.e.,

$$\tan\left(\frac{n+2}{2}\varphi\right)=-f(\varphi)\tanh\left(\frac{n+2}{2}\beta(\varphi)\right)$$

Similarly, substituting $\xi(\varphi)=-i\beta(\varphi)$ in (3.2) and since $\sin(-ix)=-i\sinh x$, we arrive at the equation

$$\frac{\sin((p+\frac{3}{2})\varphi)}{\sin((p+\frac{1}{2})\varphi)}=\frac{\sinh((p+\frac{3}{2})\beta(\varphi))}{\sinh((p+\frac{1}{2})\beta(\varphi))}. \tag{3.6}$$

Again, notice that (3.2) and (3.6) are equivalent. Furthermore,

$$\sin\left(\left(p+\frac{3}{2}\right)\varphi\right)=\sin\left(\left(p+\frac{1}{2}\right)\varphi\right)\cos\varphi+\cos\left(\left(p+\frac{1}{2}\right)\varphi\right)\sin\varphi.$$

Hence, if $\sin((p+\frac{1}{2})\varphi)=0$, then $\varphi=\frac{2j\pi}{2p+1}$, for some $j\in\mathbb{Z}$. Thus, $\sin((p+\frac{3}{2})\varphi)=(-1)^j\sin\frac{2j\pi}{2p+1}$. Since $j\neq 0$ it follows that $\cos((p+\frac{3}{2})\varphi)\neq 0$.

Equation (3.6) can be rewritten as follows

$$\frac{\sin((p+1)\varphi+\frac{\varphi}{2})}{\sin((p+1)\varphi-\frac{\varphi}{2})}=\frac{\sinh((p+1)\beta(\varphi)+\frac{\beta(\varphi)}{2})}{\sinh((p+1)\beta(\varphi)-\frac{\beta(\varphi)}{2})}.$$

Simplifying this equation and using $p=\frac{n}{2}$, we obtain (2.2), i.e.,

$$\tan\left(\frac{n+2}{2}\varphi\right)=\frac{1}{f(\varphi)}\tanh\left(\frac{n+2}{2}\beta(\varphi)\right).$$

Once again, from Lemma 3.2 we find that $\det T_{2p+1}(a-\lambda)=0$ is equivalent to

$$\cos((p+1)\varphi)\cos((p+2)\xi(\varphi))-\cos((p+1)\xi(\varphi))\cos((p+2)\varphi)=0 \qquad (3.7)$$

or

$$\sin((p+1)\varphi)\sin((p+2)\xi(\varphi))-\sin((p+1)\xi(\varphi))\sin((p+2)\varphi)=0. \qquad (3.8)$$

Substituting $\xi(\varphi)=-i\beta(\varphi)$ in (3.7) and since $\cos(-ix)=\cosh x$, we arrive at the equation

$$\frac{\cos((p+2)\varphi)}{\cos((p+1)\varphi)}=\frac{\cosh((p+2)\beta(\varphi))}{\cosh((p+1)\beta(\varphi))}.$$

Again, if $\cos((p+1)\varphi)=0$ then, $\cos((p+2)\varphi)\neq 0$.

The last equation can be rewritten as follows

$$\frac{\cos((p+\frac{3}{2})\varphi+\frac{\varphi}{2})}{\cos((p+\frac{3}{2})\varphi-\frac{\varphi}{2})}=\frac{\cosh((p+\frac{3}{2})\beta(\varphi)+\frac{\beta(\varphi)}{2})}{\cosh((p+\frac{3}{2})\beta(\varphi)-\frac{\beta(\varphi)}{2})}.$$

Simplifying this equation and using $p=\frac{n}{2}-\frac{1}{2}$, we obtain (2.1).

Similarly, substituting $\xi(\varphi)=-i\beta(\varphi)$ in (3.8) with $\sin(-ix)=-i\sinh x$, we arrive at the equation

$$\frac{\sin((p+2)\varphi)}{\sin((p+1)\varphi)}=\frac{\sinh((p+2)\beta(\varphi))}{\sinh((p+1)\beta(\varphi))}.$$

Once again, if $\sin((p+1)\varphi)=0$, then $\sin((p+2)\varphi)\neq 0$.

The last equation can be rewritten as follows

$$\frac{\sin((p+\frac{3}{2})\varphi+\frac{\varphi}{2})}{\sin((p+\frac{3}{2})\varphi-\frac{\varphi}{2})}=\frac{\sinh((p+\frac{3}{2})\beta(\varphi)+\frac{\beta(\varphi)}{2})}{\sinh((p+\frac{3}{2})\beta(\varphi)-\frac{\beta(\varphi)}{2})}.$$

Finally, simplifying this equation and using $p=\frac{n}{2}-\frac{1}{2}$, we get (2.2). □

4. Proof of the main results

Proof of Theorem 2.1. This theorem follows directly from Lemma 3.3. □

Now, we introduce two functions $F:[0,\pi]\times\mathbb{N}\to\mathbb{R}$ and $G:[0,\pi]\times\mathbb{N}\to\mathbb{R}$ given by

$$F(\varphi,n):=\frac{n+2}{2}\varphi+\arctan\left(f(\varphi)\tanh\left(\frac{n+2}{2}\beta(\varphi)\right)\right)$$

and

$$G(\varphi,n) := \frac{n+2}{2}\varphi - \arctan\left(\frac{1}{f(\varphi)}\tanh\left(\frac{n+2}{2}\beta(\varphi)\right)\right).$$

The functions F and G will have an important role in this work, the solutions of

$$F(\varphi,n) = \pi j \quad \text{and} \quad G(\varphi,n) = \pi j \quad (j \in \mathbb{N}),$$

will allow us to find asymptotic formulas for the eigenvalues of $T_n(a)$.

Lemma 4.1. *If n is sufficiently large, then*

i) *F is increasing on $[0,\pi]$.*

ii) *G is increasing on $[\frac{\pi}{2(n+2)},\pi]$.*

Proof. i) For $\varphi \in [0,\pi]$, we have

$$F'(\varphi,n) = \frac{n+2}{2} + \frac{\frac{n+2}{2} f^2(\varphi)\operatorname{sech}^2(\frac{n+2}{2}\beta(\varphi))}{1+f^2(\varphi)\tanh^2(\frac{n+2}{2}\beta(\varphi))} + \frac{f'(\varphi)\tanh(\frac{n+2}{2}\beta(\varphi))}{1+f^2(\varphi)\tanh^2(\frac{n+2}{2}\beta(\varphi))},$$

where $f'(\varphi) = -\frac{\sin\frac{\varphi}{2}}{2\sqrt{1+\sin^2\frac{\varphi}{2}}}\left(1+f^2(\varphi)\right)$.

Since $\frac{\frac{n+2}{2} f^2(\varphi)\operatorname{sech}^2(\frac{n+2}{2}\beta(\varphi))}{1+f^2(\varphi)\tanh^2(\frac{n+2}{2}\beta(\varphi))} \geq 0$ for $\varphi \in [0,\pi]$, it follows that

$$F'(\varphi,n) \geq \frac{n+2}{2} + \frac{f'(\varphi)\tanh(\frac{n+2}{2}\beta(\varphi))}{1+f^2(\varphi)\tanh^2(\frac{n+2}{2}\beta(\varphi))}. \tag{4.1}$$

Furthermore, since f decreases on $[0,\pi]$, we see that $0 \leq f(\varphi) \leq 1$. Thus, we arrive at the estimate

$$|f'(\varphi)| \leq 1, \ \varphi \in [0,\pi].$$

Additionally, given that β increases on $[0,\pi]$, we get that $0 \leq \beta(\varphi) \leq 2\log\left(1+\sqrt{2}\right)$. Hence, if n is large enough, then we have

$$\left|\frac{f'(\varphi)\tanh\left(\frac{n+2}{2}\beta(\varphi)\right)}{1+f^2(\varphi)\tanh^2\left(\frac{n+2}{2}\beta(\varphi)\right)}\right| < \tanh\left(\frac{n+2}{2}\beta(\varphi)\right) \leq \tanh\left(\log\left(1+\sqrt{2}\right)(n+2)\right)$$
$$= 1 + O\left(e^{-2\log(1+\sqrt{2})(n+2)}\right) < 1.2.$$

Substituting this inequality in (4.1) we arrive at the estimate

$$F'(\varphi,n) > \frac{n}{2} + 1 - 1.2 = \frac{n}{2} - \frac{1}{5}.$$

Finally, if n satisfies $\frac{n}{2} > \frac{1}{5}$, then $F'(\varphi,n) > 0$.

ii) Similarly, for $\varphi \in [\frac{\pi}{2(n+2)},\pi]$ we obtain that

$$G'(\varphi,n) = \frac{n+2}{2} - \frac{\frac{n+2}{2}\operatorname{sech}^2(\frac{n+2}{2}\beta(\varphi))}{1+\frac{1}{f^2(\varphi)}\tanh^2(\frac{n+2}{2}\beta(\varphi))} + \frac{f'(\varphi)\tanh(\frac{n+2}{2}\beta(\varphi))}{f^2(\varphi)+\tanh^2(\frac{n+2}{2}\beta(\varphi))}.$$

Since $|f'(\varphi)| \leq 1$ for $\varphi \in [0,\pi]$, we arrive at the estimate

$$G'(\varphi,n) > \frac{n+2}{2} - \frac{n+2}{2\cosh^2(\frac{n+2}{2}\beta(\varphi))} - \coth\left(\frac{n+2}{2}\beta(\varphi)\right). \tag{4.2}$$

Since β increases on $[0,\pi]$ with $\frac{\pi}{2(n+2)} \le \varphi \le \pi$, it follows that $\frac{n+2}{2}\beta\left(\frac{\pi}{2(n+2)}\right) \le \frac{n+2}{2}\beta(\varphi)$. Furthermore, $\beta(\varphi) = \varphi + O(\varphi^3)$ as $\varphi \to 0^+$. Therefore, if n is large enough, then we get

$$\frac{1}{\cosh^2(\frac{n+2}{2}\beta(\varphi))} \le \left(\frac{2}{e^{\frac{1}{2}(n+2)(\frac{\pi}{2(n+2)}+O((\frac{\pi}{2(n+2)})^3))} + e^{-\frac{1}{2}(n+2)(\frac{\pi}{2(n+2)}+O((\frac{\pi}{2(n+2)})^3))}}\right)^2$$

$$= \left(\frac{2}{e^{\frac{\pi}{4}+O\left(\frac{1}{(n+2)^2}\right)} + e^{-\frac{\pi}{4}+O\left(\frac{1}{(n+2)^2}\right)}}\right)^2 \le \left(\frac{2.3}{e^{\frac{\pi}{4}} + e^{-\frac{\pi}{4}}}\right)^2 < 0.8$$

and

$$\coth\left(\frac{n+2}{2}\beta(\varphi)\right) \le \coth\left(\frac{n+2}{2}\beta\left(\frac{\pi}{2(n+2)}\right)\right) = \coth\left(\frac{\pi}{4} + O\left(\frac{1}{(n+2)^2}\right)\right)$$

$$= \frac{e^{\frac{\pi}{4}} + e^{-\frac{\pi}{4}}}{e^{\frac{\pi}{4}} - e^{-\frac{\pi}{4}}} + \frac{c_3}{(n+2)^2} < 1.6,$$

where c_3 does not depend on n.

Substituting these inequalities in (4.2) we arrive at the estimate

$$G'(\varphi, n) > \frac{n}{10} - \frac{7}{5}.$$

Finally, if n satisfies $\frac{n}{10} > \frac{7}{5}$, then $G'(\varphi, n) > 0$. □

Proof of Theorem 2.3. Let n be an even number.

i) For each j, F is continuous on the segment $\left[\frac{(2j-1)\pi}{n+2}, \frac{2\pi j}{n+2}\right]$. Given that

$$\arctan\left(f\left(\frac{(2j-1)\pi}{n+2}\right)\tanh\left(\frac{n+2}{2}\beta\left(\frac{(2j-1)\pi}{n+2}\right)\right)\right) - \frac{\pi}{2} < 0$$

and

$$\arctan\left(f\left(\frac{2\pi j}{n+2}\right)\tanh\left(\frac{n+2}{2}\beta\left(\frac{2\pi j}{n+2}\right)\right)\right) > 0,$$

we easily see that

$$F\left(\frac{(2j-1)\pi}{n+2}, n\right) < \pi j, \quad F\left(\frac{2\pi j}{n+2}, n\right) > \pi j.$$

Thus, the intermediate value theorem tells us that (2.3) have solutions, i.e., for $1 \le j \le \frac{n}{2}$ there is at least one $\varphi_{2j-1}^{(n)}$ with $F(\varphi_{2j-1}^{(n)}, n) = \pi j$. By Lemma 4.1, F is a strictly increasing function on $[0,\pi]$ and we conclude that these solutions are unique.

ii) Similarly, for each j the function G is continuous on the segment $\left[\frac{2\pi j}{n+2}, \frac{(2j+1)\pi}{n+2}\right]$. Since

$$\arctan\left(\frac{1}{f\left(\frac{2\pi j}{n+2}\right)}\tanh\left(\frac{n+2}{2}\beta\left(\frac{2\pi j}{n+2}\right)\right)\right) > 0,$$

and

$$\frac{\pi}{2} - \arctan\left(\frac{1}{f\left(\frac{(2j+1)\pi}{n+2}\right)}\tanh\left(\frac{n+2}{2}\beta\left(\frac{(2j+1)\pi}{n+2}\right)\right)\right) > 0,$$

we easily see that

$$G\left(\frac{2\pi j}{n+2}, n\right) < \pi j, \quad G\left(\frac{(2j+1)\pi}{n+2}, n\right) > \pi j.$$

Once again, the intermediate value theorem tells us that(2.4) have solutions, i.e., for $1 \leq j \leq \frac{n}{2}$ there is at least one $\varphi_{2j}^{(n)}$ with $G(\varphi_{2j}^{(n)}, n) = \pi j$. By Lemma 4.1, G is a strictly increasing function on $[\frac{\pi}{2(n+2)}, \pi]$ and we conclude that these solutions are unique.

iii) For example, when n is an even number, from i) and ii) it follows that

$$\varphi_1^{(n)} < \varphi_2^{(n)} < \cdots < \varphi_{2j-1}^{(n)} < \varphi_{2j}^{(n)} < \cdots < \varphi_{n-1}^{(n)} < \varphi_n^{(n)}.$$

Furthermore, given that g is a strictly increasing function on $[0, \pi]$, we obtain

$$g(\varphi_1^{(n)}) < g(\varphi_2^{(n)}) < \cdots < g(\varphi_{2j-1}^{(n)}) < g(\varphi_{2j}^{(n)}) < \cdots < g(\varphi_{n-1}^{(n)}) < g(\varphi_n^{(n)}).$$

Since $\lambda = g(\varphi)$, we see that

$$\lambda_1^{(n)} < \lambda_2^{(n)} < \cdots < \lambda_{n-1}^{(n)} < \lambda_n^{(n)}. \qquad \square$$

Proof of Theorem 2.5. Let $q := \frac{n+2}{2}$.

i) Making the variable change $\frac{u}{q} = \varphi - d_-$ in (2.3), we arrive at the equation

$$u = -\arctan\left(f\left(d_- + \frac{u}{q}\right)\tanh\left(q\beta\left(d_- + \frac{u}{q}\right)\right)\right), \tag{4.3}$$

where $u \in \left(-\frac{\pi}{2}, 0\right)$.

By Taylor's theorem, we have

$$f\left(d_- + \frac{u}{q}\right) = f(d_-) + \frac{f'(d_-)u}{q} + O\left(\frac{u^2}{q^2}\right), \quad q \to \infty \tag{4.4}$$

and

$$q\beta\left(d_- + \frac{u}{q}\right) = q\beta(d_-) + \beta'(d_-)u + \frac{\beta''(d_-)u^2}{2q} + O\left(\frac{u^3}{q^2}\right), \quad q \to \infty. \tag{4.5}$$

Replacing (4.4) and (4.5) in (4.3), we obtain

$$\begin{aligned} u = & -\arctan\left[\left(f(d_-) + \frac{f'(d_-)u}{q} + O\left(\frac{u^2}{q^2}\right)\right)\right. \\ & \left.\times \tanh\left(q\beta(d_-) + \beta'(d_-)u + \frac{\beta''(d_-)u^2}{2q} + O\left(\frac{u^3}{q^2}\right)\right)\right], \quad q \to \infty. \end{aligned} \tag{4.6}$$

We will seek for an asymptotic formula for the solution $u_{*,j}$ of (4.3) with the form

$$u_{*,j} := u_{1,j} + \frac{u_{2,j}}{q} + O\left(\frac{1}{q^2}\right), \quad q \to \infty \tag{4.7}$$

where the coefficients $u_{1,j}$ and $u_{2,j}$ will be properly defined. Substituting (4.7) in (4.4) and (4.5), as $q \to \infty$, we arrive at

$$f(d_-) + \frac{f'(d_-)u_{*,j}}{q} + O\left(\frac{u_{*,j}^2}{q^2}\right) = f(d_-) + \frac{f'(d_-)u_{1,j}}{q} + O\left(\frac{1}{q^2}\right) \tag{4.8}$$

and

$$\begin{aligned} q\beta(d_-) + \beta'(d_-)u_{*,j} + \frac{\beta''(d_-)u_{*,j}^2}{2q} + O\left(\frac{u_{*,j}^3}{q^2}\right) \\ = \Psi_j^- + \frac{\beta''(d_-)u_{1,j}^2 + 2\beta'(d_-)u_{2,j}}{2q} + O\left(\frac{1}{q^2}\right), \end{aligned} \tag{4.9}$$

where $\Psi_j^- := q\beta(d_-) + \beta'(d_-)u_{1,j}$.

Replacing (4.7), (4.8), and (4.9) in (4.6), it follows that

$$\begin{aligned} u_{1,j} + \frac{u_{2,j}}{q} = &-\left(\frac{f(d_-)[\beta''(d_-)u_{1,j}^2 + 2\beta'(d_-)u_{2,j}] + f'(d_-)u_{1,j}\sinh(2\Psi_j^-)}{2\cosh^2\Psi_j^-[1 + f^2(d_-)\tanh^2\Psi_j^-]}\right)\frac{1}{q} \\ &- \arctan\left(f(d_-)\tanh\Psi_j^-\right) + O\left(1/q^2\right), \quad q \to \infty. \end{aligned}$$

Notice that $u_{1,j}$ can be defined as the solution of the equation

$$u = -\arctan(f(d_-)\tanh\Psi_j^-(u)),$$

where $\Psi_j^-(u) := q\beta(d_-) + \beta'(d_-)u$.

Indeed, if we define $F_1(u) := u + \arctan(f(d_-)\tanh\Psi_j^-(u))$ for $u \in [-\frac{\pi}{2}, 0]$, then

$$F_1'(u) = 1 + \frac{f^2(d_-)}{\cosh^2\Psi_j^-(u) + f^2(d_-)\sinh^2\Psi_j^-(u)} > 0.$$

This implies that F_1 is a strictly increasing function on $(-\frac{\pi}{2}, 0)$. Since F_1 is continuous on $[-\frac{\pi}{2}, 0]$ with $F_1(-\frac{\pi}{2}) < 0$ and $F_1(0) > 0$, there is a unique $u_{1,j}$ such that

$$u_{1,j} = -\arctan(f(d_-)\tanh\Psi_j^-(u_{1,j})). \tag{4.10}$$

Additionally, $u_{2,j}$ should satisfy

$$u_{2,j} = -\frac{f(d_-)[\beta''(d_-)u_{1,j}^2 + 2\beta'(d_-)u_{2,j}] + f'(d_-)u_{1,j}\sinh(2\Psi_j^-)}{2\left[\cosh^2\Psi_j^- + f^2(d_-)\sinh^2\Psi_j^-\right]}.$$

A simple calculation reveals that

$$u_{2,j} = -\frac{f(d_-)\beta''(d_-)u_{1,j}^2 + f'(d_-)u_{1,j}\sinh(2\Psi_j^-)}{2\left[1 + f^2(d_-)\right]\cosh^2\Psi_j^-}. \tag{4.11}$$

Given $u_{*,j} = q(\varphi_{2j-1}^{(n)} - d_-)$ with $q = \frac{n+2}{2}$ we get the asymptotic formula (2.7), i.e.,

$$\varphi_{2j-1}^{(n)} = d_- + \frac{2u_{1,j}}{n+2} + \frac{4u_{2,j}}{(n+2)^2} + O\left(\frac{1}{n^3}\right), \quad n \to \infty$$

for $j = 1, \dots, [\frac{n+1}{2}]$.

ii) Let $h(\varphi) := \frac{1}{f(\varphi)}$. The variable change $\frac{w}{q} = \varphi - d_+$ in (2.4), leads us to

$$w = -\operatorname{arccot}\left(h\left(d_+ + \frac{w}{q}\right)\tanh\left(q\beta\left(d_+ + \frac{w}{q}\right)\right)\right), \tag{4.12}$$

where $w \in \left(-\frac{\pi}{2}, 0\right)$.

Substituting (4.4) and (4.5) in (4.12), with d_+ and h instead of d_- and f, it follows that

$$\begin{aligned} w = & -\operatorname{arccot}\left[\left(h(d_+) + \frac{h'(d_+)w}{q} + O\left(\frac{w^2}{q^2}\right)\right)\right. \\ & \left.\times \tanh\left(q\beta(d_+) + \beta'(d_+)w + \frac{\beta''(d_+)w^2}{2q} + O\left(\frac{w^3}{q^2}\right)\right)\right], \quad q \to \infty. \end{aligned} \tag{4.13}$$

We also seek for an asymptotic formula for the solution $w_{*,j}$ of (4.12) with the form

$$w_{*,j} := w_{1,j} + \frac{w_{2,j}}{q} + O\left(\frac{1}{q^2}\right), \quad q \to \infty, \tag{4.14}$$

where the coefficients $w_{1,j}$ and $w_{2,j}$ will be properly defined.

Now, as $q \to \infty$, using (4.14) we arrive at

$$h(d_+) + \frac{h'(d_+)w_{*,j}}{q} + O\left(\frac{w_{*,j}^2}{q^2}\right) = h(d_+) + \frac{h'(d_+)w_{1,j}}{q} + O\left(\frac{1}{q^2}\right) \tag{4.15}$$

and

$$\begin{aligned} & q\beta(d_+) + \beta'(d_+)w_{*,j} + \frac{\beta''(d_+)w_{*,j}^2}{2q} + O\left(\frac{w_{*,j}^3}{q^2}\right) \\ & = \Psi_j^+ + \frac{\beta''(d_+)w_{1,j}^2 + 2\beta'(d_+)w_{2,j}}{2q} + O\left(\frac{1}{q^2}\right), \end{aligned} \tag{4.16}$$

where $\Psi_j^+ := q\beta(d_+) + \beta'(d_+)w_{1,j}$.

Substituting (4.14), (4.15), and (4.16) in (4.13), it follows that

$$\begin{aligned} w_{1,j} + \frac{w_{2,j}}{q} = & \left(\frac{h(d_+)[\beta''(d_+)w_{1,j}^2 + 2\beta'(d_+)w_{2,j}] + h'(d_+)w_{1,j}\sinh(2\Psi_j^+)}{2\cosh^2\Psi_j^+[1 + h^2(d_+)\tanh^2\Psi_j^+]}\right)\frac{1}{q} \\ & - \operatorname{arccot}\left(h(d_+)\tanh\Psi_j^+\right) + O\left(\frac{1}{q^2}\right), \quad q \to \infty. \end{aligned}$$

Similarly, notice that $w_{1,j}$ can be defined as the solution of the equation

$$w = -\operatorname{arccot}(h(d_+)\tanh\Psi_j^+(w)),$$

where $\Psi_j^+(w) := q\beta(d_+) + \beta'(d_+)w$.

Indeed, if we define $F_2(w) := w + \operatorname{arccot}(h(d_+)\tanh\Psi_j^+(w))$ for $w \in [-\frac{\pi}{2}, 0]$, then

$$F_2'(w) = 1 - \frac{1}{\cosh^2\Psi_j^+(w)[1 + h^2(d_+)\tanh^2\Psi_j^+(w)]}.$$

Additionally, we have

$$q\beta(d_+) - \frac{\beta'(d_+)\pi}{2} < \Psi_j^+(w) < q\beta(d_+),$$

for every $w \in (-\frac{\pi}{2}, 0)$.
Since $d_+ > \frac{\pi}{q}$ and $\beta'(d_+) < 1$, we see that

$$q\beta\left(\frac{\pi}{q}\right) - \frac{\pi}{2} < q\beta(d_+) - \frac{\beta'(d_+)\pi}{2} < \Psi_j^+(w). \tag{4.17}$$

Using the inequality (4.17) we obtain

$$\frac{1}{\cosh^2\Psi_j^+(w)[1 + h^2(d_+)\tanh^2\Psi_j^+(w)]} < \frac{1}{\cosh^2\Psi_j^+(w)} < \frac{1}{\cosh^2(q\beta(\frac{\pi}{q}) - \frac{\pi}{2})}$$
$$= \left(\frac{2}{e^{q\beta(\frac{\pi}{q})-\frac{\pi}{2}} + e^{-q\beta(\frac{\pi}{q})+\frac{\pi}{2}}}\right)^2 < \left(\frac{4}{e^{\frac{\pi}{2}} + e^{-\frac{\pi}{2}}}\right)^2 < 0.8.$$

This implies that F_2 is a strictly increasing function on $(-\frac{\pi}{2}, 0)$. Given that F_2 is continuous on $[-\frac{\pi}{2}, 0]$ with $F_2(-\frac{\pi}{2}) < 0$ and $F_2(0) > 0$, there is a unique $w_{1,j}$ such that

$$w_{1,j} = -\operatorname{arccot}(h(d_+)\tanh\Psi_j^+(w_{1,j})). \tag{4.18}$$

Now, $w_{2,j}$ should satisfy the equation

$$w_{2,j} = \frac{h'(d_+)w_{1,j}\sinh(2\Psi_j^+) + h(d_+)[\beta''(d_+)w_{1,j}^2 + 2\beta'(d_+)w_{2,j}]}{2\left[\cosh^2\Psi_j^+ + h^2(d_+)\sinh^2\Psi_j^+\right]}.$$

A simple calculation reveals that

$$w_{2,j} = \frac{f(d_+)\beta''(d_+)w_{1,j}^2 - f'(d_+)w_{1,j}\sinh(2\Psi_j^+)}{2\left[1 + f^2(d_+)\right]\sinh^2\Psi_j^+}. \tag{4.19}$$

Given $w_{*,j} = q(\varphi_{2j}^{(n)} - d_+)$ with $q = \frac{n+2}{2}$ we get the asymptotic formula (2.8), i.e.,

$$\varphi_{2j}^{(n)} = d_+ + \frac{2w_{1,j}}{n+2} + \frac{4w_{2,j}}{(n+2)^2} + O\left(\frac{1}{n^3}\right), \quad n \to \infty,$$

for $j = 1, \ldots, [\frac{n}{2}]$. □

The asymptotic formulas for the eigenvalues will need $\lambda_{2j-1}^{(n)} = g(\varphi_{2j-1}^{(n)})$ and $\lambda_{2j}^{(n)} = g(\varphi_{2j}^{(n)})$. By Taylor's theorem applied to (2.7) and (2.8), as $q = \frac{n+2}{2} \to \infty$, there are points $\zeta_1 \in (d_- - \frac{\pi}{2q}, d_-)$ and $\zeta_2 \in (d_+ - \frac{\pi}{2q}, d_+)$ such that

$$\lambda_{2j-1}^{(n)} = g(\varphi_{2j-1}^{(n)}) = g(d_-) + g'(d_-)\left(\frac{u_{1,j}}{q} + \frac{u_{2,j}}{q^2} + O\left(\frac{1}{q^3}\right)\right) \tag{4.20}$$
$$+ \frac{g''(d_-)}{2}\left(\frac{u_{1,j}}{q} + \frac{u_{2,j}}{q^2} + O\left(\frac{1}{q^3}\right)\right)^2 + \frac{g'''(\zeta_1)}{6}\left(\frac{u_{1,j}}{q} + \frac{u_{2,j}}{q^2} + O\left(\frac{1}{q^3}\right)\right)^3$$

and

$$\lambda_{2j}^{(n)} = g(\varphi_{2j}^{(n)}) = g(d_+) + g'(d_+)\left(\frac{w_{1,j}}{q} + \frac{w_{2,j}}{q^2} + O\left(\frac{1}{q^3}\right)\right) \tag{4.21}$$
$$+ \frac{g''(d_+)}{2}\left(\frac{w_{1,j}}{q} + \frac{w_{2,j}}{q^2} + O\left(\frac{1}{q^3}\right)\right)^2 + \frac{g'''(\zeta_2)}{6}\left(\frac{w_{1,j}}{q} + \frac{w_{2,j}}{q^2} + O\left(\frac{1}{q^3}\right)\right)^3.$$

Proof of Theorem 2.6. i) Notice that g', g'', and g''' are bounded. The relation $q = \frac{n+2}{2}$ in (4.20) shows that

$$\lambda_{2j-1}^{(n)} = g(d_-) + \frac{2u_{1,j}g'(d_-)}{n+2} + \frac{2(u_{1,j})^2 g''(d_-) + 4u_{2,j}g'(d_-)}{(n+2)^2} + O\left(\frac{1}{n^3}\right), \quad n \to \infty.$$

Now, we analyze $u_{1,j}$ and $u_{2,j}$. From (4.10) we know that

$$u_{1,j} = -\arctan(f(d_-)\tanh \Psi_j^-(u_{1,j})),$$

where $\Psi_j^-(u_{1,j}) = q\beta(d_-) + \beta'(d_-)u_{1,j}$. Let $c_4 := \beta(\pi\epsilon)$.

Given that $\pi\epsilon < d_- < (1-\epsilon)\pi$ and that β is an increasing function, we obtain $-2q\beta(d_-) < -2q\beta(\pi\epsilon)$. Furthermore, if q is sufficiently large, then we see that

$$f(d_-)\tanh \Psi_j^-(u_{1,j}) = f(d_-)(1 + O(e^{(-2\beta(\pi\epsilon)q)})) = f(d_-) + O(e^{-2c_4 q}), \quad q \to \infty.$$

Hence, we get

$$u_{1,j} = -\arctan f(d_-) + O\left(e^{-c_4(n+2)}\right), \quad n \to \infty.$$

Similarly, from (4.11) we arrive at the representation

$$u_{2,j} = \frac{f'(d_-)\arctan f(d_-)}{1 + f^2(d_-)} + O\left(e^{-c_4(n+2)}\right), \quad n \to \infty.$$

Now, if $u_{1,j}^* := -\arctan f(d_-)$ and $u_{2,j}^* := \frac{f'(d_-)\arctan f(d_-)}{1+f^2(d_-)}$, then it follows that

$$|u_{1,j} - u_{1,j}^*| = O\left(e^{-c_4(n+2)}\right), \quad |u_{2,j} - u_{2,j}^*| = O\left(e^{-c_4(n+2)}\right), \quad n \to \infty.$$

ii) Analogously, using (4.21) we arrive at the formula (2.10).

We analyze $w_{1,j}$ and $w_{2,j}$, as we did with $u_{1,j}$ and $u_{2,j}$ in the first part of the proof. Therefore, as $n \to \infty$, from (4.18) and (4.19) $w_{1,j}$ and $w_{2,j}$ satisfy

$$w_{1,j} = -\operatorname{arccot}\frac{1}{f(d_+)} + O\left(e^{-c_4(n+2)}\right)$$

and

$$w_{2,j} = \frac{f'(d_+)\operatorname{arccot}\frac{1}{f(d_+)}}{1+f^2(d_+)} + O\left(e^{-c_4(n+2)}\right).$$

Similarly, if $w^*_{1,j} := -\operatorname{arccot}\frac{1}{f(d_+)}$ and $w^*_{2,j} := \frac{f'(d_+)\operatorname{arccot}\frac{1}{f(d_+)}}{1+f^2(d_+)}$, then it follows that

$$|w_{1,j} - w^*_{1,j}| = O\left(e^{-c_4(n+2)}\right), \quad |w_{2,j} - w^*_{2,j}| = O\left(e^{-c_4(n+2)}\right), \quad n \to \infty. \quad \square$$

Proof of Theorem 2.7. Recall that $g(\varphi) := M\sin^4(\varphi/2)$, for $\varphi \in [0,\pi]$.
i) Given that $\lambda^{(n)}_{2j-1} = g(\varphi^{(n)}_{2j-1})$, where $\varphi^{(n)}_{2j-1}$ is given by (2.7), we arrive at

$$\lambda^{(n)}_{2j-1} = M\sin^4\left(\frac{d_-}{2} + \frac{u_{1,j}}{2q} + \frac{u_{2,j}}{2q^2} + O\left(\frac{1}{q^3}\right)\right), \quad q \to \infty.$$

Let $C_j := 1 + \frac{u_{1,j}}{j\pi}$. Since $\sin x = x - \frac{1}{6}x^3 + O(x^5)$, $x \to 0$, a simple calculation shows that $\lambda^{(n)}_{2j-1}$ equals

$$C_j^4 d_-^4\left[1+\frac{u_{2,j}}{d_-q^2C_j}+O\left(\frac{1}{d_-q^3}\right)-\frac{d_-^2C_j^2}{24}\left(1+\frac{u_{2,j}}{d_-q^2C_j}+O\left(\frac{1}{d_-q^3}\right)\right)^3+O\left(d_-^4\right)\right]^4$$

$$= C_j^4 d_-^4\left[1+\frac{u_{2,j}}{d_-q^2C_j}+O\left(\frac{1}{d_-q^3}\right)+O\left(d_-^4\right)\right]^4$$

$$= C_j^4 d_-^4 + \frac{4u_{2,j}C_j^3d_-^3}{q^2} + O\left(\frac{d_-^3}{q^3}\right) + O\left(d_-^8\right), \quad q \to \infty.$$

Substituting our expressions for C_j and q we arrive at (2.11) when $d_- \to 0$, i.e.,

$$\lambda^{(n)}_{2j-1} = \frac{(2\pi j + 2u_{1,j})^4}{(n+2)^4} + \frac{16u_{2,j}(2\pi j + 2u_{1,j})^3}{(n+2)^5} + O\left(\frac{d_-^3}{n^3}\right) + O\left(d_-^8\right),$$

where $u_{1,j}$ and $u_{2,j}$ are given by (4.10) and (4.11), respectively.
ii) To deduce the formula (2.12), we proceed as we did with (2.11). In this case $w_{1,j}$ and $w_{2,j}$ are given by (4.18) and (4.19), respectively. $\square$

In Theorem 2.7, the constants $u_{1,j}$ and $w_{1,j}$ are obtained as the unique solution of the equations (4.10) and (4.18), respectively. In order to simplify the calculations, we present the following result.

Corollary 4.2. *Let $a(t) = \left(\frac{1}{t} + t - 2\right)^2$ and suppose that $d_- \to 0$ and $d_+ \to 0$ as $n \to \infty$. Then,*

i) *If $\pi j \gg 1$ as $j \to \infty$, then*

$$|u_{1,j} - u^*_{1,j}| = O\left(e^{-\pi j}\right),$$

where $u^*_{1,j} := -\arctan f(d_-)$.

ii) *If* $(j+\frac{1}{2})\pi \gg 1$ *as* $j \to \infty$, *then*

$$|w_{1,j} - w^*_{1,j}| = O\left(e^{-\pi j}\right),$$

where $w^*_{1,j} := -\operatorname{arccot}\frac{1}{f(d_+)}$.

iii) *If* j *is bounded, as* $n \to \infty$, *then*

$$u_{1,j} = -\arctan(\tanh(\pi j + u_{1,j})) + O\left(\frac{1}{n^2}\right),$$

$$w_{1,j} = -\arctan\left(\coth\left(\left(j+\frac{1}{2}\right)\pi + w_{1,j}\right)\right) + O\left(\frac{1}{n^2}\right).$$

Proof. i) From (4.10) we know that

$$u_{1,j} = -\arctan\left(f(d_-)\tanh\left(\frac{n+2}{2}\beta(d_-) + \beta'(d_-)u_{1,j}\right)\right). \tag{4.22}$$

Given that $d_- \to 0$ it follows that $\beta(d_-) = d_- + O(d_-^3)$ and $\beta'(d_-) = 1 + O(d_-^2)$. Substituting these equalities in (4.22) we arrive at the representation

$$u_{1,j} = -\arctan\left(f(d_-)\tanh\left(\pi j + u_{1,j} + O\left(\frac{j^3}{n^2}\right)\right)\right), \quad n \to \infty. \tag{4.23}$$

Now, if $\pi j \gg 1$ as $j \to \infty$, then we find that

$$\tanh\left(\pi j + u_{1,j} + O\left(\frac{j^3}{n^2}\right)\right) = 1 + O\left(e^{-\pi j}\right). \tag{4.24}$$

Substituting (4.24) in (4.23) we see that

$$u_{1,j} = -\arctan f(d_-) + O\left(e^{-\pi j}\right), \quad j \to \infty.$$

Once again, if $u^*_{1,j} := -\arctan f(d_-)$, then we obtain that

$$|u_{1,j} - u^*_{1,j}| = O\left(e^{-\pi j}\right), \quad j \to \infty.$$

ii) Similarly, from (4.18) and given that $d_+ \to 0$ we arrive at the representation

$$w_{1,j} = -\operatorname{arccot}\left(\frac{1}{f(d_+)}\tanh\left(\left(j+\frac{1}{2}\right)\pi + w_{1,j} + O\left(\frac{j^3}{n^2}\right)\right)\right).$$

Analogously, if $\left(j+\frac{1}{2}\right)\pi \gg 1$ as $j \to \infty$, then we get that

$$|w_{1,j} - w^*_{1,j}| = O\left(e^{-\pi j}\right),$$

where $w^*_{1,j} := -\operatorname{arccot}\frac{1}{f(d_+)}$.

iii) Given that j is bounded and $d_- \to 0$ as $n \to \infty$, we have

$$f(d_-) = 1 + O\left(\frac{1}{n^2}\right), \quad n \to \infty.$$

Substituting this equality in (4.22) it follows that

$$u_{1,j} = -\arctan(\tanh(\pi j + u_{1,j})) + O\left(\frac{1}{n^2}\right), \quad n \to \infty.$$

Similarly, since j is bounded and $d_+ \to 0$ as $n \to \infty$, using (4.18) we see that

$$w_{1,j} = -\arctan\left(\coth\left(\left(j+\frac{1}{2}\right)\pi + w_{1,j}\right)\right) + O\left(\frac{1}{n^2}\right), \quad n \to \infty.$$

□

Proof of Theorem 2.8. i) Notice that, as $n \to \infty$,

$$g(d_{2j}^{(n)}) = M - \frac{M}{2}(d_{2j}^{(n)} - \pi)^2 + O((d_{2j}^{(n)} - \pi)^4).$$

Therefore, $g'(d_{2j}^{(n)}) = -M(d_{2j}^{(n)} - \pi) + O((d_{2j}^{(n)} - \pi)^3)$ and $g''(d_{2j}^{(n)}) = -M + O((d_{2j}^{(n)} - \pi)^2)$.

Substituting these equalities in (2.9) we get

$$\lambda_{2j-1}^{(n)} = M - \frac{M(d_{2j}^{(n)} - \pi)^2}{2} - \frac{Mu_{1,j}^*(d_{2j}^{(n)} - \pi)}{q} - \frac{Mu_{2,j}^*(d_{2j}^{(n)} - \pi)}{q^2} \tag{4.25}$$
$$- \frac{M(u_{1,j}^*)^2}{2q^2} + O\left((d_{2j}^{(n)} - \pi)^4\right) + O\left(\frac{1}{n^3}\right), \quad n \to \infty.$$

From Theorem 2.6 we know that $u_{1,j}^* = -\arctan f(d_{2j}^{(n)})$. Furthermore, as $n \to \infty$, $f(d_{2j}^{(n)}) = f'(\pi)(d_{2j}^{(n)} - \pi) + O((d_{2j}^{(n)} - \pi)^3)$ and $f'(d_{2j}^{(n)}) = f'(\pi) + O((d_{2j}^{(n)} - \pi)^2)$. Hence,

$u_{1,j}^* = -\arctan(f'(\pi)(d_{2j}^{(n)} - \pi) + O((d_{2j}^{(n)} - \pi)^3)) = -f'(\pi)(d_{2j}^{(n)} - \pi) + O((d_{2j}^{(n)} - \pi)^3)$,

i.e., $u_{1,j}^*$ is given by

$$u_{1,j}^* = -f'(\pi)(d_{2j}^{(n)} - \pi) + O((d_{2j}^{(n)} - \pi)^3), \quad n \to \infty. \tag{4.26}$$

Once again, by Theorem 2.6 we have

$$u_{2,j}^* = \frac{f'(d_{2j}^{(n)}) \arctan f(d_{2j}^{(n)})}{1 + f^2(d_{2j}^{(n)})}.$$

Similarly, it follows that $u_{2,j}^*$ is given by

$$u_{2,j}^* = (f'(\pi))^2(d_{2j}^{(n)} - \pi) + O((d_{2j}^{(n)} - \pi)^3), \quad n \to \infty. \tag{4.27}$$

Substituting (4.26) and (4.27) in (4.25) with $q = \frac{n+2}{2}$, we arrive at (2.13), i.e.,

$$\lambda_{2j-1}^{(n)} = M - \frac{M(d_{2j}^{(n)} - \pi)^2}{2} + \frac{2Mf'(\pi)(d_{2j}^{(n)} - \pi)^2}{n+2} + O\left((d_{2j}^{(n)} - \pi)^4\right) + O\left(\frac{1}{n^3}\right), \quad n \to \infty.$$

ii) (2.14) can be obtained as we did with (2.13). □

5. Extreme eigenvalues

In this section, the well-known results of [11] and [21] are compared with our Theorems 2.7 and 2.8.

5.1. Smallest eigenvalues

In [11] the author considered the class of functions g satisfying:

a1) g is real, continuous, and periodic with period 2π; $\min g = g(0) = m$ and $\varphi = 0$ is the only value of $\varphi \pmod{2\pi}$ for which this minimum is attained.

a2) If g satisfies a1) then, it has continuous derivatives of order 2α ($\alpha \in \mathbb{N}$) in some neighborhood of $\varphi = 0$ and $g^{(2\alpha)}(0) = \sigma^2 > 0$ is the first non-vanishing derivative of g at $\varphi = 0$.

Theorem 5.1 ([11] Theorem 3). *Let g be a function satisfying* a1) *and* a2) *with $\alpha = 2$. Let $\lambda_{j,n}$ be the eigenvalues of $T_n(a)$ ordered in non-decreasing order. Then, for fixed $j = 1, 2, \ldots$, as $n \to \infty$, we have*

$$\lambda_{j,n} = m + \frac{1}{4!}\left(\frac{(2j+1)\pi + E_j}{2(n+3)}\right)^4 g^{(4)}(0) + o(n^{-4}), \quad n \to \infty, \tag{5.1}$$

where the E_j is determined by the equation

$$\tan\left(\frac{(2j+1)\pi + E_j}{4}\right) = (-1)^j \tanh\left(\frac{(2j+1)\pi + E_j}{4}\right).$$

For the eigenvalues near zero, we compare the first asymptotic term in (2.11) and (2.12) with (5.1).

Since $a(t) = \left(\frac{1}{t} + t - 2\right)^2$ it follows that $g(\varphi) = a(e^{i\varphi}) = 4(\cos\varphi - 1)^2$ satisfies a1) and a2). Moreover, as $n \to \infty$, using $m = g(0) = 0$ and $g^{(4)}(0) = 4!$ we can rewrite (5.1) as

$$\lambda_{2k-1,n} = \left(\frac{2k\pi + \frac{E_{2k-1}-\pi}{2}}{n+3}\right)^4 + o(n^{-4}), \tag{5.2}$$

and

$$\lambda_{2k,n} = \left(\frac{(2k+1)\pi + \frac{E_{2k}-\pi}{2}}{n+3}\right)^4 + o(n^{-4}), \tag{5.3}$$

where $k = 1, 2, \ldots$. E_{2k} and E_{2k-1} satisfy the following equations

$$\tan\left(\frac{E_{2k-1} - \pi}{4}\right) = -\tanh\left(\frac{(4k-1)\pi + E_{2k-1}}{4}\right) \tag{5.4}$$

and

$$\tan\left(\frac{E_{2k} + \pi}{4}\right) = \tanh\left(\frac{(4k+1)\pi + E_{2k}}{4}\right). \tag{5.5}$$

By Theorem 2.7, for every fixed k and $n \to \infty$, we get

$$\lambda_{2k-1}^{(n)} = \frac{(2k\pi + 2u_{1,k})^4}{(n+2)^4} + O\left(\frac{1}{n^5}\right) \tag{5.6}$$

and

$$\lambda_{2k}^{(n)} = \frac{((2k+1)\pi + 2w_{1,k})^4}{(n+2)^4} + O\left(\frac{1}{n^5}\right), \tag{5.7}$$

where $u_{1,k}$ and $w_{1,k}$ satisfy

$$u_{1,k} = -\arctan\left(f(d_-)\tanh\left(\frac{n+2}{2}\beta(d_-) + \beta'(d_-)u_{1,k}\right)\right),$$

$$w_{1,k} = -\arctan\left(f(d_+)\coth\left(\frac{n+2}{2}\beta(d_+) + \beta'(d_+)w_{1,k}\right)\right),$$

where $d_- = \frac{2k\pi}{n+2}$ and $d_+ = \frac{(2k+1)\pi}{n+2}$.

Notice that the first term in (5.2) and (5.3) coincides with the first term in (5.6) and (5.7), respectively, if and only if $u_{1,k} = \frac{E_{2k-1}-\pi}{4}$ and $w_{1,k} = \frac{E_{2k}-\pi}{4}$. Hence, if $u_{1,k} = \frac{E_{2k-1}-\pi}{4}$ and $w_{1,k} = \frac{E_{2k}-\pi}{4}$, (5.4) and (5.5) can be rewritten as

$$\tan u_{1,k} = -\tanh(k\pi + u_{1,k}) \quad \text{and} \quad \tan w_{1,k} = -\coth\left(\left(k+\frac{1}{2}\right)\pi + w_{1,k}\right),$$

respectively.

Moreover, since $d_- \to 0$, $d_+ \to 0$, and k is fixed, by Corollary 4.2, as $n \to \infty$, we know that

$$\tan(u_{1,k}) = -\tanh(\pi k + u_{1,k}) + O\left(\frac{1}{n^2}\right),$$

$$\tan(w_{1,k}) = -\coth\left(\left(k+\frac{1}{2}\right)\pi + w_{1,k}\right) + O\left(\frac{1}{n^2}\right).$$

Therefore, Theorems 2.7 and 5.1 coincide when k is fixed.

5.2. Largest eigenvalues

In [21] the author considered the class of functions g satisfying:

b) g is continuous and periodic with period 2π; $\max g = g(0) = M$ and $\varphi = 0$ is the only value of $\varphi \pmod{2\pi}$ for which this maximum is reached. Assume also that g is even, has continuous derivatives up to the fourth order in some neighborhood of $\varphi = 0$, and $\sigma^2 = -g^{(2)}(0) \neq 0$.

Theorem 5.2 ([21] Theorem 2.1). *Let g be a function satisfying* b). *Let $\lambda_{j,n}$ be the eigenvalues of $T_n(a)$ ordered in decreasing order. Then, as $n \to \infty$, for fixed $j = 1, 2, \dots,$ we have*

$$\lambda_{j,n} = M - \frac{\sigma^2\pi^2 j^2}{2(n+1)^2}\left(1 + \frac{\rho}{n+1}\right) + o(n^{-3}), \tag{5.8}$$

where

$$\rho = 2 + \frac{1}{\pi}\int_{-\pi}^{\pi}\left(\frac{g'(\varphi)}{g(\varphi) - M} - 2\csc\varphi\right)\cot\frac{\varphi}{2}d\varphi.$$

Remark 5.3. In [21], there is a misprint when calculating ρ, i.e., the constant 2 was not included. Hence, we make the respective correction here.

For the eigenvalues near M, we compare now the asymptotic terms of (2.13) and (2.14) with (5.8).

In order to satisfy b), we need to modify the symbol a. Therefore, we consider $a_1(t) := a(-t) = \left(\frac{1}{t}+t+2\right)^2$ and $g(\varphi) := a_1(e^{i\varphi})$ for $\varphi \in [-\pi,\pi]$. Notice that $T_n(a)$ and $T_n(a_1)$ are similar matrices with same eigenvalues. Furthermore, notice that $g'(\varphi) = 2i(e^{-i\varphi}+e^{i\varphi}+2)(e^{i\varphi}-e^{-i\varphi})$ for $\varphi \in [-\pi,\pi]$. Therefore, $\rho-2$ equals

$$\frac{1}{\pi}\int_{\mathbb{T}}\left(\frac{\frac{2i(t-1)(t+1)^3}{t^2}}{(t^{-1}+t+2)^2-16}-\frac{4i}{t-t^{-1}}\right)\frac{t+1}{t(t-1)}dt = \frac{1}{2\pi i}\int_{\mathbb{T}}\frac{-4(t^2+4t+1)}{t(t^2+6t+1)}dt.$$

Let $h_2(t) := \frac{-4(t^2+4t+1)}{t(t^2+6t+1)}$ for $t \in \mathbb{T}$. Using the residue theorem we get

$$\rho = 2+\operatorname{Res}(h_2,0)+\operatorname{Res}(h_2,-3+2\sqrt{2}) = -2+\sqrt{2}.$$

For every $k = 1,2,\ldots$ with $n \to \infty$, (5.8) can be written as

$$\lambda_{2k,n} = 16-\frac{8(2k\pi)^2}{(n+1)^2}+\frac{8(2-\sqrt{2})(2k\pi)^2}{(n+1)^3}+o(n^{-3}), \tag{5.9}$$

and

$$\lambda_{2k-1,n} = 16-\frac{8((2k-1)\pi)^2}{(n+1)^2}+\frac{8(2-\sqrt{2})((2k-1)\pi)^2}{(n+1)^3}+o(n^{-3}). \tag{5.10}$$

From Theorem 2.8, as $n \to \infty$, we have

$$\lambda_{2j-1}^{(n)} = 16-8(d_{2j}^{(n)}-\pi)^2+\frac{32f'(\pi)(d_{2j}^{(n)}-\pi)^2}{n+2}+O\left((d_{2j}^{(n)}-\pi)^4\right) \tag{5.11}$$

and

$$\lambda_{2j}^{(n)} = 16-8(d_{2j+1}^{(n)}-\pi)^2+\frac{32f'(\pi)(d_{2j+1}^{(n)}-\pi)^2}{n+2}+O\left((d_{2j+1}^{(n)}-\pi)^4\right), \tag{5.12}$$

where $d_{2j}^{(n)} = \frac{2\pi j}{n+2}$, $d_{2j+1}^{(n)} = \frac{(2j+1)\pi}{n+2}$ and $f'(\pi) = -\frac{\sqrt{2}}{4}$.
Notice that if n is an even number, then $\pi - d_{2j}^{(n)} = \frac{(n-2j+2)\pi}{n+2}$ for $j = \frac{n}{2}, \frac{n}{2}-1, \ldots$. Take $k = 1,2,\ldots$ with $n \to \infty$. Making the variable change $2k = n-2j+2$ in (5.11), we arrive at the equation

$$\lambda_{n-2k+1}^{(n)} = 16-\frac{8(2k\pi)^2}{(n+2)^2}+\frac{32f'(\pi)(2k\pi)^2}{(n+2)^3}+O\left(\frac{k^4}{n^4}\right).$$

Since $(1+\frac{1}{n+1})^{-p} = 1-\frac{p}{n+1}+O\left(\frac{1}{n^2}\right)$ $(p \in \mathbb{R})$, we obtain

$$\lambda_{n-2k+1}^{(n)} = 16-\frac{8(2k\pi)^2}{(n+1)^2}+\frac{(16+32f'(\pi))(2k\pi)^2}{(n+1)^3}+O\left(\frac{k^4}{n^4}\right).$$

Since $f'(\pi) = -\frac{\sqrt{2}}{4}$ it follows that $16+32f'(\pi) = 8(2-\sqrt{2})$. Therefore,

$$\lambda_{n-2k+1}^{(n)} = 16-\frac{8(2k\pi)^2}{(n+1)^2}+\frac{8(2-\sqrt{2})(2k\pi)^2}{(n+1)^3}+O\left(\frac{k^4}{n^4}\right),$$

which coincides with (5.9) when k is fixed.

Similarly, from (5.12) it follows that

$$\lambda^{(2)}_{n-2k+2} = 16 - \frac{8((2k-1)\pi)^2}{(n+1)^2} + \frac{8(2-\sqrt{2})((2k-1)\pi)^2}{(n+1)^3} + O\left(\frac{k^4}{n^4}\right),$$

which coincides with (5.10) when k is fixed.

Now, if n is an odd number, then we can proceed as we did for an even number.

Remark 5.4. The formulas obtained in [11] and [21] work for fixed eigenvalues (j is fixed). Our formulas work for fixed j as well, but we can consider new cases, for example $j \ll n$.

6. Numerical tests

The main idea of the theorems in Section 2 is to provide us asymptotic expansions which reveal the fine structure of the eigenvalues bulk and show that this structure is essentially independent of the matrix dimension n. Furthermore, our asymptotic formulas also have the potential to be of use for design of numerical algorithms for the computation of the eigenvalues. For instance, we make all numerical tests of the eigenvalues of $T_n(a)$ for several moderately sized n by *Matlab R2015a*. Here, all exact eigenvalues are determined by means of the Matlab function *eig*.

The numerical tests in this section allow us to check that we calculated the coefficients in the asymptotic formulas correctly and we want to demonstrate that the asymptotic formulas deliver acceptable approximations not only for astronomically large n, but already for n in the early hundreds.

Denote by $\lambda^{(n,k)}_{2j-1}$ and $\lambda^{(n,k)}_{2j}$ the approximations of $\lambda^{(n)}_{2j-1}$ and $\lambda^{(n)}_{2j}$ with k terms, respectively. These approximations are obtained from our formulas (2.9), (2.10), (2.11), (2.12), (2.13), and (2.14). For example, from (2.9), we have

$$\lambda^{(n,2)}_{2j-1} = g(d_-) + \frac{2u^*_{1,j}g'(d_-)}{n+2}, \quad \lambda^{(n,3)}_{2j-1} = g(d_-) + \frac{2u^*_{1,j}g'(d_-)}{n+2} + \frac{2(u^*_{1,j})^2 g''(d_-) + 4u^*_{2,j}g'(d_-)}{(n+2)^2}.$$

For each j, we put $\epsilon^{(n,k)}_{2j-1} := |\lambda^{(n)}_{2j-1} - \lambda^{(n,k)}_{2j-1}|$, $\epsilon^{(n,k)}_{2j} := |\lambda^{(n)}_{2j} - \lambda^{(n,k)}_{2j}|$.

To use the formulas (2.7) and (2.8), we denote by $\tilde{\lambda}^{(n,k)}_{2j-1} := g(\varphi^{(n,k)}_{2j-1,*})$ and $\tilde{\lambda}^{(n,k)}_{2j} := g(\varphi^{(n,k)}_{2j,*})$ the approximations of $\lambda^{(n)}_{2j-1}$ and $\lambda^{(n)}_{2j}$, where $\varphi^{(n,k)}_{2j-1,*}$ and $\varphi^{(n,k)}_{2j,*}$ contain k terms. For example, when taking (2.7) and (2.8), we have

$$\tilde{\lambda}^{(n,2)}_{2j-1} = g\left(d_- + \frac{2u_{1,j}}{n+2}\right), \quad \tilde{\lambda}^{(n,3)}_{2j-1} = g\left(d_- + \frac{2u_{1,j}}{n+2} + \frac{4u_{2,j}}{(n+2)^2}\right), \tag{6.1}$$

$$\tilde{\lambda}^{(n,2)}_{2j} = g\left(d_+ + \frac{2w_{1,j}}{n+2}\right), \quad \tilde{\lambda}^{(n,3)}_{2j} = g\left(d_+ + \frac{2w_{1,j}}{n+2} + \frac{4w_{2,j}}{(n+2)^2}\right). \tag{6.2}$$

Furthermore, if n is an even number, then we denote by $\epsilon^{(n,k)}$ the corresponding maximal error, i.e.,

$$\epsilon^{(n,k)} := \max_{j\in\{1,2,\cdots,\frac{n}{2}\}} \{\epsilon^{(n,k)}_{2j-1}, \epsilon^{(n,k)}_{2j}\}.$$

n	64	128	512	1024
$\epsilon^{(n,2)}$	$1.23 \cdot 10^{-3}$	$3.18 \cdot 10^{-4}$	$2.03 \cdot 10^{-5}$	$5.11 \cdot 10^{-6}$
$\epsilon^{(n,3)}$	$1.03 \cdot 10^{-5}$	$1.35 \cdot 10^{-6}$	$1.77 \cdot 10^{-8}$	$3.25 \cdot 10^{-10}$

TABLE 1. Maximal absolute errors for the eigenvalues of $T_n(a)$ obtained with our formulas (6.1) and (6.2).

n	64	128	512	1024
$\epsilon_1^{(n,1)}$	$4.38 \cdot 10^{-8}$	$7.49 \cdot 10^{-10}$	$1.96 \cdot 10^{-13}$	$3.19 \cdot 10^{-15}$
$\epsilon_1^{(n,2)}$	$1.96 \cdot 10^{-8}$	$3.35 \cdot 10^{-10}$	$8.81 \cdot 10^{-14}$	$1.48 \cdot 10^{-15}$

TABLE 2. Maximal absolute errors for the eigenvalues of $T_n(a)$ obtained with our formula (2.11), i.e., $j = 1$.

n	64	128	512	1024
$\epsilon_2^{(n,1)}$	$6.32 \cdot 10^{-7}$	$1.08 \cdot 10^{-8}$	$2.83 \cdot 10^{-12}$	$4.45 \cdot 10^{-14}$
$\epsilon_2^{(n,2)}$	$4.59 \cdot 10^{-7}$	$7.85 \cdot 10^{-9}$	$2.05 \cdot 10^{-12}$	$3.22 \cdot 10^{-14}$

TABLE 3. Maximal absolute errors for the eigenvalues of $T_n(a)$ obtained with our formula (2.12), i.e., $j = 1$.

n	64	128	512	1024
$\epsilon^{(n,1)}$	$7.47 \cdot 10^{-4}$	$7.76 \cdot 10^{-5}$	$1.13 \cdot 10^{-6}$	$1.62 \cdot 10^{-7}$
$\epsilon^{(n,2)}$	$7.28 \cdot 10^{-4}$	$7.65 \cdot 10^{-5}$	$1.12 \cdot 10^{-6}$	$1.61 \cdot 10^{-7}$

TABLE 4. Maximal absolute errors for the eigenvalues of $T_n(a)$ near zero, obtained with our formulas (2.11) and (2.12), where

$$\epsilon^{(n,k)} := \max\{\epsilon_{2j-1}^{(n,k)}, \epsilon_{2j}^{(n,k)} : j \in \{1, \dots, [\sqrt{n}]\}\}.$$

These errors show that we can consider the case $j \ll n$.

n	64	128	512	1024	2048
$\epsilon^{(n,2)}$	$1.0 \cdot 10^{-3}$	$3.06 \cdot 10^{-4}$	$1.95 \cdot 10^{-5}$	$4.91 \cdot 10^{-6}$	$1.23 \cdot 10^{-6}$
$\epsilon^{(n,3)}$	$2.01 \cdot 10^{-5}$	$2.63 \cdot 10^{-6}$	$4.24 \cdot 10^{-8}$	$5.33 \cdot 10^{-9}$	$6.68 \cdot 10^{-10}$

TABLE 5. Maximal absolute errors for the inner eigenvalues of $T_n(a)$ obtained with our formulas (2.9) and (2.10), where

$$\epsilon^{(n,k)} := \max\{\epsilon_{2j-1}^{(n,k)}, \epsilon_{2j}^{(n,k)} : j \in \{[\tfrac{n}{6}] + 1, \dots, n - [\tfrac{n}{6}]\}\}.$$

n	64	128	512	1024	2048
$\epsilon_{n-1}^{(n,2)}$	$1.43 \cdot 10^{-3}$	$1.95 \cdot 10^{-4}$	$3.25 \cdot 10^{-6}$	$4.11 \cdot 10^{-7}$	$5.17 \cdot 10^{-8}$
$\epsilon_{n-1}^{(n,3)}$	$1.17 \cdot 10^{-4}$	$7.63 \cdot 10^{-6}$	$3.06 \cdot 10^{-8}$	$1.92 \cdot 10^{-9}$	$1.2 \cdot 10^{-10}$

TABLE 6. Maximal absolute errors for the penultimate eigenvalue of $T_n(a)$ obtained with our formula (2.13), i.e., $j = \frac{n}{2}$.

n	64	128	512	1024	2048
$\epsilon_{n}^{(n,2)}$	$3.85 \cdot 10^{-4}$	$5.06 \cdot 10^{-5}$	$8.21 \cdot 10^{-7}$	$1.03 \cdot 10^{-7}$	$1.29 \cdot 10^{-8}$
$\epsilon_{n}^{(n,3)}$	$2.61 \cdot 10^{-6}$	$1.63 \cdot 10^{-7}$	$6.39 \cdot 10^{-10}$	$3.99 \cdot 10^{-11}$	$2.49 \cdot 10^{-12}$

TABLE 7. Maximal absolute errors for the very last eigenvalue of $T_n(a)$ obtained with our formula (2.14), i.e., $j = \frac{n}{2}$.

References

[1] J.M. Bogoya, A. Böttcher, S.M. Grudsky, and E.A. Maximenko, *Eigenvalues of Hermitian Toeplitz matrices with smooth simple-loop symbols.* Journal of Mathematical Analysis and Applications **422** (2015), 1308–1334.

[2] A. Böttcher, S.M. Grudsky, and E.A. Maksimenko, *Inside the eigenvalues of certain Hermitian Toeplitz band matrices.* J. Comput. Appl. Math. **233** (2010), 2245–2264.

[3] A. Böttcher, S. Grudsky, E.A. Maksimenko, J. Unterberger, *The first-order asymptotics of the extreme eigenvectors of certain Hermitian Toeplitz matrices.* Integral Equations and Operator Theory **63** (2009), 165–180.

[4] A. Böttcher, S. Grudsky, *Spectral Properties of Banded Toeplitz Matrices.* SIAM, Philadelphia, 2005.

[5] A. Böttcher, B. Silbermann, *Introduction to Large Truncated Toeplitz Matrices.* Universitext, Springer-Verlag, New York, 1999.

[6] P. Deift, A. Its, I. Krasovsky, *Eigenvalues of Toeplitz matrices in the bulk of the spectrum.* Bull. Inst. Math. Acad. Sin. (N. S.) **7** (2012), 437–461.

[7] M. Elouafi, *On a relationship between Chebyshev polynomials and Toeplitz determinants.* Appl. Math. Comput. **229** (2014), 27–33.

[8] U. Grenander, G. Szegő, *Toeplitz Forms and Their Applications.* University of California Press, Berkeley, 1958.

[9] M. Kac, W.L. Murdock, and G. Szegő, *On the eigenvalues of certain Hermitian forms.* J. Rational Mech. Anal. **2** (1953), 767–800.

[10] A.Yu. Novosel'tsev, I.B. Simonenko, *Dependence of the asymptotics of extreme eigenvalues of truncated Toeplitz matrices on the rate of attaining the extremum by the symbol.* St. Petersburg Math. J. **16** (2005), 713–718.

[11] S.V. Parter, *Extreme eigenvalues of Toeplitz forms and applications to elliptic difference equations.* Trans. Amer. Math. Soc. **99** (1961), 153–192.

[12] S.V. Parter, *On the extreme eigenvalues of Toeplitz matrices.* Trans. Amer. Math. Soc. **100** (1961), 263–276.

[13] S. Serra Capizzano, *On the extreme spectral properties of Toeplitz matrices generated by L^1 functions with several minima/maxima.* BIT **36** (1996), 135–142.

[14] S. Serra Capizzano, *On the extreme eigenvalues of Hermitian (block) Toeplitz matrices.* Linear Algebra Appl. **270** (1998), 109–129.

[15] S. Serra Capizzano, P. Tilli, *Extreme singular values and eigenvalues of non-Hermitian block Toeplitz matrices.* J. Comput. Appl. Math. **108** (1999), 113–130.

[16] S. Serra Capizzano, C. Tablino Possio, *Spectral and structural analysis of high precision finite difference matrices for elliptic operators.* Linear Algebra Appl. **293** (1999), 85–131.

[17] P. Schmidt, F. Spitzer, *The Toeplitz matrices of an arbitrary Laurent polynomial.* Math. Scand. **8** (1960), 15–28.

[18] W.F. Trench, *Asymptotic distribution of the spectra of a class of generalized Kac–Murdock–Szegő matrices.* Linear Algebra Appl. **294** (1999), 181–192.

[19] P. Tilli, *Some results on complex Toeplitz eigenvalues.* J. Comput. Appl. Math. **239** (1999), 390–401.

[20] E.E. Tyrtyshnikov, N.L Zamarashkin, *Toeplitz eigenvalues for Radon measures.* Linear Algebra Appl. **343/344** (2002), 345–354.

[21] H. Widom, *On the eigenvalues of certain Hermitian operators.* Trans. Amer. Math. Soc. **88** (1958), 491–522.

[22] H. Widom, *Eigenvalue distribution of nonselfadjoint Toeplitz matrices and the asymptotics of Toeplitz determinants in the case of nonvanishing index.* Oper. Theory Adv. Appl. **48** (1990), 387–421.

[23] N.L Zamarashkin, E.E Tyrtyshnikov, *Distribution of eigenvalues and singular values of Toeplitz matrices under weakened conditions on the generating function.* Sb. Math. **188** (1997), 1191–1201.

[24] P. Zizler, R.A. Zuidwijk, K.F. Taylor, and S. Arimoto, *A finer aspect of eigenvalue distribution of selfadjoint band Toeplitz matrices.* SIAM J.Matrix Anal. Appl. **24** (2002), 59–67.

M. Barrera and S.M. Grudsky
Departamento de Matemáticas
CINVESTAV del I.P.N.
Apartado Postal 14-740
07000 México, D.F. México
e-mail: `mabarrera@math.cinvestav.mx`
`grudsky@math.cinvestav.mx`

Operator Theory:
Advances and Applications, Vol. 259, 79–124

Echelon Type Canonical Forms in Upper Triangular Matrix Algebras

Harm Bart, Torsten Ehrhardt and Bernd Silbermann

Dedicated to our colleague and friend Albrecht Böttcher,
on the occasion of his sixtieth birthday

Abstract. It is common knowledge that matrices can be brought in echelon form by Gaussian elimination and that the reduced echelon form of a matrix is canonical (also) in the sense that it is unique. A crucial auxiliary result in [BW] suggests a generalization of the standard echelon form. For square matrices, some new canonical forms of echelon type are introduced. One of them (suggested by observations made in [Lay] and [SW]) has the important property of being an upper triangular idempotent. The others come up when working exclusively in the context of $\mathbb{C}^{n\times n}_{\text{upper}}$, the algebra of upper triangular $n \times n$ matrices. Subalgebras of $\mathbb{C}^{n\times n}_{\text{upper}}$ determined by a pattern of zeros are considered too. The issue there is whether or not the canonical forms referred to above belong to the subalgebras in question. In general they do not, but affirmative answers are obtained under certain conditions on the given preorder which allow for a large class of examples and that also came up in [BES4]. Similar results hold for canonical generalized diagonal forms involving matrices for which all columns and rows contain at most one nonzero entry. The new canonical forms are used to study left, right and left/right equivalence in zero pattern algebras. For the archetypical full upper triangular case a connection with the Stirling numbers (of the second kind) and with the Bell numbers is made.

Mathematics Subject Classification (2010). 15A21, 11B73.

Keywords. Echelon (canonical) form, idempotent (canonical) form, generalized diagonal (canonical) form, lean structure, zero pattern matrix algebra, Stirling numbers of the second kind, Bell numbers, logarithmic residue, sum of idempotents.

1. Introduction: motivation, background and preliminaries

It is well known that a matrix can be brought in echelon form by Gaussian elimination. More precisely, a row echelon form, respectively a column echelon form, of a matrix is obtained by applying Gaussian elimination on the rows, respectively the columns, of the matrix in question. Such echelon forms are generally not unique. Uniqueness does hold, however, for the so-called reduced echelon forms. The reduced row echelon form, respectively the reduced column echelon form, of a given matrix M will be denoted by M_{rEch}, respectively M_{cEch}. Note that $M_{\mathrm{cEch}} = E\big((EM^\top E)_{\mathrm{rEch}}\big)^\top E$. Here, as in the sequel, E is the $n \times n$ reverse-identity matrix. The symbol $^\top$ signals the operation of taking the transpose.

Let $M \in \mathbb{C}^{n\times n}$, where n is a positive integer. Then there exist invertible matrices L and R in $\mathbb{C}^{n\times n}$ such that $LM = M_{\mathrm{rEch}}$ and $MR = M_{\mathrm{cEch}}$. What is the situation when we restrict ourselves to working with matrices coming from a given matrix subalgebra of $\mathbb{C}^{n\times n}$, say $\mathcal{A}$? Of course we cannot expect the matrices M_{rEch} and M_{cEch} to belong to $\mathcal{A}$. And if they don't, then, *a fortiori*, the matrices L and R bringing about the reduced echelon forms cannot be chosen to belong to $\mathcal{A}$ either.

On the basis of a lead in [BW], a property (shared by echelon type matrices) will be introduced for which the situation is better. Again let $M \in \mathbb{C}^{n\times n}$, and write $M = [m_{k,l}]_{k,l=1}^n$. By a *diagonal block* in M we mean a submatrix of M of the form $M[s,t] = [m_{k,l}]_{k,l=s}^t$ with $s, t \in \{1, \ldots, n\}$, $s \le t$. The matrix M is said to have *lean row structure* if each diagonal block $M[s,t]$ in M has rank equal to the number of nonzero rows in $M[s,t]$. In other words, a matrix has lean row structure if in each of its diagonal blocks the nonzero rows are linearly independent. Similarly M is said to have *lean column structure* if each diagonal block $M[s,t]$ in M has rank equal to the number of nonzero columns in $M[s,t]$. Clearly M has lean column structure if and only if $M^\top$ has lean row structure.

Suppose $H \in \mathbb{C}^{n\times n}$ is in row echelon form. Then the rank of H is equal to the number of nonzero rows in H. Also, along with H, each diagonal block of H is in row echelon form. Hence a row echelon form of a square matrix has lean row structure. Similarly, a column echelon form of a square matrix has lean column structure. Thus, given $M \in \mathbb{C}^{n\times n}$, there exist invertible matrices L and R in $\mathbb{C}^{n\times n}$ such that LM has lean row structure and MR has lean column structure. For L and R one can take, e.g., invertible matrices such that $LM = M_{\mathrm{rEch}}$ and $MR = M_{\mathrm{cEch}}$.

We now modify our earlier question as follows: what is the situation with respect to lean row or lean column structure when we restrict ourselves to working with matrices coming from a given matrix subalgebra $\mathcal{A}$ of $\mathbb{C}^{n\times n}$? More precisely, the questions are these:

- given $M \in \mathcal{A}$, does there always exist an invertible matrix $L \in \mathcal{A}$ such that LM (which then necessarily belongs to $\mathcal{A}$) has lean row structure?
- given $M \in \mathcal{A}$, does there always exist an invertible matrix $R \in \mathcal{A}$ such that MR (which then necessarily belongs to $\mathcal{A}$) has lean column structure?

For a first affirmative answer to both questions we turn to the subalgebra $\mathbb{C}^{n\times n}_{\text{upper}}$ of $\mathbb{C}^{n\times n}$ consisting of all upper triangular $n \times n$ matrices.

Theorem 1.1. *Let n be a positive integer. Then, given $M \in \mathbb{C}^{n\times n}_{\text{upper}}$, there exist invertible matrices L and R in $\mathbb{C}^{n\times n}_{\text{upper}}$ such that LM has lean row structure and MR has lean column structure.*

Of course there is a counterpart of the theorem in which $\mathbb{C}^{n\times n}_{\text{upper}}$ is replaced by $\mathbb{C}^{n\times n}_{\text{lower}}$, the subalgebra of $\mathbb{C}^{n\times n}$ consisting of all lower triangular $n \times n$ matrices.

Proof. For the existence of R we refer to Proposition 2.1 in [BW], specialized to the upper triangular case. To establish that of L we argue as follows. With E as above, the matrix $EM^\top E$ is upper triangular. Hence there exists an invertible matrix R in $\mathbb{C}^{n\times n}_{\text{upper}}$ such that $C = EM^\top ER$ has lean column structure. Now put $L = ER^\top E$. Then L is an invertible matrix in $\mathbb{C}^{n\times n}_{\text{upper}}$ and $LM = EC^\top E$ has lean row structure. □

Motivated by Theorem 1.1 we will turn our attention to subalgebras $\mathcal{A}$ of $\mathbb{C}^{n\times n}_{\text{upper}}$ determined by a pattern of zeros. This is done in Section 7. Prior to that, however, and (also) as a motivation for what follows later on, permuted echelon forms are considered in Section 2. Following leads to be found in [Lay] and [SW], a canonical idempotent echelon form is introduced. After having taken preparatory steps in Section 3, the full upper triangular case $\mathcal{A} = \mathbb{C}^{n\times n}_{\text{upper}}$ is dealt with in Sections 4 and 5, the considerations leading to new canonical echelon type forms for upper triangular matrices. Canonical generalized diagonal forms are considered too. As it turns out, there are strong relations between the various canonical forms. These are explored in Section 6. Section 7 is concerned with the question of whether or not the canonical forms introduced in this paper belong to a given zero pattern subalgebra of $\mathbb{C}^{n\times n}_{\text{upper}}$ in case the original matrix does. In general the answer is negative, but positive results are obtained under certain conditions on the zero pattern in question which allow for a large class of examples and that also came up in [BES4]. Canonical forms can often play the role of representatives of equivalence classes. Here that is the case too, in fact for left, right and left/right equivalence classes in the zero pattern algebras under discussion (the archetypical algebra of upper triangular matrices among them). Sections 8 and 9 contain the material in question. More specifically, the latter section is concerned with determining the number of left/right equivalence classes. For the full upper triangular case a connection with the Stirling numbers of the second kind and with the Bell numbers is made. Section 10 contains a number of illuminating examples. In Section 11, the final section of the paper, a connection is made with sums of idempotents and logarithmic residues.

We close the introduction by giving a little more background for this connection. Here we draw on [BES1] and [BES2]; cf. also [BES3] and the references given there. Roughly speaking, a logarithmic residue is a contour integral of the logarithmic derivative of an analytic vector-valued function. The paper [BES2] deals with logarithmic residues in matrix algebras. In the block upper triangular case, it characterizes those in terms of rank/trace conditions. For the upper triangular case (blocks of size 1), the characterization comes down to this.

Theorem 1.2. *For an upper triangular $n \times n$ matrix A, the following four statements are equivalent:*

(1) *A is a sum of upper triangular idempotents;*
(2) *A is a sum of rank one upper triangular idempotents;*
(3) *A is a logarithmic residue in the algebra of upper triangular $n \times n$ matrices;*
(4) *Each diagonal block in A has integer trace larger than or equal to its rank.*

The proof of Theorem 1.2, as given in [BES2], relies heavily on the material presented in [BW]. In turn, the considerations there involve Farkas' Lemma (cf. [FP]) and elements from Integer Programming (in particular material on totally unimodular systems of equations). In an essential way they are also based on (the part of) Theorem 1.1 (dealing with lean column structure).

2. Permuted echelon matrices

In dealing with echelon type constructions and echelon type matrices, one has a choice between working with rows (corresponding to multiplication on the left) or working with columns (corresponding to multiplication on the right). In this paper, as a rule, and especially in proofs, the emphasis will be on the first of the two.

Let $A \in \mathbb{C}^{n\times n}$ with n a positive integer. If a row of A is nonzero, its *leading entry* is its first nonzero entry. Recall that A is in *row echelon form* if the zero rows of A are at the bottom of the matrix, the leading entry of any nonzero row of A is a 1, and the leading entries occur in a left to right staircase pattern. When, in addition, all other entries in the column of a leading entry are zero, then A is said to be in *reduced row echelon form.*

Given $M \in \mathbb{C}^{n\times n}$, there exists a matrix $A \in \mathbb{C}^{n\times n}$ such that A is in reduced row echelon form and $A = LM$ for some invertible $L \in \mathbb{C}^{n\times n}$. The matrix A is uniquely determined by M and called *the reduced row echelon form* of M – written M_{rEch} (as in the introduction).

There is some arbitrariness in the definitions given above. Why should one insist of having all zero rows at the bottom of the matrix? Or why having the leading entries in a left to right staircase pattern? With this point in mind we expand our conceptual framework.

We say that $A \in \mathbb{C}^{n\times n}$ is in *permuted row echelon form*, respectively *permuted reduced row echelon form* if there is a permutation matrix P such that PA is in row echelon form, respectively reduced row echelon form. Given $M \in \mathbb{C}^{n\times n}$, a matrix A is called a *permuted row echelon form* of M, respectively a *permuted reduced row echelon form* of M, if A is in permuted row echelon form, respectively permuted reduced row echelon form and, in addition, there is an invertible $L \in \mathbb{C}^{n\times n}$ such that $A = LM$. Clearly A is a permuted row echelon form of M, respectively a permuted reduced row echelon form of M, if and only if A can be obtained from a row echelon form of M, respectively the reduced row echelon form of M, by a multiplication on the left with a permutation matrix.

That it does make sense to consider permuted echelon forms becomes manifest when we look at the paper [Lay] by David C. Lay and the article [SW] having as authors David Surowski and Yuhua Wang.

The paper by Lay was published in 1993. It is educationally oriented and seems to be absent from both MathSciNet and Zentralblatt. Here is what is pertinent to our present considerations. Take $M \in \mathbb{C}^{n\times n}$ and consider the (unique) reduced row echelon form M_{rEch} of M. Let V be the $n \times n$ matrix obtained from M_{rEch} by inserting or deleting rows of zeros so as to make the leading entries of M_{rEch} occur on the diagonal of V. Then V is obviously upper triangular. But, and this is what counts, V is an idempotent. (As explained by Lay, this is of interest for solving $Mx = 0$. Indeed, the solution can be obtained through solving $Vx = 0$. The latter, however, is particularly easy. Indeed, consider $V - I$, where I is the $n \times n$ identity matrix, and observe that, V being an idempotent, the null space of V coincides with the range of $V - I$. So the desired solution space has the nonzero columns of $V - I$ as a basis. In this way one circumvents the more cumbersome backward substitution.) The matrix V can be written as the product PM_{rEch} with P an $n \times n$ permutation matrix. Thus the idempotent V is a permuted (reduced) echelon form of M. Due to the possible presence of zero rows in M_{rEch}, the permutation matrix P is generally not uniquely determined. However, its rows corresponding to the positions of the leading entries in M_{rEch} are, and it follows that (along with M_{rEch}) the upper triangular idempotent PM_{rEch} is uniquely determined by M. We will call it *the canonical row echelon idempotent* of M – written M_{rEchId}.

In 2003, ten years after [Lay], permuted reduced echelon forms having their leading entries on the diagonal (implying upper triangularity) again appeared in the paper [SW] by David Surowski and Yuhua Wang. The terminology adopted there is *Hesse normal form.* Their use of this term is not standard, however (cf., for instance, [HJ], page 11). This ambiguity is the reason why in the preceding paragraph we coined our own term (in line, by the way, with our terminology in the rest of this paper). Surowski and Wang use the Hesse normal form to give an elegant alternative proof of the uniqueness of the reduced (they say: strong) echelon form. As one of the steps in their argument, they prove uniqueness of the Hesse form (without employing that of the reduced echelon form). Along the way, they (re)establish the idempotency of the Hesse normal form, earlier observed by Lay.

The permuted echelon forms considered above (following the leads from [Lay] and [SW]) are of the reduced type and have their leading entries on the diagonal (implying upper triangularity). What is the situation when one relaxes on these two aspects? This issue is addressed in the next two results.

Theorem 2.1. *Let $A \in \mathbb{C}^{n\times n}$ be upper triangular and in permuted row echelon form. Suppose, in addition, that A is idempotent. Then A is in permuted reduced row echelon form and has its leading entries on the diagonal.*

In the (non-standard) terminology of [SW], the conclusion of the theorem can be read as: *A is in Hesse normal form* (cf. [Lay], Algorithm III.)

Proof. As A is in permuted row echelon form, the rank of A is equal to the number of nonzero rows in A. Now, A being an idempotent, the rank and trace of A coincide. Thus the trace of A is equal to the number of nonzero rows in A. Since A is an upper triangular idempotent, the diagonal entries of A can only have the values 0 and 1. Hence the diagonal entries in the nonzero rows of A are all equal to 1. But then, using again that A is upper triangular, the leading entries in the nonzero rows of A must be on the diagonal of A. These leading entries have the value 1.

Next we show that the matrix A is in permuted reduced row echelon form. Assuming it is not we shall reason towards a contradiction.

Write $A = [a_{k,l}]_{k,l=1}^{n}$. By our assumption, there must be a leading entry such that above it, in the corresponding column, nonzero entries are present. Thus there exists $k \in \{2,\dots,n\}$ such that $a_{k,k} = 1$ is a leading entry and there exists $m < k$ with $a_{m,k} \neq 0$. Taking for m the largest possible value, we have $a_{r,k} = 0$, $r = m+1,\dots,k-1$. Using $A^2 = A$ and the upper triangularity of A, we get

$$\begin{aligned} a_{m,k} &= \sum_{s=1}^{n} a_{m,s}a_{s,k} = \sum_{s=m}^{k} a_{m,s}a_{s,k} \\ &= a_{m,m}a_{m,k} + a_{m,k}a_{k,k} = a_{m,m}a_{m,k} + a_{m,k}. \end{aligned}$$

Clearly the mth row of A is a nonzero row of A. Hence its leading entry is a 1 on the diagonal. So $a_{m,m} = 1$. But then $a_{m,k} = a_{m,m}a_{m,k} + a_{m,k} = 2a_{m,k}$. It follows that $a_{m,k} = 0$, and we have arrived at the desired contradiction. □

Corollary 2.2. *Let M be an $n \times n$ matrix. Then the canonical row echelon idempotent M_{rEchId} of M is the unique permuted row echelon form of M which is upper triangular and idempotent.*

As will become clear in Example A in Section 10, it may happen that among the permuted row echelon forms of M – even the reduced ones – there are idempotents failing to be upper triangular.

Proof. The canonical row echelon idempotent M_{rEchId} is a permuted (reduced) row echelon form of M which is both upper triangular and idempotent. This settles the existence issue. With regard to uniqueness we reason as follows. Let A be a permuted row echelon form of M which is upper triangular and idempotent. Theorem 2.1 then gives that A is a permuted reduced row echelon form of M having its leading entries on the diagonal. Uniqueness is now clear from that of the reduced row echelon form of M. Alternatively one can take recourse to [SW] because in the (non standard) terminology used there, A is the (unique) Hesse normal form of M. □

Of course there is a column version of Theorem 2.1. Correspondingly there is a counterpart of Corollary 2.2 involving the *canonical column echelon idempotent* $M_{\text{cEchId}} = E\big((EM^\top E)_{\text{rEchId}}\big)^\top E$.

3. Generalized echelon matrices and generalized diagonals

Above we indicated two elements of arbitrariness in the definition of echelon forms. Another such a point is the stipulation that leading entries should be 1. In view of this we expand our conceptual framework a bit further.

We say that $A \in \mathbb{C}^{n\times n}$ is in *generalized row echelon form* if the leading entries of any two different nonzero rows of A are in different columns of A. Clearly this is the case when A is in permuted row echelon form. When the matrix $A \in \mathbb{C}^{n\times n}$ is in generalized row echelon form and, in addition, it has the property that in the column of a leading entry all other entries are zero, then A is said to be in *reduced generalized row echelon form.* Evidently A is in reduced generalized row echelon form whenever it is in permuted reduced row echelon form. The difference between permuted and generalized row echelon forms lies solely in the fact that in the latter the leading entries are not required to be 1. Hence these forms can be obtained from one another by multiplication on the left by an invertible diagonal matrix.

Given $M \in \mathbb{C}^{n\times n}$, a matrix A is called a *generalized row echelon form* of M, respectively *reduced generalized row echelon form* of M if A can be obtained from M by a multiplication on the left with an invertible $n \times n$ matrix while, in addition, A is in generalized row echelon form, respectively reduced generalized row echelon form. Of course there is a simple relationship between generalized row echelon forms of M, respectively reduced generalized row echelon forms of M, on the one hand and row echelon forms of M, respectively the reduced row echelon form of M on the other. This relationship involves multiplication on the left with permutation matrices and invertible diagonal matrices. The precise formulation is left to the reader.

In the above, the focus was on row structure. As is to be expected (and as was indicated before), there are analogues in which rows are replaced by columns. So we have the notions of *generalized column echelon form* and *reduced generalized column echelon form.* In this the role of the leading entry in a row is played by what we will call the *trailing entry* (i.e., the last) nonzero entry in a column. A matrix $A \in \mathbb{C}^{n\times n}$ is in generalized column echelon form, respectively in reduced generalized column echelon form, if and only if $EA^\top E$ is in generalized row echelon form, respectively in reduced generalized row echelon form. If desired, this observation can also be taken as the basis for a formal definition.

Given $M \in \mathbb{C}^{n\times n}$, we will also use expressions as *generalized column echelon form* of M, respectively *reduced generalized column echelon form* of M. The formal definitions are analogous to those given in the third paragraph of this section. Of course multiplication on the left is replaced by multiplication on the right. Note that A is a generalized column echelon form of M, respectively, a reduced generalized column echelon form of M, if and only if $EA^\top E$ is a generalized row echelon form of $EM^\top E$, respectively, a reduced generalized column echelon form of $EM^\top E$.

Anticipating on what we shall encounter in the context of working with upper triangular matrices, we go on refining our conceptual apparatus. Let $A \in \mathbb{C}^{n\times n}$. We say that A is in *upward reduced generalized row echelon form* or, for short, *upward echelon form,* if A is in generalized row echelon form and, when A has a leading entry in a (nonzero) row of A, the entries above it in the column in question are all zero. Similarly, A is in *starboard reduced generalized column echelon form* or, for short, *starboard echelon form,* if A is in generalized column echelon form and, when A has a trailing entry in a (nonzero) column of A, the entries to the right of it in the row in question are all zero. For illustrations of these definitions, see Example C in Section 10. Note that A is in starboard echelon form if and only if $EA^\top E$ is in upward echelon form.

Next we turn to the introduction of generalized diagonal matrices. Here is some preparation. Let M be an $n \times n$ matrix. It is well known that there exist invertible $L, R \in \mathbb{C}^{n\times n}$ such that LMR is a diagonal matrix having the nonzero entries on its diagonal in the left upper corner. The number of these nonzero entries is obviously equal to the rank of M. We can also see to it that the nonzero entries are equal to 1. Under that extra requirement the diagonal matrix is uniquely determined by M (see, for instance [LT], Section 2.7).

Here, again, there is a natural possibility for generalization. We will say that $A \in \mathbb{C}^{n\times n}$ is a *generalized diagonal matrix* when all rows and columns of A contain at most one nonzero entry. We end this section with a characterization which will play a role when we will be dealing with the upper triangular case.

Theorem 3.1. *Let $A \in \mathbb{C}^{n\times n}$. Then A is a generalized diagonal matrix if and only if A is both in upward and in starboard echelon form.*

Proof. Suppose A is a generalized diagonal matrix. Then the nonzero entries in A are precisely the leading entries in A and also precisely the trailing entries in A. As all other entries (in the corresponding rows and columns) vanish, it is immediate that A is in upward echelon form as well as in starboard echelon form.

This covers the "only if" part of the theorem. So we turn to the "if" part and from now on assume that A is both in upward and in starboard echelon form.

We need to show that all rows and columns of A contain at most one nonzero entry. Suppose this is not the case. Then, either there is a row of A featuring more than one nonzero entry, or there is a column of A featuring more than one nonzero entry. It suffices to consider the situation where there is such a row. Let s be the largest integer for which the sth row of A has at least two nonzero entries. The leading entry in this row cannot be the trailing entry for the column it is in, say the tth. The reason is that A is in starboard echelon form. Let the $\hat{s}$th row of A contain the trailing entry in the tth column of A. Then $\hat{s} > s$ and, as A is in upward echelon form, the trailing entry in question cannot be the leading entry in the $\hat{s}$th row of A. But then this $\hat{s}$th row contains at least two nonzero entries, conflicting with the maximality of s. □

4. Canonical generalized echelon forms – upper triangular case

The set of upper triangular $n \times n$ matrices will be denoted by $\mathbb{C}^{n\times n}_{\text{upper}}$. Here, as before, n is a positive integer. As an aid to the reader, we begin by recalling some items from the introduction.

Let $M \in \mathbb{C}^{n\times n}$, and write $M = [m_{k,l}]_{k,l=1}^{n}$. By a *diagonal block* in M we mean a submatrix of M of the form $M[s,t] = [m_{k,l}]_{k,l=s}^{t}$ with s and t in $\{1,\dots,n\}$, $s \le t$. Now M is said to have *lean row structure* if each diagonal block $M[s,t]$ in M has rank equal to the number of nonzero rows in $M[s,t]$. In other words, a matrix has lean row structure if in each of its diagonal blocks the nonzero rows are linearly independent.

The first result in this section makes the connection with generalized and permuted row echelon forms.

Theorem 4.1. *Let $A \in \mathbb{C}^{n\times n}_{\text{upper}}$. Then A has lean row structure if and only if A is in generalized row echelon form.*

Proof. Suppose A has lean structure. Let S be the set of $s \in \{1,\dots,n\}$ for which the sth row of A is nonzero. For $s \in S$, denote the position of the leading entry in the sth row by $t(s)$. Assume now that A is not in generalized echelon form. Thus there exist two different s_1 and s_2 in S such that $t(s_1)$ and $t(s_2)$ have the same value, t say. Then the diagonal block $A[1,t]$ violates the requirement that its nonzero rows are linearly independent. To see this, look at its s_1th and s_2th row: these are a nonzero multiple of each other. Note here that $s_1, s_2 \le t$ (by upper triangularity).

Next suppose A is in generalized echelon form. Take $s,t \in \{1,\dots,n\}$ with $s \le t$. The leading entries of any two different nonzero rows in $A[s,t]$ are in different columns of $A[s,t]$. (Here, again, the upper triangularity of A plays a role.) But then the nonzero rows of $A[s,t]$ are linearly independent, and we conclude that A has lean row structure. □

Let $M \in \mathbb{C}^{n\times n}_{\text{upper}}$. According to Theorem 1.1 there exists an invertible $L \in \mathbb{C}^{n\times n}_{\text{upper}}$ such that LM has lean row structure or, what amounts here to the same (Theorem 4.1), LM is in generalized row echelon form. The matrices L and LM are not not uniquely determined by M. This appears from Example C in Section 10. The second matrix in the expression (15) featuring in the example possesses a property that the first matrix in (15) does not enjoy: for every leading entry, the positions above it on the column in question are occupied by zeros, i.e., the matrix is in upward echelon form. As the next theorem shows, with this extra condition uniqueness can be assured.

A matrix will be called *monic* if it is upper triangular and has only ones on its diagonal. Such a matrix is of course invertible and its inverse is monic again.

Theorem 4.2. *Let $M \in \mathbb{C}^{n\times n}_{\text{upper}}$. Then there is precisely one $B \in \mathbb{C}^{n\times n}_{\text{upper}}$ such that B is in upward echelon form and $B = LM$ for some monic matrix $L \in \mathbb{C}^{n\times n}_{\text{upper}}$.*

The (upper triangular) matrix in question will be called *the canonical upward reduced generalized row echelon form* of M or, for short, *the canonical upward echelon form* of M – written $M^{\uparrow}_{\text{rEch}}$.

Proof. For the existence proof we rely on that of Theorem 7.1 in Section 7 below (just take the partial order there to be standard linear order on $\{1,\dots,n\}$). So we turn to uniqueness.

Let L_1 and L_2 be monic $n\times n$ matrices such that both $B_1 = L_1M$ and $B_2 = L_2M$ are in upward echelon form. Put $L = L_2L_1^{-1}$. Then L is monic and $B_2 = LB_1$. It is our task to deduce from this that $B_2 = B_1$.

In case $n = 1$, the matrix L can be identified with the scalar 1, and the identity $B_2 = B_1$ is immediate.

Suppose $n = 2$. Write

$$B_1 = \begin{bmatrix} b^{(1)}_{1,1} & b^{(1)}_{1,2} \\ 0 & b^{(1)}_{2,2} \end{bmatrix}, \qquad B_2 = \begin{bmatrix} b^{(2)}_{1,1} & b^{(2)}_{1,2} \\ 0 & b^{(2)}_{2,2} \end{bmatrix}, \qquad L = \begin{bmatrix} 1 & l \\ 0 & 1 \end{bmatrix}.$$

From $B_2 = LB_1$, we get

$$b^{(2)}_{1,1} = b^{(1)}_{1,1}, \qquad b^{(2)}_{2,2} = b^{(1)}_{2,2}, \qquad b^{(2)}_{1,2} = b^{(1)}_{1,2} + lb^{(1)}_{2,2}.$$

If $b^{(1)}_{2,2} = 0$, we immediately get $b^{(2)}_{1,2} = b^{(1)}_{1,2}$, so $B_2 = B_1$ which is the desired result. Assume $b^{(1)}_{2,2} \neq 0$. Then $b^{(2)}_{2,2} \neq 0$ as well. Hence both B_2 and B_1 have leading entries on position $(2,2)$. It follows that $b^{(2)}_{1,2} = b^{(1)}_{1,2} = 0$. But then, again, $B_2 = B_1$.

As a further preparation for the induction argument ahead, it is illuminating to also briefly sketch the argument for the case $n = 3$. Write

$$B_1 = \begin{bmatrix} b^{(1)}_{1,1} & b^{(1)}_{1,2} & b^{(1)}_{1,3} \\ 0 & b^{(1)}_{2,2} & b^{(1)}_{2,3} \\ 0 & 0 & b^{(1)}_{3,3} \end{bmatrix}, \qquad B_2 = \begin{bmatrix} b^{(2)}_{1,1} & b^{(2)}_{1,2} & b^{(2)}_{1,3} \\ 0 & b^{(2)}_{2,2} & b^{(2)}_{2,3} \\ 0 & 0 & b^{(2)}_{3,3} \end{bmatrix}, \qquad L = \begin{bmatrix} 1 & l_1 & l_0 \\ 0 & 1 & l_2 \\ 0 & 0 & 1 \end{bmatrix}.$$

Noting that $B_2[2,3] = L[2,3]B_1[2,3]$ and applying the result of the previous paragraph, one sees that $B_2[2,3] = B_1[2,3]$, i.e.,

$$b^{(2)}_{2,2} = b^{(1)}_{2,2}, \quad b^{(2)}_{2,3} = b^{(1)}_{2,3}, \quad b^{(2)}_{3,3} = b^{(1)}_{3,3}.$$

Similarly, by looking at $B_2[1,2]$ and $B_1[1,2]$, we get

$$b_{1,1}^{(2)} = b_{1,1}^{(1)}, \quad b_{1,2}^{(2)} = b_{1,2}^{(1)}, \quad b_{2,2}^{(2)} = b_{2,2}^{(1)}.$$

It remains to prove that $b_{1,3}^{(2)} = b_{1,3}^{(1)}$. Here is the reasoning.

Suppose $b_{2,3}^{(1)}$ is a leading entry in B_1. Thus $b_{2,3}^{(1)} \neq 0$ and $b_{2,2}^{(1)} = 0$. But $b_{2,3}^{(2)} = b_{2,3}^{(1)}$ and $b_{2,2}^{(2)} = b_{2,2}^{(1)}$. Hence $b_{2,3}^{(2)}$ is a leading entry in B_2. It follows that both $b_{1,3}^{(1)}$ and $b_{1,3}^{(2)}$ vanish. The same conclusion holds when $b_{3,3}^{(1)}$ is a leading entry in B_1 or, what amounts to the same $b_{3,3}^{(2)}$ is a leading entry in B_2.

In the final step to cover the case $n = 3$ we assume that neither $b_{2,3}^{(1)}$ nor $b_{3,3}^{(1)}$ is a leading entry in B_1. In particular $b_{3,3}^{(1)}$ vanishes, and so does $b_{3,3}^{(2)} = b_{3,3}^{(1)}$. If $b_{2,3}^{(1)} = 0$ we get $b_{1,3}^{(2)} = b_{1,3}^{(1)}$ directly from $B_2 = LB_1$. Thus we are ready once we have dealt with the case $b_{2,3}^{(1)} \neq 0$. Since $b_{2,3}^{(1)}$ is not a leading entry of B_1, we must have that $b_{2,2}^{(1)} \neq 0$ too. Further $b_{1,2}^{(1)} = b_{1,2}^{(2)} = b_{1,2}^{(1)} + l_1 b_{2,2}^{(1)}$. It follows that $l_1 = 0$ and $b_{1,3}^{(2)} = b_{1,3}^{(1)} + l_1 b_{2,3}^{(1)} + l_0\, b_{3,3}^{(1)} = b_{1,3}^{(1)} + 0 + 0 = b_{1,3}^{(1)}$, as desired.

We now proceed by induction with n as parameter. Having already dealt with the cases $n = 1$, $n = 2$ and $n = 3$, we take n larger than 3. Consider the diagonal blocks $B_1[2,n]$, $B_2[2,n]$ and $L[2,n]$. These are upper triangular matrices of order $n-1$ and $B_2[2,n] = L[2,n]B_1[2,n]$. Clearly $L[2,n]$ is monic. Also, for $j = 1,2$, the matrix $B_j[2,n]$ has lean structure while, in addition, if $B_j[2,n]$ has a leading entry in a nonzero row of $B_j[2,n]$, the positions above it in the column in question are all occupied by zeros. Indeed, if $B_j[2,n]$ has a leading entry on position (s,t), then B_j has a leading entry on position $(s+1, t+1)$. (Note that upper triangularity plays a role here.) It follows (by the induction hypothesis) that $B_2[2,n] = B_1[2,n]$. Analogously we have $B_2[1,n-1] = B_1[1,n-1]$.

Write $B_1 = [b_{k,m}^{(1)}]_{k,m=1}^n$ and $B_2 = [b_{k,m}^{(2)}]_{k,m=1}^n$. By upper triangularity we have $b_{n,1}^{(1)} = b_{n,1}^{(2)} = 0$. This, together with the two identities established in the preceding paragraph, gives that the entries of B_1 and B_2 are the same in all positions (k,m) except maybe for those in the upper right corner. Thus, in order to get the identity $B_1 = B_2$, it remains to prove that $b_{1,n}^{(2)} = b_{1,n}^{(1)}$. Here is how this goes.

First, let $s \in \{2,\dots,n\}$ and suppose B_1 has a leading entry at position (s,n). Then B_2 also has a leading entry at (s,n). Hence $b_{1,n}^{(2)} = b_{1,n}^{(1)} = 0$.

Thus we may assume that there is no leading entry of B_1 on any of the positions $(2,n),\dots,(n,n)$. Then, in particular, $b_{n,n}^{(1)} = 0$ bringing with it that $b_{n,n}^{(2)} = 0$ as well. If the entries $b_{2,n}^{(1)},\dots,b_{n-1,n}^{(1)}$ vanish as well, then $B_2 = LB_1$ immediately ensues $b_{1,n}^{(2)} = b_{1,n}^{(1)}$. Therefore we suppose, additionally, that among $b_{2,n}^{(1)},\dots,b_{n-1,n}^{(1)}$ there are nonzero entries, i.e., the set J_1 of all $m \in \{2,\dots,n-1\}$ with $b_{m,n}^{(1)} \neq 0$ is nonempty. On account of the identities $B_2[1,n-1] = B_1[1,n-1]$

and $B_2[2,n] = B_1[2,n]$, we can represent B_1 and B_2 as

$$B_1 = \begin{bmatrix} b & R_1 & b^{(1)}_{1,n} \\ 0 & R_2 & b_2 \\ \vdots & \vdots & \vdots \\ 0 & R_{n-1} & b_{n-1} \\ 0 & 0 & 0 \end{bmatrix}, \qquad B_2 = \begin{bmatrix} b & R_1 & b^{(2)}_{1,n} \\ 0 & R_2 & b_2 \\ \vdots & \vdots & \vdots \\ 0 & R_{n-1} & b_{n-1} \\ 0 & 0 & 0 \end{bmatrix},$$

where $R_2, \ldots, R_{n-1}$ are the rows of the identical $(n-2)\times(n-2)$ matrices $B_1[2,n-1]$ and $B_2[2,n-1]$ while, moreover, $b_m = b^{(1)}_{m,n} = b^{(2)}_{m,n}$, $m = 2,\ldots,n-1$. Let J be the set of all $m \in \{2,\ldots,n-1\}$ for which R_m is a nonzero row of $B_1[2,n-1] = B_2[2,n-1]$. Take $m \in J_1$. Then B_1 does not have a leading entry at the position (m,n). Also $b_m = b^{(1)}_{m,n} \neq 0$. Hence R_m must be nonzero, i.e., $m \in J$. This proves that $J_1 \subset J$.

Write the first row of L as $[1,\ l_2,\ldots,l_n]$. Then, from $B_2 = LB_1$,

$$R_1 = R_1 + \sum_{m=2}^{n-1} l_m R_m,$$

and so, taking into account the definition of J,

$$\sum_{m\in J} l_m R_m = \sum_{m=2}^{n-1} l_m R_m = R_1 - R_1 = 0.$$

Now the rows R_m with $m \in J$ are the nonzero rows of $B_1[2,n-1]$. As B_1 has lean structure (Theorem 4.2), these rows are linearly independent, and it follows that l_m vanishes for all $m \in J$. But then $l_m = 0$ for all $m \in J_1$ too. For $b^{(2)}_{1,n}$ we have, again from $B_2 = LB_1$,

$$b^{(2)}_{1,n} = b^{(1)}_{1,n} + \sum_{m=2}^{n-1} l_m b_m.$$

Employing the definition of J_1, this becomes

$$b^{(2)}_{1,n} = b^{(1)}_{1,n} + \sum_{m\in J_1} l_m b_m.$$

However, $l_m = 0$ for all $m \in J_1$, hence $b^{(2)}_{1,n} = b^{(1)}_{1,n}$, as desired. □

Of course there is a counterpart of Theorem 4.2 involving column echelon forms.

Theorem 4.3. *Let $M \in \mathbb{C}^{n\times n}_{\text{upper}}$. Then there is precisely one $B \in \mathbb{C}^{n\times n}_{\text{upper}}$ such that B is in starboard echelon form and $B = MR$ for some monic matrix $R \in \mathbb{C}^{n\times n}_{\text{upper}}$.*

The (upper triangular) matrix in question will be called *the canonical starboard reduced generalized column echelon form* of M or, for short, *the canonical starboard echelon form* of M – written $M_{\mathrm{cEch}}^{\rightarrow}$. There is a close relationship between $M_{\mathrm{cEch}}^{\rightarrow}$ and $M_{\mathrm{rEch}}^{\uparrow}$:

$$M_{\mathrm{cEch}}^{\rightarrow} = E\big((EM^{\top}E)_{\mathrm{rEch}}^{\uparrow}\big)^{\top}E, \qquad M_{\mathrm{rEch}}^{\uparrow} = E\big((EM^{\top}E)_{\mathrm{cEch}}^{\rightarrow}\big)^{\top}E. \tag{1}$$

The verification of these identities is left to the reader.

We now come back to the point of the leading entries being equal to 1. Let $A \in \mathbb{C}_{\mathrm{upper}}^{n\times n}$. We say that A is in *monic upward reduced generalized row echelon form* or, for short, *monic upward echelon form* if A is in upward echelon form and, in addition, all leading entries in A are equal to 1.

Corollary 4.4. *Let $M \in \mathbb{C}_{\mathrm{upper}}^{n\times n}$. Then there exists precisely one $B \in \mathbb{C}_{\mathrm{upper}}^{n\times n}$ such that B is in monic upward echelon form and $B = LM$ for some invertible $L \in \mathbb{C}_{\mathrm{upper}}^{n\times n}$.*

The (upper triangular) matrix in question will be called *the canonical monic upward reduced generalized row echelon form* of M or, for short, *the canonical monic upward echelon form* of M – written $M_{\mathrm{mrEch}}^{\uparrow}$.

Proof. By Theorem 4.2 there exists a monic $\widehat{L} \in \mathbb{C}_{\mathrm{upper}}^{n\times n}$ for which $\widehat{L}M = M_{\mathrm{rEch}}^{\uparrow}$ is in upward echelon form. Now choose an invertible diagonal matric D such that the leading entries in $DM_{\mathrm{rEch}}^{\uparrow}$ are all equal to 1. Then $L = D\widehat{L}$ is an invertible upper triangular matrix and $LM = D\widehat{L}M = DM_{\mathrm{rEch}}^{\uparrow}$ is indeed in monic upward echelon form. As an additional observation, we note that in general the diagonal matrix D is not unique. The freedom is in the diagonal entries corresponding to the zero rows of $M_{\mathrm{rEch}}^{\uparrow}$.

This covers the existence part of the corollary. Next we deal with the uniqueness claim.

Suppose L_1 and L_2 are invertible upper triangular matrices such that $B_1 = L_1M$ and $B_2 = L_2M$ are in monic upward echelon form. Choose invertible diagonal matrices D_1 and D_2 such that D_1L_1 and D_2L_2 are monic. Clearly $D_1L_1M = D_1B_1$ and $D_2L_2M = D_2B_2$ are in upward echelon form. But then Theorem 4.2 gives $D_1B_1 = D_2B_2$. Put $D = D_2^{-1}D_1$. Then D is an invertible diagonal matrix and $B_2 = DB_1$. This implies that the nonzero entries of B_1 and those of B_2 are in the same positions. Take $k \in \{1,\dots,n\}$, and consider the kth row R_1 of B_1 and the kth row R_2 of B_2. If R_1 has zero entries only, then so has R_2. Assume R_1 has at least one nonzero entry. Then this is also the case for R_2. So R_1 contains a leading entry of B_1 and R_2 contains one for B_2. These leading entries are in the same position. Also they are both equal to 1. Hence the kth diagonal entry of D must be 1 too, and it follows that $R_1 = R_2$. □

We say that A is in *monic starboard reduced generalized column echelon form* or, for short, *monic starboard echelon form* if A is in starboard echelon form and, in addition, all trailing entries in A are equal to 1.

Corollary 4.5. *Let* $M \in \mathbb{C}^{n\times n}_{\text{upper}}$. *Then there exists precisely one* $B \in \mathbb{C}^{n\times n}_{\text{upper}}$ *such that* B *is in monic starboard echelon form and* $B = MR$ *for some invertible* $R \in \mathbb{C}^{n\times n}_{\text{upper}}$.

The (upper triangular) matrix in question will be called *the canonical monic starboard reduced generalized column echelon form* of M or, for short, *the canonical monic starboard echelon form* of M – written $M^{\rightarrow}_{\text{mcEch}}$.

Here we have that $M^{\rightarrow}_{\text{mcEch}} = M^{\rightarrow}_{\text{cEch}} D$ for a suitable, generally not unique, invertible diagonal matrix D. The situation is analogous to the one indicated in the first paragraph of the proof of Corollary 4.4 with the columns and trailing entries taking over the role of the rows and the leading entries. The following identities are the counterparts of those in (1):

$$M^{\rightarrow}_{\text{mcEch}} = E\big((EM^{\top}E)^{\uparrow}_{\text{mrEch}}\big)^{\top}E, \qquad M^{\uparrow}_{\text{mcEch}} = E\big((EM^{\top}E)^{\rightarrow}_{\text{mrEch}}\big)^{\top}E. \tag{2}$$

5. Canonical generalized diagonal forms – upper triangular case

After having discussed canonical reduced generalized echelon forms in the previous section, we now turn to considering canonical forms involving generalized diagonal upper triangular matrices.

Theorem 5.1. *Let* $M \in \mathbb{C}^{n\times n}_{\text{upper}}$. *Then there is precisely one generalized diagonal matrix* $B \in \mathbb{C}^{n\times n}_{\text{upper}}$ *such that* $B = LMR$ *for some monic matrices* $L, R \in \mathbb{C}^{n\times n}_{\text{upper}}$.

The (upper triangular) matrix in question will be called *the canonical generalized diagonal form* of M – written M_{gDiag}.

Proof. For the existence proof we rely on that of Theorem 7.10 in Section 7 below (just take the partial order there to be standard linear order on $\{1,\dots,n\}$). So we turn to uniqueness.

Let L_1, R_1, L_2 and R_2 be monic $n\times n$ matrices such that both $B_1 = L_1MR_1$ and $B_2 = L_2MR_2$ are generalized diagonal matrices. Put $L = L_2L_1^{-1}$ and $R = R_1^{-1}R_2$. Then the matrix L is monic and $B_2 = LB_1R$. It is our task to derive from this that $B_2 = B_1$.

In case $n = 1$, the matrix L can be identified with the scalar 1, and the identity $B_2 = B_1$ is immediate.

Suppose $n = 2$. Write

$$B_1 = \begin{bmatrix} b^{(1)}_{1,1} & b^{(1)}_{1,2} \\ 0 & b^{(1)}_{2,2} \end{bmatrix}, \quad B_2 = \begin{bmatrix} b^{(2)}_{1,1} & b^{(2)}_{1,2} \\ 0 & b^{(2)}_{2,2} \end{bmatrix}, \quad L = \begin{bmatrix} 1 & l \\ 0 & 1 \end{bmatrix}, \quad R = \begin{bmatrix} 1 & r \\ 0 & 1 \end{bmatrix}.$$

From $B_2 = LB_1R$, we get $b^{(2)}_{1,1} = b^{(1)}_{1,1}$ and $b^{(2)}_{2,2} = b^{(1)}_{2,2}$. So what we have to establish is that $b^{(2)}_{1,2} = b^{(1)}_{1,2}$. If both of these numbers vanish there is nothing to prove. So assume at least one of them is nonzero, $b^{(1)}_{1,2}$ say. But then $b^{(1)}_{1,1}$, $b^{(1)}_{2,2}$, $b^{(2)}_{1,1}$ and $b^{(2)}_{2,2}$ all vanish, and it follows from $B_2 = LB_1R$ that $b^{(2)}_{1,2} = b^{(1)}_{1,2}$, as desired.

We now proceed by induction with n as parameter. Having already dealt with the cases $n = 1$ and $n = 2$, we take n larger than 2. Consider the diagonal blocks $B_1[2, n]$, $B_2[2, n]$, $L[2, n]$ and $R[2, n]$. These are upper triangular matrices of order $n - 1$ and $B_2[2, n] = L[2, n]B_1[2, n]R[2, n]$. Clearly $L[2, n]$ and $R[2, n]$ are monic. Also, $B_1[2, n]$ and $B_2[2, n]$ are generalized diagonal matrices. It follows (by the induction hypothesis) that $B_2[2, n] = B_1[2, n]$. Analogously we have $B_2[1, n-1] = B_1[1, n-1]$.

Write $B_1 = [b^{(1)}_{k,m}]^n_{k,m=1}$ and $B_2 = [b^{(2)}_{k,m}]^n_{k,m=1}$. By upper triangularity we have $b^{(1)}_{n,1} = b^{(2)}_{n,1} = 0$. This, together with the two identities established in the preceding paragraph, gives that the entries of B_1 and B_2 are the same in all positions (k, m) except maybe for those in the upper right corner. Thus, in order to get the identity $B_1 = B_2$, it remains to prove that $b^{(2)}_{1,n} = b^{(1)}_{1,n}$. Here is how this goes.

We may suppose that $b^{(1)}_{1,n}$ and $b^{(2)}_{1,n}$ do not both vanish. In other words that either $b^{(1)}_{1,n} \neq 0$ or $b^{(2)}_{1,n} \neq 0$. It is sufficient to consider the first of the two cases. So suppose $b^{(1)}_{1,n} \neq 0$. We then have that

$$b^{(j)}_{1,l} = b^{(j)}_{l+1,n} = 0, \qquad l = 1, \ldots, n-1;\ j = 1, 2.$$

Now write

$$B_1 = \begin{bmatrix} 0 & 0 & b^{(1)}_{1,n} \\ 0 & B_1[2, n-1] & 0 \\ 0 & 0 & 0 \end{bmatrix}, \qquad B_2 = \begin{bmatrix} 0 & 0 & b^{(2)}_{1,n} \\ 0 & B_2[2, n-1] & 0 \\ 0 & 0 & 0 \end{bmatrix},$$

(where, by the way, $B_1[2, n-1] = B_2[2, n-1]$). Also introduce

$$L = \begin{bmatrix} 1 & L_1 & l \\ 0 & L[2, n-1] & L_2 \\ 0 & 0 & 1 \end{bmatrix}, \qquad R = \begin{bmatrix} 1 & R_1 & r \\ 0 & R[2, n-1] & R_2 \\ 0 & 0 & 1 \end{bmatrix}.$$

With these representations, the identity $B_2 = LB_1R$ becomes

$$B_2 = \begin{bmatrix} 0 & L_1B_1[2, n-1]R[2, n-1] & b^{(1)}_{1,n} + L_1B_1[2, n-1]R_2 \\ 0 & L[2, n-2]B_1[2, n-1]R[2, n-2] & L[2, n-2]B_1[2, n-2]R_2 \\ 0 & 0 & 0 \end{bmatrix},$$

and it follows that $L_1B_1[2, n-1]R[2, n-1] = 0$. As $R[2, n-1]$ is invertible, this implies $L_1B_1[2, n-1] = 0$. But then $b^{(2)}_{1,n} = b^{(1)}_{1,n} + L_1B_1[2, n-1]R_2 = b^{(1)}_{1,n}$, as desired. □

Again we come back to the point of the leading/trailing entries being equal to 1. Let $A \in \mathbb{C}^{n\times n}_{\text{upper}}$. We say that A is a *monic generalized diagonal matrix* if A is a generalized diagonal matrix and all nonzero entries in A are equal to 1.

Corollary 5.2. *Let $M \in \mathbb{C}^{n\times n}_{\text{upper}}$. Then there is precisely one monic generalized diagonal matrix $B \in \mathbb{C}^{n\times n}_{\text{upper}}$ such $B = LMR$ for some invertible matrices $L, R \in \mathbb{C}^{n\times n}_{\text{upper}}$.*

The (upper triangular) matrix in question will be called *the canonical monic generalized diagonal form* of M – written M_{mgDiag}.

Proof. By Theorem 5.1 there exist monic matrices $\widehat{L}$ and $\widehat{R}$ for which $\widehat{L}M\widehat{R} = M_{\text{gDiag}}$ is a generalized diagonal matrix. Now choose invertible diagonal matrices G and H such that all nonzero entries in $GM_{\text{gDiag}}H$ are equal to 1. Clearly $L = G\widehat{L}$ and $R = \widehat{R}H$ are invertible upper triangular matrix. Also $LMR = G\widehat{L}M\widehat{R}H = GM_{\text{gDiag}}H$ is indeed a monic generalized diagonal matrix. For completeness we note that in general there one has some freedom in choosing the diagonal matrices G and H; see the proofs of Corollaries 4.4 and 4.5. In fact one of them may always be taken to be the $n \times n$ identity matrix.

This covers the existence part of the corollary. Next we deal with the uniqueness claim.

Suppose L_1, L_2, R_1 and R_2 are invertible upper triangular matrices such that $B_1 = L_1MR_1$ and $B_2 = L_2MR_2$ are monic generalized diagonal matrices. Choose invertible diagonal matrices G_1, H_1, G_2, H_2 and H_2 such that G_1L_1, R_1H_1, G_2L_2 and R_2H_2 are monic. Clearly $G_1B_1H_1$ and $G_2B_2H_2$ are generalized diagonal matrices. Also $(G_1L_1)M(R_1H_1) = G_1B_1H_1$ and $(G_2L_2)M(R_2H_2) = G_2B_2H_2$. But then Theorem 5.1 gives $G_1B_1H_1 = G_2B_2H_2$. This implies that the nonzero entries of B_1 and those of B_2 are in the same positions. As these nonzero entries are all equal to 1, we can conclude that $B_1 = B_2$, as desired. □

6. Relations between the canonical forms

Let $A \in \mathbb{C}^{n\times n}$. The matrix obtained from A by leaving the leading entries in A untouched but replacing all other entries by zeros will be called the *leading entry reduction* of A. Similarly, we will use the term *trailing entry reduction* of A for the matrix obtained from A by leaving the trailing entries in A untouched and replacing all other entries by zeros. By the *leading entry graph* of A – written $\mathcal{L}(A)$ – we mean the directed graph on the set of nodes $\boldsymbol{N} = \{1, \ldots, n\}$ given by $k \to l$ if and only if A has a leading entry at the position (k, l). Similarly, the *trailing entry graph* of A – written $\mathcal{T}(A)$ – is defined to be the directed graph on $\boldsymbol{N}$ given by $k \to l$ if and only if A has a trailing entry at the position (k, l). So $\mathcal{L}(A)$, respectively $\mathcal{T}(A)$, is what is often called the directed graph of the leading entry reduction of A, respectively the trailing entry reduction of A (see, for instance, [HJ]). In the language of relations, $\mathcal{L}(A)$, respectively $\mathcal{T}(A)$, is the set of ordered pairs $(k, l) \in \boldsymbol{N} \times \boldsymbol{N}$ with A having a leading, respectively a trailing, entry at (k, l).

Clearly, the leading entry graph of the leading entry reduction of A coincides with the leading entry graph $\mathcal{L}(A)$ of A. Similarly, the trailing entry graph of the trailing entry reduction of A coincides with the trailing entry graph $\mathcal{T}(A)$ of A. We also note that $\mathcal{L}(A) = \mathcal{T}(A)$ whenever A is a generalized diagonal matrix.

The identities (1) and (2) already exhibited some relations between the canonical forms introduced in the previous two sections. These, coming directly along with the definitions, are of a rather straightforward character. In the remainder of this section, we will present results of a less straightforward nature.

Theorem 6.1. *Let $M \in \mathbb{C}^{n\times n}_{\text{upper}}$. Then*

$$\begin{aligned} M_{\text{gDiag}} &= (M^{\uparrow}_{\text{rEch}})_{\text{gDiag}} = (M^{\uparrow}_{\text{rEch}})^{\rightarrow}_{\text{cEch}} \\ &= (M^{\rightarrow}_{\text{cEch}})_{\text{gDiag}} = (M^{\rightarrow}_{\text{cEch}})^{\uparrow}_{\text{rEch}}, \end{aligned}$$

$$\mathcal{L}(M_{\text{gDiag}}) = \mathcal{T}(M_{\text{gDiag}}) = \mathcal{L}(M^{\uparrow}_{\text{rEch}}) = \mathcal{T}(M^{\rightarrow}_{\text{cEch}}).$$

Moreover, the matrix M_{gDiag} is the leading entry reduction of $M^{\uparrow}_{\text{rEch}}$ and it is the trailing entry reduction of $M^{\rightarrow}_{\text{cEch}}$.

Implicit in the theorem is the fact – worth noticing as such – that the leading entry reduction of $M^{\uparrow}_{\text{rEch}}$ and the trailing entry reduction of $M^{\rightarrow}_{\text{cEch}}$ coincide. Informally speaking, the leading entry structure of $M^{\uparrow}_{\text{rEch}}$ and the trailing entry structure of $M^{\rightarrow}_{\text{cEch}}$ are identical, both with regard to positions and entry values.

Theorem 6.1 shows that there are very strong relations between the possibly non-monic canonical forms discussed in the previous two sections. In fact we can do even better. Indeed, one may replace $M^{\uparrow}_{\text{rEch}}$ by any generalized row echelon form M_r of M (upward reduced or not) which can be obtained from M via multiplication on the left with a monic matrix and, likewise, one may replace $M^{\rightarrow}_{\text{cEch}}$ by any generalized column echelon form M_c of M (starboard reduced or not) which can be obtained from M via multiplication on the right with a monic matrix. It is this enhanced version of Theorem 6.1 that we will now prove.

Proof. Let M_r and M_c be as above.

Apply Lemma 7.9 below (with the partial order $\mathcal{Z}$ featuring there taken to be standard linear order on $\{1, \dots, n\}$). This yields the existence of a monic $R \in \mathbb{C}^{n\times n}_{\text{upper}}$ such that $M_r R$ is a generalized diagonal matrix while, in addition, $M_r R$ is the leading entry reduction of M_r. Thus $\mathcal{L}(M_r R) = \mathcal{T}(M_r R) = \mathcal{L}(M_r)$. Theorem 5.1 gives $M_r R = (M_r)_{\text{gDiag}}$.

Write $M_r = LM$ with $L \in \mathbb{C}^{n\times n}_{\text{upper}}$ monic. Then $LMR = M_r R$ is a generalized diagonal matrix. Hence $M_{\text{gDiag}} = LMR = M_r R$, again by Theorem 5.1. From Theorem 3.1 we know that $M_r R$ (being a generalized diagonal matrix) is in starboard echelon form. But then $M_r R = (M_r)^{\rightarrow}_{\text{cEch}}$ by Theorem 4.3.

From what we got so far, we conclude that M_{gDiag} is the leading entry reduction of M_r and

$$M_{\text{gDiag}} = (M_r)_{\text{gDiag}} = (M_r)^{\rightarrow}_{\text{cEch}}, \quad \mathcal{L}(M_{\text{gDiag}}) = \mathcal{T}(M_{\text{gDiag}}) = \mathcal{L}(M_r). \tag{3}$$

Turning our attention to M_c, we obtain in a similar fashion that M_{gDiag} is the trailing entry reduction of M_c while

$$M_{\mathrm{gDiag}} = (M_c)_{\mathrm{gDiag}} = (M_c)^{\uparrow}_{\mathrm{rEch}}, \quad \mathcal{L}(M_{\mathrm{gDiag}}) = \mathcal{T}(M_{\mathrm{gDiag}}) = \mathcal{T}(M_c). \tag{4}$$

Combining (3) and (4), yields

$$M_{\mathrm{gDiag}} = (M_r)_{\mathrm{gDiag}} = (M_r)^{\rightarrow}_{\mathrm{cEch}} = (M_c)_{\mathrm{gDiag}} = (M_c)^{\uparrow}_{\mathrm{rEch}},$$
$$\mathcal{L}(M_{\mathrm{gDiag}}) = \mathcal{T}(M_{\mathrm{gDiag}}) = \mathcal{L}(M_r) = \mathcal{T}(M_c),$$

corresponding to the two series of identities in the theorem, respectively. □

Theorem 6.2. *Let $M \in \mathbb{C}^{n\times n}_{\mathrm{upper}}$. Then*

$$\begin{aligned} M_{\mathrm{mgDiag}} &= (M^{\uparrow}_{\mathrm{mrEch}})_{\mathrm{gDiag}} = (M^{\uparrow}_{\mathrm{mrEch}})^{\rightarrow}_{\mathrm{cEch}} \\ &= (M^{\rightarrow}_{\mathrm{mcEch}})_{\mathrm{gDiag}} = (M^{\rightarrow}_{\mathrm{mcEch}})^{\uparrow}_{\mathrm{rEch}}, \end{aligned}$$

$$\mathcal{L}(M_{\mathrm{mgDiag}}) = \mathcal{T}(M_{\mathrm{mgDiag}}) = \mathcal{L}(M^{\uparrow}_{\mathrm{mrEch}}) = \mathcal{T}(M^{\rightarrow}_{\mathrm{mcEch}}).$$

Moreover, the matrix M_{mgDiag} is the leading entry reduction of $M^{\uparrow}_{\mathrm{mrEch}}$ and it is the trailing entry reduction of $M^{\rightarrow}_{\mathrm{mcEch}}$.

Note that the leading entry reduction of $M^{\uparrow}_{\mathrm{mrEch}}$ and the trailing entry reduction of $M^{\rightarrow}_{\mathrm{mcEch}}$ coincide. Loosely speaking, the leading entry structure of $M^{\uparrow}_{\mathrm{mrEch}}$ and the trailing entry structure of $M^{\rightarrow}_{\mathrm{mcEch}}$ are identical, both with regard to positions and (but this is trivial now) entry values.

Theorem 6.2, just as was the case for Theorem 6.1, allows for a stronger version. In fact, one may replace $M^{\uparrow}_{\mathrm{mrEch}}$ by any generalized row echelon form M_{mr} of M (upward reduced or not) which can be obtained from M via multiplication on the left with an invertible upper triangular matrix and has the additional property that the leading entries of M_{mr} are all equal to 1. Likewise, one may replace $M^{\rightarrow}_{\mathrm{mcEch}}$ by any generalized column echelon form M_{mc} of M (starboard reduced or not) which can be obtained from M via multiplication on the right with an invertible upper triangular matrix and which has trailing entries all equal to 1. It is on this stronger version of Theorem 6.2 that we will focus in the argument given below.

Proof. Take M_{mr} and M_{mc} as above.

Apply Lemma 7.9 below (with the partial order $\mathcal{Z}$ featuring there taken to be standard linear order on $\{1, \ldots, n\}$). This yields the existence of a monic $R \in \mathbb{C}^{n\times n}_{\mathrm{upper}}$ such that $M_{mr}R$ is a generalized diagonal matrix while, in addition, $M_{mr}R$ is the leading entry reduction of M_{mr}. As the leading entries of M_{mr} are all equal to one, the matrix $M_{mr}R$ is a generalized diagonal matrix. Theorem 5.1 now gives $M_{mr}R = (M_{mr})_{\mathrm{gDiag}}$. We also have $\mathcal{L}(M_{mr}R) = \mathcal{T}(M_{mr}R) = \mathcal{L}(M_{mr})$.

Write $M_{mr} = LM$ with $L \in \mathbb{C}^{n\times n}_{\mathrm{upper}}$ invertible. Then $LMR = M_{mr}R$ is a generalized diagonal matrix. Hence $M_{\mathrm{mgDiag}} = LMR = M_{mr}R$ by Corollary 5.2. From Theorem 3.1 we know that $M_{mr}R$ (being a generalized diagonal matrix) is in starboard echelon form. But then $M_{mr}R = (M_{mr})^{\rightarrow}_{\mathrm{cEch}}$ by Theorem 4.3.

From what we got so far, we conclude that M_{mgDiag} is the leading entry reduction of $M_{mr}H$ and

$$M_{\text{mgDiag}} = (M_{mr})_{\text{gDiag}} = (M_{mr})^{\rightarrow}_{\text{cEch}}, \tag{5}$$

$$\mathcal{L}(M_{\text{mgDiag}}) = \mathcal{T}(M_{\text{mgDiag}}) = \mathcal{L}(M_{mr}). \tag{6}$$

Turning our attention to M_{mc}, we obtain in a similar fashion that M_{mgDiag} is the trailing entry reduction of M_{mc} while

$$M_{\text{mgDiag}} = (M_{mc})_{\text{gDiag}} = (M_{mc})^{\uparrow}_{\text{rEch}}, \tag{7}$$

$$\mathcal{L}(M_{\text{mgDiag}}) = \mathcal{T}(M_{\text{mgDiag}}) = \mathcal{T}(M_{mc}). \tag{8}$$

Combining (5)–(8), yields

$$M_{\text{mgDiag}} = (M_{mr})_{\text{gDiag}} = (M_{mr})^{\rightarrow}_{\text{cEch}} = (M_{mc})_{\text{gDiag}} = (M_{mc})^{\uparrow}_{\text{rEch}},$$

$$\mathcal{L}(M_{\text{mgDiag}}) = \mathcal{T}(M_{\text{mgDiag}}) = \mathcal{L}(M_{mr}) = \mathcal{T}(M_{mc}),$$

corresponding to the two series of identities in the theorem, respectively. □

Theorem 6.3. *Let $M \in \mathbb{C}^{n\times n}_{\text{upper}}$. Then*

$$\begin{aligned} M_{\text{mgDiag}} &= (M^{\uparrow}_{\text{rEch}})_{\text{mgDiag}} = (M^{\uparrow}_{\text{rEch}})^{\rightarrow}_{\text{mcEch}} \\ &= (M^{\rightarrow}_{\text{cEch}})_{\text{mgDiag}} = (M^{\rightarrow}_{\text{cEch}})^{\uparrow}_{\text{mrEch}}, \end{aligned}$$

$$\mathcal{L}(M_{\text{mgDiag}}) = \mathcal{T}(M_{\text{mgDiag}}) = \mathcal{L}(M^{\uparrow}_{\text{rEch}}) = \mathcal{T}(M^{\rightarrow}_{\text{cEch}}).$$

Theorem 6.3 allows for an enhanced version too. Actually, one may replace $M^{\uparrow}_{\text{rEch}}$ by any generalized row echelon form M_r of M (upward reduced or not) which can be obtained from M via multiplication on the left with an invertible upper triangular matrix and, similarly, one may replace $M^{\rightarrow}_{\text{cEch}}$ by any generalized column echelon form M_c of M (starboard reduced or not) which can be obtained from M via multiplication on the right with an invertible upper triangular matrix matrix. Again we focus on the enhanced version of the theorem.

Proof. Take M_r and M_c as above.

Let H be an invertible diagonal matrix such that all leading entries of M_rH are equal to 1. Apply Lemma 7.9 below (with the partial order $\mathcal{Z}$ featuring there taken to be standard linear order on $\{1, \ldots, n\}$). This yields the existence of a monic $R \in \mathbb{C}^{n\times n}_{\text{upper}}$ such that M_rHR is of generalized diagonal type while, in addition, M_rHR is the leading entry reduction of M_rH. As the leading entries of M_rH are all equal to one, M_rHR is a monic generalized diagonal matrix. Corollary 5.2 now gives $M_rHR = (M_r)_{\text{mgDiag}}$. We also have $\mathcal{L}(M_rH) = \mathcal{T}(M_rH) = \mathcal{L}(M_r)$.

Write $M_r = LM$ with $L \in \mathbb{C}^{n\times n}_{\text{upper}}$ invertible. Then $LMHR = M_rHR$ is a monic generalized diagonal matrix. Hence $M_{\text{mgDiag}} = LMHR = M_rHR$, again by Corollary 5.2. From Theorem 3.1 we know that M_rHR (being a monic generalized diagonal matrix) is in monic starboard echelon form. But then $M_rHR = (M_r)^{\rightarrow}_{\text{mcEch}}$ by Corollary 4.5.

From what we got so far, we conclude that $M_{\mathrm{mgDiag}} = (M_r)_{\mathrm{mgDiag}} = (M_r)^{\rightarrow}_{\mathrm{mcEch}}$ and $\mathcal{L}(M_{\mathrm{mgDiag}}) = \mathcal{T}(M_{\mathrm{mgDiag}}) = \mathcal{L}(M_r)$. Turning our attention to M_c, and introducing an invertible diagonal matrix G such that GM_c has trailing entries all equal to 1, we obtain in a similar fashion that $M_{\mathrm{mgDiag}} = (M_c)_{\mathrm{mgDiag}} = (M_c)^{\uparrow}_{\mathrm{mrEch}}$ and $\mathcal{L}(M_{\mathrm{mgDiag}}) = \mathcal{T}(M_{\mathrm{mgDiag}}) = \mathcal{T}(M_c)$. It follows that

$$M_{\mathrm{mgDiag}} = (M_r)_{\mathrm{mgDiag}} = (M_r)^{\rightarrow}_{\mathrm{mcEch}} = (M_c)_{\mathrm{mgDiag}} = (M_c)^{\uparrow}_{\mathrm{mrEch}},$$
$$\mathcal{L}(M_{\mathrm{mgDiag}}) = \mathcal{T}(M_{\mathrm{mgDiag}}) = \mathcal{L}(M_r) = \mathcal{T}(M_c),$$

and with this the argument comes to an end. □

Theorem 6.4. *Let $M \in \mathbb{C}^{n\times n}_{\mathrm{upper}}$. Then*

$$\begin{aligned} M_{\mathrm{mgDiag}} &= (M^{\uparrow}_{\mathrm{mrEch}})_{\mathrm{mgDiag}} = (M^{\uparrow}_{\mathrm{mrEch}})^{\rightarrow}_{\mathrm{mcEch}} \\ &= (M^{\rightarrow}_{\mathrm{mcEch}})_{\mathrm{mgDiag}} = (M^{\rightarrow}_{\mathrm{mcEch}})^{\uparrow}_{\mathrm{mrEch}}. \end{aligned}$$

Proof. This is a special case of the enhanced version of Theorem 6.3. Indeed, for M_r and M_c featuring in the proof of that theorem, one can take $M^{\uparrow}_{\mathrm{mrEch}}$ and $M^{\rightarrow}_{\mathrm{mcEch}}$, respectively. □

Obviously the directed graphs

$$\mathcal{L}(M_{\mathrm{gDiag}}),\ \mathcal{T}(M_{\mathrm{gDiag}}),\ \mathcal{L}(M_{\mathrm{mgDiag}}),\ \mathcal{T}(M_{\mathrm{mgDiag}}),$$
$$\mathcal{L}(M^{\uparrow}_{\mathrm{rEch}}),\ \ \mathcal{T}(M^{\rightarrow}_{\mathrm{cEch}}),\ \ \mathcal{L}(M^{\uparrow}_{\mathrm{mrEch}}),\ \ \mathcal{T}(M^{\rightarrow}_{\mathrm{mcEch}})$$

all coincide. This suggests that they carry some essential information about the given upper triangular matrix M. In Section 9 we shall come back to this.

7. The case of zero pattern algebras

We begin by presenting some preliminaries on matrix algebras determined by a pattern of zeros. Throughout this section n is a positive integer and $\boldsymbol{N}$ stands for the set $\{1, \dots, n\}$.

Let $\mathcal{Z}$ be a relation with ground set $\boldsymbol{N}$ or, if one prefers, let $\mathcal{Z}$ be a directed graph on the set of nodes $\{1, \dots, n\}$. The notation $h \rightarrow_{\mathcal{Z}} k$ will often be used as an alternative for $(h, k) \in \mathcal{Z}$. In the same vein, $h \not\rightarrow_{\mathcal{Z}} k$ signals that $(h, k) \notin \mathcal{Z}$.

We introduce $\mathcal{A}_{\mathcal{Z}}$ as the collection of all $A = [a_{k,m}]_{k,m=1}^{n} \in \mathbb{C}^{n\times n}$ with $a_{k,m} = 0$ whenever $k \not\rightarrow_{\mathcal{Z}} m$. Evidently, $\mathcal{A}_{\mathcal{Z}}$ is closed under scalar multiplication and addition. So, regardless of additional properties of $\mathcal{Z}$, the set $\mathcal{A}_{\mathcal{Z}}$ is a linear subspace of $\mathbb{C}^{n\times n}$. For being an algebra more is needed. Indeed, $\mathcal{A}_{\mathcal{Z}}$ is a subalgebra of $\mathbb{C}^{n\times n}$ if and only if the relation $\mathcal{Z}$ is transitive (see [D]; cf. also [Laf]). Clearly $\mathcal{A}_{\mathcal{Z}}$ contains the identity matrix I_n, if and only if $\mathcal{Z}$ is reflexive. Thus $\mathcal{A}_{\mathcal{Z}}$ is a subalgebra of $\mathbb{C}^{n\times n}$ containing the unit element in $\mathbb{C}^{n\times n}$ if and only if $\mathcal{Z}$ is a preorder (i.e., both transitive and reflexive).

In passing, we mention that Theorem 1.2 on logarithmic residues and sums of idempotents in $\mathbb{C}^{n\times n}_{\mathrm{upper}}$ suggests to look for a similar result where the underlying algebra is a zero pattern algebra of the type introduced above. In [BES3] this challenge has been taken up.

Next we specialize to the situation where $\mathcal{Z}$ is a partial order. So besides being reflexive and transitive, $\mathcal{Z}$ is antisymmetric. It is common knowledge that this implies the existence of a linear order (also called total order) $\mathcal{L}$ on $\boldsymbol{N}$ which is compatible with $\mathcal{Z}$, i.e., for which $\mathcal{Z} \subset \mathcal{L}$. (In the present situation with a finite ground set this is not difficult to see. When the ground set is infinite, things are more involved; in particular the axiom of choice is needed then. For details, see [S(M)].)

From now on we shall assume that $\mathcal{Z}$ is of *upper triangular type*, meaning that $\mathcal{L}$ can be taken to be the standard order on $\boldsymbol{N}$, so $k \leq m$ whenever $k \to_{\mathcal{Z}} m$. To put this assumption in perspective: note that linear orders on $\boldsymbol{N}$ are permutation similar to the standard (natural) order on $\boldsymbol{N}$. Our assumption implies that $\mathcal{A}_{\mathcal{Z}}$ consists of upper triangular matrices, in other words, $\mathcal{A}_{\mathcal{Z}}$ is contained in $\mathbb{C}^{n\times n}_{\text{upper}}$. Recall here that the canonical forms considered in this paper all involve upper triangular matrices.

Let $M \in \mathbb{C}^{n\times n}_{\text{upper}}$ be in $\mathcal{A}_{\mathcal{Z}}$. A moment of reflection makes clear that one cannot expect the reduced row echelon form M_{rEch} of M to be in $\mathcal{A}_{\mathcal{Z}}$, and the same is true for or the canonical echelon idempotent M_{rEchId} of M. Indeed getting those forms from M generally requires the interchanging of rows. But what about the canonical upward echelon form $M^{\uparrow}_{\text{rEch}}$ of M? Getting that one from M does not involve any displacement of rows. Here is the rough answer. In general the matrix $M^{\uparrow}_{\text{rEch}}$ is not in $\mathcal{A}_{\mathcal{Z}}$; it is, however, under certain conditions on the given preorder $\mathcal{Z}$ allowing for a large class of examples. For an illustration, see Example C in Section 10.

The extra condition meant above is that the partial order $\mathcal{Z}$ is *in-ultra transitive*. This amounts to requiring the following:

$$k \to_{\mathcal{Z}} m \text{ and } l \to_{\mathcal{Z}} m \quad \Rightarrow \quad k \to_{\mathcal{Z}} l \text{ or } l \to_{\mathcal{Z}} k.$$

When k and l are different, the 'or' has to be understood in the exclusive sense (because of antisymmetry). For concrete instances of a partial order that are or are not in-ultra transitive, see Example B in Section 10. In-ultra transitivity can be characterized in terms of rooted trees. For details, see [BES4]; cf. also Example B in Section 10.

Theorem 7.1. *Let $M \in \mathcal{A}_{\mathcal{Z}}$ where $\mathcal{Z}$ is a partial order on $\boldsymbol{N}$ of upper triangular type. Suppose $\mathcal{Z}$ is in-ultra transitive. Then the canonical upward echelon form $M^{\uparrow}_{\text{rEch}}$ of M belongs to $\mathcal{A}_{\mathcal{Z}}$. In fact, $M^{\uparrow}_{\text{rEch}} = LM$ for some monic matrix $L \in \mathcal{A}_{\mathcal{Z}}$.*

Example C in Section 10 shows that the conclusion of the theorem need not hold when the in-ultra transitivity condition is dropped. On the other hand, it may happen that $M^{\uparrow}_{\text{rEch}} \in \mathcal{A}_{\mathcal{Z}}$ in spite of the circumstance that $\mathcal{Z}$ is not in-ultra transitive. An instance where this occurs is Example D in Section 10. An additional feature of the example is that there does not exist any invertible $K \in \mathcal{A}_{\mathcal{Z}}$ such that $KM = M^{\uparrow}_{\text{rEch}}$.

The proof of Theorem 7.1 given below is constructive in the sense that it (implicitly) provides an algorithm for computing L and the canonical upward echelon form $M^{\uparrow}_{\text{rEch}}$ of M. The algorithm is a modification of the standard one for determining the reduced echelon form M_{rEch} of M, and we could have left it with this were it not for the claim that the matrices involved belong to the zero pattern algebra $\mathcal{A}_{\mathcal{Z}}$. This makes it necessary to give the details.

Proof. The argument employs induction with the positive integer n as parameter. If $n = 1$ there is nothing to prove: take for L the scalar 1 (more precisely the one by one identity matrix). So assume n is at least two.

Let $\mathcal{N} = \mathcal{Z} \cap (\{1, \ldots, n-1\} \times \{1, \ldots, n-1\})$. Then $\mathcal{N}$ is a partial order with ground set $\{1, \ldots, n-1\}$. Also $\mathcal{N}$ is of upper triangular type and in-ultra transitive. Clearly the diagonal block $M[1, n-1]$ belongs to $\mathcal{A}_{\mathcal{N}}$. So (by induction hypothesis) there exists an invertible matrix L_0 in $\mathcal{A}_{\mathcal{N}}$, having only zeros on its diagonal, such that $M_0 = L_0 M[1, n-1]$ belongs to $\mathcal{A}_{\mathcal{N}}$ and is in upward echelon form.

Introduce the invertible (upper triangular) matrix

$$L_1 = \begin{bmatrix} L_0 & 0 \\ 0 & 1 \end{bmatrix}.$$

Along with L_0 it has only ones on its diagonal. Further $L_1 \in \mathcal{A}_{\mathcal{Z}}$, hence $L_1 M$ belongs to $\mathcal{A}_{\mathcal{Z}}$ too. Write $L_1 M$ in the form

$$L_1 M = \begin{bmatrix} M_0 & R \\ 0 & r_n \end{bmatrix},$$

where $R = [r_1 \ldots r_{n-1}]^\top$ is an $(n-1) \times 1$ matrix (column) and $r_n \in \mathbb{C}$. Let $R_1, \ldots, R_{n-1}$ be the rows of the $(n-1) \times (n-1)$ matrix M_0. Then the rows of $L_1 M$ are $[R_1\ r_1], \ldots, [R_{n-1}\ r_{n-1}], [R_n\ r_n]$, with $R_n^\top$ the zero vector in $\mathbb{C}^{n-1}$ (i.e., $[R_n\ r_n] = [0, \ldots, 0, r_n]$).

The $(n-1) \times (n-1)$ matrix M_0 is in upward echelon form. Thus the first up to the last column of M_0 contain at most one leading entry. Also the entries above such a leading entry are all zero. But then the same can be said for the first up to the $(n-1)$th column of the $n \times n$ matrix $L_1 M$. If there are no leading entries in the last column of $L_1 M$, it immediately follows that $L_1 M$ is in upward echelon form, and we can take $L = L_1$.

So assume the last column of $L_1 M$ does feature at least one leading entry. Let s be the largest of all $t \in \{1, \ldots, n\}$ such that r_t is the leading entry in the tth row of $L_1 M$. (Clearly $s = n$ if and only if $r_n \neq 0$.) Also, write C for the (possibly empty) set of all k in $\{1, \ldots, s-1\}$ with $r_k \neq 0$. Now let the matrix L_2 be obtained from the $n \times n$ identity matrix I_n as follows: for $k \in C$, subtract $r_k r_s^{-1}$ times the sth row of I_n from the kth row of I_n. (Thus $L_2 = I_n - \sum_{k \in C} r_k r_s^{-1} u_k u_s^\top$, where $u_1, \ldots, u_n$ stand for the standard unit vectors in $\mathbb{C}^{n \times n}$.) Then L_2 is monic. Multiplying $L_1 M$ on the left with L_2 has no effect on $L_1 M[1, n-1]$, the reason being that the sth row of M_0 starts with $n-1$ zeros. The effect on the last column of $L_1 M$ is that all entries in the positions (k, n) with $k \in \{1, \ldots, s-1\}$ become (or stay) zero while the other entries in the column are unaffected. In particular the entry in the position (s, n) remains equal to $r_s \neq 0$. It follows that $L_2(L_1M)$ is in upward echelon form.

Put $L = L_2L_1$. Then $LM = L_2(L_1M)$ is the canonical upward echelon form $M^{\uparrow}_{\mathrm{rEch}}$ of M. It remains to be seen that L_2 belongs to $\mathcal{A}_{\mathcal{Z}}$ (so that $L = L_2L_1 \in \mathcal{A}_{\mathcal{Z}}$ too). Here is the argument, heavily relying on the in-ultra transitivity of $\mathcal{Z}$.

A nonzero off-diagonal entry in L_2 can only occur in a position of the form (k, s) with $k \in C$. The latter implies that $k < s$ and $k \rightarrow_{\mathcal{Z}} n$. Indeed, the entry of M_0 on the (k, n)th position is $r_k \neq 0$ and $M_0 \in \mathcal{A}_{\mathcal{Z}}$. As r_s is the leading entry in the sth row of $L_1M \in \mathcal{A}_{\mathcal{Z}}$, we also have that $s \rightarrow_{\mathcal{Z}} n$. The in-ultra transitivity of $\mathcal{Z}$ now gives that either $k \rightarrow_{\mathcal{Z}} s$ or $s \rightarrow_{\mathcal{Z}} k$. By assumption $\mathcal{Z}$ is of upper triangular type. Thus $k \rightarrow_{\mathcal{Z}} s$ implies $k \leq s$ and $s \rightarrow_{\mathcal{Z}} k$ gives $s \leq k$. From $k < s$ we now get $k \rightarrow_{\mathcal{Z}} s$, and the proof is complete. □

Corollary 7.2. *Let $M \in \mathcal{A}_{\mathcal{Z}}$ where $\mathcal{Z}$ is a partial order on $\boldsymbol{N}$ of upper triangular type. Suppose $\mathcal{Z}$ is in-ultra transitive. Then the canonical monic upward echelon form $M^{\uparrow}_{\mathrm{mrEch}}$ of M belongs to $\mathcal{A}_{\mathcal{Z}}$. In fact, $M^{\uparrow}_{\mathrm{mrEch}} = LM$ for some invertible (upper triangular) matrix $L \in \mathcal{A}_{\mathcal{Z}}$.*

For how to obtain Corollary 7.2 from Theorem 7.1, consult the first part of proof of Corollary 4.4. Thus along with the argument for Theorem 7.1, that for the corollary is (implicitly) constructive too.

Let $\mathcal{Z}$ be a partial order on $\boldsymbol{N} = \{1, \ldots, n\}$ of upper triangular type. The standard (natural) order on $\boldsymbol{N}$ is of upper triangular type (of course) and in-ultra transitive. So the collection $\boldsymbol{P}$ of in-ultra transitive partial orders of upper triangular type containing $\mathcal{Z}$ is nonempty. Write

$$\mathcal{Z}_{\mathrm{in}} = \bigcap_{\mathcal{P} \in \boldsymbol{P}} \mathcal{P}.$$

Then $\mathcal{Z}_{\mathrm{in}}$ is a partial order of upper triangular type and $\mathcal{Z} \subset \mathcal{Z}_{\mathrm{in}}$. Also $\mathcal{Z}_{\mathrm{in}}$ is in-ultra transitive. This can be seen as follows. Suppose $k \rightarrow_{\mathcal{Z}_{\mathrm{in}}} m$ and $l \rightarrow_{\mathcal{Z}_{\mathrm{in}}} m$. We need to show that $k \rightarrow_{\mathcal{Z}_{\mathrm{in}}} l$ or $l \rightarrow_{\mathcal{Z}_{\mathrm{in}}} k$. Without loss of generality it may be assumed that $k < l$. Take $\mathcal{P} \in \boldsymbol{P}$. Then, $\mathcal{P}$ being in-ultra transitive, $k \rightarrow_{\mathcal{P}} l$ or $l \rightarrow_{\mathcal{P}} k$. Here the outcome of the logical operator 'or' in principle depends on $\mathcal{P}$. However, $\mathcal{P}$ is of upper triangular type and $k < l$, hence $k \rightarrow_{\mathcal{P}} l$. It follows that $k \rightarrow_{\mathcal{Z}_{\mathrm{in}}} l$.

Observe that $\mathcal{Z}_{\mathrm{in}}$ is the smallest partial order containing $\mathcal{Z}$ which is of upper triangular type and in-ultra transitive. We call $\mathcal{Z}_{\mathrm{in}}$ the *in-ultra transitive closure* of $\mathcal{Z}$. Evidently $\mathcal{Z}$ is in-ultra transitive if and only if $\mathcal{Z}_{\mathrm{in}} = \mathcal{Z}$.

In practice $\mathcal{Z}_{\mathrm{in}}$ can be obtained from $\mathcal{Z}$ via the following procedure. If there exist two pairs (k, m) and (l, m) in $\mathcal{Z}$ such that $k < l$ but with $(k, l) \notin \mathcal{Z}$, add the pair (k, l) to $\mathcal{Z}$ and take the transitive closure. This gives the relation $\mathcal{Z}'$. Now proceed with $\mathcal{Z}$ replaced by $\mathcal{Z}'$. Carrying on in this way one arrives in a finite number of steps at $\mathcal{Z}_{\mathrm{in}}$.

The next two results are immediate consequences of Theorem 7.1 and Corollary 7.2.

Corollary 7.3. *Let $M \in \mathcal{A}_{\mathcal{Z}}$ where $\mathcal{Z}$ is a partial order on $\boldsymbol{N}$ of upper triangular type. Then the canonical upward echelon form $M^{\uparrow}_{\mathrm{rEch}}$ of M belongs to $\mathcal{A}_{\mathcal{Z}_{\mathrm{in}}}$. In fact, $M^{\uparrow}_{\mathrm{rEch}} = LM$ for some monic matrix $L \in \mathcal{A}_{\mathcal{Z}_{\mathrm{in}}}$.*

Corollary 7.4. *Let $M \in \mathcal{A}_{\mathcal{Z}}$ where $\mathcal{Z}$ is a partial order on $\boldsymbol{N}$ of upper triangular type. Then the canonical monic upward echelon form $M^{\uparrow}_{\mathrm{mrEch}}$ of M belongs to $\mathcal{A}_{\mathcal{Z}_{\mathrm{in}}}$. In fact, $M^{\uparrow}_{\mathrm{mrEch}} = LM$ for some invertible (upper triangular) matrix $L \in \mathcal{A}_{\mathcal{Z}_{\mathrm{in}}}$.*

Of course there are analogues to the four results obtained so far in this section in which row echelon forms are replaced by column echelon forms and, accordingly, left multiplication is replaced by right multiplication. The counterpart of in-ultra transitivity needed in this context is out-ultra transitivity. We call $\mathcal{Z}$ *out-ultra transitive* if

$$k \to_{\mathcal{Z}} l \text{ and } k \to_{\mathcal{Z}} m \quad \Rightarrow \quad l \to_{\mathcal{Z}} m \text{ or } m \to_{\mathcal{Z}} l.$$

where (because of antisymmetry) the 'or' has to be understood in the exclusive sense. Just as in-ultra transitivity, out-ultra transitivity can be characterized in terms of rooted trees. For details we refer again to [BES4].

Corresponding to the in-ultra transitive closure $\mathcal{Z}_{\mathrm{in}}$ of a partial order $\mathcal{Z}$ of upper triangular type, we have $\mathcal{Z}_{\mathrm{out}}$, the *out-ultra transitive closure* of $\mathcal{Z}$. It is the smallest partial order containing $\mathcal{Z}$ which is of upper triangular type and out-ultra transitive. Here are the analogues of Theorem 7.1 and Corollaries 7.2, 7.3 and 7.4.

Theorem 7.5. *Let $M \in \mathcal{A}_{\mathcal{Z}}$ where $\mathcal{Z}$ is a partial order on $\boldsymbol{N}$ of upper triangular type. Suppose $\mathcal{Z}$ is out-ultra transitive. Then the canonical starboard echelon form $M^{\to}_{\mathrm{cEch}}$ of M belongs to $\mathcal{A}_{\mathcal{Z}}$. In fact, $M^{\to}_{\mathrm{cEch}} = MR$ for some monic matrix $R \in \mathcal{A}_{\mathcal{Z}}$.*

Corollary 7.6. *Let $M \in \mathcal{A}_{\mathcal{Z}}$ where $\mathcal{Z}$ is a partial order on $\boldsymbol{N}$ of upper triangular type. Suppose $\mathcal{Z}$ is out-ultra transitive. Then the canonical monic starboard echelon form $M^{\to}_{\mathrm{mcEch}}$ of M belongs to $\mathcal{A}_{\mathcal{Z}}$. In fact, $M^{\to}_{\mathrm{mcEch}} = MR$ for some invertible (upper triangular) matrix $R \in \mathcal{A}_{\mathcal{Z}}$.*

Corollary 7.7. *Let $M \in \mathcal{A}_{\mathcal{Z}}$ where $\mathcal{Z}$ is a partial order on $\boldsymbol{N}$ of upper triangular type. Then the canonical starboard echelon form $M^{\to}_{\mathrm{cEch}}$ of M belongs to $\mathcal{A}_{\mathcal{Z}_{\mathrm{out}}}$. In fact, $M^{\to}_{\mathrm{cEch}} = MR$ for some monic matrix $R \in \mathcal{A}_{\mathcal{Z}_{\mathrm{out}}}$.*

Corollary 7.8. *Let $M \in \mathcal{A}_{\mathcal{Z}}$ where $\mathcal{Z}$ is a partial order on $\boldsymbol{N}$ of upper triangular type. Then the canonical monic starboard echelon form $M^{\to}_{\mathrm{mcEch}}$ of M belongs to $\mathcal{A}_{\mathcal{Z}_{\mathrm{out}}}$. In fact, $M^{\to}_{\mathrm{mcEch}} = MR$ for some invertible (upper triangular) matrix $R \in \mathcal{A}_{\mathcal{Z}_{\mathrm{out}}}$.*

In the remainder of this section, we deal with canonical generalized diagonal forms. In line with what we did for $M^{\uparrow}_{\mathrm{rEch}}$ we now ask: if M is in $\mathcal{A}_{\mathcal{Z}}$, do the canonical generalized diagonal form M_{gDiag} of M and the canonical generalized monic diagonal form M_{gDiag} of M belong to $\mathcal{A}_{\mathcal{Z}}$ too? Again the answer is negative (see Example C in Section 10). However, similar to what we encountered before, a positive result can be obtained under a relatively mild extra condition on $\mathcal{Z}$.

It is convenient to have available the following auxiliary result.

Lemma 7.9. *Let $M \in \mathcal{A}_{\mathcal{Z}}$ where $\mathcal{Z}$ is a partial order on $\boldsymbol{N}$ of upper triangular type. Suppose M is in generalized row echelon form and $\mathcal{Z}$ is out-ultra transitive. Then there is a monic matrix $R \in \mathcal{A}_{\mathcal{Z}}$ such that MR is a generalized diagonal matrix and MR is the leading entry reduction of M.*

Proof. Let M have its leading entries at the positions $(s_1, t_1), \ldots, (s_m, t_m)$ with $s_1, \ldots, s_m$ different elements of $\boldsymbol{N}$. Then $t_1, \ldots, t_m$ are different integers (because M is in generalized row echelon form) and each t_j is larger than or equal to s_j (upper triangularity). For $s \in \boldsymbol{N}$, the sth row of M is nonzero if and only if $s \in \{s_1, \ldots, s_m\}$. Let S be the set of all $\sigma \in \{1, \ldots, m\}$ such that the σth row has at least two nonzero entries (so at least one more than the leading one). If S is empty, then M is already a generalized diagonal matrix with the nonzero entries in the positions (s_j, t_j) $j = 1, \ldots, m$. So assume S is nonempty.

Take $k \in S$ such that t_k is minimal among the integers t_σ, $\sigma \in S$ and put $\tau = t_k$. Further, write the kth row of M, which starts with $\tau - 1$ zeros, as $[0, \ldots, 0, c_\tau, \ldots, c_n]$. Then $c_\tau \neq 0$ and not all of $c_{\tau+1}, \ldots, c_n$ vanish. Now let the matrix R_1 be obtained from the $n \times n$ identity matrix I_n as follows: for $l \in \{\tau + 1, \ldots, n\}$ with $c_l \neq 0$, subtract $c_l c_\tau^{-1}$ times the τth column of I_n from the lth column of I_n. (Thus $R_1 = I_n - \sum_{l \in \{\tau+1, \ldots, n\},\, c_l \neq 0} c_l c_\tau^{-1} u_\tau u_l^\top$, where again $u_1, \ldots, u_n$ stand for the standard unit vectors in $\mathbb{C}^{n \times n}$.) Then R_1 is monic. Also $R_1 \in \mathcal{A}_{\mathcal{Z}}$. Here is the argument, using the out-ultra transitivity of $\mathcal{Z}$.

A nonzero off-diagonal entry in R_1 can only occur in a position of the form (τ, l) with $l \in \{\tau + 1, \ldots, n\}$, $c_l \neq 0$. Thus $\tau < l$ and $k \rightarrow_{\mathcal{Z}} l$. Indeed, the entry of M on the (k, l)th position is $c_l \neq 0$ and $M \in \mathcal{A}_{\mathcal{Z}}$. As c_τ is the leading entry in the kth row of $M \in \mathcal{A}_{\mathcal{Z}}$, we also have that $k \rightarrow_{\mathcal{Z}} \tau$. The in-ultra transitivity of $\mathcal{Z}$ now gives that either $l \rightarrow_{\mathcal{Z}} \tau$ or $\tau \rightarrow_{\mathcal{Z}} l$. By assumption $\mathcal{Z}$ is of upper triangular type. Thus $l \rightarrow_{\mathcal{Z}} \tau$ implies $l \leq \tau$ and $\tau \rightarrow_{\mathcal{Z}} l$ gives $\tau \leq l$. From $\tau < l$ we now get $\tau \rightarrow_{\mathcal{Z}} l$, as desired.

Put $M_1 = MR_1$. As $R_1 \in \mathcal{A}_{\mathcal{Z}}$, we have $M_1 \in \mathcal{A}_{\mathcal{Z}}$. Multiplying M on the right with R_1 (to get M_1) has only an effect on the kth row of M, the reason being that the τth column of M has just one nonzero entry which occurs at the position (k, τ). The effect on the kth row is that it changes from $[0, \ldots, 0, c_\tau, \ldots, c_n]$ into $[0, \ldots, 0, c_\tau, 0 \ldots, 0]$. It follows that M_1 is in generalized echelon form. In fact M_1 has the same leading entry structure as M (same positions, same entries).

Define S_1 for M_1 in the same as S was defined for M. Then $S_1 = S \setminus \{\tau\}$. We can now proceed with M_1 in the same way as we did with M. And so on. Ultimately this leads to a matrix in $\mathcal{A}_{\mathcal{Z}}$ for which the corresponding analogue of the set S is empty, and which is therefore a generalized diagonal matrix. Also this matrix has the same leading entry structure as M (same positions, same entries). Therefore it is the leading entry reduction of M.

We leave the formalization of all of this by (finite) induction to the reader. □

A partial order is called *ultra transitive* if it is both in-ultra transitive and out-ultra transitive. The archetypical example of an ultra transitive partial order is the standard linear order on $\boldsymbol{N} = \{1, \ldots, n\}$ where n is a positive integer.

For a characterization of ultra transitivity in terms of standard linear orders, see Proposition 9.4 below.

Theorem 7.10. *Let $M \in \mathcal{A}_{\mathcal{Z}}$ where $\mathcal{Z}$ is a partial order on $\boldsymbol{N}$ of upper triangular type. Suppose $\mathcal{Z}$ is ultra transitive. Then the canonical generalized diagonal form $M_{\rm gDiag}$ of M belongs to $\mathcal{A}_{\mathcal{Z}}$. In fact, $M_{\rm gDiag} = LMR$ for some monic matrices $L, R \in \mathcal{A}_{\mathcal{Z}}$.*

Proof. By Theorem 7.1, there exists a monic $L \in \mathcal{A}_{\mathcal{Z}}$ for which $LM = M^{\uparrow}_{\rm rEch}$. Applying Lemma 7.9 to LM, we see that there is a monic $R \in \mathcal{A}_{\mathcal{Z}}$ such that LMR is a generalized diagonal matrix. The uniqueness part of Theorem 5.1 now gives that $LMR = M_{\rm gDiag}$. As both L and R are in $\mathcal{A}_{\mathcal{Z}}$, so is $M_{\rm gDiag} = LMR$. □

Corollary 7.11. *Let $M \in \mathcal{A}_{\mathcal{Z}}$ where $\mathcal{Z}$ is a partial order on $\boldsymbol{N}$ of upper triangular type. Suppose $\mathcal{Z}$ is ultra transitive. Then the canonical generalized monic diagonal form $M_{\rm mgDiag}$ of M belongs to $\mathcal{A}_{\mathcal{Z}}$. In fact, $M_{\rm mgDiag} = LMR$ for some invertible (upper triangular) matrices $L, R \in \mathcal{A}_{\mathcal{Z}}$.*

For how to obtain Corollary 7.11 from Theorem 7.10, consult the first part of proof of Corollary 5.2.

Given a partial order $\mathcal{Z}$ of upper triangular type, we can introduce the *ultra transitive closure* $\mathcal{Z}_{\rm ultra}$ of $\mathcal{Z}$ as the smallest partial order containing $\mathcal{Z}$ which is of upper triangular type and ultra transitive. It can be constructed in practice, starting from $\mathcal{Z}$, in a way similar to that of getting $\mathcal{Z}_{\rm in}$ and $\mathcal{Z}_{\rm out}$ from $\mathcal{Z}$ (see above).

Clearly, $\mathcal{Z} \subset \mathcal{Z}_{\rm in} \subset \mathcal{Z}_{\rm ultra}$ and $\mathcal{Z} \subset \mathcal{Z}_{\rm out} \subset \mathcal{Z}_{\rm ultra}$. Also $\mathcal{Z}_{\rm in} \cap \mathcal{Z}_{\rm out}$ is a partial order of upper triangular type. Thus $\mathcal{Z}_{\rm in} \cap \mathcal{Z}_{\rm out}$ is a partial order of upper triangular type for which $\mathcal{Z} \subset \mathcal{Z}_{\rm in} \cap \mathcal{Z}_{\rm out} \subset \mathcal{Z}_{\rm ultra}$. As can be seen from Example B in Section 10, these inclusions are generally proper. For the associated matrix algebras we have $\mathcal{A}_{\mathcal{Z}} \subset \mathcal{A}_{\mathcal{Z}_{\rm in} \cap \mathcal{Z}_{\rm out}} = \mathcal{A}_{\mathcal{Z}_{\rm in}} \cap \mathcal{A}_{\mathcal{Z}_{\rm out}} \subset \mathcal{A}_{\mathcal{Z}_{\rm ultra}}$, where, again, it may happen that the inclusions are proper.

Corollary 7.12. *Let $M \in \mathcal{A}_{\mathcal{Z}}$ where $\mathcal{Z}$ is a partial order on $\boldsymbol{N}$ of upper triangular type. Then the canonical generalized diagonal form $M_{\rm gDiag}$ of M belongs to $\mathcal{A}_{\mathcal{Z}_{\rm in} \cap \mathcal{Z}_{\rm out}} = \mathcal{A}_{\mathcal{Z}_{\rm in}} \cap \mathcal{A}_{\mathcal{Z}_{\rm out}}$. Also, $M_{\rm gDiag} \in \mathcal{A}_{\mathcal{Z}_{\rm ultra}}$ and, in fact, $M_{\rm gDiag} = LMR$ for some monic matrices $L, R \in \mathcal{A}_{\mathcal{Z}_{\rm ultra}}$.*

Proof. The second part of the result is immediate from Theorem 7.10. So let us concentrate on the first part.

By Corollaries 7.3 and 7.7, we have $M^{\uparrow}_{\rm rEch} \in \mathcal{A}_{\mathcal{Z}_{\rm in}}$ and $M^{\rightarrow}_{\rm cEch} \in \mathcal{A}_{\mathcal{Z}_{\rm out}}$. From Theorem 6.1 we know that the matrix $M_{\rm gDiag}$ is the leading entry reduction of $M^{\uparrow}_{\rm rEch}$, and that it is the trailing entry reduction of $M^{\rightarrow}_{\rm cEch}$ as well. It follows that $M_{\rm gDiag} \in \mathcal{A}_{\mathcal{Z}_{\rm in}}$ and $M_{\rm gDiag} \in \mathcal{A}_{\mathcal{Z}_{\rm out}}$. So $M_{\rm gDiag} \in \mathcal{A}_{\mathcal{Z}_{\rm in}} \cap \mathcal{A}_{\mathcal{Z}_{\rm out}} = \mathcal{A}_{\mathcal{Z}_{\rm in} \cap \mathcal{Z}_{\rm out}}$. □

Corollary 7.13. *Let $M \in \mathcal{A}_{\mathcal{Z}}$ where $\mathcal{Z}$ is a partial order on $\boldsymbol{N}$ of upper triangular type. Then the canonical generalized monic diagonal form $M_{\rm mgDiag}$ of M*

belongs to $\mathcal{A}_{\mathcal{Z}_{\text{in}} \cap \mathcal{Z}_{\text{out}}} = \mathcal{A}_{\mathcal{Z}_{\text{in}}} \cap \mathcal{A}_{\mathcal{Z}_{\text{out}}}$. *Moreover,* $M_{\text{mgDiag}} \in \mathcal{A}_{\mathcal{Z}_{\text{ultra}}}$ *and, in fact,* $M_{\text{mgDiag}} = LMR$ *for some invertible (upper triangular) matrices* $L, R \in \mathcal{A}_{\mathcal{Z}_{\text{ultra}}}$.

Proof. The argument is similar to that given for Corollary 7.12. Instead of the theorems and corollaries referred to there, utilize Corollaries 7.11, 7.4, 7.8 and Theorem 6.2. □

Part of the conclusion of the last two corollaries is that M_{gDiag} and M_{mgDiag} always belong to $\mathcal{A}_{\mathcal{Z}_{\text{in}} \cap \mathcal{Z}_{\text{out}}} = \mathcal{A}_{\mathcal{Z}_{\text{in}}} \cap \mathcal{A}_{\mathcal{Z}_{\text{out}}}$. It may happen, however, that M_{gDiag} and M_{mgDiag} cannot be written in the form LMR with L and R invertible matrices in $\mathcal{A}_{\mathcal{Z}_{\text{in}} \cap \mathcal{Z}_{\text{out}}} = \mathcal{A}_{\mathcal{Z}_{\text{in}}} \cap \mathcal{A}_{\mathcal{Z}_{\text{out}}}$. An instance where this occurs is Example D in Section 10.

8. Equivalence in zero pattern algebras

Let n be a positive integer, and let $\mathcal{A}$ be a matrix subalgebra of $\mathbb{C}^{n\times n}$ containing as its unit element the $n \times n$ identity matrix. Two matrices $\widehat{M}$ and $\widetilde{M}$ in $\mathcal{A}$ are said to be *left equivalent in* $\mathcal{A}$, respectively *right equivalent in* $\mathcal{A}$, if there exists an invertible matrix $L \in \mathcal{A}$, respectively $R \in \mathcal{A}$, such that $\widetilde{M} = L\widehat{M}$, respectively $\widetilde{M} = \widehat{M}R$. Also $\widehat{M}, \widetilde{M} \in \mathcal{A}$ are called *left/right equivalent in* $\mathcal{A}$ if there exist invertible matrices $L, R \in \mathcal{A}$ such that $\widetilde{M} = L\widehat{M}R$. Left equivalence, right equivalence and left/right equivalence thus defined determine equivalence relations in $\mathcal{A}$ and corresponding equivalence classes. Issues that arise naturally are: when do two matrices belong to the same equivalence class and how many equivalence classes are there? The ground work for dealing with these issues in some specific cases has been laid in the previous section by identifying certain canonical forms.

First let us consider the case $\mathcal{A} = \mathbb{C}^{n\times n}$. For left equivalence and right equivalence in $\mathbb{C}^{n\times n}$, the canonical forms that matter are the reduced row echelon form and the reduced column echelon form, respectively. Indeed, the $n \times n$ matrices $\widehat{M}$ and $\widetilde{M}$ are left equivalent in $\mathbb{C}^{n\times n}$, respectively right equivalent in $\mathbb{C}^{n\times n}$, if and only if $\widetilde{M}_{\text{rEch}} = \widehat{M}_{\text{rEch}}$, respectively $\widetilde{M}_{\text{cEch}} = \widehat{M}_{\text{cEch}}$. Other canonical forms are the canonical row echelon idempotent and, respectively, the canonical column echelon idempotent. Thus $\widehat{M}$ and $\widetilde{M}$ are left equivalent, respectively right equivalent, if and only if $\widetilde{M}_{\text{rEchId}} = \widehat{M}_{\text{rEchId}}$, respectively $\widetilde{M}_{\text{cEchId}} = \widehat{M}_{\text{cEchId}}$ (see Corollary 2.2). When $n > 1$, the number of left equivalence classes in $\mathbb{C}^{n\times n}$, respectively right equivalence classes in $\mathbb{C}^{n\times n}$, is infinite.

For the left/right equivalence classes in $\mathbb{C}^{n\times n}$, the situation is different. Their number is always finite and, in fact, equal to $n+1$. Indeed, as is well known (see, for instance, [LT], Section 2.7), two matrices are left/right equivalent if and only if they have the same rank, and this rank is one of the numbers $0, \dots, n$.

Next we turn to the discussion of equivalence in matrix subalgebras of $\mathbb{C}^{n\times n}$ determined by a pattern of zeros. Let n be a positive integer, and let $\mathcal{Z}$ be a partial order on $\boldsymbol{N}$ (which, as always in this paper, stands for $\{1,\dots,n\}$). As before, we will assume that $\mathcal{Z}$ is of upper triangular type. The algebra $\mathcal{A}_{\mathcal{Z}}$ is then a subalgebra of $\mathbb{C}^{n\times n}_{\text{upper}}$, the algebra of upper triangular $n\times n$ matrices. Recall that monic $n\times n$ matrices by definition belong to $\mathbb{C}^{n\times n}_{\text{upper}}$.

Presently we will discuss left, right and left/right equivalence in $\mathcal{A}_{\mathcal{Z}}$. But first we want to consider three somewhat more restrictive notions. Two matrices $\widehat{M}$ and $\widetilde{M}$ in $\mathcal{A}_{\mathcal{Z}}$ are said to be *left monic-equivalent in* $\mathcal{A}_{\mathcal{Z}}$, respectively *right monic-equivalent in* $\mathcal{A}_{\mathcal{Z}}$, if there exists a monic matrix $L\in\mathcal{A}_{\mathcal{Z}}$, respectively a monic matrix $R\in\mathcal{A}_{\mathcal{Z}}$, such that $\widetilde{M}=L\widehat{M}$, respectively $\widetilde{M}=\widehat{M}R$. Also $\widehat{M},\widetilde{M}\in\mathcal{A}_{\mathcal{Z}}$ are called *left/right monic-equivalent in* $\mathcal{A}_{\mathcal{Z}}$ if there exist monic matrices $L,R\in\mathcal{A}_{\mathcal{Z}}$ such that $\widetilde{M}=L\widehat{M}R$.

Theorem 8.1. *Let $\widehat{M},\widetilde{M}\in\mathcal{A}_{\mathcal{Z}}$, where $\mathcal{Z}$ is a partial order on $\boldsymbol{N}$ of upper triangular type. Suppose $\mathcal{Z}$ is in-ultra transitive. Then $\widehat{M}$ and $\widetilde{M}$ are left monic-equivalent in $\mathcal{A}_{\mathcal{Z}}$ if and only $\widehat{M}^{\uparrow}_{\text{rEch}}=\widetilde{M}^{\uparrow}_{\text{rEch}}$.*

Proof. First assume that $\widehat{M}$ and $\widetilde{M}$ are left monic-equivalent in $\mathcal{A}_{\mathcal{Z}}$. So there exists a monic matrix $G\in\mathcal{A}_{\mathcal{Z}}$ such that $\widehat{M}=G\widetilde{M}$. On account of Theorem 7.1, we can write $\widetilde{M}$ in the form $H\widetilde{M}^{\uparrow}_{\text{rEch}}$ with H a monic matrix in $\mathcal{A}_{\mathcal{Z}}$. Put $L=GH$. Then L is a monic matrix in $\mathcal{A}_{\mathcal{Z}}$ and $\widehat{M}=L\widetilde{M}^{\uparrow}_{\text{rEch}}$. Theorem 4.2 now gives that $\widehat{M}^{\uparrow}_{\text{rEch}}=\widetilde{M}^{\uparrow}_{\text{rEch}}$.

Next suppose the latter is the case. Again applying Theorem 7.1, we write $\widehat{M}$ and $\widetilde{M}$ in the form $G\widehat{M}^{\uparrow}_{\text{rEch}}$ and $H\widetilde{M}^{\uparrow}_{\text{rEch}}$ with G and H monic matrices in $\mathcal{A}_{\mathcal{Z}}$. Put $L=GH^{-1}$. Then L is a monic matrix in $\mathcal{A}_{\mathcal{Z}}$ and $\widehat{M}=G\widehat{M}^{\uparrow}_{\text{rEch}}=G\widetilde{M}^{\uparrow}_{\text{rEch}}=GH^{-1}\widetilde{M}=L\widetilde{M}$. Thus $\widehat{M}$ and $\widetilde{M}$ are left monic-equivalent in $\mathcal{A}_{\mathcal{Z}}$, as desired. □

A similar argument, using Theorems 7.5 and 4.3, yields the following counterpart of Theorem 8.1.

Theorem 8.2. *Let $\widehat{M},\widetilde{M}\in\mathcal{A}_{\mathcal{Z}}$, where $\mathcal{Z}$ is a partial order on $\boldsymbol{N}$ of upper triangular type. Suppose $\mathcal{Z}$ is out-ultra transitive. Then $\widehat{M}$ and $\widetilde{M}$ are right monic-equivalent in $\mathcal{A}_{\mathcal{Z}}$ if and only if $\widehat{M}^{\rightarrow}_{\text{cEch}}=\widetilde{M}^{\rightarrow}_{\text{cEch}}$.*

For left/right monic-equivalence in $\mathcal{A}_{\mathcal{Z}}$ we have the following result.

Theorem 8.3. *Let $\widehat{M},\widetilde{M}\in\mathcal{A}_{\mathcal{Z}}$, where $\mathcal{Z}$ is a partial order on $\boldsymbol{N}$ of upper triangular type. Suppose $\mathcal{Z}$ is ultra transitive. Then $\widehat{M}$ and $\widetilde{M}$ are left/right monic-equivalent in $\mathcal{A}_{\mathcal{Z}}$ if and only if $\widehat{M}_{\text{gDiag}}=\widetilde{M}_{\text{gDiag}}$.*

Proof. The argument is similar to that employed to establish Theorem 8.1. Instead of Theorems 7.1 and 4.2, use Theorems 7.10 and 5.1. □

With regard to numbers of equivalence classes, we can say the following. In all situations considered in the above three theorems, the number of corresponding equivalence classes is infinite.

Now we come back to left, right and left/right equivalence in $\mathcal{A}_{\mathcal{Z}}$. Here the monic echelon forms play a role.

Theorem 8.4. *Let $\widehat{M}, \widetilde{M} \in \mathcal{A}_{\mathcal{Z}}$, where $\mathcal{Z}$ is a partial order on $\boldsymbol{N}$ of upper triangular type. Suppose $\mathcal{Z}$ is in-ultra transitive. Then $\widehat{M}$ and $\widetilde{M}$ are left equivalent in $\mathcal{A}_{\mathcal{Z}}$ if and only $\widehat{M}^{\uparrow}_{\mathrm{mrEch}} = \widetilde{M}^{\uparrow}_{\mathrm{mrEch}}$.*

Theorem 8.5. *Let $\widehat{M}, \widetilde{M} \in \mathcal{A}_{\mathcal{Z}}$, where $\mathcal{Z}$ is a partial order on $\boldsymbol{N}$ of upper triangular type. Suppose $\mathcal{Z}$ is out-ultra transitive. Then $\widehat{M}$ and $\widetilde{M}$ are right equivalent in $\mathcal{A}_{\mathcal{Z}}$ if and only if $\widehat{M}^{\rightarrow}_{\mathrm{mcEch}} = \widetilde{M}^{\rightarrow}_{\mathrm{mcEch}}$.*

Theorem 8.6. *Let $\widehat{M}, \widetilde{M} \in \mathcal{A}_{\mathcal{Z}}$, where $\mathcal{Z}$ is a partial order on $\boldsymbol{N}$ of upper triangular type. Suppose $\mathcal{Z}$ is ultra transitive. Then $\widehat{M}$ and $\widetilde{M}$ are left/right equivalent in $\mathcal{A}_{\mathcal{Z}}$ if and only if $\widehat{M}_{\mathrm{mgDiag}} = \widetilde{M}_{\mathrm{mgDiag}}$.*

The arguments for these results are completely analogous to those employed for establishing Theorems 8.1, 8.2 and 8.3, respectively. For Theorem 8.4, use Corollaries 7.2 and 4.4, for Theorem 8.5, use Corollaries 7.6 and 4.5, and for Theorem 8.6, employ Corollaries 7.11 and 5.2.

In the in-ultra transitive case, the number of left equivalence classes in $\mathcal{A}_{\mathcal{Z}}$ is generally infinite (Theorem 8.4). In the out-ultra transitive situation, the number of right equivalence classes in $\mathcal{A}_{\mathcal{Z}}$ is as a rule infinite too (Theorem 8.5). Finiteness occurs only for rather specific partial orders $\mathcal{Z}$. For left/right equivalence classes, the situation is different: when $\mathcal{Z}$ is ultra transitive, their number is always finite. This is a direct consequence of Theorem 8.6 in view of the fact that monic generalized diagonal matrices are zero/one matrices, i.e., their entries are 0 or 1. This leads to the question: how many of these left/right equivalence classes are there? This issue will be explored in the next section.

9. Counting left/right equivalence classes

Let n be a positive integer, and let $M \in \mathbb{C}^{n\times n}_{\mathrm{upper}}$. By the *frame* of M we mean the canonical monic generalized diagonal form of M, i.e., the upper triangular zero/one matrix M_{mgDiag}. Obviously the frame of M can be identified with the directed graph $\mathcal{L}(M_{\mathrm{gDiag}})$ or, what comes down to the same, $\mathcal{T}(M_{\mathrm{gDiag}})$. Actually, as we know from the last paragraph of Section 6, it may just as well be identified with each of the coinciding leading or trailing entry graphs $\mathcal{L}(M_{\mathrm{gDiag}})$, $\mathcal{T}(M_{\mathrm{gDiag}})$, $\mathcal{L}(M_{\mathrm{mgDiag}})$, $\mathcal{T}(M_{\mathrm{mgDiag}})$, $\mathcal{L}(M^{\uparrow}_{\mathrm{rEch}})$, $\mathcal{T}(M^{\rightarrow}_{\mathrm{cEch}})$, $\mathcal{L}(M^{\uparrow}_{\mathrm{mrEch}}$ and $\mathcal{T}(M^{\rightarrow}_{\mathrm{mcEch}})$.

From Corollary 7.11 we know that the frame of M belongs to $\mathcal{A}_{\mathcal{Z}}$ whenever $M \in \mathcal{A}_{\mathcal{Z}}$, where $\mathcal{Z}$ is a partial order on $\boldsymbol{N} = \{1, \dots, n\}$ which is of upper triangular type and ultra transitive. Theorem 8.6 can now be reformulated as follows.

Theorem 9.1. *Let $\mathcal{Z}$ be a partial order on $\boldsymbol{N}$. Suppose $\mathcal{Z}$ is of upper triangular type and ultra transitive. Then two matrices in $\mathcal{A}_{\mathcal{Z}}$ are left/right equivalent in $\mathcal{A}_{\mathcal{Z}}$ if and only if they have the same frame.*

Suppose $\mathcal{Z}$ is as in the theorem. To count the number of left/right equivalence classes in $\mathcal{A}_{\mathcal{Z}}$, one has to count the number of possible frames belonging to $\mathcal{A}_{\mathcal{Z}}$. In other words, one needs to count the number of monic generalized diagonal matrices in $\mathcal{A}_{\mathcal{Z}}$.

Let W be such a matrix. Then all rows and columns of W contain at most one nonzero entry (which is then equal to 1). This enables us to identify W with a 'word' $\lfloor w_1 w_2 \dots w_n \rceil$ having 'letters' from the 'alphabet' $0, 1, \dots, n$. Here is the definition. Let $j = 1, \dots, m$. If the jth column of W contains only zeros, we take the letter w_j to be 0. If the jth column of W does contain a nonzero entry (in fact a 1), then there is precisely one $k \in \boldsymbol{N}$ such that this entry is in the kth row of W, and we take w_j to be equal to this k. We record the following three properties:

$$w_l = w_m \neq 0 \;\Rightarrow\; l = m, \qquad l, m = 1, \dots, n,$$

because each row of W has at most one nonzero entry,

$$w_j \to_{\mathcal{Z}} j, \qquad j = 1, \dots, n,$$

because $W \in \mathcal{A}_{\mathcal{Z}}$, and

$$w_j \leq j, \qquad j = 1, \dots, n,$$

because of the upper triangularity of W (or, if one prefers, of $\mathcal{Z}$).

The identification of the matrix W with the word $\lfloor w_1 w_2 \dots w_n \rceil$ brings with it that we can speak of the *rank* of $\lfloor w_1 w_2 \dots w_n \rceil$. It simply is the rank of W. The rank of the word $\lfloor w_1 w_2 \dots w_n \rceil$ can be determined by counting the nonzero letters in it: $\operatorname{rank}\lfloor w_1 w_2 \dots w_n \rceil = \sharp\{j \in \boldsymbol{N} \mid w_j \neq 0\}$.

Our problem of counting the number of left/right equivalence classes in $\mathcal{A}_{\mathcal{Z}}$, is now shifted to that of counting the number of words as introduced above.

For $r = 0, \dots, n$, let $D_{\mathcal{Z}}(n; r)$ be the set of all words having rank r. Evidently $D_{\mathcal{Z}}(n; 0)$ is the singleton set consisting of the 'zero word' $\lfloor 0\,0 \,\dots\, 0 \rceil$ only. Also, a moment of reflection gives that $D_{\mathcal{Z}}(n; n)$ is a singleton set too, this time having the word $\lfloor 1\,2 \,\dots\, n \rceil$ as it single element. This 'identity word' corresponds, of course, to the $n \times n$ identity matrix which is indeed a monic generalized diagonal matrix and belongs to $\mathcal{A}_{\mathcal{Z}}$ (because of the reflexivity of $\mathcal{Z}$). The cardinality of $D_{\mathcal{Z}}(n; r)$ will be denoted by $\sigma_{\mathcal{Z}}(n; r)$. So $\sigma_{\mathcal{Z}}(n; r)$ is the number of words (as introduced above) having rank r. As $D_{\mathcal{Z}}(n; 0)$ and $D_{\mathcal{Z}}(n; n)$ are singletons, we have $\sigma_{\mathcal{Z}}(n; 0) = \sigma_{\mathcal{Z}}(n; n) = 1$. Anticipating on a connection with the so-called Stirling numbers of the second kind to be made below, we stipulate that $\sigma_0(0) = 1$ too.

We now proceed via a kind of induction process. In fact, for $n \geq 2$, we will relate the $n+1$ positive integers $\sigma_{\mathcal{Z}}(n; 0), \dots, \sigma_{\mathcal{Z}}(n; n)$ with the n positive integers $\sigma_{\mathcal{Z}'}(n-1; 0), \dots, \sigma_{\mathcal{Z}'}(n-1; n-1)$. Here $\mathcal{Z}'$ is the restriction of the relation $\mathcal{Z}$ to $\{1, \dots, n-1\}$, i.e., $\mathcal{Z}' = \mathcal{Z} \cap (\{1, \dots, n-1\} \times \{1, \dots, m-1\})$. (Of course symbols as $\sigma_{\mathcal{Z}'}(n-1; r)$ and $D_{\mathcal{Z}'}(n-1; r)$ have the meaning to be expected from the above.)

Let $r \in \{1, \dots, n-1\}$ and let $\lfloor w_1 w_2 \dots w_{n-1} w_n \rceil$ be a word in $D_{\mathcal{Z}}(n; r)$. As its rank is r, the word has r nonzero letters. But then the number of nonzero letters in the word $\lfloor w_1 w_2 \dots w_{n-1} \rceil$ is r or $r-1$, depending on whether w_n is zero or not. It follows that $\lfloor w_1 w_2 \dots w_{n-1} \rceil$ belongs to either $D_{\mathcal{Z}'}(n-1; r)$ or $D_{\mathcal{Z}'}(n-1; r-1)$

(exclusive). Thus the words in the set $D_{\mathcal{Z}}(n;r)$ are of one of the following two types: either a word from $D_{\mathcal{Z}'}(n-1;r)$ augmented by the letter 0, or a word from $D_{\mathcal{Z}'}(n-1;r-1)$ augmented by one of the letters $1,\ldots,n$. Here the augmentation should respect the requirement that there are no repetitions of nonzero letters. Also the specifics of the partial order $\mathcal{Z}$ should be taken into account, meaning that from $1,\ldots,n$ only those letters k can appear for which $k \to_{\mathcal{Z}} n$.

From the analysis given so far we can conclude that $D_{\mathcal{Z}}(n;r)$ consists of the words from $D_{\mathcal{Z}'}(n-1;r)$ augmented by the letter 0 and of the words from $D_{\mathcal{Z}'}(n-1;r-1)$ augmented by 'admissible' letters from the alphabet $1,\ldots,n$. The number of these admissible letters is larger than or equal to 1 and less than or equal to $n+1-r$. Indeed, the letter n is always admissible, and there are always al least $r-1$ letters that are not admissible (no repetition of nonzero letters and restrictions due to the specifics of $\mathcal{Z}$). It follows that

$$\begin{aligned}\sigma_{\mathcal{Z}'}(n-1;r)+\sigma_{\mathcal{Z}'}(n-1;r-1) &\leq \sigma_{\mathcal{Z}}(n;r)\\ &\leq \sigma_{\mathcal{Z}'}(n-1;r)+(n+1-r)\sigma_{\mathcal{Z}'}(n-1;r-1).\end{aligned}$$

These considerations lead to a Pascal type triangular scheme determining the (positive) integers $\sigma(n;0),\ldots,\sigma(n,n)$. In the scheme, for reasons of exposition, the reference to the given partial order is suppressed. Also the number 1 – which we interpret as $\sigma(0;0)$ – is put at the top. The scheme in question is

$$\begin{array}{ccccccccccccc}
 & & & & & & 1 & & & & & & \\
 & & & & & 1 & & 1 & & & & & \\
 & & & & 1 & & \sigma(1;1) & & 1 & & & & \\
 & & & 1 & & \sigma(2;1) & & \sigma(2;2) & & 1 & & & \\
 & & 1 & & \sigma(3;1) & & \sigma(3;2) & & \sigma(3;3) & & 1 & & \\
 & 1 & & \sigma(4;1) & & \sigma(4;2) & & \sigma(4;3) & & \sigma(4;4) & & 1 & \\
 & & & & & & \vdots & & & & & & \\
1 & \sigma(n;1) & & \sigma(n;2) & & & \ldots & & \sigma(n;n-2) & & \sigma(n;n-1) & & 1
\end{array}$$

where, for $s=1,\ldots,k-1$ and $k=2,\ldots,n$,

$$\sigma(k;s)=\sigma(k-1;s)+\alpha(k;s)\sigma(k-1;s-1), \tag{9}$$

with $\alpha(k;s)$ a rational number between 1 and $k+1-s$ (inclusive) depending on the specifics of the partial order $\mathcal{Z}$.

For given $\mathcal{Z}$ with small n, the above considerations make it possible to compute the numbers $\sigma(k,s)$. We do not have a general description of the $\alpha(k,s)$, but in some special cases involving an unspecified n they can be determined, and with them the numbers $\sigma(k,s)$.

One of these special cases is concerned with the rather simple situation where $\mathcal{Z} = \{(k,k) \mid k = 1,\dots,n\}$ is just the diagonal in $\boldsymbol{N}\times\boldsymbol{N}$. Then, trivially, all $\alpha(k,s)$ are equal to 1. So (9) becomes $\sigma(k;s) = \sigma(k-1;s) + \sigma(k-1;s-1)$, and the above scheme reduces to the celebrated Pascal triangle. Hence $\sigma(k,s) = \binom{k}{s}$ and the number of words that we are interested in is

$$\sum_{s=0}^{n} \binom{k}{s} = 2^n.$$

Returning to left/right equivalence classes in $\mathcal{A}_{\mathcal{Z}}$, we conclude that there are 2^n of them. The number of such classes consisting of rank r matrices is $\binom{n}{r}$. Here, of course, r is a nonnegative integer not exceeding n.

This information on the diagonal partial order could easily have been obtained directly. Things become already a bit more complicated when we turn to the next special case where n is at least 2 and $\mathcal{Z}$ is the diagonal partial order augmented with the single pair $(1,n)$. So for $n=5$, the (ultra transitive) partial order in question is given by the scheme

$$\begin{pmatrix} & 1 & 2 & 3 & 4 & 5 \\ 1 & * & 0 & 0 & 0 & * \\ 2 & 0 & * & 0 & 0 & 0 \\ 3 & 0 & 0 & * & 0 & 0 \\ 4 & 0 & 0 & 0 & * & 0 \\ 5 & 0 & 0 & 0 & 0 & * \end{pmatrix}. \tag{10}$$

Returning to arbitrary $n\,(\geq 2)$, we note that

$$\sigma(k,s) = \binom{k}{s}, \qquad s = 0,\dots,k;\ k = 0,\dots,n-1.$$

We also have (evidently)

$$\alpha(k;s) = 1, \qquad s = 1,\dots,k-1;\ k = 2,\dots,n-1,$$

and (requiring a little more effort)

$$\alpha(n;s) = \frac{2n-1-s}{n-1}, \quad s = 1,\dots,n-1.$$

It follows that, for $r = 1, \dots, n-1$,

$$\begin{aligned}
\sigma(n;r) &= \sigma(n-1;r) + \alpha(n;r)\sigma(n-1,r-1) \\
&= \binom{n-1}{r} + \frac{2n-1-r}{n-1}\binom{n-1}{r-1} \\
&= \binom{n-1}{r} + 2\binom{n-1}{r-1} - \frac{r-1}{n-1}\binom{n-1}{r-1} \\
&= \binom{n-1}{r} + 2\binom{n-1}{r-1} - \binom{n-2}{r-2}.
\end{aligned}$$

With the help of the standard relations between the binomial coefficients, we now arrive at $\sigma(n;r) = \binom{n}{r} + \binom{n-2}{r-1}$, also holding for $r = 0$ and $r = n$ when we adhere to the usual interpretation of $\binom{n-2}{-1}$ and $\binom{n-2}{n-1}$ as being equal to zero.

It can be concluded that the number of left/right equivalence classes in $\mathcal{A}_{\mathcal{Z}}$ consisting of rank r matrices is $\binom{n}{r} + \binom{n-2}{r-1}$. Also the total number of left/right equivalence classes in $\mathcal{A}_{\mathcal{Z}}$ can be computed as

$$\sum_{r=0}^{n} \sigma(n,r) = \sum_{r=0}^{n} \binom{n}{r} + \sum_{r=1}^{n-1} \binom{n-2}{r-1} = 2^n + 2^{n-2} = 5 \cdot 2^{n-2}. \tag{11}$$

For the special case (10), where $n = 5$, this gives 40 for the number of left/right equivalence classes in $\mathcal{A}_{\mathcal{Z}}$. Also the number of such classes consisting of rank r matrices is $\binom{5}{r} + \binom{3}{r-1}$. For $r = 0, \dots, 5$, we get the numbers $1, 6, 13, 13, 6$ and 1, respectively.

In the remainder of this section we focus on the archetypical situation where $\mathcal{Z}$ is the standard linear order on $\boldsymbol{N} = \{1, \dots, n\}$. Here the numbers $\alpha(k;s)$ have the largest possible value, i.e., $\alpha(k;s) = k + 1 - s$ (the reason being that the exclusion of repetitions of nonzero letters is the only restrictive factor). With this, the identity in (9) becomes $\sigma(k;s) = \sigma(k-1;s) + (k+1-s)\sigma(k-1;s-1)$.

For $l = 1, \dots, n+1$ and $m = 1, \dots, l$, introduce $S(l,m) = \sigma(l-1; l-m)$. Clearly $S(l,1) = S(l-1,l-1) = 1$ and $S(l,l) = S(l-1,0) = 1$. Also

$$S(l,m) = S(l-1,m-1) + mS(l-1,m), \qquad m = 2, \dots, l; l = 3, \dots, n+1.$$

These recurrence relations, together with the initial conditions $S(l,1) = S(l,l) = 1$, determine the so-called *Stirling numbers*, more specifically, the ones *of the second kind.* For information about these Stirling numbers, we refer to the extensive pertinent literature (cf., for instance, [C], [R] or [S]). Employing the so-called Knuth notation $\{{\cdot \atop \cdot}\}$ for them, we have

$$\sigma(k,s) = S(k+1,k+1-s) = \left\{ {k+1 \atop k+1-s} \right\}, \qquad s = 0, \dots, k;\ k = 0, \dots, n,$$

and, for the sum of the numbers $\sigma(n;0),\dots,\sigma(n,n)$,

$$\sum_{r=0}^{n}\sigma(n,r)=\sum_{s=1}^{n+1}\left\{ {n+1 \atop s} \right\}=B_{n+1}.$$

Here, again using standard notation, B_{n+1} stands for the $(n+1)$th *Bell number* (see again [C], [R] or [S]).

The following theorem is now immediate.

Theorem 9.2. *Let n be positive integer and $r\in\{0,\dots,n\}$. Then the number of left/right equivalence classes in $\mathbb{C}^{n\times n}_{\text{upper}}$ consisting of rank r matrices is the Stirling number $\left\{ {n+1 \atop n+1-r} \right\}$. Also, the total number of left/right equivalence classes in $\mathbb{C}^{n\times n}_{\text{upper}}$ is the Bell number B_{n+1}.*

There are explicit expressions for the Stirling and Bell numbers. For instance

$$\left\{ {l \atop m} \right\}=\frac{1}{m!}\sum_{s=0}^{m}(-1)^{m-s}\binom{m}{s}s^l,$$

and (consequently)

$$B_l=\sum_{m=0}^{l}\sum_{s=0}^{m}\frac{(-1)^{m-s}}{m!}\binom{m}{s}s^l.$$

A wealth of other expressions is available in the literature.

The Bell number counts the number of partitions of a set with l elements. Also the Stirling number $\left\{ {l \atop m} \right\}$ is equal to the number of ways a set with l elements can be partitioned into m disjoint, non-empty subsets. Without going into the details, we mention that these facts may be used to give an alternative proof of Theorem 9.2.

Specializing to the case $n=5$, we have the following triangular scheme for the first 21 Stirling numbers of the second kind:

$$\begin{array}{ccccccccccc}
 & & & & & 1 & & & & & \\
 & & & & 1 & & 1 & & & & \\
 & & & 1 & & 3 & & 1 & & & \\
 & & 1 & & 7 & & 6 & & 1 & & \\
 & 1 & & 15 & & 25 & & 10 & & 1 & \\
1 & & 31 & & 90 & & 65 & & 15 & & 1
\end{array}$$

giving, for instance, that the number of left/right equivalence classes in $\mathbb{C}^{5\times 5}_{\text{upper}}$, being the sum of the integers in the bottom row, is equal to $B_6 = 203$.

We close this section with the following result.

Theorem 9.3. *Let n be a positive integer, and suppose $\mathcal{Z}$ is an ultra transitive partial order on $\boldsymbol{N}$. Then the total number of left/right equivalence classes in $\mathcal{A}_{\mathcal{Z}}$ is a product of Bell numbers.*

From Theorem 9.2 it is clear that the conclusion of Theorem 9.3 holds for partial orders $\mathcal{Z}$ that are permutation similar to a disjoint union of standard linear orders. So we are ready once it is shown that the ultra transitive partial orders are of that type. In fact we shall prove a bit more.

Proposition 9.4. *Let n be a positive integer, and let $\mathcal{Z}$ be a partial order on $\boldsymbol{N}$. Then $\mathcal{Z}$ is ultra transitive if and only if it is permutation similar to a disjoint union of linear orders.*

For the precise meaning in which the terms 'permutation similar' and 'disjoint union' are employed here – which, by the way, is pretty much standard (and can even be guessed) – we refer to [BES3], Subsections 3.1 and 3.3. The proposition can be proved by combining the characterizations (in terms of rooted trees) of in-ultra transitivity and out-ultra transitivity obtained in [BES4]. We prefer, however, to give a more direct argument here.

Proof. The "if" part is practically immediate from the simple observation that linear orders are ultra transitive. So we concentrate on the "only if" part.

Suppose $\mathcal{Z}$ is ultra transitive. Let p be a positive integer. By a transitivity path (for $\mathcal{Z}$) of length p we mean a finite sequence $\langle k_1, k_2, \dots, k_p\rangle$ such that

$$k_s \neq k_t, \quad k_s \to_{\mathcal{Z}} k_t, \qquad s,t = 1,\dots,p;\ s < t.$$

Obviously this implies that $p \leq n$. The collection of transitivity paths is non-empty. Indeed, for each $k \in \{1,\dots,n\}$ it contains the path $\langle k\rangle$ (of length 1).

Let m be the largest (positive) integer that occurs as the length of a transitivity path. Also, let $\langle r_1, r_2, \dots, r_m\rangle$ be a transitivity path of that length m. Modulo a permutation similarity, we may assume that $\mathcal{Z}$ is such that $r_j = n - m + j$, $j = 1,\dots,m$. Thus

$$s \to_{\mathcal{Z}} t, \qquad s,t = n-m+1,\dots,n;\ s \leq t.$$

The restriction $\mathcal{Z}_0$ of $\mathcal{Z}$ to $\{n-m+1,\dots,n\}$ is given by

$$\mathcal{Z}_0 = \mathcal{Z} \cap (\{n-m+1,\dots,n\} \times \{n-m+1,\dots,n\}).$$

Note that $\mathcal{Z}_0$ consists of the pairs (s,t), $s,t = n-m+1,\dots,n$, $s \leq t$ and that $\mathcal{Z}_0$ is a linear order on $\{n-m+1,\dots,n\}$.

Write $\mathcal{Z}_1$ for the restriction of $\mathcal{Z}$ to $\{1,\dots,n-m\}$. Then $\mathcal{Z}_1$ is a partial order on $\{1,\dots,n-m\}$ which is easily seen to be ultra transitive. Passing to an induction argument, we may assume that $\mathcal{Z}_1$ is permutation similar to a disjoint union of linear orders. (Note here that the statement in the theorem is trivially all right when $n = 1$.) It now suffices to prove that $\mathcal{Z}$ is the disjoint union of $\mathcal{Z}_1$ and $\mathcal{Z}_0$. For this we need to show that

$$k \not\rightarrow_{\mathcal{Z}} l, \qquad (k,l) \in \{1,\dots,n-m\} \times \{n-m+1,\dots,n\}, \tag{12}$$

as well as

$$k \not\rightarrow_{\mathcal{Z}} l, \qquad (k,l) \in \{n-m+1,\dots,n\} \times \{1,\dots,n-m\}. \tag{13}$$

Here is the argument.

Suppose (12) does not hold. Then there exist pairs in the set

$$\{1,\dots,n-m\} \times \{n-m+1,\dots,n\}$$

that do belong to $\mathcal{Z}$. Now consider such a pair with the second coordinate having the smallest possible value, (k_1,l_1) say. Assume $l_1 = n-m+1$, implying that $(k_1 \rightarrow_{\mathcal{Z}} n-m+1)$. Then $\langle k_1, n-m+1,\dots,n\rangle$ is a transitivity path which has length $m+1$. However, a transitivity path of that length (larger than m) does not exist, and it follows that l_1 must belong to $\{n-m+2,\dots,n\}$. We now have $k_1 \rightarrow_{\mathcal{Z}} l_1$ and $l_1 - 1 \rightarrow_{\mathcal{Z}} l_1$. Hence, by the ultra transitivity (and in fact by the in-ultra transitivity) of $\mathcal{Z}$, either $k_1 \rightarrow_{\mathcal{Z}} l_1 - 1$ or $l_1 - 1 \rightarrow_{\mathcal{Z}} k_1$. The first of these is impossible by the minimality condition on l_1. So $l_1 - 1 \rightarrow_{\mathcal{Z}} k_1$. But then $\langle n-m+1,\dots,l_1-1,k_1,l_1,\dots,n\rangle$ is a transitivity path. Its length is $m+1$, violating the choice of m.

Next suppose (13) does not hold. Then there exist pairs in the set

$$\{n-m+1,\dots,n\} \times \{1,\dots,n-m\}$$

that do belong to $\mathcal{Z}$. Now consider such a pair with the first coordinate having the largest possible value, (k_2,l_2) say. Assume $k_2 = n$, so that $n \rightarrow_{\mathcal{Z}} l_2$. Then $\langle n-m+1,\dots,n,l_2\rangle$ is a transitivity path of length $m+1$, which is impossible. It follows that k_2 must belong to $\{n-m+1,\dots,n-1\}$. We now have $k_2 \rightarrow_{\mathcal{Z}} l_2$ and $k_2 \rightarrow_{\mathcal{Z}} k_2+1$. Hence, by the ultra transitivity (and in fact by the out-ultra transitivity) of $\mathcal{Z}$, either $k_2+1 \rightarrow_{\mathcal{Z}} l_2$ or $l_2 \rightarrow_{\mathcal{Z}} k_2+1$. The first of these is impossible by the maximality requirement on k_2. So $l_2 \rightarrow_{\mathcal{Z}} k_2+1$. But then $\langle n-m+1,\dots,k_2,l_2,k_2+1,\dots,n\rangle$ is a transitivity path. Its length is $m+1$, again contradicting the choice of m. □

Theorem 9.3 is corroborated by the observations presented earlier in this section pertaining to partial orders of the type described in the passage prior to (10), the specific partial order given by (10) corresponding to the case $n = 5$. Indeed, the expression $5 \cdot 2^{n-2}$ featuring in (11) can be written as $B_3 B_2^{n-2}$. We complete this comment by noting that partial order given by (10) is permutation similar to the one represented by the scheme

$$\begin{pmatrix} & 1 & 2 & 3 & 4 & 5 \\ 1 & * & 0 & 0 & 0 & 0 \\ 2 & 0 & * & 0 & 0 & 0 \\ 3 & 0 & 0 & * & 0 & 0 \\ 4 & 0 & 0 & 0 & * & * \\ 5 & 0 & 0 & 0 & 0 & * \end{pmatrix},$$

in line with Proposition 9.4.

10. Examples

In this section we present the examples that have been referred to in the above.

Example A. Take $n = 4$, and let

$$M = \begin{bmatrix} 1 & 1 & 0 & -2 \\ 1 & 1 & 0 & -1 \\ 0 & 0 & 0 & -1 \\ 0 & 0 & 0 & 3 \end{bmatrix}.$$

Then the reduced row echelon form of M is given by

$$M_{\mathrm{rEch}} = \begin{bmatrix} 1 & 1 & 0 & 0 \\ 0 & 0 & 0 & 1 \\ 0 & 0 & 0 & 0 \\ 0 & 0 & 0 & 0 \end{bmatrix}.$$

Multiplying M_{rEch} on the left with appropriate permutation matrices, one gets

$$\begin{bmatrix} 1 & 1 & 0 & 0 \\ 0 & 0 & 0 & 0 \\ 0 & 0 & 0 & 0 \\ 0 & 0 & 0 & 1 \end{bmatrix}, \qquad \begin{bmatrix} 0 & 0 & 0 & 0 \\ 1 & 1 & 0 & 0 \\ 0 & 0 & 0 & 0 \\ 0 & 0 & 0 & 1 \end{bmatrix}.$$

The first of these is the canonical row echelon idempotent M_{rEchId} of M, the second is a permuted reduced row echelon form of M which has the additional property of being an idempotent, though not an upper triangular idempotent (cf. Corollary 2.2 and the paragraph following its proof). □

Example B. Let $\mathcal{Z}$ be the partial order on $\{1,\ldots,7\}$ given by the scheme

$$\mathcal{Z}=\begin{pmatrix} & 1 & 2 & 3 & 4 & 5 & 6 & 7 \\ 1 & * & 0 & 0 & * & * & * & 0 \\ 2 & 0 & * & 0 & * & * & * & 0 \\ 3 & 0 & 0 & * & * & * & * & * \\ 4 & 0 & 0 & 0 & * & * & * & 0 \\ 5 & 0 & 0 & 0 & 0 & * & * & 0 \\ 6 & 0 & 0 & 0 & 0 & 0 & * & 0 \\ 7 & 0 & 0 & 0 & 0 & 0 & 0 & * \end{pmatrix}. \tag{14}$$

Then $\mathcal{Z}$ is of upper triangular type. As $1 \to_{\mathcal{Z}} 4$ and $2 \to_{\mathcal{Z}} 4$ but $1 \not\to_{\mathcal{Z}} 2$, the partial order $\mathcal{Z}$ is not in-ultra transitive. Neither is it out-ultra transitive. This one sees from $3 \to_{\mathcal{Z}} 4$ and $3 \to_{\mathcal{Z}} 7$ while $4 \not\to_{\mathcal{Z}} 7$. Following the procedure that has been outlined for obtaining $\mathcal{Z}_{\text{in}}$, and the analogous ones for $\mathcal{Z}_{\text{out}}$ and $\mathcal{Z}_{\text{ultra}}$, we arrive at the respective 'ultra closures'

$$\mathcal{Z}_{\text{in}}=\begin{pmatrix} & 1 & 2 & 3 & 4 & 5 & 6 & 7 \\ 1 & * & * & * & * & * & * & * \\ 2 & 0 & * & * & * & * & * & * \\ 3 & 0 & 0 & * & * & * & * & * \\ 4 & 0 & 0 & 0 & * & * & * & 0 \\ 5 & 0 & 0 & 0 & 0 & * & * & 0 \\ 6 & 0 & 0 & 0 & 0 & 0 & * & 0 \\ 7 & 0 & 0 & 0 & 0 & 0 & 0 & * \end{pmatrix}, \quad \mathcal{Z}_{\text{out}}=\begin{pmatrix} & 1 & 2 & 3 & 4 & 5 & 6 & 7 \\ 1 & * & 0 & 0 & * & * & * & * \\ 2 & 0 & * & 0 & * & * & * & * \\ 3 & 0 & 0 & * & * & * & * & * \\ 4 & 0 & 0 & 0 & * & * & * & * \\ 5 & 0 & 0 & 0 & 0 & * & * & * \\ 6 & 0 & 0 & 0 & 0 & 0 & * & * \\ 7 & 0 & 0 & 0 & 0 & 0 & 0 & * \end{pmatrix},$$

and

$$\mathcal{Z}_{\text{ultra}}=\begin{pmatrix} & 1 & 2 & 3 & 4 & 5 & 6 & 7 \\ 1 & * & * & * & * & * & * & * \\ 2 & 0 & * & * & * & * & * & * \\ 3 & 0 & 0 & * & * & * & * & * \\ 4 & 0 & 0 & 0 & * & * & * & * \\ 5 & 0 & 0 & 0 & 0 & * & * & * \\ 6 & 0 & 0 & 0 & 0 & 0 & * & * \\ 7 & 0 & 0 & 0 & 0 & 0 & 0 & * \end{pmatrix}.$$

It follows that

$$\mathcal{Z}_{\mathrm{in}} \cap \mathcal{Z}_{\mathrm{out}} = \begin{pmatrix} & 1 & 2 & 3 & 4 & 5 & 6 & 7 \\ 1 & * & 0 & 0 & * & * & * & * \\ 2 & 0 & * & 0 & * & * & * & * \\ 3 & 0 & 0 & * & * & * & * & * \\ 4 & 0 & 0 & 0 & * & * & * & 0 \\ 5 & 0 & 0 & 0 & 0 & * & * & 0 \\ 6 & 0 & 0 & 0 & 0 & 0 & * & 0 \\ 7 & 0 & 0 & 0 & 0 & 0 & 0 & * \end{pmatrix},$$

and we see that the inclusions $\mathcal{Z} \subset \mathcal{Z}_{\mathrm{in}} \cap \mathcal{Z}_{\mathrm{out}} \subset \mathcal{Z}_{\mathrm{ultra}}$ are proper (cf. the remark just before Corollary 7.12).

To finish this example, we return to $\mathcal{Z}$ and $\mathcal{Z}_{\mathrm{in}}$. Above these partial orders were given by means of matrix schemes. They can also be represented by the familiar arrow diagrams. With the reflexivity loops suppressed, the diagram for the given partial order $\mathcal{Z}$ is

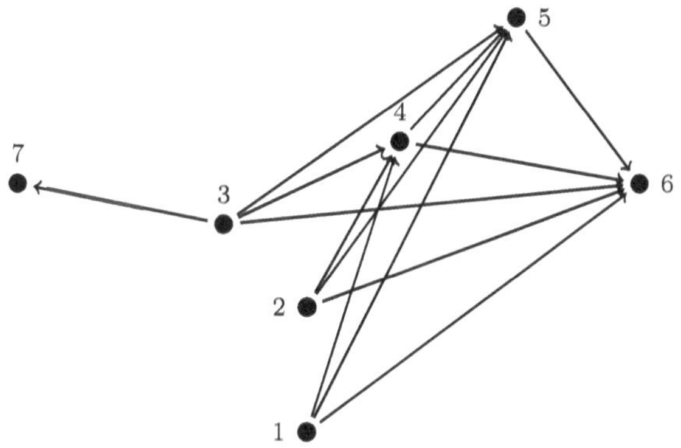

while

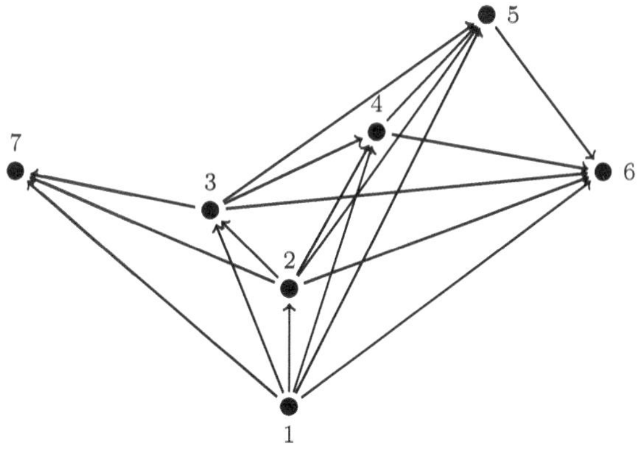

is the one for the in-ultra transitive closure $\mathcal{Z}_{\text{in}}$ of $\mathcal{Z}$. The Hasse diagram of $\mathcal{Z}_{\text{in}}$ – obtained from $\mathcal{Z}_{\text{in}}$ by omitting the arrows induced by transitivity ('short cuts') and, by definition, without reflexivity loops – is

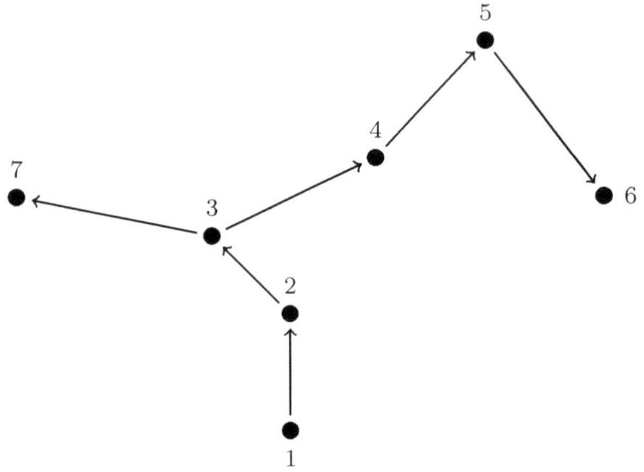

which is a rooted tree. This fact corroborates the characterization of in-ultra transitivity given in [BES4]. □

Example C. Introduce $M \in \mathbb{C}^{7\times 7}_{\text{upper}}$ via

$$M = \begin{bmatrix} 1 & 0 & 0 & 2 & 0 & 0 & 0 \\ 0 & 0 & 0 & 6 & -3 & 9 & 0 \\ 0 & 0 & 0 & 4 & -2 & -1 & 1 \\ 0 & 0 & 0 & -2 & 1 & -1 & 0 \\ 0 & 0 & 0 & 0 & 0 & 0 & 0 \\ 0 & 0 & 0 & 0 & 0 & 0 & 0 \\ 0 & 0 & 0 & 0 & 0 & 0 & 0 \end{bmatrix},$$

and introduce the monic matrices

$$G = \begin{bmatrix} 1 & 0 & 0 & 0 & 0 & 0 & 0 \\ 0 & 1 & 2 & 7 & 0 & 0 & 0 \\ 0 & 0 & 1 & 2 & 0 & 0 & 0 \\ 0 & 0 & 0 & 1 & 0 & 0 & 0 \\ 0 & 0 & 0 & 0 & 1 & 0 & 0 \\ 0 & 0 & 0 & 0 & 0 & 1 & 0 \\ 0 & 0 & 0 & 0 & 0 & 0 & 1 \end{bmatrix}, \quad L = \begin{bmatrix} 1 & \frac{1}{6} & 0 & \frac{3}{2} & 0 & 0 & 0 \\ 0 & 1 & 2 & 7 & 0 & 0 & 0 \\ 0 & 0 & 1 & 2 & 0 & 0 & 0 \\ 0 & 0 & 0 & 1 & 0 & 0 & 0 \\ 0 & 0 & 0 & 0 & 1 & 0 & 0 \\ 0 & 0 & 0 & 0 & 0 & 1 & 0 \\ 0 & 0 & 0 & 0 & 0 & 0 & 1 \end{bmatrix}.$$

Computing GM and LM we get

$$GM = \begin{bmatrix} \mathbf{1} & 0 & 0 & 2 & 0 & 0 & 0 \\ 0 & 0 & 0 & 0 & 0 & 0 & \mathbf{2} \\ 0 & 0 & 0 & 0 & 0 & \mathbf{-3} & 1 \\ 0 & 0 & 0 & \mathbf{-2} & 1 & -1 & 0 \\ 0 & 0 & 0 & 0 & 0 & 0 & 0 \\ 0 & 0 & 0 & 0 & 0 & 0 & 0 \\ 0 & 0 & 0 & 0 & 0 & 0 & 0 \end{bmatrix}, \quad LM = \begin{bmatrix} \mathbf{1} & 0 & 0 & 0 & 1 & 0 & 0 \\ 0 & 0 & 0 & 0 & 0 & 0 & \mathbf{2} \\ 0 & 0 & 0 & 0 & 0 & \mathbf{-3} & 1 \\ 0 & 0 & 0 & \mathbf{-2} & 1 & -1 & 0 \\ 0 & 0 & 0 & 0 & 0 & 0 & 0 \\ 0 & 0 & 0 & 0 & 0 & 0 & 0 \\ 0 & 0 & 0 & 0 & 0 & 0 & 0 \end{bmatrix}. \tag{15}$$

Both these matrices are in generalized row echelon form, as can be seen by looking at the positions of the leading entries (here, and also below, marked boldface for easy recognition). They are not identical, however (cf. the paragraph after the proof of Theorem 4.1). Clearly, LM is in upward echelon form. Thus LM is the (uniquely determined) canonical upward echelon form of M, i.e., $LM = M^{\uparrow}_{\mathrm{rEch}}$ (cf. Theorem 4.2 and the remark made after the proof of Theorem 4.1). The matrix

$$\begin{bmatrix} \mathbf{1} & 0 & 0 & 0 & 1 & 0 & 0 \\ 0 & 0 & 0 & 0 & 0 & 0 & \mathbf{1} \\ 0 & 0 & 0 & 0 & 0 & \mathbf{1} & -\frac{1}{3} \\ 0 & 0 & 0 & \mathbf{1} & -\frac{1}{2} & \frac{1}{2} & 0 \\ 0 & 0 & 0 & 0 & 0 & 0 & 0 \\ 0 & 0 & 0 & 0 & 0 & 0 & 0 \\ 0 & 0 & 0 & 0 & 0 & 0 & 0 \end{bmatrix},$$

is the canonical monic upward echelon form $M^{\uparrow}_{\mathrm{mrEch}}$ of M (cf. Corollary 4.4).

Let R be the monic matrix

$$R = \begin{bmatrix} 1 & 0 & 0 & -2 & -1 & 1 & \frac{1}{3} \\ 0 & 1 & 0 & 0 & 0 & 0 & 0 \\ 0 & 0 & 1 & 0 & 0 & 0 & 0 \\ 0 & 0 & 0 & 1 & \frac{1}{2} & -\frac{1}{2} & -\frac{1}{6} \\ 0 & 0 & 0 & 0 & 1 & 0 & 0 \\ 0 & 0 & 0 & 0 & 0 & 1 & \frac{1}{3} \\ 0 & 0 & 0 & 0 & 0 & 0 & 1 \end{bmatrix}.$$

Multiplying M on the right with R, we get

$$MR = \begin{bmatrix} \mathbf{1} & 0 & 0 & 0 & 0 & 0 & 0 \\ 0 & 0 & 0 & 6 & 0 & 6 & \mathbf{2} \\ 0 & 0 & 0 & 4 & 0 & \mathbf{-3} & 0 \\ 0 & 0 & 0 & \mathbf{-2} & 0 & 0 & 0 \\ 0 & 0 & 0 & 0 & 0 & 0 & 0 \\ 0 & 0 & 0 & 0 & 0 & 0 & 0 \\ 0 & 0 & 0 & 0 & 0 & 0 & 0 \end{bmatrix},$$

and this is the (uniquely determined) canonical starboard echelon form of M, i.e., $MR = M_{\mathrm{cEch}}^{\rightarrow}$ (cf. Theorem 4.3). The trailing entry structure of $MR = M_{\mathrm{cEch}}^{\rightarrow}$ is the same as the leading entry structure of $LM = M_{\mathrm{rEch}}^{\uparrow}$, both with regard to positions and entry values (cf. the remark made after Theorem 6.1). The matrix

$$\begin{bmatrix} \mathbf{1} & 0 & 0 & 0 & 0 & 0 & 0 \\ 0 & 0 & 0 & 3 & 0 & 3 & \mathbf{1} \\ 0 & 0 & 0 & -\frac{4}{3} & 0 & \mathbf{1} & 0 \\ 0 & 0 & 0 & \mathbf{1} & 0 & 0 & 0 \\ 0 & 0 & 0 & 0 & 0 & 0 & 0 \\ 0 & 0 & 0 & 0 & 0 & 0 & 0 \\ 0 & 0 & 0 & 0 & 0 & 0 & 0 \end{bmatrix},$$

is the canonical monic starboard echelon form $M_{\mathrm{mcEch}}^{\rightarrow}$ of M (cf. Corollary 4.5). The trailing entry structure of $M_{\mathrm{mcEch}}^{\rightarrow}$ is the same as the leading entry structure of $M_{\mathrm{mrEch}}^{\uparrow}$, both with regard to positions and (but this is trivial now) entry values (cf. the remark made after Theorem 6.2).

Next we turn to generalized diagonal forms. A quick way to obtain the canonical generalized diagonal form M_{gDiag} of M goes via applying Theorem 6.1. Indeed, M_{gDiag} is the leading entry reduction of $LM = M_{\mathrm{rEch}}^{\uparrow}$ (as well as the trailing entry reduction of $MR = M_{\mathrm{cEch}}^{\rightarrow}$):

$$M_{\mathrm{gDiag}} = \begin{bmatrix} 1 & 0 & 0 & 0 & 0 & 0 & 0 \\ 0 & 0 & 0 & 0 & 0 & 0 & 2 \\ 0 & 0 & 0 & 0 & 0 & -3 & 0 \\ 0 & 0 & 0 & -2 & 0 & 0 & 0 \\ 0 & 0 & 0 & 0 & 0 & 0 & 0 \\ 0 & 0 & 0 & 0 & 0 & 0 & 0 \\ 0 & 0 & 0 & 0 & 0 & 0 & 0 \end{bmatrix}.$$

The various identities featuring M_{gDiag} in Theorem 6.1 are easily seen to be valid. In the same vein one gets via Theorem 6.2 the canonical monic generalized diagonal form M_{mgDiag} of M, i.e., what we also called the frame of M:

$$M_{\text{mgDiag}} = \begin{bmatrix} 1 & 0 & 0 & 0 & 0 & 0 & 0 \\ 0 & 0 & 0 & 0 & 0 & 0 & 1 \\ 0 & 0 & 0 & 0 & 0 & 1 & 0 \\ 0 & 0 & 0 & 1 & 0 & 0 & 0 \\ 0 & 0 & 0 & 0 & 0 & 0 & 0 \\ 0 & 0 & 0 & 0 & 0 & 0 & 0 \\ 0 & 0 & 0 & 0 & 0 & 0 & 0 \end{bmatrix}.$$

Theorems 6.2–6.4 contain a number of identities for M_{mgDiag} which can easily be verified in the present situation.

Let $\mathcal{Z}$ be the partial order given by the scheme (14). Recall that $\mathcal{Z}$ is neither in-ultra transitive nor out-ultra transitive. Its in-ultra transitive closure $\mathcal{Z}_{\text{in}}$ and out-ultra transitive closure $\mathcal{Z}_{\text{out}}$ are determined in Example B. The canonical forms of M presented above are not in $\mathcal{A}_{\mathcal{Z}}$ (cf. the sixth paragraph in Section 7, the comment right after Theorem 7.1, and the paragraph after Corollary 7.8). They do belong to $\mathcal{A}_{\mathcal{Z}_{\text{in}}}$, however (cf. Corollaries 7.3 and Corollary 7.4, 7.12 and 7.13). The canonical diagonal forms (including the frame of M) even belong to $\mathcal{A}_{\mathcal{Z}_{\text{in}}} \cap \mathcal{A}_{\mathcal{Z}_{\text{out}}} = \mathcal{A}_{\mathcal{Z}_{\text{in}} \cap \mathcal{Z}_{\text{out}}}$. The scheme for $\mathcal{Z}_{\text{in}} \cap \mathcal{Z}_{\text{out}}$ can be found in Example B. In line with Corollary 7.3, the matrix L for which $LM = M^{\uparrow}_{\text{rEch}}$ is in $\mathcal{A}_{\mathcal{Z}_{\text{in}}}$. It is not in $\mathcal{A}_{\mathcal{Z}}$. □

Example D. Here $n = 3$ and $\mathcal{Z}$ is the partial order on $\{1, 2, 3\}$ given by the scheme

$$\mathcal{Z} = \begin{pmatrix} & 1 & 2 & 3 \\ 1 & * & 0 & * \\ 2 & 0 & * & * \\ 3 & 0 & 0 & * \end{pmatrix}.$$

Clearly $\mathcal{Z}$ is of upper triangular type and out-ultra transitive (trivially). Thus $\mathcal{Z}_{\text{out}} = \mathcal{Z}$. From $1 \to_{\mathcal{Z}} 3$, $2 \to_{\mathcal{Z}} 3$ and $1 \not\to_{\mathcal{Z}} 2$, we see that $\mathcal{Z}$ is not in-ultra transitive and so, *a fortiori*, not ultra transitive. The in-ultra transitive closure $\mathcal{Z}_{\text{in}}$ and the ultra transitive closure $\mathcal{Z}_{\text{ultra}}$ of $\mathcal{Z}$ coincide and are given by

$$\mathcal{Z}_{\text{in}} = \mathcal{Z}_{\text{ultra}} = \begin{pmatrix} & 1 & 2 & 3 \\ 1 & * & * & * \\ 2 & 0 & * & * \\ 3 & 0 & 0 & * \end{pmatrix}.$$

It follows that

$$\mathcal{Z}_{\text{in}} \cap \mathcal{Z}_{\text{out}} = \mathcal{Z}_{\text{out}} = \mathcal{Z} = \begin{pmatrix} & 1 & 2 & 3 \\ 1 & * & 0 & * \\ 2 & 0 & * & * \\ 3 & 0 & 0 & * \end{pmatrix}.$$

Define M and L by

$$M = \begin{bmatrix} 0 & 0 & 1 \\ 0 & 0 & 1 \\ 0 & 0 & 0 \end{bmatrix}, \qquad L = \begin{bmatrix} 1 & -1 & 0 \\ 0 & 1 & 0 \\ 0 & 0 & 1 \end{bmatrix}.$$

Then $M \in \mathcal{A}_{\mathcal{Z}}$, the matrix L is monic, and

$$LM = \begin{bmatrix} 0 & 0 & 0 \\ 0 & 0 & 1 \\ 0 & 0 & 0 \end{bmatrix} = M^{\uparrow}_{\text{rEch}} = M^{\uparrow}_{\text{mrEch}} = M_{\text{gDiag}} = M_{\text{mgDiag}}.$$

In spite of the circumstance that $\mathcal{Z}$ is not in-ultra transitive, $LM = M^{\uparrow}_{\text{rEch}}$ belongs to $\mathcal{A}_{\mathcal{Z}}$ (cf. Theorem 7.1, and the paragraph just below it). Also M_{gDiag} and M_{mgDiag} belong to $\mathcal{Z}_{\text{in}} \cap \mathcal{Z}_{\text{out}}$ corroborating Corollaries 7.12 and 7.13.

The (monic) matrix L does not belong to $\mathcal{A}_{\mathcal{Z}}$ (although the product LM does). In fact there do not exist invertible (upper triangular) G and H in $\mathcal{A}_{\mathcal{Z}} = \mathcal{A}_{\mathcal{Z}_{\text{in}} \cap \mathcal{Z}_{\text{out}}} = \mathcal{A}_{\mathcal{Z}_{\text{in}}} \cap \mathcal{A}_{\mathcal{Z}_{\text{out}}}$ for which $GMH = M^{\uparrow}_{\text{rEch}}$ or, what here amounts to the same, $GMH = M_{\text{gDiag}} = M_{\text{mgDiag}}$ (cf. the paragraphs just below Theorem 7.1 and Corollary 7.13).

This is the argument. Suppose such matrices G and H do exist. Write

$$G = \begin{bmatrix} g_{1,1} & 0 & g_{1,3} \\ 0 & g_{2,2} & g_{2,3} \\ 0 & 0 & g_{3,3} \end{bmatrix}, \quad H = \begin{bmatrix} h_{1,1} & 0 & h_{1,3} \\ 0 & h_{2,2} & h_{2,3} \\ 0 & 0 & h_{3,3} \end{bmatrix}, \quad \text{so that } GMH = \begin{bmatrix} 0 & 0 & g_{1,1}h_{3,3} \\ 0 & 0 & g_{2,2}h_{3,3} \\ 0 & 0 & 0 \end{bmatrix}.$$

Then $GMR = M^{\uparrow}_{\text{rEch}} = LM$ gives $g_{1,1}h_{3,3} = 0$. Thus, either $g_{1,1} = 0$ conflicting with the invertibility of G, or $h_{3,3} = 0$ violating the invertibility of H. □

11. Concluding remark

The ideas for the present paper came up in our study of logarithmic residues and sums of idempotents. In the introduction we gave some references while Theorem 1.2 served as background information. Here we note that, as a rather straightforward consequence of Proposition 9.4 and material from [BES3], we have the result formulated below. In terms of [BES3], it says that ultra transitive partial orders are rank/trace complete. In fact, this even holds for ultra-transitive preorders.

Note here that the concepts of in-ultra transitivity, out-ultra transitivity and ultra transitivity allow for an immediate extension to preorders.

Theorem 11.1. *Let $\mathcal{Z}$ be an ultra transitive preorder on $\{1,\dots,n\}$, where n is a positive integer. Then, for $A \in \mathcal{A}_{\mathcal{Z}}$, the following statements are equivalent:*

(i) *A is a sum of idempotents in $\mathcal{A}_{\mathcal{Z}}$;*
(ii) *A is a sum of rank one idempotents in $\mathcal{A}_{\mathcal{Z}}$;*
(iii) *A is a logarithmic residue in $\mathcal{A}_{\mathcal{Z}}$;*
(iv) *A satisfies the $\mathcal{Z}$-rank/trace conditions.*

For the meaning of (iv), we refer to [BES3], Section 2. In the case when $\mathcal{Z}$ is the standard linear order on $\{1,\dots,n\}$, the statement in question comes down to (4) in Theorem 1.2. In the more general setting, the diagonal blocks mentioned there are replaced by submatrices of A corresponding to $\mathcal{Z}$-convex subsets of $\{1,\dots,n\}$.

Proof. Evidently (ii) implies (i). For the implications (i) $\Rightarrow$ (iii) and (iii) $\Rightarrow$ (iv), see [BES1], Section 2 and [BES3], Theorem 2.4, respectively. It remains to prove that (iv) implies (ii). Here is the reasoning.

Consider the so-called condensation $\mathcal{Z}_{\text{cond}}$ of $\mathcal{Z}$ (obtained via the identification of nodes k and l for which both $k \to_{\mathcal{Z}} l$ and $l \to_{\mathcal{Z}} k$; cf. [BES3], Subsection 3.6). This is a partial order. As $\mathcal{Z}$ is ultra transitive, so is $\mathcal{Z}_{\text{cond}}$. Hence Proposition 9.4 applies, and we may conclude that $\mathcal{Z}_{\text{cond}}$ is permutation similar to a disjoint union of linear orders. But then, modulo a permutation similarity, the preorder $\mathcal{Z}$ is a disjoint union of preorders of block upper triangular type in the sense of [BES2], Subsection 4.3. From the material presented there (heavily based on [BW] and [BES2]) we know that these preorders are rank/trace complete in the sense of [BES3]. Theorem 6.3 in that paper gives that so is their disjoint union. The desired result is now immediate from [BES3], Lemma 3.1. Indeed, this lemma says that rank/trace completeness is preserved under permutation similarity. □

It is an open problem whether or not in Theorem 11.1, the requirement of ultra transitivity can be relaxed to in-ultra transitivity or out-ultra transitivity. For more on this issue, see Subsection 5.2 in [BES4].

References

[BES1] H. Bart, T. Ehrhardt, B. Silbermann, Logarithmic residues in Banach algebras, *Integral Equations and Operator Theory* 19 (1994), 135–152.

[BES2] H. Bart, T. Ehrhardt, B. Silbermann, Sums of idempotents and logarithmic residues in matrix algebras, *Operator Theory: Advances and Applications*, Vol. 122, Birkhäuser Verlag, Basel 2001, 139–168.

[BES3] H. Bart, T. Ehrhardt, B. Silbermann, Sums of idempotents and logarithmic residues in zero pattern matrix algebras, *Linear Algebra Appl.* 498 (2016), 262–316.

[BES4] H. Bart, T. Ehrhardt, B. Silbermann, Rank decomposition in zero pattern matrix algebras, *Czechoslovak Mathematical Journal* 66 (2016), 987–1005.

[BW] H. Bart, A.P.M. Wagelmans, An integer programming problem and rank decomposition of block upper triangular matrices, *Linear Algebra Appl.* 305 (2000), 107–129.

[C] C.A. Charalambides, *Enumerative combinatorics.* CRC Press Series on Discrete Mathematics and its Applications, Chapman and Hall/CRC, Boca Raton, FL, 2002.

[D] R.L. Davis, Algebras defined by patterns of zeros, *J. Combinatorial Theory* 9 (1970), 257–260.

[FP] S.-C. Fang, S. Puthenpura, *Linear Optimization and Extensions: Theory and Algorithms*, Englewood Cliffs, NJ: Prentice-Hall, 1993.

[HJ] R.A. Horn, C.R. Johnson, *Matrix Analysis, second edition*, Cambridge University Press, Cambridge, Melbourne etc., 2013.

[Laf] T.J. Laffey, A structure theorem for some matrix algebras, *Linear Algebra Appl.* 162–164 (1992), 205–215.

[Lay] D.C. Lay, Subspaces and echelon Forms, *The College Mathematics Journal*, 24(1) (1993), 57–62.

[LT] P. Lancaster, M. Tismenetsky, *The Theory of Matrices, second edition*, Academic Press, Orlando etc., 1985 **24(1)** (1993), 57–62.

[R] J. Riordan, *Combinatorial identities*, John Wiley and Sons, Inc., New York-London-Sydney, 1968.

[S] R.P. Stanley, *Enumerative combinatorics. Vol.* 1, Cambridge University Press, Cambridge, 1997.

[S(M)] E. Szpilrajn (later called E. Marczewski), Sur l'extension de l'ordre partiel, *Fundamenta Mathematicae* 16 (1930), 386–389 [in French]. (Available at `http://matwbn.icm.edu.pl/ksiazki/fm/fm16/fm16125.pdf`.)

[SW] D. Surowski, Y. Wang, The uniqueness of strong row echelon form, *Missouri J. Math. Sciences*, 15, 2003, 36–39.

[T] Ph.Thijsse, Upper triangular similarity of upper triangular matrices. *Linear Algebra Appl.* 260 (1997), 119–149.

Harm Bart
Econometric Institute
Erasmus University Rotterdam
P.O. Box 1738
NL-3000 DR Rotterdam, The Netherlands
e-mail: `bart@ese.eur.nl`

Torsten Ehrhardt
Mathematics Department
University of California
Santa Cruz, CA-95064, USA
e-mail: `tehrhard@ucsc.edu`

Bernd Silbermann
Fakultät für Mathematik
Technische Universität Chemnitz
D-09107 Chemnitz, Germany
e-mail: `silbermn.toeplitz@googlemail.com`

Operator Theory:
Advances and Applications, Vol. 259, 125–154

Asymptotic Formulas for Determinants of a Special Class of Toeplitz + Hankel Matrices

Estelle Basor and Torsten Ehrhardt

Dedicated to Albrecht Böttcher on the occasion of his sixtieth birthday

Abstract. We compute the asymptotics of the determinants of certain $n \times n$ Toeplitz + Hankel matrices $T_n(a) + H_n(b)$ as $n \to \infty$ with symbols of Fisher–Hartwig type. More specifically we consider the case where a has zeros and poles and where b is related to a in specific ways. Previous results of Deift, Its and Krasovsky dealt with the case where a is even. We are generalizing this in a mild way to certain non-even symbols.

Mathematics Subject Classification (2010). 47B35, 47A20, 15B52, 82B.

Keywords. Toeplitz-plus-Hankel determinant, Toeplitz-plus-Hankel operator, Fisher–Hartwig asymptotics.

1. Introduction

For many recent applications, an asymptotic formula for determinants of the sum of finite Toeplitz and Hankel matrices has been of interest. For example, if we let a be in $L^1(\mathbb{T})$ and denote the kth Fourier coefficients of a by a_k then understanding the behavior of

$$\det \left(a_{j-k} + a_{j+k+1}\right)_{j,k=0,\dots,n-1}$$

as $n \to \infty$ is important in random matrix theory. It has been shown in [5] that the above determinant behaves asymptotically like $G^n E$ with certain explicitly given constants G and E if a is a sufficiently well-behaved function. Such a result is an analogue of the classical Szegő–Widom limit theorem [18] for Toeplitz determinants.

The above determinant is a special case of more general determinants,

$$\det \left(a_{j-k} + b_{j+k+1}\right)_{j,k=0,\dots,n-1},$$

where both a and b are in $L^1(\mathbb{T})$. We refer to the functions a and b as symbols. The goal is to find the asymptotics in the case of well-behaved a and b and also for singular Fisher–Hartwig type symbols (symbols, with say, jump discontinuities or zeros). While an asymptotic formula in such a general case (with an explicit description of the constants) is probably not doable, much recent progress has been made in some special cases.

To be more precise a Fisher–Hartwig symbol is one of the form

$$a(e^{i\theta}) = c(e^{i\theta}) \prod_{r=1}^{R} v_{\tau_r,\alpha_r}(e^{i\theta}) u_{\tau_r,\beta_r}(e^{i\theta}) \tag{1}$$

where c is a sufficiently well-behaved function (i.e., sufficiently smooth, nonvanishing, and with winding number zero) and for $\tau = e^{i\phi}$,

$$u_{\tau,\beta}(e^{i\theta}) = \exp(i\beta(\theta - \phi - \pi)), \quad 0 < \theta - \phi < 2\pi,$$
$$v_{\tau,\alpha}(e^{i\theta}) = (2 - 2\cos(\theta - \phi))^{\alpha}.$$

The symbol $u_{\tau,\beta}$ has a jump at the point τ on the unit circle and the function $v_{\tau.\alpha}$ can be singular (say if α has negative real part,) or be zero at τ. We will generally refer to this last factor as having a singularity of "zero" type. Furthermore, α_r and β_r are complex parameters (where we assume $\operatorname{Re}\alpha_r > -1/2$) and $\tau_1, \ldots, \tau_R$ are distinct points on the unit circle $\mathbb{T}$.

In the case of smooth symbols, we cite the results in [7] where the case of

$$\det\left(a_{j-k} + b_{j+k+1}\right)_{j,k=0,\ldots,n-1}$$

with $b(e^{i\theta}) = \pm e^{i\ell\theta} a(e^{i\theta})$ and ℓ fixed is considered. It is worth mentioning that among those cases, there are four special cases of particular interest,

(i) $b(e^{i\theta}) = a(e^{i\theta})$,
(ii) $b(e^{i\theta}) = -a(e^{i\theta})$,
(iii) $b(e^{i\theta}) = e^{i\theta} a(e^{i\theta})$,
(iv) $b(e^{i\theta}) = -e^{-i\theta} a(e^{i\theta})$,

in which the asymptotics have the form $G^n E$ with non-zero E. In the case of even symbols, i.e., $a(e^{i\theta}) = a(e^{-i\theta})$, these four cases are also related to quantities that arise in the study of the random matrices taken from the classical groups [2, 14]. Furthermore, these Toeplitz + Hankel determinants are expressible as Hankel determinants as well. For other instances of these determinants and applications of such determinants, see also [4, 15].

In the case of $b = a$, two earlier papers of the authors consider the case of jump discontinuities [5, 6].

In the above four cases where in addition a is even, the results of Deift, Its, and Krasovsky [12] are quite complete and impressive. They allow quite general Fisher–Hartwig symbols with both zeros and jumps. In [12] the asymptotics are of the form

$$G^n \, n^p \, E$$

where G, p, and E are explicitly given constants. However, none of the earlier mentioned papers covers the case where the symbol is allowed to be non-even and with singularities of the zero type. So this is the focus of this paper, a non-even symbol with certain specified types of Fisher–Hartwig symbols. We prove an asymptotic formula of the same form as above. This is a step in the ultimate goal of asymptotics for non-even symbols with general Fisher–Hartwig singularities.

In order to briefly sketch the main ideas of the paper, let $T(a)$ and $H(b)$ stand for the Toeplitz and Hankel operators with symbols a and b acting on ℓ^2, and let P_n stand for the finite section projection on ℓ^2. The precise definition of these operators will be given in the next section. The above determinants can be understood as the determinant of $P_n(T(a)+H(b))P_n$ for certain symbols a, b of Fisher–Hartwig type. Since we will allow not only for jumps, but also for zeros and poles, the underlying operator (or its inverse) is generally not bounded. Hence the first step is to reformulate the problem as one which involves only bounded operators. This will be done by establishing an identity of the kind

$$\det P_n(T(a)+H(b))P_n = \det\left(P_nT^{-1}(\psi)(T(c)+H(cd\phi))T^{-1}(\psi^{-1})P_n\right) \tag{2}$$

where the functions c and d are smooth and ϕ and ψ have only jump discontinuities. The next major step is a separation theorem, which allows to "remove" the smooth functions c and d, i.e.,

$$\frac{\det\left(P_nT^{-1}(\psi)(T(c)+H(cd\phi))T^{-1}(\psi^{-1})P_n\right)}{\det\left(P_nT^{-1}(\psi)(I+H(\phi))T^{-1}(\psi^{-1})P_n\right)} \sim G^nE, \qquad n\to\infty,$$

with explicit constant G and E. Then by using the first identity again we relate the last determinant back to a Toeplitz + Hankel determinant,

$$\det P_n(T(a_0)+H(b_0))P_n = \det\left(P_nT^{-1}(\psi)(I+H(\phi))T^{-1}(\psi^{-1})P_n\right),$$

where, as it turns out, a_0 and b_0 are Fisher–Hartwig symbols which in their product representation (1) do not have the smooth part. Of course, the whole procedure is only as useful as far as we are able to obtain the asymptotics of the determinants $\det P_n(T(a_0)+H(b_0))P_n$. Here we apply the results of Deift, Its, and Krasovsky mentioned above [12] to identify this asymptotics in four special cases. The relation between our symbols a and b, and the symbols a_0 and b_0 to which we apply [12], is in fact given by

$$a = c\,a_0, \qquad b = c\,d\,b_0,$$

where a_0 is even and of Fisher–Hartwig type and b_0 relates to a_0 as in (i)–(iv). The functions c and d, while required to be sufficiently well behaved, do not have to be even. The function d is required to satisfy $d(e^{i\theta})d(e^{-i\theta})=1$, $d(\pm 1)=1$. In this sense, our results generalize some of the results of [12].

Note that in the most general case (for which we are able to do the separation theorem), the asymptotics of the corresponding $\det P_n(T(a_0)+H(b_0))P_n$ is not known.

The idea for establishing an identity of the kind (2) and proving a separation theorem is due to Böttcher and Silbermann. In 1985 they proved the Fisher–Hartwig conjecture for symbols with small parameters (i.e., symbols (1) with $|\mathrm{Re}\,\alpha_r| < 1/2$ and $|\mathrm{Re}\,\beta_r| < 1/2$), which was considered a major breakthrough at the time (see [10] and [11, Sect. 10.10]). Although they considered bounded invertible operators acting between different weighted L^2-spaces, the essential point of their analysis can be expressed as an identity of the kind

$$\det T_n(a) = \det P_n T^{-1}(\psi) T(\phi) T^{-1}(\psi^{-1}) P_n\,,$$

where a is a Fisher–Hartwig symbol with jumps and zeros/poles, while ϕ and ψ are Fisher–Hartwig symbols with jumps only (the corresponding Toeplitz operators being bounded and invertible under certain conditions).

Here is an outline of the paper. We begin with the operator theoretic preliminaries. This is done is Section 2. In Section 3, we reformulate the problem so that we can consider determinants of bounded operators only. In Section 4, some additional operator theoretic results are given that are particularly useful for our situation. We then prove, in Section 5, a "separation" theorem, that is, a theorem that allows us to compute the asymptotics from a combination of the smooth symbols and some specific cases of singular symbols where the results can be computed by other means. This is done in more generality than is needed for our final results, but it may prove to be useful in the future if other specific cases of singular symbols are obtained.

Section 6 is devoted to infinite determinant computations that are required to describe constants explicitly, and Section 7 contains the known results for the specific singular symbols. Everything is collected in Section 8 where the final asymptotics are computed.

Finally, the last section contains some additional results. In the course of the computations for the main results of this paper, we discovered that certain Identity plus Hankel operators had inverses that could be described using Toeplitz operators, their inverses and Hankel operators. So the inverse expressions may be of independent interest and are also included.

2. Preliminaries

We denote by ℓ^2 the space of all complex-valued sequences $\{x_n\}_{n=0}^{\infty}$ equipped with usual 2-norm. The set $\mathcal{L}(\ell^2)$ is the set of bounded operators on ℓ^2 and $\mathcal{C}_1(\ell^2)$ is the set of trace class operators on ℓ^2.

The Toeplitz operator $T(a)$ and Hankel operator $H(a)$ with symbol $a \in L^\infty(\mathbb{T})$ are the bounded linear operators defined on ℓ^2 with matrix representations

$$T(a) = (a_{j-k}), \quad 0 \le j,k < \infty,$$

and

$$H(a) = (a_{j+k+1}), \quad 0 \le j,k < \infty.$$

It is well known and not difficult to prove that Toeplitz and Hankel operators satisfy the fundamental identities

$$T(ab) = T(a)T(b) + H(a)H(\tilde{b}) \tag{3}$$

and

$$H(ab) = T(a)H(b) + H(a)T(\tilde{b}). \tag{4}$$

In the last two identities and throughout the paper we are using the notation

$$\tilde{b}(e^{i\theta}) := b(e^{-i\theta}).$$

It is worthwhile to point out that these identities imply that

$$T(abc) = T(a)T(b)T(c), \qquad H(ab\tilde{c}) = T(a)H(b)T(c) \tag{5}$$

for $a, b, c, \in L^\infty(\mathbb{T})$ if $a_n = c_{-n} = 0$ for all $n > 0$.

We define the (finite section) projection P_n by

$$P_n : \{x_k\}_{k=0}^\infty \in \ell^2 \mapsto \{y_k\}_{k=0}^\infty \in \ell^2, \qquad y_k = \begin{cases} x_k, & \text{if } k < n, \\ 0, & \text{if } k \geq n. \end{cases}$$

Using P_n we can view our determinants of interest as determinants of truncations of infinite matrices,

$$\det(T_n(a) + H_n(b)) = \det P_n(T(a) + H(b))P_n.$$

For bounded a and b this is the truncation of a sum of bounded operators, but it holds even more generally for $a, b \in L^1(\mathbb{T})$ providing we view the operators as being defined on the space of sequences with only a finite number of non-zero terms.

In the next sections we will be mostly concerned with functions a that are products of continuous functions times those with certain specific types of singularities. It will be convenient for the continuous function factors to satisfy certain properties. To describe this, we consider the Banach algebra called the Besov class B_1^1. This is the algebra of all functions a defined on the unit circle for which

$$\|a\|_{B_1^1} := \int_{-\pi}^{\pi} \frac{1}{y^2} \int_{-\pi}^{\pi} |a(e^{ix+iy}) + a(e^{ix-iy}) - 2a(e^{ix})|\, dx\, dy < \infty.$$

A function a is in B_1^1 if and only if the Hankel operators $H(a)$ and $H(\tilde{a})$ are both trace class. Moreover, the Riesz projection is bounded on B_1^1, and an equivalent norm is given by

$$|a_0| + \|H(a)\|_{\mathcal{C}_1} + \|H(\tilde{a})\|_{\mathcal{C}_1},$$

where $\|A\|_{\mathcal{C}_1}$ is the trace norm of the operator A.

Let us also recall the notion of Wiener–Hopf factorization. There are several versions of it. We say that $c \in L^\infty(\mathbb{T})$ has a bounded (canonical) factorization if we can write $c = c_- c_+$ with $c_+, c_+^{-1} \in H_+^\infty(\mathbb{T})$ and $c_-, c_-^{-1} \in H_-^\infty(\mathbb{T})$, where

$$H_\pm^\infty(\mathbb{T}) = \{f \in L^\infty(\mathbb{T}) \,:\, f_n = 0 \text{ for all } \pm n < 0\}.$$

We say that $c \in B_1^1$ has a canonical factorization in B_1^1 if we can write $c = c_- c_+$ with $c_+, c_+^{-1} \in H_+^\infty(\mathbb{T}) \cap B_1^1$ and $c_-, c_-^{-1} \in H_-^\infty(\mathbb{T}) \cap B_1^1$. It is well known (see, e.g.,

[11, Sect. 10.24]) that c admits a canonical factorization in B_1^1 if and only if the function c does not vanish on $\mathbb{T}$ and has winding number zero. In this case, the logarithm exists, $\log c \in B_1^1$, and one can define normalized factors,

$$c_{\pm}(t) = \exp\left(\sum_{k=1}^{\infty} t^{\pm k}[\log c]_{\pm k}\right), \tag{6}$$

which yield a factorization $c(t) = c_-(t)G[c]c_+(t)$, $t \in \mathbb{T}$, with the constant

$$G[c] := \exp([\log c]_0) \tag{7}$$

representing the geometric mean.

For our purposes it is also important to consider a factorization of the kind

$$d = \tilde{d}_+^{-1} d_+ \quad \text{with } d_+, d_+^{-1} \in B_1^1 \cap H_+^{\infty}(\mathbb{T}), \tag{8}$$

in which the "minus" factor $\tilde{d}_+^{-1}$ is given by the "plus" factor d_+. It is not too difficult to show (using the existence of a factorization in B_1^1 and the uniqueness of a factorization up to multiplicative constants) that $d \in B_1^1$ possesses a factorization of the above kind if and only if d does not vanish on $\mathbb{T}$, has winding number zero and satisfies the conditions $d\tilde{d} = 1$ and $d(\pm 1) = 1$. Notice that in this case $\log d \in B_1^1$ is an odd function and thus $G[d] = 1$.

3. Reformulating the problem

As described in the introduction, we are interested in determinants of Toeplitz plus Hankel matrices with singular symbols. Let us denote the corresponding (infinite) operator by

$$M(a,b) := T(a) + H(b).$$

Notice that when the symbols involve zeros or poles, then either $M(a,b)$ or its inverse are in general not bounded operators anymore. The purpose of this section is to reformulate the problem about the asymptotics of $\det(P_n M(a,b) P_n)$ as one for $\det(P_n A P_n)$ where A is a bounded (and invertible) operator on ℓ^2. More precisely, we are going to prove a formula

$$\det P_n M(a,b) P_n = \det P_n T^{-1}(\psi) M(c, c\, d\, \phi) T^{-1}(\psi^{-1}) P_n, \tag{9}$$

where a and b are certain functions of Fisher–Hartwig type (allowing in particular for zeros and jumps) while on the right-hand side ψ and ϕ are functions with jump discontinuities only and with ranges of parameters such that $T(\psi)$ and $T(\psi^{-1})$ are invertible Toeplitz operators. The functions c and d are smooth and nonvanishing functions with winding number zero.

Since the above formula involves inverses of Toeplitz operators, let us first recall a well-known sufficient invertibility criterion (see, e.g., [17] or [11]).

Theorem 3.1. *Let c be a continuous and nonvanishing function on $\mathbb{T}$ with winding number zero, let $\tau_1, \ldots, \tau_R \in \mathbb{T}$ be distinct, and*

$$\psi(e^{i\theta}) = c(e^{i\theta}) \prod_{r=1}^{R} u_{\tau_r,\beta_r}(e^{i\theta}) .$$

If $|\mathrm{Re}\, \beta_r| < 1/2$ for all $1 \le r \le R$, then $T(\psi)$ is invertible on ℓ^2.

Let us now introduce the functions for which identity (9) will be proved. For these functions the separation theorem will be proved later on as mentioned in the introduction. We consider

$$a = c\, v_{1,\alpha^+}\, v_{-1,\alpha^-} \prod_{r=1}^{R} v_{\tau_r,\alpha_r^+}\, v_{\bar{\tau}_r,\alpha_r^-} , \tag{10}$$

$$b = c\, d\, v_{1,\alpha^+}\, u_{1,\beta^+}\, v_{-1,\alpha^-}\, u_{-1,\beta^-} \prod_{r=1}^{R} v_{\tau_r,\alpha_r} u_{\tau_r,\beta_r}\, v_{\bar{\tau}_r,\alpha_r} u_{\bar{\tau}_r,\beta_r} . \tag{11}$$

The functions c and d are smooth nonvanishing functions with winding number zero. In addition, we will require that $d\tilde{d} = 1$ and $d(\pm 1) = 1$. We also assume that $\tau_1, \ldots, \tau_R \in \mathbb{T}_+$ are distinct, where

$$\mathbb{T}_+ := \{\, t \in \mathbb{T} : \mathrm{Im}\,(t) > 0 \,\} ,$$

and that

$$\alpha^{\pm},\ \ \beta^{\pm},\ \ \alpha_r^{\pm},\ \ \beta_r$$

are complex parameters satisfying the conditions (16) and (17) stated below, whereas

$$\alpha_r := \frac{\alpha_r^+ + \alpha_r^-}{2} \qquad \text{for } 1 \le r \le R. \tag{12}$$

The functions ψ and ϕ that will appear in the identity are

$$\psi = u_{1,\alpha^+}\, u_{-1,\alpha^-} \prod_{r=1}^{R} u_{\tau_r,\alpha_r^+}\, u_{\bar{\tau}_r,\alpha_r^-} , \tag{13}$$

$$\phi = u_{1,\gamma^+}\, u_{-1,\gamma^-} \prod_{r=1}^{R} u_{\tau_r,\gamma_r}\, u_{\bar{\tau}_r,\gamma_r} , \tag{14}$$

where

$$\gamma^{\pm} := \alpha^{\pm} + \beta^{\pm} , \qquad \gamma_r := \alpha_r + \beta_r . \tag{15}$$

The restrictions which we are going to impose on the parameters are the following:

$$|\mathrm{Re}\, \alpha^{\pm}| < 1/2, \qquad |\mathrm{Re}\, \alpha_r^{\pm}| < 1/2, \tag{16}$$

which guarantee the invertibility of $T(\psi)$ and $T(\psi^{-1})$, and

$$-3/2 < \mathrm{Re}\, \gamma^+ < 1/2, \qquad -1/2 < \mathrm{Re}\, \gamma^- < 3/2, \qquad |\mathrm{Re}\, \gamma_r| < 1/2. \tag{17}$$

The last conditions are needed later on.

Theorem 3.2. *Let a, b, c, d, ϕ, ψ be as above with* (16) *being assumed. Then*

$$\det P_n M(a,b) P_n = \det\left(P_n T^{-1}(\psi) M(c, c\, d\, \phi) T^{-1}(\psi^{-1}) P_n\right).$$

Proof. We first notice that $a, b \in L^1(\mathbb{T})$. Hence the $P_n M(a,b) P_n$ is a well-defined matrix, although $M(a,b)$ may be an unbounded operator. The proof of the identity is based on the Wiener–Hopf factorization of the underlying generating functions. In order to avoid unbounded factors, let us assume for the time being that all the parameters $\alpha^{\pm}, \beta^{\pm}, \alpha_r^{\pm}, \beta_r$ are purely imaginary. The general case follows by observing that both sides of the identity are analytic in each of these parameters.

In order to obtain the factorization introduce the functions

$$\eta_{\tau,\gamma}(t) = (1 - t/\tau)^{\gamma}, \qquad \xi_{\tau,\delta}(t) = (1 - \tau/t)^{\delta},$$

where the branches of η (analytic inside the unit circle) and ξ (analytic outside the unit circle) are chosen so that $\eta_{\tau,\gamma}(0) = \xi_{\tau,\delta}(\infty) = 1$. Using the above definitions we can produce the well-known Wiener–Hopf factorizations for

$$u_{\tau,\beta} = \xi_{\tau,-\beta}\, \eta_{\tau,\beta}\,, \qquad v_{\tau,\alpha} = \xi_{\tau,\alpha}\, \eta_{\tau,\alpha}\,.$$

Now put

$$\psi_- = \xi_{1,-\alpha^+}\, \xi_{-1,-\alpha^-} \prod_{r=1}^{R} \xi_{\tau_r,-\alpha_r^+}\, \xi_{\bar{\tau}_r,-\alpha_r^-}\,,$$

$$\psi_+ = \eta_{1,\alpha^+}\, \eta_{-1,\alpha^-} \prod_{r=1}^{R} \eta_{\tau_r,\alpha_r^+}\, \eta_{\bar{\tau}_r,\alpha_r^-}\,.$$

Then, indeed, $\psi = \psi_- \psi_+$. Furthermore,

$$\psi_-^{-1} \psi_+ = v_{1,\alpha^+}\, v_{-1,\alpha^-} \prod_{r=1}^{R} v_{\tau_r,\alpha_r^+}\, v_{\bar{\tau}_r,\alpha_r^-}\,,$$

$$\psi_-^{-1} \tilde{\psi}_+ = \xi_{1,2\alpha^+}\, \xi_{-1,2\alpha^-} \prod_{r=1}^{R} \xi_{\tau_r,\alpha_r^+ + \alpha_r^-}\, \xi_{\bar{\tau}_r,\alpha_r^+ + \alpha_r^-}\,.$$

Here notice that $\tilde{\eta}_{\tau,\alpha} = \xi_{\bar{\tau},\alpha}$. The latter can be written as the product of

$$v_{1,\alpha^+}\, v_{-1,\alpha^-} \prod v_{\tau_r,\alpha_r}\, v_{\bar{\tau}_r,\alpha_r} \quad \text{and} \quad u_{1,-\alpha^+}\, u_{-1,-\alpha^-} \prod u_{\tau_r,-\alpha_r}\, u_{\bar{\tau}_r,-\alpha_r}$$

with $\alpha_r = (\alpha_r^+ + \alpha_r^-)/2$. Thus we see that

$$\psi = \psi_- \psi_+\,, \quad a = c\, \psi_-^{-1}\, \psi_+\,, \quad b = c\, d\, \phi\, \psi_-^{-1}\, \tilde{\psi}_+\,.$$

It follows that

$$\det P_n M(a,b) P_n = \det P_n M(\psi_-^{-1}\, c\, \psi_+, \psi_-^{-1} c\, d\, \phi\, \tilde{\psi}_+) P_n$$

which equals

$$\det P_n T(\psi_-^{-1}) M(c, c\, d\, \phi) T(\psi_+) P_n$$

by using (5). Also, notice that the determinants of $P_nT(\psi_\pm)P_n$ and $P_nT(\psi_\pm^{-1})P_n$ are one since they are either upper or lower triangular matrices with ones on the diagonal. Using this and the observation that

$$P_nT(\psi_+^{-1})P_n = P_nT(\psi_+^{-1}), \qquad P_nT(\psi_-)P_n = T(\psi_-)P_n,$$

the above equals

$$\det P_nT(\psi_+^{-1})T(\psi_-^{-1})M(c, c\,d\,\psi)T(\psi_+)T(\psi_-)P_n.$$

Applying the following formulas for the inverses,

$$T^{-1}(\psi) = T(\psi_+^{-1})T(\psi_-^{-1}), \qquad T^{-1}(\psi^{-1}) = T(\psi_+)T(\psi_+),$$

concludes the proof of the identity. □

It is interesting to consider certain special cases. What we have in mind is the case where the Fisher–Hartwig part of a (i.e., the product without the function c) is even. This happens if

$$\alpha_r^+ = \alpha_r^- = \alpha_r\,.$$

If in addition, we put $\beta_r = 0$, then

$$b = u_{1,\beta^+}\, u_{-1,\beta^-}\, d\, a\,.$$

There are four specific choices of parameters $\beta^\pm$ where the factor $\phi_0 := u_{1,\beta^+}\, u_{\beta^-}$ is actually continuous:

(1) $\beta^+ = \beta^- = 0$, $\phi_0(t) = 1$;
(2) $\beta^+ = -1$, $\beta^- = 1$, $\phi_0(t) = -1$;
(3) $\beta^+ = 0$, $\beta^- = 1$, $\phi_0(t) = t$;
(4) $\beta^+ = -1$, $\beta^- = 0$, $\phi_0(t) = -1/t$.

Notice that the conditions (16) and (17) on the parameters $\alpha^\pm$ and α_r amount to the following

$$|\mathrm{Re}\,\alpha^\pm| < 1/2, \qquad |\mathrm{Re}\,\alpha_r| < 1/2. \tag{18}$$

To summarize, in these special cases we have

$$a = c\, v_{1,\alpha^+}\, v_{-1,\alpha^-} \prod_{r=1}^{R} v_{\tau_r,\alpha_r}\, v_{\bar\tau_r,\alpha_r}\,, \qquad b = \phi_0\, d\, a. \tag{19}$$

Notice that the cases (1)–(4) correspond to the cases (i)–(iv) considered in the introduction, but are slightly more general due to the factor d. The reason why we single out these four special cases, is because for the computations that are made later in this paper, it is in these cases that we can actually determine the asymptotics, whereas in the more general case we can only reduce the asymptotics to a simplified determinant problem for which an answer is unknown.

4. Additional operator theoretic results

We need some results about Toeplitz operators and Hankel operators (see [5] and [11] for the general theory). First of all, in addition to the projections P_n, and $Q_n = I - P_n$ we define

$$W_n(f_0, f_1, \dots) = (f_{n-1}, f_{n-2}, \dots, f_1, f_0, 0, 0, \dots),$$
$$V_n(f_0, f_1, \dots) = (0, 0, \dots, 0, 0, f_0, f_1, f_2, \dots),$$
$$V_{-n}(f_0, f_1, \dots) = (f_n, f_{n+1}, f_{n+2}, \dots).$$

Here V_n is the forward shift by n positions. It is easily seen that $W_n^2 = P_n$, $W_n = W_n P_n = P_n W_n$, $V_n V_{-n} = Q_n$ and $V_{-n} V_n = I$. Note also that

$$P_n T(a) V_n = W_n H(\tilde{a}), \qquad V_{-n} T(a) P_n = H(a) W_n. \tag{20}$$

Moreover, we have

$$V_{-n} T(a) V_n = T(a), \quad V_{-n} H(a) = H(a) V_n, \quad W_n T(a) W_n = P_n T(\tilde{a}) P_n. \tag{21}$$

In the proofs that follow we need the notions of stability and strong convergence and we describe those now.

Let A_n be a sequence of operators. This sequence is said to be stable if there exists an n_0 such that the operators A_n are invertible for each $n \geq n_0$ and $\sup_{n \geq n_0} \|A_n^{-1}\| < \infty$. Moreover, we say that A_n converges strongly on ℓ^2 to an operator A as $n \to \infty$ if $A_n x \to Ax$ in the norm of ℓ^2 for each $x \in \ell^2$. When dealing with finite matrices A_n, we identify the matrices and their inverses with operators acting on the image of P_n. It is well known (see [11, Th. 4.15] and worthy to note that stability is related to strong convergence of the inverses (and their adjoints) in the following sense.

Lemma 4.1. *Suppose that A_n is a stable sequence such that $A_n \to A$ and $A_n^* \to A^*$ strongly. Then A is invertible, and $A_n^{-1} \to A^{-1}$ and $(A_n^{-1})^* \to (A^{-1})^*$ strongly.*

Recall that for trace class operators, the trace "trace A" and the operator determinant "$\det(I + A)$" are well defined and continuous with respect to A in the trace class norm. The following well-known result shows the connection with strong convergence.

Lemma 4.2. *Let B be a trace class operator and suppose that A_n and C_n are sequences such that $A_n \to A$ and $C_n^* \to C^*$ strongly. Then $A_n B C_n \to ABC$ in the trace class norm.*

In what follows we will write $A = \text{s-lim} A_n$ if both $A_n \to A$ strongly and $A_n^* \to A^*$ strongly.

We can use the first lemma to obtain information about the strong convergence of the inverses of Toeplitz matrices.

Proposition 4.3. *Let $\phi \in L^\infty(\mathbb{T})$. If $T_n(\phi)$ is stable, then $T(\phi)$ is invertible and*

$$T^{-1}(\phi) = \text{s-lim}\, T_n^{-1}(\phi), \qquad T^{-1}(\tilde{\phi}) = \text{s-lim}\, W_n T_n^{-1}(\phi) W_n.$$

Proof. Since $P_n^* = P_n \to I$ strongly, it follows that $T_n(\phi) \to T(\phi)$ strongly and the same holds for the adjoints. Lemma 4.1 implies the first statement. For the second one, observe that $W_n T_n(\phi) W_n = T_n(\tilde{\phi})$ and proceed similarly. Also note that $T(\tilde{\phi})$ is the transpose of $T(\phi)$, thus also invertible. □

We will need a new definition and additional results about strong convergence in what follows (see also [11, Thm. 7.13]). Let $\mathcal{A}$ equal the set of all bounded operators A defined on ℓ^2 such that the operators

$$\begin{pmatrix} W_n \\ V_{-n} \end{pmatrix} A \begin{pmatrix} W_n & V_n \end{pmatrix} = \begin{pmatrix} W_n A W_n & W_n A V_n \\ V_{-n} A W_n & V_{-n} A V_n \end{pmatrix}$$

along with their adjoints (in which A is replaced with A^*) converge strongly to operators defined on $\ell^2 \oplus \ell^2$. In other words

$$\pi(A) := \text{s-lim}\left(\begin{pmatrix} W_n \\ V_{-n} \end{pmatrix} A \begin{pmatrix} W_n & V_n \end{pmatrix}\right)$$

exists.

Lemma 4.4. *The set $\mathcal{A}$ is a (closed in the operator topology) C^*-subalgebra of $\mathcal{L}(\ell^2)$, and the map $\pi : \mathcal{A} \to \mathcal{L}(\ell^2 \oplus \ell^2)$ is a *-homomorphism.*

Proof. It is easy to see that the sum, the product and the involution are closed operations in $\mathcal{A}$ and at the same time that π is a *-homomorphism. Using that the norms of W_n and $V_{\pm n}$ are one, one can conclude the map π is bounded. The fact that $\mathcal{A}$ is closed can be shown straightforwardly using a Cauchy sequence argument (see also [11, Thm. 7.13]). □

We now relate Toeplitz and Hankel operators to $\mathcal{A}$ and π.

Lemma 4.5. *For ϕ in $L^\infty(\mathbb{T})$ the operators $T(\phi)$ and $H(\phi)$ belong to $\mathcal{A}$. Moreover,*

$$\pi(T(\phi)) = \begin{pmatrix} T(\tilde{\phi}) & H(\tilde{\phi}) \\ H(\phi) & T(\phi) \end{pmatrix}, \qquad \pi(H(\phi)) = \begin{pmatrix} 0 & 0 \\ 0 & 0 \end{pmatrix}.$$

Proof. We consider first the Toeplitz operator. We use the identities

$$W_n T(\phi) W_n = P_n T(\tilde{\phi}) P_n, \qquad W_n T(\phi) V_n = P_n H(\tilde{\phi}),$$
$$V_{-n} T(\phi) W_n = H(\phi) P_n, \qquad V_{-n} T(\phi) V_n = T(\phi),$$

stated at the beginning of the section, to show the strong convergence. For the Hankel operator we consider

$$W_n H(\phi) W_n = W_n V_{-n} T(\phi) P_n, \qquad W_n H(\phi) V_n = W_n V_{-n} H(\phi),$$
$$V_{-n} H(\phi) W_n = V_{-2n} T(\phi) P_n, \qquad V_{-n} H(\phi) V_n = V_{-2n} H(\phi),$$

and the strong convergence follows because $V_{-n} \to 0$ strongly. For the adjoints the argumentation is analogous. □

If we abbreviate

$$L(\phi) := \begin{pmatrix} T(\tilde{\phi}) & H(\tilde{\phi}) \\ H(\phi) & T(\phi) \end{pmatrix},$$

then (3) and (4) imply that

$$L(\phi_1\phi_2) = L(\phi_1)L(\phi_2). \tag{22}$$

This is not surprising since by an appropriate identification of $\ell^2 \oplus \ell^2$ with $\ell^2(\mathbb{Z})$ it is easily seen that $L(\phi)$ is the Laurent operator with symbol ϕ.

The following result is what will be needed in the next section. Notice that in the case of $c = 1$ we have that $\pi(A)$ is the identity operator on $\ell^2 \oplus \ell^2$.

Proposition 4.6. *Let $A = T^{-1}(\psi)(T(c) + H(\phi))T^{-1}(\psi^{-1})$ where $c, \phi, \psi \in L^\infty(\mathbb{T})$ are such that $T(\psi^{\pm 1})$ are invertible. Then $A \in \mathcal{A}$ and*

$$\pi(A) = \begin{pmatrix} T(\tilde{c}) & H(\tilde{c}) \\ H(c) & T(c) \end{pmatrix}.$$

Proof. Since $\mathcal{A}$ is a C*-algebra (hence inverse closed) and π is a *-homomorphism, it follows that $T^{-1}(\psi) \in \mathcal{A}$ and

$$\pi(T^{-1}(\psi)) = (\pi(T(\psi)))^{-1} = L(\psi)^{-1} = L(\psi^{-1}).$$

Here we use (22) for the inversion. Similarly, we derive

$$\pi(T^{-1}(\psi^{-1})) = L(\psi) \quad \text{and} \quad \pi(T(c) + H(\phi)) = L(c).$$

Using again (22) we obtain $\pi(A) = L(\psi^{-1})L(c)L(\psi) = L(c)$, which is the formula for $\pi(A)$. □

5. Separation theorems

We now establish a separation theorem, which we formulate in a quite general setting.

Theorem 5.1. *Let $\psi, \phi \in L^\infty(\mathbb{T})$ with $\phi\tilde{\phi} = 1$ be such that $T(\psi)$ is invertible on ℓ^2 and such that the sequence*

$$A_n = P_n T^{-1}(\psi) M(1, \phi) T^{-1}(\psi^{-1}) P_n \tag{23}$$

is stable. Moreover, assume that $c \in B_1^1$ is nonvanishing and has winding number zero and that $d \in B_1^1$ has a Wiener–Hopf factorization $d = d_+ \tilde{d}_+^{-1}$ in B_1^1. Then

$$\lim_{n\to\infty} \frac{\det\Big(P_n T^{-1}(\psi) M(c, cd\phi) T^{-1}(\psi^{-1}) P_n\Big)}{G[c]^n \det\Big(P_n T^{-1}(\psi) M(1, \phi) T^{-1}(\psi^{-1}) P_n\Big)} = E,$$

where $G[c] = \exp([\log c]_0)$ and

$$E = \det\Big(T^{-1}(\tilde{c}\tilde{d}_+) T(\tilde{c}) T(\tilde{d}_+)\Big) \times \det\Big(T(cd_+) T(c^{-1} d_+^{-1})\Big) \tag{24}$$
$$\times \det\Big(T^{-1}(cd_+) T^{-1}(\psi) M(c, cd\phi) T^{-1}(\psi^{-1}) T(d_+) T(\psi^{-1}) M^{-1}(1, \phi) T(\psi)\Big).$$

Proof. We note that the conditions on c and d_+ imply the invertibility of $T(cd_+)$ and $T(\tilde{c}\tilde{d}_+)$ and the stability of $T_n(cd_+)$. Because $T(\psi)$ is invertible (and hence $\psi^{-1} \in L^\infty(\mathbb{T})$) and one can conclude that $T(\tilde{\psi}^{-1})$ is invertible, too. Indeed, the formula

$$T^{-1}(\tilde{\psi}^{-1}) = T(\tilde{\psi}) - H(\tilde{\psi})T^{-1}(\psi)H(\psi)$$

can be verified straightforwardly using (4) and (5). Note that $T(\tilde{\psi}^{-1})$ is the transpose of $T(\psi^{-1})$, which thus is also invertible. Furthermore the stability of A_n implies the invertibility of $T^{-1}(\psi)M(1,\phi)T^{-1}(\psi^{-1})$. Hence $M(1,\phi)$ is invertible. From the proof below it will follow that the operator determinants in (24) are well defined, by which we mean that the underlying operator is identity plus a trace class operator.

We start by looking at $M(c, cd\phi)$ modulo trace class operators. It equals

$$\begin{aligned} T(c) + H(cd_+\tilde{d}_+^{-1}\phi) &= (T(cd_+) + H(cd_+\phi))T(d_+^{-1}) \\ &= \left(T(cd_+) + T(cd_+)H(\phi) + H(cd_+)T(\tilde{\phi})\right) T(d_+^{-1}) \\ &= T(cd_+)M(1,\phi)T(d_+^{-1}) + \text{ trace class.} \end{aligned}$$

Hence modulo trace class, $T^{-1}(\psi)M(c, cd\phi)T^{-1}(\psi^{-1})$ equals

$$T^{-1}(\psi)T(cd_+)M(1,\phi)T(d_+^{-1})T^{-1}(\psi^{-1}).$$

Since the commutators $[T^{-1}(\psi), T(cd_+)]$ and $[T(d_+^{-1}), T^{-1}(\psi^{-1})]$ are trace class, it follows that

$$T^{-1}(\psi)M(c, cd\phi)T^{-1}(\psi^{-1}) = T(cd_+)T^{-1}(\psi)M(1,\phi)T^{-1}(\psi^{-1})T(d_+^{-1}) + K_1$$

with a certain trace class operator K_1.

Now multiply with P_n from the left- and the right-hand side and write

$$\begin{aligned} &P_nT(cd_+)AT(d_+^{-1})P_n \\ &\quad = P_nT(cd_+)P_nAP_nT(d_+^{-1})P_n + P_nT(cd_+)Q_nAQ_nT(d_+^{-1})P_n \\ &\qquad + P_nT(cd_+)Q_nAP_nT(d_+^{-1})P_n + P_nT(cd_+)P_nAQ_nT(d_+^{-1})P_n \end{aligned}$$

with $A := T^{-1}(\psi)M(1,\phi)T^{-1}(\psi^{-1})$. We are going to analyse the last three terms. First, using $P_n = W_n^2$, $Q_n = V_nV_{-n}$, we see that

$$P_nAQ_nT(d_+^{-1})P_n = W_n\big(W_nAV_n\big)H(d_+^{-1})W_n$$

tends to zero in trace norm because $W_nAV_n \to 0$ strongly (see Proposition 4.6). Secondly,

$$P_nT(cd_+)Q_nAP_n = W_nH(\tilde{c}\tilde{d}_+)\big(V_{-n}AW_n\big)W_n$$

tends also to zero in trace norm because $(V_{-n}AW_n)^* \to 0$ strongly (again by Proposition 4.6). This implies that the last two terms of the above expressions tend to zero in trace norm. Finally, write

$$P_nT(cd_+)Q_nAQ_nT(d_+^{-1})P_n = W_nH(\tilde{c}\tilde{d}_+)\big(V_{-n}AV_n\big)H(d_+^{-1})W_n.$$

Notice that $V_{-n}AV_n \to I$ converges strongly (by Proposition 4.6). Therefore, the latter is $W_n H(\tilde{c}\tilde{d}_+)H(d_+^{-1})W_n$ plus a sequence tending to zero in trace norm.

Summarizing the previous two paragraphs, we can conclude that

$$\begin{aligned} B_n &:= P_n T^{-1}(\psi)M(c, cd\phi)T^{-1}(\psi^{-1})P_n \\ &= T_n(cd_+)P_n A P_n T_n(d_+^{-1}) + P_n K_1 P_n + W_n L_1 W_n + D_n^{(1)} \end{aligned}$$

with K_1 and $L_1 = H(\tilde{c}\tilde{d}_+)H(d_+^{-1})$ being trace class operators and $D_n^{(1)}$ being a sequence tending to zero in trace norm.

Next multiply B_n with the inverses of

$$T_n(cd_+), \qquad P_n A P_n =: A_n, \quad \text{and} \quad T_n(d_+^{-1})$$

to arrive at

$$\begin{aligned} &T_n^{-1}(cd_+)B_n T_n(d_+^{-1})A_n^{-1} \\ &\quad = P_n + T_n^{-1}(cd_+)P_n K_1 P_n T_n(d_+^{-1})A_n^{-1} + T_n^{-1}(cd_+)W_n L_1 W_n T_n(d_+^{-1})A_n^{-1} + D_n^{(2)} \end{aligned}$$

with $D_n^{(2)} \to 0$ in trace norm due to stability. Using stability and the strong convergence of the above three sequences and their adjoints it follows from Lemmas 4.1 and 4.2 that

$$T_n^{-1}(cd_+)P_n K_1 P_n T_n(d_+^{-1})A_n^{-1} = P_n K P_n + D_n^{(3)}$$

and

$$T_n^{-1}(cd_+)W_n L_1 W_n T_n(d_+^{-1})A_n^{-1} = W_n L W_n + D_n^{(4)}$$

with $D_n^{(j)} \to 0$ in trace norm and K, L being trace class. In the latter we use that the strong limits of

$$W_n T_n(cd_+)W_n, \quad W_n A W_n = W_n A_n W_n, \quad \text{and} \quad W_n T_n(d_+^{-1})W_n$$

(and their adjoints) exist. Indeed, this can be seen by applying Proposition 4.6. Also, due to stability their inverses have a strong limit.

Thus

$$T_n^{-1}(cd_+)B_n T_n^{-1}(d_+^{-1})A_n^{-1} = P_n + P_n K P_n + W_n L W_n + D_n \tag{25}$$

with $D_n \to 0$ in trace norm. Write

$$P_n + P_n K P_n + W_n L W_n = (P_n + P_n K P_n)(P_n + W_n L W_n) - P_n K W_n L W_n$$

with the last term tending to zero in trace norm as $W_n \to 0$ weakly. Now take determinants and it follows that

$$\lim_{n\to\infty} \frac{\det B_n}{\det T_n(cd_+) \cdot \det A_n \cdot \det T_n(d_+^{-1})} = \det(I+K) \cdot \det(I+L).$$

From the standard Szegő Limit Theorem we get

$$\det T_n(cd_+) \sim G[cd_+]^n \cdot \det T(cd_+)T(c^{-1}d_+^{-1}), \qquad n \to \infty,$$

while $\det T_n(d_+^{-1}) = G[d_+^{-1}]^n$. Together we get the exponential factor $G[c] = G[cd_+] \cdot G[d_+^{-1}]$.

It remains to identify the trace class operators K and L. This is most conveniently done by passing to strong limits (and the strong limits after applying W_n from both sides) in (25). We obtain

$$T^{-1}(cd_+)BT^{-1}(d_+^{-1})A^{-1} = I + K$$

with $B = T^{-1}(\psi)M(c, cd\phi)T^{-1}(\psi^{-1})$ and $A = T^{-1}(\psi)M(1, \phi)T^{-1}(\psi^{-1})$, i.e.,

$$I + K = T^{-1}(cd_+)T^{-1}(\psi)M(c, cd\phi)T^{-1}(\psi^{-1})T(d_+)T(\psi^{-1})M(1, \phi)^{-1}T(\psi).$$

This gives one of the operator determinants in (24). As for the W_n-limits we obtain

$$T^{-1}(\tilde{c}\tilde{d}_+)T(\tilde{c})T^{-1}(\tilde{d}_+^{-1}) = I + L.$$

Here notice that $W_n B_n W_n \to T(\tilde{c})$ and $W_n A_n W_n \to I$ (again by Proposition 4.6). Thus,

$$\det(I + L) = \det T^{-1}(\tilde{c}\tilde{d}_+)T(\tilde{c})T(\tilde{d}_+),$$

which is the remaining term in (24) along with the constant term from the Szegő Limit Theorem above. □

In order to use the previous theorem we have to know the stability of the sequence A_n defined in (23). This is a non-trivial issue and is addressed in [9], where the following two theorems are proved. These results include certain "local" operators, which we are not going to define here, but instead refer to [9].

Theorem 5.2. *Let ϕ and ψ be of the form* (13) *and* (14). *Assume that conditions* (16) *are satisfied. Then the sequence*

$$A_n = P_n T^{-1}(\psi)M(1, \phi)T^{-1}(\psi^{-1})P_n$$

is stable if and only if the following conditions are satisfied:

(i) *the operator $M(1, \phi)$ is invertible on ℓ^2,*
(ii) $\operatorname{Re}\gamma^+ \notin 2\mathbb{Z} + 1/2$ *and* $\operatorname{Re}\gamma^- \notin 2\mathbb{Z} - 1/2$,
(iii) *for each $1 \le r \le R$, a certain the "local" operator $B(\alpha_r^+, \alpha_r^-, \gamma_r)$ is invertible.*

This theorem is proved in [9] by using general stability results of [16]. These general stability results imply that A_n is stable if and only if a certain collection of operators is invertible. Among these operators is the strong limit of A_n, i.e., the operator $A = T^{-1}(\psi)M(1, \phi)T^{-1}(\psi^{-1})$. Thus it is necessary for stability that $M(1, \phi)$ is invertible. In addition, there occur "local" operators (associated to each point where ψ or ϕ have jump discontinuities). Invertibility of the local operators at $t = \pm 1$ lead to conditions (ii). For the jumps at $t = \tau_r$ and $t = \bar{\tau}_r$ the local operators are Mellin convolution operators $B(\alpha_r^+, \alpha_r^-, \gamma_r)$ with 2×2 matrix-valued symbol defined in terms of the three parameters $\alpha_r^+, \alpha_r^-, \gamma_r \in \mathbb{C}$. As is well known, the invertibility of such operators is equivalent to a Wiener–Hopf factorization of the matrix symbol, which in general is not easily obtainable. Therefore, we have only the following results available [9]. On the positive side, part (c) covers the special cases we are particularly interested in.

Theorem 5.3. *Let* $\operatorname{Re}\alpha_r^{\pm} \notin \mathbb{Z}+1/2$.

(a) *If* $B(\alpha_r^+,\alpha_r^-,\gamma_r)$ *is invertible, then* $\operatorname{Re}\gamma_r \notin \mathbb{Z}+1/2$;
(b) *If* $\operatorname{Re}\gamma_r \notin \mathbb{Z}+1/2$, *then* $B(\alpha_r^+,\alpha_r^-,\gamma_r)$ *is Fredholm with index zero;*
(c) *If* $\alpha_r=\alpha_r^+=\alpha_r^-$ *and* $\operatorname{Re}\gamma_r \notin \mathbb{Z}+1/2$, *then* $B(\alpha_r,\alpha_r,\gamma_r)$ *is invertible.*

Let us return to the invertibility of the operator $M(1,\phi)$. For general Toeplitz + Hankel operators (with jump discontinuities) invertibility is a delicate issue. In [8] necessary and sufficient conditions for invertibility conditions are established for Toeplitz + Hankel operator $T(a)+H(b)$ with piecewise continuous a, b satisfying the additional condition $a\tilde{a}=b\tilde{b}$. Since $\phi\tilde{\phi}=1$ our operator $M(1,\phi)=I+H(\phi)$ falls into this class and we cite the corresponding result (Corollary 5.5 of [8]). For sake of simple presentation we only state it in the form that provides us a sufficient condition. Note that $\widetilde{u_{\tau,\gamma}}=u_{\bar{\tau},-\gamma}$, which ensures $\phi\tilde{\phi}=1$.

Theorem 5.4. *Let* ϕ *be of the form*

$$\phi = u_{1,\gamma^+}\, u_{-1,\gamma^-} \prod_{r=1}^{R} u_{\tau_r,\gamma_r}\, u_{\bar{\tau}_r,\gamma_r}$$

with distinct $\tau_1,\dots,\tau_R \in \mathbb{T}_+$, *and assume*

$$-3/2 < \operatorname{Re}\gamma^+ < 1/2, \quad -1/2 < \operatorname{Re}\gamma^- < 3/2, \quad -1/2 < \operatorname{Re}\gamma_r < 1/2.$$

Then $M(1,\phi)=I+H(\phi)$ *is invertible on* ℓ^2.

We will make further remarks on the invertibility of $M(1,\phi)$ in Section 9.

As a conclusion of the previous three results we can give a sufficient condition for stability. Notice that if (b) in Theorem 5.3 would imply invertibility (as is the case in (c)), we would not need the extra condition $\alpha_r^+=\alpha_r^-$.

Corollary 5.5. *Let* ϕ *and* ψ *be of the form* (13) *and* (14). *Assume that conditions* (16) *and* (17) *are satisfied and that in addition* $\alpha_r=\alpha_r^+=\alpha_r^-$. *Then* A_n *is stable.*

6. Determinant computations

In view of our separation theorem, we need to do two things. One is to evaluate the constant (24) and the other is to compute the asymptotics of the determinant of A_n. The goal of this section is to do the first, that is, evaluate the constant (24) which is given in terms of operator determinants. Some of the factors have been computed before, and the complicated one can be reduced to simpler ones, which also have been computed before.

We start with the following definitions and observations. For $A, B \in L^\infty(\mathbb{T})$ for which $H(A)H(\tilde{B})$ is trace class and for which $T(A)$ and $T(B)$ are invertible, let

$$E[A,B] = \det\Big(T^{-1}(A)T(AB)T^{-1}(B)\Big).$$

Note that $H(A)H(\tilde{B})$ is trace class if one of the functions is in B_1^1. If both functions are smooth (and have a continuous logarithm), then it has been shown that

$$E[A,B] = \exp\Big(\operatorname{trace} H(\log A)H(\log \tilde{B})\Big) = \exp\Big(\sum_{k\geq 1} k[\log A]_k[\log B]_{-k}\Big).$$

From this formula it follows that

$$E[A,B] = E[\tilde{B},\tilde{A}] = E[A_+,B_-],$$

where $A = A_-A_+$ and $B = B_-B_+$ are Wiener–Hopf factorizations of functions in B_1^1.

This constant is related to the constant

$$E[C] = \det T(C)T(C^{-1}) = \exp\Big(\sum_{k\geq 1} k[\log C]_k[\log C]_{-k}\Big).$$

appearing in the Szegő–Widom limit theorem. In fact, we have $E[C] = E[C,C] = E[C_+,C_-]$.

Finally, for a nonvanishing function $C \in B_1^1$ with winding number zero put

$$F[C] = \det(I + T^{-1}(C)H(C)).$$

It was computed in [7] that

$$\begin{aligned} F[C] &= \exp\Big(-\frac{1}{2}\operatorname{trace} H(\log C)^2 + \operatorname{trace} H(\log C)\Big) \\ &= \exp\Big(-\frac{1}{2}\sum_{k\geq 1} k[\log C]_k^2 + \sum_{k\geq 1}[\log C]_{2k-1}\Big). \end{aligned} \tag{26}$$

The previous determinant relates to a slightly more complicated determinant.

Lemma 6.1. *Let $C \in B_1^1$ be nonvanishing and have winding number zero. Assume that ϕ has a bounded Wiener–Hopf factorization $\phi = \phi_-\tilde{\phi}_-^{-1}$. Then*

$$\det(I + H(C)T^{-1}(C\phi)) = \det(I + H(C)T^{-1}(C)) = F[C].$$

Proof. We have

$$T(C\phi) = T(\phi_-)T(C)T(\tilde{\phi}_-^{-1})$$

with the factors being bounded invertible operators. Using

$$H(C)T(\tilde{\phi}_-) = H(C\phi_-) = T(\phi_-)H(C)$$

we obtain

$$H(C)T^{-1}(C\phi) = H(C)T(\tilde{\phi}_-)T^{-1}(C)T(\phi_-^{-1}) = T(\phi_-)H(C)T^{-1}(C)T(\phi_-^{-1}),$$

which gives the assertion. □

This next lemma illustrates some properties of the constant $E[A,B]$ which will be used later to simplify determinants.

Lemma 6.2. *Let $A, B, C \in L^\infty(\mathbb{T})$ be such that $T(A)$ is invertible and B and C admit bounded factorizations.*

(a) *If* $H(A)H(\tilde{B})$ *and* $H(A)H(\tilde{C})$ *are trace class, then*

$$E[A, BC] = E[A, B] \cdot E[A, C].$$

(b) *If* $H(B)H(\tilde{A})$ *and* $H(C)H(\tilde{A})$ *are trace class, then*

$$E[BC, A] = E[B, A] \cdot E[C, A].$$

Proof. Let $B = B_- B_+$ and $C = C_- C_+$ be the bounded factorizations. Using (3)–(5) in what follows, we remark that

$$H(A)H(\tilde{B}_-) = H(A)H(\tilde{B})T(B_+^{-1}) \quad \text{and} \quad H(A)H(\tilde{C}_-) = H(A)H(\tilde{C})T(C_+^{-1})$$

are trace class. Analogously, $H(A)H(\tilde{B}\tilde{C}) = H(A)H(\tilde{B}_-\tilde{C}_-)T(B_+^{-1}C_+^{-1})$. Observe that

$$\begin{aligned} H(A)H(\tilde{B}_-\tilde{C}_-) &= H(A)H(\tilde{B}_-)T(C_-) + H(A)T(\tilde{B}_-)H(\tilde{C}_-) \\ &= H(A)H(\tilde{B}_-)T(C_-) + T(B_-)H(A)H(\tilde{C}_-) \end{aligned}$$

where we used $T(B_-)H(A) = H(B_-A) = H(AB_-) = H(A)T(\tilde{B}_-)$. Therefore we conclude that $H(A)H(\tilde{B}\tilde{C})$ is trace class, too. The Toeplitz operators $T(B)$, $T(C)$, and $T(BC)$ are invertible due to the bounded factorizations. This implies that the three operator determinants are well defined. A straightforward computation using the factorizations yields that

$$E[A, B] = \det T^{-1}(A)T(AB)T^{-1}(B) = \det T^{-1}(A)T(AB_-)T(B_-^{-1}) = E[A, B_-]$$

and similar statements for the other two determinants. In fact, we can write

$$E[A, B] = \det T(B_-^{-1})T^{-1}(A)T(AB_-), \qquad E[A, C] = \det T^{-1}(A)T(AC_-)T(C_-^{-1})$$

and multiplication yields

$$\begin{aligned} &\det T(B_-^{-1})T^{-1}(A)T(B_-)T(A)T^{-1}(A)T(AC_-)T(C_-^{-1}) \\ &\quad = \det T(B_-^{-1})T^{-1}(A)T(B_-)T(AC_-)T(C_-^{-1}) \\ &\quad = \det T^{-1}(A)T(AB_-C_-)T(C_-^{-1}B_-^{-1}), \end{aligned}$$

which is $E[A, BC]$. This proves (a). The proof of (b) is analogous. □

The next theorem states that the operator determinant occurring in (24) is well defined under certain conditions (invertibility of $M(1, \phi)$), and that it can be expressed in terms of above constants in case of a slightly stronger condition (invertibility of $T(\phi)$). Afterwards, when we specialize to the functions ψ and ϕ with jump discontinuities we will see that the stronger condition is redundant for the evaluation of the constant. Notice that we have already shown in Theorem 5.1 that operator determinant is well defined, but under the (perhaps stronger) assumption of the stability of a certain sequence (which implies invertibility of $M(1, \phi)$).

Theorem 6.3. *Let* $\psi, \phi \in L^\infty(\mathbb{T})$ *with* $\phi\tilde{\phi} = 1$ *such that* $T(\psi)$ *is invertible on* ℓ^2. *Assume that* $c \in B_1^1$ *is nonvanishing and has winding number zero and that* $d \in B_1^1$ *has a factorization* $d = d_+\tilde{d}_+^{-1}$ *in* B_1^1.

(a) *If $M(1,\phi)$ is invertible on ℓ^2, then the following operator determinant is well defined:*

$$E_1 = \det\Big(T^{-1}(cd_+)T^{-1}(\psi)M(c,cd\phi)T^{-1}(\psi^{-1})T(d_+)T(\psi^{-1})M^{-1}(1,\phi)T(\psi)\Big).$$

(b) *If $T(\phi)$ is invertible on ℓ^2, then $M(1,\phi)$ is invertible, and*

$$E_1 = \frac{E[\psi,cd_+]}{E[cd_+,\psi]} \times E[d_+^{-1},\psi^{-1}] \times E[cd_+,\phi] \times \det\Big((T(cd_+\phi)+H(cd_+))T^{-1}(cd_+\phi)\Big).$$

(c) *If ϕ and ψ have a bounded factorization, then*

$$E_1 = \frac{E[\psi,c]}{E[c,\psi]} \times E[cd_+,\phi] \times F[cd_+].$$

Proof. (a): Abbreviate $e = cd_+$ and write

$$M(e,e\phi) = T(e)+H(e\phi) = T(e)+T(e)H(\phi)+H(e)T(\tilde{\phi}) = T(e)M(1,\phi)+K_1$$

with K_1 being trace class. This implies that

$$\det M(cd_+,cd_+\phi)M^{-1}(1,\phi)T^{-1}(cd_+) = \det M^{-1}(1,\phi)T^{-1}(cd_+)M(cd_+,cd_+\phi)$$

is well defined. Multiplying this determinant with the well-defined determinant

$$\begin{aligned} E[d_+^{-1},\psi^{-1}] &= \det T^{-1}(d_+^{-1})T(\psi^{-1}d_+^{-1})T^{-1}(\psi^{-1}) \\ &= \det T(d_+^{-1})T^{-1}(\psi^{-1})T(d_+)T(\psi^{-1}) \end{aligned}$$

and observing $M(cd_+,cd_+\phi)T(d_+^{-1}) = M(c,cd\phi)$ yields the well-defined determinant

$$\det M^{-1}(1,\phi)T^{-1}(cd_+)M(c,cd\phi)T^{-1}(\psi^{-1})T(d_+)T(\psi^{-1}),$$

which we can also write as

$$\det M(c,cd\phi)T^{-1}(\psi^{-1})T(d_+)T(\psi^{-1})M^{-1}(1,\phi)T^{-1}(cd_+).$$

Next observe that

$$\begin{aligned} \frac{E[\psi,cd_+]}{E[cd_+,\psi]} &= \det T(cd_+)T(\psi)T^{-1}(\psi cd_+)\cdot \det T(\psi cd_+)T^{-1}(cd_+)T^{-1}(\psi). \\ &= \det T(cd_+)T(\psi)T^{-1}(cd_+)T^{-1}(\psi). \end{aligned}$$

We multiply this well-defined determinant to the above one to obtain

$$\det M(c,cd\phi)T^{-1}(\psi^{-1})T(d_+)T(\psi^{-1})M^{-1}(1,\phi)T(\psi)T^{-1}(cd_+)T^{-1}(\psi),$$

which is E_1. Summarizing, besides the issue of E_1 being well defined we have shown that

$$E_1 = \frac{E[\psi,cd_+]}{E[cd_+,\psi]} \times E[d_+^{-1},\psi^{-1}] \times \det M(cd_+,cd_+\phi)M^{-1}(1,\phi)T^{-1}(cd_+).$$

(b): Now we show that $M(1,\phi)$ is invertible if $T(\phi)$ is, and express the inverse. Then we will compute the remaining determinant. The identity

$$(I+H(\phi))\,(I-H(\phi)) = I - H(\phi)H(\tilde{\phi}^{-1}) = T(\phi)\,T(\phi^{-1})$$

implies that $M(1,\phi) = I + H(\phi)$ is invertible. Moreover,

$$M^{-1}(1,\phi) = (I+H(\phi))^{-1} = (I-H(\phi))\,T^{-1}(\phi^{-1})\,T^{-1}(\phi).$$

Next observe that

$$\begin{aligned}
&M(cd_+, cd_+\phi)(I-H(\phi))\\
&\quad = T(cd_+) + H(cd_+\phi) - T(cd_+)H(\phi) - H(cd_+\phi)H(\phi)\\
&\quad = T(cd_+) + H(cd_+\phi) - H(cd_+\phi) + H(cd_+)T(\tilde{\phi}) - T(cd_+\phi\tilde{\phi}) + T(cd_+\phi)T(\tilde{\phi})\\
&\quad = H(cd_+)T(\tilde{\phi}) + T(cd_+\phi)T(\tilde{\phi})\\
&\quad = (T(cd_+\phi) + H(cd_+))\,T(\phi^{-1}).
\end{aligned}$$

Therefore,

$$\begin{aligned}
M(cd_+, cd_+\phi)M^{-1}(1,\phi) &= M(cd_+, cd_+\phi)(I-H(\phi))T^{-1}(\phi^{-1})\,T^{-1}(\phi)\\
&= (T(cd_+\phi) + H(cd_+))T^{-1}(\phi),
\end{aligned}$$

and thus

$$\begin{aligned}
&\det M(cd_+, cd_+\phi)M^{-1}(1,\phi)T^{-1}(cd_+)\\
&\qquad = \det(T(cd_+\phi) + H(cd_+))T^{-1}(\phi)T^{-1}(cd_+).
\end{aligned}$$

We split this into

$$\det\Big((T(cd_+\phi) + H(cd_+))T^{-1}(cd_+\phi)\Big) \times \det\Big(T(cd_+\phi)T^{-1}(\phi)T^{-1}(cd_+)\Big)$$

with the last determinant equal to $E[cd_+, \phi]$.

For part (c), we apply Lemma 6.1 and Lemma 6.2. □

Now we specialize to the functions we are interested in.

Corollary 6.4. *Let ψ and ϕ be given by* (13) *and* (14) *and assume that the condition* (16) *and* (17) *hold. Moreover, let $c = c_- G[c] c_+$ and $d = \tilde{d}_+^{-1} d_+$ be factorizations in B_1^1 (see* (6)–(8)*). Then the operator determinant E_1 of Theorem* 6.3 *is well defined and given by*

$$\begin{aligned}
E_1 &= \hat{E}_1 \cdot F[e_+],\\
\hat{E}_1 &= c_0(1)^{-\alpha^+} c_0(-1)^{-\alpha^-} \prod_{r=1}^{R} c_0(\tau_r)^{-\alpha_r^+} c_0(\bar{\tau}_r)^{-\alpha_r^-}\\
&\quad \times e_+(1)^{\alpha^+ + \beta^+} e_+(-1)^{\alpha^- + \beta^-} \prod_{r=1}^{R} e_+(\tau_r)^{\alpha_r^+ + \beta_r} e_+(\bar{\tau}_r)^{\alpha_r^- + \beta_r}
\end{aligned}$$

with $e_+ = c_+ d_+$ and $c_0 = c_+ c_- = c/G[c]$.

Proof. The assumptions imply that $T(\psi)$ and $M(1,\phi)$ are invertible. Hence by the previous theorem, part (a), the operator determinant E_1 is well defined. It also depends analytically on the parameters $\alpha^\pm, \beta^\pm, \alpha_r^\pm, \beta_r$. Therefore it is sufficient to prove the identity under the assumption that the real parts of all parameters vanish. This implies that $T(\phi)$ is invertible and hence part (c) of the previous theorem can be applied. Moreover, $F[cd_+] = F[c_+d_+]$ as can be easily seen from (26).

To carry out the computation of the various $E[\cdot,\cdot]$ terms, we observe that for functions C admitting a factorization in B_1^1 (see (6)) the following general formulas were established in [11, Sect. 10.62]),

$$E[u_{\tau,\beta}, C] = C_-(\tau)^{-\beta}, \qquad E[C, u_{\tau,\beta}] = C_+(\tau)^{\beta}.$$

Recalling the definition of ψ and ϕ,

$$\psi = u_{1,\alpha^+}\, u_{-1,\alpha^-} \prod_{r=1}^{R} u_{\tau_r,\alpha_r^+}\, u_{\bar\tau_r,\alpha_r^-}\,,$$

$$\phi = u_{1,\alpha^++\beta^+}\, u_{-1,\alpha^-+\beta^-} \prod_{r=1}^{R} u_{\tau_r,\alpha_r+\beta_r}\, u_{\bar\tau_r,\alpha_r+\beta_r}\,,$$

we get

$$E[\psi, c_-] = c_-(1)^{-\alpha^+} c_-(-1)^{-\alpha^-} \prod_{r=1}^{R} c_-(\tau_r)^{-\alpha_r^+} c_-(\bar\tau)^{-\alpha_r^-},$$

$$E[c_+, \psi]^{-1} = c_+(1)^{-\alpha^+} c_+(-1)^{-\alpha^-} \prod_{r=1}^{R} c_+(\tau_r)^{-\alpha_r^+} c_+(\bar\tau)^{-\alpha_r^-},$$

$$E[c_+d_+, \phi] = e_+(1)^{\alpha^++\beta^+} e_+(-1)^{\alpha^-+\beta^-} \prod_{r=1}^{R} e_+(\tau_r)^{\alpha_r^++\beta_r} e_+(\bar\tau)^{\alpha_r^-+\beta_r},$$

from which the formula follows. Here we use Lemma 6.2. □

We remark that in the special case of $\alpha_r = \alpha_r^\pm$ and $\beta_r = 0$, which will be considered later, the constant

$$\hat{E}_1 = c_+(1)^{\beta^+} c_+(-1)^{\beta^-} c_-(1)^{-\alpha^+} c_-(-1)^{-\alpha^-} \prod_{r=1}^{R} c_-(\tau_r)^{-\alpha_r} c_-(\bar\tau_r)^{-\alpha_r}$$
$$\times\, d_+(1)^{\alpha^++\beta^+} d_+(-1)^{\alpha^-+\beta^-} \prod_{r=1}^{R} d_+(\tau_r)^{\alpha_r} d_+(\bar\tau_r)^{\alpha_r}\,. \qquad (27)$$

Let us now turn to the constant E appearing in Theorem 5.1.

Corollary 6.5. *Let* ψ *and* ϕ *be given by* (13) *and* (14) *and assume that the condition* (16) *and* (17) *hold. Moreover, let* $c = c_-G[c]c_+$ *and* $d = \tilde{d}_+^{-1}d_+$ *be factorizations in* B_1^1 *(see* (6)–(8)*). Then the constant* E *in* (24) *is well defined and given by*

$$E = \hat{E}_1 \times \left(\frac{c_+(1)d_+(1)}{c_+(-1)d_+(-1)}\right)^{1/2} \times \exp\Big(\sum_{k\geq 1} k[\log c]_k[\log c]_{-k} - \frac{1}{2}\sum_{k\geq 1} k([\log c]_k + [\log d]_k)^2\Big),$$

where $\hat{E}_1$ *is the expression given in the previous corollary.*

Proof. We have to identify the additional constants,

$$\det\Big(T^{-1}(\tilde{c}\tilde{d}_+)T(\tilde{c})T(\tilde{d}_+)\Big) \times \det\Big(T(cd_+)T(c^{-1}d_+^{-1})\Big)$$

which are

$$E^{-1}[\tilde{c}, \tilde{d}_+] \cdot E[cd_+] = \frac{E[cd_+, cd_+]}{E[d_+, c]} = E[c_+, c_-].$$

This we combine with $F[cd_+] = F[e_+]$ and the previous corollary. □

Putting all this together, we have the following.

Corollary 6.6. *Let* ψ *and* ϕ *be given by* (13) *and* (14) *and assume that the condition* (16) *and* (17) *hold. Moreover, let* $c = c_-G[c]c_+$ *and* $d = \tilde{d}_+^{-1}d_+$ *be factorizations in* B_1^1 *(see* (6)–(8)*). Finally, suppose that* (iii) *of Theorem* 5.2 *holds. Then*

$$\lim_{n\to\infty} \frac{\det\Big(P_nT^{-1}(\psi)M(c, cd\phi)T^{-1}(\psi^{-1})P_n\Big)}{G[c]^n \times \det\Big(P_nT^{-1}(\psi)M(1, \phi)T^{-1}(\psi^{-1})P_n\Big)} = E,$$

where E *is as in the previous corollary.*

7. Known asymptotics

In the previous separation theorem and the constant computation we have reduced the asymptotics of

$$\det\Big(P_nT^{-1}(\psi)M(c, cd\phi)T^{-1}(\psi^{-1})P_n\Big)$$

to the asymptotics of

$$\det\Big(P_nT^{-1}(\psi)M(1, \phi)T^{-1}(\psi^{-1})P_n\Big).$$

The separation theorem is of course only as useful as far as we are able to obtain this last asymptotics. We can reverse the considerations of Section 3 and Theorem 3.2 and obtain

$$\det\Big(P_nT^{-1}(\psi)M(1, \phi)T^{-1}(\psi^{-1})P_n\Big) = \det M_n(a_0, b_0)$$

where

$$a_0 = v_{1,\alpha^+}\, v_{-1,\alpha^-} \prod_{r=1}^{R} v_{\tau_r,\alpha_r^+}\, v_{\bar\tau_r,\alpha_r^-}\,, \tag{28}$$

$$b_0 = v_{1,\alpha^+}\, u_{1,\beta^+}\, v_{-1,\alpha^-}\, u_{-1,\beta^-} \prod_{r=1}^{R} v_{\tau_r,\alpha_r} u_{\tau_r,\beta_r}\, v_{\bar\tau_r,\alpha_r} u_{\bar\tau_r,\beta_r}\,. \tag{29}$$

These are the original functions a and b without the c and d terms.

The asymptotics of $\det M_n(a_0, b_0)$ are known in the cases of $\beta_r = 0$ and $\alpha_r = \alpha_r^+ = \alpha_r^-$ and $\beta^+ \in \{0,-1\}$, $\beta^- \in \{0,1\}$. We remark that in these cases a_0 is an even function which is a product of pure Fisher–Hartwig type functions with zeros/poles only (no jumps). Furthermore, depending on the values of $\beta^\pm$, we have four cases,

$$b_0(t) = \pm a_0(t), \qquad b_0(t) = t a_0(t), \qquad b_0(t) = -t^{-1} a_0(t),$$

which are precisely the cases (i)–(iv) described in the introduction. We list them here and note that $G(1+z)$ is the Barnes G-function, an entire function satisfying $G(1+z) = \Gamma(z)G(z)$ (see [3]).

The asymptotics of $\det M_n(a_0, b_0)$ are given by the following expressions.

(1) $b_0(t) = a_0(t)$, $\beta_r = 0$ and $\alpha_r = \alpha_r^+ = \alpha_r^-$, and $\beta^+ = 0$, $\beta^- = 0$.

$$\begin{aligned} &n^{\{\frac12((\alpha^+)^2+(\alpha^-)^2-\alpha^++\alpha^-)+\sum\alpha_r^2\}}\, 2^{-\frac12(\alpha^++\alpha^-)^2+\frac12(\alpha^++\alpha^-)+\sum\alpha_r^2} \\ &\times \prod_r |1-\tau_r^2|^{-\alpha_r^2} |1-\tau_r|^{-2\alpha_r(\alpha^+-1/2)} |1+\tau_r|^{-2\alpha_r(\alpha^-+1/2)} \\ &\times \prod_{j<k} |\tau_k-\tau_j|^{-2\alpha_k\alpha_j} |\tau_k - 1/\tau_j|^{-2\alpha_k\alpha_j} \\ &\times \frac{\pi^{\frac12(\alpha^++\alpha^-)} G(1/2)G(3/2)}{G(1/2+\alpha^+)G(3/2+\alpha^-)} \prod_r \frac{G(1+\alpha_r)^2}{G(1+2\alpha_r)}\,. \end{aligned}$$

(2) $b_0(t) = -a_0(t)$, $\beta_r = 0$ and $\alpha_r = \alpha_r^+ = \alpha_r^-$, and $\beta^+ = -1$, $\beta^- = 1$.

$$\begin{aligned} &n^{\{\frac12((\alpha^+)^2+(\alpha^-)^2+\alpha^+-\alpha^-)+\sum\alpha_r^2\}}\, 2^{-\frac12(\alpha^++\alpha^-)^2+\frac12(\alpha^++\alpha^-)+\sum\alpha_r^2} \\ &\times \prod_r |1-\tau_r^2|^{-\alpha_r^2} |1-\tau_r|^{-2\alpha_r(\alpha^++1/2)} |1+\tau_r|^{-2\alpha_r(\alpha^--1/2)} \\ &\times \prod_{j<k} |\tau_k-\tau_j|^{-2\alpha_k\alpha_j} |\tau_k - 1/\tau_j|^{-2\alpha_k\alpha_j} \\ &\times \frac{\pi^{\frac12(\alpha^++\alpha^-)} G(3/2)G(1/2)}{G(3/2+\alpha^+)G(1/2+\alpha^-)} \prod_r \frac{G(1+\alpha_r)^2}{G(1+2\alpha_r)}\,. \end{aligned}$$

(3) $b_0(t) = ta_0(t)$, $\beta_r = 0$ and $\alpha_r = \alpha_r^+ = \alpha_r^-$, and $\beta^+ = 0$, $\beta^- = 1$.

$$n^{\{\frac{1}{2}((\alpha^+)^2+(\alpha^-)^2-\alpha^+-\alpha^-)+\sum\alpha_r^2\}}\, 2^{\{2-\frac{1}{2}(\alpha^++\alpha^--1)^2+\frac{1}{2}(\alpha^++\alpha^--1)+\sum\alpha_r^2\}}$$
$$\times \prod_r |1-\tau_r^2|^{-\alpha_r^2}|1-\tau_r|^{-2\alpha_r(\alpha^+-1/2)}|1+\tau_r|^{-2\alpha_r(\alpha^--1/2)}$$
$$\times \prod_{j<k} |\tau_k-\tau_j|^{-2\alpha_k\alpha_j}\,|\tau_k-1/\tau_j|^{-2\alpha_k\alpha_j}$$
$$\times \frac{\pi^{\frac{1}{2}(\alpha^++\alpha^-)}G(1/2)^2}{G(1/2+\alpha^+)G(1/2+\alpha^-)}\prod_r \frac{G(1+\alpha_r)^2}{G(1+2\alpha_r)}.$$

(4) $b_0(t) = -t^{-1}a_0(t)$, $\beta_r = 0$ and $\alpha_r = \alpha_r^+ = \alpha_r^-$, and $\beta^+ = -1$, $\beta^- = 0$.

$$n^{\{\frac{1}{2}((\alpha^+)^2+(\alpha^-)^2+\alpha^++\alpha^-)+\sum\alpha_r^2\}}\, 2^{\{-\frac{1}{2}(\alpha^++\alpha^-+1)^2+\frac{1}{2}(\alpha^++\alpha^-+1)+\sum\alpha_r^2\}}$$
$$\times \prod_r |1-\tau_r^2|^{-\alpha_r^2}|1-\tau_r|^{-2\alpha_r(\alpha^++1/2)}|1+\tau_r|^{-2\alpha_r(\alpha^-+1/2)}$$
$$\times \prod_{j<k} |\tau_k-\tau_j|^{-2\alpha_k\alpha_j}\,|\tau_k-1/\tau_j|^{-2\alpha_k\alpha_j}$$
$$\times \frac{\pi^{\frac{1}{2}(\alpha^++\alpha^-)}G(3/2)^2}{G(3/2+\alpha^+)G(3/2+\alpha^-)}\prod_r \frac{G(1+\alpha_r)^2}{G(1+2\alpha_r)}.$$

The above asymptotics have been proved in [12, Theorem 1.25]. In the special case of $\alpha^+ = \alpha^- = 0$ and case (1), see also [6]. On the other hand, if all $\alpha_r = 0$, then the determinants can be evaluated explicitly because they are related to Hankel determinants constructed from the moments of classical Jacobi orthogonal polynomials. In this later case the Hankel determinant asymptotics can also be described as Toeplitz determinants [6].

8. Final computations for the four important cases

Using the previous section and then our computations for the constant E we have the final four answers. We assume in what follows that we have the factorizations $c = c_-G[c]c_+$ and $d = d_+\tilde{d}_+^{-1}$ (see (6)–(8)) and that condition (18) holds. Then the asymptotics of $\det M_n(a,b)$ are given by the following products.

(1) $b(t) = a(t)$, $\beta_r = 0$ and $\alpha_r = \alpha_r^+ = \alpha_r^-$, and $\beta^+ = 0$, $\beta^- = 0$.

$$G[c]^n\, n^{\{\frac{1}{2}((\alpha^+)^2+(\alpha^-)^2-\alpha^++\alpha^-)+\sum\alpha_r^2\}}\, 2^{-\frac{1}{2}(\alpha^++\alpha^-)^2+\frac{1}{2}(\alpha^++\alpha^-)+\sum\alpha_r^2}$$
$$\times \exp\left(\sum_{k\geq 1}\left(k[\log c]_k[\log c]_{-k} - \frac{1}{2}k([\log c]_k+[\log d]_k)^2\right)\right)$$
$$\times\ c_+(1)^{1/2}c_+(-1)^{-1/2}c_-(1)^{-\alpha^+}c_-(-1)^{-\alpha^-}\prod_r c_-(\tau_r)^{-\alpha_r}c_-(\bar{\tau}_r)^{-\alpha_r}$$

$$\times\ d_+(1)^{\alpha^++1/2} d_+(-1)^{\alpha^--1/2} \prod_r d_+(\tau_r)^{\alpha_r} d_+(\bar{\tau}_r)^{\alpha_r}$$

$$\times \prod_r |1-\tau_r^2|^{-\alpha_r^2} |1-\tau_r|^{-2\alpha_r(\alpha^+-1/2)} |1+\tau_r|^{-2\alpha_r(\alpha^-+1/2)}$$

$$\times \prod_{j<k} |\tau_k-\tau_j|^{-2\alpha_k\alpha_j} |\tau_k - 1/\tau_j|^{-2\alpha_k\alpha_j}$$

$$\times \frac{\pi^{\frac{1}{2}(\alpha^++\alpha^-)} G(1/2)G(3/2)}{G(1/2+\alpha^+)G(3/2+\alpha^-)} \prod_r \frac{G(1+\alpha_r)^2}{G(1+2\alpha_r)}\,.$$

(2) $b(t) = -a(t)$, $\beta_r = 0$ and $\alpha_r = \alpha_r^+ = \alpha_r^-$, and $\beta^+ = -1$, $\beta^- = 1$.

$$G[c]^n\, n^{\{\frac{1}{2}((\alpha^+)^2+(\alpha^-)^2+\alpha^+-\alpha^-)+\sum\alpha_r^2\}}\, 2^{-\frac{1}{2}(\alpha^++\alpha^-)^2+\frac{1}{2}(\alpha^++\alpha^-)+\sum\alpha_r^2}$$

$$\times \exp\left(\sum_{k\geq 1}\left(k[\log c]_k[\log c]_{-k} - \frac{1}{2}k([\log c]_k + [\log d]_k)^2\right)\right)$$

$$\times\ c_+(1)^{-1/2} c_+(-1)^{1/2} c_-(1)^{-\alpha^+} c_-(-1)^{-\alpha^-} \prod_r c_-(\tau_r)^{-\alpha_r} c_-(\bar{\tau}_r)^{-\alpha_r}$$

$$\times\ d_+(1)^{\alpha^+-1/2} d_+(-1)^{\alpha^-+1/2} \prod_r d_+(\tau_r)^{\alpha_r} d_+(\bar{\tau}_r)^{\alpha_r}$$

$$\times \prod_r |1-\tau_r^2|^{-\alpha_r^2} |1-\tau_r|^{-2\alpha_r(\alpha^++1/2)} |1+\tau_r|^{-2\alpha_r(\alpha^--1/2)}$$

$$\times \prod_{j<k} |\tau_k-\tau_j|^{-2\alpha_k\alpha_j} |\tau_k - 1/\tau_j|^{-2\alpha_k\alpha_j}$$

$$\times \frac{\pi^{\frac{1}{2}(\alpha^++\alpha^-)} G(3/2)G(1/2)}{G(3/2+\alpha^+)G(1/2+\alpha^-)} \prod_r \frac{G(1+\alpha_r)^2}{G(1+2\alpha_r)}\,.$$

(3) $b(t) = ta(t)$, $\beta_r = 0$ and $\alpha_r = \alpha_r^+ = \alpha_r^-$, and $\beta^+ = 0$, $\beta^- = 1$.

$$G[c]^n\, n^{\{\frac{1}{2}((\alpha^+)^2+(\alpha^-)^2-\alpha^+-\alpha^-)+\sum\alpha_r^2\}}\, 2^{\{2-\frac{1}{2}(\alpha^++\alpha^--1)^2+\frac{1}{2}(\alpha^++\alpha^--1)+\sum\alpha_r^2\}}$$

$$\times \exp\left(\sum_{k\geq 1}\left(k[\log c]_k[\log c]_{-k} - \frac{1}{2}k([\log c]_k + [\log d]_k)^2\right)\right)$$

$$\times\, c_+(1)^{1/2} c_+(-1)^{1/2} c_-(1)^{-\alpha^+} c_-(-1)^{-\alpha^-} \prod_r c_-(\tau_r)^{-\alpha_r} c_-(\bar{\tau}_r)^{-\alpha_r}$$

$$\times\ d_+(1)^{\alpha^++1/2} d_+(-1)^{\alpha^-+1/2} \prod_r d_+(\tau_r)^{\alpha_r} d_+(\bar{\tau}_r)^{\alpha_r}$$

$$\times \prod_r |1-\tau_r^2|^{-\alpha_r^2} |1-\tau_r|^{-2\alpha_r(\alpha^+-1/2)} |1+\tau_r|^{-2\alpha_r(\alpha^--1/2)}$$

$$\times \prod_{j<k} |\tau_k-\tau_j|^{-2\alpha_k\alpha_j} |\tau_k - 1/\tau_j|^{-2\alpha_k\alpha_j}$$

$$\times \frac{\pi^{\frac{1}{2}(\alpha^++\alpha^-)} G(1/2)^2}{G(1/2+\alpha^+)G(1/2+\alpha^-)} \prod_r \frac{G(1+\alpha_r)^2}{G(1+2\alpha_r)}\,.$$

(4) $b(t) = -t^{-1}a(t)$, $\beta_r = 0$ and $\alpha_r = \alpha_r^+ = \alpha_r^-$, and $\beta^+ = -1$, $\beta^- = 0$.

$$G[c]^n\, n^{\{\frac{1}{2}((\alpha^+)^2+(\alpha^-)^2+\alpha^++\alpha^-)+\sum\alpha_r^2\}}\, 2^{\{-\frac{1}{2}(\alpha^++\alpha^-+1)^2+\frac{1}{2}(\alpha^++\alpha^-+1)+\sum\alpha_r^2\}}$$
$$\times \exp\left(\sum_{k\geq 1}\left(k[\log c]_k[\log c]_{-k} - \frac{1}{2}k([\log c]_k + [\log d]_k)^2\right)\right)$$
$$\times c_+(1)^{-1/2}c_+(-1)^{-1/2}c_-(1)^{-\alpha^+}c_-(-1)^{-\alpha^-}\prod_r c_-(\tau_r)^{-\alpha_r}c_-(\bar{\tau}_r)^{-\alpha_r}$$
$$\times d_+(1)^{\alpha^+-1/2}d_+(-1)^{\alpha^--1/2}\prod_r d_+(\tau_r)^{\alpha_r}d_+(\bar{\tau}_r)^{\alpha_r}$$
$$\times \prod_r |1-\tau_r^2|^{-\alpha_r^2}|1-\tau_r|^{-2\alpha_r(\alpha^++1/2)}|1+\tau_r|^{-2\alpha_r(\alpha^-+1/2)}$$
$$\times \prod_{j<k}|\tau_k-\tau_j|^{-2\alpha_k\alpha_j}|\tau_k - 1/\tau_j|^{-2\alpha_k\alpha_j}$$
$$\times \frac{\pi^{\frac{1}{2}(\alpha^++\alpha^-)}G(3/2)^2}{G(3/2+\alpha^+)G(3/2+\alpha^-)}\prod_r\frac{G(1+\alpha_r)^2}{G(1+2\alpha_r)}.$$

These formulas follow in a straightforward manner from simply using the known results quoted in the previous section, along with our separation theorem formulas for the constant and the evaluation of those for the specific values of β^+ and β^-. More specifically, use (27) and Corollary 6.5.

9. The inverse of $I + H(\psi)$

In the course of the computations in the previous sections, we discovered that when $\psi^{-1} = \tilde{\psi}$, explicit forms of the inverse of $I + H(\phi_0\psi)$ can be found in the four cases $\phi_0 = \pm 1$, $\phi_0(z) = z$, $\phi_0(z) = -z^{-1}$ considered previously. The inverse will be expressed in terms of the inverse of $T(\psi)$ and $T(\psi^{-1})$.

The first two cases are rather simple.

Proposition 9.1. *Suppose that $\psi^{-1} = \tilde{\psi}$ and that the operator $T(\psi)$ is invertible. Then $I \pm H(\psi)$ is invertible and*
$$(I \pm H(\psi))^{-1} = T(\psi^{-1})^{-1}(I \pm H(\psi^{-1}))\,T(\psi)^{-1}.$$

Proof. From $\psi^{-1} = \tilde{\psi}$ and thus $T(\psi)T(\psi^{-1}) = I - H(\psi)^2$, we know that
$$(T(\psi)T(\psi^{-1}))^{-1} = T(\psi^{-1})^{-1}T(\psi)^{-1} = (I - H(\psi)^2)^{-1}.$$
This means that
$$T(\psi^{-1})^{-1}(I \pm H(\psi^{-1}))\,T(\psi)^{-1} = (I - H(\psi)^2)^{-1}T(\psi)\,(I \pm H(\psi^{-1}))\,T(\psi)^{-1}.$$
Now from (4) we know that $T(\psi)H(\psi^{-1}) = -H(\psi)T(\psi)$. So the term
$$T(\psi)\,(I \pm H(\psi^{-1}))\,T(\psi)^{-1} = I \mp H(\psi)$$
and this yields the result. □

In order to continue with the other two cases, we need to make a connection with factorization. If $T(\psi)$ is invertible then there exists a Wiener–Hopf factorization of ψ such that $\psi = \psi_-\psi_+$ and $\psi_\pm$ and $\psi_\pm^{-1}$ are all in L^2. If in addition $\psi^{-1} = \tilde{\psi}$, then using the uniqueness of the Wiener–Hopf factorization (up to multiplicative constants) it is not difficult to see that the factors are related to each other either by

$$\psi_+ = \tilde{\psi}_-^{-1} \quad \text{or} \quad \psi_+ = -\tilde{\psi}_-^{-1}.$$

This means that we have either

$$\psi = \tilde{\psi}_+^{-1}\psi_+ \quad \text{or} \quad \psi = -\tilde{\psi}_+^{-1}\psi_+.$$

We can equivalently express this dichotomy as

$$P_1T^{-1}(\psi)P_1 = 1 \quad \text{or} \quad P_1T^{-1}(\psi)P_1 = -1$$

by using that $T^{-1}(\psi) = T(\psi_+^{-1})T(\psi_-^{-1})$. We need one basic result before we proceed.

Lemma 9.2. *Assume that $\psi^{-1} = \tilde{\psi}$. If the Toeplitz operator $T(\psi)$ is invertible and if $P_1T^{-1}(\psi)P_1 = 1$, then*

$$\Big(H(\tilde{\psi}) + T(\psi)^{-1}H(\psi) - H(\tilde{\psi}/z)T(\psi)^{-1}H(\psi)\Big)H(z) = 0.$$

Proof. If $T(\psi)$ is invertible then there exists a factorization of ψ such that $\psi = \psi_-\psi_+$ where $\psi_- = \tilde{\psi_+}^{-1}$ and $\psi_\pm$ and $\psi_\pm^{-1}$ are all in L^2. The operator $H(z)$ is the rank one projection P_1, so the above will be zero if the operator above-applied to the constant function $f(z) = 1$ is zero. The operator $T(\psi)^{-1}H(\psi)$ is the same as the operator $T(\psi_+^{-1})H(\psi_+)$ since it agrees with this operator on all trigonometric polynomials. Hence

$$T(\psi)^{-1}H(\psi)\,f = T(\psi_+^{-1})H(\psi_+)\,f = \psi_+^{-1}(\psi_+ - (\psi_+)_0)/z = (1 - \psi_+^{-1}(\psi_+)_0)/z.$$

We also have $H(\tilde{\psi})\,f = P(\tilde{\psi}/z)$. Finally

$$\begin{aligned} H(\tilde{\psi}/z)[(1 - \psi_+^{-1}(\psi_+)_0)/z] &= P[\tilde{\psi}/z\,(1 - \tilde{\psi}_+^{-1}(\psi_+)_0)] \\ &= P(\tilde{\psi}/z) - P[\psi_+^{-1}(\psi_+)_0/z]. \\ &= P(\tilde{\psi}/z) + (1 - \psi_+^{-1}(\psi_+)_0)/z. \end{aligned}$$

Putting the three terms together yields the desired result. □

Theorem 9.3. *Suppose that $\psi^{-1} = \tilde{\psi}$, that the operator $T(\psi)$ is invertible, and that $P_1T^{-1}(\psi)P_1 = 1$. Then $I - H(z^{-1}\psi)$ is also invertible and its inverse is given by*

$$T(\psi^{-1})^{-1}(I - H(z^{-1}\psi^{-1}))T(\psi)^{-1}.$$

Proof. The invertibility of $T(\psi)$ implies the invertibility of $T(\psi^{-1})$ (see the beginning of the proof of Theorem 5.1). Hence the operator $T(\psi)T(\psi^{-1}) = I - H(\psi)^2$ is invertible.

We do the same steps as in the first two cases, except that this time we are dealing with the term

$$T(\psi)\,(I-H(\psi^{-1}/z)\,T(\psi)^{-1}=T(\psi)\,(I-H(\tilde{\psi}/z))\,T(\psi)^{-1}$$

in a slightly different manner. First

$$T(\psi)\,(I-H(\tilde{\psi}/z))\,T(\psi)^{-1}=I+H(\psi)T(z)+H(\psi)H(z)H(\tilde{\psi})T(\psi)^{-1}.$$

Multiplying on the right by $I-H(\psi/z)$ we have that

$$T(\psi)\,(I-H(\tilde{\psi}/z))\,T(\psi)^{-1}(I-H(\psi/z))$$

is the same as

$$I+H(\psi)H(z)H(\tilde{\psi})T(\psi)^{-1}-H(\psi)T(z)\,H(\psi/z)-H(\psi)H(z)H(\tilde{\psi})T(\psi)^{-1}H(\psi/z).$$

The above uses the identity $H(\psi/z)=H(\psi)\,T(z)=T(1/z)\,H(\psi)$. Now

$$H(\psi)\,T(z)\,H(\psi/z)=H(\psi)\,T(z)\,T(1/z)H(\psi)=H(\psi)\,Q_1\,H(\psi).$$

This means the above is the same as

$$\begin{aligned}I-H(\psi)H(\psi)+H(\psi)H(z)H(\psi)+H(\psi)H(z)H(\tilde{\psi})T(\psi)^{-1}\\ -H(\psi)H(z)H(\tilde{\psi})T(\psi)^{-1}H(\psi/z).\end{aligned}$$

Thus we will have our result if we can show that the last three terms of this operator sum to zero. To do this we consider the transpose of

$$H(\psi)+H(\tilde{\psi})T(\psi)^{-1}-H(\tilde{\psi})T(\psi)^{-1}H(\psi/z),$$

that is,

$$H(\psi)+T(\tilde{\psi})^{-1}H(\tilde{\psi})-H(\psi/z)T(\tilde{\psi})^{-1}H(\tilde{\psi}).$$

Using Lemma 9.2, with ψ replaced by $\tilde{\psi}$ the statement is proved. □

Theorem 9.4. *Suppose that $\psi^{-1}=\tilde{\psi}$, that the operator $T(\psi)$ is invertible, and that $P_1T^{-1}(\psi)P_1=1$. Then $I+H(z\psi)$ is also invertible and the inverse is given by*

$$(T(\psi^{-1})+H(z))^{-1}(I+H(z\psi^{-1}))(T(\psi)+H(z))^{-1}.$$

Proof. The inverse of $(T(\psi)+H(z))^{-1}$ is the same as $T(\psi)^{-1}(I-\frac{1}{2}H(z)T(\psi)^{-1})$ which can be verified using basic linear algebra. Here $P_1T^{-1}(\psi)P_1=1$ is applied. Now consider the product

$$T(\psi)^{-1}\Big(I-\frac{1}{2}H(z)T(\psi)^{-1}\Big)(I+H(z\psi)).$$

Using the factorization of ψ we see this is the same as the operator

$$T(\psi_+^{-1})\Big(I-\frac{1}{2}H(z)\Big)(T(\tilde{\psi}_+)+H(z\psi_+)).$$

Note that $T(\psi_+^{-1})T(\tilde{\psi}_+)=T(\psi)^{-1}$ and that

$$T(\psi_+^{-1})H(z\psi_+)=T(\psi_+^{-1})T(\tilde{\psi}_+)H(z\psi_+\tilde{\psi}_+^{-1})=T(\psi)^{-1}H(z\psi)$$

and thus is also a bounded operator. Now multiplying by $I + H(z\tilde{\psi})$ from the left and doing a similar computation yields

$$(I + H(z\tilde{\psi}))T(\psi)^{-1}\left(I - \frac{1}{2}H(z)T(\psi)^{-1}\right)(I + H(z\psi)) = T(\tilde{\psi}) + H(z),$$

which gives the desired result. $\square$

A concise way to express both of the previous theorems is: The inverse of $M(1, \phi_0\psi)$ is given by

$$(T(\psi^{-1}) + H(\phi_0))^{-1}M(1, \phi_0\psi^{-1})(T(\psi) + H(\phi_0))^{-1}. \tag{30}$$

References

[1] J.C. Andrade, *Random Matrix Theory and L-functions in function fields*, PhD Thesis (2012), University of Bristol.

[2] J. Baik, E.M. Rains, Algebraic aspects of increasing subsequences, *Duke Math. J.* **109**, no. 1 (2001), 1–65.

[3] E.W. Barnes, The theory of the *G*-function, *Quart. J. Pure Appl. Math.* **31** (1900), 264–313.

[4] E.L. Basor, Y. Chen, T. Ehrhardt, Painlevé V and time-dependent Jacobi polynomials, *J. Phys. A: Math. Theor.* **43** (2010), 01524 (25 pages).

[5] E.L. Basor, T. Ehrhardt, Asymptotic formulas for determinants of a sum of finite Toeplitz and Hankel matrices, *Math. Nachr.* **122** (2001), 5–45.

[6] E.L. Basor, T. Ehrhardt, Asymptotic formulas for the determinants of symmetric Toeplitz plus Hankel matrices, in: Toeplitz matrices and singular integral equations (Pobershau, 2001), 61–90, *Oper. Theory Adv. Appl.*, **135**, Birkhäuser, Basel, 2002.

[7] E.L. Basor, T. Ehrhardt, Determinant computations for some classes of Toeplitz–Hankel matrices, *Oper. Matrices* **3**, no. 2 (2009), 167–186.

[8] E.L. Basor, T. Ehrhardt, Fredholm and invertibility theory for a special class of Toeplitz + Hankel operators, *J. Spectral Theory* **3** (2013), 171–214.

[9] E.L. Basor, T. Ehrhardt, Stability of finite sections of Toeplitz–Hankel type operators, in preparation.

[10] A. Böttcher, B. Silbermann, Toeplitz matrices and determinants with Fisher–Hartwig symbols, *J. Funct. Anal.* **63**, no. 2 (1985), 178–214.

[11] A. Böttcher, B. Silbermann, *Analysis of Toeplitz operators*, Springer, Berlin, 1990; also: 2nd ed., Springer, 2006.

[12] P. Deift, A. Its, I. Krasovsky, Asymptotics of Toeplitz, Hankel, and Toeplitz + Hankel determinants with Fisher–Hartwig singularities, *Annals of Mathematics*, **174** (2011), 1243–1299.

[13] T. Ehrhardt, A status report on the asymptotic behavior of Toeplitz determinants with Fisher–Hartwig singularities, in: Recent advances in operator theory (Groningen, 1998), 217–241, *Oper. Theory Adv. Appl.*, **124**, Birkhäuser, Basel, 2001.

[14] P.J. Forrester, N.E. Frankel, Applications and generalizations of Fisher–Hartwig asymptotics, *J. Math. Phys.* **45**, no. 5 (2004), 2003–2028.

[15] J.P. Keating, F. Mezzadri, Random matrix theory and entanglement in quantum spin chains, *Comm. Math. Phys.* **252** (2004), 543–579.

[16] S. Roch, Finite Sections of operators generated by singular integrals with Carleman shift, Preprint Nr. 52, November 1987, Sektion Mathematik, Technische Universität Karl-Marx-Stadt.

[17] H. Widom, Inversion of Toeplitz Matrices. II, *Illinois J. Math.* **4** (1960), 88–99.

[18] H. Widom, Asymptotic behavior of block Toeplitz determinants. II, *Adv. Math.* **21** (1976), 1–29.

Estelle Basor
American Institute of Mathematics
San Jose, CA-95112, USA
e-mail: ebasor@aimath.org

Torsten Ehrhardt
Mathematics Department
University of California
Santa Cruz, CA-95064, USA
e-mail: tehrhard@ucsc.edu

Operator Theory:
Advances and Applications, Vol. 259, 155–178

Generalization of the Brauer Theorem to Matrix Polynomials and Matrix Laurent Series

D.A. Bini and B. Meini

Dedicated to our friend and colleague Albrecht Böttcher on the occasion of his 60th birthday

Abstract. Given a square matrix A, Brauer's theorem [Duke Math. J. 19 (1952), 75–91] shows how to modify one single eigenvalue of A via a rank-one perturbation, without changing any of the remaining eigenvalues. We reformulate Brauer's theorem in functional form and provide extensions to matrix polynomials and to matrix Laurent series $A(z)$ together with generalizations to shifting a set of eigenvalues. We provide conditions under which the modified function $\widetilde{A}(z)$ has a canonical factorization $\widetilde{A}(z) = \widetilde{U}(z)\widetilde{L}(z^{-1})$ and we provide explicit expressions of the factors $\widetilde{U}(z)$ and $\widetilde{L}(z)$. Similar conditions and expressions are given for the factorization of $\widetilde{A}(z^{-1})$. Some applications are discussed.

Mathematics Subject Classification (2010). Primary 15A24, 47A68; Secondary 60J22.

Keywords. Brauer theorem, canonical factorizations, matrix polynomials, matrix Laurent series, matrix equations.

1. Introduction

Brauer's theorem [9] relates the eigenvalues of an $n \times n$ matrix A to the eigenvalues of the modified matrix $\widetilde{A} = A + uv^*$ when either u or v coincides with a right or with a left eigenvector of A, where v^* denotes the transpose conjugate of the vector v. Its original formulation can be stated in the following form.

Work supported by Gruppo Nazionale di Calcolo Scientifico (GNCS) of INdAM, and by the PRA project "Mathematical models and computational methods for complex networks" funded by the University of Pisa.

Theorem 1.1. *Let A be an $n \times n$ complex matrix with eigenvalues $\lambda_1, \dots, \lambda_n$. Let u_k be an eigenvector of A associated with the eigenvalue λ_k, and let v be any n-dimensional vector. Then the matrix $\widetilde{A} = A + u_k v^*$ has eigenvalues $\lambda_1, \dots, \lambda_{k-1}$, $\lambda_k + v^* u_k, \lambda_{k+1}, \dots, \lambda_n$.*

This elementary result has been applied in different contexts, more specifically to the spectral analysis of the PageRank matrix [11], [19], to the nonnegative inverse eigenvalue problem [10], [22], to accelerating the convergence of iterative methods for solving matrix equations in particular quadratic equations [21], equations related to QBD Markov chains [17], [14], to M/G/1-type Markov chains [5], [3], or for algebraic Riccati equations in [15], [16], [20], and [2].

In this paper we revisit Brauer's theorem in functional form and generalize it to a non-degenerate $n \times n$ matrix function $A(z)$ analytic in the open annulus $\mathbb{A}_{r_1,r_2} = \{z \in \mathbb{C} : \quad r_1 < |z| < r_2\}$, where $0 \le r_1 < r_2$, so that it can be described by a matrix Laurent series $A(z) = \sum_{i\in\mathbb{Z}} z^i A_i$ convergent for $z \in \mathbb{A}_{r_1,r_2}$. Here, non-degenerate means that $\det A(z)$ is not identically zero. Throughout the paper we assume that all the matrix functions are nondegenerate.

Let $\lambda \in \mathbb{A}_{r_1,r_2}$ be an eigenvalue of $A(z)$, i.e., such that $\det A(\lambda) = 0$ and $u \in \mathbb{C}^n$, $u \neq 0$, be a corresponding eigenvector, i.e., such that $A(\lambda)u = 0$. We prove that for any $v \in \mathbb{C}^n$ such that $v^* u = 1$ and for any $\mu \in \mathbb{C}$, the function $\widetilde{A}(z) = A(z)(I + \frac{\lambda-\mu}{z-\lambda} uv^*)$ is still analytic in $\mathbb{A}_{r_1,r_2}$. Moreover, $\widetilde{A}(z)$ has all the eigenvalues of $A(z)$, except for λ which, if $\mu \in \mathbb{A}_{r_1,r_2}$, is replaced by μ, otherwise is removed. We provide explicit expressions which relate the coefficients $\widetilde{A}_i$ of $\widetilde{A}(z) = \sum_{i\in\mathbb{Z}} z^i \widetilde{A}_i$ to those of $A(z)$.

This transformation, called *right shift*, is obtained by using a right eigenvector u of $A(z)$. A similar version, called *left shift*, is given by using a left eigenvector v of $A(z)$. Similarly, a combination of the left and the right shift, leads to a transformation called *double shift*, which enables one to shift a pair of eigenvalues by operating to the right and to the left of $A(z)$.

This result, restricted to a matrix Laurent polynomial $A(z) = \sum_{-h}^{k} z^i A_i$, provides a shifted function which is still a matrix Laurent polynomial $\widetilde{A}(z) = \sum_{i=-h}^{k} z^i \widetilde{A}_i$. In the particular case where $A(z) = zI - A$, we find that $\widetilde{A}(z) = zI - \widetilde{A}$ is such that $\widetilde{A}$ is the matrix given in the classical theorem of Brauer [9]. Therefore, these results provide an extension of Brauer's Theorem 1.1 to matrix Laurent series and to matrix polynomials.

A further generalization is obtained by performing the right, left, or double shift simultaneously to a packet of eigenvalues once we are given an invariant (right or left) subspace associated with this set of eigenvalues.

Assume that $A(z)$ admits a canonical factorization $A(z) = U(z)L(z^{-1})$, where $U(z) = \sum_{i=0}^{\infty} z^i U_i$, $L(z) = \sum_{i=0}^{\infty} z^i L_i$ are analytic and nonsingular for $|z| \le 1$. An interesting issue is to find out under which assumptions the function $\widetilde{A}(z)$, obtained after any kind of the above transformations, still admits a canonical factorization $\widetilde{A}(z) = \widetilde{U}(z)\widetilde{L}(z^{-1})$. We investigate this issue and provide explicit expressions to the functions $\widetilde{U}(z)$ and $\widetilde{L}(z)$. We examine also the more complicated problem to determine, if they exist, the canonical factorizations of both the functions $\widetilde{A}(z)$ and $\widetilde{A}(z^{-1})$.

These latter results find an interesting application to determining the explicit solutions of matrix difference equations, by relying on the theory of standard triples of [13]. In fact, in the case where the matrix polynomial $A(z) = \sum_{i=0}^{d} z^i A_i$, which defines the matrix difference equation, has some multiple eigenvalues, the tool of standard triples cannot be always applied. However, by shifting away the multiple eigenvalues into a set of pairwise different eigenvalues, we can transform $A(z)$ into a new matrix polynomial $\widetilde{A}(z)$, which has all simple eigenvalues. This way, the matrix difference equation associated with $\widetilde{A}(z)$ can be solved with the tool of standard triples and we can easily reconstruct the solution of the original matrix difference equation, associated with $A(z)$, from that associated with $\widetilde{A}(z)$. This fact enables us to provide formal solutions to the Poisson problem for QBD Markov chains also in the null-recurrent case [4].

Another important issue related to the shift technique concerns solving matrix equations. In fact, for quadratic matrix polynomials, computing the canonical factorization corresponds to solving a pair of matrix equations. The same property holds, to a certain extent, also for general matrix polynomials or for certain matrix Laurent series [3]. If the matrix polynomial has a repeated eigenvalue, as it happens in null recurrent stochastic processes, then the convergence of numerical algorithms slows down and the conditioning of the sought solution deteriorates. Shifting multiple eigenvalues provides matrix functions, where the associated computational problem is much easier to solve and where the solution is better conditioned. Moreover the solution of the original equation can be obtained by means of the formulas, that we provide in this paper, relating the canonical factorizations of $A(z)$, $A(z^{-1})$ to the canonical factorizations of $\widetilde{A}(z)$ and $\widetilde{A}(z^{-1})$.

An interesting application concerns manipulating a matrix polynomial $A(z)$ having a singular leading coefficient. In fact, in this case $A(z)$ has some eigenvalues at infinity. By using the shift technique it is simple to construct a new matrix polynomial $\widetilde{A}(z)$ having the same eigenvalues of $A(z)$ except for the eigenvalues at infinity which are replaced by some finite value, say 1. A nonsingular leading coefficient allows one to apply specific algorithms that require this condition, see for instance [7], [23], [24].

The paper is organized as follows. Section 2 provides a functional formulation of Brauer's theorem, followed by its extension to matrix polynomials and to matrix Laurent series. Section 3 describes a formulation where a set of selected eigenvalues is shifted somewhere in the complex plane. Section 4 deals with the transformations that the canonical factorization of a matrix function has after some eigenvalue has been shifted away. Section 5 deals with some specific cases and applications. In particular, we examine the possibility of shifting eigenvalues from/to infinity, we analyze right and left shifts together with their combination, that is, the double shift, and apply them to manipulating palindromic matrix polynomials. Through some examples we show how the degree of a matrix polynomial can be decreased by shifting eigenvalues to the infinity; we show also how the precision of the approximations of the eigenvalues can be increased by separating clustered eigenvalues by means of the shift. Finally we analyze quadratic matrix polynomials and the associated matrix equations. In particular, we relate the solutions of the shifted equations to the ones of the original equations.

2. Generalizations and extensions

In this section, we provide some generalizations and extensions of Brauer's Theorem 1.1. In particular, we give a functional formulation, from which we obtain an extension to matrix polynomials and to matrix (Laurent) power series.

2.1. Functional formulation and extension to matrix polynomials

Let A be an $n\times n$ matrix, let λ and $u\in\mathbb{C}^n$, $u\neq 0$, be such that $Au=\lambda u$. Consider the rational function $I+\frac{\lambda-\mu}{z-\lambda}Q$, where $Q=uv^*$ and v is any nonzero vector such that $v^*u=1$ and μ is any complex number. We have the following

Theorem 2.1. *The rational function $\widetilde{A}(z)=(zI-A)(I+\frac{\lambda-\mu}{z-\lambda}Q)$ coincides with the matrix polynomial $\widetilde{A}(z)=zI-\widetilde{A}$, $\widetilde{A}=A+(\mu-\lambda)Q$. Moreover the eigenvalues of $\widetilde{A}$ coincide with the eigenvalues of A except for λ which is replaced by μ.*

Proof. We have $\widetilde{A}(z)=(zI-A)(I+\frac{\lambda-\mu}{z-\lambda}Q)=zI-A+\frac{\lambda-\mu}{z-\lambda}(zI-A)uv^*=zI-A+\frac{\lambda-\mu}{z-\lambda}(z-\lambda)uv^*=zI-(A-(\lambda-\mu)Q)$. Moreover, $\det(zI-\widetilde{A})=\det(zI-A)\det(I+\frac{\lambda-\mu}{z-\lambda}Q)=\det(zI-A)(1+\frac{\lambda-\mu}{z-\lambda}v^*u)=\det(zI-A)\frac{z-\mu}{z-\lambda}$. This implies that the eigenvalue λ of A is replaced by μ. □

Relying on the above functional formulation of Brauer's theorem, we may prove the following extension to matrix polynomials:

Theorem 2.2. *Let $A(z)=\sum_{i=0}^d z^iA_i$ be an $n\times n$ matrix polynomial. Let $\lambda\in\mathbb{C}$ and $u\in\mathbb{C}^n$, $u\neq 0$, be such that $A(\lambda)u=0$. For any $\mu\in\mathbb{C}$ and for any $v\in\mathbb{C}^n$ such that $v^*u=1$, the rational function $\widetilde{A}(z)=A(z)(I+\frac{\lambda-\mu}{z-\lambda}Q)$, with $Q=uv^*$, coincides with the matrix polynomial $\widetilde{A}(z)=\sum_{i=0}^d z^i\widetilde{A}_i$, where*

$$\widetilde{A}_i=A_i+(\lambda-\mu)\sum_{k=0}^{d-i-1}\lambda^kA_{k+i+1}Q,\ i=0,\dots,d-1,\quad \widetilde{A}_d=A_d.$$

Moreover, the eigenvalues of $\widetilde{A}(z)$ coincide with those of $A(z)$ except for λ which is shifted to μ.

Proof. Since $A(\lambda)u=0$ we find that $\widetilde{A}(z)=A(z)(I+\frac{\lambda-\mu}{z-\lambda}Q)=A(z)+\frac{\lambda-\mu}{z-\lambda}(A(z)-A(\lambda))uv^*$. Moreover, since $A(z)-A(\lambda)=\sum_{i=1}^d(z^i-\lambda^i)A_i$ and $z^i-\lambda^i=(z-$

$\lambda)\sum_{j=0}^{i-1}\lambda^{i-j-1}z^j$ we get

$$\widetilde{A}(z)=A(z)+(\lambda-\mu)\sum_{i=1}^{d}\sum_{j=0}^{i-1}\lambda^{i-j-1}z^jA_iQ=A(z)+(\lambda-\mu)\sum_{i=0}^{d-1}z^i\sum_{k=0}^{d-i-1}\lambda^kA_{k+i+1}Q.$$

Whence we deduce that $\widetilde{A}_i=A_i+(\lambda-\mu)\sum_{k=0}^{d-i-1}\lambda^kA_{k+i+1}Q$, for $i=0,\dots,d-1$. Concerning the eigenvalues, by taking determinants in both sides of $\widetilde{A}(z)=A(z)(I+\frac{\lambda-\mu}{z-\lambda}Q)$, one obtains $\det\widetilde{A}(z)=\det A(z)\det(I+\frac{\lambda-\mu}{z-\lambda}v^*u)=\det A(z)\frac{z-\mu}{z-\lambda}$, which completes the proof. □

It is interesting to observe that the function obtained by applying the shift to a matrix polynomial is still a matrix polynomial with the same degree.

A different way of proving this result for a matrix polynomial consists in applying the following three steps: 1) extend Brauer's theorem to a linear pencil, that is, to a function of the kind $zB+C$; 2) apply this extension to the $nd\times nd$ linear pencil $\mathcal{A}(z)=z\mathcal{B}+\mathcal{C}$ obtained by means of a Frobenius companion linearization of the matrix polynomial; in fact, it is possible to choose the vector v in such a way that the shifted pencil still keeps the Frobenius structure; 3) reformulate the problem in terms of matrix polynomial and get the extension of Brauer's theorem. In this approach we have to rely on different companion forms if we want to apply either the right or the left shift.

This approach, based on companion linearizations, does not work in the more general case of matrix Laurent series, while the functional formulation still applies as explained in the next section.

As an example of application of this extension consider the quadratic matrix polynomial $A(z)=A_0+zA_1+z^2A_2$ where

$$A_0=-\begin{bmatrix}1&1\\0&1\end{bmatrix},\quad A_1=\begin{bmatrix}4&3\\1&4\end{bmatrix},\quad A_2=-\begin{bmatrix}3&0\\1&2\end{bmatrix}.$$

Its eigenvalues are $1/3,1/2,1,1$. Moreover $u=[1,0]^*$ is an eigenvector corresponding to the eigenvalue 1.

Performing a shift of $\lambda=1$ to $\mu=0$, with $v=u$, so that $Q=\left[\begin{smallmatrix}1&0\\0&0\end{smallmatrix}\right]$, one obtains

$$\widetilde{A}_0=-\begin{bmatrix}0&1\\0&1\end{bmatrix},\quad \widetilde{A}_1=\begin{bmatrix}1&3\\0&4\end{bmatrix},\quad \widetilde{A}_2=-\begin{bmatrix}3&0\\1&2\end{bmatrix}.$$

The polynomial $\widetilde{A}(z)=\widetilde{A}_0+z\widetilde{A}_1+z^2\widetilde{A}_2$ has eigenvalues $0,1/3,1/2,1$.

Applying the Matlab function `polyeig` to $A(z)$ provides the following approximations to the eigenvalues

```
0.333333333333333, 0.500000000000002,
0.999999974684400, 1.000000025315599.
```

The large relative error in the approximation of the eigenvalue 1 is due to the ill conditioning of this double eigenvalue. Applying the Matlab function `polyeig` to $\widetilde{A}(z)$ provides the following approximations to the eigenvalues

```
0.000000000000000, 0.333333333333333,
0.500000000000000, 1.000000000000000.
```

In this case, the approximations are accurate to the machine precision since the new polynomial has simple well-conditioned eigenvalues. This transformation has required the knowledge of the double eigenvalue and of the corresponding eigenvector. For many problems, typically for matrix polynomials encountered in the analysis of stochastic processes, this information is readily available.

2.2. The case of matrix (Laurent) power series

The shift based on the functional approach, described for a matrix polynomial, can be applied to deal with nondegenerate matrix functions which are analytic on a certain region of the complex plane. Let us start with the case of a matrix power series.

Theorem 2.3. *Let* $A(z) = \sum_{i=0}^{\infty} z^i A_i$ *be an* $n \times n$ *matrix power series which is analytic in the disk* $\mathbb{D}_r = \{z \in \mathbb{C} : |z| < r\}$. *Let* $\lambda \in \mathbb{D}_r$ *and* $u \in \mathbb{C}^n$, $u \neq 0$, *be such that* $A(\lambda)u = 0$. *For any any* $\mu \in \mathbb{C}$ *and* $v \in \mathbb{C}^n$ *such that* $v^*u = 1$, *the function* $\widetilde{A}(z) = A(z)(I + \frac{\lambda-\mu}{z-\lambda}Q)$, *with* $Q = uv^*$, *is analytic for* $z \in \mathbb{D}_r$ *and coincides with the matrix power series*

$$\widetilde{A}(z) = \sum_{i=0}^{\infty} z^i \widetilde{A}_i, \quad \widetilde{A}_i = A_i + (\lambda - \mu) \sum_{k=0}^{\infty} \lambda^k A_{k+i+1} Q, \quad i = 0, 1 \ldots.$$

Moreover, $\det \widetilde{A}(z) = \det A(z) \frac{z-\mu}{z-\lambda}$.

Proof. Since $A(\lambda)u = 0$, we find that $\widetilde{A}(z) = A(z)(I + \frac{\lambda-\mu}{z-\lambda}Q) = A(z) + \frac{\lambda-\mu}{z-\lambda}(A(z) - A(\lambda))uv^*$. Moreover, since $A(z) - A(\lambda) = \sum_{i=1}^{\infty}(z^i - \lambda^i)A_i$ and $z^i - \lambda^i = (z - \lambda)\sum_{j=0}^{i-1} \lambda^{i-j-1} z^j$ we get

$$\widetilde{A}(z) = A(z) + (\lambda - \mu) \sum_{i=1}^{\infty} \sum_{j=0}^{i-1} \lambda^{i-j-1} z^j A_i uv^*, \tag{2.1}$$

provided that the series $S(z) = \sum_{i=1}^{\infty} \sum_{j=0}^{i-1} \lambda^{i-j-1} z^j A_i$ is convergent in $\mathbb{D}_r$. We prove this latter property. Since $A(z)$ is convergent in $\mathbb{D}_r$, then for any $0 < \sigma < r$ there exists a positive matrix Γ such that $|A_i| \leq \Gamma \sigma^{-i}$ [18, Theorem 4.4c], [3, Theorem 3.6], where $|A|$ denotes the matrix whose entries are the absolute values of the entries of A and the inequality holds component-wise. Let z be such that $|\lambda| \leq |z| < r$ and let σ be such that $|z| < \sigma < r$. We have $\sum_{i=1}^{\infty} \sum_{j=0}^{i-1} |\lambda^{i-j-1} z^j| \, |A_i| \leq \sum_{i=1}^{\infty} \sum_{j=0}^{i-1} |z|^{i-1} |A_i| \leq \sigma^{-1} \sum_{i=1}^{\infty} i(|z|/\sigma)^{i-1} \Gamma$, and the latter series is convergent since $0 \leq |z|/\sigma < 1$. Therefore the series $S(z)$ is absolutely convergent, and hence convergent, for $z \in \mathbb{D}_r$. Since the series is absolutely convergent, we may exchange the order of the summations so that we obtain $S(z) = \sum_{i=0}^{\infty} z^i \sum_{k=0}^{\infty} \lambda^k A_{k+i+1}$. Whence, from (2.1), we deduce that $\widetilde{A}_i = A_i + (\lambda - \mu) \sum_{k=0}^{\infty} \lambda^k A_{k+i+1}$, for $i =$

$0,1,\ldots$. By taking determinants in both sides of $\widetilde{A}(z) = A(z)(I + \frac{\lambda-\mu}{z-\lambda}Q)$, one obtains $\det \widetilde{A}(z) = \det A(z) \det(I + \frac{\lambda-\mu}{z-\lambda}v^*u) = \det A(z)(1 + \frac{\lambda-\mu}{z-\mu}) = \det A(z)\frac{z-\mu}{z-\lambda}$, which completes the proof. □

We observe that, if in the above theorem μ does not belong to $\mathbb{D}_r$, the eigenvalues of the function $\widetilde{A}(z)$ coincide with the eigenvalues of $A(z)$, together with their multiplicities, except for λ. More specifically, if λ is a simple eigenvalue of $A(z)$, then it is not any more eigenvalue of $\widetilde{A}(z)$; if λ is an eigenvalue of multiplicity $\ell > 1$ for $A(z)$, then λ is an eigenvalue of multiplicity $\ell - 1$ for $\widetilde{A}(z)$. On the other hand, if $\mu \in \mathbb{D}_r$ then in $\widetilde{A}(z)$ the eigenvalue λ is replaced by μ.

A similar result can be stated for matrix Laurent series, i.e., matrix functions of the form $A(z) = \sum_{i\in\mathbb{Z}} z^i A_i$ which are analytic over $\mathbb{A}_{r_1,r_2} = \{z \in \mathbb{C} : \quad r_1 < |z| < r_2\}$ for $0 < r_1 < r_2$.

Theorem 2.4. *Let $A(z) = \sum_{i\in\mathbb{Z}} z^i A_i$ be an $n \times n$ matrix Laurent series which is analytic in the annulus $\mathbb{A}_{r_1,r_2}$. Let $\lambda \in \mathbb{A}_{r_1,r_2}$ and $u \in \mathbb{C}^n$, $u \neq 0$, be such that $A(\lambda)u = 0$. For any $v \in \mathbb{C}^n$ such that $v^*u = 1$ and for any $\mu \in \mathbb{C}$, the function*

$$\widetilde{A}(z) = A(z)\left(I + \frac{\lambda-\mu}{z-\lambda}Q\right), \quad Q = uv^*, \tag{2.2}$$

is analytic for $z \in \mathbb{A}_{r_1,r_2}$ and coincides with the matrix Laurent series $\widetilde{A}(z) = \sum_{i\in\mathbb{Z}} z^i \widetilde{A}_i$, where

$$\begin{aligned} \widetilde{A}_i &= A_i + (\lambda-\mu)\sum_{k=0}^{\infty}\lambda^k A_{k+i+1}Q, && \text{for } i \geq 0, \\ \widetilde{A}_i &= A_i - (\lambda-\mu)\sum_{k=0}^{\infty}\lambda^{-k-1} A_{-k+i}Q, && \text{for } i < 0. \end{aligned} \tag{2.3}$$

Moreover, $\det \widetilde{A}(z) = \det A(z)\frac{z-\mu}{z-\lambda}$.

Proof. We proceed as in the proof of Theorem 2.3. Since $A(\lambda)u = 0$, we find that $\widetilde{A}(z) = A(z)(I + \frac{\lambda-\mu}{z-\lambda}Q) = A(z) + \frac{\lambda-\mu}{z-\lambda}(A(z) - A(\lambda))uv^*$. Observe that $A(z) - A(\lambda) = \sum_{i=1}^{\infty}(z^i - \lambda^i)A_i + \sum_{i=1}^{\infty}(z^{-i} - \lambda^{-i})A_{-i}$. Moreover, we have $z^i - \lambda^i = (z-\lambda)\sum_{j=0}^{i-1}\lambda^{i-j-1}z^j$ and $z^{-i} - \lambda^{-i} = -\frac{1}{z\lambda}(z-\lambda)\sum_{j=0}^{i-1}\lambda^{-(i-j-1)}z^{-j}$. Thus we get

$$\begin{aligned} &\widetilde{A}(z) = A(z) + (\lambda-\mu)S_+(z)uv^* - \frac{\lambda-\mu}{z\lambda}S_-(z)uv^*, \\ &S_+(z) = \sum_{i=1}^{\infty}\sum_{j=0}^{i-1}\lambda^{i-j-1}z^j A_i, \quad S_-(z) = \sum_{i=1}^{\infty}\sum_{j=0}^{i-1}\lambda^{-(i-j-1)}z^{-j}A_{-i}, \end{aligned} \tag{2.4}$$

provided that the series $S_+(z)$ and $S_-(z)$ are convergent in $\mathbb{A}_{r_1,r_2}$. According to [3, Theorem 3.6], [18, Theorem 4.4c], for any $r_1 < \sigma < r_2$ there exists a positive matrix Γ such that $|A_i| \leq \Gamma\sigma^{-i}$ for $i \in \mathbb{Z}$. Thus we may proceed as in the proof of Theorem

2.3 and conclude that the series $S_+(z)$ and $S_-(z)$ are absolutely convergent, so that we may exchange the order of the summations and get

$$S_+(z) = \sum_{i=0}^{\infty} z^i \sum_{k=0}^{\infty} \lambda^k A_{k+i+1}, \quad S_-(z) = \sum_{i=0}^{\infty} z^{-i} \sum_{k=0}^{\infty} \lambda^{-k} A_{-(k+i+1)}.$$

Whence from the first equation in (2.4) we deduce that

$$\widetilde{A}_i = A_i + (\lambda - \mu) \sum_{k=0}^{\infty} \lambda^k A_{k+i+1} uv^*, \qquad \text{for } i \geq 0,$$

$$\widetilde{A}_i = A_i - (\lambda - \mu) \sum_{k=0}^{\infty} \lambda^{-k-1} A_{-k+i} uv^*, \qquad \text{for } i < 0.$$

By taking determinants in both sides of $\widetilde{A}(z) = A(z)(I + \frac{\lambda-\mu}{z-\lambda} Q)$, one obtains $\det \widetilde{A}(z) = \det A(z) \det(I + \frac{\lambda-\mu}{z-\lambda} v^* u) = \det A(z)(1 + \frac{\lambda-\mu}{z-\mu}) = \det A(z) \frac{z-\mu}{z-\lambda}$, which completes the proof. □

Observe that, if $\mu \notin \mathbb{A}_{r_1,r_2}$ then the eigenvalues of $\widetilde{A}(z)$ coincide with those of $A(z)$ with their multiplicities except for λ. More specifically, if λ is a simple eigenvalue of $A(z)$, then it is not any more eigenvalue of $\widetilde{A}(z)$; if λ is an eigenvalue of multiplicity $\ell > 1$ for $A(z)$, then λ is an eigenvalue of multiplicity $\ell - 1$ for $\widetilde{A}(z)$. On the other hand, if $\mu \in \mathbb{A}_{r_1,r_2}$ then in $\widetilde{A}(z)$ the eigenvalue λ is replaced by μ.

Observe also that if $A(z) = \sum_{i=-h}^{k} z^i A_i$ then from (2.3) it follows that $\widetilde{A}(z) = \sum_{i=-h}^{k} z^i \widetilde{A}_i$.

3. Shifting a set of eigenvalues

In this section we provide a generalization of Brauer's theorem, where we can simultaneously shift a set of eigenvalues. We will treat the case of a matrix, of a matrix polynomial and of a matrix Laurent series.

Let $A \in \mathbb{C}^{n\times n}$, $U \in \mathbb{C}^{n\times m}$ and $\Lambda \in \mathbb{C}^{m\times m}$, where $m < n$, be such that U has full rank and $AU = U\Lambda$. In other words, the eigenvalues of Λ are a subset of the eigenvalues of A and the columns of U span an invariant subspace for A corresponding to the eigenvalues of Λ. Without loss of generality we may assume that the eigenvalues of Λ are $\lambda_1, \ldots, \lambda_m$ where λ_i for $i = 1, \ldots, n$ are the eigenvalues of A.

We have the following generalization of Brauer's theorem

Theorem 3.1. *Let A be an $n \times n$ matrix. Let $U \in \mathbb{C}^{n\times m}$ and $\Lambda \in \mathbb{C}^{m\times m}$, where $m < n$, be such that U has full rank and $AU = U\Lambda$. Let $V \in \mathbb{C}^{n\times m}$ be a full rank matrix such that $V^*U = I$. Let $S \in \mathbb{C}^{m\times m}$ be any matrix. Then the rational matrix function*

$$\widetilde{A}(z) = (zI - A)(I + U(zI - \Lambda)^{-1}(\Lambda - S)V^*)$$

coincides with the linear pencil $\widetilde{A}(z) = zI - \widetilde{A}_0$, *where* $\widetilde{A}_0 = A - U(\Lambda - S)V^*$. *Moreover,* $\det \widetilde{A}(z) = \det(zI - S)\frac{\det A(z)}{\det(zI-\Lambda)}$, *that is, the eigenvalues of* $\widetilde{A}_0$ *are given by* $\mu_1, \ldots, \mu_m, \lambda_{m+1}, \ldots, \lambda_n$, *where* $\mu_1, \ldots, \mu_m$ *are the eigenvalues of the matrix* S.

Proof. Denote $X(z) = (zI - \Lambda)^{-1}(\Lambda - S)$. Since $AU = U\Lambda$, we have $\widetilde{A}(z) = zI - A + zUX(z)V^* - U\Lambda X(z)V^* = zI - A + U(zI - \Lambda)X(z)V^*$. Whence we get $\widetilde{A}(z) = zI - (A - U(\Lambda - S)V^*)$. Concerning the eigenvalues, taking determinants we have $\det \widetilde{A}(z) = \det A(z) \det(I + UX(z)V^*) = \det A(z) \det(I + X(z)V^*U) = \det A(z) \det(I + X(z))$. Since $I + X(z) = (zI - \Lambda)^{-1}(zI - \Lambda + \Lambda - S)$, we obtain $\det \widetilde{A}(z) = \frac{\det A(z)}{\det(zI-\Lambda)} \det(zI - S)$, which completes the proof. □

The above results immediately extend to matrix polynomials. Assume that we are given a matrix $U \in \mathbb{C}^{n\times m}$ and $\Lambda \in \mathbb{C}^{m\times m}$, where $m < n$, U has full rank and $\sum_{i=0}^{d} A_i U \Lambda^i = 0$. Then if w is an eigenvector of Λ corresponding to the eigenvalue λ, i.e., $\Lambda w = \lambda w$ then $0 = \sum_{i=0}^{d} A_i U \Lambda^i w = \sum_{i=0}^{d} A_i U \lambda^i w$, that is, $A(\lambda)Uw = 0$ so that λ is eigenvalue of $A(z)$ and Uw is eigenvector. We have the following result.

Theorem 3.2. *Let* $A(z) = \sum_{i=0}^{d} z^i A_i$ *be an* $n\times n$ *matrix polynomial. Let* $U \in \mathbb{C}^{n\times m}$ *and* $\Lambda \in \mathbb{C}^{m\times m}$, *where* $m < n$, *be such that* U *has full rank and* $\sum_{i=0}^{d} A_i U \Lambda^i = 0$. *Let* $V \in \mathbb{C}^{n\times m}$ *be a full rank matrix such that* $V^*U = I$. *Let* $S \in \mathbb{C}^{m\times m}$ *be any matrix. Then the rational function*

$$\widetilde{A}(z) = A(z)(I + U(zI - \Lambda)^{-1}(\Lambda - S)V^*)$$

coincides with the matrix polynomial $\widetilde{A}(z) = \sum_{i=0}^{d} z^i \widetilde{A}_i$, *where*

$$\widetilde{A}_i = A_i + \sum_{k=0}^{d-i-1} A_{k+i+1} U \Lambda^k (\Lambda - S)V^*, \quad i = 0, \ldots, d-1, \quad \widetilde{A}_d = A_d.$$

Moreover, $\det \widetilde{A}(z) = \det(zI - S)\frac{\det A(z)}{\det(zI-\Lambda)}$, *that is, the eigenvalues of* $\widetilde{A}(z)$ *coincide with those of* $A(z)$ *except for those coinciding with the eigenvalues of* Λ *which are replaced by the eigenvalues of* S.

Proof. Denote $X(z) = (zI - \Lambda)^{-1}(\Lambda - S)$. Since $\sum_{i=0}^{d} A_i U \Lambda^i = 0$ we find that

$$\begin{aligned}\widetilde{A}(z) &= A(z)(I + UX(z)V^*) = A(z) + A(z)UX(z)V^* \\ &= A(z) + (A(z)U - \sum_{i=0}^{d} A_i U \Lambda^i)X(z)V^*.\end{aligned}$$

Moreover, since

$$A(z)U - \sum_{i=0}^{d} A_i U \Lambda^i = \sum_{i=1}^{d} A_i U(z^i I - \Lambda^i)$$

and

$$z^iI - \Lambda^i = (zI - \Lambda)\sum_{j=0}^{i-1} z^j\Lambda^{i-j-1}$$

we get

$$\begin{aligned}\widetilde{A}(z) =& A(z) + \left(\sum_{i=1}^{d} A_iU(zI-\Lambda)\sum_{j=0}^{i-1} z^j\Lambda^{i-j-1}\right)X(z)V^* \\ =& A(z) + \left(\sum_{i=1}^{d} A_iU\sum_{j=0}^{i-1} z^j\Lambda^{i-j-1}\right)(\Lambda - S)V^* \\ =& A(z) + \sum_{j=0}^{d-1} z^j \sum_{k=0}^{d-j-1} A_{k+j+1}U\Lambda^k(\Lambda - S)V^*.\end{aligned}$$

Whence we deduce that $\widetilde{A}_i = A_i + \sum_{k=0}^{d-i-1} A_{k+i+1}U\Lambda^k(\Lambda - S)V^*$, $i = 0,\dots,d-1$, $\widetilde{A}_d = A_d$. Concerning the eigenvalues, taking determinants we have $\det\widetilde{A}(z) = \det A(z)\det(I+UX(z)V^*) = \det A(z)\det(I+X(z)V^*U) = \det A(z)\det(I+X(z))$. Since $I+X(z) = (zI-\Lambda)^{-1}(zI-\Lambda+\Lambda-S)$, we obtain $\det\widetilde{A}(z) = \frac{\det A(z)}{\det(zI-\Lambda)}\det(zI-S)$, which completes the proof. □

Similarly, we may prove the following extensions of the multishift to matrix power series and to matrix Laurent series. Below, we report only the case of a matrix Laurent series.

Theorem 3.3. *Let $A(z) = \sum_{i\in\mathbb{Z}} z^iA_i$ be a matrix Laurent series analytic over the annulus $\mathbb{A}_{r_1,r_2}$ where $r_1 < r_2$. Let $U \in \mathbb{C}^{n\times m}$ and $\Lambda \in \mathbb{C}^{m\times m}$, where $m < n$, such that U has full rank and $\sum_{i\in\mathbb{Z}} A_iU\Lambda^i = 0$. Let $V \in \mathbb{C}^{n\times m}$ be a full rank matrix such that $V^*U = I$. Let $S \in \mathbb{C}^{m\times m}$ be any matrix. Then the matrix function*

$$\widetilde{A}(z) = A(z)(I + U(zI-\Lambda)^{-1}(\Lambda - S)V^*)$$

coincides with the matrix Laurent series $\widetilde{A}(z) = \sum_{i\in\mathbb{Z}} z^i\widetilde{A}_i$,

$$\widetilde{A}_i = A_i + \sum_{k=0}^{\infty} A_{k+i+1}U\Lambda^k(\Lambda - S)V^*, \qquad i \ge 0,$$

$$\widetilde{A}_i = A_i - \sum_{k=0}^{\infty} A_{-k+i}U\Lambda^{-k+1}(\Lambda - S)V^*, \quad i < 0.$$

Moreover, $\det\widetilde{A}(z) = \det(zI - S)\frac{\det A(z)}{\det(zI-\Lambda)}$.

Observe that if the matrix S in the above theorem has eigenvalues $\mu_1,\dots,\mu_m$ where $\mu_1,\dots,\mu_k \in \mathbb{A}_{r_1,r_2}$ and $\mu_{k+1},\dots,\mu_m \notin \mathbb{A}_{r_1,r_2}$, then the matrix function $\widetilde{A}(z)$ is still defined and analytic in $\mathbb{A}_{r_1,r_2}$. The eigenvalues of $\widetilde{A}(z)$ coincide with those of $A(z)$ except for $\lambda_1,\dots,\lambda_m$ whose multiplicity is decreased by 1, moreover, $\widetilde{A}(z)$ has the additional eigenvalues $\mu_1,\dots,\mu_k \in \mathbb{A}_{r_1,r_2}$.

4. Brauer's theorem and canonical factorizations

Assume we are given a matrix Laurent series $A(z) = \sum_{i\in\mathbb{Z}} z^i A_i$, analytic over $\mathbb{A}_{r_1,r_2}$, $0 < r_1 < 1 < r_2$, which admits the canonical factorization [8]

$$A(z) = U(z)L(z^{-1}), \quad U(z) = \sum_{i=0}^{\infty} z^i U_i, \quad L(z) = \sum_{i=0}^{\infty} z^i L_i, \tag{4.1}$$

where $U(z)$ and $L(z)$ are invertible for $|z| \le 1$. We observe that from the invertibility of $U(z)$ and $L(z)$ for $|z| \le 1$ it follows that $\det L_0 \ne 0$, $\det U_0 \ne 0$. Therefore, the canonical factorization can be written in the form $A(z) = \check{U}(z)K\check{L}(z^{-1})$ where $K = U_0L_0$ and $\check{U}(z) = U(z)U_0^{-1}$, $\check{L}(z) = L_0^{-1}L(z)$. The factorization in this form is unique [12].

The goal of this section is to find out if, and under which conditions, this canonical factorization is preserved under the shift transformation, and to relate the canonical factorization of $\widetilde{A}(z)$ to that of $A(z)$.

This problem has been recently analyzed in the framework of stochastic processes in [4] for a quadratic matrix polynomial and used to provide an analytic solution of the Poisson problem.

We need a preliminary result whose proof is given in [4].

Theorem 4.1. *Let $A(z) = z^{-1}A_{-1} + A_0 + zA_1$ be an $n \times n$ matrix Laurent polynomial, invertible in the annulus $\mathbb{A}_{r_1,r_2}$ having the canonical factorization*

$$A(z) = (I - zR_+)K_+(I - z^{-1}G_+).$$

We have the following properties:

1. *Let $H(z) := A(z)^{-1}$. Then, for $z \in \mathbb{A}_{r_1,r_2}$ it holds that $H(z) = \sum_{i=-\infty}^{+\infty} z^i H_i$ where*

$$H_i = \begin{cases} G_+^{-i}H_0, & \text{for } i < 0, \\ \sum_{j=0}^{+\infty} G_+^j K_+^{-1} R_+^j, & \text{for } i = 0, \\ H_0 R_+^i, & \text{for } i > 0. \end{cases}$$

2. *If H_0 is nonsingular, then $A(z^{-1})$ has the canonical factorization $A(z^{-1}) = (I - zR_-)K_-(I - z^{-1}G_-)$, where $K_- = A_0 + A_{-1}G_- = A_0 + R_-A_1$ and $G_- = H_0R_+H_0^{-1}$, $R_- = H_0^{-1}G_+H_0$.*

Now we can prove the following result concerning the canonical factorization of $\widetilde{A}(z)$.

Theorem 4.2. *Assume that $A(z) = \sum_{i\in\mathbb{Z}} z^i A_i$ admits the canonical factorization (4.1). Let $\lambda \in \mathbb{A}_{r_1,r_2}$ and $u \in \mathbb{C}^n$, $u \ne 0$, be such that $A(\lambda)u = 0$ and $|\lambda| < 1$. Let $v \in \mathbb{C}^n$ be such that $v^*u = 1$ and let $\mu \in \mathbb{C}$ with $|\mu| < 1$. Then the matrix Laurent polynomial $\widetilde{A}(z) = A(z)(I + \frac{\lambda-\mu}{z-\lambda}Q)$, where $Q = uv^*$, admits the following canonical factorization*

$$\widetilde{A}(z) = \widetilde{U}(z)\widetilde{L}(z^{-1})$$

where $\widetilde{U}(z) = U(z)$, $\widetilde{L}(z) = \sum_{i=0}^{\infty} z^i \widetilde{L}_i$, *with*

$$\widetilde{L}_0 = L_0, \ \widetilde{L}_i = L_i - (\lambda - \mu) \sum_{j=1}^{\infty} \lambda^{-j} L_{j+i-1} Q, \quad i \geq 1.$$

Moreover, $\det \widetilde{L}(z^{-1}) = \det L(z^{-1}) \frac{z-\mu}{z-\lambda}$.

Proof. Observe that the condition $A(\lambda)u = 0$ implies that $U(\lambda)L(\lambda^{-1})u = 0$. Since $|\lambda| < 1$ then $U(\lambda)$ is nonsingular, therefore $L(\lambda^{-1})u = 0$. Define $\widetilde{L}(z^{-1}) = L(z^{-1})(I + \frac{\lambda-\mu}{z-\lambda} Q)$. In view of Theorem 2.4, the matrix function $\widetilde{L}(z^{-1})$ is a matrix power series in z^{-1}, obtained by applying the right shift to $L(z^{-1})$. Moreover, the expression for its matrix coefficients follows from (2.3). By taking the determinant of $\widetilde{L}(z)$ one obtains $\det \widetilde{L}(z^{-1}) = \det L(z^{-1}) \det(I + \frac{\lambda-\mu}{z-\lambda} Q) = \det L(z^{-1}) \frac{z-\mu}{z-\lambda}$. Therefore, since $\det \widetilde{L}(z) = \det L(z) \frac{z^{-1}-\mu}{z^{-1}-\lambda}$, the function $\widetilde{L}(z)$ is nonsingular for $|z| \leq 1$, so that $\widetilde{A}(z) = \widetilde{U}(z)\widetilde{L}(z^{-1})$ is a canonical factorization. □

The above theorem relates the canonical factorization of $A(z)$ with the canonical factorization of the shifted function $\widetilde{A}(z)$. Assume we are given a canonical factorization of $A(z^{-1})$, it is a natural question to figure out if also $\widetilde{A}(z^{-1})$ admits a canonical factorization, and if it is related to the one of $A(z^{-1})$. This issue is motivated by applications in the solution of the Poisson problem in stochastic models, where both the canonical factorizations of $\widetilde{A}(z)$ and $\widetilde{A}(z^{-1})$ are needed for the existence of the solution and for providing its explicit expression [1].

We give answer to this question in the case of matrix Laurent polynomials of the kind $A(z) = z^{-1}A_{-1} + A_0 + zA_1$.

Theorem 4.3. *Let* $A(z) = z^{-1}A_{-1} + A_0 + zA_1$ *be analytic in the annulus* $\mathbb{A}_{r_1,r_2}$. *Assume that* $A(z)$ *and* $A(z^{-1})$ *admit the canonical factorizations*

$$\begin{aligned} A(z) &= (I - zR_+)K_+(I - z^{-1}G_+), \\ A(z^{-1}) &= (I - zR_-)K_-(I - z^{-1}G_-). \end{aligned}$$

Let $\lambda \in \mathbb{A}_{r_1,r_2}$, $|\lambda| < 1$, *and* $u \in \mathbb{C}^n$, $u \neq 0$, *be such that* $A(\lambda)u = 0$. *Let* $\mu \in \mathbb{C}$, $|\mu| < 1$, *and* v *be any vector such that:* $v^*u = 1$, $(\lambda - \mu)v^*G_-u \neq 1$. *Set* $Q = uv^*$, *and define the matrix Laurent polynomial* $\widetilde{A}(z) = z^{-1}\widetilde{A}_{-1} + \widetilde{A}_0 + z\widetilde{A}_1$ *as in* (2.2). *Then* $\widetilde{A}_{-1} = A_{-1} + (\lambda - \mu)(A_0 + \lambda A_1)Q$, $\widetilde{A}_0 = A_0 + (\lambda - \mu)A_1 Q$, $\widetilde{A}_1 = A_1$, *moreover* $\widetilde{A}(z)$ *and* $\widetilde{A}(z^{-1})$ *have the canonical factorizations*

$$\begin{aligned} \widetilde{A}(z) &= (I - z\widetilde{R}_+)\widetilde{K}_+(I - z^{-1}\widetilde{G}_+), \\ \widetilde{A}(z^{-1}) &= (I - z\widetilde{R}_-)\widetilde{K}_-(I - z^{-1}\widetilde{G}_-), \end{aligned} \tag{4.2}$$

where $\widetilde{R}_+ = R_+$, $\widetilde{K}_+ = K_+$, $\widetilde{G}_+ = G_+ + (\mu - \lambda)Q$, $\widetilde{R}_- = \widetilde{W}^{-1}\widetilde{G}_+\widetilde{W}$, $\widetilde{G}_- = \widetilde{W}R_+\widetilde{W}^{-1}$, $\widetilde{K}_- = \widetilde{A}_0 + \widetilde{A}_{-1}\widetilde{G}_- = \widetilde{A}_0 + \widetilde{R}_-A_1$, *with* $\widetilde{W} = W + (\mu - \lambda)QWR_+$, $W = \sum_{i=0}^{\infty} G_+^i K_+^{-1} R_+^i$.

Proof. The expressions for the matrix coefficients $\widetilde{A}_i$, $i = -1, 0, 1$, follow from Theorem 2.4. Observe that, since $|\lambda| < 1$, from the definition of canonical factorization, it follows that the matrix $I - \lambda R_+$ is not singular. Therefore, since $A(\lambda)u = 0$ and also K_+ is nonsingular, from the canonical factorization of $A(z)$, we have $(I - \lambda^{-1}G_+)u = 0$. The existence and the expression of the canonical factorization of $\widetilde{A}(z)$ can be obtained from Theorem 4.2 by setting $U(z) = (I - zR_+)K_+$ and $L(z) = I - zG_+$. In fact, Theorem 4.2 implies that $\widetilde{A}(z)$ has the canonical factorization $\widetilde{A}(z) = \widetilde{U}(z)\widetilde{L}(z)$ with $\widetilde{U}(z) = U(z)$ and $\widetilde{L}(z) = \widetilde{L}_0 + z\widetilde{L}_1$, where $\widetilde{L}_0 = L_0 = I$ and $\widetilde{L}_1 = L_1 - (\lambda - \mu)\lambda^{-1}L_1 Q = -G_+ + (\lambda - \mu)\lambda^{-1}G_+Q$. Since $\lambda^{-1}G_+Q = Q$, we find that $\widetilde{L}(z) = I - z(G_+ + (\mu - \lambda)Q)$. To prove that $\widetilde{A}(z^{-1})$ has a canonical factorization, we apply Theorem 4.1 to $\widetilde{A}(z)$. To this purpose, we show that the matrix $\widetilde{W} = \sum_{i=0}^{+\infty} \widetilde{G}_+^i \widetilde{K}_+^{-1} \widetilde{R}_+^i$ is nonsingular. Observe that $\widetilde{G}_+^i = G_+^i + (\mu - \lambda)QG_+^{i-1}$, for $i \geq 1$. Therefore, since $\widetilde{R}_+ = R_+$ and $\widetilde{K}_+ = K_+$, we may write

$$\begin{aligned}
\widetilde{W} &= K_+^{-1} + \sum_{i=1}^{+\infty}(G_+^i + (\mu - \lambda)QG_+^{i-1})K_+^{-1}R_+^i \\
&= K_+^{-1} + \sum_{i=1}^{+\infty} G_+^i K_+^{-1} R_+^i + (\mu - \lambda)Q\left(\sum_{i=0}^{+\infty} G_+^i K_+^{-1} R_+^i\right) R_+ \\
&= W - (\lambda - \mu)QWR_+.
\end{aligned}$$

By Theorem 3.22 of [3] applied to $A(z)$, we have $\det W \neq 0$. Therefore $\det \widetilde{W} = \det(I - (\lambda - \mu)QWR_+W^{-1}) \det W$. Moreover, since $Q = uv^*$, then the matrix $I - (\lambda - \mu)QWRW^{-1}$ is nonsingular if and only if

$$(\lambda - \mu)v^*WR_+W^{-1}u \neq 1. \tag{4.3}$$

Since $G_- = WR_+W^{-1}$, for Theorem 4.1 where $W = H_0$, condition (4.3) holds if $(\lambda - \mu)v^*G_-u \neq 1$, which is satisfied by assumption. Therefore the matrix $\widetilde{W}$ is nonsingular, so that from Theorem 4.1 applied to the matrix Laurent polynomial $\widetilde{A}(z)$, we deduce that $\widetilde{A}(z^{-1})$ has the canonical factorization $\widetilde{A}(z^{-1}) = (I - z\widetilde{R}_-)\widetilde{K}_-(I - z^{-1}\widetilde{G}_-)$ with $\widetilde{K}_- = \widetilde{A}_0 + \widetilde{A}_{-1}\widetilde{G}_- = \widetilde{A}_0 + \widetilde{R}_-A_1$ and $\widetilde{G}_- = \widetilde{W}R_+\widetilde{W}^{-1}$, $\widetilde{R}_- = \widetilde{W}^{-1}\widetilde{G}_+\widetilde{W}$. □

5. Some specific cases and applications

5.1. Left, right, and double shift

Observe that the Brauer theorem and its extensions can be applied to the matrix A^T or to the matrix polynomial $A(z)^T$, where A^T denotes the transpose of A. In this case we need to know a left eigenvector or a left invariant subspace of $A(z)$. The result is that we may shift a group of eigenvalues relying on the knowledge of a left invariant subspace. We refer to this procedure as to *left shift*. It is interesting to point out that right shift and left shift can be combined together. This combination is particularly convenient when dealing with palindromic polynomials. In fact, by shifting a group of eigenvalues by means of the right shift and the reciprocals of these eigenvalues by means of a left shift, we may preserve the palindromic structure.

Performing the double shift is also necessary when we are given a canonical factorization of $A(z)$ and we have to shift both an eigenvalue of modulus greater than 1 and an eigenvalue of modulus less than 1. This situation is encountered in the analysis of certain stochastic processes.

We describe the left shift as follows.

Theorem 5.1. *Let $A(z) = \sum_{i\in\mathbb{Z}} z^i A_i$ be a matrix Laurent series analytic in the annulus $\mathbb{A}_{r_1,r_2}$. Let $\lambda \in \mathbb{A}_{r_1,r_2}$, be an eigenvalue of $A(z)$ such that $v^*A(\lambda) = 0$ for a nonzero vector $v \in \mathbb{C}^n$. Given $\mu \in \mathbb{C}$ define the matrix function*

$$\widetilde{A}(z) = \left(I + \frac{\lambda - \mu}{z - \lambda} S\right) A(z), \quad S = yv^*,$$

where $y \in \mathbb{C}^n$ is such that $v^*y = 1$. Then $\widetilde{A}(z)$ admits the power series expansion $\widetilde{A}(z) = \sum_{i\in\mathbb{Z}} z^i \widetilde{A}_i$, convergent for $z \in \mathbb{A}_{r_1,r_2}$, where

$$\widetilde{A}_i = A_i + (\lambda - \mu) \sum_{j=0}^{\infty} \lambda^j S A_{i+j+1}, \qquad \text{for } i \geq 0,$$

$$\widetilde{A}_i = A_i - (\lambda - \mu) \sum_{j=0}^{\infty} \lambda^{-j-1} S A_{i-j}, \qquad \text{for } i < 0.$$

Moreover, we have $\det \widetilde{A}(z) = \det A(z) \frac{z-\mu}{z-\lambda}$.

Proof. It is sufficient to apply Theorem 2.4 to the matrix function $A(z)^T$. □

We provide a brief description of the double shift where two eigenvalues λ_1 and λ_2 are shifted by means of a right and a left shift respectively. We restrict our attention to the case where $\lambda_1 \neq \lambda_2$. The general case requires a more accurate analysis and is not treated here.

Theorem 5.2. *Let $A(z) = \sum_{i\in\mathbb{Z}} z^i A_i$ be a matrix Laurent series analytic in the annulus $\mathbb{A}_{r_1,r_2}$. Let $\lambda_1, \lambda_2 \in \mathbb{A}_{r_1,r_2}$, $\lambda_1 \neq \lambda_2$, be two eigenvalues of $A(z)$ such that $A(\lambda_1)u = 0$, $v^*A(\lambda_2) = 0$ for nonzero vectors $u, v \in \mathbb{C}^n$. Given $\mu_1, \mu_2 \in \mathbb{C}$ define the matrix function*

$$\widetilde{A}(z) = \left(I + \frac{\lambda_2 - \mu_2}{z - \lambda_2} S\right) A(z) \left(I + \frac{\lambda_1 - \mu_1}{z - \lambda_1} Q\right), \quad Q = uw^*, \quad S = yv^* \tag{5.1}$$

where $w, y \in \mathbb{C}^n$ *are such that* $w^*u = 1$, $v^*y = 1$. *Then* $\widetilde{A}(z)$ *admits the power series expansion* $\widetilde{A}(z) = \sum_{i\in\mathbb{Z}} z^i \widetilde{A}_i$, *convergent for* $z \in \mathbb{A}_{r_1,r_2}$, *where*

$$\begin{aligned} \widetilde{A}_i &= \widehat{A}_i + (\lambda_2 - \mu_2)\sum_{j=0}^{\infty} \lambda_2^j S\widehat{A}_{i+j+1}, && \textit{for } i \geq 0, \\ \widetilde{A}_i &= \widehat{A}_i - (\lambda_2 - \mu_2)\sum_{j=0}^{\infty} \lambda_2^{-j-1} S\widehat{A}_{i-j}, && \textit{for } i < 0, \end{aligned} \tag{5.2}$$

with

$$\begin{aligned} \widehat{A}_j &= A_j + (\lambda_1 - \mu_1)\sum_{k=0}^{\infty} \lambda_1^k A_{k+j+1}Q, && \textit{for } j \geq 0, \\ \widehat{A}_j &= A_j - (\lambda_1 - \mu_1)\sum_{k=0}^{\infty} \lambda_1^{-k-1} A_{-k+j}Q, && \textit{for } j < 0. \end{aligned} \tag{5.3}$$

Moreover, we have $\det \widetilde{A}(z) = \det A(z) \frac{z-\mu_1}{z-\lambda_1}\frac{z-\mu_2}{z-\lambda_2}$.

Proof. Denote $\widehat{A}(z) = A(z)(I + \frac{\lambda_1-\mu_1}{z-\lambda_1}Q)$, $Q = uw^*$. In view of Theorem 2.4, the matrix function $\widehat{A}(z)$ is analytic in $\mathbb{A}_{r_1,r_2}$, moreover $\widehat{A}(z) = \sum_{i\in\mathbb{Z}} z^i \widehat{A}_i$ where $\widehat{A}_i$ are defined in (5.3). We observe that $v^*\widehat{A}(\lambda_2) = v^*A(\lambda_2)(I + \frac{\lambda_1-\mu_1}{\lambda_2-\lambda_1}Q) = 0$ since $v^*A(\lambda_2) = 0$ and $\lambda_1 \neq \lambda_2$. Thus, we may apply Theorem 5.1 to the matrix function $\widehat{A}(z)$ to shift the eigenvalue λ_2 to μ_2 and get the shifted function $\widetilde{A}(z) = \sum_{i\in\mathbb{Z}} z^i \widetilde{A}_i$ where the coefficients $\widetilde{A}_i$ are given in (5.2). The expression for the determinant follows by computing the determinant of both sides of (5.1). □

In the above theorem, one can prove that applying the left shift followed by the right shift provides the same matrix Laurent series.

Observe that, if $A(z)$ is a matrix Laurent polynomial, i.e., $A(z) = \sum_{i=-h}^{k} z^i A_i$ then $\widetilde{A}_i = 0$ for $i > k$ and $i < -h$, i.e., also $\widetilde{A}(z)$ is a matrix Laurent polynomial. In particular, if $A(z) = z^{-1}A_{-1} + A_0 + zA_1$, then $\widetilde{A}(z) = \sum_{i=-1}^{1} z^i \widetilde{A}_i$ with

$$\begin{aligned} \widetilde{A}_1 &= A_1, \\ \widetilde{A}_0 &= A_0 + (\lambda_1 - \mu_1)A_1Q + (\lambda_2 - \mu_2)SA_1, \\ \widetilde{A}_{-1} &= A_{-1} - (\lambda_1 - \mu_1)\lambda_1^{-1}A_{-1}Q - (\lambda_2 - \mu_2)\lambda_2^{-1}S(A_{-1} - (\lambda_1 - \mu_1)\lambda_1^{-1}A_{-1}Q). \end{aligned}$$

If the matrix function admits a canonical factorization $A(z) = U(z)L(z^{-1})$ then we can prove that also the double shifted function $\widetilde{A}(z)$ admits a canonical factorization. This result is synthesized in the next theorem.

Theorem 5.3. *Under the assumptions of Theorem* 5.2, *if* $A(z)$ *admits a canonical factorization* $A(z) = U(z)L(z^{-1})$, *with* $U(z) = \sum_{i=0}^{\infty} z^i U_i$, $L(z) = \sum_{i=0}^{\infty} z^i L_i$, *and if* $|\lambda_1|, |\mu_1| < 1$, $|\lambda_2|, |\mu_2| > 1$, *then* $\widetilde{A}(z)$ *has the canonical factorization*

$\widetilde{A}(z) = \widetilde{U}(z)\widetilde{L}(z^{-1})$, *with* $\widetilde{U}(z) = \sum_{i=0}^{\infty} z^i \widetilde{U}_i$, $\widetilde{L}(z) = \sum_{i=0}^{\infty} z^i \widetilde{L}_i$, *where*

$$\widetilde{L}_0 = L_0, \ \widetilde{L}_i = L_i - (\lambda_1 - \mu_1) \sum_{j=1}^{\infty} \lambda_1^{-j} L_{j+i-1} Q \quad i \geq 1,$$

$$\widetilde{U}_i = U_i + (\lambda_2 - \mu_2) \sum_{j=0}^{\infty} \lambda_2^j S U_{i+j+1}, \quad i \geq 0.$$

Proof. Replacing $A(z)$ with $U(z)L(z^{-1})$ in equation (5.1) yields the factorization $\widetilde{A}(z) = \widetilde{U}(z)\widetilde{L}(z^{-1})$ where $\widetilde{U}(z) = (I + \frac{\lambda_2 - \mu_2}{z - \lambda_2} S)U(z)$, and $\widetilde{L}(z^{-1}) = L(z^{-1})(I + \frac{\lambda_1 - \mu_1}{z - \lambda_1} Q)$. Since $v^* U(\lambda_2) = 0$ and $L(\lambda_1^{-1})u = 0$, we have that $\widetilde{U}(z)$ and $\widetilde{L}(z)$ are obtained by applying the left shift and the right shift to $U(z)$ and $L(z^{-1})$ respectively. Thus the expressions for $\widetilde{U}(z)$ and $\widetilde{L}(z)$ follow from Theorems 5.1 and 2.4, respectively. □

5.2. Shifting eigenvalues from/to infinity

Recall that for a matrix polynomial having a singular leading coefficient, some eigenvalues are at infinity. These eigenvalues may cause numerical difficulties in certain algorithms for the polynomial eigenvalue problem. In other situations, having eigenvalues far from the unit circle, possibly at infinity, may increase the convergence speed of certain iterative algorithms. With the shift technique it is possible to move eigenvalues from/to infinity in order to overcome the difficulties encountered in solving specific problems. In this section we examine this possibility.

We restrict our attention to the case of matrix polynomials. Assume we are given $A(z) = \sum_{i=0}^{d} z^i A_i$ where A_d is singular, and a vector $u \neq 0$ such that $A_d u = 0$. Consider the reverse polynomial $A_{\mathcal{R}}(z) = \sum_{i=0}^{d} z^i A_{d-i}$, observe that $A_{\mathcal{R}}(0)u = A_d u = 0$ and that the eigenvalues of $A_{\mathcal{R}}(z)$ are the reciprocals of the eigenvalues of $A(z)$, with the convention that $1/0 = \infty$ and $1/\infty = 0$. Apply to $A_{\mathcal{R}}(z)$ a right shift of $\lambda = 0$ to a value $\sigma \neq 0$, and obtain a shifted polynomial $\widetilde{A}_{\mathcal{R}}(z) = A_{\mathcal{R}}(z)(I - z^{-1}\sigma Q)$, with $Q = uv^*$, $v^* u = 1$, having the same eigenvalues of $A_{\mathcal{R}}(z)$ except for the null eigenvalue which is replaced by σ. Revert again the coefficients of the polynomial $\widetilde{A}_{\mathcal{R}}(z)$ and obtain a new matrix polynomial $\widetilde{A}(z) = \sum_{i=0}^{d} z^i \widetilde{A}_i = z^d A_{\mathcal{R}}(z^{-1}) = A(z)(I - z\sigma Q)$. This polynomial has the same eigenvalues of $A(z)$ except for the eigenvalue at infinity which is replaced by $\mu = \sigma^{-1}$. A direct analysis shows that the coefficients of $\widetilde{A}(z)$ are given by

$$\begin{aligned} &\widetilde{A}_0 = A_0, \\ &\widetilde{A}_i = A_i - \mu^{-1} A_{i-1} Q, \quad i = 1, \ldots, d. \end{aligned} \tag{5.4}$$

A similar formula can be obtained for the left shift.

If the kernel of A_d has dimension $k \geq 1$ and we know a basis of the kernel, then we may apply the above shifting technique k times in order to remove k eigenvalues at infinity. This transformation does not guarantee that the leading coefficient of $\widetilde{A}(z)$ is nonsingular. However, if $\det \widetilde{A}_d = 0$ we may repeat the shift.

An example of this transformation is given by the following quadratic matrix polynomial

$$A(z) = z^2 \begin{bmatrix} 0 & 0 & 0 \\ 0 & 1 & 0 \\ 0 & 0 & 1 \end{bmatrix} + z \begin{bmatrix} 0 & 1 & 1 \\ 0 & 1 & 1 \\ 0 & 1 & 1 \end{bmatrix} + \begin{bmatrix} 1 & 0 & -1 \\ 1 & 2 & 0 \\ 1 & 1 & 1 \end{bmatrix},$$

which has two eigenvalues at infinity, and 4 eigenvalues equal to $\pm\mathbf{i}\sqrt{3}$, $\pm\mathbf{i}$, where $\mathbf{i}$ is the imaginary unit. The Matlab function `polyeig` applied to this polynomial yields the following approximations to the eigenvalues

```
                   Inf + 0.000000000000000e+00i
 1.450878662283769e-15 + 1.732050807568876e+00i
 1.450878662283769e-15 - 1.732050807568876e+00i
-8.679677645252271e-17 + 9.999999999999998e-01i
-8.679677645252271e-17 - 9.999999999999998e-01i
                   Inf + 0.000000000000000e+00i.
```

Since $e_1 = (1, 0, 0)^*$ is such that $A_2 e_1 = 0$, we may apply (5.4) with $\mu = 1$ and $Q = e_1 e_1^*$ yielding the transformed matrix polynomial

$$z^2 \begin{bmatrix} 0 & 0 & 0 \\ 0 & 1 & 0 \\ 0 & 0 & 1 \end{bmatrix} + z \begin{bmatrix} -1 & 1 & 1 \\ -1 & 1 & 1 \\ -1 & 1 & 1 \end{bmatrix} + \begin{bmatrix} 1 & 0 & -1 \\ 1 & 2 & 0 \\ 1 & 1 & 1 \end{bmatrix},$$

which still has the same eigenvalues of $A(z)$ except one eigenvalue at infinity which is replaced by 1. Since e_1 is still in the kernel of the leading coefficient, we may perform again the shift to the above matrix polynomial with $Q = e_1 e_1^*$ and $\mu = 1/2$ yielding the matrix polynomial

$$z^2 \begin{bmatrix} 2 & 0 & 0 \\ 2 & 1 & 0 \\ 2 & 0 & 1 \end{bmatrix} + z \begin{bmatrix} -3 & 1 & 1 \\ -3 & 1 & 1 \\ -3 & 1 & 1 \end{bmatrix} + \begin{bmatrix} 1 & 0 & -1 \\ 1 & 2 & 0 \\ 1 & 1 & 1 \end{bmatrix},$$

which has no eigenvalues at infinity. Applying again `polyeig` to this latter polynomial yields

```
 5.000000000000000e-01 + 0.000000000000000e+00i
 9.999999999999993e-01 + 0.000000000000000e+00i
-1.456365117861229e-16 + 1.732050807568876e+00i
-1.456365117861229e-16 - 1.732050807568876e+00i
-4.885759970651636e-17 + 9.999999999999996e-01i
-4.885759970651636e-17 - 9.999999999999996e-01i.
```

It is evident that eigenvalues $\pm\mathbf{i}$ and $\pm\mathbf{i}\sqrt{3}$ are left unchanged, while the two eigenvalues of the original polynomial at infinity are replaced by 1 and by 1/2.

A similar technique can be applied to shift a finite eigenvalue of $A(z)$ to infinity. Assume that $A(\lambda)u = 0$ for $u \neq 0$ where $\lambda \neq 0$. Consider the reversed polynomial $A_{\mathcal{R}}(z) = z^d A(z^{-1})$ and apply the right shift to $A_{\mathcal{R}}(z)$ to move λ^{-1} to 0. This way we obtain the polynomial $\widetilde{A}_{\mathcal{R}}(z) = A_{\mathcal{R}}(z)(I + \frac{\lambda^{-1}}{z-\lambda^{-1}}Q)$, $Q = uv^*$, with v any vector such that $v^*u = 1$. Revert again the coefficients of the polynomial

$\widetilde{A}_{\mathcal{R}}(z)$ and obtain a new matrix polynomial $\widetilde{A}(z) = \sum_{i=0}^{d} z^i \widetilde{A}_i = z^d A_{\mathcal{R}}(z^{-1}) = A(z)(I + \frac{\lambda^{-1}}{z^{-1}-\lambda^{-1}} Q)$. The equations which relate the coefficients of $A(z)$ and $\widetilde{A}(z)$ can be obtained by reverting the coefficients of $\widetilde{A}_{\mathcal{R}}(z)$ and relying on Theorem 2.2. This way we get

$$\begin{aligned} \widetilde{A}_0 &= A_0, \\ \widetilde{A}_i &= A_i + \lambda^{-1} \sum_{k=0}^{i-1} \lambda^{-k} A_{i-k-1} Q, \quad i = 1, \dots, d. \end{aligned} \tag{5.5}$$

A similar expression can be obtained with the left shift. In the transformed matrix polynomial, the eigenvalue λ is shifted to infinity.

An example of this transformation, where a quadratic polynomial, is transformed into a linear polynomial is given by

$$A(z) = z^2 \begin{bmatrix} 0 & 0 & 0 \\ 0 & 0 & 0 \\ 0 & 0 & 1 \end{bmatrix} + z \begin{bmatrix} 1 & 2 & 3 \\ 1 & 1 & 1 \\ 1 & 0 & 1 \end{bmatrix} + \begin{bmatrix} -1 & -2 & -3 \\ -3 & 1 & -1 \\ -1 & -1 & -1 \end{bmatrix}.$$

This polynomial has 2 eigenvalues at infinity and the eigenvalue $\lambda = 1$ corresponding to the eigenvector $u = (1,1,1)^*$. Shifting $\lambda = 1$ to infinity with $Q = uv^*$, $v^* = [0,0,1]$, yields

$$\widetilde{A}(z) = z^2 \begin{bmatrix} 0 & 0 & 0 \\ 0 & 0 & 0 \\ 0 & 0 & 0 \end{bmatrix} + z \begin{bmatrix} 1 & 2 & -3 \\ 1 & 1 & -2 \\ 1 & 0 & -2 \end{bmatrix} + \begin{bmatrix} -1 & -2 & -3 \\ -3 & 1 & -1 \\ -1 & -1 & -1 \end{bmatrix}$$

that is a linear polynomial.

5.3. Improving the conditioning of eigenvalues

Let $A(z) = A_0 + z(A_1 - D) + z^2 A_2$ be an $n \times n$ matrix polynomial with $n = 20$ and

$$A_2 = \begin{bmatrix} 1 & & & & \\ 2 & 1 & & & \\ & \ddots & \ddots & & \\ & & 2 & 1 \end{bmatrix}, \quad A_1 = \begin{bmatrix} 1 & 1 & 1 & 1 & & & \\ & 1 & \ddots & \ddots & \ddots & & \\ & & \ddots & \ddots & \ddots & 1 \\ & & & 1 & 1 & 1 \\ & & & & 1 & 1 \\ & & & & & 1 \end{bmatrix}, \quad A_0 = \begin{bmatrix} 2 & 1 & & & \\ & \ddots & \ddots & & \\ & & \ddots & 1 \\ \alpha & & \beta & 2 \end{bmatrix},$$

where blanks denote null entries, $\alpha = 3$, $\beta = 15/1000$, and

$$D = \mathrm{diag}(d), \quad d = (A_0 + A_1 + A_2)e, \quad e = [1, \dots, 1]^*.$$

This matrix polynomial has an eigenvalue $\lambda_1 = 1$ corresponding to the eigenvector $e = [1, \dots, 1]^*$ and an eigenvalue very close to 1, whose floating point approxima-

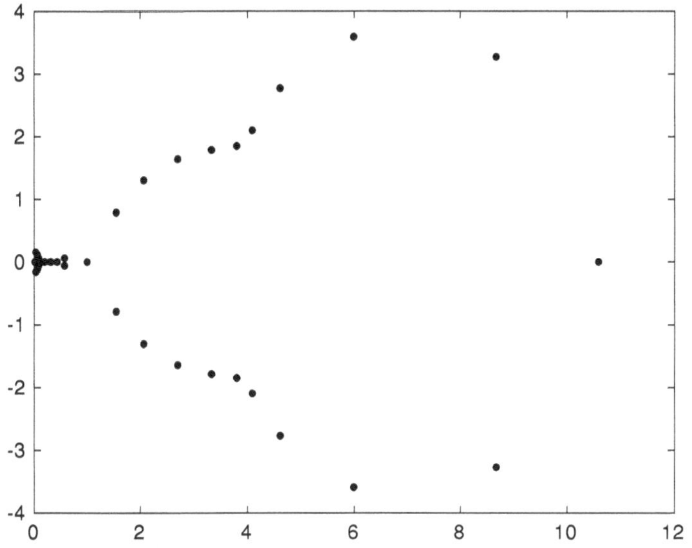

FIGURE 1. Eigenvalues of $A(z) = A_0 + z(A_1 - D) + z^2 A_2$.

tion in high precision is $\lambda_2 = 0.999868223123910$. The eigenvalues are plotted in Figure 1.

The Matlab function `polyeig` computes the two following approximations 0.999999999993444, 0.999868223130466 for λ_1 and λ_2 where the error is pretty large.

Shifting $\lambda = 1$ to $\mu = 10/9$ with $v = [1, 0, \ldots, 0]^*$ and $u = e$ yields a new matrix polynomial $\widetilde{A}(z)$. Applying `polyeig` to $\widetilde{A}(z)$ provides better approximations. In particular, the approximation to the eigenvalue λ_2 is 0.999868223123910. Moreover, almost all the other eigenvalues are approximated more accurately. Figure 2 shows the approximation errors obtained with and without the shift, where we have compared the approximations of the eigenvalues to the values obtained by running `polyeig` with 100 floating point digits. Computations have been performed with Matlab R2015b.

5.4. Palindromic polynomials

Consider a $*$-palindromic matrix polynomial $A(z) = \sum_{i=0}^{d} z^i A_i$, that is, such that $A_i = A_{d-i}^*$ for $i = 0, \ldots, d$, and observe that $A(z)^* = \bar{z}^d A(\bar{z}^{-1})$. This way, one has $A(\lambda)u = 0$ for some $\lambda \in \mathbb{C}$ and some vector $u \neq 0$, if and only if $u^* A(\bar{\lambda}^{-1}) = 0$. Thus the eigenvalues of $A(z)$ come into pairs $(\lambda, \bar{\lambda}^{-1})$, where we assume that $1/0 = \infty$ and $1/\infty = 0$.

If we are given λ and u such that $A(\lambda)u = 0$, we can apply the right shift to move λ to some value μ and the left shift to the polynomial $A(z)^*$, thus shifting

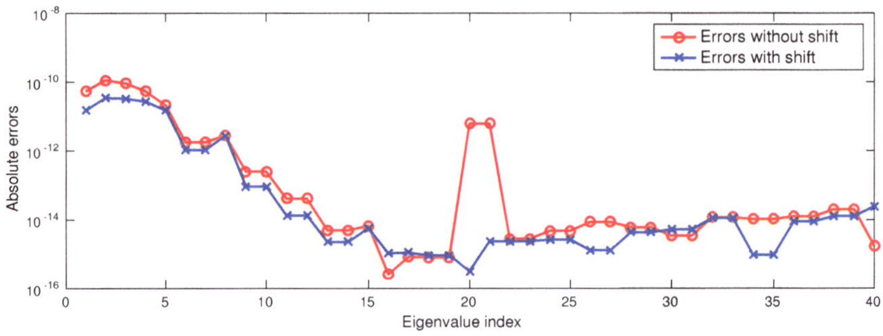

FIGURE 2. Errors with shift (crosses) and without shift (circles).

$\bar{\lambda}$ to $\bar{\mu}$. For $\mu \in \mathbb{C}$, the functional expression of this double shift is

$$\widetilde{A}(z) = \left(I + \frac{\bar{\lambda} - \bar{\mu}}{z^{-1} - \bar{\lambda}} Q\right) A(z) \left(I + \frac{\lambda - \mu}{z - \lambda} Q\right),$$

where $Q = uu^*/(u^*u)$. We may easily check that $\widetilde{A}(z)^* = \bar{z}^d \widetilde{A}(\bar{z}^{-1})$, i.e., $\widetilde{A}(z)$ is $*$-palindromic. Moreover, one has

$$\det \widetilde{A}(z) = \left(1 + \frac{\bar{\lambda} - \bar{\mu}}{z^{-1} - \bar{\lambda}}\right) \det A(z) \left(1 + \frac{\lambda - \mu}{z - \lambda}\right) = \det A(z) \frac{(z - \mu)(1 - \bar{\mu} z)}{(z - \lambda)(1 - \bar{\lambda} z)},$$

hence the pair of eigenvalues $(\lambda, \bar{\lambda}^{-1})$ is moved to the pair $(\mu, \bar{\mu}^{-1})$. The coefficients of the shifted matrix polynomial $\widetilde{A}(z) = \sum_{i=0}^{d} \widetilde{z}^i A_i$ are given by

$$\widetilde{A}_0 = \widehat{A}_0, \quad \widetilde{A}_i = \widehat{A}_i + (\bar{\lambda} - \bar{\mu}) \sum_{k=0}^{i-1} \bar{\lambda}^k Q \widehat{A}_{i-k-1}, \qquad i = 1, \ldots, d,$$

where

$$\widehat{A}_d = A_d, \quad \widehat{A}_j = A_j + (\lambda - \mu) \sum_{k=0}^{d-j-1} \lambda^k A_{k+j+1} Q, \quad j = 0, \ldots, d-1.$$

In the particular case where $\lambda = 0$, the expression for the coefficients simplifies to

$$\widetilde{A}_i = A_i - \mu A_{i+1} Q - \bar{\mu} Q A_{i-1} + |\mu|^2 Q A_i Q, \quad i = 0, \ldots, d,$$

with the convention that $A_{-1} = A_{d+1} = 0$.

In the specific case where $A(z)$ is a matrix polynomial of degree 2, the coefficients of $\widetilde{A}(z)$ have the simpler expression:

$$\begin{aligned}
\widetilde{A}_0 &= A_0 + (\lambda - \mu)(A_1 + \lambda A_2) Q, \\
\widetilde{A}_1 &= A_1 + (\lambda - \mu) A_2 Q + (\bar{\lambda} - \bar{\mu}) Q A_0 + |\lambda - \mu|^2 Q (A_1 + \lambda A_2) Q, \\
\widetilde{A}_2 &= \widetilde{A}_0^*.
\end{aligned}$$

In the above formulas, since $(A_1 + \lambda A_2) Q = -\lambda^{-1} A_0 Q$ and $Q(A_1 + \bar{\lambda} A_0) = -\bar{\lambda}^{-1} Q A_2$, we may easily check that $\widetilde{A}_1^* = \widetilde{A}_1$.

Concerning canonical factorizations, we may prove that if $d = 2m$ and

$$z^{-m}A(z) = U(z)L(z^{-1}), \quad U(z) = L(\bar{z})^*,$$

is a canonical factorization of $z^{-m}A(z)$, and if $|\lambda| < 1, |\mu| < 1$, then the shifted function $z^{-m}\widetilde{A}(z)$ has the canonical factorization

$$z^{-m}\widetilde{A}(z) = \widetilde{U}(z)\widetilde{L}(z^{-1}), \quad \widetilde{U}(z) = \widetilde{L}(\bar{z})^*, \quad \widetilde{L}(z^{-1}) = L(z^{-1})\left(I + \frac{\lambda - \mu}{z - \lambda}Q\right).$$

5.5. Quadratic matrix polynomials and matrix equations

Recall that [3] a matrix Laurent polynomial of the kind $A(z) = z^{-1}A_{-1} + A_0 + zA_1$ has a canonical factorization

$$A(z) = (I - zR_+)K_+(I - z^{-1}G_+)$$

if and only if the matrices R_+ and G_+ are the solutions of the matrix equations

$$X^2A_{-1} + XA_0 + A_1 = 0, \quad A_{-1} + A_0X + A_1X^2 = 0,$$

respectively, with spectral radius less than 1. From Theorem 4.3 it follows that the matrices $\widetilde{R}_+ = R_+$ and $\widetilde{G}_+ = G_+ + (\lambda - \mu)Q$ are the solutions of spectral radius less than 1 to the following matrix equations

$$X^2\widetilde{A}_{-1} + X\widetilde{A}_0 + \widetilde{A}_1 = 0, \quad \widetilde{A}_{-1} + \widetilde{A}_0X + \widetilde{A}_1X^2 = 0,$$

respectively, where $\widetilde{A}(z) = A(z)(I + \frac{\lambda-\mu}{z-\lambda}Q)$ is the matrix Laurent polynomial obtained with the right shift with $|\lambda|, |\mu| < 1$.

If in addition $A(z^{-1})$ has the canonical factorization

$$A(z^{-1}) = (I - zR_-)K_-(I - z^{-1}G_-),$$

and the assumptions of Theorem 4.3 are satisfied, then the matrices $\widetilde{R}_-$ and $\widetilde{G}_-$ defined in Theorem 4.3 are the solution of spectral radius less than 1 to the matrix equations

$$X^2\widetilde{A}_1 + X\widetilde{A}_0 + \widetilde{A}_{-1} = 0, \quad \widetilde{A}_1 + \widetilde{A}_0X + \widetilde{A}_{-1}X^2 = 0,$$

respectively.

For a general matrix polynomial $A(z) = \sum_{i=0}^{d} z^iA_i$ the existence of the canonical factorization $z^{-1}A(z) = U(z)(I - z^{-1}G)$ implies that G is the solution of minimal spectral radius of the equation

$$\sum_{i=0}^{d} A_iX^i = 0.$$

Let λ be an eigenvalue of $A(z)$ of modulus less than 1 such that $A(\lambda)u = 0$, $u \in \mathbb{C}^n$, $u \neq 0$ and let $\mu \in \mathbb{C}$ be such that $|\mu| < 1$. Consider the polynomial $\widetilde{A}(z)$ defined in Theorem 2.2 obtained by means of the right shift. According to Theorem 4.2 the function $z^{-1}\widetilde{A}(z)$ has the canonical factorization $z^{-1}\widetilde{A}(z) = U(z)(I - z^{-1}\widetilde{G})$ where $\widetilde{G} = G - (\lambda - \mu)Q$. Therefore the matrix $\widetilde{G}$ is the solution of minimal spectral radius of the matrix equation

$$\sum_{i=0}^{d} \widetilde{A}_i X^i = 0.$$

As an example of application of this result, consider the $n \times n$ matrix polynomial $A(z) = zI - B(z)$ where $B(z) = \sum_{i=0}^{d} z^i B_i$ with nonnegative coefficients such that $(\sum_{i=0}^{d} B_i)e = e$, $e = [1, \ldots, 1]^*$. Choose $n = 5$, $d = 4$ and construct the coefficients B_i as follows. Let $E = (e_{i,j})$, $e_{i,j} = 1$, $T = (t_{i,j})$ be the upper triangular matrix such that $t_{i,j} = 1$ for $i \leq j$, $D = \mathrm{diag}(57, 49, 41, 33, 25)$, set $B_0 = 9D^{-1}T$, $B_1 = D^{-1}T^*$, $B_2 = D^{-1}E$, $B_3 = D^{-1}E$, $B_4 = D^{-1}$. Then $B(1)e = e$ so that $A(1)e = 0$. The matrix polynomial $A(z)$ has 5 eigenvalues of modulus less than or equal to 1, that is

```
0.15140 - 0.01978i
0.15140 + 0.01978i
0.20936 - 0.09002i
0.20936 + 0.09002i
1.00000 + 0.00000i.
```

Moreover, the eigenvalue of smallest modulus among those which lie outside the unit disk is `1.01258`. Applying cyclic reduction to approximate the minimal solution of the matrix equation $\sum_{i=0}^{4} A_i X^i = 0$, relying on the software of [6], requires 12 iterations. Applying the same method with the same software to the equation $\sum_{i=0}^{4} \widetilde{A}_i X^i = 0$ obtained by shifting the eigenvalue 1 to zero provides the solution in just 6 iterations. In fact the approximation error at step k goes to zero as σ^{2^k} where σ is the ratio between the eigenvalue of largest modulus, among those in the unit disk, and the eigenvalue of smallest modulus, among those out of the unit disk. This ratio is `0.98758` for the original polynomial and `0.20676` for the polynomial obtained after the shift. This explains the difference in the number of iterations.

6. Conclusions and open problems

By means of a functional interpretation, we have extended Brauer's theorem to matrix (Laurent) polynomials, matrix (Laurent) power series and have related the canonical factorizations of these matrix functions to the corresponding factorizations of the shifted functions.

One point that has not been treated in this article is the analysis of the Jordan chains of a matrix polynomial or matrix power series under the right/left or double shift. This is the subject of next investigation.

Another interesting issue which deserves further study is the formal analysis of the conditioning of the eigenvalues of a matrix polynomial under the shift, together with the analysis of the conditioning of the shift operation itself and of the numerical stability of the algorithms for its implementation.

References

[1] D.A. Bini, S. Dendievel, G. Latouche, and B. Meini. General solution of the Poisson equation for Quasi-Birth-and-Death processes. To appear on SIAM J. Appl. Math., 2016.

[2] D.A. Bini, B. Iannazzo, and B. Meini. *Numerical solution of algebraic Riccati equations*, volume 9 of *Fundamentals of Algorithms*. Society for Industrial and Applied Mathematics (SIAM), Philadelphia, PA, 2012.

[3] D.A. Bini, G. Latouche, and B. Meini. *Numerical methods for structured Markov chains*. Numerical Mathematics and Scientific Computation. Oxford University Press, New York, 2005. Oxford Science Publications.

[4] D.A. Bini, G. Latouche, and B. Meini. Shift techniques for Quasi-Birth and Death processes: canonical factorizations and matrix equations. Applied Numerical Mathematics, Available online 8 September 2016, http://dx.doi.org/10.1016/j.apnum.2016.09.001.

[5] D.A. Bini, B. Meini, and I.M. Spitkovsky. Shift techniques and canonical factorizations in the solution of $M/G/1$-type Markov chains. *Stoch. Models*, 21(2-3):279–302, 2005.

[6] D.A. Bini, B. Meini, S. Steffé, and B. Van Houdt. Structured Markov chains solver: tool extension. In Proceedings of the Fourth International ICST Conference on Performance Evaluation Methodologies and Tools (VALUETOOLS '09). Article 20. http://dx.doi.org/10.4108/ICST.VALUETOOLS2009.7443

[7] D.A. Bini and L. Robol. On a Class of Matrix Pencils and ℓ-ifications Equivalent to a Given Matrix Polynomial. *Linear Algebra Appl.*, 502:275–298 (2016).

[8] A. Böttcher and B. Silbermann. *Introduction to large truncated Toeplitz matrices.* Universitext. Springer-Verlag, New York, 1999.

[9] A. Brauer. Limits for the characteristic roots of a matrix. IV. Applications to stochastic matrices. *Duke Math. J.*, 19:75–91, 1952.

[10] R. Bru, R. Cantó, R.L. Soto, and A.M. Urbano. A Brauer's theorem and related results. *Cent. Eur. J. Math.*, 10(1):312–321, 2012.

[11] A. Cicone and S. Serra-Capizzano. Google PageRanking problem: the model and the analysis. *J. Comput. Appl. Math.*, 234(11):3140–3169, 2010.

[12] K.F. Clancey and I. Gohberg. *Factorization of matrix functions and singular integral operators*, volume 3 of *Operator Theory: Advances and Applications*. Birkhäuser Verlag, Basel-Boston, Mass., 1981.

[13] I. Gohberg, P. Lancaster, and L. Rodman. *Matrix Polynomials*. Academic Press Inc. [Harcourt Brace Jovanovich Publishers], New York, 1982. Computer Science and Applied Mathematics.

[14] C.-H. Guo. Comments on a shifted cyclic reduction algorithm for quasi-birth-death problems. *SIAM J. Matrix Anal. Appl.*, 24(4):1161–1166 (electronic), 2003.

[15] C.-H. Guo and N.J. Higham. Iterative solution of a nonsymmetric algebraic Riccati equation. *SIAM J. Matrix Anal. Appl.*, 29(2):396–412, 2007.

[16] C.-H. Guo, B. Iannazzo, and B. Meini. On the doubling algorithm for a (shifted) nonsymmetric algebraic Riccati equation. *SIAM J. Matrix Anal. Appl.*, 29(4):1083–1100, 2007.

[17] C. He, B. Meini, and N.H. Rhee. A shifted cyclic reduction algorithm for quasi-birth-death problems. *SIAM J. Matrix Anal. Appl.*, 23(3):673–691 (electronic), 2001/02.

[18] P. Henrici. *Applied and computational complex analysis. Vol.* 1. Wiley Classics Library. John Wiley & Sons, Inc., New York, 1988. Power series – integration – conformal mapping – location of zeros, Reprint of the 1974 original, A Wiley-Interscience Publication.

[19] R. Horn, S. Serra-Capizzano. A general setting for the parametric Google matrix. *Internet Mathematics*, 3(4):385–411, 2006.

[20] B. Iannazzo and F. Poloni. A subspace shift technique for nonsymmetric algebraic Riccati equations associated with an M-matrix. *Numer. Linear Algebra Appl.*, 20(3):440–452, 2013.

[21] B. Meini. A "shift-and-deflate" technique for quadratic matrix polynomials. *Linear Algebra Appl.*, 438(4):1946–1961, 2013. 16th ILAS Conference Proceedings, Pisa 2010.

[22] R.L. Soto and O. Rojo. Applications of a Brauer theorem in the nonnegative inverse eigenvalue problem. *Linear Algebra Appl.*, 416(2-3):844–856, 2006.

[23] L. Taslaman, F. Tisseur, and I. Zaballa. Triangularizing matrix polynomials. *Linear Algebra Appl.*, 439(7):1679–1699, 2013.

[24] F. Tisseur and I. Zaballa. Triangularizing quadratic matrix polynomials. *SIAM J. Matrix Anal. Appl.*, 34(2):312–337, 2013.

D.A. Bini and B. Meini
Dipartimento di Matematica
Università di Pisa
Largo Bruno Pontecorvo, 5
I-56127 Pisa, Italy
e-mail: `dario.bini@unipi.it`
`beatrice.meini@unipi.it`

Operator Theory:
Advances and Applications, Vol. 259, 179–212

Eigenvalues of Hermitian Toeplitz Matrices Generated by Simple-loop Symbols with Relaxed Smoothness

J.M. Bogoya, S.M. Grudsky and E.A. Maximenko

To Albrecht Böttcher, who has involved the three of us in his passionate research, on the occasion of his 60th birthday

Abstract. In a sequence of previous works with Albrecht Böttcher, we established higher-order uniform individual asymptotic formulas for the eigenvalues and eigenvectors of large Hermitian Toeplitz matrices generated by symbols satisfying the so-called simple-loop condition, which means that the symbol has only two intervals of monotonicity, its first derivative does not vanish on these intervals, and the second derivative is different from zero at the minimum and maximum points. Moreover, in previous works it was supposed that the symbol belongs to the weighted Wiener algebra W^α for $\alpha \geq 4$, or satisfies even stronger smoothness conditions. We now use a different technique, which allows us to extend previous results to the case $\alpha \geq 1$ with additional smoothness at the minimum and maximum points.

MSC 2010. Primary 47B35, Secondary 15A18, 15B05, 65F15, 82B20

Keywords. Toeplitz matrix, eigenvector, spectral asymptotics, asymptotic expansion

1. Introduction

Given a Lebesgue integrable function a on the complex unit circle $\mathbb{T}$, we denote by $T_n(a)$ the $n \times n$ Toeplitz matrix defined by $T_n(a) := (a_{j-k})_{j,k=0}^{n-1}$ ($n = 1, 2, \ldots$), where a_ℓ is the ℓth Fourier coefficient of a. The function a is referred to as the *symbol* of the sequence $\{T_n(a)\}_{n=1}^{\infty}$.

Research of the first author is supported by PUJ project 7023.
Research of the second author is partially supported by CONACYT grants 180049 and 238630.
Research of the third author is partially supported by IPN-SIP project 20160733.

Toeplitz matrices and asymptotics of their spectral characteristics, as n goes to infinity, are extensively studied [5, 7, 13] and have applications in various areas, including stochastic processes and time series analysis [8, 19], signal processing [15, 21, 22], numerical solutions of differential and integral equations [17, 20, 25], image processing [9, 14], and quantum mechanics [11, 12]. See also the bibliographies of [16], [18], and [24].

In this paper we study the individual asymptotic behavior of the eigenvalues of $T_n(a)$, as n goes to infinity, assuming that the function a is sufficiently smooth, takes only real values, and satisfies an additional *simple-loop* condition (explained in the next section), which implies that the function $g(\sigma) := a(e^{i\sigma})$ defined on $[0, 2\pi]$ has only two intervals of monotonicity and the second derivative of g at the maximum and minimum points is non-zero.

This work improves some results from the articles [2, 6, 10] also studying the eigenvalues of the Toeplitz matrices generated by simple-loop symbols with different smoothness conditions. The symbols considered in [6] are Laurent polynomials, in [10] they are C^∞-functions, and [2] deals with the Wiener algebra W^α ($\alpha \geq 4$). In this article we generalize the results of [2]. We consider the case $a \in \mathrm{SL}^\alpha$ with $\alpha \geq 1$ under some additional conditions (see (2.1)). We think that these conditions are the most general in order to guarantee the existence of "normal" asymptotic expansions for the eigenvalues of $T_n(a)$ (see Theorem 2.3 part (i)). Indeed, the conditions (2.2) guarantee that the symbol a has zeros of order 2 at the extreme points with minimal additional smoothness requirements. If we change these conditions by the existence of zeros of another order we will obtain an essentially different asymptotic behavior of the eigenvalues (see for example [1]). Moreover, in previous works we proved that the eigenvalues of $T_n(a)$ are simple for every sufficiently large n, but now we are able to prove it for $n \geq 1$ (this result is a simple generalization of [23]).

Paper [2] is based on the variable change $\lambda = \psi(s)$, related with the concept of the *quantile function* (the role of the quantile function is also explained in [4]), which makes $b(t, \psi(s))$ a differentiable function of s. A considerable part of this calculation was devoted to a careful analysis of the first two derivatives of b with respect to s. In the present paper we transfer our attention from derivatives to the modulus of continuity, which allows us to avoid some technical difficulties from [2] and to relax the smoothness requirement for the symbol a.

2. Main results

For $\alpha \geq 0$, we denote by W^α the weighted Wiener algebra of all functions $a\colon \mathbb{T} \to \mathbb{C}$ that can be represented in the form

$$a(t) = \sum_{j=-\infty}^{\infty} a_j t^j \quad (t \in \mathbb{T}),$$

with Fourier coefficients satisfying

$$\|a\|_\alpha := \sum_{j=-\infty}^{\infty} |a_j|(1+|j|)^\alpha < \infty.$$

For a symbol $a \in W^\alpha$, we consider the real-valued function $g\colon [0,2\pi] \to \mathbb{R}$ given by $g(\sigma) := a(e^{i\sigma})$. As in [2] and [3], for every $\alpha \geq 1$, we define the class SL^α as the set of all symbols a in W^α satisfying:

(i) a is real-valued;

(ii) the range of g is a closed interval $[0,\mu]$ with $\mu > 0$, $g(0) = g(2\pi) = 0$, $g''(0) = g''(2\pi) > 0$, there is a $\varphi_0 \in (0,2\pi)$ such that $g(\varphi_0) = \mu$, $g'(\sigma) > 0$ for $\sigma \in (0,\varphi_0)$, $g'(\sigma) < 0$ for $\sigma \in (\varphi_0, 2\pi)$, and $g''(\varphi_0) < 0$.

Symbols in the class SL^α are known as *simple-loop symbols*. (i) is equivalent to the condition that all matrices $T_n(a)$ $(n \in \mathbb{Z}_+)$ are Hermitian (self-adjoint). If $a \in W^\alpha$, then $g \in C^{\lfloor \alpha \rfloor}[0,2\pi]$ where $\lfloor \alpha \rfloor$ is the integer part of α. So, the condition $a \in \mathrm{SL}^\alpha$ with $\alpha \geq 1$ implies, in particular, that g belongs to $C^1[0,2\pi]$. Moreover, in (ii) we require g to have second derivatives at $\sigma = 0$, φ_0, and 2π.

In this work, for every $\alpha \geq 1$, we introduce a new class of symbols MSL^α (the modified simple loop class). Namely, $a \in \mathrm{MSL}^\alpha$ if $a \in \mathrm{SL}^\alpha$ and

(iii) there exist functions $q_1, q_2 \in W^\alpha$ satisfying

$$a(t) = (t-1)q_1(t) \quad \text{and} \quad a(t) - a(e^{i\varphi_0}) = (t - e^{i\varphi_0})q_2(t). \tag{2.1}$$

According to Lemma 3.1, if $a \in W^\alpha$, then q_1 and q_2 both belong to $W^{\alpha-1}$, but we require the stronger condition (iii) instead. Note that $q_1(1) = a'(1) = 0$ and $q_2(e^{i\varphi_0}) = a'(e^{i\varphi_0}) = 0$. Lemma 3.1 will show that there exist $q_3, q_4 \in W^{\alpha-1}$ such that

$$a(t) = (t-1)^2 q_3(t) \quad \text{and} \quad a(t) - a(e^{i\varphi_0}) = (t - e^{i\varphi_0})^2 q_4(t). \tag{2.2}$$

Moreover, $q_3(1) = \frac{a''(1)}{2} \neq 0$ and $q_4(e^{i\varphi_0}) = \frac{a''(e^{i\varphi_0})}{2} \neq 0$.

For every $\alpha \geq 1$ the class $\mathrm{SL}^{\alpha+1}$ is contained in MSL^α. So every statement in Theorems 2.1, 2.2, and 2.3 formulated for $a \in \mathrm{MSL}^\alpha$ is true for all $a \in \mathrm{SL}^{\alpha+1}$.

For each $\lambda \in [0,\mu]$, there are exactly one $\varphi_1(\lambda)$ belonging to $[0,\varphi_0]$ such that $g(\varphi_1(\lambda)) = \lambda$ and exactly one $\varphi_2(\lambda) \in [\varphi_0, 2\pi]$ such that $g(\varphi_2(\lambda)) = \lambda$. In other words: φ_1, φ_2 are the inverse functions of g restricted to the intervals $[0,\varphi_0]$ and $[\varphi_0, 2\pi]$, respectively. For each $\lambda \in [0,\mu]$, the function g takes values less than or equal to λ on the segments $[0,\varphi_1(\lambda)]$ and $[\varphi_2(\lambda), 2\pi]$. Denote by $\varphi(\lambda)$ the arithmetic mean of the lengths of these two segments:

$$\varphi(\lambda) := \frac{1}{2}(\varphi_1(\lambda) - \varphi_2(\lambda)) + \pi = \frac{1}{2}|\{\sigma \in [0,2\pi] \colon g(\sigma) \leq \lambda\}|,$$

where $|\cdot|$ is the Lebesgue measure on $[0,2\pi]$. The function $\varphi\colon [0,\mu] \to [0,\pi]$ is continuous and bijective. We let $\psi\colon [0,\pi] \to [0,\mu]$ stand for the inverse function. Up to a linear change of variables, φ is the cumulative distribution function associated with a, and ψ is the corresponding quantile function.

In this paper we will use the notation of [3] and express all the main objects in terms of $s = \varphi(\lambda)$ rather than λ itself. In particular we put

$$\sigma_1(s) = \varphi_1(\psi(s)) = \varphi_1(\lambda) \quad \text{and} \quad \sigma_2(s) = \varphi_2(\psi(s)) = \varphi_2(\lambda).$$

The derivatives of the functions φ_1 and φ_2 are unbounded near the points 0 and M, whereas the functions σ_1 and σ_2 have continuous first derivatives on the whole segment $[0,\pi]$ (see Lemma 4.3 below). Additionally, we have

$$g(\sigma_1(s)) = g(\sigma_2(s)) = \psi(s) = \lambda.$$

For each $s \in [0,\pi]$ the symbol $a - \psi(s)$ has 2 zeros: one at $t = e^{i\sigma_1(s)}$ and one at $t = e^{i\sigma_2(s)}$. Consequently $T(a-\psi(s))$ is not invertible. In order to get an invertible Toeplitz operator, we define the function b by

$$b(t,s) := \frac{(a(t)-\psi(s))e^{is}}{(t-e^{i\sigma_1(s)})(t^{-1}-e^{-i\sigma_2(s)})} \qquad (t \in \mathbb{T},\ s \in [0,\pi]). \tag{2.3}$$

For each $s \in [0,\pi]$, Lemma 4.4 guarantees the existence of a Wiener–Hopf factorization $b_\pm(\cdot,s)$ of $b(\cdot,s)$ and the invertibility of $T(b(\cdot,s))$. Therefore the finite section method (see [7] for example) is applicable and the related Toeplitz matrices $T_n(b(\cdot,s))$ are also invertible. Moreover, their inverses are uniformly bounded with respect to $n \in \mathbb{Z}_+$ and $s \in [0,\pi]$, which is crucial in our forthcoming calculations. It is well known that $b(t,s)$ can be represented as $b_-(t,s)b_+(t,s)$ with

$$b_\pm(t,s) = \exp\left(\frac{1}{2}\log b(t,s) \pm \frac{1}{2\pi i}\int_{\mathbb{T}} \frac{\log b(\tau,s)}{\tau - t}d\tau\right). \tag{2.4}$$

We define the function $\eta\colon [0,\pi] \to \mathbb{R}$ by

$$\eta(s) := \frac{1}{4\pi}\int_0^{2\pi} \frac{\log b(e^{i\sigma},s)}{\tan\frac{\sigma-\sigma_2(s)}{2}}d\sigma - \frac{1}{4\pi}\int_0^{2\pi} \frac{\log b(e^{i\sigma},s)}{\tan\frac{\sigma-\sigma_1(s)}{2}}d\sigma. \tag{2.5}$$

The singular integrals above are understood in the Cauchy principal-value sense. For a symbol a let $\lambda_1^{(n)},\dots,\lambda_n^{(n)}$ stand for the eigenvalues of $T_n(a)$. Now we are ready to formulate our main results.

Theorem 2.1. *Let $\alpha \geq 1$ and $a \in \mathrm{MSL}^\alpha$. Then for every $n \geq 1$:*

(i) *the eigenvalues of $T_n(a)$ are all distinct: $\lambda_1^{(n)} < \cdots < \lambda_n^{(n)}$;*

(ii) *the numbers $s_j^{(n)} := \varphi(\lambda_j^{(n)})$ $(j = 1,\dots,n)$ satisfy*

$$(n+1)s_j^{(n)} + \eta(s_j^{(n)}) = \pi j + \Delta_1^{(n)}(j), \tag{2.6}$$

where $\Delta_1^{(n)}(j) = o\left(\frac{1}{n^{\alpha-1}}\right)$ as $n \to \infty$, uniformly in j;

(iii) *for every sufficiently large n, (2.6) has exactly one solution $s_j^{(n)} \in [0,\pi]$ for each $j = 1,\dots,n$.*

Let $d_j^{(n)} = \dfrac{\pi j}{n+1}$. The following two theorems are more general and more precise versions of Theorems 2.2 and 2.3 from [2].

Theorem 2.2. *Under the conditions of Theorem* 2.1*,*

$$s_j^{(n)} = d_j^{(n)} + \sum_{k=1}^{\lfloor\alpha\rfloor} \frac{p_k(d_j^{(n)})}{(n+1)^k} + \Delta_2^{(n)}(j),$$

where $\Delta_2^{(n)}(j) = o\left(\frac{1}{n^\alpha}\right)$ *as* $n \to \infty$*, uniformly in* j*. The coefficients* p_k *can be calculated explicitly; in particular,*

$$p_1(s) = -\eta(s) \quad \text{and} \quad p_2(s) = \eta(s)\eta'(s).$$

Theorem 2.3. *Under the conditions of Theorem* 2.1*,*

$$\lambda_j^{(n)} = \psi(d_j^{(n)}) + \sum_{k=1}^{\lfloor\alpha\rfloor} \frac{r_k(d_j^{(n)})}{(n+1)^k} + \Delta_3^{(n)}(j),$$

where $\Delta_3^{(n)}$ *is* $o\Big(\frac{1}{n^\alpha}(d_j^{(n)}(\pi - d_j^{(n)}))^{\alpha-1}\Big)$ *if* $1 \le \alpha < 2$ *and* $o\Big(\frac{d_j^{(n)}}{n^\alpha}(\pi - d_j^{(n)})\Big)$ *if* $\alpha \ge 2$ *as* $n \to \infty$*, uniformly in* j*. The coefficients* r_k *can be calculated explicitly; in particular,*

$$r_1(s) = -\psi'(s)\eta(s) \quad \text{and} \quad r_2(s) = \frac{1}{2}\psi''(s)\eta^2(s) + \psi'(s)\eta(s)\eta'(s).$$

3. Some general lemmas about Wiener and small Hölder algebras

In this section we gather some statements that are more general than the paper context. The following "Division Lemma" is just Lemma 3.1 from [2] (see a proof there). We denote by $\chi_m(t)$ the monomial t^m. In particular, χ_1 is a generating element of the involutive Banach algebras W^κ, and χ_0 is the identity element of these algebras. In Lemmas 3.1, 3.2, and 3.9, for f in W^κ, we study some properties of the quotient

$$q(t,s) = \frac{f(t) - f(e^{is})}{t - e^{is}}. \tag{3.1}$$

Usually we will treat q as a function of its second argument s, with values in some Wiener algebras: $q(\cdot,s) = \dfrac{f - f(e^{is})}{\chi_1 - e^{is}}$.

Lemma 3.1. *If* $\kappa \ge 1$*,* $f \in W^\kappa$*, and* $s \in \mathbb{R}$*, then*

$$q(\cdot,s) \in W^{\kappa-1} \quad \text{and} \quad \|q(\cdot,s)\|_{\kappa-1} \le \|f\|_\kappa.$$

In the following lemma we show that the quotient $q(\cdot,s)$, considered as an element of $W^{\kappa-1}$, depends continuously on s. Moreover, for $\kappa \ge 2$ this quotient is a continuously differentiable function of s, but each derivation decreases the weight parameter of the Wiener algebra. Given a function of several arguments, we denote by D_j its partial derivative with respect to the jth argument. If $p \in \mathbb{N}$, then D_j^p denotes the partial derivative of order p with respect to the jth argument.

Lemma 3.2. *Let $\kappa \geq 1$, $f \in W^\kappa$, and q be defined by (3.1). Then for every $p \in \{0, \dots, \lfloor \kappa \rfloor - 1\}$ the function $s \mapsto D_2^p q(\cdot, s)$ takes values in $W^{\kappa-1-p}$ and is continuous.*

Proof. As it was shown in the proof of Lemma 3.1 in [2],

$$q(t,\tau) = \sum_{j=0}^{\infty} \Big(\sum_{k=j+1}^{\infty} f_k \tau^{k-j-1} \Big) t^j - \sum_{j=-\infty}^{-1} \Big(\sum_{k=-\infty}^{j} f_k \tau^{k-j-1} \Big) t^j. \tag{3.2}$$

Take $s \in \mathbb{R}$ and $\tau = e^{is}$, and differentiate (3.2) p times with respect to s to obtain:

$$\begin{aligned} D_2^p q(t,s) = &\sum_{j=0}^{\infty} \Big(\sum_{k=j+1}^{\infty} (k-j-1)^p i^p f_k \tau^{k-j-1} \Big) t^j \\ &- \sum_{j=-\infty}^{-1} \Big(\sum_{k=-\infty}^{j} (k-j-1)^p i^p f_k \tau^{k-j-1} \Big) t^j. \end{aligned}$$

Let $\delta > 0$, $s_1, s_2 \in \mathbb{R}$, $\tau_1 = e^{is_1}$, $\tau_2 = e^{is_2}$, and $|\tau_1 - \tau_2| \leq \delta$. Then

$$\begin{aligned} D_2^p q(t,s_1) - D_2^p q(t,s_2) = &\sum_{j=0}^{\infty} \Big(\sum_{k=j+1}^{\infty} (k-j-1)^p i^p f_k (\tau_1^{k-j-1} - \tau_2^{k-j-1}) \Big) t^j \\ &- \sum_{j=-\infty}^{-1} \Big(\sum_{k=-\infty}^{j} (k-j-1)^p i^p f_k (\tau_1^{k-j-1} - \tau_2^{k-j-1}) \Big) t^j. \end{aligned}$$

Passing to the norm of $W^{\kappa-1-p}$:

$$\begin{aligned} &\|D_2^p q(\cdot, s_1) - D_2^p q(\cdot, s_2)\|_{\kappa-1-p} \\ &\quad \leq \sum_{j=0}^{\infty} \sum_{k=j+1}^{\infty} (1+j)^{\kappa-1-p} (k-j-1)^p |f_k| \, |\tau_1^{k-j-1} - \tau_2^{k-j-1}| \\ &\quad\quad + \sum_{j=-\infty}^{-1} \sum_{k=-\infty}^{j} (1+|j|)^{\kappa-1-p} |k-j-1|^p |f_k| \, |\tau_1^{k-j-1} - \tau_2^{k-j-1}|. \end{aligned}$$

If we separate N summands in the sums over k, then we obtain

$$\|q(\cdot, \tau_1) - q(\cdot, \tau_2)\|_{\kappa-1-p} \leq S_1 + S_2 + S_3 + S_4,$$

where

$$S_1 := \sum_{j=0}^{\infty} \sum_{k=j+1}^{j+N} (1+|j|)^{\kappa-1-p} (k-j-1)^p |f_k| \, |\tau_1^{k-j-1} - \tau_2^{k-j-1}|,$$

$$S_2 := \sum_{j=0}^{\infty} \sum_{k=j+1+N}^{\infty} \dots, \quad S_3 := \sum_{j=-\infty}^{-1} \sum_{k=-\infty}^{j-N} \dots, \quad S_4 := \sum_{j=-\infty}^{-1} \sum_{k=j-N+1}^{j} \dots.$$

In the expression S_1 use the inequality $|\tau_1^{k-j-1} - \tau_2^{k-j-1}| \le \delta(k-j-1) \le \delta N$ and change the order of summation:

$$\begin{aligned} S_1 &\le \delta N \sum_{j=0}^{\infty} \sum_{k=j+1}^{j+N} (1+j)^{\kappa-1-p} (k-j-1)^p |f_k| \\ &\le \delta N \sum_{k=1}^{\infty} |f_k| \sum_{k=0}^{k-1} (1+j)^{\kappa-1-p} (k-j-1)^p \\ &\le \delta N \sum_{k=1}^{\infty} (1+k)^{\kappa} |f_k| = \delta N \|f\|_{\kappa}. \end{aligned}$$

In the expression S_2 apply the upper bound $|\tau_1^{k-j-1} - \tau_2^{k-j-1}| \le 2$ and change the order of summation:

$$\begin{aligned} S_2 &\le 2 \sum_{j=0}^{\infty} \sum_{k=j+1+N}^{\infty} (1+j)^{\kappa-1-p} (k-j-1)^p |f_k| \\ &\le 2 \sum_{k=N+1}^{\infty} |f_k| \sum_{j=0}^{k-N-1} (1+j)^{\kappa-1-p} (k-j-1)^p \\ &\le 2 \sum_{k=N+1}^{\infty} (1+k)^{\kappa} |f_k|. \end{aligned}$$

After adding similar upper bounds for S_3 and S_4 we obtain

$$\|q(\cdot, \tau_1) - q(\cdot, \tau_2)\|_{\kappa-1} \le 2N\delta \|f\|_{\kappa} + 4 \sum_{|k| \ge N+1} (1+|k|)^{\kappa} |f_k|. \tag{3.3}$$

Given $\varepsilon > 0$, we apply the condition $f \in W^{\kappa}$ and choose N such that

$$\sum_{|k| \ge N+1} (1+|k|)^{\kappa} |f_k| < \frac{\varepsilon}{8}.$$

Put $\delta := \frac{\varepsilon}{4N\|f\|_{\kappa}}$. Then for every $s_1, s_2 \in \mathbb{R}$ with $|e^{is_1} - e^{is_2}| \le \delta$ the right-hand side of (3.3) is less than ε. □

As we already mentioned in the introduction, if $f \in W^{\kappa}$ ($\kappa \ge 1$), then $f' \in W^{\kappa-1}$. Moreover, $\|f'\|_{\kappa-1} \le \|f\|_{\kappa}$, thus the function $f \mapsto f'$ is a bounded linear operator from W^{κ} to $W^{\kappa-1}$.

In order to use efficiently the condition $f \in W^{\kappa}$, we need to work with some classes of entire and non-entire smoothness. Since we want to prove upper estimates with "small o" in Theorems 2.2 and 2.3, instead of classical Hölder classes we prefer to work with the following "small Hölder classes". For a function h defined on a segment S of the real line, we denote by ω_h the *modulus of continuity* of h:

$$\omega_h(\delta) := \sup\{|h(x_1) - h(x_2)| : x_1, x_2 \in S,\ |x_1 - x_2| \le \delta\}.$$

Given a segment S in $\mathbb{R}$ and a number $0 \le \kappa < 1$, we define $\mathcal{H}^\kappa(S)$ as the set of all functions $h\colon S \to \mathbb{C}$ such that

$$\omega_h(\delta) = o(\delta^\kappa) \quad (\delta \to 0^+). \tag{3.4}$$

The condition (3.4) is equivalent to

$$\lim_{\delta\to 0^+} \sup_{|x_1-x_2|\le\delta} \frac{|h(x_1)-h(x_2)|}{|x_1-x_2|^\kappa} = 0.$$

The set $\mathcal{H}^0(S)$ coincides with $C(S)$. If $0 < \kappa < 1$, then $\mathcal{H}^\kappa(S)$ is a closed subspace of the usual Hölder class with exponent κ. For $\kappa \ge 1$ we define $\mathcal{H}^\kappa(S)$ as the set of all functions $h \in C^{\lfloor\kappa\rfloor}(S)$ such that $h^{(\lfloor\kappa\rfloor)} \in \mathcal{H}^{\kappa-\lfloor\kappa\rfloor}(S)$. We also define a norm on $\mathcal{H}^\kappa(S)$ by

$$\|h\|_{\mathcal{H}^\kappa(S)} := \sum_{p=0}^{\lfloor\kappa\rfloor} \frac{\|h^{(p)}\|_{L_\infty}}{p!} + \frac{1}{\lfloor\kappa\rfloor!} \sup_{\substack{x_1,x_2\in S\\ x_1\ne x_2}} \frac{|h^{(\lfloor\kappa\rfloor)}(x_1)-h^{(\lfloor\kappa\rfloor)}(x_2)|}{|x_1-x_2|^{\kappa-\lfloor\kappa\rfloor}}. \tag{3.5}$$

The usual $\frac{\varepsilon}{3}$-argument shows that $\mathcal{H}^\kappa(S)$ is a Banach space with respect to this norm. Moreover, one can verify directly that the norm (3.5) is submultiplicative. The constant function 1 is the identity element of $\mathcal{H}^\kappa(S)$. So $\mathcal{H}^\kappa(S)$ is a unital Banach algebra. It is easy to see that the (multiplicative) invertibility of an element h of $\mathcal{H}^\kappa(S)$ is equivalent to the condition $h(s) \ne 0$ for every $s \in S$. The character space of $\mathcal{H}^\kappa(S)$ can be identified with S in the natural way.

If $\mathcal{A}$ is a unital Banach algebra, then we can define ω_h for $h\colon S \to \mathcal{A}$ using the norm of $\mathcal{A}$ instead of the absolute value in $\mathbb{C}$, and denote by $\mathcal{H}^\kappa(S,\mathcal{A})$ the corresponding generalization of $\mathcal{H}^\kappa(S)$. We need the three following simple results about compositions and inverse functions.

Lemma 3.3. *Let S and R be two segments of $\mathbb{R}$, $\kappa > 0$, $h \in \mathcal{H}^\gamma(S)$, $h(S) \subseteq R$, $\gamma \ge 1$, $\gamma \ge \kappa$, and $f \in \mathcal{H}^\kappa(R)$. Then $f \circ h \in \mathcal{H}^\kappa(S)$.*

Proof. The result can be verified directly in the case $0 \le \kappa < 1$. Assume now that $\kappa \ge 1$ and consider the $\lfloor\kappa\rfloor$th derivative of $f \circ h$. It is a sum of products of $f^{(k)} \circ h$ ($k \le \lfloor\kappa\rfloor$) times $h^{(k)}$ ($k \le \lfloor\kappa\rfloor$), and all these functions belong to the class $\mathcal{H}^{\kappa-\lfloor\kappa\rfloor}(S)$. Therefore $(f\circ h)^{(\lfloor\kappa\rfloor)} \in \mathcal{H}^{\kappa-\lfloor\kappa\rfloor}(S)$ and $f \circ h \in \mathcal{H}^\kappa(S)$. □

Lemma 3.4. *Let S be a segment in $\mathbb{R}$, $\kappa \ge 1$, and let $f\colon S \to \mathbb{R}$ be a function in $\mathcal{H}^\kappa(S)$. Suppose that $f' \ne 0$ on S. Then the inverse function $h := f^{-1}$ belongs to the class $\mathcal{H}^\kappa(f(S))$.*

Proof. The condition $f \in \mathcal{H}^\kappa(S)$ implies that $f \in C^{\lfloor\kappa\rfloor}(S)$, and therefore we have $h \in C^{\lfloor\kappa\rfloor}(f(S)) = \mathcal{H}^{\lfloor\kappa\rfloor}(f(S))$. The derivative of h is

$$h'(y) = \frac{1}{f'(h(y))}.$$

Since $f' \in \mathcal{H}^{\kappa-1}(S)$, by Lemma 3.3 the composition $f' \circ h$ belongs to the class $\mathcal{H}^{\kappa-1}(f(S))$. Furthermore, the function $f' \circ h$ does not vanish. Being its reciprocal, $h' \in \mathcal{H}^{\kappa-1}(f(S))$, which means that h belongs to $\mathcal{H}^\kappa(f(S))$. □

Lemma 3.5. *Let $\kappa \geq 0$, $\delta > 0$, and $f\colon [0,\delta] \to \mathbb{R}$ be a strictly increasing function in $C^{\lfloor\kappa\rfloor+1}[0,\delta]$ such that $f(x) = xq(x)$, with $q \in \mathcal{H}^\kappa[0,\delta]$ and $q(0) \neq 0$. Then the inverse function $h := f^{-1}$ defined on $[0, f(\delta)]$ can be written as $h(y) = yr(y)$, with $r \in \mathcal{H}^\kappa[0, f(\delta)]$.*

Proof. Since $f \in C^{\lfloor\kappa\rfloor+1}[0,\delta]$ it follows that $h \in C^{\lfloor\kappa\rfloor+1}[0, f(\delta)]$. Inserting h in the expansion of f and solving for $h(y)$ we get

$$h(y) = \frac{y}{q(h(y))}.$$

We see that h has the required form $h(y) = yr(y)$ with r defined on $[0, f(\delta)]$ by

$$r(y) := \frac{1}{q(h(y))}.$$

Because $q \in \mathcal{H}^\kappa[0,\delta]$ and $h \in C^{\lfloor\kappa\rfloor+1}[0, f(\delta)]$ ($\lfloor\kappa\rfloor + 1 > \kappa$), $q \circ h$ belongs to $\mathcal{H}^\kappa[0, f(\delta)]$. Since $q \circ h$ doesn't vanish on the segment $[0, f(\delta)]$, it is bounded from below and $r \in \mathcal{H}^\kappa[0, f(\delta)]$. □

The following lemma says that W^κ can be embedded into $\mathcal{H}^\kappa[0, 2\pi]$.

Lemma 3.6. *Let $\kappa \geq 0$, $f \in W^\kappa$, and $h\colon [0, 2\pi] \to \mathbb{C}$ be defined by $h(x) := f(e^{ix})$. Then $h \in \mathcal{H}^\kappa[0, 2\pi]$.*

Proof. If $f \in W^\kappa$, then $f^{(\lfloor\kappa\rfloor)} \in W^{\kappa-\lfloor\kappa\rfloor}$. The definition of $\mathcal{H}^\kappa[0, 2\pi]$ says that if $h^{(\lfloor\kappa\rfloor)} \in \mathcal{H}^{\kappa-\lfloor\kappa\rfloor}[0, 2\pi]$, then $h \in \mathcal{H}^\kappa[0, 2\pi]$. Therefore it is sufficient to consider the case $0 \leq \kappa < 1$. When $\kappa = 0$ we have $h \in C[0, 2\pi] = \mathcal{H}^0[0, 2\pi]$. Suppose that $0 < \kappa < 1$ and define $\rho_f\colon [0, \infty) \to [0, \infty)$ by

$$\rho_f(N) := \sum_{|k|>N} (1 + |k|)^\kappa |f_k|.$$

The condition $f \in W^\kappa$ implies that $\rho_f(N) \to 0$ as $N \to \infty$. In order to simplify the final part of the proof we prefer to work with positive numbers N without assuming that N is an integer.

For every $\delta > 0$, every $x_1, x_2 \in [0, 2\pi]$ with $|x_1 - x_2| = \delta$, and every $N > 0$ we have

$$|h(x_2) - h(x_1)| = \left| \sum_{j=-\infty}^{\infty} f_j (e^{ijx_2} - e^{ijx_1}) \right| \leq \sum_{|j|\leq N} |f_j|\, |e^{ij(x_2-x_1)} - 1| + 2 \sum_{|j|>N} |f_j|. \tag{3.6}$$

If $|j| \le N$, then $|e^{ij(x_2-x_1)} - 1| = 2|\sin\frac{j}{2}(x_2 - x_1)| \le \delta|j| \le \delta(1 + |j|)$. Thus from (3.6) we get

$$\begin{aligned}|h(x_2) - h(x_1)| &\le \delta \sum_{|j|\le N} (1 + |j|)|f_j| + 2 \sum_{|j|>N} |f_j| \\ &\le \delta \sup_{|j|\le N} \{(1 + |j|)^{1-\kappa}\} \sum_{|j|\le\kappa} |f_j|(1 + |j|)^{\kappa} \\ &\quad + 2 \sup_{|j|>N} \left\{\frac{1}{(1 + |j|)^{\kappa}}\right\} \sum_{|j|>N} |f_j|(1 + |j|)^{\kappa}.\end{aligned}$$

Since the functions $x \mapsto (1 + x)^{1-\kappa}$ and $x \mapsto \frac{1}{(1+x)^{\kappa}}$ for $x \in (0, \infty)$, are strictly increasing and decreasing, respectively, we deduce

$$|h(x_2) - h(x_1)| \le N^{1-\kappa}\delta\|f\|_{\kappa} + 2N^{-\kappa}\rho_f(N).$$

For every $\delta > 0$ and $N > 0$ it follows that

$$\frac{\omega_h(\delta)}{\delta^{\kappa}} \le (\delta N)^{1-\kappa}\|f\|_{\kappa} + 2(\delta N)^{-\kappa}\rho_f(N). \tag{3.7}$$

We will choose $N = N(\delta)$ in such a manner that $\delta N(\delta) \to 0$, $N(\delta) \to \infty$ and the second term in (3.7) tends to 0 as δ tends to 0. Put

$$\varepsilon(\delta) := \max\left\{\sqrt{\delta}, \rho_f^{\frac{1}{2\kappa}}\left(\frac{1}{\sqrt{\delta}}\right)\right\}.$$

Then $\varepsilon(\delta) \to 0$ as $\delta \to 0^+$. Furthermore, since $\varepsilon(\delta) \ge \sqrt{\delta}$ and ρ_f is non-decreasing,

$$\rho_f\left(\frac{\varepsilon(\delta)}{\delta}\right) \le \rho_f\left(\frac{1}{\sqrt{\delta}}\right) \le \varepsilon^{2\kappa}(\delta).$$

Applying (3.7) with $N := N(\delta) := \frac{\varepsilon(\delta)}{\delta}$ we obtain

$$\frac{\omega_h(\delta)}{\delta^{\kappa}} \le \varepsilon^{1-\kappa}(\delta)\|f\|_{\kappa} + 2\varepsilon^{-\kappa}(\delta)\,\varepsilon^{2\kappa}(\delta) = \varepsilon^{1-\kappa}(\delta)\|f\|_{\kappa} + 2\varepsilon^{\kappa}(\delta).$$

This expression tends to 0 as δ tends to 0. □

Given a segment S in $\mathbb{R}$ and a number $\kappa \ge 0$, we denote by $\mathcal{H}_W^{\kappa}(S)$ the set of all functions $h\colon \mathbb{T} \times S \to \mathbb{C}$ such that:

(i) for every $p = 0, \dots, \lfloor\kappa\rfloor$ the function $s \mapsto D_2^p h(\cdot, s)$ is a continuous function from S to $W^{\kappa-p}$;

(ii) for every $p = 0, \dots, \lfloor\kappa\rfloor$ the function $s \mapsto D_2^p h(\cdot, s)$ belongs to the class $\mathcal{H}^{\kappa-\lfloor\kappa\rfloor}(S, W^{\lfloor\kappa\rfloor-p})$:

$$\lim_{\delta\to 0^+}\left(\frac{1}{\delta^{\kappa-\lfloor\kappa\rfloor}} \sup_{\substack{s_1,s_2\in S\\ |s_1-s_2|\le\delta}} \|D_2^p h(\cdot, s_1) - D_2^p h(\cdot, s_2)\|_{\lfloor\kappa\rfloor-p}\right) = 0.$$

We define a norm on $\mathcal{H}_W^\kappa(S)$ by

$$\begin{aligned}\|h\|_{\mathcal{H}_W^\kappa(S)} := & \sum_{p=0}^{\lfloor\kappa\rfloor} \frac{1}{p!} \sup_{s\in S} \|D_2^p h(\cdot,s)\|_{\kappa-p} \\ & + \sum_{p=0}^{\lfloor\kappa\rfloor} \frac{1}{p!} \sup_{\substack{s_1,s_2\in S\\ s_1\neq s_2}} \frac{\|D_2^p h(\cdot,s_1) - D_2^p h(\cdot,s_2)\|_{\lfloor\kappa\rfloor-p}}{|s_1-s_2|^{\kappa-\lfloor\kappa\rfloor}}.\end{aligned} \tag{3.8}$$

In the first sum we use norms in $W^{\kappa-p}$ while in the second sum we use norms in $W^{\lfloor\kappa\rfloor-p}$. We state without a proof the following result which follows directly from the properties of the Wiener algebras W^κ.

Lemma 3.7. *The set of functions $\mathcal{H}_W^\kappa(S)$ provided with pointwise operations and with the norm* (3.8) *is a unital Banach algebra. Its character space is naturally homeomorphic to $\mathbb{T}\times S$. If $f\in\mathcal{H}_W^\kappa(S)$ and $f(t,s)>0$ for every $t\in\mathbb{T}$ and $s\in S$, then f belongs to the exponential group* $\exp(\mathcal{H}_W^\kappa(S))$.

The application of $\mathcal{H}_W^\kappa(S)$ in this paper is based on the following property.

Lemma 3.8. *Let S be a segment in $\mathbb{R}$, $\kappa\geq 0$, $h\in\mathcal{H}_W^\kappa(S)$, and $k,p\in\{0,\dots,\lfloor\kappa\rfloor\}$ such that $k+p\leq\lfloor\kappa\rfloor$. Then the mixed partial derivative $D_1^kD_2^ph$ has the following properties:*

(i) *for every $s\in S$, the function $x\mapsto D_1^kD_2^ph(e^{ix},s)$ belongs to $\mathcal{H}^{\kappa-\lfloor\kappa\rfloor}[0,2\pi]$;*
(ii) *the function $s\mapsto D_1^kD_2^ph(\cdot,s)$ belongs to the class $\mathcal{H}^{\kappa-\lfloor\kappa\rfloor}(S,W^0)$.*

Proof. (i) Fix $s\in S$ and consider the function $f(t):=D_2^ph(t,s)$. By the definition of $\mathcal{H}_W^\kappa(S)$, it belongs to the algebra $W^{\kappa-p}$, and $W^{\kappa-p}$ is contained in $W^{\kappa-\lfloor\kappa\rfloor+k}$. Applying Lemma 3.6 we obtain (i).

(ii) By the definition of $\mathcal{H}_W^\kappa(S)$, the function $s\mapsto D_2^ph(\cdot,s)$ belongs to the class $\mathcal{H}^{\kappa-\lfloor\kappa\rfloor}(S,W^{\lfloor\kappa\rfloor-p})$. Furthermore, $W^{\lfloor\kappa\rfloor-p}$ is embedded in W^k, and the differentiation of order k is a bounded linear operator from W^k to W^0. □

Lemma 3.2 shows that for $p\in\{0,\dots,\lfloor\kappa\rfloor-1\}$ the function $s\mapsto D_2^pq(\cdot,s)$, considered from S to $W^{\kappa-p-1}$, is continuous. For non-entire values of κ we need also some non-entire smoothness with respect to s. The following lemma works as a "heat machine", converting the "temperature descent" from $W^{\kappa-p-1}$ to $W^{\lfloor\kappa\rfloor-p-1}$ into the small Hölderness with exponent $\kappa-\lfloor\kappa\rfloor$, with respect to the second argument.

Lemma 3.9. *Let $\kappa>1$, $f\in W^\kappa$, and q be defined as in Lemma* 3.2. *Then q belongs to $\mathcal{H}_W^{\kappa-1}[0,2\pi]$.*

Proof. We have to show that for $p\in\{0,\dots,\lfloor\kappa\rfloor-1\}$ the function $s\mapsto D_2^pq(\cdot,s)$, considered as a function from S to $W^{\lfloor\kappa\rfloor-p-1}$, belongs to the small Hölder class. For the sake of simplicity, we will consider only the case $p=\lfloor\kappa\rfloor-1$; for other values of p the proof is similar. We will combine ideas from the proofs of Lemmas 3.2

and 3.6. Let $\delta > 0$, $s_1, s_2 \in [0, 2\pi]$ with $|s_1 - s_2| \le \delta$, $\tau_1 = e^{is_1}$, $\tau_2 = e^{is_2}$, and $N > 0$. Similarly to the proof of Lemma 3.2,

$$\begin{aligned}
&\|D_2^{\lfloor\kappa\rfloor-1}q(\cdot,s_1) - D_2^{\lfloor\kappa\rfloor-1}q(\cdot,s_2)\|_0 \\
&\quad\le \sum_{j=0}^{\infty}\sum_{k=j+1}^{\infty}(k-j-1)^{\lfloor\kappa\rfloor-1}|f_k|\,|\tau_1^{k-j-1} - \tau_2^{k-j-1}| \\
&\quad\quad + \sum_{j=-\infty}^{-1}\sum_{k=-\infty}^{j}|k-j-1|^{\lfloor\kappa\rfloor-1}\,|f_k|\,|\tau_1^{k-j-1} - \tau_2^{k-j-1}|.
\end{aligned}$$

Taking out $\lfloor N\rfloor$ summands of the sums over k we obtain

$$\|D_2^{\lfloor\kappa\rfloor-1}q(\cdot,s_1) - D_2^{\lfloor\kappa\rfloor-1}q(\cdot,s_2)\|_0 \le S_1 + S_2 + S_3 + S_4,$$

where

$$S_1 := \sum_{j=0}^{\infty}\sum_{k=j+1}^{j+\lfloor N\rfloor}(k-j-1)^{\lfloor\kappa\rfloor-1}|f_k|\,|\tau_1^{k-j-1} - \tau_2^{k-j-1}|,$$

$$S_2 := \sum_{j=0}^{\infty}\sum_{k=j+\lfloor N\rfloor+1}^{\infty}\ldots,\qquad S_3 := \sum_{j=-\infty}^{-1}\sum_{k=-\infty}^{j-\lfloor N\rfloor}\ldots,\qquad S_4 := \sum_{j=-\infty}^{-1}\sum_{k=j-\lfloor N\rfloor+1}^{j}\ldots.$$

In the expression S_1 we use the inequality $|\tau_1^{k-j-1} - \tau_2^{k-j-1}| \le \delta(k-j-1)$ and change the order of summation:

$$S_1 \le \delta\sum_{j=0}^{\infty}\sum_{k=j+1}^{j+\lfloor N\rfloor}(k-j-1)^{\lfloor\kappa\rfloor}|f_k| = \delta\sum_{k=1}^{\infty}S_5(k)|f_k|,$$

where

$$S_5(k) := \sum_{j=\max\{0,k-\lfloor N\rfloor-1\}}^{k-1}(k-j-1)^{\lfloor\kappa\rfloor} = \sum_{j=0}^{\min\{k-1,\lfloor N\rfloor-1\}} j^{\lfloor\kappa\rfloor} < \begin{cases}(k+1)^{\lfloor\kappa\rfloor+1}, & \text{if } k \le N;\\ N^{\lfloor\kappa\rfloor+1}, & \text{if } k > N.\end{cases}$$

Therefore

$$\begin{aligned}
S_1 &\le \delta\sum_{k\le N}(k+1)^{\lfloor\kappa\rfloor+1}|f_k| + \delta N^{\lfloor\kappa\rfloor+1}\sum_{k>N}^{\infty}|f_k| \\
&\le \delta\max_{k\le N+1}\{k^{\lfloor\kappa\rfloor+1-\kappa}\}\sum_{k\le N+1}k^{\kappa}|f_{k-1}| + \delta N^{\lfloor\kappa\rfloor+1}\max_{k>N+1}\left\{\frac{1}{k^{\kappa}}\right\}\sum_{k>N+1}^{\infty}k^{\kappa}|f_{k-1}| \\
&\le \delta N^{\lfloor\kappa\rfloor+1-\kappa}\sum_{k=1}^{\infty}(k+1)^{\kappa}|f_k|.
\end{aligned}$$

In the expression S_2 we apply the upper bound $|\tau_1^{k-j-1} - \tau_2^{k-j-1}| \le 2$ and change the order of summation:

$$S_2 \le 2\sum_{j=0}^{\infty}\sum_{k>j+N}^{\infty}(k-j-1)^{\lfloor\kappa\rfloor-1}|f_k| = 2\sum_{k>N}^{\infty}\sum_{0\le j<k-N}(k-j-1)^{\lfloor\kappa\rfloor-1}|f_k|$$

$$\le 2\sum_{k>N}^{\infty}\sum_{N\le j\le k-1} j^{\lfloor\kappa\rfloor}|f_k| \le 2\sum_{k>N}^{\infty}(k+1)^{\lfloor\kappa\rfloor}|f_k| = 2\sum_{k>N}^{\infty}\frac{(k+1)^{\kappa}}{(k+1)^{\kappa-\lfloor\kappa\rfloor}}|f_k|$$

$$\le 2\sup_{k>N}\left\{\frac{1}{(k+1)^{\kappa-\lfloor\kappa\rfloor}}\right\}\sum_{k>N}(k+1)^{\kappa}|f_k| \le 2N^{\lfloor\kappa\rfloor-\kappa}\sum_{k>N}(k+1)^{\kappa}|f_k|.$$

The expressions S_3 and S_4 can be bounded from above readily but with sums over negative values of k. Joining all the parts together we obtain

$$\|D_2^{\lfloor\kappa\rfloor-1}q(\cdot,s_1) - D_2^{\lfloor\kappa\rfloor-1}q(\cdot,s_2)\|_0 \le \delta N^{\lfloor\kappa\rfloor+1-\kappa}\|f\|_\kappa + 2N^{\lfloor\kappa\rfloor-\kappa}\rho_f(N). \quad (3.9)$$

Selecting N as in the proof of Lemma 3.6, with $\kappa - \lfloor\kappa\rfloor$ instead of κ, we see that the expression in the right-hand side of (3.9) is $o(\delta^{\kappa-\lfloor\kappa\rfloor})$ as $\delta \to 0^+$. □

4. Removing the zeros of $a - \lambda$

In this section we suppose that $a \in \mathrm{MSL}^\alpha$ with $\alpha \ge 1$ and use the notation of Section 2. We define the functions $z_1, z_2\colon [0,\pi] \to \mathbb{T}$ by

$$z_1(s) := e^{i\sigma_1(s)} \quad \text{and} \quad z_2(s) := e^{i\sigma_2(s)}.$$

The purpose of this section is to prove that the function b in (2.3) belongs to some weighted Wiener algebra and that the function η in (2.5) belongs to some small Hölder class. We start with a well-known fact about the asymptotic expansion of the inverse function.

Lemma 4.1. *Let $\delta > 0$, $\varepsilon > 0$, and $f\colon [0,\delta) \to [0,\varepsilon)$ be a strictly increasing function with the following asymptotic expansion:*

$$f(x) = \sum_{k=1}^{\lfloor\kappa\rfloor} f_k x^k + o(x^\kappa) \quad (x \to 0^+),$$

where $\kappa \ge 1$, $f_1 \ne 0$, $f_2, \dots, f_{\lfloor\kappa\rfloor} \in \mathbb{R}$. Then the inverse function has the asymptotic expansion:

$$f^{-1}(y) = \sum_{k=1}^{\lfloor\kappa\rfloor} h_k y^k + o(y^\kappa) \quad (y \to 0^+),$$

with $h_1 = \dfrac{1}{f_1}$ and $h_2, \dots, h_{\lfloor\kappa\rfloor} \in \mathbb{R}$.

Lemma 4.2. *Let $\alpha \geq 1$ and $a \in \mathrm{MSL}^\alpha$. Then the functions σ_1, σ_2, and ψ have the following asymptotic expansions:*

$$\psi(s) = \sum_{k\in[2,\lfloor\alpha\rfloor+1]\cap 2\mathbb{Z}} c_{\psi,k} s^k + o(s^{\alpha+1}) \quad (s \to 0^+); \tag{4.1}$$

$$\sigma_1(s) = \sum_{k=1}^{\lfloor\alpha\rfloor} c_{\sigma,k} s^k + o(s^\alpha) \quad (s \to 0^+); \tag{4.2}$$

$$\sigma_2(s) = 2\pi + \sum_{k=1}^{\lfloor\alpha\rfloor} (-1)^k c_{\sigma,k} s^k + o(s^\alpha) \quad (s \to 0^+). \tag{4.3}$$

Here $c_{\psi,k}$ and $c_{\sigma,k}$ are some real coefficients; in particular,

$$c_{\psi,2} = \frac{g''(0)}{2}, \quad c_{\psi,4} = \frac{3g''(0)g^{(4)}(0) - 5g'''(0)}{72g''(0)}, \quad c_{\sigma,1} = 1, \quad c_{\sigma,2} = -\frac{g'''(0)}{6g''(0)}.$$

Similarly to (4.1), (4.2), and (4.3), the functions σ_1, σ_2, and ψ expand by powers of $s - \pi$ about the point π.

Proof. From equation (2.2) and Lemma 3.6, g can be written as $g(x) = x^2 q(x)$, where $q \in \mathcal{H}^{\alpha-1}[0, 2\pi]$ and $q(0) \neq 0$. Therefore the function F defined on $[0, \varphi_0]$ by $F(x) := g^{\frac{1}{2}}(x) = xq^{\frac{1}{2}}(x)$ has the following expansion:

$$F(x) = \sum_{k=1}^{\lfloor\alpha\rfloor} c_{F,k} x^k + o(x^\alpha) \quad (x \to 0^+),$$

with $c_{F,1} = \left(\frac{g''(0)}{2}\right)^{\frac{1}{2}}$. Applying Lemma 4.1 we can write the inverse function as

$$F^{-1}(u) = \sum_{k=1}^{\lfloor\alpha\rfloor} c_{\varphi,k} u^k + o(u^\alpha) \quad (u \to 0^+),$$

with $c_{\varphi,1} = \left(\frac{2}{g''(0)}\right)^{\frac{1}{2}}$. Furthermore, $\varphi_1(\lambda) = F^{-1}(\lambda^{\frac{1}{2}})$, thus φ_1 expands in half-integer powers of λ:

$$\varphi_1(\lambda) = \sum_{k=1}^{\lfloor\alpha\rfloor} c_{\varphi,k} \lambda^{\frac{k}{2}} + o(\lambda^{\frac{\alpha}{2}}) \quad (\lambda \to 0^+). \tag{4.4}$$

Since the function $x \mapsto a(e^{ix})$ is 2π-periodic, the expansion of φ_2 has almost the same coefficients, except for the shift 2π and the alternating sign:

$$\varphi_2(\lambda) = 2\pi + \sum_{k=1}^{\lfloor\alpha\rfloor} (-1)^k c_{\varphi,k} \lambda^{\frac{k}{2}} + o(\lambda^{\frac{\alpha}{2}}) \quad (\lambda \to 0^+). \tag{4.5}$$

Subtracting (4.5) from (4.4) and applying the formula $\varphi(\lambda) = \frac{1}{2}(\varphi_1(\lambda)-\varphi_2(\lambda)+2\pi)$ we obtain an expansion of φ in powers of $\lambda^{\frac{k}{2}}$ with odd integers k:

$$\varphi(\lambda) = \sum_{k\in[1,\lfloor\alpha\rfloor]\cap(2\mathbb{Z}+1)} c_{\varphi,k}\lambda^{\frac{k}{2}} + o(\lambda^{\frac{\alpha}{2}}) \quad (\lambda \to 0^+).$$

In other words, $\varphi(\lambda) = G(\lambda^{\frac{1}{2}})$, where

$$G(u) = \sum_{k\in[1,\lfloor\alpha\rfloor]\cap(2\mathbb{Z}+1)} c_{\varphi,k}u^k + o(u^\alpha) \quad (u \to 0^+).$$

Let $H = G^{-1}$. Another application of Lemma 4.1 yields

$$H(s) = \sum_{k\in[1,\lfloor\alpha\rfloor]\cap(2\mathbb{Z}+1)} c_{H,k}s^k + o(s^\alpha) \quad (s \to 0^+),$$

with $c_{H,1} = \left(\frac{g''(0)}{2}\right)^{\frac{1}{2}}$. The function ψ is the square of H: $\psi = H^2$, which explains that the expansion (4.1) has only even powers of s and an error term of the order $o(s^{\alpha+1})$. Finally note that $\sigma_1(s) = \varphi_1(\psi(s)) = F^{-1}(\psi^{\frac{1}{2}}(s)) = F^{-1}(H(s))$, which leads to (4.2). Formula (4.3) is obtained in a similar manner. □

Lemma 4.3. *Let $\alpha \geq 1$ and $a \in \mathrm{MSL}^\alpha$. Then σ_1, σ_2, and $\psi \in \mathcal{H}^\alpha[0,\pi]$. Moreover, the functions*

$$s \mapsto \frac{\sigma_1(s)}{s}, \quad s \mapsto \frac{\sigma_2(s)}{s}, \quad s \mapsto \frac{\sigma_1(s)}{\pi - s}, \quad s \mapsto \frac{\sigma_2(s)}{\pi - s}$$

belong to the class $\mathcal{H}^{\alpha-1}[0,\pi]$.

Proof. From definition we know that σ_1, σ_2, and ψ belong to the class $C^{\lfloor\alpha\rfloor}(0,\pi)$. Their first derivatives are:

$$\psi'(s) = \frac{2g'(\sigma_1(s))(-g'(\sigma_2(s)))}{g'(\sigma_1(s)) - g'(\sigma_2(s))}, \quad \sigma_1'(s) = \frac{-2g'(\sigma_2(s))}{g'(\sigma_1(s)) - g'(\sigma_2(s))},$$

$$\sigma_2'(s) = \frac{2g'(\sigma_1(s))}{g'(\sigma_1(s)) - g'(\sigma_2(s))}.$$

Here $g' \in \mathcal{H}^{\alpha-1}(0,\pi)$ and $\sigma_1, \sigma_2 \in C^{\lfloor\alpha\rfloor}(0,\pi)$. Since $\lfloor\alpha\rfloor \geq 1$, by Lemma 3.3 we conclude that $g' \circ \sigma_1, g' \circ \sigma_2 \in \mathcal{H}^{\alpha-1}(0,\pi)$. So all numerators and denominators are functions of the class $\mathcal{H}^{\alpha-1}(0,\pi)$. Moreover, on every interval of the form $[\delta, \pi - \delta]$ the denominators are bounded from below. Thus the restrictions of ψ', σ_1', and σ_2' to $[\delta, \pi - \delta]$ belong to the class $\mathcal{H}^{\alpha-1}[\delta, \pi - \delta]$, and the restrictions of ψ, σ_1, and σ_2 to $[\delta, \pi - \delta]$ belong to the class $\mathcal{H}^\alpha[\delta, \pi - \delta]$.

In order to analyze the situation near the point 0, consider the function $F(x) = g^{\frac{1}{2}}(x) = xq^{\frac{1}{2}}(x)$ from the proof of Lemma 4.2. We already know that $q \in \mathcal{H}^{\alpha-1}[0,\varphi_0]$. The values of q are positive and separated from zero, therefore $q^{\frac{1}{2}} \in \mathcal{H}^{\alpha-1}[0,\varphi_0]$.

Recall that by equations (2.1) and (2.2) we have $a(t) = (t-1)q_1(t)$ with $q_1 \in W^\alpha$ and $q_1(t) = (t-1)q_3(t)$ with $q_3 \in W^{\alpha-1}$. Therefore a' satisfies the relation $a'(t) = q_1(t) + (t-1)q_1'(t) = (t-1)(q_3(t) + q_1'(t))$, where $q_3 + q_1' \in W^{\alpha-1}$, and the function $x \mapsto \frac{g'(x)}{x}$ belongs to the class $\mathcal{H}^{\alpha-1}[0, \varphi_0]$. From the identities $g(x) = F^2(x)$ and $g'(x) = 2F(x)F'(x)$ we obtain

$$F'(x) = \frac{g'(x)}{2F(x)} = \frac{g'(x)}{2x} \frac{x}{F(x)} = \frac{g'(x)}{2x} \frac{1}{q^{\frac{1}{2}}(x)},$$

thus $F' \in \mathcal{H}^{\alpha-1}[0, \varphi_0]$ and $F \in \mathcal{H}^{\alpha}[0, \varphi_0]$. Repeating the steps of the proof of Lemma 4.2 and applying Lemma 3.4 we see that G, H, ψ, σ_1, and σ_2 belong to the class $\mathcal{H}^{\alpha}[0, \varphi_0]$.

Since $F(x) = xq^{\frac{1}{2}}(x)$ with $q \in \mathcal{H}^{\alpha-1}[0, \varphi_0]$, by Lemma 3.5 we conclude that the quotient $u \mapsto \frac{F^{-1}(u)}{u}$ belongs to $\mathcal{H}^{\alpha-1}[0, \varphi_0]$. Consequently we obtain that $s \mapsto \frac{\sigma_1(s)}{s}$ and $s \mapsto \frac{\sigma_2(s)}{s}$ belong to the class $\mathcal{H}^{\alpha-1}[0, \varphi_0]$. The situation near the right extreme is considered in a similar manner. □

The following lemma shows some good properties of b, $\log b$, and $b_\pm$, considered as functions from $[0, \pi]$ to $W^{\alpha-1}$.

Lemma 4.4. *Let $\alpha \geq 1$ and $a \in \mathrm{MSL}^\alpha$. Then b has the following properties:*

(i) *the function $s \mapsto b(\cdot, s)$ acts from $[0, \pi]$ to $W^{\alpha-1}$ and is continuous;*
(ii) *the function $b\colon \mathbb{T} \times [0, \pi] \to \mathbb{R}$ is positive, bounded away from zero, and continuous;*
(iii) *the function $s \mapsto \log b(\cdot, s)$, defined as the principal (positive) logarithm of $b(\cdot, s)$, acts from $[0, \pi]$ to $W^{\alpha-1}$ and is continuous;*
(iv) *the functions $s \mapsto b_\pm(\cdot, s)$ act from $[0, \pi]$ to $W^{\alpha-1}$ and are continuous;*
(v) $\sup\limits_{s \in [0,\pi]} \|b(\cdot, s)\|_{\alpha-1} < \infty$, $\sup\limits_{s \in [0,\pi]} \|b_\pm(\cdot, s)\|_{\alpha-1} < \infty$.

Proof. (i) Starting with (2.3) we rewrite b in the form

$$b(t, s) = z_0(s)t \, \frac{a(t) - \psi(s)}{(t - z_1(s))(t - z_2(s))},$$

where $z_0(s) := e^{\frac{i}{2}(\sigma_1(s) + \sigma_2(s))}$. Using the relation $\psi(s) = a(z_1(s)) = a(z_2(s))$, for $s \in (0, \pi)$ we obtain the following "partial fraction decomposition" of b:

$$b(t, s) = \frac{z_0(s)t}{z_1(s) - z_2(s)} \left(\frac{a(t) - a(z_1(s))}{t - z_1(s)} - \frac{a(t) - a(z_2(s))}{t - z_2(s)} \right).$$

By Lemmas 3.1 and 3.2, the function $s \mapsto b(\cdot, s)$ takes values in $W^{\alpha-1}$ and is continuous on $(0, \pi)$. Recall that $a(t) = (t-1)q_1(t)$ where $q_1 \in W^\alpha$ (see the definition of the class MSL^α). For $s \in [0, \frac{\pi}{2}]$, an elementary computation shows that

$$b(t, s) = t(c_1(s)b_1(t, s) + c_2(s)b_2(t, s)), \tag{4.6}$$

where

$$b_1(t,s) := \frac{q_1(t) - q_1(z_1(s))}{t - z_1(s)}, \quad c_1(s) := \frac{z_0(s)(z_1(s)-1)}{z_1(s) - z_2(s)} = -\frac{\sin\frac{\sigma_1(s)}{2}}{\sin\frac{s}{2}}\, e^{\frac{i}{2}\sigma_1(s)},$$

$$b_2(t,s) := \frac{q_1(t) - q_1(z_2(s))}{t - z_2(s)}, \quad c_2(s) := \frac{z_0(s)(1-z_2(s))}{z_1(s) - z_2(s)} = \frac{\sin\frac{\sigma_2(s)}{2}}{\sin\frac{s}{2}}\, e^{\frac{i}{2}\sigma_2(s)}.$$

According to Lemmas 3.1 and 3.2, $b_1(\cdot,s)$ and $b_2(\cdot,s)$ belong to $W^{\alpha-1}$ and depend continuously on s. Lemma 4.3 tells us that c_1 and c_2 are continuous, and the multiplication by the monomial $\chi_1(t) = t$ is a continuous operation in $W^{\alpha-1}$. Thus the function $b(\cdot,s)$ belongs to $W^{\alpha-1}$ and depends continuously on s. These arguments work in $[0,\frac{\pi}{2}]$. The situation on $[\frac{\pi}{2},\pi]$ can be studied in a similar manner using the function q_2 from the definition of MSL^α.

(ii) It follows from (i) that the function b is well defined and continuous on $\mathbb{T}\times[0,\pi]$. A simple calculation shows that

$$b(e^{i\sigma},s) = \frac{\psi(s) - a(e^{i\sigma})}{4\sin\frac{\sigma-\sigma_1(s)}{2}\sin\frac{\sigma-\sigma_2(s)}{2}}.$$

Using the simple-loop condition one easily verifies that b is always positive. Applying the continuity we conclude that b is bounded away from zero.

(iii) and (iv) This follows from the fact that the logarithm operation is continuous on the exponential group of the algebra $W^{\alpha-1}$, and the singular integral operator is continuous on $W^{\alpha-1}$.

(v) This follows from (i) and (iv), because the continuity over the compact $[0,\pi]$ implies boundedness. □

Now we will prove that b and η belong to some small Hölder classes.

Lemma 4.5. *Let $\alpha \ge 1$ and $a \in \mathrm{MSL}^\alpha$. Define $\Xi\colon \mathbb{T}\times[0,\pi]\to\mathbb{C}$ by*

$$\Xi(t,s) := \frac{1}{4\pi}\int_0^{2\pi} \frac{\log b(e^{i\sigma},s)}{\tan\frac{\sigma-\theta}{2}}\, d\sigma \quad (t = e^{i\theta}). \tag{4.7}$$

Then b, $\log b$, and Ξ belong to $\mathcal{H}_W^{\alpha-1}[0,\pi]$.

Proof. Recall that b can be decomposed in the form (4.6) for $s\in[0,\frac{\pi}{2}]$. Using the Lemmas 3.9 and 4.3 we obtain that b_1 and b_2 belong to the class $\mathcal{H}_W^{\alpha-1}[0,\frac{\pi}{2}]$. The quotients c_1 and c_2 can be written as

$$c_1(s) = -\frac{\sigma_1(s)}{s}\,\frac{\operatorname{sinc}\frac{\sigma_1(s)}{2}}{\operatorname{sinc}\frac{s}{2}}\, e^{\frac{i}{2}\sigma_1(s)} \quad \text{and} \quad c_2(s) = \frac{\sigma_2(s)}{s}\,\frac{\operatorname{sinc}\frac{\sigma_2(s)}{2}}{\operatorname{sinc}\frac{s}{2}}\, e^{\frac{i}{2}\sigma_2(s)}.$$

Using Lemma 4.3 and the analyticity of the cardinal sine function we see that $c_1, c_2 \in \mathcal{H}^{\alpha-1}[0,\frac{\pi}{2}]$. Therefore b belongs to the class $\mathcal{H}_W^{\alpha-1}[0,\frac{\pi}{2}]$. A similar reasoning is applied on $[\frac{\pi}{2},\pi]$. So $b\in\mathcal{H}_W^{\alpha-1}[0,\pi]$.

Since b is positive on $\mathbb{T}\times[0,\pi]$, by Lemma 3.7 we have $\log b\in\mathcal{H}_W^{\alpha-1}[0,\pi]$. The singular integral operator is bounded and linear on the Wiener classes, therefore $\Xi\in\mathcal{H}_W^{\alpha-1}[0,\pi]$. □

Lemma 4.6. *Let $a \in \mathrm{MSL}^\alpha$ with $\alpha \geq 1$. Then η belongs to $\mathcal{H}^{\alpha-1}[0,\pi]$. Moreover, $\eta(0) = \eta(\pi) = 0$.*

Proof. By Lemma 4.5 we know that the function Ξ given by (4.7) belongs to $\mathcal{H}_W^{\alpha-1}[0,\pi]$. Define $\Omega\colon [0,2\pi] \times [0,\pi] \to \mathbb{C}$ by

$$\Omega(x,s) := \Xi(e^{ix}, s).$$

Given a pair $k, p \in \{0,\dots,\lfloor\alpha\rfloor - 1\}$ with $k + p \leq \lfloor\alpha\rfloor - 1$, consider the functions

$$f_{1,k,p}(s) := D_1^k D_2^p \Omega(\sigma_1(s), s), \qquad f_{2,k,p}(s) := D_1^k D_2^p \Omega(\sigma_2(s), s). \tag{4.8}$$

Using the Lemma 3.8, the Lipschitz continuity of σ_1, and the upper bound

$$\begin{aligned}|f_{1,k,p}(s_1) - f_{1,k,p}(s_2)| &\leq |D_1^k D_2^p \Omega(\sigma_1(s_1), s_1) - D_1^k D_2^p \Omega(\sigma_1(s_2), s_1)| \\ &\quad + |D_1^k D_2^p \Omega(\sigma_1(s_2), s_1) - D_1^k D_2^p \Omega(\sigma_1(s_2), s_2)| \\ &\leq |D_1^k D_2^p \Omega(\sigma_1(s_1), s_1) - D_1^k D_2^p \Omega(\sigma_1(s_2), s_1)| \\ &\quad + \|D_1^k D_2^p \Omega(\cdot, s_1) - D_1^k D_2^p \Omega(\cdot, s_2)\|_{L_\infty},\end{aligned}$$

we see that $f_{1,k,p} \in \mathcal{H}^{\alpha-\lfloor\alpha\rfloor}[0,2\pi]$. A similar reasoning works for $f_{2,k,p}$. Finally, η (given by (2.5)) can be written as

$$\eta(s) = \Omega(\sigma_2(s), s) - \Omega(\sigma_1(s), s).$$

The derivative $\eta^{(\lfloor\alpha\rfloor-1)}$ can be written as a sum of products of functions (4.8) times some derivatives of σ_1 and σ_2 with orders $\leq \lfloor\alpha\rfloor - 1$. All these functions belong to the class $\mathcal{H}^{\alpha-\lfloor\alpha\rfloor}[0,\pi]$, so $\eta^{(\lfloor\alpha\rfloor-1)} \in \mathcal{H}^{\alpha-\lfloor\alpha\rfloor}[0,\pi]$ and $\eta \in \mathcal{H}^{\alpha-1}[0,\pi]$. □

5. Derivation of the equation for the eigenvalues

Let $f(t) = \sum_{j=-\infty}^{\infty} f_j t^j$ ($t \in \mathbb{T}$) be a function in the Lebesgue space $L_2(\mathbb{T})$. Recall the function $\chi_n(t) := t^n$ and the familiar operators

$$[P_n f](t) := \sum_{j=0}^{n-1} f_j t^j, \quad [Q_n f](t) := \sum_{j=n}^{\infty} f_j t^j, \quad [Pf](t) := \sum_{j=0}^{\infty} f_j t^j,$$

$$[Qf](t) := \sum_{j=-\infty}^{-1} f_j t^j, \quad \text{and} \quad [W_n f](t) := \sum_{j=0}^{n-1} f_{n-1-j} t^j.$$

The image $L_2^{(n)}(\mathbb{T})$ of the operator P_n is naturally isomorphic to $\mathbb{C}^n$. For a symbol a in $L_\infty(\mathbb{T})$, the Toeplitz matrix $T_n(a)$ may be identified with the operator $T_n(a)\colon L_2^{(n)}(\mathbb{T}) \to L_2^{(n)}(\mathbb{T})$ given by $T_n(a)f := P_n(af)$. The following result generalizes to the Hermitian case a result stated by Trench [23] for the real symmetric case. The symbol is supposed to be continuous and to have two intervals of monotonicity, but there are no requirements on the smoothness. So the class of symbols in Lemma 5.1 is larger than our classes MSL^α.

Lemma 5.1. *Suppose that a is a continuous real-valued function on $\mathbb{T}$, the function $g(x) := a(e^{ix})$ is non-decreasing on $[0, \varphi_0]$ and non-increasing on $[\varphi_0, 2\pi]$, with $0 = g(0) < g(\varphi_0) = \mu$. Then for every n the matrix $T_n(a)$ has n distinct eigenvalues in $(0, \mu)$.*

Proof. We repeat the proof of Lemma 1 from [23], with small modifications. Let λ be an eigenvalue of $T_n(a)$. It is known that $\lambda \in (0, \mu)$, see [13]. Define $h: [0, 2\pi] \to \mathbb{R}$ by $h(\sigma) = g(\sigma) - \lambda$. Take $\sigma_1 \in [0, \varphi_0]$ and $\sigma_2 \in [\varphi_0, 2\pi]$ satisfying the relation $g(\sigma_1) = g(\sigma_2) = \lambda$. Then $h(\sigma) \geq 0$ for every $\sigma \in [\sigma_1, \sigma_2]$ and $h(\sigma) \leq 0$ for every $\sigma \in [0, \sigma_1] \cup [\sigma_2, 2\pi]$.

If $X \in L_2^{(n)}(\mathbb{T})$ is an eigenfunction associated with the eigenvalue λ, then for every polynomial $Q \in L_2^{(n)}(\mathbb{T})$ we have

$$\langle (a - \lambda)X, Q \rangle = \langle (a - \lambda)X, P_n Q \rangle = \langle P_n (a - \lambda)X, Q \rangle = 0,$$

where $\langle \cdot, \cdot \rangle$ is the inner product of $L_2(\mathbb{T})$. In other words,

$$\int_0^{2\pi} h(\sigma) X(e^{i\sigma}) \overline{Q(e^{i\sigma})} \, d\sigma = 0. \tag{5.1}$$

Suppose that there are two linearly independent eigenfunctions U and V in $L_2^{(n)}(\mathbb{T})$ associated with λ. Let X be their non-trivial linear combination satisfying the relation $X(e^{i\sigma_1}) = 0$. Consider the following polynomial Q:

$$Q(z) = X(z) \frac{z - e^{i\sigma_2}}{z - e^{i\sigma_1}}.$$

Substituting Q into (5.1) yields

$$\int_0^{2\pi} h(\sigma) |X(e^{i\sigma})|^2 \frac{e^{-i\sigma} - e^{-i\sigma_2}}{e^{-i\sigma} - e^{-i\sigma_1}} \, d\sigma = 0,$$

or, equivalently,

$$\int_0^{2\pi} h_1(\sigma)(e^{i\sigma} - e^{i\sigma_1})(e^{-i\sigma} - e^{-i\sigma_2}) \, d\sigma = 0,$$

where

$$h_1(\sigma) := h(\sigma) \left| \frac{X(e^{i\sigma})}{e^{i\sigma} - e^{i\sigma_1}} \right|^2.$$

Since U and V are linearly independent polynomials, the second factor in the definition of h_1 can vanish only on a finite set of points, and the factor h takes non-zero values near 0, φ_0, and 2π. Thus h_1 is different from zero on a set of positive measure. Note that

$$(e^{i\sigma} - e^{i\sigma_1})(e^{-i\sigma} - e^{-i\sigma_2}) = 4e^{\frac{i}{2}(\sigma_1 - \sigma_2)} \sin \frac{\sigma - \sigma_1}{2} \sin \frac{\sigma - \sigma_2}{2}.$$

Hence we obtain

$$\int_0^{2\pi} h_1(\sigma) \sin \frac{\sigma - \sigma_1}{2} \sin \frac{\sigma - \sigma_2}{2} \, d\sigma = 0. \tag{5.2}$$

On the other hand, the definitions of h, σ_1, and σ_2 imply that for every $\sigma \in [0, 2\pi]$

$$h_1(\sigma) \sin \frac{\sigma - \sigma_1}{2} \sin \frac{\sigma - \sigma_2}{2} \leq 0,$$

and this inequality is strict on a set of positive measure. So we obtain a contradiction to (5.2). □

Now we return to the assumption $a \in \mathrm{MSL}^\alpha$ ($\alpha \geq 1$) and recall some tools from [2] that are necessary to formulate our exact equation for the eigenvalues. Since the function $b(\cdot, s)$ is positive, the operator $T_n(b(\cdot, s))$ is invertible for every $n \geq 1$. Consider the polynomial Θ_k defined for $k \geq 1$ by

$$\Theta_k(t, s) := [T_k^{-1}(b(\cdot, s))\chi_0](t). \tag{5.3}$$

Note that Θ_k is a polynomial of degree $k - 1$ in the variable t.

The following Lemma 5.2 can be proved analogously to Lemma 4.1 in [2], using also Lemma 3.1 in [3]. Furthermore, Lemma 5.3 is just Lemma 4.2 from [2]. We state these results here without a proof. Recall that $z_k(s) = e^{i\sigma_k(s)}$ ($k = 1, 2$). For $z_k^m(s)$ take the branch given by $e^{im\sigma_k(s)}$.

Lemma 5.2. *Let $a \in \mathrm{MSL}^\alpha$ with $\alpha \geq 1$ and $n \geq 1$. A number $\lambda = \psi(s)$ is an eigenvalue of $T_n(a)$ if and only if*

$$z_2^{n+1}(s)\Theta_{n+2}(z_1(s), s)\overline{\Theta_{n+2}(z_2(s), s)} = z_1^{n+1}(s)\Theta_{n+2}(z_2(s), s)\overline{\Theta_{n+2}(z_1(s), s)}. \tag{5.4}$$

Lemma 5.3. *If $f \in W^\kappa$ with $\kappa \geq 0$, then*

$$|[Q_n f](t)| \leq \frac{\|Q_n f\|_\kappa}{n^\kappa} \quad \textit{for all} \quad t \in \mathbb{T}.$$

The previous lemma, in particular, says that

$$\|Q_n f\|_{L_\infty} = o\left(\frac{1}{n^\kappa}\right) \quad (n \to \infty).$$

At this point asymptotic analysis makes his entrance. Relation (5.4) is an exact equation for the eigenvalues of $T_n(a)$, but it is too complicated to be solved for s. Thus we need information about the asymptotic behavior of Θ_n as $n \to \infty$.

The Toeplitz operator $T(b(\cdot, s))$ related to the infinite matrix $(b_{k-j}(s))_{k,j=0}^\infty$ is the bounded linear operator on $L_2^+(\mathbb{T}) := PL_2(\mathbb{T})$ acting by the rule

$$T(b(\cdot, s))f := Pb(\cdot, s)f.$$

Recall that $b_\pm(\cdot, s)$ are the Wiener–Hopf factors of $b(\cdot, s)$; see (2.4). They have expansions of the form

$$b_+(t, s) = \sum_{j=0}^\infty u_j(s)t^j \quad \text{and} \quad b_-(t, s) = \sum_{j=0}^\infty v_j(s)t^{-j},$$

and are analytic functions of t inside and outside the complex unit circle, respectively. In particular, the values $b_+(0, s) = u_0(s)$ and $b_-(\infty, s) = v_0(s)$ are well defined. Formula (2.4) also ensures the standard normalization $b_-(\infty, s) = 1$.

It is well known, see, e.g., [5, p. 10] or [7, p. 18], that the inverse of the Toeplitz operator $T(b(\cdot,s))$ is

$$T^{-1}(b(\cdot,s)) = b_+^{-1}(\cdot,s)Pb_-^{-1}(\cdot,s).$$

Because of $Pb_-(t,s) = b_-(\infty,s) = 1$ and $Pb_-^{-1}(t,s) = b_-^{-1}(\infty,s) = 1$, we get

$$T^{-1}(b(\cdot,s))\chi_0 = b_+^{-1}(\cdot,s)Pb_-^{-1}(\cdot,s)\chi_0 = b_+^{-1}(\cdot,s). \tag{5.5}$$

The following lemma coincides with a part of Lemma 4.3 in [2]. We present it here with a different proof.

Lemma 5.4. *Let $a \in \mathrm{MSL}^\alpha$ with $\alpha \geq 1$. We then have the asymptotic expansion*

$$\Theta_n(t,s) = b_+^{-1}(t,s) + R_1^{(n)}(t,s),$$

where $\sup\{|R_1^{(n)}(t,s)|\colon (t,s) \in \mathbb{T}\times[0,\pi]\} = o\left(\frac{1}{n^{\alpha-1}}\right) \quad (n\to\infty).$

Proof. It follows from the definitions in the beginning of this section that P_n converges strongly to P, and $T_n(b(\cdot,s))$ converges strongly to $T(b(\cdot,s))$ as $n\to\infty$. From (5.3) and (5.5) we have $\Theta_n(t,s) = b_+^{-1}(t,s) + R_1^{(n)}(t,s)$, where

$$\begin{aligned} R_1^{(n)}(\cdot,s) =& T_n^{-1}(b(\cdot,s))\chi_0 - T^{-1}(b(\cdot,s))\chi_0 \\ =& T_n^{-1}(b(\cdot,s))\{T(b(\cdot,s)) - T_n(b(\cdot,s))\}T^{-1}(b(\cdot,s))\chi_0 - Q_nT^{-1}(b(\cdot,s))\chi_0. \end{aligned}$$

Using the identities $T(b) = PbP = P_nbP_n + P_nbQ_n + Q_nbP$, $Q_nbb_+^{-1} = Q_nb_- = 0$, $T_n(b) = P_nbP_n$, and (5.5), we obtain

$$R_1^{(n)}(\cdot,s) = T_n^{-1}(b(\cdot,s))P_nb(\cdot,s)Q_nb_+^{-1}(\cdot,s) - Q_nb_+^{-1}(\cdot,s). \tag{5.6}$$

Thanks to Lemma 4.4 the finite section method is applicable to $T(b(\cdot,s))$, thus the ℓ_1-operator norms of the matrices $T_n^{-1}(b(\cdot,s))$ are uniformly bounded with respect to n and s, i.e., there exists $C>0$ satisfying

$$\|T_n^{-1}(b(\cdot,s))y\|_{\ell_1} \leq C\|y\|_{\ell_1} \quad (y\in\mathbb{C}^n, n\in\mathbb{Z}_+, s\in[0,\pi]).$$

Note that $\|\cdot\|_{\ell_1} = \|\cdot\|_0$, $\|\cdot\|_{L_\infty} \leq \|\cdot\|_0$, $\|\cdot\|_\beta \leq \|\cdot\|_\gamma$ for $0\leq\beta\leq\gamma$, and that

$$\|Q_nf\|_0 \leq \sum_{j=n}^{\infty}|f_j|\left(\frac{1+j}{1+n}\right)^{\alpha-1} = \frac{\|Q_nf\|_{\alpha-1}}{(n+1)^{\alpha-1}} = o\left(\frac{1}{n^{\alpha-1}}\right) \quad (n\to\infty).$$

Thus Lemma 4.4 and (5.6) give us

$$\begin{aligned} \|R_1^{(n)}(\cdot,s)\|_{L_\infty} &\leq \|R_1^{(n)}(\cdot,s)\|_0 \\ &\leq \|T_n^{-1}(b(\cdot,s))P_nb(\cdot,s)Q_nb_+^{-1}(\cdot,s)\|_0 + \|Q_nb_+^{-1}(\cdot,s)\|_0 \\ &\leq C\|P_nb(\cdot,s)Q_nb_+^{-1}(\cdot,s)\|_0 + \|Q_nb_+^{-1}(\cdot,s)\|_0 \\ &\leq (C\|P_nb(\cdot,s)\|_0 + 1)\|Q_nb_+^{-1}(\cdot,s)\|_0 \\ &\leq (C\|b(\cdot,s)\|_{\alpha-1} + 1)\frac{\|Q_nb_+^{-1}(\cdot,s)\|_{\alpha-1}}{(n+1)^{\alpha-1}} = o\left(\frac{1}{n^{\alpha-1}}\right) \ (n\to\infty). \quad \square \end{aligned}$$

For every $n \geq 1$ the matrix $T_n(b(\cdot, s))$ is positive definite and invertible; moreover, these matrices and their inverses depend continuously on s. Therefore the functions Θ_n and $R_1^{(n)}$ are well defined and continuous for every $n \geq 1$. The following lemma gives us an implicit equation for the eigenvalues of $T_n(a)$. We will use this equation in the following section to find the corresponding asymptotic expansion.

Lemma 5.5. *Let $a \in \mathrm{MSL}^\alpha$ with $\alpha \geq 1$. Then for every sufficiently large natural number n there exists a real-valued function $R_2^{(n)} \in C[0, \pi]$ with the following properties:*

(i) *a number $\lambda = \psi(s)$ is an eigenvalue of $T_n(a)$ if and only if there is a $j \in \mathbb{Z}$ such that*

$$(n+1)s + \eta(s) - R_2^{(n)}(s) = j\pi; \tag{5.7}$$

(ii) $R_2^{(n)}(0) = R_2^{(n)}(\pi) = 0$;

(iii) $\|R_2^{(n)}\|_{L_\infty} = o\left(\frac{1}{n^{\alpha-1}}\right)$ $(n \to \infty)$.

Proof. We will deduce (5.7) from (5.4). First rewrite equation (5.4) in the form

$$\frac{z_1^{n+1}(s)}{z_2^{n+1}(s)} = \frac{\Theta_{n+2}(z_1(s), s)\overline{\Theta_{n+2}(z_2(s), s)}}{\Theta_{n+2}(z_2(s), s)\overline{\Theta_{n+2}(z_1(s), s)}}. \tag{5.8}$$

The left-hand side of (5.8) is just

$$\frac{z_1^{n+1}}{z_2^{n+1}} = e^{(n+1)i(\sigma_1(s)-\sigma_2(s))} = e^{i(n+1)(2s-2\pi)} = e^{2i(n+1)s}. \tag{5.9}$$

On the other hand, Lemma 5.4 and the definition of η imply (similarly to Lemma 4.4 from [2]) that

$$\frac{\Theta_{n+2}(z_1(s), s)\overline{\Theta_{n+2}(z_2(s), s)}}{\Theta_{n+2}(z_2(s), s)\overline{\Theta_{n+2}(z_1(s), s)}} = e^{2i(-\eta(s)+R_2^{(n)}(s))}, \tag{5.10}$$

where $R_2^{(n)} \in C[0, \pi]$ and $\|R_2^{(n)}\|_{L_\infty} = o\left(\frac{1}{n^{\alpha-1}}\right)$ $(n \to \infty)$. Since $z_1(0) = z_2(0)$ and $z_1(\pi) = z_2(\pi)$, the left-hand side of (5.10) takes the value 1 at the points 0 and π. Therefore $-\eta + R_2^{(n)}$ takes integer values at these points. Note also that $\eta(0) = 0$ and $\eta(\pi) = 0$. For n large enough, $\|R_2^{(n)}\|_{L_\infty}$ is small, thus $R_2^{(n)}(0) = R_2^{(n)}(\pi) = 0$. Using identities (5.9) and (5.10) we rewrite equation (5.8) in the form

$$e^{2i(n+1)s} = e^{2i(-\eta(s)+R_2^{(n)}(s))},$$

which is equivalent to (5.7) with some $j \in \mathbb{Z}$. □

Proof of Theorem 2.1. By Lemma 5.1 we already know that for every positive natural n the matrix $T_n(a)$ has n simple eigenvalues.

Suppose that n is large enough, such that $R_2^{(n)}$ is well defined and has the properties of Lemma 5.5. As in [2], let $F_n(s):=(n+1)s+\eta(s)$. Denote the left-hand side of (5.7) by G_n:

$$G_n(s):=F_n(s)-R_2^{(n)}(s).$$

Then (5.7) can be written as $G_n(s)=j\pi$. The function G_n is continuous on $[0,\pi]$, $G_n(0)=0$, and $G_n(\pi)=(n+1)\pi$. By the Intermediate Value Theorem, for each $j\in\{1,\dots,n\}$ the equation $G_n(s)=j\pi$ has *at least one* solution in $(0,\pi)$. We don't know if G_n is strictly increasing for every n, but we can prove the uniqueness of the solution with the following arguments.

By Lemma 5.5 there is a correspondence between solutions of (5.7) with integer j and eigenvalues of $T_n(a)$, and this correspondence is given by the strictly increasing function ψ. Therefore there are *exactly* n points belonging to the interval $(0,\pi)$ where the function G_n takes values $\pi,2\pi,\dots,n\pi$. So, for each $j\in\{1,\dots,n\}$ the equation $G_n(s)=j\pi$ has one and only one solution denoted by $s_j^{(n)}$. Since the function G_n is continuous, $G_n(0)=0$ and $\pi<\cdots<n\pi$, another application of the Intermediate Value Theorem yields the inequalities $s_1^{(n)}<\cdots<s_n^{(n)}$. This implies the relation $\psi(s_1^{(n)})<\cdots<\psi(s_n^{(n)})$, i.e., $\psi(s^{(n)})=\lambda_j^{(n)}$. The conclusion of Theorem 2.1 is obtained with $\Delta_1^{(n)}(j)=R_2^{(n)}(s_j^{(n)})$, for n large enough. □

6. Solution of the equation for the eigenvalues

In this section we mainly work with asymptotic expansions of functions depending on n and j. From now on, all asymptotic expansions will refer to $n\to\infty$, and the estimates of the residue terms will be uniform with respect to j. We introduce the following notation:

$$d_j^{(n)}:=\frac{\pi j}{n+1},\quad e_j^{(n)}:=d_j^{(n)}-\frac{\eta(d_j^{(n)})}{n+1},\quad\text{and}\quad I_j^{(n)}:=\left[e_j^{(n)}-\frac{r(n)}{n+1},e_j^{(n)}+\frac{r(n)}{n+1}\right],$$

where $r(n):=2\|R_2^{(n)}\|_{L_\infty}+\omega_\eta\left(\frac{4\|\eta\|_{L_\infty}}{n+1}\right)$. Thus there exists $C_1>0$ such that for every sufficiently large n

$$e_{j+1}^{(n)}-e_j^{(n)}\geq\frac{C_1}{n}. \tag{6.1}$$

If $a\in\mathrm{MSL}^\alpha$ with $\alpha\geq 1$, then Lemmas 4.6 and 5.5 part (iii) show that,

$$r(n)=\begin{cases}o\left(\frac{1}{n^{\alpha-1}}\right)+o\left(\frac{1}{n^{\alpha-1}}\right)=o\left(\frac{1}{n^{\alpha-1}}\right), & 1\leq\alpha<2;\\ o\left(\frac{1}{n^{\alpha-1}}\right)+O\left(\frac{1}{n}\right)=O\left(\frac{1}{n}\right), & \alpha\geq 2\end{cases} \tag{6.2}$$

and

$$\operatorname{diam}I_j^{(n)}=\begin{cases}o\left(\frac{1}{n^\alpha}\right), & 1\leq\alpha<2;\\ O\left(\frac{1}{n^2}\right), & \alpha\geq 2.\end{cases} \tag{6.3}$$

Thus for every sufficiently large n, formulas (6.1) and (6.3) show that the intervals $I_j^{(n)}$ are disjoint.

Lemma 6.1. *Let $a \in \mathrm{MSL}^\alpha$ with $\alpha \geq 1$. Then for every sufficiently large n the equation (5.7) has a unique solution $s_j^{(n)} \in I_j^{(n)}$, for every $j = 1, \dots, n$.*

Proof. From Lemma 5.5 we know that for every $j = 1, \dots, n$ the equation (5.7) has a unique solution in $[0, \pi]$; now we have to prove that this solution belongs to $I_j^{(n)}$. A simple calculation shows that

$$\begin{aligned} F_n\left(e_j^{(n)} - \frac{r(n)}{n+1}\right) &= (n+1)\left(e_j^{(n)} - \frac{r(n)}{n+1}\right) + \eta\left(e_j^{(n)} - \frac{r(n)}{n+1}\right) \\ &= e_j^{(n)}(n+1) - r(n) + \eta\left(e_j^{(n)} - \frac{r(n)}{n+1}\right) \\ &= \pi j - r(n) + \eta\left(e_j^{(n)} - \frac{r(n)}{n+1}\right) - \eta(d_j^{(n)}). \end{aligned} \tag{6.4}$$

Additionally, from (6.2) we obtain $r(n) = o(1)$, thus for every sufficiently large n

$$\left| e_j^{(n)} - \frac{r(n)}{n+1} - d_j^{(n)} \right| = \frac{|\eta(d_j^{(n)}) + r(n)|}{n+1} \leq \frac{2\|\eta\|_{L_\infty}}{n+1}. \tag{6.5}$$

Thus putting (6.4) and (6.5) together, and noting that ω_η is an increasing function, we obtain

$$\begin{aligned} &F_n\left(e_j^{(n)} - \frac{r(n)}{n+1}\right) - \pi j - R_2^{(n)}\left(e_j^{(n)} - \frac{r(n)}{n+1}\right) \\ &\leq -r(n) + \omega_\eta\left(\frac{2\|\eta\|_{L_\infty}}{n+1}\right) - R_2^{(n)}\left(e_j^{(n)} - \frac{r(n)}{n+1}\right) \\ &= -2\|R_2^{(n)}\|_{L_\infty} - \omega_\eta\left(\frac{4\|\eta\|_{L_\infty}}{n+1}\right) + \omega_\eta\left(\frac{2\|\eta\|_{L_\infty}}{n+1}\right) - R_2^{(n)}\left(e_j^{(n)} - \frac{r(n)}{n+1}\right) \leq 0. \end{aligned}$$

That is,

$$F_n\left(e_j^{(n)} - \frac{r(n)}{n+1}\right) - R_2^{(n)}\left(e_j^{(n)} - \frac{r(n)}{n+1}\right) \leq \pi j. \tag{6.6}$$

In a similar fashion, it is possible to show that

$$F_n\left(e_j^{(n)} + \frac{r(n)}{n+1}\right) - R_2^{(n)}\left(e_j^{(n)} + \frac{r(n)}{n+1}\right) \geq \pi j. \tag{6.7}$$

Relations (6.6) and (6.7) together with the Intermediate Value Theorem show that in the interval $I_j^{(n)}$ there exists a solution of the equation (5.7). □

Lemma 6.2. *Let $a \in \mathrm{MSL}^\alpha$ with $\alpha \geq 2$. For every sufficiently large n, we have:*

(i) *for each $j = 1, \dots, n$, the following truncated version of (5.7) has a unique solution $\hat{s}_j^{(n)} \in I_j^{(n)}$:*

$$F_n(s) = \pi j; \tag{6.8}$$

(ii) $|s_j^{(n)} - \hat{s}_j^{(n)}| = o\left(\frac{1}{n^\alpha}\right)$;

(iii) *for each* $j = 1, \ldots, n$, *the following function is a contraction on* $[0, \pi]$*:*

$$\Phi_j^{(n)}(s) := d_j^{(n)} - \frac{\eta(s)}{n+1}; \tag{6.9}$$

(iv) *The sequence defined by*

$$\hat{s}_{j,0}^{(n)} := d_j^{(n)} \quad \text{and} \quad \hat{s}_{j,\ell}^{(n)} := \Phi_j^{(n)}(\hat{s}_{j,\ell-1}^{(n)}) \quad (\ell \geq 1)$$

satisfies $|\hat{s}_j^{(n)} - \hat{s}_{j,\ell}^{(n)}| = O\left(\frac{1}{n^{\ell+1}}\right)$.

Proof. (i) and (iii) The equation (6.8) is equivalent to $\Phi_j^{(n)}(s) = s$. The existence of a solution of (6.8) in $I_j^{(n)}$ can be proved as in the previous lemma, just take 0 instead of $R_2^{(n)}(s)$. By Lemma 4.6 we have $\eta' \in C[0, \pi]$, thus we obtain the following upper bound for the derivative of $\Phi_j^{(n)}$:

$$|(\Phi_j^{(n)})'(s)| \leq \Lambda_n := \frac{\|\eta'\|_{L_\infty}}{n+1}. \tag{6.10}$$

If n is large enough, then $\Lambda_n < 1$. Moreover, we know that $\Phi_j^{(n)}$ has *at least* one fixed point. It follows that the function $\Phi_j^{(n)}$ is a contraction on $[0, \pi]$ and has *exactly one* fixed point which we denote by $\hat{s}_j^{(n)}$.

(ii) Since $s^{(n)}$ and $\hat{s}_j^{(n)}$ are the solutions of (6.8) and (5.7), respectively,

$$F_n(s_j^{(n)}) - F_n(\hat{s}_j^{(n)}) = (n+1)(s_j^{(n)} - \hat{s}_j^{(n)}) + \eta(s_j^{(n)}) - \eta(\hat{s}_j^{(n)}) = R_2^{(n)}(s_j^{(n)}).$$

Thus

$$|s_j^{(n)} - \hat{s}_j^{(n)}| \leq \frac{|\eta(s_j^{(n)}) - \eta(\hat{s}_j^{(n)})| + |R_2^{(n)}(s_j^{(n)})|}{n+1} \leq \Lambda_n |s_j^{(n)} - \hat{s}_j^{(n)}| + \frac{\|R_2^{(n)}\|_{L_\infty}}{n+1}.$$

With the help of Lemma 5.5 part (iii) we obtain

$$|s_j^{(n)} - \hat{s}_j^{(n)}| \leq (1 - \Lambda_n)^{-1} \frac{\|R_2^{(n)}\|_{L_\infty}}{n+1} = o\left(\frac{1}{n^\alpha}\right).$$

(iv) Note that

$$|\hat{s}_{j,1}^{(n)} - \hat{s}_{j,0}^{(n)}| = |e_j^{(n)} - d_j^{(n)}| = \frac{|\eta(d_j^{(n)})|}{n+1} \leq \frac{\|\eta\|_{L_\infty}}{n+1}. \tag{6.11}$$

Recall the following standard upper bound to estimate the speed of convergence of the fixed point iterations:

$$|\hat{s}_{j,\ell}^{(n)} - \hat{s}_j^{(n)}| \leq \frac{\Lambda_n^\ell}{1 - \Lambda_n} |\hat{s}_{j,1}^{(n)} - \hat{s}_{j,0}^{(n)}|.$$

Combining this with (6.11) we get

$$|\hat{s}_{j,\ell}^{(n)} - \hat{s}_j^{(n)}| \le \frac{n+1}{n+1-\|\eta'\|_{L_\infty}} \frac{\|\eta\|_{L_\infty}}{\|\eta'\|_{L_\infty}} \left(\frac{\|\eta'\|_{L_\infty}}{n+1}\right)^{\ell+1} = O\left(\frac{1}{n^{\ell+1}}\right),$$

proving (iv). □

Proof of Theorem 2.2. Suppose first that $a \in \mathrm{MSL}^\alpha$ with $1 \le \alpha < 2$. Recall that $s_j^{(n)}$ satisfies (5.7) and $e_j^{(n)}$ is given by $d_j^{(n)} - \frac{\eta(d_j^{(n)})}{n+1}$. From Lemma 6.1 we know that $s_j^{(n)} \in I_j^{(n)}$. Moreover, $e_j^{(n)} \in I_j^{(n)}$ and $\operatorname{diam} I_j^{(n)} = o\left(\frac{1}{n^\alpha}\right)$ by (6.3). Thus

$$s_j^{(n)} = d_j^{(n)} - \frac{\eta(d_j^{(n)})}{n+1} + \Delta_2^{(n)}(j), \tag{6.12}$$

where $\Delta_2^{(n)}(j) = |s_j^{(n)} - e_j^{(n)}| = o\left(\frac{1}{n^\alpha}\right)$. Assume now that $2 \le \alpha < 3$. From Lemma 6.2, for each $j = 1, \dots, n$, we have

$$\hat{s}_j^{(n)} = d_j^{(n)} - \frac{\eta(\hat{s}_j^{(n)})}{n+1}. \tag{6.13}$$

We will iterate over (6.13). By Lemma 6.2 we know that $|\hat{s}_j^{(n)} - \hat{s}_{j,\ell}^{(n)}| = O\left(\frac{1}{n^{\ell+1}}\right)$. Since $|s_j^{(n)} - \hat{s}_j^{(n)}| = o\left(\frac{1}{n^\alpha}\right)$, it is enough to consider $\ell = 2$. By Taylor's theorem

$$\begin{aligned} \hat{s}_{j,2}^{(n)} &= \Phi_j^{(n)}(\hat{s}_{j,1}^{(n)}) = \Phi_j^{(n)}\left(d_j^{(n)} - \frac{\eta(d_j^{(n)})}{n+1}\right) = d_j^{(n)} - \frac{1}{n+1}\eta\left(d_j^{(n)} - \frac{\eta(d_j^{(n)})}{n+1}\right) \\ &= d_j^{(n)} - \frac{\eta(d_j^{(n)})}{n+1} + \frac{\eta(d_j^{(n)})\eta'(d_j^{(n)})}{(n+1)^2} + \frac{\eta(d_j^{(n)})}{(n+1)^2}(\eta'(\xi_1) - \eta'(d_j^{(n)})), \end{aligned} \tag{6.14}$$

for some ξ_1 between $d_j^{(n)}$ and $\Phi_j^{(n)}(d_j^{(n)})$. Estimate $|\xi_1 - d_j^{(n)}|$ using (6.11):

$$|\xi_1 - d_j^{(n)}| \le |\hat{s}_{j,1}^{(n)} - d_j^{(n)}| \le \frac{\|\eta(d^{(n)})\|_{L_\infty}}{n+1} = O\left(\frac{1}{n}\right).$$

Combining this with the property $\eta' \in \mathcal{H}^{\alpha-2}[0,\pi]$ we obtain

$$|\eta'(\xi_1) - \eta'(d_j^{(n)})| = o\left(\frac{1}{n^{\alpha-2}}\right).$$

Thus (6.14) mixed with Lemma 6.2 part (ii), gives us

$$\begin{aligned} s_j^{(n)} &= \hat{s}_j^{(n)} + o\left(\frac{1}{n^\alpha}\right) = \hat{s}_{j,2}^{(n)} + O\left(\frac{1}{n^3}\right) + o\left(\frac{1}{n^\alpha}\right) \\ &= d_j^{(n)} - \frac{\eta(d_j^{(n)})}{n+1} + \frac{\eta(d_j^{(n)})\eta'(d_j^{(n)})}{(n+1)^2} + o\left(\frac{1}{n^\alpha}\right) + O\left(\frac{1}{n^3}\right) + o\left(\frac{1}{n^\alpha}\right) \\ &= d_j^{(n)} - \frac{\eta(d_j^{(n)})}{n+1} + \frac{\eta(d_j^{(n)})\eta'(d_j^{(n)})}{(n+1)^2} + o\left(\frac{1}{n^\alpha}\right), \end{aligned}$$

which proves the theorem when $2 \le \alpha < 3$. The remaining cases, $k \le \alpha < k+1$ with $k \in \mathbb{N}$ and $k \ge 3$, follow readily by approximating $\hat{s}_{j,k-1}^{(n)}$ by iteration over (6.9). □

We need some auxiliary lemmas before the proof of Theorem 2.3. In order to simplify our forthcoming calculations we define $\varrho_{j,n} := d_j^{(n)}(\pi - d_j^{(n)})$.

Lemma 6.3. *Let $a \in \mathrm{MSL}^\alpha$ with $\alpha \ge 1$.*

(i) *If $1 \le \alpha < 2$, then there exists $C > 0$ such that for every $s \in [0, \pi]$*

$$|\eta(s)| \le C(s(\pi - s))^{\alpha-1} \quad \text{and} \quad |\psi'(s)| \le C(s(\pi - s))^{\alpha-1}. \tag{6.15}$$

(ii) *If $\alpha \ge 2$, then there exists $C > 0$ such that for every $s \in [0, \pi]$*

$$|\eta(s)| \le Cs(\pi - s) \quad \text{and} \quad |\psi'(s)| \le Cs(\pi - s). \tag{6.16}$$

Proof. This lemma follows from the fact that the functions η and ψ' belong to the space $\mathcal{H}^{\alpha-1}[0, \pi]$ and vanish at the points 0 and π, see Lemmas 4.3 and 4.6. □

Lemma 6.4. *Let $a \in \mathrm{MSL}^\alpha$, then*

$$|s_j^{(n)} - d_j^{(n)}| = \begin{cases} O\left(\dfrac{\varrho_{j,n}^{\alpha-1}}{n}\right), & 1 \le \alpha < 2; \qquad (6.17) \\[2ex] O\left(\dfrac{\varrho_{j,n}}{n}\right), & \alpha \ge 2. \qquad (6.18) \end{cases}$$

Proof. From the equation (5.7), $s_j^{(n)} = d_j^{(n)} + \frac{1}{n+1}(R_2^{(n)}(s_j^{(n)}) - \eta(s_j^{(n)}))$. This implies immediately the upper estimate $|s_j^{(n)} - d_j^{(n)}| = O\left(\frac{1}{n}\right)$. Moreover,

$$\begin{aligned} |s_j^{(n)} - d_j^{(n)}| &= \frac{|\eta(s_j^{(n)})| + |R_2^{(n)}(s)|}{n+1} \\ &\le \frac{1}{n+1}(|\eta(d_j^{(n)})| + |\eta(d_j^{(n)}) - \eta(s_j^{(n)})| + \|R_2^{(n)}\|_{L_\infty}). \end{aligned}$$

We will use the following inequality: for every $n \ge 1$ and every $j \in \{1, \ldots, n\}$,

$$\varrho_{j,n} \ge \frac{\pi^2}{2(n+1)}. \tag{6.19}$$

In the case $1 \le \alpha < 2$ we apply (6.15), the property $\eta \in \mathcal{H}^{\alpha-1}[0, \pi]$, the Lemma 5.5 part (iii), and (6.19) to get

$$|s_j^{(n)} - d_j^{(n)}| \le \frac{1}{n+1}\left(O(\varrho_{j,n}^{\alpha-1}) + o\left(\frac{1}{n^{\alpha-1}}\right) + o\left(\frac{1}{n^{\alpha-1}}\right)\right) = O\left(\frac{\varrho_{j,n}^{\alpha-1}}{n}\right)$$

and (6.17) is proved. The case $\alpha \ge 2$ is considered in a similar manner, using (6.16) and the Lipschitz continuity of η we obtain

$$|s_j^{(n)} - d_j^{(n)}| = \frac{1}{n+1}\left(O(\varrho_{j,n}) + O\left(\frac{1}{n}\right) + o\left(\frac{1}{n^{\alpha-1}}\right)\right)$$

and applying (6.19) again, we obtain (6.18). □

Proof of Theorem 2.3. The idea is to use the formula $\lambda_j^{(n)} = \psi(s_j^{(n)})$. Apply Taylor's theorem to ψ around the point $d_j^{(n)}$ and expand $s_j^{(n)} - d_j^{(n)}$ using Theorem 2.2. Suppose that $1 \le \alpha < 2$. In this case $\psi \in C^1[0,\pi]$, and there is a point ξ_2 between $d_j^{(n)}$ and $s_j^{(n)}$ such that

$$\lambda_j^{(n)} = \psi(s_j^{(n)}) = \psi(d_j^{(n)}) + \psi'(\xi_2)(s_j^{(n)} - d_j^{(n)}).$$

Write $\psi'(\xi_2)$ as $\psi'(d_j^{(n)}) + (\psi'(\xi_2) - \psi'(d_j^{(n)}))$ and use (6.12):

$$\lambda_j^{(n)} = \psi(d_j^{(n)}) - \frac{\psi'(d_j^{(n)})\eta(d_j^{(n)})}{n+1} + \Delta_3^{(n)}(j)$$

with $\Delta_3^{(n)}(j) := \psi'(d_j^{(n)})\Delta_2^{(n)}(j) + (\psi'(\xi_2) - \psi'(d_j^{(n)}))(s_j^{(n)} - d_j^{(n)})$. The first term in this expression can be estimated with the help of (6.15):

$$|\psi'(d_j^{(n)})\Delta_2^{(n)}(j)| \le C\varrho_{j,n}^{\alpha-1} o\Big(\frac{1}{n^\alpha}\Big) = o\Big(\frac{\varrho_{j,n}^{\alpha-1}}{n^\alpha}\Big).$$

For the second term we use the property $\psi' \in \mathcal{H}^{\alpha-1}[0,\pi]$ and (6.17):

$$|\psi'(\xi_2) - \psi'(d_j^{(n)})|\,|s_j^{(n)} - d_j^{(n)}| = o(|s_j^{(n)} - d_j^{(n)}|^\alpha) = o\Big(\frac{\varrho_{j,n}^{\alpha(\alpha-1)}}{n^\alpha}\Big) = o\Big(\frac{\varrho_{j,n}^{\alpha-1}}{n^\alpha}\Big).$$

So $\Delta_3^{(n)}(j) = o\Big(\frac{\varrho_{j,n}^{\alpha-1}}{n^\alpha}\Big)$ proving the theorem when $1 \le \alpha < 2$. Assume now that $2 \le \alpha < 3$. In this case Theorem 2.2 tells us

$$s_j^{(n)} - d_j^{(n)} = -\frac{\eta(d_j^{(n)})}{n+1} + \frac{\eta(d_j^{(n)})\eta'(d_j^{(n)})}{(n+1)^2} + o\Big(\frac{1}{n^\alpha}\Big). \tag{6.20}$$

Squaring this expression we obtain

$$(s_j^{(n)} - d_j^{(n)})^2 = \frac{\eta^2(d_j^{(n)})}{(n+1)^2} + o\Big(\frac{\varrho_{j,n}^2}{n^3}\Big) = \frac{\eta^2(d_j^{(n)})}{(n+1)^2} + o\Big(\frac{\varrho_{j,n}}{n^\alpha}\Big). \tag{6.21}$$

Applying (6.16) and (6.19) we get

$$o\Big(\Big(\frac{1}{n^\alpha}\Big)^2\Big) = o\Big(\frac{1}{n}\Big)o\Big(\frac{1}{n^{2\alpha-1}}\Big) = o\Big(\frac{\varrho_{j,n}}{n^3}\Big).$$

By Lemma 4.3 we know that $\psi \in \mathcal{H}^\alpha[0,\pi]$. In particular, $\psi \in C^2[0,\pi]$. By Taylor's theorem, there exists a point ξ_3 between $s_j^{(n)}$ and $d_j^{(n)}$ such that

$$\lambda_j^{(n)} = \psi(s_j^{(n)}) = \psi(d_j^{(n)}) + \psi'(d_j^{(n)})(s_j^{(n)} - d_j^{(n)}) + \frac{\psi''(\xi_3)}{2}(s_j^{(n)} - d_j^{(n)})^2.$$

Write $\psi''(\xi_3)$ as $\psi''(d_j^{(n)}) + (\psi''(\xi_3) - \psi''(d_j^{(n)}))$ and substitute (6.20) and (6.21):

$$\begin{aligned}\lambda_j^{(n)} &= \psi(d_j^{(n)}) - \frac{\psi'(d_j^{(n)})\eta(d_j^{(n)})}{n+1} \\ &\quad + \frac{\psi'(d_j^{(n)})\eta(d_j^{(n)})\eta'(d_j^{(n)})}{(n+1)^2} + \frac{\psi''(d_j^{(n)})\eta^2(d_j^{(n)})}{2(n+1)^2} + \Delta_3^{(n)}(j),\end{aligned}$$

where, in this case, the residue term is

$$\begin{aligned}\Delta_3^{(n)}(j) &:= \psi'(d_j^{(n)})o\Big(\frac{1}{n^\alpha}\Big) + \psi''(d_j^{(n)})o\Big(\frac{\varrho_{j,n}}{n^\alpha}\Big) \\ &\quad + \frac{1}{2}(\psi''(\xi_3) - \psi''(d_j^{(n)}))(s_j^{(n)} - d_j^{(n)})^2.\end{aligned}$$

The first term in this sum can be bounded from above using (6.16):

$$\psi'(d_j^{(n)})o\Big(\frac{1}{n^\alpha}\Big) = O(\varrho_{j,n})o\Big(\frac{1}{n^\alpha}\Big) = o\Big(\frac{\varrho_{j,n}}{n^\alpha}\Big),$$

the situation with the second one is trivial, and for the third one we apply the property $\psi'' \in \mathcal{H}^{\alpha-2}[0,\pi]$ and the upper bound (6.18):

$$(\psi''(\xi_3) - \psi''(d_j^{(n)}))(s_j^{(n)} - d_j^{(n)})^2 = o(|s_j^{(n)} - d_j^{(n)}|^\alpha) = o\Big(\frac{\varrho_{j,n}^\alpha}{n^\alpha}\Big) = o\Big(\frac{\varrho_{j,n}}{n^\alpha}\Big).$$

Thereby we have proved the theorem when $2 \le \alpha < 3$. The remaining cases, $k \le \alpha < k+1$ ($k = 3, 4, \dots$), follow readily. □

7. Numerical tests

In order to test Theorem 2.3 numerically, we proceed as follows: (i) Invert g in the intervals $[0, \varphi_0]$ and $[\varphi_0, 2\pi]$ to obtain φ_1 and φ_2, respectively. This can be done with the Newton–Raphson algorithm or any other method (for example FindRoot in Wolfram Mathematica). (ii) With φ_1 and φ_2 we define φ; inverting again we get ψ. (iii) In order to avoid numerical derivatives, we compute the first and second derivatives of φ and ψ directly from g', g'', φ_1, and φ_2. (iv) With the aid of φ_1, φ_2, and ψ we define σ_1, σ_2, b, and η. The singular integrals in formula (2.5) are hard to tackle numerically; we use the standard regularization trick described in [2, sect. 7]. (v) Now it is easy to get the asymptotic terms in Theorem 2.3.

Denote by $\lambda_j^{(n,p)}$ the approximation of $\lambda_j^{(n)}$ resulting from our Theorem 2.3 with p terms. For example,

$$\lambda_j^{(n,1)} := \psi(d_j^{(n)}), \quad \lambda_j^{(n,2)} := \psi(d_j^{(n)}) - \frac{\psi'(d_j^{(n)})\eta(d_j^{(n)})}{n+1}.$$

For each $j = 1, \dots, n$, put $\varepsilon_j^{(n,p)} := |\lambda_j^{(n)} - \lambda_j^{(n,p)}|$ and let $\varepsilon^{(n,p)}$ be the corresponding maximal error,

$$\varepsilon^{(n,p)} := \max\{\varepsilon_j^{(n,p)} : 1 \le j \le n\}.$$

In order to find symbols in MSL^α we recall the well-known facts that if f is a continuous function with bounded variation, then it belongs to the Wiener algebra W, and that if $f' \in W^\beta$ for some $\beta \in \mathbb{R}$, then $f \in W^{\beta+1}$. It is easy to find rational symbols in MSL^α, but to detect explicit non-rational symbols, one has to be a little bit tricky. The following examples illustrate the situation.

Example 7.1 (A non-symmetric symbol). Consider the symbol

$$a_\beta(e^{i\sigma}) = g(\sigma) = g_2\sigma^2 + g_\beta\sigma^{2+\beta} + g_3\sigma^3 + g_4\sigma^4 + g_5\sigma^5, \quad \sigma \in [0, 2\pi],$$

where $\beta \in [0,1)$ and the coefficients g_2, g_β, g_3, g_4, and g_5 are chosen in such a manner that

$$g_\beta \neq 0, \quad g_2 > 0, \quad \text{and} \quad g^{(k)}(0) = g^{(k)}(2\pi) \quad \text{for} \quad k = 0,1,2.$$

Elementary computations yield

$$g_2 = \frac{1}{(2\pi)^2}(\beta^2 - 4\beta + 4), \quad g_\beta = \frac{-2}{(2\pi)^{2+\beta}},$$

$$g_3 = \frac{1}{(2\pi)^3}(-\beta^2 + 3\beta - 2), \quad g_4 = \frac{1}{(2\pi)^4}(-\beta^2 + 4\beta - 2), \quad g_5 = \frac{1}{(2\pi)^5}(\beta^2 - 3\beta + 2).$$

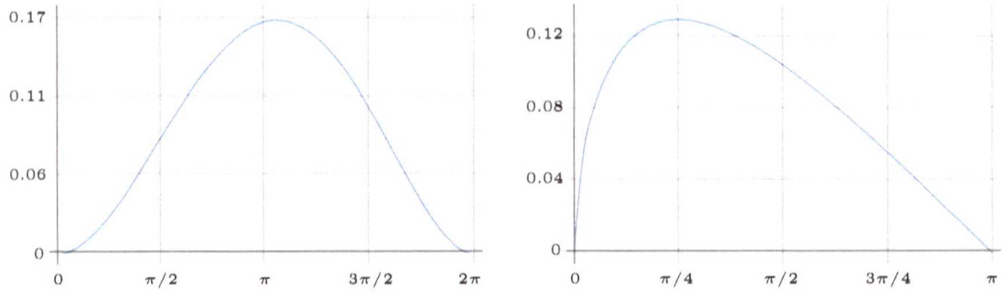

FIGURE 1. Graph of $g(\sigma) = a_\beta(e^{i\sigma})$ (left) and η (right) for $\beta = \frac{1}{7}$.

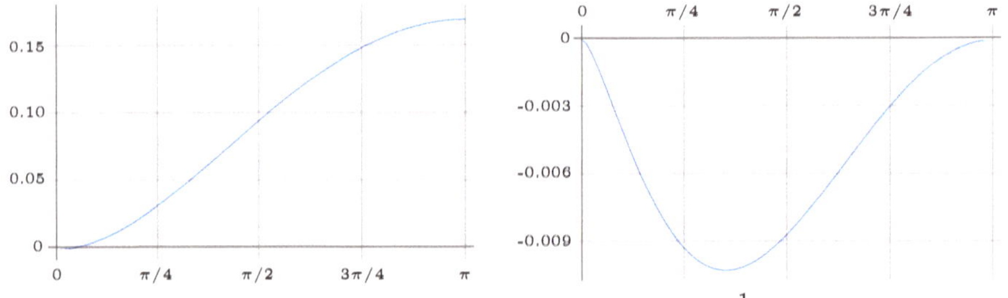

FIGURE 2. Graph of ψ (left) and r_1 (right) for $\beta = \frac{1}{7}$, where r_1 is the function in Theorem 2.3.

The symbol a_β belongs to $\mathrm{MSL}^{1+\beta-\varepsilon}$ for arbitrary $\varepsilon \in (0,1)$. Indeed, note that $a \in W^2$ because $a_\beta^{(2)}$ is in $C(\mathbb{T})$ and has bounded variation. Additionally, with the aid of standard asymptotic methods we can see that the kth Fourier coefficient of g has order $O(k^{-3-\beta})$ as $k \to \infty$, which means that $a_\beta \in W^{2+\beta-\varepsilon}$

for arbitrary $\varepsilon \in (0,1)$. The remaining simple-loop conditions are easily checked. So $a_\beta \in \mathrm{SL}^{2+\beta-\varepsilon} \subset \mathrm{MSL}^{1+\beta-\varepsilon}$. Figures 1 and 2, and Table 1 present numerical results for this example. The data in this and the following example were obtained by comparison with the eigenvalues given by *Wolfram Mathematica.*

n	128	256	512	1024	2048	4096
$\varepsilon^{(n,1)}$	$7.9\cdot10^{-5}$	$4.0\cdot10^{-5}$	$2.0\cdot10^{-5}$	$9.9\cdot10^{-6}$	$5.0\cdot10^{-6}$	$2.5\cdot10^{-6}$
$\varepsilon^{(n,2)}$	$6.3\cdot10^{-8}$	$1.6\cdot10^{-8}$	$4.0\cdot10^{-9}$	$9.9\cdot10^{-10}$	$2.5\cdot10^{-10}$	$6.2\cdot10^{-11}$

TABLE 1. Maximum errors for the eigenvalues of $T_n(a_\beta)$ obtained with our Theorem 2.3, $\varepsilon^{(n,1)}$.

Example 7.2 (A symmetric symbol). In this example we consider a simple-loop symbol resulting from gluing two parabolas:

$$a_{sym}(e^{i\sigma}) = g(\sigma) = \begin{cases} \frac{2}{\pi^2}\sigma^2 & 0 \le \sigma \le \pi/2, \\ 1 - \frac{2}{\pi^2}(\sigma-\pi)^2 & \pi/2 \le \sigma \le 3\pi/2, \\ \frac{2}{\pi^2}(\sigma-2\pi)^2 & 3\pi/2 \le \sigma \le 2\pi. \end{cases}$$

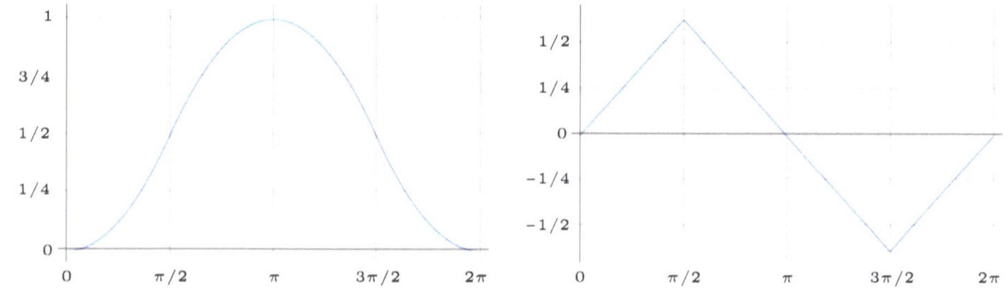

FIGURE 3. Graph of $g(\sigma) = a_{sym}(e^{i\sigma})$ (left) and g' (right).

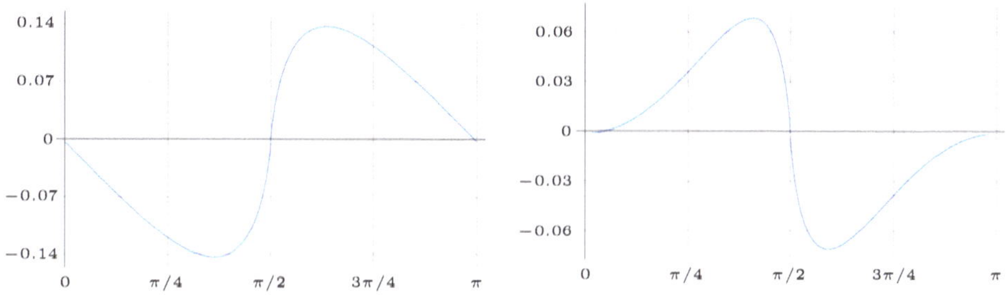

FIGURE 4. Graph of η (left) and r_1 (right), where r_1 is the function in Theorem 2.3.

A simple computation shows that $(a_{sym})_k = \frac{2i}{(k\pi)^3}((-1)^k - 1)e^{\frac{i}{2}k\pi}$ for $k \neq 0$ and $(a_{sym})_0 = \frac{1}{2}$. This implies that $a_{sym} \in W^\alpha$ for every $\alpha < 2$. The remaining simple-loop conditions are easily seen in Figure 3. The symbol a is analytic near ± 1, and its smoothness is left unchanged when dividing by $t \mp 1$; thus we conclude that $a \in \mathrm{MSL}^\alpha$ ($\alpha < 2$). The symbol a_{sym} is symmetric (i.e., $g(\sigma) = g(2\pi - \sigma)$). In this case we have

$$\varphi = \varphi_1, \quad \psi = g, \quad \sigma_1(s) = s, \quad \sigma_2(s) = 2\pi - s, \quad b(e^{i\sigma}, s) = \frac{g(s) - g(\sigma)}{2(\cos s - \cos\sigma)}.$$

Figure 4 and Table 2 present numerical results for this example.

n	128	256	512	1024	2048	4096
$\varepsilon^{(n,1)}$	$5.4\cdot10^{-4}$	$2.7\cdot10^{-4}$	$1.4\cdot10^{-4}$	$6.8\cdot10^{-5}$	$3.4\cdot10^{-5}$	$1.7\cdot10^{-5}$
$\varepsilon^{(n,2)}$	$4.5\cdot10^{-7}$	$1.1\cdot10^{-7}$	$2.9\cdot10^{-8}$	$7.2\cdot10^{-9}$	$1.8\cdot10^{-9}$	$4.6\cdot10^{-10}$

TABLE 2. Maximum errors for the eigenvalues of $T_n(a_{sym})$ obtained with our Theorem 2.3, $\varepsilon^{(n,p)}$ with $p = 1, 2$.

We are grateful to the referee for his positive review and corrections. We have taken seriously his words about "reliable mathematicians who take care of each epsilon" and revised every detail, corrected various errors, and filled some gaps in the proofs. We also simplified and improved the statements of the main theorems.

References

[1] J.M. Bogoya, A. Böttcher, and S.M. Grudsky: Asymptotics of individual eigenvalues of a class of large Hessenberg Toeplitz matrices. Oper. Theory Adv. Appl. 220, 207–224 (2012).

[2] J.M. Bogoya, A. Böttcher, S.M. Grudsky, and E.A. Maximenko: Eigenvalues of Hermitian Toeplitz matrices with smooth simple-loop symbols. J. Math. Analysis Appl. 422, 1308–1334 (2015).

[3] J.M. Bogoya, A. Böttcher, S.M. Grudsky, and E.A. Maximenko: Eigenvectors of Hermitian Toeplitz matrices with smooth simple-loop symbols. Linear Algebra Appl. 493, 606–637 (2016).

[4] J.M. Bogoya, A. Böttcher, S.M. Grudsky, and E.A. Maximenko: Maximum norm versions of the Szegő and Avram–Parter theorems for Toeplitz matrices. J. Approx. Theory 196, 79–100 (2015).

[5] A. Böttcher and S.M. Grudsky: Spectral properties of banded Toeplitz matrices. SIAM, Philadelphia 2005.

[6] A. Böttcher, S.M. Grudsky, and E.A. Maksimenko: Inside the eigenvalues of certain Hermitian Toeplitz band matrices. J. Comput. Appl. Math. 233, 2245–2264 (2010).

[7] A. Böttcher and B. Silbermann: Introduction to large truncated Toeplitz matrices. Universitext, Springer-Verlag, New York 1999.

[8] W.W. Chen, C.M. Hurvich, and Y. Lu: On the correlation matrix of the discrete Fourier transform and the fast solution of large Toeplitz systems for long-memory time series. J. Amer. Statist. Assoc. 101, 812–822 (2006).

[9] W.K. Cochran, R.J. Plemmons, and T.C. Torgersen: Exploiting Toeplitz structure in atmospheric image restoration. Contemp. Math. 280, 177–189 (2001).

[10] P. Deift, A. Its, and I. Krasovsky: Eigenvalues of Toeplitz matrices in the bulk of the spectrum. Bull. Inst. Math. Acad. Sin. (N.S.) 7, 437–461 (2012).

[11] P. Deift, A. Its, and I. Krasovsky: Toeplitz matrices and Toeplitz determinants under the impetus of the Ising model. Some history and some recent results. Comm. Pure Appl. Math. 66, 1360–1438 (2013).

[12] E. Eisenberg, A. Baram, and M. Baer: Calculation of the density of states using discrete variable representation and Toeplitz matrices. J. Phys. A. 28, 433–438 (1995).

[13] U. Grenander and G. Szegő: Toeplitz forms and their applications. Chelsea Publishing, New York 1984.

[14] P.C. Hansen, J.G. Nagy, D.P. O'Leary: Deblurring images: matrices, spectra, and filtering. SIAM, Philadelphia 2006.

[15] J. Haupt, W.U. Bajwa, G. Raz, and R. Nowak: Toeplitz compressed sensing matrices with applications to sparse channel estimation. IEEE Trans. Inform. Theory. 56, 5862–5875 (2010).

[16] K. Ye and L.-H. Lim: Every matrix is a product of Toeplitz matrices. Found. Comput. Math. 16, 577–598 (2016).

[17] R. Lin, M.K. Ng, and R.H. Chan: Preconditioners for Wiener–Hopf equations with high-order quadrature rules. SIAM J. Numer. Anal. 34, 1418–1431 (1997).

[18] M.K. Ng: Iterative methods for Toeplitz systems. Oxford University Press, 2004.

[19] D. Poland: Toeplitz matrices and random walks with memory. Phys. A. 223, 113–124 (1996).

[20] S. Serra: The rate of convergence of Toeplitz based PCG methods for second-order nonlinear boundary value problems. Numer. Math. 81, 461–495 (1999).

[21] U. Steimel: Fast computation of Toeplitz forms under narrowband conditions with applications to statistical signal processing. Signal Process. 1, 141–158 (1979).

[22] G. Strang: The discrete cosine transform, block Toeplitz matrices, and wavelets. Adv. Comput. Math. 202, 517–536 (1999).

[23] W.F. Trench: Interlacement of the even and odd spectra of real symmetric Toeplitz matrices. Linear Alg. Appl. 195, 59–68 (1993).

[24] W.F. Trench: Spectral decomposition of Kac–Murdock–Szegő matrices. The Selected Works of William F. Trench, available at: http://works.bepress.com/william trench/133, 2010.

[25] E.E. Tyrtyshnikov: Fast computation of Toeplitz forms and some multidimensional integrals. Russian J. Numer. Anal. Math. Modelling. 20, 383–390 (2005).

J.M. Bogoya
Pontificia Universidad Javeriana
Departamento de Matemáticas
Apartado Postal 01110
Bogotá D.C., Colombia
e-mail: jbogoya@javeriana.edu.co

S.M. Grudsky
CINVESTAV
Departamento de Matemáticas
Apartado Postal 07360
Ciudad de México, Mexico
e-mail: grudsky@math.cinvestav.mx

E.A. Maximenko
Instituto Politécnico Nacional
Escuela Superior de Física y Matemáticas
Apartado Postal 07730
Ciudad de México, Mexico
ORCID: 0000-0002-1497-4338
e-mail: emaximenko@ipn.mx

Operator Theory:
Advances and Applications, Vol. 259, 213–234

On the Asymptotic Behavior of a Log Gas in the Bulk Scaling Limit in the Presence of a Varying External Potential II

Thomas Bothner, Percy Deift, Alexander Its and Igor Krasovsky

Dedicated to Albrecht Böttcher on the occasion of his 60th birthday

Abstract. In this paper we continue our analysis [3] of the determinant $\det(I-\gamma K_s), \gamma \in (0,1)$ where K_s is the trace class operator acting in $L^2(-1,1)$ with kernel $K_s(\lambda,\mu) = \frac{\sin s(\lambda-\mu)}{\pi(\lambda-\mu)}$. In [3] various key asymptotic results were stated and utilized, but without proof: Here we provide the proofs (see Theorem 1.2 and Proposition 1.3 below).

Mathematics Subject Classification (2010). 82B23; Secondary 47B35, 34E05, 34M50.

Keywords. Sine kernel determinant, Toeplitz determinant, transition asymptotics, Riemann–Hilbert problem, Deift–Zhou nonlinear steepest descent method.

1. Introduction and statement of results

In this paper we consider the asymptotic behavior of the determinant $\det(I-\gamma K_s)$ as $s \to +\infty$ and $\gamma \in [0,1]$. This problem has received considerable attention over the past 40 years (see, e.g., [3] for more discussion) and, in particular, the behavior for fixed

$$v = -\frac{1}{2}\ln(1-\gamma) \in [0,+\infty] \tag{1.1}$$

P. Deift acknowledges support of NSF Grant DMS-1300965. A. Its acknowledges support of NSF Grants DMS-1001777 and DMS-1361856, and of SPbGU grant N 11.38.215.2014. I. Krasovsky acknowledges support of the European Community Seventh Framework grant "Random and Integrable Models in Mathematical Physics" and of the Leverhulme Trust research fellowship RF-2015-243.

is known: when $v=+\infty$, i.e., $\gamma=1$, we have from [14, 18, 16, 11], as $s\to+\infty$,

$$\ln\det(I-K_s)=-\frac{s^2}{2}-\frac{1}{4}\ln s+\ln c_0+\mathcal{O}\left(s^{-1}\right);\quad \ln c_0=\frac{1}{12}\ln 2+3\zeta'(-1), \tag{1.2}$$

which is the classical expansion for the gap probability in the bulk scaling limit for the eigenvalues of Hermitian matrices chosen from the Gaussian Unitary Ensemble (GUE). On the other hand, when $v\in[0,+\infty)$ is fixed, we have from [2, 6], as $s\to+\infty$,

$$\ln\det(I-\gamma K_s)=-\frac{4v}{\pi}s+\frac{2v^2}{\pi^2}\ln(4s)+2\ln G\left(1+\frac{\mathrm{i}v}{\pi}\right)G\left(1-\frac{\mathrm{i}v}{\pi}\right)+\mathcal{O}\left(s^{-1}\right) \tag{1.3}$$

in terms of Barnes G-function, cf. [13]. For $\gamma<1$, $-\ln\det(I-\gamma K_s)$ gives the free energy for the so-called log-gas of eigenvalues of GUE Hermitian matrices in the above bulk scaling limit, but now in the presence of an external field $\mathcal{V}(x)=\ln(1-\gamma)$ for $-\frac{2s}{\pi}<x<\frac{2s}{\pi}$, and zero otherwise (see [3]). The mechanism behind the transformation of the exponential decay in (1.3) to (1.2) as $v\to+\infty$ was first analyzed by Dyson in [15], although on a heuristical level. Nevertheless Dyson's computations served as a guideline in the Riemann–Hilbert based approach chosen in [3] which led to the following result: let $a=a(\kappa)\in(0,1)$ be the unique solution of the equation

$$(0,1)\ni\kappa\equiv\frac{v}{s}=\int_a^1\sqrt{\frac{\mu^2-a^2}{1-\mu^2}}\,\mathrm{d}\mu, \tag{1.4}$$

and set

$$V(\kappa)=-\frac{2}{\pi}\Big(E(a)-(1-a^2)K(a)\Big);\quad \tau(\kappa)=2\mathrm{i}\frac{K(a)}{K(a')}; \tag{1.5}$$

$$\theta_3(z|\tau)=1+2\sum_{k=1}^{\infty}\mathrm{e}^{\mathrm{i}\pi\tau k^2}\cos(2\pi kz),\ z\in\mathbb{C};$$

where $K=K(a)$ and $E=E(a)$ are the standard complete elliptic integrals (cf. [13]) with modulus $a=a(\kappa)\in(0,1)$ and complementary modulus $a'=\sqrt{1-a^2}$.

Theorem 1.1 ([3], Theorem 1.4). *Given $\delta\in(0,1)$, there exist positive $s_0=s_0(\delta), c=c(\delta)$ and $C_0=C_0(\delta)$ such that*

$$\begin{aligned}\ln\det(I-\gamma K_s)=&-\frac{s^2}{2}(1-a^2)+vsV+\ln\theta_3(sV|\tau)+A(v)\\&+\int_s^{\infty}M\Big(tV(vt^{-1}),vt^{-1}\Big)\frac{\mathrm{d}t}{t}+J(s,v)\end{aligned} \tag{1.6}$$

with

$$A(v)=2\ln G\left(1+\frac{\mathrm{i}v}{\pi}\right)G\left(1-\frac{\mathrm{i}v}{\pi}\right)-\frac{v^2}{\pi^2}\left(3+2\ln\left(\frac{\pi}{v}\right)\right),$$

and

$$\left|\int_s^{\infty}M\Big(tV(vt^{-1}),vt^{-1}\Big)\frac{\mathrm{d}t}{t}\right|\leq C_0,\qquad |J(s,v)|\leq cs^{-\frac{1}{4}}\ln s,$$

for $s \geq s_0$ *and for* γ *as in* (1.1) *such that* $0 < v \leq s(1-\delta)$. *Here, the function* $M(z,\kappa)$ *is explicitly stated in* (1.24) *of* [3] *as well as in the Appendix,* (A.1) *in terms of Jacobi theta functions.*

The derivation of the bound for $J(s,v)$ in [3] used the following extension of (1.3) which will be the first result of the current paper.

Theorem 1.2. *There exist positive constants* s_0, c_1, c_2 *such that for* $s \geq s_0$ *and for* γ *in* (1.1) *such that* $0 \leq v < s^{\frac{1}{3}}$,

$$\ln\det(I-\gamma K_s) = -\frac{4v}{\pi}s+\frac{2v^2}{\pi^2}\ln(4s)+2\ln G\left(1+\frac{\mathrm{i}v}{\pi}\right)G\left(1-\frac{\mathrm{i}v}{\pi}\right)+r(s,v). \quad (1.7)$$

The error term $r(s,v)$ *is differentiable with respect to* s *and*

$$\left|r(s,v)\right| \leq c_1\frac{v}{s}+c_2\frac{v^3}{s}. \quad (1.8)$$

Note that (1.7) reduces to (1.3) if v is fixed. However (1.7), (1.8) is a considerably stronger result than (1.3) since v is also allowed to grow at a certain rate. As noted in [3], Theorem 1.2 is needed, in particular, to make rigorous a very interesting argument of Bohigas and Pato [7] which interprets the transition (1.3) to (1.2) as a transition from a system with Poisson statistics to a system with random matrix theory GUE statistics. We will provide two derivations for Theorem 1.2: one based on a nonlinear steepest descent analysis of Riemann–Hilbert Problem 1.3 in [3], see also RHP 2.1 below, but using very different techniques than those employed in [3]. This approach will provide us with expansion (1.7) involving a slightly weaker estimate for $r(s,v)$ than (1.8) (see (2.11) below). Still, the result of this first derivation is sufficient for the arguments used in [3], Section 5.2 in the derivation of the bound for $J(s,v)$ in Theorem 1.1. Second, Theorem 1.2 will also follow by the application of recent results from [9] using the connection of $\det(I-\gamma K_s)$ to a Toeplitz determinant with Fisher–Hartwig singularities. We confirm in this way (1.7) with the stated error estimate (1.8).

Our next result will provide further insight into the integral term involving $M(z,\kappa)$.

Proposition 1.3. *Let* $a_0(u) = \int_0^1 M(x,u)\mathrm{d}x$ *denote an average of* $M(x,\kappa)$ *over the "fast" variable* x. *Given* $\delta \in (0,1)$, *we have*

$$\int_s^{\infty} M\Big(tV(vt^{-1}), vt^{-1}\Big)\frac{\mathrm{d}t}{t} = \int_0^{\kappa} a_0(u)\frac{\mathrm{d}u}{u}+\mathcal{O}\left(s^{-1}\right)$$

uniformly for $s \geq s_0 > 0$ *and* $0 < v \leq s(1-\delta)$.

It is worth noticing that Proposition 1.3 implies that the integral term in the right-hand side of (1.6) depends, up to $\mathcal{O}(s^{-1})$, on the "slow" variable $\kappa = \frac{v}{s}$ only. Hence, Theorem 1.1 captures in an explicit way all principal features of the leading asymptotic behavior of the determinant $\det(I-\gamma K_s)$ in the bulk scaling limit.

1.1. Outline of the paper

We complete the introduction with a short outline for the remainder. As mentioned above, the leading terms in Theorem 1.2 will be derived in two ways: First, in Section 2, we apply Riemann–Hilbert techniques related to the integrable structure of $K_s(\lambda,\mu)$, cf. [17]. The s derivative of $\ln\det(I-\gamma K_s)$ is expressible in terms of the solution of RHP 2.1 below which allows us to compute its asymptotic by application of nonlinear steepest descent techniques [12]. Integrating the asymptotic series indefinitely with respect to s and comparing the result to (1.3) we obtain (1.7) with a slightly weaker error estimate for $r(s,v)$. Second, in Section 3, we use the representation of $\det(I-\gamma K_s)$ as a limit of a Toeplitz determinant with Fisher–Hartwig singularities. We strengthen the results obtained in [9] and derive Theorem 1.2 with the stated error estimate (1.8). In Section 4 we provide a proof of Proposition 1.3 which uses a Fourier series representation of $M(z,\kappa)$. Several theta function expressions which are used in the definition of $M(z,\kappa)$ are summarized in the Appendix.

2. Extending 1.3 by integrable operator techniques

We first recall the central Riemann–Hilbert problem related to the integrable structure of the sine kernel determinant $\det(I-\gamma K_s)$, compare [17].

Riemann–Hilbert Problem 2.1. Determine $Y(\lambda)=Y(\lambda;s,\gamma)\in\mathbb{C}^{2\times 2}$ such that

1. $Y(\lambda)$ is analytic for $\lambda\in\mathbb{C}\backslash[-1,1]$ with square integrable limiting values
$$Y_{\pm}(\lambda)=\lim_{\varepsilon\downarrow 0}Y(\lambda\pm \mathrm{i}\varepsilon),\quad \lambda\in(-1,1).$$
2. The boundary values $Y_{\pm}(\lambda)$ are related by the jump condition
$$Y_{+}(\lambda)=Y_{-}(\lambda)\begin{pmatrix}1-\gamma & \gamma\mathrm{e}^{2\mathrm{i}\lambda s}\\ -\gamma\mathrm{e}^{-2\mathrm{i}\lambda s} & 1+\gamma\end{pmatrix},\quad \lambda\in(-1,1).$$
3. Near the endpoints $\lambda=\pm1$, we have
$$Y(\lambda)=\check{Y}(\lambda)\left[I+\frac{\gamma}{2\pi\mathrm{i}}\begin{pmatrix}-1 & 1\\ -1 & 1\end{pmatrix}\ln\left(\frac{\lambda-1}{\lambda+1}\right)\right]\mathrm{e}^{-\mathrm{i}s\lambda\sigma_3},\quad \lambda\to\pm1;$$
where $\check{Y}(\lambda)$ is analytic at $\lambda=\pm1$, ln denotes the principal branch for the logarithm and $\sigma_3=\left(\begin{smallmatrix}1 & 0\\ 0 & -1\end{smallmatrix}\right)$.
4. Near $\lambda=\infty$,
$$Y(\lambda)=I+Y_1\lambda^{-1}+\mathcal{O}\left(\lambda^{-2}\right),\quad \lambda\to\infty;\qquad Y_1=\left(Y_1^{jk}\right).$$

Our goal is to solve this problem asymptotically for sufficiently large $s\geq s_0$ and $0\leq v<s^{\frac{1}{3}}$. This will be achieved by an application of the nonlinear steepest descent method of Deift and Zhou [12] to RHP 2.1. Our approach is somewhat similar to [4].

2.1. Matrix factorization and opening of lens

We make use of a factorization of the jump matrix $G_Y(\lambda)$ occurring in RHP 2.1,

$$G_Y(\lambda) = \begin{pmatrix} 1-\gamma & \gamma e^{2i\lambda s} \\ -\gamma e^{-2i\lambda s} & 1+\gamma \end{pmatrix} = \begin{pmatrix} 1 & 0 \\ -\gamma e^{2s(\kappa - i\lambda)} & 1 \end{pmatrix} e^{-2\kappa s \sigma_3} \begin{pmatrix} 1 & \gamma e^{2s(\kappa+i\lambda)} \\ 0 & 1 \end{pmatrix}$$
$$\equiv S_L(\lambda) S_D S_U(\lambda).$$

Since $S_L(\lambda)$ and $S_U(\lambda)$ clearly admit analytical continuations to the lower and upper λ-planes respectively, we can perform a first transformation of the initial RHP 2.1: introduce

$$S(\lambda) = Y(\lambda) \begin{cases} S_U^{-1}(\lambda), & \lambda \in \Omega_1 \\ S_L(\lambda), & \lambda \in \Omega_2 \\ I, & \text{else} \end{cases}$$

where the domains $\Omega_j \subset \mathbb{C}$ are sketched in Figure 1 below. This leads to the following problem:

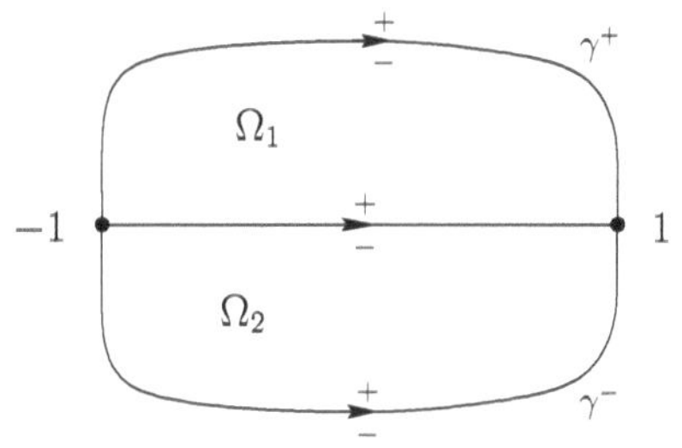

FIGURE 1. The oriented jump contour Σ_S for the function $S(\lambda)$ in the complex λ-plane.

Riemann–Hilbert Problem 2.2. Determine $S(\lambda) = S(\lambda; s, \gamma) \in \mathbb{C}^{2\times 2}$ such that

1. $S(\lambda)$ is analytic for $\lambda \in \mathbb{C}\backslash\Sigma_S$ where the jump contour Σ_S is shown in Figure 1.
2. The limiting values $S_\pm(\lambda), \lambda \in \Sigma_S$ are square integrable and satisfy the jump conditions
$$S_+(\lambda) = S_-(\lambda) S_U(\lambda), \quad \lambda \in \gamma^+; \quad S_+(\lambda) = S_-(\lambda) S_L(\lambda), \quad \lambda \in \gamma^-;$$
$$S_+(\lambda) = S_-(\lambda) S_D, \quad \lambda \in (-1, 1).$$
3. Near $\lambda = \pm 1$,
$$S(\lambda) = \check{Y}(\lambda) \left[I + \frac{\gamma}{2\pi i} \begin{pmatrix} -1 & 1 \\ -1 & 1 \end{pmatrix} \ln\left(\frac{\lambda - 1}{\lambda + 1} \right) \right] e^{-is\lambda\sigma_3} \begin{cases} S_U^{-1}(\lambda), & \lambda \in \Omega_1 \\ S_L(\lambda), & \lambda \in \Omega_2 \\ I, & \text{else} \end{cases}.$$
4. As $\lambda \to \infty$, we have $S(\lambda) \to I$.

If $v \in (0,\infty)$ were kept fixed throughout we would already know at this point that the major contribution to the asymptotic solution of RHP 2.2 arises from the segment $(-1,1)$ and two small neighborhoods of the endpoints $\lambda=\pm 1$. Indeed we have

$$\left|e^{2s(\kappa\pm i\lambda)}\right| = e^{2v}e^{\mp 2s\Im\lambda},$$

and thus exponentially fast decay to zero on $\gamma^+\backslash D(\pm 1,r)$ in the upper half-plane and $\gamma^-\backslash D(\pm 1,r)$ in the lower half-plane. Here, $D(\pm 1,r)=\{\lambda\in\mathbb{C}:\ |\lambda\mp 1|<r\}$. We will return to this observation after the next subsection.

2.2. Local Riemann–Hilbert analysis

We require three explicit model functions among which

$$P^{(\infty)}(\lambda) = \left(\frac{\lambda-1}{\lambda+1}\right)^{\frac{iv}{\pi}\sigma_3},\ \lambda\in\mathbb{C}\backslash[-1,1]:\ P^{(\infty)}(\lambda)\to I,\quad \lambda\to+\infty, \tag{2.1}$$

will serve as the outer parametrix. This function reproduces exactly the jump behavior of $S(\lambda)$ on the segment $(-1,1)$. For the local parametrices near the endpoints we follow and adapt the constructions in [4], Sections 9 and 10. First we define on the punctured plane $\zeta\in\mathbb{C}\backslash\{0\}$ with $-\pi<\arg\zeta\le\pi$,

$$P(\zeta) = \begin{pmatrix} U(-\nu;e^{-i\frac{\pi}{2}}\zeta) & -U(1+\nu;e^{i\frac{\pi}{2}}\zeta)ie^{i\pi\nu}\frac{\Gamma(1+\nu)}{\Gamma(-\nu)} \\ U(1-\nu;e^{-i\frac{\pi}{2}}\zeta)ie^{i\pi\nu}\frac{\Gamma(1-\nu)}{\Gamma(\nu)} & U(\nu;e^{i\frac{\pi}{2}}\zeta)e^{2\pi i\nu}\end{pmatrix}$$
$$\times\ e^{\frac{i}{2}\zeta\sigma_3}e^{i\frac{\pi}{2}(\frac{1}{2}-\nu)\sigma_3}\mathcal{S}(\zeta) \tag{2.2}$$

with

$$\nu=\frac{iv}{\pi};\qquad \mathcal{S}(\zeta)=\begin{cases}\begin{pmatrix}1 & \frac{2\pi i e^{i\pi\nu}}{\Gamma(\nu)\Gamma(1-\nu)}\\ 0 & 1\end{pmatrix}\begin{pmatrix}e^{i\frac{\pi}{2}\nu} & 0\\ 0 & e^{-i\frac{3\pi}{2}\nu}\end{pmatrix}, & \arg\zeta\in(\frac{\pi}{2},\pi)\\ \begin{pmatrix}1 & 0\\ \frac{2\pi i e^{-3\pi i\nu}}{\Gamma(\nu)\Gamma(1-\nu)} & 1\end{pmatrix}\begin{pmatrix}e^{i\frac{\pi}{2}\nu} & 0\\ 0 & e^{-i\frac{3\pi}{2}\nu}\end{pmatrix}, & \arg\zeta\in(-\pi,-\frac{\pi}{2})\\ \begin{pmatrix}e^{i\frac{\pi}{2}\nu} & 0\\ 0 & e^{-i\frac{3\pi}{2}\nu}\end{pmatrix}, & \arg\zeta\in(-\frac{\pi}{2},\frac{\pi}{2})\end{cases}$$

and $U(a;\zeta)=U(a,1;\zeta)\sim\zeta^{-a}$ as $\zeta\to\infty, \arg\zeta\in(-\frac{3\pi}{2},\frac{3\pi}{2})$, is the confluent hypergeometric function, cf. [13]. Using standard asymptotic and monodromy properties of $U(a;\zeta)$, see [13], one can check that the model function $P(\zeta)$ is a solution to the "bare" model problem below (cf. [10]).

Riemann–Hilbert Problem 2.3. The function $P(\zeta)$ defined in (2.2) has the following properties:

1. $P(\zeta)$ is analytic for $\zeta\in\mathbb{C}\backslash\{\arg\zeta=-\pi,-\frac{\pi}{2},\frac{\pi}{2}\}$ and the orientation of the three rays is fixed as indicated in Figure 2 below.
2. $P(\zeta)$ has ζ-independent jumps,

$$P_+(\zeta)=P_-(\zeta)\begin{pmatrix}1 & \gamma e^{2\kappa s}\\ 0 & 1\end{pmatrix},\quad \arg\zeta=\frac{\pi}{2};$$
$$P_+(\zeta)=P_-(\zeta)\begin{pmatrix}1 & 0\\ -\gamma e^{2\kappa s} & 1\end{pmatrix},\quad \arg\zeta=-\frac{\pi}{2},$$

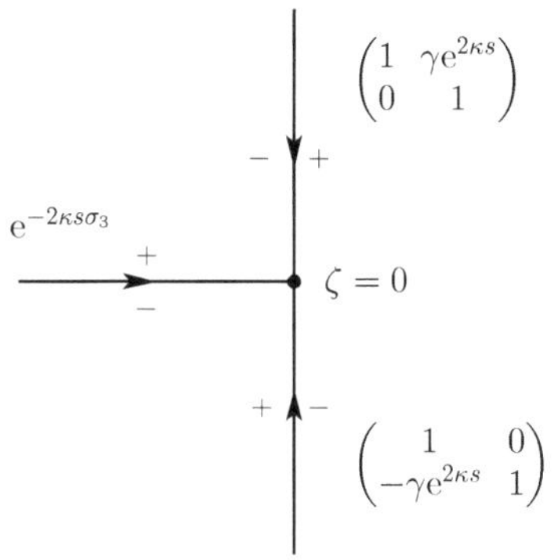

FIGURE 2. Jump behavior of $P(\zeta)$ in the complex ζ-plane.

and,

$$P_+(\zeta) = P_-(\zeta)e^{-2\kappa s\sigma_3}, \qquad \arg\zeta = -\pi.$$

3. Near $\zeta = 0$, with $-\pi < \arg\zeta \le \pi$,

$$P(\zeta) = \check{P}(\zeta)\left[I + \frac{\gamma}{2\pi \mathrm{i}}\begin{pmatrix}-1 & 1\\ -1 & 1\end{pmatrix}\ln\zeta\right]\begin{cases}\begin{pmatrix}1 & -\gamma e^{2\kappa s}\\ 0 & 1\end{pmatrix}, & \arg\zeta \in (\frac{\pi}{2}, \pi)\\ \begin{pmatrix}1 & 0\\ -\gamma e^{2\kappa s} & 1\end{pmatrix}, & \arg\zeta \in (-\pi, -\frac{\pi}{2})\\ I, & \arg\zeta \in (-\frac{\pi}{2}, \frac{\pi}{2})\end{cases} \tag{2.3}$$

where $\check{P}(\zeta)$ is analytic at $\zeta = 0$.

4. As $\zeta \to \infty$,

$$P(\zeta) \sim \left[I + \sum_{k=1}^{\infty} P_k(\nu)\frac{\zeta^{-k}}{k!}\right]\zeta^{\nu\sigma_3}e^{\frac{\mathrm{i}}{2}\zeta\sigma_3}e^{\mathrm{i}\frac{\pi}{2}(\frac{1}{2}-\nu)\sigma_3}$$

with

$$P_k(\nu) = \begin{pmatrix}\big((-\nu)_k\big)^2 e^{-\mathrm{i}\frac{\pi}{2}k} & \mathrm{i}\big((1+\nu)_{k-1}\big)^2 k\, e^{\mathrm{i}\frac{\pi}{2}k - \mathrm{i}\pi\nu}\frac{\Gamma(1+\nu)}{\Gamma(-\nu)}\\ -\mathrm{i}\big((1-\nu)_{k-1}\big)^2 k\, e^{-\mathrm{i}\frac{\pi}{2}k+\mathrm{i}\pi\nu}\frac{\Gamma(1-\nu)}{\Gamma(\nu)} & \big((\nu)_k\big)^2 e^{\mathrm{i}\frac{\pi}{2}k}\end{pmatrix}$$

where $(a)_k = a(a+k)(a+2)\cdot\ldots\cdot(a+k-1)$ is Pochhammer's symbol. Observe that in our situation $\gamma \in [0,1]$ and hence we have $\nu \in \mathrm{i}\mathbb{R}$.

In terms of $P(\zeta)$ the actual parametrix near $\lambda = 1$ is then defined as follows,

$$P^{(1)}(\lambda) = \big(2s(\lambda+1)\big)^{-\nu\sigma_3}e^{-\mathrm{i}\frac{\pi}{2}(\frac{1}{2}-\nu)\sigma_3}e^{\mathrm{i}s\sigma_3}P\big(\zeta(\lambda)\big)e^{-\frac{\mathrm{i}}{2}(\zeta(\lambda)+2s)\sigma_3}, \quad 0 < |\lambda - 1| < \frac{1}{4} \tag{2.4}$$

where $\zeta = \zeta(\lambda) = 2s(\lambda - 1)$ denotes the locally conformal change of variables from the λ- to the ζ-plane. Using the properties listed in RHP 2.3 we immediately establish that the initial function $S(\lambda)$ in RHP 2.2 and (2.4) are related by an analytic left multiplier $N_1(\lambda), |\lambda-1|<\frac{1}{4}$ via

$$S(\lambda)=N_1(\lambda)P^{(1)}(\lambda),\quad 0<|\lambda-1|<\frac{1}{4}.$$

Moreover, as $s\to\infty$, and hence $|\zeta|\to\infty$, with $0<r_1\leq|\lambda-1|\leq r_2<\frac{1}{4}$, we have asymptotic matching of the model functions $P^{(\infty)}(\lambda)$ and $P^{(1)}(\lambda)$: with $\beta(\lambda)=\big(2s(\lambda+1)\big)^{\nu}$,

$$P^{(1)}(\lambda)\sim\left[I+\sum_{k=1}^{\infty}P_k^{(1)}(\nu,\lambda)\frac{\zeta(\lambda)^{-k}}{k!}\right]P^{(\infty)}(\lambda), \tag{2.5}$$

where

$$P_k^{(1)}(\nu,\lambda)=\big(\beta(\lambda)\big)^{-\sigma_3}\mathrm{e}^{-\mathrm{i}\frac{\pi}{2}(\frac{1}{2}-\nu)\sigma_3}\mathrm{e}^{\mathrm{i}s\sigma_3}P_k(\nu)\mathrm{e}^{-\mathrm{i}s\sigma_3}\mathrm{e}^{\mathrm{i}\frac{\pi}{2}(\frac{1}{2}-\nu)\sigma_3}\big(\beta(\lambda)\big)^{\sigma_3}.$$

Near the remaining endpoint $\lambda=-1$ we can either carry out explicitly a similar construction or simply use symmetry: for $0<|\lambda+1|<\frac{1}{4}$, introduce the parametrix as

$$P^{(-1)}(\lambda)=\sigma_1P^{(1)}(-\lambda)\sigma_1,\qquad \sigma_1=\left(\begin{smallmatrix}0&1\\1&0\end{smallmatrix}\right) \tag{2.6}$$

and obtain at once with analytic $N_2(\lambda), |\lambda+1|<\frac{1}{4}$,

$$S(\lambda)=N_2(\lambda)P^{(-1)}(\lambda),\quad 0<|\lambda+1|<\frac{1}{4},$$

as well as for $s\to\infty$ with $0<r_1\leq|\lambda+1|\leq r_2<\frac{1}{4}$,

$$P^{(-1)}(\lambda)\sim\left[I+\sum_{k=1}^{\infty}\sigma_1P_k^{(1)}(\nu,-\lambda)\sigma_1\frac{\zeta(-\lambda)^{-k}}{k!}\right]P^{(\infty)}(\lambda). \tag{2.7}$$

This completes the construction of the explicit model functions $P^{(\infty)}(\lambda), P^{(1)}(\lambda)$ and $P^{(-1)}(\lambda)$.

2.3. Ratio transformation and small norm estimations

We use (2.1), (2.4), (2.6) and define in this step

$$R(\lambda)=S(\lambda)\begin{cases}\big(P^{(1)}(\lambda)\big)^{-1}, & |\lambda-1|<r\\ \big(P^{(-1)}(\lambda)\big)^{-1}, & |\lambda+1|<r\\ \big(P^{(\infty)}(\lambda)\big)^{-1}, & |\lambda\mp1|>r\end{cases} \tag{2.8}$$

with $0<r<\frac{1}{4}$. Recalling the results of the previous subsection we are led to the problem below.

Riemann–Hilbert Problem 2.4. The ratio function $R(\lambda)=R(\lambda;s,\gamma)\in\mathbb{C}^{2\times2}$ is determined by the following Riemann–Hilbert problem $(\Sigma_R,G_R(\cdot;s,\gamma))$:

1. $R(\lambda)$ is analytic for $\lambda\in\mathbb{C}\backslash\Sigma_R$ where the oriented jump contour Σ_R is depicted in Figure 3.

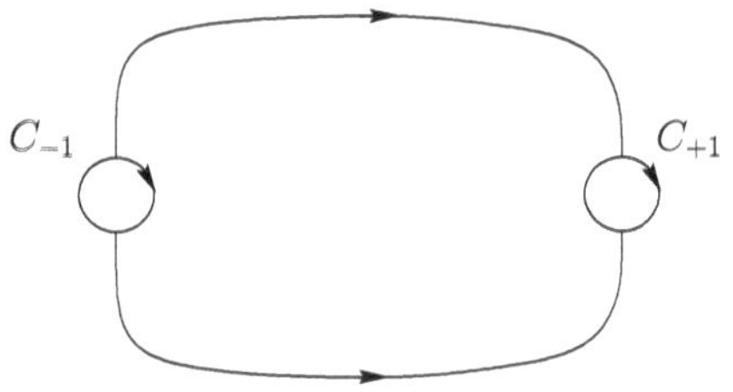

FIGURE 3. The oriented jump contour Σ_R in the ratio RHP 2.4.

2. The square integrable boundary values $R_\pm(\lambda), \lambda \in \Sigma_R$ are related by the jump conditions

$$R_+(\lambda) = R_-(\lambda)\left[I + j_{12}(\lambda; s,\gamma)\left(\begin{smallmatrix}0&1\\0&0\end{smallmatrix}\right)\right], \quad \lambda \in (\Sigma_R\backslash C_{\pm 1}) \cap \{\Im\lambda > 0\};$$
$$R_+(\lambda) = R_-(\lambda)\left[I + j_{21}(\lambda; s,\gamma)\left(\begin{smallmatrix}0&0\\1&0\end{smallmatrix}\right)\right], \quad \lambda \in (\Sigma_R\backslash C_{\pm 1}) \cap \{\Im\lambda < 0\}$$

with

$$j_{12}(\lambda; s,\gamma) = \gamma e^{2s(\kappa+i\lambda)}\left(\frac{\lambda-1}{\lambda+1}\right)^{\frac{2iv}{\pi}}, \quad j_{21}(\lambda) = -\gamma e^{2s(\kappa-i\lambda)}\left(\frac{\lambda-1}{\lambda+1}\right)^{-\frac{2iv}{\pi}}.$$

Also, along the clockwise oriented circle boundaries $C_{\pm 1}$,

$$R_+(\lambda) = R_-(\lambda)P^{(1)}(\lambda)\big(P^{(\infty)}(\lambda)\big)^{-1}, \quad \lambda \in C_{+1};$$
$$R_+(\lambda) = R_-(\lambda)P^{(-1)}(\lambda)\big(P^{(\infty)}(\lambda)\big)^{-1}, \quad \lambda \in C_{-1}.$$

3. As $\lambda \to \infty$, we have that $R(\lambda) \to I$.

On the way to small norm estimations for the jump matrix $G_R(\lambda; s,\gamma), \lambda \in \Sigma_R$ we first note that

$$\big|j_{12}(\pm 1 + ir; s,\gamma)\big| \le \exp\big[2v - v\alpha(r) - 2rs\big];$$
$$\big|j_{21}(\pm 1 - ir; s,\gamma)\big| \le \exp\big[2v - v\alpha(r) - 2rs\big]$$

with

$$\alpha(r) = 1 - \frac{2}{\pi}\arctan\left(\frac{r}{2}\right) > 1 - \frac{2}{\pi}\arctan\left(\frac{1}{8}\right) > 0.8.$$

In order to obtain an estimate in the scaling region

$$s \ge s_0, \quad 0 \le v < s^{\frac{1}{3}}$$

we have to choose a contracting radius in (2.8). In fact we will work with $r = r(s) = s^{-\frac{1}{3}}$ which leads to

Proposition 2.5. *There exists $s_0 > 0$ such that*

$$\|G_R(\cdot; s,\gamma) - I\|_{L^2\cap L^\infty(\Sigma_R\backslash C_{\pm 1})} \le e^{-s^{\frac{2}{3}}}, \qquad \forall\, s \ge s_0,\ 0 \le v < s^{\frac{1}{3}}.$$

For $C_{\pm 1}$ we go back to (2.5) and (2.7), first for $\lambda \in C_{+1}$,

$$\mathrm{e}^{\nu \ln|2s(\lambda+1)|\sigma_3}\mathrm{e}^{-\frac{\mathrm{i}}{2}\arg(\frac{\Gamma(\nu)}{\Gamma(-\nu)})\sigma_3}\mathrm{e}^{-\mathrm{i}s\sigma_3}\Big(G_R(\lambda;s,\gamma)-I\Big)\mathrm{e}^{\mathrm{i}s\sigma_3}\mathrm{e}^{\frac{\mathrm{i}}{2}\arg(\frac{\Gamma(\nu)}{\Gamma(-\nu)})\sigma_3}$$
$$\times\,\mathrm{e}^{-\nu\ln|2s(\lambda+1)|\sigma_3}\sim\sum_{k=1}^{\infty}\widehat{P}_k(\nu)\mathrm{e}^{-\mathrm{i}\frac{\pi}{2}k\sigma_3}\frac{\zeta^{-k}}{k!}$$

where

$$\widehat{P}_k(\nu)=\begin{pmatrix}\big((-\nu)_k\big)^2 & \nu\big((1+\nu)_{k-1}\big)^2k\,\mathrm{e}^{-2\mathrm{i}\nu\arg(\lambda+1)}\\ -\nu\big((1-\nu)_{k-1}\big)^2k\,\mathrm{e}^{2\mathrm{i}\nu\arg(\lambda+1)} & \big((\nu)_k\big)^2\end{pmatrix}.$$

Thus for $\lambda \in C_{+1}$,

$$\|G_R(\lambda;s,\gamma)-I\|\leq c\left\{\frac{v}{sr}\left\|\begin{pmatrix} v & \mathrm{e}^{\frac{2v}{\pi}\arg(\lambda+1)}\\ \mathrm{e}^{-\frac{2v}{\pi}\arg(\lambda+1)} & v\end{pmatrix}\right\|+\mathcal{E}(\lambda;s,\gamma)\right\},$$

and estimations for the error term $\mathcal{E}(\lambda;s,\gamma)$ follow from known error-term estimates for the confluent hypergeometric functions which are, for instance, given in [13]: there exists $s_0>0$ and constants $c_j>0$ such that

$$\|\mathcal{E}(\lambda;s,\gamma)\|\leq c_1\mathrm{e}^{c_2v-c_3sr}\leq c_1\mathrm{e}^{-c_4s^{\frac{2}{3}}},\quad \forall\, s\geq s_0,\ \ 0\leq v<s^{\frac{1}{3}}.$$

Combining this estimate with

$$\exp\left[\frac{2v}{\pi}\big|\arg(\lambda+1)\big|\right]\leq \mathrm{e}^{\frac{2v}{\pi}\delta(r)},\quad \delta(r)=\arctan\left(\frac{r}{2}\right),\quad \lambda\in C_{+1},$$

we obtain (with similar methods for the estimates on C_{-1})

Proposition 2.6. *For every $0<\epsilon<\frac{1}{3}$ there exists $s_0>0$ and $c>0$ such that*

$$\|G_R(\cdot;s,\gamma)-I\|_{L^2\cap L^\infty(C_{\pm1})}\leq c\,v^2s^{-\frac{2}{3}},\quad \forall\, s\geq s_0,\ \ 0\leq v<s^{\frac{1}{3}-\epsilon}.$$

Combining Propositions 2.5 and 2.6 we obtain by general theory [12],

Proposition 2.7. *For every $0<\epsilon<\frac{1}{3}$ there exists $s_0>0$ and $c>0$ such that the ratio RHP 2.4 is uniquely solvable for all $s\geq s_0$ and $0\leq v<s^{\frac{1}{3}-\epsilon}$. The solution can be computed iteratively through the integral equation*

$$R(\lambda)=I+\frac{1}{2\pi\mathrm{i}}\int_{\Sigma_R}R_-(w)\big(G_R(w)-I\big)\frac{\mathrm{d}w}{w-\lambda},\quad \lambda\in\mathbb{C}\backslash\Sigma_R,$$

where we use that

$$\|R_-(\cdot;s,\gamma)-I\|_{L^2(\Sigma_R)}\leq c\,v^2s^{-\frac{2}{3}},\quad \forall\, s\geq s_0,\ \ 0\leq v<s^{\frac{1}{3}-\epsilon}.$$

This concludes the nonlinear steepest descent analysis of RHP 2.1.

2.4. Proof of expansion 1.7

In order to obtain the statement in (1.7) we make use of the identity (see, e.g., [3], equation (2.1)),

$$\frac{\partial}{\partial s}\ln\det(I-\gamma K_s) = -2\mathrm{i}Y_1^{11} \tag{2.9}$$

where Y_1 appeared in RHP 2.1 and the derivative is taken with γ fixed. Tracing back the transformations $Y(\lambda) \mapsto S(\lambda) \mapsto R(\lambda)$, we have

$$Y_1 = -\frac{2\mathrm{i}v}{\pi}\sigma_3 + \frac{\mathrm{i}}{2\pi}\int_{\Sigma_R} R_-(w)\big(G_R(w)-I\big)\mathrm{d}w$$

and with the help of an explicit residue computation as well as Proposition 2.7,

$$\begin{aligned} Y_1^{11} &= -\frac{2\mathrm{i}v}{\pi} + \frac{\mathrm{i}v^2}{s\pi^2} + \frac{\mathrm{i}}{2\pi}\oint_{C_{\pm 1}}\Big[\big(R_-(w)-I\big)\big(G_R(w)-I\big)\Big]_{11}\mathrm{d}w \\ &= -\frac{2\mathrm{i}v}{\pi} + \frac{\mathrm{i}v^2}{s\pi^2} + \frac{\mathrm{i}v^2}{4\pi^2 s^2}\sin\big(\phi(s,v)\big) + \mathcal{O}\left(v^3 s^{-2}\right); \\ \phi(s,v) &= 4s - \frac{4v}{\pi}\ln(4s) + 2\arg\left(\frac{\Gamma(\nu)}{\Gamma(-\nu)}\right). \end{aligned}$$

With this we go back to (2.9) and perform an indefinite integration with respect to s,

$$\ln\det(I-\gamma K_s) = -\frac{4v}{\pi}s + \frac{2v^2}{\pi^2}\ln s + C(v) + r(s,v) \tag{2.10}$$

and the error term $r(s,v)$ is differentiable with respect to s and for any $0<\epsilon<\frac{1}{3}$ there exist $s_0>0, c>0$ such that

$$\big|r(s,v)\big| \le c\frac{v^3}{s}, \qquad \forall\, s\ge s_0,\ \ 0\le v< s^{\frac{1}{3}-\epsilon}. \tag{2.11}$$

The term $C(v)$ appearing in (2.10) is s-independent and can therefore be determined by comparison with (1.3), i.e., we have

$$C(v) = \frac{2v^2}{\pi^2}\ln 4 + 2\ln G\left(1+\frac{\mathrm{i}v}{\pi}\right)G\left(1-\frac{\mathrm{i}v}{\pi}\right),$$

and this completes the section.

Remark 2.8. Estimate (2.10), (2.11) is weaker than the one stated in Theorem 1.2. However, it is enough for the needs of [3]. The full statement of Theorem 1.2, that is the extension of (2.10) to the whole range $0\le v<s^{\frac{1}{3}}$ with uniform constants s_0, c_1 and c_2 will be given in the next section with the use of an alternative method based on the recent result [9].

3. Extending 1.3 by Toeplitz determinant techniques

Consider the following function on the unit circle,

$$f^t(\mathrm{e}^{\mathrm{i}\theta}) = \begin{cases} \mathrm{e}^{-2v}, & \theta \in [0,t) \cup [2\pi - t, 2\pi) \\ 1, & \theta \in [t, 2\pi - t) \end{cases}; \qquad 0 < t < 2\pi,$$

with Fourier coefficients $\{f_k^t\}_{k\in\mathbb{Z}}$ and the associated Toeplitz determinant $D_n(f^t)$,

$$f_k^t = \frac{1}{2\pi}\int_0^{2\pi} f^t(\mathrm{e}^{\mathrm{i}\theta})\mathrm{e}^{-\mathrm{i}k\theta}\mathrm{d}\theta, \qquad D_n(f^t) = \det\left[f_{j-k}^t\right]_{j,k=0}^{n-1}.$$

We first observe

Lemma 3.1. *For any fixed $s > 0$,*

$$\lim_{n\to\infty} D_n(f^{\frac{2s}{n}}) = \det(I - \gamma K_s)\big|_{L^2(-1,1)}; \qquad K_s(\lambda,\mu) = \frac{\sin s(\lambda-\mu)}{\pi(\lambda-\mu)}.$$

Proof. A straightforward computation of the Fourier coefficients $f_k^{\frac{2s}{n}}$ gives

$$f_k^{\frac{2s}{n}} = -\gamma\frac{\sin(k\frac{2s}{n})}{\pi k}, \quad k \neq 0; \qquad f_0^{\frac{2s}{n}} = 1 - \gamma\frac{2s}{n\pi}$$

and therefore

$$nf_{j-k}^{\frac{2s}{n}} \to I(\lambda,\mu) - \gamma K_s(\lambda,\mu), \quad \text{if} \quad \frac{j}{n}\to\lambda, \ \frac{k}{n}\to\mu.$$

This, together with the translation invariance of $K_s(\lambda,\mu)$, implies the Lemma by standard properties of trace class operators. □

We will now obtain asymptotics of $D_n(f^t)$ as $n\to\infty$, for $s > s_0$ with $s_0 > 0$ sufficiently large and $0 \leq v < s^{\frac{1}{3}}$. Note to this end that the function $f^t(z)$ is of Fisher–Hartwig type, see, for example, [10], with two jump-type singularities at $z_1 = \mathrm{e}^{\mathrm{i}t}, z_2 = \mathrm{e}^{\mathrm{i}(2\pi-t)}$ and parameters

$$\beta = \beta_1 = -\beta_2 = \frac{\mathrm{i}v}{\pi}.$$

In more detail, we have

$$f^t(z) = \mathrm{e}^{-\frac{2vt}{\pi}} g_{z_1,\beta_1}(z) g_{z_2,\beta_2}(z) z_1^{-\beta_1} z_2^{-\beta_2}, \quad z = \mathrm{e}^{\mathrm{i}\theta} \tag{3.1}$$

where

$$g_{z_j,\beta_j}(\mathrm{e}^{\mathrm{i}\theta}) = \begin{cases} \mathrm{e}^{\mathrm{i}\pi\beta_j}, & \theta \in [0, \arg z_j) \\ \mathrm{e}^{-\mathrm{i}\pi\beta_j}, & \theta \in [\arg z_j, 2\pi) \end{cases}.$$

The asymptotics of $D_n(f^t)$ as $n\to\infty$ for any *fixed* t is a classical result going back to the works of Widom, Basor, Böttcher and Silbermann [19, 1, 5]. As is shown in

[9], Theorem 1.11, this result still holds in the case of $\frac{2s}{n} < t < t_0$ for sufficiently small t_0 with adjusted error estimate:

$$\ln D_n(f^t) = -\frac{4v}{\pi}s + \frac{2v^2}{\pi^2}\ln\left(2n\sin\frac{2s}{n}\right) + 2\ln G\left(1+\frac{iv}{\pi}\right)G\left(1-\frac{iv}{\pi}\right) + r(s,v,n), \tag{3.2}$$

and there exists $n_0(v), s_0(v), C_0(v) > 0$ such that[1]

$$\forall n > n_0,\ s > s_0,\ v \geq 0 \text{ fixed}: \quad |r(s,v,n)| < C_0(v)s^{-1}.$$

We will now extend the argument of [9] to the case of $0 \leq v < s^{\frac{1}{3}}$ and see that the error term remains small (for large s) in this region of the (v,s)-plane. In order to carry out this approach we have to track the dependence of the error term [9, (7.58)], the error term [9, (8.20)] at z_1 and a similar term at z_2 on $\beta = \frac{iv}{\pi}$.

First, we obtain by a straightforward calculation that the crucial constants $\widehat{E}_j(z_j)e^{-i\pi\beta_j\sigma_3}$ in [9, (7.47), (7.52)] are bounded in $n, t > 0$ uniformly for any purely imaginary β_j and $\frac{2s}{n} < t < t_0$.

Second, we consider the auxiliary M-RHP of Section 4 in [9]:

Riemann–Hilbert Problem 3.2 (see [9, Section 4] or [8, Section 4.2.1]). Determine $M(\zeta) = M(\zeta;\beta) \in \mathbb{C}^{2\times 2}$ such that

1. $M(\zeta)$ is analytic for $\zeta \in \mathbb{C}\setminus(e^{\pm i\frac{\pi}{4}}\mathbb{R} \cup [0,\infty))$. We orient the five jump rays as shown in Figure 4 below.

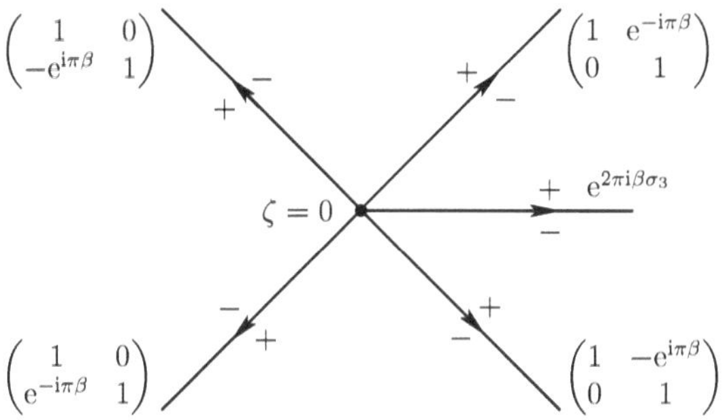

FIGURE 4. Jump behavior of $M(\zeta)$ in the complex ζ-plane.

2. The boundary values $M_\pm(\zeta)$ on $e^{\pm i\frac{\pi}{4}}\mathbb{R} \cup (0,\infty)$ are continuous and related by the following jump conditions:

$$M_+(\zeta) = M_-(\zeta)\begin{pmatrix}1 & e^{-i\pi\beta}\\ 0 & 1\end{pmatrix}, \quad \zeta \in e^{i\frac{\pi}{4}}(0,\infty);$$
$$M_+(\zeta) = M_-(\zeta)\begin{pmatrix}1 & 0\\ -e^{i\pi\beta} & 1\end{pmatrix}, \quad \zeta \in e^{i\frac{3\pi}{4}}(0,\infty)$$

[1] note that β_j in [9], and hence v, are fixed.

as well as

$$M_+(\zeta) = M_-(\zeta)\begin{pmatrix} 1 & 0 \\ e^{-i\pi\beta} & 1 \end{pmatrix}, \quad \zeta \in e^{i\frac{5\pi}{4}}(0,\infty);$$

$$M_+(\zeta) = M_-(\zeta)\begin{pmatrix} 1 & -e^{i\pi\beta} \\ 0 & 1 \end{pmatrix}, \quad \zeta \in e^{i\frac{7\pi}{4}}(0,\infty)$$

and

$$M_+(\zeta) = M_-(\zeta)e^{2\pi i\beta\sigma_3}, \qquad \zeta \in (0,\infty).$$

3. As $\zeta \to \infty$, valid in a full vicinity of $\zeta = \infty$,

$$M(\zeta) = \Big(I + \frac{M_1}{\zeta} + \mathcal{O}\left(\zeta^{-2}\right)\Big)\zeta^{-\beta\sigma_3}e^{-\frac{1}{2}\zeta\sigma_3}, \qquad \arg\zeta \in (0, 2\pi).$$

with

$$M_1 = \begin{pmatrix} -\beta^2 & -e^{-2\pi i\beta}\frac{\Gamma(1-\beta)}{\Gamma(\beta)} \\ e^{2\pi i\beta}\frac{\Gamma(1+\beta)}{\Gamma(-\beta)} & \beta^2 \end{pmatrix}.$$

This problem is solved explicitly in terms of the confluent hypergeometric function $U(a;\zeta) = U(a,1;\zeta)$, compare RHP 2.3 and (2.2) above. We define

$$M(z) = \begin{pmatrix} U(\beta,z) & U(1-\beta, e^{-i\pi}z)e^{-i\pi\beta}\frac{\Gamma(1-\beta)}{\Gamma(\beta)} \\ U(1+\beta,z)e^{2\pi i\beta}\frac{\Gamma(1+\beta)}{\Gamma(-\beta)} & U(-\beta, e^{-i\pi}z)e^{i\pi\beta} \end{pmatrix}$$
$$\times\, e^{-\frac{1}{2}z\sigma_3}\mathcal{M}(z), \quad \arg z \in (0, 2\pi)$$

with

$$\mathcal{M}(z) = \begin{cases} \begin{pmatrix} 1 & -e^{-i\pi\beta} \\ 0 & 1 \end{pmatrix}, & \arg z \in (0, \frac{\pi}{4}) \\ I, & \arg z \in (\frac{\pi}{4}, \frac{3\pi}{4}) \\ \begin{pmatrix} 1 & 0 \\ -e^{i\pi\beta} & 1 \end{pmatrix}, & \arg z \in (\frac{3\pi}{4}, \frac{5\pi}{4}) \end{cases},$$

$$\mathcal{M}(z) = \begin{cases} \begin{pmatrix} 1 & 0 \\ -2i\sin\pi\beta & 1 \end{pmatrix}, & \arg z \in (\frac{5\pi}{4}, \frac{7\pi}{4}) \\ \begin{pmatrix} 1 & -e^{i\pi\beta} \\ -2i\sin\pi\beta & e^{2\pi i\beta} \end{pmatrix}, & \arg z \in (\frac{7\pi}{4}, 2\pi) \end{cases}.$$

From the standard asymptotics of the confluent hypergeometric function (cf. [13]), we conclude that for purely imaginary β,

$$M(z) = \left[I + e^{-i\pi\beta\sigma_3}\left\{\mathcal{O}\left(\frac{\beta}{z}\right) + \mathcal{O}\left(\frac{\beta^2}{z}\right)\right\}e^{-i\pi\beta\sigma_3}\right]z^{-\beta\sigma_3}e^{-\frac{z}{2}\sigma_3}, \quad z \to \infty. \tag{3.3}$$

Here we used the identity (cf. [13])

$$\left|\frac{\Gamma(1-\beta)}{\Gamma(\beta)}\right| = |\beta|, \quad \beta \in i\mathbb{R}.$$

Now we are ready to estimate the jumps in the final Υ-RHP in the region $\frac{2s}{n} < t < t_0$, see [9, (7.55)]. The function $\Upsilon(\lambda)$ is the final ratio function for the underlying nonlinear steepest descent analysis in [9]. We will not explicitly define $\Upsilon(z)$ at this point as it would involve several other model functions which we do not need for our purposes. Instead we refer to [9, (7.55), (7.56) and (7.57)] for details. The jump contour Σ_Υ in the Υ-RHP consists of two contracting circles of radii $\mathcal{O}(t)$ and the lens boundary around the larger arc outside the circles. The jump on the outer lens lip equals identity plus a constant matrix multiplied by $z^{-n}(\mathcal{D}_{\mathrm{in},t}(z)\mathcal{D}_{\mathrm{out},t}(z))^{-1}$, defined in [9, (7.3), (7.4)]. Subject to the condition that, say, $v^2 < s$, $n > s > s_0$, $\frac{2s}{n} < t < t_0$, we obtain the estimate

$$\left|z^{-n}\left(\mathcal{D}_{\mathrm{in},t}(z)\mathcal{D}_{\mathrm{out},t}(z)\right)^{-1}\right| < c_0 \mathrm{e}^{-\epsilon nt}\left|\left(\frac{z-z_1}{z-z_2}\right)^{-2\beta}\left(\frac{z_1}{z_2}\right)^{\beta}\right|$$
$$< c_0 \mathrm{e}^{-\epsilon nt + C_0 v} < c_0 \mathrm{e}^{-\epsilon_1 nt}$$

for some $c_0, C_0, \epsilon, \epsilon_1 > 0$ on the outer lip of the lens. Thus the jump is exponentially (in s) close to the identity. The corresponding estimate on the inner lens lip is similar. For the jump on the circle centered at z_1 we use (3.3) and the properties of $\widehat{E}_1(z)$. This leads us to an estimate of the form

$$I + \mathcal{O}\left(\frac{\beta}{nt}\right) + \mathcal{O}\left(\frac{\beta^2}{nt}\right)$$

and a similar one also holds on the circle centered at z_2. From standard theory [12] it follows now that the problem for $\Upsilon(z)$ is solvable and we obtain the estimates (which replace [9, (7.58)]),

$$\Upsilon(z) = I + \mathcal{O}\left(\frac{\beta}{nt}\right) + \mathcal{O}\left(\frac{\beta^2}{nt}\right), \quad \frac{\mathrm{d}}{\mathrm{d}z}\Upsilon(z) = \frac{1}{t}\left\{\mathcal{O}\left(\frac{\beta}{nt}\right) + \mathcal{O}\left(\frac{\beta^2}{nt}\right)\right\} \tag{3.4}$$

as $s, n \to \infty, n > s > s_0, v^2 < s$, uniformly for z off the jump contour Σ_Υ and uniformly in s, v such that $\frac{2s}{n} < t < t_0$. Using (3.4) we can then estimate the error term in [9](8.20).

Remark 3.3. For simplicity of the derivation, it is assumed in [9, (8.20)] that there also exist $\alpha > 0$ Fisher–Hartwig singularities at the same points:

$$f^t(z) = \mathrm{e}^{-\frac{2vt}{\pi}} \prod_{j=1}^{2} |z - z_j|^{2\alpha} g_{z_j,\beta_j}(z) z_j^{-\beta_j},$$

as opposed to (3.1). In the end of the derivation, to obtain the actual error term in the differential identity for the determinant $D_n(f^t)$, we have to multiply [9](8.20) by α and take the limit $\alpha \to 0$.

By a straightforward computation we obtain

$$\frac{1}{t}\left\{\mathcal{O}\left(\frac{\beta}{nt}\right) + \mathcal{O}\left(\frac{\beta^3}{nt}\right)\right\}$$

as error term in the differential identity for $D_n(f^t)$, uniformly for $\frac{2s}{n} < t < t_0$, the rest of the identity for $D_n(f^t)$ is the same as in [9]. Integrating this identity over $\frac{2s}{n} < t < t_0$ for some fixed t_0 and using the well-known large n asymptotics for the determinant with two fixed Fisher–Hartwig singularities at $z_1 = e^{it_0}, z_2 = e^{i(2\pi - t_0)}$ (see, for instance, [9, (1.8)]), we derive (3.2), where now, for some $s_0, C_1, C_2 > 0$,

$$n > s > s_0, \qquad 0 \leq v < s^{\frac{1}{3}}, \qquad |r(s,v,n)| < C_1 \frac{v}{s} + C_2 \frac{v^3}{s}.$$

At this point we take the limit $n \to \infty$, use Lemma 3.1 and obtain Theorem 1.2.

4. More on the integral term – proof of Proposition 1.3

We first analyze all objects involved in the formulæ for $M(x,\kappa)$ in the limit $\kappa \downarrow 0$. Observe that the defining equation (1.4) for the branch point $a = a(\kappa)$ can be written as

$$F(a,\kappa) \equiv \kappa + \frac{i}{2} \oint_{\Sigma} \left(\frac{\mu^2 - a^2}{\mu^2 - 1} \right)^{\frac{1}{2}} d\mu = 0 \tag{4.1}$$

where $\Sigma \subset \mathbb{C}$ is a simple counterclockwise oriented Jordan curve around the interval $[a, 1]$ and we fix

$$-\pi < \arg \left(\frac{\mu^2 - a^2}{\mu^2 - 1} \right) \leq \pi \quad \text{with} \quad \sqrt{\frac{\mu^2 - a^2}{\mu^2 - 1}} > 0 \ \text{ for } \ \mu > 1.$$

Since $F(a,\kappa)$ is analytic at $(a,\kappa) = (1,0)$ and $F_a(1,0) = \frac{\pi}{2} > 0$ we have unique solvability of equation (4.1) in a neighborhood of $(a,\kappa) = (1,0)$ and the solution $a = a(\kappa)$ is analytic at $\kappa = 0$, i.e., [3, (2.5)] extends to a full Taylor series

$$a(\kappa) = 1 - \frac{2\kappa}{\pi} - \frac{\kappa^2}{\pi^2} + \sum_{j=3}^{\infty} a_j \kappa^j, \qquad \kappa \to 0. \tag{4.2}$$

Next we extend the small κ expansions listed in [3, (2.10), (2.11)] for the frequency $V = V(\kappa)$ and nome $\tau = \tau(\kappa)$. Both, compare (1.5), are expressed in terms of complete elliptic integrals and thus hypergeometric functions, cf. [13],

$$K(a) = \frac{\pi}{2} \, {}_2F_1 \left(\frac{1}{2}, \frac{1}{2}; 1; a^2 \right), \qquad E(a) = \frac{\pi}{2} \, {}_2F_1 \left(-\frac{1}{2}, \frac{1}{2}; 1; a^2 \right).$$

Hence, using expansion of the hypergeometric functions at unity (cf. [13]) and combining these with (4.2) we obtain

$$V(\kappa) = -\frac{2}{\pi} \left(1 + \ln \kappa \sum_{j=1}^{\infty} u_j \kappa^j + \sum_{j=1}^{\infty} v_j \kappa^j \right), \quad \kappa \downarrow 0; \tag{4.3}$$

with $u_1 = \frac{1}{\pi}, v_1 = -\frac{1}{\pi}(1 + \ln 4\pi)$, as well as

$$\tau(\kappa) = -\frac{2i}{\pi} \ln \kappa + \ln \kappa \sum_{j=1}^{\infty} b_j \kappa^j + \sum_{j=0}^{\infty} c_j \kappa^j, \quad \kappa \downarrow 0; \qquad c_0 = \frac{2i}{\pi} \ln 4\pi. \tag{4.4}$$

The series in the right-hand sides of (4.3) and (4.4) are convergent series.

Remark 4.1. Another parameter which appears in [3, (1.16)] and which serves as normalization for the elliptic nome $\tau = \tau(\kappa)$ is given by

$$c = c(\kappa) = \frac{\mathrm{i}}{2}\frac{1}{K(a')}, \qquad a' = \sqrt{1-a^2}.$$

Hence analyticity of $a(\kappa)$ at $\kappa = 0$ implies at once analyticity of $c(\kappa)$ at the same point and [3], Corollary 2.2 extends to a full Taylor series

$$c(\kappa) = \frac{\mathrm{i}}{\pi}\left(1 - \frac{\kappa}{\pi} + \sum_{j=2}^{\infty} d_j\kappa^j\right), \qquad \kappa \to 0.$$

At this point we recall [3], Proposition 4.2: It was shown that $M(x,\kappa)$, given by [3, (1.24)] (see also (A.1)) and defined for $x \in \mathbb{R}, \kappa \in (0, 1-\delta]$ is smooth in both its arguments and one-periodic in the first,

$$M(x+1,\kappa) = M(x,\kappa).$$

Hence we can expand $M(x,\kappa)$ in a Fourier series

$$M(x,\kappa) = \sum_{n\in\mathbb{Z}} a_n(\kappa)\mathrm{e}^{2\pi\mathrm{i}nx}; \quad a_n(\kappa) = \frac{1}{2\pi}\int_0^1 M(x,\kappa)\mathrm{e}^{-2\pi\mathrm{i}nx}\mathrm{d}x \tag{4.5}$$

and the series converges uniformly in $x \in \mathbb{R}, \kappa \in (0, 1-\delta]$ to $M(x,\kappa)$. Note that for $n \neq 0$,

$$a_n(\kappa) = -\frac{1}{8\pi^2 n^2}\int_0^1 \left[\frac{\partial^2}{\partial x^2}M(x,\kappa)\right]\mathrm{e}^{-2\pi\mathrm{i}nx}\mathrm{d}x;$$

$$\frac{\mathrm{d}}{\mathrm{d}\kappa}a_n(\kappa) = -\frac{1}{8\pi^3 n^2}\int_0^1 \left[\frac{\partial^3}{\partial x^2\partial\kappa}M(x,\kappa)\right]\mathrm{e}^{-2\pi\mathrm{i}nx}\mathrm{d}x.$$

In the derivation of Proposition 1.3 we will use the series representation (4.5), i.e., we need estimates for $a_n(\kappa)$ and its derivative.

The building blocks of $M(x,\kappa)$ are summarized in [3, (4.10), (4.11), (4.12)] (see also (A.1), (A.2) and (A.3)) and these are all functions involving the third Jacobi theta function

$$\theta_3(x|\tau(\kappa)) = \sum_{m\in\mathbb{Z}} \mathrm{e}^{\mathrm{i}\pi\tau m^2 + 2\pi\mathrm{i}mx}.$$

The roots of this function are located at $\frac{1}{2} + \frac{\tau}{2} + \mathbb{Z} + \tau\mathbb{Z}$ and from (4.4) we see that uniformly in $x \in \mathbb{R}$,

$$\theta_3(x|\tau) = 1 + \mathcal{O}\left(\kappa^2\right).$$

Hence, for any $m, \ell \in \mathbb{Z}_{\geq 0}$ there exist $C_{m,\ell}, C_m > 0$[2] such that

$$\left|\frac{\partial^{m+\ell}}{\partial x^m\partial\tau^\ell}\theta_3(x|\tau) - \delta_{m0}\delta_{\ell 0}\right| \leq C_{m,\ell}\kappa^2; \quad \left|\frac{\partial^{m+1}}{\partial x^m\partial\kappa}\theta_3(x|\tau(\kappa))\right| \leq C_m\kappa, \tag{4.6}$$

[2]In what follows, C with or without indices denotes a positive constant independent of x and κ whose value may be different in different estimates.

for all $x \in \mathbb{R}, \kappa \in (0, 1-\delta]$ and in particular

$$\begin{aligned}
\left|\frac{\partial^m}{\partial x^m}\left[\frac{\mathrm{i}}{4\pi}\frac{\kappa}{\theta_3(x|\tau)}\frac{\partial^2}{\partial y^2}\theta_3(y|\tau)\bigg|_{y=x}\frac{\mathrm{d}\tau}{\mathrm{d}\kappa}\right]\right| &\leq C_m\kappa^2,\\
\left|\frac{\partial^{m+1}}{\partial x^m\partial\kappa}\left[\frac{\mathrm{i}}{4\pi}\frac{\kappa}{\theta_3(x|\tau)}\frac{\partial^2}{\partial y^2}\theta_3(y|\tau)\bigg|_{y=x}\frac{\mathrm{d}\tau}{\mathrm{d}\kappa}\right]\right| &\leq C_m\kappa.
\end{aligned}\tag{4.7}$$

The last two estimates are needed for the second summand in [3, (1.24)] (compare (A.1)). For the first term in loc. cit. we collect from (4.2), the uniform bounds which are valid for all $x \in \mathbb{R}, \kappa \in (0, 1-\delta]$

$$\begin{aligned}
C_{1,m} &\leq \left|\frac{\mathrm{d}^m}{\mathrm{d}\kappa^m}a(\kappa)\right| \leq C_{2,m},\\
C_{1,m} &\leq \left|\frac{\mathrm{d}^m}{\mathrm{d}\kappa^m}\frac{1}{a(\kappa)(1+a(\kappa))}\right| \leq C_{2,m}.
\end{aligned}\tag{4.8}$$

Next we analyze $\Xi_0(x,\kappa), \Xi_2(x,\kappa)$ and $\Theta_0(x,\kappa)$: The functions $\Xi_j(x,\kappa)$, see [3, (4.10)] and (A.2), are combinations of $\theta_j(x|\tau)$ and thus, following the logic which lead to (4.6), we obtain similarly

$$\left|\frac{\partial^{m+\ell}}{\partial x^m\partial\tau^\ell}\theta_0(x\pm d|\tau) - \delta_{m0}\delta_{\ell 0}\right| \leq C_{m,\ell}\kappa;$$
$$\left|\frac{\partial^{m+1}}{\partial x^m\partial\kappa}\theta_0(x\pm d|\tau)\right| \leq C_m$$

for all $x \in \mathbb{R}, \kappa \in (0, 1-\delta]$. The same estimates also hold for $\theta_2(x \pm d|\tau)$ which yield

$$\begin{aligned}
\big|\Xi_j(x,\kappa)\big| \leq C, \qquad &\left|\frac{\partial^m}{\partial x^m}\Xi_j(x,\kappa)\right| \leq C_m\kappa,\\
&\left|\frac{\partial^{m+1}}{\partial x^m\partial\kappa}\Xi_j(x,\kappa)\right| \leq C_m, \qquad j = 0, 2,
\end{aligned}\tag{4.9}$$

valid for all $x \in \mathbb{R}, \kappa \in (0, 1-\delta]$. Since the remaining function $\Theta_0(x,\kappa)$ satisfies estimates of the same type as (4.9), we can combine (4.7), (4.8) and (4.9) to derive

Proposition 4.2. *Given $m \in \mathbb{Z}_{\geq 1}$ there exist positive C, C_m such that $M(x,\kappa)$ given in [3](1.24) satisfies*

$$\begin{aligned}
\big|M(x,\kappa)\big| \leq C, \qquad &\left|\frac{\partial^m}{\partial x^m}M(x,\kappa)\right| \leq C_m\kappa,\\
&\left|\frac{\partial^{m+1}}{\partial x^m\partial\kappa}M(x,\kappa)\right| \leq C_m
\end{aligned}$$

for all $x \in \mathbb{R}, \kappa \in (0, 1-\delta]$.

This proposition allows us to estimate the Fourier coefficients in (4.5).

Corollary 4.3. *For all $n \in \mathbb{Z}, \kappa \in (0, 1-\delta]$,*

$$\left|a_n(\kappa)\right| \le \frac{C}{1+n^2}, \qquad \left|\frac{\mathrm{d}}{\mathrm{d}\kappa}a_n(\kappa)\right| \le \frac{C}{1+n^2}, \qquad C > 0.$$

In order to complete the proof of Proposition 1.3 we use (4.5), [3], (4.4), Proposition 4.2 and integration by parts

$$\begin{aligned}\int_s^\infty M\Big(tV(vt^{-1}), vt^{-1}\Big)\frac{\mathrm{d}t}{t} &= \int_0^\kappa a_0(u)\frac{\mathrm{d}u}{u} + \sum_{n\neq 0}\int_s^\infty a_n(vt^{-1})\mathrm{e}^{-2\pi i n t V\left(vt^{-1}\right)}\frac{\mathrm{d}t}{t} \\ &= \int_0^\kappa a_0(u)\frac{\mathrm{d}u}{u} - \frac{1}{4\pi s c(\kappa)}\sum_{n\neq 0}\frac{a_n(\kappa)}{n}\mathrm{e}^{-2\pi i n s V(\kappa)} \\ &\quad - \frac{1}{4\pi}\sum_{n\neq 0}\frac{1}{n}\int_s^\infty \frac{\partial}{\partial t}\left[\frac{a_n(vt^{-1})}{tc(vt^{-1})}\right]\mathrm{e}^{-2\pi i n t V(vt^{-1})}\mathrm{d}t.\end{aligned}$$

We estimate the infinite sums as follows: For the first sum we use, compare Remark 4.1, the fact that

$$\left|\frac{1}{c(\kappa)}\right| \le C, \quad \forall \kappa \in (0, 1-\delta],$$

together with Corollary 4.3. Thus

$$\left|\frac{1}{sc(\kappa)}\sum_{n\neq 0}\frac{a_n(\kappa)}{n}\mathrm{e}^{-2\pi i n s V(\kappa)}\right| \le \frac{C}{s}, \quad \kappa = \frac{v}{s} \in (0, 1-\delta]. \tag{4.10}$$

For the second sum we differentiate,

$$\frac{\partial}{\partial t}\left[\frac{a_n(vt^{-1})}{tc(vt^{-1})}\right] = -\frac{\kappa}{t^2 c(\kappa)}\frac{\mathrm{d}}{\mathrm{d}\kappa}a_n(\kappa) - \frac{a_n(\kappa)}{(tc(\kappa))^2}\left(c(\kappa) - \kappa\frac{\mathrm{d}}{\mathrm{d}\kappa}c(\kappa)\right), \quad \kappa = \frac{v}{t}$$

and since in the integral $0 < \frac{v}{t} \le \frac{v}{s} \le 1-\delta$ we obtain using Corollary 4.3 and Remark 4.1 the bound

$$\left|\frac{\partial}{\partial t}\left[\frac{a_n(vt^{-1})}{tc(vt^{-1})}\right]\right| \le \frac{C}{t^2(1+n^2)}.$$

Thus,

$$\left|\sum_{n\neq 0}\frac{1}{n}\int_s^\infty \frac{\partial}{\partial t}\left[\frac{a_n(vt^{-1})}{tc(vt^{-1})}\right]\mathrm{e}^{-2\pi i n t V(vt^{-1})}\mathrm{d}t\right| \le \frac{C}{s},$$

which, together with (4.10), completes the proof of Proposition 1.3, i.e.,

$$\int_s^\infty M\Big(tV(vt^{-1}), vt^{-1}\Big)\frac{\mathrm{d}t}{t} = \int_0^\kappa a_0(u)\frac{\mathrm{d}u}{u} + \mathcal{O}\left(s^{-1}\right)$$

uniformly for $s \ge s_0 > 0$ and $0 < v \le s(1-\delta)$.

Appendix: Explicit form of $M(z,\kappa)$

We fix $a = a(\kappa), \tau = \tau(\kappa)$ and $V = V(\kappa)$ as in (1.4), (1.5) throughout. Note that in terms of $\theta_3(z|\tau)$ defined in (1.5) the remaining three Jacobi theta functions $\theta_0(z), \theta_1(z), \theta_2(z)$ are given by (cf. [13])

$$\theta_0(z|\tau) = \theta_3\left(z + \frac{1}{2}\middle|\tau\right),$$

$$\theta_1(z|\tau) = -\mathrm{i}\mathrm{e}^{\mathrm{i}\frac{\pi}{4}\tau + \mathrm{i}\pi z}\theta_3\left(z + \frac{1}{2} + \frac{\tau}{2}\middle|\tau\right).$$

$$\theta_2(z|\tau) = \mathrm{e}^{\mathrm{i}\frac{\pi}{4}\tau + \mathrm{i}\pi z}\theta_3\left(z + \frac{\tau}{2}\middle|\tau\right),$$

The integral term $M(x,\kappa)$, defined in [3, (1.24)], equals

$$M(x,\kappa) = \frac{\Xi_0(x,\kappa)\Theta_0(x,\kappa) + 6a(\kappa)\Xi_2(x,\kappa)}{48a(\kappa)(1+a(\kappa))} + \frac{\mathrm{i}}{4\pi}\frac{\kappa}{\theta_3(x|\tau)}\frac{\partial^2}{\partial y^2}\theta_3(y|\tau)\bigg|_{y=x}\frac{\mathrm{d}\tau}{\mathrm{d}\kappa} \tag{A.1}$$

with $\Xi_j(x,\kappa)$ and $\Theta_0(x,\kappa)$ equal to the following functions:

$$\Xi_j(x,\kappa) = 2\frac{\theta_3^2(0|\tau)}{\theta_3^2(x|\tau)}\frac{\theta_j(x+d|\tau)\theta_j(x-d|\tau)}{\theta_j^2(d|\tau)}, \quad j = 0,2; \qquad d = -\frac{\tau}{4}, \tag{A.2}$$

and

$$\begin{aligned}\Theta_0(x,\kappa) = {} & 5c^2\left\{\frac{\theta_0''(x+d|\tau)}{\theta_0(x+d|\tau)} - 2\left(\frac{\theta_0'(d|\tau)}{\theta_0(d|\tau)}\right)' + \frac{\theta_0''(x-d|\tau)}{\theta_0(x-d|\tau)}\right\} \\ & +14c^2\frac{\theta_0'(x-d|\tau)}{\theta_0(x-d|\tau)}\frac{\theta_0'(x+d|\tau)}{\theta_0(x+d|\tau)} \\ & -4c^2\left\{\frac{\theta_0'(x-d|\tau)}{\theta_0(x-d|\tau)} + \frac{\theta_0'(d|\tau)}{\theta_0(d|\tau)} - \frac{\theta_0'(x+d|\tau)}{\theta_0(x+d|\tau)}\right\}\frac{\theta_0'(d|\tau)}{\theta_0(d|\tau)} \\ & -2c(1+a)\left\{\frac{\theta_0'(x+d|\tau)}{\theta_0(x+d|\tau)} - 2\frac{\theta_0'(d|\tau)}{\theta_0(d|\tau)} - \frac{\theta_0'(x-d|\tau)}{\theta_0(x-d|\tau)}\right\} \\ & -2(2+a); \qquad \theta'(x|\tau) = \frac{\mathrm{d}}{\mathrm{d}x}\theta(x|\tau).\end{aligned} \tag{A.3}$$

Here, compare Remark 4.1,

$$c = c(\kappa) = \frac{\mathrm{i}}{2}\frac{1}{K(a')}, \quad a' = \sqrt{1-a^2}.$$

References

[1] E. Basor, *Asymptotic formulas for Toeplitz determinants*, Trans. Amer. Math. Soc. **239** (1978), 33–65.

[2] E. Basor, H. Widom, *Toeplitz and Wiener–Hopf determinants with piecewise continuous symbols*, Journal of Functional Analysis **50** (1983), 387–413.

[3] T. Bothner, P. Deift, A. Its, I. Krasovsky, *On the asymptotic behavior of a log gas in the bulk scaling limit in the presence of a varying external potential I.* Commun. Math. Phys. **337** (2015), 1397–1463.

[4] T. Bothner, A. Its, *Asymptotics of a cubic sine kernel determinant*, St. Petersburg Math. J. **26** (2014), 515–565.

[5] A. Böttcher, B. Silbermann, *Toeplitz matrices and determinants with Fisher–Hartwig symbols*, J. Func. Anal. **63** (1985), 178–214.

[6] A. Budylin, V. Buslaev, *Quasiclassical asymptotics of the resolvent of an integral convolution operator with a sine kernel on a finite interval*, Algebra i Analiz **7** (1995), 79–103.

[7] O. Bohigas, M. Pato, *Randomly incomplete spectra and intermediate statistics*, Phys. Rev. E **74** (2006), 036212, (6pp.).

[8] T. Claeys, A. Its, I. Krasovsky, *Emergence of a singularity for Toeplitz determinants and Painlevé V*, Duke Math. Journal **160** (2011), 207–262.

[9] T. Claeys, I. Krasovsky, *Toeplitz determinants with merging singularities*, Duke Math. Journal **164** (2015), 2897–2987.

[10] P. Deift, A. Its, and I. Krasovsky, *Asymptotics of Toeplitz, Hankel, and Toeplitz + Hankel determinants with Fisher–Hartwig singularities* Annals of Mathematics **174** (2011), 1243–1299.

[11] P. Deift, A. Its, I. Krasovsky, and X. Zhou, *The Widom–Dyson constant for the gap probability in random matrix theory*, J. Comput. Appl. Math. **202** (2007), 26–47.

[12] P. Deift and X. Zhou, *A steepest descent method for oscillatory Riemann–Hilbert problems. Asymptotics for the MKdV equation*, Annals of Mathematics **137** (1993), 295–368.

[13] NIST Digital Library of Mathematical Functions, http://dlmf.nist.gov/

[14] F. Dyson, *Fredholm determinants and inverse scattering problems*, Commun. Math. Phys. **47** (1976), 171–183.

[15] F. Dyson, *The Coulomb fluid and the fifth Painlevé transcendent*, appearing in Chen Ning Yang: A Great Physicist of the Twentieth Century, eds. C.S. Liu and S.-T. Yau, International Press, Cambridge, MA, (1995), 131–146.

[16] T. Ehrhardt, *Dyson's constant in the asymptotics of the Fredholm determinant of the sine kernel*, Comm. Math. Phys. **262** (2006), 317–341.

[17] A. Its, A. Izergin, V. Korepin, and N. Slavnov, *Differential equations for quantum correlation functions*, Int. J. Mod. Physics **B4** (1990), 1003–1037.

[18] I. Krasovsky, *Gap probability in the spectrum of random matrices and asymptotics of polynomials orthogonal on an arc of the unit circle*, Int. Math. Res. Not. **2004** (2004), 1249–1272.

[19] H. Widom, *Toeplitz determinants with singular generating functions*, Amer. J. Math **95** (1973), 333–383.

Thomas Bothner
Department of Mathematics
University of Michigan
2074 East Hall
530 Church Street
Ann Arbor, MI 48109-1043, USA
e-mail: bothner@umich.edu

Percy Deift
Courant Institute of Mathematical Sciences
251 Mercer St.
New York, NY 10012, USA
e-mail: deift@cims.nyu.edu

Alexander Its
Department of Mathematical Sciences
Indiana University – Purdue University Indianapolis
402 N. Blackford St.
Indianapolis, IN 46202, USA
e-mail: aits@iupui.edu

Igor Krasovsky
Department of Mathematics
Imperial College
London SW7 2AZ, UK
e-mail: i.krasovsky@imperial.ac.uk

Operator Theory:
Advances and Applications, Vol. 259, 235–265

Useful Bounds on the Extreme Eigenvalues and Vectors of Matrices for Harper's Operators

Daniel Bump, Persi Diaconis, Angela Hicks,
Laurent Miclo and Harold Widom

Dedicated to Albrecht Böttcher
on the occasion of his sixtieth birthday

Abstract. In analyzing a simple random walk on the Heisenberg group we encounter the problem of bounding the extreme eigenvalues of an $n \times n$ matrix of the form $M = C + D$ where C is a circulant and D a diagonal matrix. The discrete Schrödinger operators are an interesting special case. The Weyl and Horn bounds are not useful here. This paper develops three different approaches to getting good bounds. The first uses the geometry of the eigenspaces of C and D, applying a discrete version of the uncertainty principle. The second shows that, in a useful limit, the matrix M tends to the harmonic oscillator on $L^2(\mathbb{R})$ and the known eigenstructure can be transferred back. The third approach is purely probabilistic, extending M to an absorbing Markov chain and using hitting time arguments to bound the Dirichlet eigenvalues. The approaches allow generalization to other walks on other groups.

Mathematics Subject Classification (2010). 60B15; 20P05.

Keywords. Heisenberg group, almost Mathieu operator, Fourier analysis, random walk.

The first, second, third, and fifth authors would like to acknowledge partial support from NSF grants DMS 1001079, DMS 08-04324, DMS 1303761, and DMS 1400248 (respectively). The remaining author would like to acknowledge partial support from ANR grant number ANR-12-BS01-0019.

1. Introduction

Consider the $n \times n$ matrix

$$M_n = \frac{1}{4}\begin{pmatrix} 2 & 1 & & & & 1 \\ 1 & \ddots & \ddots & & & \\ & \ddots & \ddots & \ddots & & \\ & & 1 & 1 & \ddots & \\ & & & \ddots & \ddots & 1 \\ 1 & & & & 1 & \ddots \end{pmatrix} \quad 2\cos\left(\tfrac{2\pi j}{n}\right), \quad 0 \le j \le n-1. \tag{1}$$

As explained in [5] and summarized in Section 2, this matrix arises as the Fourier transform of a simple random walk on the Heisenberg group, as a discrete approximation to Harper's operator in solid state physics and in understanding the Fast Fourier Transform. Write $M = C + D$ with C a circulant, (having $\frac{1}{4}$ on the diagonals just above and below the main diagonal and in the corners) and D a diagonal matrix (with diagonal entries $\frac{1}{2}\cos\left(\frac{2\pi j}{n}\right)$ for $0 \le j \le n-1$). The Weyl bounds [20] and Horn's extensions [2] yield that the largest eigenvalue $\lambda_1(M) \le \lambda_1(C) + \lambda_1(D)$. Here $\lambda_1(C) = \lambda_1(D) = \frac{1}{2}$ giving $\lambda_1(M) \le 1$. This was not useful in our application; in particular, we need $\lambda_1(M) \le 1 - \frac{\text{const}}{n}$. This paper presents three different approaches to proving such bounds. The first approach uses the geometry of the eigenvectors and a discrete version of the Heisenberg uncertainty principle. It works for general Hermitian circulants:

Theorem 1. *Let C be an $n \times n$ Hermitian circulant with eigenvalues $\lambda_1(C) \ge \cdots \ge \lambda_n(C)$. Let D be an $n \times n$ real diagonal matrix with eigenvalues $\lambda_1(D) \ge \cdots \ge \lambda_n(D)$. If k, k' satisfy $1 \le k, k' \le n, kk' < n$, then*

$$\lambda_1(C+D) \le \lambda_1(C) + \lambda_1(D) - \frac{1}{2}\min\{\lambda_1(D) - \lambda_k(D), \lambda_1(C) - \lambda_{k'}(C)\}\left(1 - \sqrt{\frac{kk'}{n}}\right)^2.$$

Example 1. For the matrix M_n in (1), the eigenvalues of C and D are real and equal to $\left\{\frac{1}{2}\cos\left(\frac{2\pi j}{n}\right)\right\}_{0 \le j \le n-1}$. For simplicity, take n odd. Then, writing $\lambda_j = \lambda_j(C) = \lambda_j(D)$, $\lambda_1 = \frac{1}{2}$, $\lambda_2 = \lambda_3 = \frac{1}{2}\cos\left(\frac{2\pi}{n}\right)$, and $\lambda_{2j+1} = \lambda_{2j} = \frac{1}{2}\cos\left(\frac{2\pi j}{n}\right)$ for $1 \le j \le \frac{n-1}{2}$. Choose $k = k' = \lfloor c\sqrt{n} \rfloor$ for a fixed $0 < c < 1$. Then

$$\lambda_k = \frac{1}{2}\left(1 - \frac{1}{2}\left(\frac{\pi c}{n^{1/2}}\right)^2 + O\left(\frac{1}{n^{3/2}}\right)\right), \quad \left(1 - \sqrt{\frac{kk'}{n}}\right)^2 \ge (1-c)^2$$

and the bound in Theorem 1 becomes

$$\lambda_1(M_n) \le 1 - \frac{\pi^2}{8}\frac{c^2(1-c)^2}{n} + O\left(\frac{1}{n^{3/2}}\right).$$

The choice $c = \frac{1}{2}$ gives the best result. Very sharp inequalities for the largest and smallest eigenvalues of M_n follow from [3]. They get better constants than we have in this example. Their techniques make sustained careful use of the exact form of the matrix entries while the techniques in Theorem 1 work for general circulants.

The second approach passes to the large n limit, showing that the largest eigenvalues of M_n from (1) tend to suitably scaled eigenvalues of the harmonic oscillator $L = -\frac{1}{4}\frac{d^2}{dx^2} + \pi^2 x^2$.

Theorem 2. *For a fixed $k \geq 1$, the kth largest eigenvalue of M_n equals*

$$1 - \frac{\mu_k}{n} + o\left(\frac{1}{n}\right)$$

with $\mu_k = \frac{(2k-1)\pi}{2}$, the kth smallest eigenvalue of L.

Theorem 2 gets higher eigenvalues with sharp constants for a restricted family of matrices. The argument also gives a useful approximation to the kth eigenvector. Similar results (with very different proofs) are in [28].

There are many techniques available for bounding the eigenvalues of stochastic matrices ([24], [13], and [7]). We initially thought that some of these would adapt to M_n. However, M_n is far from stochastic: the row sums of M_n are not constant and the entries are sometimes negative. Our third approach is to let $M'_n = \frac{1}{3}I + \frac{2}{3}M_n$. This is substochastic (having non-negative entries and row sums at most 1). If $a_i = 1 - \sum_j M'_n(i,j)$, consider the $(n+1) \times (n+1)$ stochastic matrix:

$$M''_n = \begin{pmatrix} 1 & 0 \; 0 \; \ldots \; 0 \\ a_1 & \\ a_2 & M'_n \\ \vdots & \\ a_n & \end{pmatrix}. \tag{2}$$

This has the interpretation of an absorbing Markov chain (0 is the absorbing state) and the Dirichlet eigenvalues of M''_n (namely those whose eigenvalues vanish at 0) are the eigenvalues of M'_n. In [5] path and other geometric techniques are used to bound these Dirichlet eigenvalues. This results in bounds of the form $1 - \frac{\text{const.}}{n^{4/3}}$ for $\lambda_1(M_n)$. While sufficient for the application, it is natural to want an improvement that gets the right order. Our third approach introduces a purely probabilistic technique which works to give bounds of the right order for a variety of similar matrices.

Theorem 3. *There is a $c > 0$ such that, for all $n \geq 1$ and M_n defined at* (1)*, the largest eigenvalue satisfies $\lambda_1(M_n) \leq 1 - \frac{c}{n}$.*

Section 2 gives background and motivation. Theorems 1, 2, and 3 are proved in Sections 3, 4, and 5. Section 6 treats a simple random walk on the affine group mod p. It uses the analytic bounds to show that order p^2 steps are necessary and sufficient for convergence. It may be consulted now for further motivation. The final section gives the limiting distribution of the bulk of the spectrum of $M_n(a)$ using the Kac–Murdock–Szegő theorem.

2. Background

Our work in this area starts with the finite Heisenberg group:

$$H_1(n) = \left\{ \begin{pmatrix} 1 & x & z \\ 0 & 1 & y \\ 0 & 0 & 1 \end{pmatrix} : \qquad x, y, z \in \mathbb{Z}/n\mathbb{Z} \right\}.$$

Write such a matrix as (x, y, z), so

$$(x, y, z)(x', y', z') = (x + x', y + y', z + z' + xy').$$

Let

$$S = \{(1,0,0), (-1,0,0), (0,1,0), (0,-1,0)\} \tag{3}$$

and

$$Q(g) = \begin{cases} \frac{1}{4} & g \in S, \\ 0 & \text{otherwise}. \end{cases} \tag{4}$$

Thus S is a minimal symmetric generating set for $H_1(n)$ and Q is the probability measure associated with 'pick an element in S at random and multiply.' Repeated steps of this walk correspond to convolution. For $(x, y, z) \in H_1(n)$,

$$Q^{*^k}(x, y, z) = \sum_{(x',y',z') \in H_1(n)} Q(x', y', z') Q^{*^{k-1}}((x, y, z)(x', y', z')^{-1}).$$

When k is large, Q^{*^k} converges to the uniform distribution $U(x, y, z) = \frac{1}{n^3}$ The rate of convergence of Q^{*^k} to U can be measured by the chi-squared distance:

$$\sum_{(x,y,z) \in H} |Q^{*^k}(x, y, z) - U(x, y, z)|^2 / (U(x, y, z)) = \sum_{\substack{\rho \in \hat{H}_1 \\ \rho \neq 1}} d_\rho \|\hat{Q}(\rho)^k\|^2. \tag{5}$$

On the right, the sum is over nontrivial irreducible representations of $H_1(n)$ with ρ of dimension d_ρ and $\hat{Q}(\rho)^k = \sum_{(x,y,z)} Q^{*^k}(x, y, z)\rho(x, y, z)$. For background on the Fourier analysis approach to bounded convergence see [8], Chapter 3.

For simplicity, (see [5] for the general case) take $n = p$ a prime. Then $H_1(p)$ has p^2 1-dimensional representations $\rho_{a,b}(x, y, z) = e^{\frac{2\pi i}{p}(ax+by)}$ for a, b in $\mathbb{Z}_p$. It has $p - 1$ p-dimensional representations. These act on $V = \{f : \mathbb{Z}_p \to \mathbb{C}\}$ via

$$\rho_a(x, y, z) f(w) = e^{\frac{2\pi i a}{p}(yw+z)} f(x + w), \; 0 \leq a \leq p - 1.$$

The Fourier transform of Q at ρ_a is the matrix $M_n(a)$ as in (1) with $\cos\left(\frac{2\pi j}{p}\right)$ replaced by $\cos\left(\frac{2\pi a j}{p}\right)$ for $0 \leq j \leq p - 1$.

The chi-squared norm in (5) is the sum of the $(2k)$th power of the eigenvalues so proceeding needs bounds on these. The details are carried out in [5]. The main results show that k of order n^2 steps are necessary and sufficient for convergence. That paper also summarizes other appearances of the matrices $M_n(a)$. They occur in discrete approximations of the 'almost Mathieu' operator in solid state physics.

In particular, see [31], [3], and [1]. If $\mathcal{F}_n$ is the discrete Fourier transform matrix $\left((\mathcal{F}_n)_{jk} = \frac{1}{\sqrt{n}} e^{\frac{2\pi ijk}{n}}\right)$; it is easy to see that $\mathcal{F}_n M_n(1) = M_n(1)\mathcal{F}_n$. Diagonalizing $\mathcal{F}_n$ has engineering applications and having a 'nice' commuting matrix should help. For this reason, there is engineering interest in the eigenvalues and vectors of $M_n(1)$. See [14] and [25].

3. Proof of Theorem 1

Throughout this section C is an $n \times n$ Hermitian circulant with eigenvalues $\lambda_1(C) \geq \lambda_2(C) \geq \cdots \geq \lambda_n(C)$ and D is a real diagonal matrix with eigenvalues $\lambda_1(D) \geq \lambda_2(D) \geq \cdots \geq \lambda_n(D)$. Let x be an eigenvector of $C+D$ corresponding to $\lambda_1(C+D)$. Recall that $\left((\mathcal{F}_n)_{jk} = \frac{1}{\sqrt{n}} e^{\frac{2\pi ijk}{n}}\right)$ for $j, k \in \mathbb{Z}/n\mathbb{Z}$. This has rows or columns which simultaneously diagonalize all circulants. Write $\hat{x} = \mathcal{F}_n x$ and x^h for the conjugate transpose. We use $\|x\|^2 = x^h x$.

Our aim is to prove that for $kk' < n$,

$$\begin{aligned} \lambda_1(C+D) &\leq \lambda_1(C) + \lambda_1(D) \\ &\quad - \frac{1}{2}\min\{\lambda_1(D) - \lambda_{k+1}(D), \lambda_1(C) - \lambda_{k'+1}(C)\}\left(1 - \sqrt{\frac{kk'}{n}}\right)^2. \end{aligned} \tag{6}$$

The first step is to write $x^h C x$ in terms of a Fourier transform pair $\hat{x} = \mathcal{F}_n x$. A subtle point is that although $\mathcal{F}_n$ diagonalizes C, the resulting diagonal matrix does not necessarily have entries in decreasing order, necessitating a permutation indexing in the following lemma.

Lemma 1. *Define a permutation σ such that*

$$\frac{e^{\frac{2\pi i}{n}\sigma_j^{-1} b}}{\sqrt{n}}, \quad 0 \leq b \leq n-1$$

is the eigenvector corresponding to $\lambda_j(C)$. Then

$$x^h C x = \hat{x}^h D' \hat{x} \tag{7}$$

with $D' = \operatorname{diag}(\lambda_{\sigma_1}(C), \ldots, \lambda_{\sigma_n}(C))$.

Proof. Since C is diagonalized by $\mathcal{F}_n^h$, $\mathcal{F}_n C \mathcal{F}_n^h = D'$. Thus

$$x^h C x = x^h \mathcal{F}_n^h \mathcal{F}_n C \mathcal{F}_n^h \mathcal{F}_n x = \hat{x}^h D' \hat{x}.$$ □

A key tool is the Donoho–Stark [15] version of the Heisenberg Uncertainty Principle. For this, call a vector y 'ϵ-concentrated on a set $S \subset [n]$' if $|x_i| < \epsilon$ for $i \notin S$.

Theorem 4 (Donoho–Stark). *Let y, $\hat{y}$ be a unit norm Fourier Transform pair with y ϵ_S-concentrated on S and $\hat{y}$ ϵ_T-concentrated on T. Then*

$$|S||T| \geq n(1 - (\epsilon_S + \epsilon_T))^2. \tag{8}$$

Let $(y)_S$ be the projection onto the subspace vanishing off S:

$$((y)_S)_i = \begin{cases} y_i, & i \in S, \\ 0, & \text{otherwise.} \end{cases}$$

A simple consequence of the bound (8) is

Corollary 1. *If $kk' < n$, $z, \hat{z}$ a unit norm Fourier transform pair and S and T are sets of size k, k', then*

$$\|(z)_{S^c}\|^2 + \|(\hat{z})_{T^c}\|^2 \geq \frac{1}{2}\left(1 - \sqrt{\frac{kk'}{n}}\right)^2.$$

Proof. Let $\epsilon_S = \|(z)_{S^c}\|$ and $\epsilon_T = \|(\hat{z})_{T^c}\|$. Then $\|z - (z)_S\| = \|(z)_{S^c}\|$ and $\|\hat{z} - (\hat{z})_T\| = \|(\hat{z})_{T^c}\|$. Thus z is ϵ_S concentrated on S and $\hat{z}$ is ϵ_T concentrated on T. Thus (8) gives

$$\frac{|S||T|}{n} = \frac{kk'}{n} \geq (1 - (\epsilon_S + \epsilon_t))^2 \text{ or } (\epsilon_S + \epsilon_T) \geq 1 - \sqrt{\frac{kk'}{n}},$$

so if $kk' \leq n$

$$\|(z)_{S^c}\|^2 + \|(\hat{z})_{T^c}\|^2 = (\epsilon_S^2 + \epsilon_T^2) \geq \frac{1}{2}(\epsilon_s + \epsilon_T)^2 \geq \frac{1}{2}\left(1 - \sqrt{\frac{kk'}{n}}\right)^2. \qquad \square$$

Proof of Theorem 1. With notation as above,

$$\lambda_1(C + D) = x^h(C + D)x = x^h Cx + x^h Dx = \hat{x}^h D'\hat{x} + x^h Dx =: *.$$

Let $\overline{D} = D - \lambda_1(D)I$ and $\overline{D'} = D' - \lambda_1(C)I$. Then

$$\begin{aligned} * &= \hat{x}^h \lambda_1(C)I\hat{x} + \hat{x}^h \overline{D'}\hat{x} + x^h \lambda_1(D)Ix + x^h \overline{D}x \\ &= \lambda_1(C) + \hat{x}^h \overline{D'}\hat{x} + \lambda_1(D) + x^h \overline{D}x. \end{aligned}$$

Now $\overline{D}$ and $\overline{D'}$ have non-positive eigenvalues so our improvement over the Weyl bounds will follow by showing that x or $\hat{x}$ have support on suitably negative entries of $\overline{D}'$ or $\overline{D}$.

Let S and T correspond to the largest k, k' entries of $\overline{D}$, $\overline{D}'$, respectively. Then $x = (x)_S + (x)_{S^c}$, $\hat{x} = (\hat{x})_T + (\hat{x})_{T^c}$. Each of those decompositions is into orthogonal pieces. Multiplying any of the four pieces by an arbitrary diagonal matrix preserves this orthogonality. Thus

$$* = \lambda_1(C) + \lambda_1(D) + (\hat{x})_T^h \overline{D'}(\hat{x})_T + (\hat{x})_{T^c}^h \overline{D'}(\hat{x})_{T^c} + (x)_S^h \overline{D}(x)_S + (x)_{S^c}^h \overline{D}(x)_{S^c}.$$

For the last four terms on the right, terms 1 and 3 are bounded above by zero and 2 and 4 contribute with the following bounds:

$$\begin{aligned} * &\leq \lambda_1(C) + \lambda_1(D) + (\lambda_{k+1}(D) - \lambda_1(D))\|(x)_{S^c}\|^2 + (\lambda_{k'+1}(C) - \lambda_1(C)\|(\hat{x})_{T^c}\|^2 \\ &\leq \lambda_1(C) + \lambda_1(D) \\ &\quad + \min\{(\lambda_{k+1}(D) - \lambda_1(D)), (\lambda_{k'+1}(C) - \lambda_1(C)\} \left(\frac{1}{2}\left(1 - \sqrt{\frac{kk'}{n}}\right)^2\right) \end{aligned}$$

where the last line follows from the corollary. □

Remarks.

1. These arguments work to give the smallest eigenvalue as well, so in fact we also have for $ll' < n$:

$$\begin{aligned} \lambda_n(C + D) &\geq \lambda_n(C) + \lambda_n(D) \\ &\quad + \frac{1}{2}\min\{\lambda_{n-l}(D) - \lambda_n(D), \lambda_{n-l'}(C) - \lambda_n(C)\}\left(1 - \sqrt{\frac{ll'}{n}}\right)^2. \end{aligned} \tag{9}$$

2. Our thanks to a thoughtful anonymous reviewer, who pointed out that Corollary 1 can be improved using Cauchy–Schwarz to show that for $0 < a, b \leq 1$,

$$\left(1 - \sqrt{\frac{kk'}{n}}\right)^2 \leq (\epsilon_s + \epsilon_T)^2 \leq (a\epsilon_S^2 + b\epsilon_T^2)\left(\frac{1}{a} + \frac{1}{b}\right).$$

   Setting $a = \lambda_{k+1}(D) - \lambda_1(D)$ and $b = \lambda_{k+1}(D) - \lambda_1(D)$, one can improve the previous theorem:

$$\begin{aligned} \lambda_1(C + D) &\leq \lambda_1(C) + \lambda_1(D) \\ &\quad - \frac{(\lambda_1(D) - \lambda_{k+1}(D))(\lambda_1(C) - \lambda_{k'+1}(C))}{(\lambda_1(D) - \lambda_{k+1}(D)) + (\lambda_1(C) - \lambda_{k'+1}(C))}\left(1 - \sqrt{\frac{kk'}{n}}\right)^2. \end{aligned}$$

   In our case, the result is the same, since the eigenvalues of C and D are identical.
3. Donoho and Stark [15] give many variations on their uncertainty principle suitable for other transforms. The techniques above should generalize, at least to the G-circulants of [9].
4. There should be similar theorems with C and D replaced by general Hermitian matrices and perhaps extensions to higher Weyl and Horn inequalities (see [2] and [18]).
5. Further applications/examples are in Section 6.

4. The harmonic oscillator as a limit

We prove Theorem 2, that for $k \geq 1$ the kth largest eigenvalue of M_n is equal to $1 - \mu_k/n + o(1/n)$ and the kth smallest eigenvalue of M_n is equal to $-1 + \mu_k/n + o(1/n)$, where μ_k is the kth smallest eigenvalue of

$$L = -\frac{1}{4}\frac{d^2}{dx^2} + \pi^2 x^2$$

on $(-\infty, \infty)$. By a classical computation (see [19]), $\mu_k = (2k-1)/(2\pi)$.

The $n \times n$ matrix M_n has j, k-entry

$$\frac{1}{4}\left[\delta(j-k-1) + \delta(j-k+1)\right] + \frac{1}{2}\cos(2\pi k/n)\,\delta(j-k),$$

where $j, k \in \mathbb{Z}_n = \mathbb{Z}/n\mathbb{Z}$.

We define

$$\tilde{M_n} = n\,(I - M_n).$$

This has j, k entry $m_1(j,k) + m_2(j,k)$, where

$$m_1(j,k) = \frac{n}{2}\left(\delta(j-k) - \frac{1}{2}\left[\delta(j-k-1) + \delta(j-k+1)\right]\right) \tag{10}$$

$$m_2(j,k) = \frac{n}{2}\,(1 - \cos(2\pi k/n))\,\delta(j-k). \tag{11}$$

We will show first that if μ is any limit of eigenvalues of $\tilde{M_n}$ then μ is an eigenvalue of L; and, second, that any eigenvalue μ of L has a neighborhood that contains exactly one eigenvalue, counting multiplicity, of $\tilde{M_n}$ for n sufficiently large. These imply the stated result.

These will be accomplished as follows. Give each point of $\mathbb{Z}_n$ measure $1/\sqrt{n}$, so the total measure equals $\sqrt{n}$. We then define an isometry T from $L^2(\mathbb{Z}_n)$ to $L^2(-\sqrt{n}/2, \sqrt{n}/2)$ (thought of as a subspace of $L^2(\mathbb{R})$ with Lebesgue measure) for which the following hold:

Proposition 1. *Suppose $\{u_n\}$ is a sequence of functions of norm one in $L^2(\mathbb{Z}_n)$ such that the sequence $\{(\tilde{M_n}\,u_n, u_n)\}$ of inner products is bounded. Then $\{Tu_n\}$ has a strongly (i.e., in norm) convergent subsequence.*

Proposition 2. *If ϕ is a Schwartz function on $\mathbb{R}$ then $T\tilde{M_n}T^*\phi \to L\phi$ strongly.*[1]

These will easily give the desired results. (See Propositions 3 and 4 near the end.) The final Section 4.2 treats the smallest eigenvalues.

[1]The operator T^* acts on $L^2(-\sqrt{n}/2, \sqrt{n}/2)$, so ϕ is first to be restricted to $(-\sqrt{n}/2, \sqrt{n}/2)$.

4.1. Proofs for the largest eigenvalues

We use two transforms (with, confusingly, the same notation). First, for ϕ in $L^2(-\sqrt{n}/2, \sqrt{n}/2)$ we define

$$\widehat{\phi}(\ell) = \int_{-\sqrt{n}/2}^{\sqrt{n}/2} e^{-2\pi i \ell x/\sqrt{n}}\, \phi(x)\, dx, \qquad (\ell \in \mathbb{Z}),$$

and we have by Parseval (after making the substitution $x \mapsto x\sqrt{n}$ in the integral)

$$\|\widehat{\phi}\| = n^{1/4}\, \|\phi\|. \tag{12}$$

Here $\|\widehat{\phi}\|^2 = \sum_{\ell \in \mathbb{Z}} |\hat{\phi}(\ell)|^2$.

For $u \in L^2(\mathbb{Z}_n)$ we have its finite Fourier transform

$$\widehat{u}(\ell) = \sum_k e^{-2\pi i \ell k/n}\, u(k), \qquad (\ell \in \mathbb{Z}_n),$$

and we compute below that

$$\|\widehat{u}\| = n^{3/4}\, \|u\|. \tag{13}$$

Here

$$\|\widehat{u}\|^2 = \sum |\widehat{u}(\ell)|^2,$$

the sum over any integer interval of length n. To show (13), we have

$$|\widehat{u}(\ell)|^2 = \sum_{j,k} e^{-2\pi i \ell (j-k)/n}\, u(j)\, \overline{u(k)}.$$

Since $e^{-2\pi i(j-k)/n}$ is an nth root of unity, equal to 1 only when $j = k$ in $\mathbb{Z}_n$, we get

$$\|\widehat{u}\|^2 = \sum_\ell |\widehat{u}(\ell)|^2 = n \sum_k |u(k)|^2 = n^{3/2}\, \|u\|^2.$$

Now we define the operator T. Let J be an interval of integers of length n (which later will be specified further) and set

$$D_n(x) = \sum_{\ell \in J} e^{2\pi i \ell x}.$$

Then T is defined by

$$(Tu)(x) = \frac{1}{n} \sum_k D_n\left(\frac{x}{\sqrt{n}} - \frac{k}{n}\right) u(k).$$

Thus T has kernel

$$T(x,k) = \frac{1}{n}\, D_n\left(\frac{x}{\sqrt{n}} - \frac{k}{n}\right).$$

By the definition of the inner product on $L^2(\mathbb{Z}_n)$ we find that

$$T^* : L^2(-\sqrt{n}/2, \sqrt{n}/2) \to L^2(\mathbb{Z}_n)$$

has kernel

$$T^*(k,x) = \frac{1}{\sqrt{n}}\, D_n\left(\frac{k}{n} - \frac{x}{\sqrt{n}}\right).$$

In terms of the transforms we have the following:

Lemma 2.

(a) *For* $u \in L^2(\mathbb{Z}_n)$,

$$\widehat{Tu}(\ell) = \begin{cases} \frac{1}{\sqrt{n}}\,\widehat{u}(\ell), & \text{if } \ell \in J, \\ 0, & \text{if } \ell \notin J. \end{cases}$$

(b) *For* $\phi \in L^2(-\sqrt{n}/2, \sqrt{n}/2)$,

$$\widehat{T^*\phi}(\ell) = \sqrt{n}\,\widehat{\phi}(\ell) \quad \text{when } \ell \in J.$$

Proof. For (a), we have

$$\widehat{Tu}(\ell) = \frac{1}{n}\int \sum_{k,\,\ell' \in J} e^{-2\pi i \ell x/\sqrt{n}}\, e^{2\pi i \ell'(x/\sqrt{n}-k/n)}\, u(k)\, dx$$
$$= \frac{1}{n}\int \sum_{\ell' \in J} e^{2\pi i(\ell'-\ell)x/\sqrt{n}}\, \widehat{u}(\ell')\, dx.$$

The result follows.

For (b), we have when $\ell \in J$,

$$\widehat{T^*\phi}(\ell) = \frac{1}{\sqrt{n}} \sum_{k,\,\ell' \in J} \int e^{-2\pi i \ell k/n}\, e^{2\pi i \ell'(k/n - x/\sqrt{n})}\, \phi(x)\, dx$$
$$= \sqrt{n} \int e^{-2\pi i \ell x/\sqrt{n}}\, \phi(x)\, dx = \sqrt{n}\,\widehat{\phi}(\ell). \qquad \square$$

We show two things about T. For the second we shall assume now and hereafter that the end-points of J are $\pm n/2 + O(1)$, although this is a lot stronger than necessary.

Lemma 3. (a) $T^*T = I$. (b) $TT^* \to I$ *strongly as* $n \to \infty$.

Proof. By Lemma 2b,

$$\widehat{T^*Tu}(\ell) = \sqrt{n}\,\widehat{Tu}(\ell'),$$

where $\ell' \in J$ and $\ell' - \ell \in n\mathbb{Z}$. By Lemma 2(a) this in turn equals $\widehat{u}(\ell')$, which equals $\widehat{u}(\ell)$ since $\widehat{u}$ is n-periodic. This gives (a).

For (b) observe that TT^* is self-adjoint. Since $(TT^*)^2 = TT^*TT^* = TT^*$, it is a (nonzero) projection and so has norm one. Therefore it suffices to show that if ϕ is a Schwartz function then $TT^*\phi \to \phi$. We have from Lemma 2(a) that

$$\widehat{TT^*\phi}(\ell) = \frac{1}{\sqrt{n}}\,\widehat{T^*\phi}(\ell)$$

if $\ell \in J$ and equals zero otherwise. If $\ell \in J$ then by Lemma 2(b) it equals $\widehat{\phi}(\ell)$. It follows that

$$\|\widehat{TT^*\phi} - \widehat{\phi}\|^2 = \sum_{\ell \notin J} |\widehat{\phi}(\ell)|^2.$$

Integrating by parts shows that

$$\widehat{\phi}(\ell) = O(\sqrt{n}/\ell),$$

and so, by our assumption on J, the sum on the right side is $O(1)$. Then by (12) we get

$$\|T T^{*} \phi - \phi\| = O(n^{-1/4}). \qquad \square$$

Now the work begins. First, an identity. We introduce the notations

$$\overline{C}(\xi) = 1 - \cos(2\pi\xi), \quad S(\xi) = \sin(\pi\xi),$$

and observe that $\overline{C}(\xi) = 2\, S(\xi)^2$.

Lemma 4. *For $u \in L^2(\mathbb{Z}_n)$,*

$$(\tilde{M}_n\, u,\, u) = n^{-1/2}\, \|S(\ell/n)\, \widehat{u}(\ell)\|^2 + n\, \|S(k/n)\, u(k)\|^2.$$

Note. Here and below we display "k" as the variable in the ambient space $\mathbb{Z}_n$ and "ℓ" as the variable in the space $\mathbb{Z}_n$ of the Fourier transform. We abuse notation and, for example, the "$u(k)$" above denotes the function $k \to u(k)$.

Proof. We consider first the contribution of (10) to the inner product. If we define the operators $A_{\pm}$ by

$$(A_{\pm}u)(k) = u(k \pm 1),$$

we see that the contribution to the inner product is

$$\frac{n}{2}\,(u - [A_{+}u + A_{-}u]/2, u).$$

Now

$$\widehat{A_{\pm}u}(\ell) = e^{\pm 2\pi i \ell/n}\, \widehat{u}(\ell),$$

so if we use (13) we see that the above is equal to

$$\frac{1}{2} n^{-1/2}\, (\overline{C}(\ell/n)\, \widehat{u}, \widehat{u}(\ell)) = n^{-1/2}\, \|S(\ell/n)\, \widehat{u}\|^2.$$

To complete the proof of the lemma we note that the contribution to the inner product of (11) is clearly

$$\frac{n}{2}\, (\overline{C}(k/n)\, u(k), u(k)) = n\, \|S(k/n)\, u(k)\|^2. \qquad \square$$

Lemma 5. *Suppose u_n satisfy $(\tilde{M}_n u_n, u_n) = O(1)$. Then* (a) $\|x\, T u_n(x)\| = O(1)$, *and* (b) $\|(T u_n)'\| = O(1)$.

Proof of (a). We have

$$\widehat{u}_n(\ell) - \widehat{u}_n(\ell + 1) = \sum_k e^{-2\pi \ell k/n} (1 - e^{-2\pi i k/n})\, u_n(k),$$

the finite Fourier transform of $(1 - e^{-2\pi i k/n})\, u_n(k)$. We have,

$$|(1 - e^{-2\pi i k/n})\, u_n(k)| = 2 |S(k/n)\, u_n(k)|.$$

Therefore from (13) and

$$\|S(k/n)\, u_n(k)\| = O(n^{-1/2}),$$

which follows from Lemma 4, we get

$$\|\widehat{u}_n(\ell) - \widehat{u}_n(\ell+1)\| = O(n^{1/4}).$$

It follows from Lemma 2 (a) that $\widehat{Tu_n}(\ell) = \widehat{Tu_n}(\ell+1) = 0$ if both $\ell, \ell+1 \notin J$ and

$$\sum_{\ell,\, \ell+1 \in J} |\widehat{Tu_n}(\ell) - \widehat{Tu_n}(\ell+1)|^2 = O(n^{-1/2}). \tag{14}$$

If $\ell \in J$ but $\ell+1 \notin J$ then $\widehat{Tu_n}(\ell+1) = 0$ and ℓ is the right end-point of J and therefore $n/2 + O(1)$. From

$$\|S(\ell/n)\, \widehat{u}_n(\ell)\| = O(n^{1/4}), \tag{15}$$

which also follows from Lemma 4, and that $|S(\ell/n)|$ is bounded below for $\ell = n/2 + \ O(1)$, we have in particular that $\widehat{u}_n(\ell) = O(n^{1/4})$, Therefore $\widehat{Tu_n}(\ell) = O(n^{-1/4})$.

So the bound in (14) holds when the sum is taken over all $\ell \in \mathbb{Z}$. Since

$$\widehat{Tu_n}(\ell) - \widehat{Tu_n}(\ell+1) = \int e^{-2\pi i x\ell/\sqrt{n}}(1 - e^{-2\pi i x/\sqrt{n}})\, Tu_n(x)\, dx,$$

it follows from (12) that

$$\|S(x/\sqrt{n})\, Tu_n(x)\| = O(n^{-1/2}).$$

In the interval of integration $|x| < \sqrt{n}/2$, so $|S(x/\sqrt{n})|$ is bounded below by a constant times $|x|/\sqrt{n}$. This gives (a). □

Proof of (b). We have

$$(Tu_n)'(x) = \frac{2\pi i}{n^{3/2}} \sum_{k,\, \ell \in J} \ell\, e^{2\pi i \ell(x/\sqrt{n} - k/n)}\, u_n(k) = \frac{2\pi i}{n^{3/2}} \sum_{\ell \in J} \ell\, e^{2\pi i \ell x/\sqrt{n}}\, \widehat{u_n}(\ell).$$

Thus $\widehat{Tu_n'}(\ell) = 2\pi i \ell\, \widehat{u}_n(\ell)/n$ for $\ell \in J$, and it follows from (12) that

$$\|(Tu_n)'\|^2 = \frac{4\pi^2}{n^{5/2}} \sum_{\ell \in J} \ell^2\, |\widehat{u_n}(\ell)|^2.$$

Now $|S(\ell/n)|$ is bounded below by a constant times $|\ell/n|$ for $\ell \in J$, so (15) implies that

$$\sum_{\ell \in J} \ell^2\, |\widehat{u_n}(\ell)|^2 = O(n^{5/2}),$$

which gives the result. □

Proof of Proposition 1. Since T is an isometry each $\|Tu_n\| = 1$, and by passing to a subsequence we may assume $\{Tu_n\}$ converges weakly to some $f \in L^2(\mathbb{R})$. We use the fact that strong convergence will follow if we can show that $\|f\| \geq 1$. (In general, if $\|f_n\| = 1$ and $f_n \to f$ weakly, then $\|f\| \geq 1$ implies that $f_n \to f$ strongly. Here is the argument. We have that

$$\|f_n - f\|^2 = \|f_n\|^2 + \|f\|^2 - 2\,\mathrm{Re}\,(f_n, f).$$

By weak convergence, $(f_n, f) \to \|f\|^2$. Therefore

$$\|f_n - f\|^2 \to 1 - \|f\|^2 \leq 0,$$

so $\|f_n - f\| \to 0$.)

The hypothesis of Lemma 5 is satisfied. It follows from Lemma 5 (a) that for each $\varepsilon > 0$ there is a bounded interval A such that

$$\|(1 - \chi_A)\,Tu_n\| \leq \varepsilon$$

for all n. So $\|\chi_A\,Tu_n\| \geq 1 - \varepsilon$. It follows from Lemma 5 (b) that $\{Tu_n\}$ is equicontinuous, and this combined with $\|\chi_A\,Tu_n\| \leq 1$ shows that a subsequence of $\{Tu_n\}$ converges uniformly on A (to $\chi_A\,f$), and so $\|\chi_A\,f\| \geq 1 - \varepsilon$. Thus, $\|f\| \geq 1$. □

Proof of Proposition 2. Consider first the operator corresponding to m_1 in (10). We call it $n\Delta_k^2$. (The subscript indicates that it acts on functions of k.) We show first that

$$n\,T\,\Delta_k^2\,T^*\phi \to -\phi''/4$$

in $L^2(\mathbb{R})$. We have

$$n\,(T\,\Delta_k^2\,T^*\phi)(x)$$
$$= \frac{1}{2n^{1/2}} \sum_{\ell,\ell',k} e^{2\pi i\,\ell((x/\sqrt{n}-k/n)} \int_{-\sqrt{n}/2}^{\sqrt{n}/2} \Delta_k^2\, e^{2\pi i\,\ell'((k/n-y/\sqrt{n})}\,\phi(y)\,dy.$$

The exponent in the integral is a function of $k/\sqrt{n} - y$. So taking the second difference Δ_k^2 in k is the same as taking the second difference Δ_y^2 in y as long as the differences in the y-variable are $1/\sqrt{n}$. With this understanding, the above equals

$$\frac{1}{2n^{1/2}} \sum_{\ell,\ell',k} e^{2\pi i\,\ell((x/\sqrt{n}-k/n)} \int_{-\sqrt{n}/2}^{\sqrt{n}/2} \Delta_y^2\, e^{2\pi i\,\ell'((k/n-y/\sqrt{n})}\,\phi(y)\,dy.$$

By changing variables in two of the three summands from Δ_y^2 we can put the Δ_y^2 in front of the $\phi(y)$. There is an error because of the little change of integration domains but (for ϕ a Schwartz function) this is a rapidly decreasing function of n, and so can be ignored. After this what we get is $nT\,T^*\,\Delta^2\,\phi$. Taylor's theorem gives

$$n(\Delta_x^2\,\phi)(x) = -\phi''(x)/4 + O\left(n^{-1/2} \max_{|y-x|<1/\sqrt{n}} |\phi'''(y)|\right),$$

from which it follows that $n\,(\Delta_x^2\,\phi)(x) \to -\phi''(x)/4$ strongly. Since $T\,T^* \to I$ strongly we deduce that $n\,T\,\Delta_k^2\,T^*\phi \to -\phi''/2$ strongly.

Lastly, consider the operator corresponding to (11), which is multiplication by $n\,\overline{C}(k/n)/2 = n(1-\cos(2\pi k/n))/2$. For convenience we call this operator $\overline{C}_n/2$.

By Lemma 2(b) we know that $\widehat{T^*\phi}(\ell) = \sqrt{n}\,\widehat{\phi}(\ell)$ when $\ell \in J$. In general,

$$\widehat{\overline{C}_n\,u}(\ell) = n\sum_k (1-\cos(2\pi k/n))\,e^{-2\pi i k\ell/n}\,u(k) = n\,(\Delta^2\,\widehat{u})(\ell),$$

where here

$$(\Delta^2\,\widehat{u})(\ell) = \widehat{u}(\ell) - [\widehat{u}(\ell-1)+\widehat{u}(\ell+1)]/2.$$

Applying this to $T^*\phi$ gives

$$\widehat{\overline{C}_n\,T^*\phi}(\ell) = n\,\Delta^2\,\widehat{T^*\phi}(\ell) = n^{3/2}\,\Delta^2\,\widehat{\phi}(\ell)$$

as long as $\ell \pm 1$ are also in J. If ℓ is in J but one of $\ell \pm 1$ is not in J then ℓ is near an end-point of J and the error committed will be rapidly decreasing as $n \to \infty$. For the right side is $n^{3/2}$ times a linear combination of integrals like

$$\int_{-\sqrt{n}/2}^{\sqrt{n}/2} e^{\pm\pi i x\sqrt{n}}\,\phi(x)\,dx,$$

and integration by parts many time shows this is rapidly decreasing. For the left side, if for example $\ell+1 \notin J$ then it is the same as the value at $\ell+1-n \in J$, which is rapidly decreasing. So we ignore this little error and use

$$\Delta^2\,\widehat{\phi}(\ell) = \int_{-\sqrt{n}/2}^{\sqrt{n}/2} e^{-2\pi i x\ell/\sqrt{n}}\,\overline{C}(x/\sqrt{n})\,\phi(x)\,dx.$$

We also have $\widehat{T^*x^2\phi}(\ell) = \sqrt{n}\,\widehat{x^2\phi}(\ell)$ for $\ell \in J$. Thus (ignoring the error),

$$\widehat{\overline{C}_n\,T^*\phi}(\ell) - 2\pi^2\,\widehat{T^*x^2\phi}(\ell) = \sqrt{n}\int_{-\sqrt{n}/2}^{\sqrt{n}/2} e^{-2\pi i x\ell/\sqrt{n}}\,(n\,\overline{C}(x/\sqrt{n}) - 2\pi^2x^2)\,\phi(x)\,dx.$$

From this we get

$$\|\widehat{\overline{C}_n\,T^*\phi} - 2\pi^2\,\widehat{T^*x^2\phi}\| \le n^{3/4}\,\|(n\,\overline{C}(x/\sqrt{n}) - 2\pi^2x^2)\,\phi(x)\| = O(n^{-1/4}),$$

since $n\,\overline{C}(x/\sqrt{n}) - 2\pi^2x^2 = O(x^4/n)$ and $x^4\,\phi(x) \in L^2(\mathbb{R})$. Then we get from (13)

$$\|\overline{C}_n\,T^*\phi - 2\pi^2\,T^*x^2\phi\| = O(n^{-1}).$$

Equivalently,

$$\|T\,\overline{C}_n\,T^*\phi - 2\pi^2\,TT^*x^2\phi\| = O(n^{-1}).$$

Since $TT^* \to I$ strongly, this gives

$$\|T\,\overline{C}_n\,T^*\phi - 2\pi^2\,x^2\phi\| \to 0.$$

Since the operator corresponding to (11) is multiplication by $\overline{C}_n/2$, this completes the proof of Proposition 2. □

Now we go back to the eigenvalues of $\tilde{M}_n$ and easy consequences of Propositions 1 and 2.

Proposition 3.

(a) *If λ_n are eigenvalues of $\tilde{M}_n$ and $\lambda_n \to \mu$, then μ is an eigenvalue of L.*

(b) *Any eigenvalue μ of L has a neighborhood that contains at most one eigenvalue (counting multiplicity) of $\tilde{M}_n$ for sufficiently large n.*

Proof of (a). Suppose that u_n is an eigenvector of $\tilde{M}_n$ of norm one with eigenvalue λ_n. In particular $(\tilde{M}_n u_n, u_n) = \lambda_n$. By Proposition 1 there is a subsequence of $\{Tu_n\}$ that converges strongly to some $f \in L^2(\mathbb{R})$. For a Schwartz function ϕ we have

$$\begin{aligned}(\mu f, \phi) &= \lim (\lambda_n T u_n, \phi) = \lim (T\tilde{M}_n u_n, \phi) \\ &= \lim (u_n, \tilde{M}_n T^* \phi) = \lim (T u_n, T\tilde{M}_n T^* \phi),\end{aligned}$$

since T is an isometry. By Proposition 2, $T\tilde{M}_n T^*\phi$ converges strongly to $L\phi$. Therefore[2] the limit equals $(f, L\phi)$, and we have shown

$$\mu(f, \phi) = (f, L\phi).$$

It follows that f is an eigenfunction of L with corresponding eigenvalue μ. Here is why. The eigenfunctions of L are the harmonic oscillator wave functions ϕ_i, and therefore Schwartz functions. If the corresponding eigenvalues are μ_i, then

$$(\mu - \mu_i)(f, \phi_i) = \mu (f, \phi_i) - (f, L\phi_i) = 0.$$

Since the ϕ_i are complete and $f \neq 0$, some $\mu - \mu_k = 0$. And f, being orthogonal to the ϕ_i with $i \neq k$, must be a multiple of ϕ_k and therefore a corresponding eigenfunction. □

Proof of (b). Suppose the contrary were true. Then there would be sequences of eigenvalues $\{\lambda_n\}$ and $\{\lambda_n'\}$ of $\tilde{M}_n$, both converging to μ, and corresponding orthogonal (since $\tilde{M}_n$ is self-adjoint) eigenfunctions u_n and u_n'. The strong (sub)limits f and f' of Tu_n and Tu_n' would be mutually orthogonal eigenfunctions of L corresponding to the same eigenvalue μ of L. Since the eigenvalues of L are simple, this cannot happen. □

Proposition 4. *For each eigenvalue μ of L there is a sequence of eigenvalues λ_n of $\tilde{M}_n$ that converges to μ.*

Proof. With corresponding eigenfunction ϕ of L we have, by Proposition 2,

$$\|T\tilde{M}_n T^*\phi - \mu\, \phi\| = \|T\tilde{M}_n T^*\phi - L\, \phi\| = o(1).$$

From this, and that $\|\phi - T\,T^*\, \phi\| = o(1)$ by Lemma 3 (b), we get

$$\|T\, (\tilde{M}_n T^*\phi - \mu\, T^*\phi)\| = o(1).$$

[2]For this we need only weak convergence of one and strong convergence of the other. But we also need that $f \neq 0$, which is no easier to show than strong convergence of Tu_n, and we shall need strong convergence of $T\tilde{M}_n T^*\phi$ for Proposition 4.

Since T is an isometry this is the same as

$$\|\tilde{M}_n T^* \phi - \mu T^* \phi\| = o(1).$$

Since $\|T^*\phi\| = \|T T^*\phi\| \to \|\phi\| \neq 0$, and the other eigenvalues of L are bounded away from μ, this implies that μ is within $o(1)$ of an eigenvalue of $\tilde{M}_n$ and $T^* \phi$ within $o(1)$ of an eigenvector. See [11]. □

Combining Propositions 3 and 4 shows that the kth largest eigenvalue of M_n equals $1 - \mu_k/n + o(1/n)$.

4.2. The bottom eigenvalues of M_n

We shall find a unitary operator U on $L^2(\mathbb{Z}_n)$ such that the quadratic form for $n(I + U_n M_n U_n^*)$ is the same as for $n(I - M_n)$ when n is even and close to it when n is odd. From that it will follow that the kth bottom eigenvalue of M_n equals $-1 + \mu_k/n + o(1/n)$.

Recall that Lemma 4 says that

$$(n(I - M_n)\, u,\, u) = n^{-1/2}\, \|S(\ell/n)\, \widehat{u}\|^2 + n\, \|S(k/n)\, u(k)\|^2 =: Q(u),$$

where $S(\xi) = \sin(\pi\xi)$. For this we used the identity $1 - \cos\xi = 2\sin^2(\xi/2)$. For $n(I + M_n)$ this gets replaced by $1 + \cos\xi = 2\cos^2(\xi/2)$. So now we define

$$C(\xi) = \cos(\pi\xi),$$

and get

$$(n(I + M_n)\, u,\, u) = n^{-1/2}\, \|C(\ell/n)\, \widehat{u}\|^2 + n\, \|C(k/n)\, u(k)\|^2.$$

We consider first the less straightforward case of n odd and define

$$v(k) = e^{2\pi i\alpha k}\, u\left(k - (n+1)/2\right),$$

where α (real) will be determined below. We have

$$\|C(k/n)\, u(k)\| = \Big\|C(k/n + 1/2 + 1/2n)\, v(k)\Big\| = \|S(k/n + 1/2n)\, v(k)\|.$$

Next,

$$\begin{aligned}
\widehat{v}(\ell) &= \sum_k e^{-2\pi i k\ell/n}\, e^{2\pi i\alpha k}\, u\left(k - (n+1)/2\right) \\
&= \sum_k e^{-2\pi i(k+(n+1)/2)\ell/n}\, e^{2\pi i\alpha(k+(n+1)/2)}\, u(k) \\
&= (-1)^\ell\, e^{-\pi i\ell/n}\, e^{\pi i\alpha(n+1)} \sum_k e^{-2\pi i k\ell/n}\, e^{2\pi i\alpha k}\, u(k).
\end{aligned}$$

We choose $\alpha = -(n+1)/2n$. The factor outside the sum has absolute value 1 while the sum becomes $\widehat{u}(\ell + (n+1)/2)$. Alternatively,

$$|\widehat{u}(\ell)| = |\widehat{v}(\ell - (n+1)/2)|.$$

Therefore

$$\|C(\ell/n)\, \widehat{u}(\ell)\| = \|C(\ell/n + 1/2 + 1/2n)\, \widehat{v}(\ell)\| = \|S(\ell/n + 1/2n)\, \widehat{v}(\ell)\|.$$

The map $U : u \to v$ is unitary and we have shown

Lemma 6. *For n odd we have,*

$$(n(I + UM_nU^*)\, v,\, v) = n^{-1/2}\, \|S(\ell/n + 1/2n)\, \widehat{v}(\ell)\|^2 + n\, \|S(k/n + 1/2n)\, v(k)\|^2.$$

Remark. When n is even we replace the shift $(n+1)/2$ by $n/2$ and α by $-1/2$, and the extra $1/2n$'s do not appear in the arguments of the S's. The quadratic form becomes $Q(v)$ exactly, so $U_n M_n U_n^* = -M_n$.[3]

Proposition 5. $(n(I + UM_nU^*)\, v,\, v) = (1 + O(n^{-1/2}))\, Q(v) + O(n^{-1/2}\|v\|^2)$.

Proof. Since $dS/d\xi$ is bounded,

$$\|S(k/n + 1/2n)\, v(k)\| \le \|S(k/n)\, v(k)\| + O(n^{-1}\|v(k)\|).$$

It follows from the arithmetic-geometric mean inequality that for any $\varepsilon > 0$

$$|a + b|^2 \le (1 + \varepsilon)\, |a|^2 + (1 + \varepsilon^{-1})\, |b|^2.$$

We will take $\varepsilon \to 0$ as $n \to \infty$, so we obtain

$$\|S(k/n + 1/2n)\, v(k)\|^2 \le (1 + \varepsilon)\, \|S(k/n)\, v(k)\|^2 + O(\varepsilon^{-1} n^{-2}\|v(k)\|^2).$$

Similarly,

$$\begin{aligned}\|S(\ell/n + 1/2n)\, \widehat{v}\|^2 &\le (1 + \varepsilon)\, \|S(\ell/n)\, \widehat{v}\|^2 + O(\varepsilon^{-1} n^{-2}\|\widehat{v})\|^2) \\ &= (1 + \varepsilon)\, \|S(\ell/n)\, \widehat{v}\|^2 + O(\varepsilon^{-1} n^{-1/2}\|v\|^2),\end{aligned}$$

where we used $\|\widehat{v}\| = n^{3/4}\, \|v\|$.

Thus,

$$(n(I + UM_nU^*)\, v,\, v) \le (1 + \varepsilon)\, Q(v) + O(\varepsilon^{-1} n^{-1}\|v\|^2).$$

Similarly,

$$(n(I + UM_nU^*)\, v,\, v) \ge (1 + \varepsilon)^{-1}\, Q(v) - O(\varepsilon^{-1} n^{-1}\|v\|^2).$$

We set $\varepsilon = n^{-1/2}$ and put the inequalities together to get the statement of the proposition. □

Recall that $Q(v) = (n(I - M_n)\, v, v)$. If in the statement of the proposition we take the minimum of both sides over all v with $\|v\| = 1$ we deduce that

$$n + n\lambda_n = (1 + O(n^{-1/2}))(n - n\lambda_1) + O(n^{-1/2}),$$

where λ_n is the bottom eigenvalue of M_n and λ_1 the top eigenvalue. Since $\lambda_1 = 1 - \mu_1/n + o(1/n)$, we have $n - n\lambda_1 = \mu_1 + o(1)$, and then $n + n\lambda_n = \mu_1 + o(1)$, and then $\lambda_n = -1 + \mu_1/n + o(1)$.

Using the minimax characterization of the eigenvalues we show similarly that $\lambda_{n-k+1} = -1 + \mu_k/n + o(1)$ for each k.

---

[3]This is easy to see directly.

5. A stochastic argument

This section gives a bound on the largest eigenvalue of the matrix M_n using a probabilistic argument. By inspection,

$$M_n' = \frac{1}{3}I + \frac{2}{3}M_n$$

is a sub-stochastic matrix (with non-negative entries and row sums at most 1). Take M_n'' as in (2), an $(n+1) \times (n+1)$ stochastic matrix corresponding to a Markov chain absorbing at 0. The first (Dirichlet) eigenvector has first entry 0 and its corresponding eigenvalue β^* is the top eigenvalue of M_n'. Thus

$$\beta = \frac{(\beta^* - \frac{1}{3})}{(2/3)} = \frac{3}{2}\beta^* - \frac{1}{2}$$

is the top eigenvector of M_n.

We will work in continuous time, thus, for any transition matrix M,

$$M_t = \sum_{j=0}^{\infty} \frac{e^{-t}M^j t^j}{j!} = e^{t(M-I)}.$$

The matrix $L = I - M$, the opposite of the generator of the semigroup $\{M_t\}_{t\geq 0}$, has row sums zero, and non-positive off diagonal entries.[4] If $\mathbf{v}$ is a right eigenvector of M with eigenvalue β, then $\mathbf{v}$ is an eigenvector of L with eigenvalue $1-\beta$. A lower bound for the non trivial eigenvalues of L gives an upper bound for the eigenvalues of M. Throughout, we specialize to $L = I - M_n$, let λ^* be the lowest non-zero eigenvalue of L, and β the highest eigenvector of M_n.

Standard theory for absorbing Markov chains with all non-absorbing states connected shows that if τ is the time to first absorption, for any non absorbing state ξ, as t tends to infinity,

$$\lim_{t\to\infty} \frac{-\log P_\xi(\tau > t)}{t} = \lambda^*.$$

Thus an upper bound on β will follow from an upper bound on $P_\xi(\tau > t)$. Here is an outline of the proof. Begin by coupling the absorbing chain of interest with a simple random walk on $\mathcal{C}_n = \mathbb{Z}/(n\mathbb{Z})$. For a fixed b, let τ_b be the first time that the simple random walk travels $\pm b$ from its start. We derive the bound $P_\xi(\tau > \tau_b) \leq G_b$, where $G_b < 1$ is a particular constant described below. Define a sequence of stopping times τ_b^i as follows. $\tau_b^1 = \tau_b$, τ_b^2 is the first time following τ_b^1

[4]Since some of our readers (indeed some of our authors) may not be probabilists we insert the following note; given any matrix $L(x,y)$ with row sums zero and non-positive off diagonal entries one may construct a continuous time Markov process $W = (W_t)_{t\geq 0}$ as follows. Suppose $W_0 = w_0$ is fixed. The process stays at w_0 for an exponential time σ_0 with mean $1/|L(w_0,w_0)|$. (Thus $P\{\sigma_0 \geq t\} = e^{-tL(w_0,w_0)}$.) Then, choose $w_1 \neq w_0$ with probability $|L(w_0,w_1)|/L(w_0,w_0)$. Stay at w_1 for an exponential time σ_1 (with mean $1/L(w_1,w_1)$). Continue, choosing from $L(w_1,.)/L(w_1,w_1)$.

that the walk travels $\pm b$, similarly define τ_b^m. By the strong law of large numbers, $\tau_b^m/m \to \mu_b = E(\tau_b^1)$ almost surely. Thus

$$P\{\tau > m\mu_b\} \cong P\{\tau > \tau_b^m\}.$$

Using the Markov property, $P\{\tau > \tau_b^m\} \leq G_b^m$. This implies there are positive c_1, c_2 with

$$P\{\tau > c_1 m\mu_b\} \leq G_b^m + e^{-c_2 m}.$$

In our problem, classical random walk estimates show $\mu_b \sim b^2$. We show, for $b = \sqrt{n}$, $G_{\sqrt{n}}$ is bounded away from one. Thus

$$P\{\tau > c_1 m\mu_b\} \leq 2\max(G_b^m, e^{-c_2 m})$$

and

$$\frac{\log P\{\tau > c_1 m\mu_b\}}{m\mu_b} \leq \frac{c'}{\mu_b} = \frac{c''}{n}$$

for some $c', c'' < 0$. Backtracking gives the claimed bound in Theorem 3.

The argument is fairly robust – it works for a variety of diagonal entries. At the end of the proof, some additions are suggested which should give the right constant multiplying $\frac{1}{n}$.

We begin by constructing two processes. For as long as possible, general absorption rates will be used. Let $X = (x_t)_{t\geq 0}$ be the standard continuous time random walk on $\mathbb{Z}$ with jump rates 1 between neighbors. Take $x_0 = 0$. Fix $b \in \mathbb{Z}_+$ and let τ_b be the first hitting time of $\{-b-1, b+1\}$:

$$\tau_b = \inf\{t \geq 0 : |x_t| = b+1\}. \tag{16}$$

Let $\{u_x\}_{x\in\mathbb{Z}}$ be killing rates, e.g., arbitrary non-negative real numbers. Add a cemetery state ∞ to $\mathbb{Z}$. An absorbed process $\overline{x} = (\overline{x}_t)_{t\geq 0}$, behaving as x until it is absorbed at ∞ with the rates $\{u_x\}_{x\in\mathbb{Z}}$ can be constructed as follows: Let $\mathcal{E}$ be an independent exponential random variable with mean 1. Define an absorption time $\overline{\tau} \in [0, \infty]$ by

$$\overline{\tau} = \inf_{t\geq 0}\left\{\int_0^t \mu_{x_s} ds \geq \mathcal{E}\right\}.$$

As soon as $\{u_x\}_{x\in\mathbb{Z}}$ does not vanish identically, $\overline{\tau}$ is characterized by

$$\int_0^{\overline{\tau}} u_{x_s} ds = \mathcal{E}. \tag{17}$$

More simply, $\overline{x}_t = \begin{cases} x_t & \text{if } t < \overline{\tau}, \\ \infty & \text{otherwise,} \end{cases}$ for $0 \leq t < \infty$.

The two processes are defined on the same probability space as are $\overline{\tau}$ and τ_b. The first goal is to estimate the chance that $\overline{\tau} > \tau_b$ in terms of the given rates. Our bounds are crude but suffice for Theorem 3.

Proposition 6. *With notation as above, for any $b \geq 1$,*

$$P\{\overline{\tau} > \tau_b\} \leq \left(\frac{1}{1+(b+1)^2 v_0/2} \prod_{k=1}^{b} \frac{1}{1+(b+1)(b+1-k)v_k} \right)^{1/(b+1)} \tag{18}$$

with $v_k = \min(u_{-k}, u_k)$.

Note that the bound is achievable; if all $v_y = 0$ then both sides equal 1.

Proof. For any $k \in \mathbb{Z}$, $v_k \leq u_k$. Thus if τ is the stopping time defined in (17) with u_k replaced by v_k, $\tau \geq \overline{\tau}$. Therefore it is sufficient to bound $P\{\tau > \tau_b\}$ from above. Now, everything is symmetric about zero. Consider the process $Y = (Y_t)_{t \geq 0} = (|x_t|)_{t \geq 0}$. This is Markov with jump rates:

$$J(y, y') = \begin{cases} 2 & \text{if } y = 0, y' = 1, \\ 1 & \text{if } y = \mathbb{Z}_+ \text{ and } |y' - y| = 1, \\ 0 & \text{otherwise.} \end{cases}$$

Clearly $\tau_b = \inf_{t \geq 0}\{Y_t = b+1\}$. Define the family of local times associated to Y:

$$L_y(t) = \int_0^t \delta_y(Y_s)ds \text{ for } y \in \mathbb{Z}_+,\ t \geq 0,$$

where δ_y is the indicator function of y. For any $t \geq 0$,

$$\int_0^t v_{x_s} ds = \sum_{0 \leq y \leq b} v_y L_y(t).$$

This gives

$$\{\tau > \tau_b\} = \left\{ \sum_{0 \leq y \leq b} v_y L_y(\tau_b) < \mathcal{E} \right\}.$$

Taking expectations of both sides with respect to $\mathcal{E}$

$$P\{\tau > \tau_b\} = E\left\{ \exp\left(- \sum_{0 \leq y \leq b} v_y L_y(\tau_b) \right) \right\} \tag{19}$$

$$\leq \prod_{y=1}^{b} E\left\{ \exp\left(-(b+1) v_y L_y(\tau_b) \right) \right\}^{\frac{1}{b+1}}. \tag{20}$$

The last bound follows from Hölder's inequality (with $b+1$ functions).

It is well known (see [23] or Claim 2.4 of [26] for the discrete time version) that for any y, $1 \leq y \leq b$, $L_y(\tau_b)$ is distributed as an exponential random variable with mean $(b+1-y)$ and $L_0(\tau_b)$ is exponential with mean $\frac{b+1}{2}$. (The process leaves zero twice as fast as it leaves other points.) Thus, for $1 \le y \le b$,

$$E\{\exp(-(b+1)v_y L_y(\tau_b)\} = \frac{(b+1-y)^{-1}}{(b+1-y)^{-1} + (b+1)v_y}.$$

$$E\{\exp(-(b+1)v_0 L_0(\tau_b)\} = \frac{1}{1+(b+1)^2 v_0/2}.$$

This completes the proof of Proposition 6. □

The bound of Proposition 6 suggests introducing functions F_b, G_b on $\mathbb{R}_+^{b+1}$. Given by

$$F_b(\mathbf{v}) = \left\{ \frac{1}{1+(b+1)^2 v_0/2} \prod_{l=1}^{b} \frac{1}{1+(b+1)(b+1-i)v_l} \right\}^{\frac{1}{l+1}} \tag{21}$$

$$G_b(\mathbf{v}) = \left\{ \prod_{k=0}^{b} \frac{1}{1+(b+1)(b+1-k)v_k/2} \right\}^{\frac{1}{(b+1)}}. \tag{22}$$

They have the following crucial monotonicity properties: say that $\mathbf{v}, \mathbf{v}' \in \mathbb{R}^{b+1}$ satisfies $\mathbf{v} \le \mathbf{v}'$ if this is true coordinate-wise. For $\mathbf{v} \in \mathbb{R}^{b+1}$, let $\overline{\mathbf{v}}$ be the non-decreasing rearrangement of v. Then

$$F_b(\mathbf{v}) \le G_b(\mathbf{v}) \tag{23}$$

$$\mathbf{v} \le \mathbf{v}' \Rightarrow G_b(\mathbf{v}) \ge G_b(\mathbf{v}') \tag{24}$$

$$G_b(\mathbf{v}) \le G_b(\overline{\mathbf{v}}). \tag{25}$$

Return now to the process underlying Theorem 3 (still keeping the extinction rates general.) Let $z = (z_t)_{t \ge 0}$ be defined on $\mathbb{Z}/n\mathbb{Z}$; it jumps to nearest neighbors at rate l and is killed with rates $u = (u_\xi)_{\xi \in \mathbb{Z}/n\mathbb{Z}}$. Suppose $z_o = \xi$. Let v_p, $0 \le p \le n-1$ denote the non-decreasing rearrangement of $\mathbf{u}$. Let τ be the absorption time of z. Fix b, $0 \le b \le n/2 - 1$ and let

$$\tau_b = \inf_{t>0} \{z_\tau \in \{\xi - b - 1, \xi + b + 1\}\}.$$

Proposition 6 in conjunction with properties (23), (24), and (25), imply that for any $\mathbf{u}$, with $G_b(\mathbf{u})$ depending only on the first b coordinates of $\mathbf{u}$,

$$P_\xi[\tau > \tau_b] \le G_b(\mathbf{u}). \tag{26}$$

Note that the upper bound is independent of ξ.

Introduce a sequence ξ_i of further stopping times: $\xi_1 = \xi_b$, and if ξ_m has been constructed,

$$\xi_{m+1} = \inf\{t > \xi_m : z_t \in \{z_{\xi_m} - b - 1, z_{\xi_m} + b + 1\}\}. \tag{27}$$

Informally speaking these stopping times end up being good. Because they cannot be larger than τ, as in the previous treatment of a random walk on $\mathbb{Z}/n\mathbb{Z}$ coinciding with z_t up to the absorption time, then ξ_m are (almost surely) finite for all m and the strong law of large numbers gives:

$$\lim_{m\to\alpha} \frac{\xi_m}{m} \to \mu_b = (b+1)^2$$

where $\mu_b = E\{\xi_1\} = (b+1)^2$ from the Classical Gambler's Ruin (see Chapter 14 of [16]).

This suggests that, for m large, the quantities

$$P_\xi[\tau > m\mu_b] \text{ and } P_\xi[\tau > \xi_m]$$

should behave similarly. Of course, care must be taken because τ and ξ_m are not independent. To proceed, we use a large deviations bound for ξ_m.

Proposition 7. *For ξ_m defined in (27), there are positive constants, c_1, c_2, independent of b and n such that for all $m \geq 1$,*

$$P[\xi_m > c_1 m\mu_b] \leq e^{-c_2 m}.$$

Proof. Observe first that this is simply a large deviations bound for the first hitting time of the simple random walk (16) so that n does not enter. The law of ξ_1 is well known (see [22], [12], and [17]). It can be represented as a sum of $b+1$ independent exponential variables with means $a_1, a_2, \ldots, a_{b+1}$ given by:

$$a_k^{-1} = 2\left(1 - \cos\left(\frac{\pi(2k-1)}{2(b+1)}\right)\right).$$

Thus for $\theta \in (0, a_1)$

$$E[e^{\theta\xi_1}] = \prod_{k=1}^{b+1} \frac{a_k}{a_k - \theta}.$$

By simple calculus, there is $c > 0$ such that for all $a \in (0, \frac{1}{2}]$, $-\log(1-a) \leq ca$. Thus, for $\theta \in (0, a_1/2]$,

$$E(e^{\theta\xi_1}) \leq e^{c\theta\sum_{n=1}^{b+1}\frac{1}{a_n}}.$$

Taking $\theta = \frac{a_1}{2}$,

$$E[e^{a_1\xi_1/2}] \leq e^{c\sum_{n=0}^{b+1}\frac{a_1}{2a_n}}.$$

Note that a_n is of order $(n/b)^2$ so the right side of the last inequality is bounded uniformly in b, say by $k > 1$. Now, ξ_n is a sum of m i.i.d. random variables so for any $c_1 > 0$

$$P[\xi_m > c_1 m\mu_b] \leq e^{\frac{-c_1 a_1 m\mu_b}{2}} E[e^{a_1\xi_1/2}]^m \leq e^{-m(c_1 a_1\mu_b)/2 - \log k}.$$

Since $\mu_b = (b+1)^2$, $a_1\mu_b$ can be bounded below by a constant $\epsilon > 0$, uniformly in $b \in \mathbb{N}$. Thus if $c_1 = 4\log k/\epsilon$, the claimed bound holds with $c_2 = \log(k)$. $\square$

We can now set up a bound for the top eigenvalue. Working on $\mathbb{Z}/n\mathbb{Z}$ but still with general absorption rates:

$$\begin{aligned} P_\xi[\tau > c_1 m\mu_b] &= P_\xi[\tau > c_1 m\mu_b; \xi_m \le c_1 m\mu_b] \\ &\quad + P_\xi[\tau > c_1 m\mu_d, \xi_m > c_1 m\mu_b] && (28) \\ &\le P_\xi[\tau > \xi_m] + P_\xi[\xi_m > c_1 m\mu_b] && (29) \\ &\le G_b^m(\mathbf{v}) + e^{-c_2 m}. && (30) \end{aligned}$$

It follows that

$$\lambda^* = -\lim_{m\to\infty} \frac{1}{c_1 m\tau_b} \log(P_\xi[\tau > c_1 m\tau_b]) \ge \frac{1}{c_1\tau_b} \min\{c_2, -\log(G_b(x))\}.$$

Since $\mu_b = (b+1)^2$, proving that with b of order $\sqrt{n}$, $-\log(G_b(x))$ is bounded below by a positive constant, uniformly in n, will complete the proof.

Up to now, the kill rates $\mathbf{u}$ have been general. Specialize now to the rates for the matrix M'' with any scrambling of its diagonal. The vector $\overline{\mathbf{v}}$ is given by the $b+1$ entries of:

$$0, \frac{1}{3}\left(1-\cos\left(\frac{2\pi}{n}\right)\right), \frac{1}{3}\left(1-\cos\left(\frac{2\pi}{n}\right)\right), \frac{1}{3}\left(1-\cos\left(\frac{4\pi}{n}\right)\right),$$
$$\frac{1}{3}\left(1-\cos\left(\frac{4\pi}{n}\right)\right), \dots, \frac{1}{3}\left(1-\cos\left(\frac{2\pi\lfloor\frac{n}{2}\rfloor}{n}\right)\right).$$

From the definition of G_b at (21) with $b = \lfloor\sqrt{n}\rfloor$, a Riemann sum approximation gives

$$\lim_{n\to\infty} G_b(\mathbf{v}) = e^{\frac{-\pi^2}{24}}.$$

Indeed,

$$\begin{aligned} -\log(G_b(\mathbf{v})) &= \frac{1}{b+1}\sum_{y=0}^{b-1} \log\left(1 + (b+1)(b+1-y)\left(1-\cos\left(\frac{2\pi}{n}\left\lfloor\frac{y+1}{2}\right\rfloor\right)\right)\right) \\ &\sim \frac{2\pi^2}{n^2}\frac{1}{b}\sum_{y=0}^{b-1}(b+1)(b+1-y)\left\lfloor\frac{y+1}{2}\right\rfloor^2 \\ &\sim \frac{\pi^2 b^4}{2n^2}\frac{1}{b}\sum_{y=0}^{b-1}\left(1-\frac{y}{b}\right)\left(\frac{y}{b}\right)^2 \sim \frac{\pi^2 b^4}{2n^2}\int_0^1 (1-y)y^2 dy \sim \frac{\pi^2}{24}. \end{aligned}$$

Combining the pieces, we use β for the highest eigenvalue of M_n and thus $\beta = \frac{3}{2}\beta^* - \frac{1}{2}$. Using this notation, we have shown that $\frac{c}{n} \le \lambda^* = 1 - \left(\frac{1}{3} + \frac{2}{3}\beta\right)$. Thus $\beta \le 1 - \frac{3c}{2n}$. This completes the argument and ends the proof of Theorem 3.

Remarks. The above argument can be modified to handle quite general diagonal elements (in particular $\cos\left(\frac{2\pi a j}{n}\right), 0 \le j \le N-1$, needed for the application to the Heisenberg random walk). Indeed, for $a = o(n)$, the argument goes through with no essential change with $b = \sqrt{\frac{n}{a}}$ to show that with diagonal entries $\cos\left(\frac{2\pi a j}{n}\right)$,

$0 \le j \le n-1$, the eigenvalue bound $1 - \frac{ca}{n}$ holds (with $c > 0$ independent of n and a).

The use of Hölder's inequality in (19) is crude. The joint distribution of the local times of birth and death processes is accessible (see [23]). We hope this can be used to give sharp results for the constant. Finally we note that the approach to bound β via an associated absorbing Markov chain was used in [5]. There, a geometric path argument was used to complete the analysis. This gave cruder bounds ($\beta \le 1 - \frac{c}{n^{\frac{4}{3}}}$) but the argument worked for diagonal entries $\cos\left(\frac{2\pi a \xi}{n}\right)$ for any $1 \le a \le \frac{n}{2}$ as well as negative eigenvalues.

6. A random walk on the affine group (mod p)

Let $\mathcal{A}_p$ be the affine group (mod p). Here, p is prime and elements of A_p can be represented as pairs $(a,b), 1 \le a \le p-1,\ 0 \le b \le p-1$

$$(a_1,b_1)(a_2,b_2) = (a_1a_2, a_1b_2 + b_1).$$

All entries are taken mod p. Fix a generator g of the multiplicative group. Let

$$S = \{(1,0),(1,1),(1,-1),(g,0),(g^{-1},0)\}.$$

Set

$$Q(h) = \begin{cases} \frac{1}{5}, & \text{if } h \in S, \\ 0, & \text{else.} \end{cases} \tag{31}$$

Convolution powers of Q converge to the uniform distribution $U(h) = \frac{1}{p(p-1)}$. We use the representation theory of A_p and the analytic results of previous sections to show that order p^2 steps are necessary and sufficient for convergence.

Theorem 5. *With definitions above, there are positive universal constants c_1,c_2, and c_3 such that for all primes p and $k \ge 1$*

$$c_1 e^{-c_2\frac{k}{p^2}} \le \|Q^{*^k} - U\|_{TV} \le c_3 e^{-\frac{c_2 k}{p^2}}.$$

Proof. By the usual Upper Bound Lemma (see [8], Chapter 3):

$$4\|Q^{*^k} - U\|_{TV} \le \sum_{\rho \ne 1} d_\rho \|\hat{Q}(\rho)^k\|^2.$$

Here, the sum is over nontrivial irreducible representations ρ of A_p, d_ρ is the dimension of ρ, $\hat{Q}(\rho) = \sum_h Q(h)\rho(h)$ and the norm on the right is the trace norm. There are $p-1$ one-dimensional irreducible representations indexed by $\alpha \in \{1,2,\dots,p-1\}$.

$$\rho_\alpha(a,b) = e^{2\pi i \alpha \sigma(a)/(p-1)}. \tag{32}$$

where $\sigma : \mathbf{Z}_p^* \to \mathbf{Z}_{p-1}$ is the group morphism such that $\sigma(g) = 1$. Then

$$\hat{Q}(\rho_\alpha) = \frac{3}{5} + \frac{2}{5}\cos\left(\frac{2\pi\alpha}{p-1}\right).$$

There is one $(p-1)$-dimensional representation ρ. This may be realized on

$$V = \{f : \{1, 2, \dots, p-1\} \to \mathbb{C}\}$$

with

$$\rho(a,b)f(j) = e^{\frac{2\pi i j b}{p}} f(aj), \quad 1 \le j \le p-1.$$

It is easy to check directly that ρ is a representation with character

$$\chi(a,b) = \begin{cases} 0, & a \neq 1, \\ -1, & a = 1,\, b \neq 0, \\ p-1, & a = 1,\, b = 0. \end{cases}$$

A further simple check shows that $\langle \chi | \chi \rangle = \frac{1}{p(p-1)} \sum_{a,b} |\chi(a,b)|^2 = 1$ and that χ is orthogonal to the characters ρ_α in (32). It follows that $\{\rho_\alpha\}_{\alpha=1}^{p-1}, \rho$ is a full set of irreducible representations. Choose a basis $\delta_{g^a}(\cdot)$ for V, $0 \le a \le p-2$. Then, for $\mathcal{Q}$ in (31),

$$\hat{\mathcal{Q}}(\rho) = \frac{1}{5} \begin{pmatrix} & 1 & & & 1 \\ 1 & \ddots & \ddots & & \\ & \ddots & 1 & 1 & \\ & & \ddots & \ddots & 1 \\ 1 & & & 1 & \ddots \end{pmatrix} \longleftarrow 1 + 2\cos\left(\frac{2\pi j}{p}\right), \quad 1 \le j \le p-1 .$$

Using any of the three techniques above, there is a constant $c > 0$ such that the largest and smallest eigenvalues of $\hat{\mathcal{Q}}(\rho)$ (in absolute value) are bounded above by $1 - \frac{c}{p}$. Combining bounds

$$4\|Q^{*^k} - U\|_{TV}^2 \le \sum_{j=1}^{p-1} \left(\frac{3}{5} + \frac{2}{5}\cos\left(\frac{2\pi j}{p}\right)\right)^{2k} + (p-1)^2 \left(1 - \frac{c}{p}\right)^{2k}.$$

Using $\cos(x) = 1 + \frac{x^2}{2} + O(x^4)$, the sum is at most $c_1' e^{-c_2' \frac{k}{p^2}}$ for universal c_1', c_2'. The final term is exponentially smaller proving the upper bound. The lower bound follows from the usual second moment method. (See [8] Chapter 3 Theorem 2 for details.) Further details are omitted. □

Remark. In this example, the matrix $\hat{\mathcal{Q}}(\rho)$ is again the sum of a circulant and a diagonal matrix. Here, the circulant has eigenvalues $\frac{2}{5}\cos\left(\frac{2\pi j}{p-1}\right)$, $0 \le j \le p-2$ and the diagonal matrix has entries $\frac{3}{5} + \frac{2}{5}\cos\left(\frac{2\pi j}{p}\right)$, $1 \le j \le p-1$. The Weyl bounds show that the largest and smallest eigenvalues are bounded in absolute value by $1 - \frac{\theta}{p^2}$ for some fixed $\theta > 0$. Using this to bound the final term in the upper bound gives $(p-1)^2\left(1 - \frac{c}{p^2}\right)^{2k}$. This shows that the walk is close to random after order $p^2 \log(p)$ steps. In the Heisenberg examples the Weyl bounds give a bound of 1 which is useless.

The methods above can be applied to other walks on other groups. While we won't carry out the details here, we briefly describe two further examples and point to our companion paper [4] for more.

Example 2 (Borel Subgroup of $SL_2(\mathbb{F}_p)$). Let G be the 2×2 matrices of the form:

$$\begin{bmatrix} a & b \\ 0 & a^{-1} \end{bmatrix} \quad a \in \mathbb{F}_p^*, b \in \mathbb{F}_p \leftrightarrow (a, b).$$

A minimal generating set (with the identity) is

$$S = \{\mathrm{id}, (g, 0), (g^{-1}, 0), (1, 1), (1, -1)\}, \ g \text{ a generator of } \mathbb{F}_p^*.$$

The group has order $p(p-1)$ with $p-1$ 1-dimensional representations and 4 representations of dimension $(p-1)/2$. They are explicitly described in [6] p. 67. The Fourier analysis of the measure Q supported on S is almost the same as the analysis for the affine group. The results are that order p^2 steps are necessary and sufficient for convergence to the uniform distribution.

Example 3 ($M(p^3)$). There are two non-Abelian groups of order p^3: the Heisenberg group discussed above and $M(p^3)$. See [29], Chapter 4, Section 4. One description of the latter is:

$$M(p^3) = \{(a, b) : a \in \mathbb{Z}_p, b \in \mathbb{Z}_p^2\}, \ (a, b)(a', b') = (a + a', a * b' + b)$$

with $a * b = (1 + ap)b \pmod{p^2}$. This group has the same character table as $H_1(p)$. It thus has p^2 1-dimensional representations and $p-1$ representations of dimension p. A minimal generating set (for odd p the identity is not needed to take care of parity problems) is

$$S = \{(1, 0)(-1, 0)(0, 1), (0, -1)\}.$$

The Fourier transforms of the associated Q at the p-dimensional representations have the same form as the matrices in (1) with diagonal elements

$$2 \cos\left(\frac{2\pi c}{p^2}(1 + jp)\right), \ 0 \le j \le p - 1$$

where $1 \le c \le p-1$ is fixed (for the cth representation). We have not carried out the details, but, as shown in [10], it is known that order p^2 steps are necessary and sufficient for convergence.

7. Eigenvalues in the bulk

Consider the matrix $M_n(a)$ as in (1) with

$$\cos\left(\frac{2\pi j a}{m}\right), \ 0 \le j \le n - 1$$

as the diagonal elements. The sections above give bounds on the largest and smallest eigenvalues. It is natural to give bounds for the empirical measure of all the eigenvalues. This is straightforward, using a theorem of Kac–Murdock–Szegő from [21]. We use the elegant form of Trotter [30]. If $\lambda_1 \ge \lambda_2 \ge \cdots \ge \lambda_n$ are the eigenvalues of $M_n(a)$, let

$$\Lambda_n = \frac{1}{n}\sum_{i=1}^{n}\delta_{\lambda_i}$$

be the associated empirical measure. To describe the limit let

$$f_2(x) = \begin{cases} \frac{2}{\pi(1+|x|)} F_{2,1}\left(\frac{1}{2}, \frac{1}{2}; 1; \left(\frac{1-|x|}{1+|x|}\right)^2\right), & -1 \le x \le 1, \\ 0, & \text{else}, \end{cases} \tag{33}$$

where $F_{2,1}$ is the hypergeometric function. Let μ_2 be the associated measure. Distance between Λ_n and μ_2 is measured in the d_2 Wasserstein distance:

$$d_2^2(\Lambda_n, \mu_2) = \sup E|W - Z|^2 \text{ with } W \sim \Lambda_n,\ Z \sim \mu_2.$$

Theorem 6. *Let Λ_n be the empirical measure of the matrix $M_n(a)$ with $1 \le a \le n-1$. Let μ_2 be defined by (33). Then, with a fixed, as $n \to \infty$,*

$$d_2(\Lambda_n, \mu_2) \to 0.$$

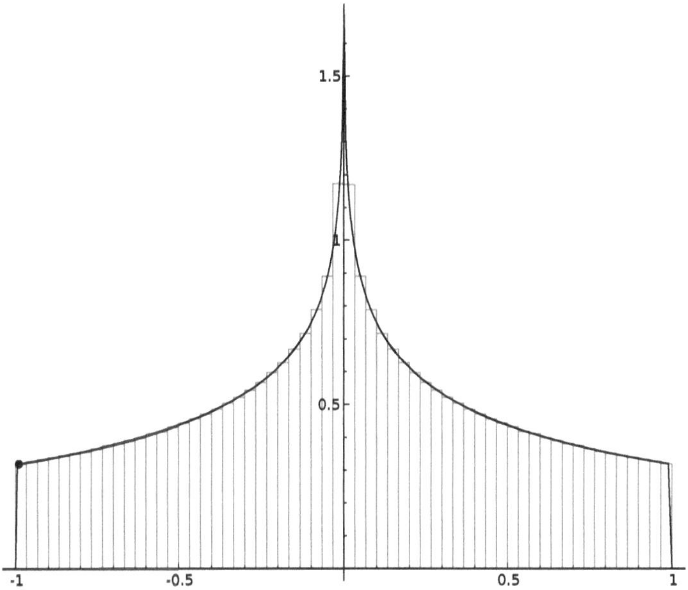

FIGURE 1. The curve shows the eigenvalues predicted by f_2, while the histogram gives the distribution of the actual eigenvalues of $M_{10,000}$ ($= M_{10,000}(1)$). Note that the curve has a very extreme, but finite slope around -1 and 1.
For example, although it is clear that $f_2(-1) = 0$, the small point on the left of the picture corresponds to $(-.99, f_2(-.99)) \approx .32$

See Figure 1 for an example.

Remark. We have not seen a way to use this kind of asymptotics to bound the rate of convergence of a random walk. Indeed our limit theorem shows that the distribution of the bulk does not depend on a while previous results show the extreme eigenvalues crucially depend on a.

Proof. Trotter's version of the Kac–Murdock–Szegő theorem applies to M_n. If

$$\sigma(x,y) = \cos(2\pi a x) + \cos(2\pi y)\ 0 \leq x, y \leq 1,$$

consider σ as a random variable on $[0,1]^2$, endowed with the Lebesgue measure. This has distribution $\cos(2\pi a U_1) + \cos(2\pi U_2)$ where U_1 and U_2 are independent uniform on $[0,1]$. An elementary calculation shows that $\cos(2\pi a U)$ has an arc-sine density $f(x)$ no matter what the integer a is.

$$f(x) = \begin{cases} \frac{1}{\pi\sqrt{1-x^2}}, & -1 \leq x \leq 1, \\ 0, & \text{else.} \end{cases} \tag{34}$$

Trotter shows that the empirical measure is close to μ_2, the distribution of σ. It follows that the empirical measure of the eigenvalues has limiting distribution the law of $(X+Y)/2$ where X and Y are independent with density $f(x)$. This convolution has density

$$f_2(x) = \begin{cases} \frac{2}{\pi^2}\int_{\max\{-1,2x-1\}}^{\min\{1,2x+1\}} \frac{1}{\sqrt{((2x-y)^2-1)(y^2-1)}}\,dy, & -1 \leq x \leq 1, \\ 0, & \text{else.} \end{cases} \tag{35}$$

The argument below shows that this integral is in fact

$$f_2(x) = \begin{cases} \frac{2}{\pi(1+|x|)} F_{2,1}\left(\frac{1}{2}, \frac{1}{2}; 1; \left(\frac{1-|x|}{1+|x|}\right)^2\right), & -1 \leq x \leq 1, \\ 0, & \text{else.} \end{cases} \tag{36}$$

The integral in (34) is in fact a well-known integral in a different guise. Let $0 \leqslant k \leqslant 1$. Define

$$K(k) = \int_0^1 \frac{dt}{\sqrt{(1-t^2)(1-k^2t^2)}}. \tag{37}$$

This is a complete elliptic integral and equals

$$\frac{\pi}{2} F_{2,1}\left(\frac{1}{2}, \frac{1}{2}; 1, k^2\right).$$

(See Section 22.301 of [32].) For ease of notation, we will evaluate

$$f_3(x) = \int_{\max(-1,x-1)}^{\min(1,x+1)} \frac{dt}{\sqrt{(1-t^2)(1-(x-t)^2)}}$$

for $|x| \leqslant 2$ Making the variable change $t \to t + \frac{x}{2}$ it becomes

$$\int_{\max(-\frac{x}{2}-1,\frac{x}{2}-1)}^{\min(-\frac{x}{2}+1,\frac{x}{2}+1)} \frac{dt}{\sqrt{h(t)}}$$

where

$$h(t) = (a^2 - t^2)(b^2 - t^2), \qquad a = 1 + \frac{x}{2}, \quad b = 1 - \frac{x}{2}.$$

This is an even function of x so it is enough to consider when $x \geqslant 0$. Then we need to evaluate

$$\int_{-b}^{b} \frac{dt}{\sqrt{h(t)}}.$$

Make the variable change $t \to bt$ and the integral becomes

$$\int_{-1}^{1} \frac{dt}{\sqrt{(1-t^2)(a^2-b^2t^2)}} = \frac{2}{a}K(k), \qquad k = \frac{b}{a}.$$

The factor of 2 comes from the fact that we are integrating an even function from -1 to 1, whereas in (37) the integral is from 0 to 1. Thus

$$f_3(x) = \frac{2\pi}{2+|x|} F_{2,1}\left(\frac{1}{2}, \frac{1}{2}; 1; \left(\frac{2-|x|}{2+|x|}\right)^2\right).$$

Sending $x \to 2x$ and multiplying by the appropriate constant, we have that the integral in (34) is in fact

$$\frac{2}{\pi(1+|x|)} F_{2,1}\left(\frac{1}{2}, \frac{1}{2}; 1; \left(\frac{1-|x|}{1+|x|}\right)^2\right). \qquad \square$$

Remark. [27] gives a similar expression for the sum of two general beta variables.

Acknowledgment. As this work progressed, we received useful comments from Florin Boca, Ben Bond, Bob Guralnick, Susan Holmes, Marty Isaacs, Evita Nestoridi, Jim Pitman, Laurent Saloff-Coste, and Thomas Strohmer. We offer thanks for the remarks of our very helpful reviewer.

References

[1] Cédric Béguin, Alain Valette, and Andrzej Zuk. On the spectrum of a random walk on the discrete Heisenberg group and the norm of Harper's operator. *J. Geom. Phys.*, 21(4):337–356, 1997.

[2] Rajendra Bhatia. Linear algebra to quantum cohomology: the story of Alfred Horn's inequalities. *Amer. Math. Monthly*, 108(4):289–318, 2001.

[3] Florin P. Boca and Alexandru Zaharescu. Norm estimates of almost Mathieu operators. *J. Funct. Anal.*, 220(1):76–96, 2005.

[4] D. Bump, P. Diaconis, A. Hicks, L. Miclo, and H. Widom. Characters and super characters for step two nilpotent groups with applications to random walks. to appear.

[5] D. Bump, P. Diaconis, A. Hicks, L. Miclo, and H. Widom. An Exercise (?) in Fourier Analysis on the Heisenberg Group. *ArXiv e-prints*, February 2015.

[6] Charles W. Curtis and Irving Reiner. *Representation theory of finite groups and associative algebras.* AMS Chelsea Publishing, Providence, RI, 2006. Reprint of the 1962 original.

[7] P. Diaconis and L. Saloff-Coste. Logarithmic Sobolev inequalities for finite Markov chains. *Ann. Appl. Probab.*, 6(3):695–750, 1996.

[8] Persi Diaconis. *Group representations in probability and statistics.* Institute of Mathematical Statistics Lecture Notes – Monograph Series, 11. Institute of Mathematical Statistics, Hayward, CA, 1988.

[9] Persi Diaconis. Patterned matrices. In *Matrix theory and applications* (*Phoenix, AZ,* 1989), volume 40 of *Proc. Sympos. Appl. Math.*, pages 37–58. Amer. Math. Soc., Providence, RI, 1990.

[10] Persi Diaconis. Threads through group theory. In *Character theory of finite groups*, volume 524 of *Contemp. Math.*, pages 33–47. Amer. Math. Soc., Providence, RI, 2010.

[11] Persi Diaconis, Sharad Goel, and Susan Holmes. Horseshoes in multidimensional scaling and local kernel methods. *Ann. Appl. Stat.*, 2(3):777–807, 2008.

[12] Persi Diaconis and Laurent Miclo. On times to quasi-stationarity for birth and death processes. *J. Theoret. Probab.*, 22(3):558–586, 2009.

[13] Persi Diaconis and Daniel Stroock. Geometric bounds for eigenvalues of Markov chains. *Ann. Appl. Probab.*, 1(1):36–61, 1991.

[14] Bradley W. Dickinson and Kenneth Steiglitz. Eigenvectors and functions of the discrete Fourier transform. *IEEE Trans. Acoust. Speech Signal Process.*, 30(1):25–31, 1982.

[15] David L. Donoho and Philip B. Stark. Uncertainty principles and signal recovery. *SIAM J. Appl. Math.*, 49(3):906–931, 1989.

[16] William Feller. *An introduction to probability theory and its applications. Vol. I.* Third edition. John Wiley & Sons, Inc., New York-London-Sydney, 1968.

[17] James Allen Fill. The passage time distribution for a birth-and-death chain: strong stationary duality gives a first stochastic proof. *J. Theoret. Probab.*, 22(3):543–557, 2009.

[18] William Fulton. Eigenvalues, invariant factors, highest weights, and Schubert calculus. *Bull. Amer. Math. Soc.* (*N.S.*), 37(3):209–249 (electronic), 2000.

[19] D.J. Griffiths. *Introduction to Quantum Mechanics.* Pearson international edition. Pearson Prentice Hall, 2005.

[20] Roger A. Horn and Charles R. Johnson. *Matrix analysis.* Cambridge University Press, Cambridge, second edition, 2013. p. 181.

[21] M. Kac, W.L. Murdock, and G. Szegő. On the eigenvalues of certain Hermitian forms. *J. Rational Mech. Anal.*, 2:767–800, 1953.

[22] Julian Keilson. Log-concavity and log-convexity in passage time densities of diffusion and birth-death processes. *J. Appl. Probability*, 8:391–398, 1971.

[23] John T. Kent. The appearance of a multivariate exponential distribution in sojourn times for birth-death and diffusion processes. In *Probability, statistics and analysis*, volume 79 of *London Math. Soc. Lecture Note Ser.*, pages 161–179. Cambridge Univ. Press, Cambridge, 1983.

[24] Gregory F. Lawler and Alan D. Sokal. Bounds on the L^2 spectrum for Markov chains and Markov processes: a generalization of Cheeger's inequality. *Trans. Amer. Math. Soc.*, 309(2):557–580, 1988.

[25] M.L. Mehta. Eigenvalues and eigenvectors of the finite Fourier transform. *J. Math. Phys.*, 28(4):781–785, 1987.

[26] Y. Peres and P. Sousi. Total variation cutoff in a tree. *ArXiv e-prints*, July 2013.

[27] T.G. Pham and N. Turkkan. Reliability of a standby system with beta-distributed component lives. *Reliability, IEEE Transactions on*, 43(1):71–75, Mar 1994.

[28] T. Strohmer and T. Wertz. Almost Eigenvalues and Eigenvectors of Almost Mathieu Operators. *ArXiv e-prints*, January 2015.

[29] Michio Suzuki. *Group theory. II*, volume 248 of *Grundlehren der Mathematischen Wissenschaften [Fundamental Principles of Mathematical Sciences]*. Springer-Verlag, New York, 1986. Translated from the Japanese.

[30] Hale F. Trotter. Eigenvalue distributions of large Hermitian matrices; Wigner's semi-circle law and a theorem of Kac, Murdock, and Szegő. *Adv. in Math.*, 54(1):67–82, 1984.

[31] Wang, Y.Y., Pannetier, B., and Rammal, R. Quasiclassical approximations for almost-Mathieu equations. *J. Phys. France*, 48(12):2067–2079, 1987.

[32] E.T. Whittaker and G.N. Watson. *A course of modern analysis*. Cambridge Mathematical Library. Cambridge University Press, Cambridge, 1996. An introduction to the general theory of infinite processes and of analytic functions; with an account of the principal transcendental functions, Reprint of the fourth (1927) edition.

Daniel Bump, Persi Diaconis
Department of Mathematics
Stanford University
450 Serra Mall, Bldg. 380
Stanford, CA 94305-2125, USA
e-mails:
`bump@math.stanford.edu`
`diaconis@math.stanford.edu`

Angela Hicks
Lehigh University
Mathematics Department
Christmas-Saucon Hall
14 E. Packer Ave
Bethlehem, PA 18015, USA
e-mail: `anh316@lehigh.edu`

Laurent Miclo
Institut de Mathématiques de Toulouse
Université Paul Sabatier
118 route de Narbonne
F-31062 Toulouse Cedex 9, France
e-mail:
`laurent.miclo@math.univ-toulouse.fr`

Harold Widom
UC Santa Cruz
Department of Mathematics
Santa Cruz, CA 95064, USA
e-mail: `widom@ucsc.edu`

Operator Theory:
Advances and Applications, Vol. 259, 267–300

Fast Inversion of Centrosymmetric Toeplitz-plus-Hankel Bezoutians

Torsten Ehrhardt and Karla Rost

Dedicated in admiration to Albrecht Böttcher on the occasion of his 60th birthday

Abstract. This paper establishes an algorithm for the computation of the inverse of a nonsingular, centrosymmetric Toeplitz-plus-Hankel Bezoutian B of order n. The algorithm has $O(n^2)$ computational complexity. In comparison with a previous paper on this topic the main key here is the reduction to the inversion of two symmetric Toeplitz Bezoutians of order n. This approach leads to a simpler algorithm, but it requires an additional assumption in one case. Furthermore, we obtain an explicit representation of B^{-1} as a sum of a Toeplitz and a Hankel matrix.

Mathematics Subject Classification (2010). Primary 15A09; Secondary 15B05, 65F05.

Keywords. Bezoutian matrix, Toeplitz matrix, Hankel matrix, Toeplitz-plus-Hankel matrix, matrix inversion.

1. Introduction

The present paper is devoted to the inversion of special types of structured matrices whose entries belong to a field $\mathbb{F}$ with characteristic not equal to 2. It can be considered as a continuation of papers dedicated to the inversion of Bezoutians. This series of papers started with [4], where Toeplitz Bezoutians and Hankel Bezoutians were considered. The inversion algorithms presented there relied on results about the nullspace of generalized resultant matrices [3]. The main key was that this nullspace can be described by means of solutions of Bézout equations. In [5] and [6] the inversion of Toeplitz-plus-Hankel Bezoutians was discussed for the first time.

A matrix $B = [b_{ij}]_{i,j=0}^{n-1}$ is called a *Toeplitz-plus-Hankel Bezoutian* (briefly, $T + H$*-Bezoutian*) if there exist eight polynomials $\mathbf{u}_i(t), \mathbf{v}_i(t)$ $(i = 1, 2, 3, 4)$ with coefficients in $\mathbb{F}$ and of degree at most $n+1$ such that, in polynomial language,

$$\sum_{i,j=0}^{n-1} b_{ij} t^i s^j = \frac{\sum_{i=1}^{4} \mathbf{u}_i(t)\mathbf{v}_i(s)}{(t-s)(1-ts)} . \tag{1.1}$$

In 1974 Lander [19] observed that inverses of Toeplitz or Hankel Bezoutians are Toeplitz or Hankel matrices, respectively. In 1988 Heinig and Rost [13] discovered that inverses of $T+H$-Bezoutians are *Toeplitz-plus-Hankel matrices*, i.e., matrices which are the sum of a Toeplitz matrix $[\, a_{i-j}\,]_{i,j=0}^{n-1}$ and a Hankel matrix $[\, s_{i+j}\,]_{i,j=0}^{n-1}$.

Let us mention that there is a vast literature dedicated to the inversion of Toeplitz, Hankel, and also Toeplitz-plus-Hankel matrices. The starting point were the papers [22], [7], [20], and [12]. On the other hand, the converse problem – the inversion of Bezoutians – has received little attention in the past (see [11], [9], [8]).

Our goal is to invert also $T+H$-Bezoutians. In the papers [5] and [6], the $T+H$-Bezoutians are assumed to be centrosymmetric or centroskewsymmetric, respectively. Recall that an $n \times n$ matrix A is called *centrosymmetric* if $J_n A J_n = A$ and *centroskewsymmetric* if $J_n A J_n = -A$, where J_n denotes the flip matrix of order n,

$$J_n := \begin{bmatrix} 0 & & 1 \\ & \cdot^{\cdot^{\cdot}} & \\ 1 & & 0 \end{bmatrix} . \tag{1.2}$$

In both cases a splitting of the $T+H$-Bezoutian into two special, highly structured matrices, called *split-Bezoutians*, (introduced in [14], see also [16]) is possible, and this enabled us to design fast $O(n^2)$ algorithms. Furthermore, these split-Bezoutians are related to Toeplitz Bezoutians and Hankel Bezoutians. Consequently, the approaches in [5] and [6] relied on the paper [4], which deals with the inversion of Toeplitz Bezoutians and Hankel Bezoutians.

In the present paper we consider centrosymmetric $T+H$-Bezoutians again. Whereas in [5] the main idea behind the algorithm was to reduce the problem to the inversion of two Hankel Bezoutians of about half order (via a certain transformation whose computational cost is $O(n^2)$), the algorithm designed here is based on relating both splitting parts of the $T+H$-Bezoutian B to Toeplitz Bezoutians more directly (without the use of a transformation). Both algorithms lead to a representation of the inverse B^{-1} as a Toeplitz-plus-Hankel matrix. Note that while the algorithm established here is simpler than that of [5], the proof is not much shorter.

The analogues of both kinds of algorithms have been already established in the centroskewsymmetric case in [6]. However, whereas the first kind of algorithm (reduction to Hankel Bezoutians of half the size via a transformation) is equally complicated in the centroskewsymmetric and in the centrosymmetric case, the second kind of algorithm (direct reduction to Toeplitz Bezoutians) is – surprisingly – much easier in the centroskewsymmetric case than in the centrosymmetric. One difference is that in the centrosymmetric case the resulting Toeplitz Bezoutians are not always nonsingular. As a consequence, in the centrosymmetric case with n odd we need an additional (somewhat unnatural) assumption. We conjecture that this condition is generally redundant, but we are able to remove it only in the case of fields $\mathbb{F}$ equal to $\mathbb{R}$ or $\mathbb{C}$.

The paper is organized as follows. After preliminaries in Section 2 we introduce the various kinds of structured matrices and discuss some basic facts in Section 3. At that point we are able to describe the main ideas of the inversion of centrosymmetric $T+H$-Bezoutians pursued in this paper in more detail. The results of the centroskewsymmetric case [6] will be recalled for sake of comparison in Section 5. Earlier, in Section 4 we deal with the splitting of centrosymmetric $T+H$-Bezoutians, whereas Section 6 is dedicated to the connections between centrosymmetric Toeplitz Bezoutians and split-Bezoutians.

In Section 7 first possibilities for the inversion of centrosymmetric $T+H$-Bezoutians via inversion of Toeplitz Bezoutians are deduced, which require additional assumptions. Then, in Section 8, we discuss how to eliminate the extra conditions. In Section 9 further simplifications are made and the corresponding fast $O(n^2)$ algorithms are designed.

We conclude the paper by presenting two examples in Section 10. In the first example the entries of the $n \times n$ nonsingular, centrosymmetric $T+H$-Bezoutian are in $\mathbb{F}=\mathbb{R}$ and n is odd. In the other example they are taken from the finite field $\mathbb{F}=\mathbb{Z}/(7\mathbb{Z})$ and n is even.

2. Preliminaries

Throughout this paper we consider vectors or matrices whose entries belong to a field $\mathbb{F}$ with a characteristic not equal to 2. By $\mathbb{F}^n$ we denote the linear space of all vectors of length n, by $\mathbb{F}^{m\times n}$ the linear space of all $m \times n$ matrices, and I_n denotes the identity matrix in $\mathbb{F}^{n\times n}$.

In what follows we often use polynomial language. We associate with a matrix $A = [\, a_{ij}\,]_{i,j=0}^{n-1} \in \mathbb{F}^{n\times n}$ the bivariate polynomial

$$A(t,s) := \sum_{i,j=0}^{n-1} a_{ij}\, t^i s^j \,, \tag{2.1}$$

and call it the *generating polynomial of* A.

Similarly, with a vector $\mathbf{x} = (x_j)_{j=0}^{m-1} \in \mathbb{F}^m$ we associate the polynomial

$$\mathbf{x}(t) := \sum_{j=0}^{m-1} x_j t^j \in \mathbb{F}^m[t]\,, \tag{2.2}$$

where $\mathbb{F}^m[t]$ denotes the linear space of all polynomials in t of degree less than m with coefficients in $\mathbb{F}$. Sometimes, it is more convenient to use a different way of assigning the indices, namely, $\mathbf{x} = (x_j)_{j=-n+1}^{n-1} \in \mathbb{F}^{2n-1}$. In this case, we define the corresponding polynomial by

$$\mathbf{x}(t) := t^{n-1} \sum_{j=-n+1}^{n-1} x_j t^j \in \mathbb{F}^{2n-1}[t]. \tag{2.3}$$

For a vector $\mathbf{x} \in \mathbb{F}^n$ we put

$$\mathbf{x}^J := J_n \mathbf{x}\,,$$

where J_n was introduced in (1.2). In polynomial language this means

$$\mathbf{x}^J(t) = \mathbf{x}(t^{-1})t^{n-1}\,.$$

A vector $\mathbf{x} \in \mathbb{F}^n$, or its corresponding polynomial $\mathbf{x}(t) \in \mathbb{F}^n[t]$, is said to be *symmetric* if $\mathbf{x} = \mathbf{x}^J$ and *skewsymmetric* if $\mathbf{x} = -\mathbf{x}^J$. The subspaces of $\mathbb{F}^n$ consisting of all symmetric or skewsymmetric vectors, respectively, are denoted by

$$\mathbb{F}^n_\pm := \left\{ \mathbf{x} \in \mathbb{F}^n \;:\; \mathbf{x}^J = \pm \mathbf{x} \right\}. \tag{2.4}$$

The corresponding subspaces of polynomials are abbreviated by $\mathbb{F}^n_\pm[t]$. The matrices

$$P_\pm := \tfrac{1}{2}(I_n \pm J_n) \tag{2.5}$$

are the projections from $\mathbb{F}^n$ onto $\mathbb{F}^n_\pm$. Note that here we use the assumption that the characteristic of $\mathbb{F}$ is not equal to 2.

Obviously, $\mathbf{x}_\pm(t) \in \mathbb{F}^n_\pm[t]$ has a zero at ∓ 1 in the case where n is even. In case n is odd we have $\mathbf{x}_-(\pm 1) = 0$. In other words,

$$\begin{aligned} \mathbb{F}^{2\ell}_\pm[t] &= \left\{ (t \pm 1)\,\mathbf{x}(t) \;:\; \mathbf{x}(t) \in \mathbb{F}^{2\ell-1}_+[t] \right\}, \\ \mathbb{F}^{2\ell+1}_-[t] &= \left\{ (t^2 - 1)\,\mathbf{x}(t) \;:\; \mathbf{x}(t) \in \mathbb{F}^{2\ell-1}_+[t] \right\}. \end{aligned} \tag{2.6}$$

Recall that a matrix A of order n is called *centrosymmetric* if $A = J_n A J_n$. It is easily seen that a matrix A is centrosymmetric if and only if

$$P_- A P_+ = P_+ A P_- = \mathbf{0}. \tag{2.7}$$

In particular, a centrosymmetric matrix A maps $\mathbb{F}^n_\pm$ to $\mathbb{F}^n_\pm$, i.e., $AP_\pm = P_\pm A P_\pm$. Hereafter we need the following simple fact.

Lemma 2.1. *Let $A = P_+ C_1 P_+ + P_- C_2 P_-$, where C_1 and C_2 are both centrosymmetric and nonsingular matrices. Then A is nonsingular and*

$$A^{-1} = P_+ C_1^{-1} P_+ + P_- C_2^{-1} P_-\,.$$

Proof. Using (2.7), we obtain

$$(P_+ C_1^{-1} P_+ + P_- C_2^{-1} P_-)(P_+ C_1 P_+ + P_- C_2 P_-) = P_+ + P_- = I_n\,,$$

which completes the proof. □

3. Matrices under considerations

3.1. Toeplitz-plus-Hankel matrices

First of all let us recall the definition of Toeplitz and Hankel matrices. An $n \times n$ *Toeplitz matrix* generated by the vector $\mathbf{a} = (a_i)_{i=-n+1}^{n-1} \in \mathbb{F}^{2n-1}$ is the matrix

$$T_n(\mathbf{a}) = [\, a_{i-j}\,]_{i,j=0}^{n-1} .$$

We will use (2.3) in order to assign its (polynomial) symbol. An $n \times n$ *Hankel matrix* generated by $\mathbf{s} = (s_i)_{i=0}^{2n-2} \in \mathbb{F}^{2n-1}$ is the matrix

$$H_n(\mathbf{s}) = [\, s_{i+j}\,]_{i,j=0}^{n-1} ,$$

where (2.2) with $m = 2n-1$ is used to denote its symbol. Clearly, $H_n(\mathbf{s})$ is symmetric.

For a Toeplitz matrix we have $T_n(\mathbf{a})^T = J_n T_n(\mathbf{a}) J_n$ which means that $T_n(\mathbf{a})$ is *persymmetric*. Since, moreover,

$$T_n(\mathbf{a})^T = T_n(\mathbf{a}^J),$$

a Toeplitz matrix is symmetric if and only if it is centrosymmetric, or, equivalently, if its symbol is a symmetric vector.

Recall that *Toeplitz-plus-Hankel matrices* (shortly, *$T+H$ matrices*) are matrices which are a sum of a Toeplitz and a Hankel matrix. Since $T_n(\mathbf{b})J_n$ is a Hankel matrix it is possible to represent any $T+H$ matrix by means of two Toeplitz matrices,

$$R_n = T_n(\mathbf{a}) + T_n(\mathbf{b})J_n \qquad (\mathbf{a}, \mathbf{b} \in \mathbb{F}^{2n-1}). \tag{3.1}$$

Related to this representation is another one, using the projections (2.5) and the symbols $\mathbf{c} = \mathbf{a} + \mathbf{b}$ and $\mathbf{d} = \mathbf{a} - \mathbf{b}$,

$$R_n = T_n(\mathbf{c})P_+ + T_n(\mathbf{d})P_- . \tag{3.2}$$

The following result describes the symbols of centrosymmetric $T+H$ matrices (see [16]).

Proposition 3.1. *The $T+H$ matrix R_n is centrosymmetric if and only if the symbols $\mathbf{a}, \mathbf{b}$ and $\mathbf{c}, \mathbf{d}$ of the Toeplitz matrices in (3.1) and (3.2), resp., are symmetric vectors.*

Moreover, $T_n(\mathbf{c})P_+ + T_n(\mathbf{d})P_- = \mathbf{0}$ if and only if

$$\mathbf{c} = \begin{cases} 0 & \text{if } n \text{ is odd,} \\ \mathbf{e}_{\alpha,-\alpha} & \text{if } n \text{ is even,} \end{cases} \qquad \mathbf{d} = \begin{cases} \mathbf{e}_{\alpha,\beta} & \text{if } n \text{ is odd,} \\ \mathbf{e}_{\alpha,\alpha} & \text{if } n \text{ is even,} \end{cases}$$

where $\mathbf{e}_{\alpha,\beta} = (\alpha, \beta, \alpha, \ldots, \beta, \alpha) \in \mathbb{F}_+^{2n-1}$.

Proof. Let R_n be given by (3.1). Using centrosymmetry, it follows that

$$T_n(\mathbf{a} - \mathbf{a}^J) + T_n(\mathbf{b} - \mathbf{b}^J)J_n = \mathbf{0}.$$

Passing to the transpose we get

$$-T_n(\mathbf{a} - \mathbf{a}^J) + T_n(\mathbf{b} - \mathbf{b}^J)J_n = \mathbf{0},$$

from which $\mathbf{a} = \mathbf{a}^J$ and $\mathbf{b} = \mathbf{b}^J$ follows. This also implies $\mathbf{c} = \mathbf{c}^J$ and $\mathbf{d} = \mathbf{d}^J$. The other direction of the proof can be verified easily.

As for the uniqueness issue, note that $T_n(\mathbf{c})P_+ + T_n(\mathbf{d})P_- = \mathbf{0}$ implies $T_n(\mathbf{c})P_+ = T_n(\mathbf{d})P_- = \mathbf{0}$. Hence $T_n(\mathbf{c}) = -T_n(\mathbf{c})J_n$ and $T_n(\mathbf{d}) = T_n(\mathbf{d})J_n$, which can be easily analyzed by observing that the only matrices which are both Toeplitz and Hankel matrices are of the form $T_n(\mathbf{e}_{\alpha,\beta})$. □

As a consequence, a centrosymmetric $T+H$ matrix is also symmetric. A centrosymmetric $T+H$ matrix (3.2) can also be written as

$$R_n = P_+T_n(\mathbf{c})P_+ + P_-T_n(\mathbf{d})P_- \,. \tag{3.3}$$

This follows from the previous proposition and (2.7) since the underlying Toeplitz matrices are symmetric and centrosymmetric. Moreover, from Lemma 2.1 the following is clear.

Proposition 3.2. *The centrosymmetric $T+H$ matrix $R_n = T_n(\mathbf{c})P_+ + T_n(\mathbf{d})P_-$ is nonsingular if $T_n(\mathbf{c})$ and $T_n(\mathbf{d})$ are both nonsingular. In this case*

$$R_n^{-1} = T_n(\mathbf{c})^{-1}P_+ + T_n(\mathbf{d})^{-1}P_- . \tag{3.4}$$

According to [16], both matrices $T_n(\mathbf{c})$ and $T_n(\mathbf{d})$ are nonsingular if and only if both R_n and

$$R_n^- := T_n(\mathbf{c})P_- + T_n(\mathbf{d})P_+$$

are nonsingular. This is easy to see from the equality (compare [6])

$$\begin{bmatrix} R_n & \mathbf{0} \\ \mathbf{0} & R_n^- \end{bmatrix} = \begin{bmatrix} P_+ & P_- \\ P_- & P_+ \end{bmatrix} \begin{bmatrix} T_n(\mathbf{c}) & \mathbf{0} \\ \mathbf{0} & T_n(\mathbf{d}) \end{bmatrix} \begin{bmatrix} P_+ & P_- \\ P_- & P_+ \end{bmatrix} . \tag{3.5}$$

As the following example shows, it can happen that R_n is nonsingular while R_n^- is singular (or, equivalently, at least one of $T_n(\mathbf{c})$ or $T_n(\mathbf{d})$ is singular). This indicates some limitations of computing the inverse of R_n by using inverses of Toeplitz matrices.

Example 3.3. Consider $n = 3$ and $\mathbf{c} = [1,0,1,0,1]^T$, $\mathbf{d} = [0,0,1,0,0]^T$, i.e.,

$$R_3 = \begin{bmatrix} 1 & 0 & 1 \\ 0 & 1 & 0 \\ 1 & 0 & 1 \end{bmatrix} + \frac{1}{2}\begin{bmatrix} 1 & 0 & -1 \\ 0 & 0 & 0 \\ -1 & 0 & 1 \end{bmatrix} = \begin{bmatrix} \frac{3}{2} & 0 & \frac{1}{2} \\ 0 & 1 & 0 \\ \frac{1}{2} & 0 & \frac{3}{2} \end{bmatrix}$$

and

$$R_3^- = \begin{bmatrix} 0 & 0 & 0 \\ 0 & 0 & 0 \\ 0 & 0 & 0 \end{bmatrix} + \begin{bmatrix} \frac{1}{2} & 0 & \frac{1}{2} \\ 0 & 1 & 0 \\ \frac{1}{2} & 0 & \frac{1}{2} \end{bmatrix} .$$

Here R_3 and $T_3(\mathbf{d}) = I_3$ are nonsingular, while R_3^- and $T_3(\mathbf{c}) = T_3(\mathbf{c})P_+$ are not. Furthermore, for R_3 given as above it is impossible to find a (different) representation $R_3 = T_3(\mathbf{c})P_+ + T_3(\mathbf{d})P_-$ such that $T_3(\mathbf{c})$ is nonsingular. Indeed, by Proposition 3.1, the vector $\mathbf{c}$ is unique. □

The inverses $T_n(\mathbf{c})^{-1}$ and $T_n(\mathbf{d})^{-1}$ of the Toeplitz matrices in (3.3) are so-called Toeplitz Bezoutians, which together with their Hankel counterparts are introduced now.

3.2. Toeplitz and Hankel Bezoutians

A matrix $B_T \in \mathbb{F}^{n\times n}$ is called a *Toeplitz Bezoutian* (shortly, *T-Bezoutian*) if there exist vectors $\mathbf{u}, \mathbf{v} \in \mathbb{F}^{n+1}$ such that, in polynomial language,

$$B_T(t,s) = \frac{\mathbf{u}(t)\mathbf{v}^J(s) - \mathbf{v}(t)\mathbf{u}^J(s)}{1 - ts}.$$

In this case we write $B_T = \mathrm{Bez}_T(\mathbf{u}, \mathbf{v})$. Notice that $\mathrm{Bez}_T(\mathbf{u}, \mathbf{v}) = \mathrm{Bez}_T(\hat{\mathbf{u}}, \hat{\mathbf{v}})$ if and only if there is $\Phi \in \mathbb{F}^{2\times 2}$ with $\det \Phi = 1$ such that

$$[\,\mathbf{u}, \mathbf{v}\,] = [\,\hat{\mathbf{u}}, \hat{\mathbf{v}}\,]\,\Phi\,.$$

Analogously, a matrix $B_H \in \mathbb{F}^{n\times n}$ is called an *Hankel Bezoutian* if there exist vectors $\mathbf{u}, \mathbf{v} \in \mathbb{F}^{n+1}$ such that

$$B_H(t,s) = \frac{\mathbf{u}(t)\mathbf{v}(s) - \mathbf{v}(t)\mathbf{u}(s)}{t - s}.$$

One can define these Bezoutians also by means of suitable displacement transformations as introduced in [17] and [10] (see also [11]). Hankel Bezoutians are always symmetric, while T-Bezoutians are always persymmetric. The two kinds of Bezoutians are related to each other by $\mathrm{Bez}_H(\mathbf{u}, \mathbf{v}) = -\mathrm{Bez}_T(\mathbf{u}, \mathbf{v})J_n$. Therefore, the results stated below for T-Bezoutians hold correspondingly also for Hankel Bezoutians.

It is well known (see [11]) that a T-Bezoutian $\mathrm{Bez}_T(\mathbf{u}, \mathbf{v})$ with $\mathbf{u}, \mathbf{v} \in \mathbb{F}^{n+1}$ is nonsingular if and only if $\mathbf{u}(t)$ and $\mathbf{v}(t)$ are *generalized coprime*, which means that the polynomials $\mathbf{u}(t)$ and $\mathbf{v}(t)$ are coprime in the usual sense and $\deg \mathbf{u}(t) = n$ or $\deg \mathbf{v}(t) = n$. Notice that if $\mathbf{u}(t)$ is symmetric or skewsymmetric, then $\deg \mathbf{u}(t) < n$ is equivalent to $\mathbf{u}(0) = 0$. Thus, two symmetric or skewsymmetric polynomials are coprime if and only if they are generalized coprime.

The following connection between Toeplitz matrices and T-Bezoutians is a classical result discovered by Lander in 1974 [19].

Theorem 3.4. *A nonsingular matrix is a T-Bezoutian if and only if its inverse is a Toeplitz matrix.*

The following problem arises naturally from this result: Given the T-Bezoutian $B_T = \mathrm{Bez}_T(\mathbf{u}, \mathbf{v})$ with generalized coprime polynomials $\mathbf{u}(t)$ and $\mathbf{v}(t)$, how can one compute the symbol $\mathbf{c}$ of its inverse, a Toeplitz matrix $T_n(\mathbf{c}) = B_T^{-1}$? The solution was given in [4].

Theorem 3.5. *Assume $\mathbf{u}(t), \mathbf{v}(t) \in \mathbb{F}^{n+1}[t]$ are generalized coprime polynomials, and let $B_T = \mathrm{Bez}_T(\mathbf{u}, \mathbf{v})$. Then B_T is nonsingular, the Bézout equations*

$$\mathbf{u}(t)\boldsymbol{\alpha}(t) + \mathbf{v}(t)\boldsymbol{\beta}(t) = 1\,, \tag{3.6}$$

$$\mathbf{u}^J(t)\boldsymbol{\gamma}^J(t) + \mathbf{v}^J(t)\boldsymbol{\delta}^J(t) = 1 \tag{3.7}$$

have unique solutions $\boldsymbol{\alpha}(t), \boldsymbol{\beta}(t), \boldsymbol{\gamma}(t), \boldsymbol{\delta}(t) \in \mathbb{F}^n[t]$, *and* $\mathbf{c} = (c_i)_{i=-n+1}^{n-1} \in \mathbb{F}^{2n-1}$ *given by*

$$\mathbf{c}(t) = t^{n-1} \sum_{i=-n+1}^{n-1} c_i t^i = \boldsymbol{\alpha}(t)\boldsymbol{\delta}(t) - \boldsymbol{\beta}(t)\boldsymbol{\gamma}(t)$$

is the symbol of the inverse of B_T, *i.e.,*

$$B_T^{-1} = T_n(\mathbf{c}) = [\, c_{i-j} \,]_{i,j=0}^{n-1} \,.$$

For our purposes it is important to specialize the previous result to the case of centrosymmetric T-Bezoutians. As shown in [15], Section 5, if the T-Bezoutian $\mathrm{Bez}_T(\mathbf{u}, \mathbf{v})$ is nonsingular and centrosymmetric, then $\mathbf{u}$ can be chosen as a symmetric vector, $\mathbf{u} \in \mathbb{F}_+^{n+1}$, and $\mathbf{v}$ as a skewsymmetric vector, $\mathbf{v} \in \mathbb{F}_-^{n+1}$. Thus we have $\boldsymbol{\alpha} = \boldsymbol{\gamma}^J$ and $\boldsymbol{\beta} = -\boldsymbol{\delta}^J$ for the (unique) solutions of (3.6) and (3.7). This implies

$$\mathbf{c}(t) = t^{n-1} \sum_{i=-n+1}^{n-1} c_i t^i = -\boldsymbol{\alpha}(t)\boldsymbol{\beta}^J(t) - \boldsymbol{\beta}(t)\boldsymbol{\alpha}^J(t). \tag{3.8}$$

Remark 3.6. Observe that $\mathbf{c} = (c_i)_{i=-n+1}^{n-1} \in \mathbb{F}_+^{2n-1}$ is a symmetric vector as expected. Therefore it suffices to compute only the last n components $(c_i)_{i=0}^{n-1}$. To that aim introduce for given $\mathbf{x} = (x_i)_{i=0}^{n-1}$ the following upper triangular Toeplitz matrix of order n,

$$U_n(\mathbf{x}) = \begin{bmatrix} x_0 & x_1 & \cdots & \cdots & x_{n-1} \\ & x_0 & x_1 & & \vdots \\ & & \ddots & \ddots & \vdots \\ & & & x_0 & x_1 \\ 0 & & & & x_0 \end{bmatrix}. \tag{3.9}$$

Now, it is easily seen and has already been stated in Section 6 of [4], that equation (3.8) becomes

$$(c_i)_{i=0}^{n-1} = -U_n(\boldsymbol{\beta})\boldsymbol{\alpha} - U_n(\boldsymbol{\alpha})\boldsymbol{\beta}\,. \tag{3.10}$$

3.3. $T + H$-Bezoutians

In order to introduce Toeplitz-plus-Hankel Bezoutians ($T + H$-Bezoutians) let us consider the following displacement transformation

$$\nabla_{T+H} : \mathbb{F}^{n\times n} \to \mathbb{F}^{(n+2)\times(n+2)}$$

defined in polynomial language by

$$(\nabla_{T+H}(B))\,(t,s) = (t-s)(1-ts)B(t,s)\,.$$

A matrix $B \in \mathbb{F}^{n\times n}$ is called a $T + H$*-Bezoutian* if

$$\operatorname{rank} \nabla_{T+H}(B) \le 4\,.$$

This condition is equivalent to the existence of eight vectors $\mathbf{u}_i, \mathbf{v}_i$ $(i = 1, 2, 3, 4)$ in $\mathbb{F}^{n+2}$ such that

$$(t-s)(1-ts)B(t,s) = \sum_{i=1}^{4} \mathbf{u}_i(t)\mathbf{v}_i(s)$$

(see also (1.1)). For the $T+H$ case we know from [13] the following important fact.

Theorem 3.7. *A nonsingular matrix is a $T+H$-Bezoutian if and only if its inverse is a $T+H$ matrix.*

Let us now briefly explain the main ideas of this paper for inverting a nonsingular, centrosymmetric $T+H$-Bezoutian B. Assume we are given such a B. Then its inverse B^{-1} is a centrosymmetric $T+H$ matrix, which by Proposition 3.1 can be written as

$$B^{-1} = T_n(\mathbf{c})P_+ + T_n(\mathbf{d})P_- \,.$$

Our goal is the computation of $\mathbf{c}$ and $\mathbf{d}$. However, they are not always unique. In view of Proposition 3.2, let us make the additional assumption that both Toeplitz matrices are nonsingular. Then

$$B = T_n^{-1}(\mathbf{c})P_+ + T_n^{-1}(\mathbf{d})P_- \,,$$

where the Toeplitz inverses are T-Bezoutians. In fact, it turns out that

$$T_n^{-1}(\mathbf{c})P_+ = \mathrm{Bez}_T(\mathbf{u}_1, \mathbf{v}_1)P_+ \quad \text{and} \quad T_n^{-1}(\mathbf{d})P_- = \mathrm{Bez}_T(\mathbf{u}_2, \mathbf{v}_2)P_-$$

are split-Bezoutians. These split-Bezoutions are easily obtained from B. Note that the notion of split-Bezoutians is introduced in Section 4 below.

Thus the idea is to determine first the T-Bezoutians from the split-Bezoutians of B. Then $\mathbf{c}$ and $\mathbf{d}$ are computed by inverting the T-Bezoutians. This program is carried out in Sections 4, 6, and 7. In Section 5 we will discuss without proofs the centroskewsymmetric case (see [6]) for sake of comparison and because it is much simpler.

In Sections 8 and 9 simplifications to the formulas obtained in Section 7 are made. Basically we will eliminate $\mathbf{u}_1, \mathbf{v}_1, \mathbf{u}_2, \mathbf{v}_2$ and compute $\mathbf{c}$ and $\mathbf{d}$ directly from the parameters occurring in the split-Bezoutians.

Above we made the assumption that both $T_n(\mathbf{c})$ and $T_n(\mathbf{d})$ are nonsingular. In what follows we will encounter this assumption not directly, but as an assumption on the parameters occurring in the split-Bezoutians (see Lemma 6.3). The simplifications that will be made in Section 8 allow us to drop the extra assumption in the case of n even and considerably weaken it in case of n odd.

4. Splitting of centrosymmetric $T + H$-Bezoutians

Nonsingular, centrosymmetric $T+H$-Bezoutians have a nice property, namely, they admit a certain splitting which was discovered in Section 8 of [14] (see also [16]).

Theorem 4.1. *A nonsingular, centrosymmetric $T + H$-Bezoutian B of order n allows the following (unique) splitting*

$$B = B_{++} + B_{--}\,, \tag{4.1}$$

where $B_{++} = P_+ B$ and $B_{--} = P_- B$ are special centrosymmetric $T+H$-Bezoutians,

$$B_{\pm\pm}(t,s) = \frac{\mathbf{f}_\pm(t)\mathbf{g}_\pm(s) - \mathbf{g}_\pm(t)\mathbf{f}_\pm(s)}{(t-s)(1-ts)}\,, \tag{4.2}$$

with $\mathbf{f}_\pm, \mathbf{g}_\pm \in \mathbb{F}_\pm^{n+2}$ satisfying

$$\gcd(\mathbf{f}_+(t), \mathbf{g}_+(t)) = \begin{cases} 1 & \text{if } \; n \text{ odd,} \\ t+1 & \text{if } \; n \text{ even,} \end{cases} \tag{4.3}$$

$$\gcd(\mathbf{f}_-(t), \mathbf{g}_-(t)) = \begin{cases} t^2-1 & \text{if } \; n \text{ odd,} \\ t-1 & \text{if } \; n \text{ even.} \end{cases} \tag{4.4}$$

Note that in view of (2.6) the conditions (4.3) and (4.4) mean that the greatest common divisors (gcd's) are minimal.

We call B_{++} and B_{--} the *split-Bezoutian of* (+) *and of* (−)*type*, resp., and write

$$B_{\pm\pm} = \mathrm{Bez}_{\mathrm{sp}}(\mathbf{f}_\pm, \mathbf{g}_\pm). \tag{4.5}$$

The matrices B_{++} and B_{--} are not only centrosymmetric, but have the following additional symmetries,

$$B_{\pm\pm}P_\pm = P_\pm B_{\pm\pm} = \pm B_{\pm\pm}\,. \tag{4.6}$$

This means that all the rows and columns of B_{++} are symmetric vectors, whereas all the rows and columns of B_{--} are skewsymmetric vectors. Additionally, B_{++} and B_{--} are symmetric matrices.

Remark 4.2. A natural problem is the following: Given a centrosymmetric matrix B of order n, how can one decide whether B is a nonsingular $T + H$-Bezoutian? Furthermore, if this is the case, how can one determine the vectors $\mathbf{f}_+, \mathbf{g}_+, \mathbf{f}_-, \mathbf{g}_-$ occurring in (4.2)? A complete answer was given in [5], Remark 4.10.

Remark 4.3. Different pairs of linearly independent symmetric vectors $\{\mathbf{f}_+, \mathbf{g}_+\}$ and $\{\hat{\mathbf{f}}_+, \hat{\mathbf{g}}_+\}$ produce the same split-Bezoutian of (+)type, i.e.,

$$\mathrm{Bez}_{\mathrm{sp}}(\mathbf{f}_+, \mathbf{g}_+) = \mathrm{Bez}_{\mathrm{sp}}(\hat{\mathbf{f}}_+, \hat{\mathbf{g}}_+)\,,$$

if and only if there is $\Phi \in \mathbb{F}^{2\times 2}$ with $\det \Phi = 1$ such that

$$[\,\mathbf{f}_+, \mathbf{g}_+\,] = [\,\hat{\mathbf{f}}_+, \hat{\mathbf{g}}_+\,]\,\Phi\,.$$

The same, of course, is true for split-Bezoutians of (−)type.

For our purposes here we need a deeper inside into the structure of the two splitting parts. To that aim let us introduce the $n \times (n-1)$ matrices $M_{n-1}^{t\pm 1}$ and the $n \times (n-2)$ matrix $M_{n-2}^{t^2-1}$,

$$M_{n-1}^{t\pm 1} = \begin{bmatrix} \pm 1 & & & 0 \\ 1 & \pm 1 & & \\ & 1 & \ddots & \\ & & \ddots & \pm 1 \\ 0 & & & 1 \end{bmatrix}, \qquad M_{n-2}^{t^2-1} = \begin{bmatrix} -1 & & & 0 \\ 0 & -1 & & \\ 1 & 0 & \ddots & \\ & 1 & \ddots & -1 \\ & & \ddots & 0 \\ 0 & & & 1 \end{bmatrix}. \tag{4.7}$$

These matrices are matrices of multiplication operators (in the corresponding polynomial spaces with respect to the canonical bases) with the polynomials being indicated in their superscripts.

There are two cases to consider, n even and n odd. The following theorems were established in [16].

Theorem 4.4. *Let n be even. Then $B \in \mathbb{F}^{n\times n}$ is a nonsingular, centrosymmetric $T+H$-Bezoutian if and only if it can be represented in the form*

$$B = M_{n-1}^{t+1}\mathrm{Bez}_{\mathrm{sp}}(\mathbf{f}_1, \mathbf{g}_1)\,(M_{n-1}^{t+1})^T + M_{n-1}^{t-1}\mathrm{Bez}_{\mathrm{sp}}(\mathbf{f}_2, \mathbf{g}_2)\,(M_{n-1}^{t-1})^T, \tag{4.8}$$

where $\{\mathbf{f}_i(t), \mathbf{g}_i(t)\}$ are two pairs of coprime polynomials in $\mathbb{F}_+^{n+1}[t]$, $i = 1, 2$.

Theorem 4.5. *Let n be odd. Then $B \in \mathbb{F}^{n\times n}$ is a nonsingular, centrosymmetric $T+H$-Bezoutian if and only if it can be represented in the form*

$$B = \mathrm{Bez}_{\mathrm{sp}}(\mathbf{f}_1, \mathbf{g}_1) + M_{n-2}^{t^2-1}\mathrm{Bez}_{\mathrm{sp}}(\mathbf{f}_2, \mathbf{g}_2)(M_{n-2}^{t^2-1})^T, \tag{4.9}$$

where $\{\mathbf{f}_i(t), \mathbf{g}_i(t)\}$ are two pairs of coprime polynomials in $\mathbb{F}_+^{n+2}[t]$ for $i = 1$ and in $\mathbb{F}_+^n[t]$ for $i = 2$.

Remark 4.6. Both theorems are consequences of Theorem 4.1. The first term of the sum in (4.8) or (4.9) is equal to B_{++}, the second one equals B_{--}. In particular, taking (4.3) and (4.4) into account the relationship of (4.8) and (4.9) to (4.1) and (4.2) is given by

$$\begin{aligned} \mathbf{f}_+(t) &= (t+1)\mathbf{f}_1(t), & \mathbf{f}_-(t) &= (t-1)\mathbf{f}_2(t), \\ \mathbf{g}_+(t) &= (t+1)\mathbf{g}_1(t), & \mathbf{g}_-(t) &= (t-1)\mathbf{g}_2(t), \end{aligned} \tag{4.10}$$

for (4.8), n even, and by

$$\begin{aligned} \mathbf{f}_+(t) &= \mathbf{f}_1(t), & \mathbf{f}_-(t) &= (t^2-1)\mathbf{f}_2(t), \\ \mathbf{g}_+(t) &= \mathbf{g}_1(t), & \mathbf{g}_-(t) &= (t^2-1)\mathbf{g}_2(t), \end{aligned} \tag{4.11}$$

for (4.9), n odd.

Remark 4.7. Let B be a nonsingular, centrosymmetric $T+H$-Bezoutian with splitting (4.1), i.e., $B = B_{++} + B_{--}$, $B_{\pm\pm} = BP_{\pm}$. From Theorem 3.7 we know that B^{-1} is a $T+H$ matrix, which by Proposition 3.1 can be represented in the form

$$B^{-1} = T_n(\mathbf{c})P_+ + T_n(\mathbf{d})P_- \,.$$

Using the symmetries of the terms in both splittings, it follows that

$$T_n(\mathbf{c})P_+B_{++} = B_{++}T_n(\mathbf{c})P_+ = P_+$$

and

$$T_n(\mathbf{d})P_-B_{--} = B_{--}T_n(\mathbf{d})P_- = P_-$$

(see also Lemma 2.1). In other word, $T_n(\mathbf{c})P_+$ and $T_n(\mathbf{d})P_-$ are the inverses of the split-Bezoutians B_{++} and B_{--} restricted to $\mathbb{F}^n_+$ and $\mathbb{F}^n_-$, respectively.

5. Intermezzo: Inversion of centroskewsymmetric $T + H$-Bezoutians

For sake of comparison, let us present the result obtained for centroskewsymmetric $T+H$-Bezoutians B, i.e., where $B = -J_nBJ_n$. We omit the proofs and most details as they can be found in our paper [6].

Surprisingly, the centroskewsymmetric case is much simpler than the centrosymmetric case. Note that, obviously, the order n of a nonsingular, centroskewsymmetric $T+H$-Bezoutian is always even. The following theorem is similar to Theorem 4.4 and was first established in [16].

Theorem 5.1. *Let n be even. Then $B \in \mathbb{F}^{n\times n}$ is a nonsingular, centroskewsymmetric $T+H$-Bezoutian if and only if it can be represented in the form*

$$B = M^{t+1}_{n-1}\mathrm{Bez}_{\mathrm{sp}}(\mathbf{f}_1,\mathbf{g}_1)(M^{t-1}_{n-1})^T + M^{t-1}_{n-1}\mathrm{Bez}_{\mathrm{sp}}(\mathbf{f}_2,\mathbf{g}_2)(M^{t+1}_{n-1})^T \,, \tag{5.1}$$

where $\{\mathbf{f}_i(t),\mathbf{g}_i(t)\}$ are two pairs of coprime polynomials in $\mathbb{F}^{n+1}_+[t]$, $i=1,2$.

Taking into account that

$$\frac{1}{1-ts} \pm \frac{1}{t-s} = \pm\frac{(1\pm t)(1\mp s)}{(1-ts)(t-s)}$$

we obtain

$$-2\,\mathrm{Bez}_T(\mathbf{f}_1,\mathbf{g}_1)P_- = M^{t+1}_{n-1}\mathrm{Bez}_{\mathrm{sp}}(\mathbf{f}_1,\mathbf{g}_1)(M^{t-1}_{n-1})^T$$

as well as

$$2\,\mathrm{Bez}_T(\mathbf{f}_2,\mathbf{g}_2)P_+ = M^{t-1}_{n-1}\mathrm{Bez}_{\mathrm{sp}}(\mathbf{f}_2,\mathbf{g}_2)(M^{t+1}_{n-1})^T \,.$$

These identities show that the vectors occurring in the split-Bezoutians and in the corresponding T-Bezoutians are the same.

Theorem 5.2. *Let n be even and $B \in \mathbb{F}^{n\times n}$ be given by (5.1) with $\mathbf{f}_i,\mathbf{g}_i \in \mathbb{F}^{n+1}_+$. Then*

$$B = -2\,\mathrm{Bez}_T(\mathbf{f}_1,\mathbf{g}_1)P_- + 2\,\mathrm{Bez}_T(\mathbf{f}_2,\mathbf{g}_2)P_+ \,. \tag{5.2}$$

It follows from the definition of T-Bezoutians that for the symmetric vectors $\mathbf{f}_i, \mathbf{g}_i$, the matrices

$$B_1 := \mathrm{Bez}_T(\mathbf{f}_1, \mathbf{g}_1) \quad \text{and} \quad B_2 := \mathrm{Bez}_T(\mathbf{f}_2, \mathbf{g}_2)$$

are centroskewsymmetric and skewsymmetric T-Bezoutians, i.e., $B_i^T = J_n B_i J_n = -B_i$. Moreover, using the coprimeness conditions for the nonsingularity of the T- and $T+H$-Bezoutians, it follows that $B = -2\,B_1 P_- + 2\,B_2 P_+$ is nonsingular if and only if both B_1 and B_2 are nonsingular. This contrasts the centrosymmetric case and is one of the reasons why the centroskewsymmetric case is easier.

Theorem 5.3. *Let $B \in \mathbb{F}^{n\times n}$ be a centroskewsymmetric $T+H$-Bezoutian given by (5.1) or (5.2) with pairs $\{\mathbf{f}_i(t), \mathbf{g}_i(t)\}$ of symmetric coprime polynomials in $\mathbb{F}_+^{n+1}[t]$. Then n is even, B is nonsingular, and*

$$B^{-1} = \frac{1}{2}(T_n(\mathbf{c}_1)P_+ + T_n(\mathbf{c}_2)P_-), \tag{5.3}$$

where $\mathbf{c}_1, \mathbf{c}_2 \in \mathbb{F}_-^{2n-1}$ are given by

$$\mathbf{c}_i(t) = (-1)^i \Big(\boldsymbol{\alpha}_i(t)\boldsymbol{\beta}_i^J(t) - \boldsymbol{\beta}_i(t)\boldsymbol{\alpha}_i^J(t)\Big) \tag{5.4}$$

and $\boldsymbol{\alpha}_i, \boldsymbol{\beta}_i \in \mathbb{F}^n$ are the unique solutions of the Bézout equations

$$\mathbf{f}_i(t)\boldsymbol{\alpha}_i(t) + \mathbf{g}_i(t)\boldsymbol{\beta}_i(t) = 1\,. \tag{5.5}$$

This is Theorem 6.3 of [6]. Its proof uses an analogue of Lemma 2.1, which implies $B^{-1} = -\frac{1}{2}B_1^{-1}P_+ + \frac{1}{2}B_2^{-1}P_-$, as well as formulas for the inversion of centroskewsymmetric T-Bezoutians, analogous to the centrosymmetric case (3.8), but differing by a minus sign.

In [6] a corresponding inversion algorithm was given.

Algorithm 5.4. We are given a centroskewsymmetric $T+H$-Bezoutian B of even order n in the form (5.1) with pairs $\{\mathbf{f}_i(t), \mathbf{g}_i(t)\}$ of symmetric coprime polynomials in $\mathbb{F}_+^{n+1}[t]$ $\quad (i = 1, 2)$.

1. Solve the Bézout equations (5.5) by the extended Euclidean algorithm.
2. Determine the skewsymmetric symbols $\mathbf{c}_i = (c_j^{(i)})_{j=-n+1}^{n-1}$ by (5.4). Equivalently, in matrix language (compare (3.9)–(3.10)) this means computing the last n components of $\mathbf{c}_i$ by

$$(c_j^{(i)})_{j=0}^{n-1} = (-1)^i \Big(U_n(\boldsymbol{\beta}_i)\boldsymbol{\alpha}_i - U_n(\boldsymbol{\alpha}_i)\boldsymbol{\beta}_i\Big)\,, \qquad i = 1, 2.$$

3. Compute the matrices

$$A_1 := T_n(\mathbf{c}_1)P_+ \quad \text{and} \quad A_2 := T_n(\mathbf{c}_2)P_-\,.$$

4. Then the inverse of B is given by

$$B^{-1} = \frac{1}{2}(A_1 + A_2).$$

6. Connection between centrosymmetric T-Bezoutians and split-Bezoutians

Following the program outlined at the end of Section 3, the goal of this section is to represent a nonsingular centrosymmetric $T+H$-Bezoutian in the form

$$B = \mathrm{Bez}_T(\mathbf{u}_1, \mathbf{v}_1)P_+ + \mathrm{Bez}_T(\mathbf{u}_2, \mathbf{v}_2)P_- ,$$

where $\mathbf{u}_i \in \mathbb{F}_+^{n+1}$ and $\mathbf{v}_i \in \mathbb{F}_-^{n+1}$. Therein the T-Bezoutians are symmetric and centrosymmetric. We also want to achieve that they are invertible.

We start with the following simple identity, which is crucial for what follows.

Lemma 6.1. *Let* $\mathbf{u} \in \mathbb{F}_+^{n+1}$, $\mathbf{v} \in \mathbb{F}_-^{n+1}$. *Then*

$$\mathrm{Bez}_T(\mathbf{u}, \mathbf{v})P_+ = \frac{1}{2}\mathrm{Bez}_{\mathrm{sp}}(\tilde{\mathbf{u}}, \tilde{\mathbf{v}}) \quad \text{with} \quad \begin{aligned} \tilde{\mathbf{u}}(t) &= (t+1)\mathbf{u}(t) \in \mathbb{F}_+^{n+2}[t] \\ \tilde{\mathbf{v}}(t) &= (t-1)\mathbf{v}(t) \in \mathbb{F}_+^{n+2}[t] \end{aligned} \tag{6.1}$$

and

$$\mathrm{Bez}_T(\mathbf{u}, \mathbf{v})P_- = -\frac{1}{2}\mathrm{Bez}_{\mathrm{sp}}(\tilde{\mathbf{u}}, \tilde{\mathbf{v}}) \quad \text{with} \quad \begin{aligned} \tilde{\mathbf{u}}(t) &= (t-1)\mathbf{u}(t) \in \mathbb{F}_-^{n+2}[t] \\ \tilde{\mathbf{v}}(t) &= (t+1)\mathbf{v}(t) \in \mathbb{F}_-^{n+2}[t]. \end{aligned} \tag{6.2}$$

Proof. With $B_T = \mathrm{Bez}_T(\mathbf{u}, \mathbf{v})$, using that $\mathbf{u} = \mathbf{u}^J$ and $\mathbf{v} = -\mathbf{v}^J$, we have

$$B_T(t,s) = \frac{-\mathbf{u}(t)\mathbf{v}(s) - \mathbf{v}(t)\mathbf{u}(s)}{1-ts}, \quad (B_T J_n)(t,s) = \frac{-\mathbf{u}(t)\mathbf{v}(s) + \mathbf{v}(t)\mathbf{u}(s)}{t-s}.$$

Hence

$$(B_T(I_n \pm J_n))(t,s) = \mathbf{u}(t)\mathbf{v}(s)\left[-\frac{1}{1-ts} \mp \frac{1}{t-s}\right] + \mathbf{v}(t)\mathbf{u}(s)\left[-\frac{1}{1-ts} \pm \frac{1}{t-s}\right].$$

Taking into account

$$-\frac{1}{1-ts} + \frac{1}{t-s} = -\frac{(t-1)(s+1)}{(1-ts)(t-s)} \quad \text{and} \quad -\frac{1}{1-ts} - \frac{1}{t-s} = \frac{(t+1)(s-1)}{(1-ts)(t-s)}$$

we obtain

$$2(B_T P_+)(t,s) = \frac{(t+1)\mathbf{u}(t)(s-1)\mathbf{v}(s) - (t-1)\mathbf{v}(t)(s+1)\mathbf{u}(s)}{(1-ts)(t-s)}$$

and

$$2(B_T P_-)(t,s) = -\frac{(t-1)\mathbf{u}(t)(s+1)\mathbf{v}(s) - (t+1)\mathbf{v}(t)(s-1)\mathbf{u}(s)}{(1-ts)(t-s)},$$

which implies the identities (6.1) and (6.2). □

A consequence of this lemma is the following result, where we obtain expressions analogous to those in Theorems 4.4 and 4.5.

Proposition 6.2. *Let* $\mathbf{u} \in \mathbb{F}_+^{n+1}$, $\mathbf{v} \in \mathbb{F}_-^{n+1}$.
If n *is even, then*

$$\mathrm{Bez}_T(\mathbf{u},\mathbf{v})P_+ = \frac{1}{2} M_{n-1}^{t+1} \mathrm{Bez}_{\mathrm{sp}}(\mathbf{f},\mathbf{g}) (M_{n-1}^{t+1})^T \quad \textit{with} \quad \begin{array}{l} \mathbf{f}(t) := \mathbf{u}(t) \\ \mathbf{g}(t) := \frac{t-1}{t+1}\mathbf{v}(t) \end{array} \tag{6.3}$$

and

$$\mathrm{Bez}_T(\mathbf{u},\mathbf{v})P_- = -\frac{1}{2} M_{n-1}^{t-1} \mathrm{Bez}_{\mathrm{sp}}(\mathbf{f},\mathbf{g}) (M_{n-1}^{t-1})^T \quad \textit{with} \quad \begin{array}{l} \mathbf{f}(t) := \mathbf{u}(t) \\ \mathbf{g}(t) := \frac{t+1}{t-1}\mathbf{v}(t). \end{array} \tag{6.4}$$

If n *is odd, then*

$$\mathrm{Bez}_T(\mathbf{u},\mathbf{v})P_+ = \frac{1}{2} \mathrm{Bez}_{\mathrm{sp}}(\mathbf{f},\mathbf{g}) \quad \textit{with} \quad \begin{array}{l} \mathbf{f}(t) := (t+1)\mathbf{u}(t) \\ \mathbf{g}(t) := (t-1)\mathbf{v}(t) \end{array} \tag{6.5}$$

and

$$\mathrm{Bez}_T(\mathbf{u},\mathbf{v})P_- = -\frac{1}{2} M_{n-2}^{t^2-1} \mathrm{Bez}_{\mathrm{sp}}(\mathbf{f},\mathbf{g}) (M_{n-2}^{t^2-1})^T \quad \textit{with} \quad \begin{array}{l} \mathbf{f}(t) := \frac{1}{t+1}\mathbf{u}(t) \\ \mathbf{g}(t) := \frac{1}{t-1}\mathbf{v}(t). \end{array} \tag{6.6}$$

Note that it follows from (2.6) that $\mathbf{f}(t)$ and $\mathbf{g}(t)$ are well-defined polynomials in all four cases. Furthermore, they are symmetric and hence the split-Bezoutians are of odd order and (+)type. In the cases (6.3) and (6.4) we have $\mathbf{f},\mathbf{g} \in \mathbb{F}_+^{n+1}$, in the case (6.5) we have $\mathbf{f},\mathbf{g} \in \mathbb{F}_+^{n+2}$, and in the case (6.6) we have $\mathbf{f},\mathbf{g} \in \mathbb{F}_+^{n}$.

Proof. The proof is done by appropriate substitutions in combination with Lemma 6.1. For instance, in order to prove (6.3), observe that

$$(t+1)\mathbf{f}(t) = \tilde{\mathbf{u}}(t) = (t+1)\mathbf{u}(t), \qquad (t+1)\mathbf{g}(t) = \tilde{\mathbf{v}}(t) = (t-1)\mathbf{v}(t)$$

holds. Then by Lemma 6.1 one has

$$2\,\mathrm{Bez}_T(\mathbf{u},\mathbf{v})P_+ = \mathrm{Bez}_{\mathrm{sp}}(\tilde{\mathbf{u}},\tilde{\mathbf{v}}).$$

The generating polynomial of the matrix on the right-hand side equals

$$\frac{(t+1)\mathbf{f}(t)\mathbf{g}(s)(s+1) - (t+1)\mathbf{g}(t)\mathbf{f}(s)(s+1)}{(t-s)(1-ts)}$$

and thus coincides the generating polynomial of

$$M_{n-1}^{t+1} \mathrm{Bez}_{\mathrm{sp}}(\mathbf{f},\mathbf{g}) (M_{n-1}^{t+1})^T .$$

Hence (6.3) is verified. The proof of (6.4)–(6.6) is similar. □

The last proposition implies that every centrosymmetric T-Bezoutian is related to a split-Bezoutian via the above formulas. We are actually interested in the converse problem. Namely, we would like to express a given split-Bezoutain as a T-Bezoutian. As can be seen in the following lemma, this is not always immediately possible unless $\mathbf{f}$ and $\mathbf{g}$ satisfy certain conditions.

In what follows the four cases above, (6.3)–(6.6), are referred to as (e^+) and (e^-) when n is even, and as (o^+) and (o^-) when n is odd.

Lemma 6.3. *Let* $\mathbf{f}, \mathbf{g}$ *be given symmetric vectors such that* $\{\mathbf{f}(t), \mathbf{g}(t)\}$ *are (generalized) coprime. Let* $\mathbf{u}(t)$ *and* $\mathbf{v}(t)$ *be defined by one of the conditions in* (6.3)–(6.6). *Then* $\mathbf{u}(t)$ *and* $\mathbf{v}(t)$ *are well-defined (generalized) coprime polynomials if and only if the following condition is fulfilled:*

in the case (e^+): $\mathbf{f}(-1) \neq 0$ *and* $\mathbf{g}(1) = 0$,
in the case (e^-): $\mathbf{f}(1) \neq 0$ *and* $\mathbf{g}(-1) = 0$,
in the case (o^+): $\mathbf{f}(-1) = 0$ *and* $\mathbf{g}(1) = 0$,
in the case (o^-): $\mathbf{f}(1) \neq 0$ *and* $\mathbf{g}(-1) \neq 0$.

Proof. Two issues play a role. One is the question whether $\mathbf{u}(t)$ and $\mathbf{v}(t)$ are polynomials at all, the other is their coprimeness.

Consider the case (e^+). Following (6.3), for the pair $\{\mathbf{f}(t), \mathbf{g}(t)\}$ of coprime polynomials in $\mathbb{F}_+^{n+1}[t]$ we define

$$\mathbf{u}(t) = \mathbf{f}(t) \quad \text{and} \quad \mathbf{v}(t) = \frac{t+1}{t-1}\mathbf{g}(t),$$

and note that $\mathbf{v}(t)$ is a polynomial if and only if $\mathbf{g}(1) = 0$. Since $\mathbf{v}(-1) = 0$, the coprimeness of $\{\mathbf{u}(t), \mathbf{v}(t)\}$ requires that $\mathbf{f}(-1) \neq 0$. Conversely, assume that $\mathbf{f}(-1) \neq 0$. We show that the pair $\{\mathbf{u}(t), \mathbf{v}(t)\}$ is coprime by contradiction and assume that there exists an irreducible (nonconstant) polynomial $\varphi(t)$ such that $\varphi(t) \mid \mathbf{u}(t)$ and $\varphi(t) \mid \mathbf{v}(t)$. It follows that $\varphi(t) \mid (t+1)\mathbf{g}(t)$. But $\varphi(t) \nmid (t+1)$ since otherwise $\varphi(t) = c(t+1)$, $c \neq 0$, contradicting $\varphi(t) \mid \mathbf{f}(t)$ and $\mathbf{f}(-1) \neq 0$. Thus we must have $\varphi(t) \mid \mathbf{g}(t)$, which implies that the pair $\{\mathbf{f}(t), \mathbf{g}(t)\}$ is not coprime, again a contradiction.

The case (e^-) is similar, interchange the roles of $+1$ and -1.

In the case (o^+) we have

$$\mathbf{f}(t) = (t+1)\mathbf{u}(t), \qquad \mathbf{g}(t) = (t-1)\mathbf{v}(t),$$

so that $\mathbf{u}(t)$ and $\mathbf{v}(t)$ are polynomials if and only if $\mathbf{f}(-1) = \mathbf{g}(1) = 0$. In this case the coprimeness of $\{\mathbf{f}(t), \mathbf{g}(t)\}$ implies the coprimeness of $\{\mathbf{u}(t), \mathbf{v}(t)\}$.

It remains to consider the case (o^-). Here we define

$$\mathbf{u}(t) = (t+1)\mathbf{f}(t) \in \mathbb{F}_+^{n+1}, \quad \mathbf{v}(t) = (t-1)\mathbf{g}(t) \in \mathbb{F}_-^{n+1}.$$

Thus $\{\mathbf{u}(t), \mathbf{v}(t)\}$ are coprime only if $\mathbf{f}(1) \neq 0$ and $\mathbf{g}(-1) \neq 0$. Conversely, if $\mathbf{f}(1) \neq 0$ and $\mathbf{g}(-1) \neq 0$, then $\{\mathbf{u}(t), \mathbf{v}(t)\}$ are coprime, which can be seen as follows. Take an irreducible (nonconstant) polynomial $\varphi(t)$ dividing both $\mathbf{u}(t)$ and $\mathbf{v}(t)$. If $\varphi(t) = c(t+1)$ or $\varphi(t) = c(t-1)$, we get a contradiction to $\mathbf{f}(1) \neq 0$ and $\mathbf{g}(-1) \neq 0$. If $\varphi(t)$ is not of this form, then the assumption that $\varphi(t)$ divides both $\mathbf{u}(t)$ and $\mathbf{v}(t)$ implies that $\varphi(t)$ divides both $\mathbf{f}(t)$ and $\mathbf{g}(t)$, contradicting the coprimeness assumption. This completes the proof. □

We know from Remark 4.3 that if $[\,\mathbf{f}, \mathbf{g}\,] = [\,\hat{\mathbf{f}}, \hat{\mathbf{g}}\,]\,\Phi$ with $\det \Phi = 1$, then

$$\mathrm{Bez}_{\mathrm{sp}}(\mathbf{f}, \mathbf{g}) = \mathrm{Bez}_{\mathrm{sp}}(\hat{\mathbf{f}}, \hat{\mathbf{g}}).$$

Consequently, we have some freedom to modify the vectors appearing in the split-Bezoutians in order to satisfy the conditions above for the cases ($e^\pm$) and ($o^\pm$).

Let $m_{\mathbf{f},\mathbf{g}}$ be the following matrix of order two,

$$m_{\mathbf{f},\mathbf{g}} = \begin{bmatrix} \mathbf{f}(1) & \mathbf{g}(1) \\ \mathbf{f}(-1) & \mathbf{g}(-1) \end{bmatrix}. \tag{6.7}$$

Lemma 6.4. *Let $\hat{\mathbf{f}}, \hat{\mathbf{g}}$ be given symmetric vectors such that $\{\hat{\mathbf{f}}(t), \hat{\mathbf{g}}(t)\}$ are (generalized) coprime. Consider any of the cases* (e$^{\pm}$) *and* (o$^{\pm}$). *In case* (o^{+}) *assume in addition that*

$$\operatorname{rank} m_{\hat{\mathbf{f}},\hat{\mathbf{g}}} = 2. \tag{6.8}$$

Then there exists $\Phi \in \mathbb{F}^{2\times 2}$ with $\det \Phi = 1$ such that for $\mathbf{f}, \mathbf{g}$ given by

$$[\,\mathbf{f}, \mathbf{g}\,] = [\,\hat{\mathbf{f}}, \hat{\mathbf{g}}\,]\Phi$$

the corresponding conditions for the cases (e$^{\pm}$) *or* (o$^{\pm}$), *respectively, of Lemma* 6.3 *are satisfied.*

Proof. It follows from the coprimeness of $\{\hat{\mathbf{f}}(t), \hat{\mathbf{g}}(t)\}$ that $m_{\hat{\mathbf{f}},\hat{\mathbf{g}}}$ cannot have a zero row. In particular, $m_{\hat{\mathbf{f}},\hat{\mathbf{g}}} = \mathbf{0}$ is impossible. Note that

$$m_{\mathbf{f},\mathbf{g}} = m_{\hat{\mathbf{f}},\hat{\mathbf{g}}}\Phi.$$

If $\operatorname{rank} m_{\hat{\mathbf{f}},\hat{\mathbf{g}}} = 2$, it is easily seen that we can find a matrix Φ such that the entries of $m_{\mathbf{f},\mathbf{g}}$ are in accordance with the conditions of Lemma 6.3 for each of the four cases (e$^{\pm}$) and (o$^{\pm}$).

If $\operatorname{rank} m_{\hat{\mathbf{f}},\hat{\mathbf{g}}} = 1$ we only need to consider the cases (e$^{\pm}$) and (o^{-}). In the cases (e$^{\pm}$) our aim is to define Φ such that

$$m_{\hat{\mathbf{f}},\hat{\mathbf{g}}}\Phi = \begin{bmatrix} \mathbf{f}(1) & 0 \\ \mathbf{f}(-1) & 0 \end{bmatrix}.$$

Since coprimeness is preserved by a transformation Φ, we then have $\mathbf{f}(\pm 1) \neq 0$ as required in the cases (e$^{\pm}$). If $\hat{\mathbf{g}}(1) = \hat{\mathbf{g}}(-1) = 0$ we simply choose $\Phi = I_2$ so that $\mathbf{f} = \hat{\mathbf{f}}$, $\mathbf{g} = \hat{\mathbf{g}}$. If $\hat{a} := \hat{\mathbf{g}}(1) \neq 0$ or $\hat{b} := \hat{\mathbf{g}}(-1) \neq 0$, then the rank one condition implies that

$$\hat{\mathbf{f}}(1) = c\,\hat{a}, \quad \hat{\mathbf{f}}(-1) = c\,\hat{b}$$

for some constant $c \in \mathbb{F}$ and

$$\begin{bmatrix} c\,\hat{a} & \hat{a} \\ c\,\hat{b} & \hat{b} \end{bmatrix} \begin{bmatrix} 0 & -1 \\ 1 & c \end{bmatrix} = \begin{bmatrix} \hat{a} & 0 \\ \hat{b} & 0 \end{bmatrix}.$$

This means $\mathbf{f}(t) = \hat{\mathbf{g}}(t)$ and $\mathbf{g}(t) = c\,\hat{\mathbf{g}}(t) - \hat{\mathbf{f}}(t)$.

Now let us consider the case (o^{-}), where we need to achieve $\mathbf{f}(1) \neq 0$ and $\mathbf{g}(-1) \neq 0$. First notice that $\hat{\mathbf{f}}(1) = \hat{\mathbf{g}}(-1) = 0$ is impossible due to the rank condition and coprimeness. In case $\hat{\mathbf{f}}(1) \neq 0$ and $\hat{\mathbf{g}}(-1) = 0$ (thus $\hat{\mathbf{f}}(-1) \neq 0$ due to coprimeness) we choose $\Phi = \begin{bmatrix} 1 & 1 \\ 0 & 1 \end{bmatrix}$, i.e.,

$$\mathbf{f}(t) = \hat{\mathbf{f}}(t), \quad \mathbf{g}(t) = \hat{\mathbf{g}}(t) + \hat{\mathbf{f}}(t).$$

In case $\hat{\mathbf{f}}(1) = 0$ and $\hat{\mathbf{g}}(-1) \neq 0$ (thus $\hat{\mathbf{g}}(1) \neq 0$) we choose $\Phi = \begin{bmatrix} 1 & 0 \\ 1 & 1 \end{bmatrix}$. Finally, in case $\hat{\mathbf{f}}(1) \neq 0$ and $\hat{\mathbf{g}}(-1) \neq 0$ we take $\Phi = I_2$. This completes the proof. □

7. Inversion of centrosymmetric $T + H$-Bezoutians – first formulas

As was observed in Section 4, any nonsingular centrosymmetric $T + H$-Bezoutian B can be represented in the form (4.8) or (4.9) depending on whether n is even or odd. We will take this representation of B as our starting point here.

In the theorems below we will assume extra conditions (7.1) and (7.3) regarding the polynomials $\mathbf{f}_i(t)$ and $\mathbf{g}_i(t)$, which arise from the conditions in Lemma 6.3. Note that regarding Lemma 6.4 one could take the point of view that these conditions can be assumed without loss of generality (in case (o$^+$) only if the weaker condition (6.8) is assumed). Namely, as Lemma 6.4 and its proof show, it is always possible to transform arbitrary pairs of polynomials into ones satisfying the conditions (7.1) and (7.3).

However, later on (Section 8) we will remove the extra conditions (7.1) and (7.3) and hence also the necessity of a transformation via some Φ. Therefore, the results of this section can be viewed as intermediate steps.

Theorem 7.1. *Let n be even and let*

$$B = M_{n-1}^{t+1} \operatorname{Bez}_{\mathrm{sp}}(\mathbf{f}_1, \mathbf{g}_1) \, (M_{n-1}^{t+1})^T + M_{n-1}^{t-1} \operatorname{Bez}_{\mathrm{sp}}(\mathbf{f}_2, \mathbf{g}_2) \, (M_{n-1}^{t-1})^T ,$$

where $\mathbf{f}_1, \mathbf{f}_2, \mathbf{g}_1, \mathbf{g}_2 \in \mathbb{F}_+^{n+1}$ are such that $\{\mathbf{f}_1(t), \mathbf{g}_1(t)\}$ and $\{\mathbf{f}_2(t), \mathbf{g}_2(t)\}$ are two pairs of coprime polynomials. Assume in addition

$$\mathbf{f}_1(-1) \neq 0, \quad \mathbf{f}_2(1) \neq 0, \quad \mathbf{g}_1(1) = \mathbf{g}_2(-1) = 0, \tag{7.1}$$

and define

$$\mathbf{u}_1(t) = \mathbf{f}_1(t), \; \mathbf{v}_1(t) = \frac{t+1}{t-1}\, \mathbf{g}_1(t), \; \mathbf{u}_2 = \mathbf{f}_2(t), \; \mathbf{v}_2(t) = \frac{t-1}{t+1}\, \mathbf{g}_2(t). \tag{7.2}$$

Then the Bézout equations

$$\mathbf{u}_i(t)\tilde{\boldsymbol{\alpha}}_i(t) + \mathbf{v}_i(t)\tilde{\boldsymbol{\beta}}_i(t) = 1, \qquad i = 1, 2,$$

admit unique solutions $\tilde{\boldsymbol{\alpha}}_i, \tilde{\boldsymbol{\beta}}_i \in \mathbb{F}^n$, and B is invertible. The inverse has the representation

$$2\, B^{-1} = T_n(\mathbf{c}_1)P_+ + T_n(\mathbf{c}_2)P_- \,,$$

where the symbols are given by

$$\mathbf{c}_i(t) = t^{n-1} \sum_{k=-n+1}^{n-1} c_k^{(i)} t^k = (-1)^i \Big(\tilde{\boldsymbol{\alpha}}_i(t)\tilde{\boldsymbol{\beta}}_i^J(t) + \tilde{\boldsymbol{\beta}}_i(t)\tilde{\boldsymbol{\alpha}}_i^J(t) \Big).$$

Proof. Using identities (6.3) and (6.4) of Proposition 6.2 we obtain

$$\frac{1}{2}B = \mathrm{Bez}_T(\mathbf{u}_1, \mathbf{v}_1)P_+ - \mathrm{Bez}_T(\mathbf{u}_2, \mathbf{v}_2)P_- .$$

Notice that due to the assumptions (7.1) we can apply Lemma 6.3, part (e^+) and (e^-). Hence $\{\mathbf{u}_i(t), \mathbf{v}_i(t)\}$ are pairs of polynomials which are (generalized) coprime. Therefore the Bézout equations have a unique solution. It is also clear that $\mathbf{u}_i \in \mathbb{F}_+^{n+1}$ and $\mathbf{v}_i \in \mathbb{F}_-^{n+1}$. Thus the T-Bezoutians are (centro)symmetric. Again due to coprimeness, they are invertible and as has been stated before (see Theorem 3.5 and (3.8)), the inverses can be obtained from the solutions of the Bézout equations,

$$\mathrm{Bez}_T(\mathbf{u}_i, \mathbf{v}_i)^{-1} = (-1)^{i+1} T_n(\mathbf{c}_i).$$

(Here we change the sign in the case $i = 2$ for sake of convenience.) Taking Lemma 2.1 into account it remains to note

$$2B^{-1} = \mathrm{Bez}_T(\mathbf{u}_1, \mathbf{v}_1)^{-1}P_+ - \mathrm{Bez}_T(\mathbf{u}_2, \mathbf{v}_2)^{-1}P_-.$$

This completes the proof. □

Theorem 7.2. *Let n be odd and let*

$$B = \mathrm{Bez}_{\mathrm{sp}}(\mathbf{f}_1, \mathbf{g}_1) + M_{n-2}^{t^2-1}\,\mathrm{Bez}_{\mathrm{sp}}(\mathbf{f}_2, \mathbf{g}_2)\,(M_{n-2}^{t^2-1})^T ,$$

where $\mathbf{f}_1, \mathbf{g}_1 \in \mathbb{F}_+^{n+2}$, $\mathbf{f}_2, \mathbf{g}_2 \in \mathbb{F}_+^n$ are such that $\{\mathbf{f}_1(t), \mathbf{g}_1(t)\}$ and $\{\mathbf{f}_2(t), \mathbf{g}_2(t)\}$ are two pairs of coprime polynomials. Assume in addition

$$\mathbf{f}_1(-1) = \mathbf{g}_1(1) = 0, \quad \mathbf{f}_2(1) \neq 0, \quad \mathbf{g}_2(-1) \neq 0, \tag{7.3}$$

and define

$$\mathbf{u}_1(t) = \frac{\mathbf{f}_1(t)}{t+1}, \ \mathbf{v}_1(t) = \frac{\mathbf{g}_1(t)}{t-1}, \ \mathbf{u}_2(t) = (t+1)\mathbf{f}_2(t), \ \mathbf{v}_2(t) = (t-1)\mathbf{g}_2(t). \tag{7.4}$$

Then the Bézout equations

$$\mathbf{u}_i(t)\tilde{\boldsymbol{\alpha}}_i(t) + \mathbf{v}_i(t)\tilde{\boldsymbol{\beta}}_i(t) = 1, \qquad i = 1, 2,$$

admit unique solutions $\tilde{\boldsymbol{\alpha}}_i, \tilde{\boldsymbol{\beta}}_i \in \mathbb{F}^n$, and B is invertible. The inverse has the representation

$$2\,B^{-1} = T_n(\mathbf{c}_1)P_+ + T_n(\mathbf{c}_2)P_- ,$$

where the symbols are given by

$$\mathbf{c}_i(t) = t^{n-1} \sum_{k=-n+1}^{n-1} c_k^{(i)} t^k = (-1)^i \Big(\tilde{\boldsymbol{\alpha}}_i(t)\tilde{\boldsymbol{\beta}}_i^J(t) + \tilde{\boldsymbol{\beta}}_i(t)\tilde{\boldsymbol{\alpha}}_i^J(t)\Big).$$

The proof is analogous to the one of Theorem 7.1.

8. Elimination of the extra conditions in the inversion formulas

In this section we simplify the inversion formulas established in Theorems 7.1 and 7.2 by eliminating the polynomials $\mathbf{u}_i(t)$ and $\mathbf{v}_i(t)$. At the same time, and even more importantly, we remove the extra conditions (7.1) and (7.3), and keep only one weaker condition in the case (o^+). Furthermore, we show that it is not necessary to apply appropriate transformations to the vectors that define the split-Bezoutians. On the other hand, the immediate connection to T-Bezoutians is lost.

Theorem 8.1. *Let n be even and let*

$$B = M_{n-1}^{t+1}\,\mathrm{Bez}_{\mathrm{sp}}(\mathbf{f}_1,\mathbf{g}_1)\,(M_{n-1}^{t+1})^T + M_{n-1}^{t-1}\,\mathrm{Bez}_{\mathrm{sp}}(\mathbf{f}_2,\mathbf{g}_2)\,(M_{n-1}^{t-1})^T ,$$

where $\mathbf{f}_1,\mathbf{f}_2,\mathbf{g}_1,\mathbf{g}_2 \in \mathbb{F}_+^{n+1}$ are such that $\{\mathbf{f}_1(t),\mathbf{g}_1(t)\}$ and $\{\mathbf{f}_2(t),\mathbf{g}_2(t)\}$ are two pairs of (generalized) coprime polynomials. Then the Bézout equations

$$\mathbf{f}_1(t)\boldsymbol{\alpha}_1(t) + \mathbf{g}_1(t)\boldsymbol{\beta}_1(t) = t-1, \tag{8.1}$$

$$\mathbf{f}_2(t)\boldsymbol{\alpha}_2(t) + \mathbf{g}_2(t)\boldsymbol{\beta}_2(t) = t+1, \tag{8.2}$$

admit solutions $\boldsymbol{\alpha}_i,\boldsymbol{\beta}_i \in \mathbb{F}^{n+1}$, and B is invertible. The inverse has the representation

$$2\,B^{-1} = T_n(\mathbf{c}_1)P_+ + T_n(\mathbf{c}_2)P_- ,$$

where

$$\mathbf{c}_i(t) = \frac{\boldsymbol{\beta}_i(t)\boldsymbol{\alpha}_i^J(t) - \boldsymbol{\alpha}_i(t)\boldsymbol{\beta}_i^J(t)}{t^2-1}, \qquad i=1,2, \tag{8.3}$$

and $\boldsymbol{\alpha}_i,\boldsymbol{\beta}_i$ are any solutions of (8.1) *and* (8.2).

Proof. Before we start with the actual proof, let us remark that $\mathbf{c}_i(t)$ are well-defined symmetric polynomials. Consider $\mathbf{y}_i(t) = \boldsymbol{\beta}_i(t)\boldsymbol{\alpha}_i^J(t) - \boldsymbol{\alpha}_i(t)\boldsymbol{\beta}_i^J(t)$. Since $\boldsymbol{\alpha}_i,\boldsymbol{\beta}_i \in \mathbb{F}^{n+1}$ it follows that $\mathbf{y}_i \in \mathbb{F}^{2n+1}$. Moreover, the definition implies $\mathbf{y}_i^J = -\mathbf{y}_i$. From (2.6) it is clear that $\mathbf{y}_i(t) = (t^2-1)\mathbf{c}_i(t)$ with some $\mathbf{c}_i \in \mathbb{F}_+^{2n-1}$. Note that the vector $\mathbf{c}_i$ has size $2n-1$, and hence $\mathbf{c}_i$ can be the symbol of an $n\times n$ Toeplitz matrix.

Furthermore, notice that the Bézout equations (8.1) and (8.2) admit solutions because the pairs $\{\mathbf{f}_i(t),\mathbf{g}_i(t)\}$ are coprime (hence generalized coprime). Since we are seeking $\boldsymbol{\alpha}_i,\boldsymbol{\beta}_i$ in $\mathbb{F}^{n+1}$ (rather than in $\mathbb{F}^n$) the solutions are not unique. Instead we have a one-parameter family of solutions.

We now prove the formula for B^{-1} in several steps.

Step 1*:* We first assume that $\mathbf{f}_1,\mathbf{f}_2,\mathbf{g}_1,\mathbf{g}_2$ satisfy the conditions stated in (7.1). Then we can proceed as in Theorem 7.1, introduce $\mathbf{u}_i$ and $\mathbf{v}_i$ as in (7.2), and obtain the unique solutions of the Bézout equations

$$\mathbf{u}_i(t)\tilde{\boldsymbol{\alpha}}_i(t) + \mathbf{v}_i(t)\tilde{\boldsymbol{\beta}}_i(t) = 1.$$

We substitute $\mathbf{u}_i$ and $\mathbf{v}_i$ and get the two equations

$$(t-1)\mathbf{f}_1(t)\tilde{\boldsymbol{\alpha}}_1(t) + (t+1)\mathbf{g}_1(t)\tilde{\boldsymbol{\beta}}_1(t) = t-1,$$

$$(t+1)\mathbf{f}_2(t)\tilde{\boldsymbol{\alpha}}_2(t) + (t-1)\mathbf{g}_2(t)\tilde{\boldsymbol{\beta}}_2(t) = t+1.$$

Now we make the substitutions

$$\boldsymbol{\alpha}_1(t) = (t-1)\tilde{\boldsymbol{\alpha}}_1(t), \qquad \boldsymbol{\beta}_1(t) = (t+1)\tilde{\boldsymbol{\beta}}_1(t),$$
$$\boldsymbol{\alpha}_2(t) = (t+1)\tilde{\boldsymbol{\alpha}}_2(t), \qquad \boldsymbol{\beta}_2(t) = (t-1)\tilde{\boldsymbol{\beta}}_2(t),$$

and obtain the corresponding equations (8.1) and (8.2). Next we express $\mathbf{c}_i$ in terms of $\boldsymbol{\alpha}_i$ and $\boldsymbol{\beta}_i$. Notice that $\tilde{\boldsymbol{\alpha}}_i, \tilde{\boldsymbol{\beta}}_i \in \mathbb{F}^n$ while $\boldsymbol{\alpha}_i, \boldsymbol{\beta}_i \in \mathbb{F}^{n+1}$. Moreover,

$$\boldsymbol{\alpha}_1^J(t) = -(t-1)\tilde{\boldsymbol{\alpha}}_1^J(t), \qquad \boldsymbol{\beta}_1^J(t) = (t+1)\tilde{\boldsymbol{\beta}}_1^J(t),$$
$$\boldsymbol{\alpha}_2^J(t) = (t+1)\tilde{\boldsymbol{\alpha}}_2^J(t), \qquad \boldsymbol{\beta}_2^J(t) = -(t-1)\tilde{\boldsymbol{\beta}}_2^J(t).$$

This can be checked straightforwardly by noting that $(t \pm 1)^J = \pm(t \pm 1)$. From

$$\mathbf{c}_i(t) = (-1)^i \Big(\tilde{\boldsymbol{\alpha}}_i(t)\tilde{\boldsymbol{\beta}}_i^J(t) + \tilde{\boldsymbol{\beta}}_i(t)\tilde{\boldsymbol{\alpha}}_i^J(t)\Big)$$

equation (8.3) with the appropriate changes in sign follows. To summarize, at this point we have shown that for specific $\mathbf{f}_i, \mathbf{g}_i$ there exist solutions $\{\boldsymbol{\alpha}_i, \boldsymbol{\beta}_i\}$ such that the inversion formula holds.

Step 2: Now we prove that if the inversion formula holds for some solution of the Bézout equations (8.1) and (8.2), then it also holds for any other solution of these equations. As we will see shortly, for different $\{\boldsymbol{\alpha}_i, \boldsymbol{\beta}_i\}$ we may obtain different $\mathbf{c}_i$. However, $T_n(\mathbf{c}_1)P_+$ and $T_n(\mathbf{c}_2)P_-$ remain unchanged.

Now assume that the pair $\{\boldsymbol{\alpha}_i, \boldsymbol{\beta}_i\}$ is some solution to the Bézout equation (8.1) or (8.2). Then, using the (generalized) coprimeness of $\{\mathbf{g}_i(t), \mathbf{f}_i(t)\}$ and taking into account the degrees of the polynomials under consideration, it follows that any other solution of (8.1) or (8.2) is given by

$$\boldsymbol{\alpha}_i'(t) = \boldsymbol{\alpha}_i(t) + \mu\mathbf{g}_i(t), \quad \boldsymbol{\beta}_i'(t) = \boldsymbol{\beta}_i(t) - \mu\mathbf{f}_i(t),$$

where $\mu \in \mathbb{F}$ is an arbitrary parameter. Now compute the corresponding $\mathbf{c}_i'$ as follows:

$$\begin{aligned}
(t^2-1)\mathbf{c}_i'(t) &= \boldsymbol{\beta}_i'(t)(\boldsymbol{\alpha}_i')^J(t) - \boldsymbol{\alpha}_i'(t)(\boldsymbol{\beta}_i')^J(t) \\
&= \Big(\boldsymbol{\beta}_i(t) - \mu\mathbf{f}_i(t)\Big)\Big(\boldsymbol{\alpha}_i^J(t) + \mu\mathbf{g}_i(t)\Big) \\
&\quad - \Big(\boldsymbol{\alpha}_i(t) + \mu\mathbf{g}_i(t)\Big)\Big(\boldsymbol{\beta}_i^J(t) - \mu\mathbf{f}_i(t)\Big) \\
&= (t^2-1)\mathbf{c}_i(t) \\
&\quad + \mu\Big(\mathbf{f}_i(t)\boldsymbol{\alpha}_i(t) + \mathbf{g}_i(t)\boldsymbol{\beta}_i(t) - \mathbf{f}_i^J(t)\boldsymbol{\alpha}_i^J(t) - \mathbf{g}_i^J(t)\boldsymbol{\beta}_i^J(t)\Big) \\
&= (t^2-1)\mathbf{c}_i(t) + \mu(t + (-1)^i) - \mu(t^{2n-1} + (-1)^i t^{2n}).
\end{aligned}$$

Here we used that $\mathbf{f}_i^J = \mathbf{f}_i$ and $\mathbf{g}_i^J = \mathbf{g}_i$. Hence

$$\mathbf{c}_1'(t) = \mathbf{c}_1(t) + \mu\,\frac{t^{2n} - t^{2n-1} + t - 1}{t^2-1}, \quad \mathbf{c}_2'(t) = \mathbf{c}_2(t) - \mu\,\frac{t^{2n} + t^{2n-1} - t - 1}{t^2-1},$$

i.e.,

$$\mathbf{c}_1'(t) = \mathbf{c}_1(t) + \mu \sum_{k=0}^{2n} (-t)^k, \qquad \mathbf{c}_2'(t) = \mathbf{c}_2(t) - \mu \sum_{k=0}^{2n} t^k.$$

Now $T_n(\mathbf{c}_1')P_+ = T_n(\mathbf{c}_1)P_+$ and $T_n(\mathbf{c}_2')P_- = T_n(\mathbf{c}_2)P_-$ due to Proposition 3.1 or via direct inspection.

Step 3*:* We are going to show that the inversion formula holds for any coprime pairs $\{\mathbf{f}_i(t), \mathbf{g}_i(t)\}$, even if they do not satisfy the additional condition (7.1) (see Step 1), and for any solution $\{\boldsymbol{\alpha}_i, \boldsymbol{\beta}_i\}$ of the Bézout equations. The key is Lemma 6.4. Let us start with arbitrary pairs $\{\hat{\mathbf{f}}_i(t), \hat{\mathbf{g}}_i(t)\}$ satisfying merely the coprimeness conditions, and let $\{\hat{\boldsymbol{\alpha}}_i, \hat{\boldsymbol{\beta}}_i\}$ be any solutions to the Bézout equations (8.1) or (8.2). Also denote the corresponding polynomial (8.3) by $\hat{\mathbf{c}}_i(t)$.

Because of Lemma 6.4, we find $\Phi_i \in \mathbb{F}^{2\times 2}$ with $\det \Phi_i = 1$ such that the polynomials $\{\mathbf{f}_i(t), \mathbf{g}_i(t)\}$ given by

$$[\,\mathbf{f}_i, \mathbf{g}_i\,] = [\,\hat{\mathbf{f}}_i, \hat{\mathbf{g}}_i\,]\,\Phi_i$$

satisfy the additional conditions (7.1). For these pairs of polynomials we have proved the inversion formula in Step 1 and 2.

Then solutions of the Bézout equations for $\{\mathbf{f}_i, \mathbf{g}_i\}$ are given by

$$[\,\boldsymbol{\alpha}_i, \boldsymbol{\beta}_i\,] = [\,\hat{\boldsymbol{\alpha}}_i, \hat{\boldsymbol{\beta}}_i\,]\,\Phi_i^{-T}$$

where Φ_i^{-T} stands for the inverse of the transpose. To see this it suffices to note that

$$\mathbf{f}_i(t)\boldsymbol{\alpha}_i(t) + \mathbf{g}_i(t)\boldsymbol{\beta}_i(t) = \Big[\mathbf{f}_i(t), \mathbf{g}_i(t)\Big] \begin{bmatrix} \boldsymbol{\alpha}_i(t) \\ \boldsymbol{\beta}_i(t) \end{bmatrix},$$

which equals

$$\Big[\hat{\mathbf{f}}_i(t), \hat{\mathbf{g}}_i(t)\Big] \begin{bmatrix} \hat{\boldsymbol{\alpha}}_i(t) \\ \hat{\boldsymbol{\beta}}_i(t) \end{bmatrix} = \hat{\mathbf{f}}_i(t)\hat{\boldsymbol{\alpha}}_i(t) + \hat{\mathbf{g}}_i(t)\hat{\boldsymbol{\beta}}_i(t)$$

as $\Phi_i \Phi_i^{-1} = I_2$.

Finally, we remark that

$$(t^2-1)\mathbf{c}_i(t) = \Big[\boldsymbol{\alpha}_i(t), \boldsymbol{\beta}_i(t)\Big] \begin{bmatrix} 0 & -1 \\ 1 & 0 \end{bmatrix} \begin{bmatrix} \boldsymbol{\alpha}_i^J(t) \\ \boldsymbol{\beta}_i^J(t) \end{bmatrix}$$

and

$$(t^2-1)\hat{\mathbf{c}}_i(t) = \Big[\hat{\boldsymbol{\alpha}}_i(t), \hat{\boldsymbol{\beta}}_i(t)\Big] \begin{bmatrix} 0 & -1 \\ 1 & 0 \end{bmatrix} \begin{bmatrix} \hat{\boldsymbol{\alpha}}_i^J(t) \\ \hat{\boldsymbol{\beta}}_i^J(t) \end{bmatrix}.$$

These two expressions are equal since

$$\Phi_i^T \begin{bmatrix} 0 & -1 \\ 1 & 0 \end{bmatrix} \Phi_i = \begin{bmatrix} 0 & -1 \\ 1 & 0 \end{bmatrix}$$

as is easy to check. Thus we obtain that $\mathbf{c}_i = \hat{\mathbf{c}}_i$. □

Theorem 8.2. *Let n be odd and let*

$$B = \mathrm{Bez}_{\mathrm{sp}}(\mathbf{f}_1, \mathbf{g}_1) + M_{n-2}^{t^2-1} \mathrm{Bez}_{\mathrm{sp}}(\mathbf{f}_2, \mathbf{g}_2) \, (M_{n-2}^{t^2-1})^T ,$$

where $\mathbf{f}_1, \mathbf{g}_1 \in \mathbb{F}_+^{n+2}$, $\mathbf{f}_2, \mathbf{g}_2 \in \mathbb{F}_+^n$ are such that $\{\mathbf{f}_1(t), \mathbf{g}_1(t)\}$ and $\{\mathbf{f}_2(t), \mathbf{g}_2(t)\}$ are two pairs of (generalized) coprime polynomials. Assume in addition

$$\operatorname{rank} \begin{bmatrix} \mathbf{f}_1(1) & \mathbf{g}_1(1) \\ \mathbf{f}_1(-1) & \mathbf{g}_1(-1) \end{bmatrix} = 2. \tag{8.4}$$

Then the Bézout equations

$$\mathbf{f}_1(t)\boldsymbol{\alpha}_1(t) + \mathbf{g}_1(t)\boldsymbol{\beta}_1(t) = t^2 - 1, \tag{8.5}$$

$$\mathbf{f}_2(t)\boldsymbol{\alpha}_2(t) + \mathbf{g}_2(t)\boldsymbol{\beta}_2(t) = 1, \tag{8.6}$$

admit solutions $\boldsymbol{\alpha}_i, \boldsymbol{\beta}_i \in \mathbb{F}^{n+1}$, and B is invertible. The inverse has the representation

$$2\,B^{-1} = T_n(\mathbf{c}_1)P_+ + T_n(\mathbf{c}_2)P_- ,$$

where

$$\mathbf{c}_i(t) = \frac{\boldsymbol{\beta}_i(t)\boldsymbol{\alpha}_i^J(t) - \boldsymbol{\alpha}_i(t)\boldsymbol{\beta}_i^J(t)}{t^2 - 1}, \qquad i = 1, 2, \tag{8.7}$$

and $\boldsymbol{\alpha}_i, \boldsymbol{\beta}_i$ are any solutions of (8.5) *and* (8.6).

Proof. Again, before we start with the actual proof let us remark that the Bézout equation (8.5) is uniquely solvable, while the solution to (8.6) is not unique. Indeed, the allowed degree of $\boldsymbol{\alpha}_1(t)$ and $\boldsymbol{\beta}_1(t)$ is one less than the maximal degree of $\mathbf{f}_1(t)$ and $\mathbf{g}_1(t)$ (hence uniqueness), and the allowed degree of $\boldsymbol{\alpha}_2(t)$ and $\boldsymbol{\beta}_2(t)$ is one more than the maximal degree of $\mathbf{f}_2(t)$ and $\mathbf{g}_2(t)$ (hence two degrees of freedom). Solutions exist irrespective of the right-hand side.

As in the proof of Theorem 8.1 we show the formula for B^{-1} in three steps.

Step 1*:* Again we first assume that $\mathbf{f}_1, \mathbf{f}_2, \mathbf{g}_1, \mathbf{g}_2$ satisfy the conditions stated in (7.3). Then we can proceed as in Theorem 7.2, introduce $\mathbf{u}_i$ and $\mathbf{v}_i$ as in (7.4) and obtain unique solutions of the Bézout equations

$$\mathbf{u}_i(t)\tilde{\boldsymbol{\alpha}}_i(t) + \mathbf{v}_i(t)\tilde{\boldsymbol{\beta}}_i(t) = 1.$$

We substitute $\mathbf{u}_i$ and $\mathbf{v}_i$ and get the two equations

$$(t-1)\mathbf{f}_1(t)\tilde{\boldsymbol{\alpha}}_1(t) + (t+1)\mathbf{g}_1(t)\tilde{\boldsymbol{\beta}}_1(t) = t^2 - 1,$$

$$(t+1)\mathbf{f}_2(t)\tilde{\boldsymbol{\alpha}}_2(t) + (t-1)\mathbf{g}_2(t)\tilde{\boldsymbol{\beta}}_2(t) = 1.$$

The remaining is similar to the corresponding part of the proof of Theorem 8.1 and is therefore omitted. We can conclude that for specific $\mathbf{f}_i, \mathbf{g}_i$, there exist solutions $\{\boldsymbol{\alpha}_i, \boldsymbol{\beta}_i\}$ such that the inversion formula holds.

Step 2*:* We only need to consider $\{\mathbf{f}_2(t), \mathbf{g}_2(t)\}$ and the corresponding solutions $\{\boldsymbol{\alpha}_2, \boldsymbol{\beta}_2\}$. Any other solution is given by

$$\boldsymbol{\alpha}_2'(t) = \boldsymbol{\alpha}_2(t) + (\mu + \lambda t)\mathbf{g}_2(t), \quad \boldsymbol{\beta}_2'(t) = \boldsymbol{\beta}_2(t) - (\mu + \lambda t)\mathbf{f}_2(t).$$

with arbitrary parameters $\lambda, \mu \in \mathbb{F}$. Then the corresponding $\mathbf{c}_i'$ evaluates to

$$\begin{aligned}(t^2-1)\mathbf{c}_2'(t) &= \boldsymbol{\beta}_2'(t)(\boldsymbol{\alpha}_2')^J(t) - \boldsymbol{\alpha}_2'(t)(\boldsymbol{\beta}_2')^J(t)\\ &= \Big(\boldsymbol{\beta}_2(t) - (\mu+t\lambda)\mathbf{f}_2(t)\Big)\Big(\boldsymbol{\alpha}_2^J(t) + (\mu+\lambda t)^J\mathbf{g}_2(t)\Big)\\ &\qquad - \Big(\boldsymbol{\alpha}_2(t) + (\mu+t\lambda)\mathbf{g}_2(t)\Big)\Big(\boldsymbol{\beta}_2^J(t) - (\mu+t\lambda)^J\mathbf{f}_2(t)\Big)\\ &= (t^2-1)\mathbf{c}_2(t) + (t\mu+\lambda)\Big(\mathbf{f}_2(t)\boldsymbol{\alpha}_2(t) + \mathbf{g}_2(t)\boldsymbol{\beta}_2(t)\Big)\\ &\qquad - (\mu+t\lambda)\Big(\mathbf{f}_2^J(t)\boldsymbol{\alpha}_2^J(t) + \mathbf{g}_2^J(t)\boldsymbol{\beta}_2^J(t)\Big)\\ &= (t^2-1)\mathbf{c}_2(t) + (t\mu+\lambda) - (\mu+t\lambda)t^{2n-1}\,.\end{aligned}$$

Here we used that $\mathbf{f}_i^J = \mathbf{f}_i$ and $\mathbf{g}_i^J = \mathbf{g}_i$ and $(\mu+\lambda t)^J = \mu t + \lambda$. Hence

$$\mathbf{c}_2'(t) = \mathbf{c}_2(t) - \mu\,\frac{t^{2n-1}-t}{t^2-1} - \lambda\,\frac{t^{2n}-1}{t^2-1}\,,$$

i.e.,

$$\mathbf{c}_2'(t) = \mathbf{c}_2(t) - \mu\sum_{k=0}^{n-2} t^{2k+1} - \lambda\sum_{k=0}^{n-1} t^{2k}\,.$$

It follows that $T_n(\mathbf{c}_2')P_- = T_n(\mathbf{c}_2)P_-$ in view of Proposition 3.1.

Step 3*:* This is analogous to Step 3 in the proof of Theorem 8.1. □

Finally, let us show that if $\mathbb{F} = \mathbb{R}$ or $\mathbb{F} = \mathbb{C}$, then the rank condition (8.4) in the case (o^+) can be eliminated. We will use an approximation argument.

Theorem 8.3. *In case* $\mathbb{F} = \mathbb{R}$ *or* $\mathbb{F} = \mathbb{C}$*, the results of Theorem* 8.2 *are true even without the rank condition* (8.4).

Proof. Introduce the following sets

$$\mathcal{S} = \Big\{\, [\mathbf{f},\mathbf{g}] \in \mathbb{F}_+^{n+2} \times \mathbb{F}_+^{n+2} \,:\, \{\mathbf{f}(t),\mathbf{g}(t)\} \text{ are coprime} \,\Big\}$$

and

$$\mathcal{S}_0 = \Big\{\, [\mathbf{f},\mathbf{g}] \in \mathcal{S} \,:\, \operatorname{rank} m_{\mathbf{f},\mathbf{g}} = 2 \,\Big\}.$$

The first observation is that $\mathcal{S}_0$ is dense in $\mathcal{S}$. This can be seen as follows.

Let $[\mathbf{f},\mathbf{g}] \in \mathcal{S} \setminus \mathcal{S}_0$. This means $\operatorname{rank} m_{\mathbf{f},\mathbf{g}} = 1$ with $m_{\mathbf{f},\mathbf{g}}$ being defined by (6.7). We can assume without loss of generality that $\mathbf{g}(\pm 1) = 0$ because this can be achieved by a suitable transformation Φ. Now, for $\gamma, \delta \in \mathbb{F}$ we consider

$$\mathbf{q}(t) = \gamma\,\frac{t^{n+1}+1}{2} + \delta\,\frac{t^n+t}{2} \in \mathbb{F}_+^{n+2}[t].$$

Note that $\mathbf{q}(\pm 1) = \gamma \pm \delta$ since n is odd. By choosing γ, δ suitably we can thus achieve that $m_{\mathbf{f},\mathbf{q}}$ has rank 2. Denote $\mathbf{g}_\lambda := \mathbf{g} + \lambda\mathbf{q}$ and consider $[\mathbf{f},\mathbf{g}_\lambda]$. It follows that

$$\det m_{\mathbf{f},\mathbf{g}_\lambda} = \det m_{\mathbf{f},\mathbf{g}} + \lambda \det m_{\mathbf{f},\mathbf{q}} = \lambda \det m_{\mathbf{f},\mathbf{q}}\,.$$

Hence for $\lambda \neq 0$, the rank of $m_{\mathbf{f},\mathbf{g}_\lambda}$ equals two. Since small perturbations of coprime polynomials are coprime again, $\{\mathbf{f}(t), \mathbf{g}_\lambda(t)\}$ are coprime for sufficiently small λ. This proves that $\mathcal{S}_0$ is dense in $\mathcal{S}$.

Now assume that $[\mathbf{f}_1, \mathbf{g}_1] =: [\mathbf{f}, \mathbf{g}]$ and $[\mathbf{f}_2, \mathbf{g}_2]$ satisfy the assumption of Theorem 8.2 except for (8.4). We then consider a sequence $[\mathbf{f}, \mathbf{g}_\lambda] \in \mathcal{S}_0$ which approximates $[\mathbf{f}, \mathbf{g}] \in \mathcal{S} \setminus \mathcal{S}_0$ as $\lambda \to 0$.

The Bézout equation (8.5) with $[\mathbf{f}_1, \mathbf{g}_1]$ replaced by $[\mathbf{f}, \mathbf{g}]$ and $[\mathbf{f}, \mathbf{g}_\lambda]$, respectively, can be written in matrix form, where the matrix involved is a resultant matrix generated by $[\mathbf{f}, \mathbf{g}]$ and $[\mathbf{f}, \mathbf{g}_\lambda]$, respectively. From this kind of argument we see that the solutions of

$$\mathbf{f}(t)\boldsymbol{\alpha}_\lambda(t) + \mathbf{g}_\lambda(t)\boldsymbol{\beta}_\lambda(t) = t^2 - 1$$

converge to the solutions of

$$\mathbf{f}(t)\boldsymbol{\alpha}(t) + \mathbf{g}(t)\boldsymbol{\beta}(t) = t^2 - 1,$$

i.e., $[\boldsymbol{\alpha}_\lambda, \boldsymbol{\beta}_\lambda] \to [\boldsymbol{\alpha}, \boldsymbol{\beta}]$ as $\lambda \to 0$. This implies that the corresponding

$$\mathbf{c}_{1,\lambda}(t) := \frac{\boldsymbol{\beta}_\lambda(t)\boldsymbol{\alpha}_\lambda^J(t) - \boldsymbol{\alpha}_\lambda(t)\boldsymbol{\beta}_\lambda^J(t)}{t^2 - 1}$$

converges to

$$\mathbf{c}_1(t) = \frac{\boldsymbol{\beta}(t)\boldsymbol{\alpha}^J(t) - \boldsymbol{\alpha}(t)\boldsymbol{\beta}^J(t)}{t^2 - 1}$$

(compare (8.7)). □

Remark 8.4. The four Bézout equation (8.1), (8.2), (8.5) and (8.6), have in common that the given polynomials $\{\mathbf{f}_i(t), \mathbf{g}_i(t)\}$ are coprime although their degrees are different in each case. The coprimeness makes sure that solutions exist. They are non-standard in the sense that they have different right-hand sides.

Remark 8.5. We can write these equations in a more uniform way by replacing $\{\mathbf{f}_i(t), \mathbf{g}_i(t)\}$ with $\{\mathbf{f}_\pm(t), \mathbf{g}_\pm(t)\}$, which occur in the original split-Bezoutains (see (4.10) and (4.11) of Remark 4.6). Then, in both cases of n even and odd, we start with the decomposition

$$B = \mathrm{Bez}_{\mathrm{sp}}(\mathbf{f}_+, \mathbf{g}_+) + \mathrm{Bez}_{\mathrm{sp}}(\mathbf{f}_-, \mathbf{g}_-)$$

with $\mathbf{f}_\pm, \mathbf{g}_\pm \in \mathbb{F}_\pm^{n+2}$ (see (4.1) and (4.2)). The Bézout equations take the form

$$\begin{aligned} \mathbf{f}_+(t)\boldsymbol{\alpha}_1(t) + \mathbf{g}_+(t)\boldsymbol{\beta}_1(t) &= t^2 - 1, \\ \mathbf{f}_-(t)\boldsymbol{\alpha}_2(t) + \mathbf{g}_-(t)\boldsymbol{\beta}_2(t) &= t^2 - 1. \end{aligned} \tag{8.8}$$

This looks perhaps more esthetic, but one looses the advantage that the polynomials $\{\mathbf{f}_\pm(t), \mathbf{g}_\pm(t)\}$ are coprime (except in the case (o^+)). Notice also that $\boldsymbol{\alpha}_i, \boldsymbol{\beta}_i \in \mathbb{F}^{n+1}$ in all cases.

9. Further simplifications and algorithms

Let an $n \times n$ nonsingular, centrosymmetric $T + H$-Bezoutian be given by (4.8) in case n even and by (4.9) in case n odd. The corresponding Bézout equations (8.1)–(8.2) and (8.5)–(8.6) have right-hand sides not always equal to one, and the solutions are not always unique. Since in all cases the pairs $\{\mathbf{f}_i(t), \mathbf{g}_i(t)\}$ are coprime, the standard Bézout equations (i.e., the ones with right-hand side equal to one) are uniquely solvable.

In this section we show that the solutions of (8.1)–(8.2) and (8.5)–(8.6), respectively, can be obtained from the (unique) solutions of the standard Bézout equations. Furthermore, we will express the symbols $\mathbf{c}_i$ in terms of the latter solutions.

Proposition 9.1. *For n even, let $\{\mathbf{f}_i(t), \mathbf{g}_i(t)\}$ be two pairs of coprime polynomials in $\mathbb{F}_+^{n+1}[t]$, $i = 1, 2$, and let $\tilde{\boldsymbol{\alpha}}_i, \tilde{\boldsymbol{\beta}}_i \in \mathbb{F}^n$ be the unique solutions of the Bézout equation*

$$\mathbf{f}_i(t)\tilde{\boldsymbol{\alpha}}_i(t) + \mathbf{g}_i(t)\tilde{\boldsymbol{\beta}}_i(t) = 1, \qquad i = 1, 2. \tag{9.1}$$

Then one of the solutions $\mathbf{c}_i$ determined by (8.1), (8.2), and (8.3) is given by

$$\mathbf{c}_1(t) = -\frac{t-1}{t+1}\left(\tilde{\boldsymbol{\beta}}_1(t)\tilde{\boldsymbol{\alpha}}_1^J(t) - \tilde{\boldsymbol{\alpha}}_1(t)\tilde{\boldsymbol{\beta}}_1^J(t)\right), \tag{9.2}$$

$$\mathbf{c}_2(t) = \frac{t+1}{t-1}\left(\tilde{\boldsymbol{\beta}}_2(t)\tilde{\boldsymbol{\alpha}}_2^J(t) - \tilde{\boldsymbol{\alpha}}_2(t)\tilde{\boldsymbol{\beta}}_2^J(t)\right). \tag{9.3}$$

Proof. Given $\tilde{\boldsymbol{\alpha}}_i, \tilde{\boldsymbol{\beta}}_i$ define

$$\boldsymbol{\alpha}_1(t) = (t-1)\tilde{\boldsymbol{\alpha}}_1(t), \qquad \boldsymbol{\beta}_1(t) = (t-1)\tilde{\boldsymbol{\beta}}_1(t),$$
$$\boldsymbol{\alpha}_2(t) = (t+1)\tilde{\boldsymbol{\alpha}}_2(t), \qquad \boldsymbol{\beta}_2(t) = (t+1)\tilde{\boldsymbol{\beta}}_2(t).$$

These are solutions of (8.1) and (8.2). Plugging this into (8.3) implies (9.2) and (9.3). □

It is perhaps interesting to compare the above with the kind of substitutions done in the proof of Theorem 8.1, Step 1. The $\tilde{\boldsymbol{\alpha}}_i$ and $\tilde{\boldsymbol{\beta}}_i$ occurring there are different. Thus the symbols $\mathbf{c}_i$ obtained in Theorem 7.1 and here are generally different from each other, too. However, the resulting $T + H$ matrices are the same (see also Proposition 3.1). We are not going to elaborate further on the relationship.

Algorithm 9.2. Let n be even and let $B \in \mathbb{F}^{n\times n}$ be a nonsingular, centrosymmetric $T + H$-Bezoutian given by

$$B = M_{n-1}^{t+1}\mathrm{Bez}_{\mathrm{sp}}(\mathbf{f}_1, \mathbf{g}_1)\,(M_{n-1}^{t+1})^T + M_{n-1}^{t-1}\mathrm{Bez}_{\mathrm{sp}}(\mathbf{f}_2, \mathbf{g}_2)\,(M_{n-1}^{t-1})^T,$$

where $\{\mathbf{f}_i(t), \mathbf{g}_i(t)\}$ are two pairs of coprime polynomials in $\mathbb{F}_+^{n+1}[t]$, $(i = 1, 2)$.

1. Compute the unique solutions $\tilde{\boldsymbol{\alpha}}_i, \tilde{\boldsymbol{\beta}}_i \in \mathbb{F}^n$ of (9.1).
2. Compute $\mathbf{c}_i = (c_k^{(i)})_{k=-n+1}^{n-1} \in \mathbb{F}_+^{2n-1}$ $(i = 1, 2)$ by using (9.2) and (9.3).

3. Then B^{-1} is given by

$$2\,B^{-1} = T_n(\mathbf{c}_1)P_+ + T_n(\mathbf{c}_2)P_- \,,$$

i.e.,

$$B^{-1} = \frac{1}{4}\left[c_{i-j}^{(1)} + c_{i-j}^{(2)}\right]_{i,j=0}^{n-1} + \frac{1}{4}\left[c_{i+j-n+1}^{(1)} - c_{i+j-n+1}^{(2)}\right]_{i,j=0}^{n-1} .$$

Remark 9.3. Instead of using (9.2) and (9.3) to compute $\mathbf{c}_i \in \mathbb{F}_+^{2n-1}$ one can compute $\mathbf{z}_i = (\mathbf{z}_k^{(i)})_{k=-n+1}^{n-1} \in \mathbb{F}_-^{2n-1}$, $\mathbf{z}_i(t) := \tilde{\boldsymbol{\beta}}_i(t)\tilde{\boldsymbol{\alpha}}_i^J(t) - \tilde{\boldsymbol{\alpha}}_i(t)\tilde{\boldsymbol{\beta}}_i^J(t)$, via

$$(z_k^{(i)})_{k=0}^{n-1} = U_n(\tilde{\boldsymbol{\alpha}}_i)\tilde{\boldsymbol{\beta}}_i - U_n(\tilde{\boldsymbol{\beta}}_i)\tilde{\boldsymbol{\alpha}}_i$$

(see (3.9)–(3.10)) and obtain $\mathbf{c}_i(t)$ from $\mathbf{z}_i(t)$ through elementary operations.

The case of n odd is somewhat more difficult.

Proposition 9.4. *For n odd, let $\{\mathbf{f}_i(t), \mathbf{g}_i(t)\}$ be two pairs of coprime polynomials in $\mathbb{F}_+^{n+2}[t]$ for $i = 1$ and in $\mathbb{F}_+^{n}[t]$ for $i = 2$, respectively. Let $\tilde{\boldsymbol{\alpha}}_1, \tilde{\boldsymbol{\beta}}_1 \in \mathbb{F}^{n+1}$ and $\tilde{\boldsymbol{\alpha}}_2, \tilde{\boldsymbol{\beta}}_2 \in \mathbb{F}^{n-1}$ be the unique solutions of the Bézout equation*

$$\mathbf{f}_i(t)\tilde{\boldsymbol{\alpha}}_i(t) + \mathbf{g}_i(t)\tilde{\boldsymbol{\beta}}_i(t) = 1, \qquad i = 1, 2. \tag{9.4}$$

Then one of the solutions $\mathbf{c}_i$ determined by (8.5), (8.6), and (8.7) is given by

$$\mathbf{c}_1 = \begin{bmatrix} 0 & 0 & 1 & & 0 & 0 & 0 \\ \vdots & \vdots & & \ddots & & \vdots & \vdots \\ 0 & 0 & 0 & & 1 & 0 & 0 \end{bmatrix} \mathbf{y}_1 \in \mathbb{F}_+^{2n-1}, \tag{9.5}$$

where

$$\mathbf{y}_1(t) = -(t^2-1)\left(\tilde{\boldsymbol{\beta}}_1(t)\tilde{\boldsymbol{\alpha}}_1^J(t) - \tilde{\boldsymbol{\alpha}}_1(t)\tilde{\boldsymbol{\beta}}_1^J(t)\right) \in \mathbb{F}_+^{2n+3}[t]\,, \tag{9.6}$$

and by

$$\mathbf{c}_2 = \begin{bmatrix} 0 & \dots & 0 \\ 0 & \dots & 0 \\ 1 & & 0 \\ & \ddots & \\ 0 & & 1 \\ 0 & \dots & 0 \\ 0 & \dots & 0 \end{bmatrix} \mathbf{y}_2 \in \mathbb{F}_+^{2n-1}, \tag{9.7}$$

where

$$\mathbf{y}_2(t) = \frac{\tilde{\boldsymbol{\beta}}_2(t)\tilde{\boldsymbol{\alpha}}_2^J(t) - \tilde{\boldsymbol{\alpha}}_2(t)\tilde{\boldsymbol{\beta}}_2^J(t)}{t^2-1} \in \mathbb{F}_+^{2n-5}[t]. \tag{9.8}$$

Proof. Let us consider the first case ($i = 1$) of (8.5). We start with a solution of (9.4), where $\mathbf{f}_1, \mathbf{g}_1 \in \mathbb{F}_+^{n+2}$ and $\tilde{\boldsymbol{\alpha}}_1, \tilde{\boldsymbol{\beta}}_1 \in \mathbb{F}^{n+1}$. Comparing

$$\mathbf{f}_1(t)\tilde{\boldsymbol{\alpha}}_1(t)(t^2-1) + \mathbf{g}_1(t)\tilde{\boldsymbol{\beta}}_1(t)(t^2-1) = t^2 - 1$$

with (8.5),

$$\mathbf{f}_1(t)\boldsymbol{\alpha}_1(t) + \mathbf{g}_1(t)\boldsymbol{\beta}_1(t) = t^2 - 1,$$

where $\boldsymbol{\alpha}_1, \boldsymbol{\beta}_1 \in \mathbb{F}^{n+1}$, by taking the difference and using generalized coprimeness of $\{\mathbf{f}_1(t), \mathbf{g}_1(t)\}$ we obtain

$$\boldsymbol{\alpha}_1(t) = \tilde{\boldsymbol{\alpha}}_1(t)(t^2-1) + (\mu+\lambda t)\mathbf{g}_1(t),$$
$$\boldsymbol{\beta}_1(t) = \tilde{\boldsymbol{\beta}}_1(t)(t^2-1) - (\mu+\lambda t)\mathbf{f}_1(t),$$

with certain (unique) $\lambda, \mu \in \mathbb{F}$. Notice the degrees of the various polynomials herein. In particular, the coefficients of the two highest-order terms t^{n+1} and t^{n+2} on the right-hand side must be zero. Accordingly,

$$\boldsymbol{\alpha}_1^J(t) = \tilde{\boldsymbol{\alpha}}_1^J(t)(t^{-2}-1) + (\mu t^{-1}+\lambda t^{-2})\mathbf{g}_1(t),$$
$$\boldsymbol{\beta}_1^J(t) = \tilde{\boldsymbol{\beta}}_1^J(t)(t^{-2}-1) - (\mu t^{-1}+\lambda t^{-2})\mathbf{f}_1(t).$$

Now the vector $\mathbf{c}_1 \in \mathbb{F}_+^{2n-1}$ of (8.7) evaluates to

$$\begin{aligned}(t^2-1)\mathbf{c}_1(t) &= \boldsymbol{\beta}_1(t)\boldsymbol{\alpha}_1^J(t) - \boldsymbol{\alpha}_1(t)\boldsymbol{\beta}_1^J(t)\\
&= (t^{-2}-1)(t^2-1)\left(\tilde{\boldsymbol{\beta}}_1(t)\tilde{\boldsymbol{\alpha}}_1^J(t) - \tilde{\boldsymbol{\alpha}}_1(t)\tilde{\boldsymbol{\beta}}_1^J(t)\right)\\
&\quad + (\mu t^{-1}+\lambda t^{-2})(t^2-1)\left(\mathbf{f}_1(t)\tilde{\boldsymbol{\alpha}}_1(t) + \mathbf{g}_1(t)\tilde{\boldsymbol{\beta}}_1(t)\right)\\
&\quad - (\mu+\lambda t)(t^{-2}-1)\left(\mathbf{f}_1^J(t)\tilde{\boldsymbol{\alpha}}_1^J(t) + \mathbf{g}_1^J(t)\tilde{\boldsymbol{\beta}}_1^J(t)\right)\\
&= (t^{-2}-1)(t^2-1)\left(\tilde{\boldsymbol{\beta}}_1(t)\tilde{\boldsymbol{\alpha}}_1^J(t) - \tilde{\boldsymbol{\alpha}}_1(t)\tilde{\boldsymbol{\beta}}_1^J(t)\right)\\
&\quad + (\mu t^{-1}+\lambda t^{-2})(t^2-1) - (\mu+\lambda t)(t^{-2}-1)t^{2n+1}.\end{aligned}$$

Here we used that $\mathbf{f}_i^J = \mathbf{f}_i$ and $\mathbf{g}_i^J = \mathbf{g}_i$. It follows that

$$\begin{aligned}\mathbf{c}_1(t) &= -t^{-2}(t^2-1)\left(\tilde{\boldsymbol{\beta}}_1(t)\tilde{\boldsymbol{\alpha}}_1^J(t) - \tilde{\boldsymbol{\alpha}}_1(t)\tilde{\boldsymbol{\beta}}_1^J(t)\right)\\
&\qquad + \lambda t^{-2} + \mu t^{-1} + \mu t^{2n-1} + \lambda t^{2n}.\end{aligned}$$

Therefore, we have to drop the two lowest and two highest entries in the vector corresponding to the polynomial

$$\mathbf{y}_1(t) = -(t^2-1)\left(\tilde{\boldsymbol{\beta}}_1(t)\tilde{\boldsymbol{\alpha}}_1^J(t) - \tilde{\boldsymbol{\alpha}}_1(t)\tilde{\boldsymbol{\beta}}_1^J(t)\right) \in \mathbb{F}_+^{2n+3}[t].$$

But this just means (9.6). Thus the case $i=1$ is proved.

Let us now consider the case $(i=2)$ of (8.6). Again we start with a solution of (9.4), where $\mathbf{f}_2, \mathbf{g}_2 \in \mathbb{F}_+^n$ and $\tilde{\boldsymbol{\alpha}}_2, \tilde{\boldsymbol{\beta}}_2 \in \mathbb{F}^{n-1}$. As (9.1) is the same equation as (8.6) we put

$$\boldsymbol{\alpha}_2(t) = \tilde{\boldsymbol{\alpha}}_2(t), \qquad \boldsymbol{\beta}_2(t) = \tilde{\boldsymbol{\beta}}_2(t).$$

But note that $\boldsymbol{\alpha}_2, \boldsymbol{\beta}_2 \in \mathbb{F}^{n+1}$. Thus

$$\boldsymbol{\alpha}_2^J(t) = t^2\tilde{\boldsymbol{\alpha}}_2^J(t), \qquad \boldsymbol{\beta}_2^J(t) = t^2\tilde{\boldsymbol{\beta}}_2^J(t).$$

It follows that

$$\mathbf{c}_2(t) = \frac{t^2}{t^2-1}\left(\tilde{\boldsymbol{\beta}}_2(t)\tilde{\boldsymbol{\alpha}}_2^J(t) - \tilde{\boldsymbol{\alpha}}_2(t)\tilde{\boldsymbol{\beta}}_2^J(t)\right).$$

This means that the first two (and the last two) entries of the vector $\mathbf{c}_2 \in \mathbb{F}_+^{2n-1}$ are zero. Therefore, we can write this formula also as (9.7) and (9.8). □

Algorithm 9.5. Let n be odd and let $B \in \mathbb{F}^{n\times n}$ be a nonsingular, centrosymmetric $T+H$-Bezoutian given by

$$B = \mathrm{Bez}_{\mathrm{sp}}(\mathbf{f}_1, \mathbf{g}_1) + M_{n-2}^{t^2-1}\, \mathrm{Bez}_{\mathrm{sp}}(\mathbf{f}_2, \mathbf{g}_2)\, (M_{n-2}^{t^2-1})^T$$

with $\mathbf{f}_1, \mathbf{g}_1 \in \mathbb{F}_+^{n+2}$, $\mathbf{f}_2, \mathbf{g}_2 \in \mathbb{F}_+^n$, such that $\{\mathbf{f}_1(t), \mathbf{g}_1(t)\}$ and $\{\mathbf{f}_2(t), \mathbf{g}_2(t)\}$ are two pairs of coprime polynomials. Assume in addition

$$\mathrm{rank}\begin{bmatrix} \mathbf{f}_1(1) & \mathbf{g}_1(1) \\ \mathbf{f}_1(-1) & \mathbf{g}_1(-1) \end{bmatrix} = 2.$$

1. Compute the unique solutions $\tilde{\boldsymbol{\alpha}}_i, \tilde{\boldsymbol{\beta}}_i \in \mathbb{F}^{m_i}$ of (9.4), where
$$m_i = \begin{cases} n-1 & \text{for } i=1 \\ n+1 & \text{for } i=2. \end{cases}$$
2. Compute $\mathbf{c}_i = (c_k^{(i)})_{k=-n+1}^{n-1} \in \mathbb{F}_+^{2n-1}$ by using (9.5)–(9.8).
3. Then B^{-1} is given by
$$2\, B^{-1} = T_n(\mathbf{c}_1)P_+ + T_n(\mathbf{c}_2)P_- \,,$$
i.e.,
$$B^{-1} = \frac{1}{4}\left[c_{i-j}^{(1)} + c_{i-j}^{(2)}\right]_{i,j=0}^{n-1} + \frac{1}{4}\left[c_{i+j-n+1}^{(1)} - c_{i+j-n+1}^{(2)}\right]_{i,j=0}^{n-1}.$$

Remark 9.6. As before, one can use

$$(z_k^{(i)})_{k=0}^{m_i} = U_{m_i+1}(\tilde{\boldsymbol{\alpha}}_i)\tilde{\boldsymbol{\beta}}_i - U_{m_i+1}(\tilde{\boldsymbol{\beta}}_i)\tilde{\boldsymbol{\alpha}}_i$$

to compute $\mathbf{z}_i = (z_k^{(i)})_{k=-m_i}^{m_i} \in \mathbb{F}_-^{2m_i-1}$, $\mathbf{z}_i(t) := \tilde{\boldsymbol{\beta}}_i(t)\tilde{\boldsymbol{\alpha}}_i^J(t) - \tilde{\boldsymbol{\alpha}}_i(t)\tilde{\boldsymbol{\beta}}_i^J(t)$, and obtain $\mathbf{c}_1$ and $\mathbf{c}_2$ through elementary operations.

Remark 9.7. Algorithms 9.2 and 9.5 have complexity $O(n^2)$. Indeed, the solution of the Bézout equations in Step 1 requires $O(n^2)$ operations using the extended Euclidean algorithm. In case of $\mathbb{F}=\mathbb{C}$ or $\mathbb{F}=\mathbb{R}$ the complexity can be reduced to $O(n\log^2 n)$ (see [1], [18], [2]).

The computation in Step 2, discrete convolution or polynomial multiplication, can be done with $O(n(\log n)\log\log n)$ complexity (see [21], Section 2.4), which can be accelerated to $O(n\log n)$ complexity using FFT (see [2] and references therein). Note that polynomial division by linear factors $(t\pm 1)$ can be done using the Horner scheme with linear complexity $O(n)$.

Regarding the solution of a linear system $B\mathbf{x} = \mathbf{y}$ with a coefficient matrix B being a nonsingular, centrosymmetric $T+H$-Bezoutian of order n, it is recommendable to use our algorithms and compute the symbols $\mathbf{c}_1$ and $\mathbf{c}_2$ of the $T+H$ matrix B^{-1}. Indeed, $\mathbf{x} = B^{-1}\mathbf{y}$ can be computed by convolving $\mathbf{c}_1$ and $\mathbf{c}_2$ with $\mathbf{y}$. For more details see Section 8 of [6].

10. Examples

Let us now discuss two examples.

Example 10.1. We consider the case where $\mathbb{F} = \mathbb{R}$, n is odd, and where condition (8.4) is not satisfied. Let B be the following nonsingular, centrosymmetric $T+H$-Bezoutian of order three,

$$B = \begin{bmatrix} 3 & 0 & -1 \\ 0 & 4 & 0 \\ -1 & 0 & 3 \end{bmatrix}.$$

The splitting $B = B_{++} + B_{--}$ is given by

$$B_{++} = BP_+ = \begin{bmatrix} 1 & 0 & 1 \\ 0 & 4 & 0 \\ 1 & 0 & 1 \end{bmatrix}, \quad B_{--} = BP_- = \begin{bmatrix} 2 & 0 & -2 \\ 0 & 0 & 0 \\ -2 & 0 & 2 \end{bmatrix}.$$

Since $B_{++}(t,s) = 1 + t^2 + s^2 + 4ts + t^2s^2$ we observe that

$$B_{++}(t,s) = (\mathrm{Bez}_{\mathrm{sp}}(\mathbf{f}_1, \mathbf{g}_1))(t,s) = \frac{\mathbf{f}_1(t)\mathbf{g}_1(s) - \mathbf{g}_1(t)\mathbf{f}_1(s)}{(t-s)(1-ts)},$$

where $\mathbf{f}_1(t) = t + t^3$, $\mathbf{g}_1(t) = (1-t^2)^2$. Obviously, $\mathbf{g}_1(\pm 1) = 0$, $\mathbf{f}_1(\pm 1) \neq 0$, so that

$$m_{\mathbf{f}_1,\mathbf{g}_1} = \begin{bmatrix} \mathbf{f}_1(1) & 0 \\ \mathbf{f}_1(-1) & 0 \end{bmatrix}.$$

In other words, this is the case (o^+) with $\operatorname{rank} m_{\mathbf{f}_1,\mathbf{g}_1} = 1$. Note that the additional condition (8.4) is not satisfied. Nevertheless, we solve the Bézout equation (8.5),

$$\mathbf{f}_1(t)\boldsymbol{\alpha}_1(t) + \mathbf{g}_1(t)\boldsymbol{\beta}_1(t) = t^2 - 1,$$

and obtain $\boldsymbol{\alpha}_1(t) = \frac{1}{2}t(t^2-1)$, $\boldsymbol{\beta}_1(t) = -1 - \frac{1}{2}t^2$, i.e.,

$$\boldsymbol{\alpha}_1 = \frac{1}{2}\begin{bmatrix} 0 \\ -1 \\ 0 \\ 1 \end{bmatrix}, \quad \boldsymbol{\beta}_1 = \frac{1}{2}\begin{bmatrix} -2 \\ 0 \\ -1 \\ 0 \end{bmatrix}.$$

Consequently, in view of Theorem 8.2 and (8.7),

$$\mathbf{c}_1(t) = \frac{1}{t^2-1}\left(-(1+\frac{t^2}{2})\frac{1-t^2}{2} + \frac{t}{2}(t^2-1)(t^3+\frac{t}{2})\right) = \frac{1}{2}(1+t^2+t^4) \in \mathbb{R}^5[t].$$

Thus,

$$T_3(\mathbf{c}_1) = \frac{1}{2}\begin{bmatrix} 1 & 0 & 1 \\ 0 & 1 & 0 \\ 1 & 0 & 1 \end{bmatrix} = T_3(\mathbf{c}_1)P_+.$$

For verification, let us note that we get the same result using (9.4), (9.5), and (9.6). Indeed, the solution of $\mathbf{f}_1(t)\tilde{\boldsymbol{\alpha}}_1(t) + \mathbf{g}_1(t)\tilde{\boldsymbol{\beta}}_1(t) = 1$ is

$$\tilde{\boldsymbol{\alpha}}_1(t) = \frac{1}{4}(5t - 3t^3), \quad \tilde{\boldsymbol{\beta}}_1(t) = \frac{1}{4}(4+3t^2),$$

which yields

$$\begin{aligned}\mathbf{y}_1(t) &= \frac{1-t^2}{16}\left((4+3t^2)(5t^2-3)-(5t-3t^3)(4t^3+3t)\right)\\ &= -\frac{3}{4}+\frac{1}{2}t^2+\frac{1}{2}t^4+\frac{1}{2}t^8-\frac{3t^8}{4} \in \mathbb{R}_+^9[t].\end{aligned}$$

After removing the first two and the last two entries of the corresponding vector $\mathbf{y}_1$, we obtain the same $\mathbf{c}_1 = \frac{1}{2}[1,0,1,0,1]^T$.

Analogously, the second splitting part can be written as

$$\begin{aligned}B_{--}(t,s) &= \frac{(t-t^3)(1-s^4)-(1-t^4)(s-s^3)}{(t-s)(1-ts)}\\ &= (t^2-1)\frac{\mathbf{f}_2(t)\mathbf{g}_2(s)-\mathbf{g}_2(t)\mathbf{f}_2(s)}{(t-s)(1-ts)}(s^2-1),\end{aligned}$$

where $\mathbf{f}_2(t) = t$, $\mathbf{g}_2(t) = t^2+1$. Clearly, $\mathbf{g}_1(1) \neq 0$, $\mathbf{f}_2(-1) \neq 0$, and the condition of (o^-) in Lemma 6.3 is satisfied. Solving the Bézout equations (8.6) or (9.4), and combining it with (8.7) or (9.7)–(9.8), we finally end up with $\mathbf{c}_2(t) = \frac{t^2}{2} \in \mathbb{R}_+^5[t]$. This leads to

$$T_3(\mathbf{c}_2) = \frac{1}{2}I_3, \quad \text{hence} \quad T_3(\mathbf{c}_2)P_- = \frac{1}{4}\begin{bmatrix} 1 & 0 & -1\\ 0 & 0 & 0\\ -1 & 0 & 1\end{bmatrix}.$$

Putting all things together, the inverse $B^{-1} = \frac{1}{2}(T_3(\mathbf{c}_1)P_+ + T_3(\mathbf{c}_2)P_-)$ is given by

$$B^{-1} = \frac{1}{4}\begin{bmatrix} 1 & 0 & 1\\ 0 & 1 & 0\\ 1 & 0 & 1\end{bmatrix} + \frac{1}{8}\begin{bmatrix} 1 & 0 & -1\\ 0 & 0 & 0\\ -1 & 0 & 1\end{bmatrix} = \frac{1}{8}\begin{bmatrix} 3 & 0 & 1\\ 0 & 2 & 0\\ 1 & 0 & 3\end{bmatrix}.$$

We can represent B^{-1} also as sum of a Toeplitz and a Hankel matrix, e.g.,

$$B^{-1} = \frac{1}{8}\begin{bmatrix} 2 & 0 & 1\\ 0 & 2 & 0\\ 1 & 0 & 2\end{bmatrix} + \frac{1}{8}\begin{bmatrix} 1 & 0 & 0\\ 0 & 0 & 0\\ 0 & 0 & 1\end{bmatrix}.$$

Note $4B^{-1}$ is equal to the matrix R_n of Example 3.3.

Although the condition (8.4) is not satisfied and $T_3(\mathbf{c}_1)$ is singular, we observe

$$\frac{1}{2}T_3(\mathbf{c}_1)P_+BP_+ = P_+ \quad \text{and} \quad \frac{1}{2}T_3(\mathbf{c}_2)P_-BP_- = P_-$$

(compare Remark 4.7). □

Example 10.2. Consider the finite field $\mathbb{F} = \mathbb{Z}/(7\mathbb{Z})$ and let $n = 4$ even. We start with

$$B = \begin{bmatrix} 2 & 3 & 6 & 5\\ 3 & 1 & 0 & 6\\ 6 & 0 & 1 & 3\\ 5 & 6 & 3 & 2\end{bmatrix},$$

which decomposes into

$$B_{++} = \begin{bmatrix} 0 & 1 & 1 & 0 \\ 1 & 4 & 4 & 1 \\ 1 & 4 & 4 & 1 \\ 0 & 1 & 1 & 0 \end{bmatrix}, \quad B_{--} = \begin{bmatrix} 2 & 2 & 5 & 5 \\ 2 & 4 & 3 & 5 \\ 5 & 3 & 4 & 2 \\ 5 & 5 & 2 & 2 \end{bmatrix}.$$

We can represent these split-Bezoutians as

$$B_{++}(t,s) = (t+1)\frac{\mathbf{f}_1(t)\mathbf{g}_1(s) - \mathbf{g}_1(t)\mathbf{f}_1(s)}{(t-s)(1-ts)}(s+1)$$

with $\mathbf{f}_1(t) = t^2$, $\mathbf{g}_1(t) = 1 + 2t + 2t^3 + t^4$, and

$$B_{--}(t,s) = (t-1)\frac{\mathbf{f}_2(t)\mathbf{g}_2(s) - \mathbf{g}_2(t)\mathbf{f}_2(s)}{(t-s)(1-ts)}(s-1)$$

with $\mathbf{f}_2(t) = t + 2t^2 + t^3$, $\mathbf{g}_2(t) = 2 + t^2 + 2t^4$.

The computation of the solutions of $\mathbf{f}_i(t)\tilde{\boldsymbol{\alpha}}_i(t) + \mathbf{g}_i(t)\tilde{\boldsymbol{\beta}}_i(t) = 1$ in $\mathbb{F}^4[t]$ gives

$$\tilde{\boldsymbol{\alpha}}_1(t) = 4 + 5t + 3t^3 + 2t^3\,, \qquad \tilde{\boldsymbol{\beta}}_1(t) = 1 + 5t\,,$$
$$\tilde{\boldsymbol{\alpha}}_2(t) = 1 + 4t + 2t^2 + 3t^3\,, \qquad \tilde{\boldsymbol{\beta}}_2(t) = 4 + 3t + 2t^2\,.$$

Using (9.2) and (9.3), we obtain by a straightforward calculation

$$\mathbf{c}_1(t) = 2+2t+6t^2+t^3+6t^4+2t^5+2t^6, \quad \mathbf{c}_2(t) = 2+3t+6t^2+3t^3+6t^4+3t^5+2t^6.$$

Thus

$$T_4(\mathbf{c}_1) = \begin{bmatrix} 1 & 6 & 2 & 2 \\ 6 & 1 & 6 & 2 \\ 2 & 6 & 1 & 6 \\ 2 & 2 & 6 & 1 \end{bmatrix}, \quad T_4(\mathbf{c}_2) = \begin{bmatrix} 3 & 6 & 3 & 2 \\ 6 & 3 & 6 & 3 \\ 3 & 6 & 3 & 6 \\ 2 & 3 & 6 & 3 \end{bmatrix}.$$

Taking $\frac{1}{4} = 2 \mod 7$ into account, we see that

$$\frac{1}{2}T_4(\mathbf{c}_1)P_+ = \begin{bmatrix} 6 & 2 & 2 & 6 \\ 2 & 0 & 0 & 2 \\ 2 & 0 & 0 & 2 \\ 6 & 2 & 2 & 6 \end{bmatrix}, \quad \frac{1}{2}T_4(\mathbf{c}_2)P_- = \begin{bmatrix} 2 & 6 & 1 & 5 \\ 6 & 1 & 6 & 1 \\ 1 & 6 & 1 & 6 \\ 5 & 1 & 6 & 2 \end{bmatrix},$$

and thus

$$B^{-1} = \begin{bmatrix} 1 & 1 & 3 & 4 \\ 1 & 1 & 6 & 3 \\ 3 & 6 & 1 & 1 \\ 4 & 3 & 1 & 1 \end{bmatrix}.$$

This is indeed the inverse of B in $\mathbb{Z}/(7\mathbb{Z})$. □

References

[1] A.V. Aho, J.E. Hopcroft, and J.D. Ullman. *The design and analysis of computer algorithms.* Addison-Wesley Publishing Co., Reading, Mass.-London-Amsterdam, 1975. Second printing, Addison-Wesley Series in Computer Science and Information Processing.

[2] D. Bini and V.Y. Pan. *Polynomial and matrix computations. Vol.* 1. Progress in Theoretical Computer Science. Birkhäuser Boston Inc., Boston, MA, 1994. Fundamental algorithms.

[3] T. Ehrhardt and K. Rost. On the kernel structure of generalized resultant matrices. *Indagationes Mathematicae*, 23:1053–1069, 2012.

[4] T. Ehrhardt and K. Rost. Resultant matrices and inversion of Bezoutians. *Linear Algebra Appl.*, 439:621–639, 2013.

[5] T. Ehrhardt and K. Rost. Inversion of centrosymmetric Toeplitz-plus-Hankel Bezoutians. *Electron. Trans. Numer. Anal.*, 42:106–135, 2014.

[6] T. Ehrhardt and K. Rost. Inversion of centroskewsymmetric Toeplitz-plus-Hankel Bezoutians. *Electron. J. Linear Algebra*, 30:336–359, 2015.

[7] I.C. Gohberg and A.A. Semencul. The inversion of finite Toeplitz matrices and their continuous analogues. *Mat. Issled.*, 7(2(24)):201–223, 290, 1972.

[8] M.C. Gouveia. Group and Moore–Penrose invertibility of Bezoutians. *Linear Algebra Appl.*, 197/198:495–509, 1994. Second Conference of the International Linear Algebra Society (ILAS) (Lisbon, 1992).

[9] G. Heinig and U. Jungnickel. Hankel matrices generated by the Markov parameters of rational functions. *Linear Algebra Appl.*, 76:121–135, 1986.

[10] G. Heinig and K. Rost. *Invertierung einiger Klassen von Matrizen und Operatoren. I. Endliche Toeplitzmatrizen und ihre Verallgemeinerungen*, volume 12 of *Wissenschaftliche Informationen [Scientific Information]*. Technische Hochschule Karl-Marx-Stadt Sektion Mathematik, Karl-Marx-Stadt, 1979.

[11] G. Heinig and K. Rost. *Algebraic methods for Toeplitz-like matrices and operators*, volume 13 of *Operator Theory: Advances and Applications.* Birkhäuser Verlag, Basel, 1984.

[12] G. Heinig and K. Rost. Fast inversion of Toeplitz-plus-Hankel matrices. *Wiss. Z. Tech. Hochsch. Karl-Marx-Stadt*, 27(1):66–71, 1985.

[13] G. Heinig and K. Rost. On the inverses of Toeplitz-plus-Hankel matrices. *Linear Algebra Appl.*, 106:39–52, 1988.

[14] G. Heinig and K. Rost. Hartley transform representations of inverses of real Toeplitz-plus-Hankel matrices. In *Proceedings of the International Conference on Fourier Analysis and Applications (Kuwait,* 1998), volume 21, pages 175–189, 2000.

[15] G. Heinig and K. Rost. Centro-symmetric and centro-skewsymmetric Toeplitz matrices and Bezoutians. *Linear Algebra Appl.*, 343/344:195–209, 2002. Special issue on structured and infinite systems of linear equations.

[16] G. Heinig and K. Rost. Centrosymmetric and centro-skewsymmetric Toeplitz-plus-Hankel matrices and Bezoutians. *Linear Algebra Appl.*, 366:257–281, 2003. Special issue on structured matrices: analysis, algorithms and applications (Cortona, 2000).

[17] T. Kailath, S.Y. Kung, and M. Morf. Displacement ranks of matrices and linear equations. *J. Math. Anal. Appl.*, 68(2):395–407, 1979.

[18] D.E. Knuth. *The art of computer programming. Vol.* 2. Addison-Wesley, Reading, MA, 1998. Seminumerical algorithms, Third edition [of MR0286318].

[19] F.I. Lander. The Bezoutian and the inversion of Hankel and Toeplitz matrices (in Russian). *Mat. Issled.*, 9(2 (32)):69–87, 249–250, 1974.

[20] A.B. Nersesyan and A.A. Papoyan. Construction of a matrix inverse to the sum of Toeplitz and Hankel matrices. *Izv. Akad. Nauk Armyan. SSR Ser. Mat.*, 18(2):150–160, 1983.

[21] V.Y. Pan. *Structured matrices and polynomials.* Birkhäuser Boston Inc., Boston, MA, 2001. Unified superfast algorithms.

[22] W.F. Trench. An algorithm for the inversion of finite Toeplitz matrices. *J. Soc. Indust. Appl. Math.*, 12:515–522, 1964.

Torsten Ehrhardt
Mathematics Department
University of California
Santa Cruz, CA-95064, USA
e-mail: tehrhard@ucsc.edu

Karla Rost
Faculty of Mathematics
Technische Universität Chemnitz
Reichenhainer Straße 39
D-09126 Chemnitz, Germany
e-mail: karla.rost@mathematik.tu-chemnitz.de

Operator Theory:
Advances and Applications, Vol. 259, 301–352

On Matrix-valued Stieltjes Functions with an Emphasis on Particular Subclasses

Bernd Fritzsche, Bernd Kirstein and Conrad Mädler

Dedicated to our friend and colleague Albrecht Böttcher on the occasion of his 60th birthday

Abstract. The paper deals with particular classes of $q \times q$ matrix-valued functions which are holomorphic in $\mathbb{C} \setminus [\alpha, +\infty)$, where α is an arbitrary real number. These classes are generalizations of classes of holomorphic complex-valued functions studied by Kats and Krein [17] and by Krein and Nudelman [19]. The functions are closely related to truncated matricial Stieltjes problems on the interval $[\alpha, +\infty)$. Characterizations of these classes via integral representations are presented. Particular emphasis is placed on the discussion of the Moore–Penrose inverse of these matrix-valued functions.

Mathematics Subject Classification (2010). Primary 30E05, 47A57; Secondary 44A60.

Keywords. Matrix-valued Stieltjes functions, integral representations, Moore–Penrose pseudoinverse.

1. Introduction

In their papers [11, 13, 14], the authors developed a simultaneous approach to the even and odd truncated matricial Hamburger moment problems. This approach was based on three cornerstones. One of them, namely the paper [11] is devoted to several function-theoretical aspects concerning special subclasses of matrix-valued Herglotz–Nevanlinna functions. Now we are going to work out a similar simultaneous approach to the even and odd truncated matricial Stieltjes moment problems. Our approach is again subdivided into three steps. This paper is concerned with the first step which is aimed at a closer analysis of several classes of holomorphic matrix-valued functions in the complex plane which turn out to be closely related to Stieltjes type matrix moment problems. In the scalar case, the corresponding classes were carefully studied by Kats/Krein [17] and Krein/Nudelman [19, Appendix]. What concerns the treatment of several matricial and operatorial generalizations, we refer the reader to the monograph Arlinskii/Belyi/Tsekanovskii [1] and the references therein.

In order to give a precise formulation of the matricial moment problem standing in the background of our investigations, we introduce some notation. Throughout this paper, let p and q be positive integers. Let $\mathbb{C}$ and $\mathbb{R}$ be the set of all complex numbers and the set of all real numbers, respectively. Furthermore, let $\mathbb{N}_0$ and $\mathbb{N}$ be the set of all non-negative integers and the set of all positive integers, respectively. Further, for every choice of $\alpha, \beta \in \mathbb{R} \cup \{-\infty, +\infty\}$, let $\mathbb{Z}_{\alpha,\beta}$ be the set of all integers k for which $\alpha \leq k \leq \beta$ holds. If $\mathcal{X}$ is a non-empty set, then let $\mathcal{X}^{p\times q}$ be the set of all $p \times q$ matrices each entry of which belongs to $\mathcal{X}$, and $\mathcal{X}^p$ is short for $\mathcal{X}^{p\times 1}$. The notations $\mathbb{C}_{\mathrm{H}}^{q\times q}$ and $\mathbb{C}_{\geq}^{q\times q}$ stand for the subsets of $\mathbb{C}^{q\times q}$ which are formed by the sets of Hermitian and non-negative Hermitian matrices, respectively. If $(\Omega, \mathfrak{A})$ is a measurable space, then each countably additive mapping whose domain is $\mathfrak{A}$ and whose values belong to $\mathbb{C}_{\geq}^{q\times q}$ is called a non-negative Hermitian $q \times q$ measure on $(\Omega, \mathfrak{A})$. Let $\mathfrak{B}_{\mathbb{R}}$ (resp. $\mathfrak{B}_{\mathbb{C}}$) be the σ-algebra of all Borel subsets of $\mathbb{R}$ (resp. $\mathbb{C}$). For each $\Omega \in \mathfrak{B}_{\mathbb{R}} \setminus \{\emptyset\}$, let $\mathfrak{B}_{\Omega} := \mathfrak{B}_{\mathbb{R}} \cap \Omega$, and let $\mathcal{M}_{\geq}^{q}(\Omega)$ be the set of all non-negative Hermitian $q \times q$ measures on $(\Omega, \mathfrak{B}_{\Omega})$. Furthermore, for each $\Omega \in \mathfrak{B}_{\mathbb{R}} \setminus \{\emptyset\}$ and every $\kappa \in \mathbb{N}_0 \cup \{+\infty\}$, let $\mathcal{M}_{\geq,\kappa}^{q}(\Omega)$ be the set of all $\sigma \in \mathcal{M}_{\geq}^{q}(\Omega)$ such that the integral

$$s_j^{(\sigma)} := \int_{\Omega} t^j \sigma(\mathrm{d}t)$$

exists for all $j \in \mathbb{Z}_{0,\kappa}$. In this paper, for an arbitrarily fixed $\alpha \in \mathbb{R}$, we study classes of $q \times q$ matrix-valued functions which are holomorphic in $\mathbb{C} \setminus [\alpha, +\infty)$. These classes turn out to be closely related via the Stieltjes transform with the following truncated matricial Stieltjes type moment problem:

$\mathsf{M}[[\alpha, +\infty); (s_j)_{j=0}^{m}, =]$: Let $\alpha \in \mathbb{R}$, let $m \in \mathbb{N}_0$, and let $(s_j)_{j=0}^{m}$ be a sequence of complex $q \times q$ matrices. Describe the set $\mathcal{M}_{\geq}^{q}[[\alpha, +\infty); (s_j)_{j=0}^{m}, =]$ of all $\sigma \in \mathcal{M}_{\geq,m}^{q}([\alpha, +\infty))$ for which $s_j^{(\sigma)} = s_j$ is fulfilled for each $j \in \mathbb{Z}_{0,m}$.

In a forthcoming paper, we will indicate how classes of holomorphic functions studied in this paper can be used to parametrize the set $\mathcal{M}_{\geq}^{q}[[\alpha, +\infty); (s_j)_{j=0}^{m}, =]$.

This paper is organized as follows. In Section 2, we introduce several classes of holomorphic $q \times q$ matrix-valued functions. A particular important role will be played by the class $\mathcal{S}_{q;[\alpha,+\infty)}$ of $[\alpha, +\infty)$-Stieltjes functions of order q, which was considered in the special case $q = 1$ and $\alpha = 0$ by I.S. Kats and M.G. Krein in [17] (see Definition 2.1). In Section 3, we derive several integral representations for functions belonging to $\mathcal{S}_{q;[\alpha,+\infty)}$ (see Theorems 3.1 and 3.6). Furthermore we analyse the structure of ranges and null spaces of the values of functions belonging to $\mathcal{S}_{q;[\alpha,+\infty)}$ (see Proposition 3.15). In Section 4, we state characterizations of the membership of a function to the class $\mathcal{S}_{q;[\alpha,+\infty)}$. In Section 5, we investigate the subclass $\mathcal{S}_{0,q;[\alpha,+\infty)}$ (see Notation 2.7 below) of $\mathcal{S}_{q;[\alpha,+\infty)}$. It is shown in Theorem 5.1 that this class is formed exactly by those $q \times q$ matrix-valued functions defined in $\mathbb{C} \setminus [\alpha, +\infty)$ which can be written as the Stieltjes transform of a non-negative Hermitian $q \times q$ measure defined on the Borelian σ-algebra of the interval $[\alpha, +\infty)$. In Section 6, we investigate the Moore–Penrose inverses of the functions belonging to $\mathcal{S}_{q;[\alpha,+\infty)}$. In particular, we show that the Moore–Penrose inverse $F^\dagger$ of a function $F \in \mathcal{S}_{q;[\alpha,+\infty)}$ is holomorphic in $\mathbb{C} \setminus [\alpha, +\infty)$ and that the function $G : \mathbb{C} \setminus [\alpha, +\infty) \to \mathbb{C}^{q\times q}$ defined by $G(z) := -(z-\alpha)^{-1}[F(z)]^\dagger$ belongs to $\mathcal{S}_{q;[\alpha,+\infty)}$ as well (see Theorem 6.3). The second main theme of Section 6 is concerned with the investigation of the class $\mathcal{S}^{[-1]}_{q;[\alpha,+\infty)}$ (see Notation 6.5), which was considered in the special case $q = 1$ and $\alpha = 0$ by I.S. Kats and M.G. Krein in [17]. The main result on the class $\mathcal{S}^{[-1]}_{q;[\alpha,+\infty)}$ is Theorem 6.10 which contains an integral representation which is even new for the case $q = 1$ and $\alpha = 0$. The application of Theorem 6.10 enables us to obtain much information about ranges, null spaces and Moore–Penrose inverses of the functions belonging to the class $\mathcal{S}^{[-1]}_{q;[\alpha,+\infty)}$ (see Proposition 6.14 and Theorem 6.18). In the remaining sections of this paper, we carry out corresponding investigations for dual classes of $q \times q$ matrix-valued functions which are related to an interval $(-\infty, \beta]$. These classes occur in the treatment of a matrix moment problem $\mathsf{M}[(-\infty, \beta]; (s_j)_{j=0}^m, =]$, which is analogous to Problem $\mathsf{M}[[\alpha, +\infty); (s_j)_{j=0}^m, =]$ formulated above. In Appendix 10, we summarize some facts from the integration theory with respect to non-negative Hermitian $q \times q$ measures.

2. On several classes of holomorphic matrix-valued functions

In this section, we introduce those classes of holomorphic $q \times q$ matrix-valued functions which form the central objects of this paper. For each $A \in \mathbb{C}^{q\times q}$, let $\operatorname{Re} A := \frac{1}{2}(A + A^*)$ and $\operatorname{Im} A := \frac{1}{2\mathrm{i}}(A - A^*)$ be the real part of A and the imaginary part of A, respectively. Let $\Pi_+ := \{z \in \mathbb{C} : \operatorname{Im} z > 0\}$ and $\Pi_- := \{z \in \mathbb{C} : \operatorname{Im} z < 0\}$ be the open upper half-plane and the open lower half-plane of $\mathbb{C}$, respectively. The first two dual classes of holomorphic matrix-valued functions, which are particularly important for this paper, are the following.

Definition 2.1. Let $\alpha \in \mathbb{R}$ and let $F : \mathbb{C} \setminus [\alpha, +\infty) \to \mathbb{C}^{q\times q}$. Then F is called an $[\alpha, +\infty)$-*Stieltjes function of order* q if F satisfies the following three conditions:

(i) F is holomorphic in $\mathbb{C} \setminus [\alpha, +\infty)$;
(ii) for all $w \in \Pi_+$, the matrix $\operatorname{Im}[F(w)]$ is non-negative Hermitian;
(iii) for all $w \in (-\infty, \alpha)$, the matrix $F(w)$ is non-negative Hermitian.

We denote by $\mathcal{S}_{q;[\alpha,+\infty)}$ the set of all $[\alpha, +\infty)$-Stieltjes functions of order q.

Example 2.2. Let $\alpha \in \mathbb{R}$ and let $A, B \in \mathbb{C}^{q\times q}_{\geq}$. Let $F : \mathbb{C} \setminus [\alpha, +\infty) \to \mathbb{C}^{q\times q}$ be defined by $F(z) := A + \frac{1}{\alpha - z}B$. Since $\operatorname{Im} F(z) = \frac{\operatorname{Im} z}{|\alpha - z|^2}B$ holds true for each $z \in \mathbb{C} \setminus [\alpha, +\infty)$, we have $F \in \mathcal{S}_{q;[\alpha,+\infty)}$.

Definition 2.3. Let $\beta \in \mathbb{R}$ and let $G : \mathbb{C} \setminus (-\infty, \beta] \to \mathbb{C}^{q\times q}$. Then G is called a $(-\infty, \beta]$-*Stieltjes function of order* q if G fulfills the following three conditions:

(i) G is holomorphic in $\mathbb{C} \setminus (-\infty, \beta]$;
(ii) for all $w \in \Pi_+$, the matrix $\operatorname{Im}[G(w)]$ is non-negative Hermitian;
(iii) for all $w \in (\beta, +\infty)$, the matrix $-G(w)$ is non-negative Hermitian.

We denote by $\mathcal{S}_{q;(-\infty,\beta]}$ the set of all $(-\infty, \beta]$-Stieltjes functions of order q.

Example 2.4. Let $\beta \in \mathbb{R}$ and let $A, B \in \mathbb{C}^{q\times q}_{\geq}$. Let $G : \mathbb{C} \setminus (-\infty, \beta] \to \mathbb{C}^{q\times q}$ be defined by $G(z) := -A + \frac{1}{\beta - z} B$. Since $\operatorname{Im} G(z) = \frac{\operatorname{Im} z}{|\beta - z|^2} B$ holds true for each $z \in \mathbb{C} \setminus (-\infty, \beta]$, we have $G \in \mathcal{S}_{q;(-\infty,\beta]}$.

The particular functions belonging to the class $\mathcal{S}_{q;[\alpha,+\infty)}$ (resp. $\mathcal{S}_{q;(-\infty,\beta]}$), which were introduced in Example 2.2 (resp. Example 2.4), were called by V.E. Katsnelson [18] *special* functions belonging to $\mathcal{S}_{q;[\alpha,+\infty)}$ (resp. $\mathcal{S}_{q;(-\infty,\beta]}$), whereas all remaining functions contained in $\mathcal{S}_{q;[\alpha,+\infty)}$ (resp. $\mathcal{S}_{q;(-\infty,\beta]}$) were called *generic* functions belonging to $\mathcal{S}_{q;[\alpha,+\infty)}$ (resp. $\mathcal{S}_{q;(-\infty,\beta]}$). In the case $q = 1$ and $\alpha = 0$ the general theory of multiplicative representations of functions belonging to $\mathcal{S}_{1;[0,+\infty)}$ was treated in detail by Aronszajn/Donoghue [2].

It should be mentioned that Yu.M. Dyukarev and V.E. Katsnelson studied in [6–8] an interpolation problem for functions belonging to the class $\mathcal{S}_{q;(-\infty,0]}$. Their approach was based on V.P. Potapov's method of fundamental matrix inequalities. V.E. Katsnelson [18] used the class $\mathcal{S}_{1;(-\infty,0]}$ to construct Hurwitz stable entire functions.

First more or less obvious interrelations between the above-introduced classes of functions can be described by the following result. For each $\alpha \in \mathbb{R}$, let the mapping $T_\alpha : \mathbb{R} \to \mathbb{R}$ be defined by $T_\alpha(x) := x + \alpha$. If $\mathcal{X}$, $\mathcal{Y}$, and $\mathcal{Z}$ are non-empty sets with $\mathcal{Z} \subseteq \mathcal{X}$ and if $f : \mathcal{X} \to \mathcal{Y}$ is a mapping, then we will use $\operatorname{Rstr}_{\mathcal{Z}} f$ to denote the restriction of f onto $\mathcal{Z}$.

Remark 2.5.

(a) If $\alpha \in \mathbb{R}$ and if $F : \mathbb{C} \setminus [\alpha, +\infty) \to \mathbb{C}^{q\times q}$, then F belongs to $\mathcal{S}_{q;[\alpha,+\infty)}$ if and only if the function $F \circ \operatorname{Rstr}_{\mathbb{C}\setminus[0,+\infty)} T_\alpha$ belongs to the class $\mathcal{S}_{q;[0,+\infty)}$.
(b) If $\alpha \in \mathbb{R}$ and if $F : \mathbb{C} \setminus [0, +\infty) \to \mathbb{C}^{q\times q}$, then $F \in \mathcal{S}_{q;[0,+\infty)}$ if and only if $F \circ \operatorname{Rstr}_{\mathbb{C}\setminus[\alpha,+\infty)} T_{-\alpha} \in \mathcal{S}_{q;[\alpha,+\infty)}$.

Remark 2.6.

(a) If $\beta \in \mathbb{R}$ and if $G : \mathbb{C} \setminus (-\infty, \beta] \to \mathbb{C}^{q\times q}$, then G belongs to $\mathcal{S}_{q;(-\infty,\beta]}$ if and only if the function $G \circ \operatorname{Rstr}_{\mathbb{C}\setminus(-\infty,0]} T_\beta$ belongs to the class $\mathcal{S}_{q;(-\infty,0]}$.
(b) If $\beta \in \mathbb{R}$ and if $G : \mathbb{C} \setminus (-\infty, 0] \to \mathbb{C}^{q\times q}$, then $F \in \mathcal{S}_{q;(-\infty,0]}$ if and only if $F \circ \operatorname{Rstr}_{\mathbb{C}\setminus(-\infty,\beta]} T_{-\beta} \in \mathcal{S}_{q;(-\infty,\beta]}$.

For each $A \in \mathbb{C}^{q\times q}$, let $\operatorname{tr} A$ be the trace of A. If $A \in \mathbb{C}^{p\times q}$, then we denote by $\|A\|_{\mathrm{E}} := \sqrt{\operatorname{tr}(A^* A)}$ the Euclidean norm of A.

Now we introduce particular subclasses of $\mathcal{S}_{q;[\alpha,+\infty)}$ and $\mathcal{S}_{q;(-\infty,\beta]}$, which will turn out to be important in studying matricial versions of the Stieltjes moment problem. Special attention will be put in the sequel also to the following subclasses of the classes of holomorphic matrix-valued functions introduced in Definitions 2.1 and 2.3.

Notation 2.7. Let $\alpha \in \mathbb{R}$. Then let $\mathcal{S}_{0,q;[\alpha,+\infty)}$ be the class of all S which belong to $\mathcal{S}_{q;[\alpha,+\infty)}$ and which satisfy

$$\sup_{y\in[1,+\infty)} y\|S(\mathrm{i}y)\|_{\mathrm{E}} < +\infty. \tag{2.1}$$

Furthermore, let $\mathcal{S}_{0,q;(-\infty,\alpha]}$ be the class of all $S \in \mathcal{S}_{q;(-\infty,\alpha]}$ which satisfy (2.1).

Remark 2.8. Let $\alpha \in \mathbb{R}$. If $S \in \mathcal{S}_{0,q;[\alpha,+\infty)}$ or if $S \in \mathcal{S}_{0,q;(-\infty,\alpha]}$, then

$$\lim_{y\to+\infty} S(\mathrm{i}y) = O_{q\times q}.$$

Remark 2.8 leads us to the following classes, which will play an important role in the framework of truncated matricial Stieltjes moment problems.

Notation 2.9. Let $\alpha \in \mathbb{R}$. Then by $\mathcal{S}^{\diamond}_{q;[\alpha,+\infty)}$ (resp. $\mathcal{S}^{\diamond}_{q;(-\infty,\alpha]}$) we denote the set of all $F \in \mathcal{S}_{q;[\alpha,+\infty)}$ (resp. $\mathcal{S}_{q;(-\infty,\alpha]}$) which satisfy

$$\lim_{y\to+\infty} F(\mathrm{i}y) = O_{q\times q}.$$

Remark 2.10. If $\alpha \in \mathbb{R}$, then Remark 2.8 shows that $\mathcal{S}_{0,q;[\alpha,+\infty)} \subseteq \mathcal{S}^{\diamond}_{q;[\alpha,+\infty)}$ and $\mathcal{S}_{0,q;(-\infty,\alpha]} \subseteq \mathcal{S}^{\diamond}_{q;(-\infty,\alpha]}$.

Remark 2.11. Let $\alpha \in \mathbb{R}$ and let

$$\mathcal{S} \in \{\mathcal{S}_{q;[\alpha,+\infty)}, \mathcal{S}_{0,q;[\alpha,+\infty)}, \mathcal{S}^{\diamond}_{q;[\alpha,+\infty)}, \mathcal{S}_{q;(-\infty,\alpha]}, \mathcal{S}_{0,q;(-\infty,\alpha]}, \mathcal{S}^{\diamond}_{q;(-\infty,\alpha]}\}.$$

Then $F \in \mathcal{S}$ if and only if $F^{\mathrm{T}} \in \mathcal{S}$.

Remark 2.12.

(a) Let $\alpha \in \mathbb{R}$ and let $F : \mathbb{C} \setminus [\alpha, +\infty) \to \mathbb{C}^{q\times q}$ be a matrix-valued function. Then it is readily checked that F belongs to one of the classes $\mathcal{S}_{q;[\alpha,+\infty)}$, $\mathcal{S}_{0,q;[\alpha,+\infty)}$, and $\mathcal{S}^{\diamond}_{q;[\alpha,+\infty)}$ if and only if for each $u \in \mathbb{C}^q$ the function u^*Fu belongs to the corresponding classes $\mathcal{S}_{1;[\alpha,+\infty)}$, $\mathcal{S}_{0,1;[\alpha,+\infty)}$, and $\mathcal{S}^{\diamond}_{1;[\alpha,+\infty)}$, respectively.

(b) Let $\beta \in \mathbb{R}$ and let $G : \mathbb{C}\setminus(-\infty, \beta] \to \mathbb{C}^{q\times q}$ be a matrix-valued function. Then it is readily checked that G belongs to one of the classes $\mathcal{S}_{q;(-\infty,\beta]}$, $\mathcal{S}_{0,q;(-\infty,\beta]}$, and $\mathcal{S}^{\diamond}_{q;(-\infty,\beta]}$ if and only if for each $u \in \mathbb{C}^q$ the function u^*Gu belongs to the corresponding classes $\mathcal{S}_{1;(-\infty,\beta]}$, $\mathcal{S}_{0,1;(-\infty,\beta]}$, and $\mathcal{S}^{\diamond}_{1;(-\infty,\beta]}$, respectively.

In order to get integral representations and other useful information about the one-sided Stieltjes functions of order q, we exploit the fact that these classes of functions can be embedded via restriction to the upper half-plane Π_+ to the well-studied Herglotz–Nevanlinna class $\mathcal{R}_q(\Pi_+)$ of all Herglotz–Nevanlinna functions in Π_+. A matrix-valued function $F : \Pi_+ \to \mathbb{C}^{q\times q}$ is called a $q \times q$ *Herglotz–Nevanlinna function in* Π_+ if F is holomorphic in Π_+ and satisfies the condition $\mathrm{Im}[F(w)] \in \mathbb{C}^{q\times q}_{\geq}$ for all $w \in \Pi_+$. For a comprehensive study on the class $\mathcal{R}_q(\Pi_+)$, we refer the reader to the paper [16] by F. Gesztesy and E.R. Tsekanovskii and to the paper [11]. In particular, in [11] one can find a detailed discussion of the holomorphicity properties of the Moore–Penrose pseudoinverse of matrix-valued Herglotz–Nevanlinna functions. Before we recall the well-known characterization of the class $\mathcal{R}_q(\Pi_+)$ via Nevanlinna parametrization, we observe that, for each $\nu \in \mathcal{M}^q_{\geq}(\mathbb{R})$ and each $z \in \mathbb{C} \setminus \mathbb{R}$, the function $h_z : \mathbb{R} \to \mathbb{C}$ defined by

$$h_z(t) := (1 + tz)/(t - z)$$

belongs to $\mathcal{L}^1(\mathbb{R}, \mathfrak{B}_{\mathbb{R}}, \nu; \mathbb{C})$.

Theorem 2.13.

(a) *Let $F \in \mathcal{R}_q(\Pi_+)$. Then there are unique matrices $A \in \mathbb{C}^{q\times q}_{\mathrm{H}}$ and $B \in \mathbb{C}^{q\times q}_{\geq}$ and a unique non-negative Hermitian measure $\nu \in \mathcal{M}^q_{\geq}(\mathbb{R})$ such that*

$$F(z) = A + zB + \int_{\mathbb{R}} \frac{1 + tz}{t - z} \nu(\mathrm{d}t) \quad \text{for each} \quad z \in \Pi_+. \tag{2.2}$$

(b) *If $A \in \mathbb{C}^{q\times q}_{\mathrm{H}}$, if $B \in \mathbb{C}^{q\times q}_{\geq}$, and if $\nu \in \mathcal{M}^q_{\geq}(\mathbb{R})$, then $F : \Pi_+ \to \mathbb{C}^{q\times q}$ defined by (2.2) belongs to the class $\mathcal{R}_q(\Pi_+)$.*

For each $F \in \mathcal{R}_q(\Pi_+)$, the unique triple $(A, B, \nu) \in \mathbb{C}^{q\times q}_{\mathrm{H}} \times \mathbb{C}^{q\times q}_{\geq} \times \mathcal{M}^q_{\geq}(\mathbb{R})$ for which the representation (2.2) holds true is called the *Nevanlinna parametrization of F* and we will also write (A_F, B_F, ν_F) for (A, B, ν).

Remark 2.14. Let $F \in \mathcal{R}_q(\Pi_+)$ with Nevanlinna parametrization (A_F, B_F, ν_F). Then it is immediately seen that F^{T} belongs to $\mathcal{R}_q(\Pi_+)$ and that

$$(A_{F^{\mathrm{T}}}, B_{F^{\mathrm{T}}}, \nu_{F^{\mathrm{T}}}) = (A_F^{\mathrm{T}}, B_F^{\mathrm{T}}, \nu_F^{\mathrm{T}}).$$

Remark 2.15. Let $F \in \mathcal{R}_q(\Pi_+)$ with Nevanlinna parametrization (A, B, ν). In view of Theorem 2.13, we have then $-\overline{z} \in \Pi_+$ and

$$-[F(-\overline{z})]^* = -A + zB + \int_{\mathbb{R}} \frac{1 - tz}{-t - z} \nu(\mathrm{d}t) = -A + zB + \int_{\mathbb{R}} \frac{1 + tz}{t - z} \theta(\mathrm{d}t)$$

for all $z \in \Pi_+$, where θ is the image measure of ν under the reflection $t \mapsto -t$ on $\mathbb{R}$. Because of $-A \in \mathbb{C}^{q\times q}_{\mathrm{H}}$ and $B \in \mathbb{C}^{q\times q}_{\geq}$, Theorem 2.13 yields then that $G : \Pi_+ \to \mathbb{C}^{q\times q}$ defined by $G(z) := -[F(-\overline{z})]^*$ belongs to $\mathcal{R}_q(\Pi_+)$ and that the Nevanlinna parametrization of G is given by $(-A, B, \theta)$.

From Definitions 2.1 and 2.3 we see immediately that

$$\{\mathrm{Rstr}_{\Pi_+} F : F \in \mathcal{S}_{q;[\alpha,+\infty)}\} \subseteq \mathcal{R}_q(\Pi_+), \quad \{\mathrm{Rstr}_{\Pi_+} G : G \in \mathcal{S}_{q;(-\infty,\beta]}\} \subseteq \mathcal{R}_q(\Pi_+).$$

Now we analyse the Nevanlinna parametrizations of the restrictions on Π_+ of the members of these classes of functions.

Proposition 2.16. *Let $\alpha \in \mathbb{R}$ and let $F \in \mathcal{S}_{q;[\alpha,+\infty)}$. Then the Nevanlinna parametrization (A, B, ν) of $\mathrm{Rstr}_{\Pi_+} F$ fulfills*

$$\nu((-\infty, \alpha)) = O_{q\times q}, \quad B = O_{q\times q}, \quad \nu \in \mathcal{M}^q_{\geq,1}(\mathbb{R}). \tag{2.3}$$

In particular, for each $z \in \mathbb{C} \setminus [\alpha, +\infty)$, then

$$F(z) = A + \int_{[\alpha,+\infty)} \frac{1 + tz}{t - z} \nu(\mathrm{d}t). \tag{2.4}$$

Proof. From [4, Proposition 8.3] and its proof we obtain $\nu((-\infty,\alpha)) = O_{q\times q}$ and

$$F(z) = A + zB + \int_{[\alpha,+\infty)} \frac{1+tz}{t-z}\nu(\mathrm{d}t) \tag{2.5}$$

for all $z \in \mathbb{C} \setminus [\alpha,+\infty)$. Let $u \in \mathbb{C}^q$. Then $u^*\nu u \in \mathcal{M}^1_{\geq}(\mathbb{R})$ and (2.5) yields

$$u^*F(x)u = u^*Au + xu^*Bu + \int_{[\alpha,+\infty)} \frac{1+tx}{t-x}(u^*\nu u)(\mathrm{d}t) \tag{2.6}$$

for all $x \in (-\infty,\alpha)$. Let $\alpha_1 := \min\{\alpha-1,-1\}$ and $\alpha_2 := \max\{\alpha+1,1\}$. Since $F(x) \in \mathbb{C}^{q\times q}_{\geq}$ for all $x \in (-\infty,\alpha_1)$, and furthermore $(1+tx)/(t-x) < 0$ for all $x \in (-\infty,\alpha_1)$ and all $t \in (\alpha_2,+\infty)$, we conclude from (2.6) then

$$-xu^*Bu \leq u^*Au + \int_{[\alpha,\alpha_2]} \frac{1+tx}{t-x}(u^*\nu u)(\mathrm{d}t)$$

for all $x \in (-\infty,\alpha_1)$. One can easily check that there exists a constant $L_\alpha \in \mathbb{R}$ depending only on α such that $|(1+tx)/(t-x)| \leq L_\alpha$ for all $x \in (-\infty,\alpha_1)$ and all $t \in [\alpha,\alpha_2]$. For all $x \in (-\infty,\alpha_1)$, hence

$$\left|\int_{[\alpha,\alpha_2]} \frac{1+tx}{t-x}(u^*\nu u)(\mathrm{d}t)\right| \leq L_\alpha \cdot (u^*\nu u)([\alpha,\alpha_2]) < +\infty. \tag{2.7}$$

Setting $K := u^*Au + L_\alpha \cdot (u^*\nu u)([\alpha,\alpha_2])$, we get then $-xu^*Bu \leq K < +\infty$. In view of $B \in \mathbb{C}^{q\times q}_{\geq}$, we have $u^*Bu \geq 0$, where $u^*Bu > 0$ is impossible, since $-xu^*Bu$ would then tend to $+\infty$ as x tends to $-\infty$. Thus, $u^*Bu = 0$. Since $u \in \mathbb{C}^q$ was arbitrarily chosen, we get $B = O_{q\times q}$ and, in view of (2.5), thus (2.4) holds true for all $z \in \mathbb{C} \setminus [\alpha,+\infty)$. Taking additionally into account $F(x) \in \mathbb{C}^{q\times q}_{\geq}$ for all $x \in (-\infty,\alpha_1)$, we conclude from (2.6) and (2.7) furthermore

$$\int_{[\alpha_2,+\infty)} \left(-\frac{1+tx}{t-x}\right)(u^*\nu u)(\mathrm{d}t) \leq K < +\infty$$

for all $x \in (-\infty,\alpha_1)$. Now we consider an arbitrary sequence $(x_n)_{n=1}^\infty$ from $(-\infty,\alpha_1)$ with $\lim_{n\to\infty} x_n = -\infty$. Then $-(1+tx_n)/(t-x_n) > 0$ for all $n \in \mathbb{N}$ and all $t \in [\alpha_2,+\infty)$ and, furthermore,

$$|t| = \liminf_{n\to\infty} \left(-\frac{1+tx_n}{t-x_n}\right)$$

for all $t \in [\alpha_2,+\infty)$. The application of Fatou's lemma then yields

$$\int_{[\alpha_2,+\infty)} |t|(u^*\nu u)(\mathrm{d}t) \leq \liminf_{n\to\infty} \int_{[\alpha_2,+\infty)} \left(-\frac{1+tx}{t-x}\right)(u^*\nu u)(\mathrm{d}t) \leq K < +\infty.$$

Since $(u^*\nu u)((-\infty,\alpha)) = 0$ and since

$$\int_{[\alpha,\alpha_2]} |t|(u^*\nu u)(\mathrm{d}t)$$

is finite, we conclude then

$$\int_{\mathbb{R}} |t|(u^* \nu u)(\mathrm{d}t) < +\infty.$$

Because $u \in \mathbb{C}^q$ was arbitrarily chosen, we get $\nu \in \mathcal{M}^q_{\geq,1}(\mathbb{R})$. □

3. Integral representations for the class $\mathcal{S}_{q;[\alpha,+\infty)}$

The main goal of this section is to derive some integral representations for $[\alpha, +\infty)$-Stieltjes functions of order q.

Theorem 3.1. *Let $\alpha \in \mathbb{R}$ and let $F : \mathbb{C} \setminus [\alpha, +\infty) \to \mathbb{C}^{q\times q}$.*

(a) *Suppose $F \in \mathcal{S}_{q;[\alpha,+\infty)}$. Denote by (A, B, ν) the Nevanlinna parametrization of $\operatorname{Rstr}_{\Pi_+} F$. Then $\tilde{\nu} := \operatorname{Rstr}_{\mathfrak{B}_{[\alpha,+\infty)}} \nu$ belongs to $\mathcal{M}^q_{\geq,1}([\alpha, +\infty))$ and there is a unique pair (C, η) belonging to $\mathbb{C}^{q\times q}_{\geq} \times \mathcal{M}^q_{\geq,1}([\alpha, +\infty))$ such that*

$$F(z) = C + \int_{[\alpha,+\infty)} \frac{1+t^2}{t-z} \eta(\mathrm{d}t) \tag{3.1}$$

for all $z \in \mathbb{C} \setminus [\alpha, +\infty)$, namely

$$C = A - \int_{[\alpha,+\infty)} t\tilde{\nu}(\mathrm{d}t)$$

and $\eta = \tilde{\nu}$. Furthermore,

$$C = \lim_{r\to+\infty} F(\alpha + r\mathrm{e}^{\mathrm{i}\phi}) \quad \textit{for all} \quad \phi \in (\pi/2, 3\pi/2).$$

(b) *Let $C \in \mathbb{C}^{q\times q}_{\geq}$ and let $\eta \in \mathcal{M}^q_{\geq,1}([\alpha, +\infty))$ be such that (3.1) holds true for all $z \in \mathbb{C} \setminus [\alpha, +\infty)$. Then F belongs to $\mathcal{S}_{q;[\alpha,+\infty)}$.*

Proof. (a) From Proposition 2.16 we conclude $\tilde{\nu} \in \mathcal{M}^q_{\geq,1}([\alpha, +\infty))$. Let $u \in \mathbb{C}^q$. Then

$$\int_{[\alpha,+\infty)} t(u^* \tilde{\nu} u)(\mathrm{d}t)$$

is finite. Obviously, for all $z \in \mathbb{C}$ and all $t \in \mathbb{R}$ with $t \neq z$, we get

$$\frac{1+tz}{t-z} + t = \frac{1+t^2}{t-z}. \tag{3.2}$$

Thus, in view of Proposition 2.16, we obtain

$$\int_{[\alpha,+\infty)} \frac{1+t^2}{t-z} (u^* \tilde{\nu} u)(\mathrm{d}t) < +\infty \tag{3.3}$$

and

$$u^* F(z) u = u^* A u + \int_{[\alpha,+\infty)} \frac{1+t^2}{t-z} (u^* \tilde{\nu} u)(\mathrm{d}t) - \int_{[\alpha,+\infty)} t(u^* \tilde{\nu} u)(\mathrm{d}t) \tag{3.4}$$

for all $z \in \mathbb{C} \setminus [\alpha, +\infty)$. Since $u \in \mathbb{C}^q$ was arbitrarily chosen, it follows (3.1) for all $z \in \mathbb{C} \setminus [\alpha, +\infty)$ with

$$C := A - \int_{[\alpha,+\infty)} t\tilde{\nu}(\mathrm{d}t)$$

and $\eta := \tilde{\nu}$. Let $\phi \in (\pi/2, 3\pi/2)$. Then $\cos\phi < 0$. To show

$$C = \lim_{r\to+\infty} F(\alpha + r\mathrm{e}^{\mathrm{i}\phi}),$$

we consider an arbitrary sequence $(r_n)_{n=1}^\infty$ from $\mathbb{R}$ with $r_n \geq 1/|\cos\phi|$ for all $n \in \mathbb{N}$ and $\lim_{n\to\infty} r_n = +\infty$. We have then

$$\lim_{n\to\infty} \frac{1+t^2}{t - \alpha - r_n\mathrm{e}^{\mathrm{i}\phi}} = 0$$

for all $t \in [\alpha, +\infty)$. For all $n \in \mathbb{N}$ and all $t \in [\alpha, +\infty)$, we get furthermore

$$|t - \alpha - r_n\mathrm{e}^{\mathrm{i}\phi}| \geq t - \alpha - r_n\cos\phi = t - \alpha + r_n|\cos\phi| \geq t - \alpha + 1 \geq 1$$

and hence

$$\left|\frac{1+t^2}{t - \alpha - r_n\mathrm{e}^{\mathrm{i}\phi}}\right| \leq \frac{1+t^2}{t - \alpha + 1}.$$

Since, because of (3.3), the integral

$$\int_{[\alpha,+\infty)} \frac{1+t^2}{t-\alpha+1}(u^*\tilde{\nu}u)(\mathrm{d}t)$$

is finite, the application of Lebesgue's dominated convergence theorem yields

$$\lim_{n\to\infty} \int_{[\alpha,+\infty)} \frac{1+t^2}{t - \alpha - r_n\mathrm{e}^{\mathrm{i}\phi}}(u^*\tilde{\nu}u)(\mathrm{d}t) = 0.$$

From (3.4) we conclude then

$$u^*Cu = \lim_{n\to\infty} u^*F(\alpha + r_n\mathrm{e}^{\mathrm{i}\phi})u.$$

Since $u \in \mathbb{C}^q$ was arbitrarily chosen, we obtain

$$C = \lim_{n\to\infty} F(\alpha + r_n\mathrm{e}^{\mathrm{i}\phi}).$$

Taking into account $F(x) \in \mathbb{C}_{\geq}^{q\times q}$ for all $x \in (-\infty, \alpha)$, we get with $\phi = \pi$ in particular $C \in \mathbb{C}_{\geq}^{q\times q}$.

Now let $C \in \mathbb{C}_{\geq}^{q\times q}$ and $\eta \in \mathcal{M}_{\geq,1}^q([\alpha, +\infty))$ be such that (3.1) holds true for all $z \in \mathbb{C} \setminus [\alpha, +\infty)$. Then $\chi : \mathfrak{B}_{\mathbb{R}} \to \mathbb{C}_{\geq}^{q\times q}$ defined by $\chi(M) := \eta(M \cap [\alpha, +\infty))$ belongs to $\mathcal{M}_{\geq,1}^q(\mathbb{R})$ and the matrix $C + \int_{\mathbb{R}} t\chi(\mathrm{d}t)$ is Hermitian. Using (3.2), we infer from (3.1) that the integral

$$\int_{\mathbb{R}} \frac{1+tz}{t-z}\chi(\mathrm{d}t)$$

exists and that

$$F(z) = C + \int_{\mathbb{R}} t\chi(\mathrm{d}t) + z \cdot O_{q\times q} + \int_{\mathbb{R}} \frac{1+tz}{t-z}\chi(\mathrm{d}t)$$

is fulfilled for all $z \in \Pi_+$. Theorem 2.13(a) yields then $C + \int_{\mathbb{R}} t\chi(\mathrm{d}t) = A$ and $\chi = \nu$. Hence $\eta = \tilde{\nu}$ and

$$C = A - \int_{[\alpha,+\infty)} t\tilde{\nu}(\mathrm{d}t).$$

(b) Let $C \in \mathbb{C}_{\geq}^{q\times q}$ and $\eta \in \mathcal{M}_{\geq,1}^{q}([\alpha,+\infty))$ be such that (3.1) holds true for all $z \in \mathbb{C} \setminus [\alpha,+\infty)$. Using a result on holomorphic dependence of an integral on a complex parameter (see, e.g., [9, Ch. IV, §5, Satz 5.8]), we conclude then that F is a matrix-valued function which is holomorphic in $\mathbb{C} \setminus [\alpha,+\infty)$. Furthermore,

$$\operatorname{Im} F(z) = \int_{[\alpha,+\infty)} \operatorname{Im}\left(\frac{1+t^2}{t-z}\right)\eta(\mathrm{d}t) = \int_{[\alpha,+\infty)} \frac{(1+t^2)\operatorname{Im} z}{|t-z|^2}\eta(\mathrm{d}t) \in \mathbb{C}_{\geq}^{q\times q}$$

for all $z \in \Pi_+$ and

$$F(x) = C + \int_{[\alpha,+\infty)} \frac{1+t^2}{t-x}\eta(\mathrm{d}t) \in \mathbb{C}_{\geq}^{q\times q}$$

for all $x \in (-\infty,\alpha)$. Thus, F belongs to $\mathcal{S}_{q;[\alpha,+\infty)}$. □

In the following, if $\alpha \in \mathbb{R}$ and $F \in \mathcal{S}_{q;[\alpha,+\infty)}$ are given, then we will write (C_F, η_F) for the unique pair (C,η) belonging to $\mathbb{C}_{\geq}^{q\times q} \times \mathcal{M}_{\geq,1}^{q}([\alpha,+\infty))$ which fulfills (3.1) for all $z \in \mathbb{C} \setminus [\alpha,+\infty)$. Furthermore, if A and B are complex $q \times q$ matrices, then we write $A \leq B$ or $B \geq A$ to indicate that the matrices A and B are Hermitian and that $B - A$ is non-negative Hermitian.

Remark 3.2. Let $\alpha \in \mathbb{R}$ and let $F \in \mathcal{S}_{q;[\alpha,+\infty)}$. For all $x_1, x_2 \in (-\infty,\alpha)$ with $x_1 \leq x_2$, then $O_{q\times q} \leq F(x_1) \leq F(x_2)$, by virtue of Theorem 3.1(a).

Remark 3.3. Let $\alpha \in \mathbb{R}$ and let $z \in \mathbb{C}\setminus[\alpha,+\infty)$. Then, for each $\mu \in \mathcal{M}_{\geq}^{q}([\alpha,+\infty))$, in view of the equation $(1+t-\alpha)/(t-z) = 1 + (1+z-\alpha)/(t-z)$, which holds for each $t \in [\alpha,+\infty)$, and Lemma A.8(a), one can easily see that the function $h_{\alpha,z} : [\alpha,+\infty) \to \mathbb{C}$ defined by $h_{\alpha,z}(t) := (1+t-\alpha)/(t-z)$ belongs to $\mathcal{L}^1([\alpha,+\infty), \mathfrak{B}_{[\alpha,+\infty)}, \mu; \mathbb{C})$.

Lemma 3.4. *Let $\alpha \in \mathbb{R}$ and let $F : \mathbb{C} \setminus [\alpha,+\infty) \to \mathbb{C}^{q\times q}$ be a continuous matrix-valued function such that $(\operatorname{Im} F)(\Pi_+) \subseteq \mathbb{C}_{\geq}^{q\times q}$ and $(-\operatorname{Im} F)(\Pi_-) \subseteq \mathbb{C}_{\geq}^{q\times q}$. Then $F(x) = \operatorname{Re} F(x)$ and $\operatorname{Im} F(x) = O_{q\times q}$ for each $x \in (-\infty,\alpha)$.*

Proof. Let $x \in (-\infty,\alpha)$. Then $(\operatorname{Im} F(x + \mathrm{i}/n))_{n=1}^{\infty}$ and $(-\operatorname{Im} F(x - \mathrm{i}/n))_{n=1}^{\infty}$ are sequences of non-negative Hermitian complex $q\times q$ matrices which converge to the non-negative Hermitian complex matrices $\operatorname{Im} F(x)$ and $-\operatorname{Im} F(x)$, respectively. Consequently, $\operatorname{Im} F(x) = O_{q\times q}$, which implies $F(x) = \operatorname{Re} F(x)$. □

Let $\mathbb{C}_{\alpha,-} := \{z \in \mathbb{C} : \operatorname{Re} z \in (-\infty,\alpha)\}$.

Lemma 3.5. *Let $\alpha \in \mathbb{R}$, let $\gamma \in \mathbb{C}_{\geq}^{q\times q}$, and let $\mu \in \mathcal{M}_{\geq}^{q}([\alpha,+\infty))$. Suppose the matrix function $F : \mathbb{C} \setminus [\alpha,+\infty) \to \mathbb{C}^{q\times q}$ is defined by*

$$F(z) = \gamma + \int_{[\alpha,+\infty)} \frac{1+t-\alpha}{t-z}\mu(\mathrm{d}t). \tag{3.5}$$

Then

$$\operatorname{Re} F(z) = \gamma + \int_{[\alpha,+\infty)} \frac{1+t-\alpha}{|t-z|^2}(t - \operatorname{Re} z)\mu(\mathrm{d}t) \tag{3.6}$$

and

$$\operatorname{Im} F(z) = (\operatorname{Im} z) \int_{[\alpha,+\infty)} \frac{1+t-\alpha}{|t-z|^2}\mu(\mathrm{d}t) \tag{3.7}$$

hold true for each $z \in \mathbb{C} \setminus [\alpha, +\infty)$. *In particular,*

$$(\operatorname{Re} F)(\mathbb{C}_{\alpha,-}) \subseteq \mathbb{C}_{\geq}^{q\times q}, \quad (\operatorname{Im} F)(\Pi_+) \subseteq \mathbb{C}_{\geq}^{q\times q}, \quad \textit{and} \quad (-\operatorname{Im} F)(\Pi_-) \subseteq \mathbb{C}_{\geq}^{q\times q}.$$

Furthermore, $F(x) = \operatorname{Re} F(x)$ *and* $F(x) \in \mathbb{C}_{\geq}^{q\times q}$ *for each* $x \in (-\infty, \alpha)$.

Proof. For each $z \in \mathbb{C} \setminus [\alpha, +\infty)$ and each $t \in [\alpha, +\infty)$, we have

$$\operatorname{Re}\left(\frac{1+t-\alpha}{t-z}\right) = \frac{1+t-\alpha}{|t-z|^2}(t - \operatorname{Re} z) \tag{3.8}$$

and

$$\operatorname{Im}\left(\frac{1+t-\alpha}{t-z}\right) = \frac{(\operatorname{Im} z)(1+t-\alpha)}{|t-z|^2}.$$

Taking into account $\gamma \in \mathbb{C}_{\geq}^{q\times q}$, thus (3.6) and (3.7) follow for each $z \in \mathbb{C} \setminus [\alpha, +\infty)$. For each $z \in \mathbb{C}_{\alpha,-}$ and each $t \in [\alpha, +\infty)$, the right-hand side of (3.8) belongs to $[0, +\infty)$. Thus, $\gamma \in \mathbb{C}_{\geq}^{q\times q}$ and (3.6) show that $\operatorname{Re} F(z)$ belongs to $\mathbb{C}_{\geq}^{q\times q}$ for each $z \in \mathbb{C}_{\alpha,-}$. Since $(1+t-\alpha)/|t-z|^2 \in [0, +\infty)$ for every choice of $z \in \mathbb{C} \setminus [\alpha, +\infty)$ and $t \in [\alpha, +\infty)$, from (3.7) we see that $\operatorname{Im} F(w) \in \mathbb{C}_{\geq}^{q\times q}$ for each $w \in \Pi_+$ and $-\operatorname{Im} F(v) \in \mathbb{C}_{\geq}^{q\times q}$ for each $v \in \Pi_-$ are fulfilled. Applying Lemma 3.4 completes the proof. □

Now we give a further integral representation of the matrix-valued functions which belong to the class $\mathcal{S}_{q;[\alpha,+\infty)}$. In the special case that $q = 1$ and $\alpha = 0$ hold, one can find this result in [19, Appendix].

Theorem 3.6. *Let* $\alpha \in \mathbb{R}$ *and let* $F : \mathbb{C} \setminus [\alpha, +\infty) \to \mathbb{C}^{q\times q}$.

(a) *If* $F \in \mathcal{S}_{q;[\alpha,+\infty)}$, *then there are a unique matrix* $\gamma \in \mathbb{C}_{\geq}^{q\times q}$ *and a unique non-negative Hermitian measure* $\mu \in \mathcal{M}_{\geq}^{q}([\alpha, +\infty))$ *such that* (3.5) *holds true for each* $z \in \mathbb{C} \setminus [\alpha, +\infty)$. *Furthermore,* $\gamma = C_F$.

(b) *If there are a matrix* $\gamma \in \mathbb{C}_{\geq}^{q\times q}$ *and a non-negative Hermitian measure* $\mu \in \mathcal{M}_{\geq}^{q}([\alpha, +\infty))$ *such that* F *can be represented via* (3.5) *for each* $z \in \mathbb{C} \setminus [\alpha, +\infty)$, *then* F *belongs to the class* $\mathcal{S}_{q;[\alpha,+\infty)}$.

Proof. Denote $f := F \circ \operatorname{Rstr}_{\mathbb{C}\setminus[0,+\infty)} T_\alpha$.

(a) Let $F \in \mathcal{S}_{q;[\alpha,+\infty)}$. According to Remark 2.5, the function f belongs to $\mathcal{S}_{q;[0,+\infty)}$. In view of Proposition 2.16, then $\operatorname{Rstr}_{\Pi_+} f$ belongs to $\mathcal{R}_q(\Pi_+)$, the

Nevanlinna parametrization (A, B, ν) of $\operatorname{Rstr}_{\Pi_+} f$ fulfills (2.3), and, for each $w \in \mathbb{C} \setminus [0, +\infty)$, we have

$$f(w) = A + \int_{[0,+\infty)} \frac{1 + xw}{x - w} \nu(\mathrm{d}x).$$

Because of (2.3), the integral

$$\int_{[0,+\infty)} x\nu(\mathrm{d}x)$$

exists and the mapping $\hat{\mu} : \mathfrak{B}_{[0,+\infty)} \to \mathbb{C}_{\geq}^{q \times q}$ given by

$$\hat{\mu}(B) := \int_B \frac{1 + x^2}{1 + x} \nu(\mathrm{d}x)$$

is well defined and belongs to $\mathcal{M}_{\geq}^{q}([0, +\infty))$. Setting

$$\gamma := A - \int_{[0,+\infty)} x\nu(\mathrm{d}x),$$

for each $w \in \mathbb{C} \setminus [0, +\infty)$, we get

$$\begin{aligned} f(w) &= A + \int_{[0,+\infty)} \left(\frac{1 + x^2}{x - w} - x \right) \nu(\mathrm{d}x) \\ &= A - \int_{[0,+\infty)} x\nu(\mathrm{d}x) + \int_{[0,+\infty)} \left(\frac{1 + x}{x - w} \cdot \frac{1 + x^2}{1 + x} \right) \nu(\mathrm{d}x) \\ &= \gamma + \int_{[0,+\infty)} \frac{1 + x}{x - w} \hat{\mu}(\mathrm{d}x). \end{aligned}$$

Obviously, $\mu := (\operatorname{Rstr}_{[0,+\infty)} T_\alpha)(\hat{\mu})$ belongs to $\mathcal{M}_{\geq}^{q}([\alpha, +\infty))$. For each $z \in \mathbb{C} \setminus [\alpha, +\infty)$, Proposition A.5 yields

$$\begin{aligned} F(z) = f(z - \alpha) &= \gamma + \int_{[0,+\infty)} \frac{1 + x + \alpha - \alpha}{x + \alpha - z} \hat{\mu}(\mathrm{d}x) \\ &= \gamma + \int_{[0,+\infty)} (h_{\alpha,z} \circ T_\alpha)(x) \hat{\mu}(\mathrm{d}x) = \gamma + \int_{T_\alpha([0,+\infty))} h_{\alpha,z} \mathrm{d}\mu \end{aligned}$$

and, hence, (3.5) holds for each $z \in \mathbb{C} \setminus [\alpha, +\infty)$.

Now we assume that γ is an arbitrary complex $q \times q$ matrix and that μ is an arbitrary non-negative Hermitian measure belonging to $\mathcal{M}_{\geq}^{q}([\alpha, +\infty))$ such that (3.5) holds for each $z \in \mathbb{C} \setminus [\alpha, +\infty)$. Observe that $\lim_{n \to +\infty} h_{\alpha,\alpha-1-n}(t) = 0$ is valid for each $t \in [\alpha, +\infty)$. Moreover, for every choice of $n \in \mathbb{N}_0$ and $t \in [\alpha, +\infty)$, one can easily check that the estimation $|h_{\alpha,\alpha-1-n}(t)| \leq 1$ holds. Consequently, a matrix generalization of Lebesgue's dominated convergence theorem (see Proposition A.6) yields

$$\lim_{n \to +\infty} \int_{[\alpha,+\infty)} \frac{1 + t - \alpha}{t - (\alpha - 1 - n)} \mu(\mathrm{d}t) = 0.$$

From (3.5) we obtain then

$$\gamma = \gamma + \lim_{n\to+\infty} \int_{[\alpha,+\infty)} \frac{1+t-\alpha}{t-(\alpha-1-n)} \mu(\mathrm{d}t) = \lim_{n\to+\infty} F(\alpha-1-n).$$

From Theorem 3.1(a) we conclude then $\gamma = C_F$ and thus $\gamma \in \mathbb{C}_{\geq}^{q\times q}$ follows. The mapping $\tilde{T}_{-\alpha} : [\alpha,+\infty) \to [0,+\infty)$ defined by $\tilde{T}_{-\alpha}(t) := t-\alpha$ is obviously bijective and $\mathfrak{B}_{[\alpha,+\infty)}$-$\mathfrak{B}_{[0,+\infty)}$-measurable. Further, the mapping $\rho : \mathfrak{B}_{\mathbb{R}} \to \mathbb{C}_{\geq}^{q\times q}$ given by

$$\rho(B) := \int_{B\cap[0,+\infty)} \frac{1+x}{1+x^2} \left[\tilde{T}_{-\alpha}(\mu)\right](\mathrm{d}x)$$

is well defined, belongs to $\mathcal{M}_{\geq}^{q}(\mathbb{R})$, and satisfies $\rho((-\infty,0)) = O_{q\times q}$. Furthermore, the integral

$$\int_{[0,+\infty)} x\rho(\mathrm{d}x)$$

exists. For each $w \in \mathbb{C} \setminus [0,+\infty)$, using (3.5), the relation $\tilde{T}_{-\alpha}([\alpha,+\infty)) = [0,+\infty)$ and Proposition A.5 provide us

$$\begin{aligned}
f(w) = F(w+\alpha) &= \gamma + \int_{[\alpha,+\infty)} \frac{1+t-\alpha}{t-(w+\alpha)} \mu(\mathrm{d}t) \\
&= \gamma + \int_{[\alpha,+\infty)} \frac{1+\tilde{T}_{-\alpha}(t)}{\tilde{T}_{-\alpha}(t)-w} \mu(\mathrm{d}t) = \gamma + \int_{[0,+\infty)} \frac{1+x}{x-w} \left[\tilde{T}_{-\alpha}(\mu)\right](\mathrm{d}x) \\
&= \gamma + \int_{[0,+\infty)} \left(\frac{1+x^2}{x-w} \cdot \frac{1+x}{1+x^2}\right) \left[\tilde{T}_{-\alpha}(\mu)\right](\mathrm{d}x) \\
&= \gamma + \int_{[0,+\infty)} \frac{1+x^2}{x-w} \rho(\mathrm{d}x) = \gamma + \int_{[0,+\infty)} \left(x + \frac{1+xw}{x-w}\right) \rho(\mathrm{d}x) \\
&= \gamma + \int_{[0,+\infty)} x\rho(\mathrm{d}x) + \int_{[0,+\infty)} \frac{1+xw}{x-w} \rho(\mathrm{d}x) \qquad (3.9)
\end{aligned}$$

and, consequently,

$$\operatorname{Rstr}_{\Pi_+} f(w) = \gamma + \int_{[0,+\infty)} x\rho(\mathrm{d}x) + w \cdot O_{q\times q} + \int_{\mathbb{R}} \frac{1+xw}{x-w} \rho(\mathrm{d}x)$$

for each $w \in \Pi_+$. Since γ is non-negative Hermitian we see that

$$A^\diamond := \gamma + \int_{[0,+\infty)} x\rho(\mathrm{d}x)$$

belongs to $\mathbb{C}_{\geq}^{q\times q}$. Thus, $(A^\diamond, O_{q\times q}, \rho)$ coincides with the Nevanlinna parametrization (A,B,ν) of the Herglotz–Nevanlinna function $\operatorname{Rstr}_{\Pi_+} f$. In particular, ρ is exactly the (unique) Nevanlinna measure ν of $\operatorname{Rstr}_{\Pi_+} f$. For each $B \in \mathfrak{B}_{[\alpha,+\infty)}$, we have $\tilde{T}_{-\alpha}(B) \in \mathfrak{B}_{[0,+\infty)}$ and hence

$$\mu(B) = \left[\tilde{T}_{-\alpha}(\mu)\right]\left(\tilde{T}_{-\alpha}(B)\right) = \int_{\tilde{T}_{-\alpha}(B)} \left(\frac{1+x^2}{1+x} \cdot \frac{1+x}{1+x^2}\right)\left[\tilde{T}_{-\alpha}(\mu)\right](\mathrm{d}x)$$
$$= \int_{\tilde{T}_{-\alpha}(B)} \frac{1+x^2}{1+x}\rho(\mathrm{d}x) = \int_{\tilde{T}_{-\alpha}(B)} \frac{1+x^2}{1+x}\nu(\mathrm{d}x).$$

In particular, μ is uniquely determined.

(b) Let $\gamma \in \mathbb{C}_{\geq}^{q\times q}$ and $\mu \in \mathcal{M}_{\geq}^{q}([\alpha,+\infty))$ be such that F can be represented via (3.5) for each $z \in \mathbb{C} \setminus [\alpha,+\infty)$. Then the mapping $\hat{\rho} : \mathfrak{B}_{[0,+\infty)} \to \mathbb{C}_{\geq}^{q\times q}$ given by

$$\hat{\rho}(B) := \int_{B} \frac{1+x}{1+x^2}\left[\tilde{T}_{-\alpha}(\mu)\right](\mathrm{d}x)$$

is well defined and belongs to $\mathcal{M}_{\geq}^{q}([0,+\infty))$. Furthermore, f satisfies (3.9) with $\hat{\rho}$ instead of ρ for each $w \in \mathbb{C}\setminus[0,+\infty)$. Hence, using a result on holomorphic dependence of an integral on a complex parameter (see, e.g., [9, Ch. IV, §5, Satz 5.8]), we conclude that f is a matrix-valued function which is holomorphic in $\mathbb{C}\setminus[0,+\infty)$. Because of $F(z) = f(z-\alpha)$ for each $z \in \mathbb{C} \setminus [\alpha,+\infty)$, we obtain then that F is holomorphic in $\mathbb{C} \setminus [\alpha,+\infty)$. From Lemma 3.5 we get $\operatorname{Im} F(w) \in \mathbb{C}_{\geq}^{q\times q}$ for each $w \in \Pi_+$ and $F(x) \in \mathbb{C}_{\geq}^{q\times q}$ for each $x \in (-\infty,\alpha)$. Thus, F belongs to $\mathcal{S}_{q;[\alpha,+\infty)}$. □

Remark 3.7. In the following, if $F \in \mathcal{S}_{q;[\alpha,+\infty)}$ is given, then we will write (γ_F, μ_F) for the unique pair $(\gamma,\mu) \in \mathbb{C}_{\geq}^{q\times q} \times \mathcal{M}_{\geq}^{q}([\alpha,+\infty))$ which realizes the integral representation (3.5) for each $z \in \mathbb{C} \setminus [\alpha,+\infty)$.

Example 3.8. Let $\alpha \in \mathbb{R}$ and let $A, B \in \mathbb{C}_{\geq}^{q\times q}$. Let $F \in \mathbb{C} \setminus [\alpha,+\infty) \to \mathbb{C}^{q\times q}$ be the function from $\mathcal{S}_{q;[\alpha,+\infty)}$ which is defined in Example 2.2. Then $\gamma_F = A$ and $\mu_F = \delta_\alpha B$ where δ_α denotes the Dirac measure on $([\alpha,+\infty), \mathfrak{B}_{[\alpha,+\infty)})$ with unit mass at α.

Example 3.9. Let $\alpha \in \mathbb{R}$, let $\gamma \in \mathbb{C}_{\geq}^{q\times q}$, and let $F : \mathbb{C} \setminus [\alpha,+\infty) \to \mathbb{C}^{q\times q}$ be defined for each $z \in \mathbb{C} \setminus [\alpha,+\infty)$ by $F(z) := \gamma$. In view of Theorem 3.6, then $F \in \mathcal{S}_{q;[\alpha,+\infty)}$, $\gamma_F = \gamma$, and μ_F is the zero measure belonging to $\mathcal{M}_{\geq}^{q}([\alpha,+\infty))$.

Now we state some observations on the arithmetic of the class $\mathcal{S}_{q;[\alpha,+\infty)}$.

Remark 3.10. If $F \in \mathcal{S}_{q;[\alpha,+\infty)}$ then $F^{\mathrm{T}} \in \mathcal{S}_{q;[\alpha,+\infty)}$ and $(\gamma_{F^{\mathrm{T}}}, \mu_{F^{\mathrm{T}}}) = (\gamma_F^{\mathrm{T}}, \mu_F^{\mathrm{T}})$.

Remark 3.11. Let $\alpha \in \mathbb{R}$, let $n \in \mathbb{N}$, and let $(q_k)_{k=1}^{n}$ be a sequence of positive integers. For each $k \in \mathbb{Z}_{1,n}$, let $F_k \in \mathcal{S}_{q_k;[\alpha,+\infty)}$. Then $F := \operatorname{diag}[F_k]_{k=1}^{n}$ belongs to $\mathcal{S}_{\sum_{k=1}^{n} q_k;[\alpha,+\infty)}$ and

$$(\gamma_F, \mu_F) = (\operatorname{diag}[\gamma_{F_k}]_{k=1}^{n}, \operatorname{diag}[\mu_{F_k}]_{k=1}^{n}).$$

Moreover, if $A_k \in \mathbb{C}^{q_k\times q}$ for each $k \in \mathbb{Z}_{1,n}$, then $G := \sum_{k=1}^{n} A_k^* F_k A_k$ belongs to $\mathcal{S}_{q;[\alpha,+\infty)}$ and

$$(\gamma_G, \mu_G) = \left(\sum_{k=1}^{n} A_k^* \gamma_{F_k} A_k, \sum_{k=1}^{n} A_k^* \mu_{F_k} A_k\right).$$

Remark 3.12. Let $\alpha \in \mathbb{R}$ and let $F \in \mathcal{S}_{q;[\alpha,+\infty)}$. For each matrix $A \in \mathbb{C}^{q\times q}$ for which the matrix $\gamma_F + A$ is non-negative Hermitian, from Theorem 3.6 one can easily see that the function $G := F + A$ belongs to $\mathcal{S}_{q;[\alpha,+\infty)}$ and that $(\gamma_G, \mu_G) = (\gamma_F + A, \mu_F)$.

Proposition 3.13. *Let $\alpha \in \mathbb{R}$ and let $F \in \mathcal{S}_{q;[\alpha,+\infty)}$. Then $\lim_{y\to+\infty} F(\mathrm{i}y) = \gamma_F$.*

Proof. Let $(y_n)_{n=1}^{\infty}$ be a sequence from $[1,+\infty)$ such that $\lim_{n\to+\infty} y_n = +\infty$. Obviously, in view of Remark 3.3, we have

$$\lim_{n\to+\infty} h_{\alpha,\mathrm{i}y_n}(t) = 0$$

for each $t \in [\alpha,+\infty)$. Furthermore, for each $t \in [\alpha,+\infty)$, we get $|h_{\alpha,\mathrm{i}y_n}(t)| \leq 3+|\alpha|$. By virtue of Proposition A.6, we obtain then

$$\lim_{n\to+\infty} \int_{[\alpha,+\infty)} h_{\alpha,\mathrm{i}y_n} \mathrm{d}\mu_F = O_{q\times q}.$$

Application of the integral representation stated in Theorem 3.6(a) completes the proof. □

Corollary 3.14. *Let $\alpha \in \mathbb{R}$. Then $\mathcal{S}^{\diamond}_{q;[\alpha,+\infty)} = \{F \in \mathcal{S}_{q;[\alpha,+\infty)} : \gamma_F = O_{q\times q}\}$.*

Proof. Combine Proposition 3.13 and Remark 3.7. □

If $\mathcal{X}$ is a non-empty subset of $\mathbb{C}^q$, then we will use $\mathcal{X}^{\perp}$ to denote the orthogonal space of $\mathcal{X}$. For each $A \in \mathbb{C}^{p\times q}$, let $\mathcal{N}(A)$ be the null space of A and let $\mathcal{R}(A)$ be the column space of A.

Recall that a complex $q\times q$ matrix A is called an *EP matrix* if $\mathcal{R}(A^*) = \mathcal{R}(A)$. The class $\mathbb{C}^{q\times q}_{\mathrm{EP}}$ of these complex $q \times q$ matrices was introduced by Schwerdtfeger [20]. For a comprehensive treatment of the class $\mathbb{C}^{q\times q}_{\mathrm{EP}}$ against to the background of this paper, we refer the reader to [10, Appendix A]. If $G \in \mathcal{R}_q(\Pi_+)$ then it was proved in [10, Lemma 9.1] that, for each $w \in \Pi_+$, the matrix $G(w)$ belongs to $\mathbb{C}^{q\times q}_{\mathrm{EP}}$. Part (b) of the following proposition shows that an analogous result is true for functions belonging to the class $\mathcal{S}_{q;[\alpha,+\infty)}$. Furthermore, the following proposition contains also extensions to the class $\mathcal{S}_{q;[\alpha,+\infty)}$ of former results (see [10, Theorem 9.4], [11, Proposition 3.7]) concerning the class $\mathcal{R}_q(\Pi_+)$.

Proposition 3.15. *Let $\alpha \in \mathbb{R}$ and let $F \in \mathcal{S}_{q;[\alpha,+\infty)}$.*

(a) *If $z \in \mathbb{C} \setminus [\alpha,+\infty)$, then $\overline{z} \in \mathbb{C} \setminus [\alpha,+\infty)$ and $F^*(z) = F(\overline{z})$.*

(b) *For all $z \in \mathbb{C} \setminus [\alpha,+\infty)$,*

$$\mathcal{N}(F(z)) = \mathcal{N}(\gamma_F) \cap \mathcal{N}(\mu_F([\alpha,+\infty))), \tag{3.10}$$

$$\mathcal{R}(F(z)) = \mathcal{R}(\gamma_F) + \mathcal{R}(\mu_F([\alpha,+\infty))), \tag{3.11}$$

and, in particular,

$$\mathcal{N}(F(z)) = \mathcal{N}(F^*(z)) = [\mathcal{R}(F(z))]^{\perp},$$

$$\mathcal{R}(F(z)) = \mathcal{R}(F^*(z)) = [\mathcal{N}(F(z))]^{\perp}.$$

(c) *Let* $r \in \mathbb{N}_0$. *Then the following statements are equivalent:*
   (i) *for each* $z \in \mathbb{C} \setminus [\alpha, +\infty)$, *the equation* $\operatorname{rank} F(z) = r$ *holds;*
   (ii) *there is some* $z_0 \in \mathbb{C} \setminus [\alpha, +\infty)$ *such that* $\operatorname{rank} F(z_0) = r$*;*
   (iii) $\dim[\mathcal{R}(\gamma_F) + \mathcal{R}(\mu_F([\alpha, +\infty)))] = r$.

(d) *If* $\det \gamma_F \neq 0$ *or* $\det[\mu_F([\alpha, +\infty))] \neq 0$, *then* $\det[F(w)] \neq 0$ *for all* $w \in \mathbb{C} \setminus [\alpha, +\infty)$.

Proof. In view of Remark 3.7, we have

$$\gamma_F \in \mathbb{C}_{\geq}^{q\times q} \quad \text{and} \quad \mu_F \in \mathcal{M}_{\geq}^{q}([\alpha, +\infty)). \tag{3.12}$$

Taking into account Remark 3.3, for all $z \in \mathbb{C} \setminus [\alpha, +\infty)$, we get then that $h_{\alpha,z}$ belongs to $\mathcal{L}^1([\alpha, +\infty), \mathfrak{B}_{[\alpha,+\infty)}, \mu_F; \mathbb{C})$.

(a) Let $z \in \mathbb{C} \setminus [\alpha, +\infty)$. From Remark 3.7 and (3.12) we obtain

$$\begin{aligned} F^*(z) &= \gamma_F^* + \left[\int_{[\alpha,+\infty)} \frac{1+t-\alpha}{t-z} \mu_F(\mathrm{d}t)\right]^* \\ &= \gamma_F + \int_{[\alpha,+\infty)} \frac{1+t-\alpha}{t-\overline{z}} \mu_F(\mathrm{d}t) = F(\overline{z}). \end{aligned}$$

(b) Let $z \in \mathbb{C} \setminus [\alpha, +\infty)$. For each $t \in [\alpha, +\infty)$, we get then

$$\operatorname{Re} h_{\alpha,z}(t) = \frac{(t - \operatorname{Re} z)(1+t-\alpha)}{|t-z|^2} \tag{3.13}$$

and

$$\operatorname{Im} h_{\alpha,z}(t) = \frac{(\operatorname{Im} z)(1+t-\alpha)}{|t-z|^2}. \tag{3.14}$$

Since $h_{\alpha,z}$ belongs to $\mathcal{L}^1([\alpha, +\infty), \mathfrak{B}_{[\alpha,+\infty)}, \mu_F; \mathbb{C})$, from Lemma A.4(a) we see that

$$\mathcal{N}(\mu_F([\alpha, +\infty))) \subseteq \mathcal{N}\left(\int_{[\alpha,+\infty)} h_{\alpha,z} \mathrm{d}\mu_F\right) \tag{3.15}$$

holds true. Now we consider an arbitrary $u \in \mathcal{N}(F(z))$. In view of the definition of the pair (γ_F, μ_F) (see Remark 3.7), we have

$$u^*\gamma_F u + \int_{[\alpha,+\infty)} h_{\alpha,z} \mathrm{d}(u^*\mu_F u) = u^*F(z)u = 0. \tag{3.16}$$

Consequently, because of (3.16), (3.12), and (3.14), then

$$0 = u^*\gamma_F u + \int_{[\alpha,+\infty)} \operatorname{Re} h_{\alpha,z} \mathrm{d}(u^*\mu_F u) \geq \int_{[\alpha,+\infty)} \operatorname{Re} h_{\alpha,z} \mathrm{d}(u^*\mu_F u) \tag{3.17}$$

and

$$0 = \operatorname{Im}\left[\int_{[\alpha,+\infty)} h_{\alpha,z} \mathrm{d}(u^*\mu_F u)\right] = \int_{[\alpha,+\infty)} \frac{(\operatorname{Im} z)(1+t-\alpha)}{|t-z|^2} (u^*\mu_F u)(\mathrm{d}t) \tag{3.18}$$

follow. In the case $\operatorname{Im} z \neq 0$, from (3.18) and (3.12) we get $(u^*\mu_F u)([\alpha, +\infty)) = 0$. If $z \in (-\infty, \alpha)$, then from (3.13) we see that $\operatorname{Re} h_{\alpha,z}(t) \in (0, +\infty)$ holds for each

$t \in [\alpha, +\infty)$, and, by virtue of (3.17) and (3.12), we obtain $(u^*\mu_F u)([\alpha, +\infty)) = 0$. Thus $(u^*\mu_F u)([\alpha, +\infty)) = 0$ is proved in each case, which, in view of (3.12), implies $u \in \mathcal{N}(\mu_F([\alpha, +\infty)))$. Taking into account a standard argument of the integration theory of non-negative Hermitian measures and (3.15), we conclude that

$$\int_{[\alpha,+\infty)} h_{\alpha,z} \mathrm{d}(u^*\mu_F u) = u^* \left(\int_{[\alpha,+\infty)} h_{\alpha,z} \mathrm{d}\mu_F \right) u = u^* \cdot O_{q\times 1} = 0.$$

Consequently, from (3.16) we infer $u^*\gamma_F u = 0$. Thus, (3.12) shows that u belongs to $\mathcal{N}(\gamma_F)$. Hence,

$$\mathcal{N}(F(z)) \subseteq \mathcal{N}(\gamma_F) \cap \mathcal{N}(\mu_F([\alpha, +\infty))) \tag{3.19}$$

is valid. Now we are going to check that

$$\mathcal{N}(\gamma_F) \cap \mathcal{N}(\mu_F([\alpha, +\infty))) \subseteq \mathcal{N}(F(z)) \tag{3.20}$$

holds. For this reason, we consider an arbitrary $u \in \mathcal{N}(\gamma_F) \cap \mathcal{N}(\mu_F([\alpha, +\infty)))$. From (3.15) we get then

$$F(z)u = \gamma_F u + \left(\int_{[\alpha,+\infty)} h_{\alpha,z} \mathrm{d}\mu_F \right) u = O_{q\times 1}$$

and therefore $u \in \mathcal{N}(F(z))$. Hence (3.20) is verified. From (3.19) and (3.20) then (3.10) follows. Keeping in mind (a), (3.10) for $\overline{z}$ instead of z, and (3.12), standard arguments of functional analysis yield then

$$\begin{aligned}
\mathcal{R}(F(z)) &= [\mathcal{N}(F(z)^*)]^\perp = [\mathcal{N}(F(\overline{z}))]^\perp = [\mathcal{N}(\gamma_F) \cap \mathcal{N}(\mu_F([\alpha, +\infty)))]^\perp \\
&= \left[\mathcal{R}(\gamma_F)^\perp \cap \mathcal{R}(\mu_F([\alpha, +\infty)))^\perp\right]^\perp \\
&= \left([\mathrm{span}(\mathcal{R}(\gamma_F) \cup \mathcal{R}(\mu_F([\alpha, +\infty))))]^\perp\right)^\perp \\
&= \mathcal{R}(\gamma_F) + \mathcal{R}(\mu_F([\alpha, +\infty))).
\end{aligned}$$

Thus, (3.11) is proved. Using (3.10) for z and for $\overline{z}$ instead of z, from (a) we obtain

$$\mathcal{N}(F(z)) = \mathcal{N}(F^*(z)) = [\mathcal{R}(F(z))]^\perp.$$

Similarly,

$$\mathcal{R}(F(z)) = \mathcal{R}(F^*(z)) = [\mathcal{N}(F(z))]^\perp$$

follows from (3.11) and (a).

(c)–(d) These are immediate consequences of (b). □

Proposition 3.15 yields a generalization of a result due to Kats and Krein [17, Corollary 5.1]:

Corollary 3.16. *Let $\alpha \in \mathbb{R}$, let $F \in \mathcal{S}_{q;[\alpha,+\infty)}$, and let $z_0 \in \mathbb{C} \setminus [\alpha, +\infty)$. Then $F(z_0) = O_{q\times q}$ if and only if $F(z) = O_{q\times q}$ for all $z \in \mathbb{C} \setminus [\alpha, +\infty)$.*

Proof. This is an immediate consequence of Proposition 3.15(c). □

Corollary 3.17. *Let $\alpha \in \mathbb{R}$, let $F \in \mathcal{S}_{q;[\alpha,+\infty)}$, and let $\lambda \in \mathbb{R}$ be such that the matrix $\gamma_F - \lambda I_q$ is non-negative Hermitian. Then*

$$\mathcal{R}(F(z) - \lambda I_q) = \mathcal{R}(F(w) - \lambda I_q), \quad \mathcal{N}(F(z) - \lambda I_q) = \mathcal{N}(F(w) - \lambda I_q) \tag{3.21}$$

for all $z, w \in \mathbb{C} \setminus [\alpha, +\infty)$. In particular, if $\lambda \leq 0$, then λ is an eigenvalue of the matrix $F(z_0)$ for some $z_0 \in \mathbb{C} \setminus [\alpha, +\infty)$ if and only if λ is an eigenvalue of the matrix $F(z)$ for all $z \in \mathbb{C} \setminus [\alpha, +\infty)$. In this case, the eigenspaces $\mathcal{N}(F(z) - \lambda I_q)$ are independent of $z \in \mathbb{C} \setminus [\alpha, +\infty)$.

Proof. In view of Remark 3.7 and Theorem 3.6, we conclude that the function $G : \mathbb{C} \setminus [\alpha, +\infty) \to \mathbb{C}^{q\times q}$ defined by $G(z) := F(z) - \lambda I_q$ belongs to $\mathcal{S}_{q;[\alpha,+\infty)}$. The application of Proposition 3.15(b) to the function G yields then (3.21). Since the matrix γ_F is non-negative Hermitian, we have $\gamma_F - \lambda I_q \in \mathbb{C}^{q\times q}_{\geq}$ if $\lambda \leq 0$. Thus, the remaining assertions are an immediate consequence of (3.21). □

At the end of this section we add a useful technical result.

Lemma 3.18. *Let $\alpha \in \mathbb{R}$, let $A \in \mathbb{C}^{p\times q}$, and let $F \in \mathcal{S}_{q;[\alpha,+\infty)}$. Then the following statements are equivalent:*

(i) *$\mathcal{N}(A) \subseteq \mathcal{N}(F(z))$ for all $z \in \mathbb{C} \setminus [\alpha, +\infty)$;*
(ii) *there is a $z_0 \in \mathbb{C} \setminus [\alpha, +\infty)$ such that $\mathcal{N}(A) \subseteq \mathcal{N}(F(z_0))$;*
(iii) *$\mathcal{N}(A) \subseteq \mathcal{N}(\gamma_F) \cap \mathcal{N}(\mu_F([\alpha, +\infty)))$;*
(iv) *$FA^\dagger A = F$;*
(v) *$[\mathcal{N}(A)]^\perp \supseteq \mathcal{R}(F(z))$ for all $z \in \mathbb{C} \setminus [\alpha, +\infty)$;*
(vi) *there is a $z_0 \in \mathbb{C} \setminus [\alpha, +\infty)$ such that $[\mathcal{N}(A)]^\perp \supseteq \mathcal{R}(F(z_0))$;*
(vii) *$[\mathcal{N}(A)]^\perp \supseteq \mathcal{R}(\gamma_F) + \mathcal{R}(\mu_F([\alpha, +\infty)))$;*
(viii) *$A^\dagger AF = F$.*

Proof. (i)⇒(ii) and (v)⇒(vi): These implications hold true obviously.

(i)⇔(iii) and (ii)⇒(iii): Use equation (3.10) in Proposition 3.15(b).

(i)⇔(iv): This equivalence follows from a well-known result for the Moore–Penrose inverse of complex matrices.

(i)⇔(v): Because of Proposition 3.15(b), we have $\mathcal{N}(F(z)) = \mathcal{R}(F(z))^\perp$ for all $z \in \mathbb{C} \setminus [\alpha, +\infty)$. Hence, (i) and (v) are equivalent.

(v)⇔(vii) and (vi)⇒(vii): Use equation (3.11) in Proposition 3.15(b).

(v)⇔(viii): Use $\mathcal{N}(A)^\perp = \mathcal{R}(A^*)$ and $A^\dagger A\mathcal{R}(A^*) = \mathcal{R}(A^*)$. □

Now we apply the preceding results to the subclass $\mathcal{S}^\diamond_{q;[\alpha,+\infty)}$ of $\mathcal{S}_{q;[\alpha,+\infty)}$ (see Notation 2.9).

Example 3.19. Let $\alpha \in \mathbb{R}$ and let $F : \mathbb{C} \setminus [\alpha, +\infty) \to \mathbb{C}^{q\times q}$ be defined by $F(z) := O_{q\times q}$. In view of Example 3.9 and Corollary 3.14, one can easily see then that F belongs to $\mathcal{S}^\diamond_{q;[\alpha,+\infty)}$ and that μ_F is the zero measure belonging to $\mathcal{M}^q_{\geq}([\alpha, +\infty))$.

Remark 3.20. Let $\alpha \in \mathbb{R}$, let $n \in \mathbb{N}$, and let $(q_k)_{k=1}^n$ be a sequence of positive integers. For each $k \in \mathbb{Z}_{1,n}$, let $F_k \in \mathcal{S}^\diamond_{q_k;[\alpha,+\infty)}$ and let $A_k \in \mathbb{C}^{q_k\times q}$. In view of Corollary 3.14 and Remark 3.11, then:

(a) the function $G := \sum_{k=1}^{n} A_k^* F_k A_k$ belongs to $\mathcal{S}^{\diamond}_{q;[\alpha,+\infty)}$ and

$$\mu_G = \sum_{k=1}^{n} A_k^* \mu_{F_k} A_k;$$

(b) the function $F := \operatorname{diag}[F_k]_{k=1}^{n}$ belongs to $\mathcal{S}^{\diamond}_{\sum_{k=1}^{n} q_k;[\alpha,+\infty)}$ and

$$\mu_F = \operatorname{diag}[\mu_{F_k}]_{k=1}^{n}.$$

Remark 3.21. Let $\alpha \in \mathbb{R}$ and let $F \in \mathcal{S}^{\diamond}_{q;[\alpha,+\infty)}$. In view of Corollary 3.14 and Proposition 3.15b, then $\mathcal{N}(F(z)) = \mathcal{N}(\mu_F([\alpha,+\infty)))$ and $\mathcal{R}(F(z)) = \mathcal{R}(\mu_F([\alpha,+\infty)))$ for all $z \in \mathbb{C} \setminus [\alpha,+\infty)$.

4. Characterizations of the class $\mathcal{S}_{q;[\alpha,+\infty)}$

In this section, we give several characterizations of the class $\mathcal{S}_{q;[\alpha,+\infty)}$.

Lemma 4.1. *Let $\alpha \in \mathbb{R}$, let $F : \mathbb{C} \setminus [\alpha,+\infty) \to \mathbb{C}^{q\times q}$ be holomorphic, and let $F^{\square} : \mathbb{C} \setminus [\alpha,+\infty) \to \mathbb{C}^{q\times q}$ be defined by*

$$F^{\square}(z) := (z-\alpha)F(z). \tag{4.1}$$

Suppose that $\operatorname{Rstr}_{\Pi_+} F$ and $\operatorname{Rstr}_{\Pi_+} F^{\square}$ both belong to $\mathcal{R}_q(\Pi_+)$. Then

$$(\operatorname{Re} F)((-\infty,\alpha)) \subseteq \mathbb{C}^{q\times q}_{\geq}.$$

Proof. We consider an arbitrary $x \in (-\infty,\alpha)$. For each $n \in \mathbb{N}$, we have then

$$\operatorname{Re} F\left(x + \frac{\mathrm{i}}{n}\right) = n \operatorname{Im} F^{\square}\left(x + \frac{\mathrm{i}}{n}\right) + n(\alpha - x)\operatorname{Im} F\left(x + \frac{\mathrm{i}}{n}\right). \tag{4.2}$$

For each $n \in \mathbb{N}$, $\operatorname{Rstr}_{\Pi_+} F^{\square} \in \mathcal{R}_q(\Pi_+)$ implies $n \operatorname{Im} F^{\square}(x + \mathrm{i}/n) \in \mathbb{C}^{q\times q}_{\geq}$, whereas $\operatorname{Rstr}_{\Pi_+} F \in \mathcal{R}_q(\Pi_+)$ yields $n(\alpha - x)\operatorname{Im} F(x + \mathrm{i}/n) \in \mathbb{C}^{q\times q}_{\geq}$. Thus, (4.2) provides us $\operatorname{Re} F(x + \mathrm{i}/n) \in \mathbb{C}^{q\times q}_{\geq}$ for each $n \in \mathbb{N}$. Since F is continuous, we get then $\operatorname{Re} F(x) = \lim_{n\to\infty} \operatorname{Re} F(x+\mathrm{i}/n)$. In particular, the matrix $\operatorname{Re} F(x)$ is non-negative Hermitian. □

In order to give further characterizations of the class $\mathcal{S}_{q;[\alpha,+\infty)}$, we state the following technical result. The proof of which uses an idea which is originated in [19, Theorem A.5].

Lemma 4.2. *Let $\alpha \in \mathbb{R}$ and let $F \in \mathcal{S}_{q;[\alpha,+\infty)}$. Then $F^{\square} : \mathbb{C} \setminus [\alpha,+\infty) \to \mathbb{C}^{q\times q}$ defined by (4.1) is holomorphic and fulfills*

$$\operatorname{Im} F^{\square}(z) = (\operatorname{Im} z)\left[\gamma_F + \int_{[\alpha,+\infty)} \frac{(1+t-\alpha)(t-\alpha)}{|t-z|^2}\mu_F(\mathrm{d}t)\right] \tag{4.3}$$

for each $z \in \mathbb{C} \setminus [\alpha,+\infty)$. Furthermore, $\frac{1}{\operatorname{Im} z} \operatorname{Im} F^{\square}(z) \in \mathbb{C}^{q\times q}_{\geq}$ for each $z \in \mathbb{C} \setminus \mathbb{R}$ and $F^{\square}(x) \in \mathbb{C}^{q\times q}_{\mathrm{H}}$ for each $x \in (-\infty,\alpha)$.

Proof. Since F is holomorphic, the matrix-valued function $F^\square$ is holomorphic as well. In view of Remark 3.7, using a well-known result on integrals with respect to non-negative Hermitian measures, we have

$$[F(z)]^* = \gamma_F + \int_{[\alpha,+\infty)} \frac{1+t-\alpha}{t-\overline{z}} \mu_F(\mathrm{d}t) \tag{4.4}$$

for each $z \in \mathbb{C} \setminus [\alpha, +\infty)$. Thus, from Remark 3.7 and (4.4) we get

$$\begin{aligned} 2\mathrm{i} \operatorname{Im} F^\square(z) &= (z-\alpha)F(z) - (\overline{z}-\alpha)[F(z)]^* \\ &= (z-\overline{z})\gamma_F + \int_{[\alpha,+\infty)} (1+t-\alpha)\left(\frac{z-\alpha}{t-z} - \frac{\overline{z}-\alpha}{t-\overline{z}}\right)\mu_F(\mathrm{d}t) \end{aligned} \tag{4.5}$$

for each $z \in \mathbb{C} \setminus [\alpha, +\infty)$. Since

$$\frac{z-\alpha}{t-z} - \frac{\overline{z}-\alpha}{t-\overline{z}} = \frac{2\mathrm{i}(\operatorname{Im} z)(t-\alpha)}{|t-z|^2}$$

holds for every choice of $z \in \mathbb{C} \setminus [\alpha, +\infty)$ and $t \in [\alpha, +\infty)$, from (4.5) it follows (4.3) for each $z \in \mathbb{C} \setminus [\alpha, +\infty)$. Since $(1+t-\alpha)(t-\alpha)/|t-z|^2 \in [0, +\infty)$ holds true for each $z \in \mathbb{C} \setminus [\alpha, +\infty)$ and each $t \in [\alpha, +\infty)$, from $\gamma_F \in \mathbb{C}_{\geq}^{q\times q}$ and (4.3) we get $\frac{1}{\operatorname{Im} z} \operatorname{Im} F^\square(z) \in \mathbb{C}_{\geq}^{q\times q}$ for each $z \in \mathbb{C} \setminus \mathbb{R}$. In view of Lemma 3.4, then $F^\square(x) = \operatorname{Re} F^\square(x)$ and hence $F^\square(x) \in \mathbb{C}_{\mathrm{H}}^{q\times q}$ for each $x \in (-\infty, \alpha)$. □

Proposition 4.3. *Let $\alpha \in \mathbb{R}$, let $F : \mathbb{C} \setminus [\alpha, +\infty) \to \mathbb{C}^{q\times q}$ be a matrix-valued function, and let $F^\square : \mathbb{C} \setminus [\alpha, +\infty) \to \mathbb{C}^{q\times q}$ be defined by* (4.1)*. Then F belongs to $\mathcal{S}_{q;[\alpha,+\infty)}$ if and only if the following two conditions hold true:*

(i) *F is holomorphic in $\mathbb{C} \setminus [\alpha, +\infty)$;*
(ii) *the matrix-valued functions $\operatorname{Rstr}_{\Pi_+} F$ and $\operatorname{Rstr}_{\Pi_+} F^\square$ both belong to $\mathcal{R}_q(\Pi_+)$.*

Proof. First suppose that F belongs to $\mathcal{S}_{q;[\alpha,+\infty)}$. Then (i) and $\operatorname{Rstr}_{\Pi_+} F \in \mathcal{R}_q(\Pi_+)$ follow from the definition of the class $\mathcal{S}_{q;[\alpha,+\infty)}$. Furthermore, Lemma 4.2 provides us $\operatorname{Rstr}_{\Pi_+} F^\square \in \mathcal{R}_q(\Pi_+)$.

Conversely, now suppose that (i) and (ii) hold true. Because of the definition of the classes $\mathcal{R}_q(\Pi_+)$ and $\mathcal{S}_{q;[\alpha,+\infty)}$, it then remains to prove that $F((-\infty, \alpha)) \subseteq \mathbb{C}_{\geq}^{q\times q}$. We consider an arbitrary $x \in (-\infty, \alpha)$. First we show that $\operatorname{Im} F(x) = O_{q\times q}$. Because of (ii), for each $n \in \mathbb{N}$, the matrices $\operatorname{Im} F(x + \mathrm{i}/n)$ and $\operatorname{Im} F^\square(x + \mathrm{i}/n)$ are non-negative Hermitian. Thus, the matrices $\operatorname{Im} F(x)$ and $\operatorname{Im} F^\square(x)$ are (as limits of the sequences $(\operatorname{Im} F(x+\mathrm{i}/n))_{n=1}^\infty$ and $(\operatorname{Im} F^\square(x+\mathrm{i}/n))_{n=1}^\infty$, respectively) non-negative Hermitian as well. Since (4.1) implies $\operatorname{Im} F^\square(x) = (x-\alpha) \operatorname{Im} F(x)$, we get then $-\operatorname{Im} F(x) = \frac{1}{\alpha - x} \operatorname{Im} F^\square(x) \in \mathbb{C}_{\geq}^{q\times q}$, which together with $\operatorname{Im} F(x) \in \mathbb{C}_{\geq}^{q\times q}$ yields $\operatorname{Im} F(x) = O_{q\times q}$. Hence, $\operatorname{Re} F(x) = F(x)$. Because of (i), (ii), and Lemma 4.1, we have $\operatorname{Re} F(x) \in \mathbb{C}_{\geq}^{q\times q}$. Thus, $F((-\infty, \alpha)) \subseteq \mathbb{C}_{\geq}^{q\times q}$. □

Proposition 4.4. *Let $\alpha \in \mathbb{R}$ and let $F : \mathbb{C} \setminus [\alpha, +\infty) \to \mathbb{C}^{q\times q}$ be a matrix-valued function. Then F belongs to $\mathcal{S}_{q;[\alpha,+\infty)}$ if and only if the following four conditions are fulfilled:*

(i) *F is holomorphic in $\mathbb{C} \setminus [\alpha, +\infty)$;*
(ii) *for each $z \in \Pi_+$, the matrix $\operatorname{Im} F(z)$ is non-negative Hermitian;*
(iii) *for each $z \in \Pi_-$, the matrix $-\operatorname{Im} F(z)$ is non-negative Hermitian;*
(iv) *for each $z \in \mathbb{C}_{\alpha,-}$, the matrix $\operatorname{Re} F(z)$ is non-negative Hermitian.*

Proof. First suppose that $F \in \mathcal{S}_{q;[\alpha,+\infty)}$. By definition of the class $\mathcal{S}_{q;[\alpha,+\infty)}$, conditions (i) and (ii) are fulfilled. From Theorem 3.6 and Lemma 3.5 we obtain (iii) and (iv).

Conversely, (i)–(iii) and Lemma 3.4 imply $\operatorname{Im} F(x) = O_{q\times q}$ for all $x \in (-\infty, \alpha)$. In view of (iv), we have then $F((-\infty,\alpha)) \subseteq \mathbb{C}_{\geq}^{q\times q}$. Together with (i) and (ii), this yields $F \in \mathcal{S}_{q;[\alpha,+\infty)}$. □

5. The class $\mathcal{S}_{0,q;[\alpha,+\infty)}$

In this section, we prove an important integral representation for functions which belong to the class $\mathcal{S}_{0,q;[\alpha,+\infty)}$. It can be considered as modified integral representation of the functions belonging to the class

$$\mathcal{R}_{0,q}(\Pi_+) := \left\{ F \in \mathcal{R}_q(\Pi_+) : \sup_{y\in[1,+\infty)} y\|F(\mathrm{i}y)\| < +\infty \right\}$$

(see [4, Theorem 8.7]). Observe that if $\alpha \in \mathbb{R}$ and if $z \in \mathbb{C} \setminus [\alpha, +\infty)$, then in view of Lemma A.8(a), for each $\sigma \in \mathcal{M}_{\geq}^q([\alpha,+\infty))$, the integral

$$\int_{[\alpha,+\infty)} \frac{1}{t-z}\sigma(\mathrm{d}t)$$

exists.

Theorem 5.1. *Let $\alpha \in \mathbb{R}$ and let $F : \mathbb{C} \setminus [\alpha, +\infty) \to \mathbb{C}^{q\times q}$.*

(a) *If $F \in \mathcal{S}_{0,q;[\alpha,+\infty)}$, then there is a unique non-negative Hermitian measure $\sigma \in \mathcal{M}_{\geq}^q([\alpha,+\infty))$ such that*

$$F(z) = \int_{[\alpha,+\infty)} \frac{1}{t-z}\sigma(\mathrm{d}t) \tag{5.1}$$

for each $z \in \mathbb{C} \setminus [\alpha, +\infty)$.

(b) *If there is a non-negative Hermitian measure $\sigma \in \mathcal{M}_{\geq}^q([\alpha,+\infty))$ such that F can be represented via (5.1) for each $z \in \mathbb{C} \setminus [\alpha,+\infty)$, then F belongs to the class $\mathcal{S}_{0,q;[\alpha,+\infty)}$.*

Proof. We modify ideas of proofs of integral representations of similar classes of holomorphic functions (see [19, Appendix]).

(a) First suppose $F \in \mathcal{S}_{0,q;[\alpha,+\infty)}$. Then $F \in \mathcal{S}_{q;[\alpha,+\infty)}$ and the function $F := \operatorname{Rstr}_{\Pi_+} F$ belongs to the class $\mathcal{R}_{0,q}(\Pi_+)$. From a matricial version of a well-known integral representation of functions belonging to $\mathcal{R}_{0,1}(\Pi_+)$ (see, e.g., [4,

Theorem 8.7]) we know that there is a unique $\mu \in \mathcal{M}^q_{\geq}(\mathbb{R})$ such that

$$F(w) = \int_{\mathbb{R}} \frac{1}{t-w} \mu(\mathrm{d}t) \tag{5.2}$$

for all $w \in \Pi_+$, namely the so-called matricial spectral measure of F, i.e., for each $B \in \mathfrak{B}_{\mathbb{R}}$, we have

$$\mu(B) = \int_B (1+t^2)\nu(\mathrm{d}t),$$

where ν denotes the Nevanlinna measure of F. Proposition 4.4 shows that F is holomorphic in $\mathbb{C} \setminus [\alpha, +\infty)$ and that $\operatorname{Im} F(z) \in \mathbb{C}^{q\times q}_{\geq}$ for all $z \in \Pi_+$ and $-\operatorname{Im} F(z) \in \mathbb{C}^{q\times q}_{\geq}$ for all $z \in \Pi_-$. Hence, for each $t \in (-\infty, \alpha)$, we have $F^*(t) = F(t)$. Applying the Stieltjes–Perron inversion formula (see, e.g., [4, Theorem 8.2]), one can verify that $\nu((-\infty, \alpha)) = O_{q\times q}$. Hence $\mu((-\infty, \alpha)) = O_{q\times q}$. Consequently, formula (5.2) shows that (5.1) holds for each $z \in \Pi_+$, where $\sigma := \operatorname{Rstr}_{\mathfrak{B}_{[\alpha,+\infty)}} \mu$. Since $[\alpha, +\infty)$ is a closed interval the function $G : \mathbb{C} \setminus [\alpha, +\infty) \to \mathbb{C}^{q\times q}$ defined by

$$G(z) := \int_{[\alpha,+\infty)} \frac{1}{t-z} \sigma(\mathrm{d}t)$$

is holomorphic (see, e.g., [9, Ch. IV, §5, Satz 5.8]). Because of $F(w) = G(w)$ for each $w \in \Pi_+$, we have $F = G$. If σ is an arbitrary measure belonging to $\mathcal{M}^q_{\geq}([\alpha, +\infty))$ such that (5.1) holds for each $z \in \mathbb{C} \setminus [\alpha, +\infty)$, then using standard arguments of measure theory and the uniqueness of the non-negative Hermitian $q \times q$ measure ν in the integral representation (2.2), one gets necessary $\sigma = \operatorname{Rstr}_{\mathfrak{B}_{[\alpha,+\infty)}} \mu$.

(b) Let $\sigma \in \mathcal{M}^q_{\geq}([\alpha, +\infty))$ be such that (5.1) holds for each $z \in \mathbb{C} \setminus [\alpha, +\infty)$. Then F is holomorphic (see, e.g., [9, Ch. IV, §5, Satz 5.8]) and, for each $z \in \mathbb{C} \setminus \mathbb{R}$, we have

$$\frac{1}{\operatorname{Im} z} \operatorname{Im} F(z) = \int_{[\alpha,+\infty)} \frac{1}{\operatorname{Im} z} \operatorname{Im} \frac{1}{t-z} \sigma(\mathrm{d}t) = \int_{[\alpha,+\infty)} \frac{1}{|t-z|^2} \sigma(\mathrm{d}t) \in \mathbb{C}^{q\times q}_{\geq}.$$

and, for each z belonging to $\mathbb{C}_{\alpha,-}$, moreover

$$\operatorname{Re} F(z) = \int_{[\alpha,+\infty)} \operatorname{Re} \frac{1}{t-z} \sigma(\mathrm{d}t) = \int_{[\alpha,+\infty)} \frac{t - \operatorname{Re} z}{|t-z|^2} \sigma(\mathrm{d}t) \in \mathbb{C}^{q\times q}_{\geq}.$$

Thus, $F \in \mathcal{S}_{q;[\alpha,+\infty)}$. From the definition of F and [4, Theorem 8.7(b)] we see that $\operatorname{Rstr}_{\Pi_+} F \in \mathcal{R}_{0,q}(\Pi_+)$. Thus, (2.1) is satisfied. Hence, $F \in \mathcal{S}_{0,q;[\alpha,+\infty)}$ holds. □

If σ is a measure belonging to $\mathcal{M}^q_{\geq}([\alpha, +\infty))$, then we will call the matrix-valued function $F : \mathbb{C} \setminus [\alpha, +\infty) \to \mathbb{C}^{q\times q}$ which is, for each $z \in \mathbb{C} \setminus [\alpha, +\infty)$, given by (5.1) the *$[\alpha, +\infty)$-Stieltjes transform of σ*. If $F \in \mathcal{S}_{0,q;[\alpha,+\infty)}$, then the unique measure σ which belongs to $\mathcal{M}^q_{\geq}([\alpha, +\infty))$ and which fulfills (5.1) for each $z \in \mathbb{C} \setminus [\alpha, +\infty)$ is said to be the *$[\alpha, +\infty)$-Stieltjes measure of F* and will be denoted by σ_F.

Note that, in view of Theorem 5.1, the matricial Stieltjes moment problem $\mathsf{M}[[\alpha, +\infty); (s_j)_{j=0}^m, =]$ can be obviously reformulated in the language of

$[\alpha, +\infty)$-Stieltjes transforms of non-negative Hermitian measures. We omit the details.

Remark 5.2. Let $\alpha \in \mathbb{R}$, let $n \in \mathbb{N}$, and let $(q_k)_{k=1}^n$ be a sequence of positive integers. For each $k \in \mathbb{Z}_{1,n}$, let $F_k \in \mathcal{S}_{0,q_k;[\alpha,+\infty)}$, and let σ_{F_k} be the $[\alpha, +\infty)$-Stieltjes measure of F_k. Then $F := \operatorname{diag}[F_k]_{k=1}^n$ belongs to $\mathcal{S}_{0,\sum_{k=1}^n q_k;[\alpha,+\infty)}$ and $\operatorname{diag}[\sigma_{F_k}]_{k=1}^n$ is the $[\alpha, +\infty)$-Stieltjes measure of F. Moreover, if $A_k \in \mathbb{C}^{q_k \times q}$ for each $k \in \mathbb{Z}_{1,n}$, then $G := \sum_{k=1}^n A_k^* F_k A_k$ belongs to $\mathcal{S}_{0,q;[\alpha,+\infty)}$ and $\sum_{k=1}^n A_k^* \sigma_{F_k} A_k$ is the $[\alpha, +\infty)$-Stieltjes measure of G.

Proposition 5.3. *Let $\alpha \in \mathbb{R}$, let $F \in \mathcal{S}_{0,q;[\alpha,+\infty)}$, and let σ_F be the $[\alpha, +\infty)$-Stieltjes measure of F. For each $z \in \mathbb{C} \setminus [\alpha, +\infty)$, then*

$$\mathcal{N}(F(z)) = \mathcal{N}(\sigma_F([\alpha, +\infty))) \qquad \text{and} \qquad \mathcal{R}(F(z)) = \mathcal{R}(\sigma_F([\alpha, +\infty))). \tag{5.3}$$

Furthermore,

$$\sigma_F([\alpha, +\infty)) = -\mathrm{i} \lim_{y \to +\infty} yF(\mathrm{i}y). \tag{5.4}$$

In particular, $\operatorname{rank} F(z) = \operatorname{rank} \sigma_F([\alpha, +\infty))$ *holds true for each* $z \in \mathbb{C} \setminus [\alpha, +\infty)$.

Proof. Let $z \in \mathbb{C} \setminus [\alpha, +\infty)$. From Theorem 5.1(a) and Lemma A.8(b) we obtain the second equation in (5.3). The first one is an immediate consequence of the second one and the equation $F(z) = F^*(\overline{z})$, which can be seen from (5.1). Because of Theorem 5.1(a) and Lemma A.8(c), the equation (5.4) holds true. □

6. Moore–Penrose inverses of functions belonging to the class $\mathcal{S}_{q;[\alpha,+\infty)}$

We start with some further notation. If $\mathcal{Z}$ is a non-empty subset of $\mathbb{C}$ and if a matrix-valued function $F : \mathcal{Z} \to \mathbb{C}^{p \times q}$ is given, then let $F^\dagger : \mathcal{Z} \to \mathbb{C}^{q \times p}$ be defined by $F^\dagger(z) := [F(z)]^\dagger$, where $[F(z)]^\dagger$ stands for the Moore–Penrose inverse of the matrix $F(z)$. In [11] (see also [10]), we investigated the Moore–Penrose inverse of an arbitrary function F belonging to the class $\mathcal{R}_q(\Pi_+)$. In particular, it turned out that $-F^\dagger$ belongs to $\mathcal{R}_q(\Pi_+)$ (see [10, Theorem 9.4]). The close relation between $\mathcal{S}_{q;[\alpha,+\infty)}$ and $\mathcal{R}_q(\Pi_+)$ suggests now to study the Moore–Penrose inverse of a function $F \in \mathcal{S}_{q;[\alpha,+\infty)}$.

Lemma 6.1. *Let $\alpha \in \mathbb{R}$ and $F \in \mathcal{S}_{q;[\alpha,+\infty)}$. Then $F^\dagger$ is holomorphic in $\mathbb{C} \setminus [\alpha, +\infty)$.*

Proof. In view of formulas (3.10) and (3.11), we obtain for all $z \in \mathbb{C} \setminus [\alpha, +\infty)$ the identities $\mathcal{N}(F(z)) = \mathcal{N}(F(\mathrm{i}))$ and $\mathcal{R}(F(z)) = \mathcal{R}(F(\mathrm{i}))$. Thus, the application of [15, Proposition 8.4] completes the proof. □

Let $\alpha \in \mathbb{R}$ and $F \in \mathcal{S}_{q;[\alpha,+\infty)}$. Then Lemma 6.1 suggests to look if there are functions closely related to $F^\dagger$ which belong again to $\mathcal{S}_{q;[\alpha,+\infty)}$. Against to this background, we are led to the function $G : \mathbb{C} \setminus [\alpha, +\infty) \to \mathbb{C}^{q \times q}$ defined by

$$G(z) := -(z - \alpha)^{-1} [F(z)]^\dagger.$$

Remark 6.2. If $A \in \mathbb{C}^{q\times q}_{\mathrm{EP}}$, i.e., if $A \in \mathbb{C}^{q\times q}$ fulfills $\mathcal{R}(A^*) = \mathcal{R}(A)$, then

$$\operatorname{Im}(A^\dagger) = -A^\dagger(\operatorname{Im} A)(A^\dagger)^*$$

(see also [10, Propostions A.5 and A.6]).

Theorem 6.3. *Let $\alpha \in \mathbb{R}$ and let $F \in \mathcal{S}_{q;[\alpha,+\infty)}$. Then $G : \mathbb{C}\setminus[\alpha,+\infty) \to \mathbb{C}^{q\times q}$ defined by $G(z) := -(z-\alpha)^{-1}[F(z)]^\dagger$ belongs to the class $\mathcal{S}_{q;[\alpha,+\infty)}$ as well.*

Proof. Lemma 6.1 yields that the function $F^\dagger$ is holomorphic in $\mathbb{C}\setminus[\alpha,+\infty)$. Consequently, the function G is holomorphic in $\mathbb{C}\setminus[\alpha,+\infty)$. Let $z \in \Pi_+$. Using Proposition 3.15(b), we have $\mathcal{R}(F^*(z)) = \mathcal{R}(F(z))$. Hence, because of Remark 6.2, the equations

$$\operatorname{Im} G(z) = G(z)(\operatorname{Im}[(z-\alpha)F(z)])G^*(z) \tag{6.1}$$

and

$$\operatorname{Im}[(z-\alpha)G(z)] = \operatorname{Im}\big[-F^\dagger(z)\big] = F^\dagger(z)[\operatorname{Im} F(z)]\big[F^\dagger(z)\big]^* \tag{6.2}$$

hold. Taking into account $F \in \mathcal{S}_{q;[\alpha,+\infty)}$, the application of Proposition 4.3 yields

$$\operatorname{Im} F(z)P \in \mathbb{C}^{q\times q}_{\geq} \quad \text{and} \quad \operatorname{Im}[(z-\alpha)F(z)] \in \mathbb{C}^{q\times q}_{\geq}. \tag{6.3}$$

Thus, combining the identities (6.1) (resp. (6.2)) and (6.3), we get $\operatorname{Im} G(z) \in \mathbb{C}^{q\times q}_{\geq}$ and $\operatorname{Im}[(z-\alpha)G(z)] \in \mathbb{C}^{q\times q}_{\geq}$. Now, the application of Proposition 4.3 yields $G \in \mathcal{S}_{q;[\alpha,+\infty)}$. □

Now we specify the result of Theorem 6.3 for functions belonging to $\mathcal{S}_{0,q;[\alpha,+\infty)}$.

Proposition 6.4. *Let $\alpha \in \mathbb{R}$, let $F \in \mathcal{S}_{0,q;[\alpha,+\infty)}$, and let σ_F be the $[\alpha,+\infty)$-Stieltjes measure of F. Then $G : \mathbb{C}\setminus[\alpha,+\infty) \to \mathbb{C}^{q\times q}$ defined by $G(z) := -(z-\alpha)^{-1}[F(z)]^\dagger$ belongs to $\mathcal{S}_{q;[\alpha,+\infty)}$ and*

$$\gamma_G = [\sigma_F([\alpha,+\infty))]^\dagger. \tag{6.4}$$

In particular, if F is not the constant function with value $O_{q\times q}$, then G belongs to $\mathcal{S}_{q;[\alpha,+\infty)} \setminus \mathcal{S}_{0,q;[\alpha,+\infty)}$.

Proof. In view of Theorem 6.3, we have $G \in \mathcal{S}_{q;[\alpha,+\infty)}$. From Proposition 3.13 we obtain

$$\gamma_G = \lim_{y\to+\infty} G(\mathrm{i}y). \tag{6.5}$$

Since F belongs to $\mathcal{S}_{0,q;[\alpha,+\infty)}$, we have $\lim_{y\to+\infty} F(\mathrm{i}y) = O_{q\times q}$. Proposition 5.3 yields (5.4). Consequently,

$$\lim_{y\to+\infty} (\alpha - \mathrm{i}y)F(\mathrm{i}y) = \sigma_F([\alpha,+\infty)).$$

In view of Proposition 5.3, we have $\mathcal{R}(F(\mathrm{i}y)) = \mathcal{R}(\sigma_F([\alpha,+\infty)))$ and, in particular, $\operatorname{rank}[(\alpha-\mathrm{i}y)F(\mathrm{i}y)] = \operatorname{rank}\sigma_F([\alpha,+\infty))$ for each $y \in (0,+\infty)$. Hence, taking into account [3, Theorem 10.4.1], we obtain

$$\lim_{y\to+\infty}\Big([(\alpha-\mathrm{i}y)F(\mathrm{i}y)]^\dagger\Big) = [\sigma_F([\alpha,+\infty))]^\dagger. \tag{6.6}$$

Since $G(\mathrm{i}y) = -(\mathrm{i}y-\alpha)^{-1}F^{\dagger}(\mathrm{i}y) = [(\alpha-\mathrm{i}y)F(\mathrm{i}y)]^{\dagger}$ holds true for each $y \in (0,+\infty)$, from (6.5) and (6.6) we get then (6.4). Now assume that G belongs to $\mathcal{S}_{0,q;[\alpha,+\infty)}$. From the definition of the class $\mathcal{S}_{0,q;[\alpha,+\infty)}$ we obtain then

$$\lim_{y\to+\infty} G(\mathrm{i}y) = O_{q\times q},$$

which, in view of (6.5) and (6.4), implies $\sigma_F([\alpha,+\infty)) = O_{q\times q}^{\dagger} = O_{q\times q}$. Proposition 5.3 yields then $\mathcal{N}(F(z)) = \mathbb{C}^q$ and hence $F(z) = O_{q\times q}$ for all $z \in \mathbb{C}\setminus[\alpha,+\infty)$. This proves $G \notin \mathcal{S}_{0,q;[\alpha,+\infty)}$ if F is not the constant function with value $O_{q\times q}$. □

For the special choice $q = 1$ and $\alpha = 0$, the following class was introduced by Kats/Krein [17, Definition D1.5.2].

Notation 6.5. Let $\alpha \in \mathbb{R}$. Then let $\mathcal{S}^{[-1]}_{q;[\alpha,+\infty)}$ be the class of all matrix-valued functions $F : \mathbb{C}\setminus[\alpha,+\infty) \to \mathbb{C}^{q\times q}$ which fulfill the following two conditions:

(i) F is holomorphic in $\mathbb{C}\setminus[\alpha,+\infty)$ with $\operatorname{Rstr}_{\Pi_+} F \in \mathcal{R}_q(\Pi_+)$;
(ii) for all $x \in (-\infty,\alpha)$, the matrix $-F(x)$ is non-negative Hermitian.

Remark 6.6. Let $\alpha \in \mathbb{R}$, then $F \in \mathbb{C}\setminus[\alpha,+\infty) \to \mathbb{C}^{q\times q}$ belongs to $\mathcal{S}^{[-1]}_{q;[\alpha,+\infty)}$ if and only if $u^*Fu \in \mathcal{S}^{[-1]}_{1;[\alpha,+\infty)}$ for all $u \in \mathbb{C}^q$.

Example 6.7. Let $\alpha \in \mathbb{R}$ and let $D, E \in \mathbb{C}^{q\times q}_{\geq}$. Then $F : \mathbb{C}\setminus[\alpha,+\infty) \to \mathbb{C}^{q\times q}$ defined by $F(z) := -D + (z-\alpha)E$ belongs to $\mathcal{S}^{[-1]}_{q;[\alpha,+\infty)}$.

Lemma 6.8. *Let $\alpha \in \mathbb{R}$ and let $f : \mathbb{C}\setminus[\alpha,+\infty) \to \mathbb{C}$ be such that there are real numbers d and e and a finite signed measure ρ on $((\alpha,+\infty), \mathfrak{B}_{(\alpha,+\infty)})$ such that*

$$f(z) = -d + (z-\alpha)\left[e + \int_{(\alpha,+\infty)} \frac{1+t-\alpha}{t-z}\rho(\mathrm{d}t)\right]$$

holds true for all $z \in \mathbb{C}\setminus[\alpha,+\infty)$. Then

$$d = -\lim_{x\to+0} f(\alpha - x) \quad \text{and} \quad e = -\lim_{x\to+\infty}[f(\alpha-x)+d]/x.$$

Furthermore, d, e, and ρ are uniquely determined.

Proof. With $z_x := \alpha - x$ we have $(1+t-\alpha)/(t-z_x) = (1+t-\alpha)/(t-\alpha+x) \geq 0$ for all $t \in (\alpha,+\infty)$ and all $x \in (0,+\infty)$, which decreases to 0 as x increases to infinity. Since the signed measure ρ is finite, its Jordan decomposition $\rho = \rho_+ - \rho_-$ consists of two finite measures. Hence,

$$\int_{(\alpha,+\infty)} \frac{1+t-\alpha}{t-z_1}\rho_\pm(\mathrm{d}t) = \rho_\pm((\alpha,+\infty)) < \infty$$

holds true. Thus, Lebesgue's monotone convergence theorem yields

$$\lim_{x\to+\infty}\int_{(\alpha,+\infty)} \frac{1+t-\alpha}{t-z_x}\rho_\pm(\mathrm{d}t) = 0,$$

which implies

$$\lim_{x\to+\infty}\frac{f(z_x)+d}{z_x-\alpha}=e.$$

Furthermore, we have

$$\frac{-(z_x-\alpha)(1+t-\alpha)}{t-z_x}=\frac{1+t-\alpha}{1+(t-\alpha)/x}\geq 0$$

for all $t\in(\alpha,+\infty)$ and all $x\in(0,+\infty)$, which decreases to 0 as x decreases to 0. Since

$$\int_{(\alpha,+\infty)}\frac{-(z_1-\alpha)(1+t-\alpha)}{t-z_1}\rho_\pm(\mathrm{d}t)=\rho_\pm((\alpha,+\infty))<\infty$$

holds true, Lebesgue's monotone convergence theorem yields

$$\lim_{x\to+0}\int_{(\alpha,+\infty)}\frac{-(z_x-\alpha)(1+t-\alpha)}{t-z_x}\rho_\pm(\mathrm{d}t)=0$$

showing $-\lim_{x\to+0}f(z_x)=d$. In particular, d and e are uniquely determined. Now let σ be an arbitrary finite signed measure on $((\alpha,+\infty),\mathfrak{B}_{(\alpha,+\infty)})$ such that

$$f(z)=-d+(z-\alpha)\left[e+\int_{(\alpha,+\infty)}\frac{1+t-\alpha}{t-z}\sigma(\mathrm{d}t)\right]$$

holds true for all $z\in\mathbb{C}\setminus[\alpha,+\infty)$. Then

$$\int_{(\alpha,+\infty)}\frac{1+t-\alpha}{t-z}\sigma(\mathrm{d}t)=\int_{(\alpha,+\infty)}\frac{1+t-\alpha}{t-z}\rho(\mathrm{d}t)$$

for all $z\in\mathbb{C}\setminus[\alpha,+\infty)$. Since the signed measure σ is finite, its Jordan decomposition $\sigma=\sigma_+-\sigma_-$ consists of two finite measures. Hence, we obtain

$$\int_{(\alpha,+\infty)}\frac{1+t-\alpha}{t-z}(\sigma_++\rho_-)(\mathrm{d}t)=\int_{(\alpha,+\infty)}\frac{1+t-\alpha}{t-z}(\rho_++\sigma_-)(\mathrm{d}t)$$

for all $z\in\mathbb{C}\setminus[\alpha,+\infty)$ with finite measures $\sigma_++\rho_-$ and $\rho_++\sigma_-$. Using Theorem 3.6, it is readily checked then that $\sigma_++\rho_-$ and $\rho_++\sigma_-$ coincide. Consequently, $\sigma=\rho$ follows. □

Lemma 6.9. *Let $\alpha\in\mathbb{R}$ and let $f\in\mathcal{S}_{1;[\alpha,+\infty)}^{[-1]}$. Then there are unique non-negative real numbers d and e and a unique measure $\rho\in\mathcal{M}_{\geq}^{1}((\alpha,+\infty))$ such that*

$$f(z)=-d+(z-\alpha)\left[e+\int_{(\alpha,+\infty)}\frac{1+t-\alpha}{t-z}\rho(\mathrm{d}t)\right]$$

holds true for all $z\in\mathbb{C}\setminus[\alpha,+\infty)$.

Proof. Obviously, the function $g : \mathbb{C} \setminus [0, +\infty) \to \mathbb{C}$ defined by $g(w) := f(w+\alpha)$ belongs to $\mathcal{S}^{[-1]}_{1;[0,+\infty)}$. Hence, by virtue of [17, Theorem S1.5.2], there exist unique numbers $a \in (-\infty, 0]$ and $b \in [0, +\infty)$ and a unique measure τ on $((0, +\infty), \mathfrak{B}_{(0,+\infty)})$ with

$$\int_{(0,+\infty)} \frac{1}{x + x^2} \tau(\mathrm{d}x) < \infty$$

such that

$$g(w) = a + bw + \int_{(0,+\infty)} \left[\frac{1}{x - w} - \frac{1}{x} \right] \tau(\mathrm{d}x)$$

for all $w \in \mathbb{C} \setminus [0, +\infty)$. Then χ defined on $\mathfrak{B}_{(0,+\infty)}$ by

$$\chi(B) := \int_B \frac{1}{x + x^2} \tau(\mathrm{d}x)$$

is a finite measure on $((0, +\infty), \mathfrak{B}_{(0,+\infty)})$ and the integral

$$\int_{(0,+\infty)} (x + x^2) \left[\frac{1}{x - w} - \frac{1}{x} \right] \chi(\mathrm{d}x)$$

exists for all $w \in \mathbb{C} \setminus [0, +\infty)$ and equals to

$$\int_{(0,+\infty)} \left[\frac{1}{x - w} - \frac{1}{x} \right] \tau(\mathrm{d}x).$$

We have $(x + x^2)[1/(x - w) - 1/x] = w(1 + x)/(x - w)$ for all $w \in \mathbb{C} \setminus [0, +\infty)$ and all $x \in (0, +\infty)$. In view of $z - \alpha \in \mathbb{C} \setminus [0, +\infty)$ for all $z \in \mathbb{C} \setminus [\alpha, +\infty)$, we obtain thus

$$\begin{aligned} f(z) = g(z - \alpha) &= a + b(z - \alpha) + \int_{(0,+\infty)} \left[\frac{1}{x - (z - \alpha)} - \frac{1}{x} \right] \tau(\mathrm{d}x) \\ &= a + (z - \alpha) \left[b + \int_{(0,+\infty)} \frac{1 + x}{x - (z - \alpha)} \chi(\mathrm{d}x) \right] \\ &= -d + (z - \alpha) \left[e + \int_{(\alpha,+\infty)} \frac{1 + t - \alpha}{t - z} \rho(\mathrm{d}t) \right] \end{aligned}$$

for all $z \in \mathbb{C} \setminus [\alpha, +\infty)$, where $d := -a$, $e := b$, and ρ is the image measure of χ under the translation $T : (0, +\infty) \to (\alpha, +\infty)$ defined by $T(x) := x + \alpha$. In particular $d, e \in [0, +\infty)$ and $\rho \in \mathcal{M}^1_{\geq}((\alpha, +\infty))$. Hence, the triple (d, e, ρ) is unique by virtue of Lemma 6.8. □

Theorem 6.10. *Let $\alpha \in \mathbb{R}$ and let $F : \mathbb{C} \setminus [\alpha, +\infty) \to \mathbb{C}^{q \times q}$.*

(a) *If $F \in \mathcal{S}^{[-1]}_{q;[\alpha,+\infty)}$, then there are unique non-negative Hermitian complex $q \times q$ matrices D and E and a unique non-negative Hermitian measure $\rho \in \mathcal{M}^q_{\geq}((\alpha, +\infty))$ such that*

$$F(z) = -D + (z - \alpha) \left[E + \int_{(\alpha,+\infty)} \frac{1 + t - \alpha}{t - z} \rho(\mathrm{d}t) \right] \tag{6.7}$$

for all $z \in \mathbb{C} \setminus [\alpha, +\infty)$. *Furthermore, the function* $P : \mathbb{C} \setminus [\alpha, +\infty) \to \mathbb{C}^{q\times q}$ *defined by* $P(z) := (z-\alpha)^{-1}F(z)$ *belongs to* $\mathcal{S}_{q;[\alpha,+\infty)}$ *with* $D = \mu_P(\{\alpha\})$ *and* $(E, \rho) = (\gamma_P, \operatorname{Rstr}_{\mathfrak{B}_{(\alpha,+\infty)}} \mu_P)$.

(b) *If* $D \in \mathbb{C}_{\geq}^{q\times q}$, $E \in \mathbb{C}_{\geq}^{q\times q}$, *and* $\rho \in \mathcal{M}_{\geq}^{q}((\alpha, +\infty))$ *are such that* F *can be represented via* (6.7) *for all* $z \in \mathbb{C} \setminus [\alpha, +\infty)$, *then* F *belongs to* $\mathcal{S}_{q;[\alpha,+\infty)}^{[-1]}$.

Proof. (a) We consider an arbitrary vector $u \in \mathbb{C}^q$. According to Remark 6.6, then $f_u := u^* F u$ belongs to $\mathcal{S}_{1;[\alpha,+\infty)}^{[-1]}$. Hence, Lemma 6.9 yields the existence of a unique triple $(d_u, e_u, \rho_u) \in [0, +\infty) \times [0, +\infty) \times \mathcal{M}_{\geq}^{1}((\alpha, +\infty))$ such that

$$f_u(z) = -d_u + (z-\alpha)\left[e_u + \int_{(\alpha,+\infty)} \frac{1+t-\alpha}{t-z}\rho_u(\mathrm{d}t)\right] \tag{6.8}$$

holds true for all $z \in \mathbb{C} \setminus [\alpha, +\infty)$. With the standard basis $(\mathbf{e}_1, \mathbf{e}_2, \dots, \mathbf{e}_q)$ of $\mathbb{C}^q$, let

$$d_{jk} := \frac{1}{4}\sum_{\ell=0}^{3}(-\mathrm{i})^{\ell} d_{\mathbf{e}_j+\mathrm{i}^{\ell}\mathbf{e}_k}, \quad e_{jk} := \frac{1}{4}\sum_{\ell=0}^{3}(-\mathrm{i})^{\ell} e_{\mathbf{e}_j+\mathrm{i}^{\ell}\mathbf{e}_k}, \quad \rho_{jk} := \frac{1}{4}\sum_{\ell=0}^{3}(-\mathrm{i})^{\ell} \rho_{\mathbf{e}_j+\mathrm{i}^{\ell}\mathbf{e}_k}$$

for all $j, k \in \mathbb{Z}_{1,q}$. We have then

$$\begin{aligned}
\mathbf{e}_j^*[F(z)]\mathbf{e}_k &= \frac{1}{4}\sum_{\ell=0}^{3}(-\mathrm{i})^{\ell} f_{\mathbf{e}_j+\mathrm{i}^{\ell}\mathbf{e}_k}(z)\\
&= \frac{1}{4}\sum_{\ell=0}^{3}(-\mathrm{i})^{\ell}\left(-d_{\mathbf{e}_j+\mathrm{i}^{\ell}\mathbf{e}_k} + (z-\alpha)\left[e_{\mathbf{e}_j+\mathrm{i}^{\ell}\mathbf{e}_k} + \int_{(\alpha,+\infty)} \frac{1+t-\alpha}{t-z}\rho_{\mathbf{e}_j+\mathrm{i}^{\ell}\mathbf{e}_k}(\mathrm{d}t)\right]\right)\\
&= -d_{jk} + (z-\alpha)\left[e_{jk} + \int_{(\alpha,+\infty)} \frac{1+t-\alpha}{t-z}\rho_{jk}(\mathrm{d}t)\right]
\end{aligned}$$

for all $j, k \in \mathbb{Z}_{1,q}$ and all $z \in \mathbb{C} \setminus [\alpha, +\infty)$. Hence, (6.7) follows for all $z \in \mathbb{C} \setminus [\alpha, +\infty)$ with $D := [d_{jk}]_{j,k=1}^{q}$, $E := [e_{jk}]_{j,k=1}^{q}$, and $\rho := [\rho_{jk}]_{j,k=1}^{q}$. For all $\zeta \in \mathbb{C}$ with $|\zeta| = 1$, we have $f_{\zeta u} = f_u$ and thus $d_{\zeta u} = d_u$, $e_{\zeta u} = e_u$, and $\rho_{\zeta u} = \rho_u$ by virtue of the uniqueness of the triple (d_u, e_u, ρ_u). A straightforward calculation yields for all $j, k \in \mathbb{Z}_{1,q}$ then $d_{kj} = \overline{d_{jk}}$, $e_{kj} = \overline{e_{jk}}$, and $\rho_{kj}(B) = \overline{\rho_{jk}(B)}$ for all $B \in \mathfrak{B}_{(\alpha,+\infty)}$. Thus, the matrices D and E are Hermitian and ρ is a σ-additive mapping defined on $\mathfrak{B}_{(\alpha,+\infty)}$ with values in $\mathbb{C}_{\mathrm{H}}^{q\times q}$. From (6.7) we obtain

$$f_u(z) = -u^*Du + (z-\alpha)\left[u^*Eu + \int_{(\alpha,+\infty)} \frac{1+t-\alpha}{t-z}(u^*\rho u)(\mathrm{d}t)\right]$$

for all $z \in \mathbb{C} \setminus [\alpha, +\infty)$, where u^*Du and u^*Eu belong to $\mathbb{R}$ and $u^*\rho u$ is a finite signed measure on $((\alpha, +\infty), \mathfrak{B}_{(\alpha,+\infty)})$. In view of (6.8), Lemma 6.8 yields $u^*Du = d_u$, $u^*Eu = e_u$, and $u^*\rho u = \rho_u$. In particular, u^*Du and u^*Eu belong to $[0, +\infty)$ and $u^*\rho u \in \mathcal{M}_{\geq}^{1}((\alpha, +\infty))$. Since $u \in \mathbb{C}^q$ was arbitrarily chosen, hence $D, E \in \mathbb{C}_{\geq}^{q\times q}$, and $\rho \in \mathcal{M}_{\geq}^{q}((\alpha, +\infty))$ follow.

Now let $D, E \in \mathbb{C}^{q\times q}_{\geq}$, and $\rho \in \mathcal{M}^q_{\geq}((\alpha,+\infty))$ be such that (6.7) holds true for all $z \in \mathbb{C} \setminus [\alpha,+\infty)$. Denote by δ_α the Dirac measure on $([\alpha,+\infty), \mathfrak{B}_{[\alpha,+\infty)})$ with unit mass at α. Then P admits the representation

$$P(z) = E + \int_{[\alpha,+\infty)} \frac{1+t-\alpha}{t-z} \theta(\mathrm{d}t)$$

for all $z \in \mathbb{C} \setminus [\alpha,+\infty)$, where $\theta : \mathfrak{B}_{[\alpha,+\infty)} \to \mathbb{C}^{q\times q}_{\geq}$ defined by $\theta(B) := \rho(B \cap (\alpha,+\infty)) + [\delta_\alpha(B)]D$ belongs to $\mathcal{M}^q_{\geq}([\alpha,+\infty))$. Hence, Theorem 3.6(b) and Remark 3.7 yield $P \in \mathcal{S}_{q;[\alpha,+\infty)}$ with $\gamma_P = E$ and $\mu_P = \theta$. In particular, $\mu_P(\{\alpha\}) = D$ and $\operatorname{Rstr}_{\mathfrak{B}_{(\alpha,+\infty)}} \mu_P = \rho$. Hence, the triple (D, E, ρ) is unique.

(b) Let $D, E \in \mathbb{C}^{q\times q}_{\geq}$ and $\rho \in \mathcal{M}^q_{\geq}((\alpha,+\infty))$ be such that (6.7) holds true for all $z \in \mathbb{C} \setminus [\alpha,+\infty)$. As explained above, then P belongs to $\mathcal{S}_{q;[\alpha,+\infty)}$. Since $F(z) = (z-\alpha)P(z)$ for all $z \in \mathbb{C} \setminus [\alpha,+\infty)$, we hence conclude with Proposition 4.3 that F belongs to $\mathcal{S}^{[-1]}_{q;[\alpha,+\infty)}$. □

In the following, if $\alpha \in \mathbb{R}$ and $F \in \mathcal{S}^{[-1]}_{q;[\alpha,+\infty)}$ are given, then we will write (D_F, E_F, ρ_F) for the unique triple (D, E, ρ) from $\mathbb{C}^{q\times q}_{\geq} \times \mathbb{C}^{q\times q}_{\geq} \times \mathcal{M}^q_{\geq}((\alpha,+\infty))$ which fulfills (6.7) for all $z \in \mathbb{C} \setminus [\alpha,+\infty)$.

Corollary 6.11. *Let $\alpha \in \mathbb{R}$. Then $F : \mathbb{C} \setminus [\alpha,+\infty) \to \mathbb{C}^{q\times q}$ belongs to $\mathcal{S}^{[-1]}_{q;[\alpha,+\infty)}$ if and only if $P : \mathbb{C} \setminus [\alpha,+\infty) \to \mathbb{C}^{q\times q}$ defined by $P(z) = (z-\alpha)^{-1}F(z)$ belongs to $\mathcal{S}_{q;[\alpha,+\infty)}$.*

Proof. If $F \in \mathcal{S}^{[-1]}_{q;[\alpha,+\infty)}$, then $P \in \mathcal{S}_{q;[\alpha,+\infty)}$, by virtue of Theorem 6.10(a).

Conversely, now suppose $P \in \mathcal{S}_{q;[\alpha,+\infty)}$. According to Theorem 3.6 and Remark 3.7, then

$$\begin{aligned} F(z) = (z-\alpha)P(z) &= (z-\alpha)\left[\gamma_P + \int_{[\alpha,+\infty)} \frac{1+t-\alpha}{t-z} \mu_P(\mathrm{d}t)\right] \\ &= -D + (z-\alpha)\left[E + \int_{(\alpha,+\infty)} \frac{1+t-\alpha}{t-z} \rho(\mathrm{d}t)\right] \end{aligned}$$

for all $z \in \mathbb{C} \setminus [\alpha,+\infty)$, where the matrices $D := \mu_P(\{\alpha\})$ and $E := \gamma_P$ are non-negative Hermitian and $\rho := \operatorname{Rstr}_{\mathfrak{B}_{(\alpha,+\infty)}} \mu_P$ belongs to $\mathcal{M}^q_{\geq}((\alpha,+\infty))$. Hence, Theorem 6.10(b) yields $F \in \mathcal{S}^{[-1]}_{q;[\alpha,+\infty)}$. □

Corollary 6.12. *Let $\alpha \in \mathbb{R}$ and let $F \in \mathcal{S}^{[-1]}_{q;[\alpha,+\infty)}$. For all $x_1, x_2 \in (-\infty,\alpha)$ with $x_1 \leq x_2$, then $F(x_1) \leq F(x_2) \leq O_{q\times q}$.*

Proof. Using Theorem 6.10, we obtain

$$F(x_2) - F(x_1) = (x_2 - x_1)\left[E_F + \int_{[\alpha,+\infty)} \frac{(1+t-\alpha)(t-\alpha)}{(t-x_2)(t-x_1)} \rho_F(\mathrm{d}t)\right]$$

for all $x_1, x_2 \in (-\infty, \alpha)$ with $x_1 \leq x_2$, by direct calculation. Since $E_F \in \mathbb{C}_{\geq}^{q\times q}$ and $-F(x) \in \mathbb{C}_{\geq}^{q\times q}$ for all $x \in (-\infty, \alpha)$, thus the proof is complete. □

Now we consider again the situation of Example 6.7.

Example 6.13. Let $\alpha \in \mathbb{R}$ and let $D, E \in \mathbb{C}_{\geq}^{q\times q}$. Then $F : \mathbb{C} \setminus [\alpha, +\infty) \to \mathbb{C}^{q\times q}$ defined by $F(z) := -D + (z - \alpha)E$ belongs to $\mathcal{S}_{q;[\alpha,+\infty)}^{[-1]}$, where $D_F = D$, $E_F = E$, and ρ_F is the constant measure with value $O_{q\times q}$.

Proposition 6.14. *Let $\alpha \in \mathbb{R}$ and let $F \in \mathcal{S}_{q;[\alpha,+\infty)}^{[-1]}$.*

(a) *If $z \in \mathbb{C} \setminus [\alpha, +\infty)$, then $\overline{z} \in \mathbb{C} \setminus [\alpha, +\infty)$ and $[F(z)]^* = F(\overline{z})$.*
(b) *For all $z \in \mathbb{C} \setminus [\alpha, +\infty)$,*

$$\mathcal{N}(F(z)) = \mathcal{N}(D_F) \cap \mathcal{N}(E_F) \cap \mathcal{N}(\rho_F((\alpha, +\infty))), \tag{6.9}$$

$$\mathcal{R}(F(z)) = \mathcal{R}(D_F) + \mathcal{R}(E_F) + \mathcal{R}(\rho_F((\alpha, +\infty))), \tag{6.10}$$

and, in particular, $\mathcal{N}([F(z)]^*) = \mathcal{N}(F(z))$ and $\mathcal{R}([F(z)]^*) = \mathcal{R}(F(z))$.
(c) *Let $r \in \mathbb{N}_0$. Then the following statements are equivalent:*
   (i) $\operatorname{rank} F(z) = r$ *for all $z \in \mathbb{C} \setminus [\alpha, +\infty)$;*
   (ii) *there is some $z_0 \in \mathbb{C} \setminus [\alpha, +\infty)$ such that* $\operatorname{rank} F(z_0) = r$;
   (iii) $\dim[\mathcal{R}(D_F) + \mathcal{R}(E_F) + \mathcal{R}(\rho_F((\alpha, +\infty)))] = r$.

Proof. (a) This can be seen from Theorem 6.10(a).

(b) According to Theorem 6.10(a), the function $P : \mathbb{C} \setminus [\alpha, +\infty) \to \mathbb{C}^{q\times q}$ defined by $P(z) := (z - \alpha)^{-1}F(z)$ belongs to $\mathcal{S}_{q;[\alpha,+\infty)}$ with $D_F = \mu_P(\{\alpha\})$ and $(E_F, \rho_F) = (\gamma_P, \operatorname{Rstr}_{\mathfrak{B}_{(\alpha,+\infty)}} \mu_P)$. In particular, $\mu_P([\alpha, +\infty)) = D_F + \rho_F((\alpha, +\infty))$. Since the two matrices on the right-hand side of the last equation are both non-negative Hermitian, we get

$$\mathcal{N}(\mu_P([\alpha, +\infty))) = \mathcal{N}(D_F) \cap \mathcal{N}(\rho_F((\alpha, +\infty))).$$

Now let $z \in \mathbb{C} \setminus [\alpha, +\infty)$. Applying Proposition 3.15(b) to P, we get

$$\mathcal{N}(P(z)) = \mathcal{N}(\gamma_P) \cap \mathcal{N}(\mu_P([\alpha, +\infty))).$$

In view of $\mathcal{N}(F(z)) = \mathcal{N}(P(z))$, then (6.9) follows. Thus, (6.9) is proved for all $z \in \mathbb{C} \setminus [\alpha, +\infty)$. From (a) and (6.9) we get $\mathcal{N}([F(z)]^*) = \mathcal{N}(F(z))$ for all $z \in \mathbb{C} \setminus [\alpha, +\infty)$. Taking additionally into account that the matrices D_F, E_F, and $\rho_F((\alpha, +\infty))$ are non-negative Hermitian, we obtain (6.10) from (6.9) in the same way as in the proof of Proposition 3.15(b). Using (a) and (6.10), we get $\mathcal{R}([F(z)]^*) = \mathcal{R}(F(z))$ for all $z \in \mathbb{C} \setminus [\alpha, +\infty)$.

(c) This is a consequence of (6.10). □

Corollary 6.15. *Let $\alpha \in \mathbb{R}$, let $F \in \mathcal{S}_{q;[\alpha,+\infty)}^{[-1]}$, and let $z_0 \in \mathbb{C} \setminus [\alpha, +\infty)$. Then $F(z_0) = O_{q\times q}$ if and only if $F(z) = O_{q\times q}$ for all $z \in \mathbb{C} \setminus [\alpha, +\infty)$.*

Proof. This is an immediate consequence of Proposition 6.14(c). □

Corollary 6.16. *Let $\alpha \in \mathbb{R}$, let $F \in \mathcal{S}^{[-1]}_{q;[\alpha,+\infty)}$, and let $\lambda \in \mathbb{R}$ be such that the matrix $D_F + \lambda I_q$ is non-negative Hermitian. Then*

$$\mathcal{R}(F(z) - \lambda I_q) = \mathcal{R}(F(w) - \lambda I_q), \quad \mathcal{N}(F(z) - \lambda I_q) = \mathcal{N}(F(w) - \lambda I_q) \quad (6.11)$$

for every choice of z and w in $\mathbb{C} \setminus [\alpha, +\infty)$. In particular, if $\lambda \geq 0$, then λ is an eigenvalue of the matrix $F(z_0)$ for some $z_0 \in \mathbb{C} \setminus [\alpha, +\infty)$ if and only if λ is an eigenvalue of the matrix $F(z)$ for all $z \in \mathbb{C} \setminus [\alpha, +\infty)$. In this case, the eigenspaces $\mathcal{N}(F(z) - \lambda I_q)$ are independent of $z \in \mathbb{C} \setminus [\alpha, +\infty)$.

Proof. In view of Theorem 6.10, we conclude that the function $G : \mathbb{C} \setminus [\alpha, +\infty) \to \mathbb{C}^{q\times q}$ defined by $G(z) := F(z) - \lambda I_q$ belongs to $\mathcal{S}^{[-1]}_{q;[\alpha,+\infty)}$. The application of Proposition 6.14(b) to the function G yields then (6.11). Since the matrix D_F is non-negative Hermitian, we have $D_F + \lambda I_q \in \mathbb{C}^{q\times q}_{\geq}$ if $\lambda \geq 0$. Thus, the remaining assertions are an immediate consequence of (6.11). □

Lemma 6.17. *Let $\alpha \in \mathbb{R}$ and $F \in \mathcal{S}^{[-1]}_{q;[\alpha,+\infty)}$. Then $F^\dagger$ is holomorphic in $\mathbb{C}\setminus[\alpha, +\infty)$.*

Proof. In view of (6.9) and (6.10), we obtain for all $z \in \mathbb{C} \setminus [\alpha, +\infty)$ the identities $\mathcal{N}(F(z)) = \mathcal{N}(F(\mathrm{i}))$ and $\mathcal{R}(F(z)) = \mathcal{R}(F(\mathrm{i}))$. Thus, the application of [15, Proposition 8.4] completes the proof. □

The following proposition is a generalization of a result due to Kats and Krein [17, Lemma D1.5.2], who considered the case $q = 1$ and $\alpha = 0$.

Theorem 6.18. *Let $\alpha \in \mathbb{R}$ and let $F : \mathbb{C} \setminus [\alpha, +\infty) \to \mathbb{C}^{q\times q}$ be a matrix-valued function. Then F belongs to $\mathcal{S}^{[-1]}_{q;[\alpha,+\infty)}$ if and only if $G := -F^\dagger$ belongs to $\mathcal{S}_{q;[\alpha,+\infty)}$.*

Proof. First suppose $F \in \mathcal{S}^{[-1]}_{q;[\alpha,+\infty)}$. Then G is holomorphic in $\mathbb{C} \setminus [\alpha, +\infty)$ by virtue of Lemma 6.17. In view of $\mathrm{Rstr}_{\Pi_+} F \in \mathcal{R}_q(\Pi_+)$, we conclude from [11, Proposition 3.8] that $\mathrm{Rstr}_{\Pi_+} G$ belongs to $\mathcal{R}_q(\Pi_+)$ as well. In particular, $\operatorname{Im} G(w) \in \mathbb{C}^{q\times q}_{\geq}$ for all $w \in \Pi_+$. Because of $-F(x) \in \mathbb{C}^{q\times q}_{\geq}$ for each $x \in (-\infty, \alpha)$, we have $G(x) = [-F(x)]^\dagger \in \mathbb{C}^{q\times q}_{\geq}$ for all $x \in (-\infty, \alpha)$ (see, e.g., [5, Lemma 1.1.5]). Hence, G belongs to $\mathcal{S}_{q;[\alpha,+\infty)}$.

Now suppose $G \in \mathcal{S}_{q;[\alpha,+\infty)}$. Then $F = -G^\dagger$. Hence, F is holomorphic in $\mathbb{C} \setminus [\alpha, +\infty)$ by virtue of Lemma 6.1. Since Proposition 4.3 yields $\mathrm{Rstr}_{\Pi_+} G \in \mathcal{R}_q(\Pi_+)$, we conclude from [11, Proposition 3.8] that $\mathrm{Rstr}_{\Pi_+} F$ belongs to $\mathcal{R}_q(\Pi_+)$ as well. Because of $G(x) \in \mathbb{C}^{q\times q}_{\geq}$ for all $x \in (-\infty, \alpha)$, we have $-F(x) = [G(x)]^\dagger \in \mathbb{C}^{q\times q}_{\geq}$ for all $x \in (-\infty, \alpha)$ (see, e.g., [5, Lemma 1.1.5]). Hence, $F \in \mathcal{S}^{[-1]}_{q;[\alpha,+\infty)}$. □

7. Integral representations for the class $\mathcal{S}_{q;(-\infty,\beta]}$

The main goal of this section is to derive some integral representations for $(-\infty, \beta]$-Stieltjes functions of order q. Our strategy is based on using the corresponding results for the class $\mathcal{S}_{q;[-\beta,+\infty)}$. The following observation provides the key to realize our aims.

Remark 7.1. Let $\alpha, \beta \in \mathbb{R}$ and let $T : \mathbb{C} \to \mathbb{C}$ be defined by $T(z) := \alpha + \beta - \overline{z}$. Then $T(\mathbb{C} \setminus [\alpha, +\infty)) = \mathbb{C} \setminus (-\infty, \beta]$, $T(\mathbb{C} \setminus (-\infty, \beta]) = \mathbb{C} \setminus [\alpha, +\infty)$, $T(\Pi_+) = \Pi_+$, $T((-\infty, \alpha)) = (\beta, +\infty)$ and $T((\beta, +\infty)) = (-\infty, \alpha)$. Consequently, in view of Proposition 4.4, one can easily check that, for each $F \in \mathcal{S}_{q;[\alpha,+\infty)}$, the function $G : \mathbb{C} \setminus (-\infty, \beta] \to \mathbb{C}^{q\times q}$ defined by $G(z) := -[F(\alpha+\beta-\overline{z})]^*$ belongs to $\mathcal{S}_{q;(-\infty,\beta]}$ and that, conversely, for any function $G \in \mathcal{S}_{q;(-\infty,\beta]}$, the function $F : \mathbb{C} \setminus [\alpha, +\infty) \to \mathbb{C}^{q\times q}$ defined by $F(z) := -[G(\alpha + \beta - \overline{z})]^*$ belongs to $\mathcal{S}_{q;[\alpha,+\infty)}$.

Proposition 7.2. *Let $\beta \in \mathbb{R}$ and let $G \in \mathcal{S}_{q;(-\infty,\beta]}$. Then the Nevanlinna parametrization (A, B, ν) of $\operatorname{Rstr}_{\Pi_+} G$ fulfills $\nu((\beta, +\infty)) = O_{q\times q}$, $B = O_{q\times q}$, and $\nu \in \mathcal{M}^q_{\geq,1}(\mathbb{R})$. In particular, for each $z \in \mathbb{C} \setminus (-\infty, \beta]$, then*

$$G(z) = A + \int_{(-\infty,\beta]} \frac{1+tz}{t-z} \nu(\mathrm{d}t).$$

Proof. According to Remark 7.1, the function $F : \mathbb{C} \setminus [-\beta, +\infty) \to \mathbb{C}^{q\times q}$ defined by $F(z) := -[G(-\overline{z})]^*$ belongs to $\mathcal{S}_{q;[-\beta,+\infty)}$. From Remark 2.15 we obtain then that the Nevanlinna parametrization of $\operatorname{Rstr}_{\Pi_+} F$ is given by $(-A, B, \theta)$, where θ is the image measure of ν under the reflection $t \mapsto -t$ on $\mathbb{R}$. Now Proposition 2.16 yields $\theta((-\infty, -\beta)) = O_{q\times q}$, $B = O_{q\times q}$, $\theta \in \mathcal{M}^q_{\geq,1}(\mathbb{R})$, and

$$F(z) = -A + \int_{[-\beta,+\infty)} \frac{1+tz}{t-z} \theta(\mathrm{d}t)$$

for all $z \in \mathbb{C} \setminus [-\beta, +\infty)$. Hence, $\nu((\beta, +\infty)) = O_{q\times q}$, $\nu \in \mathcal{M}^q_{\geq,1}(\mathbb{R})$, and

$$G(z) = -[F(-\overline{z})]^* = A + \int_{[-\beta,+\infty)} \frac{1-tz}{-t-z} \theta(\mathrm{d}t) = A + \int_{(-\infty,\beta]} \frac{1+tz}{t-z} \nu(\mathrm{d}t)$$

for all $z \in \mathbb{C} \setminus (-\infty, \beta]$. □

Theorem 7.3. *Let $\beta \in \mathbb{R}$ and let $G : \mathbb{C} \setminus (-\infty, \beta] \to \mathbb{C}^{q\times q}$.*

(a) *Suppose $G \in \mathcal{S}_{q;(-\infty,\beta]}$. Denote by (A, B, ν) the Nevanlinna parametrization of $\operatorname{Rstr}_{\Pi_+} G$ and let $\tilde{\nu} := \operatorname{Rstr}_{\mathfrak{B}_{(-\infty,\beta]}} \nu$. Then $\tilde{\nu} \in \mathcal{M}^q_{\geq,1}((-\infty, \beta])$ and there is a unique pair (C, η) from $\mathbb{C}^{q\times q}_{\geq} \times \mathcal{M}^q_{\geq,1}((-\infty, \beta])$ such that*

$$G(z) = -C + \int_{(-\infty,\beta]} \frac{1+t^2}{t-z} \eta(\mathrm{d}t) \tag{7.1}$$

for all $z \in \mathbb{C} \setminus (-\infty, \beta]$, namely

$$C = \int_{(-\infty,\beta]} t\tilde{\nu}(\mathrm{d}t) - A$$

and $\eta = \tilde{\nu}$. Furthermore,

$$C = -\lim_{r\to+\infty} G(\beta + r\mathrm{e}^{\mathrm{i}\phi}) \quad \textit{for all} \quad \phi \in (-\pi/2, \pi/2).$$

(b) *Let $C \in \mathbb{C}^{q\times q}_{\geq}$ and let $\eta \in \mathcal{M}^q_{\geq,1}((-\infty, \beta])$ be such that (7.1) holds true for all $z \in \mathbb{C} \setminus (-\infty, \beta]$. Then G belongs to $\mathcal{S}_{q;(-\infty,\beta]}$.*

Proof. (a) According to Remark 7.1, the function $F : \mathbb{C}\setminus[-\beta,+\infty) \to \mathbb{C}^{q\times q}$ defined by $F(z) := -[G(-\overline{z})]^*$ belongs to $\mathcal{S}_{q;[-\beta,+\infty)}$. From Remark 2.15 we obtain then that the Nevanlinna parametrization of $\operatorname{Rstr}_{\Pi_+} F$ is given by $(-A, B, \theta)$, where θ is the image measure of ν under the reflection $t \mapsto -t$ on $\mathbb{R}$. Now Theorem 3.1(a) yields that $\tilde{\theta} := \operatorname{Rstr}_{\mathfrak{B}_{[-\beta,+\infty)}} \theta$ belongs to $\mathcal{M}^q_{\geq,1}([-\beta,+\infty))$ and that there is a unique pair $(D,\tau) \in \mathbb{C}^{q\times q}_{\geq} \times \mathcal{M}^q_{\geq,1}([-\beta,+\infty))$ such that

$$F(z) = D + \int_{[-\beta,+\infty)} \frac{1+t^2}{t-z}\tau(\mathrm{d}t)$$

for all $z \in \mathbb{C}\setminus[-\beta,+\infty)$, namely

$$D = -A - \int_{[-\beta,+\infty)} t\tilde{\theta}(\mathrm{d}t)$$

and $\tau = \tilde{\theta}$. Since $\tilde{\theta}$ is the image measure of $\tilde{\nu}$ under the transformation $T : (-\infty,\beta] \to [-\beta,+\infty)$ defined by $T(t) := -t$, we can conclude $\tilde{\nu} \in \mathcal{M}^q_{\geq,1}((-\infty,\beta])$ and

$$\begin{aligned}
G(z) &= -[F(-\overline{z})]^* \\
&= -D^* - \left[\int_{[-\beta,+\infty)} \frac{1+t^2}{t+\overline{z}}\tau(\mathrm{d}t)\right]^* = -D - \int_{[-\beta,+\infty)} \frac{1+t^2}{t+z}\tau(\mathrm{d}t) \\
&= -\left[\int_{[-\beta,+\infty)} (-t)\tilde{\theta}(\mathrm{d}t) - A\right] + \int_{[-\beta,+\infty)} \frac{1+t^2}{-t-z}\tilde{\theta}(\mathrm{d}t) \\
&= -C + \int_{(-\infty,\beta]} \frac{1+t^2}{t-z}\eta(\mathrm{d}t)
\end{aligned}$$

for all $z \in \mathbb{C}\setminus(-\infty,\beta]$, where

$$C := \int_{(-\infty,\beta]} t\tilde{\nu}(\mathrm{d}t) - A$$

and $\eta := \tilde{\nu}$. From the above computation we see $C = D$ and hence $C \in \mathbb{C}^{q\times q}_{\geq}$ follows. Taking additionally into account Theorem 3.1(a), for all $\phi \in (-\pi/2, \pi/2)$, we get

$$\begin{aligned}
C = D^* &= \left[\lim_{r\to+\infty} F(-\beta + r\mathrm{e}^{\mathrm{i}(\pi-\phi)})\right]^* \\
&= \lim_{r\to+\infty} \left[F(-\beta - r\mathrm{e}^{-\mathrm{i}\phi})\right]^* = -\lim_{r\to+\infty} G(\beta + r\mathrm{e}^{\mathrm{i}\phi}).
\end{aligned}$$

Now let $C \in \mathbb{C}^{q\times q}_{\geq}$ and $\eta \in \mathcal{M}^q_{\geq,1}((-\infty,\beta])$ be such that (7.1) holds true for all $z \in \mathbb{C}\setminus(-\infty,\beta]$. Then $\chi : \mathfrak{B}_{\mathbb{R}} \to \mathbb{C}^{q\times q}_{\geq}$ defined by $\chi(M) := \eta(M \cap (-\infty,\beta])$ belongs to $\mathcal{M}^q_{\geq,1}(\mathbb{R})$ and the matrix $-C + \int_{\mathbb{R}} t\chi(\mathrm{d}t)$ is Hermitian. Using (3.2), we conclude from (7.1) then that the integral

$$\int_{\mathbb{R}} \frac{1+tz}{t-z} \chi(\mathrm{d}t)$$

exists and that

$$G(z) = -C + \int_{\mathbb{R}} t\chi(\mathrm{d}t) + z \cdot O_{q\times q} + \int_{\mathbb{R}} \frac{1+tz}{t-z} \chi(\mathrm{d}t)$$

for all $z \in \Pi_+$. Theorem 2.13(a) yields then

$$-C + \int_{\mathbb{R}} t\chi(\mathrm{d}t) = A$$

and $\chi = \nu$. Hence $\eta = \tilde{\nu}$ and

$$C = \int_{(-\infty,\beta]} t\tilde{\nu}(\mathrm{d}t) - A.$$

(b) Let $C \in \mathbb{C}_{\geq}^{q\times q}$ and $\eta \in \mathcal{M}_{\geq,1}^{q}((-\infty,\beta])$ be such that (7.1) holds true for all $z \in \mathbb{C} \setminus (-\infty,\beta]$. Using a result on holomorphic dependence of an integral on a complex parameter (see, e.g., [9, Ch. IV, §5, Satz 5.8]), we conclude then that G is a matrix-valued function which is holomorphic in $\mathbb{C} \setminus (-\infty,\beta]$. Furthermore,

$$\operatorname{Im} G(z) = \int_{(-\infty,\beta]} \operatorname{Im}\left(\frac{1+t^2}{t-z}\right)\eta(\mathrm{d}t) = \int_{(-\infty,\beta]} \frac{(1+t^2)\operatorname{Im} z}{|t-z|^2}\eta(\mathrm{d}t) \in \mathbb{C}_{\geq}^{q\times q}$$

for all $z \in \Pi_+$ and

$$-G(x) = C + \int_{(-\infty,\beta]} \frac{1+t^2}{x-t}\eta(\mathrm{d}t) \in \mathbb{C}_{\geq}^{q\times q}$$

for all $x \in (\beta,+\infty)$. Thus, G belongs to $\mathcal{S}_{q;(-\infty,\beta]}$. □

In the following, if $\beta \in \mathbb{R}$ and $G \in \mathcal{S}_{q;(-\infty,\beta]}$ are given, then we will write (C_G,η_G) for the unique pair (C,η) from $\mathbb{C}_{\geq}^{q\times q} \times \mathcal{M}_{\geq,1}^{q}((-\infty,\beta])$ which fulfills (7.1) for all $z \in \mathbb{C} \setminus (-\infty,\beta]$.

Remark 7.4. Let $\beta \in \mathbb{R}$ and let $G \in \mathcal{S}_{q;(-\infty,\beta]}$. For all $x_1, x_2 \in (\beta,+\infty)$ with $x_1 \leq x_2$, then $G(x_1) \leq G(x_2) \leq O_{q\times q}$, by virtue of Theorem 7.3(a).

Remark 7.5. (a) Let $\alpha \in \mathbb{R}$ and let $F \in \mathcal{S}_{q;[\alpha,+\infty)}$. In view of Theorem 3.1, we have then

$$-[F(-\overline{z})]^* = -C_F + \int_{[\alpha,+\infty)} \frac{1+t^2}{-t-z}\eta_F(\mathrm{d}t) = -C_F + \int_{(-\infty,-\alpha]} \frac{1+t^2}{t-z}\hat{\theta}(\mathrm{d}t)$$

for all $z \in \mathbb{C} \setminus (-\infty,-\alpha]$, where $\hat{\theta}$ is the image measure of η_F under the transformation $R : [\alpha,+\infty) \to (-\infty,-\alpha]$ defined by $R(t) := -t$. Because of $C_F \in \mathbb{C}_{\geq}^{q\times q}$ and $\hat{\theta} \in \mathcal{M}_{\geq,1}^{q}((-\infty,\alpha])$, Theorem 7.3 yields then that $G : \mathbb{C} \setminus (-\infty,-\alpha] \to \mathbb{C}^{q\times q}$ defined by $G(z) := -[F(-\overline{z})]^*$ belongs to $\mathcal{S}_{q;(-\infty,-\alpha]}$ and that $(C_G,\eta_G) = (C_F,\hat{\theta})$.

(b) Let $\beta \in \mathbb{R}$ and let $G \in \mathcal{S}_{q;(-\infty,\beta]}$. In view of Theorem 7.3 and Proposition A.5, we have then

$$-[G(-\overline{z})]^* = C_G + \int_{(-\infty,\beta]} \frac{1+t^2}{-t-z}\eta_G(\mathrm{d}t) = C_G + \int_{[-\beta,+\infty)} \frac{1+t^2}{t-z}\tilde{\theta}(\mathrm{d}t)$$

for all $z \in \mathbb{C} \setminus [-\beta,+\infty)$, where $\tilde{\theta}$ is the image measure of η_G under the transformation $T : (-\infty,\beta] \to [-\beta,+\infty)$ defined by $T(t) := -t$. Because of $C_G \in \mathbb{C}^{q\times q}_{\geq}$ and $\tilde{\theta} \in \mathcal{M}^q_{\geq,1}([-\beta,+\infty))$, Theorem 3.1 yields then that $F : \mathbb{C} \setminus [-\beta,+\infty) \to \mathbb{C}^{q\times q}$ defined by $F(z) := -[G(-\overline{z})]^*$ belongs to $\mathcal{S}_{q;[-\beta,+\infty)}$ and that $(C_F,\eta_F) = (C_G,\tilde{\theta})$.

Now we get an integral representation for functions which belong to the class $\mathcal{S}_{q;(-\infty,\beta]}$.

Theorem 7.6. *Let $\beta \in \mathbb{R}$ and let $G : \mathbb{C} \setminus (-\infty,\beta] \to \mathbb{C}^{q\times q}$.*

(a) *If $G \in \mathcal{S}_{q;(-\infty,\beta]}$, then there are a unique matrix $\gamma \in \mathbb{C}^{q\times q}_{\geq}$ and a unique non-negative Hermitian measure $\mu \in \mathcal{M}^q_{\geq}((-\infty,\beta])$ such that*

$$G(z) = -\gamma + \int_{(-\infty,\beta]} \frac{1+\beta-t}{t-z}\mu(\mathrm{d}t). \tag{7.2}$$

holds for each $z \in \mathbb{C} \setminus (-\infty,\beta]$. Furthermore, $\gamma = C_G$ and

$$\gamma = -\lim_{y\to+\infty} G(\mathrm{i}y).$$

(b) *If there exist a matrix $\gamma \in \mathbb{C}^{q\times q}_{\geq}$ and a non-negative Hermitian measure $\mu \in \mathcal{M}^q_{\geq}((-\infty,\beta])$ such that G can be represented via (7.2) for each $z \in \mathbb{C} \setminus (-\infty,\beta]$, then G belongs to the class $\mathcal{S}_{q;(-\infty,\beta]}$.*

Proof. (a) According to Remark 7.5(b), the function $F : \mathbb{C} \setminus [-\beta,+\infty) \to \mathbb{C}^{q\times q}$ defined by $F(z) := -[G(-\overline{z})]^*$ belongs to $\mathcal{S}_{q;[-\beta,+\infty)}$ and $C_F = C_G$. Theorem 3.6(a) yields then that there is a unique pair (δ,ρ) from $\mathbb{C}^{q\times q}_{\geq} \times \mathcal{M}^q_{\geq}([-\beta,+\infty))$ such that

$$F(z) = \delta + \int_{[-\beta,+\infty)} \frac{1+t+\beta}{t-z}\rho(\mathrm{d}t) \tag{7.3}$$

for all $z \in \mathbb{C}\setminus[-\beta,+\infty)$ and that $\delta = C_F$. Applying Proposition A.5, we now infer

$$\begin{aligned} G(z) = -[F(-\overline{z})]^* &= -C_F - \int_{[-\beta,+\infty)} \frac{1+t+\beta}{t+z}\rho(\mathrm{d}t) \\ &= -C_G + \int_{[-\beta,+\infty)} \frac{1+\beta-(-t)}{-t-z}\rho(\mathrm{d}t) = -\gamma + \int_{(-\infty,\beta]} \frac{1+\beta-t}{t-z}\mu(\mathrm{d}t) \end{aligned}$$

for all $z \in \mathbb{C} \setminus (-\infty,\beta]$, where $\gamma := C_G$ and μ is the image measure of ρ under the transformation $R : [-\beta,+\infty) \to (-\infty,\beta]$ defined by $R(t) := -t$. Since Proposition 3.13 yields $\lim_{y\to+\infty} F(\mathrm{i}y) = \delta$, we conclude furthermore

$$\gamma = C_G = C_F = \delta = \delta^* = \left[\lim_{y\to+\infty} F(\mathrm{i}y)\right]^* = -\lim_{y\to+\infty} G(\mathrm{i}y).$$

Now let $\gamma \in \mathbb{C}_{\geq}^{q\times q}$ and $\mu \in \mathcal{M}_{\geq}^{q}((-\infty,\beta])$ be arbitrary such that (7.2) holds true for all $z \in \mathbb{C}\setminus(-\infty,\beta]$. Then using Proposition A.5 we get

$$F(z) = -[G(-\overline{z})]^* = \gamma - \int_{(-\infty,\beta]} \frac{1+\beta-t}{t+z}\mu(\mathrm{d}t)$$

$$= \gamma + \int_{(-\infty,\beta]} \frac{1-t+\beta}{-t-z}\mu(\mathrm{d}t) = \gamma + \int_{[-\beta,+\infty)} \frac{1+t+\beta}{t-z}\tilde{\theta}(\mathrm{d}t) \tag{7.4}$$

for all $z \in \mathbb{C}\setminus[-\beta,+\infty)$, where $\tilde{\theta}$ is the image measure of μ under the transformation $T : (-\infty,\beta] \to [-\beta,+\infty)$ defined by $T(t) := -t$. Since we know from Theorem 3.6(a) that the pair $(\delta,\rho) \in \mathbb{C}_{\geq}^{q\times q} \times \mathcal{M}_{\geq}^{q}([-\beta,+\infty))$ with (7.3) for all $z \in \mathbb{C}\setminus[-\beta,+\infty)$ is unique, comparing with (7.4), we conclude $\gamma = \delta$ and $\tilde{\theta} = \rho$. Hence, $\gamma = C_F = C_G$ and μ is the image measure of ρ under the transformation R.

(b) Let $\gamma \in \mathbb{C}_{\geq}^{q\times q}$ and $\mu \in \mathcal{M}_{\geq}^{q}((-\infty,\beta])$ be such that (7.2) holds true for all $z \in \mathbb{C}\setminus(-\infty,\beta]$. Then $F : \mathbb{C}\setminus[-\beta,+\infty) \to \mathbb{C}^{q\times q}$ defined by $F(z) := -[G(-\overline{z})]^*$ fulfills (7.4) for all $z \in \mathbb{C}\setminus[-\beta,+\infty)$, where $\tilde{\theta}$ is the image measure of μ under the transformation T. Theorem 3.6(b) yields then $F \in \mathcal{S}_{q;[-\beta,+\infty)}$. In view of $G(z) = -[F(-\overline{z})]^*$ for all $z \in \mathbb{C}\setminus(-\infty,\beta]$, hence G belongs to $\mathcal{S}_{q;(-\infty,\beta]}$ by virtue of Remark 7.1. □

In the following, if $\beta \in \mathbb{R}$ and $G \in \mathcal{S}_{q;(-\infty,\beta]}$ are given, then we will write (γ_G,μ_G) for the unique pair (γ,μ) from $\mathbb{C}_{\geq}^{q\times q} \times \mathcal{M}_{\geq}^{q}((-\infty,\beta])$ which fulfills (7.2) for all $z \in \mathbb{C}\setminus(-\infty,\beta]$.

Remark 7.7. (a) Let $\alpha \in \mathbb{R}$ and let $F \in \mathcal{S}_{q;[\alpha,+\infty)}$. In view of Theorem 3.6 and Proposition A.5, we have then

$$-[F(-\overline{z})]^* = -\gamma_F + \int_{[\alpha,+\infty)} \frac{1-(-t)-\alpha}{-t-z}\mu_F(\mathrm{d}t) = -\gamma_F + \int_{(-\infty,-\alpha]} \frac{1-\alpha-t}{t-z}\hat{\theta}(\mathrm{d}t)$$

for all $z \in \mathbb{C}\setminus(-\infty,-\alpha]$, where $\hat{\theta}$ is the image measure of μ_F under the transformation $R : [\alpha,+\infty) \to (-\infty,-\alpha]$ defined by $R(t) := -t$. Because of $\gamma_F \in \mathbb{C}_{\geq}^{q\times q}$, Theorem 7.6 yields then that $G : \mathbb{C}\setminus(-\infty,-\alpha] \to \mathbb{C}^{q\times q}$ defined by $G(z) := -[F(-\overline{z})]^*$ belongs to $\mathcal{S}_{q;(-\infty,-\alpha]}$ and that $(\gamma_G,\mu_G) = (\gamma_F,\hat{\theta})$.

(b) Let $\beta \in \mathbb{R}$ and let $G \in \mathcal{S}_{q;(-\infty,\beta]}$. In view of Theorem 7.6, we have then

$$-[G(-\overline{z})]^* = \gamma_G + \int_{(-\infty,\beta]} \frac{1+\beta-t}{-t-z}\mu_G(\mathrm{d}t) = \gamma_G + \int_{[-\beta,+\infty)} \frac{1+t+\beta}{t-z}\tilde{\theta}(\mathrm{d}t)$$

for all $z \in \mathbb{C}\setminus[-\beta,+\infty)$, where $\tilde{\theta}$ is the image measure of μ_G under the transformation $T : (-\infty,\beta] \to [-\beta,+\infty)$ defined by $T(t) := -t$. Because of $\gamma_G \in \mathbb{C}_{\geq}^{q\times q}$, Theorem 3.6 yields then that $F : \mathbb{C}\setminus[-\beta,+\infty) \to \mathbb{C}^{q\times q}$ defined by $F(z) := -[G(-\overline{z})]^*$ belongs to $\mathcal{S}_{q;[-\beta,+\infty)}$ and that $(\gamma_F,\mu_F) = (\gamma_G,\tilde{\theta})$.

Proposition 7.8. *Let $\beta \in \mathbb{R}$ and let $G \in \mathcal{S}_{q;(-\infty,\beta]}$.*

(a) *Let $z \in \mathbb{C}\setminus(-\infty,\beta]$. Then $\overline{z} \in \mathbb{C}\setminus(-\infty,\beta]$ and $[G(z)]^* = G(\overline{z})$.*

(b) *For all* $z \in \mathbb{C} \setminus (-\infty, \beta]$,

$$\begin{aligned} \mathcal{R}(G(z)) &= \mathcal{R}(\gamma_G) + \mathcal{R}(\mu_G((-\infty, \beta])), \\ \mathcal{N}(G(z)) &= \mathcal{N}(\gamma_G) \cap \mathcal{N}(\mu_G((-\infty, \beta])), \end{aligned} \tag{7.5}$$

and, in particular, $\mathcal{R}([G(z)]^*) = \mathcal{R}(G(z))$ *and* $\mathcal{N}([G(z)]^*) = \mathcal{N}(G(z))$.

(c) *Let* $r \in \mathbb{N}_0$. *Then the following statements are equivalent:*

(i) $\operatorname{rank} G(z) = r$ *for all* $z \in \mathbb{C} \setminus (-\infty, \beta]$;

(ii) *there is some* $z_0 \in \mathbb{C} \setminus (-\infty, \beta]$ *such that* $\operatorname{rank} G(z_0) = r$;

(iii) $\dim[\mathcal{R}(\gamma_G) + \mathcal{R}(\mu_G((-\infty, \beta]))] = r$.

Proof. (a) This can be concluded from the representation (7.2) in Theorem 7.6(a).

(b) According to Remark 7.7(b), the function $F : \mathbb{C} \setminus [-\beta, +\infty) \to \mathbb{C}^{q\times q}$ defined by $F(z) := -[G(-\overline{z})]^*$ belongs to $\mathcal{S}_{q;[-\beta,+\infty)}$ and $(\gamma_F, \mu_F) = (\gamma_G, \tilde{\theta})$, where $\tilde{\theta}$ is the image measure of μ_G under the transformation $T : (-\infty, \beta] \to [-\beta, +\infty)$ defined by $T(t) := -t$. In particular, $\mu_F([-\beta, +\infty)) = \mu_G((-\infty, \beta])$. Proposition 3.15(b) yields

$$\mathcal{R}([F(w)]^*) = \mathcal{R}(F(w)) = \mathcal{R}(\gamma_F) + \mathcal{R}(\mu_F([-\beta, +\infty)))$$

and

$$\mathcal{N}([F(w)]^*) = \mathcal{N}(F(w)) = \mathcal{N}(\gamma_F) \cap \mathcal{N}(\mu_F([-\beta, +\infty)))$$

for all $w \in \mathbb{C} \setminus [-\beta, +\infty)$. We now infer

$$\begin{aligned} \mathcal{R}(G(z)) &= \mathcal{R}([F(-\overline{z})]^*) \\ &= \mathcal{R}(\gamma_F) + \mathcal{R}(\mu_F([-\beta, +\infty))) = \mathcal{R}(\gamma_G) + \mathcal{R}(\mu_G((-\infty, \beta])) \end{aligned}$$

and

$$\begin{aligned} \mathcal{N}(G(z)) &= \mathcal{N}([F(-\overline{z})]^*) \\ &= \mathcal{N}(\gamma_F) \cap \mathcal{N}(\mu_F([-\beta, +\infty))) = \mathcal{N}(\gamma_G) \cap \mathcal{N}(\mu_G((-\infty, \beta])) \end{aligned}$$

for all $z \in \mathbb{C} \setminus (-\infty, \beta]$. From (a) and (7.5) we get $\mathcal{R}([G(z)]^*) = \mathcal{R}(G(z))$ and $\mathcal{N}([G(z)]^*) = \mathcal{N}(G(z))$.

(c) This is an immediate consequence of (b). □

Corollary 7.9. *Let* $\beta \in \mathbb{R}$, *let* $G \in \mathcal{S}_{q;(-\infty,\beta]}$, *and let* $z_0 \in \mathbb{C} \setminus (-\infty, \beta]$. *Then* $G(z_0) = O_{q\times q}$ *if and only if* G *is identically* $O_{q\times q}$.

Proof. This is an immediate consequence of Proposition 7.8(c). □

Corollary 7.10. *Let* $\beta \in \mathbb{R}$, *let* $G \in \mathcal{S}_{q;(-\infty,\beta]}$, *and let* $\lambda \in \mathbb{R}$ *be such that the matrix* $\gamma_G + \lambda I_q$ *is non-negative Hermitian. Then*

$$\mathcal{R}(G(z) - \lambda I_q) = \mathcal{R}(G(w) - \lambda I_q), \quad \mathcal{N}(G(z) - \lambda I_q) = \mathcal{N}(G(w) - \lambda I_q) \tag{7.6}$$

for all $z, w \in \mathbb{C} \setminus [\alpha, +\infty)$. *In particular, if* $\lambda \geq 0$, *then* λ *is an eigenvalue of the matrix* $G(z_0)$ *for some* $z_0 \in \mathbb{C} \setminus [\alpha, +\infty)$ *if and only if* λ *is an eigenvalue of the matrix* $G(z)$ *for all* $z \in \mathbb{C} \setminus [\alpha, +\infty)$. *In this case, the eigenspaces* $\mathcal{N}(G(z) - \lambda I_q)$ *are independent of* $z \in \mathbb{C} \setminus [\alpha, +\infty)$.

Proof. In view of Theorem 7.6, we conclude that the function $F : \mathbb{C} \setminus [\alpha, +\infty) \to \mathbb{C}^{q\times q}$ defined by $F(z) := G(z) - \lambda I_q$ belongs to $\mathcal{S}_{q;(-\infty,\beta]}$. The application of Proposition 7.8(b) to the function F yields then (7.6). Since the matrix γ_G is non-negative Hermitian, we have $\gamma_G + \lambda I_q \in \mathbb{C}^{q\times q}_{\geq}$ if $\lambda \geq 0$. Thus, the remaining assertions are an immediate consequence of (7.6). □

Now we apply the preceding results to the subclass $\mathcal{S}^{\diamond}_{q;(-\infty,\beta]}$ of $\mathcal{S}_{q;(-\infty,\beta]}$.

Remark 7.11. Let $\beta \in \mathbb{R}$. From Theorem 7.6(a) we see that the class $\mathcal{S}^{\diamond}_{q;(-\infty,\beta]}$ consists of all $G \in \mathcal{S}_{q;(-\infty,\beta]}$ with $\gamma_G = O_{q\times q}$.

Remark 7.12. Let $\beta \in \mathbb{R}$ and let $G \in \mathcal{S}^{\diamond}_{q;(-\infty,\beta]}$. In view of Remark 7.11 and Proposition 7.8(b), then $\mathcal{N}(G(z)) = \mathcal{N}(\mu_G((-\infty, \beta]))$ and $\mathcal{R}(G(z)) = \mathcal{R}(\mu_G((-\infty, \beta]))$ for all $z \in \mathbb{C} \setminus (-\infty, \beta]$.

8. Characterizations of the class $\mathcal{S}_{q;(-\infty,\beta]}$

While we have discussed the class $\mathcal{S}_{q;[\alpha,+\infty)}$ in Section 4, we give here now the corresponding results for the class $\mathcal{S}_{q;(-\infty,\beta]}$.

Proposition 8.1. *Let $\beta \in \mathbb{R}$, let $G : \mathbb{C}\setminus(-\infty, \beta] \to \mathbb{C}^{q\times q}$ be a matrix-valued function, and let $G^{\square} : \mathbb{C} \setminus (-\infty, \beta] \to \mathbb{C}^{q\times q}$ be defined by $G^{\square}(z) := (\beta - z)G(z)$. Then G belongs to $\mathcal{S}_{q;(-\infty,\beta]}$ if and only if the following two conditions hold true:*

(i) *G is holomorphic in $\mathbb{C} \setminus (-\infty, \beta]$;*
(ii) *the matrix-valued functions $\operatorname{Rstr}_{\Pi_+} G$ and $\operatorname{Rstr}_{\Pi_+} G^{\square}$ both belong to $\mathcal{R}_q(\Pi_+)$.*

Proof. According to Remark 7.1, the function G belongs to $\mathcal{S}_{q;(-\infty,\beta]}$ if and only if the function $F : \mathbb{C} \setminus [-\beta, +\infty) \to \mathbb{C}^{q\times q}$ defined by $F(z) := -[G(-\overline{z})]^*$ belongs to $\mathcal{S}_{q;[-\beta,+\infty)}$. Furthermore, G is holomorphic in $\mathbb{C} \setminus (-\infty, \beta]$ if and only if F is holomorphic in $\mathbb{C} \setminus [-\beta, +\infty)$. By virtue of Remark 2.15, the function $\operatorname{Rstr}_{\Pi_+} G$ belongs to $\mathcal{R}_q(\Pi_+)$ if and only if $\operatorname{Rstr}_{\Pi_+} F$ belongs to $\mathcal{R}_q(\Pi_+)$. Let $F^{\square} : \mathbb{C} \setminus [-\beta, +\infty) \to \mathbb{C}^{q\times q}$ be defined by $F^{\square}(z) := (z + \beta)F(z)$. Then

$$-\left[G^{\square}(-\overline{z})\right]^* = -[(\beta + \overline{z})G(-\overline{z})]^* = (z + \beta)\left(-[G(-\overline{z})]^*\right) = F^{\square}(z)$$

for all $z \in \mathbb{C} \setminus [-\beta, +\infty)$. According to Remark 2.15, hence $\operatorname{Rstr}_{\Pi_+} G^{\square}$ belong to $\mathcal{R}_q(\Pi_+)$ if and only if $\operatorname{Rstr}_{\Pi_+} F^{\square}$ belong to $\mathcal{R}_q(\Pi_+)$. The application of Proposition 4.3 completes the proof. □

For each $\beta \in \mathbb{R}$, let $\mathbb{C}_{\beta,+} := \{z \in \mathbb{C} : \operatorname{Re} z > \beta\}$.

Proposition 8.2. *Let $\beta \in \mathbb{R}$ and let $G : \mathbb{C} \setminus (-\infty, \beta] \to \mathbb{C}^{q\times q}$ be a matrix-valued function. Then G belongs to $\mathcal{S}_{q;(-\infty,\beta]}$ if and only if the following four conditions are fulfilled:*

(i) *G is holomorphic in $\mathbb{C} \setminus (-\infty, \beta]$;*
(ii) *$\operatorname{Im} G(z)$ is non-negative Hermitian for all $z \in \Pi_+$;*

(iii) $-\operatorname{Im} G(z)$ *is non-negative Hermitian for all* $z \in \Pi_-$;
(iv) $-\operatorname{Re} G(z)$ *is non-negative Hermitian for all* $z \in \mathbb{C}_{\beta,+}$.

Proof. According to Remark 7.1, the function G belongs to $\mathcal{S}_{q;(-\infty,\beta]}$ if and only if the function $F : \mathbb{C}\setminus[-\beta,+\infty) \to \mathbb{C}^{q\times q}$ defined by $F(z) := -[G(-\overline{z})]^*$ belongs to $\mathcal{S}_{q;[-\beta,+\infty)}$. Furthermore, G is holomorphic in $\mathbb{C}\setminus(-\infty,\beta]$ if and only if F is holomorphic in $\mathbb{C}\setminus[-\beta,+\infty)$. For all $z \in \mathbb{C}\setminus[-\beta,+\infty)$, we have $\operatorname{Re} F(z) = -\operatorname{Re} G(-\overline{z})$ and $\operatorname{Im} F(z) = \operatorname{Im} G(-\overline{z})$. Hence, $\operatorname{Im} F(\Pi_+) = \operatorname{Im} G(\Pi_+)$, $\operatorname{Im} F(\Pi_-) = \operatorname{Im} G(\Pi_-)$, and $\operatorname{Re} F(\mathbb{C}_{-\beta,-}) = -\operatorname{Re} G(\mathbb{C}_{\beta,+})$. The application of Proposition 4.4 completes the proof. □

9. The class $\mathcal{S}_{0,q;(-\infty,\beta]}$

In Section 5, we have studied the class $\mathcal{S}_{0,q;[\alpha,+\infty)}$. The aim of this section is to derive corresponding results for the dual class $\mathcal{S}_{0,q;(-\infty,\beta]}$. The following observation establishes the bridge to Section 5.

Remark 9.1. (a) Let $\alpha \in \mathbb{R}$, let $F \in \mathcal{S}_{0,q;[\alpha,+\infty)}$, and let $G : \mathbb{C}\setminus(-\infty,-\alpha] \to \mathbb{C}^{q\times q}$ be defined by $G(z) := -[F(-\overline{z})]^*$. We have then $\|G(\mathrm{i}y)\|_{\mathrm{E}} = \|F(\mathrm{i}y)\|_{\mathrm{E}}$ for all $y \in [1,+\infty)$. Taking additionally into account Remark 7.1, one can see that G belongs to $\mathcal{S}_{0,q;(-\infty,-\alpha]}$.
(b) Let $\beta \in \mathbb{R}$, let $G \in \mathcal{S}_{0,q;(-\infty,\beta]}$, and let $F : \mathbb{C}\setminus[-\beta,+\infty) \to \mathbb{C}^{q\times q}$ be defined by $F(z) := -[G(-\overline{z})]^*$. We have then $\|F(\mathrm{i}y)\|_{\mathrm{E}} = \|G(\mathrm{i}y)\|_{\mathrm{E}}$ for all $y \in [1,+\infty)$. Taking additionally into account Remark 7.1, one can see that F belongs to $\mathcal{S}_{0,q;[-\beta,+\infty)}$.

Theorem 9.2. *Let* $\beta \in \mathbb{R}$ *and let* $G : \mathbb{C}\setminus(-\infty,\beta] \to \mathbb{C}^{q\times q}$.
(a) *If* $G \in \mathcal{S}_{0,q;(-\infty,\beta]}$, *then there is a unique measure* $\sigma \in \mathcal{M}^q_{\geq}((-\infty,\beta])$ *such that*
$$G(z) = \int_{(-\infty,\beta]} \frac{1}{t-z}\sigma(\mathrm{d}t) \tag{9.1}$$
for each $z \in \mathbb{C}\setminus(-\infty,\beta]$. *Furthermore,* $\sigma((-\infty,\beta]) = -\mathrm{i}\lim_{y\to+\infty} yG(\mathrm{i}y)$.
(b) *If there is a measure* $\sigma \in \mathcal{M}^q_{\geq}((-\infty,\beta])$ *such that* G *can be represented via* (9.1) *for each* $z \in \mathbb{C}\setminus(-\infty,\beta]$, *then* G *belongs to the class* $\mathcal{S}_{0,q;(-\infty,\beta]}$.

Proof. (a) According to Remark 9.1(b), the function $F : \mathbb{C}\setminus[-\beta,+\infty) \to \mathbb{C}^{q\times q}$ defined by $F(z) := -[G(-\overline{z})]^*$ belongs to $\mathcal{S}_{0,q;[-\beta,+\infty)}$. Theorem 5.1(a) yields then the existence of a unique non-negative Hermitian measure $\tau \in \mathcal{M}^q_{\geq}([-\beta,+\infty))$ such that
$$F(z) = \int_{[-\beta,+\infty)} \frac{1}{t-z}\tau(\mathrm{d}t) \tag{9.2}$$
for all $z \in \mathbb{C}\setminus[-\beta,+\infty)$. Using Proposition A.5, we now infer
$$G(z) = -[F(-\overline{z})]^* = \int_{[-\beta,+\infty)} \frac{1}{-t-z}\tau(\mathrm{d}t) = \int_{(-\infty,\beta]} \frac{1}{t-z}\sigma(\mathrm{d}t)$$

for all $z \in \mathbb{C} \setminus (-\infty, \beta]$, where σ is the image measure of τ under the transformation $R : [-\beta, +\infty) \to (-\infty, \beta]$ defined by $R(t) := -t$. Since Proposition 5.3 yields $\tau([-\beta, +\infty)) = -\mathrm{i} \lim_{y \to +\infty} yF(\mathrm{i}y)$, we conclude furthermore

$$\begin{aligned}\sigma((-\infty, \beta]) &= [\sigma((-\infty, \beta])]^* = [\tau([-\beta, +\infty))]^* \\ &= \mathrm{i} \lim_{y \to +\infty} y[F(\mathrm{i}y)]^* = -\mathrm{i} \lim_{y \to +\infty} yG(\mathrm{i}y).\end{aligned}$$

Now let $\sigma \in \mathcal{M}^q_{\geq}((-\infty, \beta])$ be such that (9.1) holds true for all $z \in \mathbb{C} \setminus (-\infty, \beta]$. Using Proposition A.5, we get then

$$F(z) = -[G(-\overline{z})]^* = \int_{(-\infty, \beta]} \frac{1}{-t - z} \sigma(\mathrm{d}t) = \int_{[-\beta, +\infty)} \frac{1}{t - z} \tilde{\theta}(\mathrm{d}t) \tag{9.3}$$

for all $z \in \mathbb{C} \setminus [-\beta, +\infty)$, where $\tilde{\theta}$ is the image measure of σ under the transformation $T : (-\infty, \beta] \to [-\beta, +\infty)$ defined by $T(t) := -t$. Since we know from Theorem 5.1(a) that the measure $\tau \in \mathcal{M}^q_{\geq}([-\beta, +\infty))$ with (9.2) for all $z \in \mathbb{C} \setminus [-\beta, +\infty)$ is unique, we obtain $\tilde{\theta} = \tau$. Hence, σ is the image measure of τ under the transformation R.

(b) Let $\sigma \in \mathcal{M}^q_{\geq}((-\infty, \beta])$ be such that (9.1) holds true for all $z \in \mathbb{C} \setminus (-\infty, \beta]$. Thus, in view of Proposition A.5, the function $F : \mathbb{C} \setminus [-\beta, +\infty) \to \mathbb{C}^{q \times q}$ defined by $F(z) := -[G(-\overline{z})]^*$ fulfills (9.3) for all $z \in \mathbb{C} \setminus [-\beta, +\infty)$, where $\tilde{\theta}$ is the image measure of σ under the transformation T. Theorem 5.1(b) yields then $F \in \mathcal{S}_{0,q;[-\beta,+\infty)}$. Because of $G(z) = -[F(-\overline{z})]^*$ for all $z \in \mathbb{C} \setminus (-\infty, \beta]$, the function G belongs to $\mathcal{S}_{0,q;(-\infty,\beta]}$ by virtue of Remark 9.1(a). □

If $\beta \in \mathbb{R}$ and σ is a measure belonging to $\mathcal{M}^q_{\geq}((-\infty, \beta])$, then we will call the matrix-valued function $G : \mathbb{C} \setminus (-\infty, \beta] \to \mathbb{C}^{q \times q}$ which is, for each $z \in \mathbb{C} \setminus (-\infty, \beta]$, given by (9.1) the *$(-\infty, \beta]$-Stieltjes transform of σ*. If $G \in \mathcal{S}_{0,q;(-\infty,\beta]}$, then the unique measure $\sigma \in \mathcal{M}^q_{\geq}((-\infty, \beta])$ which fulfills (9.1) for each $z \in \mathbb{C} \setminus (-\infty, \beta]$ is said to be the *$(-\infty, \beta]$-Stieltjes measure of G* and will be denoted by σ_G.

Remark 9.3. (a) Let $\alpha \in \mathbb{R}$ and let $F \in \mathcal{S}_{0,q;[\alpha,+\infty)}$. In view of Theorem 5.1 and Proposition A.5, we have then

$$-[F(-\overline{z})]^* = \int_{[\alpha, +\infty)} \frac{1}{-t - z} \sigma_F(\mathrm{d}t) = \int_{(-\infty, -\alpha]} \frac{1}{t - z} \hat{\theta}(\mathrm{d}t)$$

for all $z \in \mathbb{C} \setminus (-\infty, -\alpha]$, where $\hat{\theta}$ is the image measure of σ_F under the transformation $R : [\alpha, +\infty) \to (-\infty, -\alpha]$ defined by $R(t) := -t$. Theorem 9.2 yields then that $G : \mathbb{C} \setminus (-\infty, -\alpha] \to \mathbb{C}^{q \times q}$ defined by $G(z) := -[F(-\overline{z})]^*$ belongs to $\mathcal{S}_{0,q;(-\infty,-\alpha]}$ and that $\sigma_G = \hat{\theta}$.

(b) Let $\beta \in \mathbb{R}$ and let $G \in \mathcal{S}_{0,q;(-\infty,\beta]}$. In view of Theorem 9.2 and Proposition A.5, we have then

$$-[G(-\overline{z})]^* = \int_{(-\infty, \beta]} \frac{1}{-t - z} \sigma_G(\mathrm{d}t) = \int_{[-\beta, +\infty)} \frac{1}{t - z} \tilde{\theta}(\mathrm{d}t)$$

for all $z \in \mathbb{C} \setminus [-\beta, +\infty)$, where $\tilde{\theta}$ is the image measure of σ_G under the transformation $T : (-\infty, \beta] \to [-\beta, +\infty)$ defined by $T(t) := -t$. Theorem 5.1 yields then that $F : \mathbb{C} \setminus [-\beta, +\infty) \to \mathbb{C}^{q\times q}$ defined by $F(z) := -[G(-\overline{z})]^*$ belongs to $\mathcal{S}_{0,q;[-\beta,+\infty)}$ and that $\sigma_F = \tilde{\theta}$.

Proposition 9.4. *Let $\beta \in \mathbb{R}$ and let $G \in \mathcal{S}_{0,q;(-\infty,\beta]}$. Then*

$$\mathcal{R}(G(z)) = \mathcal{R}(\sigma_G((-\infty, \beta])), \quad \mathcal{N}(G(z)) = \mathcal{N}(\sigma_G((-\infty, \beta]))$$

for all $z \in \mathbb{C} \setminus (-\infty, \beta]$.

Proof. According to Remark 9.3(b), the function $F : \mathbb{C}\setminus[-\beta, +\infty) \to \mathbb{C}^{q\times q}$ defined by $F(z) := -[G(-\overline{z})]^*$ belongs to $\mathcal{S}_{0,q;[-\beta,+\infty)}$ and σ_F is the image measure of σ_G under the transformation $T : (-\infty, \beta] \to [-\beta, +\infty)$ defined by $T(t) := -t$. In particular, $\sigma_F([-\beta, +\infty)) = \sigma_G((-\infty, \beta])$. Proposition 5.3 yields

$$\mathcal{R}(F(w)) = \mathcal{R}(\sigma_F([-\beta, +\infty))), \quad \mathcal{N}(F(w)) = \mathcal{N}(\sigma_F([-\beta, +\infty)))$$

for all $w \in \mathbb{C}\setminus[-\beta, +\infty)$. Furthermore, $\mathcal{R}([F(w)]^*) = \mathcal{R}(F(w))$ and $\mathcal{N}([F(w)]^*) = \mathcal{N}(F(w))$ follow for all $w \in \mathbb{C} \setminus [-\beta, +\infty)$ from Proposition 3.15(b). Finally for all $z \in \mathbb{C} \setminus (-\infty, \beta]$, we infer

$$\mathcal{R}(G(z)) = \mathcal{R}([F(-\overline{z})]^*) = \mathcal{R}(F(-\overline{z})) = \mathcal{R}(\sigma_F([-\beta, +\infty))) = \mathcal{R}(\sigma_G((-\infty, \beta]))$$

and

$$\mathcal{N}(G(z)) = \mathcal{N}([F(-\overline{z})]^*) = \mathcal{N}(F(-\overline{z})) = \mathcal{N}(\sigma_F([-\beta, +\infty))) = \mathcal{N}(\sigma_G((-\infty, \beta])),$$

which completes the proof. □

10. Moore–Penrose inverses of functions belonging to the class $\mathcal{S}_{q;(-\infty,\beta]}$

This section is the dual counterpart to Section 6.

Proposition 10.1. *Let $\beta \in \mathbb{R}$ and let $G \in \mathcal{S}_{q;(-\infty,\beta]}$. Then $G^\dagger$ is holomorphic in $\mathbb{C} \setminus (-\infty, \beta]$.*

Proof. In view of Proposition 7.8(b), we obtain the identities $\mathcal{N}(G(z)) = \mathcal{N}(G(\mathrm{i}))$ and $\mathcal{R}(G(z)) = \mathcal{R}(G(\mathrm{i}))$ for all $z \in \mathbb{C} \setminus (-\infty, \beta]$. Thus, the application of [15, Proposition 8.4] completes the proof. □

Let $\beta \in \mathbb{R}$ and $G \in \mathcal{S}_{q;(-\infty,\beta]}$. Then Proposition 10.1 suggests to look if there are functions closely related to $G^\dagger$ which belong again to $\mathcal{S}_{q;(-\infty,\beta]}$. Against to this background, we are led to the function $F : \mathbb{C} \setminus (-\infty, \beta] \to \mathbb{C}^{q\times q}$ defined by

$$F(z) := -(\beta - z)^{-1}[G(z)]^\dagger.$$

Theorem 10.2. *Let $\beta \in \mathbb{R}$ and let $G \in \mathcal{S}_{q;(-\infty,\beta]}$. Then $F : \mathbb{C} \setminus (-\infty, \beta] \to \mathbb{C}^{q\times q}$ defined by $F(z) := -(\beta - z)^{-1}[G(z)]^\dagger$ belongs to $\mathcal{S}_{q;(-\infty,\beta]}$.*

Proof. According to Remark 7.1, the function $P : \mathbb{C} \setminus [-\beta, +\infty) \to \mathbb{C}^{q\times q}$ defined by $P(z) := -[G(-\overline{z})]^*$ belongs to $\mathcal{S}_{q;[-\beta,+\infty)}$. Theorem 6.3 yields then that the function $Q : \mathbb{C} \setminus [-\beta, +\infty) \to \mathbb{C}^{q\times q}$ defined by $Q(z) := -(z+\beta)^{-1}[P(z)]^{\dagger}$ belongs to $\mathcal{S}_{q;[-\beta,+\infty)}$ as well. We now infer

$$\begin{aligned} -[Q(-\overline{z})]^* &= -(-z+\beta)^{-1}\Big(-[P(-\overline{z})]^{\dagger}\Big)^* \\ &= -(\beta - z)^{-1}\big(-[P(-\overline{z})]^*\big)^{\dagger} = F(z) \end{aligned} \tag{10.1}$$

for all $z \in \mathbb{C} \setminus (-\infty, \beta]$. Hence, $F \in \mathcal{S}_{q;(-\infty,\beta]}$ by virtue of Remark 7.1. □

Now we specify the result of Theorem 10.2 for functions belonging to the class $\mathcal{S}_{0,q;(-\infty,\beta]}$.

Proposition 10.3. *Let $\beta \in \mathbb{R}$ and let $G \in \mathcal{S}_{0,q;(-\infty,\beta]}$. Then $F : \mathbb{C}\setminus(-\infty, \beta] \to \mathbb{C}^{q\times q}$ defined by $F(z) := -(\beta - z)^{-1}[G(z)]^{\dagger}$ belongs to $\mathcal{S}_{q;(-\infty,\beta]}$ and $\gamma_F = [\sigma_G((-\infty,\beta])]^{\dagger}$. If G is not the constant function with value $O_{q\times q}$, then $F \in \mathcal{S}_{q;(-\infty,\beta]} \setminus \mathcal{S}_{0,q;(-\infty,\beta]}$.*

Proof. According to Remark 9.3(b), the function $P : \mathbb{C}\setminus[-\beta, +\infty) \to \mathbb{C}^{q\times q}$ defined by $P(z) := -[G(-\overline{z})]^*$ belongs to $\mathcal{S}_{0,q;[-\beta,+\infty)}$ and σ_P is the image measure of σ_G under the transformation $T : (-\infty, \beta] \to [-\beta, +\infty)$ defined by $T(t) := -t$. In particular, $\sigma_P([-\beta, +\infty)) = \sigma_G((-\infty, \beta])$. Proposition 6.4 yields that the function $Q : \mathbb{C} \setminus [-\beta, +\infty) \to \mathbb{C}^{q\times q}$ defined by $Q(z) := -(z + \beta)^{-1}[P(z)]^{\dagger}$ belongs to $\mathcal{S}_{q;[-\beta,+\infty)}$, that $\gamma_Q = [\sigma_P([-\beta, +\infty))]^{\dagger}$, and that $Q \notin \mathcal{S}_{0,q;[-\beta,+\infty)}$ if P is not the constant function with value $O_{q\times q}$. Furthermore, we have (10.1) for all $z \in \mathbb{C} \setminus (-\infty, \beta]$. Hence, $F \in \mathcal{S}_{q;(-\infty,\beta]}$ and $\gamma_F = \gamma_Q$ by virtue of Remark 7.7(a). We now infer $\gamma_F = [\sigma_G((-\infty, \beta])]^{\dagger}$. From (10.1) we conclude $Q(z) = -[F(-\overline{z})]^*$ for all $z \in \mathbb{C} \setminus [-\beta, +\infty)$. Since $G \not\equiv O_{q\times q}$ implies $P \not\equiv O_{q\times q}$, the proof is complete in view of Remark 9.1(b). □

Now we introduce the dual counterpart of the class introduced in Notation 6.5.

Notation 10.4. Let $\beta \in \mathbb{R}$. Then let $\mathcal{S}^{[-1]}_{q;(-\infty,\beta]}$ be the class of all matrix-valued functions $G : \mathbb{C} \setminus (-\infty, \beta] \to \mathbb{C}^{q\times q}$ which fulfill the following two conditions:

(i) G is holomorphic in $\mathbb{C} \setminus (-\infty, \beta]$ with $\operatorname{Rstr}_{\Pi_+} G \in \mathcal{R}_q(\Pi_+)$;
(ii) for all $x \in (\beta, +\infty)$, the matrix $G(x)$ is non-negative Hermitian.

For the special case $q = 1$ and $\beta = 0$ the class introduced in Notation 10.4 was studied by Katsnelson [18].

Example 10.5. Let $\beta \in \mathbb{R}$ and let $D, E \in \mathbb{C}^{q\times q}_{\geq}$. Then $G : \mathbb{C} \setminus (-\infty, \beta] \to \mathbb{C}^{q\times q}$ defined by $G(z) := D - (\beta - z)E$ belongs to $\mathcal{S}^{[-1]}_{q;(-\infty,\beta]}$.

Remark 10.6. Let $\alpha, \beta \in \mathbb{R}$.

(a) If $F \in \mathcal{S}^{[-1]}_{q;[\alpha,+\infty)}$, then $G : \mathbb{C} \setminus (-\infty, \beta] \to \mathbb{C}^{q\times q}$ defined by

$$G(z) := -[F(\alpha + \beta - \overline{z})]^*$$

belongs to $\mathcal{S}^{[-1]}_{q;(-\infty,\beta]}$.

(b) If $G \in \mathcal{S}^{[-1]}_{q;(-\infty,\beta]}$, then $F : \mathbb{C} \setminus [\alpha, +\infty) \to \mathbb{C}^{q\times q}$ defined by
$$F(z) := -[G(\alpha + \beta - \overline{z})]^*$$
belongs to $\mathcal{S}^{[-1]}_{q;[\alpha,+\infty)}$.

Theorem 10.7. *Let $\beta \in \mathbb{R}$ and let $G : \mathbb{C} \setminus (-\infty, \beta] \to \mathbb{C}^{q\times q}$.*

(a) *If $G \in \mathcal{S}^{[-1]}_{q;(-\infty,\beta]}$, then there are unique non-negative Hermitian complex $q \times q$ matrices D and E and a unique non-negative Hermitian measure $\rho \in \mathcal{M}^q_{\geq}((-\infty, \beta))$ such that*
$$G(z) = D + (\beta - z)\left[-E + \int_{(-\infty,\beta)} \frac{1 + \beta - t}{t - z} \rho(\mathrm{d}t)\right] \tag{10.2}$$
for all $z \in \mathbb{C} \setminus (-\infty, \beta]$. Furthermore, the function $Q : \mathbb{C} \setminus (-\infty, \beta] \to \mathbb{C}^{q\times q}$ defined by $Q(z) := (\beta - z)^{-1}G(z)$ belongs to $\mathcal{S}_{q;(-\infty,\beta]}$ with $D = \mu_Q(\{\beta\})$ and $(E, \rho) = (\gamma_Q, \mathrm{Rstr}_{\mathfrak{B}_{(-\infty,\beta)}} \mu_Q)$.

(b) *If $D \in \mathbb{C}^{q\times q}_{\geq}$, $E \in \mathbb{C}^{q\times q}_{\geq}$ and $\rho \in \mathcal{M}^q_{\geq}((-\infty, \beta))$ are such that G can be represented via (10.2) for all $z \in \mathbb{C} \setminus (-\infty, \beta]$, then G belongs to $\mathcal{S}^{[-1]}_{q;(-\infty,\beta]}$.*

Proof. (a) Let $G \in \mathcal{S}^{[-1]}_{q;(-\infty,\beta]}$. According to Remark 10.6(b), the matrix-valued function $F : \mathbb{C} \setminus [-\beta, +\infty) \to \mathbb{C}^{q\times q}$ defined by $F(z) := -[G(-\overline{z})]^*$ belongs to $\mathcal{S}^{[-1]}_{q;[-\beta,+\infty)}$. Theorem 6.10(a) yields then that there is a unique triple (M, N, ω) from $\mathbb{C}^{q\times q}_{\geq} \times \mathbb{C}^{q\times q}_{\geq} \times \mathcal{M}^q_{\geq}((-\beta, +\infty))$ such that
$$F(z) = -M + (z + \beta)\left[N + \int_{(-\beta,+\infty)} \frac{1 + t + \beta}{t - z} \omega(\mathrm{d}t)\right]$$
for all $z \in \mathbb{C} \setminus [-\beta, +\infty)$ and that the function $P : \mathbb{C} \setminus [-\beta, +\infty) \to \mathbb{C}^{q\times q}$ defined by $P(z) := (z + \beta)^{-1}F(z)$ belongs to $\mathcal{S}_{q;[-\beta,+\infty)}$ with $M = \mu_P(\{-\beta\})$ and $(N, \omega) = (\gamma_P, \mathrm{Rstr}_{\mathfrak{B}_{(-\beta,+\infty)}} \mu_P)$. Applying Proposition A.5, we now infer
$$\begin{aligned} G(z) = -[F(-\overline{z})]^* &= M - (-z + \beta)\left[N + \int_{(-\beta,+\infty)} \frac{1 + t + \beta}{t + z} \omega(\mathrm{d}t)\right] \\ &= M + (\beta - z)\left[-N + \int_{(-\beta,+\infty)} \frac{1 + \beta - (-t)}{-t - z} \omega(\mathrm{d}t)\right] \\ &= D + (\beta - z)\left[-E + \int_{(-\infty,\beta)} \frac{1 + \beta - t}{t - z} \rho(\mathrm{d}t)\right] \end{aligned}$$
for all $z \in \mathbb{C} \setminus (-\infty, \beta]$, where $D := M$, where $E := N$, and where ρ is the image measure of ω under the transformation $R_0 : (-\beta, +\infty) \to (-\infty, \beta)$ defined by $R_0(t) := -t$. The function Q fulfills $Q(z) = -[P(-\overline{z})]^*$ for all $z \in \mathbb{C} \setminus (-\infty, \beta]$. Remark 7.7(a) yields then $Q \in \mathcal{S}_{q;(-\infty,\beta]}$ with $\gamma_Q = \gamma_P$ and μ_Q being the image measure of μ_P under the transformation $R : [-\beta, +\infty) \to (-\infty, \beta]$ defined by

$R(t) := -t$. Hence, $D = M = \mu_P(\{-\beta\}) = \mu_Q(\{\beta\})$ and $E = N = \gamma_P = \gamma_Q$ follow. Furthermore, $\operatorname{Rstr}_{\mathfrak{B}_{(-\infty,\beta)}} \mu_Q$ is the image measure of $\operatorname{Rstr}_{\mathfrak{B}_{(-\beta,+\infty)}} \mu_P$ under the transformation R_0, implying $\rho = \operatorname{Rstr}_{\mathfrak{B}_{(-\infty,\beta)}} \mu_Q$. In particular, the triple (D, E, ρ) is unique.

(b) Let $D, E \in \mathbb{C}_{\geq}^{q\times q}$ and let $\rho \in \mathcal{M}_{\geq}^{q}((-\infty, \beta))$ be such that (10.2) holds true for all $z \in \mathbb{C} \setminus (-\infty, \beta]$. Let $F : \mathbb{C} \setminus [-\beta, +\infty) \to \mathbb{C}^{q\times q}$ be defined by $F(z) := -[G(-\overline{z})]^*$. Then, using Proposition A.5, we get

$$\begin{aligned} F(z) = -[G(-\overline{z})]^* &= -D - (\beta + z)\left[-E + \int_{(-\infty,\beta)} \frac{1+\beta-t}{t+z}\rho(\mathrm{d}t)\right] \\ &= -D + (z+\beta)\left[E + \int_{(-\infty,\beta)} \frac{1-t+\beta}{-t-z}\rho(\mathrm{d}t)\right] \\ &= -D + (z+\beta)\left[E + \int_{(-\beta,\infty)} \frac{1+t+\beta}{t-z}\tilde{\theta}(\mathrm{d}t)\right] \end{aligned}$$

for all $z \in \mathbb{C} \setminus [-\beta, +\infty)$, where $\tilde{\theta}$ is the image measure of ρ under the transformation $T_0 : (-\infty, \beta) \to (-\beta, +\infty)$ defined by $T_0(t) := -t$. Theorem 6.10(b) yields then $F \in \mathcal{S}_{q;[-\beta,+\infty)}^{[-1]}$. In view of $G(z) = -[F(-\overline{z})]^*$ for all $z \in \mathbb{C} \setminus (-\infty, \beta]$, hence G belongs to $\mathcal{S}_{q;(-\infty,\beta]}^{[-1]}$ by virtue of Remark 10.6(a). □

In the following, if $\beta \in \mathbb{R}$ and $G \in \mathcal{S}_{q;(-\infty,\beta]}^{[-1]}$ are given, then we will write (D_G, E_G, ρ_G) for the unique triple $(D, E, \rho) \in \mathbb{C}_{\geq}^{q\times q} \times \mathbb{C}_{\geq}^{q\times q} \times \mathcal{M}_{\geq}^{q}((-\infty, \beta))$ which fulfills (10.2) for all $z \in \mathbb{C} \setminus (-\infty, \beta]$.

Remark 10.8.

(a) Let $\alpha \in \mathbb{R}$ and let $F \in \mathcal{S}_{q;[\alpha,+\infty)}^{[-1]}$. In view of Theorem 6.10(a) and Proposition A.5, we have then

$$\begin{aligned} -[F(-\overline{z})]^* &= D_F - (-z-\alpha)\left[E_F + \int_{(\alpha,+\infty)} \frac{1+t-\alpha}{t+z}\rho_F(\mathrm{d}t)\right] \\ &= D_F + (-\alpha - z)\left[-E_F + \int_{(\alpha,+\infty)} \frac{1-\alpha-(-t)}{-t-z}\rho_F(\mathrm{d}t)\right] \\ &= D_F + (-\alpha - z)\left[-E_F + \int_{(-\infty,-\alpha)} \frac{1-\alpha-t}{t-z}\hat{\theta}(\mathrm{d}t)\right] \end{aligned}$$

for all $z \in \mathbb{C} \setminus (-\infty, -\alpha]$, where $\hat{\theta}$ is the image measure of ρ_F under the transformation $R_0 : (\alpha, +\infty) \to (-\infty, -\alpha)$ defined by $R_0(t) := -t$. Because of $D_F \in \mathbb{C}_{\geq}^{q\times q}$ and $E_F \in \mathbb{C}_{\geq}^{q\times q}$, Theorem 10.7 yields then that $G : \mathbb{C} \setminus (-\infty, -\alpha] \to \mathbb{C}^{q\times q}$ defined by $G(z) := -[F(-\overline{z})]^*$ belongs to $\mathcal{S}_{q;(-\infty,-\alpha]}^{[-1]}$ and that $(D_G, E_G, \rho_G) = (D_F, E_F, \hat{\theta})$.

(b) Let $\beta \in \mathbb{R}$ and let $G \in \mathcal{S}^{[-1]}_{q;(-\infty,\beta]}$. In view of Theorem 10.7(a) and Proposition A.5, we have then

$$\begin{aligned}
-[G(-\bar{z})]^* &= -D_G - (\beta + z)\left[-E_G + \int_{(-\infty,\beta)} \frac{1+\beta-t}{t+z}\rho_G(\mathrm{d}t)\right] \\
&= -D_G + (z+\beta)\left[E_G + \int_{(-\infty,\beta)} \frac{1-t+\beta}{-t-z}\rho_G(\mathrm{d}t)\right] \\
&= -D_G + (z+\beta)\left[E_G + \int_{(-\beta,+\infty)} \frac{1+t+\beta}{t-z}\tilde{\theta}(\mathrm{d}t)\right]
\end{aligned}$$

for all $z \in \mathbb{C} \setminus [-\beta, +\infty)$, where $\tilde{\theta}$ is the image measure of ρ_G under the transformation $T_0 : (-\infty, \beta) \to (-\beta, +\infty)$ defined by $T_0(t) := -t$. Because of $D_G \in \mathbb{C}^{q\times q}_{\geq}$ and $E_G \in \mathbb{C}^{q\times q}_{\geq}$, Theorem 6.10 yields then that $F : \mathbb{C} \setminus [-\beta, +\infty) \to \mathbb{C}^{q\times q}$ defined by $F(z) := -[G(-\bar{z})]^*$ belongs to $\mathcal{S}^{[-1]}_{q;[-\beta,+\infty)}$ and that $(D_F, E_F, \rho_F) = (D_G, E_G, \tilde{\theta})$.

Corollary 10.9. *Let $\beta \in \mathbb{R}$. Then $G : \mathbb{C} \setminus (-\infty, \beta] \to \mathbb{C}^{q\times q}$ belongs to $\mathcal{S}^{[-1]}_{q;(-\infty,\beta]}$ if and only if $Q : \mathbb{C} \setminus (-\infty, \beta] \to \mathbb{C}^{q\times q}$ defined by $Q(z) = (\beta - z)^{-1}G(z)$ belongs to $\mathcal{S}_{q;(-\infty,\beta]}$.*

Proof. From Remark 10.8 we conclude that G belongs to $\mathcal{S}^{[-1]}_{q;(-\infty,\beta]}$ if and only if $F : \mathbb{C} \setminus [-\beta, +\infty) \to \mathbb{C}^{q\times q}$ defined by $F(z) := -[G(-\bar{z})]^*$ belongs to $\mathcal{S}^{[-1]}_{q;[-\beta,+\infty)}$. Corollary 6.11 yields that F belongs to $\mathcal{S}^{[-1]}_{q;[-\beta,+\infty)}$ if and only if $P : \mathbb{C}\setminus[-\beta, +\infty) \to \mathbb{C}^{q\times q}$ defined by $P(z) = (z + \beta)^{-1}F(z)$ belongs to $\mathcal{S}_{q;[-\beta,+\infty)}$. Since $Q(z) = -[P(-\bar{z})]^*$ holds true for all $z \in \mathbb{C} \setminus (-\infty, \beta]$, we see by virtue of Remark 7.7 that P belongs to $\mathcal{S}_{q;[-\beta,+\infty)}$ if and only if Q belongs to $\mathcal{S}_{q;(-\infty,\beta]}$, which completes the proof. □

Corollary 10.10. *Let $\beta \in \mathbb{R}$ and let $G \in \mathcal{S}^{[-1]}_{q;(-\infty,\beta]}$. For all $x_1, x_2 \in (\beta, +\infty)$ with $x_1 \leq x_2$, then $O_{q\times q} \leq G(x_1) \leq G(x_2)$.*

Proof. Using Theorem 10.7, we obtain

$$G(x_2) - G(x_1) = (x_2 - x_1)\left[E_G + \int_{(-\infty,\beta]} \frac{(1+\beta-t)(\beta-t)}{(t-x_2)(t-x_1)}\rho_G(\mathrm{d}t)\right]$$

for all $x_1, x_2 \in (\beta, +\infty)$ with $x_1 \leq x_2$, by direct calculation. Since $E_G \in \mathbb{C}^{q\times q}_{\geq}$ and $G(x) \in \mathbb{C}^{q\times q}_{\geq}$ for all $x \in (\beta, +\infty)$, thus the proof is complete. □

Proposition 10.11. *Let $\beta \in \mathbb{R}$ and let $G \in \mathcal{S}^{[-1]}_{q;(-\infty,\beta]}$.*

(a) *If $z \in \mathbb{C} \setminus (-\infty, \beta]$, then $[G(z)]^* = G(\bar{z})$.*

(b) *For all* $z \in \mathbb{C} \setminus (-\infty, \beta]$,

$$\mathcal{R}(G(z)) = \mathcal{R}(D_G) + \mathcal{R}(E_G) + \mathcal{R}(\rho_G((-\infty, \beta))), \tag{10.3}$$

$$\mathcal{N}(G(z)) = \mathcal{N}(D_G) \cap \mathcal{N}(E_G) \cap \mathcal{N}(\rho_G((-\infty, \beta))), \tag{10.4}$$

and, in particular, $\mathcal{R}([G(z)]^*) = \mathcal{R}(G(z))$ *and* $\mathcal{N}([G(z)]^*) = \mathcal{N}(G(z))$.

(c) *Let* $r \in \mathbb{N}_0$. *Then the following statements are equivalent:*
  (i) $\operatorname{rank} G(z) = r$ *for all* $z \in \mathbb{C} \setminus (-\infty, \beta]$;
  (ii) *there is some* $z_0 \in \mathbb{C} \setminus (-\infty, \beta]$ *such that* $\operatorname{rank} G(z_0) = r$;
  (iii) $\dim[\mathcal{R}(D_G) + \mathcal{R}(E_G) + \mathcal{R}(\rho_G((-\infty, \beta)))] = r$.

Proof. (a) This can be concluded from the representation (10.2) in Theorem 10.7(a).

(b) According to Remark 10.8(b), the function $F : \mathbb{C} \setminus [-\beta, +\infty) \to \mathbb{C}^{q\times q}$ defined by $F(z) := -[G(-\overline{z})]^*$ belongs to $\mathcal{S}^{[-1]}_{q;[-\beta,+\infty)}$ and $(D_F, E_F, \rho_F) = (D_G, E_G, \tilde{\theta})$, where $\tilde{\theta}$ is the image measure of ρ_G under the transformation $T_0 : (-\infty, \beta) \to (-\beta, +\infty)$ defined by $T_0(t) := -t$. In particular, $\rho_F((-\beta, +\infty)) = \rho_G((-\infty, \beta))$. Proposition 6.14(b) yields

$$\mathcal{R}([F(w)]^*) = \mathcal{R}(F(w)) = \mathcal{R}(D_F) + \mathcal{R}(E_F) + \mathcal{R}(\rho_F((-\beta, +\infty)))$$

and

$$\mathcal{N}([F(w)]^*) = \mathcal{N}(F(w)) = \mathcal{N}(D_F) \cap \mathcal{N}(E_F) \cap \mathcal{N}(\rho_F((-\beta, +\infty)))$$

for all $w \in \mathbb{C} \setminus [-\beta, +\infty)$. We infer

$$\begin{aligned}\mathcal{R}(G(z)) = \mathcal{R}([F(-\overline{z})]^*) &= \mathcal{R}(D_F) + \mathcal{R}(E_F) + \mathcal{R}(\rho_F((-\beta, +\infty)))\\ &= \mathcal{R}(D_G) + \mathcal{R}(E_G) + \mathcal{R}(\rho_G((-\infty, \beta)))\end{aligned}$$

and, analogously,

$$\mathcal{N}(G(z)) = \mathcal{N}(D_G) \cap \mathcal{N}(E_G) \cap \mathcal{N}(\rho_G((-\infty, \beta)))$$

for all $z \in \mathbb{C} \setminus (-\infty, \beta]$. In view of (a), part (b) is proved.

(c) This is an immediate consequence of (b). □

Corollary 10.12. *Let* $\beta \in \mathbb{R}$, *let* $G \in \mathcal{S}^{[-1]}_{q;(-\infty,\beta]}$, *and let* $z_0 \in \mathbb{C} \setminus (-\infty, \beta]$. *Then* $G(z_0) = O_{q\times q}$ *if and only if* $G(z) = O_{q\times q}$ *for all* $z \in \mathbb{C} \setminus (-\infty, \beta]$.

Proof. This is an immediate consequence of Proposition 10.11(c). □

Corollary 10.13. *Let* $\beta \in \mathbb{R}$, *let* $G \in \mathcal{S}^{[-1]}_{q;(-\infty,\beta]}$, *and let* $\lambda \in \mathbb{R}$ *be such that the matrix* $D_G - \lambda I_q$ *is non-negative Hermitian. Then*

$$\mathcal{R}(G(z) - \lambda I_q) = \mathcal{R}(G(w) - \lambda I_q), \quad \mathcal{N}(G(z) - \lambda I_q) = \mathcal{N}(G(w) - \lambda I_q) \tag{10.5}$$

for every choice of z *and* w *in* $\mathbb{C} \setminus (-\infty, \beta]$. *In particular, if* $\lambda \leq 0$, *then* λ *is an eigenvalue of the matrix* $G(z_0)$ *for some* $z_0 \in \mathbb{C} \setminus (-\infty, \beta]$ *if and only if* λ *is an eigenvalue of the matrix* $G(z)$ *for all* $z \in \mathbb{C} \setminus (-\infty, \beta]$. *In this case, the eigenspaces* $\mathcal{N}(G(z) - \lambda I_q)$ *are independent of* $z \in \mathbb{C} \setminus (-\infty, \beta]$.

Proof. In view of Theorem 10.7, we conclude that the function $F : \mathbb{C} \setminus (-\infty, \beta] \to \mathbb{C}^{q\times q}$ defined by $F(z) := G(z) - \lambda I_q$ belongs to $\mathcal{S}^{[-1]}_{q;(-\infty,\beta]}$. The application of Proposition 10.11(b) to the function F yields then (10.5). Since the matrix D_G is non-negative Hermitian, we have $D_G - \lambda I_q \in \mathbb{C}^{q\times q}_{\geq}$ if $\lambda \leq 0$. Thus, the remaining assertions are an immediate consequence of (10.5). □

Lemma 10.14. *Let $\beta \in \mathbb{R}$ and $G \in \mathcal{S}^{[-1]}_{q;(-\infty,\beta]}$. Then the matrix function $G^\dagger$ is holomorphic in $\mathbb{C} \setminus (-\infty, \beta]$.*

Proof. In view of formulas (10.3) and (10.4), we obtain $\mathcal{N}(G(z)) = \mathcal{N}(G(\mathrm{i}))$ and $\mathcal{R}(G(z)) = \mathcal{R}(G(\mathrm{i}))$ for all $z \in \mathbb{C} \setminus (-\infty, \beta]$. Thus, the application of [15, Proposition 8.4] completes the proof. □

The following result is an analogue of Theorem 6.18.

Theorem 10.15. *Let $\beta \in \mathbb{R}$ and $G : \mathbb{C}\setminus(-\infty, \beta] \to \mathbb{C}^{q\times q}$ be a matrix-valued function. Then G belongs to $\mathcal{S}^{[-1]}_{q;(-\infty,\beta]}$ if and only if $F := -G^\dagger$ belongs to $\mathcal{S}_{q;(-\infty,\beta]}$.*

Proof. From Remark 10.8 we get that G belongs to $\mathcal{S}^{[-1]}_{q;(-\infty,\beta]}$ if and only if $Q : \mathbb{C} \setminus [-\beta, +\infty) \to \mathbb{C}^{q\times q}$ defined by $Q(z) := -[G(-\overline{z})]^*$ belongs to $\mathcal{S}^{[-1]}_{q;[-\beta,+\infty)}$. Theorem 6.18 yields that Q belongs to $\mathcal{S}^{[-1]}_{q;[-\beta,+\infty)}$ if and only if $P := -Q^\dagger$ belongs to $\mathcal{S}_{q;[-\beta,+\infty)}$. Since $F(z) = -[P(-\overline{z})]^*$ is true for all $z \in \mathbb{C} \setminus (-\infty, \beta]$, we see from Remark 7.7 that P belongs to $\mathcal{S}_{q;[-\beta,+\infty)}$ if and only if F belongs to $\mathcal{S}_{q;(-\infty,\beta]}$, which completes the proof. □

Appendix
Some considerations on non-negative Hermitian measures

In this appendix, we summarize some facts on integration with respect to non-negative Hermitian measures. For each non-negative Hermitian $q \times q$ measure $\mu = (\mu_{jk})_{j,k=1}^q$ on a measurable space $(\Omega, \mathfrak{A})$, we denote by $\mathcal{L}^1(\Omega, \mathfrak{A}, \mu; \mathbb{C})$ the set of all $\mathfrak{A}$-$\mathfrak{B}_\mathbb{C}$-measurable functions $f : \Omega \to \mathbb{C}$ such that the integral $\int_\Omega f \mathrm{d}\mu$ exists, i.e., that $\int_\Omega |f| \mathrm{d}\tilde{\mu}_{jk} < \infty$ for every choice of $j, k \in \mathbb{Z}_{1,q}$, where $\tilde{\mu}_{jk}$ is the variation of the complex measure μ_{jk}.

Remark A.1. Let μ be a non-negative Hermitian measure on a measurable space $(\Omega, \mathfrak{A})$, let $\tau := \operatorname{tr} \mu$ be the trace measure of μ, and let $f : \Omega \to \mathbb{C}$ be an $\mathfrak{A}$-$\mathfrak{B}_\mathbb{C}$-measurable function. Then f belongs to $\mathcal{L}^1(\Omega, \mathfrak{A}, \mu; \mathbb{C})$ if and only if f belongs to $\mathcal{L}^1(\Omega, \mathfrak{A}, \tau; \mathbb{C})$.

Remark A.2. Let μ be a non-negative Hermitian $q \times q$ measure on a measurable space $(\Omega, \mathfrak{A})$ and let $u \in \mathbb{C}^q$. Then $\nu := u^*\mu u$ is a finite measure on $(\Omega, \mathfrak{A})$ which is absolutely continuous with respect to the trace measure of μ. If f belongs to $\mathcal{L}^1(\Omega, \mathfrak{A}, \mu; \mathbb{C})$, then $\int_\Omega |f| \mathrm{d}\nu < \infty$ and

$$\int_\Omega f \mathrm{d}\nu = u^* \left(\int_\Omega f \mathrm{d}\mu \right) u.$$

Remark A.3. Let μ be a non-negative Hermitian $q \times q$ measure on a measurable space $(\Omega, \mathfrak{A})$. An $\mathfrak{A}$-$\mathfrak{B}_\mathbb{C}$-measurable function $f : \Omega \to \mathbb{C}$ belongs to $\mathcal{L}^1(\Omega, \mathfrak{A}, \mu; \mathbb{C})$ if and only if $\int_\Omega |f| \mathrm{d}(u^* \mu u) < \infty$ for all $u \in \mathbb{C}^q$.

Lemma A.4 (cf. [11, Lemma B.2]). *Let $(\Omega, \mathfrak{A})$ be a measurable space, let σ be a non-negative Hermitian $q \times q$ measure on $(\Omega, \mathfrak{A})$, and let τ be the trace measure of σ.*

(a) *If $f \in \mathcal{L}^1(\Omega, \mathfrak{A}, \sigma; \mathbb{C})$, then*

$$\mathcal{R}\left(\int_\Omega f \mathrm{d}\sigma \right) \subseteq \mathcal{R}(\sigma(\Omega)), \quad \mathcal{N}(\sigma(\Omega)) \subseteq \mathcal{N}\left(\int_\Omega f \mathrm{d}\sigma \right).$$

(b) *If $f \in \mathcal{L}^1(\Omega, \mathfrak{A}, \sigma; \mathbb{C})$ fulfills $\tau(\{f \notin (0, +\infty)\}) = 0$, then*

$$\mathcal{R}\left(\int_\Omega f \mathrm{d}\sigma \right) = \mathcal{R}(\sigma(\Omega)), \quad \mathcal{N}(\sigma(\Omega)) = \mathcal{N}\left(\int_\Omega f \mathrm{d}\sigma \right).$$

Proposition A.5 ([12, Proposition B.1]). *Let $(\Omega, \mathfrak{A})$ and $(\tilde{\Omega}, \tilde{\mathfrak{A}})$ be measurable spaces and let μ be a non-negative Hermitian $q \times q$ measure on $(\Omega, \mathfrak{A})$. Further, let $T : \Omega \to \tilde{\Omega}$ be an $\mathfrak{A}$-$\tilde{\mathfrak{A}}$-measurable mapping. Then $T(\mu) : \tilde{\mathfrak{A}} \to \mathbb{C}^{q \times q}$ defined by $[T(\mu)](\tilde{A}) := \mu(T^{-1}(\tilde{A}))$ is a non-negative Hermitian $q \times q$ measure on $(\tilde{\Omega}, \tilde{\mathfrak{A}})$. Furthermore, if $\tilde{f} : \tilde{\Omega} \to \mathbb{C}$ is an $\tilde{\mathfrak{A}}$-$\mathfrak{B}_\mathbb{C}$-measurable mapping, then $\tilde{f} \in \mathcal{L}^1(\tilde{\Omega}, \tilde{\mathfrak{A}}, T(\mu); \mathbb{C})$ if and only if $\tilde{f} \circ T \in \mathcal{L}^1(\Omega, \mathfrak{A}, \mu; \mathbb{C})$. If $\tilde{f}$ belongs to $\mathcal{L}^1(\tilde{\Omega}, \tilde{\mathfrak{A}}, T(\mu); \mathbb{C})$, then*

$$\int_{\tilde{A}} \tilde{f} \mathrm{d}[T(\mu)] = \int_{T^{-1}(\tilde{A})} (\tilde{f} \circ T) \mathrm{d}\mu$$

for all $\tilde{A} \in \tilde{\mathfrak{A}}$.

Proposition A.6 (Lebesgue's dominated convergence for non-negative Hermitian measures). *Let μ be a non-negative Hermitian $q \times q$ measure on a measurable space $(\Omega, \mathfrak{A})$ with trace measure τ. For all $n \in \mathbb{N}$, let $f_n : \Omega \to \mathbb{C}$ be an $\mathfrak{A}$-$\mathfrak{B}_\mathbb{C}$-measurable function. Let $f : \Omega \to \mathbb{C}$ be an $\mathfrak{A}$-$\mathfrak{B}_\mathbb{C}$-measurable function and let $g \in \mathcal{L}^1(\Omega, \mathfrak{A}, \mu; \mathbb{C})$ be such that $\lim_{n \to +\infty} f_n(\omega) = f(\omega)$ for τ-a.a. $\omega \in \Omega$ and that $|f_n(\omega)| \leq |g(\omega)|$ for all $n \in \mathbb{N}$ and τ-a.a. $\omega \in \Omega$. Then $f \in \mathcal{L}^1(\Omega, \mathfrak{A}, \mu; \mathbb{C})$, $f_n \in \mathcal{L}^1(\Omega, \mathfrak{A}, \mu; \mathbb{C})$ for all $n \in \mathbb{N}$, and*

$$\lim_{n \to +\infty} \int_\Omega f_n \mathrm{d}\mu = \int_\Omega f \mathrm{d}\mu.$$

Proof. Let $u \in \mathbb{C}^q$ and let $\nu := u^* \mu u$. According to Remark A.2, then ν is a finite measure on $(\Omega, \mathfrak{A})$ which is absolutely continuous with respect to τ and $\int_\Omega |g| \mathrm{d}\nu < \infty$. In particular, $\lim_{n \to +\infty} f_n(\omega) = f(\omega)$ for ν-a.a. $\omega \in \Omega$ and $|f_n(\omega)| \leq |g(\omega)|$

for all $n \in \mathbb{N}$ and ν-a.a. $\omega \in \Omega$. Thus, Lebesgue's dominated convergence theorem provides us $\int_\Omega |f| \mathrm{d}\nu < \infty$, $\int_\Omega |f_n| \mathrm{d}\nu < \infty$ for all $n \in \mathbb{N}$, and

$$\lim_{n\to+\infty} \int_\Omega f_n \mathrm{d}\nu = \int_\Omega f \mathrm{d}\nu.$$

Since $u \in \mathbb{C}^q$ was arbitrarily chosen, $f \in \mathcal{L}^1(\Omega, \mathfrak{A}, \mu; \mathbb{C})$ and $f_n \in \mathcal{L}^1(\Omega, \mathfrak{A}, \mu; \mathbb{C})$ for all $n \in \mathbb{N}$ follow by virtue of Remark A.3. Thus, using Remark A.2, we get

$$\lim_{n\to+\infty} u^* \left(\int_\Omega f_n \mathrm{d}\mu \right) u = u^* \left(\int_\Omega f \mathrm{d}\mu \right) u$$

for all $u \in \mathbb{C}^q$. Hence,

$$\lim_{n\to+\infty} \int_\Omega f_n \mathrm{d}\mu = \int_\Omega f \mathrm{d}\mu$$

holds true. □

Remark A.7. If $A \in \mathbb{C}^{q\times q}$ is such that $\operatorname{Im} A \in \mathbb{C}^{q\times q}_{\geq}$, then $\mathcal{N}(A) \subseteq \mathcal{N}(\operatorname{Im} A)$ (see, e.g., [10, Lemma A.10]).

Lemma A.8. *Let Ω be a non-empty closed subset of $\mathbb{R}$, let $\sigma \in \mathcal{M}^q_{\geq}(\Omega)$, and let τ be the trace measure of σ. Then:*

(a) *For each $w \in \mathbb{C} \setminus \Omega$, the function $g_w : \Omega \to \mathbb{C}$ defined by $g_w(t) := 1/(t - w)$ belongs to $\mathcal{L}^1(\Omega, \mathfrak{B}_\Omega, \sigma; \mathbb{C})$.*

(b) *The matrix-valued function $S : \mathbb{C} \setminus \Omega \to \mathbb{C}^{q\times q}$ given by $S(w) := \int_\Omega g_w \mathrm{d}\sigma$ satisfies $\mathcal{R}(S(z)) = \mathcal{R}(\sigma(\Omega))$ and, in particular, $\operatorname{rank} S(z) = \operatorname{rank} \sigma(\Omega)$ for each $z \in \mathbb{C} \setminus [\inf \Omega, \sup \Omega]$.*

(c) $-\mathrm{i} \lim_{y\to+\infty} y S(\mathrm{i}y) = \sigma(\Omega)$.

Proof. Part a is readily checked. In particular, the function S is well defined. Let $z \in \mathbb{C} \setminus [\inf \Omega, \sup \Omega]$. From Lemma A.4(a) we get then

$$\mathcal{R}(S(z)) = \mathcal{R}\left(\int_\Omega g_z \mathrm{d}\sigma \right) \subseteq \mathcal{R}(\sigma(\Omega)).$$

Thus, in order to prove part (b), it remains to verify that $\mathcal{R}(\sigma(\Omega)) \subseteq \mathcal{R}(S(z))$, i.e., that

$$\mathcal{N}(S^*(z)) \subseteq \mathcal{N}(\sigma(\Omega)) \tag{A.1}$$

is true. Part (a) implies

$$\operatorname{Im} g_z \in \mathcal{L}^1(\Omega, \mathfrak{B}_\Omega, \sigma; \mathbb{C}). \tag{A.2}$$

For each $t \in \Omega$, one can easily check the equations

$$g_z(t) = \frac{t - \overline{z}}{|t - z|^2}, \qquad \overline{g_z}(t) = \frac{t - z}{|t - z|^2}, \qquad \text{and} \qquad \operatorname{Im} g_z(t) = \frac{\operatorname{Im} z}{|t - z|^2}. \tag{A.3}$$

First we consider the case that z belongs to Π_+. In view of

$$\operatorname{Im}(-S^*(z)) = \operatorname{Im} S(z) = \int_\Omega \operatorname{Im} g_z \mathrm{d}\sigma$$

and (A.3), we see that the matrix $\operatorname{Im}(-S^*(z))$ is non-negative Hermitian. Remark A.7 and (A.3) yield then

$$\mathcal{N}(S^*(z)) = \mathcal{N}(-S^*(z)) \subseteq \mathcal{N}(\operatorname{Im}(-S^*(z))) = \mathcal{N}\left(\int_\Omega \operatorname{Im} g_z \mathrm{d}\sigma\right). \tag{A.4}$$

From (A.3) we know that $\tau(\{\operatorname{Im} g_z \in (-\infty, 0]\}) = \tau(\emptyset) = 0$. Taking into account (A.2) and Lemma A.4(b), then

$$\mathcal{N}\left(\int_\Omega \operatorname{Im} g_z \mathrm{d}\sigma\right) = \mathcal{N}(\sigma(\Omega)) \tag{A.5}$$

follows. Combining (A.4) and (A.5) we obtain (A.1).

Now we study the case that z belongs to Π_-. By virtue of

$$\operatorname{Im} S^*(z) = -\operatorname{Im} S(z) = \int_\Omega (-\operatorname{Im} g_z)\mathrm{d}\sigma \tag{A.6}$$

and (A.3), the matrix $\operatorname{Im} S^*(z)$ belongs to $\mathbb{C}_{\geq}^{q\times q}$. Hence, Remark A.7 and (A.6) yield

$$\mathcal{N}(S^*(z)) \subseteq \mathcal{N}(\operatorname{Im} S^*(z)) = \mathcal{N}\left(\int_\Omega (-\operatorname{Im} g_z)\mathrm{d}\sigma\right). \tag{A.7}$$

Because of (A.3), we have $\tau(\{-\operatorname{Im} g_z \in (-\infty, 0]\}) = \tau(\emptyset) = 0$. Thus, using (A.2) and Lemma A.4(b), we have

$$\mathcal{N}\left(\int_\Omega (-\operatorname{Im} g_z)\mathrm{d}\sigma\right) = \mathcal{N}(\sigma(\Omega)).$$

Taking into account (A.7), this implies (A.1).

Now we discuss the case that $\inf\Omega > -\infty$ and that $z \in (-\infty, \inf\Omega)$. In view of $t \in \Omega$ and (A.3), we have then $\tau(\{\overline{g_z} \in (-\infty, 0]\}) = \tau(\emptyset) = 0$. Since one can see from part (a) that $\overline{g_z}$ belongs to $\mathcal{L}^1(\Omega, \mathfrak{B}_\Omega, \sigma; \mathbb{C})$, application of Lemma A.4(b) provides us then

$$\mathcal{N}(S^*(z)) = \mathcal{N}\left(\int_\Omega \overline{g_z}\mathrm{d}\sigma\right) = \mathcal{N}(\sigma(\Omega)).$$

Hence, (A.1) is valid. Similarly, in the case that $\sup\Omega < +\infty$ and $z \in (\sup\Omega, +\infty)$ hold, we get that (A.1) is fulfilled. Thus (A.1) is verified for each $z \in \mathbb{C} \setminus [\inf\Omega, \sup\Omega]$.

(c) For each $y \in (0, +\infty)$ and each $t \in \Omega$, we have

$$\operatorname{Re}\left(\frac{-\mathrm{i}y}{t - \mathrm{i}y}\right) = \frac{y^2}{t^2 + y^2}, \quad \operatorname{Im}\left(\frac{-\mathrm{i}y}{t - \mathrm{i}y}\right) = \frac{-yt}{t^2 + y^2}, \quad \left|\frac{-\mathrm{i}y}{t - \mathrm{i}y}\right| \leq 1.$$

From Proposition A.6 we obtain then

$$\sigma(\Omega) = \int_\Omega 1\mathrm{d}\sigma = \lim_{y\to+\infty} \int_\Omega \frac{-\mathrm{i}y}{t - \mathrm{i}y}\sigma(\mathrm{d}t) = -\mathrm{i}\lim_{y\to+\infty} yS(\mathrm{i}y). \qquad \square$$

References

[1] Yu.M. Arlinskii, S.V. Belyi, and E.R. Tsekanovskii. *Conservative realizations of Herglotz–Nevanlinna functions*, volume 217 of *Operator Theory: Advances and Applications*. Birkhäuser/Springer Basel AG, Basel, 2011.

[2] N. Aronszajn and W.F. Donoghue. A supplement to the paper on exponential representations of analytic functions in the upper half-plane with positive imaginary part. *J. Analyse Math.*, 12:113–127, 1964.

[3] S.L. Campbell and C.D. Meyer, Jr. *Generalized inverses of linear transformations*. Dover Publications Inc., New York, 1991. Corrected reprint of the 1979 original.

[4] A.E. Choque Rivero, Yu. M. Dyukarev, B. Fritzsche, and B. Kirstein. A truncated matricial moment problem on a finite interval. In *Interpolation, Schur functions and moment problems*, volume 165 of *Oper. Theory Adv. Appl.*, pages 121–173. Birkhäuser, Basel, 2006.

[5] V.K. Dubovoj, B. Fritzsche, and B. Kirstein. *Matricial version of the classical Schur problem*, volume 129 of *Teubner-Texte zur Mathematik* [*Teubner Texts in Mathematics*]. B.G. Teubner Verlagsgesellschaft mbH, Stuttgart, 1992. With German, French and Russian summaries.

[6] Yu.M. Dyukarev. Multiplicative and additive Stieltjes classes of analytic matrix-valued functions and interpolation problems connected with them. II. *Teor. Funktsiĭ Funktsional. Anal. i Prilozhen.*, (38):40–48, 127, 1982.

[7] Yu.M. Dyukarev and V.È. Katsnel′son. Multiplicative and additive Stieltjes classes of analytic matrix-valued functions and interpolation problems connected with them. I. *Teor. Funktsiĭ Funktsional. Anal. i Prilozhen.*, (36):13–27, 126, 1981.

[8] Yu.M. Dyukarev and V.È. Katsnel′son. Multiplicative and additive Stieltjes classes of analytic matrix-valued functions, and interpolation problems connected with them. III. *Teor. Funktsiĭ Funktsional. Anal. i Prilozhen.*, (41):64–70, 1984.

[9] J. Elstrodt. *Maß- und Integrationstheorie*. Springer-Lehrbuch. [Springer Textbook]. Springer-Verlag, Berlin, fourth edition, 2005. Grundwissen Mathematik. [Basic Knowledge in Mathematics].

[10] B. Fritzsche, B. Kirstein, A. Lasarow, and A. Rahn. On reciprocal sequences of matricial Carathéodory sequences and associated matrix functions. In *Interpolation, Schur functions and moment problems. II*, volume 226 of *Oper. Theory Adv. Appl.*, pages 57–115. Birkhäuser/Springer Basel AG, Basel, 2012.

[11] B. Fritzsche, B. Kirstein, and C. Mädler. On matrix-valued Herglotz–Nevanlinna functions with an emphasis on particular subclasses. *Math. Nachr.*, 285(14-15):1770–1790, 2012.

[12] B. Fritzsche, B. Kirstein, and C. Mädler. Transformations of matricial α-Stieltjes non-negative definite sequences. *Linear Algebra Appl.*, 439(12):3893–3933, 2013.

[13] B. Fritzsche, B. Kirstein, and C. Mädler. On a simultaneous approach to the even and odd truncated matricial Hamburger moment problems. In *Recent Advances in Inverse Scattering, Schur Analysis and Stochastic Processes*, volume 244 of *Oper. Theory Adv. Appl.*, pages 181–285. Birkhäuser/Springer Basel AG, Basel, 2015.

[14] B. Fritzsche, B. Kirstein, C. Mädler, and T. Schwarz. On a Schur-type algorithm for sequences of complex $p \times q$-matrices and its interrelations with the canonical

Hankel parametrization. In *Interpolation, Schur functions and moment problems. II*, volume 226 of *Oper. Theory Adv. Appl.*, pages 117–192. Birkhäuser/Springer Basel AG, Basel, 2012.

[15] B. Fritzsche, B. Kirstein, C. Mädler, and T. Schwarz. On the concept of invertibility for sequences of complex $p \times q$-matrices and its application to holomorphic $p \times q$-matrix-valued functions. In *Interpolation, Schur functions and moment problems. II*, volume 226 of *Oper. Theory Adv. Appl.*, pages 9–56. Birkhäuser/Springer Basel AG, Basel, 2012.

[16] F. Gesztesy and E.R. Tsekanovskii. On matrix-valued Herglotz functions. *Math. Nachr.*, 218:61–138, 2000.

[17] I.S. Kats and M.G. Kreĭn. R-functions – analytic functions mapping the upper half-plane into itself (Russian). Appendix I in F.V. Atkinson. *Diskretnye i nepreryvnye granichnye zadachi.* Translated from the English by I.S. Iohvidov and G.A. Karal′nik. Edited and supplemented by I.S. Kats and M.G. Kreĭn. Izdat. "Mir", Moscow, 1968. English translation in *American Mathematical Society Translations, Series* 2, *Vol.* 103*: Nine papers in analysis.* American Mathematical Society, Providence, R.I., 1974, pages 1–18.

[18] V.E. Katsnelson. Stieltjes functions and Hurwitz stable entire functions. *Complex Anal. Oper. Theory*, 5(2):611–630, 2011.

[19] M.G. Kreĭn and A.A. Nudel′man. *The Markov moment problem and extremal problems.* American Mathematical Society, Providence, R.I., 1977. Ideas and problems of P.L. Čebyšev and A.A. Markov and their further development, translated from the Russian by D. Louvish, Translations of Mathematical Monographs, Vol. 50.

[20] H. Schwerdtfeger. *Introduction to Linear Algebra and the Theory of Matrices.* P. Noordhoff, Groningen, 1950.

Bernd Fritzsche, Bernd Kirstein and Conrad Mädler
Universität Leipzig
Fakultät für Mathematik und Informatik
PF 10 09 20
D-04009 Leipzig, Germany
e-mail: **fritzsche@math.uni-leipzig.de**
kirstein@math.uni-leipzig.de
maedler@math.uni-leipzig.de

Operator Theory:
Advances and Applications, Vol. 259, 353–394

The Theory of Generalized Locally Toeplitz Sequences: a Review, an Extension, and a Few Representative Applications

Carlo Garoni and Stefano Serra-Capizzano

Dedicated to Albrecht Böttcher on his 60th birthday:
from him we learned beauty and elegance of the theory of Toeplitz operators

Abstract. We review and extend the theory of Generalized Locally Toeplitz (GLT) sequences, which goes back to Tilli's work on Locally Toeplitz sequences and was developed by the second author during the last decade. Informally speaking, a GLT sequence $\{A_n\}_n$ is a sequence of matrices with increasing size equipped with a function κ (the so-called symbol). We write $\{A_n\}_n \sim_{\rm GLT} \kappa$ to indicate that $\{A_n\}_n$ is a GLT sequence with symbol κ. This symbol characterizes the asymptotic singular value distribution of $\{A_n\}_n$; if the matrices A_n are Hermitian, it also characterizes the asymptotic eigenvalue distribution of $\{A_n\}_n$. Three fundamental examples of GLT sequences are: (i) the sequence of Toeplitz matrices generated by a function f in L^1; (ii) the sequence of diagonal sampling matrices containing the samples of a Riemann-integrable function a over equispaced grids; (iii) any zero-distributed sequence, i.e., any sequence of matrices with an asymptotic singular value distribution characterized by 0. The symbol of the GLT sequence (i) is f, the symbol of the GLT sequence (ii) is a, and the symbol of the GLT sequences (iii) is 0. The set of GLT sequences is a *-algebra. More precisely, suppose that $\{A_n^{(i)}\}_n \sim_{\rm GLT} \kappa_i$ for $i = 1,\ldots,r$, and let $A_n = {\rm ops}(A_n^{(1)},\ldots,A_n^{(r)})$ be a matrix obtained from $A_n^{(1)},\ldots,A_n^{(r)}$ by means of certain algebraic operations "ops", such as linear combinations, products, inversions and conjugate transpositions; then $\{A_n\}_n \sim_{\rm GLT} \kappa = {\rm ops}(\kappa_1,\ldots,\kappa_r)$.

The theory of GLT sequences is a powerful apparatus for computing the asymptotic singular value and eigenvalue distribution of the discretization ma-

---

This work was supported by the Italian MIUR Program FIR 2013 through the Project DREAMS, by the INdAM GNCS (Gruppo Nazionale per il Calcolo Scientifico), and by the Donation KAW 2013.0341 from the Knut & Alice Wallenberg Foundation in collaboration with the Royal Swedish Academy of Sciences.

trices A_n arising from the numerical approximation of continuous problems, such as integral equations and, especially, partial differential equations. Indeed, when the discretization parameter n tends to infinity, the matrices A_n give rise to a sequence $\{A_n\}_n$, which often turns out to be a GLT sequence.

Nevertheless, this work is not primarily concerned with the applicative interest of the theory of GLT sequences. Although we will provide some illustrative applications at the end, the attention is focused on the mathematical foundations of the theory. We first propose a modification of the original definition of GLT sequences. With the new definition, we are able to enlarge the applicability of the theory, by generalizing/simplifying a lot of key results. In particular, we remove the Riemann-integrability assumption from the main spectral distribution and algebraic results for GLT sequences. As a final step, we extend the theory. We first prove an approximation result, which is useful to show that a given sequence of matrices is a GLT sequence. By using this result, we provide a new and easier proof of the fact that $\{A_n^{-1}\}_n \sim_{\rm GLT} \kappa^{-1}$ whenever $\{A_n\}_n \sim_{\rm GLT} \kappa$, the matrices A_n are invertible, and $\kappa \neq 0$ almost everywhere. Finally, using again the approximation result, we prove that $\{f(A_n)\}_n \sim_{\rm GLT} f(\kappa)$ whenever $\{A_n\}_n \sim_{\rm GLT} \kappa$, the matrices A_n are Hermitian, and $f : \mathbb{R} \to \mathbb{R}$ is continuous.

Mathematics Subject Classification (2010). Primary 47B35, 47B06, 15A18, 15A60; Secondary 65N06, 65N30.

Keywords. Generalized Locally Toeplitz sequences, singular value and eigenvalue distribution, approximating class of sequences, discretization of partial differential equations, Finite Differences, Finite Elements, Schur complement.

1. Introduction

We review and extend the theory of Generalized Locally Toeplitz (GLT) sequences, which stems from Tilli's work on Locally Toeplitz sequences [46] and from the theory of classical Toeplitz operators [3, 8, 9, 10, 11, 31, 34, 45, 48, 49, 50], and was developed by the second author in [39, 40]. In Section 1.1 we mention some of the main applications of this theory. In Section 1.2 we summarize its main features. In Section 1.3 we describe the main contributions of this work.

Why the name "GLT sequences"? Before proceeding further, let us discuss the motivations behind the choice of the name "Generalized Locally Toeplitz sequences". A "Toeplitz sequence" is a sequence of Toeplitz matrices $T_n(f)$ generated by some function $f \in L^1([-\pi, \pi])$. Recall that $T_n(f)$ is a matrix of size n whose coefficients along each diagonal are constant. More precisely, the coefficients of $T_n(f)$ along the kth diagonal are equal to the kth Fourier coefficient of f.

For example,

$$T_n(2-2\cos\theta)=\begin{bmatrix}2 & -1 & & & \\ -1 & 2 & -1 & & \\ & \ddots & \ddots & \ddots & \\ & & -1 & 2 & -1 \\ & & & -1 & 2\end{bmatrix}. \tag{1.1}$$

A "Locally Toeplitz sequence" is a sequence of matrices A_n possessing a local Toeplitz structure. This means that the coefficients of A_n vary continuously along each diagonal, so that the difference between two consecutive entries in a fixed diagonal goes to zero as n tends to infinity – just like the difference between the samples of a continuous function over two consecutive points in a uniform grid with step size $h=\frac{1}{n+1}$. For example, a Locally Toeplitz version of (1.1) is

$$A_n=\begin{bmatrix}2a(x_1) & -a(x_1) & & & \\ -a(x_1) & 2a(x_2) & -a(x_2) & & \\ & \ddots & \ddots & \ddots & \\ & & -a(x_{n-2}) & 2a(x_{n-1}) & -a(x_{n-1}) \\ & & & -a(x_{n-1}) & 2a(x_n)\end{bmatrix}, \tag{1.2}$$

where $a:[0,1]\to\mathbb{C}$ is a continuous function and $x_j=\frac{j}{n+1}$, $j=1,\dots,n$. Looking at a relatively small submatrix of A_n (according to a "local" perspective), one easily recognizes an approximate Toeplitz structure. For instance, if $B_{\lfloor\sqrt{n}\rfloor}$ is a submatrix of size $\lfloor\sqrt{n}\rfloor$, obtained by taking $\lfloor\sqrt{n}\rfloor$ consecutive rows and columns of A_n, one easily sees that $B_{\lfloor\sqrt{n}\rfloor}=a(x_j)T_{\lfloor\sqrt{n}\rfloor}(2-2\cos\theta)+E_{\lfloor\sqrt{n}\rfloor}$, where $a(x_j)$ is one of the evaluations of $a(x)$ appearing in $B_{\lfloor\sqrt{n}\rfloor}$, and $E_{\lfloor\sqrt{n}\rfloor}$ tends to zero in spectral norm as n tends to infinity (the spectral norm $\|E_{\lfloor\sqrt{n}\rfloor}\|$ is proportional to the modulus of continuity of a evaluated at $\lfloor\sqrt{n}\rfloor/n$). The latter assertion remains true if $\lfloor\sqrt{n}\rfloor$ is replaced by any integer s_n such that $s_n=o(n)$. In other words, if we explore "locally" the matrix (1.2), using an ideal microscope and considering a large value of n, then we realize that the "local" structure of A_n is approximately a Toeplitz structure "weighted" through the function $a(x)$. This is the key idea that led to the definition of Locally Toeplitz sequences; see also the introduction of Tilli's paper [46] and Section 5.1 of [24]. The notion of Locally Toeplitz sequences was then generalized in [39, 40], so as to include Riemann-integrable "weight functions" $a(x)$, algebraic combinations of Toeplitz matrices, multilevel Toeplitz structures, and many other concepts that led to the birth of "Generalized Locally Toeplitz sequences".

1.1. Applications of the theory of GLT sequences

The theory of GLT sequences is a powerful apparatus for computing/analyzing the asymptotic spectral distribution of the discretization matrices arising from the numerical approximation of continuous problems, such as Integral Equations (IEs) and, especially, Partial Differential Equations (PDEs). Let us explain this point in more detail. When discretizing a linear PDE by means of a linear numerical method, the computation of the numerical solution reduces to solving a linear system $A_n\mathbf{u}_n = \mathbf{b}_n$. The size d_n of this linear system increases when the discretization parameter n tends to infinity. Hence, what we actually have is not just a single linear system, but an whole sequence of linear systems with increasing size; and what is often observed in practice is that the sequence of discretization matrices $\{A_n\}_n$ enjoys an asymptotic spectral distribution, which is somehow related to the spectrum of the differential operator associated with the considered PDE. More precisely, it often happens that, for a large set of test functions F (usually, for all continuous functions F with bounded support), the following limit relation holds:

$$\lim_{n\to\infty}\frac{1}{d_n}\sum_{j=1}^{d_n}F(\lambda_j(A_n)) = \frac{1}{\mu_k(D)}\int_D F(f(\mathbf{x}))d\mathbf{x},$$

where $\lambda_j(A_n)$, $j = 1,\ldots,d_n$, are the eigenvalues of A_n, μ_k is the Lebesgue measure in $\mathbb{R}^k$, and $f : D \subset \mathbb{R}^k \to \mathbb{C}$ is a measurable function. In this situation, f is referred to as the spectral symbol of the sequence $\{A_n\}_n$. The theory of GLT sequences, in combination with the results of [26, 29] concerning the asymptotic spectral distribution of sequences of perturbed Hermitian matrices, is one of the most powerful tool for computing the spectral symbol f. Indeed, the sequence of discretization matrices $\{A_n\}_n$ turns out to be a GLT sequence for many classes of PDEs and numerical methods, especially when the numerical method belongs to the class of the so-called "local methods" (locality again!). Local methods are, for example, Finite Difference (FD) methods, Finite Element (FE) methods with "locally supported" basis functions, and collocation methods; in short, all standard numerical methods for the approximation of PDEs. We refer the reader to [39, 40] for applications of the theory of GLT sequences in the context of FD discretizations of PDEs; to [5, 20, 40, 43] for the FE and collocation settings; and to [16, 21, 22, 23] for recent applications to the case of B-spline Isogeometric Analysis (IgA) approximations of PDEs, both in the collocation and Galerkin frameworks.[1] We also refer the reader to [1, 37] for a look at the GLT approach to deal with sequences of matrices coming from the approximation of IEs.

It is worth emphasizing that the discretization matrices A_n arising from the numerical approximation of PDEs are often ill-conditioned for large n. In fact, their condition number diverges when $n \to \infty$. The knowledge of the spectral symbol f, which can be attained through the theory of GLT sequences, is not only interesting in itself, but could also be exploited in two different ways:

(a) to analyze/predict the convergence rate of iterative methods, such as preconditioned Krylov and multigrid methods, when they are applied to the ill-conditioned linear systems with coefficient matrix A_n;
(b) to design effective preconditioners and iterative solvers for these linear systems.

[1] IgA is a modern paradigm introduced in [2, 12, 32] for the numerical approximation of PDEs.

The reason is clear: the convergence properties of general purpose iterative methods depend on the spectral features of the matrix to which they are applied. Hence, the spectral information provided by f can be conveniently used for designing fast iterative solvers and/or analyzing their convergence properties. In this respect, we recall that recent estimates on the superlinear convergence of the Conjugate Gradient method are strictly related to the asymptotic spectral distribution of the matrices to which the method is applied; see [4]. We also refer the reader to [13, 14, 15] for recent developments in the IgA framework, where the spectral symbol was exploited to design ad hoc iterative solvers for the IgA discretization matrices.

Despite the numerous applications of the theory of GLT sequences, we should say at this point that this work is not primarily concerned with the applications. On the contrary, we focus on the mathematical foundations of the theory, as explained in Section 1.3. Nevertheless, in order to give an idea of the applicative interest, we will outline some applications at the end, in Section 6.

1.2. Summary of the theory of GLT sequences

Informally speaking, a GLT sequence $\{A_n\}_n$ is a sequence of matrices with increasing size, equipped with a Lebesgue-measurable (complex-valued) function κ. This function is referred to as the symbol (or kernel) of $\{A_n\}_n$ and it is defined over a domain D of the form $[0,1]^d \times [-\pi,\pi]^d$, $d \geq 1$. Due to the experience coming from the applications and to the analogy with the Fourier Analysis,[2] a point of $D = [0,1]^d \times [-\pi,\pi]^d$ is usually denoted by $(\mathbf{x}, \boldsymbol{\theta})$, where $\mathbf{x} = (x_1, \ldots, x_d)$ are the so-called "physical variables", while $\boldsymbol{\theta} = (\theta_1, \ldots, \theta_d)$ are the "Fourier variables".

The main properties of GLT sequences are summarized in **GLT 1**–**GLT 9**, and will be made more precise in Section 5.5. In the following, we write $\{A_n\}_n \sim_{\mathrm{GLT}} \kappa$ to indicate that $\{A_n\}_n$ is a GLT sequence with symbol κ.

GLT 1. The symbol $\kappa : [0,1]^d \times [-\pi,\pi]^d \to \mathbb{C}$ of a GLT sequence $\{A_n\}_n$ characterizes the asymptotic singular value distribution of $\{A_n\}_n$. This means that, for all continuous functions F with bounded support, we have

$$\lim_{n\to\infty} \frac{1}{d_n} \sum_{j=1}^{d_n} F(\sigma_j(A_n)) = \frac{1}{(2\pi)^d} \int_{[0,1]^d \times [-\pi,\pi]^d} F(|\kappa(\mathbf{x}, \boldsymbol{\theta})|) d\mathbf{x} d\boldsymbol{\theta}, \tag{1.3}$$

where d_n is the size of A_n and $\sigma_j(A_n)$, $j = 1, \ldots, d_n$, are the singular values of A_n. If moreover the matrices A_n are Hermitian, then the symbol κ also characterizes the asymptotic spectral distribution of $\{A_n\}_n$. This means that, for all continuous functions F with bounded support, we have

$$\lim_{n\to\infty} \frac{1}{d_n} \sum_{j=1}^{d_n} F(\lambda_j(A_n)) = \frac{1}{(2\pi)^d} \int_{[0,1]^d \times [-\pi,\pi]^d} F(\kappa(\mathbf{x}, \boldsymbol{\theta})) d\mathbf{x} d\boldsymbol{\theta}, \tag{1.4}$$

where $\lambda_j(A_n)$, $j = 1, \ldots, d_n$, are the eigenvalues of A_n.

[2]The theory of GLT sequences can be seen as a generalized Fourier Analysis [40].

GLT 2. Any sequence of (multilevel) Toeplitz matrices $\{T_n(f)\}_n$ generated by a function f in $L^1([-\pi,\pi]^d)$ is a GLT sequence with symbol $\kappa(\mathbf{x},\boldsymbol{\theta}) = f(\boldsymbol{\theta})$.

GLT 3. Any sequence of (multilevel) diagonal sampling matrices $\{D_n(a)\}_n$ containing the samples of a Riemann-integrable function $a : [0,1]^d \to \mathbb{C}$ over equispaced grids is a GLT sequence with symbol $\kappa(\mathbf{x},\boldsymbol{\theta}) = a(\mathbf{x})$.

GLT 4. Any zero-distributed sequence $\{Z_n\}_n$, i.e., any sequence of matrices possessing an asymptotic singular value distribution characterized by the identically zero function, in the sense of eq. (1.3), is a GLT sequence with symbol $\kappa(\mathbf{x},\boldsymbol{\theta}) = 0$.

GLT 5. If $\{A_n\}_n \sim_{\rm GLT} \kappa$ then $\{A_n^*\}_n \sim_{\rm GLT} \overline{\kappa}$, where A_n^* is the conjugate transpose of A_n.

GLT 6. If $A_n = \sum_{i=1}^r \alpha_i \prod_{j=1}^{q_i} A_n^{(i,j)}$, where $r, q_1,\ldots,q_r \in \mathbb{N}$, $\alpha_1,\ldots,\alpha_r \in \mathbb{C}$, and $\{A_n^{(i,j)}\}_n \sim_{\rm GLT} \kappa_{ij}$, then $\{A_n\}_n \sim_{\rm GLT} \kappa = \sum_{i=1}^r \alpha_i \prod_{j=1}^{q_i} \kappa_{ij}$.

GLT 7. If $\{A_n\}_n \sim_{\rm GLT} \kappa$ and $\kappa \neq 0$ almost everywhere, then $\{A_n^\dagger\}_n \sim_{\rm GLT} \kappa^{-1}$, where $A_n^\dagger$ is the (Moore–Penrose) pseudoinverse of A_n.

GLT 8. If $\{A_n\}_n \sim_{\rm GLT} \kappa$ and each A_n is Hermitian, then $\{f(A_n)\}_n \sim_{\rm GLT} f(\kappa)$ for all continuous functions $f : \mathbb{R} \to \mathbb{R}$.

GLT 9. $\{A_n\}_n \sim_{\rm GLT} \kappa$ if and only if there exist GLT sequences $\{B_{n,m}\}_n \sim_{\rm GLT} \kappa_m$ such that κ_m converges to κ in measure and $\{B_{n,m}\}_n$ "converges" to $\{A_n\}_n$.

At this stage, we cannot be more precise about the "convergence" of $\{B_{n,m}\}_n$ to $\{A_n\}_n$. We only anticipate that, by saying «$\{B_{n,m}\}_n$ "converges" to $\{A_n\}_n$», we mean that «$\{\{B_{n,m}\}_n\}_m$ is an approximating class of sequences for $\{A_n\}_n$ (as $m \to \infty$)». Things will become more clear in Section 3, where the notion of approximating classes of sequences is introduced. We note that **GLT 5**–**GLT 7** can be summarized by saying that the set of GLT sequences is a *-algebra. Roughly speaking, we may rephrase them as follows: suppose that $\{A_n^{(1)}\}_n,\ldots,\{A_n^{(r)}\}_n$ are GLT sequences with symbols $\kappa_1,\ldots,\kappa_r$, and let $A_n = \text{ops}(A_n^{(1)},\ldots,A_n^{(r)})$ be a matrix obtained from $A_n^{(1)},\ldots,A_n^{(r)}$ by means of certain algebraic operations "ops", such as linear combinations, products, (pseudo)inversions and conjugate transpositions; then, $\{A_n\}_n$ is a GLT sequence with symbol $\kappa = \text{ops}(\kappa_1,\ldots,\kappa_r)$.

1.3. Contributions of this work

In this work, we review and extend the theory of GLT sequences. We first propose a modification of the original definition appeared in [39, Definition 2.3] and [40, Definition 1.5]. The new definition of GLT sequences is based on the revised notions of Locally Toeplitz (LT) and separable Locally Toeplitz (sLT) sequences, which have been recently introduced and studied in [24]. Using the new definition and the work carried out in [24], we are able to enlarge the applicability of the theory, by generalizing and/or simplifying a lot of key results. In particular, we remove the Riemann-integrability assumption of the so-called "weight functions" from the main spectral distribution and algebraic results, i.e., [39, Theorems 4.5 and 4.8] and [40, Theorem 2.2]. This implies that **GLT 1** and **GLT 6**–**GLT 7** can be stated as in Section 1.2, without any Riemann-integrability hypothesis. As a final step, we extend the theory of GLT sequences. This is the completely new part of the paper. We first prove an approximation result in Section 5.2, which is essentially the content of **GLT 9** and is particularly useful to show that a given sequence of matrices is a GLT sequence. By using this result, we provide in Section 5.4 a new and easier proof of **GLT 7** (the original proof appeared in [40, Theorem 2.2]). Finally, using again the approximation result in **GLT 9**, we prove **GLT 8**.

This paper completes the work started by the authors in [24], where the theory of LT sequences was presented. As we shall see herein, LT sequences are specific examples of GLT sequences. However, their role in developing the theory of GLT sequences is so fundamental that they deserve a special attention. Note that the GLT sequences in **GLT 2**–**GLT 4** are actually LT sequences [24, Section 4.3].

The paper is organized as follows. In Section 2 we collect all the necessary preliminaries. Section 3 focuses on the notion of approximating classes of sequences. In Section 4 we report the main results obtained in [24] about LT sequences. In Section 5 we develop the theory of GLT sequences. In Section 6 we present some applications. Section 7 concludes the work and outlines future lines of research.

2. Mathematical background

2.1. Notation and terminology

- O_m and I_m are the $m \times m$ zero matrix and the $m \times m$ identity matrix.
- Given a $m \times m$ matrix $X \in \mathbb{C}^{m\times m}$, the eigenvalues and the singular values of X are denoted by $\lambda_j(X)$, $j = 1, \dots, m$, and $\sigma_j(X)$, $j = 1, \dots, m$.
- If $1 \le p \le \infty$, the symbol $|\cdot|_p$ denotes the p-norm of both vectors and matrices. The 2-norm $|\cdot|_2$ is the spectral (Euclidean) norm and will be denoted by $\|\cdot\|$.
- If $1 \le p \le \infty$ and $X \in \mathbb{C}^{m\times m}$, $\|X\|_p$ is the Schatten p-norm of X, i.e., the p-norm of the vector $(\sigma_1(X), \dots, \sigma_m(X))$; see [7]. The norm $\|\cdot\|_1$ is also known as the trace-norm.
- $\Re(X)$ is the real part of the (square) matrix X, i.e., $\Re(X) = \frac{X+X^*}{2}$.
- Let $\mathbb{K}$ be either $\mathbb{R}$ or $\mathbb{C}$. We denote by $C_c(\mathbb{K})$ the space of complex-valued continuous functions with bounded support defined on $\mathbb{K}$.
- A "functional" ϕ is any complex-valued function defined on some vector space.
- If $w_i : D_i \to \mathbb{C}$, $i = 1, \dots, d$, we denote by $w_1 \otimes \cdots \otimes w_d : D_1 \times \cdots \times D_d \to \mathbb{C}$ the tensor-product function

$$(w_1 \otimes \cdots \otimes w_d)(\xi_1, \dots, \xi_d) = w_1(\xi_1) \cdots w_d(\xi_d), \qquad (\xi_1, \dots, \xi_d) \in D_1 \times \cdots \times D_d.$$

- We say that $a : [0,1]^d \to \mathbb{C}$ is Riemann-integrable if its real and imaginary parts $\Re(a), \Im(a) : [0,1]^d \to \mathbb{R}$ are Riemann-integrable in the classical sense.
- If $g : D \to \mathbb{C}$, we set $\|g\|_\infty = \sup_{\xi\in D} |g(\xi)|$.
- μ_k is the Lebesgue measure in $\mathbb{R}^k$. Throughout this paper, all the terminology from measure theory (such as "measure", "measurable", "a.e.", "in L^p", etc.) is always referred to the Lebesgue measure.

- χ_E is the characteristic (indicator) function of the set E, so $\chi_E(\xi) = 1$ if $\xi \in E$, and $\chi_E(\xi) = 0$ otherwise.
- We use a notation borrowed from probability theory to indicate sets. For example, if $f, g : D \subseteq \mathbb{R}^k \to \mathbb{C}$, then $\{f \le 1,\ g > 2\} = \{\mathbf{x} \in D :\ f(\mathbf{x}) \le 1,\ g(\mathbf{x}) > 2\}$, $\mu_k\{f > 0\}$ is the measure of $\{\mathbf{x} \in D :\ f(\mathbf{x}) > 0\}$, $\chi_{\{f=0\}}$ is the characteristic function of the set where f vanishes, and so on.
- If $E_1, \dots, E_d \subseteq \mathbb{R}$ are measurable sets and $f : E_1 \times \cdots \times E_d \to \mathbb{C}$, we say that f is separable if there exist measurable functions $f_i : E_i \to \mathbb{C}$, $i = 1, \dots, d$, such that $f = f_1 \otimes \cdots \otimes f_d$.
- A d-variate trigonometric polynomial is a finite linear combination of the Fourier frequencies $\mathrm{e}^{\mathrm{i}\boldsymbol{j}\cdot\boldsymbol{\theta}}$, $\boldsymbol{j} \in \mathbb{Z}^d$. Here, $\boldsymbol{\theta} = (\theta_1, \dots, \theta_d) \in \mathbb{R}^d$ and $\boldsymbol{j}\cdot\boldsymbol{\theta} = j_1\theta_1 + \ldots + j_d\theta_d$.

2.1.1. Multi-index notation. A multi-index $\boldsymbol{i} \in \mathbb{Z}^d$, also called a d-index, is just a vector in $\mathbb{Z}^d$; its components are denoted by $i_1, \dots, i_d$. $\mathbf{0}$ and $\mathbf{1}$ are the vectors of all zeros and all ones (their size will be clear from the context). For any d-index $\boldsymbol{m}$, we set $N(\boldsymbol{m}) = \prod_{j=1}^d m_j$ and we write $\boldsymbol{m} \to \infty$ to indicate that $\min(\boldsymbol{m}) \to \infty$. If $\boldsymbol{h}, \boldsymbol{k}$ are d-indices, $\boldsymbol{h} \le \boldsymbol{k}$ means that $h_r \le k_r$ for all $r = 1, \dots, d$. In this case, we define the multi-index range $\boldsymbol{h}, \dots, \boldsymbol{k}$ as the set $\{\boldsymbol{j} \in \mathbb{Z}^d :\ \boldsymbol{h} \le \boldsymbol{j} \le \boldsymbol{k}\}$. We assume for $\boldsymbol{h}, \dots, \boldsymbol{k}$ the lexicographic ordering:

$$\Big[\cdots \Big[\, [\, (j_1, \dots, j_d)\,]_{j_d = h_d, \dots, k_d} \Big]_{j_{d-1} = h_{d-1}, \dots, k_{d-1}} \cdots \Big]_{j_1 = h_1, \dots, k_1}. \tag{2.1}$$

For instance, if $d = 2$ the ordering (2.1) is

$$(h_1, h_2),\ (h_1, h_2 + 1),\ \dots,\ (h_1, k_2),\ (h_1 + 1, h_2),\ (h_1 + 1, h_2 + 1),\ \dots,\ (h_1 + 1, k_2),$$
$$\dots\dots\dots,\ (k_1, h_2),\ (k_1, h_2 + 1),\ \dots,\ (k_1, k_2).$$

When a multi-index $\boldsymbol{j}$ varies over $\boldsymbol{h}, \dots, \boldsymbol{k}$ (this is often written as $\boldsymbol{j} = \boldsymbol{h}, \dots, \boldsymbol{k}$), it is understood that $\boldsymbol{j}$ varies from $\boldsymbol{h}$ to $\boldsymbol{k}$ following the ordering (2.1). For instance, if $\boldsymbol{m} \in \mathbb{N}^d$ and $\mathbf{x} = [x_{\boldsymbol{i}}]_{\boldsymbol{i}=\mathbf{1}}^{\boldsymbol{m}}$, then $\mathbf{x}$ is a vector of size $N(\boldsymbol{m})$ whose components $x_{\boldsymbol{i}}$, $\boldsymbol{i} = \mathbf{1}, \dots, \boldsymbol{m}$, are ordered in accordance with (2.1). Similarly, if $X = [x_{\boldsymbol{i}\boldsymbol{j}}]_{\boldsymbol{i},\boldsymbol{j}=\mathbf{1}}^{\boldsymbol{m}}$, then X is a $N(\boldsymbol{m}) \times N(\boldsymbol{m})$ matrix whose components are indexed by two d-indices $\boldsymbol{i}, \boldsymbol{j}$, both varying from $\mathbf{1}$ to $\boldsymbol{m}$ according to (2.1). If $\boldsymbol{h} \le \boldsymbol{k}$, $\sum_{\boldsymbol{j}=\boldsymbol{h}}^{\boldsymbol{k}}$ denotes the summation over all $\boldsymbol{j} = \boldsymbol{h}, \dots, \boldsymbol{k}$. Operations involving multi-indices that have no meaning in the vector space $\mathbb{Z}^d$ must always be interpreted in the componentwise sense. For instance, $\boldsymbol{i}\boldsymbol{j} = (i_1 j_1, \dots, i_d j_d)$, $\boldsymbol{i}/\boldsymbol{j} = (i_1/j_1, \dots, i_d/j_d)$, etc.

2.1.2. Multilevel diagonal sampling matrices. For $\boldsymbol{n} \in \mathbb{N}^d$ and $a : [0,1]^d \to \mathbb{C}$, we define the d-level diagonal sampling matrix $D_{\boldsymbol{n}}(a)$ as the following diagonal matrix of size $N(\boldsymbol{n})$:

$$D_{\boldsymbol{n}}(a) = \operatorname*{diag}_{\boldsymbol{i}=\mathbf{1},\dots,\boldsymbol{n}} a\Big(\frac{\boldsymbol{i}}{\boldsymbol{n}}\Big),$$

where $\boldsymbol{i}$ varies from $\mathbf{1}$ to $\boldsymbol{n}$ according to the lexicographic ordering (2.1).

2.1.3. Matrix-sequences. In all this paper, by a sequence of matrices (or matrix-sequence) we mean a sequence of the form $\{A_{\boldsymbol{n}}\}_n$, where:

- n varies in some infinite subset of $\mathbb{N}$;
- $\boldsymbol{n} = \boldsymbol{n}(n)$ is a d-index in $\mathbb{N}^d$ which depends on n, and $\boldsymbol{n} \to \infty$ as $n \to \infty$;
- $A_{\boldsymbol{n}}$ is a square matrix of size $N(\boldsymbol{n})$.

Recall that $\boldsymbol{n} \to \infty$ means $\min(\boldsymbol{n}) \to \infty$. The multi-index that parameterizes a matrix-sequence is always assumed to be a d-index.

2.2. Preliminaries on linear algebra and matrix analysis

Let $X \in \mathbb{C}^{m\times m}$. We know from the Singular Value Decomposition (SVD) that $\mathrm{rank}(X)$ is the number of nonzero singular values of X and $\|X\| = \sigma_{\max}(X)$. Hence,

$$\|X\|_1 = \sum_{i=1}^{m} \sigma_i(X) \le \mathrm{rank}(X)\|X\| \le m\|X\|, \qquad X \in \mathbb{C}^{m\times m}.$$

Another useful trace-norm inequality is the following [24, p. 6]:

$$\|X\|_1 \le \sum_{i,j=1}^{m} |x_{ij}|, \qquad X \in \mathbb{C}^{m\times m}. \tag{2.2}$$

We also recall the classical spectral norm inequality in [30, Corollary 2.3.2]:

$$\|X\| \le \sqrt{|X|_1 |X|_\infty}, \qquad X \in \mathbb{C}^{m\times m}. \tag{2.3}$$

This inequality is particularly useful to estimate the spectral norm of a matrix when we have upper bounds for its components. Indeed, we recall that $|X|_1$ (resp., $|X|_\infty$) is the maximum among the 1-norms of the columns (resp., rows) of X.

2.3. Preliminaries on measure theory

Let $\mathbb{K}$ be either $\mathbb{R}$ or $\mathbb{C}$ and let $g : D \subset \mathbb{R}^k \to \mathbb{K}$ be a measurable function defined on a set D with $0 < \mu_k(D) < \infty$. Consider the functional

$$\phi_{[g]} : C_c(\mathbb{K}) \to \mathbb{C}, \qquad \phi_{[g]}(F) = \frac{1}{\mu_k(D)} \int_D F(g(\mathbf{x}))d\mathbf{x}. \tag{2.4}$$

$\phi_{[g]}$ is a continuous linear functional on the normed vector space $(C_c(\mathbb{K}), \|\cdot\|_\infty)$, and $\|\phi_{[g]}\| \le 1$. Indeed, the linearity is obvious and the continuity, as well as the bound $\|\phi_{[g]}\| \le 1$, follows from the observation that $|\phi_{[g]}(F)| \le \|F\|_\infty$ for all $F \in C_c(\mathbb{K})$. If g is constant, say $g = \gamma$ a.e., then $\phi_{[g]} = \phi_{[\gamma]}$ is the evaluation functional at γ.

In the following lemmas we collect some well-known results from measure theory that will be used later on. The proofs of the lemmas can be found, e.g., in [25, Section 2.3].

Lemma 2.1. *Let $\mathbb{K}$ be either $\mathbb{R}$ or $\mathbb{C}$, and let $g_m, g : D \subset \mathbb{R}^k \to \mathbb{K}$ be measurable functions defined on a set D with $0 < \mu_k(D) < \infty$. If $g_m \to g$ in measure, then $F \circ g_m \to F \circ g$ in $L^1(D)$ for all $F \in C_c(\mathbb{K})$ and $\phi_{[g_m]} \to \phi_{[g]}$ pointwise over $C_c(\mathbb{K})$.*

Lemma 2.2. *Let $\mathbb{K}$ be either $\mathbb{R}$ or $\mathbb{C}$, and let $g_m : D \subset \mathbb{R}^k \to \mathbb{K}$ be measurable functions defined on a set D with $0 < \mu_k(D) < \infty$. If $\phi_{[g_m]} \to \phi_{[0]}$ pointwise over $C_c(\mathbb{K})$, then $g_m \to 0$ in measure.*

Remark 2.3. Let $\phi_{[g]}$ be defined as in (2.4) and assume that $\phi_{[g]} = \phi_{[0]}$; then $g = 0$ a.e. Indeed, if $\phi_{[g]} = \phi_{[0]}$, the constant sequence $\{\phi_{[g]}\}_m$ converges pointwise to $\phi_{[0]}$ over $C_c(\mathbb{K})$. By Lemma 2.2, this implies that $g \to 0$ in measure, i.e., $\mu_k\{|g| \geq \epsilon\} = 0$ for every $\epsilon > 0$. Hence, $g = 0$ a.e.

Lemma 2.4. *Let $\kappa : [0,1]^d \times [-\pi,\pi]^d \to \mathbb{C}$ be a measurable function. Then, there exists a sequence of functions $\kappa_m : [0,1]^d \times [-\pi,\pi]^d \to \mathbb{C}$ such that $\kappa_m \to \kappa$ a.e. and κ_m is of the form*

$$\kappa_m(\mathbf{x}, \boldsymbol{\theta}) = \sum_{\boldsymbol{j}=-\boldsymbol{N}_m}^{\boldsymbol{N}_m} a_{\boldsymbol{j}}^{(m)}(\mathbf{x})\, \mathrm{e}^{\mathrm{i}\boldsymbol{j}\cdot\boldsymbol{\theta}}, \qquad a_{\boldsymbol{j}}^{(m)} \in C^\infty([0,1]^d), \qquad \boldsymbol{N}_m \in \mathbb{N}^d. \quad (2.5)$$

2.4. Singular value and eigenvalue distribution of a matrix-sequence

Definition 2.5. Let $\{A_{\boldsymbol{n}}\}_n$ be a matrix-sequence.

- We say that $\{A_{\boldsymbol{n}}\}_n$ has an asymptotic singular value distribution described by a functional $\phi : C_c(\mathbb{R}) \to \mathbb{C}$, and we write $\{A_{\boldsymbol{n}}\}_n \sim_\sigma \phi$, if, for all $F \in C_c(\mathbb{R})$,

$$\lim_{n\to\infty} \frac{1}{N(\boldsymbol{n})} \sum_{j=1}^{N(\boldsymbol{n})} F(\sigma_j(A_{\boldsymbol{n}})) = \phi(F).$$

  In the case where $\phi = \phi_{[|f|]}$ for a measurable $f : D \subset \mathbb{R}^k \to \mathbb{C}$ defined on a set D with $0 < \mu_k(D) < \infty$, the function f is called the singular value symbol of $\{A_{\boldsymbol{n}}\}_n$, and we write $\{A_{\boldsymbol{n}}\}_n \sim_\sigma f$ as an alternative to $\{A_{\boldsymbol{n}}\}_n \sim_\sigma \phi_{[|f|]}$.
- We say that $\{A_{\boldsymbol{n}}\}_n$ has an asymptotic eigenvalue (or spectral) distribution described by a functional $\phi : C_c(\mathbb{C}) \to \mathbb{C}$, and we write $\{A_{\boldsymbol{n}}\}_n \sim_\lambda \phi$, if, for all $F \in C_c(\mathbb{C})$,

$$\lim_{n\to\infty} \frac{1}{N(\boldsymbol{n})} \sum_{j=1}^{N(\boldsymbol{n})} F(\lambda_j(A_{\boldsymbol{n}})) = \phi(F).$$

  In the case where $\phi = \phi_{[f]}$ for a function $f : D \subset \mathbb{R}^k \to \mathbb{C}$ defined on a set D with $0 < \mu_k(D) < \infty$, the function f is called the eigenvalue (or spectral) symbol of $\{A_{\boldsymbol{n}}\}_n$, and we write $\{A_{\boldsymbol{n}}\}_n \sim_\lambda f$ as an alternative to $\{A_{\boldsymbol{n}}\}_n \sim_\lambda \phi_{[f]}$.

When we write a relation such as $\{A_{\boldsymbol{n}}\}_n \sim_\sigma \phi$ (resp., $\{A_{\boldsymbol{n}}\}_n \sim_\lambda \phi$), it is understood that ϕ is a functional on $C_c(\mathbb{R})$ (resp., $C_c(\mathbb{C})$), as in Definition 2.5.

Similarly, when we write $\{A_{\boldsymbol{n}}\}_n \sim_\sigma f$ or $\{A_{\boldsymbol{n}}\}_n \sim_\lambda f$, it is understood that f is as in Definition 2.5; that is, f is a measurable function defined on a subset D of some $\mathbb{R}^k$ with $0 < \mu_k(D) < \infty$. We will use the notation $\{A_{\boldsymbol{n}}\}_n \sim_{\sigma,\lambda} f$ to indicate that $\{A_{\boldsymbol{n}}\}_n \sim_\sigma f$ and $\{A_{\boldsymbol{n}}\}_n \sim_\lambda f$.

By (2.4), the relation $\{A_{\boldsymbol{n}}\}_n \sim_\sigma f$ means that, for all $F \in C_c(\mathbb{R})$,

$$\lim_{n\to\infty} \frac{1}{N(\boldsymbol{n})} \sum_{j=1}^{N(\boldsymbol{n})} F(\sigma_j(A_{\boldsymbol{n}})) = \frac{1}{\mu_k(D)} \int_D F(|f(\mathbf{x})|)d\mathbf{x}.$$

Similarly, the relation $\{A_{\boldsymbol{n}}\}_n \sim_\lambda f$ means that, for all $F \in C_c(\mathbb{C})$,

$$\lim_{n\to\infty} \frac{1}{N(\boldsymbol{n})} \sum_{j=1}^{N(\boldsymbol{n})} F(\lambda_j(A_{\boldsymbol{n}})) = \frac{1}{\mu_k(D)} \int_D F(f(\mathbf{x}))d\mathbf{x}.$$

2.4.1. Zero-distributed sequences. A class of matrix-sequences that plays a central role in the theory of GLT sequences is the class of zero-distributed sequences. A zero-distributed sequence is a matrix-sequence $\{Z_{\boldsymbol{n}}\}_n$ such that $\{Z_{\boldsymbol{n}}\}_n \sim_\sigma 0$.

Theorem 2.6. *Let $\{Z_{\boldsymbol{n}}\}_n$ be a matrix-sequence. Then, the following properties hold.*

- $\{Z_{\boldsymbol{n}}\}_n \sim_\sigma 0$ *if and only if* $Z_{\boldsymbol{n}} = R_{\boldsymbol{n}} + N_{\boldsymbol{n}}$ *with* $\lim_{n\to\infty} \frac{\mathrm{rank}(R_{\boldsymbol{n}})}{N(\boldsymbol{n})} = \lim_{n\to\infty} \|N_{\boldsymbol{n}}\| = 0.$
- *If* $\lim_{n\to\infty} \frac{\|Z_{\boldsymbol{n}}\|_p}{N(\boldsymbol{n})^{1/p}} = 0$ *for some* $p \in [1,\infty]$, *then* $\{Z_{\boldsymbol{n}}\}_n \sim_\sigma 0$.

Note that in the statement of the second property we used the convention $1/\infty = 0$. For the proof of Theorem 2.6, see [25, Theorems 2.5–2.6].

2.5. Multilevel Toeplitz matrices

Given $\boldsymbol{n} \in \mathbb{N}^d$, a matrix of the form $[a_{\boldsymbol{i}-\boldsymbol{j}}]_{\boldsymbol{i},\boldsymbol{j}=\boldsymbol{1}}^{\boldsymbol{n}} \in \mathbb{C}^{N(\boldsymbol{n})\times N(\boldsymbol{n})}$, whose $(\boldsymbol{i},\boldsymbol{j})$ entry depends only on the difference between the d-indices $\boldsymbol{i}$ and $\boldsymbol{j}$, is called a multilevel Toeplitz matrix (or, more precisely, a d-level Toeplitz matrix). If $f : [-\pi,\pi]^d \to \mathbb{C}$ is a function in $L^1([-\pi,\pi]^d)$, we denote its Fourier coefficients by

$$f_{\boldsymbol{k}} = \frac{1}{(2\pi)^d} \int_{[-\pi,\pi]^d} f(\boldsymbol{\theta})\mathrm{e}^{-\mathrm{i}\boldsymbol{k}\cdot\boldsymbol{\theta}} d\boldsymbol{\theta}, \qquad \boldsymbol{k} \in \mathbb{Z}^d. \tag{2.6}$$

The $\boldsymbol{n}$th multilevel Toeplitz matrix associated with f is defined as

$$T_{\boldsymbol{n}}(f) = [f_{\boldsymbol{i}-\boldsymbol{j}}]_{\boldsymbol{i},\boldsymbol{j}=\boldsymbol{1}}^{\boldsymbol{n}}. \tag{2.7}$$

We call $\{T_{\boldsymbol{n}}(f)\}_{\boldsymbol{n}\in\mathbb{N}^d}$ the family of multilevel Toeplitz matrices associated with f, which, in turn, is called the generating function of $\{T_{\boldsymbol{n}}(f)\}_{\boldsymbol{n}\in\mathbb{N}^d}$. For each fixed $\boldsymbol{n} \in \mathbb{N}^d$, the application $T_{\boldsymbol{n}}(\cdot) : L^1([-\pi,\pi]^d) \to \mathbb{C}^{N(\boldsymbol{n})\times N(\boldsymbol{n})}$ is linear:

$$T_{\boldsymbol{n}}(\alpha f + \beta g) = \alpha T_{\boldsymbol{n}}(f) + \beta T_{\boldsymbol{n}}(g), \qquad \alpha,\beta \in \mathbb{C}, \qquad f,g \in L^1([-\pi,\pi]^d).$$

This follows from the relation $(\alpha f + \beta g)_{\boldsymbol{k}} = \alpha f_{\boldsymbol{k}} + \beta g_{\boldsymbol{k}}$, $\boldsymbol{k} \in \mathbb{Z}^d$, which is a consequence of the linearity of the integral in (2.6). Another nice property of the Toeplitz operator $T_{\boldsymbol{n}}(\cdot)$ is that $T_{\boldsymbol{n}}(1) = I_{N(\boldsymbol{n})}$. For every $f \in L^1([-\pi,\pi]^d)$ and

$\boldsymbol{n} \in \mathbb{N}^d$, it can be shown that $T_{\boldsymbol{n}}(f)^* = T_{\boldsymbol{n}}(\overline{f})$. Hence, if f is real (a.e.), then all the matrices $T_{\boldsymbol{n}}(f)$ are Hermitian.

Theorem 2.7 is the multilevel version of Szegő's first limit theorem and of the Avram–Parter theorem, and it is a fundamental result concerning multilevel Toeplitz matrices. For the eigenvalues it goes back to Szegő [31], and for the singular values it was established by Avram [3] and Parter [34]. They assumed that $d = 1$ and $f \in L^\infty([-\pi, \pi])$; see [10, Sections 5 and 6] and also [11, Section 10.14] for more on the subject in the case of L^∞ generating functions. The extension to any $d \geq 1$ and $f \in L^1([-\pi, \pi]^d)$ was performed by Tyrtyshnikov and Zamarashkin [48, 49, 50] and Tilli [45]. We also refer the reader to [28] for a proof of Theorem 2.7 based on the notion of approximating classes of sequences (see Section 3); the proof in [28] is made in the case of eigenvalues for $d = 1$, but the argument is general and can be extended to singular values and to higher dimensionalities.

Theorem 2.7. *If $f \in L^1([-\pi, \pi]^d)$, then $\{T_{\boldsymbol{n}}(f)\}_n \sim_\sigma f$. If moreover f is real a.e., then $\{T_{\boldsymbol{n}}(f)\}_n \sim_\lambda f$.*

We stress that, in Theorem 2.7, $\{T_{\boldsymbol{n}}(f)\}_n$ is any matrix-sequence extracted from the family $\{T_{\boldsymbol{n}}(f)\}_{\boldsymbol{n} \in \mathbb{N}^d}$.

3. Approximating classes of sequences

In this section, we introduce the fundamental definition on which the theory of GLT sequences is based: the notion of approximating classes of sequences, originally appeared in [38].

Definition 3.1 (a.c.s.). Let $\{A_{\boldsymbol{n}}\}_n$ be a matrix-sequence. An approximating class of sequences (a.c.s.) for $\{A_{\boldsymbol{n}}\}_n$ is a sequence of matrix-sequences $\{\{B_{\boldsymbol{n},m}\}_n\}_m$ with the following property: for every m there exists n_m such that, for $n \geq n_m$,

$$\begin{aligned} A_{\boldsymbol{n}} &= B_{\boldsymbol{n},m} + R_{\boldsymbol{n},m} + N_{\boldsymbol{n},m}, \\ \operatorname{rank}(R_{\boldsymbol{n},m}) &\leq c(m) N(\boldsymbol{n}), \qquad \|N_{\boldsymbol{n},m}\| \leq \omega(m), \end{aligned} \tag{3.1}$$

where n_m, $c(m)$, $\omega(m)$ depend only on m, and $\lim_{m\to\infty} c(m) = \lim_{m\to\infty} \omega(m) = 0$.

Roughly speaking, $\{\{B_{\boldsymbol{n},m}\}_n\}_m$ is an a.c.s. for $\{A_{\boldsymbol{n}}\}_n$ if $A_{\boldsymbol{n}}$ equals $B_{\boldsymbol{n},m}$ up to a small-norm matrix plus a small-rank matrix (compared to the matrix size $N(\boldsymbol{n})$). We will use "a.c.s.es" as an abbreviation of "approximating classes of sequences".

3.1. The a.c.s. machinery as a tool for computing singular value and eigenvalue distributions

The importance of a.c.s.es resides in Theorems 3.2–3.3, which appeared in [28, 38]. These theorems provide general tools for computing the singular value (resp., eigenvalue) distribution of a "difficult" matrix-sequence $\{A_{\boldsymbol{n}}\}_n$ from the singular value (resp., eigenvalue) distributions of "simpler" matrix-sequences $\{B_{\boldsymbol{n},m}\}_n$ that "approximate" $\{A_{\boldsymbol{n}}\}_n$ in the sense of Definition 3.1. For the proof of Theorems 3.2–3.3, see [25, Section 3.1].

Theorem 3.2. *Let $\{A_{\boldsymbol{n}}\}_n$ be a matrix-sequence and let ϕ be a functional on $C_c(\mathbb{R})$. Assume that:*

1. *$\{\{B_{\boldsymbol{n},m}\}_n\}_m$ is an a.c.s. for $\{A_{\boldsymbol{n}}\}_n$;*
2. *$\{B_{\boldsymbol{n},m}\}_n \sim_\sigma \phi_m$ for every m;*
3. *$\phi_m \to \phi$ pointwise over $C_c(\mathbb{R})$.*

Then $\{A_{\boldsymbol{n}}\}_n \sim_\sigma \phi$.

Theorem 3.3. *Let $\{A_{\boldsymbol{n}}\}_n$ be a sequence of Hermitian matrices and let ϕ be a functional on $C_c(\mathbb{C})$. Assume that:*

1. *$\{\{B_{\boldsymbol{n},m}\}_n\}_m$ is an a.c.s. for $\{A_{\boldsymbol{n}}\}_n$ formed by Hermitian matrices;*
2. *$\{B_{\boldsymbol{n},m}\}_n \sim_\lambda \phi_m$ for every m;*
3. *$\phi_m \to \phi$ pointwise over $C_c(\mathbb{C})$.*

Then $\{A_{\boldsymbol{n}}\}_n \sim_\lambda \phi$.

We provide below two important corollaries of Theorems 3.2–3.3.

Corollary 3.4. *Let $\{A_{\boldsymbol{n}}\}_n$ be a matrix-sequence. Assume that:*

1. *$\{\{B_{\boldsymbol{n},m}\}_n\}_m$ is an a.c.s. for $\{A_{\boldsymbol{n}}\}_n$;*
2. *for every m, $\{B_{\boldsymbol{n},m}\}_n \sim_\sigma f_m$ for some measurable $f_m : D \subset \mathbb{R}^k \to \mathbb{C}$;*
3. *$|f_m| \to |f|$ in measure over D, being $f : D \to \mathbb{C}$ another measurable function.*

Then $\{A_{\boldsymbol{n}}\}_n \sim_\sigma f$.

Proof. Apply Theorem 3.2 with $\phi_m = \phi_{[|f_m|]}$ and $\phi = \phi_{[|f|]}$. Note that $\phi_{[|f_m|]} \to \phi_{[|f|]}$ pointwise over $C_c(\mathbb{R})$ by Lemma 2.1. $\square$

Corollary 3.5. *Let $\{A_{\boldsymbol{n}}\}_n$ be a sequence of Hermitian matrices. Assume that:*

1. *$\{\{B_{\boldsymbol{n},m}\}_n\}_m$ is an a.c.s. for $\{A_{\boldsymbol{n}}\}_n$ formed by Hermitian matrices;*
2. *for every m, $\{B_{\boldsymbol{n},m}\}_n \sim_\lambda f_m$ for some measurable $f_m : D \subset \mathbb{R}^k \to \mathbb{C}$;*
3. *$f_m \to f$ in measure over D, being $f : D \to \mathbb{C}$ another measurable function.*

Then $\{A_{\boldsymbol{n}}\}_n \sim_\lambda f$.

Proof. Apply Theorem 3.3 with $\phi_m = \phi_{[f_m]}$ and $\phi = \phi_{[f]}$. Note that $\phi_{[f_m]} \to \phi_{[f]}$ pointwise over $C_c(\mathbb{C})$ by Lemma 2.1. $\square$

Theorem 3.2 admits an interesting converse, which we report for future use. For the proof, see [25, Theorem 3.2].

Theorem 3.6. *Let $\{A_{\boldsymbol{n}}\}_n$ be a matrix-sequence. Assume that:*

1. *$\{A_{\boldsymbol{n}}\}_n \sim_\sigma \phi$;*
2. *$\{\{B_{\boldsymbol{n},m}\}_n\}_m$ is an a.c.s. for $\{A_{\boldsymbol{n}}\}_n$;*
3. *$\{B_{\boldsymbol{n},m}\}_n \sim_\sigma \phi_m$ for every m;*

Then $\phi_m \to \phi$ pointwise over $C_c(\mathbb{R})$.

Remark 3.7. Let $\{A_{\boldsymbol{n}}\}_n$, $\{B_{\boldsymbol{n},m}\}_n$ be matrix-sequences and let ϕ, $\phi_m : C_c(\mathbb{R}) \to \mathbb{C}$ be functionals. Consider the following conditions:

(1) $\{A_{\boldsymbol{n}}\}_n \sim_\sigma \phi$;
(2) $\{B_{\boldsymbol{n},m}\}_n \sim_\sigma \phi_m$ for every m;
(3) $\{\{B_{\boldsymbol{n},m}\}_n\}_m$ is an a.c.s. for $\{A_{\boldsymbol{n}}\}_n$;
(4) $\phi_m \to \phi$ pointwise over $C_c(\mathbb{R})$.

Theorems 3.2 and 3.6 show that

$$(1) \wedge (2) \wedge (3) \Rightarrow (4), \qquad (2) \wedge (3) \wedge (4) \Rightarrow (1).$$

The implication $(1) \wedge (2) \wedge (4) \Rightarrow (3)$ is false. As a counterexample, take $A_n = I_n$ and $B_{n,m} = \mathrm{diag}_{i=1,\ldots,n}(-1)^i$. Then (1)–(2) and (4) are satisfied with $\phi = \phi_m = \phi_{[1]}$, $\phi_{[1]}(F) = F(1)$. However, (3) cannot be satisfied, because otherwise $\{\{A_n - B_{n,m}\}_n\}_m$ would be an a.c.s. of $\{O_n\}_n$ and so, since $\{O_n\}_n \sim_\sigma \phi_{[0]}(F) = F(0)$ and $\{A_n - B_{n,m}\}_n \sim_\sigma \varphi(F) = \frac{1}{2}\big(F(0) + F(2)\big)$, Theorem 3.6 would imply that $\varphi \to \phi_{[0]}$ pointwise.

The implication $(1) \wedge (3) \wedge (4) \Rightarrow (2)$, written in this way, is meaningless. However, a natural modification reads as follows: $(1) \wedge (3) \Rightarrow$ there exists a functional ϕ_m such that $\phi_m \to \phi$ pointwise over $C_c(\mathbb{R})$ and $\{B_{\boldsymbol{n},m}\}_n \sim_\sigma \phi_m$ for all sufficiently large m. This statement is false in general. As a counterexample, take $A_n = O_n$ and $B_{n,m} = (1 + (-1)^n)\frac{1}{m} I_n$. Then, $\|B_{n,m}\| \le \frac{2}{m}$, so $\{\{B_{n,m}\}_n\}_m$ is an a.c.s. of $\{O_n\}_n$. However, the limit

$$\lim_{n\to\infty} \frac{1}{n} \sum_{j=1}^{n} F(\sigma_j(B_{n,m})) = \lim_{n\to\infty} F((1 + (-1)^n)\tfrac{1}{m})$$

does not exist for any $F \in C_c(\mathbb{R})$ such that $F(0) \neq F(\frac{2}{m})$. Therefore, the relation $\{B_{n,m}\}_n \sim_\sigma \phi_m$ cannot hold for any functional ϕ_m.

3.2. The a.c.s. algebra

In this section, we report the algebraic properties possessed by the a.c.s.es. The properties in Proposition 3.8 are direct consequences of Definition 3.1.

Proposition 3.8. *Let $\{\{B_{\boldsymbol{n},m}\}_n\}_m$, $\{\{B'_{\boldsymbol{n},m}\}_n\}_m$ be a.c.s.es for $\{A_{\boldsymbol{n}}\}_n$, $\{A'_{\boldsymbol{n}}\}_n$, respectively. Then:*

1. *$\{\{B^*_{\boldsymbol{n},m}\}_n\}_m$ is an a.c.s. for $\{A^*_{\boldsymbol{n}}\}_n$;*
2. *$\{\{\alpha B_{\boldsymbol{n},m} + \beta B'_{\boldsymbol{n},m}\}_n\}_m$ is an a.c.s. for $\{\alpha A_{\boldsymbol{n}} + \beta A'_{\boldsymbol{n}}\}_n$, for all $\alpha, \beta \in \mathbb{C}$.*

Proposition 3.8 addresses the case of a linear combination $\{\alpha A_{\boldsymbol{n}} + \beta A'_{\boldsymbol{n}}\}_n$. The analog of Proposition 3.8 for the product $\{A_{\boldsymbol{n}} A'_{\boldsymbol{n}}\}_n$ requires the additional assumption that $\{A_{\boldsymbol{n}}\}_n$, $\{A'_{\boldsymbol{n}}\}_n$ are sparsely unbounded (s.u.).

Definition 3.9 (s.u. matrix-sequence). We say that a matrix-sequence $\{A_{\boldsymbol{n}}\}_n$ is sparsely unbounded (s.u.) if for every $M > 0$ there exists n_M such that, for $n \ge n_M$,

$$\frac{\#\{i \in \{1, \ldots, N(\boldsymbol{n})\} : \ \sigma_i(A_{\boldsymbol{n}}) > M\}}{N(\boldsymbol{n})} \le r(M),$$

where $\lim_{M\to\infty} r(M) = 0$.

Simple characterizations of s.u. matrix-sequences are given in the following proposition [25, Proposition 3.2].

Proposition 3.10. *Let $\{A_{\boldsymbol{n}}\}_n$ be a matrix-sequence. Then, the following conditions are equivalent.*

1. *$\{A_{\boldsymbol{n}}\}_n$ is s.u.*
2. $\displaystyle \lim_{M\to\infty} \limsup_{n\to\infty} \frac{\#\{i \in \{1,\dots,N(\boldsymbol{n})\} : \sigma_i(A_{\boldsymbol{n}}) > M\}}{N(\boldsymbol{n})} = 0.$
3. *For every $M > 0$ there exists n_M such that, for $n \geq n_M$,*
$$A_{\boldsymbol{n}} = \hat{A}_{\boldsymbol{n},M} + \tilde{A}_{\boldsymbol{n},M},$$
$$\operatorname{rank}(\hat{A}_{\boldsymbol{n},M}) \leq r(M)N(\boldsymbol{n}), \qquad \|\tilde{A}_{\boldsymbol{n},M}\| \leq M,$$
where $\lim_{M\to\infty} r(M) = 0$.

It is not difficult to see that any matrix-sequence enjoying an asymptotic singular value distribution is s.u. For a formal proof, see [25, Proposition 3.3].

Proposition 3.11. *If $\{A_{\boldsymbol{n}}\}_n \sim_\sigma f$ then $\{A_{\boldsymbol{n}}\}_n$ is s.u.*

Proposition 3.12 is the analog of Proposition 3.8 for the case of the product $\{A_{\boldsymbol{n}}A'_{\boldsymbol{n}}\}_n$. This important result appeared in [38, Proposition 2.4].

Proposition 3.12. *Let $\{\{B_{\boldsymbol{n},m}\}_n\}_m, \{\{B'_{\boldsymbol{n},m}\}_n\}_m$ be a.c.s.es for $\{A_{\boldsymbol{n}}\}_n, \{A'_{\boldsymbol{n}}\}_n$, respectively. Assume that $\{A_{\boldsymbol{n}}\}_n, \{A'_{\boldsymbol{n}}\}_n$ are s.u. Then $\{\{B_{\boldsymbol{n},m}B'_{\boldsymbol{n},m}\}_n\}_m$ is an a.c.s. for $\{A_{\boldsymbol{n}}A'_{\boldsymbol{n}}\}_n$.*

3.3. Some criterions to identify a.c.s.es

In practical applications, it happens that a matrix-sequence $\{A_{\boldsymbol{n}}\}_n$ is given together with a sequence of matrix-sequences $\{\{B_{\boldsymbol{n},m}\}_n\}_m$, and one would like to prove that $\{\{B_{\boldsymbol{n},m}\}_n\}_m$ is an a.c.s. for $\{A_{\boldsymbol{n}}\}_n$, without constructing the splitting (3.1). In this section, we provide two useful criterions to solve this problem.

Theorem 3.13. *Let $\{A_{\boldsymbol{n}}\}_n$ be a matrix-sequence, let $\{\{B_{\boldsymbol{n},m}\}_n\}_m$ be a sequence of matrix-sequences, and let $1 \leq p < \infty$. Suppose that for every m there exists n_m such that, for $n \geq n_m$,*
$$\|A_{\boldsymbol{n}} - B_{\boldsymbol{n},m}\|_p^p \leq \epsilon(m,\boldsymbol{n})N(\boldsymbol{n}),$$
where $\displaystyle\lim_{m\to\infty}\limsup_{n\to\infty} \epsilon(m,\boldsymbol{n}) = 0$. Then $\{\{B_{\boldsymbol{n},m}\}_n\}_m$ is an a.c.s. for $\{A_{\boldsymbol{n}}\}_n$.

Proof. See [24, Section 3.3]. □

Theorem 3.14. *Let $\{A_{\boldsymbol{n}}\}_n$ be a matrix-sequence and let $\{\{B_{\boldsymbol{n},m}\}_n\}_m$ be a sequence of matrix-sequences. Assume $\{A_{\boldsymbol{n}} - B_{\boldsymbol{n},m}\}_n \sim_\sigma g_m$ for some $g_m : D \subset \mathbb{R}^k \to \mathbb{C}$ such that $g_m \to 0$ in measure. Then $\{\{B_{\boldsymbol{n},m}\}_n\}_m$ is an a.c.s. for $\{A_{\boldsymbol{n}}\}_n$.*

Proof. Let $C_{\boldsymbol{n},m} = A_{\boldsymbol{n}} - B_{\boldsymbol{n},m}$. For any $h \in \mathbb{N}$, let $F_h \in C_c(\mathbb{R})$ such that $F_h = 1$ over $[0, 1/2h]$, $F_h = 0$ over $[1/h, \infty)$, and $0 \le F_h \le 1$ over $[0, \infty)$. Note that $F_h \le \chi_{[0,1/h]}$ over $[0, \infty)$. For every m, h, we have

$$\begin{aligned}
&\frac{\#\{i \in \{1, \dots, N(\boldsymbol{n})\} : \ \sigma_i(C_{\boldsymbol{n},m}) > 1/h\}}{N(\boldsymbol{n})} \\
&\quad = 1 - \frac{\#\{i \in \{1, \dots, N(\boldsymbol{n})\} : \ \sigma_i(C_{\boldsymbol{n},m}) \le 1/h\}}{N(\boldsymbol{n})} \\
&\quad = 1 - \frac{1}{N(\boldsymbol{n})} \sum_{i=1}^{N(\boldsymbol{n})} \chi_{[0,1/h]}(\sigma_i(C_{\boldsymbol{n},m})) \le 1 - \frac{1}{N(\boldsymbol{n})} \sum_{i=1}^{N(\boldsymbol{n})} F_h(\sigma_i(C_{\boldsymbol{n},m})), \qquad (3.2)
\end{aligned}$$

and the latter tends to $c(m, h)$ as $n \to \infty$, where

$$c(m, h) = 1 - \frac{1}{\mu_k(D)} \int_D F_h(|g_m(\mathbf{x})|) d\mathbf{x}.$$

Since $F_h(0) = 1$ and $g_m \to 0$ in measure, Lemma 2.1 implies that, for every h, $c(m, h) \to 0$ as $m \to \infty$. Hence, there exists a sequence $\{h_m\}_m \subseteq \mathbb{N}$ such that

$$\lim_{m \to \infty} h_m = \infty, \qquad \lim_{m \to \infty} c(m, h_m) = 0.$$

In view of (3.2), for every m we have

$$\limsup_{n \to \infty} \frac{\#\{i \in \{1, \dots, N(\boldsymbol{n})\} : \ \sigma_i(C_{\boldsymbol{n},m}) > 1/h_m\}}{N(\boldsymbol{n})} \le c(m, h_m). \qquad (3.3)$$

Let $C_{\boldsymbol{n},m} = U_{\boldsymbol{n},m} \Sigma_{\boldsymbol{n},m} V_{\boldsymbol{n},m}^*$ be an SVD of $C_{\boldsymbol{n},m}$. Let $\hat{\Sigma}_{\boldsymbol{n},m}$ be the matrix obtained from $\Sigma_{\boldsymbol{n},m}$ by setting to 0 all the singular values that are less than or equal to $1/h_m$, and let $\tilde{\Sigma}_{\boldsymbol{n},m} = \Sigma_{\boldsymbol{n},m} - \hat{\Sigma}_{\boldsymbol{n},m}$ be the matrix obtained from $\Sigma_{\boldsymbol{n},m}$ by setting to 0 all the singular values that exceed $1/h_m$. Then we can write

$$A_{\boldsymbol{n}} - B_{\boldsymbol{n},m} = C_{\boldsymbol{n},m} = R_{\boldsymbol{n},m} + N_{\boldsymbol{n},m},$$

where $R_{\boldsymbol{n},m} = U_{\boldsymbol{n},m} \hat{\Sigma}_{\boldsymbol{n},m} V_{\boldsymbol{n},m}^*$ and $N_{\boldsymbol{n},m} = U_{\boldsymbol{n},m} \tilde{\Sigma}_{\boldsymbol{n},m} V_{\boldsymbol{n},m}^*$. By definition, we have $\|N_{\boldsymbol{n},m}\| \le 1/h_m$. Moreover, (3.3) says that

$$\limsup_{n \to \infty} \frac{\operatorname{rank}(R_{\boldsymbol{n},m})}{N(\boldsymbol{n})} \le c(m, h_m),$$

implying the existence of a n_m such that, for $n \ge n_m$,

$$\operatorname{rank}(R_{\boldsymbol{n},m}) \le \big(c(m, h_m) + 1/m\big) N(\boldsymbol{n}).$$

This shows that $\{\{B_{\boldsymbol{n},m}\}_n\}_m$ is an a.c.s. of $\{A_{\boldsymbol{n}}\}_n$. □

3.4. An extension of the concept of a.c.s.es

We provide in this section an extension of the a.c.s. notion that will be used to define GLT sequences in Section 5. The extension is plain. The underlying idea is that, in Definition 3.1, one could choose to approximate $\{A_{\boldsymbol{n}}\}_n$ by a class of sequences $\{\{B_{\boldsymbol{n},\alpha}\}_n\}_{\alpha\in\mathcal{A}}$ parameterized by a not necessarily integer parameter α. For example, one may want to use a parameter $\epsilon > 0$ and to claim that a given class of sequences $\{\{B_{\boldsymbol{n},\epsilon}\}_n\}_{\epsilon>0}$ is an a.c.s. of $\{A_{\boldsymbol{n}}\}_n$ for $\epsilon \to 0$. As it is intuitive, this assertion has the following meaning.

Definition 3.15 (a.c.s. for $\epsilon \to 0$). Let $\{A_{\boldsymbol{n}}\}_n$ be a matrix-sequence. We say that $\{\{B_{\boldsymbol{n},\epsilon}\}_n\}_{\epsilon>0}$ is an a.c.s. of $\{A_{\boldsymbol{n}}\}_n$ for $\epsilon \to 0$ if the following property holds: for every $\epsilon > 0$ there exists n_ϵ such that, for $n \geq n_\epsilon$,

$$A_{\boldsymbol{n}} = B_{\boldsymbol{n},\epsilon} + R_{\boldsymbol{n},\epsilon} + N_{\boldsymbol{n},\epsilon},$$
$$\operatorname{rank}(R_{\boldsymbol{n},\epsilon}) \leq c(\epsilon)N(\boldsymbol{n}), \qquad \|N_{\boldsymbol{n},\epsilon}\| \leq \omega(\epsilon),$$

where n_ϵ, $c(\epsilon)$, $\omega(\epsilon)$ depend only on ϵ and $\lim_{\epsilon\to 0} c(\epsilon) = \lim_{\epsilon\to 0}\omega(\epsilon) = 0$.

It is easy to see that, if $\{\{B_{\boldsymbol{n},\epsilon}\}_n\}_{\epsilon>0}$ is an a.c.s. of $\{A_{\boldsymbol{n}}\}_n$ for $\epsilon \to 0$, then $\{\{B_{\boldsymbol{n},\epsilon(m)}\}_n\}_m$ is an a.c.s. for $\{A_{\boldsymbol{n}}\}_n$ (in the sense of the classical Definition 3.1) for all sequences $\{\epsilon(m)\}_m \subset (0,\infty)$ such that $\epsilon(m) \to 0$.

4. LT and sLT sequences

In this section, we provide the essentials of the theory of LT sequences. This theory was the subject of paper [24]. To avoid too many repetitions from [24], as well as to keep the presentation as concise as possible, we only report a summary of the results that are necessary to develop the theory of GLT sequences.

A LT sequence $\{A_{\boldsymbol{n}}\}_n$ is a matrix-sequence equipped with a function of the form $a \otimes f : [0,1]^d \times [-\pi,\pi]^d \to \mathbb{C}$, where $a : [0,1]^d \to \mathbb{C}$ is Riemann-integrable and $f \in L^1([-\pi,\pi]^d)$. The function $a \otimes f$ is referred to as the symbol (or kernel) of $\{A_{\boldsymbol{n}}\}_n$, a is the weight function and f is the generating function.[3] Throughout this paper, we write $\{A_{\boldsymbol{n}}\}_n \sim_{\rm LT} a \otimes f$ to indicate that $\{A_{\boldsymbol{n}}\}_n$ is a LT sequence with symbol $a\otimes f$; it is understood that $a : [0,1]^d \to \mathbb{C}$ is Riemann-integrable and $f \in L^1([-\pi,\pi]^d)$. If $\{A_{\boldsymbol{n}}\}_n \sim_{\rm LT} a \otimes f$ and f is separable, then $\{A_{\boldsymbol{n}}\}_n$ is said to be a sLT sequence with symbol $a \otimes f$, and we write $\{A_{\boldsymbol{n}}\}_n \sim_{\rm sLT} a \otimes f$. In other words, a sLT sequence is a LT sequence with separable generating function.

The formal definitions of LT and sLT sequences are given in [24, Section 4.2]. They are formulated in terms of a.c.s.es and, in particular, they are different from the original ones appeared in [39, 40, 46]. However, more than the definitions (which are rather cumbersome and not especially helpful), what we really need is the following list of properties [24]. For more details, as well as for the motivations that led to the birth of LT sequences, we refer the reader to Section 1 and to

[3]See the introduction of Tilli's paper [46] for the origin and the meaning of this terminology.

[24, 46]; see in particular the first part of Section 1, the introduction of [46] and Section 5.1 of [24] for the justification of the name "Locally Toeplitz sequences".

*LT*1. If $\{A_{\boldsymbol{n}}^{(i)}\}_n \sim_{\rm LT} a_i \otimes f_i$, $i = 1, \ldots, p$, then $\{\sum_{i=1}^p A_{\boldsymbol{n}}^{(i)}\}_n \sim_\sigma \sum_{i=1}^p a_i \otimes f_i$ and $\{\Re(\sum_{i=1}^p A_{\boldsymbol{n}}^{(i)})\}_n \sim_\lambda \Re(\sum_{i=1}^p a_i \otimes f_i)$.

*LT*2. $\{T_{\boldsymbol{n}}(f)\}_n \sim_{\rm LT} 1 \otimes f$ for every $f \in L^1([-\pi,\pi]^d)$.

*LT*3. $\{D_{\boldsymbol{n}}(a)\}_n \sim_{\rm LT} a \otimes 1$ for every Riemann-integrable $a : [0,1]^d \to \mathbb{C}$.

*LT*4. $\{Z_{\boldsymbol{n}}\}_n \sim_{\rm LT} 0$ if and only if $\{Z_{\boldsymbol{n}}\}_n \sim_\sigma 0$.

*LT*5. If $\{A_{\boldsymbol{n}}\}_n \sim_{\rm LT} a \otimes f$ then $\{A_{\boldsymbol{n}}^*\}_n \sim_{\rm LT} \overline{a} \otimes \overline{f}$.

*LT*6. If $\{A_{\boldsymbol{n}}\}_n \sim_{\rm LT} a \otimes f$ then $\{\alpha A_{\boldsymbol{n}}\}_n \sim_{\rm LT} \alpha a \otimes f$ for any $\alpha \in \mathbb{C}$.

*LT*7. If $\{A_{\boldsymbol{n}}^{(i)}\}_n \sim_{\rm LT} a \otimes f_i$, $i = 1, \ldots, r$, then $\{\sum_{i=1}^r A_{\boldsymbol{n}}^{(i)}\}_n \sim_{\rm LT} a \otimes (\sum_{i=1}^r f_i)$; and if $\{A_{\boldsymbol{n}}^{(i)}\}_n \sim_{\rm LT} a_i \otimes f$, $i = 1, \ldots, r$, then $\{\sum_{i=1}^r A_{\boldsymbol{n}}^{(i)}\}_n \sim_{\rm LT} (\sum_{i=1}^r a_i) \otimes f$.

*LT*8. If $\{A_{\boldsymbol{n}}\}_n \sim_{\rm LT} a \otimes f$ and $\{\tilde{A}_{\boldsymbol{n}}\}_n \sim_{\rm LT} \tilde{a} \otimes \tilde{f}$, where $f \in L^p([-\pi,\pi]^d)$, $\tilde{f} \in L^q([-\pi,\pi]^d)$, and $1 \le p, q \le \infty$ are conjugate exponents, then $\{A_{\boldsymbol{n}}\tilde{A}_{\boldsymbol{n}}\}_n \sim_{\rm LT} a\tilde{a} \otimes f\tilde{f}$.

*LT*9. Let $\{A_{\boldsymbol{n}}\}_n \sim_{\rm LT} a \otimes f$. If $\{A_{\boldsymbol{n}}^{(m)}\}_n \sim_{\rm LT} a_m \otimes f_m$, where $a_m \to a$ in $L^1([0,1]^d)$ and $f_m \to f$ in $L^1([-\pi,\pi]^d)$, then $\{\{A_{\boldsymbol{n}}^{(m)}\}_n\}_m$ is an a.c.s. for $\{A_{\boldsymbol{n}}\}_n$.

The next proposition provides a relation between LT and sLT sequences, which will be used in Section 5 to show that any LT sequence is a GLT sequence.

Proposition 4.1. *Let $\{A_{\boldsymbol{n}}\}_n \sim_{\rm LT} a \otimes f$. Then, for any $m \in \mathbb{N}$ there exist matrix-sequences $\{A_{\boldsymbol{n}}^{(i,m)}\}_n \sim_{\rm sLT} a \otimes f_{i,m}$, $i = 1, \ldots, N_m$, such that $\sum_{i=1}^{N_m} f_{i,m} \to f$ in $L^1([-\pi,\pi]^d)$ and $\{\{\sum_{i=1}^{N_m} A_{\boldsymbol{n}}^{(i,m)}\}_n\}_m$ is an a.c.s. for $\{A_{\boldsymbol{n}}\}_n$.*

Proof. Take any sequence of d-variate trigonometric polynomials f_m such that $f_m \to f$ in $L^1([-\pi,\pi]^d)$. Note that such a sequence exists because the set of d-variate trigonometric polynomials is dense in $L^1([-\pi,\pi]^d)$. By definition, any d-variate trigonometric polynomial is a finite sum of separable d-variate trigonometric polynomials. Hence, we can write $f_m = \sum_{i=1}^{N_m} f_{i,m}$, for some separable d-variate trigonometric polynomials $f_{i,m}$, $i = 1, \ldots, N_m$. Take arbitrary matrix-sequences $\{A_{\boldsymbol{n}}^{(i,m)}\}_n \sim_{\rm sLT} a \otimes f_{i,m}$, $i = 1, \ldots, N_m$. In view of *LT*2–*LT*3 and *LT*8, we can choose $A_{\boldsymbol{n}}^{(i,m)} = D_{\boldsymbol{n}}(a)T_{\boldsymbol{n}}(f_{i,m})$. By *LT*7, $\{\sum_{i=1}^{N_m} A_{\boldsymbol{n}}^{(i,m)}\}_n \sim_{\rm LT} a \otimes (\sum_{i=1}^{N_m} f_{i,m}) = a \otimes f_m$. Hence, $\{\{\sum_{i=1}^{N_m} A_{\boldsymbol{n}}^{(i,m)}\}_n\}_m$ is an a.c.s. for $\{A_{\boldsymbol{n}}\}_n$ by *LT*9. □

5. GLT sequences

In this section, we develop the theory of GLT sequences. We first report in Definition 5.1 a "corrected" version of the original notion of GLT sequences appeared in [39, Definition 2.3] and [40, Definition 1.5]. The new definition looks formally the same as the original one, but it is formulated in terms of the new concepts of LT and sLT sequences introduced in [24] and discussed in Section 4.

Definition 5.1 (GLT sequence). Let $\{A_{\boldsymbol{n}}\}_n$ be a matrix-sequence and let $\kappa : [0,1]^d \times [-\pi,\pi]^d \to \mathbb{C}$ be a measurable function. We say that $\{A_{\boldsymbol{n}}\}_n$ is a Generalized

Locally Toeplitz (GLT) sequence with symbol κ, and we write $\{A_{\boldsymbol{n}}\}_n \sim_{\mathrm{GLT}} \kappa$, if the following condition is met.

For every $\epsilon > 0$ there is a finite number of sLT sequences $\{A_{\boldsymbol{n}}^{(i,\epsilon)}\}_n \sim_{\mathrm{sLT}} a_{i,\epsilon} \otimes f_{i,\epsilon}$, $i = 1, \ldots, N_\epsilon$, such that:

- $\sum_{i=1}^{N_\epsilon} a_{i,\epsilon} \otimes f_{i,\epsilon} \to \kappa$ in measure over $[0,1]^d \times [-\pi,\pi]^d$ as $\epsilon \to 0$;
- $\big\{\{\sum_{i=1}^{N_\epsilon} A_{\boldsymbol{n}}^{(i,\epsilon)}\}_n\big\}_{\epsilon>0}$ is an a.c.s. of $\{A_{\boldsymbol{n}}\}_n$ for $\epsilon \to 0$.

From now on, if we write $\{A_{\boldsymbol{n}}\}_n \sim_{\mathrm{GLT}} \kappa$, it is understood that $\kappa : [0,1]^d \times [-\pi,\pi]^d \to \mathbb{C}$ is measurable. Note that any sLT sequence is a GLT sequence, because $\{A_{\boldsymbol{n}}\}_n \sim_{\mathrm{sLT}} a \otimes f$ immediately implies $\{A_{\boldsymbol{n}}\}_n \sim_{\mathrm{GLT}} a \otimes f$. The following proposition provides a first characterization of GLT sequences.

Proposition 5.2. *We have $\{A_{\boldsymbol{n}}\}_n \sim_{\mathrm{GLT}} \kappa$ if and only if the following condition is met.*

For every m varying in some infinite subset of $\mathbb{N}$ there is a finite number of sLT sequences $\{A_{\boldsymbol{n}}^{(i,m)}\}_n \sim_{\mathrm{sLT}} a_{i,m} \otimes f_{i,m}$, $i = 1, \ldots, N_m$, such that:

- *$\sum_{i=1}^{N_m} a_{i,m} \otimes f_{i,m} \to \kappa$ in measure over $[0,1]^d \times [-\pi,\pi]^d$ as $m \to \infty$;*
- *$\big\{\{\sum_{i=1}^{N_m} A_{\boldsymbol{n}}^{(i,m)}\}_n\big\}_m$ is an a.c.s. for $\{A_{\boldsymbol{n}}\}_n$.*

Proof. If $\{A_{\boldsymbol{n}}\}_n \sim_{\mathrm{GLT}} \kappa$, then the condition of the proposition holds with

$$a_{i,m} = a_{i,\epsilon(m)}, \qquad f_{i,m} = f_{i,\epsilon(m)}, \qquad \{A_{\boldsymbol{n}}^{(i,m)}\}_n = \{A_{\boldsymbol{n}}^{(i,\epsilon(m))}\}_n, \qquad N_m = N_{\epsilon(m)},$$

where $\{\epsilon(m)\}_m$ is any sequence of positive numbers such that $\epsilon(m) \to 0$ as $m \to \infty$, and $a_{i,\epsilon}$, $f_{i,\epsilon}$, $\{A_{\boldsymbol{n}}^{(i,\epsilon)}\}_n$, N_ϵ are as in Definition 5.1.

Conversely, suppose that the condition of the proposition holds. Let $\mathcal{M} \subseteq \mathbb{N}$ be the infinite subset of $\mathbb{N}$ where m varies. Then, the condition of Definition 5.1 holds with

$$a_{i,\epsilon} = a_{i,m(\epsilon)}, \qquad f_{i,\epsilon} = f_{i,m(\epsilon)}, \qquad \{A_{\boldsymbol{n}}^{(i,\epsilon)}\}_n = \{A_{\boldsymbol{n}}^{(i,m(\epsilon))}\}_n, \qquad N_\epsilon = N_{m(\epsilon)},$$

where $\{m(\epsilon)\}_{\epsilon>0} \subseteq \mathcal{M}$ is any family of indices such that $m(\epsilon) \to \infty$ as $\epsilon \to 0$, and $a_{i,m}$, $f_{i,m}$, $\{A_{\boldsymbol{n}}^{(i,m)}\}_n$, N_m are as in the statement of the proposition. It follows that $\{A_{\boldsymbol{n}}\}_n \sim_{\mathrm{GLT}} \kappa$. □

Proposition 5.2 is essentially the same as Definition 5.1, but it is easier to handle, because it is based on the standard notion of a.c.s.es (Definition 3.1). It follows from Propositions 4.1 and 5.2 that any LT sequence is a GLT sequence. More precisely,

$$\{A_{\boldsymbol{n}}\}_n \sim_{\mathrm{LT}} a \otimes f \quad \Rightarrow \quad \{A_{\boldsymbol{n}}\}_n \sim_{\mathrm{GLT}} a \otimes f. \tag{5.1}$$

5.1. Singular value and eigenvalue distribution of GLT sequences

This section collects the main asymptotic distribution results for GLT sequences, namely Theorems 5.3 and 5.7.

Theorem 5.3 (Serra-Capizzano). *If $\{A_{\boldsymbol{n}}\}_n \sim_{\mathrm{GLT}} \kappa$ then $\{A_{\boldsymbol{n}}\}_n \sim_\sigma \kappa$.*

Proof. By Proposition 5.2, there exist LT sequences $\{A_{\boldsymbol{n}}^{(i,m)}\}_n \sim_{\mathrm{LT}} a_{i,m} \otimes f_{i,m}$, $i = 1,\ldots,N_m$, such that $\sum_{i=1}^{N_m} a_{i,m} \otimes f_{i,m} \to \kappa$ in measure and $\{\{\sum_{i=1}^{N_m} A_{\boldsymbol{n}}^{(i,m)}\}_n\}_m$ is an a.c.s. for $\{A_{\boldsymbol{n}}\}_n$. By $LT1$, we have $\{\sum_{i=1}^{N_m} A_{\boldsymbol{n}}^{(i,m)}\}_n \sim_\sigma \sum_{i=1}^{N_m} a_{i,m} \otimes f_{i,m}$. Since $\sum_{i=1}^{N_m} a_{i,m} \otimes f_{i,m} \to \kappa$ in measure, all the assumptions of Corollary 3.4 are satisfied and so $\{A_{\boldsymbol{n}}\}_n \sim_\sigma \kappa$. □

As a consequence of Theorem 5.3, every GLT sequence is s.u. (see Proposition 3.11). Using Theorem 5.3, we show in Proposition 5.5 that the symbol of a GLT sequence is unique. For the proof of Proposition 5.5, we point out that any linear combination of GLT sequences is again a GLT sequence with symbol given by the same linear combination of the symbols. This is one of the most elementary results in the world of the algebraic properties possessed by GLT sequences. These properties will be investigated in Section 5.4 and give rise to the so-called GLT algebra.

Proposition 5.4. *Let $\{A_{\boldsymbol{n}}\}_n \sim_{\mathrm{GLT}} \kappa$ and $\{B_{\boldsymbol{n}}\}_n \sim_{\mathrm{GLT}} \xi$. Then, $\{A_{\boldsymbol{n}}^*\}_n \sim_{\mathrm{GLT}} \overline{\kappa}$ and $\{\alpha A_{\boldsymbol{n}} + \beta B_{\boldsymbol{n}}\}_n \sim_{\mathrm{GLT}} \alpha\kappa + \beta\xi$ for all $\alpha, \beta \in \mathbb{C}$.*

The proof of Proposition 5.4 is easy: it suffices to write down the meaning of $\{A_{\boldsymbol{n}}\}_n \sim_{\mathrm{GLT}} \kappa$ and $\{B_{\boldsymbol{n}}\}_n \sim_{\mathrm{GLT}} \xi$ (using the characterization of Proposition 5.2), and to apply Proposition 3.8. The details are left to the reader.

Proposition 5.5. *Assume that $\{A_{\boldsymbol{n}}\}_n \sim_{\mathrm{GLT}} \kappa$ and $\{A_{\boldsymbol{n}}\}_n \sim_{\mathrm{GLT}} \xi$. Then $\kappa = \xi$ a.e. in $[0,1]^d \times [-\pi,\pi]^d$.*

Proof. By Proposition 5.4, $\{O_{N(\boldsymbol{n})}\}_n \sim_{\mathrm{GLT}} \kappa - \xi$. Therefore, by Theorem 5.3, for all test functions $F \in C_c(\mathbb{R})$ we have

$$F(0) = \frac{1}{(2\pi)^d} \int_{[0,1]^d \times [-\pi,\pi]^d} F(|\kappa(\mathbf{x},\boldsymbol{\theta}) - \xi(\mathbf{x},\boldsymbol{\theta})|) d\mathbf{x} d\boldsymbol{\theta}.$$

This means that $\phi_{[|\kappa-\xi|]} = \phi_{[0]}$ and so, by Remark 2.3, $|\kappa - \xi| = 0$ a.e. □

Proposition 5.6. *Let $\{A_{\boldsymbol{n}}\}_n \sim_{\mathrm{GLT}} \kappa$ and assume that the matrices $A_{\boldsymbol{n}}$ are Hermitian. Then $\kappa \in \mathbb{R}$ a.e.*

Proof. Since the matrices $A_{\boldsymbol{n}}$ are Hermitian, Proposition 5.4 yields $\{A_{\boldsymbol{n}}\}_n \sim_{\mathrm{GLT}} \overline{\kappa}$. Thus, by Proposition 5.5, $\kappa = \overline{\kappa}$ a.e., i.e., $\kappa \in \mathbb{R}$ a.e. □

Theorem 5.7 (Serra-Capizzano). *If $\{A_{\boldsymbol{n}}\}_n \sim_{\mathrm{GLT}} \kappa$ and the matrices $A_{\boldsymbol{n}}$ are Hermitian, then $\{A_{\boldsymbol{n}}\}_n \sim_\lambda \kappa$.*

Proof. By Proposition 5.2, there exist LT sequences $\{A_{\boldsymbol{n}}^{(i,m)}\}_n \sim_{\mathrm{LT}} a_{i,m} \otimes f_{i,m}$, $i = 1,\ldots,N_m$, such that $\sum_{i=1}^{N_m} a_{i,m} \otimes f_{i,m} \to \kappa$ in measure and $\{\{\sum_{i=1}^{N_m} A_{\boldsymbol{n}}^{(i,m)}\}_n\}_m$ is an a.c.s. for $\{A_{\boldsymbol{n}}\}_n$. Since the matrices $A_{\boldsymbol{n}}$ are Hermitian, $\{\{\Re(\sum_{i=1}^{N_m} A_{\boldsymbol{n}}^{(i,m)})\}_n\}_m$ is another a.c.s. for $\{A_{\boldsymbol{n}} = \Re(A_{\boldsymbol{n}})\}_n$, and it is formed by Hermitian matrices. By $LT1$, $\{\Re(\sum_{i=1}^{N_m} A_{\boldsymbol{n}}^{(i,m)})\}_n \sim_\lambda \Re(\sum_{i=1}^{N_m} a_{i,m} \otimes f_{i,m})$. The function κ is real a.e. by Proposition 5.6, and so from $\sum_{i=1}^{N_m} a_{i,m} \otimes f_{i,m} \to \kappa$ (in measure) we get

$\Re(\sum_{i=1}^{N_m} a_{i,m} \otimes f_{i,m}) \to \kappa$ (in measure). All the assumptions of Corollary 3.5 are then satisfied, and it follows that $\{A_{\boldsymbol{n}}\}_n \sim_\lambda \kappa$. □

5.2. Approximation results for GLT sequences

Theorem 5.8 is the main approximation result for GLT sequences. It is formally the same as Corollaries 3.4–3.5 with "$\sim_\sigma$" and "$\sim_\lambda$" replaced by "$\sim_{\mathrm{GLT}}$", and it is particularly useful to show that a given matrix-sequence $\{A_{\boldsymbol{n}}\}_n$ is a GLT sequence.

Theorem 5.8 (Garoni). *Let $\{A_{\boldsymbol{n}}\}_n$ be a matrix-sequence and let $\kappa : [0,1]^d \times [-\pi,\pi]^d \to \mathbb{C}$ be a measurable function. Suppose that:*

1. *$\{\{B_{\boldsymbol{n},m}\}_n\}_m$ is an a.c.s. for $\{A_{\boldsymbol{n}}\}_n$;*
2. *$\{B_{\boldsymbol{n},m}\}_n \sim_{\mathrm{GLT}} \kappa_m$ for every m;*
3. *$\kappa_m \to \kappa$ in measure.*

Then $\{A_{\boldsymbol{n}}\}_n \sim_{\mathrm{GLT}} \kappa$.

Proof. Since $\{B_{\boldsymbol{n},m}\}_n \sim_{\mathrm{GLT}} \kappa_m$ for every m, Proposition 5.2 implies that the following condition is satisfied for each m: for every h varying in some infinite subset $\mathcal{H} \subseteq \mathbb{N}$, there exists a finite number of sLT sequences $\{A_{\boldsymbol{n},m}^{(i,h)}\}_n \sim_{\mathrm{sLT}} a_{i,h,m} \otimes f_{i,h,m}$, $i = 1,\dots,N_{h,m}$, such that:

- $\sum_{i=1}^{N_{h,m}} a_{i,h,m} \otimes f_{i,h,m} \to \kappa_m$ in measure over $[0,1]^d \times [-\pi,\pi]^d$ as $h \to \infty$;
- $\{\{\sum_{i=1}^{N_{h,m}} A_{\boldsymbol{n},m}^{(i,h)}\}_n\}_h$ is an a.c.s. for $\{B_{\boldsymbol{n},m}\}_n$.

In particular, for every m and every h there exists $n_{h,m}$ such that, for $n \geq n_{h,m}$,

$$B_{\boldsymbol{n},m} = \sum_{i=1}^{N_{h,m}} A_{\boldsymbol{n},m}^{(i,h)} + R_{\boldsymbol{n},h,m} + N_{\boldsymbol{n},h,m},$$

$$\mathrm{rank}(R_{\boldsymbol{n},h,m}) \leq c(h,m)N(\boldsymbol{n}), \qquad \|N_{\boldsymbol{n},h,m}\| \leq \omega(h,m),$$

where

$$\lim_{h\to\infty} c(h,m) = \lim_{h\to\infty} \omega(h,m) = 0.$$

Let $\{\delta_m\}_m$ be a sequence of positive numbers such that $\delta_m \to 0$. For each fixed m, the function $\sum_{i=1}^{N_{h,m}} a_{i,h,m} \otimes f_{i,h,m}$ tends to κ_m in measure as $h \to \infty$, and so

$$\mu(m,h,\delta_m) = \mu_{2d}\left\{\left|\sum_{i=1}^{N_{h,m}} a_{i,h,m} \otimes f_{i,h,m} - \kappa_m\right| \geq \delta_m\right\} \to 0 \quad \text{as} \quad h \to \infty.$$

Now we recall that $\{\{B_{\boldsymbol{n},m}\}_n\}_m$ is an a.c.s. for $\{A_{\boldsymbol{n}}\}_n$, i.e., for every m there exists n_m such that, for $n \geq n_m$,

$$A_{\boldsymbol{n}} = B_{\boldsymbol{n},m} + R_{\boldsymbol{n},m} + N_{\boldsymbol{n},m},$$

$$\mathrm{rank}(R_{\boldsymbol{n},m}) \leq c(m)N(\boldsymbol{n}), \qquad \|N_{\boldsymbol{n},m}\| \leq \omega(m),$$

where

$$\lim_{m\to\infty} c(m) = \lim_{m\to\infty} \omega(m) = 0.$$

It follows that, for every m, every h, and every $n \geq \max(n_m, n_{h,m})$,

$$A_{\boldsymbol{n}} = \sum_{i=1}^{N_{h,m}} A_{\boldsymbol{n},m}^{(i,h)} + (R_{\boldsymbol{n},h,m} + R_{\boldsymbol{n},m}) + (N_{\boldsymbol{n},h,m} + N_{\boldsymbol{n},m}),$$

$$\operatorname{rank}(R_{\boldsymbol{n},h,m} + R_{\boldsymbol{n},m}) \leq (c(h,m) + c(m))\, N(\boldsymbol{n}),$$

$$\|N_{\boldsymbol{n},h,m} + N_{\boldsymbol{n},m}\| \leq \omega(h,m) + \omega(m).$$

Choose a sequence $\{h_m\}_m \subseteq \mathcal{H}$ such that $h_m \to \infty$ and

$$\lim_{m\to\infty} c(h_m, m) = \lim_{m\to\infty} \omega(h_m, m) = \lim_{m\to\infty} \mu(m, h_m, \delta_m) = 0.$$

Then, for every m and every $n \geq \max(n_m, n_{h_m,m})$,

$$A_{\boldsymbol{n}} = \sum_{i=1}^{N_{h_m,m}} A_{\boldsymbol{n},m}^{(i,h_m)} + (R_{\boldsymbol{n},h_m,m} + R_{\boldsymbol{n},m}) + (N_{\boldsymbol{n},h_m,m} + N_{\boldsymbol{n},m}),$$

$$\operatorname{rank}(R_{\boldsymbol{n},h_m,m} + R_{\boldsymbol{n},m}) \leq (c(h_m,m) + c(m))\, N(\boldsymbol{n}),$$

$$\|N_{\boldsymbol{n},h_m,m} + N_{\boldsymbol{n},m}\| \leq \omega(h_m,m) + \omega(m).$$

As a consequence, $\{\{\sum_{i=1}^{N_{h_m,m}} A_{\boldsymbol{n},m}^{(i,h_m)}\}_n\}_m$ is an a.c.s. for $\{A_{\boldsymbol{n}}\}_n$. Moreover, for all m and $i = 1, \ldots, N_{h_m,m}$, we have $\{A_{\boldsymbol{n},m}^{(i,h_m)}\}_n \sim_{\rm sLT} a_{i,h_m,m} \otimes f_{i,h_m,m}$. Finally, $\sum_{i=1}^{N_{h_m,m}} a_{i,h_m,m} \otimes f_{i,h_m,m} \to \kappa$ in measure over $[0,1]^d \times [-\pi,\pi]^d$ when $m \to \infty$. Indeed, for any $\delta > 0$,

$$\mu_{2d}\left\{\left|\sum_{i=1}^{N_{h_m,m}} a_{i,h_m,m} \otimes f_{i,h_m,m} - \kappa\right| \geq \delta\right\}$$

$$\leq \mu_{2d}\left\{\left|\sum_{i=1}^{N_{h_m,m}} a_{i,h_m,m} \otimes f_{i,h_m,m} - \kappa_m\right| \geq \delta/2\right\} + \mu_{2d}\left\{|\kappa_m - \kappa| \geq \delta/2\right\}.$$

The term $\mu_{2d}\{|\kappa_m - \kappa| \geq \delta/2\}$ tends to 0 because $\kappa_m \to \kappa$ in measure, and

$$\mu_{2d}\left\{\left|\sum_{i=1}^{N_{h_m,m}} a_{i,h_m,m} \otimes f_{i,h_m,m} - \kappa_m\right| \geq \delta/2\right\} = \mu(m, h_m, \delta/2)$$

tends to 0 because it is eventually less than $\mu(m, h_m, \delta_m)$. Thus, $\{A_{\boldsymbol{n}}\}_n \sim_{\rm GLT} \kappa$ by Proposition 5.2. □

The approximation result stated in Theorem 5.8 admits the following converse, which can be interpreted as another approximation result for GLT sequences.

Theorem 5.9 (Garoni, Serra-Capizzano). *Let $\{A_{\boldsymbol{n}}\}_n$ be a matrix-sequence and let $\{\{B_{\boldsymbol{n},m}\}_n\}_m$ be a sequence of matrix-sequences. Suppose that:*

1. $\{A_{\boldsymbol{n}}\}_n \sim_{\rm GLT} \kappa$;
2. $\{B_{\boldsymbol{n},m}\}_n \sim_{\rm GLT} \kappa_m$ *for every* m.

Then, $\{\{B_{\boldsymbol{n},m}\}_n\}_m$ is an a.c.s. for $\{A_{\boldsymbol{n}}\}_n$ if and only if $\kappa_m \to \kappa$ in measure.

Proof. Assume that 1–2 hold. Then, for each m, $\{A_{\boldsymbol{n}} - B_{\boldsymbol{n},m}\}_n \sim_{\mathrm{GLT}} \kappa - \kappa_m$ (by Proposition 5.4) and $\{A_{\boldsymbol{n}} - B_{\boldsymbol{n},m}\}_n \sim_\sigma \kappa - \kappa_m$ (by Theorem 5.3). Therefore, if $\kappa_m \to \kappa$ in measure then $\{B_{\boldsymbol{n},m}\}_n$ is an a.c.s. for $\{A_{\boldsymbol{n}}\}_n$ by Theorem 3.14. Conversely, if $\{\{B_{\boldsymbol{n},m}\}_n\}_m$ is an a.c.s. for $\{A_{\boldsymbol{n}}\}_n$ then $\{\{A_{\boldsymbol{n}} - B_{\boldsymbol{n},m}\}_n\}_m$ is an a.c.s. of $\{O_{N(\boldsymbol{n})}\}_n$ with $\{O_{N(\boldsymbol{n})}\}_n \sim_\sigma 0$; hence, $\phi_{[|\kappa-\kappa_m|]} \to \phi_{[0]}$ pointwise over $C_c(\mathbb{R})$ (by Theorem 3.6) and $|\kappa - \kappa_m| \to 0$ in measure (by Lemma 2.2). □

Corollary 5.10. *Let $\{A_{\boldsymbol{n}}\}_n \sim_{\mathrm{GLT}} \kappa$. Then, for all $a_{i,m}$, $f_{i,m}$, $i = 1, \dots, N_m$, with the following properties:*

- *$a_{i,m} : [0,1]^d \to \mathbb{C}$ is Riemann-integrable and $f_{i,m} \in L^1([-\pi,\pi]^d)$;*
- *$\sum_{i=1}^{N_m} a_{i,m} \otimes f_{i,m} \to \kappa$ in measure when $m \to \infty$;*

it holds that $\{\{\sum_{i=1}^{N_m} D_{\boldsymbol{n}}(a_{i,m}) T_{\boldsymbol{n}}(f_{i,m})\}_n\}_m$ is an a.c.s. for $\{A_{\boldsymbol{n}}\}_n$. In particular, $\{A_{\boldsymbol{n}}\}_n$ admits an a.c.s. of the form

$$\left\{\left\{\sum_{\boldsymbol{j}=-\boldsymbol{N}_m}^{\boldsymbol{N}_m} D_{\boldsymbol{n}}(a_{\boldsymbol{j}}^{(m)}) T_{\boldsymbol{n}}(\mathrm{e}^{\mathrm{i}\boldsymbol{j}\cdot\boldsymbol{\theta}})\right\}_n\right\}_m, \qquad a_{\boldsymbol{j}}^{(m)} \in C^\infty([0,1]^d), \qquad \boldsymbol{N}_m \in \mathbb{N}^d, \tag{5.2}$$

where $\sum_{\boldsymbol{j}=-\boldsymbol{N}_m}^{\boldsymbol{N}_m} a_{\boldsymbol{j}}^{(m)}(\mathbf{x}) \mathrm{e}^{\mathrm{i}\boldsymbol{j}\cdot\boldsymbol{\theta}} \to \kappa(\mathbf{x},\boldsymbol{\theta})$ a.e. in $[0,1]^d \times [-\pi,\pi]^d$.

Proof. For every m, we have $\{\sum_{i=1}^{N_m} D_{\boldsymbol{n}}(a_{i,m}) T_{\boldsymbol{n}}(f_{i,m})\}_n \sim_{\mathrm{GLT}} \sum_{i=1}^{N_m} a_{i,m} \otimes f_{i,m}$ by *LT*2–*LT*3, *LT*8, (5.1), and Proposition 5.4. The thesis follows from Theorem 5.9 applied with $B_{\boldsymbol{n},m} = \sum_{i=1}^{N_m} D_{\boldsymbol{n}}(a_{i,m}) T_{\boldsymbol{n}}(f_{i,m})$ and $\kappa_m = \sum_{i=1}^{N_m} a_{i,m} \otimes f_{i,m}$. To obtain for $\{A_{\boldsymbol{n}}\}_n$ an a.c.s. of the form (5.2), it suffices to use the result of the corollary in combination with Lemma 2.4. □

Remark 5.11. Let $\{A_{\boldsymbol{n}}\}_n$, $\{B_{\boldsymbol{n},m}\}_n$ be matrix-sequences and let κ, $\kappa_m : [0,1]^d \times [-\pi,\pi]^d \to \mathbb{C}$ be measurable functions. Consider the following conditions:

(1) $\{A_{\boldsymbol{n}}\}_n \sim_{\mathrm{GLT}} \kappa$;
(2) $\{B_{\boldsymbol{n},m}\}_n \sim_{\mathrm{GLT}} \kappa_m$ for every m;
(3) $\{\{B_{\boldsymbol{n},m}\}_n\}_m$ is an a.c.s. for $\{A_{\boldsymbol{n}}\}_n$;
(4) $\kappa_m \to \kappa$ in measure.

Theorems 5.8–5.9 show that

$$(1) \wedge (2) \wedge (3) \Rightarrow (4), \qquad (1) \wedge (2) \wedge (4) \Rightarrow (3), \qquad (2) \wedge (3) \wedge (4) \Rightarrow (1).$$

The implication $(1) \wedge (3) \wedge (4) \Rightarrow (2)$, written in this way, is meaningless. However, a natural modification reads as follows: $(1) \wedge (3) \Rightarrow$ there exists a measurable function κ_m, tending to κ in measure, such that $\{B_{\boldsymbol{n},m}\}_n \sim_{\mathrm{GLT}} \kappa_m$ for all sufficiently large m. This statement is false in general. As a counterexample, take $A_n = O_n$ and $B_{n,m} = (1 + (-1)^n)\frac{1}{m} I_n$, as in Remark 3.7. Since we have seen in Remark 3.7 that the relation $\{B_{n,m}\}_n \sim_\sigma \phi_m$ cannot hold for any functional ϕ_m, in particular there is no κ_m such that $\{B_{n,m}\}_n \sim_{\mathrm{GLT}} \kappa_m$.

5.3. Characterizations of GLT sequences

As a first application of Theorem 5.8, we show in Proposition 5.12 that GLT sequences can be defined in terms of LT sequences instead of sLT sequences. Proposition 5.12 is then a characterization of GLT sequences in terms of LT sequences and, specifically, it is the same as Proposition 5.2 with "sLT" replaced by "LT".

Proposition 5.12. *We have* $\{A_{\boldsymbol{n}}\}_n \sim_{\rm GLT} \kappa$ *if and only if the following condition is met.*

For every m *varying in some infinite subset of* $\mathbb{N}$ *there is a finite number of LT sequences* $\{A_{\boldsymbol{n}}^{(i,m)}\}_n \sim_{\rm LT} a_{i,m} \otimes f_{i,m}$, $i = 1, \ldots, N_m$, *such that:*

- $\sum_{i=1}^{N_m} a_{i,m} \otimes f_{i,m} \to \kappa$ *in measure over* $[0,1]^d \times [-\pi,\pi]^d$ *as* $m \to \infty$;
- $\{\{\sum_{i=1}^{N_m} A_{\boldsymbol{n}}^{(i,m)}\}_n\}_m$ *is an a.c.s. for* $\{A_{\boldsymbol{n}}\}_n$.

Proof. It is clear that, if $\{A_{\boldsymbol{n}}\}_n \sim_{\rm GLT} \kappa$, then the condition holds by Proposition 5.2. Conversely, suppose that the condition holds. Then, $\{\{\sum_{i=1}^{N_m} A_{\boldsymbol{n}}^{(i,m)}\}_n\}_m$ is an a.c.s. for $\{A_{\boldsymbol{n}}\}_n$ by hypothesis, $\{\sum_{i=1}^{N_m} A_{\boldsymbol{n}}^{(i,m)}\}_n \sim_{\rm GLT} \sum_{i=1}^{N_m} a_{i,m} \otimes f_{i,m}$ by (5.1) and Proposition 5.4, and $\sum_{i=1}^{N_m} a_{i,m} \otimes f_{i,m} \to \kappa$ in measure by hypothesis. Hence, the thesis follows from Theorem 5.8. □

The next result is a characterization theorem for GLT sequences. All the provided characterizations have already been proved before, but it is anyway useful to collect them in a single statement.

Theorem 5.13. *Let* $\{A_{\boldsymbol{n}}\}_n$ *be a matrix-sequence and let* $\kappa : [0,1]^d \times [-\pi,\pi]^d \to \mathbb{C}$ *be a measurable function. Then, the following conditions are equivalent.*

1. $\{A_{\boldsymbol{n}}\}_n \sim_{\rm GLT} \kappa$.
2. *For all sequences* $\{\kappa_m\}_m$, $\{\{B_{\boldsymbol{n},m}\}_n\}_m$ *with the following properties:*
   - $\{B_{\boldsymbol{n},m}\}_n \sim_{\rm GLT} \kappa_m$ *for every* m;
   - $\kappa_m \to \kappa$ *in measure;*

   it holds that $\{\{B_{\boldsymbol{n},m}\}_n\}_m$ *is an a.c.s. for* $\{A_{\boldsymbol{n}}\}_n$.
3. *There exist functions* $a_{i,m}$, $f_{i,m}$, $i = 1, \ldots, N_m$, $m \in \mathbb{N}$, *such that:*
   - $a_{i,m} : [0,1]^d \to \mathbb{C}$ *belongs to* $C^\infty([0,1]^d)$ *and* $f_{i,m}$ *is a trigonometric monomial belonging to* $\{\mathrm{e}^{\mathrm{i}\boldsymbol{j}\cdot\boldsymbol{\theta}} : \boldsymbol{j} \in \mathbb{Z}^d\}$;
   - $\sum_{i=1}^{N_m} a_{i,m} \otimes f_{i,m} \to \kappa$ *a.e.;*
   - $\{\{\sum_{i=1}^{N_m} D_{\boldsymbol{n}}(a_{i,m}) T_{\boldsymbol{n}}(f_{i,m})\}_n\}_m$ *is an a.c.s. for* $\{A_{\boldsymbol{n}}\}_n$.
4. *There exist sequences* $\{\kappa_m\}_m$, $\{\{B_{\boldsymbol{n},m}\}_n\}_m$ *such that:*
   - $\{B_{\boldsymbol{n},m}\}_n \sim_{\rm GLT} \kappa_m$ *for every* m;
   - $\kappa_m \to \kappa$ *in measure;*
   - $\{\{B_{\boldsymbol{n},m}\}_n\}_m$ *is an a.c.s. for* $\{A_{\boldsymbol{n}}\}_n$.

Proof. The implication $(1 \Rightarrow 2)$ follows from Theorem 5.9. The implication $(2 \Rightarrow 3)$ follows from the observation that, by Lemma 2.4, we can find functions $a_{i,m}$, $f_{i,m}$, $i = 1, \ldots, N_m$, $m \in \mathbb{N}$, with the first two properties specified in item 3, and, moreover, $\{\sum_{i=1}^{N_m} D_{\boldsymbol{n}}(a_{i,m}) T_{\boldsymbol{n}}(f_{i,m})\}_n \sim_{\rm GLT} \sum_{i=1}^{N_m} a_{i,m} \otimes f_{i,m}$. The implication

$(3 \Rightarrow 4)$ is obvious (take $B_{\boldsymbol{n},m} = \sum_{i=1}^{N_m} D_{\boldsymbol{n}}(a_{i,m})T_{\boldsymbol{n}}(f_{i,m})$ and $\kappa_m = \sum_{i=1}^{N_m} a_{i,m} \otimes f_{i,m}$). Finally, the implication $(4 \Rightarrow 1)$ is Theorem 5.8. □

5.4. The GLT algebra

We investigate in this section the important algebraic properties possessed by GLT sequences, which give rise to the so-called GLT algebra. These properties establish that, if $\{A_{\boldsymbol{n}}^{(1)}\}_n, \ldots, \{A_{\boldsymbol{n}}^{(r)}\}_n$ are given GLT sequences with symbols $\kappa_1, \ldots, \kappa_r$, and if $A_{\boldsymbol{n}} = \text{ops}(A_{\boldsymbol{n}}^{(1)}, \ldots, A_{\boldsymbol{n}}^{(r)})$ is obtained from $A_{\boldsymbol{n}}^{(1)}, \ldots, A_{\boldsymbol{n}}^{(r)}$ by means of certain operations "ops", then $\{A_{\boldsymbol{n}}\}_n$ is a GLT sequence with symbol $\kappa = \text{ops}(\kappa_1, \ldots, \kappa_r)$.

Theorem 5.14 (Serra-Capizzano). *If $\{A_{\boldsymbol{n}}\}_n \sim_{\rm GLT} \kappa$ and $\{B_{\boldsymbol{n}}\}_n \sim_{\rm GLT} \xi$, then:*

1. $\{A_{\boldsymbol{n}}^*\}_n \sim_{\rm GLT} \overline{\kappa}$;
2. $\{\alpha A_{\boldsymbol{n}} + \beta B_{\boldsymbol{n}}\}_n \sim_{\rm GLT} \alpha\kappa + \beta\xi$, *for all* $\alpha, \beta \in \mathbb{C}$;
3. $\{A_{\boldsymbol{n}} B_{\boldsymbol{n}}\}_n \sim_{\rm GLT} \kappa\xi$.

Proof. The first two statements have already been settled before (see Proposition 5.4). We prove the third statement. By assumption and Proposition 5.12, there exist LT sequences $\{A_{\boldsymbol{n}}^{(i,m)}\}_n \sim_{\rm LT} a_{i,m} \otimes f_{i,m}$, $i = 1, \ldots, N_m$, and $\{B_{\boldsymbol{n}}^{(j,m)}\}_n \sim_{\rm LT} b_{j,m} \otimes g_{j,m}$, $j = 1, \ldots, M_m$, such that:

- $\sum_{i=1}^{N_m} a_{i,m} \otimes f_{i,m} \to \kappa$ in measure and $\sum_{j=1}^{M_m} b_{j,m} \otimes g_{j,m} \to \xi$ in measure;
- $\{\{\sum_{i=1}^{N_m} A_{\boldsymbol{n}}^{(i,m)}\}_n\}_m$ is a.c.s. of $\{A_{\boldsymbol{n}}\}_n$ and $\{\{\sum_{j=1}^{M_m} B_{\boldsymbol{n}}^{(j,m)}\}_n\}_m$ is a.c.s. of $\{B_{\boldsymbol{n}}\}_n$.

Thanks to Theorem 5.13 (item 3), the functions $f_{i,m}$, $g_{j,m}$ may be supposed to be in $L^\infty([-\pi, \pi]^d)$.[4] By Theorem 5.3, $\{A_{\boldsymbol{n}}\}_n \sim_\sigma \kappa$ and $\{B_{\boldsymbol{n}}\}_n \sim_\sigma \xi$, which implies, by Proposition 3.11, that $\{A_{\boldsymbol{n}}\}_n$ and $\{B_{\boldsymbol{n}}\}_n$ are s.u. Thus, by Proposition 3.12,

$$\left\{\left\{\left(\sum_{i=1}^{N_m} A_{\boldsymbol{n}}^{(i,m)}\right)\left(\sum_{j=1}^{M_m} B_{\boldsymbol{n}}^{(j,m)}\right)\right\}_n\right\}_m = \left\{\left\{\sum_{i=1}^{N_m}\sum_{j=1}^{M_m} A_{\boldsymbol{n}}^{(i,m)} B_{\boldsymbol{n}}^{(j,m)}\right\}_n\right\}_m$$

is an a.c.s. for $\{A_{\boldsymbol{n}} B_{\boldsymbol{n}}\}_n$. Since $f_{i,m}, g_{j,m} \in L^\infty([-\pi, \pi]^d)$, it follows from $LT8$ that $\{A_{\boldsymbol{n}}^{(i,m)} B_{\boldsymbol{n}}^{(j,m)}\}_n \sim_{\rm LT} a_{i,m} b_{j,m} \otimes f_{i,m} g_{j,m}$, $i = 1, \ldots, N_m$, $j = 1, \ldots, M_m$. Finally,

$$\sum_{i=1}^{N_m}\sum_{j=1}^{M_m} a_{i,m} b_{j,m} \otimes f_{i,m} g_{j,m} = \left(\sum_{i=1}^{N_m} a_{i,m} \otimes f_{i,m}\right)\left(\sum_{j=1}^{M_m} b_{j,m} \otimes g_{j,m}\right) \to \kappa\xi$$

in measure, because it is known that $\varphi_m \psi_m \to \varphi\psi$ in measure over Ω whenever $\varphi_m \to \varphi$ in measure over Ω, $\psi_m \to \psi$ in measure over Ω, and the measure of Ω is finite. The application of Proposition 5.12 concludes the proof. □

Corollary 5.15. *Let $r, q_1, \ldots, q_r \in \mathbb{N}$, $\alpha_1, \ldots, \alpha_r \in \mathbb{C}$, and $\{A_{\boldsymbol{n}}^{(i,j)}\}_n \sim_{\rm GLT} \kappa_{ij}$ for $i = 1, \ldots, r$, $j = 1, \ldots, q_i$. Then $\left\{\sum_{i=1}^r \alpha_i \prod_{j=1}^{q_i} A_{\boldsymbol{n}}^{(i,j)}\right\}_n \sim_{\rm GLT} \sum_{i=1}^r \alpha_i \prod_{j=1}^{q_i} \kappa_{ij}$.*

---

[4] Actually, they may be supposed to be trigonometric monomials, the functions $a_{i,m}$, $b_{j,m}$ may be supposed to belong to $C^\infty([0,1]^d)$, and $\{A_{\boldsymbol{n}}^{(i,m)}\}_n$, $\{B_{\boldsymbol{n}}^{(j,m)}\}_n$ may be chosen of the form $\{D_{\boldsymbol{n}}(a_{i,m})T_{\boldsymbol{n}}(f_{i,m})\}_n$, $\{D_{\boldsymbol{n}}(b_{j,m})T_{\boldsymbol{n}}(g_{j,m})\}_n$.

The results we have seen so far are enough to conclude that the set of GLT sequences is a *-algebra over the complex field $\mathbb{C}$. More precisely, fix any sequence of d-indices $\{\boldsymbol{n}=\boldsymbol{n}(n)\}_n \subseteq \mathbb{N}^d$ such that $\boldsymbol{n}\to\infty$ when $n\to\infty$; then,

$$\mathscr{A} = \big\{\{A_{\boldsymbol{n}}\}_n : \{A_{\boldsymbol{n}}\}_n \sim_{\mathrm{GLT}} \kappa \quad \text{for some measurable function } \kappa : [0,1]^d \times [-\pi,\pi]^d \to \mathbb{C}\big\} \tag{5.3}$$

is a *-algebra over $\mathbb{C}$, with respect to the natural operations of conjugate transposition, addition, scalar-multiplication and product of matrix-sequences: $\{A_{\boldsymbol{n}}\}_n^* = \{A_{\boldsymbol{n}}^*\}_n$, $\{A_{\boldsymbol{n}}\}_n + \{B_{\boldsymbol{n}}\}_n = \{A_{\boldsymbol{n}} + B_{\boldsymbol{n}}\}_n$, $\alpha\{A_{\boldsymbol{n}}\}_n = \{\alpha A_{\boldsymbol{n}}\}_n$, $\{A_{\boldsymbol{n}}\}_n\{B_{\boldsymbol{n}}\}_n = \{A_{\boldsymbol{n}}B_{\boldsymbol{n}}\}_n$. We call $\mathscr{A}$ the GLT algebra. $\mathscr{A}$ contains the algebra generated by zero-distributed sequences, Toeplitz sequences and sequences of diagonal sampling matrices, because these matrix-sequences fall in the class of GLT sequences (by $LT2$–$LT4$ and (5.1)). To be precise, let

$$\mathscr{B} = \left\{\left\{\sum_{i=1}^{r}\prod_{j=1}^{q_i} X_{\boldsymbol{n}}^{(i,j)}\right\} : r, q_1,\ldots,q_r \in \mathbb{N}, \quad \{X_{\boldsymbol{n}}^{(i,j)}\}_n \in \mathcal{B} \text{ for all } i=1,\ldots,r \text{ and } j=1,\ldots,q_i\right\},$$

where

$$\mathcal{B} = \big\{\{Z_{\boldsymbol{n}}\}_n :\ \{Z_{\boldsymbol{n}}\}_n \sim_\sigma 0\big\} \cup \big\{\{T_{\boldsymbol{n}}(g)\}_n :\ g \in L^1([-\pi,\pi]^d)\big\} \cup \big\{\{D_{\boldsymbol{n}}(a)\}_n :\ a : [0,1]^d \to \mathbb{C} \text{ is Riemann-integrable}\big\}.$$

Then, $\mathscr{B}$ is the algebra generated by $\mathcal{B}$ and $\mathscr{B} \subseteq \mathscr{A}$. We are going to see in Theorems 5.16 and 5.19 that the GLT algebra enjoys other nice properties, in addition to those of Theorem 5.14, which make it look like a "big container", closed under any type of "regular" operation.

Theorem 5.16 provides a positive answer to a question raised in [41]. Incidentally, we note that in [41] the authors proved that $\{\{f(B_{\boldsymbol{n},m})\}_n\}_m$ is an a.c.s. for $\{f(A_{\boldsymbol{n}})\}_n$ whenever $\{\{B_{\boldsymbol{n},m}\}_n\}_m$ is an a.c.s. for $\{A_{\boldsymbol{n}}\}_n$ and $f:\mathbb{R}\to\mathbb{R}$ is continuous (and other mild assumptions are met); this result enlarges the algebraic properties of a.c.s.es studied in Section 3.2.

Theorem 5.16 (Garoni). *Let $\{A_{\boldsymbol{n}}\}_n \sim_{\mathrm{GLT}} \kappa$ and suppose that the matrices $A_{\boldsymbol{n}}$ are Hermitian. Then $\{f(A_{\boldsymbol{n}})\}_n \sim_{\mathrm{GLT}} f(\kappa)$ for any continuous function $f:\mathbb{R}\to\mathbb{R}$.*[5]

Proof. For each $M>0$, let $\{p_{m,M}\}_m$ be a sequence of polynomials that converges uniformly to f over the compact interval $[-M,M]$:

$$\lim_{m\to\infty} \|f - p_{m,M}\|_{\infty,[-M,M]} = 0.$$

For every $M>0$ and every m,n, write

$$f(A_{\boldsymbol{n}}) = p_{m,M}(A_{\boldsymbol{n}}) + f(A_{\boldsymbol{n}}) - p_{m,M}(A_{\boldsymbol{n}}). \tag{5.4}$$

[5]Proposition 5.6 ensures that $\kappa \in \mathbb{R}$ a.e., so $f(\kappa)$ is well defined.

Since any GLT sequence is s.u. (by Theorem 5.3 and Proposition 3.11), the sequence $\{A_{\boldsymbol{n}}\}_n$ is s.u. Hence, by Proposition 3.10, for all $M > 0$ there exists n_M such that, for $n \geq n_M$,

$$A_{\boldsymbol{n}} = \hat{A}_{\boldsymbol{n},M} + \tilde{A}_{\boldsymbol{n},M}, \qquad \operatorname{rank}(\hat{A}_{\boldsymbol{n},M}) \leq r(M)N(\boldsymbol{n}), \qquad \|\tilde{A}_{\boldsymbol{n},M}\| \leq M, \quad (5.5)$$

where $\lim_{M\to\infty} r(M) = 0$. However, for the purposes of this proof we need a splitting of the form (5.5) such that $g(\hat{A}_{\boldsymbol{n},M} + \tilde{A}_{\boldsymbol{n},M}) = g(\hat{A}_{\boldsymbol{n},M}) + g(\tilde{A}_{\boldsymbol{n},M})$ for all functions $g : \mathbb{R} \to \mathbb{R}$. Luckily, the matrices $A_{\boldsymbol{n}}$ are Hermitian and, consequently, such a splitting can be constructed by following the same argument that is used for proving Proposition 3.10. For the reader's convenience, we include the details of the construction. By definition, since $\{A_{\boldsymbol{n}}\}_n$ is s.u. and formed by Hermitian matrices, for every $M > 0$ there exists n_M such that, for $n \geq n_M$,

$$\frac{\#\{i \in \{1,\ldots,N(\boldsymbol{n})\} : \ |\lambda_i(A_{\boldsymbol{n}})| > M\}}{N(\boldsymbol{n})} \leq r(M),$$

where $\lim_{M\to\infty} r(M) = 0$. Let $A_{\boldsymbol{n}} = U_{\boldsymbol{n}}\Lambda_{\boldsymbol{n}}U_{\boldsymbol{n}}^*$ be a spectral decomposition of $A_{\boldsymbol{n}}$. Let $\hat{\Lambda}_{\boldsymbol{n},M}$ be the matrix obtained from $\Lambda_{\boldsymbol{n}}$ by setting to 0 all the eigenvalues of $A_{\boldsymbol{n}}$ whose absolute value is less than or equal to M, and let $\tilde{\Lambda}_{\boldsymbol{n},M} = \Lambda_{\boldsymbol{n}} - \hat{\Lambda}_{\boldsymbol{n},M}$ be the matrix obtained from $\Lambda_{\boldsymbol{n}}$ by setting to 0 all the eigenvalues of $A_{\boldsymbol{n}}$ whose absolute value is greater than M. Then, for $M > 0$ and $n \geq n_M$,

$$A_{\boldsymbol{n}} = U_{\boldsymbol{n}}\Lambda_{\boldsymbol{n}}U_{\boldsymbol{n}}^* = U_{\boldsymbol{n}}\hat{\Lambda}_{\boldsymbol{n},M}U_{\boldsymbol{n}}^* + U_{\boldsymbol{n}}\tilde{\Lambda}_{\boldsymbol{n},M}U_{\boldsymbol{n}}^* = \hat{A}_{\boldsymbol{n},M} + \tilde{A}_{\boldsymbol{n},M},$$

where $\hat{A}_{\boldsymbol{n},M} = U_{\boldsymbol{n}}\hat{\Lambda}_{\boldsymbol{n},M}U_{\boldsymbol{n}}^*$ and $\tilde{A}_{\boldsymbol{n},M} = U_{\boldsymbol{n}}\tilde{\Lambda}_{\boldsymbol{n},M}U_{\boldsymbol{n}}^*$. The matrices $\hat{A}_{\boldsymbol{n},M}$, $\tilde{A}_{\boldsymbol{n},M}$ constructed in this way are Hermitian, satisfy the properties in (5.5) and, moreover,

$$g(\hat{A}_{\boldsymbol{n},M} + \tilde{A}_{\boldsymbol{n},M}) = g(\hat{A}_{\boldsymbol{n},M}) + g(\tilde{A}_{\boldsymbol{n},M}) = U_{\boldsymbol{n}}\, g(\hat{\Lambda}_{\boldsymbol{n},M})U_{\boldsymbol{n}}^* + U_{\boldsymbol{n}}\, g(\tilde{\Lambda}_{\boldsymbol{n},M})U_{\boldsymbol{n}}^*$$

for all functions $g : \mathbb{R} \to \mathbb{R}$.

Going back to (5.4), for each $M > 0$, each m and each $n \geq n_M$ we can write

$$\begin{aligned} f(A_{\boldsymbol{n}}) &= p_{m,M}(A_{\boldsymbol{n}}) + f(\hat{A}_{\boldsymbol{n},M}) + f(\tilde{A}_{\boldsymbol{n},M}) - p_{m,M}(\hat{A}_{\boldsymbol{n},M}) - p_{m,M}(\tilde{A}_{\boldsymbol{n},M}) \\ &= p_{m,M}(A_{\boldsymbol{n}}) + (f - p_{m,M})(\hat{A}_{\boldsymbol{n},M}) + (f - p_{m,M})(\tilde{A}_{\boldsymbol{n},M}). \end{aligned} \quad (5.6)$$

The term $(f - p_{m,M})(\hat{A}_{\boldsymbol{n},M})$ can be split in the sum of two terms $R_{\boldsymbol{n},m,M} + N'_{\boldsymbol{n},m,M}$: $R_{\boldsymbol{n},m,M}$ is obtained from $(f - p_{m,M})(\hat{A}_{\boldsymbol{n},M})$ by setting to 0 all the eigenvalues that are equal to $(f - p_{m,M_m})(0)$, so that $\operatorname{rank}(R_{\boldsymbol{n},m,M}) \leq \operatorname{rank}(\hat{A}_{\boldsymbol{n},M})$; $N'_{\boldsymbol{n},m,M}$ is obtained from $(f - p_{m,M})(\hat{A}_{\boldsymbol{n},M})$ by setting to 0 all the eigenvalues that are different from $(f - p_{m,M_m})(0)$. Let $N''_{\boldsymbol{n},m,M} = (f - p_{m,M})(\tilde{A}_{\boldsymbol{n},M})$ and $N_{\boldsymbol{n},m,M} = N'_{\boldsymbol{n},m,M} + N''_{\boldsymbol{n},m,M}$. From (5.6), for each $M > 0$, each m and each $n \geq n_M$ we have

$$f(A_{\boldsymbol{n}}) = p_{m,M}(A_{\boldsymbol{n}}) + R_{\boldsymbol{n},m,M} + N_{\boldsymbol{n},m,M},$$

and, by our construction,

$$\operatorname{rank}(R_{\boldsymbol{n},m,M}) \le \operatorname{rank}(\hat{A}_{\boldsymbol{n},M}) \le r(M)N(\boldsymbol{n}),$$
$$\|N_{\boldsymbol{n},m,M}\| \le |f(0) - p_{m,M}(0)| + \|f - p_{m,M}\|_{\infty,[-M,M]} \le 2\|f - p_{m,M}\|_{\infty,[-M,M]}.$$

Choose a sequence $\{M_m\}_m$ such that, when $m \to \infty$,

$$M_m \to \infty, \qquad \|f - p_{m,M_m}\|_{\infty,[-M_m,M_m]} \to 0. \tag{5.7}$$

Then, for every m and every $n \ge n_{M_m}$,

$$f(A_{\boldsymbol{n}}) = p_{m,M_m}(A_{\boldsymbol{n}}) + R_{\boldsymbol{n},m,M_m} + N_{\boldsymbol{n},m,M_m},$$

$$\operatorname{rank}(R_{\boldsymbol{n},m,M_m}) \le r(M_m)N(\boldsymbol{n}), \qquad \|N_{\boldsymbol{n},m,M_m}\| \le 2\|f - p_{m,M_m}\|_{\infty,[-M_m,M_m]},$$

which implies that $\{\{p_{m,M_m}(A_{\boldsymbol{n}})\}_n\}_m$ is an a.c.s. for $\{f(A_{\boldsymbol{n}})\}_n$. Moreover, we have $\{p_{m,M_m}(A_{\boldsymbol{n}})\}_n \sim_{\mathrm{GLT}} p_{m,M_m}(\kappa)$ by Corollary 5.15. Finally, $p_{m,M_m}(\kappa) \to f(\kappa)$ a.e. in $[0,1]^d \times [-\pi,\pi]^d$, due to (5.7). In conclusion, all the hypotheses of Theorem 5.8 are satisfied and $\{f(A_{\boldsymbol{n}})\}_n \sim_{\mathrm{GLT}} f(\kappa)$. □

The last issue we are interested in is to know if $\{A_{\boldsymbol{n}}^{-1}\}_n \sim_{\mathrm{GLT}} \kappa^{-1}$ in the case where $\{A_{\boldsymbol{n}}\}_n \sim_{\mathrm{GLT}} \kappa$, each $A_{\boldsymbol{n}}$ is invertible, and $\kappa \ne 0$ a.e. (so that κ^{-1} is a well-defined measurable function). More in general, we may ask if $\{A_{\boldsymbol{n}}^{\dagger}\}_n \sim_{\mathrm{GLT}} \kappa^{-1}$ when $\{A_{\boldsymbol{n}}\} \sim_{\mathrm{GLT}} \kappa$ and $\kappa \ne 0$ a.e., where $A_{\boldsymbol{n}}^{\dagger}$ is the (Moore–Penrose) pseudoinverse of $\{A_{\boldsymbol{n}}\}_n$. The answer to both the previous questions is affirmative, but some work is needed to bring out the related proofs. Note that these results cannot be inferred from Theorem 5.16, because the matrices $A_{\boldsymbol{n}}$ may fail to be Hermitian and, moreover, $f(x) = x^{-1}$ is not continuous on $\mathbb{R}$. We begin by introducing the concept of sparsely vanishing (s.v.) matrix-sequences.

Definition 5.17 (s.v. matrix-sequence). We say that a matrix-sequence $\{A_{\boldsymbol{n}}\}_n$ is sparsely vanishing (s.v.) if for every $M > 0$ there exists n_M such that, for $n \ge n_M$,

$$\frac{\#\{i \in \{1,\dots,N(\boldsymbol{n})\} : \sigma_i(A_{\boldsymbol{n}}) < 1/M\}}{N(\boldsymbol{n})} \le r(M),$$

where $\lim_{M\to\infty} r(M) = 0$.

Note that if $\{A_{\boldsymbol{n}}\}_n$ is s.v. then $\{A_{\boldsymbol{n}}^{\dagger}\}_n$ is s.u. To see this, it suffices to recall that the singular values of $A^{\dagger}$ are $1/\sigma_1(A),\dots,1/\sigma_r(A),0,\dots,0$, where $\sigma_1(A)\dots\sigma_r(A)$ are the nonzero singular values of A ($r = \operatorname{rank}(A)$).

Proposition 5.18 is the analog of Proposition 3.11 for s.v. matrix-sequences. For the proof, we refer the reader to [25, Proposition 5.6].

Proposition 5.18. *If $\{A_{\boldsymbol{n}}\}_n \sim_\sigma f$ then $\{A_{\boldsymbol{n}}\}_n$ is s.v. if and only if $f \ne 0$ a.e.*

Theorem 5.19 (Garoni, Serra-Capizzano). *If $\{A_{\boldsymbol{n}}\}_n \sim_{\mathrm{GLT}} \kappa$ and $\kappa \ne 0$ a.e., then $\{A_{\boldsymbol{n}}^{\dagger}\}_n \sim_{\mathrm{GLT}} \kappa^{-1}$.*

Proof. Fix a sequence of matrix-sequences $\{\{B_{\boldsymbol{n},m}\}_n\}_m$ such that $\{B_{\boldsymbol{n},m}\}_m \sim_{\rm GLT} \xi_m$ for every m and $\xi_m \to \kappa^{-1}$ a.e. Note that a sequence with these properties exists. Indeed, by Lemma 2.4 there is a sequence $\{\xi_m\}_m$, with ξ_m of the form

$$\xi_m(\mathbf{x},\boldsymbol{\theta}) = \sum_{\boldsymbol{j}=-\boldsymbol{N}_m}^{\boldsymbol{N}_m} a_{\boldsymbol{j}}^{(m)}(\mathbf{x})\mathrm{e}^{\mathrm{i}\boldsymbol{j}\cdot\boldsymbol{\theta}}, \qquad a_{\boldsymbol{j}}^{(m)} \in C^\infty([0,1]^d), \qquad \boldsymbol{N}_m \in \mathbb{N}^d,$$

such that $\xi_m \to \kappa^{-1}$ a.e. Take $B_{\boldsymbol{n},m} = \sum_{\boldsymbol{j}=-\boldsymbol{N}_m}^{\boldsymbol{N}_m} D_{\boldsymbol{n}}(a_{\boldsymbol{j}}^{(m)})T_{\boldsymbol{n}}(\mathrm{e}^{\mathrm{i}\boldsymbol{j}\cdot\boldsymbol{\theta}})$ and observe that $\{B_{\boldsymbol{n},m}\}_n \sim_{\rm GLT} \xi_m$ by *LT*2–*LT*3, (5.1), and Theorem 5.14.

By Theorem 5.14, we have $\{I_{N(\boldsymbol{n})} - B_{\boldsymbol{n},m}A_{\boldsymbol{n}}\}_n \sim_{\rm GLT} 1 - \xi_m\kappa$ for every m, and $1 - \xi_m\kappa \to 0$ a.e. (and hence also in measure). Therefore, by Theorem 3.14, for every m there exists n_m such that, for $n \ge n_m$,

$$\begin{aligned} B_{\boldsymbol{n},m}A_{\boldsymbol{n}} &= I_{N(\boldsymbol{n})} + R_{\boldsymbol{n},m} + N_{\boldsymbol{n},m}, \\ \mathrm{rank}(R_{\boldsymbol{n},m}) &\le c(m)N(\boldsymbol{n}), \qquad \|N_{\boldsymbol{n},m}\| \le \omega(m), \end{aligned} \tag{5.8}$$

where $\lim_{m\to\infty} c(m) = \lim_{m\to\infty}\omega(m) = 0$. Multiplying (5.8) by $A_{\boldsymbol{n}}^\dagger$, we obtain that, for every m and every $n \ge n_m$,

$$B_{\boldsymbol{n},m}\, A_{\boldsymbol{n}}A_{\boldsymbol{n}}^\dagger = A_{\boldsymbol{n}}^\dagger + (R_{\boldsymbol{n},m} + N_{\boldsymbol{n},m})A_{\boldsymbol{n}}^\dagger. \tag{5.9}$$

Since $\kappa \ne 0$ a.e. by hypothesis, $\{A_{\boldsymbol{n}}\}_n$ is s.v. (by Theorem 5.3 and Proposition 5.18). It follows that $\{A_{\boldsymbol{n}}^\dagger\}_n$ is s.u. and so, by Proposition 3.10, for all $M > 0$ there is $\overline{n}_M$ such that, for $n \ge \overline{n}_M$,

$$\begin{aligned} A_{\boldsymbol{n}}^\dagger &= \hat{A}_{\boldsymbol{n},M}^\dagger + \tilde{A}_{\boldsymbol{n},M}^\dagger, \\ \mathrm{rank}(\hat{A}_{\boldsymbol{n},M}^\dagger) &\le r(M)N(\boldsymbol{n}), \qquad \|\tilde{A}_{\boldsymbol{n},M}^\dagger\| \le M, \end{aligned}$$

where $\lim_{M\to\infty} r(M) = 0$. Choosing $M_m = \omega(m)^{-1/2}$, from (5.9) we see that, for every m and every $n \ge \max(n_m, \overline{n}_{M_m})$,

$$\begin{aligned} B_{\boldsymbol{n},m}\, A_{\boldsymbol{n}}A_{\boldsymbol{n}}^\dagger &= A_{\boldsymbol{n}}^\dagger + R'_{\boldsymbol{n},m} + N'_{\boldsymbol{n},m}, \\ \mathrm{rank}(R'_{\boldsymbol{n},m}) &\le \big(c(m) + r(M_m)\big)N(\boldsymbol{n}), \qquad \|N'_{\boldsymbol{n},m}\| \le \omega(m)^{1/2}, \end{aligned} \tag{5.10}$$

where we have set $R'_{\boldsymbol{n},m} = R_{\boldsymbol{n},m}A_{\boldsymbol{n}}^\dagger + N_{\boldsymbol{n},m}\hat{A}_{\boldsymbol{n},M_m}^\dagger$ and $N'_{\boldsymbol{n},m} = N_{\boldsymbol{n},m}\tilde{A}_{\boldsymbol{n},M_m}^\dagger$.

If the matrices $A_{\boldsymbol{n}}$ were invertible, then $A_{\boldsymbol{n}}^\dagger = A_{\boldsymbol{n}}^{-1}$ and (5.10) would imply that $\{\{B_{\boldsymbol{n},m}\}_n\}_m$ is an a.c.s. for $\{A_{\boldsymbol{n}}^{-1}\}_n$; this, in combination with Theorem 5.8, would conclude the proof. In the general case where the matrices $A_{\boldsymbol{n}}$ are not invertible, the thesis will follow again from (5.10) and Theorem 5.8 as soon as we have proved the following: for every m there exists $\hat{n}_m$ such that, for $n \ge \hat{n}_m$,

$$A_{\boldsymbol{n}}A_{\boldsymbol{n}}^\dagger = I_{N(\boldsymbol{n})} + S_{\boldsymbol{n}}, \qquad \mathrm{rank}(S_{\boldsymbol{n}}) \le \theta(m)N(\boldsymbol{n}),$$

where $\lim_{m\to\infty}\theta(m) = 0$. This is easy, because, by definition of $A_{\boldsymbol{n}}^\dagger$, the rank of $S_{\boldsymbol{n}} = A_{\boldsymbol{n}}A_{\boldsymbol{n}}^\dagger - I_{N(\boldsymbol{n})}$ is given by $\mathrm{rank}(S_{\boldsymbol{n}}) = \#\{i \in \{1,\ldots,N(\boldsymbol{n})\} : \sigma_i(A_{\boldsymbol{n}}) = 0\}$. Hence, the previous claim follows directly from the fact that $\{A_{\boldsymbol{n}}\}_n$ is s.v. □

5.5. Summary of the theory of GLT sequences

After developing the theory of GLT sequences, we note at this point that we have proved all the properties **GLT 1**–**GLT 9** of Section 1.2. In particular, **GLT 1** was proved in Theorems 5.3 and 5.7, **GLT 2**–**GLT 4** follow from *LT*2–*LT*4 and (5.1), **GLT 5**–**GLT 8** were proved in Theorems 5.14, 5.16, 5.19, and **GLT 9** is the equivalence ($1 \Leftrightarrow 4$) of Theorem 5.13.

We note however that, in Section 1.2, we adopted for simplicity a not completely precise notation and we used "n" instead of "$\boldsymbol{n}$". For the reader's convenience, we report again **GLT 1**–**GLT 9** in the correct notation. We also simplify a little bit their statements and add further details.

GLT 1. If $\{A_{\boldsymbol{n}}\}_n \sim_{\rm GLT} \kappa$, then $\{A_{\boldsymbol{n}}\}_n \sim_\sigma \kappa$. If moreover the matrices $A_{\boldsymbol{n}}$ are Hermitian, then $\{A_{\boldsymbol{n}}\}_n \sim_\lambda \kappa$.
GLT 2. $\{T_{\boldsymbol{n}}(f)\}_n \sim_{\rm GLT} 1 \otimes f$ for every $f \in L^1([-\pi,\pi]^d)$.
GLT 3. $\{D_{\boldsymbol{n}}(a)\}_n \sim_{\rm GLT} a \otimes 1$ for every Riemann-integrable $a : [0,1]^d \to \mathbb{C}$.
GLT 4. $\{Z_{\boldsymbol{n}}\}_n \sim_\sigma 0$ if and only if $\{Z_{\boldsymbol{n}}\}_n \sim_{\rm GLT} 0$.
GLT 5. If $\{A_{\boldsymbol{n}}\}_n \sim_{\rm GLT} \kappa$ then $\{A_{\boldsymbol{n}}^*\}_n \sim_{\rm GLT} \overline{\kappa}$.
GLT 6. If $A_{\boldsymbol{n}} = \sum_{i=1}^r \alpha_i \prod_{j=1}^{q_i} A_{\boldsymbol{n}}^{(i,j)}$, where $r, q_1, \ldots, q_r \in \mathbb{N}$, $\alpha_1, \ldots, \alpha_r \in \mathbb{C}$ and $\{A_{\boldsymbol{n}}^{(i,j)}\}_n \sim_{\rm GLT} \kappa_{ij}$, then $\{A_{\boldsymbol{n}}\}_n \sim_{\rm GLT} \kappa = \sum_{i=1}^r \alpha_i \prod_{j=1}^{q_i} \kappa_{ij}$.
GLT 7. If $\{A_{\boldsymbol{n}}\}_n \sim_{\rm GLT} \kappa$ and $\kappa \neq 0$ a.e., then $\{A_{\boldsymbol{n}}^\dagger\}_n \sim_{\rm GLT} \kappa^{-1}$.
GLT 8. If $\{A_{\boldsymbol{n}}\}_n \sim_{\rm GLT} \kappa$ and each $A_{\boldsymbol{n}}$ is Hermitian, then $\{f(A_{\boldsymbol{n}})\}_n \sim_{\rm GLT} f(\kappa)$ for all continuous functions $f : \mathbb{R} \to \mathbb{R}$.
GLT 9. $\{A_{\boldsymbol{n}}\}_n \sim_{\rm GLT} \kappa$ if and only if there exist GLT sequences $\{B_{\boldsymbol{n},m}\}_n \sim_{\rm GLT} \kappa_m$ such that $\kappa_m \to \kappa$ in measure and $\{\{B_{\boldsymbol{n},m}\}_n\}_m$ is an a.c.s. for $\{A_{\boldsymbol{n}}\}_n$.

6. Applications

In this section, we present a few representative applications of the theory of GLT sequences. For more applications, refer to the literature cited in Section 1.1.

6.1. The algebra generated by Toeplitz sequences

Fix a sequence of d-indices $\{\boldsymbol{n} = \boldsymbol{n}(n)\}_n \subseteq \mathbb{N}^d$ such that $\boldsymbol{n} \to \infty$ as $n \to \infty$. Let $\mathscr{T}$ be the algebra over $\mathbb{C}$ generated by the Toeplitz sequences $\{T_{\boldsymbol{n}}(g)\}_n$, $g \in L^1([-\pi,\pi]^d)$, i.e.,

$$\mathscr{T} = \left\{ \left\{ \sum_{i=1}^r \prod_{j=1}^{q_i} T_{\boldsymbol{n}}(g_{ij}) \right\}_n : r, q_1, \ldots, q_r \in \mathbb{N}, \right. \\ \left. g_{ij} \in L^1([-\pi,\pi]^d) \text{ for all } i = 1, \ldots, r \text{ and } j = 1, \ldots, q_i \right\}. \tag{6.1}$$

The algebra $\mathscr{T}$ is a sub-algebra of the GLT algebra $\mathscr{A}$ defined in (5.3). Indeed, by **GLT 2** and **GLT 6**, we have

$$\left\{\sum_{i=1}^{r}\prod_{j=1}^{q_i}T_{\boldsymbol{n}}(g_{ij})\right\}_{\boldsymbol{n}} \sim_{\rm GLT} \sum_{i=1}^{r}\prod_{j=1}^{q_i} 1\otimes g_{ij} = 1\otimes\sum_{i=1}^{r}\prod_{j=1}^{q_i} g_{ij}.$$

Taking into account Definition 2.5, **GLT 1** yields

$$\left\{\sum_{i=1}^{r}\prod_{j=1}^{q_i}T_{\boldsymbol{n}}(g_{ij})\right\}_{\boldsymbol{n}} \sim_{\sigma} \sum_{i=1}^{r}\prod_{j=1}^{q_i} g_{ij}. \qquad (6.2)$$

Moreover, if the matrices $\sum_{i=1}^{r}\prod_{j=1}^{q_i}T_{\boldsymbol{n}}(g_{ij})$ are Hermitian, **GLT 1** gives

$$\left\{\sum_{i=1}^{r}\prod_{j=1}^{q_i}T_{\boldsymbol{n}}(g_{ij})\right\}_{\boldsymbol{n}} \sim_{\lambda} \sum_{i=1}^{r}\prod_{j=1}^{q_i} g_{ij}. \qquad (6.3)$$

The result in (6.2) was originally obtained in [38]. In the case where $d = 1$ and $g_{ij} \in L^\infty([-\pi,\pi])$ for all $i, j = 1,\ldots,d$, both (6.2) and (6.3) already appeared in [10, Section 5.7]. Clearly, the distribution relations (6.2)–(6.3) extend Theorem 2.7.

The extension of the spectral distribution relation (6.3) to the case where the matrices $\sum_{i=1}^{r}\prod_{j=1}^{q_i}T_{\boldsymbol{n}}(g_{ij})$ are not Hermitian has been the subject of recent researches [18, 42]. Note that, if we remove the hypothesis of Hermitianess, then we necessarily have to add some additional assumption. Indeed, (6.3) does not hold in general; a simple counterexample is provided, e.g., by the sequence of Toeplitz matrices $\{T_n(\mathrm{e}^{\mathrm{i}j\theta})\}_n$. The hypothesis added in [18, 42] is a topological assumption on the range of the functions g_{ij}. A completely analogous hypothesis was already used in [17, 19] and, especially, in the pioneering work by Tilli [47], in order to extend the spectral distribution relation expressed in Theorem 2.7 to the case where the generating function f is not real (and hence the related Toeplitz matrices $T_{\boldsymbol{n}}(f)$ are not Hermitian).

We conclude this section by observing that the theory of GLT sequences allows us to manage matrix-sequences obtained from more complicated operations on Toeplitz sequences than sums and products. For example, in a context of Toeplitz preconditioning, one is interested in the singular value and eigenvalue distribution of a sequence of preconditioned matrices of the form $\{T_{\boldsymbol{n}}(g)^{-1}T_{\boldsymbol{n}}(f)\}_{\boldsymbol{n}}$. If $g \neq 0$ a.e., then **GLT 2** and **GLT 6**–**GLT 7** give $\{T_{\boldsymbol{n}}(g)^{-1}T_{\boldsymbol{n}}(f)\}_{\boldsymbol{n}} \sim_{\rm GLT} 1\otimes g^{-1}f$. As a consequence, by **GLT 1** we have $\{T_{\boldsymbol{n}}(g)^{-1}T_{\boldsymbol{n}}(f)\}_{\boldsymbol{n}} \sim_\sigma g^{-1}f$ and, if the matrices $T_{\boldsymbol{n}}(g)^{-1}T_{\boldsymbol{n}}(f)$ are Hermitian, $\{T_{\boldsymbol{n}}(g)^{-1}T_{\boldsymbol{n}}(f)\}_{\boldsymbol{n}} \sim_\lambda g^{-1}f$. In the case of classical CG (Conjugate Gradient) preconditioning, the functions f, g are nonnegative and the matrices $T_{\boldsymbol{n}}(f), T_{\boldsymbol{n}}(g)$ are Hermitian positive definite. In this situation, the spectral distribution $\{T_{\boldsymbol{n}}(g)^{-1}T_{\boldsymbol{n}}(f)\}_{\boldsymbol{n}} \sim_\lambda g^{-1}f$ holds even if the preconditioned matrices $T_{\boldsymbol{n}}(g)^{-1}T_{\boldsymbol{n}}(f)$ are not Hermitian. Indeed, $T_{\boldsymbol{n}}(g)^{-1}T_{\boldsymbol{n}}(f)$ is similar to the Hermitian matrix $T_{\boldsymbol{n}}(g)^{-1/2}T_{\boldsymbol{n}}(f)T_{\boldsymbol{n}}(g)^{-1/2}$. Assuming as before that $g \neq 0$ a.e., by **GLT 2** and **GLT 6–GLT 8** we obtain[6]

$$\{T_n(g)^{-1/2}T_n(f)T_n(g)^{-1/2}\}_n \sim_{\rm GLT} g^{-1/2}fg^{-1/2} = g^{-1}f.$$

Hence, **GLT 1** yields $\{T_n(g)^{-1/2}T_n(f)T_n(g)^{-1/2}\}_n \sim_\lambda g^{-1}f$, and, by similarity, $\{T_n(g)^{-1}T_n(f)\}_n \sim_\lambda g^{-1}f$.

6.2. PDE discretizations

The main application of the theory of GLT sequences was already described in Section 1.1. It consists in the computation of the spectral distribution of the sequences of discretization matrices arising from the approximation of PDEs by numerical methods. In fact, these matrix-sequences are often GLT sequences.

In Section 6.2.1, we present the GLT analysis of a simple model problem, approximated by standard FDs. In Section 6.2.2, we consider the matrices arising from the FE approximation of a system of PDEs; these matrices naturally show up in saddle point form, and we analyze their Schur complements through the theory of GLT sequences. The idea is to show that **GLT 1–GLT 9** are a powerful tool for computing the asymptotic singular value and eigenvalue distribution of PDE discretization matrices.

6.2.1. FD discretization of convection-diffusion-reaction equations. Consider the following convection-diffusion-reaction problem:

$$\begin{cases} -a(x)u''(x) + b(x)u'(x) + c(x)u(x) = f(x), \quad x \in (0,1), \\ u(0) = \alpha, \quad u(1) = \beta, \end{cases} \tag{6.4}$$

where $a, b, c, f : [0,1] \to \mathbb{R}$ are continuous and $\alpha, \beta \in \mathbb{R}$. We consider the discretization of (6.4) by the classical central second-order FD scheme. Let us describe it shortly; for more details on FD methods, we refer the reader to the available literature (see, e.g., [44]). We choose a discretization parameter $n \in \mathbb{N}$, we set $h = \frac{1}{n+1}$ and $x_j = jh$ for all $j = 0, \ldots, n+1$, and we note that, for $j = 1, \ldots, n$,

$$\begin{aligned} -a(x)u''(x)|_{x=x_j} &\approx a(x_j)\frac{-u(x_{j+1}) + 2u(x_j) - u(x_{j-1})}{h^2}, \\ b(x)u'(x)|_{x=x_j} &\approx b(x_j)\frac{u(x_{j+1}) - u(x_{j-1})}{2h}, \\ c(x)u(x)|_{x=x_j} &= c(x_j)u(x_j). \end{aligned}$$

Then, we approximate the solution of (6.4) by the piecewise linear function that takes the value u_j in x_j for $j = 0, \ldots, n+1$, where $u_0 = \alpha$, $u_{n+1} = \beta$, and $\mathbf{u} = (u_1, \ldots, u_n)^T$ is the solution of the linear system

$$a(x_j)(-u_{j+1} + 2u_j - u_{j-1}) + \frac{h}{2}b(x_j)(u_{j+1} - u_{j-1}) + h^2c(x_j)u_j = h^2f(x_j),$$
$$j = 1, \ldots, n.$$

The matrix A_n associated with this linear system can be decomposed as

$$A_n = K_n + Z_n, \tag{6.5}$$

[6] **GLT 8** is applied with the function $t \mapsto |t|^{1/2}$.

where:

- K_n is the diffusion matrix, i.e., the matrix arising from the discretization of the higher-order (diffusion) term $-a(x)u''(x)$;
- Z_n is the matrix resulting from the discretization of the terms with lower-order derivatives, i.e., the convection and reaction terms $b(x)u'(x)$, $c(x)u(x)$.

K_n and Z_n are given explicitly by

$$K_n = \begin{bmatrix} 2a(x_1) & -a(x_1) & & & \\ -a(x_2) & 2a(x_2) & -a(x_2) & & \\ & \ddots & \ddots & \ddots & \\ & & -a(x_{n-1}) & 2a(x_{n-1}) & -a(x_{n-1}) \\ & & & -a(x_n) & 2a(x_n) \end{bmatrix}, \tag{6.6}$$

$$Z_n = \frac{h}{2} \begin{bmatrix} 0 & b(x_1) & & \\ -b(x_2) & \ddots & \ddots & \\ & \ddots & \ddots & b(x_{n-1}) \\ & & -b(x_n) & 0 \end{bmatrix} + h^2 \begin{bmatrix} c(x_1) & & \\ & \ddots & \\ & & c(x_n) \end{bmatrix}, \tag{6.7}$$

and we immediately see that $A_n \approx K_n$ when n is large, because Z_n is negligible with respect to K_n. Indeed, we shall see that the GLT analysis of $\{A_n\}_n$ reduces to the GLT analysis of $\{K_n\}_n$. We show that

$$\{A_n\}_n \sim_{\mathrm{GLT}} a(x)(2 - 2\cos\theta) \tag{6.8}$$

and

$$\{A_n\}_n \sim_{\sigma,\lambda} a(x)(2 - 2\cos\theta). \tag{6.9}$$

Consider the matrix

$$D_n(a)T_n(2 - 2\cos\theta) = \begin{bmatrix} 2a(\frac{1}{n}) & -a(\frac{1}{n}) & & & \\ -a(\frac{2}{n}) & 2a(\frac{2}{n}) & -a(\frac{2}{n}) & & \\ & \ddots & \ddots & \ddots & \\ & & -a(\frac{n-1}{n}) & 2a(\frac{n-1}{n}) & -a(\frac{n-1}{n}) \\ & & & -a(1) & 2a(1) \end{bmatrix}. \tag{6.10}$$

In view of the inequalities $\left|x_j - \frac{j}{n}\right| \le \frac{1}{n+1} = h$, $j = 1, \ldots, n$, a direct comparison between (6.6) and (6.10) shows that the modulus of each entry of the matrix $Y_n = K_n - D_n(a)T_n(2-2\cos\theta)$ is bounded by $2\,\omega_a(h)$, where $\omega_a(\cdot)$ is the modulus of continuity of a. Therefore, the 1-norm and the ∞-norm of Y_n are bounded by $6\,\omega_a(h)$, and so, by (2.3), $\|Y_n\| \le 6\,\omega_a(h) \to 0$ as $n \to \infty$. It follows that $\{Y_n\}_n \sim_\sigma 0$. Since

$$K_n = D_n(a)T_n(2 - 2\cos\theta) + Y_n,$$

GLT 2–**GLT 4** and **GLT 6** imply

$$\{K_n\}_n \sim_{\rm GLT} a(x)(2-2\cos\theta). \tag{6.11}$$

Moreover, by (2.3), $\|Z_n\| \le h\|b\|_\infty + h^2\|c\|_\infty \to 0$ as $n\to\infty$, so $\{Z_n\}_n \sim_\sigma 0$. Thus, the decomposition (6.5), together with (6.11), **GLT 4** and **GLT 6**, yields (6.8). From (6.8) and **GLT 1** we immediately obtain the singular value distribution in (6.9). To obtain the eigenvalue distribution, the idea is to exploit the fact that A_n is "almost" symmetric. Indeed, $A_n \approx K_n$ for n large enough, and K_n is "almost" symmetric because $a(x)$ varies continuously when x ranges in $[0,1]$ (and so $a(x_j)\approx a(x_{j+1})$ for all $j=1,\dots,n-1$, when n is large). Consider the following symmetric approximation of K_n:

$$\tilde{K}_n = \begin{bmatrix} 2a(x_1) & -a(x_1) & & & \\ -a(x_1) & 2a(x_2) & -a(x_2) & & \\ & \ddots & \ddots & \ddots & \\ & & -a(x_{n-2}) & 2a(x_{n-1}) & -a(x_{n-1}) \\ & & & -a(x_{n-1}) & 2a(x_n) \end{bmatrix}.$$

We have $\|\tilde{K}_n\| \le 4\|a\|_\infty$ and $\|K_n - \tilde{K}_n\| \le \omega_a(h) \to 0$ as $n\to\infty$, so in particular $\{K_n - \tilde{K}_n\}_n \sim_\sigma 0$. From

$$A_n = \tilde{K}_n + (K_n - \tilde{K}_n) + Z_n,$$

we get $\{\tilde{K}_n\}_n \sim_{\rm GLT} a(x)(2-2\cos\theta)$ via (6.8), **GLT 4** and **GLT 6**. Considering that $\tilde{K}_n$ is symmetric, **GLT 1** gives $\{\tilde{K}_n\}_n \sim_\lambda a(x)(2-2\cos\theta)$. Hence, all the hypotheses of [29, Theorem 3.4] or of its generalized version [26, Theorem 3.3] are met, and the eigenvalue distribution in (6.9) follows.

6.2.2. Schur complements of matrices arising from the FE approximation of a system of PDEs. We consider in this section the FE approximation of a system of PDEs. The resulting discretization matrices show up in saddle point form [6], and we will describe the way to compute the asymptotic spectral distribution of their Schur complements using the theory of GLT sequences. We recall that the Schur complement is a key tool for the numerical treatment of the related linear systems [6, Section 5]. The analysis of this section is similar to the analysis in [20, Section 2], but the discretization technique considered herein is a pure FE approximation, whereas in [20] the authors adopted a mixed FD/FE technique.

Consider the system of PDEs

$$\begin{cases} -(a(x)u'(x))' + v'(x) = f(x), & x\in(0,1), \\ -u'(x) - \rho\, v(x) = g(x), & x\in(0,1), \end{cases} \tag{6.12}$$

with homogeneous Dirichlet boundary conditions: $u(0)=u(1)=0$ and $v(0)=v(1)=0$. In (6.12), ρ is a constant, $a\in L^\infty([0,1])$, and $f,g\in L^2([0,1])$. We consider the approximation of (6.12) by linear FEs on the uniform mesh in $[0,1]$ with step size $h=\frac{1}{n+1}$. Let us briefly describe this approximation technique; for more details on FE methods, see, e.g., [35, Chapter 4]. The weak form of (6.12) reads as follows: find $u, v \in H_0^1([0,1])$ such that, for all $w \in H_0^1([0,1])$,

$$\begin{cases} \int_0^1 a(x)u'(x)w'(x)dx + \int_0^1 v'(x)w(x)dx = \int_0^1 f(x)w(x)dx, \\ -\int_0^1 u'(x)w(x)dx - \rho\int_0^1 v(x)w(x)dx = \int_0^1 g(x)w(x)dx. \end{cases}$$

Set $x_j = jh$, $j = 0, \ldots, n+1$, and fix the subspace $\mathscr{W}_n = \mathrm{span}(\varphi_1, \ldots, \varphi_n) \subset H_0^1([0,1])$, where $\varphi_1, \ldots, \varphi_n$ are the so-called "hat-functions",

$$\varphi_i(x) = \frac{x - x_{i-1}}{x_i - x_{i-1}}\chi_{[x_{i-1},x_i)}(x) + \frac{x_{i+1} - x}{x_{i+1} - x_i}\chi_{[x_i,x_{i+1})}(x), \qquad i = 1, \ldots, n. \quad (6.13)$$

In the FE approach, we look for approximations $u_{\mathscr{W}_n}, v_{\mathscr{W}_n}$ of u, v by solving the following (Galerkin) problem: find $u_{\mathscr{W}_n}, v_{\mathscr{W}_n} \in \mathscr{W}_n$ such that, for all $w \in \mathscr{W}_n$,

$$\begin{cases} \int_0^1 a(x)u_{\mathscr{W}_n}'(x)w'(x)dx + \int_0^1 v_{\mathscr{W}_n}'(x)w(x)dx = \int_0^1 f(x)w(x)dx, \\ -\int_0^1 u_{\mathscr{W}_n}'(x)w(x)dx - \rho\int_0^1 v_{\mathscr{W}_n}(x)w(x)dx = \int_0^1 g(x)w(x)dx. \end{cases}$$

Since $\{\varphi_1, \ldots, \varphi_n\}$ is a basis of $\mathscr{W}_n$, we can write $u_{\mathscr{W}_n} = \sum_{j=1}^n u_j\varphi_j$ and $v_{\mathscr{W}_n} = \sum_{j=1}^n v_j\varphi_j$ for unique vectors $\mathbf{u} = (u_1, \ldots, u_n)^T$ and $\mathbf{v} = (v_1, \ldots, v_n)^T$. By linearity, the computation of $u_{\mathscr{W}_n}, v_{\mathscr{W}_n}$ (i.e., of $\mathbf{u}, \mathbf{v}$) reduces to solving the linear system

$$A_{2n}\begin{bmatrix}\mathbf{u}\\ \mathbf{v}\end{bmatrix} = \begin{bmatrix}\mathbf{f}\\ \mathbf{g}\end{bmatrix},$$

where $\mathbf{f} = \left[\int_0^1 f(x)\varphi_i(x)dx\right]_{i=1}^n$, $\mathbf{g} = \left[\int_0^1 g(x)\varphi_i(x)dx\right]_{i=1}^n$, and A_{2n} is the stiffness matrix, which admits the following saddle point structure:

$$A_{2n} = \begin{bmatrix} K_n(a) & H_n \\ H_n^T & -\rho M_n \end{bmatrix}.$$

Here, the blocks $K_n(a), H_n, M_n$ are square matrices of size n, and precisely

$$K_n(a) = \left[\int_0^1 a(x)\varphi_j'(x)\varphi_i'(x)dx\right]_{i,j=1}^n,$$

$$H_n = \left[\int_0^1 \varphi_j'(x)\varphi_i(x)dx\right]_{i,j=1}^n = -\mathrm{i}\,T_n(\sin\theta),$$

$$M_n = \left[\int_0^1 \varphi_j(x)\varphi_i(x)dx\right]_{i,j=1}^n = \frac{h}{3}\,T_n(2+\cos\theta).$$

Note that $K_n(a)$, M_n are symmetric, while H_n is skew-symmetric. In particular, $H_n^T = -H_n = \mathrm{i}\,T_n(\sin\theta)$. From now on we assume that the matrices $K_n(a)$ are invertible.[7] The (negative) Schur complement of A_{2n} is the symmetric matrix

$$\begin{aligned} S_n &= \rho M_n + H_n^T(K_n(a))^{-1}H_n \\ &= \frac{\rho h}{3}\,T_n(2+\cos\theta) + T_n(\sin\theta)\,(K_n(a))^{-1}\,T_n(\sin\theta). \end{aligned} \tag{6.14}$$

In the following, we perform the GLT analysis of the sequence of normalized Schur complements $\{(n+1)S_n\}_n$. Under the hypothesis that $a \neq 0$ a.e., we prove that

$$\{(n+1)S_n\}_n \sim_{\rm GLT} \varsigma(x,\theta) = \frac{\rho}{3}(2+\cos\theta) + \frac{\sin^2\theta}{a(x)(2-2\cos\theta)} \tag{6.15}$$

and

$$\{(n+1)S_n\}_n \sim_{\sigma,\lambda} \varsigma(x,\theta). \tag{6.16}$$

Actually, we only prove (6.15), because (6.16) follows immediately from (6.15) and **GLT 1** since the matrices S_n are symmetric.

To prove (6.15), we show that

$$\left\{\frac{1}{n+1}K_n(a)\right\}_n \sim_{\rm GLT} a(x)(2-2\cos\theta). \tag{6.17}$$

Once this is done, (6.15) follows from the expression (6.14) and from **GLT 2**, **GLT 6**, **GLT 7**, taking into account that $a(x)(2-2\cos\theta) \neq 0$ a.e. The remainder of this section is devoted to the proof of (6.17).

Step 1. We first consider the constant-coefficient case where $a = 1$ a.e. We have

$$K_n(1) = \left[\int_0^1 \varphi_j'(x)\varphi_i'(x)dx\right]_{i,j=1}^n = \frac{1}{h}\,T_n(2-2\cos\theta),$$

and (6.17) follows from **GLT 2**.

Step 2. Now we consider the case where $a \in C([0,1])$. To prove (6.17) in this case, we first illustrate the idea, and then we go into the details. The proof is based on the fact that the hat-functions (6.13) are "locally supported". Indeed, the support $[x_{i-1}, x_{i+1}]$ of the ith hat-function $\varphi_i(x)$ is localized near the point $\hat{x}_i = \frac{i}{n} \in [x_i, x_{i+1}]$, and the amplitude of the support tends to 0 as $n \to \infty$. Since $a(x)$ varies continuously over $[0,1]$, the (i,j) entry of $K_n(a)$ can be approximated as follows, for all $i,j = 1,\dots,n$:

$$\begin{aligned} (K_n(a))_{ij} &= \int_0^1 a(x)\varphi_j'(x)\varphi_i'(x)dx = \int_{x_{i-1}}^{x_{i+1}} a(x)\varphi_j'(x)\varphi_i'(x)dx \\ &\approx a(\hat{x}_i)\int_{x_{i-1}}^{x_{i+1}} \varphi_j'(x)\varphi_i'(x)dx = a(\hat{x}_i)\int_0^1 \varphi_j'(x)\varphi_i'(x)dx = a(\hat{x}_i)(K_n(1))_{ij}. \end{aligned}$$

This approximation can be rewritten in matrix form as follows:

$$K_n(a) \approx D_n(a)K_n(1). \tag{6.18}$$

[7] This is satisfied, for example, if $a > 0$ a.e., in which case the matrices $K_n(a)$ are positive definite.

We shall see that (6.18) implies that $\{\frac{1}{n+1}K_n(a) - \frac{1}{n+1}D_n(a)K_n(1)\}_n \sim_\sigma 0$, and, consequently, (6.17) follows from *Step* 1 and **GLT 2**, **GLT 3**, **GLT 4**, **GLT 6**.

Now let us go into the details. Since $|\varphi_i'(x)| \le n+1$, we have

$$|(K_n(a))_{ij} - (D_n(a)K_n(1))_{ij}| = \left|\int_0^1 [a(x) - a(\hat{x}_i)]\varphi_j'(x)\varphi_i'(x)dx\right|$$
$$\le (n+1)^2 \int_{x_{i-1}}^{x_{i+1}} |a(x) - a(\hat{x}_i)|dx \le 2(n+1)\,\omega_a\Big(\frac{2}{n+1}\Big),$$

for all $i,j = 1,\ldots,n$, where $\omega_a(\cdot)$ is the modulus of continuity of a. It follows that each component of $\frac{1}{n+1}K_n(a) - \frac{1}{n+1}D_n(a)K_n(1)$ is bounded in modulus by $2\,\omega_a(\frac{2}{n+1})$. Moreover, $\frac{1}{n+1}K_n(a) - \frac{1}{n+1}D_n(a)K_n(1)$ is banded (actually, tridiagonal), because $\mathrm{supp}(\varphi_i)\cap\mathrm{supp}(\varphi_j)$ contains at most one point whenever $|i-j| > 1$. Thus, the 1-norm and the ∞-norm of $\frac{1}{n+1}K_n(a) - \frac{1}{n+1}D_n(a)K_n(1)$ are bounded by $6\,\omega_a(\frac{2}{n+1})$, and so, by (2.3), $\|\frac{1}{n+1}K_n(a) - \frac{1}{n+1}D_n(a)K_n(1)\| \le 6\,\omega_a(\frac{2}{n+1}) \to 0$ as $n\to\infty$. By Theorem 2.6, we conclude that $\{\frac{1}{n+1}K_n(a) - \frac{1}{n+1}D_n(a)K_n(1)\}_n \sim_\sigma 0$, and this implies (6.17) by *Step* 1 and **GLT 2**, **GLT 3**, **GLT 4**, **GLT 6**.

Step 3. Finally, we consider the general case where $a \in L^\infty([0,1])$. By the Lusin theorem [36], there exists a sequence of continuous functions $a_m \in C([0,1])$ such that $\mu_1\{a_m \ne a\} \le \frac{1}{m}$ and $\|a_m\|_\infty \le \|a\|_{L^\infty}$. By *Step* 2, $\{\frac{1}{n+1}K_n(a_m)\}_n \sim_{\rm GLT} a_m(x)(2-2\cos\theta)$ and, moreover, $a_m(x)(2-2\cos\theta) \to a(x)(2-2\cos\theta)$ in measure. In addition, $\{\{\frac{1}{n+1}K_n(a_m)\}_n\}_m$ is an a.c.s. for $\{\frac{1}{n+1}K_n(a)\}_n$. Indeed, using (2.2) and observing that $\sum_{i=1}^n |\varphi_i'(x)| \le 2(n+1)$ for all $x\in[0,1]$, we obtain

$$\|K_n(a) - K_n(a_m)\|_1$$
$$\le \sum_{i,j=1}^n |(K_n(a))_{ij} - (K_n(a_m))_{ij}| = \sum_{i,j=1}^n \left|\int_0^1 [a(x) - a_m(x)]\varphi_j'(x)\varphi_i'(x)dx\right|$$
$$\le 2\|a\|_{L^\infty}\int_{\{a\ne a_m\}}\sum_{i,j=1}^n |\varphi_j'(x)|\,|\varphi_i'(x)|dx \le 8(n+1)^2\|a\|_{L^\infty}\mu_1\{a\ne a_m\},$$

and so

$$\left\|\frac{1}{n+1}K_n(a) - \frac{1}{n+1}K_n(a_m)\right\|_1 \le C\,\frac{n}{m}$$

for some constant C independent of n and m. Thus, $\{\{\frac{1}{n+1}K_n(a_m)\}_n\}_m$ is an a.c.s. for $\{\frac{1}{n+1}K_n(a)\}_n$ by Theorem 3.13, and (6.17) follows from **GLT 9**.

7. Conclusions and future works

In this work, we fully developed the theory of GLT sequences. We made a significant review of the original theory, by generalizing and/or simplifying a lot of key results. We also extended the theory: the main novelties of this work are the results of Sections 5.2–5.3, Theorem 5.16, and the new proof of Theorem 5.19. Finally, we provided a precise summary of the theory of GLT sequences in Section 5.5, with the purpose of giving to the reader an easy-to-use "GLT manual". Some hints on how to use this manual in practical applications were given in Section 6.

We conclude this paper with a list of possible future lines of research.

1. Try to understand if the a.c.s. notion is related to some topology $\tau_{\text{a.c.s.}}$ defined on the algebra of matrix-sequences. If the answer is affirmative, try to give a topological characterization of the sub-algebra of GLT sequences. Let us be more precise. Fix a sequence of d-indices $\{\boldsymbol{n} = \boldsymbol{n}(n)\}_n \subseteq \mathbb{N}^d$ such that $\boldsymbol{n} \to \infty$ as $n \to \infty$, and consider the complex *-algebra of all matrix-sequences

$$\mathscr{E} = \big\{\{A_{\boldsymbol{n}}\}_n : \ \{A_{\boldsymbol{n}}\}_n \ \text{ is a matrix-sequence}\big\}. \tag{7.1}$$

   If $\{A_{\boldsymbol{n}}\}_n$, $\{B_{\boldsymbol{n},m}\}_n \in \mathscr{E}$, we say that $\{B_{\boldsymbol{n},m}\}_n$ converges to $\{A_{\boldsymbol{n}}\}_n$ as $m \to \infty$ if $\{\{B_{\boldsymbol{n},m}\}_n\}_m$ is an a.c.s. for $\{A_{\boldsymbol{n}}\}_n$. Note that this statement could be meaningless, because no one proved that the "a.c.s. convergence" is a true notion of convergence. In other words, no one proved the existence of a topology on $\mathscr{E}$ that induces this kind of convergence. Trying to understand if such a topology exists is an interesting subject for future research. Assuming for the moment that such a topology $\tau_{\text{a.c.s.}}$ exists, it is easy to see that $\tau_{\text{a.c.s.}}$ is not "well behaved". For instance, the limit of a converging sequence $\{\{B_{\boldsymbol{n},m}\}_n\}_m$ is not unique. Indeed, any fixed sequence $\{B_{\boldsymbol{n}}\}_n$ converges to an infinite number of sequences, because $\{B_{\boldsymbol{n}}\}_n$ is an a.c.s. of $\{B_{\boldsymbol{n}} + Z_{\boldsymbol{n}}\}_n$ whenever $\{Z_{\boldsymbol{n}}\}_n$ is zero-distributed. In other words, $\{B_{\boldsymbol{n}}\}_n \to \{B_{\boldsymbol{n}} + Z_{\boldsymbol{n}}\}_n$ for all zero-distributed sequences $\{Z_{\boldsymbol{n}}\}_n$. This implies that every open set containing a sequence $\{B_{\boldsymbol{n}}\}_n$, actually contains all the sequences $\{B_{\boldsymbol{n}} + Z_{\boldsymbol{n}}\}_n$ such that $\{Z_{\boldsymbol{n}}\}_n$ is zero-distributed. In particular, $(\mathscr{E}, \tau_{\text{a.c.s.}})$ is not a Hausdorff space and so, a fortiori, it is not a topological vector space. Despite of this, if $\tau_{\text{a.c.s.}}$ exists, it would be interesting to provide a topological characterization of the GLT algebra $\mathscr{A}$ defined in (5.3). In particular, is it a sequentially closed subset of $\mathscr{E}$? Is it closed? If it is not closed, what is its closure?
2. Develop the theory of *block* GLT sequences. Multilevel block Toeplitz matrices (defined, e.g., in [45]) naturally arise in the numerical approximation of constant-coefficient systems of PDEs and, surprisingly enough, also in the FE approximation of constant-coefficient elliptic problems [27]. In the nonconstant-coefficient case, the counterpart of multilevel block Toeplitz sequences would be block GLT sequences, in the same way as GLT sequences are the counterpart of multilevel Toeplitz sequences. Hence, a theory of block GLT sequences, already mentioned in [40, Section 3.3], would be the ideal framework to deal with variable-coefficient differential problems approximated by FEs of any regularity, as well as to face the linear systems of nonconstant-coefficient PDEs. We note that the first step of such a theory has already been made. Indeed, thanks to the work of Böttcher, Silbermann, Miranda, and Tilli (see [10, 33, 45]), we know that any sequence of multilevel block Toeplitz matrices $\{T_{\boldsymbol{n}}(\mathbf{f})\}_n$, generated by a multivariate matrix-valued function $\mathbf{f} : [-\pi, \pi]^d \to \mathbb{C}^{s \times s}$ with components

$f_{ij} \in L^1([-\pi,\pi]^d)$, has an asymptotic spectral and singular value distribution described by $\mathbf{f}$ (see [45] for the precise meaning of this statement).

3. Revisit the work in [40, Section 3.1.4] concerning *reduced* GLT sequences. A suitable theory of reduced GLT sequences would allow one to deal with sequences of matrices associated with the FE approximation of PDEs defined on non-rectangular domains Ω. Actually, also the theory of GLT sequences allows one to deal with this kind of matrices, but under the additional assumption that: (a) the non-rectangular domain Ω is exactly described by a regular geometry map $\mathbf{G} : \widehat{\Omega} \to \overline{\Omega}$ defined on a rectangular domain $\widehat{\Omega}$; (b) the basis functions used in the FE approximation are defined as the "$\mathbf{G}$-deformations" of basis functions defined over $\widehat{\Omega}$. The assumptions (a) and (b) are satisfied in the IgA context [12], but not in the general FE setting [5].
4. Try to design an automatic procedure for computing the symbol of a GLT sequence of PDE discretization matrices. The idea would be to express the symbol as a function of the higher-order differential operator associated with the PDE, of the related coefficient, and of the used approximation technique. Some hints in this direction are given in [39, Section 2] and [40, Question 3.1].

References

[1] A.S. Al-Fhaid, S. Serra-Capizzano, D. Sesana, and M.Z. Ullah, *Singular-value (and eigenvalue) distribution and Krylov preconditioning of sequences of sampling matrices approximating integral operators.* Numer. Linear Algebra Appl. **21** (2014), 722–743.

[2] F. Auricchio, L. Beirão da Veiga, T.J.R. Hughes, A. Reali, and G. Sangalli, *Isogeometric collocation methods.* Math. Models Methods Appl. Sci. **20** (2010), 2075–2107.

[3] F. Avram, *On bilinear forms in Gaussian random variables and Toeplitz matrices.* Probab. Theory Related Fields **79** (1988), 37–45.

[4] B. Beckermann and A.B.J. Kuijlaars, *Superlinear convergence of Conjugate Gradients.* SIAM J. Numer. Anal. **39** (2001), 300–329.

[5] B. Beckermann and S. Serra-Capizzano, *On the asymptotic spectrum of Finite Element matrix sequences.* SIAM J. Numer. Anal. **45** (2007), 746–769.

[6] M. Benzi, G.H. Golub, and J. Liesen, *Numerical solution of saddle point problems.* Acta Numerica **14** (2005), 1–137.

[7] R. Bhatia, *Matrix Analysis.* Springer-Verlag, New York 1997.

[8] A. Böttcher and S.M. Grudsky, *Toeplitz matrices, asymptotic Linear Algebra, and Functional Analysis.* Birkhäuser Verlag, Basel–Boston–Berlin 2000.

[9] A. Böttcher and S.M. Grudsky, *Spectral properties of banded Toeplitz matrices.* SIAM, Philadelphia 2005.

[10] A. Böttcher and B. Silbermann, *Introduction to large truncated Toeplitz matrices.* Springer-Verlag, New York 1999.

[11] A. Böttcher and B. Silbermann, *Analysis of Toeplitz operators.* 2nd ed., Springer-Verlag, Berlin 2006.

[12] J.A. Cottrell, T.J.R. Hughes, and Y. Bazilevs, *Isogeometric Analysis: toward integration of CAD and FEA.* John Wiley & Sons, Chichester 2009.

[13] M. Donatelli, C. Garoni, C. Manni, S. Serra-Capizzano, and H. Speleers, *Symbol-based multigrid methods for Galerkin B-spline Isogeometric Analysis.* SIAM J. Numer. Anal. **55** (2017), 31–62.

[14] M. Donatelli, C. Garoni, C. Manni, S. Serra-Capizzano, and H. Speleers, *Robust and optimal multi-iterative techniques for IgA Galerkin linear systems.* Comput. Methods Appl. Mech. Engrg. **284** (2015), 230–264.

[15] M. Donatelli, C. Garoni, C. Manni, S. Serra-Capizzano, and H. Speleers, *Robust and optimal multi-iterative techniques for IgA collocation linear systems.* Comput. Methods Appl. Mech. Engrg. **284** (2015), 1120–1146.

[16] M. Donatelli, C. Garoni, C. Manni, S. Serra-Capizzano, and H. Speleers, *Spectral analysis and spectral symbol of matrices in isogeometric collocation methods.* Math. Comput. **85** (2016), 1639–1680.

[17] M. Donatelli, C. Garoni, M. Mazza, S. Serra-Capizzano, and D. Sesana, *Spectral behavior of preconditioned non-Hermitian multilevel block Toeplitz matrices with matrix-valued symbol.* Appl. Math. Comput. **245** (2014), 158–173.

[18] M. Donatelli, C. Garoni, M. Mazza, S. Serra-Capizzano, and D. Sesana, *Preconditioned HSS method for large multilevel block Toeplitz linear systems via the notion of matrix-valued symbol.* Numer. Linear Algebra Appl. **23** (2016), 83–119.

[19] M. Donatelli, M. Neytcheva, and S. Serra-Capizzano, *Canonical eigenvalue distribution of multilevel block Toeplitz sequences with non-Hermitian symbols.* Oper. Theory Adv. Appl. **221** (2012), 269–291.

[20] A. Dorostkar, M. Neytcheva, and S. Serra-Capizzano, *Spectral analysis of coupled PDEs and of their Schur complements via the notion of Generalized Locally Toeplitz sequences.* Technical Report 2015-008 (2015), Dept. of Information Technology, Uppsala University.

[21] C. Garoni, *Spectral distribution of PDE discretization matrices from Isogeometric Analysis: the case of L^1 coefficients and non-regular geometry.* J. Spectral Theory (in press).

[22] C. Garoni, C. Manni, S. Serra-Capizzano, D. Sesana, and H. Speleers, *Spectral analysis and spectral symbol of matrices in isogeometric Galerkin methods.* Math. Comput. (in press), DOI: 10.1090/mcom/3143.

[23] C. Garoni, C. Manni, S. Serra-Capizzano, D. Sesana, and H. Speleers, *Lusin theorem, GLT sequences and matrix computations: an application to the spectral analysis of PDE discretization matrices.* J. Math. Anal. Appl. **446** (2017), 365–382.

[24] C. Garoni and S. Serra-Capizzano, *The theory of Locally Toeplitz sequences: a review, an extension, and a few representative applications.* Bol. Soc. Mat. Mex. **22** (2016), 529–565.

[25] C. Garoni and S. Serra-Capizzano, *Generalized Locally Toeplitz sequences: theory and applications.* Book in preparation for Springer.[8]

[8]Citations throughout this paper refer to the preliminary version of this book, which is available, under a different title, as technical report: C. Garoni and S. Serra-Capizzano, *The theory of Generalized Locally Toeplitz sequences: a review, an extension, and a few representative applications.* Technical Report 2015-023 (2015), Dept. of Information Technology, Uppsala University.

[26] C. Garoni, S. Serra-Capizzano, and D. Sesana, *Tools for determining the asymptotic spectral distribution of non-Hermitian perturbations of Hermitian matrix-sequences and applications.* Integr. Equ. Oper. Theory **81** (2015), 213–225.

[27] C. Garoni, S. Serra-Capizzano, and D. Sesana, *Spectral analysis and spectral symbol of d-variate $\mathbb{Q}_p$ Lagrangian FEM stiffness matrices.* SIAM J. Matrix Anal. Appl. **36** (2015), 1100–1128.

[28] C. Garoni, S. Serra-Capizzano, and P. Vassalos, *A general tool for determining the asymptotic spectral distribution of Hermitian matrix-sequences.* Oper. Matrices **9** (2015), 549–561.

[29] L. Golinskii and S. Serra-Capizzano, *The asymptotic properties of the spectrum of nonsymmetrically perturbed Jacobi matrix sequences.* J. Approx. Theory **144** (2007), 84–102.

[30] G.H. Golub and C.F. Van Loan, *Matrix computations.* 4th ed., The Johns Hopkins University Press, Baltimore 2013.

[31] U. Grenander and G. Szegő, *Toeplitz forms and their applications.* 2nd ed., AMS Chelsea Publishing, New York 1984.

[32] T.J.R. Hughes, J.A. Cottrell, and Y. Bazilevs, *Isogeometric Analysis: CAD, Finite Elements, NURBS, exact geometry and mesh refinement.* Comput. Methods Appl. Mech. Engrg. **194** (2005), 4135–4195.

[33] M. Miranda and P. Tilli, *Asymptotic spectra of Hermitian block Toeplitz matrices and preconditioning results.* SIAM J. Matrix Anal. Appl. **21** (2000), 867–881.

[34] S.V. Parter, *On the distribution of the singular values of Toeplitz matrices.* Linear Algebra Appl. **80** (1986), 115–130.

[35] A. Quarteroni, *Numerical models for differential problems.* 2nd ed., Springer-Verlag Italia, Milan 2014.

[36] W. Rudin, *Real and complex analysis.* 3rd ed., McGraw-Hill, Singapore 1987.

[37] E. Salinelli, S. Serra-Capizzano, and D. Sesana, *Eigenvalue-eigenvector structure of Schoenmakers–Coffey matrices via Toeplitz technology and applications.* Linear Algebra Appl. **491** (2016), 138–160.

[38] S. Serra-Capizzano, *Distribution results on the algebra generated by Toeplitz sequences: a finite-dimensional approach.* Linear Algebra Appl. **328** (2001), 121–130.

[39] S. Serra-Capizzano, *Generalized Locally Toeplitz sequences: spectral analysis and applications to discretized partial differential equations.* Linear Algebra Appl. **366** (2003), 371–402.

[40] S. Serra-Capizzano, *The GLT class as a generalized Fourier Analysis and applications.* Linear Algebra Appl. **419** (2006), 180–233.

[41] S. Serra-Capizzano and D. Sesana, *Approximating classes of sequences: the Hermitian case.* Linear Algebra Appl. **434** (2011), 1163–1170.

[42] S. Serra-Capizzano, D. Sesana, and E. Strouse, *The eigenvalue distribution of products of Toeplitz matrices – Clustering and attraction.* Linear Algebra Appl. **432** (2010), 2658–2678.

[43] S. Serra-Capizzano and C. Tablino-Possio, *Analysis of preconditioning strategies for collocation linear systems.* Linear Algebra Appl. **369** (2003), 41–75.

[44] G.D. Smith, *Numerical solution of partial differential equations: Finite Difference methods.* 3rd ed., Oxford University Press, New York 1985.

[45] P. Tilli, *A note on the spectral distribution of Toeplitz matrices.* Linear and Multilinear Algebra **45** (1998), 147–159.

[46] P. Tilli, *Locally Toeplitz sequences: spectral properties and applications.* Linear Algebra Appl. **278** (1998), 91–120.

[47] P. Tilli, *Some results on complex Toeplitz eigenvalues.* J. Math. Anal. Appl. **239** (1999), 390–401.

[48] E.E. Tyrtyshnikov, *A unifying approach to some old and new theorems on distribution and clustering.* Linear Algebra Appl. **232** (1996), 1–43.

[49] E.E. Tyrtyshnikov and N.L. Zamarashkin, *Spectra of multilevel Toeplitz matrices: advanced theory via simple matrix relationships.* Linear Algebra Appl. **270** (1998), 15–27.

[50] N.L. Zamarashkin and E.E. Tyrtyshnikov, *Distribution of eigenvalues and singular values of Toeplitz matrices under weakened conditions on the generating function.* Sb. Math. **188** (1997), 1191–1201.

Carlo Garoni
Department of Mathematics
University of Rome "Tor Vergata"
00133 Rome, Italy
e-mail: garoni@mat.uniroma2.it

Stefano Serra-Capizzano
Department of Science and High Technology
University of Insubria
22100 Como, Italy
e-mail: stefano.serrac@uninsubria.it

Operator Theory:
Advances and Applications, Vol. 259, 395–411

The Bézout Equation on the Right Half-plane in a Wiener Space Setting

G.J. Groenewald, S. ter Horst and M.A. Kaashoek

Dedicated to Albrecht Böttcher on the occasion of his 60th birthday

Abstract. This paper deals with the Bézout equation $G(s)X(s) = I_m$, $\Re s \geq 0$, in the Wiener space of analytic matrix-valued functions on the right half-plane. In particular, G is an $m \times p$ matrix-valued analytic Wiener function, where $p \geq m$, and the solution X is required to be an analytic Wiener function of size $p \times m$. The set of all solutions is described explicitly in terms of a $p \times p$ matrix-valued analytic Wiener function Y, which has an inverse in the analytic Wiener space, and an associated inner function Θ defined by Y and the value of G at infinity. Among the solutions, one is identified that minimizes the H^2-norm. A Wiener space version of Tolokonnikov's lemma plays an important role in the proofs. The results presented are natural analogs of those obtained for the discrete case in [11].

Mathematics Subject Classification (2010). Primary 47A56; Secondary 47A57, 47B35, 46E40, 46E15.

Keywords. Bézout equation, corona problem, Wiener space on the line, matrix-valued functions, minimal norm solutions, Tolokonnikov's lemma.

1. Introduction and main results

In this paper we deal with the Bézout equation $G(s)X(s) = I_m$ on the closed right half-plane $\Re s \geq 0$, assuming that the given function G is of the form

$$G(s) = D + \int_0^\infty e^{-st} g(t)\, dt \quad (\Re s \geq 0),$$
$$\text{where} \quad g \in L^1_{m\times p}(\mathbb{R}_+) \cap L^2_{m\times p}(\mathbb{R}_+). \tag{1.1}$$

The third author gratefully thanks the mathematics department of North-West University, Potchefstroom campus, South Africa, for the hospitality and support during his visit from September 21–October 15, 2015.
This work is based on the research supported in part by the National Research Foundation of South Africa (Grant Number 93406).

In particular, G belongs to the analytic Wiener space $\mathcal{W}_+^{m\times p}$. We are interested in solutions $X \in \mathcal{W}_+^{p\times m}$, that is,

$$X(s) = D_X + \int_0^\infty e^{-st}x(t)\,dt \quad (\Re s \geq 0), \quad \text{where} \quad x \in L^1_{m\times p}(\mathbb{R}_+). \tag{1.2}$$

Throughout $p \geq m$. We refer to the final paragraph of this introduction for a further explanation of the notation.

With G given by (1.1) we associate the Wiener–Hopf operator T_G mapping $L^2_p(\mathbb{R}_+)$ into $L^2_m(\mathbb{R}_+)$ which is defined by

$$(T_G h)(t) = Dh(t) + \int_0^\infty g(t-\tau)h(\tau)d\tau, \quad t \geq 0 \quad (h \in L^2_p(\mathbb{R}_+)). \tag{1.3}$$

For X as in (1.2) we define the Wiener–Hopf operator T_X mapping $L^2_m(\mathbb{R}_+)$ into $L^2_p(\mathbb{R}_+)$ in a similar way, replacing D by D_X and g by x. If the Bézout equation

$$G(s)X(s) = I_m, \quad \Re s \geq 0. \tag{1.4}$$

has a solution X as in (1.2), then (using the analyticity of G and X) the theory of Wiener–Hopf operators (see [7, Section XII.2] or [2, Section 9]) tells us that $T_GT_X = T_{GX} = I$, where I stands for the identity operator on $L^2_m(\mathbb{R}_+)$. Thus for the Bézout equation (1.4) to be solvable the operator T_G must be surjective or, equivalently, $T_GT_G^*$ must be strictly positive. We shall see that this condition is also sufficient.

To state our main results, we assume that $T_GT_G^*$ is strictly positive. Then $D = G(\infty)$ is surjective, and hence DD^* is strictly positive too. We introduce two matrices D^+ and E, of sizes $p\times m$ and $p\times(p-m)$, respectively, and a $p\times p$ matrix function Y in $\mathcal{W}_+^{p\times p}$, as follows:

(i) $D^+ = D^*(DD^*)^{-1}$, where $D = G(\infty)$;
(ii) E is an isometry mapping $\mathbb{C}^{p-m}$ into $\mathbb{C}^p$ such that $\operatorname{Im} E = \operatorname{Ker} D$;
(iii) Y is the $p\times p$ matrix function given by

$$Y(s) = I_p - \int_0^\infty e^{-st}y(t)dt, \ \Re s \geq 0, \ \text{where } y = T_G^*(T_GT_G^*)^{-1}g. \tag{1.5}$$

From the definitions of D^+ and E it follows that the $p\times p$ matrix $\begin{bmatrix} D^+ & E \end{bmatrix}$ is non-singular. In fact

$$\begin{bmatrix} D \\ E^* \end{bmatrix} \begin{bmatrix} D^+ & E \end{bmatrix} = \begin{bmatrix} I_m & 0 \\ 0 & I_{p-m} \end{bmatrix}. \tag{1.6}$$

As we shall see (Proposition 2.1 in Section 2 below), the fact that the given function $g \in L^1_{m\times p}(\mathbb{R}_+) \cap L^2_{m\times p}(\mathbb{R}_+)$ implies that a similar result holds true for y. In particular, $Y \in \mathcal{W}_+^{p\times p}$. In what follows Ξ and Θ are the functions defined by

$$\Xi(s) = \Big(I_p - \int_0^\infty e^{-st}y(t)dt\Big)D^+ = Y(s)D^+, \ \Re s > 0; \tag{1.7}$$

$$\Theta(s) = \Big(I_p - \int_0^\infty e^{-st}y(t)dt\Big)E \ = Y(s)E,, \ \Re s > 0. \tag{1.8}$$

Since $Y \in \mathcal{W}_+^{p\times p}$, we have $\Xi \in \mathcal{W}_+^{p\times m}$ and $\Theta \in \mathcal{W}_+^{p\times (p-m)}$. Finally, recall that a function Ω in the analytic Wiener space $\mathcal{W}_+^{k\times r}$ is inner whenever $\Omega(s)$ is an isometry for each $s \in i\mathbb{R}$. We now state our main results.

Theorem 1.1. *Let G be the $m \times p$ matrix-valued function given by* (1.1). *Then the equation $G(s)X(s) = I_m$, $\Re s > 0$, has a solution $X \in \mathcal{W}_+^{p\times m}$ if and only if T_G is right invertible. In that case the function Ξ defined by* (1.7) *is a particular solution and the set of all solutions $X \in \mathcal{W}_+^{p\times m}$ is given by*

$$X(s) = \Xi(s) + \Theta(s)Z(s), \quad \Re s > 0, \tag{1.9}$$

where Ξ and Θ are defined by (1.7) *and* (1.8), *respectively, and the free parameter Z is an arbitrary function in $\mathcal{W}_+^{(p-m)\times m}$. Moreover, the function Θ belongs to $\mathcal{W}_+^{p\times (p-m)}$ and is inner. Furthermore, the solution Ξ is the minimal H^2 solution in the following sense*

$$\|X(\cdot)u\|^2_{H^2_p} = \|\Xi(\cdot)u\|^2_{H^2_p} + \|Z(\cdot)u\|^2_{H^2_{p-m}},$$
$$\text{where } u \in \mathbb{C}^m \text{ and } Z \in \widetilde{\mathcal{W}}_+^{(p-m)\times m}. \tag{1.10}$$

In the above theorem, for any positive integer k, $H^2_k = H^2_k(i\mathbb{R})$ is the Hardy space of $\mathbb{C}^k$-valued functions on the right half-plane given by $H^2_k(i\mathbb{R}) = JL^2_k(\mathbb{R}_+)$, where J is the unitary operator defined by

$$J = \frac{1}{\sqrt{2\pi}}\mathcal{F} : L^2_k(\mathbb{R}) \to L^2_k(i\mathbb{R}) \tag{1.11}$$

with $\mathcal{F}$ being the Fourier transform mapping $L^2_k(\mathbb{R})$ onto $L^2_k(i\mathbb{R})$. Moreover, $Z \in \widetilde{\mathcal{W}}_+^{(p-m)\times m}$ means that

$$Z(s) = D_Z + \int_0^\infty e^{-st}z(t)\,dt \quad (\Re s \geq 0), \quad \text{where}$$
$$D_Z \text{ is a } (p-m)\times p \text{ matrix and } z \in L^1_{(p-m)\times p}(\mathbb{R}_+) \cap L^2_{(p-m)\times p}(\mathbb{R}_+).$$

See the final part of this introduction for further information about the used notation, in particular, see (1.14) for the definition of the Fourier transform $\mathcal{F}$.

The second theorem is a variant of the Tolokonnikov lemma [20] in the present setting. The result emphasizes the central role of the function Y.

Theorem 1.2. *Assume $T_G T_G^*$ is strictly positive, and let Y be the matrix function defined by* (1.5). *Then Y belongs to the Wiener space $\mathcal{W}_+^{p\times p}$*, $\det Y(s) \neq 0$ *whenever*

$\Re s \geq 0$, *and hence* Y *is invertible in* $\mathcal{W}_+^{p\times p}$. *Furthermore, the* $p\times p$ *matrix function*

$$\begin{bmatrix} G(s) \\ E^*Y(s)^{-1} \end{bmatrix}, \quad \Re s \geq 0, \tag{1.12}$$

is invertible in the Wiener algebra $\mathcal{W}_+^{p\times p}$ *and its inverse is given by*

$$\begin{bmatrix} G(s) \\ E^*Y(s)^{-1} \end{bmatrix}^{-1} = Y(s)\begin{bmatrix} D^+ & E \end{bmatrix} = \begin{bmatrix} \Xi(s) & \Theta(s) \end{bmatrix}, \quad \Re s \geq 0. \tag{1.13}$$

The literature on the Bézout equation and the related corona problem is extensive, starting with Carleson's corona theorem [3] (for the case when $m = 1$) and Fuhrmann's extension to the matrix-valued case [6], both in a H^∞ setting. The topic has beautiful connections with operator theory (see the books [14], [16], [17], [18], and the more recent papers [21], [22], [23]). Rational matrix equations of the form (1.4) play an important role in solving systems and control theory problems, in particularly, in problems involving coprime factorization, see, e.g., [24, Section 4.1], [10, Section A.2], [25, Chapter 21]. For more recent work see [12] and [13], and [15, page 3] where it is proved that the scalar analytic Wiener algebra is a pre-Bézout ring. For matrix polynomials, the equation (1.4) is closely related to the Sylvester resultant; see, e.g., Section 3 in [9] and the references in that paper.

The present paper is inspired by [5] and [11]. The paper [5] deals with equation (1.4) assuming the matrix function G to be a stable rational matrix function, and the solutions are required to be stable rational matrix functions as well. The comment in the final paragraph of [5, Section 2] was the starting point for our analysis. The paper [11] deals with the discrete case (when the right half-plane is replaced by the open unit disc). Theorems 1.1 and 1.2 are the continuous analogue of Theorem 1.1 in [11]. The absence of an explicit formula for the function Y^{-1} in the present setting makes the proofs more complicated than those in [11].

The paper consists of five sections, including the present introduction and an appendix. Section 2, which deals with the right invertibility of the operator T_G, has an auxiliary character. Theorem 1.2 is proved in Section 3, and Theorem 1.1 in Section 4. The Appendix contains a number of auxiliary results involving the Lebesgue space $L^1(\mathbb{R}) \cap L^2(\mathbb{R})$ and its vector-valued counterpart, which are collected together simply for the convenience of the reader and contains no significantly new material.

Notation and terminology. We conclude this section with some notation and terminology. Throughout, a linear map $A : \mathbb{C}^r \to \mathbb{C}^k$ is identified with the $k \times r$ matrix of A relative to the standard orthonormal bases in $\mathbb{C}^r$ and $\mathbb{C}^k$. The space of all $k \times r$ matrices with entries in $L^1(\mathbb{R})$ will be denoted by $L^1_{k\times r}(\mathbb{R})$. As usual $\widehat{f}$ denotes the Fourier transform of $f \in L^1_{k\times r}(\mathbb{R})$, that is,

$$\widehat{f}(s) = (\mathcal{F}f)(s) = \int_{-\infty}^{\infty} e^{-st} f(t)dt, \quad s \in i\mathbb{R}. \tag{1.14}$$

Note that $\widehat{f}$ is continuous on the extended imaginary axis $i\mathbb{R}\cup\{\pm i\infty\}$, and is zero at $\pm i\infty$ by the Riemann–Lebesgue lemma. By $\mathcal{W}^{k\times r}$ we denote the *Wiener space* consisting of all $k \times r$ matrix functions F on the imaginary axis of the form

$$F(s) = D_F + \widehat{f}(s), \quad s \in i\mathbb{R}, \text{where } f \in L^1_{k\times r}(\mathbb{R}) \text{ and } D_F \text{ is a constant matrix.} \tag{1.15}$$

Since $\widehat{f}$ is continuous on the extended imaginary axis and is zero at $\pm i\infty$, the function F given by (1.15) is also continuous on the extended imaginary axis and the constant matrix D_F is equal to the value of F at infinity. We write $\mathcal{W}^{k\times r}_+$ for the space of all F of the form (1.15) with the additional property that f has its support in $\mathbb{R}_+ = [0,\infty)$, that is, f is equal to zero on $(-\infty, 0)$. Any function $F \in \mathcal{W}^{k\times r}_+$ is analytic and bounded on the open right half-plane. Thus any $F \in \mathcal{W}^{k\times r}_+$ is a matrix-valued H^∞ function. Finally, by $\mathcal{W}^{k\times r}_{-,0}$ we denote the Wiener space consisting of all F of the form (1.15) with the additional property that $D_F = 0$ and f has its support in $(-\infty, 0]$. Thus we have the following direct sum decomposition:

$$\mathcal{W}^{k\times r} = \mathcal{W}^{k\times r}_+ \dot{+} \mathcal{W}^{k\times r}_{-,0}. \tag{1.16}$$

We write $F \in \widetilde{\mathcal{W}}^{k\times r}_+$ if the function f in (1.15) belongs to $L^1_{k\times r}(\mathbb{R}_+) \cap L^2_{k\times r}(\mathbb{R}_+)$. Similarly, $F \in \widetilde{\mathcal{W}}^{k\times r}_{-,0}$ if $f \in L^1_{k\times r}(\mathbb{R}_-) \cap L^2_{k\times r}(\mathbb{R}_-)$ and $D_F = 0$.

Let $F \in \mathcal{W}^{k\times r}$ be given by (1.15). With F we associate the Wiener–Hopf operator T_F mapping $L^2_r(\mathbb{R}_+)$ into $L^2_k(\mathbb{R}_+)$. This operator (see [7, Section XII.2]) is defined by

$$(T_F h)(t) = D_F h(t) + \int_0^\infty f(t-\tau)h(\tau)d\tau, \quad t \geq 0 \quad (h \in L^2_r(\mathbb{R}_+)). \tag{1.17}$$

The orthogonal complement of $H^2_k(i\mathbb{R}) = JL^2_k(\mathbb{R}_+)$, with J as in (1.11), in $L^2_k(i\mathbb{R})$ will be denoted by $K^2_k(i\mathbb{R})$. If $F \in \widetilde{\mathcal{W}}^{k\times r}_+$, then for each $u \in \mathbb{C}^r$ the function $F(\cdot)u$ belongs to $H^2_k(i\mathbb{R})$. Similarly, $F(\cdot)u$ belongs to $K^2_k(i\mathbb{R})$ if $F \in \widetilde{\mathcal{W}}^{k\times r}_{-,0}$.

Finally, for $f \in L^1_{k\times r}(\mathbb{R})$ and $g \in L^1_{r\times m}(\mathbb{R})$ the *convolution product* $f \star g$ is the function in $L^1_{k\times m}(\mathbb{R})$, see [19, Section 7.13], given by

$$(f \star g)(t) = \int_{-\infty}^\infty f(t-\tau)g(\tau)\,d\tau \quad \text{a.e. on } \mathbb{R}. \tag{1.18}$$

2. Right invertibility of T_G

In this section $G \in \mathcal{W}^{m\times p}_+$, where G is given by (1.1) and $p \geq m$. We already know that the Bézout equation (1.4) having a solution X in $\mathcal{W}^{p\times m}_+$ implies that T_G is right invertible or, equivalently, $T_G T_G^*$ is strictly positive; see the paragraph containing formula (1.4).

In this section we present an auxiliary result that will be used to prove our main theorems. For this purpose we need the $m \times m$ matrix-valued function R on the imaginary axis defined by $R(s) = G(s)G(s)^*$, $s \in i\mathbb{R}$. It follows that

$R \in \mathcal{W}^{m\times m}$. By T_R we denote the corresponding Wiener–Hopf operator acting on $L^2_m(\mathbb{R}_+)$. Thus

$$(T_R f)(t) = DD^* f(t) + \int_0^\infty r(t-\tau) f(\tau)\, d\tau, \quad 0 \le t < \infty,$$

$$\text{with } r(t) = Dg^*(t) + g(t)D^* + \int_{-\infty}^\infty g(t-\tau) g^*(\tau)\, d\tau, \quad t \in \mathbb{R}.$$

Here $g^*(t) = g(-t)^*$ for $t \in \mathbb{R}$. It is well known (see, e.g., formula (24) in Section XII.2 of [7]) that

$$T_R = T_G T_G^* + H_G H_G^*. \tag{2.1}$$

Here H_G is the Hankel operator mapping $L^2_p(\mathbb{R}_+)$ into $L^2_m(\mathbb{R}_+)$ defined by G, that is,

$$(H_G f)(t) = \int_0^\infty g(t+\tau) f(\tau) d\tau, \quad f \in L^2_p(\mathbb{R}_+). \tag{2.2}$$

We shall prove the following proposition. For the case when G is a rational matrix function, the first part (of the "if and only if" part) of the proposition is covered by Lemma 2.3 in [5]. The proof given in [5] can also be used in the present setting. For the sake of completeness we include a proof of the first part.

Proposition 2.1. *Let G be given by* (1.1). *Then the operator T_G is right invertible if and only if T_R and $I - H_G^* T_R^{-1} H_G$ are both invertible operators. In that case the inverse of $T_G T_G^*$ is given by*

$$(T_G T_G^*)^{-1} = T_R^{-1} + T_R^{-1} H_G (I - H_G^* T_R^{-1} H_G)^{-1} H_G^* T_R^{-1}. \tag{2.3}$$

Furthermore,

(a) *$(T_G T_G^*)^{-1}$ maps $L^1_m(\mathbb{R}_+) \cap L^2_m(\mathbb{R}_+)$ in a one-to-one way onto itself;*
(b) *the function y defined by $y = T_G^*(T_G T_G^*)^{-1} g$ belongs to $L^1_{p\times p}(\mathbb{R}_+) \cap L^2_{p\times p}(\mathbb{R}_+)$, in particular, the function Y given by* (1.5) *is in $\widetilde{\mathcal{W}}_+^{p\times p}$.*

Proof. We split the proof into four parts. In the first part we assume that T_G is right invertible, and we show that T_R and $I - H_G^* T_R^{-1} H_G$ are both invertible operators and that the inverse of $T_G T_G^*$ is given by (2.3). The second part deals with the reverse implication. Items (a) and (b) are proved in the last two parts.

PART 1. Assume T_G is right invertible. Then the operator $T_G T_G^*$ is strictly positive. According to (2.1) we have $T_R \ge T_G T_G^*$, and hence T_R is also strictly positive. In particular, T_R is invertible. Rewriting (2.1) as $T_G T_G^* = T_R - H_G H_G^*$, and multiplying the latter identity from the left and from the right by $T_R^{-1/2}$ shows that

$$T_R^{-1/2} T_G T_G^* T_R^{-1/2} = I - T_R^{-1/2} H_G H_G^* T_R^{-1/2}. \tag{2.4}$$

Hence $I - T_R^{-1/2} H_G H_G^* T_R^{-1/2}$ is strictly positive which shows that $H_G^* T_R^{-1/2}$ is a strict contraction. But then $H_G^* T_R^{-1} H_G = \big(H_G^* T_R^{-1/2}\big)\big(H_G^* T_R^{-1/2}\big)^*$ is also a strict contraction, and thus the operator $I - H_G^* T_R^{-1} H_G$ is strictly positive. In particular, $I - H_G^* T_R^{-1} H_G$ is invertible. Finally, since $T_G T_G^* = T_R - H_G H_G^*$, a usual Schur complement type of argument (see, e.g., Section 2.2 in [1]), including the well-known inversion formula

$$(A - BC)^{-1} = A^{-1} + A^{-1}B(I - CA^{-1}B)^{-1}CA^{-1},$$

then shows that $(T_G T_G^*)^{-1}$ is given by (2.3).

PART 2. In this part we assume that T_R and $I - H_G^* T_R^{-1} H_G$ are both invertible operators, and we show that T_G is right invertible. According to (2.1) the operator T_R is positive. Since we assume T_R to be invertible, we conclude that T_R is strictly positive. Rewriting (2.1) as $T_G T_G^* = T_R - H_G H_G^*$, and multiplying the latter identity from the left and from the right by $T_R^{-1/2}$ we obtain the identity (2.4). Hence $I - T_R^{-1/2} H_G H_G^* T_R^{-1/2}$ is positive which shows that $H_G^* T_R^{-1/2}$ is a contraction. But then $H_G^* T_R^{-1} H_G = \big(H_G^* T_R^{-1/2}\big)\big(H_G^* T_R^{-1/2}\big)^*$ is also a contraction, and thus the operator $I - H_G^* T_R^{-1} H_G$ is positive. By assumption $I - H_G^* T_R^{-1} H_G$ is invertible. It follows that $I - H_G^* T_R^{-1} H_G$ is strictly positive, and hence $T_R^{-1/2} H_G$ is a strict contraction. But then the same holds true for $T_R^{-1/2} H_G H_G^* T_R^{-1/2}$. This implies that $I - T_R^{-1/2} H_G H_G^* T_R^{-1/2}$ is strictly positive, and (2.4) shows that T_G is right invertible.

PART 3. In this part we prove item (a). Observe that $g \in L^1_{m\times p}(\mathbb{R})$ implies that T_G maps $L^1_p(\mathbb{R}_+) \cap L^2_p(\mathbb{R}_+)$ into $L^1_m(\mathbb{R}_+) \cap L^2_m(\mathbb{R}_+)$. Since $g^* \in L^1_{p\times m}(\mathbb{R})$ and

$$(T_G^* f)(t) = D^* f(t) + \int_0^\infty g^*(t-\tau) f(\tau)\, d\tau, \quad 0 \le t < \infty,$$

the operator T_G^* maps $L^1_m(\mathbb{R}_+) \cap L^2_m(\mathbb{R}_+)$ into $L^1_p(\mathbb{R}_+) \cap L^2_p(\mathbb{R}_+)$. Thus $T_G T_G^*$ maps $L^1_m(\mathbb{R}_+) \cap L^2_m(\mathbb{R}_+)$ into itself. We have to show that the same holds true for its inverse. To do this we apply Lemmas A.3 and A.4.

Lemma A.3 tells us that T_R^{-1} maps $L^1_m(\mathbb{R}_+) \cap L^2_m(\mathbb{R}_+)$ in a one-to-one way onto itself. This allows us to apply Lemma A.4 with

$$Q = T_R^{-1}, \quad H = H_G \quad \text{and} \quad \tilde{H} = H_G^*.$$

Recall that H_G is a Hankel operator, see (2.2), and H_G^* is also a Hankel operator, in fact

$$(H_G^* f)(t) = \int_0^\infty g(t+\tau)^* f(\tau)\, d\tau, \quad 0 \le t < \infty.$$

Since $I - H_G^* T_R^{-1} H_G$ is invertible, Lemma A.4 then shows that $I - H_G^* T_R^{-1} H_G$ maps $L^1_p(\mathbb{R}_+) \cap L^2_p(\mathbb{R}_+)$ in a one-to-one way onto itself, and hence the same holds true for its inverse $(I - H_G^* T_R^{-1} H_G)^{-1}$. To complete the proof of item (a) note that $H_G^* T_R^{-1}$ maps $L^1_m(\mathbb{R}_+) \cap L^2_m(\mathbb{R}_+)$ into $L^1_p(\mathbb{R}_+) \cap L^2_p(\mathbb{R}_+)$, and $T_R^{-1} H_G$ maps $L^1_p(\mathbb{R}_+) \cap L^2_p(\mathbb{R}_+)$ into $L^1_m(\mathbb{R}_+) \cap L^2_m(\mathbb{R}_+)$. But then (2.3) shows that $(T_G T_G^*)^{-1}$ maps $L^1_m(\mathbb{R}_+) \cap L^2_m(\mathbb{R}_+)$ into itself. To see that $(T_G T_G^*)^{-1}$ is one-to-one on $L^1_m(\mathbb{R}_+) \cap L^2_m(\mathbb{R}_+)$ and maps $L^1_m(\mathbb{R}_+) \cap L^2_m(\mathbb{R}_+)$ onto itself, one can follow the same argumentation as in the last part of the proof of Lemma A.3.

PART 4. In this part we prove item (b). Since g belongs to $L^1_{m\times p}(\mathbb{R}_+)\cap L^2_{m\times p}(\mathbb{R}_+)$, item (a) tells us that $f := (T_G T_G^*)^{-1} g$ also belongs to $L^1_{m\times p}(\mathbb{R}_+)\cap L^2_{m\times p}(\mathbb{R}_+)$. We already have seen (in the first paragraph of the previous part) that T_G^* maps $L^1_m(\mathbb{R}_+)\cap L^2_m(\mathbb{R}_+)$ into $L^1_p(\mathbb{R}_+)\cap L^2_p(\mathbb{R}_+)$. It follows that $y = T_G^* f$ belongs to $L^1_{p\times p}(\mathbb{R}_+)\cap L^2_{p\times p}(\mathbb{R}_+)$, as desired. □

3. The functions Y and Θ, and proof of Theorem 1.2

We begin with three lemmas involving the functions Y and Θ defined by (1.5) and (1.8), respectively. From Proposition 2.1, item (b), and (1.8) we know that $Y \in \mathcal{W}_+^{p\times p}$ and $\Theta \in \mathcal{W}_+^{p\times(p-m)}$; see also the paragraph preceding Theorem 1.1.

Lemma 3.1. *Assume that T_G is right invertible, and let $Y \in \mathcal{W}_+^{p\times p}$ be the function defined by* (1.5). *Then*

$$G(s)Y(s) = D, \qquad \Re s > 0. \tag{3.1}$$

Proof. To prove (3.1) note that $T_G y = T_G T_G^*(T_G T_G^*)^{-1} g = g$. Since the functions g and y both have their support in $\mathbb{R}_+$, the identity $T_G y = g$ can be rewritten as $Dy + g\star y = g$, where $\star$ is the convolution product of matrix-valued functions with entries in $L^1(\mathbb{R})$; see (1.18). Thus

$$Dy(t) + (g\star y)(t) = Dy(t) + \int_{-\infty}^{\infty} g(t-\tau)y(\tau)dt = g(t), \quad t\in\mathbb{R}. \tag{3.2}$$

Next use that the Fourier transform of a convolution product is just the product of the Fourier transforms of the functions in the convolution product. Thus taking Fourier transforms in (3.2) yields $D\widehat{y} + \widehat{g}\widehat{y} = \widehat{g}$. The latter identity can be rewritten as $G\widehat{y} = \widehat{g}$. Hence, using the definition of Y in (1.5), we obtain

$$G(s)Y(s) = G(s)\Big(I_p - \widehat{y}(s)\Big) = G(s) - \widehat{g}(s) = D.$$

This proves (3.1). □

Lemma 3.2. *Assume that T_G is right invertible. Then the function Θ defined by* (1.8) *belongs to $\mathcal{W}_+^{p\times(p-m)}$ and is an inner function, that is, $\Theta(s)$ is an isometry for each $s\in i\mathbb{R}$ and at infinity.*

Proof. We already know that $\Theta \in \mathcal{W}_+^{p\times(p-m)}$. To prove that Θ is inner, let $y = T_G^*(T_G T_G^*)^{-1} g$ as in (1.5), and put $f = (T_G T_G^*)^{-1} g$. Thus $f \in L^1_{m\times m}(\mathbb{R}_+)$, by Proposition 2.1 (b), and $y = T_G^* f$. The latter can be rewritten as

$$y(t) = D^* f(t) + \int_0^{\infty} g^*(t-\tau) f(\tau)\, d\tau, \quad t \geq 0.$$

Note that $g^*(t) = g(-t)^*$, and hence g^* has its support in $(-\infty, 0]$. Therefore

$$y(t) = D^* f(t) + \int_{-\infty}^{\infty} g^*(t-\tau) f(\tau)\, d\tau, \quad t\in\mathbb{R}. \tag{3.3}$$

Put

$$\rho(t) = \begin{cases} 0 & \text{when } t \geq 0, \\ (g^* \star f)(t) & \text{when } t < 0. \end{cases} \tag{3.4}$$

Using the definition of the convolution product $\star$, see (1.18), we can rewrite (3.3) as

$$y(t) = D^* f(t) + (g^* \star f)(t) - \rho(t) \quad t \in \mathbb{R}.$$

Taking Fourier transforms we obtain

$$\widehat{y}(s) = D^* \widehat{f}(s) + \widehat{g}^*(s)\widehat{f}(s) - \widehat{\rho}(s) = G(s)^* \widehat{f}(s) - \widehat{\rho}(s), \quad s \in i\mathbb{R}.$$

Hence, we have

$$Y(s) = I - G(s)^* \widehat{f}(s) + \widehat{\rho}(s), \quad s \in i\mathbb{R}. \tag{3.5}$$

Now let us compute $\Theta(s)^*\Theta(s) = E^*Y(s)^*Y(s)E$ for $s \in i\mathbb{R}$. We have

$$\begin{aligned} E^*Y(s)^*Y(s)E &= \\ &= E^*Y(s)^*E - E^*Y(s)^*G(s)^*\widehat{f}(s)E + E^*Y(s)^*\widehat{\rho}(s)E \\ &= E^*Y(s)^*E + E^*Y(s)^*\widehat{\rho}(s)E \\ &\qquad\qquad (\text{because } G(s)Y(s)E = 0 \text{ by (3.1) and } DE = 0) \\ &= E^*E - E^*\widehat{y}(s)^*E + E^*Y(s)^*\widehat{\rho}(s)E \\ &= I_{p-m} - \Omega(s). \end{aligned}$$

Here $\Omega(s) = E^*\widehat{y}(s)^*E - E^*Y(s)^*\widehat{\rho}(s)E$. Note that the functions $\widehat{y}(\cdot)^*$ and $Y(\cdot)^*\widehat{\rho}(\cdot)$ belong to $\mathcal{W}_{-,0}^{p\times p}$, and thus Ω belongs to $\mathcal{W}_{-,0}^{(p-m)\times(p-m)}$. On the other hand, the function $E^*Y(\cdot)^*Y(\cdot)E$ is Hermitian on the imaginary axis, and hence the same is true for Ω. But for any positive integer k we have

$$\mathcal{W}_{-,0}^{k\times k} \cap (\mathcal{W}_{-,0}^{k\times k})^* = \{0\}.$$

Thus Ω is identically zero, and thus $\Theta(s)^*\Theta(s) = E^*Y(s)^*Y(s)E = I_{p-m}$ for any $s \in i\mathbb{R}$. Moreover, $\Theta(\infty)^*\Theta(\infty) = E^*E = I$. This proves that Θ is inner. □

Lemma 3.3. *Assume that T_G is right invertible, and let $Y \in \mathcal{W}_+^{p\times p}$ be the function defined by* (1.5)*. Then Y is invertible in $\mathcal{W}_+^{p\times p}$.*

Proof. Fix $s \in i\mathbb{R}$, and assume $u \in \mathbb{C}^p$ such that $Y(s)u = 0$. Then $G(s)Y(s) = D$ implies that $Du = 0$. By definition of E, $u = Ev$ for some $v \in \mathbb{C}^{p-m}$. Next use $\Theta(s) = Y(s)E$. It follows that $\Theta(s)v = Y(s)Ev = Y(s)u = 0$. However, $\Theta(s)$ is an isometry, by Lemma 3.2. So $v = 0$, and hence $u = 0$. We see that $\det Y(s) \neq 0$. Also $Y(\infty) = I_p$. We conclude that T_Y is a Fredholm operator; see [7, Theorem XII.3.1].

Next we prove that $\operatorname{Ker} T_Y = \{0\}$. Take $h \in \operatorname{Ker} T_Y$. Then $T_Y h = 0$, and hence $Y(s)\widehat{h}(s) = 0$ for each $s \in i\mathbb{R}$. But $\det Y(s) \neq 0$ for each $s \in i\mathbb{R}$. Hence $\widehat{h} = 0$, and therefore $h = 0$.

We want to prove that T_Y is invertible. Given the results of the preceding first two paragraphs it suffices to show that $\operatorname{ind} T_Y = 0$. This will be done in the next step by an approximation argument, using the fact, from [5], that we know the result is true for rational matrix functions.

Let g be as in (1.1). Note that g is the limit in L^1 of a sequence $g_1, g_2, \dots$ such that $G_n(s) = D + \widehat{g_n}(s)$ is a stable rational matrix function; cf., Part (v) on page 229 of [7]. Since T_G is right invertible, T_{G_n} will also be right invertible for n sufficiently large. In fact, $T_{G_n}T_{G_n}^* \to T_G T_G^*$ in operator norm. Put $y_n = T_{G_n}^*(T_{G_n}T_{G_n}^*)^{-1}g_n$. Then $y_n \to y$ in the L^1-norm. Put $Y_n(s) = I - \widehat{y_n}(s)$. Then $T_{Y_n} \to T_Y$ in operator norm. For n sufficiently large the operator T_{Y_n} is invertible (see the paragraph preceding Theorem 1.2 in [5] and formula (2.17) in [5]). In particular, the Fredholm index of T_{Y_n} is zero. But $\operatorname{ind} T_Y = \lim_{n\to\infty} \operatorname{ind} T_{Y_n} = 0$. Thus T_Y is invertible, and hence Y is invertible in $\mathcal{W}_+^{p\times p}$. □

Proof of Theorem 1.2. From Lemma 3.3 we know that $Y \in \mathcal{W}_+^{p\times p}$ and that Y is invertible in $\mathcal{W}_+^{p\times p}$. Thus we only have to prove the second part of the theorem. Since Y is invertible in $\mathcal{W}_+^{p\times p}$, the $p \times p$ matrix function given by (1.12) is well defined and belongs to $\mathcal{W}_+^{p\times p}$. Furthermore, from (1.6) we know that the $p \times p$ matrix $\begin{bmatrix} D^+ & E \end{bmatrix}$ is invertible. Hence the function defined by the right-hand side of (1.13) belongs to $\mathcal{W}_+^{p\times p}$ and is invertible in $\mathcal{W}_+^{p\times p}$. Using (3.1) and the identity (1.6) we see that

$$\begin{aligned} \begin{bmatrix} G(s) \\ E^*Y(s)^{-1} \end{bmatrix} Y(s) \begin{bmatrix} D^+ & E \end{bmatrix} &= \begin{bmatrix} G(s)Y(s) \\ E^* \end{bmatrix} \begin{bmatrix} D^+ & E \end{bmatrix} \\ &= \begin{bmatrix} D \\ E^* \end{bmatrix} \begin{bmatrix} D^+ & E \end{bmatrix} = \begin{bmatrix} I_m & 0 \\ 0 & I_{p-m} \end{bmatrix}, \quad \Re s \geq 0. \end{aligned}$$

This proves the first identity (1.13). The second identity is an immediate consequence of the definitions of Ξ and Θ in (1.7) and (1.8), respectively. □

4. Proof of Theorem 1.1

We begin with a lemma concerning the functions Ξ and Θ.

Lemma 4.1. *Assume that T_G is right invertible, and let Ξ and Θ be the functions defined by* (1.7) *and* (1.8)*, respectively. Then*

$$\operatorname{Ker} T_G = T_\Theta L^2_{p-m}(\mathbb{R}_+) \quad \textit{and} \quad \Theta^*\Xi \in \widetilde{\mathcal{W}}_{0,-}^{(p-m)\times m}. \tag{4.1}$$

Proof. We split the proof into two parts.

PART 1. In this part we prove the inclusion of (4.1). Take $s \in i\mathbb{R}$. Note that in Proposition 3.2 it was shown for $s \in i\mathbb{R}$ that

$$Y(s) = I - G(s)^*\widehat{f}(s) + \widehat{\rho}(s),$$

where $f = (T_G T_G^*)^{-1}g$ and ρ is defined by (3.4); see (3.5). From (1.7) and (1.8) we then see that

$$\Theta(s)^*\Xi(s) = E^*Y(s)^*Y(s)D^+ = E^*Y(s)^*\Big(I - G(s)^*\widehat{f}(s) + \widehat{\rho}(s)\Big)D^+.$$

Now use that $G(s)Y(s)E = DE = 0$, and hence $E^*Y(s)^*G(s)^* = 0$. The latter identity and the fact that $E^*D^+ = 0$ and $Y = I - \widehat{y}$ imply that

$$\begin{aligned}\Theta(s)^*\Xi(s) &= -E^*\widehat{y}(s)^*D^+ + E^*\widehat{\rho}(s)D^+ - E^*\widehat{y}(s)^*\widehat{\rho}(s)D^+ \\ &= -A(s) + B(s) - C(s).\end{aligned} \tag{4.2}$$

From item (b) in Proposition 4.1 we know that $y \in L^1_{p\times p}(\mathbb{R}_+) \cap L^2_{p\times p}(\mathbb{R}_+)$, and thus $\widehat{y} \in \widetilde{\mathcal{W}}_+^{p\times p}$ and $\widehat{y}(\infty) = 0$, that is, $\widehat{y} \in \widetilde{\mathcal{W}}_{0,+}^{p\times p}$. It follows that

$$A(\cdot) := E^*\widehat{y}(\cdot)^*D^+ \in \widetilde{\mathcal{W}}_{0,-}^{(p-m)\times p}. \tag{4.3}$$

Recall that ρ is given by (3.4) with $f = (T_GT_G^*)^{-1}g$. Since the function g belongs to $L^1_{m\times p}(\mathbb{R}_+) \cap L^2_{m\times p}(\mathbb{R}_+)$, item (b) in Proposition 2.1 tells us that the same holds true for f. It follows that $g^* \star f \in L^1_{p\times p}(\mathbb{R}) \cap L^2_{p\times p}(\mathbb{R})$. The latter implies that $\rho \in L^1_{p\times p}(\mathbb{R}_-) \cap L^2_{p\times p}(\mathbb{R}_-)$. We conclude that

$$B(\cdot) := E^*\widehat{\rho}(\cdot)D^+ \in \widetilde{\mathcal{W}}_{0,-}^{(p-m)\times p}. \tag{4.4}$$

Finally, note that $y^*(t) = y(-t)^*$ for $t \in \mathbb{R}$ and $\big(\widehat{y}(s)\big)^* = \widehat{y^*}(s)$ for $s \in i\mathbb{R}$. Thus

$$\widehat{y}(s)^*\widehat{\rho}(s) = \Big(\widehat{y^* \star \rho}\Big)(s), \quad s \in i\mathbb{R},$$

and

$$(y^* \star \rho)(t) = \int_{-\infty}^{\infty} y^*(t-\tau)\rho(\tau)\,d\tau = \int_{-\infty}^{0} y^*(t-\tau)\rho(\tau)\,d\tau.$$

Since both y^* and ρ belong to $L^1_{p\times p}(\mathbb{R}_-) \cap L^2_{p\times p}(\mathbb{R}_-)$, it is well known (see, e.g., Section 2 in [4]) that the same holds true for $y^* \star \rho$. But then

$$C(\cdot) := E^*\widehat{y}(\cdot)^*\widehat{\rho}(\cdot)D^+ \in \widetilde{\mathcal{W}}_{0,-}^{(p-m)\times p}. \tag{4.5}$$

From (4.3), (4.4), (4.5) and (4.2) it follows that $\Theta^*\Xi \in \widetilde{\mathcal{W}}_{0,-}^{(p-m)\times m}$.

PART 2. In this part we prove the identity of (4.1). Using (3.1) we see that

$$G(s)\Theta(s) = G(s)Y(s)E = DE = 0, \quad s \in i\mathbb{R}.$$

This implies that $T_GT_\Theta = 0$, and hence $\operatorname{Im} T_\Theta \subset \operatorname{Ker} T_G$. To prove the reverse inclusion, take $h \in \operatorname{Ker} T_G$. Thus $h \in L^2_p(\mathbb{R}_+)$ and $T_Gh = 0$. It follows that $G(s)\widehat{h}(s) = 0$ for $\Re s > 0$. Put $H(s) = \widehat{h}(s)$. Then $H(\cdot)$ belongs to $H^2_m(i\mathbb{R})$. Next we apply Theorem 1.2. Using the identities in (1.13) we see that

$$\begin{aligned}H(s) &= \begin{bmatrix}\Xi(s) & \Theta(s)\end{bmatrix}\begin{bmatrix}G(s)\\ E^*Y(s)^{-1}\end{bmatrix}H(s)\\ &= \begin{bmatrix}\Xi(s) & \Theta(s)\end{bmatrix}\begin{bmatrix}0\\ E^*Y(s)^{-1}H(s)\end{bmatrix} = \Theta(s)E^*Y(s)^{-1}H(s).\end{aligned}$$

Hence $\widehat{h}(s) = \Theta(s)\Psi(s)$, where $\Psi(s) = E^*Y(s)^{-1}\widehat{h}(s)$. Since $h \in L^2_p(\mathbb{R}_+)$ and $Y(\cdot)^{-1}$ is a matrix function with H^∞ entries, we conclude that $\Psi \in H^2_{p-m}$, and hence $\Psi = \widehat{u}$ for some $u \in L^2_{p-m}(\mathbb{R}_+)$. The identity $\widehat{h}(s) = \Theta(s)\Psi(s)$ then yields $\widehat{h}(s) = \Theta(s)\widehat{u}(s)$. This shows that $h = T_\Theta u$, and hence $\operatorname{Ker} T_G \subset \operatorname{Im} T_\Theta$. □

Proof of Theorem 1.1. From Lemma 3.2 we know that Θ is an inner function in $\mathcal{W}_+^{(p-m)\times m}$. The proof of the other statements is split into three parts.

PART 1. In this part we show that the equation $G(s)X(s) = I_m$, $\Re s > 0$, has a solution $X \in \mathcal{W}_+^{p\times m}$ if and only if T_G is right invertible. Furthermore, we show that in that case the function Ξ defined by (1.7) is a particular solution. From the one but last sentence of the paragraph containing (1.4) we know that it suffices to prove the "if part" only. Therefore, in what follows we assume that T_G is right invertible. Since $\Xi(s) = Y(s)D^+$ and $Y \in \mathcal{W}_+^{p\times p}$, we have $\Xi \in \mathcal{W}_+^{p\times m}$. Moreover, using the identity (4.1) we have

$$G(s)\Xi(s) = G(s)Y(s)D^+ = DD^+ = I_m.$$

Thus Ξ is a particular solution.

PART 2. This second part deals with the description of all in solutions in $\mathcal{W}_+^{p\times m}$. Let Z be an arbitrary function in $\mathcal{W}_+^{p\times m}$, and let $X \in \mathcal{W}_+^{p\times m}$ be defined by (1.9). Then

$$G(s)X(s) = G(s)\Xi(s) + G(s)\Theta(s)Z(s) = I_m + G(s)\Theta(s)Z(s), \quad \Re s \geq 0.$$

Recall that $G(s)\Theta(s) = G(s)Y(s)E = DE = 0$. Thus $G(s)X(s) = I_m$, $\Re s \geq 0$, and thus X is a solution.

To prove the converse implication, let $X \in \mathcal{W}_+^{p\times m}$ be a solution of the equation $G(s)X(s) = I_m$. Put $H = X - \Xi$. Then $H \in \mathcal{W}_+^{p\times m}$ and $G(s)H(s) = 0$. Using the identities in (1.13), we obtain

$$\begin{aligned} H(s) &= \begin{bmatrix} \Xi(s) & \Theta(s) \end{bmatrix} \begin{bmatrix} G(s) \\ E^*Y(s)^{-1} \end{bmatrix} H(s) \\ &= \begin{bmatrix} \Xi(s) & \Theta(s) \end{bmatrix} \begin{bmatrix} 0 \\ E^*Y(s)^{-1}H(s) \end{bmatrix} = \Theta(s)E^*Y(s)^{-1}H(s). \end{aligned}$$

Thus $H(s) = \Theta(s)Z(s)$, where $Z(s) = E^*Y(s)^{-1}H(s)$. Since Y is invertible in $\mathcal{W}_+^{p\times p}$, the function $Y(\cdot)^{-1}$ is in $\mathcal{W}_+^{p\times p}$. Together with the fact that $H \in \mathcal{W}_+^{p\times m}$, this yields $Z \in \mathcal{W}_+^{(p-m)\times m}$. It follows X has the desired representation (1.9).

PART 3. In this part we prove the identity (1.10). Assume $Z \in \widetilde{\mathcal{W}}^{(p-m)\times m}$, and let X be the function defined by (1.9). Fix $u \in \mathbb{C}^m$. Then $Z(\cdot)u \in H^2_m(i\mathbb{R})$, and

$$\Theta(\cdot)Z(\cdot)u = M_\Theta Z(\cdot)u \in H^2_{p-m}(i\mathbb{R}). \tag{4.6}$$

Here M_Θ is the operator of multiplication by $\Theta(\cdot)$ mapping $H^2_m(i\mathbb{R})$ into $H^2_{p-m}(i\mathbb{R})$. Furthermore, since M_Θ is an isometry, we also see that

$$\|Z(\cdot)u\| = \|\Theta(\cdot)Z(\cdot)u\|. \tag{4.7}$$

The fact that $y \in L^1_{p\times p}(\mathbb{R}_+) \cap L^2_{p\times p}(\mathbb{R}_+)$ implies that $Y \in \widetilde{\mathcal{W}}_+^{p\times p}$. But then $\Xi(s) = Y(s)D^+$ yields $\Xi(\cdot)u \in H^2_m(i\mathbb{R})$. Using the identity (1.9) we conclude that

$\Xi(\cdot)u$ also belongs to $H^2_m(i\mathbb{R})$. It follows that all norms in (1.10) are well defined, and in order to prove the identity (1.10) it suffices that to show that in $H^2_m(i\mathbb{R})$ the function $\Theta(\cdot)Z(\cdot)u$ is orthogonal to the function $\Xi(\cdot)v$ for any $v \in \mathbb{C}^m$. The latter fact follows from the inclusion in the second part of (4.1). Indeed, this inclusion tells us that $M^*_\Theta \Xi(\cdot)v = 0$, and hence

$$\begin{aligned}\langle \Xi(\cdot)v, \Theta(\cdot)Z(\cdot)u\rangle_{H^2_m(i\mathbb{R})} &= \langle \Xi(\cdot)v, M_\Theta Z(\cdot)u\rangle_{H^2_m(i\mathbb{R})}\\ &= \langle M^*_\Theta \Xi(\cdot)v, Z(\cdot)u\rangle_{H^2_m(i\mathbb{R})} = 0.\end{aligned}$$

This completes the proof. □

Appendix: The Lebesgue space $L^1(\mathbb{R}) \cap L^2(\mathbb{R})$

The material in this section is standard and is presented for the convenience of the reader. Throughout we deal with the Lebesgue spaces of complex-valued functions on the real line $L^1(\mathbb{R})$ and $L^2(\mathbb{R})$, their vector-valued counterparts $L^1_m(\mathbb{R})$ and $L^2_m(\mathbb{R})$, and the intersection of the latter two spaces: $L^1_m(\mathbb{R}) \cap L^2_m(\mathbb{R})$. The norms on these spaces are given by

$$\begin{aligned}
\|f\|_1 &= \int_{-\infty}^{\infty} |f(t)|\,dt && \text{for } f \in L^1(\mathbb{R}),\\
\|f\|_2 &= \Big(\int_{-\infty}^{\infty} |f(t)|^2\,dt\Big)^{1/2} && \text{for } f \in L^2(\mathbb{R}),\\
\|f\|_1 &= \Big(\sum_{i=1}^{m} \|f_i\|_1^2\Big)^{1/2} && \text{for } f = (f_1,\dots,f_m)^\top \in L^1_m(\mathbb{R}),\\
\|f\|_2 &= \Big(\sum_{i=1}^{m} \|f_i\|_2^2\Big)^{1/2} && \text{for } f = (f_1,\dots,f_m)^\top \in L^2_m(\mathbb{R}),\\
\|f\|_0 &= \max\{\|f\|_1, \|f\|_2\} && \text{for } f \in L^1_m(\mathbb{R}) \cap L^2_m(\mathbb{R}).
\end{aligned}$$

Let $k \in L^1_{m\times p}(\mathbb{R})$. Thus k is an $m \times p$ matrix function of which the (i,j)th entry $k_{ij} \in L^1(\mathbb{R})$. For each $\varphi \in L^1_p(\mathbb{R})$ and $\psi \in L^2_p(\mathbb{R})$ the convolution products $k\star\varphi$ and $k\star\psi$, see (1.18), are well defined, $k\star\varphi$ belongs to $L^1_m(\mathbb{R})$ and $k\star\psi$ belongs to $L^2_m(\mathbb{R})$. In particular, if $f \in L^1_p(\mathbb{R})\cap L^2_p(\mathbb{R})$, then $k\star f$ belongs to $L^1_m(\mathbb{R})\cap L^2_m(\mathbb{R})$. It follows that for a given $k \in L^1_{m\times p}(\mathbb{R})$ the convolution product induces linear maps from the space $L^1_p(\mathbb{R})$ into $L^1_m(\mathbb{R})$, from the space $L^2_p(\mathbb{R})$ into $L^2_m(\mathbb{R})$, and from the space $L^1_p(\mathbb{R}) \cap L^2_p(\mathbb{R})$ into $L^1_m(\mathbb{R}) \cap L^2_m(\mathbb{R})$. The resulting operators will be denoted by K_1, K_2 and K_0, respectively. The proof of the following lemma is standard (see, e.g., page 216 in [7]) and therefore it is omitted.

Lemma A.1. *The operators K_1, K_2 and K_0 are bounded linear operators, and*

$$\|K_\nu\| \le \kappa \ \ (\nu = 1,2,0), \ \ \textit{where } \kappa = \Big(\sum_{i=1}^{m}\sum_{j=1}^{p} \|k_{ij}\|_1^2\Big)^{1/2}. \tag{A1}$$

With $k \in L^1_{m\times p}(\mathbb{R}_+)$ we also associate the Wiener–Hopf operator W and the Hankel operator H defined by

$$(Wf)(t) = \int_0^\infty k(t-\tau)f(\tau)\,d\tau, \quad 0 \le t < \infty,$$

$$(Hf)(t) = \int_0^\infty k(t+\tau)f(\tau)\,d\tau, \quad 0 \le t < \infty.$$

Using the classical relation between the convolution operator defined by k and the operators W and H (see, e.g., Section XII.2 in [7]) it is easy to see that W and H map the space $L^1_p(\mathbb{R}_+)$ into $L^1_m(\mathbb{R}_+)$, the space $L^2_p(\mathbb{R}_+)$ into $L^2_m(\mathbb{R}_+)$, and the space $L^1_p(\mathbb{R}_+) \cap L^2_p(\mathbb{R}_+)$ into $L^1_m(\mathbb{R}_+) \cap L^2_m(\mathbb{R}_+)$. We denote the resulting operators by W_1, W_2, W_0, and H_1, H_2, H_0, respectively. Lemma A.1 shows that these operators are bounded and

$$\|W_\nu\| \le \kappa \text{ and } \|H_\nu\| \le \kappa \ \ (\nu = 1,2,0), \text{ where } \kappa = \left(\sum_{i=1}^m \sum_{j=1}^p \|k_{ij}\|_1^2\right)^{1/2}. \tag{A2}$$

Furthermore, using the line of reasoning in Lemma XX.2.4 in [7], we have the following corollary.

Corollary A.2. *The Hankel operators H_1, H_2, and H_0 are the limit in operator norm of finite rank operators, and hence compact.*

Next we present an auxiliary result that is used in the proof of Proposition 2.1. Put

$$R(s) = D_R + \int_{-\infty}^\infty e^{-st} r(t)\,dt \quad \text{where} \quad r \in L^1_{m\times m}(\mathbb{R}). \tag{A3}$$

By T_R we denote the Wiener–Hopf operator on $L^2_m(\mathbb{R}_+)$ defined by R, that is,

$$(T_R f)(t) = D_R f(t) + \int_0^\infty r(t-\tau)f(\tau)\,d\tau, \quad 0 \le t < \infty. \tag{A4}$$

As we know from the first paragraph of this section, the fact that $r \in L^1_{m\times m}(\mathbb{R})$ implies that T_R maps $L^1_m(\mathbb{R}_+) \cap L^2_m(\mathbb{R}_+)$ into itself.

Lemma A.3. *If T_R is invertible as an operator on $L^2_m(\mathbb{R}_+)$, then T_R^{-1} maps the space $L^1_m(\mathbb{R}_+) \cap L^2_m(\mathbb{R}_+)$ in a one-to-one way onto itself.*

Proof. Since T_R is invertible, R admits a canonical factorization (see Section XXX.10 in [8]), and hence we can write $T_R^{-1} = LU$, where L and U are Wiener–Hopf operators on $L^2_m(\mathbb{R}_+)$,

$$(Lf)(t) = D_L f(t) + \int_0^t \ell(t-\tau)f(t)\,dt, \qquad 0 \le t < \infty, \tag{A5}$$

$$(Uf)(t) = D_U f(t) + \int_t^\infty u(t-\tau)f(t)\,dt, \quad 0 \le t < \infty. \tag{A6}$$

Here ℓ and u both belong to $L^1_{m\times m}(\mathbb{R})$, with support of ℓ in $\mathbb{R}_+$ and support of u in $\mathbb{R}_-$. The fact that both ℓ and u belong to $L^1_{m\times m}(\mathbb{R})$ implies that both

L and U map $L^1_m(\mathbb{R}_+) \cap L^2_m(\mathbb{R}_+)$ into itself. Hence T_R^{-1} has the same property. Since T_R^{-1} is one-to-one on $L^2_m(\mathbb{R}_+)$, it is also one-to-one on $L^1_m(\mathbb{R}_+)\cap L^2_m(\mathbb{R}_+)$. For $f \in L^1_m(\mathbb{R}_+)\cap L^2_m(\mathbb{R}_+)$, we have $g = T_R f \in L^1_m(\mathbb{R}_+)\cap L^2_m(\mathbb{R}_+)$ and $f = T_R^{-1}T_R f = T_R^{-1}g$. This shows that T_R^{-1} maps $L^1_m(\mathbb{R}_+)\cap L^2_m(\mathbb{R}_+)$ onto $L^1_m(\mathbb{R}_+)\cap L^2_m(\mathbb{R}_+)$. □

Lemma A.4. *Let $k \in L^1_{p\times m}(\mathbb{R}_+)$ and $\tilde{k} \in L^1_{m\times p}(\mathbb{R}_+)$, and let H and $\tilde{H}$ be the corresponding Hankel operators acting from $L^2_m(\mathbb{R}_+)$ into $L^2_p(\mathbb{R}_+)$ and from $L^2_p(\mathbb{R}_+)$ into $L^2_m(\mathbb{R}_+)$, respectively. Let Q be any operator on $L^2_m(\mathbb{R}_+)$ mapping $L^1_p(\mathbb{R}_+)\cap L^2_m(\mathbb{R}_+)$ into itself, and assume that the restricted operator Q_0 acting on $L^1_m(\mathbb{R}_+)\cap L^2_m(\mathbb{R}_+)$ is bounded. If the operator $I - \tilde{H}QH$ is invertible on $L^2_m(\mathbb{R}_+)$, then $I - \tilde{H}QH$ maps the space $L^1_m(\mathbb{R}_+)\cap L^2_m(\mathbb{R}_+)$ in a one-to-one way onto itself.*

Proof. We know that H maps $L^1_p(\mathbb{R}_+)\cap L^p_2(\mathbb{R}_+)$ into $L^1_m(\mathbb{R}_+)\cap L^2_m(\mathbb{R}_+)$. Furthermore the same holds true for $\tilde{H}$ with the role of p and m interchanged. Hence our hypothesis on Q implies that $I - \tilde{H}QH$ maps the space $L^1_m(\mathbb{R}_+) \cap L^2_m(\mathbb{R}_+)$ into itself. Let M_0 be the corresponding restricted operator. We have to prove that M_0 is invertible. Note that Corollary A.2 implies that M_0 is equal to the identity operator minus a compact operator, and hence M_0 is a Fredholm operator of index zero. Therefore, in order to prove that M_0 is invertible, it suffices to show that $\operatorname{Ker} M_0$ consists of the zero element only. Assume not. Then there exists a non-zero f in $L^1_m(\mathbb{R}_+) \cap L^2_m(\mathbb{R}_+)$ such that $M_0 f = 0$. The fact that f belongs to $L^1_m(\mathbb{R}_+)\cap L^2_m(\mathbb{R}_+) \subset L^2_m(\mathbb{R}_+)$ shows that $0 = M_0 f = (I - \tilde{H}QH)f$. But $I - \tilde{H}QH$ is assumed to be invertible. Hence f must be zero. Thus M_0 is invertible. □

Acknowledgement

This work is based on the research supported in part by the National Research Foundation of South Africa. Any opinion, finding and conclusion or recommendation expressed in this material is that of the authors and the NRF does not accept any liability in this regard.

References

[1] H. Bart, I. Gohberg, M.A. Kaashoek, and A.C.M. Ran, *Factorization of matrix and operator functions: the state space method*, Oper. Theory Adv. Appl. **178**, Birkhäuser Verlag, Basel, 2008.

[2] A. Böttcher and B. Silbermann, *Analysis of Toeplitz operators*, 2nd Edition, Springer Monographs in Mathematics, Springer-Verlag, Berlin-Heidelberg, 2006.

[3] L. Carleson, Interpolation by bounded analytic functions and the corona problem, *Ann. Math.* **76** (1962), 547–559.

[4] H. Dym and I. Gohberg, On an extension problem, generalized Fourier analysis, and an entropy formula, *Integr. Equ. Oper. Theory* **3** (1980), 144–215.

[5] A.E. Frazho, M.A. Kaashoek, and A.C.M. Ran, Rational Matrix Solutions of a Bezout Type Equation on the Half-plane, in: *Advances in Structured Operator Theory and Related Areas. The Leonid Lerer Anniversary Volume*, Oper. Theory Adv. Appl. **237** (2012), 145–160.

[6] P. Fuhrmann, On the corona theorem and its applications to spectral problems in Hilbert space, *Trans. Amer. Math. Soc.* **132** (1968), 55–66.

[7] I. Gohberg, S. Goldberg, and M.A. Kaashoek, *Classes of Linear Operators*, Volume I, Oper. Theory Adv. Appl. **49**, Birkhäuser Verlag, Basel, 1990.

[8] I. Gohberg, S. Goldberg, and M.A. Kaashoek, *Classes of Linear Operators*, Volume II, Oper. Theory Adv. Appl. **63**, Birkhäuser Verlag, Basel, 1993.

[9] I. Gohberg, M.A. Kaashoek, and L. Lerer, The resultant for regular matrix polynomials and quasi commutativity, *Indiana Univ. Math. J.*, **57** (2008), 2783–2813.

[10] M. Green and D.J.N. Limebeer, *Linear Robust Control*, Prentice Hall, Englewood Cliffs, NJ, 1995.

[11] G.J. Groenewald, S. ter Horst and M.A. Kaashoek, The Bezout–Corona problem revisited: Wiener space setting, *Complex Anal. Oper. Theory* **10** (2016), 115–139.

[12] G. Gu and E.F. Badran, Optimal design for channel equalization via the filterbank approach, *IEEE Trans. Signal Proc.* **52** (2004), 536–545.

[13] G. Gu and L. Li, Worst-case design for optimal channel equalization in filterbank transceivers, *IEEE Trans. Signal Proc.* **51** (2003), 2424–2435.

[14] J.W. Helton, *Operator Theory, analytic functions, matrices and electrical engineering*, Regional Conference Series in Mathematics **68**, Amer. Math. Soc., Providence, RI, 1987.

[15] R. Mortini and A. Sasane, On the pre-Bézout property of Wiener algebras on the disc and the half-plane, *New Zealand J. Math.* **38** (2008), 45–55.

[16] N.K. Nikol'skii, *Treatise on the shift operator*, Grundlehren **273**, Springer-Verlag, Berlin 1986.

[17] N.K. Nikol'skii, *Operators, Functions and Systems*, Math. Surveys Monographs **92**, Amer. Math. Soc., Providence, RI, 2002.

[18] V.V. Peller, *Hankel Operators and their Applications*, Springer Monographs in Mathematics, Springer, 2003.

[19] W. Rudin, *Real and Complex Analysis*, McGraw-Hill, New York, 1966.

[20] V.A. Tolokonnikov, Estimates in Carleson's corona theorem. Ideals of the algebra H^∞, the problem of Szekefalvi-Nagy, *Zap. Naučn. Sem. Leningrad. Otdel. Mat. Inst. Steklov.* (LOMI) **113** (1981), 178–198 (Russian).

[21] S. Treil, Lower bounds in the matrix corona theorem and the codimension one conjecture, *GAFA* **14** (2004), 1118–1133.

[22] S. Treil and B.D. Wick, The matrix-valued H^p corona problem in the disk and polydisk, *J. Funct. Anal.* **226** (2005), 138–172.

[23] T.T. Trent and X. Zhang, A matricial corona theorem, *Proc. Amer. Math. Soc.* **134** (2006), 2549–2558.

[24] M. Vidyasagar, *Control system synthesis: a factorization approach*, The MIT Press, Cambridge, MA, 1985.

[25] K. Zhou with J.C. Doyle and K. Glover, *Robust and optimal control*, Prentice Hall, NJ, 1996.

G.J. Groenewald and S. ter Horst
Department of Mathematics
Unit for BMI
North-West University
Private Bag X6001-209
Potchefstroom 2520, South Africa
e-mail: Gilbert.Groenewald@nwu.ac.za
sanne.terhorst@nwu.ac.za

M.A. Kaashoek
Department of Mathematics
VU University Amsterdam
De Boelelaan 1081a
NL-1081 HV Amsterdam, The Netherlands
e-mail: m.a.kaashoek@vu.nl

Operator Theory:
Advances and Applications, Vol. 259, 413–462

On a Collocation-quadrature Method for the Singular Integral Equation of the Notched Half-plane Problem

Peter Junghanns and Robert Kaiser

Dedicated to our colleague Albrecht Böttcher on the occasion of his 60th Birthday

Abstract. A Cauchy singular integral equation describing the notched half-plane problem of two-dimensional elasticity theory is considered. This equation contains an additional fixed singularity represented by a Mellin convolution operator. We study a polynomial collocation-quadrature method for its numerical solution which takes into account the "natural" asymptotic of the solution at the endpoints of the integration interval and for which until now no criterion for stability is known. The present paper closes this gap. For this, we present a new technique of proving that the operator sequence of the respective collocation-quadrature method belongs to a certain C^*-algebra, in which we can study the stability of these sequences. One of the main ingredients of this technique is to show that the part of the operator sequence, associated with the Mellin part of the original equation, is "very close" to the finite section of particular operators belonging to a C^*-algebra of Toeplitz operators. Moreover, basing on these stability results numerical results are presented obtained by an implementation of the proposed method.

Mathematics Subject Classification (2010). Primary 65R20; Secondary 45E05.
Keywords. Notched half-plane problem, Cauchy singular integral operator, Mellin convolution operator, collocation-quadrature method, stability.

1. Introduction

One variant of an integral equation for the notched half-plane problem is (see [4, (1.8)])

$$\frac{1}{\pi}\int_{-1}^{1}\left[\frac{1}{y-x}-h(x,y)\right]v(y)\,dy=-\frac{1+\kappa}{2\mu}\,p\,,\quad -1<x<1\,, \tag{1.1}$$

where

$$h(x,y) = \frac{1}{2+y+x} - \frac{6(1+x)}{(2+y+x)^2} + \frac{4(1+x)^2}{(2+y+x)^3}$$

This equation is concerned with the following model situation (cf. the figure). The half-plane $\{(x,y) \in \mathbb{R}^2 : x \geq -1\}$ is the region occupied by the elastic medium and is subjected at infinity by (constant) tensile forces p perpendicular to a straight crack $\{(x,0) : -1 \leq x \leq 1\}$ of normalized length 2. The faces of the cut and the edge of the half-plane are assumed to be free of external forces. The elastic constants μ and κ are the shearing modulus and Muskhelishvili's constant of the elastic medium, respectively. The unknown function $v(x)$ is the first derivative of the crack opening displacement.

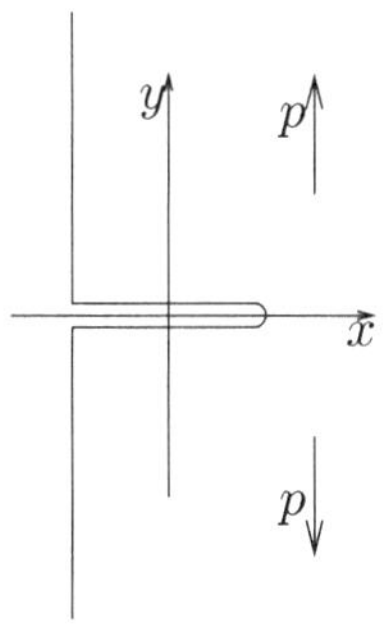

For practical purposes the so-called normalized stress intensity factor at the point $(1,0)$

$$\widetilde{k}_1(1) = -\frac{2\mu}{(\kappa+1)p} \lim_{x\to 1-0} \left[v(x)\sqrt{1-x}\right] \tag{1.2}$$

as well as the normalized crack opening displacement at the point $(-1,0)$

$$d(-1) = \frac{\mu}{(\kappa+1)p} \int_{-1}^{1} v(x)\,dx \tag{1.3}$$

are of interest. A result of Duduchava [5, Theorem 14.1] says that the solution of (1.1) can be written in the form

$$v(x) = \frac{\widetilde{v}(x)}{\sqrt{1-x}}$$

with a bounded and Hölder continuous function $\widetilde{v} : (-1,1] \longrightarrow \mathbb{C}$, which is infinitely differentiable on $(-1,1)$. For that reason we are interested in approximate solutions of (1.1) of the form

$$v_n(x) = \frac{p_n(x)}{\sqrt{1-x}}, \tag{1.4}$$

where $p_n(x)$ is a polynomial of degree less than n. Investigations on collocation methods [12, 10] (cf. also [13, 14, 9]) and collocation-quadrature methods [11] for Cauchy singular integral equations with fixed singularities of Mellin-type like in (1.1) do not cover the case (1.4). Indeed, in all these mentioned papers, we have

$v_n(x) = (1-x)^\gamma(1+x)^\delta p_n(x)$ with $\gamma \neq 0$ and $\delta \neq 0$. The present paper describes a possibility to overcome these difficulties.

A first step is to use the isometrical isomorphism $\mathcal{J}_0 : \mathbf{L}^2_\varphi \longrightarrow \mathbf{L}^2_\mu$, $f \mapsto v^{0,\frac{1}{2}} f$, where $\varphi(x) = \sqrt{1-x^2}$ and $\mu(x) = \sqrt{\dfrac{1-x}{1+x}}$ are the Chebyshev weights of second and fourth kind, respectively, and where, for a Jacobi weight $v^{\alpha,\beta}(x) = (1-x)^\alpha(1+x)^\beta$, by $\mathbf{L}^2_{v^{\alpha,\beta}}$ we denote the Hilbert space of all (classes of) functions $u : (-1,1) \longrightarrow \mathbb{C}$, which are square integrable w.r.t. the weight $v^{\alpha,\beta}$, equipped with the inner product and the norm

$$\langle u, v\rangle_{v^{\alpha,\beta}} = \int_{-1}^1 u(x)\overline{v(x)}v^{\alpha,\beta}(x)\,dx \quad \text{and} \quad \|u\|_{v^{\alpha,\beta}} = \sqrt{\langle u, u\rangle_{v^{\alpha,\beta}}}\,,$$

respectively. In more detail (cf. [4, p. 101]), write equation (1.1) in short form

$$\mathcal{A}_0 v = f\,, \tag{1.5}$$

where the operator $\mathcal{A}_0 : \mathbf{L}^2_\varphi \longrightarrow \mathbf{L}^2_\varphi$ is defined by the left-hand side of (1.1), and consider in $\mathbf{L}^2_\mu$ the equivalent equation

$$\mathcal{A}u = \widetilde{f}\,, \tag{1.6}$$

where $\mathcal{A} = \mathcal{J}_0\mathcal{A}_0\mathcal{J}_0^{-1} : \mathbf{L}^2_\mu \longrightarrow \mathbf{L}^2_\mu$, $\widetilde{f} = \mathcal{J}_0 f \in \mathbf{L}^2_\mu$, and $u = \mathcal{J}_0 v \in \mathbf{L}^2_\mu$ is looked for. If we write the kernel function of the operator $\mathcal{A}_0$ in the form

$$\frac{1}{y-x} + \mathbf{h}_0\left(\frac{1+x}{1+y}\right)\frac{1}{1+y} \quad \text{with} \quad \mathbf{h}_0(t) = -\frac{1}{1+t} + \frac{6t}{(1+t)^2} - \frac{4t^2}{(1+t)^3}\,,$$

then the kernel function of the operator $\mathcal{A}$ of equation (1.6) is equal to

$$\begin{aligned}
&\sqrt{\frac{1+x}{1+y}}\left[\frac{1}{y-x} + \mathbf{h}_0\left(\frac{1+x}{1+y}\right)\frac{1}{1+y}\right]\\
&\quad = \frac{1}{y-x} - \frac{1}{y-x}\left(1 - \sqrt{\frac{1+x}{1+y}}\right) + \sqrt{\frac{1+x}{1+y}}\,\mathbf{h}_0\left(\frac{1+x}{1+y}\right)\frac{1}{1+y} \qquad (1.7)\\
&\quad = \frac{1}{y-x} + \mathbf{h}\left(\frac{1+x}{1+y}\right)\frac{1}{1+y}
\end{aligned}$$

with

$$\mathbf{h}(t) = \sqrt{t}\,\mathbf{h}_0(t) - \frac{1}{1+\sqrt{t}}\,. \tag{1.8}$$

Note that $\mathbf{h}(t)$ is also not of the form as considered in [10, 11] and does not fulfil the conditions imposed on the Mellin kernels in [12], where it is assumed that its Mellin transform is holomorphic in a strip parallel to the imaginary axis and of wide greater than 1.

The second step is to apply to the Mellin part of the original equation the Gaussian rule associated to the collocation nodes before collocating the equation. Then, it turns out that we are able to prove uniform boundedness of the sequence of the discretized operators and to show that this sequence belongs to a C^*-algebra, in which we can study stability. While in [10] the authors used the respective properties of the sequence of the collocation method to show that the sequence of the collocation-quadrature method belongs to the considered C^*-algebra, here we have to go another way, since in the present situation these properties of the collocation method are not available. One of the main ingredients in our new approach is to show that the part of the sequence of the collocation-quadrature method concerned with the Mellin part of the operator is "very close" to the finite sections of an operator belonging to a certain C^*-algebra of Toeplitz operators (cf. Lemma 2.10 and the proof of Lemma 2.13).

The remaining part of the paper is organized as follows. In Section 2, by considering a slight generalization of equation (1.1) (see (2.1)), we describe the collocation-quadrature methods (w.r.t. Chebyshev nodes of first and third kind) under consideration, define an appropriate C^*-algebra of operator sequences, and prove that the sequences of these collocation-quadrature methods belong to this C^*-algebra. In Section 3 it is shown that the collocation-quadrature methods for the class of operator equations, we are interested in, are stable if and only if the four limit operators, associated to the method and used in the definition of the C^*-algebra, are invertible. Section 4 is devoted to the application of this stability result to the notched half-plane problem. Thereby, it turns out that, in case of using Chebyshev nodes of first kind, three limit operators are invertible and the fourth one is Fredholm with index zero. But, since we are not able to show that the null space of the fourth limit operator is trivial, we can only conjecture that the collocation-quadrature method w.r.t. the Chebyshev nodes of first kind is stable, which is confirmed by numerical results also presented in this section and compared with results obtained in [4] for the collocation method, for which no stability result is known. On the other hand, the collocation-quadrature method w.r.t. Chebyshev nodes of third kind applied to the notched half-plane problem is not stable, which is also seen by numerical results. In Section 5 we give a self-contained proof of Lemma 2.10 which is crucial for our approach as already mentioned above.

2. The limit operators of the collocation-quadrature method

As a generalization of equation (1.6) we consider the Cauchy singular integral equation with additional Mellin operators

$$a(x)u(x) + \frac{b(x)}{\pi\mathbf{i}} \int_{-1}^{1} \frac{u(y)\,dy}{y-x} + \int_{-1}^{1} \mathbf{h}\left(\frac{1+x}{1+y}\right) \frac{u(y)\,dy}{1+y} = f(x)\,, \tag{2.1}$$

$-1 < x < 1$, where $a(x)$ and $b(x)$ are piecewise continuous functions[1] on $[-1,1]$ and where $\mathbf{h} = \mathbf{h}_1 + \mathbf{h}_2$ and (comp. (1.8))

$$\mathbf{h}_1(t) = \sum_{k=1}^{m} \frac{\beta_k}{\pi\mathbf{i}} \frac{t^{k-\frac{1}{2}}}{(1+t)^k}\,, \quad \mathbf{h}_2(t) = \frac{\beta_0}{\pi\mathbf{i}} \frac{1}{1+\sqrt{t}}\,, \quad t>0\,, \quad \beta_k \in \mathbb{C}\,, \quad m \in \mathbb{N}\,.$$

[1]We call a function $a : [-1,1] \to \mathbb{C}$ piecewise continuous if it is continuous at ± 1, if the one-sided limits $a(x \pm 0)$ exist for all $x \in (-1,1)$ and at least one of them coincides with $a(x)$.

Let $a\mathcal{I} : \mathbf{L}^2_\mu \longrightarrow \mathbf{L}^2_\mu$ denote the multiplication operator by the function a,

$$\mathcal{S} : \mathbf{L}^2_\mu \to \mathbf{L}^2_\mu, \quad u \mapsto \frac{1}{\pi \mathbf{i}} \int_{-1}^{1} \frac{u(y)\,dy}{y - \cdot}$$

denote the Cauchy singular integral operator, and

$$\mathcal{B} : \mathbf{L}^2_\mu \longrightarrow \mathbf{L}^2_\mu , \quad u \mapsto \int_{-1}^{1} \mathbf{h}\left(\frac{1+\cdot}{1+y}\right) \frac{u(y)\,dy}{1+y}$$

denote the Mellin operator with a fixed singularity at -1. Then, equation (2.1) can be written in short form as

$$\mathcal{A}u := (a\mathcal{I} + b\mathcal{S} + \mathcal{B})u = f\,. \tag{2.2}$$

Note that the linear operator $\mathcal{A} : \mathbf{L}^2_\mu \longrightarrow \mathbf{L}^2_\mu$ is bounded, which can be seen from the boundedness of $\mathcal{S}$ in $\mathbf{L}^2_\varphi$ and in $\mathbf{L}^2_\mu$ (cf. [5, Theorem 1.16]), the formula $\mathcal{B} = \mathcal{J}_0\,(\mathcal{B}_0 - \beta_0\mathcal{S})\,\mathcal{J}_0^{-1} + \beta_0\mathcal{S}$ (cf. (1.7)), and the boundedness of (see [5, Remark 8.3, Theorem 9.1])

$$\mathcal{B}_0 : \mathbf{L}^2_\varphi \longrightarrow \mathbf{L}^2_\varphi , \quad u \mapsto \sum_{k=1}^{m} \frac{\beta_k}{\pi \mathbf{i}} \int_{-1}^{1} \frac{(1+\cdot)^{k-1} u(y)}{(2+y+\cdot)^k}\,dy\,, \quad k \in \mathbb{N}\,.$$

Furthermore, on the space $\mathbf{C}(-1,1)$ of continuous functions $u : (-1,1) \longrightarrow \mathbb{C}$, we define, for $n \in \mathbb{N}$ and $\tau \in \{\sigma, \nu\}$, the quadrature operators

$$\mathcal{H}^\tau_n : \mathbf{C}(-1,1) \longrightarrow \mathbf{C}[-1,1], \quad u \mapsto (\omega^\tau_n)^2 \sum_{k=1}^{n} \varphi(x^\tau_{kn})\,\mathbf{h}\left(\frac{1+\cdot}{1+x^\tau_{kn}}\right) \frac{u(x^\tau_{kn})}{1+x^\tau_{kn}} \tag{2.3}$$

with the Chebyshev nodes of first and third kind, $x^\sigma_{kn} = \cos\frac{2k-1}{2n}\pi$ and $x^\nu_{kn} = \cos\frac{2k-1}{2n+1}\pi$, respectively, as well as $\omega^\sigma_n = \sqrt{\frac{\pi}{n}}$ and $\omega^\nu_n = \sqrt{\frac{\pi}{n+\frac{1}{2}}}$. Note that $\sigma(x) = \dfrac{1}{\sqrt{1-x^2}}$ and $\nu(x) = \sqrt{\dfrac{1+x}{1-x}}$, and that the weights λ^τ_{kn} of the respective Gaussian rules are given by

$$\lambda^\sigma_{kn} = \frac{\pi}{n} \quad \text{and} \quad \lambda^\nu_{kn} = \frac{\pi(1+x^\nu_{kn})}{n+\frac{1}{2}}\,, \quad k = 1, \ldots, n\,.$$

We look for an approximate solution $u_n \in \operatorname{im} \mathcal{L}_n$ of (2.2) by solving

$$\big((a\mathcal{I} + b\mathcal{S} + \mathcal{H}^\tau_n)u_n\big)(x^\tau_{jn}) = f_n(x^\tau_{jn})\,, \tag{2.4}$$

where

$$\mathcal{L}_n : \mathbf{L}^2_\mu \longrightarrow \mathbf{L}^2_\mu , \quad u \mapsto \sum_{j=0}^{n-1} \langle u, \widetilde{p}_j \rangle_\mu\, \widetilde{p}_j$$

are the Fourier projections w.r.t. the orthonormal system $(\widetilde{p}_n)_{n=0}^\infty$ with $\widetilde{p}_j = \nu R_j$ and the normalized Chebyshev polynomials of third kind (cf., for example, [17, pp. 122, 123])

$$R_j(x) = \frac{\cos\left(j+\frac{1}{2}\right)\theta}{\sqrt{\pi}\,\cos\frac{\theta}{2}}\,, \quad x = \cos\theta\,,$$

and where $f_n \in \operatorname{im}\mathcal{L}_n$ is an approximation of $f \in \mathbf{L}^2_\mu$ in the sense that

$$\lim_{n\to\infty} \|f_n - f\|_\mu = 0\,. \tag{2.5}$$

If we denote by $\mathcal{M}^\tau_n = \nu\mathcal{L}^\tau_n\mu\mathcal{I}$ a weighted interpolation operator, where $\mathcal{L}^\tau_n$ is the Lagrange interpolation operator w.r.t. the nodes x^τ_{kn}, $k = 1,\dots,n$, then the collocation-quadrature method (2.4) can be written as

$$\mathcal{A}_n u_n = f_n\,, \quad u_n \in \operatorname{im}\mathcal{L}_n \tag{2.6}$$

with $\mathcal{A}_n = \mathcal{M}^\tau_n(a\mathcal{I} + b\mathcal{S} + \mathcal{H}^\tau_n)\mathcal{L}_n$. It is well known that, in the investigation of numerical methods for operator equations, the stability of the respective operator sequence plays an essential role.

Definition. We call the sequence $(\mathcal{A}_n)$ in (2.6) stable (in $\mathbf{L}^2_\mu$) if, for all sufficiently large n, the operators $\mathcal{A}_n : \operatorname{im}\mathcal{L}_n \longrightarrow \operatorname{im}\mathcal{L}_n$ are invertible and if the norms $\left\|\mathcal{A}_n^{-1}\mathcal{L}_n\right\|_{\mathcal{L}(\mathbf{L}^2_\mu)}$ are uniformly bounded.

Note that, if the method is stable and if $\mathcal{A}_n\mathcal{L}_n$ converges strongly to $\mathcal{A} \in \mathcal{L}(\mathbf{L}^2_\mu)$, then the operator $\mathcal{A}$ is injective. If additionally the image of $\mathcal{A}$ equals $\mathbf{L}^2_\mu$, then (2.5) implies the $\mathbf{L}^2_\mu$-convergence of the solution u_n of (2.6) to the (unique) solution $u \in \mathbf{L}^2_\mu$ of (2.2).

To investigate the stability of the operator sequence in (2.6) we follow the C^*-algebra approach already used in, for example, [9, 10, 11] (cf. also [12, 13, 14]). For this, we will consider the operator sequence under consideration as an element of a C^*-algebra, which we describe in the following. By ℓ^2 we denote the Hilbert space of all square summable sequences $\xi = (\xi_j)_{j=0}^\infty$, $\xi_j \in \mathbb{C}$ with the inner product $\langle \xi, \eta\rangle = \sum_{j=0}^\infty \xi_j\overline{\eta_j}$. Moreover, we define the operators

$$\mathcal{W}_n : \mathbf{L}^2_\mu \longrightarrow \mathbf{L}^2_\mu\,, \quad u \mapsto \sum_{j=0}^{n-1} \langle u, \widetilde{p}_{n-1-j}\rangle_\mu\,\widetilde{p}_j\,,$$

$$\mathcal{P}_n : \ell^2 \longrightarrow \ell^2\,, \quad (\xi_j)_{j=0}^\infty \mapsto (\xi_0,\cdots,\xi_{n-1},0,\dots)\,,$$

and, for $\tau \in \{\sigma,\nu\}$,

$$\mathcal{V}^\tau_n : \operatorname{im}\mathcal{L}_n \longrightarrow \operatorname{im}\mathcal{P}_n\,,$$

$$u \mapsto \left(\omega^\tau_n\sqrt{1-x^\tau_{1n}}\,u(x^\tau_{1n}),\dots,\omega^\tau_n\sqrt{1-x^\tau_{nn}}\,u(x^\tau_{nn}),0,\dots\right)\,,$$

$$\widetilde{\mathcal{V}}^\tau_n : \operatorname{im}\mathcal{L}_n \longrightarrow \operatorname{im}\mathcal{P}_n\,,$$

$$u \mapsto \left(\omega^\tau_n\sqrt{1-x^\tau_{nn}}\,u(x^\tau_{nn}),\dots,\omega^\tau_n\sqrt{1-x^\tau_{1n}}\,u(x^\tau_{1n}),0,\dots\right).$$

Let $T = \{1, 2, 3, 4\}$, set

$$\mathbf{X}^{(1)} = \mathbf{X}^{(2)} = \mathbf{L}^2_\mu, \quad \mathbf{X}^{(3)} = \mathbf{X}^{(4)} = \ell^2, \quad \mathcal{L}_n^{(1)} = \mathcal{L}_n^{(2)} = \mathcal{L}_n, \quad \mathcal{L}_n^{(3)} = \mathcal{L}_n^{(4)} = \mathcal{P}_n,$$

and define $\mathcal{E}_n^{(t)} : \operatorname{im}\mathcal{L}_n \longrightarrow \mathbf{X}_n^{(t)} := \operatorname{im}\mathcal{L}_n^{(t)}$ for $t \in T$ by

$$\mathcal{E}_n^{(1)} = \mathcal{L}_n, \qquad \mathcal{E}_n^{(2)} = \mathcal{W}_n, \qquad \mathcal{E}_n^{(3)} = \mathcal{V}_n^\tau, \qquad \mathcal{E}_n^{(4)} = \widetilde{\mathcal{V}}_n^\tau.$$

Here and at other places, we use the notion $\mathcal{L}_n$, $\mathcal{W}_n$, ... instead of $\mathcal{L}_n|_{\operatorname{im}\mathcal{L}_n}$, $\mathcal{W}_n|_{\operatorname{im}\mathcal{L}_n}$, ..., respectively. All operators $\mathcal{E}_n^{(t)}$, $t \in T$ are unitary with the inverses

$$\left(\mathcal{E}_n^{(1)}\right)^{-1} = \mathcal{E}_n^{(1)}, \quad \left(\mathcal{E}_n^{(2)}\right)^{-1} = \mathcal{E}_n^{(2)}, \quad \left(\mathcal{E}_n^{(3)}\right)^{-1} = (\mathcal{V}_n^\tau)^{-1}, \quad \left(\mathcal{E}_n^{(4)}\right)^{-1} = (\widetilde{\mathcal{V}}_n^\tau)^{-1},$$

where, for $\xi \in \operatorname{im}\mathcal{P}_n$,

$$(\mathcal{V}_n^\tau)^{-1}\xi = \sum_{k=1}^n \frac{\xi_{k-1}}{\omega_n^\tau\sqrt{1-x_{kn}^\tau}}\,\widetilde{\ell}_{kn}^\tau \quad \text{and} \quad \left(\widetilde{\mathcal{V}}_n^\tau\right)^{-1}\xi = \sum_{k=1}^n \frac{\xi_{n-k}}{\omega_n^\tau\sqrt{1-x_{kn}^\tau}}\,\widetilde{\ell}_{kn}^\tau,$$

and where

$$\widetilde{\ell}_{kn}^\tau(x) = \frac{\nu(x)}{\nu(x_{kn}^\tau)}\,\ell_{kn}^\tau(x), \quad k = 1, \ldots, n,$$

are the weighted fundamental interpolation polynomials with the usual interpolation polynomials $\ell_{kn}^\tau(x)$ w.r.t. the nodes x_{kn}^τ. It is easily seen that, for all indices $r, t \in T$ with $r \neq t$, the operators $\mathcal{E}_n^{(r)}\left(\mathcal{E}_n^{(t)}\right)^{-1}\mathcal{L}_n^{(t)}$ as well as their adjoints converge weakly to zero (cf., for example, the proof of [9, Lemma 2.1]). Now we can introduce the algebra of operator sequences we are interested in. By $\mathfrak{F}$ we denote the set of all sequences $(\mathcal{A}_n)$ of linear operators $\mathcal{A}_n : \operatorname{im}\mathcal{L}_n \longrightarrow \operatorname{im}\mathcal{L}_n$ for which the strong limits

$$\mathcal{W}^t(\mathcal{A}_n) := \lim_{n\to\infty} \mathcal{E}_n^{(t)}\mathcal{A}_n\left(\mathcal{E}_n^{(t)}\right)^{-1}\mathcal{L}_n^{(t)}$$

and

$$\left(\mathcal{W}^t(\mathcal{A}_n)\right)^* = \lim_{n\to\infty}\left(\mathcal{E}_n^{(t)}\mathcal{A}_n\left(\mathcal{E}_n^{(t)}\right)^{-1}\mathcal{L}_n^{(t)}\right)^*, \quad t \in T,$$

exist. If $\mathfrak{F}$ is provided with the supremum norm $\|(\mathcal{A}_n)\|_{\mathfrak{F}} := \sup_{n\geq 1}\|\mathcal{A}_n\mathcal{L}_n\|_{\mathfrak{L}(\mathbf{L}^2_\nu)}$ and with the operations $(\mathcal{A}_n) + (\mathcal{B}_n) := (\mathcal{A}_n + \mathcal{B}_n)$, $(\mathcal{A}_n)(\mathcal{B}_n) := (\mathcal{A}_n\mathcal{B}_n)$, and $(\mathcal{A}_n)^* := (\mathcal{A}_n^*)$, then $\mathfrak{F}$ becomes a C^*-algebra with the identity element $(\mathcal{L}_n)$. Furthermore, we introduce the set $\mathfrak{J} \subset \mathfrak{F}$ of all sequences of the form

$$\left(\sum_{t=1}^4 \left(\mathcal{E}_n^{(t)}\right)^{-1}\mathcal{L}_n^{(t)}\mathcal{T}_t\mathcal{E}_n^{(t)} + \mathcal{C}_n\right),$$

where the linear operators $\mathcal{T}_t : \mathbf{X}^{(t)} \longrightarrow \mathbf{X}^{(t)}$ are compact and where the sequence $(\mathcal{C}_n) \in \mathfrak{F}$ belongs to the closed ideal $\mathfrak{G}$ of all sequences from $\mathfrak{F}$ tending to zero in norm, i.e., $\lim_{n\to\infty}\|\mathcal{C}_n\mathcal{L}_n\|_{\mathfrak{L}(\mathbf{L}^2_\nu)} = 0$. From [18, 19, Theorem 10.33] (see also [8, Theorem 6.1]) we get the following proposition, which is the basis for our further investigations in the remaining part of the paper.

Proposition 2.1. *The set $\mathfrak{J}$ forms a two-sided closed ideal in the C^*-algebra $\mathfrak{F}$. Moreover, a sequence $(\mathcal{A}_n) \in \mathfrak{F}$ is stable if and only if the operators $\mathcal{W}^t(\mathcal{A}_n) : \mathbf{X}^{(t)} \longrightarrow \mathbf{X}^{(t)}$, $t \in T$ and the coset $(\mathcal{A}_n) + \mathfrak{J} \in \mathfrak{F}/\mathfrak{J}$ are invertible.*

As a first step, we have to show that the sequence $(\mathcal{M}_n^\tau(a\mathcal{I} + b\mathcal{S} + \mathcal{H}_n^\tau)\mathcal{L}_n)$ of the collocation-quadrature method (2.6) belongs to $\mathfrak{F}$. In particular, we have to determine the respective limit operators $\mathcal{W}^t\left(\mathcal{M}_n^\tau(a\mathcal{I} + b\mathcal{S} + \mathcal{H}_n^\tau)\mathcal{L}_n\right)$ for $t \in T$. For the part $\mathcal{M}_n^\tau(a\mathcal{I} + b\mathcal{S})\mathcal{L}_n$ the respective results are already known (cf. [9, Section 3] and [10, Section 3.2, Section 5]). For completeness, we formulate these results here. Since in [9, 10] the collocation method for $\tau \in \{\sigma, \mu\}$ in the space $\mathbf{L}^2_\nu$ is considered, we have to translate the results presented there into the case $\tau \in \{\sigma, \nu\}$ in the space $\mathbf{L}^2_\mu$. To this end, we define the isometrical isomorphism $\mathcal{J} : \mathbf{L}^2_\nu \longrightarrow \mathbf{L}^2_\mu$, $(\mathcal{J}u)(x) = u(-x)$ and the orthoprojections $\widehat{\mathcal{L}}_n : \mathbf{L}^2_\nu \longrightarrow \mathbf{L}^2_\nu$, $u \mapsto \sum_{j=0}^{n-1} \langle u, \widehat{p}_j \rangle_\nu \widehat{p}_j$, where $\widehat{p}_j(x) = \mu(x)P_j(x)$ with the Chebyshev polynomials of fourth kind $P_j(x) = \dfrac{\sin\left(j + \frac{1}{2}\right)\theta}{\sqrt{\pi}\sin\frac{\theta}{2}}$, $x = \cos\theta$. We get

$$\mathcal{J}\widehat{a}\mathcal{J}^{-1} = a\mathcal{I}\,, \ \mathcal{J}\mathcal{S}\mathcal{J}^{-1} = -\mathcal{S}\,, \ \mathcal{J}\widehat{\mathcal{L}}_n\mathcal{J}^{-1} = \mathcal{L}_n\,, \ \mathcal{J}\widehat{\mathcal{W}}_n\mathcal{J}^{-1} = (-1)^{n-1}\mathcal{W}_n\,, \tag{2.7}$$

where $\widehat{a}(x) := a(-x)$ and $\widehat{\mathcal{W}}_n u = \sum_{j=0}^{n-1} \langle u, \widehat{p}_{n-1-j} \rangle_\nu \widehat{p}_j$, and where, for the last two relations, we took into account $\widehat{p}_n(-x) = (-1)^n \widetilde{p}_n(x)$. Since $x^\sigma_{n-j+1,n} = -x^\sigma_{jn}$ and $x^\nu_{n-j+1,n} = -x^\mu_{jn}$, $j = 1, \ldots, n$, it follows

$$\ell^\sigma_{n-j+1,n}(-x) = \ell^\sigma_{j,n}(x) \quad \text{and} \quad \ell^\nu_{n-j+1,n}(-x) = \ell^\mu_{j,n}(x),$$

which implies, for every function $f : (-1,1) \to \mathbb{C}$, the relations $\mathcal{J}\mathcal{L}^\sigma_n\mathcal{J}^{-1}f = \mathcal{L}^\sigma_n f$ and $\mathcal{J}\mathcal{L}^\mu_n\mathcal{J}^{-1}f = \mathcal{L}^\nu_n f$. Consequently, using $\mathcal{J}\mu\mathcal{J}^{-1} = \nu\mathcal{I}$,

$$\mathcal{J}\widehat{\mathcal{M}}^\sigma_n\mathcal{J}^{-1}f = \mathcal{M}^\sigma_n f \quad \text{and} \quad \mathcal{J}\widehat{\mathcal{M}}^\mu_n\mathcal{J}^{-1}f = \mathcal{M}^\nu_n f\,, \tag{2.8}$$

where $\widehat{\mathcal{M}}^\tau_n = \mu\mathcal{L}^\tau_n\nu\mathcal{I}$. Moreover, we define, for $\tau \in \{\sigma, \mu\}$ and $\omega^\mu_n = \omega^\nu_n$,

$$\widehat{\mathcal{V}}^\tau_n : \operatorname{im}\widehat{\mathcal{L}}_n \longrightarrow \operatorname{im}\mathcal{P}_n\,, \quad u \mapsto \left[\ \omega^\tau_n\sqrt{1 + x^\tau_{kn}}\,u(x^\tau_{kn})\ \right]_{k=1}^n\,,$$

$$\widetilde{\widehat{\mathcal{V}}}^\tau_n : \operatorname{im}\widehat{\mathcal{L}}_n \longrightarrow \operatorname{im}\mathcal{P}_n\,, \quad u \mapsto \left[\ \omega^\tau_n\sqrt{1 + x^\tau_{n+1-k,n}}\,u(x^\tau_{n+1-k,n})\ \right]_{k=1}^n\,.$$

To describe the limit operators $\mathcal{W}^t\left(\mathcal{M}^\tau_n(a\mathcal{T} + b\mathcal{S})\mathcal{L}_n\right)$, we need the isometrical isomorphisms $\widehat{\mathcal{J}}_k : \mathbf{L}^2_\nu \longrightarrow \mathbf{L}^2_\nu$ and $\mathcal{J}_k : \mathbf{L}^2_\mu \longrightarrow \mathbf{L}^2_\mu$, $k = 1, 2, 3$, defined by

$$\widehat{\mathcal{J}}_1 u = \sum_{j=0}^\infty \langle u, \widehat{p}_j \rangle_\nu R_j\,, \quad \widehat{\mathcal{J}}_2 u = \sum_{j=0}^\infty \langle u, \widehat{p}_j \rangle_\nu \sqrt{1 - \cdot}\,U_j\,,$$

$$\widehat{\mathcal{J}}_3 u = \sum_{j=0}^\infty \langle u, \widehat{p}_j \rangle_\nu (1 + \cdot)^{-\frac{1}{2}} T_j\,,$$

and

$$\mathcal{J}_1 u = \sum_{j=0}^{\infty} \langle u, \widetilde{p}_j \rangle_\mu P_j \,, \quad \mathcal{J}_2 u = \sum_{j=0}^{\infty} \langle u, \widetilde{p}_j \rangle_\mu \sqrt{1+\cdot}\, U_j \,, \\ \mathcal{J}_3 u = \sum_{j=0}^{\infty} \langle u, \widetilde{p}_j \rangle_\mu (1-\cdot)^{-\frac{1}{2}} T_j \,, \tag{2.9}$$

where U_j and T_j denote the Chebyshev polynomials

$$U_j(x) = \frac{\sqrt{2}\sin(j+1)\theta}{\sqrt{\pi}\sin\theta} \,, \ j \in \mathbb{N}_0 \,,$$

and

$$T_0(x) = \frac{1}{\sqrt{\pi}} \,, \ T_j(x) = \frac{\sqrt{2}\,\cos j\theta}{\sqrt{\pi}} \,, \ j \in \mathbb{N} \,, \ x = \cos\theta \,,$$

of second and first kind, respectively. Finally, we have to use the shift operators

$$\widehat{\mathcal{V}} : \mathbf{L}^2_\nu \longrightarrow \mathbf{L}^2_\nu \,, \quad u \mapsto \sum_{j=0}^{\infty} \langle u, \widehat{p}_j \rangle_\nu \widehat{p}_{j+1} \,, \\ \mathcal{V} : \mathbf{L}^2_\mu \longrightarrow \mathbf{L}^2_\mu \,, \quad u \mapsto \sum_{j=0}^{\infty} \langle u, \widetilde{p}_j \rangle_\mu \widetilde{p}_{j+1} \,. \tag{2.10}$$

It is easily seen that

$$\widehat{\mathcal{J}}_k \mathcal{J}^{-1} = \mathcal{J}^{-1} \mathcal{J}_k \,, \ k = 1,2,3 \,, \quad \text{and} \quad \widehat{\mathcal{V}} \mathcal{J}^{-1} = -\mathcal{J}^{-1} \mathcal{V} \,, \tag{2.11}$$

as well as

$$\mathcal{V}_n^\tau = \widetilde{\widehat{\mathcal{V}}}{}_n^{\overline{\tau}} \mathcal{J}^{-1} \,, \quad \widetilde{\mathcal{V}}_n^\tau = \widehat{\mathcal{V}}_n^{\overline{\tau}} \mathcal{J}^{-1} \quad \text{with} \quad \overline{\sigma} = \sigma \,, \ \overline{\nu} = \mu \,. \tag{2.12}$$

By $\mathbf{I}$, $\widetilde{\mathbf{S}}$, $\mathbf{S}^\tau : \ell^2 \longrightarrow \ell^2$ we denote the operators defined by the matrices

$$\mathbf{I} = \left[\ \delta_{j,k}\ \right]_{j,k=0}^{\infty} \,, \qquad \widetilde{\mathbf{S}} = \left[\ \frac{1-(-1)^{j-k}}{\pi\mathbf{i}(j-k)} + \frac{1-(-1)^{j+k+1}}{\pi\mathbf{i}(j+k+1)}\ \right]_{j,k=0}^{\infty} \,,$$

and

$$\mathbf{S}^\tau = \begin{cases} \left[\ \dfrac{1-(-1)^{j-k}}{\pi\mathbf{i}(j-k)} - \dfrac{1-(-1)^{j+k+1}}{\pi\mathbf{i}(j+k+1)}\ \right]_{j,k=0}^{\infty} & : \ \tau = \sigma \,, \\ \left[\ \dfrac{1-(-1)^{j-k}}{\pi\mathbf{i}} \left(\dfrac{1}{j-k} - \dfrac{1}{j+k+2} \right)\ \right]_{j,k=0}^{\infty} & : \ \tau = \mu \,. \end{cases}$$

We recall the following result.

Lemma 2.2 ([9], Prop. 3.5 [10], Prop. 3.2, Remark 3.3). *If $\mathcal{A} = a\mathcal{I} + b\mathcal{S} : \mathbf{L}^2_\nu \longrightarrow \mathbf{L}^2_\nu$ with $a, b \in \mathbf{PC}$ and $\mathcal{A}_n = \widehat{\mathcal{M}}_n^\tau \mathcal{A} \widehat{\mathcal{L}}_n$, then, for $\tau \in \{\sigma, \mu\}$, we have the strong* *convergences (in $\mathbf{L}^2_\nu$)*

$$\mathcal{A}_n \longrightarrow \mathcal{A}\,,\quad \widehat{\mathcal{W}}_n\mathcal{A}_n\widehat{\mathcal{W}}_n \longrightarrow \begin{cases} \widehat{\mathcal{J}}_1^{-1}(a\widehat{\mathcal{J}}_1 + \mathbf{i}\, b\mathcal{I}) & : \ \tau = \sigma\,, \\ \widehat{\mathcal{J}}_2^{-1}(a\widehat{\mathcal{J}}_2 - \mathbf{i}\, b\widehat{\mathcal{J}}_3\widehat{\mathcal{V}}) & : \ \tau = \mu\,, \end{cases}$$

$$\widehat{\mathcal{V}}^\tau_n\mathcal{A}_n\left(\widehat{\mathcal{V}}^\tau_n\right)^{-1}\mathcal{P}_n \longrightarrow a(1)\mathbf{I} + b(1)\mathbf{S}^\tau\,,$$

$$\widetilde{\widehat{\mathcal{V}}}^\tau_n\mathcal{A}_n\left(\widetilde{\widehat{\mathcal{V}}}^\tau_n\right)^{-1}\mathcal{P}_n \longrightarrow a(-1)\mathbf{I} - b(-1)\widetilde{\mathbf{S}}\,,$$

where also the respective sequences of adjoint operators are strongly convergent.

Corollary 2.3. *Let $a, b \in \mathbf{PC}$ and consider $\mathcal{A} = a\mathcal{I} + b\mathcal{S} : \mathbf{L}^2_\mu \longrightarrow \mathbf{L}^2_\mu$ as well as $\mathcal{A}_n = \mathcal{M}^\tau_n\mathcal{A}\mathcal{L}_n$ for $\tau \in \{\sigma, \nu\}$. Then $(\mathcal{A}_n) \in \mathfrak{F}$ with*

$$\mathcal{W}^1(\mathcal{A}_n) = \mathcal{A}\,,\quad \mathcal{W}^2(\mathcal{A}_n) = \begin{cases} \mathcal{J}_1^{-1}(a\mathcal{J}_1 + \mathbf{i}\, b\mathcal{I}) & : \ \tau = \sigma\,, \\ \mathcal{J}_2^{-1}(a\mathcal{J}_2 + \mathbf{i}\, b\mathcal{J}_3\mathcal{V}) & : \ \tau = \nu\,. \end{cases}$$

$$\mathcal{W}^3(\mathcal{A}_n) = a(1)\mathbf{I} + b(1)\widetilde{\mathbf{S}}\,,\quad \mathcal{W}^4(\mathcal{A}_n) = a(-1)\mathbf{I} - b(-1)\mathbf{S}^{\overline{\tau}}\,.$$

Proof. By (2.7), (2.8), (2.11), and Lemma 2.2, we have

$$\mathcal{A}_n = \mathcal{J}\widehat{\mathcal{M}^\tau_n}(\widehat{a}\mathcal{I} - \widehat{b}\mathcal{S})\widehat{\mathcal{L}}_n\mathcal{J}^{-1} \longrightarrow \mathcal{J}(\widehat{a}\mathcal{I} - \widehat{b}\mathcal{S})\mathcal{J}^{-1} = a\mathcal{I} + b\mathcal{S}$$

and

$$\begin{aligned} \mathcal{W}_n\mathcal{A}_n\mathcal{W}_n &= \mathcal{J}\widehat{\mathcal{W}}_n\widehat{\mathcal{M}^\tau_n}(\widehat{a}\mathcal{I} - \widehat{b}\mathcal{S})\widehat{\mathcal{L}}_n\widehat{\mathcal{W}}_n\mathcal{J}^{-1} \\ &\longrightarrow \begin{cases} \mathcal{J}\widehat{\mathcal{J}}_1^{-1}(\widehat{a}\widehat{\mathcal{J}}_1 + \mathbf{i}\,\widehat{b}\mathcal{I})\mathcal{J}^{-1} & : \ \tau = \sigma \\ \mathcal{J}\widehat{\mathcal{J}}_2^{-1}(\widehat{a}\widehat{\mathcal{J}}_2 - \mathbf{i}\,\widehat{b}\widehat{\mathcal{J}}_3\widehat{\mathcal{V}})\mathcal{J}^{-1} & : \ \tau = \nu\,, \end{cases} \\ &= \begin{cases} \mathcal{J}_1^{-1}\mathcal{J}(\widehat{a}\mathcal{J}\mathcal{J}_1\mathcal{J} + \mathbf{i}\,\widehat{b}\mathcal{I})\mathcal{J}^{-1} & : \ \tau = \sigma\,, \\ \mathcal{J}_2^{-1}\mathcal{J}(\widehat{a}\mathcal{J}\mathcal{J}_2 + \mathbf{i}\,\widehat{b}\mathcal{J}^{-1}\mathcal{J}_3\mathcal{V}) & : \ \tau = \nu\,, \end{cases} \\ &= \begin{cases} \mathcal{J}_1^{-1}(a\mathcal{J}_1 + \mathbf{i}\, b\mathcal{I}) & : \ \tau = \sigma\,, \\ \mathcal{J}_2^{-1}(a\mathcal{J}_2 + \mathbf{i}\, b\mathcal{J}_3\mathcal{V}) & : \ \tau = \nu\,. \end{cases} \end{aligned}$$

Taking further into account (2.12), we conclude

$$\begin{aligned} \mathcal{V}^\tau_n\mathcal{A}_n(\mathcal{V}^\tau_n)^{-1}\mathcal{P}_n &= \widetilde{\widehat{\mathcal{V}}}^{\overline{\tau}}_n\mathcal{J}^{-1}\mathcal{A}_n\mathcal{J}\left(\widetilde{\widehat{\mathcal{V}}}^{\overline{\tau}}_n\right)^{-1}\mathcal{P}_n = \widetilde{\widehat{\mathcal{V}}}^{\overline{\tau}}_n\widehat{\mathcal{M}^{\overline{\tau}}_n}(\widehat{a}\mathcal{I} - \widehat{b}\mathcal{S})\widehat{\mathcal{L}}_n\left(\widetilde{\widehat{\mathcal{V}}}^{\overline{\tau}}_n\right)^{-1}\mathcal{P}_n \\ &\longrightarrow \widehat{a}(-1)\mathbf{I} + \widehat{b}(-1)\widetilde{\mathbf{S}} = a(1)\mathbf{I} + b(1)\widetilde{\mathbf{S}} \end{aligned}$$

and

$$\begin{aligned} \widetilde{\mathcal{V}}^\tau_n\mathcal{A}_n\left(\widetilde{\mathcal{V}}^\tau_n\right)^{-1}\mathcal{P}_n &= \widehat{\mathcal{V}}^{\overline{\tau}}_n\mathcal{J}^{-1}\mathcal{A}_n\mathcal{J}\left(\widehat{\mathcal{V}}^{\overline{\tau}}_n\right)^{-1}\mathcal{P}_n = \widehat{\mathcal{V}}^{\overline{\tau}}_n\widehat{\mathcal{M}^{\overline{\tau}}_n}(\widehat{a}\mathcal{I} - \widehat{b}\mathcal{S})\widehat{\mathcal{L}}_n\left(\widehat{\mathcal{V}}^{\overline{\tau}}_n\right)^{-1}\mathcal{P}_n \\ &\longrightarrow \widehat{a}(1)\mathbf{I} - \widehat{b}(1)\mathbf{S}^{\overline{\tau}} = a(-1)\mathbf{I} - b(-1)\mathbf{S}^{\overline{\tau}}\,. \end{aligned}$$

From these computations we can also see that the respective sequences of adjoint operators converge strongly, if we take into consideration Lemma 2.2 together with the fact that $\mathcal{J}^* : \mathbf{L}^2_\mu \longrightarrow \mathbf{L}^2_\nu$ is defined by $(\mathcal{J}^* u)(x) = u(-x)$. The corollary is proved. □

The remaining part of this section is concerned with the proof, that the sequence $(\mathcal{M}_n^\tau \mathcal{H}_n^\tau \mathcal{L}_n)$ belongs to $\mathfrak{F}$, and with the determination of its limit operators $\mathcal{W}^t(\mathcal{M}_n^\tau \mathcal{H}_n^\tau \mathcal{L}_n)$, $t \in T$. In what follows we use the notation $\alpha_{n,k,\dots} = \mathcal{O}(\beta_{n,k,\dots})$ in the sense that there is a positive constant c such that $|\alpha_{n.k.\dots}| \le c\,\beta_{n,k,\dots}$ for all parameters $n, k, \dots$ under consideration.

Lemma 2.4. *For the kernel functions* $\mathbf{h}_1, \mathbf{h}_2 : (0,\infty) \longrightarrow \mathbb{C}$, *we have*

$$\begin{aligned} &\frac{\pi}{n}\,\mathbf{h}_1\left(\frac{1+x_{jn}^\sigma}{1+x_{kn}^\sigma}\right)\frac{1}{\sqrt{1+x_{kn}^\sigma}} \\ &\quad = \mathbf{h}_1\left(\frac{\left(n-j+\frac{1}{2}\right)^2}{\left(n-k+\frac{1}{2}\right)^2}\right)\frac{\sqrt{2}}{n-k+\frac{1}{2}} + \mathcal{O}\left(\frac{n-j+1}{n^2}\right) \end{aligned} \tag{2.13}$$

and

$$\begin{aligned} &\frac{\pi}{n+\frac{1}{2}}\,\mathbf{h}_1\left(\frac{1+x_{jn}^\nu}{1+x_{kn}^\nu}\right)\frac{1}{\sqrt{1+x_{kn}^\nu}} \\ &\quad = \mathbf{h}_1\left(\frac{(n-j+1)^2}{(n-k+1)^2}\right)\frac{\sqrt{2}}{n-k+1} + \mathcal{O}\left(\frac{n-j+1}{n^2}\right), \end{aligned} \tag{2.14}$$

as well as

$$\begin{aligned} &\frac{\pi}{n}\,\mathbf{h}_2\left(\frac{1+x_{jn}^\sigma}{1+x_{kn}^\sigma}\right)\frac{1}{\sqrt{1+x_{kn}^\sigma}} \\ &\quad = \mathbf{h}_2\left(\frac{\left(n-j+\frac{1}{2}\right)^2}{\left(n-k+\frac{1}{2}\right)^2}\right)\frac{\sqrt{2}}{n-k+\frac{1}{2}} + \mathcal{O}\left(\frac{2n-j-k+1}{n^2}\right) \end{aligned} \tag{2.15}$$

and

$$\begin{aligned} &\frac{\pi}{n+\frac{1}{2}}\,\mathbf{h}_2\left(\frac{1+x_{jn}^\nu}{1+x_{kn}^\nu}\right)\frac{1}{\sqrt{1+x_{kn}^\nu}} \\ &\quad = \mathbf{h}_2\left(\frac{(n-j+1)^2}{(n-k+1)^2}\right)\frac{\sqrt{2}}{n-k+1} + \mathcal{O}\left(\frac{2n-j-k+1}{n^2}\right), \end{aligned} \tag{2.16}$$

where $1 \le j,k \le n$ *and where the constants regarding the* $\mathcal{O}$*-terms are independent of* j, k, *and* $n \in \mathbb{N}$.

Proof. Define the functions

$$h_1 : (0,\infty) \longrightarrow \mathbb{R}\,, \quad t \mapsto \frac{t^{k-\frac{1}{2}}}{(1+t)^k}\,, \quad k \in \mathbb{N}\,,$$

and

$$h_2 : (0,\infty) \longrightarrow \mathbb{R}\,, \quad t \mapsto \frac{1}{1+\sqrt{t}}.$$

At first we show that the functions

$$g_1 : \left(0, \frac{\pi}{2}\right]^2 \longrightarrow \mathbb{C}, \quad (s,t) \mapsto \frac{1}{s}\left[h_1\left(\frac{\sin^2 s}{\sin^2 t}\right)\frac{1}{\sin t} - h_1\left(\frac{s^2}{t^2}\right)\frac{1}{t}\right]$$

and

$$g_2 : \left(0, \frac{\pi}{2}\right]^2 \longrightarrow \mathbb{C}, \quad (s,t) \mapsto \frac{1}{t+s}\left[h_2\left(\frac{\sin^2 s}{\sin^2 t}\right)\frac{1}{\sin t} - h_2\left(\frac{s^2}{t^2}\right)\frac{1}{t}\right]$$

are bounded. For g_1, we have

$$g_1(s,t) = \frac{\sin s}{s}\,\frac{(\sin^2 s)^{k-1}}{(\sin^2 t + \sin^2 s)^k} - \frac{(s^2)^{k-1}}{(t^2+s^2)^k}\,.$$

Since

$$\left|\frac{\sin s}{s} - 1\right| \frac{(\sin^2 s)^{k-1}}{(\sin^2 t + \sin^2 s)^k} \le \frac{s - \sin s}{s \sin^2 s} \le \text{const}$$

holds, it suffices to verify the boundedness of

$$\frac{(\sin^2 s)^{k-1}}{(\sin^2 t + \sin^2 s)^k} - \frac{(s^2)^{k-1}}{(t^2+s^2)^k}, \quad (s,t) \in \left(0, \frac{\pi}{2}\right]^2.$$

For $k = 1$, we have

$$\left|\frac{1}{\sin^2 t + \sin^2 s} - \frac{1}{t^2+s^2}\right| \le \frac{t^2 - \sin^2 t}{t^2 \sin^2 t} + \frac{s^2 - \sin^2 s}{s^2 \sin^2 s} \le \text{const}\,.$$

Let the assertion be fulfilled for some $k \in \mathbb{N}$. With

$$\begin{aligned}
&\left|\frac{(\sin^2 s)^k}{(\sin^2 t + \sin^2 s)^{k+1}} - \frac{(s^2)^k}{(t^2+s^2)^{k+1}}\right| \\
&\quad \le \left|\frac{(\sin^2 s)^k}{(\sin^2 t + \sin^2 s)^{k+1}} - \frac{(\sin^2 s)^{k-1}s^2}{(\sin^2 t + \sin^2 s)^{k+1}}\right| \\
&\qquad + \left|\frac{(\sin^2 s)^{k-1}s^2}{(\sin^2 t + \sin^2 s)^{k+1}} - \frac{(s^2)^k}{(t^2+s^2)^{k+1}}\right| \\
&\quad \le \frac{|\sin^2 s - s^2|}{(\sin^2 t + \sin^2 s)^2} + \left|\frac{(\sin^2 s)^{k-1}s^2}{(\sin^2 t + \sin^2 s)^{k+1}} - \frac{(\sin^2 s)^{k-1}s^2}{(\sin^2 t + \sin^2 s)^k (t^2+s^2)}\right| \\
&\qquad + \left|\frac{(\sin^2 s)^{k-1}s^2}{(\sin^2 t + \sin^2 s)^k (t^2+s^2)} - \frac{(s^2)^k}{(t^2+s^2)^{k+1}}\right| \\
&\quad \le \frac{|\sin^2 s - s^2|}{(\sin^2 t + \sin^2 s)^2} + \frac{(\sin^2 s)^{k-1}s^2}{(\sin^2 t + \sin^2 s)^k}\left|\frac{1}{\sin^2 t + \sin^2 s} - \frac{1}{t^2+s^2}\right| \\
&\qquad + \left|\frac{(\sin^2 s)^{k-1}}{(\sin^2 t + \sin^2 s)^k} - \frac{(s^2)^{k-1}}{(t^2+s^2)^k}\right| \frac{s^2}{t^2+s^2} \le \text{const},
\end{aligned}$$

we get the assertion by induction. For the function g_2, we have

$$g_2(s,t) = \frac{1}{t+s}\left(\frac{1}{\sin t + \sin s} - \frac{1}{t+s}\right), \quad (s,t) \in \left(0, \frac{\pi}{2}\right]^2,$$

and

$$\left|\frac{1}{\sin t+\sin s}-\frac{1}{t+s}\right| = \left|\frac{t+s-\sin t-\sin s}{(\sin t+\sin s)(t+s)}\right|$$
$$\leq \frac{t-\sin t}{t\,\sin t}+\frac{s-\sin s}{s\,\sin s} \leq \text{const}\,(t+s)\,.$$

Since

$$\frac{\pi}{n}\,h_1\left(\frac{1+x^\sigma_{jn}}{1+x^\sigma_{kn}}\right)\frac{1}{\sqrt{1+x^\sigma_{kn}}}-h_1\left(\frac{\left(n-j+\frac{1}{2}\right)^2}{\left(n-k+\frac{1}{2}\right)^2}\right)\frac{\sqrt{2}}{n-k+\frac{1}{2}}$$
$$=\frac{\pi}{n\sqrt{2}}\left[h_1\left(\frac{\sin^2\frac{n-j+\frac{1}{2}}{2n}\pi}{\sin^2\frac{n-k+\frac{1}{2}}{2n}\pi}\right)\frac{1}{\sin\frac{n-k+\frac{1}{2}}{2n}\pi}-h_1\left(\frac{\left(n-j+\frac{1}{2}\right)^2}{\left(n-k+\frac{1}{2}\right)^2}\right)\frac{1}{\frac{n-k+\frac{1}{2}}{2n}\pi}\right]$$
$$=\frac{\pi}{n\sqrt{2}}\,g_1\left(\frac{n-j+\frac{1}{2}}{2n}\pi,\frac{n-k+\frac{1}{2}}{2n}\pi\right)\frac{n-j+\frac{1}{2}}{2n}\pi$$

and

$$\frac{\pi}{n}\,h_2\left(\frac{1+x^\sigma_{jn}}{1+x^\sigma_{kn}}\right)\frac{1}{\sqrt{1+x^\sigma_{kn}}}-h_2\left(\frac{\left(n-j+\frac{1}{2}\right)^2}{\left(n-k+\frac{1}{2}\right)^2}\right)\frac{\sqrt{2}}{n-k+\frac{1}{2}}$$
$$=\frac{\pi}{n\sqrt{2}}\left[h_2\left(\frac{\sin^2\frac{n-j+\frac{1}{2}}{2n}\pi}{\sin^2\frac{n-k+\frac{1}{2}}{2n}\pi}\right)\frac{1}{\sin\frac{n-k+\frac{1}{2}}{2n}\pi}-h_2\left(\frac{\left(n-j+\frac{1}{2}\right)^2}{\left(n-k+\frac{1}{2}\right)^2}\right)\frac{1}{\frac{n-k+\frac{1}{2}}{2n}\pi}\right]$$
$$=\frac{\pi}{n\sqrt{2}}\,g_2\left(\frac{n-j+\frac{1}{2}}{2n}\pi,\frac{n-k+\frac{1}{2}}{2n}\pi\right)\left(\frac{n-k+\frac{1}{2}}{2n}\pi+\frac{n-j+\frac{1}{2}}{2n}\pi\right),$$

we obtain (2.13) and (2.15). Relations (2.14) and (2.16) can be proved in the same manner. □

Corollary 2.5. *For the kernel functions* $\mathbf{h}_1$, $\mathbf{h}_2 : (0,\infty) \longrightarrow \mathbb{C}$, *we have*

$$\frac{\pi}{n}\,\mathbf{h}_1\left(\frac{1+x^\sigma_{n+1-j,n}}{1+x^\sigma_{n+1-k,n}}\right)\frac{1}{\sqrt{1+x^\sigma_{n+1-k,n}}}=\mathbf{h}_1\left(\frac{\left(j-\frac{1}{2}\right)^2}{\left(k-\frac{1}{2}\right)^2}\right)\frac{\sqrt{2}}{k-\frac{1}{2}}+\mathcal{O}\left(\frac{j}{n^2}\right) \tag{2.17}$$

and

$$\frac{\pi}{n+\frac{1}{2}}\,\mathbf{h}_1\left(\frac{1+x^\nu_{n+1-j,n}}{1+x^\nu_{n+1-k,n}}\right)\frac{1}{\sqrt{1+x^\nu_{n+1-k,n}}}=\mathbf{h}_1\left(\frac{j^2}{k^2}\right)\frac{\sqrt{2}}{k}+\mathcal{O}\left(\frac{j}{n^2}\right), \tag{2.18}$$

as well as

$$\frac{\pi}{n}\,\mathbf{h}_2\left(\frac{1+x^\sigma_{n+1-j,n}}{1+x^\sigma_{n+1-k,n}}\right)\frac{1}{\sqrt{1+x^\sigma_{n+1-k,n}}}=\mathbf{h}_2\left(\frac{\left(j-\frac{1}{2}\right)^2}{\left(k-\frac{1}{2}\right)^2}\right)\frac{\sqrt{2}}{k-\frac{1}{2}}+\mathcal{O}\left(\frac{j+k}{n^2}\right) \tag{2.19}$$

and

$$\frac{\pi}{n+\frac{1}{2}}\,\mathbf{h}_2\left(\frac{1+x^{\nu}_{n+1-j,n}}{1+x^{\nu}_{n+1-k,n}}\right)\frac{1}{\sqrt{1+x^{\nu}_{n+1-k,n}}}=\mathbf{h}_2\left(\frac{j^2}{k^2}\right)\frac{\sqrt{2}}{k}+\mathcal{O}\left(\frac{j+k}{n^2}\right), \tag{2.20}$$

where $1 \le j,k \le n$ and where the constants regarding the $\mathcal{O}$-terms are independent of j, k, and $n \in \mathbb{N}$.

By $\operatorname{alg}\mathcal{T}(\mathbf{PC})$ we denote the smallest C^*-subalgebra of the algebra $\mathcal{L}(\ell^2)$ of all linear and bounded operators in the Hilbert space ℓ^2 generated by the Toeplitz matrices $\mathbf{T}(g) = \left[\ \widehat{g}_{j-k}\ \right]_{j,k=0}^{\infty}$ with piecewise continuous generating functions $g(t) := \sum_{\ell\in\mathbb{Z}} \widehat{g}_\ell t^\ell$ defined on the unit circle $\mathbb{T} := \{t \in \mathbb{C} : |t| = 1\}$ and continuous on $\mathbb{T}\setminus\{\pm1\}$.

Of course, $\operatorname{alg}\mathcal{T}(\mathbf{PC})$ is a C^*-subalgebra of the C^*-algebra $\mathcal{L}_T(\ell^2) \subset \mathcal{L}(\ell^2)$ generated by all Toeplitz matrices $\mathbf{T}(f)$ with piecewise continuous generating function $f : \mathbb{T} \longrightarrow \mathbb{C}$. It is well known (see Chapter 16 in [15] or in [16]) that there exists an isometrical isomorphism smb from the quotient algebra $\mathcal{L}_T(\ell^2)/\mathcal{K}(\ell^2)$ ($\mathcal{K}(\ell^2)$ - the ideal in $\mathcal{L}(\ell^2)$ of compact operators) onto the algebra $(\mathbf{C}(\mathbb{M}), \|.\|_\infty)$ of all complex-valued and continuous functions on the compact space $\mathbb{M} = \mathbb{T}\times[0,1]$, where the topology on $\mathbb{M}$ is defined by the neighborhoods (cf. Theorem 16.1 in [15] or in [16])

$$\begin{aligned}
U_{\varepsilon,\delta}(e^{\mathbf{i}\eta_0},0)&:=\left\{(e^{\mathbf{i}\eta},\lambda):\eta_0-\delta<\eta<\eta_0,0\le\lambda\le1\right\}\cup\left\{(e^{\mathbf{i}\eta_0},\lambda):0\le\lambda<\varepsilon\right\},\\
U_{\varepsilon,\delta}(e^{\mathbf{i}\eta_0},1)&:=\left\{(e^{\mathbf{i}\eta},\lambda):\eta_0<\eta<\eta_0+\delta,0\le\lambda\le1\right\}\cup\left\{(e^{\mathbf{i}\eta_0},\lambda):\varepsilon\le\lambda\le1\right\},\\
U_{\delta_1,\delta_2}(e^{\mathbf{i}\eta_0},\lambda_0)&:=\left\{(e^{\mathbf{i}\eta_0},\lambda):\lambda_0-\delta_1<\lambda<\lambda_0+\delta_2\right\}
\end{aligned}$$

with $0 < \delta_1 < \lambda_0 < 1-\delta_2 < 1$, $0 < \delta < 2\pi$, and $0 < \varepsilon < 1$. The operator $\mathbf{T} \in \mathcal{L}_T(\ell^2)$ is Fredholm if and only if $\operatorname{smb}_{\mathbf{T}}(t,\lambda) \ne 0$ for all $(t,\lambda) \in \mathbb{T}\times[0,1]$, where $\operatorname{smb}_{\mathbf{T}} := \operatorname{smb}(\mathbf{T})$. If this is the case, then the Fredhom index of $\mathbf{T}$ is equal to the negative winding number of the closed curve $\{\operatorname{smb}_{\mathbf{T}}(t,\lambda) : t \in \mathbb{T},\ \lambda \in [0,1]\}$, oriented in a natural way by the orientation of $\mathbb{T}$ and $[0,1]$. If $g : \mathbb{T} \longrightarrow \mathbb{C}$ is piecewise continuous, then

$$\operatorname{smb}_{\mathbf{T}(g)}(t,\lambda) = g(t+0)\lambda + g(t-0)(1-\lambda)\,. \tag{2.21}$$

Moreover, if $g : \mathbb{T} \longrightarrow \mathbb{C}$ is piecewise continuous and continuous on $\mathbb{T}\setminus\{\pm1\}$, then the Hankel matrix $\mathbf{H}(g) = \left[\ \widehat{g}_{j+k+1}\ \right]_{j,k=0}^{\infty}$ also belongs to $\operatorname{alg}\mathcal{T}(\mathbf{PC})$ and ([3, Theorem 4.97])

$$\operatorname{smb}_{\mathbf{H}(g)}(t,\lambda) = \mathbf{i}\,t[g(t-0)-g(t+0)]\sqrt{\lambda(1-\lambda)}\,. \tag{2.22}$$

Note that, for $\operatorname{smb}_* : \mathcal{L}_T(\ell^2) \longrightarrow \mathbf{C}(\mathbb{M})$, $\mathbf{T} \mapsto \operatorname{smb}_{\mathbf{T}}$, we have

$$\|\operatorname{smb}_{\mathbf{T}}\|_\infty = \left\|\mathbf{T}+\mathcal{K}(\ell^2)\right\|_{\mathcal{L}(\ell^2)/\mathcal{K}(\ell^2)} \le \|\mathbf{T}\|_{\mathcal{L}(\ell^2)} \quad \forall\,\mathbf{T}\in\mathcal{L}_T(\ell^2)\,. \tag{2.23}$$

For $z \in \mathbb{C}$ and a measurable function $f : (0,\infty) \longrightarrow \mathbb{C}$, for which $x^{z-1}f(x)$ is integrable on each compact subinterval of $(0,\infty)$, the Mellin transform $\widehat{f}(z)$ is defined as

$$\widehat{f}(z) = \lim_{R\to\infty} \int_{R^{-1}}^{R} x^{z-1} f(x)\, dx\,, \tag{2.24}$$

if this limit exists. Under weaker conditions on $f(x)$ we have to modify this definition. For this we start with some well-known properties of the Fourier–Plancherel operator $\mathcal{F}$ which is well defined for $u \in \mathbf{L}^1(\mathbb{R})$ by

$$(\mathcal{F}u)(\eta) := \lim_{R\to\infty} \int_{-R}^{R} e^{\mathbf{i}\eta t} u(t)\, dt\,, \quad \eta \in \mathbb{R}\,.$$

Lemma 2.6 (cf. [5], p. 22). *We have $\mathcal{F}u \in \mathbf{C}(\mathbb{R})$ and $\lim_{\eta\to\pm\infty} (\mathcal{F}u)(\eta) = 0$ if $u \in \mathbf{L}^1(\mathbb{R})$.*

Proof. If $u = \chi_{[a,b]}$ is the characteristic function of a compact interval $[a,b] \subset \mathbb{R}$, then

$$(\mathcal{F}u)(\eta) = \int_a^b e^{\mathbf{i}\eta t}\, dt = \left\{ \begin{array}{ccc} \frac{1}{\mathbf{i}\eta}\left(e^{\mathbf{i}b\eta} - e^{\mathbf{i}a\eta}\right) & : & \eta \neq 0 \\ b-a & : & \eta = 0 \end{array} \right\} \longrightarrow 0 \quad \text{if} \quad \eta \longrightarrow \pm\infty\,.$$

Since $\|\mathcal{F}u\|_{\mathbf{L}^\infty(\mathbb{R})} \le \|u\|_{\mathbf{L}^1(\mathbb{R})}$ for all $u \in \mathbf{L}^1(\mathbb{R})$, the assertion follows by the density of the set of all piecewise constant functions in $\mathbf{L}^1(\mathbb{R})$. □

For $0 < R < \infty$ and $u \in \mathbf{L}^1(-R,R)$, set

$$(\mathcal{F}_R u)(\eta) = \int_{-R}^{R} e^{\mathbf{i}\eta t} u(t)\, dt\,, \quad \eta \in \mathbb{R}\,.$$

Lemma 2.7 ([20], Theorem 74). *Let $1 < p \le 2$, $\frac{1}{p} + \frac{1}{q} = 1$, and $u \in \mathbf{L}^p(\mathbb{R})$. Then, $\mathcal{F}_R u$ converges for $R \longrightarrow \infty$ in $\mathbf{L}^q$, say to $\mathcal{F}u \in \mathbf{L}^q$. The so-defined operator $\mathcal{F} : \mathbf{L}^p(\mathbb{R}) \longrightarrow \mathbf{L}^q(\mathbb{R})$ is linear and bounded. Moreover, for $v = \mathcal{F}u$, we have $u = \mathcal{F}^- v$, where $\mathcal{F}^- v = \lim_{R\to\infty} \mathcal{F}_R^- v$ in $\mathbf{L}^p$ and*

$$\left(\mathcal{F}_R^- v\right)(\eta) = \frac{1}{2\pi} \int_{-R}^{R} e^{-\mathbf{i}\eta t} v(t)\, dt\,, \quad \eta \in \mathbb{R}\,.$$

Define, for $1 \le p < \infty$, $0 < x < \infty$, $\xi \in \mathbb{R}$, and a function $f : (0,\infty) \longrightarrow \mathbb{C}$,

$$\omega_{\xi,p}(x) = x^{\xi - \frac{1}{p}} \quad \text{and} \quad (\mathcal{T}_\xi f)(t) = e^{\xi t} f(e^t)\,,\ t \in \mathbb{R}\,.$$

If we set $\mathbf{L}^p_\xi := \{f : \omega_{\xi,p} f \in \mathbf{L}^p(0,\infty)\}$ with $\|f\|_{\mathbf{L}^p_\xi} := \|\omega_{\xi,p} f\|_{\mathbf{L}^p(0,\infty)}$, then $\mathcal{T}_\xi : \mathbf{L}^p_\xi \longrightarrow \mathbf{L}^p(\mathbb{R})$ is an isometrical isomorphism. Note that, for $f \in \mathbf{L}^1_\xi$ and $\eta \in \mathbb{R}$,

$$\begin{aligned} \widehat{f}(\xi + \mathbf{i}\eta) &= \lim_{R\to\infty} \int_{e^{-R}}^{e^R} x^{\xi + \mathbf{i}\eta - 1} f(x)\, dx \\ &= \lim_{R\to\infty} \int_{-R}^{R} e^{\mathbf{i}\eta t} e^{\xi t} f(e^t)\, dt = \lim_{R\to\infty} (\mathcal{F}_R \mathcal{T}_\xi f)(\eta) \end{aligned}$$

and, due to Lemma 2.6, $\widehat{f}(\xi + \mathbf{i}\cdot) \in \mathbf{C}(\mathbb{R})$ with $\widehat{f}(\xi \pm \mathbf{i}\infty) = 0$.

Lemma 2.8 (cf. [20], Theorem 86). *If* $1 < p \le 2$, $\xi \in \mathbb{R}$, *and* $f \in \mathbf{L}^p_\xi$, *then*

$$\widehat{f}(\xi + \mathbf{i}\,\cdot) := \mathcal{F}\mathcal{T}_\xi f \in \mathbf{L}^q(\mathbb{R}) \tag{2.25}$$

and

$$\lim_{R\to\infty} \|f - f_R\|_{\mathbf{L}^p_\xi} = 0\,, \tag{2.26}$$

where $f_R(x) = \dfrac{1}{2\pi\mathbf{i}} \displaystyle\int_{\xi-\mathbf{i}R}^{\xi+\mathbf{i}R} x^{-z}\widehat{f}(z)\,dz$, $0 < x < \infty$. *If even* $\widehat{f}(\xi+\mathbf{i}\cdot) \in \mathbf{L}^1(\mathbb{R})$ *and* $f \in \mathbf{C}(0,\infty)$, *then*

$$\begin{aligned} f(x) &= \frac{1}{2\pi}\int_{-\infty}^{\infty} x^{-\xi-\mathbf{i}\eta}\widehat{f}(\xi+\mathbf{i}\eta)\,d\eta \\ &= \frac{1}{2\pi\mathbf{i}}\int_{\mathrm{Re}\,z=\xi} x^{-z}\widehat{f}(z)\,dz \quad \forall\, x \in (0,\infty)\,. \end{aligned} \tag{2.27}$$

Proof. Due to $f \in \mathbf{L}^p_\xi$, we have $\mathcal{T}_\xi f \in \mathbf{L}^p(\mathbb{R})$, such that, in virtue of Lemma 2.7, $\mathcal{F}\mathcal{T}_\xi f \in \mathbf{L}^q(\mathbb{R})$, and $\widehat{f}(\xi+\mathbf{i}\,\cdot) \in \mathbf{L}^q(\mathbb{R})$ is well defined. Moreover, again by Lemma 2.7,

$$f = T_\xi^{-1}\mathcal{F}^-\left[\widehat{f}(\xi+\mathbf{i}\,\cdot)\right],$$

where $\left(T_\xi^{-1}g\right)(x) = x^{-\xi}g(\ln x)$. It remains to use the relation

$$\begin{aligned} \left(\mathcal{T}_\xi^{-1}\mathcal{F}_R^-\left[\widehat{f}(\xi+\mathbf{i}\,\cdot)\right]\right)(x) &= \frac{x^{-\xi}}{2\pi}\int_{-R}^{R} e^{-\mathbf{i}(\ln x)t}\widehat{f}(\xi+\mathbf{i}t)\,dt \\ &= \frac{1}{2\pi\mathbf{i}}\int_{\xi-\mathbf{i}R}^{\xi+\mathbf{i}R} x^{-\xi-\mathbf{i}t}\widehat{f}(\xi+\mathbf{i}t)\,d(\xi+\mathbf{i}t) = f_R(x) \end{aligned}$$

to get (2.26). If $\widehat{f}(\xi+\mathbf{i}\cdot) \in \mathbf{L}^1(\mathbb{R})$, then

$$\lim_{R\to\infty} f_R(x) = \frac{1}{2\pi}\int_{-\infty}^{\infty} x^{-\xi-\mathbf{i}\eta}\widehat{f}(\xi+\mathbf{i}\eta)\,d\eta$$

is continuous on $(0,\infty)$ in virtue of Lemma 2.6. Since (2.26) implies that there exists a sequence of integers $n_1 < n_2 < \cdots$ such that $f_{n_k}(x)$ converges to $f(x)$ for almost all $x \in \mathbb{R}$, this leads to (2.27) if $f : (0,\infty) \longrightarrow \mathbb{C}$ is also continuous. □

Corollary 2.9. *Let* $1 < p \le 2$ *and* $\varepsilon > 0$. *If* $f \in \mathbf{C}(0,\infty) \cap \mathbf{L}^p_{\xi+\varepsilon} \cap \mathbf{L}^p_{\xi-\varepsilon}$, *then* $f \in \mathbf{L}^p_{\xi\pm\eta}$ *for* $0 \le \eta \le \varepsilon$. *Hence, by Lemma* 2.8, $\widehat{f}(\xi\pm\eta+\mathbf{i}\cdot) \in \mathbf{L}^q(\mathbb{R})$ *exists for all these* η. *If, additionally,* $\widehat{f}(z)$ *is holomorphic in the strip* $\{z \in \mathbb{C} : \xi - \varepsilon < \mathrm{Re}\,z < \xi + \varepsilon\}$ *with*

$$M_0 := \sup\left\{|\widehat{f}'(z)|(1+|z|)^{1+\delta} : \xi - \varepsilon < \mathrm{Re}\,z < \xi + \varepsilon\right\} < \infty \tag{2.28}$$

for some $\delta > 0$, *then*

$$f(x) = \frac{x^\mu}{2\pi\mathbf{i}(1-x^{2\mu})}\int_{\mathrm{Re}\,z=\xi} x^{-z}\left[\widehat{f}(z-\mu) - \widehat{f}(z+\mu)\right]dz \tag{2.29}$$

for all $x \in (0,\infty)$ *and* $\mu \in (0,\varepsilon)$, *where* $\widehat{f}(\xi-\mu+\mathbf{i}\cdot) - \widehat{f}(\xi+\mu+\mathbf{i}\cdot) \in \mathbf{L}^1(\mathbb{R})$ *and*

$$\int_{\mathrm{Re}\, z=\xi} \left[\widehat{f}(z-\mu) - \widehat{f}(z+\mu)\right] dz = 0\,. \tag{2.30}$$

Proof. For $0 \le \eta \le \varepsilon$, the estimate

$$\begin{aligned}
&\int_0^\infty x^{p(\xi-\eta)-1}|f(x)|^p\,dx \\
&\quad = \int_0^1 x^{p(\xi-\varepsilon)-1}x^{p(\varepsilon-\eta)}|f(x)|^p\,dx + \int_1^\infty x^{p(\xi+\varepsilon)-1}x^{-p(\varepsilon+\eta)}|f(x)|^p\,dx \\
&\quad \le \|f\|_{\mathbf{L}^p_{\xi+\varepsilon}} + \|f\|_{\mathbf{L}^p_{\xi-\varepsilon}}
\end{aligned}$$

shows that $f \in \mathbf{L}^p_{\xi-\eta}$. Analogously, one proves $f \in \mathbf{L}^p_{\xi+\eta}$. Under the additional assumption on $\widehat{f}(z)$, we have, for all $\mu \in (0,\varepsilon)$,

$$\widehat{f}(\xi+\mu+\mathbf{i}\eta) - \widehat{f}(\xi-\mu+\mathbf{i}\eta) = \int_{-\mu}^{\mu} \widehat{f}'(\xi+t+\mathbf{i}\eta)\,dt\,,$$

such that (2.28) implies $\widehat{f}(\xi+\mu+\mathbf{i}\cdot) - \widehat{f}(\xi-\mu+\mathbf{i}\cdot) \in \mathbf{L}^1(\mathbb{R})$. Hence, the continuous function $g(x) = f(x)x^{-\mu} - f(x)x^{\mu}$ is in $\mathbf{L}^p_\xi$ and its Mellin transform $\widehat{g}(\xi+\mathbf{i}\cdot) = \widehat{f}(\xi-\mu+\mathbf{i}\cdot) - \widehat{f}(\xi+\mu+\mathbf{i}\cdot)$ belongs to $\mathbf{L}^1(\mathbb{R})$. Now, relation (2.27) applied to $g(x)$ leads to (2.29), while (2.30) is a consequence of $g(1) = 0$. □

The following lemma is a modification of [13, Lemma 7.1, Part iv)] (cf. also [14, Lemma 4.5]). The proof is given in Section 5 and uses ideas from [18, 19, Section 11.27].

Lemma 2.10. *Suppose that the continuous function* $\mathbf{y} : (0,\infty) \longrightarrow \mathbb{C}$ *belongs to* $\mathbf{L}^p_{\frac12-\varepsilon} \cap \mathbf{L}^p_{\frac12+\varepsilon}$ *for some* $p \in (1,2]$ *and some* $\varepsilon \in (0,1)$ *and that* $\widehat{\mathbf{y}}(z)$ *is holomorphic in the strip (cf. Corollary* 2.9) $\left\{z \in \mathbb{C} : \frac12 - \varepsilon < \mathrm{Re}\, z < \frac12 + \varepsilon\right\}$, *such that, for some* $\delta > 0$,

$$M_k := \sup\left\{\left|\widehat{\mathbf{y}}^{(k)}(z)\right|(1+|z|)^{k+\delta} : \frac12 - \varepsilon < \mathrm{Re}\, z < \frac12 + \varepsilon\right\} < \infty \tag{2.31}$$

for $k = 0, 1$. *Then, for all* $s > 0$, *the matrix*

$$\mathbf{M}_s := \left[\ \mathbf{y}\left(\frac{j+s}{k+s}\right)\frac{1}{k+s}\ \right]_{j,k=0}^{\infty}$$

defines an operator $\mathbf{M}_s \in \mathcal{L}(\ell^2)$, *which belongs to the algebra* $\mathrm{alg}\,\mathcal{T}(\mathbf{PC})$, *and its symbol is given by*

$$\mathrm{smb}_{\mathbf{M}_s}(t,\lambda) = \begin{cases} 0 & : \quad t \in \mathbb{T}\setminus\{1\}\,, \\ \widehat{\mathbf{y}}\left(\dfrac12 + \dfrac{\mathbf{i}}{2\pi}\ln\dfrac{\lambda}{1-\lambda}\right) & : \quad t = 1\,. \end{cases}$$

For $\alpha, \beta \geq 0$, by $\mathbf{C}_{\alpha,\beta} = \mathbf{C}_{\alpha,\beta}(-1,1)$ we denote the set of all continuous functions $f : (-1,1) \to \mathbb{C}$, for which the finite limits

$$c_{\pm} := \lim_{x \to \pm 1} (1-x)^{\alpha}(1+x)^{\beta} f(x)$$

exist and where $c_+ = 0$ if $\alpha > 0$ and $c_- = 0$ if $\beta > 0$. Equipped with the norm

$$\|f\|_{\alpha,\beta,\infty} = \sup\left\{(1-x)^{\alpha}(1+x)^{\beta}|f(x)| : -1 < x < 1\right\}$$

$\mathbf{C}_{\alpha,\beta}$ becomes a Banach space. In case $\alpha = \beta = 0$, we write $(\mathbf{C}, \|.\|_{\infty})$ instead of $\left(\mathbf{C}_{0,0}, \|.\|_{0,0,\infty}\right)$.

Lemma 2.11 ([11], Lemma 4.2). *Let $\vartheta(x) = (1-x)^{\gamma}(1+x)^{\delta}$ with $\gamma, \delta > -1$.*

(a) *If $0 \leq \alpha < \gamma + 1$ and $0 \leq \beta < \delta + 1$, then*

$$\lim_{n \to \infty} \sum_{k=1}^{n} \lambda_{kn}^{\vartheta} f(x_{kn}^{\vartheta}) = \int_{-1}^{1} f(x)\vartheta(x)\, dx \quad \forall f \in \mathbf{C}_{\alpha,\beta}\,.$$

(b) *If $0 \leq \alpha < \frac{\gamma+1}{2}$ and $0 \leq \beta < \frac{\delta+1}{2}$, then*

$$\lim_{n \to \infty} \left\|f - \mathcal{L}_n^{\vartheta} f\right\|_{\vartheta} = 0 \quad \forall f \in \mathbf{C}_{\alpha,\beta}\,.$$

Corollary 2.12. *Let $0 \leq \alpha < \frac{3}{4}$, $0 \leq \beta < \frac{1}{4}$, and $f \in \mathbf{C}_{\alpha,\beta}$. Then, for $\tau = \sigma$ and $\tau = \nu$, $\mathcal{M}_n^{\tau} f \longrightarrow f$ in $\mathbf{L}_{\mu}^2$.*

Proof. From [11, Corollary 4.3] we infer that $\widehat{\mathcal{M}_n^{\overline{\tau}}} \mathcal{J}^{-1} f \longrightarrow \mathcal{J}^{-1} f$ in $\mathbf{L}_{\nu}^2$. Thus, due to (2.8), $\mathcal{M}_n^{\tau} f = \mathcal{J}\widehat{\mathcal{M}_n^{\overline{\tau}}}\mathcal{J}^{-1} f \longrightarrow f$ in $\mathbf{L}_{\mu}^2$. □

Lemma 2.13. *Let $\tau \in \{\sigma, \nu\}$ and $\beta_k \in \mathbb{C}$, $m \in \mathbb{N}$. Then*

$$\mathcal{M}_n^{\tau}\mathcal{H}_n^{\tau}\mathcal{L}_n \longrightarrow \mathcal{B} \quad \text{and} \quad (\mathcal{M}_n^{\tau}\mathcal{H}_n^{\tau}\mathcal{L}_n)^* \longrightarrow \nu\widetilde{\mathcal{B}}\mu$$

in the sense of strong convergence in $\mathbf{L}_{\mu}^2$, where

$$\widetilde{\mathcal{B}} : \mathbf{L}_{\nu}^2 \longrightarrow \mathbf{L}_{\nu}^2\,, \quad u \mapsto \int_{-1}^{1} \overline{\mathbf{h}\left(\frac{1+y}{1+\cdot}\right)\frac{u(y)\,dy}{1+\cdot}}\,.$$

Proof. First of all we show the uniform boundedness of the operator sequences under consideration. Since the operators $\widetilde{\mathcal{V}}_n^{\tau} : \operatorname{im} \mathcal{L}_n \longrightarrow \operatorname{im} \mathcal{P}_n$ are unitary operators, it is sufficient to show the uniform boundedness of the operators $\mathbf{H}_n : \operatorname{im} \mathcal{P}_n \longrightarrow \operatorname{im} \mathcal{P}_n$ with $\mathbf{H}_n$ equal to

$$\begin{aligned} &\widetilde{\mathcal{V}}_n^{\tau}\mathcal{M}_n^{\tau}\mathcal{H}_n^{\tau}\mathcal{L}_n(\widetilde{\mathcal{V}}_n^{\tau})^{-1}\mathcal{P}_n \\ &= \left[\,(\omega_n^{\tau})^2\sqrt{1-x_{n+1-j,n}^{\tau}}\,\mathbf{h}\left(\frac{1+x_{n+1-j,n}^{\tau}}{1+x_{n+1-k,n}^{\tau}}\right)\frac{1}{\sqrt{1+x_{n+1-k,n}^{\tau}}}\,\right]_{j,k=1}^{n}. \end{aligned} \tag{2.32}$$

Set $c^\sigma = \frac{1}{2}$ and $c^\nu = 0$. With the help of (2.17)–(2.20), we deliver

$$\begin{aligned}\mathbf{H}_n &= \left[\ \sqrt{1 - x^\tau_{n+1-j,n}} \left\{ \mathbf{h}\left(\frac{(j-c^\tau)^2}{(k-c^\tau)^2} \right) \frac{\sqrt{2}}{k - c^\tau} + \mathcal{O}\left(\frac{1}{n}\right) \right\} \ \right]_{j,k=1}^{n} \\ &= \mathbf{D}_n \left(\mathbf{E}^1_n + \mathbf{E}^2_n + \mathbf{F}_n \right) ,\end{aligned} \tag{2.33}$$

where

$$\begin{aligned}\mathbf{D}_n &= \left[\ \sqrt{\frac{1 - x^\tau_{n+1-j,n}}{2}}\, \delta_{j,k} \ \right]_{j,k=1}^{n} , \\ \mathbf{E}^r_n &= \left[\ \mathbf{h}_r \left(\frac{(j-c^\tau)^2}{(k-c^\tau)^2} \right) \frac{2}{k - c^\tau} \ \right]_{j,k=1}^{n} ,\end{aligned} \tag{2.34}$$

and $\mathbf{F}_n = \left[\ \mathcal{O}\left(\frac{1}{n}\right) \ \right]_{j,k=1}^{n}$. We set $\mathbf{g}_r(x) := 2\,\mathbf{h}_r(x^2)$. Then, for $p \ge 1$,

$$\mathbf{g}_1 \in \mathbf{L}^p_\xi \,, \ -1 < \xi < 1 \quad \text{and} \quad \mathbf{g}_2 \in \mathbf{L}^p_\xi \,, \ 0 < \xi < 1 \,. \tag{2.35}$$

Furthermore, the Mellin transforms of $\mathbf{g}_1$ and $\mathbf{g}_2$ are given by (see [6, 6.2.(3), (6)])

$$\begin{aligned}\widehat{\mathbf{g}}_1(z) &= 2\sum_{k=1}^{m} \frac{\beta_k}{\pi \mathbf{i}} \int_0^\infty \frac{x^{z+2k-2}}{(1+x^2)^k}\, dx = \sum_{k=1}^{m} \frac{\beta_k}{\pi \mathbf{i}} \int_0^\infty \frac{y^{\frac{z}{2}+k-\frac{3}{2}}}{(1+y)^k}\, dy \\ &= \sum_{k=1}^{m} \frac{\beta_k}{\mathbf{i}} \frac{(-1)^{k-1}\binom{\frac{z}{2}+k-\frac{3}{2}}{k-1}}{\sin\pi\left(\frac{z}{2}+k-\frac{1}{2}\right)} = \sum_{k=1}^{m} \frac{\beta_k}{\mathbf{i}} \frac{\binom{\frac{z}{2}+k-\frac{3}{2}}{k-1}}{\sin\frac{\pi(z+1)}{2}}\end{aligned} \tag{2.36}$$

and

$$\widehat{\mathbf{g}}_2(z) = 2\beta_0 \int_0^\infty \frac{x^{z-1}}{1+x}\, dx = \frac{2\beta_0}{\mathbf{i}\,\sin(\pi z)} , \tag{2.37}$$

which are holomorphic in the strips

$$\{ z \in \mathbb{C} : -1 < \operatorname{Re} z < 1 \} \quad \text{and} \quad \{ z \in \mathbb{C} : 0 < \operatorname{Re} z < 1 \} ,$$

respectively, and fulfil

$$\sup_{\frac{1}{2}-\varepsilon < \operatorname{Re} z < \frac{1}{2}+\varepsilon} \left| \frac{d^k \widehat{\mathbf{g}}_r(z)}{dz^k} (1+|z|)^{k+\delta} \right| < \infty, \quad k = 0, 1, \dots \,, \ r = 1, 2 \,, \ \delta > 0 \,,$$

with $0 < \varepsilon < \frac{1}{2}$. Hence, we can apply Lemma 2.10 to

$$\mathbf{E}^r_n = \mathcal{P}_n \left[\ \mathbf{g}_r \left(\frac{j+1-c^\tau}{k+1-c^\tau} \right) \frac{2}{k+1-c^\tau} \ \right]_{j,k=0}^{\infty} \mathcal{P}_n \,, \quad r = 1, 2 \,, \tag{2.38}$$

and get the uniform boundedness of the operators $\mathbf{E}^r_n : \ell^2 \longrightarrow \ell^2$, $r = 1, 2$. Since $\mathbf{D}_n$ and $\mathbf{F}^r_n$ are obviously uniformly bounded in ℓ^2, the uniform boundedness of the operators $\mathcal{M}^\tau_n \mathcal{H}^\tau_n \mathcal{L}_n : \mathbf{L}^2_\mu \longrightarrow \mathbf{L}^2_\mu$ is proved.

It remains to verify the convergence on a dense subset of $\mathbf{L}^2_\mu$. For this, we take the set of all functions of the form $v(x) = \nu(x)(1+x)p(x)$, where $p(x)$ is a polynomial. Corollary 2.12 implies $\sup\left\{\|\mathcal{M}_n^\tau\|_{\mathbf{C}\to\mathbf{L}^2_\mu} : n\in\mathbb{N}\right\}<\infty$. Consequently, for n sufficiently large, we get

$$\begin{aligned}\|\mathcal{M}_n^\tau\mathcal{H}_n^\tau\mathcal{L}_n v-\mathcal{B}v\|_\mu &\le \|\mathcal{M}_n^\tau\|_{\mathbf{C}\to\mathbf{L}^2_\mu}\|\mathcal{H}_n^\tau v-\mathcal{B}v\|_\infty+\|\mathcal{M}_n^\tau\mathcal{B}v-\mathcal{B}v\|_\mu\\ &\le \text{const}\,\|\mathcal{H}_n^\tau v-\mathcal{B}v\|_\infty+\|\mathcal{M}_n^\tau\mathcal{B}v-\mathcal{B}v\|_\mu\,.\end{aligned}\tag{2.39}$$

The function $(\mathcal{B}v)(x)$ is continuous on $(-1,1)$, since the kernel function of the operator $\mathcal{B}$, namely $\mathbf{h}\left(\frac{1+x}{1+y}\right)\frac{1}{1+y}$, is continuous on $(-1,1)\times[-1,1]$. Moreover,

$$|(\mathcal{B}v)(x)|\le\text{const}\int_{-1}^1\frac{|v(y)|\,dy}{1+y}=\text{const}\int_{-1}^1\nu(y)|p(y)|\,dy\,,\quad -1<x<1\,,$$

which shows that $\mathcal{B}v\in\mathbf{C}_{\varepsilon_1,\varepsilon_2}$ for all $\varepsilon_1,\varepsilon_2>0$. Thus, in (2.39) we have

$$\|\mathcal{M}_n^\tau\mathcal{B}v-\mathcal{B}v\|_\mu\longrightarrow 0$$

in view of Corollary 2.12. If $R_n(f)_\tau$ denotes the quadrature error

$$R_n(f)_\tau=\int_{-1}^1 f(x)\tau(x)\,dx-\sum_{k=1}^n\lambda_{kn}^\tau f(x_{kn}^\tau)\,,$$

then ([17, (5.1.35)])

$$|R_n(f)_\tau|\le\frac{\text{const}}{n}\int_{-1}^1|f'(x)|\tau(x)\varphi(x)\,dx\,,\tag{2.40}$$

where the constant does not depend on n and f. We have

$$(\mathcal{H}_n^\tau v)(x)-(\mathcal{B}v)(x)=R_n(f_x)_\tau$$

with $f_x(y)=\mathbf{h}\left(\frac{1+x}{1+y}\right)\frac{v(y)}{(1+y)\tau(y)}=\mathbf{h}\left(\frac{1+x}{1+y}\right)v_0(y)$ and

$$v_0(y)=\begin{cases}(1+y)p(y) & : \quad \tau=\sigma\\ p(y) & : \quad \tau=\nu\,.\end{cases}\tag{2.41}$$

With the estimates

$$\mathbf{h}(t)\le\text{const}\quad\text{and}\quad|\mathbf{h}'(t)|\le\frac{\text{const}}{\sqrt{t}}\,,\quad t\in(0,\infty)\,,$$

we get, using (2.40),

$$\begin{aligned}|R_n(f_x)_\sigma| &\le\frac{\text{const}}{n}\int_{-1}^1\left|\frac{d}{dy}\left[\mathbf{h}\left(\frac{1+x}{1+y}\right)v_0(y)\right]\right|\tau(y)\varphi(y)\,dy\\ &\le\frac{\text{const}}{n}\left[\int_{-1}^1\mathbf{h}\left(\frac{1+x}{1+y}\right)|v_0'(y)|\tau(y)\varphi(y)\,dy\right.\\ &\qquad\left.+\int_{-1}^1\left|\mathbf{h}'\left(\frac{1+x}{1+y}\right)\right|\frac{1+x}{(1+y)^2}|v_0(y)|\tau(y)\varphi(y)\,dy\right]\end{aligned}$$

$$\leq \frac{\text{const}}{n}\left[\int_{-1}^{1}|v_0'(y)|\tau(y)\varphi(y)\,dy\right.$$
$$\left.+\sqrt{1+x}\int_{-1}^{1}(1+y)^{-\frac{3}{2}}|v_0(y)|\tau(y)\varphi(y)\,dy\right].$$

This, together with (2.41), implies $\lim\limits_{n\to\infty}\left\|\mathcal{H}_n^\tau v-\mathcal{B}v\right\|_\infty=0$. Now we consider the adjoint operators. Since, for $u,v\in\mathbf{L}^2_\mu$,

$$\begin{aligned}
&\langle\mathcal{M}_n^\tau\mathcal{H}_n^\tau\mathcal{L}_n u,v\rangle_\mu\\
&=\left\langle\mathcal{L}_n^\tau\mu\mathcal{H}_n^\tau\mathcal{L}_n u,(\nu\tau^{-1})\mu\mathcal{L}_n v\right\rangle_\tau\\
&=\sum_{k=1}^{n}\lambda_{kn}^\tau\mu(x_{kn}^\tau)\left(\mathcal{H}_n^\tau\mathcal{L}_n u\right)(x_{kn}^\tau)\overline{(\tau^{-1}\mathcal{L}_n v)\,(x_{kn}^\tau)}\\
&=\sum_{k=1}^{n}\lambda_{kn}^\tau\mu(x_{kn}^\tau)(\omega_n^\tau)^2\sum_{j=1}^{n}\varphi(x_{jn}^\tau)\mathbf{h}\left(\frac{1+x_{kn}^\tau}{1+x_{jn}^\tau}\right)\frac{(\mathcal{L}_n u)(x_{jn}^\tau)}{1+x_{jn}^\tau}\frac{\overline{(\mathcal{L}_n v)(x_{kn}^\tau)}}{\tau(x_{kn}^\tau)}\\
&=\sum_{k=1}^{n}\lambda_{kn}^\tau\mu(x_{kn}^\tau)\sum_{j=1}^{n}\frac{\lambda_{jn}^\tau}{\tau(x_{jn}^\tau)}\mathbf{h}\left(\frac{1+x_{kn}^\tau}{1+x_{jn}^\tau}\right)\frac{(\mathcal{L}_n u)(x_{jn}^\tau)}{1+x_{jn}^\tau}\frac{\overline{(\mathcal{L}_n v)(x_{kn}^\tau)}}{\tau(x_{kn}^\tau)}\\
&=\sum_{j=1}^{n}\frac{\lambda_{jn}^\tau}{\tau(x_{jn}^\tau)}(\mathcal{L}_n u)(x_{jn}^\tau)\overline{\sum_{k=1}^{n}\frac{\lambda_{kn}^\tau}{\tau(x_{kn}^\tau)}\mathbf{h}\left(\frac{1+x_{kn}^\tau}{1+x_{jn}^\tau}\right)\frac{(\mu\mathcal{L}_n v)(x_{kn}^\tau)}{1+x_{jn}^\tau}}\\
&=\left\langle\mu\mathcal{L}_n u,\mathcal{L}_n^\tau(\omega_n^\tau)^2\sum_{k=1}^{n}\varphi(x_{kn}^\tau)\overline{\mathbf{h}\left(\frac{1+x_{kn}^\tau}{1+\cdot}\right)\frac{(\mu\mathcal{L}_n v)(x_{kn}^\tau)}{1+\cdot}}\right\rangle_\nu\\
&=\left\langle(\nu\tau^{-1})\mu\mathcal{L}_n u,\mathcal{L}_n^\tau\widetilde{\mathcal{H}}_n^\tau\mu\mathcal{L}_n v\right\rangle_\tau=\left\langle u,\mathcal{M}_n^\tau\nu\widetilde{\mathcal{H}}_n^\tau\mu\mathcal{L}_n v\right\rangle_\mu
\end{aligned}$$

with

$$\begin{aligned}
\left(\widetilde{\mathcal{H}}_n^\tau f\right)(x)&=(\omega_n^\tau)^2\sum_{k=1}^{n}\varphi(x_{kn}^\tau)\overline{\mathbf{h}\left(\frac{1+x_{kn}^\tau}{1+x}\right)\frac{f(x_{kn}^\tau)}{1+x}}\\
&=\sum_{k=1}^{n}\lambda_{kn}^\tau\overline{\mathbf{h}\left(\frac{1+x_{kn}^\tau}{1+x}\right)\frac{(\tau^{-1}f)(x_{kn}^\tau)}{1+x}},
\end{aligned}$$

we deliver

$$(\mathcal{M}_n^\tau\mathcal{H}_n^\tau\mathcal{L}_n)^*=\mathcal{M}_n^\tau\nu\widetilde{\mathcal{H}}_n^\tau\mu\mathcal{L}_n\,. \tag{2.42}$$

Define $v(x)=\nu(x)(1+x)p(x)$ with a polynomial $p(x)$. Then, due to $|\mathbf{h}(t)|\leq \text{const}\,t^{-\frac{1}{2}}$,

$$\left|(\widetilde{\mathcal{B}}\mu v)(x)\right|\leq\text{const}(1+x)^{-\frac{1}{2}}\int_{-1}^{1}(1+y)^{\frac{1}{2}}|p(y)|\,dy\,,$$

which, together with the continuity of $\widetilde{\mathcal{B}}\mu v:(-1,1)\longrightarrow\mathbb{C}$ (cf. the sentence after (2.39)), yields $\nu\widetilde{\mathcal{B}}\mu v\in\mathbf{C}_{\frac{1}{2}+\varepsilon,0}$ for $0<\varepsilon<\frac{1}{4}$. Corollary 2.12 implies, for n

sufficiently large,

$$\begin{aligned}&\left\|\mathcal{M}_n^\tau\nu\widetilde{\mathcal{H}}_n^\tau\mu\mathcal{L}_n v-\nu\widetilde{\mathcal{B}}\mu v\right\|_\mu\\&\qquad\le\left\|\nu\left(\widetilde{\mathcal{H}}_n^\tau-\widetilde{\mathcal{B}}\right)(1+\cdot)p\right\|_{\frac12+\varepsilon,0,\infty}+\left\|\mathcal{M}_n^\tau\nu\widetilde{\mathcal{B}}\mu v-\nu\widetilde{\mathcal{B}}\mu v\right\|_\mu\end{aligned}$$

with $\lim_{n\to\infty}\left\|\mathcal{M}_n^\tau\nu\widetilde{\mathcal{B}}\mu v-\nu\widetilde{\mathcal{B}}\mu v\right\|_\mu=0$. Moreover,

$$\left((\widetilde{\mathcal{H}}_n^\tau-\widetilde{\mathcal{B}})(1+\cdot)p\right)(x)=R_n(g_x)_\tau$$

with $g_x(y)=\overline{\mathbf{h}\left(\frac{1+y}{1+x}\right)}\frac{(1+y)p(y)}{1+x}$. Further, by using that $|\mathbf{h}'(t)|\le\operatorname{const}t^{-\frac32}$ we conclude from (2.40)

$$\begin{aligned}|R_n(g_x)_\tau|&\le\frac{\operatorname{const}}{n(1+x)}\int_{-1}^1\left|\frac{d}{dy}\left[\mathbf{h}\left(\frac{1+y}{1+x}\right)(1+y)p(y)\right]\right|\tau(y)\varphi(y)\,dy\\&\le\frac{\operatorname{const}}{n(1+x)}\left[\int_{-1}^1\mathbf{h}\left(\frac{1+y}{1+x}\right)\left|\frac{d}{dy}[(1+y)p(y)]\right|\tau(y)\varphi(y)\,dy\right.\\&\qquad\left.+\frac{1}{1+x}\int_{-1}^1\left|\mathbf{h}'\left(\frac{1+y}{1+x}\right)(1+y)p(y)\right|\tau(y)\varphi(y)\,dy\right]\\&\le\frac{\operatorname{const}}{n\sqrt{1+x}}\,.\end{aligned}$$

This shows that $\lim_{n\to\infty}\left\|\nu\widetilde{\mathcal{H}}_n^\tau\mu v-\mu\widetilde{\mathcal{B}}\mu v\right\|_{\frac12+\varepsilon,\infty}=0$, and the strong convergence of $(\mathcal{M}_n^\tau\mathcal{H}_n^\tau\mathcal{L}_n)^*$ in $\mathbf{L}_\mu^2$ is proved. □

We recall the following version of Lebesgue's dominant convergence theorem.

Lemma 2.14. *If* $\xi,\eta\in\ell^2$, $\xi^n=(\xi_k^n)_{k=1}^\infty$, $|\xi_k^n|\le|\eta_k|$ *for all* $k=0,1,2,\dots$ *and for all* $n\ge n_0$ *and if* $\lim_{n\to\infty}\xi_k^n=\xi_k$ *for all* $k=0,1,2,\dots$, *then* $\lim_{n\to\infty}\|\xi^n-\xi\|_{\ell^2}=0$.

Set

$$\mathbf{A}^\sigma=\left[\ \mathbf{h}\left(\frac{(j+\frac12)^2}{(k+\frac12)^2}\right)\frac{2}{k+\frac12}\ \right]_{j,k=0}^\infty\quad\text{and}\quad\mathbf{A}^\nu=\left[\ \mathbf{h}\left(\frac{(j+1)^2}{(k+1)^2}\right)\frac{2}{k+1}\ \right]_{j,k=0}^\infty.$$

Lemma 2.15. *Let* $\beta_k\in\mathbb{C}$ *and* $m\in\mathbb{N}$. *For* $\tau\in\{\sigma,\nu\}$ *the strong limits*

$$\mathcal{W}^3(\mathcal{M}_n^\tau\mathcal{H}_n^\tau\mathcal{L}_n)=\Theta\quad\text{and}\quad\mathcal{W}^4(\mathcal{M}_n^\tau\mathcal{H}_n^\tau\mathcal{L}_n)=\mathbf{A}^\tau\tag{2.43}$$

exist, where Θ *is the zero operator in* ℓ^2. *Moreover, the respective sequences of adjoint operators converge strongly, too.*

Proof. We have to consider the sequences

$$\left(\mathcal{V}_n^\tau\mathcal{M}_n^\tau\mathcal{H}_n^\tau\mathcal{L}_n(\mathcal{V}_n^\tau)^{-1}\mathcal{P}_n\right)\quad\text{and}\quad\left(\widetilde{\mathcal{V}}_n^\tau\mathcal{M}_n^\tau\mathcal{H}_n^\tau\mathcal{L}_n(\widetilde{\mathcal{V}}_n^\tau)^{-1}\mathcal{P}_n\right)$$

in $\mathcal{L}(\ell^2)$. Since, due to the unitarity of the operators $\mathcal{V}_n^\tau$ and $\widetilde{\mathcal{V}}_n^\tau$, the uniform boundedness of these sequences is already known from Lemma 2.13, it suffices to check the convergence on the elements $e_m = (\delta_{j,m})_{j=0}^\infty$, $m = 0, 1, 2, \ldots$ of the standard basis of ℓ^2. For $n > m = k-1 \geq 0$, we can write (cf. (2.32))

$$\begin{aligned} &\mathcal{V}_n^\tau \mathcal{M}_n^\tau \mathcal{H}_n^\tau \mathcal{L}_n (\mathcal{V}_n^\tau)^{-1} \mathcal{P}_n e_{k-1} \\ &\quad = \left[\, (\omega_n^\tau)^2 \sqrt{1 - x_{jn}^\tau} \sqrt{1 + x_{kn}^\tau}\, \mathbf{h}\left(\frac{1 + x_{jn}^\tau}{1 + x_{kn}^\tau} \right) \frac{1}{1 + x_{kn}^\tau} \, \right]_{j=1}^n \qquad (2.44) \\ &\quad =: \left[\, b_{jk}^\tau \, \right]_{j=1}^n \end{aligned}$$

as well as

$$\widetilde{\mathcal{V}}_n^\tau \mathcal{M}_n^\tau \mathcal{H}_n^\tau \mathcal{L}_n (\widetilde{\mathcal{V}}_n^\tau)^{-1} \mathcal{L}_n^{(3)} e_{k-1} = \left[\, b_{n-j+1,n-k+1}^\tau \, \right]_{j=1}^n .$$

Since $\mathbf{h}(t) \leq \mathrm{const}$ for $t > 0$, we get (for fixed k)

$$|b_{jk}^\tau| \leq \frac{\mathrm{const}}{n} ,$$

which proves the first relation in (2.43). For $j, k \geq 1$ fixed, we get, for $n \longrightarrow \infty$,

$$b_{n+1-j,n+1-k}^\sigma = \frac{\pi \cos \frac{(j-\frac{1}{2})\pi}{2n}}{n \sin \frac{(k-\frac{1}{2})\pi}{2n}} \, \mathbf{h}\left(\frac{\sin^2 \frac{(j-\frac{1}{2})\pi}{2n}}{\sin^2 \frac{(k-\frac{1}{2})\pi}{2n}} \right) \longrightarrow \mathbf{h}\left(\frac{(j-\frac{1}{2})^2}{(k-\frac{1}{2})^2} \right) \frac{2}{k-\frac{1}{2}}$$

and

$$b_{n+1-j,n+1-k}^\mu = \frac{\pi \cos \frac{j\pi}{2n+1}}{\left(n+\frac{1}{2}\right) \sin \frac{k\pi}{2n+1}} \, \mathbf{h}\left(\frac{\sin^2 \frac{j\pi}{2n+1}}{\sin^2 \frac{k\pi}{2n+1}} \right) \longrightarrow \mathbf{h}\left(\frac{j^2}{k^2} \right) \frac{2}{k}$$

Moreover, for fixed k and due to $\mathbf{h}(t) \leq \mathrm{const}\, t^{-\frac{1}{2}}$, we get

$$\left| b_{n+1-j,n+1-k}^\sigma \right| \leq \frac{\mathrm{const}}{n} \, \frac{\cos \frac{2j-1}{4n}\pi}{\sin \frac{2j-1}{4n}\pi} \leq \frac{\mathrm{const}}{j}$$

and

$$\left| b_{n+1-j,n+1-k}^\nu \right| \leq \frac{\mathrm{const}}{n} \, \frac{\cos \frac{j\pi}{2n+1}}{\sin \frac{j\pi}{2n+1}} \leq \frac{\mathrm{const}}{j} .$$

Hence, we can apply Lemma 2.14 to get the second relation in (2.43). The convergence of the sequences of the adjoint operators can be seen analogously. □

Lemma 2.16. *Let $\beta_k \in \mathbb{C}$ and $m \in \mathbb{N}$. If $\tau \in \{\sigma, \nu\}$, then $\mathcal{W}^2(\mathcal{M}_n^\tau \mathcal{H}_n^\tau \mathcal{L}_n) = \Theta$ and the sequences of the adjoint operators $\left(\mathcal{W}_n \mathcal{M}_n^\tau \mathcal{H}_n^\tau \mathcal{W}_n\right)^*$ also converge strongly in $\mathbf{L}_\mu^2$.*

Proof. It suffices to show the strong convergence on a dense subset of $\mathbf{L}_\mu^2$. The set of continuous functions $f : [-1, 1] \longrightarrow \mathbb{C}$, for which an $\varepsilon_f > 0$ exists such that $f(x) = 0$ for all $x \in [-1, -1+\varepsilon_f]$, is dense in $\mathbf{L}_\mu^2$. For every such f, there exists a continuous function $\chi : [-1, 1] \longrightarrow [0, 1]$ vanishing in $\left[-1, -1 + \frac{\varepsilon_f}{2}\right]$ such that $\chi f = f$. We have the strong convergences $\mathcal{L}_n^\sigma \chi \mathcal{L}_n^1 \longrightarrow \chi\mathcal{I}$ and $\sqrt{1+\cdot}\,\mathcal{L}_n^\nu \frac{1}{\sqrt{1+\cdot}} \chi \mathcal{L}_n^2 \longrightarrow \chi\mathcal{I}$ in $\mathbf{L}^2_\mu$, where

$$\mathcal{L}_n^1 : \mathbf{L}^2_\mu \longrightarrow \mathbf{L}^2_\mu\,, \quad u \mapsto \sum_{j=0}^{n-1} \langle u, P_j\rangle_\mu \, P_j$$

and

$$\mathcal{L}_n^2 : \mathbf{L}^2_\mu \longrightarrow \mathbf{L}^2_\mu\,, \quad u \mapsto \sum_{j=0}^{n-1} \left\langle u, \sqrt{1+\cdot}\,U_j\right\rangle_\mu \sqrt{1+\cdot}\,U_j\,.$$

Indeed, this is a consequence of the Banach–Steinhaus theorem and of the estimates

$$\begin{aligned}\left\|\mathcal{L}_n^\sigma \chi \mathcal{L}_n^1 u\right\|_\mu^2 &= \left\|\sqrt{1-\cdot}\,\mathcal{L}_n^\sigma \chi \mathcal{L}_n^1 u\right\|_\sigma^2 \\ &= \frac{\pi}{n}\sum_{k=1}^{n}(1-x_{kn}^\sigma)\left|\chi(x_{kn}^\sigma)(\mathcal{L}_n^1 u)(x_{kn}^\sigma)\right|^2 \le \left\|\mathcal{L}_n^1 u\right\|_\mu^2 \le \|u\|_\mu^2\end{aligned}$$

and

$$\begin{aligned}\left\|\sqrt{1+\cdot}\,\mathcal{L}_n^\nu \frac{1}{\sqrt{1+\cdot}}\chi\mathcal{L}_n^2 u\right\|_\mu^2 &= \left\|\sqrt{1-\cdot}\,\mathcal{L}_n^\nu \frac{1}{\sqrt{1+\cdot}}\chi\mathcal{L}_n^2 u\right\|_\nu^2 \\ &= \frac{\pi}{n+\frac{1}{2}}\sum_{k=1}^{n}(1-x_{kn}^\nu)\left|\chi(x_{kn}^\nu)(\mathcal{L}_n^2 u)(x_{kn}^\nu)\right|^2 \\ &\le \left\|\mu\mathcal{L}_n^2 u\right\|_\nu^2 = \left\|\mathcal{L}_n^2 u\right\|_\mu^2 \le \|u\|_\mu^2\,.\end{aligned}$$

Note, that, due to Lemma 2.11,(b),

$$\left\|\mathcal{L}_n^\sigma \chi u - \chi u\right\|_\mu = \left\|\sqrt{1-\cdot}\,(\mathcal{L}_n^\sigma \chi u - \chi u)\right\|_\sigma \longrightarrow 0$$

for all $u \in \bigcup_{j=1}^{\infty} \operatorname{im} \mathcal{L}_j^1$ and that

$$\left\|\sqrt{1+\cdot}\,\mathcal{L}_n^\nu \frac{1}{\sqrt{1+\cdot}}\chi u - \chi u\right\|_\mu = \left\|\sqrt{1-\cdot}\left[\mathcal{L}_n^\nu \frac{1}{\sqrt{1+\cdot}}\chi u - \frac{1}{\sqrt{1+\cdot}}\chi u\right]\right\|_\nu \longrightarrow 0$$

for all $u \in \bigcup_{j=1}^{\infty} \operatorname{im} \mathcal{L}_j^2$. The relation (cf. [9, (3.19)]) $\widehat{\mathcal{W}}_n \widehat{\mathcal{M}}_n^\sigma \chi \widehat{\mathcal{L}}_n \widehat{\mathcal{W}}_n = \widehat{\mathcal{J}}_1^{-1} \mathcal{L}_n^\sigma \chi \widehat{\mathcal{J}}_1 \widehat{\mathcal{L}}_n$ together with formulas from (2.7)–(2.11) gives us

$$\begin{aligned}\mathcal{W}_n \mathcal{M}_n^\sigma \chi \mathcal{L}_n \mathcal{W}_n &= \mathcal{J}\widehat{\mathcal{W}}_n \widehat{\mathcal{M}}_n^\sigma \mathcal{J}^{-1} \chi \mathcal{J} \widehat{\mathcal{L}}_n \widehat{\mathcal{W}}_n \mathcal{J}^{-1} \\ &= \mathcal{J}\widehat{\mathcal{J}}_1^{-1} \mathcal{L}_n^\sigma \widehat{\chi} \widehat{\mathcal{J}}_1 \widehat{\mathcal{L}}_n \mathcal{J}^{-1} = \mathcal{J}_1^{-1} \mathcal{L}_n^\sigma \chi \mathcal{J}_1 \mathcal{L}_n\,.\end{aligned}$$

This we can explore to estimate, for $\widetilde{f}_1 = \mathcal{J}_1^{-1} f$,

$$\begin{aligned}
&\left\| \mathcal{W}_n \mathcal{M}_n^\sigma \mathcal{H}_n^\sigma \mathcal{L}_n \mathcal{M}_n^\sigma \chi \mathcal{L}_n \mathcal{W}_n \widetilde{f}_1 - \mathcal{W}_n \mathcal{M}_n^\sigma \mathcal{H}_n^\sigma \mathcal{W}_n \widetilde{f}_1 \right\|_\mu \\
&\quad \le \text{const} \left\| \mathcal{W}_n \mathcal{M}_n^\sigma \chi \mathcal{L}_n \mathcal{W}_n \mathcal{J}_1^{-1} f - \mathcal{J}_1^{-1} f \right\|_\mu \\
&\quad = \text{const} \left\| \mathcal{J}_1^{-1} \mathcal{L}_n^\sigma \chi \mathcal{J}_1 \mathcal{L}_n \mathcal{J}_1^{-1} f - \mathcal{J}_1^{-1} f \right\|_\mu = \text{const} \left\| \mathcal{L}_n^\sigma \chi \mathcal{L}_n^1 f - f \right\|_\mu \longrightarrow 0 .
\end{aligned}$$

Analogously, we take the formula (cf. [9, (3.20)])

$$\widehat{\mathcal{W}}_n \widehat{\mathcal{M}}_n^\mu \chi \widehat{\mathcal{L}}_n \widehat{\mathcal{W}}_n = \widehat{\mathcal{J}}_2^{-1} \sqrt{1-\cdot}\, \mathcal{L}_n^\mu \frac{1}{\sqrt{1-\cdot}} \chi \widehat{\mathcal{J}}_2 \widehat{\mathcal{L}}_n$$

to obtain

$$\begin{aligned}
\mathcal{W}_n \mathcal{M}_n^\nu \chi \mathcal{L}_n \mathcal{W}_n &= \mathcal{J} \widehat{\mathcal{W}}_n \widehat{\mathcal{M}}_n^\mu \mathcal{J}^{-1} \chi \mathcal{J} \widehat{\mathcal{L}}_n \widehat{\mathcal{W}}_n \mathcal{J}^{-1} \\
&= \mathcal{J} \widehat{\mathcal{J}}_2^{-1} \sqrt{1-\cdot}\, \mathcal{L}_n^\mu \frac{1}{\sqrt{1-\cdot}} \widehat{\chi} \widehat{\mathcal{J}}_2 \widehat{\mathcal{L}}_n \mathcal{J}^{-1} \\
&= \mathcal{J}_2^{-1} \sqrt{1+\cdot}\, \mathcal{L}_n^\nu \frac{1}{\sqrt{1+\cdot}} \chi \mathcal{J}_2 \mathcal{L}_n .
\end{aligned}$$

Hence, in the same manner as above we get, for $\widetilde{f}_2 = \mathcal{J}_2^{-1} f$,

$$\left\| \mathcal{W}_n \mathcal{M}_n^\nu \mathcal{H}_n^\nu \mathcal{L}_n \mathcal{M}_n^\nu \chi \mathcal{L}_n \mathcal{W}_n \widetilde{f}_2 - \mathcal{W}_n \mathcal{M}_n^\nu \mathcal{H}_n^\nu \mathcal{W}_n \widetilde{f}_2 \right\|_\mu \longrightarrow 0 .$$

Consequently, it remains to prove the strong convergence

$$\mathcal{W}_n \mathcal{M}_n^\tau \mathcal{H}_n^\tau \mathcal{L}_n \mathcal{M}_n^\tau \chi \mathcal{L}_n \mathcal{W}_n \longrightarrow \Theta \quad \text{in} \quad \mathbf{L}_\mu^2 , \tag{2.45}$$

which we have only to check on all $\widetilde{p}_m$ for every fixed $m \in \mathbb{N}_0$. In virtue of (2.44),

$$\widetilde{p}_{n-1-m}(x_{kn}^\sigma) = (-1)^{k+1} P_m(x_{kn}^\sigma)$$

and

$$\widetilde{p}_{n-1-m}(x_{kn}^\nu) = (-1)^{k+1} \sqrt{1+x_{kn}^\nu}\, U_m(x_{kn}^\nu)$$

as well as the unitarity of $\mathcal{W}_n$ and $\mathcal{V}_n^\tau$, it yields, for $n > m$,

$$\begin{aligned}
&\left\| \mathcal{W}_n \mathcal{M}_n^\tau \mathcal{H}_n^\tau \mathcal{L}_n \mathcal{M}_n^\tau \chi \mathcal{L}_n \mathcal{W}_n \widetilde{p}_m \right\|_\mu \\
&= \left\| \mathcal{V}_n^\tau \mathcal{M}_n^\tau \mathcal{H}_n^\tau \mathcal{L}_n \left(\mathcal{V}_n^\tau\right)^{-1} \mathcal{V}_n^\tau \mathcal{M}_n^\tau \chi \mathcal{L}_n \left(\mathcal{V}_n^\tau\right)^{-1} \mathcal{V}_n^\tau \widetilde{p}_{n-1-m} \right\|_{\ell^2} \\
&= \Bigg\| \left[\left(\omega_n^\tau\right)^2 \sqrt{1-x_{jn}^\tau} \sqrt{1+x_{kn}^\tau}\, \mathbf{h}\left(\frac{1+x_{jn}^\tau}{1+x_{kn}^\tau} \right) \frac{\chi(x_{kn}^\tau)}{1+x_{kn}^\tau} \right]_{j,k=1}^n \\
&\qquad\qquad \cdot \left[\omega_n^\tau \sqrt{1-x_{kn}^\tau}\, \widetilde{p}_{n-1-m}(x_{kn}^\tau) \right]_{k=1}^n \Bigg\|_{\ell^2} \\
&= \left(\omega_n^\tau\right)^3 \sqrt{ \sum_{j=1}^n \left[\sum_{k=1}^n (-1)^{k+1} F_m^\tau(x_{kn}^\tau, x_{jn}^\tau) \right]^2 } ,
\end{aligned}$$

where

$$F_m^\tau(x,y) = \begin{cases} \sqrt{(1-y)(1-x^2)}\,\mathbf{h}\left(\dfrac{1+y}{1+x}\right)\dfrac{\chi(x)}{1+x}\,P_m(x) & : \ \tau=\sigma\,, \\ \sqrt{(1-y)(1-x)}(1+x)\,\mathbf{h}\left(\dfrac{1+y}{1+x}\right)\dfrac{\chi(x)}{1+x}\,U_m(x) & : \ \tau=\nu\,, \end{cases}$$

is a continuous function on $[-1,1]^2$. Let $\varepsilon > 0$ and let $n(\chi)$ denote the smallest integer with the property $\chi(x_{kn}^\tau) = 0$ for all $k > 2\,n(\chi)$. Since, for $x_{kn}^\tau = \cos\theta_{kn}^\tau$ with $\theta_{kn}^\tau \in (0,\pi)$, we have the relation $\theta_{k+1,n}^\tau - \theta_{kn}^\tau \sim_{k,n} \frac{1}{n}$, there exist an $n_0 \in \mathbb{N}$ such that $\left|F_m^\tau(x_{kn}^\tau, x_{jn}^\tau) - F_m^\tau(x_{k+1,n}^\tau, x_{jn}^\tau)\right| < \frac{\varepsilon}{\pi^{\frac{3}{2}}}$ for all $k = 1,\dots,n-1$, $j = 1,\dots,n$, $n > n_0$. This implies

$$\begin{aligned} &\|\mathcal{W}_n\mathcal{M}_n^\tau\mathcal{H}_n^\tau\mathcal{L}_n\mathcal{M}_n^\tau\chi\mathcal{L}_n\mathcal{W}_n\widetilde{p}_m\|_\mu \\ &\quad\le (\omega_n^\tau)^3\sqrt{\sum_{j=1}^n\left[\sum_{k=1}^{n(\chi)}\left|F_m^\tau(x_{2k-1,n}^\tau, x_{jn}^\tau) - F_m^\tau(x_{2k,n}^\tau, x_{jn}^\tau)\right|\right]^2} < \varepsilon \end{aligned}$$

for all $n > n_0$, and (2.45) is shown. The strong convergence of the sequence of the adjoint operators can be seen analogously by using (see (2.42))

$$(\mathcal{W}_n\mathcal{M}_n^\tau\mathcal{H}_n^\tau\mathcal{L}_n\mathcal{W}_n)^* = \mathcal{W}_n\mathcal{M}_n^\tau\nu\widetilde{\mathcal{H}}_n^\tau\mu\mathcal{L}_n\mathcal{W}_n\,. \qquad \square$$

The following proposition summarizes the results of Corollary 2.3, Lemma 2.13, Lemma 2.15, and Lemma 2.16.

Proposition 2.17. *Let* $\tau \in \{\sigma,\nu\}$ *and* $a, b \in \mathbf{PC}$. *The quadrature method (cf. (2.4) and (2.6))* $(\mathcal{A}_n^\tau) = (\mathcal{M}_n^\tau(a\mathcal{I} + b\mathcal{S} + \mathcal{H}_n^\tau)\mathcal{L}_n)$ *belongs to the C^*-algebra* $\mathfrak{F}$ *and the associated limit operators are given by*

$$\mathcal{W}^1(\mathcal{A}_n^\tau) = \mathcal{A} = a\mathcal{I} + b\mathcal{S} + \mathcal{B}\,,$$

$$\mathcal{W}^2(\mathcal{A}_n^\tau) = \left\{\begin{array}{lcl} \mathcal{J}_1^{-1}(a\mathcal{J}_1 + \mathbf{i}b\mathcal{I}) & : & \tau=\sigma\,, \\ \mathcal{J}_2^{-1}(a\mathcal{J}_2 + \mathbf{i}b\mathcal{J}_3\mathcal{V}) & : & \tau=\mu\,, \end{array}\right\}$$

$$\mathcal{W}^3(\mathcal{A}_n^\tau) = a(1)\mathbf{I} + b(1)\widetilde{\mathbf{S}}\,,$$

$$\mathcal{W}^4(\mathcal{A}_n^\tau) = a(-1)\mathbf{I} - b(-1)\mathbf{S}^\tau + \mathbf{A}^\tau\,,$$

where

$$\widetilde{\mathbf{S}} = \left[\frac{1-(-1)^{j-k}}{\pi\mathbf{i}(j-k)} + \frac{1-(-1)^{j+k+1}}{\pi\mathbf{i}(j+k+1)}\right]_{j,k=0}^\infty\,,$$

$$\mathbf{S}^\tau = \begin{cases} \left[\dfrac{1-(-1)^{j-k}}{\pi\mathbf{i}(j-k)} - \dfrac{1-(-1)^{j+k+1}}{\pi\mathbf{i}(j+k+1)}\right]_{j,k=0}^\infty & : \ \tau=\sigma\,, \\ \left[\dfrac{1-(-1)^{j-k}}{\pi\mathbf{i}}\left(\dfrac{1}{j-k} - \dfrac{1}{j+k+2}\right)\right]_{j,k=0}^\infty & : \ \tau=\nu\,, \end{cases}$$

$$\mathbf{A}^\sigma = \left[\mathbf{h}\left(\frac{(j+\frac{1}{2})^2}{(k+\frac{1}{2})^2}\right) \frac{2}{k+\frac{1}{2}} \right]_{j,k=0}^{\infty},$$

and

$$\mathbf{A}^\nu = \left[\mathbf{h}\left(\frac{(j+1)^2}{(k+1)^2}\right) \frac{2}{k+1} \right]_{j,k=0}^{\infty},$$

and where the operators $\mathcal{J}_k$ and $\mathcal{V}$ are defined in (2.9) *and* (2.10).

3. The stability theorem for the collocation-quadrature method

Since we want to apply Proposition 2.1 to find conditions for the stability of the collocation-quadrature method (2.4), we have to investigate the invertibility of the limit operators $\mathcal{W}^t(\mathcal{A}_n^\tau)$, $t \in \{1,2,3,4\}$, given by proposition 2.17 and the invertibility of the coset $(\mathcal{A}_n^\tau) + \mathfrak{J}$ in the quotient algebra $\mathfrak{F}/\mathfrak{J}$, where $\mathcal{A}_n^\tau = \mathcal{M}_n^\tau(a\mathcal{I} + b\mathcal{S} + \mathcal{H}_n^\tau)\mathcal{L}_n$. For that, we need some further operator sequences. Let $\mathbf{R}$ belong to $\operatorname{alg}\mathcal{T}(\mathbf{PC})$, define the finite sections $\mathbf{R}_n := \mathcal{P}_n\mathbf{R}\mathcal{P}_n \in \mathcal{L}(\operatorname{im}\mathcal{P}_n)$, and set $\mathbf{R}_n^t := (\mathcal{E}_n^{(t)})^{-1}\mathbf{R}_n\mathcal{E}_n^{(t)}$, $t \in \{3,4\}$.

Lemma 3.1. *For $\mathbf{R} \in \operatorname{alg}\mathcal{T}(\mathbf{PC})$ and $t \in \{3,4\}$, the sequences $(\mathbf{R}_n^t)$ belong to the algebra $\mathfrak{F}$.*

Proof. The analogous result is proved in [9, Lemma 5.4] for the algebra $\widehat{\mathfrak{F}}$ defined by the operator sequences $\widehat{\mathcal{L}}_n$, $\widehat{\mathcal{W}}_n$, $\widehat{\mathcal{V}}_n$, and $\widehat{\widehat{\mathcal{V}}}_n$ (cf. Section 2) such that the present lemma can be proved by using (2.7), (2.8), (2.11), and (2.12) (cf. the proof of Corollary 2.3). □

Let $m \in \mathbb{N}$ be fixed. We denote by $\mathfrak{A}_0 = \mathfrak{A}_0^\tau$ the smallest C^*-subalgebra of $\mathfrak{F}$ containing all sequences from $\mathfrak{J}$ and all sequences $(\mathcal{A}_n^\tau)$ with

$$\mathcal{A}_n^\tau = \mathcal{M}_n^\tau\left(a\mathcal{I} + b\mathcal{S} + \mathcal{H}_n^\tau\right)\mathcal{L}_n,$$

$a, b \in \mathbf{PC}$, $\beta_k \in \mathbb{C}$. Moreover, let $\mathfrak{A} = \mathfrak{A}^\tau$ be the smallest C^*-subalgebra of $\mathfrak{F}$ generated by all sequences from $\mathfrak{A}_0$ and all sequences $(\mathbf{R}_n^t)$ with $t \in \{3,4\}$ and $\mathbf{R} \in \operatorname{alg}\mathcal{T}(\mathbf{PC})$. For the coset $(\mathcal{A}_n) + \mathfrak{J}$ we use the abbreviation $(\mathcal{A}_n)^o$. Due to the property of inverse closedness of C^*-subalgebras, the invertibility of $(\mathcal{A}_n^\tau)^o$ in $\mathfrak{F}/\mathfrak{J}$ is equivalent to its invertibility in $\mathfrak{A}/\mathfrak{J}$. As a main tool for proving invertibility in the quotient algebra $\mathfrak{A}/\mathfrak{J}$ we will use the local principle of Allan and Douglas. For this, we have to find a C^*-subalgebra of the center of $\mathfrak{A}/\mathfrak{J}$ as well as its maximal ideal space. Completely analogous to the proof of [13, Lemma 5.1] one can prove the following lemma.

Lemma 3.2. *Let $\tau \in \{\sigma, \nu\}$ and $f \in \mathbf{C}[-1,1]$. Then, for every $a, b \in \mathbf{PC}$ and every $\mathbf{R} \in \operatorname{alg}\mathcal{T}(\mathbf{PC})$,*

$$(\mathcal{M}_n^\tau f\mathcal{L}_n)(\mathcal{M}_n^\tau(a\mathcal{I} + b\mathcal{S})\mathcal{L}_n) - (\mathcal{M}_n^\tau(a\mathcal{I} + b\mathcal{S})\mathcal{L}_n)(\mathcal{M}_n^\tau f\mathcal{L}_n) \in \mathfrak{J}$$

and

$$(\mathcal{M}_n^\tau f\mathcal{L}_n)(\mathbf{R}_n^t) - (\mathbf{R}_n^t)(\mathcal{M}_n^\tau f\mathcal{L}_n) \in \mathfrak{J}, \quad t \in \{3,4\}.$$

Let $\mathbf{k} : (-1,1) \times [-1,1] \longrightarrow \mathbb{C}$ be a given function, let $0 \le \alpha < \frac{3}{4}$ and $0 \le \beta < \frac{1}{4}$, and assume that

$$\widetilde{\mathbf{k}}(x,y) := v^{\alpha,\beta}(x)\mathbf{k}(x,y) \quad \text{is continuous on} \quad [-1,1]\times[-1,1]. \tag{3.1}$$

Define

$$(\mathcal{K}u)(x) := \int_{-1}^{1} \mathbf{k}(x,y)u(y)\,dy, \quad u \in \mathbf{L}^2_\mu,$$

and, for $n \in \mathbb{N}$ and $\tau \in \{\sigma, \nu\}$, the quadrature operators

$$\mathcal{K}_n^\tau : \mathbf{C}(-1,1) \longrightarrow \mathbf{C}(-1,1), \quad u \mapsto (\omega_n^\tau)^2 \sum_{k=1}^{n} \varphi(x_{kn}^\tau)\,\mathbf{k}(\cdot, x_{kn}^\tau)u(x_{kn}^\tau),$$

i.e., $\mathcal{K}_n^\tau u = \sum_{k=1}^{n} \frac{\lambda_{kn}^\tau}{\tau(x_{kn}^\tau)} \mathbf{k}(\cdot, x_{kn}^\tau)u(x_{kn}^\tau)$.

Lemma 3.3. *Suppose the function* $\mathbf{k} : (-1,1)\times[-1,1] \longrightarrow \mathbb{C}$ *satisfies* (3.1). *Then* $(\mathcal{M}_n^\tau \mathcal{K}_n^\tau \mathcal{L}_n) \in \mathfrak{J}$.

Proof. Note that, in virtue of the exactness of the Gaussian rule,

$$\begin{aligned}(\mathcal{K}_n^\tau u)(x) &= \int_{-1}^{1} \mathcal{L}_n^\tau[\mathbf{k}(x,\cdot)\tau^{-1}u](y)\,\tau(y)\,dy \\ &= \int_{-1}^{1} \mathcal{L}_n^\tau[\mathbf{k}(x,\cdot)\mu u](y)\,\nu(y)\,dy\end{aligned} \tag{3.2}$$

and that

$$\mathcal{M}_n^\tau \mathcal{K}_n^\tau \mathcal{L}_n = (\mathcal{V}_n^\tau)^{-1} \left[\ (\omega_n^\tau)^2 \sqrt{1 - x_{jn}^\tau}\sqrt{1 + x_{kn}^\tau}\, \mathbf{k}(x_{jn}^\tau, x_{kn}^\tau)\ \right]_{j,k=1}^{n} \mathcal{V}_n^\tau. \tag{3.3}$$

Moreover, due to the Arzelà–Ascoli theorem, the operator $\mathcal{K} : \mathbf{L}^2_\mu \longrightarrow \mathbf{C}_{\alpha,\beta}$ is compact. This implies, taking into account Corollary 2.12,

$$\|\mathcal{M}_n^\tau \mathcal{K}\mathcal{L}_n - \mathcal{L}_n\mathcal{K}\mathcal{L}_n\|_{\mathcal{L}(\mathbf{L}^2_\mu)} \longrightarrow 0,$$

and it remains to prove (cf. the definition of $\mathfrak{J}$)

$$\lim_{n\to\infty} \|\mathcal{K}_n^\tau \mathcal{L}_n - \mathcal{K}\mathcal{L}_n\|_{\mathcal{L}(\mathbf{L}^2_\mu, \mathbf{C}_{\alpha,\beta})} = 0. \tag{3.4}$$

For this, let $u \in \mathbf{L}^2_\mu$, let $\widetilde{\mathbf{k}}_n(x,y)$ be a polynomial of degree less than n in both variables with

$$\gamma_n := \int_{-1}^{1} \nu(y)\,dy\ \sup\left\{\left|\widetilde{\mathbf{k}}(x,y) - \widetilde{\mathbf{k}}_n(x,y)\right| : (x,y) \in [-1,1]^2\right\} \longrightarrow 0, \tag{3.5}$$

and write $\mathcal{L}_n u = \nu p_n$ with $p_n(x)$ being a polynomial of degree less than n. Then, again using the exactness of the Gaussian rule,

$$
\begin{aligned}
&\left|v^{\alpha,\beta}(x)\left(\mathcal{K}_n^\tau \mathcal{L}_n u - \mathcal{K}\mathcal{L}_n u\right)(x)\right| \\
&\overset{(3.2)}{=} \left|\int_{-1}^{1}\left[\mathcal{L}_n^\tau[\widetilde{\mathbf{k}}(x,\cdot)p_n](y) - \widetilde{\mathbf{k}}(x,y)p_n(y)\right]\nu(y)\,dy\right| \\
&= \left|\int_{-1}^{1}\left[\mathcal{L}_n^\tau[\widetilde{\mathbf{k}}(x,\cdot)](y) - \widetilde{\mathbf{k}}(x,y)\right]p_n(y)\nu(y)\,dy\right| \\
&\le \left|\int_{-1}^{1}\left[\mathcal{L}_n^\tau[\widetilde{\mathbf{k}}(x,\cdot)](y) - \widetilde{\mathbf{k}}_n(x,y)\right]p_n(y)\nu(y)\,dy\right| \\
&\qquad\qquad + \left|\int_{-1}^{1}\left[\widetilde{\mathbf{k}}_n(x,y) - \widetilde{\mathbf{k}}(x,y)\right]p_n(y)\nu(y)\,dy\right| \\
&= \left|\sum_{k=1}^{n}\lambda_{kn}^\tau\left[\widetilde{\mathbf{k}}(x,x_{kn}^\tau) - \widetilde{\mathbf{k}}(x,x_{kn}^\tau)\right]p_n(x_{kn}^\tau)\left(\tau^{-1}\nu\right)(x_{kn}^\tau)\right| \\
&\qquad\qquad + \left|\int_{-1}^{1}\left[\widetilde{\mathbf{k}}_n(x,y) - \widetilde{\mathbf{k}}(x,y)\right]p_n(y)\nu(y)\,dy\right| \\
&\le \gamma_n\left(\sum_{k=1}^{n}\lambda_{kn}^\tau\left|p_n(x_{kn}^\tau)\right|^2\right)^{\frac{1}{2}} + \gamma_n\left\|p_n\right\|_\nu = 2\gamma_n\left\|\mathcal{L}_n u\right\|_\mu \le 2\gamma_n\left\|u\right\|_\mu,
\end{aligned}
$$

which proves (3.4). □

Lemma 3.4. *For $f \in \mathbf{C}[-1,1]$, the coset $(\mathcal{M}_n^\tau f \mathcal{L}_n)^o$ belongs to the center of $\mathfrak{A}/\mathfrak{J}$.*

Proof. Due to Lemma 3.2 it remains to verify that

$$
(\mathcal{M}_n^\tau f \mathcal{L}_n)(\mathcal{M}_n^\tau \mathcal{H}_n^\tau \mathcal{L}_n) - (\mathcal{M}_n^\tau \mathcal{H}_n^\tau \mathcal{L}_n)(\mathcal{M}_n^\tau f \mathcal{L}_n) \in \mathfrak{J} \tag{3.6}
$$

is true. Using representation (2.33), we get

$$
\begin{aligned}
&\mathcal{M}_n^\tau f \mathcal{L}_n \mathcal{M}_n^\tau \mathcal{H}_n^\tau \mathcal{L}_n - \mathcal{M}_n^\tau \mathcal{H}_n^\tau \mathcal{L}_n \mathcal{M}_n^\tau f \mathcal{L}_n \\
&\quad = (\widetilde{\mathcal{V}}_n^\tau)^{-1}\Big(\left[\, f(x_{n+1-j,n}^\tau)\delta_{j,k} \,\right]_{j,k=1}^n \mathbf{H}_n \\
&\qquad\qquad\qquad - \mathbf{H}_n \left[\, f(x_{n+1-j,n}^\tau)\delta_{j,k} \,\right]_{j,k=1}^n\Big)\widetilde{\mathcal{V}}_n^\tau \mathcal{L}_n \\
&\quad = (\widetilde{\mathcal{V}}_n^\tau)^{-1}\left[\mathbf{F}_n^\tau \mathbf{D}_n\left(\mathbf{E}_n + \mathbf{F}_n\right) - \mathbf{D}_n\left(\mathbf{E}_n + \mathbf{F}_n\right)\mathbf{F}_n^\tau\right]\widetilde{\mathcal{V}}_n^\tau \mathcal{L}_n \\
&\quad = (\widetilde{\mathcal{V}}_n^\tau)^{-1}\mathbf{D}_n\left[\mathbf{F}_n^\tau\left(\mathbf{E}_n + \mathbf{F}_n\right) - \left(\mathbf{E}_n + \mathbf{F}_n\right)\mathbf{F}_n^\tau\right]\widetilde{\mathcal{V}}_n^\tau \mathcal{L}_n,
\end{aligned}
$$

where $\mathbf{D}_n$ and $\mathbf{E}_n = \mathbf{E}_n^1 + \mathbf{E}_n^2$ are defined in (2.34), $\mathbf{F}_n^\tau = \left[\, f(x_{n+1-j,n}^\tau)\delta_{j,k} \,\right]_{j,k=1}^n$, and $\mathbf{F}_n = \left[\, \mathcal{O}(n^{-1}) \,\right]_{j,k=1}^n$ with the constant in $\mathcal{O}(n^{-1})$ independent of j, k, and n. By (2.38) and Lemma 2.10, $\mathbf{E}_n = \mathcal{P}_n \mathbf{R} \mathcal{P}_n$ with $\mathbf{R} \in \operatorname{alg}\mathcal{T}(\mathbf{PC})$. Because of

$\mathbf{D}_n = \widetilde{\mathcal{V}}_n^\tau \mathcal{M}_n^\tau \psi \mathcal{L}_n (\widetilde{\mathcal{V}}_n^\tau)^{-1}$, where $\psi(x) = \sqrt{\frac{1-x}{2}}$, we get that

$$\left((\widetilde{\mathcal{V}}_n^\tau)^{-1}\mathbf{D}_n\left[\mathbf{F}_n^\tau \mathbf{E}_n - \mathbf{E}_n \mathbf{F}_n^\tau\right]\widetilde{\mathcal{V}}_n^\tau \mathcal{L}_n\right) = (\mathcal{M}_n^\tau \psi \mathcal{L}_n)\left(\mathcal{M}_n^\tau f \mathcal{L}_n \mathbf{R}_n^4 - \mathbf{R}_n^4 \mathcal{M}_n^\tau f \mathcal{L}_n\right)$$

belongs to $\mathfrak{J}$ in virtue of Lemma 3.2. It remains to show

$$\left((\widetilde{\mathcal{V}}_n^\tau)^{-1}\left(\mathbf{F}_n^\tau \mathbf{F}_n - \mathbf{F}_n \mathbf{F}_n^\tau\right)\widetilde{\mathcal{V}}_n^\tau\right) \in \mathfrak{J}\,.$$

Since the entries of the matrix $\mathbf{F}_n$ are bounded by $\frac{\text{const}}{n}$ with a constant independent of j, k, and n, we can replace $\mathbf{F}_n$ by

$$\widetilde{\mathbf{F}}_n := \left[\ \chi(x_{n+1-j,n}^\tau)\delta_{j,k}\ \right]_{j,k=1}^n \mathbf{F}_n \left[\ \chi(x_{n+1-j,n}^\tau)\delta_{j,k}\ \right]_{j,k=1}^n$$

with a smooth function $\chi : [-1,1] \longrightarrow [0,1]$, which is identically equal to 1 except in two small neighborhoods of the interval endpoints. Indeed, $\mathfrak{J}$ is closed and the norm of the difference of $\mathbf{F}_n$ and the modified matrix $\widetilde{\mathbf{F}}_n$ can be made arbitrarily small (uniformly w.r.t. $n \in \mathbb{N}$) due to a simple Frobenius norm estimate. Now, we additionally suppose that χ vanishes identically in a small neighborhood of the endpoints of the interval $[-1,1]$. Let $\mathbf{g} : (-1,1)\times(-1,1) \longrightarrow \mathbb{C}$ be continuous and consider the matrix

$$\begin{aligned}\mathbf{K}_n &= \left[\ \chi(x_{n+1-j,n}^\tau)\delta_{j,k}\ \right]_{j,k=1}^n \\ &\qquad \cdot \left[\ (\omega_n^\tau)^2 \mathbf{g}(x_{n+1-j,n}^\tau, x_{n+1-k,n}^\tau)\ \right]_{j,k=1}^n \left[\ \chi(x_{n+1-j,n}^\tau)\delta_{j,k}\ \right]_{j,k=1}^n \\ &= \left[\ (\omega_n^\tau)^2 \sqrt{1 - x_{n+1-j,n}^\tau}\,\sqrt{1 + x_{n+1-k,n}^\tau}\,\widetilde{\mathbf{g}}(x_{n+1-j,n}^\tau, x_{n+1-k,n}^\tau)\ \right]_{j,k=1}^n,\end{aligned}$$

where

$$\widetilde{\mathbf{g}}(x,y) = \frac{\chi(x)}{\sqrt{1-x}}\,\mathbf{g}(x,y)\,\frac{\chi(y)}{\sqrt{1+y}}$$

is continuous on $[-1,1]\times[-1,1]$. Thus, due to Lemma 3.3 and (3.3),

$$\left((\widetilde{\mathcal{V}}_n^\tau)^{-1}\mathbf{K}_n \widetilde{\mathcal{V}}_n^\tau\right) = \left((\mathcal{V}_n^\tau)^{-1}\mathbf{K}_n^0 \mathcal{V}_n^\tau\right) \in \mathfrak{J}\,, \tag{3.7}$$

where

$$\mathbf{K}_n^0 = \left[\ (\omega_n^\tau)^2 \sqrt{1 - x_{jn}^\tau}\,\sqrt{1 + x_{kn}^\tau}\,\widetilde{\mathbf{g}}(x_{jn}^\tau, x_{kn}^\tau)\ \right]_{j,k=1}^n.$$

In view of (2.32) and (2.33), we have

$$\begin{aligned}\mathbf{F}_n = (\omega_n^\tau)^2 \Bigg[\mathbf{h}\left(\frac{1+x_{n+1-j,n}^\tau}{1+x_{n+1-k,n}^\tau}\right)&\frac{\sqrt{2}}{\sqrt{1+x_{n+1-k,n}^\tau}} \\ &-\mathbf{h}\left(\frac{(j-c^\tau)^2}{(k-c^\tau)^2}\right)\frac{2}{(\omega_n^\tau)^2(k-c^\tau)}\Bigg]_{j,k=1}^n,\end{aligned}$$

which gives, together with $\dfrac{(\omega_n^\tau)^2(k-c^\tau)}{2} = \arcsin\sqrt{\dfrac{1+x_{n+1-k,n}^\tau}{2}}$,

$$\mathbf{F}_n = \left[\ (\omega_n^\tau)^2 \mathbf{g}_0(x_{n+1-j,n}^\tau, x_{n+1-k,n}^\tau)\ \right]_{j,k=1}^n$$

with

$$\mathbf{g}_0(x,y) = \mathbf{h}\left(\frac{1+x}{1+y}\right)\frac{\sqrt{2}}{\sqrt{1+y}} - \mathbf{h}\left(\frac{\arcsin^2\sqrt{\frac{1+x}{2}}}{\arcsin^2\sqrt{\frac{1+y}{2}}}\right)\frac{1}{\arcsin\sqrt{\frac{1+y}{2}}}\,.$$

Hence, we can apply (3.7) with $\mathbf{g}(x,y) = f(x)\mathbf{g}_0(x,y)$ and $\mathbf{g}(x,y) = \mathbf{g}_0(x,y)f(y)$ in order to get

$$\left((\widetilde{\mathcal{V}}_n^\tau)^{-1}\mathbf{F}_n^\tau\mathbf{F}_n\widetilde{\mathcal{V}}_n^\tau\right) \in \mathfrak{J} \quad \text{and} \quad \left((\widetilde{\mathcal{V}}_n^\tau)^{-1}\mathbf{F}_n\mathbf{F}_n^\tau\widetilde{\mathcal{V}}_n^\tau\right) \in \mathfrak{J}\,,$$

respectively. □

Lemma 3.4 says that the set $\mathfrak{C} := \{(\mathcal{M}_n^\tau f\mathcal{L}_n)^o : f \in \mathbf{C}[-1,1]\}$ forms a $\mathcal{C}^*$-subalgebra of the center of $\mathfrak{A}/\mathfrak{J}$. This subalgebra is *-isomorphic to the algebra $\mathbf{C}[-1,1]$ via the mapping $(\mathcal{M}_n^\tau f\mathcal{L}_n)^o \longrightarrow f$. The definition of this mapping is correct, since $(\mathcal{M}_n^\tau f_1\mathcal{L}_n)^o = (\mathcal{M}_n^\tau f_2\mathcal{L}_n)^o$ implies that $(f_1 - f_2)\mathcal{I} : \mathbf{L}_\nu^2 \longrightarrow \mathbf{L}_\nu^2$ is compact, i.e., $f_1 - f_2 = 0$. Consequently, the maximal ideals of $\mathfrak{C}$ are $\mathfrak{T}_\omega$, $\omega \in [-1,1]$, with

$$\mathfrak{T}_\omega := \{(\mathcal{M}_n^\tau f\mathcal{L}_n)^o : f \in \mathbf{C}[-1,1], f(\omega) = 0\}\,.$$

By $\mathfrak{J}_\omega = \mathfrak{J}_\omega^\tau$ we denote the smallest closed ideal of $\mathfrak{A}/\mathfrak{J}$, which contains $\mathfrak{T}_\omega$, i.e., $\mathfrak{J}_\omega$ is equal to

$$\operatorname{clos}_{\mathfrak{A}/\mathfrak{J}}\left\{\sum_{j=1}^m \left(\mathcal{A}_n^j\mathcal{M}_n^\tau f_j\mathcal{L}_n\right)^o : \left(\mathcal{A}_n^j\right) \in \mathfrak{A},\ f_j \in \mathbf{C}[-1,1],\ f_j(\omega) = 0,\ m \in \mathbb{N}\right\}.$$

The local principle of Allan and Douglas claims the following.

Theorem 3.5 (cf. [7], Sections 1.4.4, 1.4.6). *For all $\omega \in [-1,1]$, the ideal $\mathfrak{J}_\omega$ is a proper ideal in $\mathfrak{A}/\mathfrak{J}$. An element $(\mathcal{A}_n)^o$ of $\mathfrak{A}/\mathfrak{J}$ is invertible if and only if $(\mathcal{A}_n)^o + \mathfrak{J}_\omega$ is invertible in $(\mathfrak{A}/\mathfrak{J})/\mathfrak{J}_\omega$ for all $\omega \in [-1,1]$.*

Lemma 3.6 (cf. [11], Lemma 5.3). *For $-1 < \omega \le 1$, the coset $(\mathcal{M}_n^\tau\mathcal{H}_n^\tau\mathcal{L}_n)^o$ belongs to $\mathfrak{J}_\omega$.*

Proof. Let $-1 < \omega \le 1$ and let $\chi : [-1,1] \longrightarrow [0,1]$ be a continuous function vanishing in a neighborhood of -1 and satisfying $\chi(\omega) = 1$. Then, the kernel function of the integral operator $\mathcal{K} = \mathcal{B}\chi\mathcal{I}$ is continuous, such that, due to Lemma 3.3, $(\mathcal{M}_n^\tau\mathcal{H}_n^\tau\mathcal{L}_n)(\mathcal{M}_n^\tau\chi\mathcal{L}_n) = (\mathcal{M}_n^\tau\mathcal{K}_n^\tau\mathcal{L}_n) \in \mathfrak{J}$. Consequently, $(\mathcal{M}_n^\tau\mathcal{H}_n^\tau\mathcal{L}_n)^o = (\mathcal{M}_n^\tau\mathcal{H}_n^\tau\mathcal{L}_n\mathcal{M}_n^\tau(1-\chi)\mathcal{L}_n)^o \in \mathfrak{J}_\omega$. □

Lemma 3.7 ([9], Lemma 5.9). *If $a \in \mathbf{PC}$, $\omega \in [-1,1]$, $a(\omega) = 0$, and if a is continuous at the point ω, then $(\mathcal{M}_n^\tau a\mathcal{L}_n)^o \in \mathfrak{J}_\omega$.*

Lemma 3.8. *If $(\mathcal{A}_n) \in \mathfrak{A}_0$ and if the operator $\mathcal{W}^1(\mathcal{A}_n) : \mathbf{L}_\mu^2 \longrightarrow \mathbf{L}_\mu^2$ is Fredholm, then, for all $\omega \in (-1,1)$, the coset $(\mathcal{A}_n)^o + \mathfrak{J}_\omega$ is invertible in $(\mathfrak{A}/\mathfrak{J})/\mathfrak{J}_\omega$.*

Proof. Let $\mathfrak{G}$ be the smallest C^*-subalgebra of $\mathcal{L}(\mathbf{L}^2_\mu)$ which contains all operators $a\mathcal{I} + b\mathcal{S}$ with $a, b \in \mathbf{PC}$ and the ideal $\mathfrak{K} := \mathcal{K}(\mathbf{L}^2_\mu)$ of all compact operators from $\mathcal{L}(\mathbf{L}^2_\mu)$. By $\mathfrak{J}^{\mathfrak{G}}_\omega$, $\omega \in [-1, 1]$, we denote the smallest closed ideal of $\mathfrak{G}/\mathfrak{K}$ which contains all cosets of the form $f\mathcal{I} + \mathfrak{K}$ with $f \in \mathbf{C}[-1, 1]$ and $f(\omega) = 0$. Note that the set $\{f\mathcal{I} + \mathfrak{K} : f \in \mathbf{C}[-1, 1]\}$ forms a C^*-subalgebra of the center of $\mathfrak{G}/\mathfrak{K}$. With these notations, the local principle of Allan and Douglas states that an operator $\mathcal{A} \in \mathfrak{G}$ is Fredholm if and only if $(\mathcal{A} + \mathfrak{K}) + \mathfrak{J}^{\mathfrak{G}}_\omega$ is invertible in $(\mathfrak{G}/\mathfrak{K})/\mathfrak{J}^{\mathfrak{G}}_\omega$ for all $\omega \in [-1, 1]$.

Let $\omega \in (-1, 1)$. By Lemma 3.6, we have $(\mathcal{M}^\tau_n(a\mathcal{I} + b\mathcal{S} + \mathcal{H}^\tau_n)\mathcal{L}_n)^o + \mathfrak{J}_\omega = (\mathcal{M}^\tau_n(a\mathcal{I} + b\mathcal{S})\mathcal{L}_n)^o + \mathfrak{J}_\omega$, which implies, together with Lemma 3.7, that the C^*-algebra $(\mathfrak{A}_0/\mathfrak{J})/\mathfrak{J}_\omega$ is generated by the elements (cf. [9, p. 763])

$$(\mathcal{L}_n)^o + \mathfrak{J}_\omega\,, \quad \frac{1}{2}\left((\mathcal{L}_n)^o + (\mathcal{M}^\tau_n \mathcal{S}\mathcal{L}_n)^o\right) + \mathfrak{J}_\omega\,,$$

and

$$(\mathcal{M}^\tau_n h_\omega \mathcal{L}_n)^o + \mathfrak{J}_\omega\,,$$

where

$$h_\omega(x) = \begin{cases} 0 & : \quad -1 \le x \le \omega\,, \\ 1 & : \quad \omega < x \le 1\,. \end{cases}$$

As in [9, p. 76], one can prove that $(\mathfrak{A}_0/\mathfrak{J})\,/\mathfrak{J}_\omega$ and $(\mathfrak{G}/\mathfrak{K})/\mathfrak{J}^{\mathfrak{G}}_\omega$ are $*$-isomorphic via the isomorphism $(\mathcal{A}_n)^o + \mathfrak{J}_\omega \mapsto \left(\mathcal{W}^1(\mathcal{A}_n) + \mathfrak{K}\right) + \mathfrak{J}^{\mathfrak{G}}_\omega$, which finishes the proof of the lemma (cf. also [9, Corollary 5.13]). □

Now, let us investigate the invertibility of $(\mathcal{A}^\tau_n)^o + \mathfrak{J}_{\pm 1}$ in $(\mathfrak{A}/\mathfrak{J})/\mathfrak{J}_{\pm 1}$ for $\mathcal{A}^\tau_n = \mathcal{M}^\tau_n(a\mathcal{I} + b\mathcal{S} + \mathcal{H}^\tau_n)\mathcal{L}_n$, $a, b \in \mathbf{PC}$. To this end, we show that the invertibility of the limit operators $\mathcal{W}^3(\mathcal{A}^\tau_n)$ and $\mathcal{W}^4(\mathcal{A}^\tau_n)$ implies the invertibility of $(\mathcal{A}^\tau_n)^o + \mathfrak{J}_{+1}$ and $(\mathcal{A}^\tau_n)^o + \mathfrak{J}_{-1}$, respectively.

Referring by $\mathbf{C}_{\pm 1}$ to the set of all continuous functions $f : [-1, 1] \longrightarrow \mathbb{R}$ with $f(\pm 1) = 1$ and $0 \le f(x) \le 1$ for all $x \in [-1, 1]$, from Lemma 3.7 we conclude that, for $(\mathcal{A}_n) \in \mathfrak{A}$, we have

$$\|(\mathcal{A}_n)^o + \mathfrak{J}_{\pm 1}\|_{(\mathfrak{A}/\mathfrak{J})/\mathfrak{J}_{\pm 1}} \le \inf_{f \in \mathbf{C}_{\pm 1}} \|(\mathcal{M}^\tau_n f \mathcal{L}_n)^o (\mathcal{A}_n)^o\|_{\mathfrak{A}/\mathfrak{J}}\,. \tag{3.8}$$

Let us remark that, due to [2, Prop. 5.1], actually we can write the equality sign in (3.8).

Lemma 3.9. *Let* $\mathbf{R}, \widetilde{\mathbf{R}} \in \operatorname{alg} \mathcal{T}(\mathbf{PC})$.

(a) *We have* $\left(\mathbf{R}^3_n \widetilde{\mathbf{R}}^3_n - [\mathbf{R}\widetilde{\mathbf{R}}]^3_n\right)^o \in \mathfrak{J}_{+1}$ *and* $\left(\mathbf{R}^4_n \widetilde{\mathbf{R}}^4_n - [\mathbf{R}\widetilde{\mathbf{R}}]^4_n\right)^o \in \mathfrak{J}_{-1}$.

(b) *If* $\mathbf{R}$ *is invertible, then the coset* $\left([\mathbf{R}^{-1}]^{3/4}_n\right)^o + \mathfrak{J}_{\pm 1}$ *is the inverse of* $\left(\mathbf{R}^{3/4}_n\right)^o + \mathfrak{J}_{\pm 1}$ *in* $(\mathfrak{A}/\mathfrak{J})/\mathfrak{J}_{\pm 1}$.

Proof. The analogue to assertion (a) was already proved in [9, Lemma 5.14] for collocation using $\mathcal{M}^{\overline{\tau}}_n$ (cf. (2.8) and (2.12)), and the present case follows by symmetry reasons. Assertion (b) is an immediate consequence of (a). □

Lemma 3.10. *If* $\mathcal{A}_n^\tau = \mathcal{M}_n^\tau\left(a\mathcal{I} + b\mathcal{S} + \mathcal{H}_n^\tau\right)\mathcal{L}_n$ *with* $a, b \in \mathbf{PC}$, $\mathbf{S} := \mathcal{W}^3(\mathcal{A}_n^\tau)$, *and* $\mathbf{T} := \mathcal{W}^4(\mathcal{A}_n^\tau)$, *then* $\mathbf{S}, \mathbf{T} \in \operatorname{alg}\mathcal{T}(\mathbf{PC})$, $\left(\mathbf{S}_n^3\right)^o + \mathfrak{J}_1 = \left(\mathcal{A}_n^\tau\right)^o + \mathfrak{J}_1$, *and* $\left(\mathbf{T}_n^4\right)^o + \mathfrak{J}_{-1} = \left(\mathcal{A}_n^\tau\right)^o + \mathfrak{J}_{-1}$.

Proof. From [9, Proposition 3.4, Lemma 3.9] we infer that $\widetilde{\mathbf{S}}, \mathbf{S}^\tau \in \operatorname{alg}\mathcal{T}(\mathbf{PC})$ such that, due to Proposition 2.17 and Lemma 2.10 (cf. also the proof of Lemma 2.13, in particular (2.38)), $\mathbf{S}$ and $\mathbf{T}$ belong to $\operatorname{alg}\mathcal{T}(\mathbf{PC})$. In case $\mathcal{B}$ is the zero operator, the present lemma is the analogue to [9, Lemma 5.16]. Since the limit operator $\mathcal{W}^3(\mathcal{A}_n^\tau)$ does not depend on $\mathcal{H}_n^\tau$, we have only to show that

$$\left([\mathbf{A}^\tau]_n^4 - \mathcal{M}_n^\tau\mathcal{H}_n^\tau\mathcal{L}_n\right)^o \in \mathfrak{J}_{-1} \tag{3.9}$$

is true, where $\mathbf{A}^\tau$ is defined in Proposition 2.17. By (2.33) we get

$$[\mathbf{A}^\tau]_n^4 - \mathcal{M}_n^\tau\mathcal{H}_n^\tau\mathcal{L}_n = \left(\widetilde{\mathcal{V}}_n^\tau\right)^{-1}\left[(\mathbf{I}_n - \mathbf{D}_n)\mathbf{E}_n - \mathbf{D}_n\mathbf{F}_n\right]\widetilde{\mathcal{V}}_n^\tau\,,$$

where

$$\mathbf{I}_n - \mathbf{D}_n = \left[\ \left(1 - \sqrt{\frac{1 - x_{n+1-j,n}^\tau}{2}}\right)\delta_{j,k}\ \right]_{j,k=1}^n, \qquad \mathbf{E}_n = \mathcal{P}_n\mathbf{A}^\tau\mathcal{P}_n\,,$$

and

$$\mathbf{F}_n = \left[\ \mathcal{O}\left(\frac{1}{n}\right)\ \right]_{j,k=1}^n,$$

with the constant in $\mathcal{O}\left(\frac{1}{n}\right)$ not depending on j, k, and n. Hence, the matrices $\mathbf{E}_n$ and $\mathbf{F}_n$ are uniformly bounded and, using (3.8) and taking into account the unitarity of $\widetilde{\mathcal{V}}_n^\tau : \operatorname{im}\mathcal{L}_n \longrightarrow \operatorname{im}\mathcal{P}_n$, we can write, for some $n_0 \in \mathbb{N}$,

$$\begin{aligned}
&\left\|\left([\mathbf{A}^\tau]_n^4 - \mathcal{M}_n^\tau\mathcal{H}_n^\tau\mathcal{L}_n\right)^o + \mathfrak{J}_{-1}\right\|_{(\mathfrak{A}/\mathfrak{J})/\mathfrak{J}_{-1}} \\
&\quad\le \left\|\left(\left(\widetilde{\mathcal{V}}_n^\tau\right)^{-1}(\mathbf{I}_n - \mathbf{D}_n)\mathbf{E}_n\widetilde{\mathcal{V}}_n^\tau\right)^o + \mathfrak{J}_{-1}\right\|_{(\mathfrak{A}/\mathfrak{J})/\mathfrak{J}_{-1}} \\
&\qquad\qquad + \left\|\left(\left(\widetilde{\mathcal{V}}_n^\tau\right)^{-1}\mathbf{D}_n\mathbf{F}_n\widetilde{\mathcal{V}}_n^\tau\right)^o + \mathfrak{J}_{-1}\right\|_{(\mathfrak{A}/\mathfrak{J})/\mathfrak{J}_{-1}} \\
&\quad\le \inf_{f\in\mathbf{C}_{-1}}\sup_{n\ge n_0}\left\|\left[\ f(x_{n+1-j,n}^\tau)\delta_{j,k}\ \right]_{j,k=1}^n(\mathbf{I}_n - \mathbf{D}_n)\mathbf{E}_n\right\|_{\mathcal{L}(\ell^2)} \\
&\qquad\qquad + \inf_{f\in\mathbf{C}_{-1}}\sup_{n\ge n_0}\left\|\left[\ f(x_{n+1-j,n}^\tau)\delta_{j,k}\ \right]_{j,k=1}^n\mathbf{D}_n\mathbf{F}_n\right\|_{\mathcal{L}(\ell^2)} \\
&\quad\le \text{const}\left(\inf_{f\in\mathbf{C}_{-1}}\sup_{n\ge n_0}\left\|\left[\ f(x_{n+1-j,n}^\tau)\delta_{j,k}\ \right]_{j,k=1}^n(\mathbf{I}_n - \mathbf{D}_n)\right\|_{\mathcal{L}(\ell^2)}\right. \\
&\qquad\qquad \left. + \inf_{f\in\mathbf{C}_{-1}}\sup_{n\ge n_0}\left\|\left[\ f(x_{n+1-j,n}^\tau)\delta_{j,k}\ \right]_{j,k=1}^n\mathbf{F}_n\right\|_{\mathcal{L}(\ell^2)}\right).
\end{aligned}$$

Choosing $f \in \mathbf{C}_{-1}$ with $\operatorname{supp}(f \circ \cos) \subset [\pi(1-\varepsilon), \pi]$ with $\varepsilon \in (0,1)$, a simple Frobenius norm estimate shows that, for all $n \ge \frac{1}{2\varepsilon}$,

$$\left\| \left[\, f(x^\sigma_{n+1-j,n})\delta_{j,k} \,\right]_{j,k=1}^n \mathbf{F}_n \right\|^2_{\mathcal{L}(\ell^2)} \le \text{const} \sum_{1\le j\le n\varepsilon+\frac{1}{2}} \sum_{k=1}^n \frac{1}{n^2} \le \varepsilon + \frac{1}{2n} \le 2\varepsilon$$

and, for all $n \in \mathbb{N}$,

$$\left\| \left[\, f(x^\nu_{n+1-j,n})\delta_{j,k} \,\right]_{j,k=1}^n \mathbf{F}_n \right\|^2_{\mathcal{L}(\ell^2)} \le \text{const} \sum_{1\le j\le (n+\frac{1}{2})\varepsilon} \sum_{k=1}^n \frac{1}{n^2} \le \left(1+\frac{1}{2n}\right)\varepsilon \le 2\varepsilon$$

which implies

$$\inf_{f\in\mathbf{C}_{-1}} \sup_{n\ge n_0} \left\| \left[\, f(x^\tau_{n+1-j,n})\delta_{j,k} \,\right]_{j,k=1}^n \mathbf{F}_n \right\|_{\mathcal{L}(\ell^2)} = 0\,.$$

Moreover, since $1 - \sqrt{\dfrac{1-x^\tau_{n+1-j,n}}{2}}$ equals

$$\begin{cases} 1-\cos\dfrac{2j-1}{4n}\pi = 2\sin^2\dfrac{2j-1}{8n}\pi \le \left(\dfrac{2j-1}{2n}\right)^2\dfrac{\pi^2}{8} & : \ \tau=\sigma\,, \\ 1-\cos\dfrac{2j-1}{2(2n+1)}\pi = 2\sin^2\dfrac{2j-1}{4(2n+1)}\pi \le \left(\dfrac{2j}{2n+1}\right)^2\dfrac{\pi^2}{8} & : \ \tau=\nu\,, \end{cases}$$

we have

$$\left\| \left[\, f(x^\tau_{n+1-j,n})\delta_{j,k} \,\right]_{j,k=1}^n (\mathbf{I}_n - \mathbf{D}_n) \right\|_{\mathcal{L}(\ell^2)} \le \frac{(\varepsilon\pi)^2}{8}\,.$$

Consequently,

$$\left\| \left([\mathbf{A}^\tau]_n^4 - \mathcal{M}_n^\tau \mathcal{H}_n^\tau \mathcal{L}_n \right)^o + \mathfrak{J}_{-1} \right\|_{(\mathfrak{A}/\mathfrak{J})/\mathfrak{J}_{-1}} = 0\,.$$

Thus (3.9) is proved. □

Since we know that, for the generating sequences of $\mathfrak{A}_0$, the limit operators with $t \in \{3,4\}$ belong to $\operatorname{alg}\mathcal{T}(\mathbf{PC})$ (cf. Lemma 3.10) and since the mappings $\mathcal{W}^{3/4} : \mathfrak{F} \longrightarrow \mathfrak{L}(\ell^2)$ are continuous *-homomorphisms (see [9, Corollary 2.4]), we have $\mathcal{W}^{3/4}(\mathcal{A}_n) \in \operatorname{alg}\mathcal{T}(\mathbf{PC})$ if $(\mathcal{A}_n) \in \mathfrak{A}_0$. Thus, by Lemma 3.9, Lemma 3.10, and the closedness of $\mathfrak{J}_{\pm 1}$, we get the following corollary.

Corollary 3.11. *Let $(\mathcal{A}_n) \in \mathfrak{A}_0$. Then, the invertibility of $\mathcal{W}^3(\mathcal{A}_n)$ and $\mathcal{W}^4(\mathcal{A}_n)$ implies, respectively, the invertibility of $(\mathcal{A}_n)^o + \mathfrak{J}_{+1}$ and $(\mathcal{A}_n)^o + \mathfrak{J}_{-1}$ in $(\mathfrak{A}/\mathfrak{J})/\mathfrak{J}_{\pm 1}$.*

Now, we are able to formulate the stability theorem for sequences of the algebra $\mathfrak{A}_0$, in particular for the collocation-quadrature method $(\mathcal{A}_n^\tau)$ given by (2.6). Indeed, with the help of Proposition 2.17, Lemma 3.8, Corollary 3.11, and the local principle of Allan and Douglas we can state the following theorem.

Theorem 3.12. *A sequence $(\mathcal{A}_n) \in \mathfrak{A}_0$ is stable if and only if all limit operators $\mathcal{W}^t(\mathcal{A}_n) : \mathbf{X}^{(t)} \longrightarrow \mathbf{X}^{(t)}$, $t = 1,2,3,4$ are invertible.*

4. Application to the notched half-plane problem

In this section we apply the collocation-quadrature method to the transformed integral equation of the notched half-plane problem (cf. (1.6)–(1.8))

$$(\mathcal{S}u)(x)+(\mathcal{B}u)(x) := \frac{1}{\pi \mathbf{i}}\int_{-1}^{1}\frac{u(y)\,dy}{y-x}+\int_{-1}^{1}\mathbf{h}\left(\frac{1+x}{1+y}\right)\frac{u(y)}{1+y}\,dy = -\mathbf{i}\widetilde{f}(x)\,, \quad (4.1)$$

$-1 < x < 1$, where we take the right-hand side in (1.1) to be identically 1, i.e., $f(x) = 1$ in (1.5) such that $\widetilde{f}(x) = \sqrt{1+x}$. Further,

$$\mathbf{h}(t) = -\frac{\sqrt{t}}{\pi\mathbf{i}(1+t)}+\frac{6\,t\,\sqrt{t}}{\pi\mathbf{i}(1+t)^2}-\frac{4\,t^2\sqrt{t}}{\pi\mathbf{i}(1+t)^3}-\frac{1}{\pi\mathbf{i}(1+\sqrt{t})}\,. \quad (4.2)$$

Then, we have (cf. (1.2) and (1.3))

$$\widetilde{k}_1(1) = \lim_{x\to 1-0}\left[u(x)\sqrt{\frac{1-x}{1+x}}\right] = \frac{1}{\sqrt{2}}\lim_{x\to 1-0}\left[u(x)\sqrt{1-x}\right] \quad (4.3)$$

for the normalized stress intensity factor at the end of the crack inside the elastic half-plane and

$$d(-1) = -\frac{1}{2}\int_{-1}^{1}\frac{u(x)\,dx}{\sqrt{1+x}} \quad (4.4)$$

for the normalized crack opening displacement at the end of the crack located at the boundary of the elastic half-plane.

If we want to ask for the stability of the method

$$\mathcal{A}_n^\tau u_n := \mathcal{M}_n^\tau(\mathcal{S}+\mathcal{H}_n^\tau)\mathcal{L}_n u_n = -\mathbf{i}\mathcal{M}_n^\tau\widetilde{f}\,, \quad u_n \in \operatorname{im}\mathcal{L}_n\,,$$

where $\mathcal{H}_n^\tau$ is defined according to (2.3), then, due to Theorem 3.12, we have to check the invertibility of the four limit operators $\mathcal{W}^t\left(\mathcal{M}_n^\tau(\mathcal{S}+\mathcal{H}_n^\tau)\mathcal{L}_n\right)$, $t = 1,2,3,4$, which are given in Proposition 2.17.

The invertibility of $\mathcal{W}^1(\mathcal{A}_n^\tau) = \mathcal{S}+\mathcal{B} : \mathbf{L}^2_\mu \longrightarrow \mathbf{L}^2_\mu$ is equivalent to the invertibility of the operator $\mathcal{S}+\mathcal{B}_0 : \mathbf{L}^2_\varphi \longrightarrow \mathbf{L}^2_\varphi$, where

$$(\mathcal{B}_0 u)\,u(x) = \frac{1}{\pi\mathbf{i}}\int_{-1}^{1}\mathbf{h}_0\left(\frac{1+x}{1+y}\right)\frac{u(y)\,dy}{1+y}$$

with

$$\mathbf{h}_0(t) = -\frac{1}{\pi\mathbf{i}(1+t)}+\frac{6\,t}{\pi\mathbf{i}(1+t)^2}-\frac{4\,t^2}{\pi\mathbf{i}(1+t)^3}\,.$$

The invertibility of this operator was already mentioned in [4, Cor. 2.3].

The operator $\mathcal{W}^2(\mathcal{A}_n^\sigma) = \mathbf{i}\mathcal{J}_1^{-1} : \mathbf{L}^2_\mu \longrightarrow \mathbf{L}^2_\mu$ is obviously invertible. On the other hand, it is clear that the operator $\mathcal{W}^2(\mathcal{A}_n^\nu) = \mathbf{i}\mathcal{J}_2^{-1}\mathcal{J}_3\mathcal{V} : \mathbf{L}^2_\mu \longrightarrow \mathbf{L}^2_\mu$ is not invertible in virtue of the presence of the shift operator $\mathcal{V}$. Consequently, the collocation method $(\mathcal{A}_n^\nu)$ is not stable.

The limit operator $\mathcal{W}^3(\mathcal{A}_n^\tau) = \widetilde{\mathbf{S}}$ turns out to be equal to $\mathbf{T}(\phi) + \mathbf{H}(\phi)$ with $\phi(t) = \operatorname{sgn}(\operatorname{Im} t)$ (cf. [9, (4.1)]). Hence, in view of (2.21) and (2.22),

$$\operatorname{smb}_{\widetilde{\mathbf{S}}}(t,\lambda) = \begin{cases} 1 & : \ \operatorname{Im} t > 0\,, \\ 1 - 2\lambda - 2\mathbf{i}\sqrt{\lambda(1-\lambda)} & : \ t = -1\,, \\ -1 & : \ \operatorname{Im} t < 0\,, \\ 2\lambda - 1 - 2\mathbf{i}\sqrt{\lambda(1-\lambda)} & : \ t = 1\,, \end{cases}$$

such that the curve $\left\{\operatorname{smb}_{\widetilde{\mathbf{S}}}(t,\lambda) : t \in \mathbb{T},\ \lambda \in [0,1]\right\}$ equals the lower unit circle (from 1 to -1 and back) and has winding number zero. Taking into account the special structure of $\widetilde{\mathbf{S}} = \mathbf{T}(\phi) + \mathbf{H}(\phi)$, we get the invertibility of $\widetilde{\mathbf{S}} : \ell^2 \longrightarrow \ell^2$ (cf. [1, Cor. 2.6]).

Also, the fourth limit operators

$$\mathcal{W}^4(\mathcal{A}_n^\sigma) = -\mathbf{S}^\sigma + \mathbf{A}^\sigma = -\mathbf{T}(\phi) + \mathbf{H}(\phi) + \mathbf{A}^\sigma$$

and

$$\mathcal{W}^4(\mathcal{A}_n^\nu) = -\mathbf{S}^\nu + \mathbf{A}^\nu = -\mathbf{T}(\phi) + \mathbf{H}(\psi\phi) + \mathbf{A}^\nu\,,$$

where $\psi(t) = t^{-1}$, belong to $\operatorname{alg}\mathcal{T}(\mathbf{PC})$ and, by (2.21), (2.22), and Lemma 2.10, their symbols are given by

$$\operatorname{smb}_{\mathcal{W}^4(\mathcal{A}_n^\sigma)}(t,\lambda) = \begin{cases} -1 & : \ \operatorname{Im} t > 0\,, \\ 2\lambda - 1 - 2\mathbf{i}\sqrt{\lambda(1-\lambda)} & : \ t = -1\,, \\ 1 & : \ \operatorname{Im} t < 0\,, \\ 1 - 2\lambda - 2\mathbf{i}\sqrt{\lambda(1-\lambda)} + \widehat{\mathbf{y}}\left(\frac{1}{2} + \frac{\mathbf{i}}{2\pi}\ln\frac{\lambda}{1-\lambda}\right) & : \ t = 1\,, \end{cases}$$

and

$$\operatorname{smb}_{\mathcal{W}^4(\mathcal{A}_n^\nu)}(t,\lambda) = \begin{cases} -1 & : \ \operatorname{Im} t > 0\,, \\ 2\lambda - 1 + 2\mathbf{i}\sqrt{\lambda(1-\lambda)} & : \ t = -1\,, \\ 1 & : \ \operatorname{Im} t < 0\,, \\ 1 - 2\lambda - 2\mathbf{i}\sqrt{\lambda(1-\lambda)} + \widehat{\mathbf{y}}\left(\frac{1}{2} + \frac{\mathbf{i}}{2\pi}\ln\frac{\lambda}{1-\lambda}\right) & : \ t = 1\,, \end{cases}$$

where $\mathbf{y}(t) = 2\,\mathbf{h}(t^2)$. Taking into account (2.36), (2.37), and (4.2), we get, for $\operatorname{Im} z = \frac{1}{2}$,

$$\widehat{\mathbf{y}}(z) = \frac{-1 + 6\,\frac{z+1}{2} - 4\,\frac{1}{2}\,\frac{z+3}{2}\,\frac{z+1}{2}}{\mathbf{i}\sin\frac{\pi(z+1)}{2}} - \frac{2}{\mathbf{i}\sin(\pi z)} = \frac{(z+1)(3-z) - 2}{2\,\mathbf{i}\sin\frac{\pi(z+1)}{2}} - \frac{2}{\mathbf{i}\sin(\pi z)}\,.$$

From the following pictures we can see that $\mathcal{W}^4(\mathcal{A}_n^\sigma)$ and $\mathcal{W}^4(\mathcal{A}_n^\nu)$ are Fredholm operators with indices $\operatorname{ind}\mathcal{W}^4(\mathcal{A}_n^\sigma) = 0$ and $\operatorname{ind}\mathcal{W}^4(\mathcal{A}_n^\nu) = 1$.

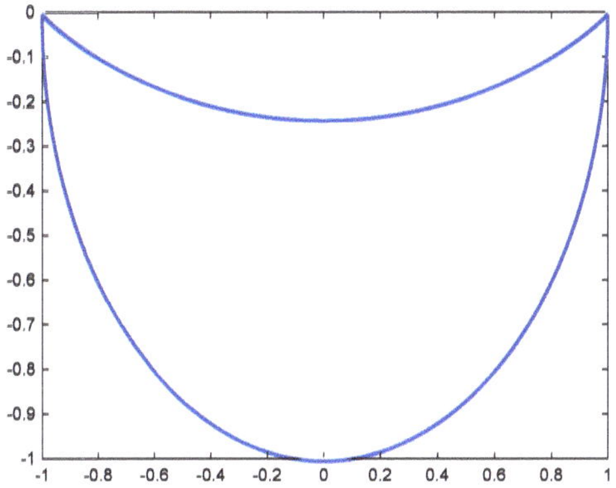

$$\left\{ \mathrm{smb}_{\mathcal{W}^4(\mathcal{A}_n^\sigma)}(t,\lambda) : (t \in \mathbb{T}, \lambda \in [0,1] \right\}$$

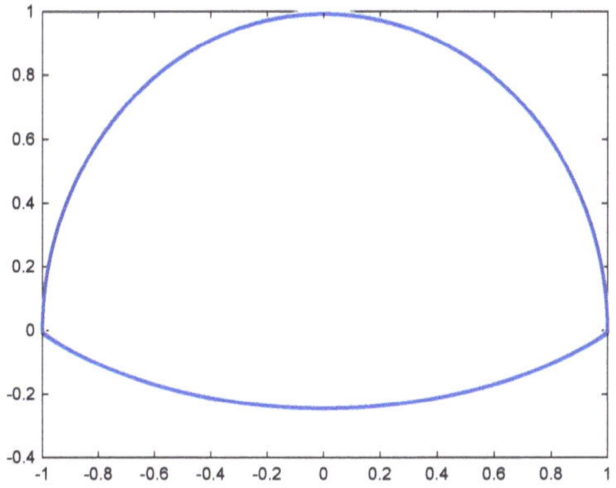

$$\left\{ \mathrm{smb}_{\mathcal{W}^4(\mathcal{A}_n^\nu)}(t,\lambda) : (t \in \mathbb{T}, \lambda \in [0,1] \right\}$$

In the following Tables 1 and 3 we present the numerical results obtained by applying the collocation-quadrature method w.r.t. the Chebyshev nodes of first and third kind to equation (4.1). The respective system of linear equations is solved with the Krylov subspace method CGNR, where the system matrix and the right-hand side are given by (cf. the definition on $\mathcal{V}_n^\tau$ and (2.32))

$$\mathbb{A}_n := \left[\sqrt{\frac{1 - x_{jn}^\tau}{1 - x_{kn}^\tau}} \left(\mathcal{S}\widetilde{\ell}_{kn}^\tau + \mathcal{H}_n^\tau \widetilde{\ell}_{kn}^\tau \right) (x_{jn}^\tau) \right]_{j,k=1}^n$$

and

$$\left[-\mathbf{i}\sqrt{1 - x_{jn}^\tau}\, \widetilde{f}(x_{jn}^\tau) \right]_{j=1}^n ,$$

respectively. The number M denotes the number of CGNR-iterations needed to obtain a residual norm which is smaller than 10^{-12} times the initial residual norm.

n	M	cond$(\mathbb{A}_n)$	$k_1(1)$	$d(-1)$	s_1	s_2	s_3	s_n
16	9	2.5434	1.12095644	1.45002768	0.3982	0.6794	0.8751	1.0127
32	11	2.7936	1.12138101	1.45297370	0.3625	0.6017	0.7977	1.0127
64	12	2.9985	1.12148696	1.45390311	0.3377	0.5422	0.7288	1.0127
128	13	3.1668	1.12151343	1.45418327	0.3198	0.4961	0.6698	1.0127
256	14	3.3061	1.12152005	1.45426521	0.3063	0.4596	0.6199	1.0127
512	15	3.4221	1.12152170	1.45428868	0.2959	0.4303	0.5776	1.0127
1024	16	3.5194	1.12152212	1.45429529	0.2877	0.4064	0.5416	1.0127
2048	17	3.6016	1.12152222	1.45429712	0.2812	0.3867	0.5109	1.0127
4096	18	3.6716	1.12152225	1.45429763	0.2758	0.3702	0.4845	1.0127
8192	19	3.7315	1.12152225	1.45429777	0.2714	0.3563	0.4616	1.0127
16384	19	3.7831	1.12152225	1.45429781	0.2677	0.3444	0.4417	1.0127

TABLE 1. Collocation-quadrature (2.4) for (4.1) with $\widetilde{f}(x) = \sqrt{1+x}$ and $\tau = \sigma$

n	M	cond$(\mathbb{A}_n)$	$k_1(1)$	$d(-1)$
8	8	2.37	1.12150794	1.45679430
16	10	2.65	1.12152262	1.45492880
32	11	2.88	1.12152227	1.45445562
64	12	3.06	1.12152226	1.45433725
128	13	3.22	1.12152226	1.45430768
256	14	3.34	1.12152226	1.45430028
512	15	3.45	1.12152226	1.45429844
1024	16	3.53	1.12152226	1.45429797
2048	17	3.61	1.12152226	1.45429786

TABLE 2. Collocation $(\mathcal{M}_n^\sigma(\mathcal{S}+\mathcal{B})\mathcal{L}_n)$ for (4.1) with $\widetilde{f}(x) = \sqrt{1+x}$ ([4])

The numerical value of $\widetilde{k}_1(1)$ is calculated by interpolating the computed approximate values of $u(x)\sqrt{1-x}$ at x_{1n}^τ, x_{2n}^τ, and x_{3n}^τ. For $d(-1)$ we use

$$d(-1) \approx -\frac{1}{2}\sum_{k=1}^{n} \frac{\lambda_{kn}^\tau u(x_{kn}^\tau)}{\tau(x_{kn}^\tau)\sqrt{1+x_{kn}^\tau}}\,.$$

In the last four columns of these tables one can see the three smallest singular values and the greatest singular value of $\mathbb{A}_n$. Unfortunately, we only know that $\mathcal{W}^4(\mathcal{A}_n^\sigma)$ is Fredholm with index zero, but we are not able to show that the null space of this operator is trivial. Also the sequence of the condition numbers in Table 1 do not surely confirm stability, it seems to be a weak instability. In comparison to Table 1, in Table 2 we recall the results, already obtained in [4, p. 114, 2nd Table]

n	M	$\mathrm{cond}(\mathbb{A}_n)$	$k_1(1)$	$d(-1)$	s_1	s_2	s_3	s_n
16	10	7.7444	1.12180201	1.50072897	0.1292	0.5031	0.7767	1.0002
32	11	10.9193	1.12159404	1.47739741	0.0916	0.4404	0.6909	1.0003
64	13	15.4050	1.12154047	1.46578686	0.0649	0.3966	0.6210	1.0003
128	14	21.7530	1.12152684	1.46001884	0.0460	0.3651	0.5646	1.0004
256	15	30.7369	1.12152341	1.45715037	0.0325	0.3418	0.5190	1.0004
512	16	43.4485	1.12152254	1.45572158	0.0230	0.3242	0.4818	1.0004
1024	17	61.4308	1.12152223	1.45500894	0.0163	0.3106	0.4512	1.0004
2048	18	86.8656	1.12152227	1.45465316	0.0115	0.2999	0.4257	1.0004
4096	19	122.8389	1.12152226	1.45447543	0.0081	0.2913	0.4043	1.0004
8192	20	173.7150	1.12152226	1.45438660	0.0058	0.2843	0.3863	1.0004
16384	22	245.6662	1.12152226	1.45434221	0.0041	0.2786	0.3708	1.0004

TABLE 3. Collocation-quadrature (2.4) for (4.1) with $\widetilde{f}(x) = \sqrt{1+x}$ and $\tau = \nu$

for the collocation method $(\mathcal{M}_n^\sigma(\mathcal{S}+\mathcal{B})\mathcal{L}_n)$, for which a stability result is not known.

In Table 3 we can observe the instability of the collocation-quadrature method w.r.t. the Chebyshev nodes of third kind and the 1-splitting property of the singular values, which means that the smallest singular value tends to zero and the others stay away from zero for n tending to infinity. For this, let us mention that, as in [14, Section 8], one can prove that the number of singular values tending to zero equals the sum of the kernel dimensions of the limit operators $\mathcal{W}^t\left(\mathcal{A}_n^\nu\right)$. Since $\mathcal{W}^1\left(\mathcal{A}_n^\nu\right)$ and $\mathcal{W}^3\left(\mathcal{A}_n^\nu\right)$ are invertible and since $\dim\ker\mathcal{W}^2\left(\mathcal{A}_n^\nu\right) = 0$, we conclude that $\dim\ker\mathcal{W}^4\left(\mathcal{A}_n^\nu\right) = 1$ seems to be true.

5. Proof of Lemma 2.10

Let us start with some properties of the C^*-algebra $\operatorname{alg}\mathcal{T}(\mathbf{PC})$.

Lemma 5.1 ([18], Lemma 11.4). *Let the generating function* $g(t) = \sum_{m=-\infty}^{\infty} g_m t^m$, $t\in\mathbb{T}$, *of the Toeplitz operator* $\mathbf{T}(g) = \left[\ g_{j-k}\ \right]_{j,k=0}^\infty$ *be piecewise continuous on* $\mathbb{T}$ *and continuous on* $\mathbb{T}\setminus\{\pm 1\}$. *Moreover, for* $z\in\mathbb{C}$, *set* $\mathbf{\Lambda}^z = \left[\ (j+1)^z\delta_{j,k}\ \right]_{j,k=0}^\infty$ *with* $(j+1)^z = e^{z\ln(j+1)}$. *Then,*

(a) *for* $-\frac{1}{2} < \operatorname{Re} z < \frac{1}{2}$, *the operator* $\mathbf{\Lambda}^{-z}\mathbf{T}(g)\mathbf{\Lambda}^z$ *belongs to* $\operatorname{alg}\mathcal{T}(\mathbf{PC})$, *and its symbol is given by*

$$\mathrm{smb}_{\mathbf{\Lambda}^{-z}\mathbf{T}(g)\mathbf{\Lambda}^z}(t,\lambda) = \begin{cases} g(t) & : \quad t\in\mathbb{T}\setminus\{\pm 1\}\,, \\ \dfrac{\lambda g(t+0) + (1-\lambda)g(t-0)e^{-2\pi\mathbf{i}z}}{\lambda + (1-\lambda)e^{-2\pi\mathbf{i}z}} & : \quad t=\pm 1\,, \end{cases}$$

(b) *the function* $\left\{z \in \mathbb{C} : -\frac{1}{2} < \operatorname{Re} z < \frac{1}{2}\right\} \longrightarrow \operatorname{alg} \mathcal{T}(\mathbf{PC})$, $z \mapsto \mathbf{\Lambda}^{-z}\mathbf{T}(g)\mathbf{\Lambda}^{z}$ *is continuous if the function* $(0,1) \longrightarrow \mathbb{C}$, $x \mapsto \widetilde{g}(x) := g\left(e^{2\pi \mathrm{i} x}\right)$ *is twice differentiable on* $(0,1)$ *and if the finite limits* $\lim_{x\to+0} \widetilde{g}''(x)$ *and* $\lim_{x\to 1-0} \widetilde{g}''(x)$ *exist,*

(c) *for arbitrary* $\mathbf{D} \in \operatorname{alg} \mathcal{T}(\mathbf{PC})$, *the function* $\mathbb{R} \longrightarrow \operatorname{alg} \mathcal{T}(\mathbf{PC})$, $\xi \mapsto \mathbf{\Lambda}^{-\mathrm{i}\xi}\mathbf{D}\mathbf{\Lambda}^{\mathrm{i}\xi}$ *is continuous.*

Corollary 5.2. *For* $s > 0$ *and* $-\frac{1}{2} < \operatorname{Re} z < \frac{1}{2}$, *set* $\mathbf{\Lambda}_s^z = \left[\ (j+s)^z \delta_{j,k}\ \right]_{j,k=0}^{\infty}$ *with* $(j+s)^z = e^{z \ln(j+s)}$. *Then, for* $g(t)$ *as in Lemma* 5.1, $\mathbf{\Lambda}_s^{-z}\mathbf{T}(g)\mathbf{\Lambda}_s^z \in \operatorname{alg} \mathcal{T}(\mathbf{PC})$ *and* $\operatorname{smb}_{\mathbf{\Lambda}_s^{-z}\mathbf{T}(g)\mathbf{\Lambda}_s^z}(t,\lambda) = \operatorname{smb}_{\mathbf{\Lambda}^{-z}\mathbf{T}(g)\mathbf{\Lambda}^z}(t,\lambda)$. *Moreover, assertion* (b) *of Lemma* 5.1 *remains true if* $\mathbf{\Lambda}^z$ *is replaced by* $\mathbf{\Lambda}_s^z$.

Proof. Note, that the ideal $\mathcal{K}(\ell^2)$ of compact operators in ℓ^2 is contained in $\operatorname{alg} \mathcal{T}(\mathbf{PC})$ and that, due to the closedness of $\mathcal{K}(\ell^2)$ in $\mathcal{L}(\ell^2)$, a diagonal operator $\left[\ a_j \delta_{j,k}\ \right]_{j,k=0}^{\infty} : \ell^2 \longrightarrow \ell^2$ belongs to $\mathcal{K}(\ell^2)$ if $\lim_{j\to\infty} a_j = 0$. Hence, the difference

$$\begin{aligned}
&\mathbf{\Lambda}_s^{-z}\mathbf{T}(g)\mathbf{\Lambda}_s^z - \mathbf{\Lambda}^{-z}\mathbf{T}(g)\mathbf{\Lambda}^z \\
&= \mathbf{\Lambda}^{-z}\mathbf{T}(g)\mathbf{\Lambda}^z \left[\ \frac{(j+s)^z - (j+1)^z}{(j+1)^z}\, \delta_{j,k}\ \right]_{j,k=0}^{\infty} \\
&\quad - \left[\ \frac{(k+1)^{-z} - (k+s)^{-z}}{(k+1)^{-z}}\, \delta_{j,k}\ \right]_{j,k=0}^{\infty} \mathbf{\Lambda}^{-z}\mathbf{T}(g)\mathbf{\Lambda}^z \left[\ \left(\frac{j+s}{j+1}\right)^z \delta_{j,k}\ \right]_{j,k=0}^{\infty}
\end{aligned}$$

is compact in ℓ^2, and the first assertion follows from the property of the symbol map being identically zero for compact operators. Finally, since

$$\mathbf{\Lambda}_s^{-z}\mathbf{T}(g)\mathbf{\Lambda}_s^z = \left[\ \left(\frac{j+s}{j+1}\right)^{-z} \delta_{j,k}\ \right]_{j,k=0}^{\infty} \mathbf{\Lambda}^{-z}\mathbf{T}(g)\mathbf{\Lambda}^z \left[\ \left(\frac{j+s}{j+1}\right)^{z} \delta_{j,k}\ \right]_{j,k=0}^{\infty}$$

and since the diagonal operators on the right-hand side of the last equation depend continuously (in operator norm) on $z \in \mathbb{C}$, the second assertion follows from Lemma 5.1,(b) and the continuity of the multiplication in $\operatorname{alg} \mathcal{T}(\mathbf{PC})$. □

Let $\xi \in \mathbb{R} \setminus \{0\}$ and $0 < \mu < 1$ be fixed and consider

$$\mathbf{y}_\xi(x) = \frac{\mu\, x^{\frac{\mu-1}{2}}}{\pi \mathbf{i}\,(1 - x^\mu)} \left(x^{-\mathrm{i}\xi} - 1\right), \quad 0 < x < \infty. \tag{5.1}$$

Note that $\mathbf{y}_\xi : (0,\infty) \longrightarrow \mathbb{C}$ is a continuous function, where $\mathbf{y}_\xi(1) = \pi^{-1}\xi$, and that $\mathbf{y}_\xi \in \mathbf{L}^p_{\frac{1}{2}}$ for all $p \in [1,\infty)$. Taking into account ([6, Chapter VI, 6.2.(4)])

$$\text{v.p.} \int_0^\infty \frac{x^{z-1}\,dx}{1-x} = \lim_{\varepsilon\to+0} \left(\int_0^{1-\varepsilon} \frac{x^{z-1}\,dx}{1-x} + \int_{1+\varepsilon}^\infty \frac{x^{z-1}\,dx}{1-x} \right) = \pi \cot(\pi z),$$

$0 < \operatorname{Re} z < 1$, we get

$$\begin{aligned}\widehat{\mathbf{y}}_\xi(z) &= \lim_{R\to\infty}\lim_{\varepsilon\to+0}\left(\int_{\frac{1}{R}}^{1-\varepsilon} x^{z-1}\mathbf{y}_\xi(x)\,dx + \int_{1+\varepsilon}^{R} x^{z-1}\mathbf{y}_\xi(x)\,dx\right)\\ &= \text{v.p.}\,\frac{\mu}{\pi\mathbf{i}}\int_0^\infty \frac{x^{z-\mathbf{i}\xi-1+\frac{\mu-1}{2}}}{1-x^\mu}\,dx - \text{v.p.}\,\frac{\mu}{\pi\mathbf{i}}\int_0^\infty \frac{x^{z-1+\frac{\mu-1}{2}}}{1-x^\mu}\,dx\\ &= \text{v.p.}\frac{1}{\pi\mathbf{i}}\int_0^\infty \frac{y^{\frac{z-\mathbf{i}\xi}{\mu}-1+\frac{1}{2}-\frac{1}{2\mu}}}{1-y}\,dy - \text{v.p.}\frac{1}{\pi\mathbf{i}}\int_0^\infty \frac{y^{\frac{z}{\mu}-1+\frac{1}{2}-\frac{1}{2\mu}}}{1-y}\,dy\,,\end{aligned}$$

i.e., for $\frac{1-\mu}{2} < \operatorname{Re} z < \frac{1+\mu}{2}$,

$$\widehat{\mathbf{y}}_\xi(z) = \mathbf{i}\left[\cot\left(\pi\frac{z+\frac{\mu-1}{2}}{\mu}\right) - \cot\left(\pi\frac{z-\mathbf{i}\xi+\frac{\mu-1}{2}}{\mu}\right)\right] \tag{5.2}$$

This Mellin transform is a meromorphic function on $\mathbb{C}$ with the (isolated) singularities (which are pols of order 1) at

$$z_{n,1} = \mathbf{i}\xi + \mu n - \frac{\mu-1}{2} \quad\text{and}\quad z_{n,2} = \mu n - \frac{\mu-1}{2}\,, \qquad n\in\mathbb{Z}\,, \tag{5.3}$$

where

$$\operatorname{res}_{z_{n,1}}\widehat{\mathbf{y}}_\xi = \frac{\mu}{\pi\mathbf{i}} \quad\text{and}\quad \operatorname{res}_{z_{n,2}}\widehat{\mathbf{y}}_\xi = -\frac{\mu}{\pi\mathbf{i}}\,, \qquad n\in\mathbb{Z}\,. \tag{5.4}$$

From $\sin z = \frac{e^{\mathbf{i}z}-e^{-\mathbf{i}z}}{2\mathrm{i}}$ we conclude, in case $\operatorname{Im} z > 0$,

$$\frac{1}{|\sin z|} = \frac{2\,e^{\operatorname{Im} z}}{|e^{2\mathrm{i}\operatorname{Re} z} - e^{2\operatorname{Im} z}|} \le \frac{2\,e^{\operatorname{Im} z}}{e^{2\operatorname{Im} z}-1}$$

and, in case $\operatorname{Im} z < 0$,

$$\frac{1}{|\sin z|} = \frac{2\,e^{-\operatorname{Im} z}}{|e^{-2\mathrm{i}\operatorname{Re} z} - e^{-2\operatorname{Im} z}|} \le \frac{2\,e^{-\operatorname{Im} z}}{e^{-2\operatorname{Im} z}-1}\,,$$

i.e.,

$$\frac{1}{|\sin z|} \le 4\,e^{-|\operatorname{Im} z|} \quad\text{for}\quad |\operatorname{Im} z| \ge \ln\sqrt{2}\,. \tag{5.5}$$

Since, due to (5.2), $|\widehat{\mathbf{y}}_\xi(z)| \le \int_0^1 |g'(t)|\,dt$ with

$$g(t) = \cot\left(\pi\,\frac{z-\mathbf{i}t\xi+\frac{\mu-1}{2}}{\mu}\right) \quad\text{and}\quad g'(t) = \frac{\pi\mathbf{i}\xi}{\mu\left[\sin\left(\pi\frac{z-\mathbf{i}t\xi+\frac{\mu-1}{2}}{\mu}\right)\right]^2}\,,$$

we conclude from (5.5)

$$|g'(t)| \le \frac{16\pi|\xi|}{\mu}\,e^{-\frac{2\pi}{\mu}|\operatorname{Im} z - t\xi|} \le \frac{16\pi|\xi|}{\mu}\,e^{\frac{2\pi|\xi|}{\mu}}\,e^{-\frac{2\pi|\operatorname{Im} z|}{\mu}}\,, \qquad |\operatorname{Im} z| \ge \ln\sqrt{2}+|\xi|\,,$$

and, consequently,

$$|\widehat{\mathbf{y}}_\xi(z)| \le \frac{16\pi|\xi|}{\mu}\,e^{\frac{2\pi|\xi|}{\mu}}\,e^{-\frac{2\pi|\operatorname{Im} z|}{\mu}}\,, \qquad |\operatorname{Im} z| \ge \ln\sqrt{2}+|\xi|\,. \tag{5.6}$$

Thus, due to Lemma 2.8, we have

$$\mathbf{y}_\xi(x) = \frac{1}{2\pi\mathbf{i}} \int_{\mathrm{Re}\, z=\frac{1}{2}} x^{-z} \widehat{\mathbf{y}}_\xi(z)\, dz\,, \quad 0 < x < \infty\,. \tag{5.7}$$

This can also be verified by direct computations. For this, let $R > 0$ and $x > 0$ and consider

$$\begin{aligned}
&\frac{1}{2\pi\mathbf{i}} \int_{\frac{1}{2}-\mathbf{i}R}^{\frac{1}{2}+\mathbf{i}R} x^{-z} \widehat{y}_\xi(z)\, dz \\
&= \frac{1}{2\pi} \int_{\frac{1}{2}-\mathbf{i}R}^{\frac{1}{2}+\mathbf{i}R} x^{-z} \left[\cot\left(\pi \frac{z + \frac{\mu-1}{2}}{\mu}\right) - \cot\left(\pi \frac{z - \mathbf{i}\xi + \frac{\mu-1}{2}}{\mu}\right)\right] dz \\
&= \frac{\mathbf{i}}{2\pi} \int_{-R}^{R} x^{-\frac{1}{2}-\mathbf{i}s} \left[\cot\left(\frac{\pi\mathbf{i}s}{\mu} + \frac{\pi}{2}\right) - \cot\left(\frac{\pi\mathbf{i}(s-\xi)}{\mu} + \frac{\pi}{2}\right)\right] ds \\
&= \frac{1}{2\pi\mathbf{i}\sqrt{x}} \int_{-R}^{R} x^{-\mathbf{i}s} \left[\tan\frac{\pi\mathbf{i}s}{\mu} - \tan\frac{\pi\mathbf{i}(s-\xi)}{\mu}\right] ds \\
&= \frac{1}{2\pi\sqrt{x}} \int_{-R}^{R} x^{-\mathbf{i}s} \left[\frac{e^{\frac{2\pi s}{\mu}} - 1}{e^{\frac{2\pi s}{\mu}} + 1} - \frac{e^{\frac{2\pi(s-\xi)}{\mu}} - 1}{e^{\frac{2\pi(s-\xi)}{\mu}} + 1}\right] ds \\
&= \frac{\mu}{4\pi^2\sqrt{x}} \int_{e^{-\frac{2\pi R}{\mu}}}^{e^{\frac{2\pi R}{\mu}}} t^{-\frac{\mathbf{i}\mu\ln x}{2\pi}} \left(\frac{t-1}{t+1} - \frac{t\, e^{-\frac{2\pi\xi}{\mu}} - 1}{t\, e^{-\frac{2\pi\xi}{\mu}} + 1}\right) \frac{dt}{t} \\
&= \frac{\mu\left(e^{\frac{2\pi\xi}{\mu}} - 1\right)}{2\pi^2\sqrt{x}} \int_{e^{-\frac{2\pi R}{\mu}}}^{e^{\frac{2\pi R}{\mu}}} \frac{t^{1-\frac{\mathbf{i}\mu\ln x}{2\pi}}}{\left(t + e^{\frac{2\pi\xi}{\mu}}\right)(t+1)} \frac{dt}{t}\,.
\end{aligned}$$

For $\alpha > 0$, $\alpha \neq 1$, and $0 < \mathrm{Re}\, z < 2$, $z \neq 1$, we have (see [6, Chapter VI, 6.2.(7)])

$$\int_0^\infty \frac{t^z}{(t+\alpha)(t+1)} \frac{dt}{t} = \frac{\pi}{\sin(\pi z)} \frac{\alpha^{z-1} - 1}{1-\alpha}\,.$$

Consequently, for $x > 0$ and $x \neq 1$,

$$\begin{aligned}
&\lim_{R\to\infty} \frac{1}{2\pi\mathbf{i}} \int_{\frac{1}{2}-\mathbf{i}R}^{\frac{1}{2}+\mathbf{i}R} x^{-z} \widehat{y}_\xi(z)\, dz \\
&= -\frac{\mu}{2\pi\sqrt{x}} \frac{e^{-\mathbf{i}\xi\ln x} - 1}{\sin\left(\pi - \frac{\mathbf{i}\mu\ln x}{2}\right)} = -\frac{\mu}{2\pi\sqrt{x}} \frac{x^{-\mathbf{i}\xi} - 1}{\sin\frac{\mathbf{i}\mu\ln x}{2}} \\
&= \frac{\mu}{\pi\mathbf{i}\sqrt{x}} \frac{x^{-\mathbf{i}\xi} - 1}{x^{-\frac{\mu}{2}} - x^{\frac{\mu}{2}}} = \frac{\mu\, x^{\frac{\mu-1}{2}}\left(x^{-\mathbf{i}\xi} - 1\right)}{\pi\mathbf{i}\,(1 - x^\mu)} = \mathbf{y}_\xi(x)\,.
\end{aligned}$$

For $x = 1$, we can use continuity arguments and we again arrive at (5.7).

Taking into account (5.6), we can apply (two times) the residue theorem together with (5.3) and (5.4), and obtain, for $0 < \mu < 1$ such that $\frac{\mu-1}{2\mu}$ and $\frac{\mu+1}{2\mu}$

are not integer,

$$\mathbf{y}_\xi(x) = \frac{1}{2\pi\mathbf{i}} \int_{\operatorname{Re} z=0} x^{-z} \widehat{\mathbf{y}}_\xi(z)\, dz + \sum_{n\in\mathbb{Z}:0<\mu\, n-\frac{\mu-1}{2}<\frac{1}{2}} \frac{\mu}{\pi\mathbf{i}} \left(x^{-\mathbf{i}\xi} - 1\right) x^{\frac{\mu-1}{2}-\mu\, n}$$

and

$$\mathbf{y}_\xi(x) = \frac{1}{2\pi\mathbf{i}} \int_{\operatorname{Re} z=1} x^{-z} \widehat{\mathbf{y}}_\xi(z)\, dz - \sum_{n\in\mathbb{Z}:\frac{1}{2}<\mu\, n-\frac{\mu-1}{2}<1} \frac{\mu}{\pi\mathbf{i}} \left(x^{-\mathbf{i}\xi} - 1\right) x^{\frac{\mu-1}{2}-\mu\, n}$$

or

$$\mathbf{y}_\xi(x)x = \frac{1}{2\pi\mathbf{i}} \int_{\operatorname{Re} z=0} x^{-z} \widehat{\mathbf{y}}_\xi(z+1)\, dz - \sum_{n\in\mathbb{Z}:\frac{1}{2}<\mu\, n-\frac{\mu-1}{2}<1} \frac{\mu}{\pi\mathbf{i}} \left(x^{-\mathbf{i}\xi} - 1\right) x^{\frac{\mu+1}{2}-\mu\, n}.$$

It follows, for $x > 0$,

$$\begin{aligned}
&\mathbf{y}_\xi(x)(1-x)\\
&= \frac{1}{2\pi\mathbf{i}} \int_{\operatorname{Re} z=0} x^{-z} \left[\widehat{\mathbf{y}}_\xi(z) - \widehat{\mathbf{y}}_\xi(z+1)\right] dz\\
&\quad + \frac{\mu}{\pi\mathbf{i}} \left(x^{-\mathbf{i}\xi} - 1\right) \left[\sum_{n\in\mathbb{Z}:\frac{\mu-1}{2\mu}<n<\frac{1}{2}} x^{\frac{\mu-1}{2}-\mu\, n} + \sum_{n\in\mathbb{Z}:\frac{1}{2}<n<\frac{\mu+1}{2\mu}} x^{\frac{\mu+1}{2\mu}-\mu\, n} \right]\\
&= \frac{1}{2\pi\mathbf{i}} \int_{\operatorname{Re} z=0} x^{-z} \left[\widehat{\mathbf{y}}_\xi(z) - \widehat{\mathbf{y}}_\xi(z+1)\right] dz\\
&\quad + \frac{\mu}{\pi\mathbf{i}} \left(x^{-\mathbf{i}\xi} - 1\right) \sum_{n\in\mathbb{Z}:\frac{\mu-1}{2\mu}<n<\frac{\mu+1}{2\mu}} x^{-z_n},
\end{aligned} \tag{5.8}$$

where

$$z_n = \begin{cases} \mu\, n - \frac{\mu-1}{2} & : \quad n < \frac{1}{2}, \\ \mu\, n - \frac{\mu+1}{2} & : \quad n > \frac{1}{2}. \end{cases}$$

In particular, for $x = 1$,

$$0 = \frac{1}{2\pi\mathbf{i}} \int_{\operatorname{Re} z=0} \left[\widehat{\mathbf{y}}_\xi(z) - \widehat{\mathbf{y}}_\xi(z+1)\right] dz. \tag{5.9}$$

For $s > 0$, define

$$\begin{aligned}
\mathbf{M}_\xi^s &:= \left[\mathbf{y}_\xi\left(\frac{j+s}{k+s}\right) \frac{1-\delta_{j,k}}{k+s} \right]_{j,k=0}^{\infty}\\
&= \left[\left(1 - \frac{j+s}{k+s}\right) \mathbf{y}_\xi\left(\frac{j+s}{k+s}\right) \frac{1-\delta_{j,k}}{k-j} \right]_{j,k=0}^{\infty},
\end{aligned} \tag{5.10}$$

where the entries on the main diagonal are zero by definition. From (5.8), we get that $\mathbf{M}_\xi^s$ equals

$$\begin{aligned} &\left[\frac{1}{2\pi\mathbf{i}} \int_{\mathrm{Re}\, z=0} \left(\frac{j+s}{k+s}\right)^{-z} [\widehat{\mathbf{y}}_\xi(z) - \widehat{\mathbf{y}}_\xi(z+1)]\, dz\, \frac{1-\delta_{j,k}}{k-j} \right]_{j,k=0}^{\infty} \\ &+ \frac{\mu}{\pi\mathbf{i}} \sum_{n\in\mathbb{Z}:\frac{\mu-1}{2\mu}<n<\frac{\mu+1}{2\mu}} \left[\left\{ \left(\frac{j+s}{k+s}\right)^{-\mathbf{i}\xi} - 1 \right\} \left(\frac{j+s}{k+s}\right)^{-z_n} \frac{1-\delta_{j,k}}{k-j} \right]_{j,k=0}^{\infty} . \end{aligned} \tag{5.11}$$

Obviously,

$$\frac{1}{\pi\mathbf{i}} \left[\frac{1-\delta_{j,k}}{k-j} \right]_{j,k=0}^{\infty} = \mathbf{T}(f) \in \mathrm{alg}\,\mathcal{T}(\mathbf{PC}) \tag{5.12}$$

with $f\left(e^{2\pi\mathbf{i}x}\right) = 2x-1$, $0 \le x < 1$. In what follows, we use the concept of Bochner's integral (cf. [21, Chapter V, Section 5]). If $\mathbf{X}$ is a Banach space, then a function $\mathbf{x} : \mathbb{R} \longrightarrow \mathbf{X}$ is called Bochner–Lebesgue-integrable, if there exists a sequence $(\mathbf{x}_n)_{n=0}^\infty$ of finitely-valued Lebesgue-measurable functions $\mathbf{x}_n : \mathbb{R} \longrightarrow \mathbf{X}$ such that $\mathbf{x}_n(\eta)$ converges in $\mathbf{X}$ to $\mathbf{x}(\eta)$ for almost all $\eta \in \mathbb{R}$ and

$$\lim_{n\to\infty} \int_{-\infty}^{\infty} \|\mathbf{x}_n(\eta) - \mathbf{x}(\eta)\|_{\mathbf{X}}\, d\eta = 0\,.$$

In that case, the Bochner–Lebesgue integral of $\mathbf{x}(\eta)$ is defined by

$$\int_{-\infty}^{\infty} \mathbf{x}(\eta)\, d\eta := \lim_{n\to\infty} \int_{-\infty}^{\infty} \mathbf{x}_n(\eta)\, d\eta \quad \text{in} \quad \mathbf{X}\,.$$

Thereby, a function $\mathbf{x} : \mathbb{R} \longrightarrow \mathbf{X}$ is called finitely-valued and measurable, if there exist $x_k \in \mathbf{X}$ and pairwise disjoint measurable sets $E_k \subset \mathbb{R}$ with Lebesgue measure $m(E_k) < \infty$, $k = 1, \ldots, m \in \mathbb{N}$, such that $\mathbf{x} = \sum_{k=1}^{m} x_k \chi_{E_k}$ with the characteristic functions χ_{E_k} of the sets E_k. In that case,

$$\int_{-\infty}^{\infty} \mathbf{x}(\eta)\, d\eta := \sum_{k=1}^{m} x_k m(E_k)\,.$$

Lemma 5.3 ([21], Chapter V, Section 5, Theorem 1). *If $\mathbf{x} : \mathbb{R} \longrightarrow \mathbf{X}$ is measurable, i.e., there exists an almost everywhere to $\mathbf{x}$ converging sequence of finitely-valued and measurable functions, then $\mathbf{x}$ is Bochner-integrable if and only if $\eta \mapsto \|\mathbf{x}(\eta)\|_{\mathbf{X}}$ is integrable.*

Corollary 5.4. *For the operator $\mathbf{M}_\xi^s : \ell^2 \longrightarrow \ell^2$ defined in (5.10), we have*

$$\begin{aligned} \mathbf{M}_\xi^s = {} & \frac{1}{2} \int_{\mathrm{Re}\, z=0} [\widehat{\mathbf{y}}_\xi(z) - \widehat{\mathbf{y}}_\xi(z+1)]\, \mathbf{\Lambda}_s^{-z} \mathbf{T}(f) \mathbf{\Lambda}_s^{z}\, dz \\ & + \mu \sum_{n\in\mathbb{Z}:\frac{\mu-1}{2\mu}<n<\frac{\mu+1}{2\mu}} \left[\mathbf{\Lambda}_s^{-z_n-\mathbf{i}\xi} \mathbf{T}(f) \mathbf{\Lambda}_s^{z_n+\mathbf{i}\xi} - \mathbf{\Lambda}_s^{-z_n} \mathbf{T}(f) \mathbf{\Lambda}_s^{z_n} \right] \end{aligned} \tag{5.13}$$

with the integral as a Bochner integral and with $\mathbf{T}(f)$ *defined in* (5.12). *Moreover,* $\mathbf{M}_\xi^s \in \operatorname{alg}\mathcal{T}(\mathbf{PC})$.

Proof. In virtue of Corollary 5.2, the function

$$\{\operatorname{Re} z = 0\} \longrightarrow \operatorname{alg}\mathcal{T}(\mathbf{PC}), \quad z \mapsto [\widehat{\mathbf{y}}(z) - \widehat{\mathbf{y}}(z+1)]\, \mathbf{\Lambda}_s^{-z}\mathbf{T}(f)\mathbf{\Lambda}_s^z$$

is continuous. Moreover, since, for $\mathbf{A} \in \mathcal{L}(\ell^2)$ and $\eta \in \mathbb{R}$,

$$\left\| \mathbf{\Lambda}_s^{-\mathrm{i}\eta}\mathbf{A}\mathbf{\Lambda}_s^{\mathrm{i}\eta}\right\|_{\mathcal{L}(\ell^2)} = \|\mathbf{A}\|_{\mathcal{L}(\ell^2)}\,, \tag{5.14}$$

the function $\mathbb{R} \longrightarrow \mathbb{R}$, $\eta \mapsto \left\| [\widehat{\mathbf{y}}_\xi(\mathrm{i}\eta) - \widehat{\mathbf{y}}_\xi(\mathrm{i}\eta+1)]\, \mathbf{\Lambda}_s^{-\mathrm{i}\eta}\mathbf{T}(f)\mathbf{\Lambda}_s^{\mathrm{i}\eta}\right\|_{\mathcal{L}(\ell^2)}$ is, due to (5.6), integrable on $\mathbb{R}$. Lemma 5.3 implies that the integral in (5.13) exists as a Bochner integral. Since the integral in (5.13) is the norm limit of a sequence of finite linear combinations of operators from $\mathcal{L}(\ell^2)$, it is easily seen that the entries of the matrix (5.11) coincide with $\left\langle \mathbf{e}_j, \mathbf{M}_\xi^s \mathbf{e}_k\right\rangle_{\ell^2}$ if for $\mathbf{M}_\xi^s$ formula (5.13) is used and where $\mathbf{e}_k = (\delta_{n,k})_{n=0}^\infty$. Finally, since $\operatorname{alg}\mathcal{T}(\mathbf{PC})$ is a closed subspace of the Banach space $\mathcal{L}(\ell^2)$ and since the integrands in (5.13) belong to $\operatorname{alg}\mathcal{T}(\mathbf{PC})$, we get $\mathbf{M}_\xi^s \in \operatorname{alg}\mathcal{T}(\mathbf{PC})$. □

Lemma 5.5 ([21], Chapter V, Section 5, Corollary 2). *If* $\mathcal{T} : \mathbf{X} \longrightarrow \mathbf{Y}$ *is a linear and bounded operator between the Banach spaces* $\mathbf{X}$ *and* $\mathbf{Y}$ *and if* $\mathbf{x} : \mathbb{R} \longrightarrow \mathbf{X}$ *is Bochner integrable, then* $\mathcal{T}\mathbf{x} : \mathbb{R} \longrightarrow \mathbf{Y}$ *is Bochner integrable and*

$$\int_{-\infty}^{\infty} (\mathcal{T}\mathbf{x})(\eta)\, d\eta = \mathcal{T}\int_{-\infty}^{\infty} \mathbf{x}(\eta)\, d\eta\,.$$

Let $a_z := \operatorname{smb}_{\mathbf{\Lambda}^{-z}\mathbf{T}(f)\mathbf{\Lambda}^z} = \operatorname{smb}_{\mathbf{\Lambda}_s^{-z}\mathbf{T}(f)\mathbf{\Lambda}_s^z}$ (cf. Corollary 5.2). Since $\operatorname{smb}_* : \operatorname{alg}\mathcal{T}(\mathbf{PC}) \longrightarrow \mathbf{C}(\mathbb{M})$, $\mathbf{T} \mapsto \operatorname{smb}_{\mathbf{T}}$ is linear and bounded (see (2.23)), Corollary 5.4 and Lemma 5.5 lead to

$$\begin{aligned}
\operatorname{smb}_{\mathbf{M}_\xi^s} &= \frac{1}{2}\int_{\operatorname{Re} z=0} [\widehat{\mathbf{y}}_\xi(z) - \widehat{\mathbf{y}}_\xi(z+1)]\, a_z\, dz \\
&\qquad + \mu \sum_{n\in\mathbb{Z}:\frac{\mu-1}{2\mu}<n<\frac{\mu+1}{2\mu}} \left[a_{z_n+\mathrm{i}\xi} - a_{z_n}\right] \\
&\overset{(5.6)}{=} \frac{1}{2}\left[\int_{\operatorname{Re} z=0} \widehat{\mathbf{y}}_\xi(z) a_z\, dz - \int_{\operatorname{Re} z=1} \widehat{\mathbf{y}}_\xi(z) a_z\, dz\right] \\
&\qquad + \mu \sum_{n\in\mathbb{Z}:\frac{\mu-1}{2\mu}<n<\frac{\mu+1}{2\mu}} \left[a_{z_n+\mathrm{i}\xi} - a_{z_n}\right].
\end{aligned}$$

Since, in virtue of Lemma 5.1(a),

$$a_z(t,\lambda) = \begin{cases} 2x-1 & : \quad 0<x<1,\, 0\le\lambda\le 1\,, \\[1ex] \dfrac{(1-\lambda)e^{-2\pi\mathrm{i}z} - \lambda}{(1-\lambda)e^{-2\pi\mathrm{i}z} + \lambda} =: b(z,\lambda) & : \quad x=0,\, 0\le\lambda\le 1\,, \end{cases}$$

where $t = e^{2\pi \mathbf{i} x}$, we get, for $t \in \mathbb{T} \setminus \{1\}$,

$$\mathrm{smb}_{\mathbf{M}_\xi^s}(t,\lambda) = \frac{2x-1}{2} \int_{\mathrm{Re}\, z=0} \left[\widehat{\mathbf{y}}_\xi(z) - \widehat{\mathbf{y}}_\xi(z+1)\right] dz \overset{(5.9)}{=} 0$$

and

$$\mathrm{smb}_{\mathbf{M}_\xi^s}(1,\lambda) = \frac{1}{2}\left[\int_{\mathrm{Re}\, z=0} - \int_{\mathrm{Re}\, z=1}\right] \widehat{\mathbf{y}}_\xi(z) b(z,\lambda)\, dz \\ + \mu \sum_{n\in\mathbb{Z}:\, \frac{\mu-1}{2\mu}<n<\frac{\mu+1}{2\mu}} \left[b(z_n + \mathbf{i}\xi,\lambda) - b(z_n,\lambda)\right].$$

In $\{0 \le \mathrm{Re}\, z \le 1\}$, the function $b(z,\lambda)$ has as singularity only the pole $z_* = \frac{1}{2} + \frac{\mathbf{i}}{2\pi} \ln \frac{\lambda}{1-\lambda}$ of order 1 with the residue $\mathrm{res}_{z_*} b = -\frac{1}{\pi \mathbf{i}}$. Consequently, by applying the residue theorem together with (5.6), we deliver

$$\begin{aligned}\mathrm{smb}_{\mathbf{M}_\xi^s}(1,\lambda) = & -\pi\mathbf{i}\left[\sum_{j=1,2:\, 0<z_{n,j}<1} \left(\mathrm{res}_{z_{n,j}} \widehat{\mathbf{y}}_\xi\right) b(z_{n,j},\lambda) + \left(\mathrm{res}_{z_*} b\right) \widehat{\mathbf{y}}_\xi(z_*)\right] \\ & + \mu \sum_{n\in\mathbb{Z}:\, \frac{\mu-1}{2\mu}<n<\frac{\mu+1}{2\mu}} \left[b(z_n+\mathbf{i}\xi,\lambda) - b(z_n,\lambda)\right] \\ \overset{(5.4)}{=} & -\mu \sum_{n\in\mathbb{Z}:0<\mu n - \frac{\mu-1}{2}<1} \left[b(z_{n,1},\lambda) - b(z_{n,2},\lambda)\right] + \widehat{\mathbf{y}}_\xi(z_*) \\ & + \mu \sum_{n\in\mathbb{Z}:\, \frac{\mu-1}{2\mu}<n<\frac{\mu+1}{2\mu}} \left[b(z_n+\mathbf{i}\xi,\lambda) - b(z_n,\lambda)\right] \\ = & \ \widehat{\mathbf{y}}_\xi(z_*)\,,\end{aligned}$$

where we took into account $z_{n,1} = z_n + \mathbf{i}\xi$ and $z_{n,2} = z_n$ if $n < \frac{1}{2}$, $z_{n,1} = z_n + 1 + \mathbf{i}\xi$ and $z_{n,2} = z_n + 1$ if $n > \frac{1}{2}$, as well as $b(z+1,\lambda) = b(z,\lambda)$. Summarizing we have, for $(t,\lambda) \in \mathbb{T} \times [0,1]$,

$$\mathrm{smb}_{\mathbf{M}_\xi^s}(t,\lambda) = \begin{cases} 0 & : \ t \in \mathbb{T}\setminus\{1\}\,, \\ \widehat{\mathbf{y}}_\xi\left(\frac{1}{2} + \frac{\mathbf{i}}{2\pi} \ln \frac{\lambda}{1-\lambda}\right) & : \ t = 1\,. \end{cases} \tag{5.15}$$

Now, assume that the continuous function $\mathbf{y} : (0,\infty) \longrightarrow \mathbb{C}$ satisfies the conditions of Lemma 2.10. Choose $\mu \in (0, \min\{\varepsilon, 1\})$ such that $\frac{\mu-1}{2\mu}$ and $\frac{\mu+1}{2\mu}$ are not integer. Then, due to Corollary 2.9 (in particular, (2.29) and (2.30)),

$$\begin{aligned}\mathbf{y}(x) &= \frac{x^{\frac{\mu}{2}}}{2\pi\mathbf{i}\,(1-x^\mu)} \int_{\mathrm{Re}\, z=0} x^{-\frac{1}{2}-z}\left[\widehat{\mathbf{y}}\left(\frac{1}{2}+z-\frac{\mu}{2}\right) - \widehat{\mathbf{y}}\left(\frac{1}{2}+z+\frac{\mu}{2}\right)\right] dz \\ &= \frac{1}{2\mu} \int_{\mathrm{Re}\, z=0} \frac{\mu\, x^{\frac{\mu-1}{2}}}{\pi\mathbf{i}\,(1-x^\mu)} \left(x^{-\mathbf{i}\,\mathrm{Im}\, z} - 1\right) \left[\widehat{\mathbf{y}}\left(\frac{1}{2}+z-\frac{\mu}{2}\right) - \widehat{\mathbf{y}}\left(\frac{1}{2}+z+\frac{\mu}{2}\right)\right] dz\,.\end{aligned}$$

Consequently,

$$\begin{aligned}\mathbf{M}_s^0 &:= \left[\ \mathbf{y}\left(\frac{j+s}{k+s}\right)\frac{1-\delta_{j,k}}{k+s}\ \right]_{j,k=0}^{\infty}\\ &= \frac{1}{2\mu}\left[\int_{\operatorname{Re} z=0}\mathbf{y}_{\operatorname{Im} z}\left(\frac{j+s}{k+s}\right)\frac{1-\delta_{j,k}}{k+s}\right.\\ &\qquad\left.\cdot\left[\widehat{\mathbf{y}}\left(\frac{1}{2}+z-\frac{\mu}{2}\right)-\widehat{\mathbf{y}}\left(\frac{1}{2}+z+\frac{\mu}{2}\right)\right]\,dz\right]_{j,k=0}^{\infty}\\ &= \frac{\mathbf{i}}{2\mu}\left[\int_{-\infty}^{\infty}\mathbf{y}_{\xi}\left(\frac{j+s}{k+s}\right)\frac{1-\delta_{j,k}}{k+s}\right.\\ &\qquad\left.\cdot\left[\widehat{\mathbf{y}}\left(\frac{1}{2}+\mathbf{i}\xi-\frac{\mu}{2}\right)-\widehat{\mathbf{y}}\left(\frac{1}{2}+\mathbf{i}\xi+\frac{\mu}{2}\right)\right]\,d\xi\right]_{j,k=0}^{\infty}\end{aligned}$$

(cf. (5.1)). Having in mind definition (5.10) of $\mathbf{M}_\xi^s$, we want to prove that

$$\mathbf{M}_s^0 = \frac{\mathbf{i}}{2\mu}\int_{-\infty}^{\infty}\mathbf{M}_\xi^s\left[\widehat{\mathbf{y}}\left(\frac{1}{2}+\mathbf{i}\xi-\frac{\mu}{2}\right)-\widehat{\mathbf{y}}\left(\frac{1}{2}+\mathbf{i}\xi+\frac{\mu}{2}\right)\right]\,d\xi \tag{5.16}$$

in the sense of a Bochner integral by using Lemma 5.3. For this, in virtue of (5.6), it is sufficient to show that the map $\mathbb{R}\longrightarrow \operatorname{alg}\mathcal{T}(\mathbf{PC})$, $\xi\mapsto\mathbf{M}_\xi^s$ is continuous and bounded. Indeed, this is a consequence of the representation (5.13) of $\mathbf{M}_\xi^s$. Firstly, the continuity and boundedness of

$$\mathbb{R}\longrightarrow \operatorname{alg}\mathcal{T}(\mathbf{PC})\,,\quad \xi\mapsto \sum_{n\in\mathbb{Z}:\,\frac{\mu-1}{2\mu}<n<\frac{\mu+1}{2\mu}}\left[\mathbf{\Lambda}_s^{-z_n-\mathbf{i}\xi}\mathbf{T}(f)\mathbf{\Lambda}_s^{z_n+\mathbf{i}\xi}-\mathbf{\Lambda}_s^{-z_n}\mathbf{T}(f)\mathbf{\Lambda}_s^{z_n}\right]$$

follows from Corollary 5.2 and (5.14). Secondly, due to (5.2) and the periodicity of $\cot z$ as well as (5.5), the integral in (5.13) can be written as

$$\begin{aligned}&\frac{\mathbf{i}}{2}\int_{\operatorname{Re} z=0}\left[\cot\left(\pi\,\frac{z+\frac{\mu-1}{2}}{\mu}\right)-\cot\left(\pi\,\frac{z+\frac{\mu+1}{2}}{\mu}\right)\right]\mathbf{\Lambda}_s^{-z}\mathbf{T}(f)\mathbf{\Lambda}_s^{z}\,dz\\ &-\frac{\mathbf{i}}{2}\int_{\operatorname{Re} z=0}\left[\cot\left(\pi\,\frac{z-\mathbf{i}\xi+\frac{\mu-1}{2}}{\mu}\right)-\cot\left(\pi\,\frac{z-\mathbf{i}\xi+\frac{\mu+1}{2}}{\mu}\right)\right]\mathbf{\Lambda}_s^{-z}\mathbf{T}(f)\mathbf{\Lambda}_s^{z}\,dz\\ &=\frac{\mathbf{i}}{2}\int_{\operatorname{Re} z=0}\left[\cot\left(\pi\,\frac{z+\frac{\mu-1}{2}}{\mu}\right)-\cot\left(\pi\,\frac{z+\frac{\mu+1}{2}}{\mu}\right)\right]\mathbf{\Lambda}_s^{-z}\mathbf{T}(f)\mathbf{\Lambda}_s^{z}\,dz\\ &-\frac{\mathbf{i}}{2}\,\mathbf{\Lambda}_s^{-\mathbf{i}\xi}\int_{\operatorname{Re} z=0}\left[\cot\left(\pi\,\frac{z+\frac{\mu-1}{2}}{\mu}\right)-\cot\left(\pi\,\frac{z+\frac{\mu+1}{2}}{\mu}\right)\right]\mathbf{\Lambda}_s^{-z}\mathbf{T}(f)\mathbf{\Lambda}_s^{z}\,dz\,\mathbf{\Lambda}_s^{\mathbf{i}\xi},\end{aligned}$$

and we can again apply Corollary 5.2, such that (5.16) is proved, which yields $\mathbf{M}_s^0\in\operatorname{alg}\mathcal{T}(\mathbf{PC})$. If we define $\mathbf{a}_{(t,\lambda)}=0$ for $(t,\lambda)\in(\mathbb{T}\setminus\{1\})\times[0,1]$ and

$$\mathbf{a}_{(1,\lambda)}(z)=\mathbf{i}\left[\cot\left(\frac{\pi}{2}+\frac{\mathbf{i}}{2\mu}\ln\frac{\lambda}{1-\lambda}\right)-\cot\left(\frac{\pi}{2}+\frac{\mathbf{i}}{2\mu}\ln\frac{\lambda}{1-\lambda}-\frac{\pi z}{\mu}\right)\right]$$

for $0 \le \lambda \le 1$, then, according to (5.15) and (5.2), $\mathrm{smb}_{\mathbf{M}_\xi^s}(t,\lambda) = \mathbf{a}_{(t,\lambda)}(\mathbf{i}\xi)$. Together with Lemma 5.5 we get, for

$$\begin{aligned}\Gamma_R := &\left\{\frac{\mu}{2} + \mathbf{i}\,t : -R \le t \le R\right\}\\ &\cup\left\{\mathbf{i}\,R - t : -\frac{\mu}{2} \le t \le \frac{\mu}{2}\right\}\\ &\cup\left\{-\frac{\mu}{2} - \mathbf{i}\,t : -R \le t \le R\right\}\\ &\cup\left\{-\mathbf{i}\,R + t : -\frac{\mu}{2} \le t \le \frac{\mu}{2}\right\}\end{aligned}$$

and by taking into account $\mathbf{a}_{(1,\lambda)}\left(z - \frac{\mu}{2}\right) = \mathbf{a}_{(1,\lambda)}\left(z + \frac{\mu}{2}\right)$, the relation

$$\begin{aligned}&\mathrm{smb}_{\mathbf{M}_s^0}(1,\lambda)\\ &= \frac{\mathbf{i}}{2\mu}\int_{-\infty}^{\infty}\mathbf{a}_{(1,\lambda)}(\mathbf{i}\xi)\left[\widehat{\mathbf{y}}\left(\frac{1}{2} + \mathbf{i}\xi - \frac{\mu}{2}\right) - \widehat{\mathbf{y}}\left(\frac{1}{2} + \mathbf{i}\xi + \frac{\mu}{2}\right)\right]d\xi\\ &= \frac{\mathbf{i}}{2\mu}\lim_{R\to\infty}\left[\int_{-R}^{R}\mathbf{a}_{(1,\lambda)}(\mathbf{i}\xi)\widehat{\mathbf{y}}\left(\frac{1}{2} + \mathbf{i}\xi - \frac{\mu}{2}\right)d\xi\right.\\ &\qquad\left. - \int_{-R}^{R}\mathbf{a}_{(1,\lambda)}(\mathbf{i}\xi)\widehat{\mathbf{y}}\left(\frac{1}{2} + \mathbf{i}\xi + \frac{\mu}{2}\right)d\xi\right]\\ &= \frac{\mathbf{i}}{2\mu}\lim_{R\to\infty}\left[-\int_{\Gamma_R}\mathbf{a}_{(1,\lambda)}\left(z + \frac{\mu}{2}\right)\widehat{\mathbf{y}}\left(\frac{1}{2} + z\right)dz\right.\\ &\quad + \int_{\left\{z=\mathbf{i}\,R-t:-\frac{\mu}{2}\le t\le\frac{\mu}{2}\right\}}\mathbf{a}_{(1,\lambda)}\left(z + \frac{\mu}{2}\right)\widehat{\mathbf{y}}\left(\frac{1}{2} + z\right)dz\\ &\quad\left. + \int_{\left\{z=-\mathbf{i}\,R+t:-\frac{\mu}{2}\le t\le\frac{\mu}{2}\right\}}\mathbf{a}_{(1,\lambda)}\left(z + \frac{\mu}{2}\right)\widehat{\mathbf{y}}\left(\frac{1}{2} + z\right)dz\right]\end{aligned} \tag{5.17}$$

For fixed $\lambda \in (0,1)$, the function $\mathbf{a}_{(1,\lambda)}\left(z + \frac{\mu}{2}\right)$ is bounded for $-\frac{\mu}{2} \le \operatorname{Re} z \le \frac{\mu}{2}$. Hence, by assumption (2.31) for $k = 0$, the last two integrals in (5.17) tend to zero if R tends to infinity, and, by applying the residue theorem to the integral over Γ_R, we obtain

$$\begin{aligned}\mathrm{smb}_{\mathbf{M}_s^0}(1,\lambda) &= -\frac{\pi\mathbf{i}}{\mu}\,\widehat{\mathbf{y}}\left(\frac{1}{2} + \frac{\mathbf{i}}{2\mu}\ln\frac{\lambda}{1-\lambda}\right)\operatorname{res}_{\frac{\mathbf{i}}{2\pi}\ln\frac{\lambda}{1-\lambda}}\mathbf{a}_{(1,\lambda)}\left(\cdot + \frac{\mu}{2}\right)\\ &= \widehat{\mathbf{y}}\left(\frac{1}{2} + \frac{\mathbf{i}}{2\mu}\ln\frac{\lambda}{1-\lambda}\right).\end{aligned}$$

It remains to note that

$$\mathbf{M}_s = \mathbf{M}_s^0 + \mathbf{y}(1)\mathbf{K}$$

with the compact operator $\mathbf{K} = \left[\begin{array}{c}\frac{\delta_{j,k}}{k+s}\end{array}\right] \in \mathcal{K}(\ell^2)$, and Lemma 2.10 is proved.

References

[1] E.L. Basor and T. Ehrhardt. On a class of Toeplitz + Hankel operators. *New York J. Math.*, 5:1–16 (electronic), 1999.

[2] A. Böttcher, N. Krupnik, and B. Silbermann. A general look at local principles with special emphasis on the norm computation aspect. *Integral Equations Operator Theory*, 11(4):455–479, 1988.

[3] A. Böttcher and B. Silbermann. *Analysis of Toeplitz operators.* Springer Monographs in Mathematics. Springer-Verlag, Berlin, second edition, 2006. Prepared jointly with Alexei Karlovich.

[4] M.R. Capobianco, G. Criscuolo, and P. Junghanns. On the numerical solution of a hypersingular integral equation with fixed singularities. In *Recent advances in operator theory and applications*, volume 187 of *Oper. Theory Adv. Appl.*, pages 95–116. Birkhäuser, Basel, 2008.

[5] R. Duduchava. *Integral equations in convolution with discontinuous presymbols, singular integral equations with fixed singularities, and their applications to some problems of mechanics.* BSB B.G. Teubner Verlagsgesellschaft, Leipzig, 1979. With German, French and Russian summaries, Teubner-Texte zur Mathematik. [Teubner Texts on Mathematics].

[6] A. Erdélyi, W. Magnus, F. Oberhettinger, and F.G. Tricomi. *Tables of integral transforms. Vol. I.* McGraw-Hill Book Company, Inc., New York-Toronto-London, 1954. Based, in part, on notes left by Harry Bateman.

[7] R. Hagen, S. Roch, and B. Silbermann. *Spectral theory of approximation methods for convolution equations*, volume 74 of *Operator Theory: Advances and Applications.* Birkhäuser Verlag, Basel, 1995.

[8] R. Hagen, S. Roch, and B. Silbermann. *C^*-algebras and numerical analysis*, volume 236 of *Monographs and Textbooks in Pure and Applied Mathematics.* Marcel Dekker Inc., New York, 2001.

[9] P. Junghanns and R. Kaiser. Collocation for Cauchy singular integral equations. *Linear Algebra Appl.*, 439(3):729–770, 2013.

[10] P. Junghanns, R. Kaiser, and G. Mastroianni. Collocation for singular integral equations with fixed singularities of particular Mellin type. *Electron. Trans. Numer. Anal.*, 41:190–248, 2014.

[11] P. Junghanns, R. Kaiser, and D. Potts. Collocation-quadrature methods and fast summation for Cauchy singular integral equations with fixed singularities. *Linear Algebra Appl.*, 491:187–238, 2016.

[12] P. Junghanns and A. Rathsfeld. On polynomial collocation for Cauchy singular integral equations with fixed singularities. *Integral Equations Operator Theory*, 43(2):155–176, 2002.

[13] P. Junghanns and A. Rathsfeld. A polynomial collocation method for Cauchy singular integral equations over the interval. *Electron. Trans. Numer. Anal.*, 14:79–126, 2002. Orthogonal polynomials, approximation theory, and harmonic analysis (Inzel, 2000).

[14] P. Junghanns and A. Rogozhin. Collocation methods for Cauchy singular integral equations on the interval. *Electron. Trans. Numer. Anal.*, 17:11–75 (electronic), 2004.

[15] N.Y. Krupnik. *Banakhovy algebry s simbolom i singulyarnye integralnye operatory.* "Shtiintsa", Kishinev, 1984.

[16] N.Y. Krupnik. *Banach algebras with symbol and singular integral operators*, volume 26 of *Operator Theory: Advances and Applications.* Birkhäuser Verlag, Basel, 1987. Translated from the Russian by A. Iacob.

[17] G. Mastroianni and G.V. Milovanović. *Interpolation processes.* Springer Monographs in Mathematics. Springer-Verlag, Berlin, 2008. Basic theory and applications.

[18] S. Prössdorf and B. Silbermann. *Numerical analysis for integral and related operator equations*, volume 84 of *Mathematische Lehrbücher und Monographien, II. Abteilung: Mathematische Monographien [Mathematical Textbooks and Monographs, Part II: Mathematical Monographs]*. Akademie-Verlag, Berlin, 1991.

[19] S. Prössdorf and B. Silbermann. *Numerical analysis for integral and related operator equations*, volume 52 of *Operator Theory: Advances and Applications.* Birkhäuser Verlag, Basel, 1991.

[20] E.C. Titchmarsh. *Introduction to the theory of Fourier integrals.* Chelsea Publishing Co., New York, third edition, 1986.

[21] K. Yosida. *Functional analysis.* Die Grundlehren der Mathematischen Wissenschaften, Band 123. Academic Press, Inc., New York; Springer-Verlag, Berlin, 1965.

Peter Junghanns and Robert Kaiser
Technische Universität Chemnitz
Reichenhainer Str. 39
D-09107 Chemnitz, Germany
e-mail: `peter.junghanns@mathematik.tu-chemnitz.de`
`robert.kaiser@mathematik.tu-chemnitz.de`

Operator Theory:
Advances and Applications, Vol. 259, 463–500

The Haseman Boundary Value Problem with Slowly Oscillating Coefficients and Shifts

Yu.I. Karlovich

To my friend and colleague Albrecht Böttcher on the occasion of his 60th birthday

Abstract. The paper is devoted to studying the Haseman boundary value problem $\Phi^+ \circ \alpha = G\Phi^- + g$ on a star-like Carleson curve Γ composed by logarithmic spirals in the setting of Lebesgue spaces, where $\Phi^\pm$ are angular boundary values of an unknown analytic function Φ on Γ, G and g are given functions, and α is an orientation-preserving homeomorphism of Γ onto itself. This problem is reduced to the equivalent singular integral operator with a shift $T = V_\alpha P_+ + GP_-$ on a Lebesgue space $L^p(\Gamma)$, where the operators $P_\pm = 2^{-1}(I \pm S_\Gamma)$ are related to the Cauchy singular integral operator S_Γ, and the shift operator V_α is given by $V_\alpha f = f \circ \alpha$. Applying the theory of Mellin pseudodifferential operators with non-regular symbols of limited smoothness and essentially decreasing the smoothness of the shift α, we establish a Fredholm criterion and an index formula for the operator T provided that the shift derivative α' and the coefficient G are slowly oscillating functions on Γ.

Mathematics Subject Classification (2010). Primary 45E05, 47A53; Secondary 30E25, 47G10, 47G30.

Keywords. Haseman boundary value problem, singular integral operator with a shift, slowly oscillating function, Lebesgue space, logarithmic star, Mellin pseudodifferential operator, non-regular symbol, Fredholmness, index.

1. Introduction

Let $\mathcal{B}(X)$ be the Banach algebra of all bounded linear operators acting on a Banach space X, and let $\mathcal{K}(X)$ be the closed two-sided ideal of all compact operators in $\mathcal{B}(X)$. An operator $A \in \mathcal{B}(X)$ is said to be *Fredholm*, if its image is closed and the spaces $\ker A$ and $\ker A^*$ are finite-dimensional. In that case the number $\operatorname{Ind} A = \dim\ker A - \dim\ker A^*$ is referred to as the *index* of A. (see, e.g., [10]).

The work was supported by the SEP-CONACYT Projects No. 168104 and No. 169496 (México).

Let Γ be a logarithmic star composed by open oriented logarithmic spirals Γ_k $(k = 1, 2, \dots, N)$ with endpoints 0 and ∞ and forming there non-zero angles, $C_b(\Gamma)$ the C^*-algebra of all bounded continuous functions on $\Gamma = \overline{\Gamma} \setminus \{0, \infty\}$, let $SO(\Gamma)$ be the C^*-algebra of all functions $a \in C_b(\Gamma)$ that slowly oscillate at 0 and ∞, that is, for every $k = 1, 2, \dots, N$ satisfy the condition

$$\lim_{r \to s} \max \Big\{ |a(t) - a(\tau)| : \ t, \tau \in \Gamma_k, \ |t|, |\tau| \in [r, 2r] \Big\} = 0 \quad (s \in \{0, \infty\}),$$

and let α be an orientation-preserving diffeomorphism of each open arc Γ_k onto itself with the derivative $\alpha' \in SO(\Gamma)$.

Given $p \in (1, \infty)$, we study the Haseman boundary value problem (see, e.g., [30]): Find a function Φ analytic in $\mathbb{C} \setminus \overline{\Gamma}$, represented by the Cauchy type integral over Γ with a density $\varphi \in L^p(\Gamma)$ and satisfying the boundary condition

$$\Phi^+(\alpha(t)) = G(t)\Phi^-(t) + g(t) \quad \text{for} \quad t \in \Gamma, \tag{1.1}$$

where $\Phi^+(t)$ and $\Phi^-(t)$ are angular boundary values of Φ on Γ from the left and from the right, respectively, $G \in SO(\Gamma)$, $g \in L^p(\Gamma)$. Taking the Cauchy type integral

$$\Phi(z) = \frac{1}{2\pi i} \int_\Gamma \frac{\varphi(\tau)}{\tau - z} \, d\tau \quad (z \in \mathbb{C} \setminus \overline{\Gamma})$$

with a density $\varphi \in L^p(\Gamma)$ and applying the Sokhotski–Plemelj formulas

$$\Phi^\pm(t) = \pm \frac{1}{2} \varphi(t) + \frac{1}{2\pi i} \int_\Gamma \frac{\varphi(\tau)}{\tau - t} \, d\tau$$

for almost all $t \in \Gamma$, we reduce the Haseman boundary value problem (1.1) to the equivalent singular integral equation with a shift

$$(V_\alpha P_+ + GP_-)\varphi = g,$$

where V_α is the shift operator given by $V_\alpha f = f \circ \alpha$, $P_\pm = 2^{-1}(I \pm S_\Gamma)$, I is the identity operator and S_Γ is the Cauchy singular integral operator defined by

$$(S_\Gamma f)(t) := \lim_{\varepsilon \to 0} \frac{1}{\pi i} \int_{\Gamma \setminus \Gamma(t, \varepsilon)} \frac{f(\tau)}{\tau - t} \, d\tau, \quad \Gamma(t, \varepsilon) := \big\{ \tau \in \Gamma : |\tau - t| < \varepsilon \big\}.$$

As is known (see, e.g., [11], [4]), the Cauchy singular integral operator S_Γ is bounded on the space $L^p(\Gamma)$. The shift operator V_α is also bounded on this space. Thus, with the Haseman problem (1.1) we associate the equivalent singular integral operator with a shift

$$T = V_\alpha P_+ + GP_- \in \mathcal{B}(L^p(\Gamma)). \tag{1.2}$$

The Haseman boundary value problem under different conditions on its data was studied by C. Haseman, D.A. Kveselava, L.I. Chibrikova, G.F. Mandzhavidze and B.V. Khvedelidze, I.B. Simonenko, and by the author in teams with V.G. Kravchenko, A.V. Aizenshtat and G.S. Litvinchuk (see [1], [20], [23], [25], [27], [29], [30], [31] and the references therein). Applications of Mellin pseudodifferential operators to singular integral operators, elaborated by V.S. Rabinovich (see [32]–

[34] and the references therein) and by A. Böttcher, V.S. Rabinovich and the author ([4]–[8]), essentially extended the class of studied operators. The theory of Mellin pseudodifferential operators with non-regular symbols was applied for the first time to studying nonlocal singular integral operators (1.2) with non-Carleman shifts α in [23] and [25].

The present paper is devoted to studying the operator (1.2) related to the Haseman boundary value problem with condition (1.1) given on the Carleson star-like contour Γ composed by logarithmic spirals under weaker conditions on slowly oscillating data: the shift derivative α' and the coefficient G. Applying the theory of Mellin pseudodifferential operators with non-regular symbols of limited smoothness, which was constructed in [23], [24], [26] on the basis of [21], generalizing results on singular integral operators with shifts obtained in [16]–[19] and following partially the scheme of [25], we establish a Fredholm criterion and an index formula for the operator T on the space $L^p(\Gamma)$ under much weaker smoothness conditions on the shift derivative α'. To obtain the index formula for T, we calculate the generalized Cauchy index of some semi-almost periodic function related to the Haseman boundary value problem (the notion of generalized Cauchy index for any invertible semi-almost periodic function was introduced in [27], also see [28] and, for a particular case, [9]).

The paper is organized as follows. In Section 2 we introduce slowly oscillating data. In Section 3 we collect necessary results from [23], [24], [26] on Mellin pseudodifferential operators with slowly oscillating $V(\mathbb{R})$-valued symbols on $\mathbb{R}_+$, where $V(\mathbb{R})$ is the Banach algebra of absolutely continuous functions of bounded total variation on $\mathbb{R}$. In Section 4 we study properties of some specific slowly oscillating functions and symbols. In Section 5 we apply the results on Mellin pseudodifferential operators presented in Section 3 to singular integral operators with a shift. The crucial role in studying singular integral operators with shifts plays the fact that the product of the shift operator V_α and a singular integral operator with point singularities at the endpoints of Γ is similar to a Mellin pseudodifferential operator to within compact operators. In Section 6, applying the results of Section 5, we study auxiliary singular integral operators with shifts $T_\pm$ that are Fredholm on the space $L^p(\Gamma)$ and have zero indices. Finally, in Sections 7 and 8 we establish a Fredholm criterion and an index formula for the operator T with a shift α on the space $L^p(\Gamma)$ in the case of slowly oscillating functions G and α'. The form of this criterion and the index formula are different of those in [25].

2. Slowly oscillating data

Slowly oscillating functions. Let $\mathbb{R}_+ = (0,\infty)$ and let $SO(\mathbb{R}_+)$ stand for the set of all functions $a \in C_b(\mathbb{R}_+) := C(\mathbb{R}_+) \cap L^\infty(\mathbb{R}_+)$ that slowly oscillate at 0 and ∞, that is (see, e.g., [36]), satisfy the condition

$$\lim_{r\to s} \max\Big\{|a(x)-a(y)| : x,y \in [r,2r]\Big\} = 0, \quad \text{for } s \in \{0,\infty\}. \tag{2.1}$$

It is clear that (2.1) is equivalent to the condition

$$\lim_{r\to s} \max\Big\{|a(r)-a(tr)| : t\in[\lambda^{-1},\lambda]\Big\}=0 \quad \text{for} \quad s\in\{0,\infty\}, \tag{2.2}$$

with any $\lambda>1$. Obviously, $SO(\mathbb{R}_+)$ is a unital C^*-subalgebra of $L^\infty(\mathbb{R}_+)$.

Let $\overline{\mathbb{R}}_+=[0,+\infty]$. Since $C(\overline{\mathbb{R}}_+)\subset SO(\mathbb{R}_+)$ and since the maximal ideal space $M(C(\overline{\mathbb{R}}_+))$ of $C(\overline{\mathbb{R}}_+)$ can be defined as $\overline{\mathbb{R}}_+$ by identifying the points $t\in\overline{\mathbb{R}}_+$ with the evaluation functionals $\delta_t(f)=f(t)$, we conclude that the maximal ideal space $M(SO(\mathbb{R}_+))$ of $SO(\mathbb{R}_+)$ can be represented in the form

$$M(SO(\mathbb{R}_+))=M_0(SO(\mathbb{R}_+))\cup\mathbb{R}_+\cup M_\infty(SO(\mathbb{R}_+)),$$

where $M_s(SO(\mathbb{R}_+))=\big\{\xi\in M(SO(\mathbb{R}_+)):\ \xi|_{C(\overline{\mathbb{R}}_+)}=s\big\}$ are fibers of $M(SO(\mathbb{R}_+))$ over the points $s\in\{0,\infty\}$, respectively.

The union of the fibers $M_0(SO(\mathbb{R}_+))$ and $M_\infty(SO(\mathbb{R}_+))$ is characterized by the following corollary of [21, Proposition 2.4] proved by analogy with [7, Proposition 4.1] and [3, Proposition 3.29].

Proposition 2.1 ([24, Proposition 2.1]). *We have*

$$\Delta:=M_0(SO(\mathbb{R}_+))\cup M_\infty(SO(\mathbb{R}_+))=\mathrm{clos}_{SO^*}\mathbb{R}_+\setminus\mathbb{R}_+, \tag{2.3}$$

where $\mathrm{clos}_{SO^*}\mathbb{R}_+$ *is the weak-star closure of* $\mathbb{R}_+$ *in* $(SO(\mathbb{R}_+))^*$, *the dual space of* $SO(\mathbb{R}_+)$.

The fibers $M_s(SO(\mathbb{R}_+))$ are related to the partial limits of functions $a\in SO(\mathbb{R}_+)$ at the points $s\in\{0,\infty\}$ as follows (cf. [7, Proposition 4.2 and Corollary 4.3] and [2, Corollary 3.3]).

Proposition 2.2 ([24, Proposition 2.2]). *If* $\{a_k\}_{k=1}^\infty$ *is a countable subset of* $SO(\mathbb{R}_+)$ *and* $s\in\{0,\infty\}$, *then for each* $\xi\in M_s(SO(\mathbb{R}_+))$ *there exists a sequence* $\{x_n\}\subset\mathbb{R}_+$ *such that* $x_n\to s$ *as* $n\to\infty$ *and*

$$\lim_{n\to\infty} a_k(x_nt)=\xi(a_k) \quad \textit{for all} \quad t\in\mathbb{R}_+ \ \textit{and all} \ \ k\in\mathbb{N}. \tag{2.4}$$

Conversely, if $\{x_n\}\subset\mathbb{R}_+$ *is a sequence such that* $x_n\to s$ *as* $n\to\infty$ *and the limits* $\lim\limits_{n\to\infty} a_k(x_n)$ *exist for all* $k\in\mathbb{N}$, *then there is a* $\xi\in M_s(SO(\mathbb{R}_+))$ *such that* (2.4) *holds.*

Note that (2.2) leads to the equality $\lim_{n\to\infty}a_k(x_nt)=\lim_{n\to\infty}a_k(x_n)$ for every $t\in\mathbb{R}_+$. In what follows we write $a(\xi):=\xi(a)$ for every $a\in SO(\mathbb{R}_+)$ and every $\xi\in\Delta$.

Logarithmic stars. Given $\delta\in\mathbb{R}$, let $\Gamma=\bigcup_{k=1}^N\Gamma_k$, where all arcs

$$\Gamma_k:=\big\{e^{i\beta_k}r^{1+i\delta}\ :\ r\in\mathbb{R}_+\big\} \tag{2.5}$$

are logarithmic spirals if $\delta\in\mathbb{R}\setminus\{0\}$ and rays if $\delta=0$, and

$$0\le\beta_1<\beta_2<\dots<\beta_N<2\pi. \tag{2.6}$$

Obviously, Γ is a Carleson curve (see, e.g., [5] or [4, Section 10.6]).

Slowly oscillating shifts. Let α be an orientation-preserving diffeomorphism of each arc Γ_k $(k = 1, 2, \dots, N)$ onto itself and such that

$$\ln|\alpha'| \in C_b(\Gamma), \tag{2.7}$$

which implies the boundedness of the shift operator V_α given by $V_\alpha f = f \circ \alpha$ and its inverse operator V_α^{-1} on the space $L^p(\Gamma)$. We call α a slowly oscillating shift (at 0 and ∞) if, for every $r \in \mathbb{R}_+$ and every $k = 1, 2, \dots, N$,

$$\alpha\big(e^{i\beta_k} r^{1+i\delta}\big) = e^{i\beta_k}\big(re^{\omega_k(r)}\big)^{1+i\delta}, \tag{2.8}$$

where ω_k are real-valued functions in $SO(\mathbb{R}_+) \cap C^1(\mathbb{R}_+)$ such that the functions $\phi_k : r \mapsto r\omega_k'(r)$ also belong to $SO(\mathbb{R}_+)$. The properties $\omega_k \in SO(\mathbb{R}_+) \cap C^1(\mathbb{R}_+)$ and $\phi_k \in SO(\mathbb{R}_+)$ imply (see, e.g., [15, Proposition 3.3]) that

$$\lim_{r \to s}(r\omega_k'(r)) = 0 \quad \text{for} \quad s \in \{0, \infty\}. \tag{2.9}$$

On the other hand, if $\omega_k \in C_b(\mathbb{R}_+) \cap C^1(\mathbb{R}_+)$ and (2.9) holds, then $\omega_k \in SO(\mathbb{R}_+)$.

If α is a slowly oscillating shift then, for every $k = 1, 2, \dots, N$,

$$\alpha'\big(e^{i\beta_k} r^{1+i\delta}\big) = (1 + r\omega_k'(r))e^{(1+i\delta)\omega_k(r)} \quad \text{for all} \quad r \in \mathbb{R}_+ \tag{2.10}$$

and, since $\ln|\alpha'| \in C_b(\Gamma)$, we conclude from (2.9) and (2.10) that

$$\inf_{r \in \mathbb{R}_+} \big(1 + r\omega_k'(r)\big) > 0 \quad \text{for all} \quad k = 1, 2, \dots, N. \tag{2.11}$$

Slowly oscillating coefficients. We denote by $SO(\Gamma)$ the set of all functions c_Γ : $\Gamma \to \mathbb{C}$ such that

$$c_\Gamma\big(e^{i\beta_k} r^{1+i\delta}\big) = c_k(r) \quad \text{for all} \quad r \in \mathbb{R}_+ \quad \text{and all} \quad k = 1, 2, \dots, N, \tag{2.12}$$

where $c_k \in SO(\mathbb{R}_+)$. Below we assume that the coefficient G in (1.1) is in $SO(\Gamma)$. By (2.10) and (2.12), $\alpha' \in SO(\Gamma)$ as well.

3. Mellin pseudodifferential operators

If a is an absolutely continuous function of finite total variation on $\mathbb{R}$, then $a' \in L^1(\mathbb{R})$ and the total variation of a is given by

$$V(a) = \int_{\mathbb{R}} |a'(x)|dx.$$

The set $V(\mathbb{R})$ of all absolutely continuous functions of finite total variation on $\mathbb{R}$ becomes a Banach algebra when equipped with the norm

$$\|a\|_V := \|a\|_{L^\infty(\mathbb{R})} + V(a) = \|a\|_{L^\infty(\mathbb{R})} + \int_{\mathbb{R}} |a'(x)|dx.$$

Following [21, 22], let $C_b(\mathbb{R}_+, V(\mathbb{R}))$ denote the Banach algebra of all bounded continuous $V(\mathbb{R})$-valued functions $r \mapsto \mathfrak{a}(r, \cdot)$ on $\mathbb{R}_+$ with the norm

$$\|\mathfrak{a}(\cdot, \cdot)\|_{C_b(\mathbb{R}_+, V(\mathbb{R}))} = \sup_{r \in \mathbb{R}_+} \|\mathfrak{a}(r, \cdot)\|_V.$$

As usual, let $C_0^\infty(\mathbb{R}_+)$ be the set of all infinitely differentiable functions of compact support on $\mathbb{R}_+$.

Let $d\mu(\varrho) = d\varrho/\varrho$ be the (normalized) invariant measure on $\mathbb{R}_+$. The following boundedness result for Mellin pseudodifferential operators was obtained in [22, Theorem 3.1] (see also [21, Theorem 3.1]).

Theorem 3.1. *If* $\mathfrak{a} \in C_b(\mathbb{R}_+, V(\mathbb{R}))$, *then the Mellin pseudodifferential operator* $\operatorname{Op}(\mathfrak{a})$, *defined for functions* $f \in C_0^\infty(\mathbb{R}_+)$ *by the iterated integral*

$$\big[\operatorname{Op}(\mathfrak{a})f\big](r) = \frac{1}{2\pi}\int_{\mathbb{R}} dx \int_{\mathbb{R}_+} \mathfrak{a}(r,x)\left(\frac{r}{\varrho}\right)^{ix} f(\varrho)\,\frac{d\varrho}{\varrho} \quad \text{for} \quad r \in \mathbb{R}_+, \tag{3.1}$$

extends to a bounded linear operator on the space $L^p(\mathbb{R}_+, d\mu)$ *and there is a number* $C_p \in (0,\infty)$ *depending only on* p *such that*

$$\|\operatorname{Op}(\mathfrak{a})\|_{\mathcal{B}(L^p(\mathbb{R}_+,d\mu))} \le C_p \|\mathfrak{a}\|_{C_b(\mathbb{R}_+,V(\mathbb{R}))}.$$

Mellin pseudodifferential operators are generalizations of Mellin convolution operators $\mathcal{M}^{-1}a\mathcal{M}$, where a belongs to the Banach algebra of Mellin multipliers on the space $L^p(\mathbb{R}_+, d\mu)$ and the Mellin transform and its inverse are defined by

$$\mathcal{M} : L^2(\mathbb{R}_+, d\mu) \to L^2(\mathbb{R}), \quad (\mathcal{M}f)(x) = \int_{\mathbb{R}_+} f(\varrho)\varrho^{-ix}\,\frac{d\varrho}{\varrho},$$

$$\mathcal{M}^{-1} : L^2(\mathbb{R}) \to L^2(\mathbb{R}_+, d\mu), \quad (\mathcal{M}^{-1}g)(r) = \frac{1}{2\pi}\int_{\mathbb{R}} g(x) r^{ix}\,dx.$$

Let $SO(\mathbb{R}_+, V(\mathbb{R}))$ denote the Banach subalgebra of $C_b(\mathbb{R}_+, V(\mathbb{R}))$ consisting of all $V(\mathbb{R})$-valued functions $\mathfrak{a}$ on $\mathbb{R}_+$ that slowly oscillate at 0 and ∞, that is,

$$\lim_{r\to 0} \operatorname{cm}_r^C(\mathfrak{a}) = \lim_{r\to\infty} \operatorname{cm}_r^C(\mathfrak{a}) = 0,$$

where

$$\operatorname{cm}_r^C(\mathfrak{a}) = \max\big\{\big\|\mathfrak{a}(\rho,\cdot) - \mathfrak{a}(\varrho,\cdot)\big\|_{L^\infty(\mathbb{R})} : \rho, \varrho \in [r, 2r]\big\}.$$

Let $\mathcal{E}(\mathbb{R}_+, V(\mathbb{R}))$ be the Banach algebra of all $V(\mathbb{R})$-valued functions $\mathfrak{a}$ belonging to $SO(\mathbb{R}_+, V(\mathbb{R}))$ and such that

$$\lim_{|h|\to 0} \sup_{r\in\mathbb{R}_+} \big\|\mathfrak{a}(r,\cdot) - \mathfrak{a}^h(r,\cdot)\big\|_V = 0 \tag{3.2}$$

where $h \in \mathbb{R}$ and $\mathfrak{a}^h(r,x) := \mathfrak{a}(r, x+h)$ for all $(r,x) \in \mathbb{R}_+ \times \mathbb{R}$.

Passing to Fourier pseudodifferential operators, we deduce from [21, Theorem 8.3] the following result.

Theorem 3.2 ([18, Theorem 3.4]). *If* $\mathfrak{a}, \mathfrak{b} \in \mathcal{E}(\mathbb{R}_+, V(\mathbb{R}))$, *then* $\operatorname{Op}(\mathfrak{a})\operatorname{Op}(\mathfrak{b}) - \operatorname{Op}(\mathfrak{a}\mathfrak{b})$ *is a compact operator on every space* $L^p(\mathbb{R}_+, d\mu)$ *with* $1 < p < \infty$.

The following result on compactness of commutators of Mellin pseudodifferential operators was obtained in [24, Theorem 3.5] (see also [21, Corollary 8.4]).

Theorem 3.3. *If* $\mathfrak{a}, \mathfrak{b} \in \mathcal{E}(\mathbb{R}_+, V(\mathbb{R}))$, *then the commutator*

$$\mathrm{Op}(\mathfrak{a})\,\mathrm{Op}(\mathfrak{b}) - \mathrm{Op}(\mathfrak{b})\,\mathrm{Op}(\mathfrak{a})$$

is compact on the space $L^p(\mathbb{R}_+, d\mu)$ *with* $1 < p < \infty$.

Theorem 3.4 ([24, Theorems 3.4]). *If* $\mathfrak{a} \in \mathcal{E}(\mathbb{R}_+, V(\mathbb{R}))$ *and*

$$\lim_{(\ln r)^2 + x^2 \to \infty} \mathfrak{a}(r, x) = 0,$$

then the Mellin pseudodifferential operator $\mathrm{Op}(\mathfrak{a})$ *is compact on every Lebesgue space* $L^p(\mathbb{R}_+, d\mu)$ *with* $1 < p < \infty$.

By [22, Section 2], the set $\widetilde{\mathcal{E}}(\mathbb{R}_+, V(\mathbb{R}))$ of all functions $\mathfrak{a} \in \mathcal{E}(\mathbb{R}_+, V(\mathbb{R}))$ possessing the property

$$\lim_{M\to\infty} \sup_{r\in\mathbb{R}_+} \int_{\mathbb{R}\setminus[-M,M]} |\partial_x \mathfrak{a}(r, x)|\, dx = 0 \tag{3.3}$$

is a Banach subalgebra of $SO(\mathbb{R}_+, V(\mathbb{R})) \subset C_b(\mathbb{R}_+, V(\mathbb{R}))$.

For any algebra $\mathcal{A}$, we will denote by $\mathcal{A}_{N\times N}$ the algebra of $N \times N$ matrices with entries in $\mathcal{A}$.

Below we need the following Fredholm criterion and index formula for Mellin pseudodifferential operators $\mathrm{Op}(\mathfrak{a})$ with symbols $\mathfrak{a} \in [\widetilde{\mathcal{E}}(\mathbb{R}_+, V(\mathbb{R}))]_{N\times N}$, which were obtained for $N = 1$ in [26, Theorem 4.3] on the basis of [21, Theorems 12.2 and 12.5]. Note that for infinite differentiable slowly oscillating matrix symbols such result was obtained earlier in [32, Theorem 2.6].

Theorem 3.5. *If* $\mathfrak{a} \in [\widetilde{\mathcal{E}}(\mathbb{R}_+, V(\mathbb{R}))]_{N\times N}$, *then the Mellin pseudodifferential operator* $\mathrm{Op}(\mathfrak{a})$ *is Fredholm on the space* $L^p_N(\mathbb{R}_+, d\mu)$ *if and only if*

$$\det \mathfrak{a}(r, \pm\infty) \neq 0 \ \ \textit{for all} \ \ r \in \mathbb{R}_+, \qquad \det \mathfrak{a}(\xi, x) \neq 0 \ \ \textit{for all} \ \ (\xi, x) \in \Delta \times \overline{\mathbb{R}}, \tag{3.4}$$

where Δ *is given by* (2.3). *In the case of Fredholmness*

$$\mathrm{Ind}\,\mathrm{Op}(\mathfrak{a}) = \lim_{m\to\infty} \frac{1}{2\pi} \big\{ \arg\det \mathfrak{a}(r, x) \big\}_{(r,x)\in\partial\Pi_m}$$

where $\Pi_m = [m^{-1}, m] \times \overline{\mathbb{R}}$ *and* $\big\{\arg\det \mathfrak{a}(r, x)\big\}_{(r,x)\in\partial\Pi_m}$ *denotes the increment of* $\arg\det \mathfrak{a}(r, x)$ *when the point* (r, x) *traces the boundary* $\partial\Pi_m$ *of* Π_m *counterclockwise.*

Proof. Let the matrix function $\mathfrak{a} = [\mathfrak{a}_{k,j}]_{k,j=1}^N$ satisfy all conditions of the theorem. Since its entries belong to $\widetilde{\mathcal{E}}(\mathbb{R}_+, V(\mathbb{R}))$, we infer from Theorem 3.3 that the entries of the operator matrix $\mathrm{Op}(\mathfrak{a}) = [\mathrm{Op}(\mathfrak{a}_{k,j})]_{k,j=1}^N$ commute to within compact operators. Hence the operator $\mathrm{Op}(\mathfrak{a})$ is Fredholm on the space $L^p_N(\mathbb{R}_+, d\mu)$ if and only if the operator $\mathrm{Op}(\det \mathfrak{a})$ is Fredholm on the space $L^p(\mathbb{R}_+, d\mu)$. Since $\det \mathfrak{a} \in \widetilde{\mathcal{E}}(\mathbb{R}_+, V(\mathbb{R}))$, the Mellin pseudodifferential operator $\mathrm{Op}(\det \mathfrak{a})$ is Fredholm on the space $L^p(\mathbb{R}_+, d\mu)$ if and only if (3.4) holds (see [24, Theorem 4.3] and [19,

Theorem 5.8]). Hence from [19, Lemma 5.4] it follows that there exists an $m > 1$ such that

$$A(m) := \sup_{(r,x)\in T_m\times\overline{\mathbb{R}}} \left|\frac{1}{\det \mathfrak{a}(r,x)}\right| < \infty, \quad A_\pm := \sup_{r\in\mathbb{R}_+} \frac{1}{|\det \mathfrak{a}(r,\pm\infty)|} < \infty, \tag{3.5}$$

where $T_m := (0, m^{-1}] \cup [m, \infty)$.

We now define the matrix function $\widetilde{\mathfrak{a}}_m \in [C(\overline{\mathbb{R}}_+ \times \overline{\mathbb{R}})]_{N\times N}$ by

$$\widetilde{\mathfrak{a}}_m(r,x) = \mathfrak{a}(m^{-1},x)\widetilde{\chi}_-(r) + \mathfrak{a}(m,x)\widetilde{\chi}_+(r) + c_+(r)\zeta_+(x) + c_-(r)\zeta_-(x),$$

where $\zeta_\pm(x) = (1 \pm \tanh x)/2$ for $x \in \overline{\mathbb{R}}$, $\widetilde{\chi}_-(r) = 1 - \widetilde{\chi}_+(r)$ for $r \in \overline{\mathbb{R}}_+$,

$$\widetilde{\chi}_+(r) = \begin{cases} 0 & \text{if} \quad r \in [0, m^{-1}], \\ (r - m^{-1})/(m - m^{-1}) & \text{if} \quad r \in [m^{-1}, m], \\ 1 & \text{if} \quad r \in [m, \infty], \end{cases}$$

$$c_\pm(r) = \begin{cases} \mathfrak{a}(r,\pm\infty) - \mathfrak{a}(m^{-1},\pm\infty)\widetilde{\chi}_-(r) - \mathfrak{a}(m,\pm\infty)\widetilde{\chi}_+(r) & \text{if} \ \ r \in (m^{-1}, m), \\ 0 & \text{if} \ \ r \in T_m. \end{cases}$$

Then $\widetilde{\mathfrak{a}}_m(r,x) = \mathfrak{a}(m^{-1},x)$ for all $(r,x) \in [0,m^{-1}] \times \overline{\mathbb{R}}$, $\widetilde{\mathfrak{a}}_m(r,x) = \mathfrak{a}(m,x)$ for all $(r,x) \in [m,\infty] \times \overline{\mathbb{R}}$, and $\widetilde{\mathfrak{a}}_m(r,\pm\infty) = \mathfrak{a}(r,\pm\infty)$ for all $r \in [m^{-1}, m]$. Hence we conclude from (3.5) that

$$\det \widetilde{\mathfrak{a}}_m(r,x) \neq 0 \quad \text{for all} \quad (r,x) \in \partial(\overline{\mathbb{R}}_+ \times \overline{\mathbb{R}}). \tag{3.6}$$

Clearly, the Mellin pseudodifferential operator $\mathrm{Op}(\widetilde{\mathfrak{a}}_m)$ belongs to the Banach subalgebra of $\mathcal{B}(L^p_N(\mathbb{R}_+, d\mu))$ generated by all operators of multiplication by matrix functions in $[C(\overline{\mathbb{R}}_+)]_{N\times N}$ and by all Mellin convolution operators $\mathcal{M}^{-1}b\mathcal{M}$ with $b \in [C_p(\overline{\mathbb{R}})]_{N\times N}$, where $C_p(\overline{\mathbb{R}})$ is the closure of $V(\mathbb{R})$ in the Banach algebra of Fourier multipliers on $L^p(\mathbb{R})$. Passing to Fourier convolution type operators, we deduce from [12], [13], [14, Theorem 7.7] and (3.6) that the Mellin convolution type operator $\mathrm{Op}(\widetilde{\mathfrak{a}}_m)$ is Fredholm on the space $L^p_N(\mathbb{R}_+, d\mu)$ and

$$\mathrm{Ind}\ \mathrm{Op}(\widetilde{\mathfrak{a}}_m) = \mathrm{Ind}\ \mathrm{Op}(\det \widetilde{\mathfrak{a}}_m) = \frac{1}{2\pi}\{\arg\det\widetilde{\mathfrak{a}}_m(r,x)\}_{(r,x)\in\partial\Pi_m}. \tag{3.7}$$

Let $[\mathrm{Op}(\widetilde{\mathfrak{a}}_m)]^{(-1)}$ be a regularizer of the operator $\mathrm{Op}(\widetilde{\mathfrak{a}}_m)$. From [19, Theorem 5.8] it follows that the operator $[\mathrm{Op}(\widetilde{\mathfrak{a}}_m)]^{(-1)}$ is of the form $\mathrm{Op}(\mathfrak{b}) + K$, where $\mathfrak{b} \in [\widetilde{\mathcal{E}}(\mathbb{R}_+, V(\mathbb{R}))]_{N\times N} \cap [C(\overline{\mathbb{R}}_+ \times \overline{\mathbb{R}})]_{N\times N}$ and K is a compact operator on the space $L^p_N(\mathbb{R}_+, d\mu)$. Moreover, by [19, Lemmas 5.5–5.6], $\mathfrak{b}(r,x) = (\widetilde{\mathfrak{a}}_m(r,x))^{-1}$ for all $(r,x) \in ([0,m^{-1}] \cup [m,\infty]) \times \overline{\mathbb{R}}$ and $\mathfrak{b}(r,\pm\infty) = (\widetilde{\mathfrak{a}}_m(r,\pm\infty))^{-1}$ for all $r \in [m^{-1}, m]$. Then

$$\mathfrak{a}(r,x)\mathfrak{b}(r,x) = \mathfrak{c}(r,x) + \mathfrak{d}(r,x) \quad \text{for all} \quad (r,x) \in \mathbb{R}_+ \times \overline{\mathbb{R}}, \tag{3.8}$$

with matrix functions $\mathfrak{c}, \mathfrak{d} \in [\widetilde{\mathcal{E}}(\mathbb{R}_+, V(\mathbb{R}))]_{N\times N}$ such that

$$\inf_{(r,x)\in T_m\times\overline{\mathbb{R}}} |\det \mathfrak{c}(r,x)| > 0, \quad \mathfrak{c}(r,x) = I_N \ \text{ for all } \ (r,x) \in [m^{-1}, m] \times \overline{\mathbb{R}}, \tag{3.9}$$

$\mathfrak{d}(r,x) = 0_N$ for all $(r,x) \in (\overline{\mathbb{R}}_+ \times \overline{\mathbb{R}}) \setminus ((m^{-1},m) \times \mathbb{R})$. It follows from (3.8) that

$$\mathrm{Op}(\mathfrak{a})\,\mathrm{Op}(\mathfrak{b}) = \mathrm{Op}(\mathfrak{c}) + K \tag{3.10}$$

where $K = \mathrm{Op}(\mathfrak{d})$ is a compact operator in view of Theorem 3.4. Hence, by (3.10), the operator $\mathrm{Op}(\mathfrak{c})$ is Fredholm on the space $L^p_N(\mathbb{R}_+, d\mu)$ along with $\mathrm{Op}(\mathfrak{a})$ and $\mathrm{Op}(\mathfrak{b})$, and

$$\mathrm{Ind}\,\mathrm{Op}(\mathfrak{a}) + \mathrm{Ind}\,\mathrm{Op}(\mathfrak{b}) = \mathrm{Ind}\,\mathrm{Op}(\mathfrak{c}). \tag{3.11}$$

It remains to prove that there exists a homotopy (in the class of Fredholm operators in $\mathcal{B}(L^p_N(\mathbb{R}_+, d\mu))$ of the operator $\mathrm{Op}(\mathfrak{c})$ to an invertible operator. The homotopy is given as follows. Let $z \in [0,1]$, $\varphi(z) := \ln[z/(1-z)]$ and

$$\mathfrak{c}_z(r,x) = \begin{cases} \mathfrak{c}(r,\varphi(z)) & \text{if } (r,x) \in \mathbb{R}_+ \times [-\infty, \varphi(z)], \\ \mathfrak{c}(r,x) & \text{if } (r,x) \in \mathbb{R}_+ \times (\varphi(z), +\infty]. \end{cases}$$

It is easily seen from the definition of $\widetilde{\mathcal{E}}(\mathbb{R}_+, V(\mathbb{R}))$, Theorem 3.1 and (3.9) that $\mathfrak{c}_z \in [\widetilde{\mathcal{E}}(\mathbb{R}_+, V(\mathbb{R}))]_{N\times N}$ for all $z \in [0,1]$ and the map $z \mapsto \mathrm{Op}(\mathfrak{c}_z)$ is the required homotopy of $\mathrm{Op}(\mathfrak{c}_0) = \mathrm{Op}(\mathfrak{c})$ to $\mathrm{Op}(\mathfrak{c}_1)$. Since the operator $\mathrm{Op}(\mathfrak{c}_1) = \mathfrak{c}(\cdot, +\infty)I$ is invertible on the space $L^p_N(\mathbb{R}_+, d\mu)$, we conclude that

$$\mathrm{Ind}\,\mathrm{Op}(\mathfrak{c}) = \mathrm{Ind}\,\mathrm{Op}(\mathfrak{c}_0) = \mathrm{Ind}\,\mathrm{Op}(\mathfrak{c}_1) = 0.$$

This implies according to (3.11) that

$$\mathrm{Ind}\,\mathrm{Op}(\mathfrak{a}) = -\mathrm{Ind}\,\mathrm{Op}(\mathfrak{b}) = \mathrm{Ind}\,\mathrm{Op}(\widetilde{\mathfrak{a}}_m),$$

which completes the proof in view of (3.7) because $\widetilde{\mathfrak{a}}_m = \mathfrak{a}$ on $\partial\Pi_m$. □

4. Important functions in $\widetilde{\mathcal{E}}(\mathbb{R}_+, V(\mathbb{R}))$

Lemma 4.1. *If $p \in (1,\infty)$, $\delta \in \mathbb{R}$ and $\beta \in (-\pi,\pi)$, then the functions*

$$\mathfrak{s}_{p,\delta}(r,x) := s_{p,\delta}(x) \quad := \coth\left(\pi\frac{x+i/p}{1+i\delta}\right), \quad (r,x) \in \mathbb{R}_+ \times \overline{\mathbb{R}}, \tag{4.1}$$

$$\mathfrak{h}_{p,\delta,\beta}(r,x) := h_{p,\delta,\beta}(x) := \frac{\exp\left(\beta\frac{x+i/p}{1+i\delta}\right)}{\sinh\left(\pi\frac{x+i/p}{1+i\delta}\right)}, \quad (r,x) \in \mathbb{R}_+ \times \overline{\mathbb{R}}, \tag{4.2}$$

belong to the Banach algebra $\widetilde{\mathcal{E}}(\mathbb{R}_+, V(\mathbb{R}))$.

Proof. Since the functions $\mathfrak{s}_{p,\delta}$ and $\mathfrak{h}_{p,\delta,\beta}$ are constants in the first variable, it only remains to prove that the first and second derivatives of the functions $s_{p,\delta}$ and $h_{p,\delta,\beta}$ belong to $L^1(\mathbb{R})$, which implies that $s_{p,\delta}$, $h_{p,\delta,\beta} \in V(\mathbb{R})$ and conditions (3.2)–(3.3) for the functions $\mathfrak{s}_{p,\delta}$ and $\mathfrak{h}_{p,\delta,\beta}$ are fulfilled. As $1/p \in (0,1)$, it follows that the function $x \mapsto \sinh\left(\pi(x+i/p)/(1+i\delta)\right)$ is separated from $z = 0$ and tends exponentially to ∞ as $x \to \infty$. Then $s_{p,\delta} \in C_b(\overline{\mathbb{R}})$ and, in view of $\beta \in (-\pi,\pi)$, the function $h_{p,\delta,\beta}$ tends exponentially to 0 as $x \to \infty$. Hence, we conclude from the equalities

$$
\begin{aligned}
s'_{p,\delta}(x) &= -\pi(1+i\delta)^{-1}(h_{p,\delta,0}(x))^2,\\
s''_{p,\delta}(x) &= 2\pi^2(1+i\delta)^{-2}s_{p,\delta}(x)(h_{p,\delta,0}(x))^2,\\
h'_{p,\delta,\beta}(x) &= (1+i\delta)^{-1}(\beta-\pi s_{p,\delta}(x))\,h_{p,\delta,\beta}(x),\\
h''_{p,\delta,\beta}(x) &= (1+i\delta)^{-2}\big[(\beta-\pi s_{p,\delta}(x))^2+(\pi\, h_{p,\delta,0}(x))^2\big]h_{p,\delta,\beta}(x)
\end{aligned}
$$

that the functions on the left of these equalities are in $L^1(\mathbb{R})$, which completes the proof. □

We start with the following obvious auxiliary statement.

Lemma 4.2. *For every $p\in(1,\infty)$, $\delta\in\mathbb{R}$, $\beta\in(-\pi,\pi)$ and $j\in\{0,1\}$, we have*

$$C_j^\infty(p,\delta,\beta):=\sup_{x\in\mathbb{R}}|x|^j|h_{p,\delta,\beta}(x)|<\infty, \tag{4.3}$$

$$C_j^1(p,\delta,\beta):=\int_{\mathbb{R}}|x|^j|h_{p,\delta,\beta}(x)|\,dx<\infty, \tag{4.4}$$

$$M_0(p,\delta,\beta):=(1+\delta^2)^{-1/2}\Big(|\beta|+\sup_{x\in\mathbb{R}}\big|\pi s_{p,\delta}(x)\big|\Big). \tag{4.5}$$

The following result generalizes [18, Lemma 4.1].

Lemma 4.3. *If $p\in(1,\infty)$, $\delta\in\mathbb{R}$, $\beta\in(-\pi,\pi)$ and the functions ω and $r\mapsto r\omega'(r)$ belong to $SO(\mathbb{R}_+)$, then the functions*

$$\mathfrak{n}_{p,\delta,\beta,\omega}(r,x):=e^{i\omega(r)x}\,h_{p,\delta,\beta}(x)=e^{i\omega(r)x}\frac{\exp\big(\beta\frac{x+i/p}{1+i\delta}\big)}{\sinh\big(\pi\frac{x+i/p}{1+i\delta}\big)},\quad (r,x)\in\mathbb{R}_+\times\overline{\mathbb{R}}, \tag{4.6}$$

belong to the Banach algebra $\widetilde{\mathcal{E}}(\mathbb{R}_+,V(\mathbb{R}))$ and there is a constant $C(p,\delta,\beta)\in(0,\infty)$ depending only on p,δ,β such that

$$\|\mathfrak{n}_{p,\delta,\beta,\omega}\|_{C_b(\mathbb{R}_+,V(\mathbb{R}))}\le C(p,\delta,\beta)\Big(1+\sup_{r\in\mathbb{R}_+}|\omega(r)|\Big). \tag{4.7}$$

Proof. First, by analogy with [16, Lemma 7.3], we will show that $\mathfrak{b}\in C_b(\mathbb{R}_+,V(\mathbb{R}))$. Through the proof we assume that $r,\varrho\in\mathbb{R}_+$ and $x\in\mathbb{R}$. From (4.6) and (4.3) it follows that

$$\|\mathfrak{n}_{p,\delta,\beta,\omega}(r,\cdot)\|_{L^\infty(\mathbb{R})}\le C_0^\infty(p,\delta,\beta). \tag{4.8}$$

Since $\omega\in SO(\mathbb{R}_+)$, we have

$$M_1(\omega):=\sup_{r\in\mathbb{R}_+}|\omega(r)|<\infty. \tag{4.9}$$

From (4.6) it easily follows that

$$\frac{\partial\mathfrak{n}_{p,\delta,\beta,\omega}}{\partial x}(r,x)=\big[i\omega(r)+(1+i\delta)^{-1}(\beta-\pi s_{p,\delta}(x))\big]\mathfrak{n}_{p,\delta,\beta,\omega}(r,x), \tag{4.10}$$

$$\frac{\partial^2\mathfrak{n}_{p,\delta,\beta,\omega}}{\partial x^2}(r,x)=\big(\pi^2(1+i\delta)^{-2}(h_{p,\delta,0}(x))^2+\big[i\omega(r)+(1+i\delta)^{-1}(\beta-\pi s_{p,\delta}(x))\big]^2\big)\times\mathfrak{n}_{p,\delta,\beta,\omega}(r,x). \tag{4.11}$$

From (4.10), (4.9) and (4.4), (4.5) we obtain

$$V(\mathfrak{n}_{p,\delta,\beta,\omega}(r,\cdot)) = \int_{\mathbb{R}} \left| \frac{\partial \mathfrak{n}_{p,\delta,\beta,\omega}(r,x)}{\partial x} \right| dx \le (M_1(\omega)+M_0(p,\delta,\beta))C_0^1(p,\delta,\beta). \quad (4.12)$$

Combining (4.8) and (4.12), we arrive at

$$\begin{aligned} \|\mathfrak{n}_{p,\delta,\beta,\omega}(r,\cdot)\|_V &\le C_0^\infty(p,\delta,\beta) + (M_1(\omega) + M_0(p,\delta,\beta))C_0^1(p,\delta,\beta) \\ &\le C(p,\delta,\beta)(1 + M_1(\omega)), \end{aligned} \quad (4.13)$$

where $C(p,\delta,\beta) := \max\left\{C_0^\infty(p,\delta,\beta) + C_0^1(p,\delta,\beta)M_0(p,\delta,\beta), C_0^1(p,\delta,\beta)\right\}$. Further from (4.6), we get

$$\begin{aligned} \left|\mathfrak{n}_{p,\delta,\beta,\omega}(r,x) - \mathfrak{n}_{p,\delta,\beta,\omega}(\varrho,x)\right| &= \left| ix \left(\int_{\omega(\varrho)}^{\omega(r)} e^{i\theta x} d\theta \right) h_{p,\delta,\beta}(x) \right| \\ &\le |\omega(r) - \omega(\varrho)|\, |x|\, |h_{p,\delta,\beta}(x)|. \end{aligned} \quad (4.14)$$

From (4.14) and (4.3) we obtain

$$\|\mathfrak{n}_{p,\delta,\beta,\omega}(r,\cdot) - \mathfrak{n}_{p,\delta,\beta,\omega}(\varrho,\cdot)\|_{L^\infty(\mathbb{R})} \le C_1^\infty(p,\delta,\beta)|\omega(r) - \omega(\varrho)|. \quad (4.15)$$

It follows from (4.10) that

$$\begin{aligned} &\frac{\partial \mathfrak{n}_{p,\delta,\beta,\omega}(r,x)}{\partial x} - \frac{\partial \mathfrak{n}_{p,\delta,\beta,\omega}(\varrho,x)}{\partial x} = i(\omega(r) - \omega(\varrho))\mathfrak{n}_{p,\delta,\beta,\omega}(r,x) \\ &+ \left[i\omega(\varrho) + (1+i\delta)^{-1}(\beta - \pi s_{p,\delta}(x))\right]\left(\mathfrak{n}_{p,\delta,\beta,\omega}(r,x) - \mathfrak{n}_{p,\delta,\beta,\omega}(\varrho,x)\right). \end{aligned} \quad (4.16)$$

Starting with (4.16) and taking into account (4.14), (4.5) and (4.9), we arrive at

$$\begin{aligned} &\left| \frac{\partial \mathfrak{n}_{p,\delta,\beta,\omega}(r,x)}{\partial x} - \frac{\partial \mathfrak{n}_{p,\delta,\beta,\omega}(\varrho,x)}{\partial x} \right| \\ &\quad \le \left(|h_{p,\delta,\beta}(x)| + (M_1(\omega) + M_0(p,\delta,\beta))|x|\,|h_{p,\delta,\beta}(x)|\right)|\omega(r) - \omega(\varrho)|. \end{aligned}$$

Therefore, taking into account (4.4), we infer from the latter inequality that

$$\begin{aligned} V(\mathfrak{n}_{p,\delta,\beta,\omega}(r,\cdot) - \mathfrak{n}_{p,\delta,\beta,\omega}(\varrho,\cdot)) &= \int_{\mathbb{R}} \left| \frac{\partial \mathfrak{n}_{p,\delta,\beta,\omega}(r,x)}{\partial x} - \frac{\partial \mathfrak{n}_{p,\delta,\beta,\omega}(\varrho,x)}{\partial x} \right| dx \\ &\le L_1(\omega,p,\delta,\beta)|\omega(r) - \omega(\varrho)|, \end{aligned} \quad (4.17)$$

where $L_1(\omega,p,\delta,\beta) := C_0^1(p,\delta,\beta) + (M_1(\omega) + M_0(p,\delta,\beta))C_1^1(p,\delta,\beta)$. Combining (4.15) and (4.17), we arrive at

$$\|\mathfrak{n}_{p,\delta,\beta,\omega}(r,\cdot) - \mathfrak{n}_{p,\delta,\beta,\omega}(\varrho,\cdot)\|_V \le L_2(\omega,p,\delta,\beta)|\omega(r) - \omega(\varrho)| \quad (4.18)$$

for $t,\tau \in \mathbb{R}_+$, where $L_2(\omega,p,\delta,\beta) := C_1^\infty(p,\delta,\beta) + L_1(\omega,p,\delta,\beta)$. From (4.13) and (4.18) it follows that $\mathfrak{n}_{p,\delta,\beta,\omega}$ is a bounded and continuous $V(\mathbb{R})$-valued function. Thus, $\mathfrak{n}_{p,\delta,\beta,\omega} \in C_b(\mathbb{R}_+, V(\mathbb{R}))$. Estimate (4.7) follows immediately from (4.13).

Further, we will show by analogy with [16, Lemma 7.4] that $\mathfrak{n}_{p,\delta,\beta,\omega}$ belongs to the algebra $\mathcal{E}(\mathbb{R}_+, V(\mathbb{R}))$. Estimate (4.15) immediately implies that

$$\mathrm{cm}_r^C(\mathfrak{n}_{p,\delta,\beta,\omega}) \le C_1^\infty(p,\delta,\beta)\,\mathrm{osc}(\omega,[r,2r]), \quad r \in \mathbb{R}_+,$$

where $\mathrm{osc}(\omega, J) := \max_{x,y\in J}|\omega(x)-\omega(y)|$ for any set $J \subset \mathbb{R}_+$. Since $\omega \in SO(\mathbb{R}_+)$, from this estimate we obtain

$$\lim_{r\to s}\mathrm{cm}_r^C(\mathfrak{n}_{p,\delta,\beta,\omega}) = \lim_{r\to s}\mathrm{osc}(\omega,[r,2r]) = 0, \quad s \in \{0,\infty\}.$$

Thus, $\mathfrak{n}_{p,\delta,\beta,\omega} \in SO(\mathbb{R}_+, V(\mathbb{R}))$.

From (4.10), (4.9) and (4.3), (4.5) we infer that

$$\left|\frac{\partial \mathfrak{n}_{p,\delta,\beta,\omega}(r,x)}{\partial x}\right| \le M_2(\omega,p,\delta,\beta),$$

where $M_2(\omega,p,\delta,\beta) := (M_1(\omega)+M_0(p,\delta,\beta))C_0^\infty(p,\delta,\beta) < \infty$.

Let $h \in \mathbb{R}$. Then

$$\begin{aligned}|\mathfrak{n}_{p,\delta,\beta,\omega}(r,x) - \mathfrak{n}^h_{p,\delta,\beta,\omega}(r,x)| &= \left|\int_x^{x+h}\frac{\partial \mathfrak{n}_{p,\delta,\beta,\omega}}{\partial y}(r,y)\,dy\right| \le \left|\int_x^{x+h}\left|\frac{\partial \mathfrak{n}_{p,\delta,\beta,\omega}}{\partial y}(r,y)\right|dy\right| \\ &\le M_2(\omega,p,\delta,\beta)|h|.\end{aligned}$$

Therefore,

$$\sup_{t\in\mathbb{R}_+}\left\|\mathfrak{n}_{p,\delta,\beta,\omega}(r,\cdot) - \mathfrak{n}^h_{p,\delta,\beta,\omega}(r,\cdot)\right\|_{L^\infty(\mathbb{R})} \le M_2(\omega,p,\delta,\beta)|h|, \quad h\in\mathbb{R}. \tag{4.19}$$

From (4.11), (4.9) and (4.3), (4.5) we obtain

$$\left|\frac{\partial^2 \mathfrak{n}_{p,\delta,\beta,\omega}(r,x)}{\partial x^2}\right| \le M_3(\omega,p,\delta,\beta)|h_{p,\delta,\beta}(x)|,$$

where $M_3(\omega,p,\delta,\beta) := (\pi C_0^\infty(p,\delta,0))^2 + (M_1(\omega)+M_0(p,\delta,\beta))^2 < \infty$. Fix $h > 0$. Then

$$\begin{aligned}V\big(\mathfrak{n}_{p,\delta,\beta,\omega}(r,\cdot) - \mathfrak{n}^h_{p,\delta,\beta,\omega}(r,\cdot)\big) &= \int_{\mathbb{R}}\left|\frac{\partial \mathfrak{n}_{p,\delta,\beta,\omega}}{\partial x}(r,x+h) - \frac{\partial \mathfrak{n}_{p,\delta,\beta,\omega}}{\partial x}(r,x)\right|dx \\ = \int_{\mathbb{R}}\left|\int_x^{x+h}\frac{\partial^2 \mathfrak{n}_{p,\delta,\beta,\omega}}{\partial y^2}(r,y)dy\right|dx &\le \int_{\mathbb{R}}\int_x^{x+h}\left|\frac{\partial^2 \mathfrak{n}_{p,\delta,\beta,\omega}}{\partial y^2}(r,y)\right|dy\,dx \\ \le M_3(\omega,p,\delta,\beta)\int_{\mathbb{R}}\int_x^{x+h}|h_{p,\delta,\beta}(y)|dy\,dx.\end{aligned} \tag{4.20}$$

Changing the order of integration in the integral on the right-hand side of (4.20) and taking into account (4.3), we get for $h \in \mathbb{R}$,

$$\int_{\mathbb{R}} \int_{x}^{x+h} |h_{p,\delta,\beta}(y)| dy\, dx = \int_{\mathbb{R}} \int_{y-h}^{y} |h_{p,\delta,\beta}(y)| dx\, dy = h \int_{\mathbb{R}} |h_{p,\delta,\beta}(y)| dy = C_0^1(p,\delta,\beta)h. \tag{4.21}$$

Combining (4.20) and (4.21), we see that

$$V\big(\mathfrak{n}_{p,\delta,\beta,\omega}(r,\cdot) - \mathfrak{n}^h_{p,\delta,\beta,\omega}(r,\cdot)\big) \le M_4(\omega,p,\delta,\beta)h \quad (h > 0), \tag{4.22}$$

where $M_4(\omega,p,\delta,\beta) := C_0^1(p,\delta,\beta)M_3(\omega,p,\delta,\beta)$. Analogously it can be shown that

$$V\big(\mathfrak{n}_{p,\delta,\beta,\omega}(r,\cdot) - \mathfrak{n}^h_{p,\delta,\beta,\omega}(r,\cdot)\big) \le M_4(\omega,p,\delta,\beta)(-h) \quad (h < 0). \tag{4.23}$$

From (4.22) and (4.23) we get for $h \in \mathbb{R}$,

$$\sup_{r\in\mathbb{R}_+} V\big(\mathfrak{n}_{p,\delta,\beta,\omega}(r,\cdot) - \mathfrak{n}^h_{p,\delta,\beta,\omega}(r,\cdot)\big) \le M_4(\omega,p,\delta,\beta)|h|. \tag{4.24}$$

Combining (4.19) with (4.24), we arrive at

$$\lim_{|h|\to 0} \sup_{t\in\mathbb{R}_+} \big\|\mathfrak{n}_{p,\delta,\beta,\omega}(r,\cdot) - \mathfrak{n}^h_{p,\delta,\beta,\omega}(r,\cdot)\big\|_V \le M_5(\omega,p,\delta,\beta) \lim_{|h|\to 0} |h| = 0,$$

where $M_5(\omega,p,\delta,\beta) := M_2(\omega,p,\delta,\beta) + M_4(\omega,p,\delta,\beta)$, which implies that the function $\mathfrak{n}_{p,\delta,\beta,\omega} \in SO(\mathbb{R}_+, V(\mathbb{R}))$ actually belongs to $\mathcal{E}(\mathbb{R}_+, V(\mathbb{R}))$.

Finally, from (4.10), (4.9) and (4.5) it follows that

$$\begin{aligned} &\lim_{m\to\infty} \sup_{r\in\mathbb{R}_+} \int_{\mathbb{R}\setminus[-m,m]} \left| \frac{\partial \mathfrak{n}_{p,\delta,\beta,\omega}}{\partial x}(r,x) \right| dx \\ &\le (M_1(\omega) + M_0(p,\delta,\beta)) \lim_{m\to\infty} \int_{\mathbb{R}\setminus[-m,m]} |h_{p,\delta,\beta}(x)|\, dx = 0, \end{aligned}$$

which in view of (3.3) implies that $\mathfrak{n}_{p,\delta,\beta,\omega} \in \widetilde{\mathcal{E}}(\mathbb{R}_+, V(\mathbb{R}))$. □

5. Reduction to Mellin pseudodifferential operators

Let $1 < p < \infty$ and let $\Gamma = \bigcup_{k=1}^N \Gamma_k$ be the logarithmic star given by (2.5). In this section we apply the results on Mellin pseudodifferential operators collected in Section 3 to singular integral operators acting on the Lebesgue space $L^p(\Gamma)$.

Slightly modifying the proof of [5, Theorem 2.2.1] and [4, Theorem 10.20], we obtain the following.

Lemma 5.1 ([25, Lemma 5.1]). *If $y > 0$, $0 < \nu < 1$, $\xi \in \mathbb{C}$ and $\operatorname{Re}\xi > \nu$, then*

$$\frac{1}{\pi i} \frac{\xi y^\nu}{1 - y^\xi} = \frac{1}{2\pi} \int_{\mathbb{R}} \coth[\pi(x + i\nu)/\xi]\, y^{ix}\, dx. \tag{5.1}$$

Given $b \in \mathbb{C}$ with $\operatorname{Re} b \in (0, 2\pi)$, let K_b be the open sector in $\mathbb{C}$ given by

$$K_b = \begin{cases} \left\{ z = re^{i\theta} : r > 0, \ \arctan \dfrac{2\pi - \operatorname{Re} b}{\operatorname{Im} b} < \theta < \arctan \dfrac{-\operatorname{Re} b}{\operatorname{Im} b} \right\} & \text{if } \operatorname{Im} b < 0, \\ \left\{ z = re^{i\theta} : r > 0, \ \arctan \dfrac{-\operatorname{Re} b}{\operatorname{Im} b} < \theta < \arctan \dfrac{2\pi - \operatorname{Re} b}{\operatorname{Im} b} \right\} & \text{if } \operatorname{Im} b > 0, \\ \left\{ z = re^{i\theta} : r > 0, \ -\pi/2 < \theta < \pi/2 \right\} \ \text{if } \operatorname{Im} b = 0, \end{cases}$$

and let

$$\mathcal{D}_{b,\nu} := \{ z \in K_b : \operatorname{Re} z > \nu \}, \quad b \in \mathbb{C}, \quad \operatorname{Re} b \in (0, 2\pi), \quad \nu \in (0,1).$$

By analogy with [5, Theorem 2.2.1], we get the following.

Lemma 5.2 ([25, Lemma 5.2]). *If* $y > 0$, $b \in \mathbb{C}$ *with* $\operatorname{Re} b \in (0, 2\pi)$, $0 < \nu < 1$ *and* $\xi \in \mathcal{D}_{b,\nu}$, *then*

$$\frac{1}{\pi i} \frac{\xi y^\nu}{1 - e^{ib} y^\xi} = \frac{1}{2\pi} \int_{\mathbb{R}} \frac{e^{(\pi - b)(x + i\nu)/\xi}}{\sinh[\pi(x + i\nu)/\xi]} y^{ix} dx. \tag{5.2}$$

Given $1 < p < \infty$ and $N \in \mathbb{N}$, we consider the Banach space $L^p_N(\mathbb{R}_+, d\mu)$ of vector-functions $\varphi = \{\varphi_k\}_{k=1}^N$ with entries $\varphi_k \in L^p(\mathbb{R}_+, d\mu)$ and the norm

$$\|\varphi\| = \Big(\sum\nolimits_{k=1}^N \|\varphi_k\|^p_{L^p(\mathbb{R}_+, d\mu)} \Big)^{1/p}.$$

By analogy with [7], we introduce the isomorphisms

$$\begin{aligned} &\Phi : L^p(\Gamma) \to L^p_N(\mathbb{R}_+, d\mu), \quad (\Phi f)(r) = \Big\{ r^{1/p} f\big(e^{i\beta_k} r^{1+i\delta}\big) \Big\}_{k=1}^N \quad (r \in \mathbb{R}_+), \\ &\Psi : \mathcal{B}(L^p(\Gamma)) \to \mathcal{B}(L^p_N(\mathbb{R}_+, d\mu)), \quad A \mapsto \Phi A \Phi^{-1}. \end{aligned} \tag{5.3}$$

Since Γ is a Carleson curve and (2.7) holds, the operators $V_\alpha^{\pm 1}$ and S_Γ are bounded on the space $L^p(\Gamma)$. As we will see below, the Ψ-image of the operator S_Γ is a Mellin convolution operator.

Theorem 5.3. *Let* $1 < p < \infty$ *and let* $\Gamma = \bigcup_{k=1}^N \Gamma_k$ *be the logarithmic star given by* (2.5) *and* (2.6). *Then*

$$\Psi(S_\Gamma) = \big[\operatorname{Op}(\varepsilon_k \mathfrak{g}_{j,k}) \big]_{j,k=1}^N, \quad \mathfrak{g}_{j,k}(r, x) := g_{j,k}(x) \ \text{for} \ (r, x) \in \mathbb{R}_+ \times \overline{\mathbb{R}}, \tag{5.4}$$

where $\varepsilon_k := 1$ *if* 0 *is the starting point of* Γ_k, $\varepsilon_k := -1$ *if* 0 *is the terminating point of* Γ_k, *the functions* $g_{j,k}$ *are given for all* $x \in \mathbb{R}$ *and all* $k, j = 1, 2, \dots, N$ *by*

$$g_{j,k}(x) := \begin{cases} \coth\Big(\pi \dfrac{x + i/p}{1 + i\delta}\Big) & \text{if } j = k, \\ \dfrac{\exp\big(\theta_{j,k} \frac{x + i/p}{1 + i\delta}\big)}{\sinh\big(\pi \frac{x + i/p}{1 + i\delta}\big)} & \text{if } j \neq k, \end{cases} \tag{5.5}$$

$$\theta_{j,k} := \pi \operatorname{sgn}(j - k) - (\beta_j - \beta_k) \in (-\pi, \pi), \tag{5.6}$$

and $g_{j,k} \in V(\mathbb{R})$, $\mathfrak{g}_{j,k} \in \widetilde{\mathcal{E}}(\mathbb{R}_+, V(\mathbb{R}))$ *for all* $j, k = 1, 2, \dots, N$.

Proof. Clearly, $\Psi(S_\Gamma) = \big([\Psi(S_\Gamma)]_{j,k}\big)_{j,k=1}^N$ with entries $[\Psi(S_\Gamma)]_{j,k} \in \mathcal{B}(L^p(\mathbb{R}_+, d\mu))$. For $r \in \mathbb{R}_+$, all $j,k = 1,2,\dots,N$ and $f \in L^p(\mathbb{R}_+, d\mu)$, we obtain

$$\begin{aligned}\big([\Psi(S_\Gamma)]_{j,k} f\big)(r) &= \frac{\varepsilon_k}{\pi i}\int_{\mathbb{R}_+} \frac{(r/\varrho)^{1/p} e^{i\beta_k}(1+i\delta)\varrho^{i\delta}}{e^{i\beta_k}\varrho^{1+i\delta} - e^{i\beta_j} r^{1+i\delta}} f(\varrho)\, d\varrho \\ &= \frac{\varepsilon_k}{\pi i}\int_{\mathbb{R}_+} \frac{(1+i\delta)(r/\varrho)^{1/p}}{1 - e^{i(\beta_j-\beta_k)}(r/\varrho)^{1+i\delta}} f(\varrho)\frac{d\varrho}{\varrho}. \end{aligned} \tag{5.7}$$

Setting $y := r/\varrho$, $\nu := 1/p$, $\xi := 1 + i\delta$ and

$$b := 2\pi + \beta_j - \beta_k \ \text{ if } \ j < k, \quad b := \beta_j - \beta_k \ \text{ if } \ j > k, \tag{5.8}$$

we deduce that $\xi \in \mathcal{D}_{b,\nu}$ for $j \ne k$ because $\operatorname{Re} b \in (0, 2\pi)$, $\operatorname{Im} b = 0$, and $\operatorname{Re}\xi > \nu$. Hence, applying (5.1) and (5.2), we infer from (5.7) that, for all $r \in \mathbb{R}_+$ and $j,k = 1,2,\dots,N$,

$$\big([\Psi(S_\Gamma)]_{j,k} f\big)(r) = \frac{\varepsilon_k}{2\pi}\int_{\mathbb{R}} dx \int_{\mathbb{R}_+} g_{j,k}(x)\Big(\frac{r}{\varrho}\Big)^{ix} f(\varrho)\frac{d\varrho}{\varrho}, \tag{5.9}$$

where the functions $g_{j,k}$ are given by (5.5)–(5.6). Hence, according to (4.1) and (4.2), $g_{k,k} = s_{p,\delta}$ and $g_{j,k} = h_{p,\delta,\theta_{j,k}}$ for all $j,k = 1,2,\dots,N$; $j \ne k$, where $\theta_{j,k} \in (-\pi,\pi)$ is given by (5.6). Since $1/p \in (0,1)$, we infer from Lemma 4.1 with $\beta = \theta_{j,k}$ that $g_{j,k} \in V(\mathbb{R})$ and $\mathfrak{g}_{j,k} \in \widetilde{\mathcal{E}}(\mathbb{R}_+, V(\mathbb{R}))$ for all $k,j = 1,2,\dots,N$. Hence, from (5.9) and (3.1) it follows that $[\Psi(S_\Gamma)]_{j,k} = \operatorname{Op}(\varepsilon_k\, \mathfrak{g}_{j,k})$, which gives (5.4). $\square$

Under the conditions of Theorem 5.3, we consider the operators $S_\delta, R_{j,k} \in \mathcal{B}(L^p(\mathbb{R}_+))$ $(k,j = 1,2,\dots,N;\ j \ne k)$ given for $r \in \mathbb{R}_+$ by

$$(S_\delta f)(r) = \frac{1}{\pi i}\int_{\mathbb{R}_+} \frac{(1+i\delta)\varrho^{i\delta}}{\varrho^{1+i\delta} - r^{1+i\delta}} f(\varrho)\, d\varrho, \tag{5.10}$$

$$(R_\delta f)(r) = \frac{1}{\pi i}\int_{\mathbb{R}_+} \frac{(1+i\delta)\varrho^{i\delta}}{\varrho^{1+i\delta} + r^{1+i\delta}} f(\varrho)\, d\varrho, \tag{5.11}$$

$$(R_{j,k} f)(r) = \frac{1}{\pi i}\int_{\mathbb{R}_+} \frac{(1+i\delta)\varrho^{i\delta}}{\varrho^{1+i\delta} - e^{i(\beta_j-\beta_k)} r^{1+i\delta}} f(\varrho)\, d\varrho. \tag{5.12}$$

Along with these operators we also consider the operators $S_{\mathbb{R}_+}, R \in \mathcal{B}(L^p(\mathbb{R}_+))$ given for $r \in \mathbb{R}_+$ by

$$(S_{\mathbb{R}_+} f)(r) = \frac{1}{\pi i}\int_{\mathbb{R}_+} \frac{f(\varrho)}{\varrho - r}\, d\varrho, \quad (Rf)(r) = \frac{1}{\pi i}\int_{\mathbb{R}_+} \frac{f(\varrho)}{\varrho + r}\, d\varrho. \tag{5.13}$$

We now define the isomorphism

$$\Upsilon:\ L^p(\mathbb{R}_+) \to L^p(\mathbb{R}_+, d\mu), \quad (\Upsilon f)(r) = r^{1/p} f(r) \quad (r \in \mathbb{R}_+). \tag{5.14}$$

Let $C_p(\overline{\mathbb{R}})$ be the closure of the set $V(\mathbb{R}) \subset C(\overline{\mathbb{R}})$ in the Banach algebra $\mathcal{M}_p(\mathbb{R})$ of Fourier multipliers on $L^p(\mathbb{R})$, and let $C_{p,0}(\overline{\mathbb{R}}) = \{a \in C_p(\overline{\mathbb{R}}) :\ a(\infty) = 0\}$ be the closed two-sided ideal in the Banach algebra $C_p(\overline{\mathbb{R}})$.

Corollary 5.4. *Under the conditions of Theorem* 5.3*, the operators* S_δ*,* R_δ *and* $R_{j,k}$ $(j,k=1,2,\dots,N;\ j\neq k)$ *given by* (5.10)*,* (5.11) *and* (5.12) *belong to the Banach algebra* $\mathcal{A}=\mathrm{alg}\{I,S_{\mathbb{R}_+}\}$ *generated by the identity operator* I *and by the Cauchy singular integral operator* $S_{\mathbb{R}_+}$*. Moreover, the operators* R_δ*,* $R_{j,k}$ *for* $j\neq k$ *and the operator* $S_\delta-S_{\mathbb{R}_+}$ *belong to the closed two-sided ideal* $\mathrm{id}\{R\}$ *of* $\mathcal{A}$ *generated by the operator* R *defined in* (5.13).

Proof. By the proof of Theorem 5.3, for all $k,j=1,2,\dots,m$ with $j\neq k$,

$$\Upsilon S_\delta\Upsilon^{-1}=\mathrm{Op}(\mathfrak{s}_{p,\delta}),\quad \Upsilon R_\delta\Upsilon^{-1}=\mathrm{Op}(\mathfrak{h}_{p,\delta,0}),\quad \Upsilon R_{j,k}\Upsilon^{-1}=\mathrm{Op}(\mathfrak{h}_{p,\delta,\theta_{j,k}}). \tag{5.15}$$

As is well known (see, e.g., [35, Sections 4.2.2–4.2.3]),

$$\Upsilon S_{\mathbb{R}_+}\Upsilon^{-1}=\mathrm{Op}(\mathfrak{s}_p),\quad \Upsilon R\Upsilon^{-1}=\mathrm{Op}(\mathfrak{h}_p), \tag{5.16}$$

where $\mathfrak{s}_p(r,x)=s_p(x)$ and $\mathfrak{h}_p(r,x)=h_p(x)$ for all $(r,x)\in\mathbb{R}_+\times\overline{\mathbb{R}}$, the functions $s_p,h_p\in C_p(\overline{\mathbb{R}})$ are given by

$$s_p(x):=\coth[\pi(x+i/p)],\quad h_p(x):=1/(\sinh[\pi(x+i/p)])\quad \text{for}\quad x\in\overline{\mathbb{R}}, \tag{5.17}$$

and the functions 1, s_p are generators of $C_p(\overline{\mathbb{R}})$, while the function h_p is a generator of the ideal $C_{p,0}(\overline{\mathbb{R}})$. Since $s_{p,\delta}=g_{k,k}$ and $h_{p,\delta,\theta_{j,k}}=g_{j,k}$ for $j\neq k$ and since all functions $g_{j,k}$ belong to $V(\mathbb{R})\subset C_p(\overline{\mathbb{R}})$ according to Theorem 5.3, from (5.15)–(5.17) it follows that $S_\delta,R_{j,k}\in\mathcal{A}=\mathrm{alg}\{I,S_{\mathbb{R}_+}\}$. Analogously, $R_\delta\in\mathcal{A}$ because $h_{p,\delta,0}\in C_p(\overline{\mathbb{R}})$. Moreover, by (4.1), (4.2) and (5.17), the functions $s_{p,\delta}-s_p$, $h_{p,\delta,0}$ and $h_{p,\delta,\theta_{j,k}}$ in $C_p(\overline{\mathbb{R}})$ vanish at ∞ and therefore belong to the ideal $C_{p,0}(\overline{\mathbb{R}})$. Hence, the operators $S_\delta-S_{\mathbb{R}_+}$, R_δ and $R_{j,k}$ for $j\neq k$ belong to the ideal $\mathrm{id}\{R\}$ of the Banach algebra $\mathcal{A}$. □

Along with isomorphisms (5.3), we introduce the isomorphisms

$$\begin{aligned}&\Phi_0: L^p(\Gamma)\to L^p_N(\mathbb{R}_+),\quad (\Phi_0 f)(r)=\left\{f\left(e^{i\beta_k}r^{1+i\delta}\right)\right\}_{k=1}^N\quad (r\in\mathbb{R}_+),\\ &\Psi_0:\mathcal{B}(L^p(\Gamma))\to\mathcal{B}(L^p_N(\mathbb{R}_+)),\quad A\mapsto\Phi_0A\Phi_0^{-1}.\end{aligned} \tag{5.18}$$

Setting $\alpha_k(r):=re^{\omega_k(r)}$, we infer that

$$\Psi_0(V_\alpha)=\mathrm{diag}\left\{V_{\alpha_k}\right\}_{k=1}^N,\quad V_{\alpha_k}\in\mathcal{B}(L^p(\mathbb{R}_+)),\quad V_{\alpha_k}f:=f\circ\alpha_k. \tag{5.19}$$

Lemma 5.5. *If* $1<p<\infty$*,* $\delta\in\mathbb{R}$ *and* $b\in SO(\mathbb{R}_+)$*, then for all* $j,s,k=1,2,\dots,N;\ j\neq s$ *the commutators*

$$[bI,S_\delta],\quad [bI,R_\delta],\quad [bI,R_{j,s}],\quad [V_{\alpha_k},S_\delta],\quad [V_{\alpha_k},R_\delta],\quad [V_{\alpha_k},R_{j,s}],$$

where the operators S_δ*,* R_δ *and* $R_{j,s}$ *are given, respectively, by* (5.10)*,* (5.11) *and* (5.12)*, are compact operators on the space* $L^p(\mathbb{R}_+)$.

Proof. By [16, Theorem 6.3], the commutators $[bI,S_{\mathbb{R}_+}]$ and $[V_{\alpha_k},S_{\mathbb{R}_+}]$ are compact on the space $L^p(\mathbb{R}_+)$ for all $k=1,2,\dots,N$. Since the operators S_δ, R_δ and $R_{j,s}$ $(j,s=1,2,\dots,N;\ j\neq s)$ belong to the Banach algebra $\mathcal{A}=\mathrm{alg}\{I,S_{\mathbb{R}_+}\}$ according to Corollary 5.4, we immediately obtain the compactness of the required commutators from the compactness of $[bI,S_{\mathbb{R}_+}]$ and $[V_{\alpha_k},S_{\mathbb{R}_+}]$. □

Theorem 5.6. *Let $1 < p < \infty$ and let $\Gamma = \bigcup_{k=1}^{N} \Gamma_k$ be the logarithmic star given by (2.5) and (2.6), and let $\alpha : \Gamma \to \Gamma$ be the slowly oscillating shift given by (2.8) and satisfying (2.7) and (2.11), where the functions ω_k and $\phi_k : r \mapsto r\omega_k'(r)$ are in $SO(\mathbb{R}_+)$ for all $k = 1, 2, \dots, N$. If $j, k = 1, 2, \dots, N$ and $j \neq k$, then*

$$[\Psi_0(V_\alpha S_\Gamma)]_{j,k} = \varepsilon_k V_{\alpha_j} R_{j,k}, \tag{5.20}$$

$$[\Psi(V_\alpha S_\Gamma)]_{j,k} = \Upsilon(\varepsilon_k V_{\alpha_j} R_{j,k})\Upsilon^{-1} = \varepsilon_k \operatorname{Op}(\mathfrak{v}_{j,k}), \tag{5.21}$$

where the operators $R_{j,k}$ and V_{α_j} are defined by (5.12) and (5.19), respectively, and the functions $\mathfrak{v}_{j,k} \in \widetilde{\mathcal{E}}(\mathbb{R}_+, V(\mathbb{R}))$ are given for $(r, x) \in \mathbb{R}_+ \times \mathbb{R}$ by

$$\mathfrak{v}_{j,k}(r, x) := e^{i\omega_j(r)(x+i/p)} \frac{\exp\left(\theta_{j,k} \frac{x+i/p}{1+i\delta}\right)}{\sinh\left(\pi \frac{x+i/p}{1+i\delta}\right)} \tag{5.22}$$

with $\theta_{j,k}$ defined by (5.6).

Proof. Let $j, k \in \{1, 2, \dots, N\}$ and $j \neq k$. Then, by (5.7), $[\Psi(S_\Gamma)]_{j,k} = \varepsilon_k \widetilde{R}_{j,k}$, where

$$(\widetilde{R}_{j,k} f)(r) = \frac{1}{\pi i} \int_{\mathbb{R}_+} \frac{(r/\varrho)^{1/p}(1 + i\delta)\varrho^{i\delta}}{\varrho^{1+i\delta} - e^{i(\beta_j - \beta_k)} r^{1+i\delta}} f(\varrho)\, d\varrho \quad \text{for } r \in \mathbb{R}_+. \tag{5.23}$$

Since $\Psi_0 = \Upsilon^{-1}\Psi\Upsilon$ by(5.3), (5.14) and (5.18), and since $R_{j,k} = \Upsilon^{-1}\widetilde{R}_{j,k}\Upsilon$, we infer that $[\Psi_0(S_\Gamma)]_{j,k} = \varepsilon_k R_{j,k}$, which in view of (5.19) implies (5.20). Further,

$$\Psi(V_\alpha) = \operatorname{diag}\left\{c_k \widetilde{V}_{\alpha_k}\right\}_{k=1}^{N}, \quad \widetilde{V}_{\alpha_k} f := f \circ \alpha_k, \quad c_k(r) := e^{-\omega_k(r)/p} \quad (r \in \mathbb{R}_+), \tag{5.24}$$

and therefore

$$[\Psi(V_\alpha S_\Gamma)]_{j,k} = \Upsilon(\varepsilon_k V_{\alpha_j} R_{j,k})\Upsilon^{-1} = \varepsilon_k c_j \widetilde{V}_{\alpha_j} \widetilde{R}_{j,k}. \tag{5.25}$$

Then

$$\begin{aligned}(c_j \widetilde{V}_{\alpha_j} \widetilde{R}_{j,k} f)(r) &= \frac{1}{\pi i} \int_{\mathbb{R}_+} \frac{e^{-\omega_j(r)/p}(re^{\omega_j(r)}/\varrho)^{1/p}(1 + i\delta)\varrho^{i\delta}}{\varrho^{1+i\delta} - e^{i(\beta_j - \beta_k)}(re^{\omega_j(r)})^{1+i\delta}} f(\varrho)\, d\varrho \\ &= \frac{1}{\pi i} \int_{\mathbb{R}_+} \frac{e^{-\omega_j(r)/p}(1 + i\delta)(re^{\omega_j(r)}/\varrho)^{1/p}}{1 - e^{i(\beta_j - \beta_k)}(re^{\omega_j(r)}/\varrho)^{1+i\delta}} f(\varrho)\, \frac{d\varrho}{\varrho}.\end{aligned} \tag{5.26}$$

Taking b given by (5.8) and

$$y := re^{\omega_j(r)}/\varrho, \quad \nu := 1/p, \quad \xi := 1 + i\delta,$$

we infer from (5.26) and the Mellin transform identity (5.2) that

$$(c_j \widetilde{V}_{\alpha_j} \widetilde{R}_{j,k} f)(r) = \frac{1}{2\pi} \int_{\mathbb{R}} dx \int_{\mathbb{R}_+} \mathfrak{v}_{j,k}(r, x) \left(\frac{r}{\varrho}\right)^{ix} f(\varrho) \frac{d\varrho}{\varrho} \quad \text{for } r \in \mathbb{R}_+, \tag{5.27}$$

where the functions $\mathfrak{v}_{j,k}$ are given by (5.22) with (5.6).

Further, from (5.22), (5.24) and (4.6) it follows that $\mathfrak{v}_{j,k} = \mathfrak{c}_j \mathfrak{n}_{p,\delta,\theta_{j,k},\omega_j}$, where $\mathfrak{n}_{p,\delta,\theta_{j,k},\omega_j} \in \widetilde{\mathcal{E}}(\mathbb{R}_+, V(\mathbb{R}))$ by Lemma 4.3, $\mathfrak{c}_j(r, x) := c_j(r)$ for $(r, x) \in \mathbb{R}_+ \times \overline{\mathbb{R}}$, $c_j = e^{-\omega_j(r)/p}$ are in $SO(\mathbb{R}_+)$ along with ω_j, and hence $\mathfrak{c}_j \in \widetilde{\mathcal{E}}(\mathbb{R}_+, V(\mathbb{R}))$. This implies that $\mathfrak{v}_{j,k} = \mathfrak{c}_j \mathfrak{n}_{p,\delta,\theta_{j,k},\omega_j} \in \widetilde{\mathcal{E}}(\mathbb{R}_+, V(\mathbb{R}))$ too, and then (5.27) gives the equality

$$c_j \widetilde{V}_{\alpha_j} \widetilde{R}_{j,k} = \mathrm{Op}(\mathfrak{v}_{j,k}). \tag{5.28}$$

Finally, (5.21) follows from (5.25) and (5.28). □

Let $U_\alpha := |\alpha'|^{1/p} V_\alpha \in \mathcal{B}(L^p(\Gamma))$ be the isometric weighted shift operator, where $|\alpha'|^{1/p} \in SO(\Gamma)$. Hence, by (5.3) and (5.24), we obtain

$$\Psi(U_\alpha) = \mathrm{diag}\left\{\widetilde{U}_{\alpha_k}\right\}_{k=1}^{N}, \quad \widetilde{U}_{\alpha_k} := \widetilde{c}_k \widetilde{V}_{\alpha_k}, \tag{5.29}$$

where $\widetilde{V}_{\alpha_k}$ is given by (5.24) and, by (2.10),

$$\widetilde{c}_k(r) := \left|\alpha'\left(e^{i\beta_k} r^{1+i\delta}\right)\right|^{1/p} c_k(r) = (1 + r\omega_k'(r))^{1/p} \quad \text{for } r \in \mathbb{R}_+. \tag{5.30}$$

Corollary 5.7. *If the conditions of Theorem* 5.6 *are fulfilled,* $j, k = 1, 2, \dots, N$ *and* $j \neq k$, *then*

$$\widetilde{U}_{\alpha_j} \widetilde{R}_{j,k} = \mathrm{Op}(\mathfrak{u}_{j,k}) + K_{j,k}, \tag{5.31}$$

where $\widetilde{U}_{\alpha_j}$ *and* $\widetilde{R}_{j,k}$ *are given by* (5.29) *and* (5.23), *respectively, and the functions* $\mathfrak{u}_{j,k} \in \widetilde{\mathcal{E}}(\mathbb{R}_+, V(\mathbb{R}))$ *are given by*

$$\mathfrak{u}_{j,k}(r,x) := \mathfrak{n}_{p,\delta,\theta_{j,k},\omega_j}(r,x) = e^{i\omega_j(r)x} \frac{\exp\left(\theta_{j,k} \frac{x+i/p}{1+i\delta}\right)}{\sinh\left(\pi \frac{x+i/p}{1+i\delta}\right)} \tag{5.32}$$

for all $(r,x) \in \mathbb{R}_+ \times \overline{\mathbb{R}}$ *and* $K_{j,k} \in \mathcal{K}(L^p(\mathbb{R}_+, d\mu))$.

Proof. For $j \neq k$, we derive from (5.28), (5.22), (5.29) and (5.30) that

$$\widetilde{U}_{\alpha_j} \widetilde{R}_{j,k} = \mathrm{Op}(\widetilde{\mathfrak{c}}_j (\mathfrak{c}_j)^{-1} \mathfrak{v}_{j,k}) = \mathrm{Op}(\widetilde{\mathfrak{c}}_j \mathfrak{u}_{j,k}), \tag{5.33}$$

where $\widetilde{\mathfrak{c}}_j(r,x) = \widetilde{c}_j(r)$ for all $(r,x) \in \mathbb{R}_+ \times \overline{\mathbb{R}}$, the functions $\widetilde{c}_j$ given by (5.30) are in $SO(\mathbb{R}_+)$, whence $\widetilde{\mathfrak{c}}_j \in \widetilde{\mathcal{E}}(\mathbb{R}_+, V(\mathbb{R}))$, and by (5.32) the functions $\widetilde{\mathfrak{c}}_j \mathfrak{u}_{j,k}$ belong to $\widetilde{\mathcal{E}}(\mathbb{R}_+, V(\mathbb{R}))$ along with $\widetilde{\mathfrak{c}}_j$ and $\mathfrak{v}_{j,k} = \mathfrak{n}_{p,\delta,\theta_{j,k},\omega_j}$ (see Lemma 4.3).

By (2.9), $\lim_{r \to s} \widetilde{c}_j(r) = 1$ for $s \in \{0, \infty\}$, and therefore

$$(\widetilde{\mathfrak{c}}_j \mathfrak{u}_{j,k})(r,x) - \mathfrak{u}_{j,k}(r,x) = 0 \quad \text{for all} \quad (r,x) \in \partial(\mathbb{R}_+ \times \mathbb{R}).$$

Consequently, from Theorem 3.4 it follows that

$$\mathrm{Op}(\widetilde{\mathfrak{c}}_j \mathfrak{u}_{j,k}) - \mathrm{Op}(\mathfrak{u}_{j,k}) \in \mathcal{K}(L^p(\mathbb{R}_+, d\mu)),$$

which together with (5.33) gives (5.31). □

6. Fredholmness and index of auxiliary operators $T_{\pm}$

Given $p \in (1, \infty)$, let the contour Γ and the shift $\alpha : \Gamma \to \Gamma$ satisfy the conditions of Theorem 5.6. Consider the function $b_\alpha \in SO(\Gamma)$ defined by

$$\Psi(b_\alpha I) = \mathcal{B}_\alpha I, \quad \mathcal{B}_\alpha(r) := \operatorname{diag}\big\{ \exp\big[i\omega_k(r)\delta/p\big]\big\}_{k=1}^{N} \quad \text{for } r \in \mathbb{R}_+, \tag{6.1}$$

where $\mathcal{B}_\alpha I \in \mathcal{B}(L^p_N(\mathbb{R}_+, d\mu))$. Then

$$\Psi_0(b_\alpha I) = \Upsilon^{-1}\Psi(b_\alpha I)\Upsilon = \Upsilon^{-1}(\mathcal{B}_\alpha I)\Upsilon = \mathcal{B}_\alpha I \in \mathcal{B}(L^p_N(\mathbb{R}_+)). \tag{6.2}$$

Since $U_\alpha = |\alpha'|^{1/p} V_\alpha$, from (6.1)–(6.2) and (5.19) it follows that

$$\Psi_0(b_\alpha U_\alpha) = \operatorname{diag}\big\{b_k V_{\alpha_k}\big\}_{k=1}^{N}, \quad \Psi_0(U_\alpha^{-1} b_\alpha^{-1} I) = \operatorname{diag}\big\{V_{\alpha_k}^{-1} b_k^{-1} I\big\}_{k=1}^{N}, \tag{6.3}$$

where

$$b_k(r) := (1 + r\omega_k'(r))^{1/p} e^{\omega_k(r)(1+i\delta)/p} \quad \text{for all } r \in \mathbb{R}_+, \tag{6.4}$$

and hence $b_k \in SO(\mathbb{R}_+)$.

Lemma 6.1. *Under the conditions of Section 2, for all $j, k = 1, 2, \dots, N$ with $j \neq k$,*

$$\Upsilon(b_j V_{\alpha_j})^{\epsilon_j} R_{j,k} (b_k V_{\alpha_k})^{\epsilon_k} \Upsilon^{-1} \simeq \operatorname{Op}(\mathfrak{b}_j^{\epsilon_j} \mathfrak{w}_j^{\epsilon_j} \mathfrak{h}_{p,\delta,\theta_{j,k}} \mathfrak{w}_k^{\epsilon_k} \mathfrak{b}_k^{\epsilon_k}), \tag{6.5}$$

$$\Upsilon(b_j V_{\alpha_j})^{\epsilon_j} R_\delta \Upsilon^{-1} \simeq \operatorname{Op}(\mathfrak{b}_j^{\epsilon_j} \mathfrak{w}_j^{\epsilon_j} \mathfrak{h}_{p,\delta,0}), \tag{6.6}$$

where b_k and V_{α_k} are given by (6.4) and (5.19), respectively, $\epsilon_k \in \{-1, 0, 1\}$, $\mathfrak{h}_{p,\delta,\beta}$ and $\theta_{j,k}$ are given by (4.2) and (5.6),

$$\mathfrak{b}_k(r, x) := e^{i\omega_k(r)\delta/p}, \quad \mathfrak{w}_k(r, x) := e^{i\omega_k(r)x}, \tag{6.7}$$

and $\mathfrak{b}_k$, $\mathfrak{w}_j^{\epsilon_j} \mathfrak{h}_{p,\delta,\theta_{j,k}} \mathfrak{w}_k^{\epsilon_k}$, $\mathfrak{w}_j^{\epsilon_j} \mathfrak{h}_{p,\delta,0} \in \widetilde{\mathcal{E}}(\mathbb{R}_+, V(\mathbb{R}))$.

Proof. Let $j, k = 1, 2, \dots, m$ and $j \neq k$. By Lemma 5.5,

$$\begin{aligned}(b_j V_{\alpha_j})^{\epsilon_j} R_{j,k} (b_k V_{\alpha_k})^{\epsilon_k} &\simeq (b_j V_{\alpha_j})^{\epsilon_j} (b_k V_{\alpha_k})^{\epsilon_k} R_{j,k} \\ &\simeq b_j^{\epsilon_j} (b_k^{\epsilon_k} \circ \alpha_j^{\epsilon_j}) V_{\alpha_k^{\epsilon_k} \circ \alpha_j^{\epsilon_j}} R_{j,k}.\end{aligned} \tag{6.8}$$

Let, for example, $\epsilon_j = -\epsilon_k = 1$. Then

$$b_j^{\epsilon_j}(r)(b_k^{\epsilon_k} \circ \alpha_j^{\epsilon_j})(r) = b_j(r) b_k^{-1}(re^{\omega_j(r)}) \quad \text{for } r \in \mathbb{R}_+, \quad V_{\alpha_k^{\epsilon_k} \circ \alpha_j^{\epsilon_j}} = V_{\alpha_k^{-1} \circ \alpha_j}.$$

Since $\alpha_k(r) = re^{\omega_k(r)}$, $\alpha_j(r) = re^{\omega_j(r)}$ and $\alpha_k^{-1}(r) = re^{\widetilde{\omega}_k(r)}$, where $\widetilde{\omega}_k \in SO(\mathbb{R}_+)$ (see [16, Lemmas 2.2 and 2.4]), we infer that

$$\alpha_k^{-1}[\alpha_j(r)] = \alpha_j(r) e^{\widetilde{\omega}_k(\alpha_j(r))} = re^{\omega_j(r) + \widetilde{\omega}_k(re^{\omega_j(r)})}.$$

By [16, Lemma 2.3] and [19, Lemma 2.6], the function $r \mapsto \omega_j(r) + \widetilde{\omega}_k(re^{\omega_j(r)})$ belongs to $SO(\mathbb{R}_+)$ along with ω_j and $\widetilde{\omega}_k$, and

$$\lim_{r \to s} [\widetilde{\omega}_k(re^{\omega_j(r)}) - \widetilde{\omega}_k(r)] = 0 \quad \text{and} \quad \lim_{r \to s} [\widetilde{\omega}_k(r) + \omega_k(r)] = 0 \quad \text{for } s \in \{0, \infty\}. \tag{6.9}$$

Analogously, by [16, Lemma 2.3], the function $r \mapsto b_j(r) b_k^{-1}(re^{\omega_j(r)})$ also belong to $SO(\mathbb{R}_+)$ and

$$\lim_{r \to s} [b_j(r) b_k^{-1}(re^{\omega_j(r)}) - b_j(r) b_k^{-1}(r)] = 0 \quad \text{for } s \in \{0, \infty\}. \tag{6.10}$$

Furthermore, due to (6.4) and (2.9),

$$\lim_{r\to s}\big[b_j(r)b_k^{-1}(r)-e^{(\omega_j(r)-\omega_k(r))(1+i\delta)/p}\big]=0\quad\text{for}\quad s\in\{0,\infty\}. \tag{6.11}$$

Applying Theorem 5.6 to the operator $V_{\alpha_k^{-1}\circ\alpha_j}$, we obtain

$$\Upsilon\big(b_j(b_k^{-1}\circ\alpha_j)V_{\alpha_k^{-1}\circ\alpha_j}R_{j,k}\big)\Upsilon^{-1}=\mathrm{Op}(\mathfrak{f}), \tag{6.12}$$

where

$$\mathfrak{f}(r,x):=b_j(r)b_k^{-1}(re^{\omega_j(r)})e^{i(\omega_j(r)+\widetilde{\omega}_k(re^{\omega_j(r)}))(x+i/p)}\,\frac{\exp\big(\theta_{j,k}\frac{x+i/p}{1+i\delta}\big)}{\sinh\big(\pi\frac{x+i/p}{1+i\delta}\big)}, \tag{6.13}$$

and therefore $\mathfrak{f}\in\widetilde{\mathcal{E}}(\mathbb{R}_+,V(\mathbb{R}))$ along with the functions $r\mapsto b_j(r)b_k^{-1}(re^{\omega_j(r)})$ and

$$(r,x)\mapsto e^{i(\omega_j(r)+\widetilde{\omega}_k(re^{\omega_j(r)}))(x+i/p)}\,\exp\Big(\theta_{j,k}\frac{x+i/p}{1+i\delta}\Big)\Big/\sinh\Big(\pi\frac{x+i/p}{1+i\delta}\Big).$$

For every $x\in\mathbb{R}$, we deduce from (6.9)–(6.11) that

$$\lim_{r\to s}\Big[b_j(r)b_k^{-1}(re^{\omega_j(r)})e^{i(\omega_j(r)+\widetilde{\omega}_k(re^{\omega_j(r)}))(x+i/p)} - e^{(\omega_j(r)-\omega_k(r))(1+i\delta)/p}e^{i(\omega_j(r)-\omega_k(r))(x+i/p)}\Big]=0, \tag{6.14}$$

where according to (6.7),

$$e^{(\omega_j(r)-\omega_k(r))(1+i\delta)/p}e^{i(\omega_j(r)-\omega_k(r))(x+i/p)}=\mathfrak{b}_j(r,x)\mathfrak{w}_j(r,x)(\mathfrak{b}_k(r,x)\mathfrak{w}_k(r,x))^{-1}.$$

Hence, we infer from (6.13), (6.14) and (6.5) that

$$\lim_{(\ln r)^2+x^2\to\infty}\big[\mathfrak{f}(r,x)-\big(\mathfrak{b}_j\mathfrak{w}_j\mathfrak{h}_{p,\delta,\theta_{j,k}}\mathfrak{w}_k^{-1}\mathfrak{b}_k^{-1}\big)(r,x)\big]=0,$$

which implies by Theorem 3.4 that

$$\mathrm{Op}(\mathfrak{f})\simeq\mathrm{Op}(\mathfrak{b}_j\mathfrak{w}_j\mathfrak{h}_{p,\delta,\theta_{j,k}}\mathfrak{w}_k^{-1}\mathfrak{b}_k^{-1}). \tag{6.15}$$

Finally, from (6.8), (6.12) and (6.15) it follows that

$$\Upsilon(b_jV_{\alpha_j}R_{j,k}(b_kV_{\alpha_k})^{-1}\Upsilon^{-1}\simeq\mathrm{Op}(\mathfrak{b}_j\mathfrak{w}_j\mathfrak{h}_{p,\delta,\theta_{j,k}}\mathfrak{w}_k^{-1}\mathfrak{b}_k^{-1}),$$

which gives (6.5) in case $\epsilon_j=-\epsilon_k=1$.

Equalities (6.5) for other values of ϵ_j and ϵ_k (as well as equalities (6.6)) are proved analogously. □

Specifying [25, Theorem 6.1] for the logarithmic star $\Gamma=\bigcup_{k=1}^N\Gamma_k$ and applying Theorems 5.3, 5.6 and Lemmas 5.5, 6.1 allowing to decrease the smoothness of the shift α, we obtain the following.

Theorem 6.2. *If α is an orientation-preserving diffeomorphism of Γ onto itself that satisfies the conditions of Section 2, then the operators $T_\pm:=(b_\alpha U_\alpha)^{\pm1}P_++P_-$, with $b_\alpha\in SO(\Gamma)$ given by (6.1), are Fredholm on every space $L^p(\Gamma)$ $(1<p<\infty)$ and* Ind $T_\pm=0$.

Proof. Here we partially follow the scheme of the proof of [25, Theorem 6.1]. Let us consider the operator

$$\begin{aligned}&\Psi_0\big[(U_\alpha^{-1}b_\alpha^{-1}P_+ + P_-)(b_\alpha U_\alpha P_+ + P_-)\big] = I+\\&\Psi_0\big[U_\alpha^{-1}b_\alpha^{-1}P_+P_- + P_-b_\alpha U_\alpha P_+ - U_\alpha^{-1}b_\alpha^{-1}P_-b_\alpha U_\alpha P_+ - P_+P_-\big].\end{aligned}\tag{6.16}$$

Since $\Psi_0(P_\pm) = \Upsilon^{-1}\Psi(P_\pm)\Upsilon$ we infer from Theorem 5.3 that

$$\Psi_0(P_\pm) = 2^{-1}\big(\delta_{j,k}(I \pm \varepsilon_k S_\delta) + \varepsilon_k R_{j,k}\big)_{j,k=1}^N,$$

where $\delta_{j,k}$ is the Kronecker symbol and $R_{k,k}$ are zero operators in $\mathcal{B}(L^p(\mathbb{R}_+))$. Hence, applying (6.3)–(6.4), Lemma 5.5 and the equality $I - S_\delta^2 = -R_\delta^2$, we conclude that

$$\begin{aligned}[\Psi_0(P_-b_\alpha U_\alpha P_+)]_{k,k} &= -\sum_{s=1,2,\dots,N;\ s\neq k}\big(2^{-1}\varepsilon_s R_{k,s} b_s V_{\alpha_s} 2^{-1}\varepsilon_k R_{s,k}\big)\\&\quad + 2^{-1}(I - \varepsilon_k S_\delta) b_k V_{\alpha_k} 2^{-1}(I + \varepsilon_k S_\delta)\\&\simeq -\sum_{s=1,2,\dots,N;\ s\neq k}\big(4^{-1}\varepsilon_s\varepsilon_k R_{k,s} b_s V_{\alpha_s} R_{s,k}\big) + 4^{-1} b_k V_{\alpha_k}(I - S_\delta^2)\\&= -\sum_{s=1,2,\dots,N;\ s\neq k}\big(4^{-1}\varepsilon_s\varepsilon_k R_{k,s} b_s V_{\alpha_s} R_{s,k}\big) - 4^{-1} b_k V_{\alpha_k} R_\delta^2,\end{aligned}$$

while for $j \neq k$ we get

$$\begin{aligned}[\Psi_0(P_-b_\alpha U_\alpha P_+)]_{j,k} &= -\sum_{s=1,2,\dots,N;\ s\neq j,k}\big(2^{-1}\varepsilon_s R_{k,s} b_s V_{\alpha_s} 2^{-1}\varepsilon_k R_{s,k}\big)\\&\quad + 4^{-1}(I - \varepsilon_j S_\delta) b_j V_{\alpha_j}\varepsilon_k R_{j,k} - 4^{-1}\varepsilon_k R_{j,k} b_k V_{\alpha_k}(I + \varepsilon_k S_\delta)\\&\simeq -\sum_{s=1,2,\dots,N;\ s\neq j,k}\big(4^{-1}\varepsilon_s\varepsilon_k R_{k,s} b_s V_{\alpha_s} R_{s,k}\big)\\&\quad + 4^{-1}\varepsilon_k(I - \varepsilon_j S_\delta) b_j V_{\alpha_j} R_{j,k} - 4^{-1}\varepsilon_k b_k V_{\alpha_k} R_{j,k}(I + \varepsilon_k S_\delta).\end{aligned}$$

Hence, applying Lemma 6.1 and Theorem 3.2, we obtain

$$\begin{aligned}[\Psi(P_-b_\alpha U_\alpha P_+)]_{k,k} &\simeq -\sum_{s=1,2,\dots,N;\ s\neq k}\big(4^{-1}\varepsilon_s\varepsilon_k \operatorname{Op}(\mathfrak{h}_{p,\delta,\theta_{k,s}}) \operatorname{Op}(\mathfrak{b}_s\mathfrak{w}_s\mathfrak{h}_{p,\delta,\theta_{s,k}})\big)\\&\quad - 4^{-1}\operatorname{Op}(\mathfrak{b}_k\mathfrak{w}_k\mathfrak{h}_{p,\delta,0})\operatorname{Op}(\mathfrak{h}_{p,\delta,0})\\&\simeq -\sum_{s=1,2,\dots,N;\ s\neq k}\operatorname{Op}\big(4^{-1}\varepsilon_s\varepsilon_k\mathfrak{h}_{p,\delta,\theta_{k,s}}\mathfrak{b}_s\mathfrak{w}_s\mathfrak{h}_{p,\delta,\theta_{s,k}}\big)\\&\quad - \operatorname{Op}\big(4^{-1}\mathfrak{b}_k\mathfrak{w}_k(\mathfrak{h}_{p,\delta,0})^2\big)\end{aligned}$$

and, analogously,

$$[\Psi(P_- b_\alpha U_\alpha P_+)]_{j,k} \simeq - \sum_{s=1,2,\dots,N;\ s\neq j,k} \mathrm{Op}\left(4^{-1}\varepsilon_s\varepsilon_k \mathfrak{h}_{p,\delta,\theta_{k,s}} \mathfrak{b}_s \mathfrak{w}_s \mathfrak{h}_{p,\delta,\theta_{s,k}}\right)$$
$$+ \mathrm{Op}\left(4^{-1}\varepsilon_k(1-\varepsilon_j \mathfrak{s}_{p,\delta})\mathfrak{b}_j\mathfrak{w}_j\mathfrak{h}_{p,\delta,\theta_{j,k}}\right)$$
$$+ \mathrm{Op}\left(4^{-1}\varepsilon_k \mathfrak{h}_{p,\delta,\theta_{j,k}}\mathfrak{b}_k\mathfrak{w}_k(1+\varepsilon_k s_{p,\delta})\right).$$

It is easy to see from these formulas and the equalities

$$\mathfrak{D}_\alpha(r,x) := \mathrm{diag}\big\{(\mathfrak{b}_k\mathfrak{w}_k)(r,x)\big\}_{k=1}^N = \mathrm{diag}\big\{e^{i\omega_k(r)(x+\delta/p)}\big\}_{k=1}^N, \tag{6.17}$$

$$\mathfrak{P}_\pm(r,x) = 2^{-1}\big(I_N \pm \big(\varepsilon_k \mathfrak{g}_{j,k}(r,x)\big)_{j,k=1}^N\big), \tag{6.18}$$

with functions $\mathfrak{g}_{j,k}$ defined by (5.5), that

$$\Psi(P_- b_\alpha U_\alpha P_+) \simeq \mathrm{Op}(\mathfrak{P}_-\mathfrak{D}_\alpha\mathfrak{P}_+),$$

where the entries of the matrix function $\mathfrak{P}_-\mathfrak{D}_\alpha\mathfrak{P}_+$ belong to $\widetilde{\mathcal{E}}(\mathbb{R}_+, V(\mathbb{R}))$.

By analogy, we obtain

$$\Psi(U_\alpha^{-1}b_\alpha^{-1}P_+P_-) \simeq \mathrm{Op}(\mathfrak{D}_\alpha^{-1}\mathfrak{P}_+\mathfrak{P}_-),$$
$$\Psi(U_\alpha^{-1}b_\alpha^{-1}P_- b_\alpha U_\alpha P_+) \simeq \mathrm{Op}(\mathfrak{D}_\alpha^{-1}\mathfrak{P}_-\mathfrak{D}_\alpha\mathfrak{P}_+),$$
$$\Psi(P_+P_-) = \mathrm{Op}(\mathfrak{P}_+\mathfrak{P}_-),$$

where the entries of the matrix symbols of these Mellin pseudodifferential operators belong to $\widetilde{\mathcal{E}}(\mathbb{R}_+, V(\mathbb{R}))$. Consequently, we deduce from (6.16) that

$$\Psi\big[(U_\alpha^{-1}b_\alpha^{-1}P_+ + P_-)(b_\alpha U_\alpha P_+ + P_-)\big]$$
$$\simeq \mathrm{Op}\big((\mathfrak{D}_\alpha^{-1}\mathfrak{P}_+ + \mathfrak{P}_-)(\mathfrak{D}_\alpha\mathfrak{P}_+ + \mathfrak{P}_-)\big) = \mathrm{Op}(A_{\alpha^{-1}}A_\alpha), \tag{6.19}$$

where the matrix functions A_α and $A_{\alpha^{-1}}$ are given by

$$A_\alpha := \mathfrak{D}_\alpha\mathfrak{P}_+ + \mathfrak{P}_- \quad \text{and} \quad A_{\alpha^{-1}} := \mathfrak{D}_\alpha^{-1}\mathfrak{P}_+ + \mathfrak{P}_-, \tag{6.20}$$

and the entries of the matrix function $A_{\alpha^{-1}}A_\alpha$ belong to $\widetilde{\mathcal{E}}(\mathbb{R}_+, V(\mathbb{R}))$.

Similarly to (6.19), we obtain

$$\Psi\big[(b_\alpha U_\alpha P_+ + P_-)(U_\alpha^{-1}b_\alpha^{-1}P_+ + P_-)\big]$$
$$\simeq \mathrm{Op}\big((\mathfrak{D}_\alpha\mathfrak{P}_+ + \mathfrak{P}_-)(\mathfrak{D}_\alpha^{-1}\mathfrak{P}_+ + \mathfrak{P}_-)\big) = \mathrm{Op}(A_\alpha A_{\alpha^{-1}}), \tag{6.21}$$

where the entries of the matrix function $A_\alpha A_{\alpha^{-1}}$ also are in $\widetilde{\mathcal{E}}(\mathbb{R}_+, V(\mathbb{R}))$.

It follows from (6.20) and formulas (6.7), (6.8), (6.14) and (6.15) in [25] that

$$\begin{aligned}\det A_\alpha &= \Big(\prod\nolimits_{\varepsilon_k=1} d_k\Big)\mathcal{P}_+ + \Big(\prod\nolimits_{\varepsilon_k=-1} d_k\Big)\mathcal{P}_-,\\ \det A_{\alpha^{-1}} &= \Big(\prod\nolimits_{\varepsilon_k=1} d_k^{-1}\Big)\mathcal{P}_+ + \Big(\prod\nolimits_{\varepsilon_k=-1} d_k^{-1}\Big)\mathcal{P}_-,\end{aligned} \tag{6.22}$$

where

$$d_k(r,x) := \exp[i\omega_k(r)(x+\delta/p)], \quad \mathcal{P}_\pm(x) := \frac{1}{2}\bigg(1 \pm \coth\Big(\pi\frac{x+i/p}{1+i\delta}\Big)\bigg). \tag{6.23}$$

Hence, we infer from (6.22) that

$$\det(A_{\alpha^{-1}}A_\alpha)=\det(A_\alpha A_{\alpha^{-1}})=1+\Big[\prod_{k=1}^{N} d_k^{\varepsilon_k}+\prod_{k=1}^{N} d_k^{-\varepsilon_k}-2\Big]\mathcal{P}_+\mathcal{P}_-, \tag{6.24}$$

where

$$\det(A_{\alpha^{-1}}A_\alpha)=\det(A_\alpha A_{\alpha^{-1}})=\det A_{\alpha^{-1}}\det A_\alpha\in\widetilde{\mathcal{E}}(\mathbb{R}_+,V(\mathbb{R})). \tag{6.25}$$

Since the entries of the operator matrices $\mathrm{Op}(A_\alpha A_{\alpha^{-1}})$ and $\mathrm{Op}(A_{\alpha^{-1}}A_\alpha)$ commute to within compact operators by Theorem 3.3, we conclude that these operators are Fredholm on the space $L^p_N(\mathbb{R}_+,d\mu)$ if and only if the operator $\mathrm{Op}(\det A_\alpha\det A_{\alpha^{-1}})$ is Fredholm on the space $L^p(\mathbb{R}_+,d\mu)$, and in the case of Fredholmness

$$\mathrm{Ind}\,\mathrm{Op}(A_\alpha A_{\alpha^{-1}})=\mathrm{Ind}\,\mathrm{Op}(A_{\alpha^{-1}}A_\alpha)=\mathrm{Ind}\,\mathrm{Op}(\det A_\alpha\det A_{\alpha^{-1}}) \tag{6.26}$$

in view of Theorem 3.5. Taking into account (6.25) and (6.23)–(6.24), we infer from Theorem 3.5 that the operators $\mathrm{Op}(A_\alpha A_{\alpha^{-1}})$ and $\mathrm{Op}(A_{\alpha^{-1}}A_\alpha)$ are Fredholm on the space $L^p_N(\mathbb{R}_+,d\mu)$ if and only if

$$\begin{aligned}&\det A_\alpha(r,+\infty)\det A_{\alpha^{-1}}(r,+\infty)\neq 0\quad\text{for all}\quad r\in\mathbb{R}_+,\\&\det A_\alpha(r,-\infty)\det A_{\alpha^{-1}}(r,-\infty)\neq 0\quad\text{for all}\quad r\in\mathbb{R}_+,\end{aligned} \tag{6.27}$$

$$\det A_\alpha(\xi,x)\det A_{\alpha^{-1}}(\xi,x)\neq 0\quad\text{for all}\quad(\xi,x)\in\Delta\times\overline{\mathbb{R}}, \tag{6.28}$$

and in the case of Fredholmness

$$\mathrm{Ind}\,\mathrm{Op}(\det A_\alpha\det A_{\alpha^{-1}})=\lim_{m\to\infty}\frac{1}{2\pi}\{\arg\det(A_\alpha A_{\alpha^{-1}})(r,x)\}_{(r,x)\in\partial\Pi_m} \tag{6.29}$$

where $\Pi_m=[m^{-1},m]\times\overline{\mathbb{R}}$. Obviously, conditions (6.27) are fulfilled because

$$\det A_\alpha(r,\pm\infty)\det A_{\alpha^{-1}}(r,\pm\infty)=1\quad\text{for all}\quad r\in\mathbb{R}_+ \tag{6.30}$$

in view of (6.24). On the other hand, from (6.24) and (6.23) it follows that

$$\begin{aligned}Q(r,x)&:=\det\big[A_\alpha(r,x)A_{\alpha^{-1}}(r,x)\big]\\&=1-\Big[\prod_{k=1}^{N}(d_k(r,x))^{\varepsilon_k}+\prod_{k=1}^{N}(d_k(r,x))^{-\varepsilon_k}-2\Big]\Big(2\sinh\Big(\pi\frac{x+i/p}{1+i\delta}\Big)\Big)^{-2}\\&=\exp\Big(-2\pi\,\frac{x+i/p}{1+i\delta}\Big)\Big(2\sinh\Big(\pi\frac{x+i/p}{1+i\delta}\Big)\Big)^{-2}f_+(r,x)f_-(r,x),\end{aligned} \tag{6.31}$$

where $d_k(r,x)$ are given by (6.23) and the functions

$$f_\pm(r,x):=\exp\Big(\pm i(x+\delta/p)\sum_{k=1}^{N}\varepsilon_k\omega_k(r)\Big)-\exp\big(2\pi(x+i/p)/(1+i\delta)\big) \tag{6.32}$$

are separated from zero for all $(r,x)\in\mathbb{R}_+\times\mathbb{R}$. Indeed, $f_\pm(r,x)$ can equal zero only for $x_r=-\delta/p$, but $f_\pm(r,x_r)=1-\exp(2\pi i/p)\neq 0$. Thus, by (6.30)–(6.32),

$$\inf_{(r,x)\in\mathbb{R}_+\times\overline{\mathbb{R}}}|Q(r,x)|>0,$$

and therefore we conclude from (6.22) that

$$\inf_{(r,x)\in\mathbb{R}_+\times\mathbb{R}}|\det A_\alpha(r,x)|>0,\qquad\inf_{(r,x)\in\mathbb{R}_+\times\mathbb{R}}|\det A_{\alpha^{-1}}(r,x)|>0 \tag{6.33}$$

for all $\delta\in\mathbb{R}$ and all functions ω_k $(k=1,2,\dots,N)$.

Thus, from (6.33) it follows that (6.28) is also fulfilled. Hence, the operators $\mathrm{Op}(A_\alpha A_{\alpha^{-1}})$ and $\mathrm{Op}(A_{\alpha^{-1}} A_\alpha)$ are Fredholm on the space $L^p_N(\mathbb{R}_+, d\mu)$. Then, in view of (6.21) and (6.19), the operators

$$(b_\alpha U_\alpha P_+ + P_-)(U_\alpha^{-1} b_\alpha^{-1} P_+ + P_-) \quad \text{and} \quad (U_\alpha^{-1} b_\alpha^{-1} P_+ + P_-)(b_\alpha U_\alpha P_+ + P_-)$$

are Fredholm on the space $L^p(\Gamma)$, which implies in its turn that both the operators $b_\alpha U_\alpha P_+ + P_-$ and $U_\alpha^{-1} b_\alpha^{-1} P_+ + P_-$ are also Fredholm on this space.

Further, from (6.32) it follows that the functions $f_\pm$ remain separated from zero under the replacement of the functions ω_k by $z\omega_k$, respectively, where $z \in [0,1]$. Accordingly, along with $Q(r,x)$ given by (6.31), we consider the functions

$$\begin{aligned} d_k(r,x,z) &:= \exp\big(iz\omega_k(r)(x+\delta/p)\big) \quad (k=1,2,\dots,N),\\ Q(r,x,z) &:= 1 - \Big[\prod\nolimits_{k=1}^{N} (d_k(r,x,z))^{\varepsilon_k} + \prod\nolimits_{k=1}^{N} (d_k(r,x,z))^{-\varepsilon_k} - 2\Big]\\ &\qquad \times \left(4\sinh^2\left(\pi\frac{x+i/p}{1+i\delta}\right)\right)^{-1}. \end{aligned}$$

As $Q(r,x,z)$ is continuous and separated from zero for all $(r,x,z) \in \mathbb{R}_+ \times \overline{\mathbb{R}} \times [0,1]$, we conclude that $\{\arg Q(r,x,z)\}_{(r,x)\in\partial\Pi_m}$ does not depend on $z \in [0,1]$. Hence,

$$\frac{1}{2\pi}\big\{\arg Q(r,x)\big\}_{(r,x)\in\partial\Pi_m} = \frac{1}{2\pi}\big\{\arg Q(r,x,0)\big\}_{(r,x)\in\partial\Pi_m} = 0,$$

and therefore, by (6.29), $\mathrm{Ind}\,\mathrm{Op}(\det A_\alpha \det A_{\alpha^{-1}}) = 0$. Consequently, applying (6.21) and (6.26), we deduce that

$$\mathrm{Ind}\,(b_\alpha U_\alpha P_+ + P_-) + \mathrm{Ind}\,(U_\alpha^{-1} b_\alpha^{-1} P_+ + P_-) = 0. \tag{6.34}$$

Let $Cf = \overline{f}$. It remains to observe that both the operators $b_\alpha U_\alpha P_+ + P_-$ and $C(P_+ + U_\alpha^{-1} b_\alpha^{-1} P_-)C$ are Fredholm on the space $L^p(\Gamma)$ by the part already proved and

$$\mathrm{Ind}\,(b_\alpha U_\alpha P_+ + P_-) = \mathrm{Ind}\,\big(C(P_+ + U_\alpha^{-1} b_\alpha^{-1} P_-)C\big). \tag{6.35}$$

It is easily seen that

$$\Psi\big((b_\alpha U_\alpha P_+ + P_-)C(P_+ + U_\alpha^{-1} b_\alpha^{-1} P_-)C\big) \simeq \mathrm{Op}(Y), \tag{6.36}$$

where

$$\begin{aligned} Y(r,x) &:= \big(\mathfrak{D}_\alpha(r,x)\mathfrak{P}_+(r,x) + \mathfrak{P}_-(r,x)\big)\overline{\big(\mathfrak{P}_+(r,-x) + \mathfrak{D}_\alpha^{-1}(r,-x)\mathfrak{P}_-(r,-x)\big)}\\ &= A_\alpha(r,x)\overline{A_{\alpha^{-1}}(r,-x)}, \end{aligned} \tag{6.37}$$

the entries of the matrix function $Y(r,x)$ belong to the Banach algebra $\widetilde{\mathcal{E}}(\mathbb{R}_+, V(\mathbb{R}))$ and, according to (6.33) and (6.37),

$$\inf_{(r,x)\in\mathbb{R}_+\times\overline{\mathbb{R}}} |\det Y(r,x)| > 0.$$

Since the operator $\mathrm{Op}(Y)$ is Fredholm on the space $L^p_N(\mathbb{R}_+, d\mu)$, we infer that the operator $\mathrm{Op}(\det Y)$ is Fredholm on the space $L^p(\mathbb{R}_+, d\mu)$ and

$$\mathrm{Ind}\,\mathrm{Op}(Y) = \mathrm{Ind}\,\mathrm{Op}(\det Y). \tag{6.38}$$

Similarly to (6.24), we deduce from (6.37) and (6.22) that

$$\begin{aligned}\det Y(r,x) &= \Big[\Big(\prod\nolimits_{\varepsilon_k=1} d_k(r,x)\Big)\mathcal{P}_+(x) + \Big(\prod\nolimits_{\varepsilon_k=-1} d_k(r,x)\Big)\mathcal{P}_-(x)\Big] \\ &\quad\times \Big[\Big(\prod\nolimits_{\varepsilon_k=1} \overline{d_k^{-1}(r,-x)}\Big)\overline{\mathcal{P}_-(-x)} + \Big(\prod\nolimits_{\varepsilon_k=-1} \overline{d_k^{-1}(r,-x)}\Big)\overline{\mathcal{P}_+(-x)}\Big] \\ &= \exp\Big(2i\sum\nolimits_{\varepsilon_k=1} \omega_k(r)\,\delta/p\Big)\mathcal{P}_+(x)\overline{\mathcal{P}_-(-x)} \\ &\quad+ \exp\Big(2i\sum\nolimits_{\varepsilon_k=-1} \omega_k(r)\,\delta/p\Big)\mathcal{P}_-(x)\overline{\mathcal{P}_+(-x)} \\ &\quad+ \exp\Big(i\sum\nolimits_{k=1}^{N} \omega_k(r)(\varepsilon_k x + \delta/p)\Big)\mathcal{P}_+(x)\overline{\mathcal{P}_+(-x)} \\ &\quad+ \exp\Big(i\sum\nolimits_{k=1}^{N} \omega_k(r)(-\varepsilon_k x + \delta/p)\Big)\mathcal{P}_-(x)\overline{\mathcal{P}_-(-x)},\end{aligned} \tag{6.39}$$

where the functions $d_k(r,x)$ and $\mathcal{P}_\pm(x)$ are given by (6.23). From (6.39) it follows that $\det Y(r,x) \in \widetilde{\mathcal{E}}(\mathbb{R}_+, V(\mathbb{R}))$, and therefore, by Theorem 3.5,

$$\operatorname{Ind}\operatorname{Op}(\det Y) = \lim_{m\to\infty} \frac{1}{2\pi}\big\{\arg\det Y(r,x)\big\}_{(r,x)\in\partial\Pi_m}. \tag{6.40}$$

Consider the function $\det Y(r,x,z)$ for $(r,x,z) \in \mathbb{R}_+ \times \overline{\mathbb{R}} \times [0,1]$ which is obtained from $\det Y(r,x)$ by replacing $\omega_k(r)$ by $z\omega_k(r)$ for $k = 1,2,\dots,N$. Since the function $\det Y(r,x,z)$ is continuous and separated from zero for all $(r,x,z) \in \mathbb{R}_+ \times \overline{\mathbb{R}} \times [0,1]$, we conclude that $\frac{1}{2\pi}\{\arg\det Y(r,x,z)\}_{(r,x)\in\partial\Pi_m}$ does not depend on $z \in [0,1]$. Consequently, the operator $\operatorname{Op}(\det Y)$ is homotopic in the class of Fredholm operators to the identity operator because $\det Y(r,x,0) = 1$, and

$$\frac{1}{2\pi}\big\{\arg\det Y(r,x)\big\}_{(r,x)\in\partial\Pi_m} = \frac{1}{2\pi}\big\{\arg\det Y(r,x,0)\big\}_{(r,x)\in\partial\Pi_m} = 0. \tag{6.41}$$

Thus, by (6.38), (6.40) and (6.41),

$$\operatorname{Ind}\operatorname{Op}(Y) = 0. \tag{6.42}$$

Finally, from (6.35), (6.36) and (6.42) it follows that $\operatorname{Ind}(b_\alpha U_\alpha P_+ + P_-) = 0$. This and (6.34) imply that $\operatorname{Ind}(U_\alpha^{-1} b_\alpha^{-1} P_+ + P_-) = 0$ as well. □

7. Fredholmness of the operator $T = V_\alpha P_+ + GP_-$

In this section we establish a Fredholm criterion for the operator $T = V_\alpha P_+ + GP_-$ on the space $L^p(\Gamma)$ under the conditions of Section 2. Modifying the proof of [25, Theorem 6.1] with the aid of Theorem 5.6 and Lemma 6.1, we obtain the following.

Theorem 7.1. *If $G \in SO(\Gamma)$ and all the conditions of Theorem 5.6 hold, then the operator $T = V_\alpha P_+ + GP_-$ is Fredholm on the space $L^p(\Gamma)$ if and only if*

$$\inf_{t\in\Gamma} |G(t)| > 0, \quad \liminf_{r\to s}\, \inf_{x\in\mathbb{R}} |\det \mathcal{T}(r,x)| > 0 \quad (s \in \{0,\infty\}), \tag{7.1}$$

where

$$\det \mathcal{T}(r,x) = \Big(\prod\nolimits_{\varepsilon_k=1} e^{i\omega_k(r)(x+i/p)}\Big)\Big(\prod\nolimits_{\varepsilon_k=-1} G_k(r)\Big)\mathcal{P}_+(x) + \Big(\prod\nolimits_{\varepsilon_k=-1} e^{i\omega_k(r)(x+i/p)}\Big)\Big(\prod\nolimits_{\varepsilon_k=1} G_k(r)\Big)\mathcal{P}_-(x) \tag{7.2}$$

and the functions $\mathcal{P}_\pm(r,x)$ *are given by* (6.23).

Proof. Setting $G_\alpha := c_\alpha G$ where $c_\alpha := |\alpha'|^{1/p} \in SO(\Gamma)$, we obtain

$$(V_\alpha P_+ + GP_-)(U_\alpha^{-1} b_\alpha^{-1} P_+ + P_-) = c_\alpha^{-1} N_\alpha, \tag{7.3}$$

where

$$N_\alpha := (U_\alpha P_+ + G_\alpha P_-)(U_\alpha^{-1} b_\alpha^{-1} P_+ + P_-) = b_\alpha^{-1} P_+ + G_\alpha P_- - \big(U_\alpha P_- U_\alpha^{-1} b_\alpha^{-1} P_+ - U_\alpha P_+ P_- - G_\alpha P_- U_\alpha^{-1} b_\alpha^{-1} P_+ + G_\alpha P_+ P_-\big). \tag{7.4}$$

Introducing the diagonal matrix function

$$\mathfrak{G}_\alpha(r) := \operatorname{diag}\big\{(1 + r\omega_k'(r))^{1/p} e^{\omega_k(r)/p} G_k(r)\big\}_{k=1}^N, \quad r \in \mathbb{R}_+, \tag{7.5}$$

$$\mathfrak{U}_\alpha(r,x) := \operatorname{diag}\big\{e^{i\omega_k(r)x}\big\}_{k=1}^N, \quad (r,x) \in \mathbb{R}_+ \times \mathbb{R}, \tag{7.6}$$

we deduce from (7.4) by analogy with (6.21) that

$$\Psi(N_\alpha) \simeq \mathrm{Op}(\mathfrak{N}_\alpha), \tag{7.7}$$

where

$$\mathfrak{N}_\alpha := (\mathfrak{U}_\alpha \mathfrak{P}_+ + \mathfrak{G}_\alpha \mathfrak{P}_-)(\mathfrak{D}_\alpha^{-1}\mathfrak{P}_+ + \mathfrak{P}_-) = \mathcal{B}_\alpha^{-1}\mathfrak{P}_+ + \mathfrak{G}_\alpha \mathfrak{P}_- + \big(-\mathfrak{U}_\alpha \mathfrak{P}_- \mathfrak{D}_\alpha^{-1}\mathfrak{P}_+ + \mathfrak{U}_\alpha \mathfrak{P}_+ \mathfrak{P}_- + \mathfrak{G}_\alpha \mathfrak{P}_- \mathfrak{D}_\alpha^{-1}\mathfrak{P}_+ - \mathfrak{G}_\alpha \mathfrak{P}_+ \mathfrak{P}_-\big), \tag{7.8}$$

the matrix functions $\mathcal{B}_\alpha$, $\mathfrak{D}_\alpha$ and $\mathfrak{P}_\pm$ are given by (6.1), (6.17) and (6.18), respectively, and from (7.8), Corollary 5.7 and Lemma 6.1 it follows by analogy with Theorem 6.2 that the entries of the matrix function $\mathfrak{N}_\alpha$ are in $\widetilde{\mathcal{E}}(\mathbb{R}_+, V(\mathbb{R}))$.

Since the operator $U_\alpha^{-1} b_\alpha^{-1} P_+ + P_-$ is Fredholm on the space $L^p(\Gamma)$ in view of Theorem 6.2, from (7.3) and (7.7) it follows that the operator $V_\alpha P_+ + GP_- = c_\alpha^{-1}(U_\alpha P_+ + G_\alpha P_-)$ is Fredholm on the space $L^p(\Gamma)$ if and only if the Mellin pseudodifferential operator $\mathrm{Op}(\mathfrak{N}_\alpha)$ is Fredholm on the space $L^p_N(\mathbb{R}_+, d\mu)$.

As the entries of the matrix function $\mathfrak{N}_\alpha$ belong to $\widetilde{\mathcal{E}}(\mathbb{R}_+, V(\mathbb{R}))$, we deduce from Theorem 3.3 that the operator $\mathrm{Op}(\mathfrak{N}_\alpha)$ is Fredholm on the space $L^p_N(\mathbb{R}_+, d\mu)$ if and only if the operator $\mathrm{Op}(\det \mathfrak{N}_\alpha)$ is Fredholm on the space $L^p(\mathbb{R}_+, d\mu)$, which is equivalent, by Theorem 3.5, to the conditions

$$\det \mathfrak{N}_\alpha(r, \pm\infty) \neq 0 \quad \text{for all} \quad r \in \mathbb{R}_+, \tag{7.9}$$

$$\det \mathfrak{N}_\alpha(\xi, x) \neq 0 \quad \text{for all} \quad (\xi, x) \in \Delta \times \overline{\mathbb{R}}. \tag{7.10}$$

Thus, it remains to prove that conditions (7.9)–(7.10) are equivalent to both conditions (7.1). Since the matrix functions $\mathfrak{N}_\alpha(r, \pm\infty)$ are diagonal, we infer from

from (6.1), (7.5), (6.18) and (5.5) that

$$\det \mathfrak{N}_\alpha(r,\pm\infty) = \det\left(\mathcal{B}_\alpha^{-1}(r,\pm\infty)\mathfrak{P}_+(r,\pm\infty) + \mathfrak{G}_\alpha(r,\pm\infty)\mathfrak{P}_-(r,\pm\infty)\right)$$
$$= \left(\prod\nolimits_{\varepsilon_k=\pm1} e^{-i\omega_k(r)\delta/p}\right)\left(\prod\nolimits_{\varepsilon_k=\mp1} (1+r\omega_k'(r))^{1/p} e^{\omega_k(r)/p}\, G_k(r)\right). \quad (7.11)$$

Hence (7.9) holds if and only if $\inf_{t\in\Gamma}|G(t)|>0$, which gives the first condition in (7.1).

On the other hand, from (7.8) and (6.20) it follows that

$$\mathfrak{N}_\alpha = (\mathfrak{U}_\alpha\mathfrak{P}_+ + \mathfrak{G}_\alpha\mathfrak{P}_-)A_{\alpha^{-1}}. \quad (7.12)$$

Along with (7.12) we introduce the matrix function

$$\mathcal{N}_\alpha := (\mathfrak{U}_\alpha\mathfrak{P}_+ + \mathcal{G}_\alpha\mathfrak{P}_-)A_{\alpha^{-1}}, \quad (7.13)$$

where $\mathfrak{U}_\alpha$ is given by (7.6) and

$$\mathcal{G}_\alpha(r) := \operatorname{diag}\left\{(\mathcal{G}_\alpha)_k(r)\right\}_{k=1}^N, \quad (\mathcal{G}_\alpha)_k(r) = e^{\omega_k(r)/p}G_k(r), \quad r\in\mathbb{R}_+. \quad (7.14)$$

Comparing (7.5) and (7.14), we conclude from (7.12) and (7.13) that for all $x\in\overline{\mathbb{R}}$,

$$\lim_{r\to s}\left[\det\mathfrak{N}_\alpha(r,x) - \det\mathcal{N}_\alpha(r,x)\right] = 0 \quad (s\in\{0,\infty\}), \quad (7.15)$$

which implies in its turn that (7.10) is equivalent to the property

$$\liminf_{r\to s}\min_{x\in\overline{\mathbb{R}}}\left|\det\mathcal{N}_\alpha(r,x)\right| > 0 \quad \text{for} \quad s\in\{0,\infty\}. \quad (7.16)$$

Introducing the matrix functions

$$\mathfrak{C}_\alpha(r) := \operatorname{diag}\left\{e^{\omega_k(r)/p}\right\}_{k=1}^N, \quad \mathcal{G}(r) := \operatorname{diag}\left\{G_k(r)\right\}_{k=1}^N \quad (r\in\mathbb{R}_+), \quad (7.17)$$
$$\mathcal{V}_\alpha(r,x) := \operatorname{diag}\left\{e^{i\omega_k(r)(x+i/p)}\right\}_{k=1}^N \quad \text{for} \quad (r,x)\in\mathbb{R}_+\times\mathbb{R}; \quad (7.18)$$

we conclude from (7.13) and (7.17)–(7.18) that

$$\mathcal{N}_\alpha = \mathfrak{C}_\alpha\mathcal{T}A_{\alpha^{-1}}, \quad \mathcal{T} := \mathcal{V}_\alpha\mathfrak{P}_+ + \mathcal{G}\mathfrak{P}_-. \quad (7.19)$$

Hence

$$\det\mathcal{N}_\alpha(r,x) = \det\mathfrak{C}_\alpha(r)\det\mathcal{T}(r,x)\det A_{\alpha^{-1}}(r,x) \quad \text{for all} \quad (r,x)\in\mathbb{R}_+\times\mathbb{R}, \quad (7.20)$$

where $\det A_{\alpha^{-1}}$ is given by (6.22) and, by analogy with (6.22) and [25, Section 6], $\det\mathcal{T}(r,x)$ is calculated by (7.2).

Since $\det\mathfrak{C}_\alpha(r) = \prod_{k=1}^N e^{\omega_k(r)/p}$ by (7.17), it is easily seen from (7.20) and (6.33) that (7.16) is equivalent to the property

$$\liminf_{r\to s}\inf_{x\in\mathbb{R}}\left|\det\mathcal{T}(r,x)\right| > 0 \quad (s\in\{0,\infty\}),$$

which completes the proof. □

Clearly, for $\mathcal{T}$ given by (7.19), the relation

$$\liminf_{r\to s}\inf_{x\in\mathbb{R}}|\det\mathcal{T}(r,x)|>0\quad\text{for all}\quad s\in\{0,\infty\}$$

is equivalent to the inequality

$$\inf_{(\xi,x)\in\Delta\times\mathbb{R}}|\det\mathcal{T}(\xi,x)|>0. \tag{7.21}$$

Lemma 7.2. *If $G_k(\xi)\neq 0$ for all $\xi\in\Delta$ and all $k=1,2,\dots,N$, then the inequality (7.21) holds if and only if*

$$\varphi_{p,\delta}(\eta(\xi),\lambda(\xi))\notin\mathbb{Z}\quad\textit{for all}\quad \xi\in\Delta, \tag{7.22}$$

where

$$\begin{aligned}\varphi_{p,\delta}(\eta,\lambda)=&\Big(\Big(\frac{\eta}{2\pi}-\frac{\delta}{1+\delta^2}\Big)\frac{1}{p}+\frac{\ln|\lambda|}{2\pi}\Big)\Big(\frac{\eta}{2\pi}-\frac{\delta}{1+\delta^2}\Big)(1+\delta^2)\\&+\frac{1}{(1+\delta^2)p}-\frac{\arg\lambda}{2\pi},\end{aligned} \tag{7.23}$$

$$\eta(\xi):=\sum_{k=1}^{N}\varepsilon_k\omega_k(\xi),\quad \lambda(\xi):=\prod_{k=1}^{N}(G_k(\xi))^{\varepsilon_k}\quad\textit{for}\ \ \xi\in\Delta. \tag{7.24}$$

Proof. For every $\xi\in\Delta$ and every $x\in\mathbb{R}$, from (7.2) it follows that

$$\begin{aligned}\det\mathcal{T}(\xi,x)=&\Big(\prod\nolimits_{\varepsilon_k=1}e^{i\omega_k(\xi)(x+i/p)}\Big)\Big(\prod\nolimits_{\varepsilon_k=-1}G_k(\xi)\Big)\mathcal{P}_+(x)\\&+\Big(\prod\nolimits_{\varepsilon_k=-1}e^{i\omega_k(\xi)(x+i/p)}\Big)\Big(\prod\nolimits_{\varepsilon_k=1}G_k(\xi)\Big)\mathcal{P}_-(x).\end{aligned} \tag{7.25}$$

If $G_k(\xi)\neq 0$ for all $\xi\in\Delta$ and all $k=1,2,\dots,N$, then (7.21) is equivalent to the condition

$$\det\mathcal{T}(\xi,x)\neq 0\quad\text{for all}\quad(\xi,x)\in\Delta\times\mathbb{R},$$

which in its turn is equivalent to the relations

$$\Big(\prod\nolimits_{k=1}^{N}e^{i\varepsilon_k\omega_k(\xi)(x+i/p)}(G_k(\xi))^{-\varepsilon_k}\Big)\mathcal{P}_+(x)+\mathcal{P}_-(x)\neq 0,\ \ (\xi,x)\in\Delta\times\mathbb{R}. \tag{7.26}$$

Applying (7.24), we conclude in view of the identities

$$\mathcal{P}_\pm(x)=\frac{\pm\exp\Big(\pm\frac{\pi}{1+\delta^2}\big[(x+\delta/p)+i(1/p-\delta x)\big]\Big)}{2\sinh\Big(\frac{\pi}{1+\delta^2}\big[(x+\delta/p)+i(1/p-\delta x)\big]\Big)},\quad x\in\mathbb{R},$$

that (7.26) is equivalent to the condition

$$e^{i\eta(\xi)(x+i/p)-\ln|\lambda(\xi)|-i\arg\lambda(\xi)}\,e^{\frac{2\pi}{1+\delta^2}[(x+\delta/p)+i(1/p-\delta x)]}\neq 1\ \text{ for all }\ (\xi,x)\in\Delta\times\mathbb{R}. \tag{7.27}$$

Clearly, (7.27) holds if and only if for any $k\in\mathbb{Z}$ the system

$$\begin{cases}-\frac{\eta(\xi)}{p}-\ln|\lambda(\xi)|+\frac{2\pi}{1+\delta^2}\big(x+\frac{\delta}{p}\big)=0,\\ \eta(\xi)x-\arg\lambda(\xi)+\frac{2\pi}{1+\delta^2}\big(\frac{1}{p}-\delta x\big)=2\pi k\end{cases}$$

does not have solutions $(\xi,x)\in\Delta\times\mathbb{R}$. But the latter is equivalent to (7.22). □

Theorem 7.1 and Lemma 7.2 immediately imply the following.

Corollary 7.3. *If $G \in SO(\Gamma)$ and all the conditions of Theorem 5.6 hold, then the operator $T = V_\alpha P_+ + GP_-$ is Fredholm on the space $L^p(\Gamma)$ if and only if*

$$\inf_{t\in\Gamma} |G(t)| > 0, \quad \varphi_{p,\delta}(\eta(\xi), \lambda(\xi)) \notin \mathbb{Z} \quad \textit{for all} \quad \xi \in \Delta,$$

where $\varphi_{p,\delta}(\eta, \lambda)$ is defined by (7.23), and $\eta(\xi)$ and $\lambda(\xi)$ are given by (7.24).

8. Index of the operator $T = V_\alpha P_+ + GP_-$

In this section we establish an index formula for the operator $T = V_\alpha P_+ + GP_-$ on the space $L^p(\Gamma)$ under the conditions of Section 2.

Let us recall the generalization of the Cauchy index obtained in [27] (also see [28] and [9]) for semi-almost periodic (SAP) functions. As is known (see, e.g., [37] and [9]), every function $f \in SAP$ is represented in the form

$$f = f_+ u_+ + f_- u_- + f_0,$$

where $f_\pm$ are (Bohr) almost periodic functions, $u_\pm$ and f_0 are continuous functions on $\overline{\mathbb{R}}$, $u_+ + u_- = 1$, and $u_+(+\infty) = 1$, $u_+(-\infty) = 0$, $f_0(\pm\infty) = 0$. If $\inf_{x\in\mathbb{R}} |f(x)| > 0$, then by [27] the generalized Cauchy index $\mathrm{Ind}_{\mathbb{R}} f$ is defined as follows:

$$\mathrm{Ind}_{\mathbb{R}} f = \frac{1}{2\pi}\big[M_+(\varphi_+) - M_-(\varphi_-)\big], \tag{8.1}$$
$$M_\pm(\varphi) = \lim_{x\to\pm\infty} \frac{1}{x}\int_0^x \varphi(y)dy,$$
$$\varphi_\pm(x) = \arg f(x) - x \lim_{y\to\pm\infty} \big[y^{-1} \arg f(y)\big]$$

and $\arg f$ is any continuous branch of argument of f. This index exists for every invertible function $f \in SAP$, is finite, independent of the branch choice, stable with respect to small perturbations and possesses the logarithmic property [28]:

$$\mathrm{Ind}_{\mathbb{R}} [f_1 f_2] = \mathrm{Ind}_{\mathbb{R}} f_1 + \mathrm{Ind}_{\mathbb{R}} f_2 \quad \text{if} \quad f_k \in SAP,\ \inf_{x\in\mathbb{R}} |f_k(x)| > 0\ (k = 1, 2). \tag{8.2}$$

Lemma 8.1. *If $G \in SO(\Gamma)$, where Γ and α satisfy the conditions of Theorem 5.6, $\prod_{k=1}^N G_k(\xi) \neq 0$ and $\varphi_{p,\delta}(\eta(\xi), \lambda(\xi)) \notin \mathbb{Z}$ for some $\xi \in \Delta$, then the function $\det \mathcal{T}(\xi, \cdot)$ is invertible in SAP and*

$$\mathrm{Ind}_{\mathbb{R}} \det \mathcal{T}(\xi, \cdot) = -E(\varphi_{p,\delta}(\eta(\xi), \lambda(\xi))) - \frac{\arg \lambda(\xi)}{2\pi}, \tag{8.3}$$

where $E(y)$ is the integer part of a real number y.

Proof. Since the function $\det \mathcal{T}(\xi, \cdot)$ belongs to SAP and is separated from the origin by the proof of Lemma 7.2, we conclude that the function $\det \mathcal{T}(\xi, \cdot)$ is invertible in SAP. Taking then any continuous branch of the argument of the function $\arg\det\mathcal{T}(\xi,\cdot)$ where $\det\mathcal{T}(\xi,\cdot)$ is given by (7.25), we infer from (8.1) that

$$\begin{aligned}\operatorname{Ind}_{\mathbb{R}}\det\mathcal{T}(\xi,\cdot) &= \lim_{x\to+\infty}\frac{1}{2\pi x}\int_0^x\Big(\arg\det\mathcal{T}(\xi,y)-y\sum_{\varepsilon_k=1}\omega_k(\xi)\Big)dy\\ &\quad-\lim_{x\to-\infty}\frac{1}{2\pi x}\int_0^x\Big(\arg\det\mathcal{T}(\xi,y)-y\sum_{\varepsilon_k=-1}\omega_k(\xi)\Big)dy\\ &=\frac{1}{2\pi}\big\{\arg\mathcal{T}_+(\xi,x)\big\}_{x\in[0,+\infty]}+\frac{1}{2\pi}\big\{\arg\mathcal{T}_-(\xi,x)\big\}_{x\in[-\infty,0]},\end{aligned}\tag{8.4}$$

where

$$\begin{aligned}\mathcal{T}_+(\xi,x) &:= \Big(\prod\nolimits_{\varepsilon_k=1}e^{-\omega_k(\xi)/p}\Big)\Big(\prod\nolimits_{\varepsilon_k=-1}G_k(\xi)\Big)\mathcal{P}_+(x)\\ &\quad+\Big(\prod\nolimits_{k=1}^N e^{-i\varepsilon_k\omega_k(\xi)x}\Big)\Big(\prod\nolimits_{\varepsilon_k=-1}e^{-\omega_k(\xi)/p}\Big)\Big(\prod\nolimits_{\varepsilon_k=1}G_k(\xi)\Big)\mathcal{P}_-(x),\\ \mathcal{T}_-(\xi,x) &:= \Big(\prod\nolimits_{\varepsilon_k=-1}e^{-\omega_k(\xi)/p}\Big)\Big(\prod\nolimits_{\varepsilon_k=1}G_k(\xi)\Big)\mathcal{P}_-(x)\\ &\quad+\Big(\prod\nolimits_{k=1}^N e^{i\varepsilon_k\omega_k(\xi)x}\Big)\Big(\prod\nolimits_{\varepsilon_k=1}e^{-\omega_k(\xi)/p}\Big)\Big(\prod\nolimits_{\varepsilon_k=-1}G_k(\xi)\Big)\mathcal{P}_+(x).\end{aligned}$$

As $\arg\mathcal{T}_+(\xi,x)=-\lambda(\xi)x+\arg\mathcal{T}_-(\xi,x)$, we infer from (8.4) for any $x_0\neq 0$ that

$$\begin{aligned}\operatorname{Ind}_{\mathbb{R}}\det\mathcal{T}(\xi,\cdot) &= \frac{1}{2\pi}\big\{\arg\mathcal{T}_+(\xi,x)\big\}_{x\in[0,+\infty]}+\frac{1}{2\pi}\big\{\arg\mathcal{T}_-(\xi,x)\big\}_{x\in[-\infty,0]}\\ &=\frac{1}{2\pi}\big\{\arg\mathcal{T}_+(\xi,x)\big\}_{x\in[x_0,+\infty]}+\frac{1}{2\pi}\big\{\arg\mathcal{T}_-(\xi,x)\big\}_{x\in[-\infty,x_0]}-\frac{\eta(\xi)x_0}{2\pi}\\ &=\frac{1}{2\pi}\{\arg\mathcal{P}_+(x)\}_{x\in[x_0,+\infty]}+\frac{1}{2\pi}\{\arg\mathcal{Y}_+(\xi,x)\}_{x\in[x_0,+\infty]}-\frac{\eta(\xi)x_0}{2\pi}\\ &\quad+\frac{1}{2\pi}\{\arg\mathcal{P}_-(x)\}_{x\in[-\infty,x_0]}+\frac{1}{2\pi}\{\arg\mathcal{Y}_-(\xi,x)\}_{x\in[-\infty,x_0]},\end{aligned}\tag{8.5}$$

where

$$\begin{aligned}\mathcal{Y}_+(\xi,x) &:= 1+\frac{\mathcal{P}_-(x)}{\mathcal{P}_+(x)}\prod_{k=1}^N\big(e^{-i\varepsilon_k\omega_k(\xi)(x+i/p)}(G_k(\xi))^{\varepsilon_k}\big),\\ \mathcal{Y}_-(\xi,x) &:= 1+\frac{\mathcal{P}_+(x)}{\mathcal{P}_-(x)}\prod_{k=1}^N\big(e^{i\varepsilon_k\omega_k(\xi)(x+i/p)}(G_k(\xi))^{-\varepsilon_k}\big).\end{aligned}\tag{8.6}$$

By (8.6), we get

$$\begin{aligned}\mathcal{Y}_+(\xi,x) &= 1-e^{-2\pi(x+\delta/p)/(1+\delta^2)+\eta(\xi)/p+\ln|\lambda(\xi)|}e^{-i[2\pi(1/p-\delta x)/(1+\delta^2)+\eta(\xi)x-\arg\lambda(\xi)]},\\ \mathcal{Y}_-(\xi,x) &= 1-e^{2\pi(x+\delta/p)/(1+\delta^2)-\eta(\xi)/p-\ln|\lambda(\xi)|}e^{i[2\pi(1/p-\delta x)/(1+\delta^2)+\eta(\xi)x-\arg\lambda(\xi)]}.\end{aligned}\tag{8.7}$$

Since

$$\mathcal{P}_\pm(x)=\Big(1-\exp\big(\mp 2\pi\big((x+\delta/p)+i(1/p-\delta x)\big)/(1+\delta^2)\big)\Big)^{-1},$$

we conclude that $\operatorname{Re}\mathcal{P}_\pm(x) > 0$ for all $\pm(x+\delta/p) > 0$, respectively. Hence

$$\frac{1}{2\pi}\big\{\arg\mathcal{P}_+(x)\big\}_{x\in[-\delta/p,+\infty]} = \frac{1}{2\pi}\arg\left(1-e^{-2\pi i/p}\right) = \frac{1}{2\pi}\left(\frac{\pi}{2}-\frac{\pi}{p}\right),$$
$$\frac{1}{2\pi}\big\{\arg\mathcal{P}_-(x)\big\}_{x\in[-\infty,-\delta/p]} = -\frac{1}{2\pi}\arg\left(1-e^{2\pi i/p}\right) = \frac{1}{2\pi}\left(\frac{\pi}{2}-\frac{\pi}{p}\right),$$

which gives

$$\frac{1}{2\pi}\Big(\big\{\arg\mathcal{P}_+(x)\big\}_{x\in[-\delta/p,+\infty]} + \big\{\arg\mathcal{P}_-(x)\big\}_{x\in[-\infty,-\delta/p]}\Big) = \frac{1}{2}-\frac{1}{p}. \tag{8.8}$$

On the other hand, by (8.8),

$$\begin{aligned}
&\frac{1}{2\pi}\{\arg\mathcal{P}_-(x)\}_{x\in[-\infty,x_0]} + \frac{1}{2\pi}\{\arg\mathcal{P}_+(x)\}_{x\in[x_0,+\infty]}\\
&\quad= \frac{1}{2\pi}\{\arg\mathcal{P}_-(x)\}_{x\in[-\infty,-\delta/p]} + \frac{1}{2\pi}\{\arg\mathcal{P}_+(x)\}_{x\in[-\delta/p,+\infty]} + \frac{\delta}{1+\delta^2}(x_0+\delta/p)\\
&\quad= \frac{1}{2}-\frac{1}{p}+\frac{\delta}{1+\delta^2}(x_0+\delta/p).
\end{aligned}\tag{8.9}$$

Hence, we infer from (8.5) and (8.9) that

$$\begin{aligned}
\operatorname{Ind}_{\mathbb{R}}\det\mathcal{T}(\xi,\cdot) &= \frac{1}{2\pi}\{\arg\mathcal{Y}_-(\xi,x)\}_{x\in[-\infty,x_0]} + \frac{1}{2\pi}\{\arg\mathcal{Y}_+(\xi,x)\}_{x\in[x_0,+\infty]}\\
&\quad - \frac{\eta(\xi)x_0}{2\pi} + \frac{1}{2} - \frac{1}{p} + \frac{\delta}{1+\delta^2}(x_0+\delta/p).
\end{aligned}\tag{8.10}$$

Let $\varphi_{p,\delta}(\eta(\xi),\lambda(\xi)) > 0$ and $\frac{\eta(\xi)}{2\pi} - \frac{\delta}{1+\delta^2} \notin \mathbb{Z}$. Choose the branches of $\eta(\xi)$ and $\arg\lambda(\xi)$ so that

$$0 < \frac{\eta(\xi)}{2\pi} - \frac{\delta}{1+\delta^2} < 1, \quad 0 \le \frac{\arg\lambda(\xi)}{2\pi} - \frac{1}{(1+\delta^2)p} < 1. \tag{8.11}$$

Let us look for the points $x_k \ge 0$ for which $\mathcal{Y}_+(\xi,x_k) < 0$ or, equivalently,

$$\begin{cases}
-\dfrac{x_k+\delta/p}{1+\delta^2} + \dfrac{\eta(\xi)}{2\pi p} + \dfrac{\ln|\lambda(\xi)|}{2\pi} > 0,\\[2ex]
-\dfrac{1/p-\delta x_k}{1+\delta^2} - \dfrac{\eta(\xi)x_k}{2\pi} + \dfrac{\arg\lambda(\xi)}{2\pi} = -k \in \mathbb{Z}.
\end{cases}\tag{8.12}$$

Since, by the second equation in (8.12),

$$x_k = \left(\frac{\arg\lambda(\xi)}{2\pi} - \frac{1}{(1+\delta^2)p} + k\right)\Big/\left(\frac{\eta(\xi)}{2\pi} - \frac{\delta}{1+\delta^2}\right), \tag{8.13}$$

we infer from (8.11) and the first inequality in (8.12) that

$$\begin{aligned}
k < \varphi_{p,\delta}(\eta(\xi),\lambda(\xi)) &= \left(\left(\frac{\eta(\xi)}{2\pi} - \frac{\delta}{1+\delta^2}\right)\frac{1}{p} + \frac{\ln|\lambda(\xi)|}{2\pi}\right)\left(\frac{\eta(\xi)}{2\pi} - \frac{\delta}{1+\delta^2}\right)(1+\delta^2)\\
&\quad + \frac{1}{(1+\delta^2)p} - \frac{\arg\lambda(\xi)}{2\pi}.
\end{aligned}$$

On the other hand, (8.11) and (8.13) imply that $x_k \ge 0$ if and only if

$$\frac{1}{(1+\delta^2)p} - \frac{\arg\lambda(\xi)}{2\pi} \le 0 \le k.$$

Thus, system (8.12) has solutions $x_k \ge 0$ for $k \in \mathbb{Z}$ such that

$$\frac{1}{(1+\delta^2)p} - \frac{\arg\lambda(\xi)}{2\pi} \le 0 \le k < \varphi_{p,\delta}(\eta(\xi), \lambda(\xi)),$$

which happens if $\varphi_{p,\delta}(\eta(\xi), \lambda(\xi)) > 0$ and in that case

$$n_+ := E(\varphi_{p,\delta}(\eta(\xi), \lambda(\xi))) \ge 0. \tag{8.14}$$

We now choose

$$x_0 := \Big(\frac{\arg\lambda(\xi)}{2\pi} - \frac{1}{(1+\delta^2)p}\Big)\Big/\Big(\frac{\eta(\xi)}{2\pi} - \frac{\delta}{1+\delta^2}\Big) \ge 0. \tag{8.15}$$

Then from (8.7) and (8.12) it follows that

$$\frac{1}{2\pi}\{\arg\mathcal{Y}_+(\xi,x)\}_{x\in[x_0,+\infty]} = -n_+ - \frac{1}{2}, \tag{8.16}$$

where the integer number n_+ is given by (8.14). Further, we infer from the first inequality in (8.12) that, for all $x \le x_0$,

$$\frac{x+\delta/p}{1+\delta^2} - \frac{\eta(\xi)}{2\pi p} - \frac{\ln|\lambda(\xi)|}{2\pi} < 0,$$

and therefore $\operatorname{Re}\mathcal{Y}_-(\xi,x) > 0$ for all $x \le x_0$, while $\mathcal{Y}_-(\xi,x_0) > 0$. Hence

$$\frac{1}{2\pi}\{\arg\mathcal{Y}_-(\xi,x)\}_{x\in[-\infty,x_0]} = 0. \tag{8.17}$$

Thus, we deduce from (8.10), (8.16), (8.17) and (8.15) that

$$\begin{aligned}\operatorname{Ind}_{\mathbb{R}} \det \mathcal{T}(\xi,\cdot) &= -\frac{1}{p} + \frac{\delta}{1+\delta^2}(x_0+\delta/p) - \frac{\eta(\xi)x_0}{2\pi} - E(\varphi_{p,\delta}(\eta(\xi),\lambda(\xi)))\\ &= -E(\varphi_{p,\delta}(\eta(\xi),\lambda(\xi))) - \frac{\arg\lambda(\xi)}{2\pi}.\end{aligned} \tag{8.18}$$

Let now $\varphi_{p,\delta}(\eta(\xi),\lambda(\xi)) < 0$ and $\frac{\eta(\xi)}{2\pi} - \frac{\delta}{1+\delta^2} \notin \mathbb{Z}$. Choose the branches of $\eta(\xi)$ and $\arg\lambda(\xi)$ so that

$$0 < \frac{\eta(\xi)}{2\pi} - \frac{\delta}{1+\delta^2} < 1, \quad -1 < \frac{\arg\lambda(\xi)}{2\pi} - \frac{1}{(1+\delta^2)p} \le 0. \tag{8.19}$$

Consider solutions $x_k \le 0$ of the system

$$\begin{cases} -\dfrac{x_k+\delta/p}{1+\delta^2} + \dfrac{\eta(\xi)}{2\pi p} + \dfrac{\ln|\lambda(\xi)|}{2\pi} < 0, \\ -\dfrac{1/p-\delta x_k}{1+\delta^2} - \dfrac{\eta(\xi)x_k}{2\pi} + \dfrac{\arg\lambda(\xi)}{2\pi} = -k \in \mathbb{Z}, \end{cases} \tag{8.20}$$

which imply that $\mathcal{Y}_-(\xi, x_k) < 0$. By the second equation in (8.20), x_k is given by (8.13). Then we infer from (8.19) and the first inequality in (8.20) that

$$k > \varphi_{p,\delta}(\eta(\xi), \lambda(\xi)) = \Big(\Big(\frac{\eta(\xi)}{2\pi} - \frac{\delta}{1+\delta^2}\Big)\frac{1}{p} + \frac{\ln|\lambda(\xi)|}{2\pi}\Big)\Big(\frac{\eta(\xi)}{2\pi} - \frac{\delta}{1+\delta^2}\Big)(1+\delta^2) + \frac{1}{(1+\delta^2)p} - \frac{\arg\lambda(\xi)}{2\pi}.$$

On the other hand, (8.19) and (8.13) imply that $x_k \le 0$ if and only if

$$\frac{1}{(1+\delta^2)p} - \frac{\arg\lambda(\xi)}{2\pi} \ge 0 \ge k.$$

Thus, system (8.12) has solutions $x_k \le 0$ for $k \in \mathbb{Z}$ such that

$$\frac{1}{(1+\delta^2)p} - \frac{\arg\lambda(\xi)}{2\pi} \ge 0 \ge k > \varphi_{p,\delta}(\eta(\xi), \lambda(\xi)),$$

which happens if $\varphi_{p,\delta}(\eta(\xi), \lambda(\xi)) < 0$ and in that case

$$n_- := E(\varphi_{p,\delta}(\eta(\xi), \lambda(\xi))) < 0. \tag{8.21}$$

Taking

$$x_0 := \Big(\frac{\arg\lambda(\xi)}{2\pi} - \frac{1}{(1+\delta^2)p}\Big)\Big/\Big(\frac{\eta(\xi)}{2\pi} - \frac{\delta}{1+\delta^2}\Big) < 0, \tag{8.22}$$

we conclude from (8.7) and (8.20) that $k \in \{n_- + 1, \ldots, -1, 0\}$ are admissible, and hence

$$\frac{1}{2\pi}\{\arg\mathcal{Y}_-(\xi, x)\}_{x\in[-\infty, x_0]} = -n_- - \frac{1}{2}, \tag{8.23}$$

where the integer number n_- is given by (8.21). Further we infer from the first inequality in (8.20) that $\operatorname{Re}\mathcal{Y}_+(\xi, x) > 0$ for all $x \ge x_0$, while $\mathcal{Y}_+(\xi, x_0) > 0$. Hence

$$\frac{1}{2\pi}\{\arg\mathcal{Y}_+(\xi, x)\}_{x\in[x_0, +\infty]} = 0. \tag{8.24}$$

Thus, we deduce from (8.10), (8.23), (8.24) and (8.22) that again

$$\begin{aligned}\operatorname{Ind}_{\mathbb{R}} \det \mathcal{T}(\xi, \cdot) &= -\frac{1}{p} + \frac{\delta}{1+\delta^2}(x_0 + \delta/p) - \frac{\eta(\xi)x_0}{2\pi} - E(\varphi_{p,\delta}(\eta(\xi), \lambda(\xi))) \\ &= -E(\varphi_{p,\delta}(\eta(\xi), \lambda(\xi))) - \frac{\arg\lambda(\xi)}{2\pi}.\end{aligned} \tag{8.25}$$

Finally, let $\frac{\eta(\xi)}{2\pi} - \frac{\delta}{1+\delta^2} \in \mathbb{Z}$. Choose the branch of $\eta(\xi)$ so that

$$\frac{\eta(\xi)}{2\pi} - \frac{\delta}{1+\delta^2} = 0. \tag{8.26}$$

In that case

$$\varphi_{p,\delta}(\eta(\xi), \lambda(\xi)) = \frac{1}{(1+\delta^2)p} - \frac{\arg\lambda(\xi)}{2\pi} \notin \mathbb{Z}, \tag{8.27}$$

and we choose the branch of $\arg\lambda(\xi)$ such that

$$-1 < \frac{\arg\lambda(\xi)}{2\pi} - \frac{1}{(1+\delta^2)p} < 0. \tag{8.28}$$

Then, by (8.26), we deduce from (8.7) that

$$\begin{aligned}\mathcal{Y}_+(\xi,x)&=1-e^{-2\pi x/(1+\delta^2)+\ln|\lambda(\xi)|}e^{-i[2\pi/((1+\delta^2)p)-\arg\lambda(\xi)]},\\ \mathcal{Y}_-(\xi,x)&=1-e^{2\pi x/(1+\delta^2)-\ln|\lambda(\xi)|}e^{i[2\pi/((1+\delta^2)p)-\arg\lambda(\xi)]}.\end{aligned}\tag{8.29}$$

Taking $x_0:=\frac{\ln|\lambda(\xi)|}{2\pi}(1+\delta^2)$, we infer from (8.10) and (8.26) that

$$\begin{aligned}\operatorname{Ind}_{\mathbb{R}}\det\mathcal{T}(\xi,\cdot)&=\frac{1}{2\pi}\{\arg\mathcal{Y}_-(\xi,x)\}_{x\in[-\infty,x_0]}+\frac{1}{2\pi}\{\arg\mathcal{Y}_+(\xi,x)\}_{x\in[x_0,+\infty]}\\ &\quad+\frac{1}{2}-\frac{1}{p}+\frac{\delta^2}{(1+\delta^2)p}.\end{aligned}\tag{8.30}$$

On the other hand, from (8.29) it follows for such x_0 that

$$\begin{aligned}\frac{1}{2\pi}\{\arg\mathcal{Y}_-(\xi,x)\}_{x\in[-\infty,x_0]}&=-\frac{1}{2}\left(\frac{1}{2}-\frac{1}{(1+\delta^2)p}+\frac{\arg\lambda(\xi)}{2\pi}\right),\\ \frac{1}{2\pi}\{\arg\mathcal{Y}_+(\xi,x)\}_{x\in[x_0,+\infty]}&=-\frac{1}{2}\left(\frac{1}{2}-\frac{1}{(1+\delta^2)p}+\frac{\arg\lambda(\xi)}{2\pi}\right).\end{aligned}\tag{8.31}$$

Combining (8.30) and (8.31), we obtain

$$\begin{aligned}\operatorname{Ind}_{\mathbb{R}}\det\mathcal{T}(\xi,\cdot)&=\frac{1}{2}-\frac{1}{p}+\frac{\delta^2}{(1+\delta^2)p}-\frac{1}{2}+\frac{1}{(1+\delta^2)p}-\frac{\arg\lambda(\xi)}{2\pi}\\ &=-\frac{\arg\lambda(\xi)}{2\pi}.\end{aligned}\tag{8.32}$$

Since $E(\varphi_{p,\delta}(\eta(\xi),\lambda(\xi)))=0$ according to (8.31) and (8.28), we conclude from (8.32) that again $\operatorname{Ind}_{\mathbb{R}}\det\mathcal{T}(\xi,\cdot)$ is calculated by (8.3). $\square$

Following the scheme of the proof of [25, Theorem 6.3] and applying Corollary 7.3 and Lemma 8.1, we obtain the index formula for the operator T.

Theorem 8.2. *If the operator $T=V_\alpha P_+ + GP_-$ with $G\in SO(\Gamma)$ is Fredholm on the space $L^p(\Gamma,w)$, then*

$$\begin{aligned}\operatorname{Ind}T=\lim_{m\to\infty}\Big(&\sum\nolimits_{k=1}^{N}\frac{\varepsilon_k}{2\pi}\{\arg G_k(r)\}_{r\in[m^{-1},m]}+\frac{\arg\lambda(m^{-1})}{2\pi}-\frac{\arg\lambda(m)}{2\pi}\\ &+E(\varphi_{p,\delta}(\eta(m^{-1}),\lambda(m^{-1})))-E(\varphi_{p,\delta}(\eta(m),\lambda(m)))\Big),\end{aligned}\tag{8.33}$$

where $\varphi_{p,\delta}(\eta,\lambda)$ is given by (7.23) and

$$\eta(r):=\sum_{k=1}^{N}\varepsilon_k\omega_k(r),\quad \lambda(r):=\prod_{k=1}^{N}(G_k(r))^{\varepsilon_k}\quad\textit{for all}\quad r\in\mathbb{R}_+.\tag{8.34}$$

Proof. By Theorem 6.2, the operator $T_-=U_\alpha^{-1}b_\alpha^{-1}P_+ + P_-$ is Fredholm on the space $L^p(\Gamma)$ and $\operatorname{Ind}T_-=0$. Hence, from (7.3) and (7.7) it follows that

$$\operatorname{Ind}T=\operatorname{Ind}N_\alpha=\operatorname{Ind}\Psi(N_\alpha)=\operatorname{Ind}\operatorname{Op}(\mathfrak{N}_\alpha).\tag{8.35}$$

Since all the entries of the matrix function $\mathfrak{N}_\alpha$ given by (7.8) belong to the Banach algebra $\widetilde{\mathcal{E}}(\mathbb{R}_+, V(\mathbb{R}))$, we infer from Theorem 3.5 that

$$\text{Ind Op}(\mathfrak{N}_\alpha) = \text{Ind Op}(\det \mathfrak{N}_\alpha) = \lim_{m\to\infty} \frac{1}{2\pi}\{\arg\det \mathfrak{N}_\alpha(r,x)\}_{(r,x)\in\partial\Pi_m}, \tag{8.36}$$

where $\Pi_m = [m^{-1}, m] \times \overline{\mathbb{R}}$.

By analogy with (7.11), we conclude from (7.8), (7.12) and (7.13) that

$$\begin{aligned}\det \mathcal{N}_\alpha(r,\pm\infty) &= \det\big(\mathcal{B}_\alpha^{-1}(r,\pm\infty)\mathfrak{P}_+(r,\pm\infty) + \mathcal{G}_\alpha(r,\pm\infty)\mathfrak{P}_-(r,\pm\infty)\big)\\ &= \Big(\prod\nolimits_{\varepsilon_k=\pm1} e^{-i\omega_k(r)\delta/p}\Big)\Big(\prod\nolimits_{\varepsilon_k=\mp1} e^{\omega_k(r)/p}\, G_k(r)\Big).\end{aligned} \tag{8.37}$$

Taking into account (7.11), (8.37) and (7.15), we infer that

$$\begin{aligned}\det \mathfrak{N}_\alpha(r,\pm\infty) &= H(r,\pm\infty)\det\mathcal{N}_\alpha(r,\pm\infty) \quad \text{for all} \quad r\in\mathbb{R}_+,\\ \det \mathfrak{N}_\alpha(\xi,x) &= H(\xi,x)\det\mathcal{N}_\alpha(\xi,x) \quad \text{for all} \quad (\xi,x)\in\Delta\times\overline{\mathbb{R}},\end{aligned} \tag{8.38}$$

where

$$H(r,x) = \Big(\prod\nolimits_{\varepsilon_k=-1}\big(1+r\omega_k'(r)\big)^{1/p}\Big)\mathcal{P}_+(x) + \Big(\prod\nolimits_{\varepsilon_k=1}\big(1+r\omega_k'(r)\big)^{1/p}\Big)\mathcal{P}_-(x).$$

As the function H is positive on $\partial(\mathbb{R}_+\times\mathbb{R})$, we deduce from (8.38) that

$$\lim_{m\to\infty}\frac{1}{2\pi}\{\arg\det\mathfrak{N}_\alpha(r,x)\}_{(r,x)\in\partial\Pi_m} = \lim_{m\to\infty}\frac{1}{2\pi}\{\arg\det\mathcal{N}_\alpha(r,x)\}_{(r,x)\in\partial\Pi_m}, \tag{8.39}$$

and therefore, by (8.36),

$$\text{Ind Op}(\mathcal{N}_\alpha) = \text{Ind Op}(\det\mathfrak{N}_\alpha) = \lim_{m\to\infty}\frac{1}{2\pi}\{\arg\det\mathcal{N}_\alpha(r,x)\}_{(r,x)\in\partial\Pi_m}. \tag{8.40}$$

By (8.37) and (8.1), we obtain

$$\begin{aligned}\frac{1}{2\pi}\{\arg\det\mathcal{N}_\alpha(r,x)\}_{(r,x)\in\partial\Pi_m} &= \text{Ind}_{\mathbb{R}}\det\mathcal{N}_\alpha(m,\cdot) - \text{Ind}_{\mathbb{R}}\det\mathcal{N}_\alpha(m^{-1},\cdot)\\ &+ \sum\nolimits_{k=1}^{N}\frac{\varepsilon_k}{2\pi}\Big(\{\arg G_k(r)\}_{r\in[m^{-1},m]} - \{\omega_k(r)\delta/p\}_{r\in[m^{-1},m]}\Big).\end{aligned} \tag{8.41}$$

Further, from (7.20) and (8.2) it follows that

$$\text{Ind}_{\mathbb{R}}\det\mathcal{N}_\alpha(r,\cdot) = \text{Ind}_{\mathbb{R}}\det\mathcal{T}(r,\cdot) + \text{Ind}_{\mathbb{R}}\det A_{\alpha^{-1}}(r,\cdot) \quad \text{for} \quad r = m^{\pm1}, \tag{8.42}$$

where $\det\mathcal{T}(r,x)$ and $\det A_{\alpha^{-1}}(r,x)$ are given by (7.2) and (6.22), respectively.

Substituting $d_k(r,x)$ in (6.39) by $d_k^{-1}(r,x)$ according to (6.22) and (6.23) and saving the notation $\det Y(r,x)$ for the obtained determinant, we infer from the modified formula (6.39) that

$$\begin{aligned}\frac{1}{4\pi}\{\arg\det Y(r,x)\}_{(r,x)\in\partial\Pi_m} &= \text{Ind}_{\mathbb{R}}\det A_{\alpha^{-1}}(m,\cdot) - \text{Ind}_{\mathbb{R}}\det A_{\alpha^{-1}}(m^{-1},\cdot)\\ &- \sum\nolimits_{k=1}^{N}\frac{\varepsilon_k}{2\pi}\{\omega_k(r)\delta/p\}_{r\in[m^{-1},m]}.\end{aligned} \tag{8.43}$$

Here, by analogy with the proof of Theorem 6.2,

$$\Psi\big((U_\alpha^{-1}b_\alpha^{-1}P_+ + P_-)C(P_+ + b_\alpha U_\alpha P_-)C\big) \simeq \text{Op}(Y),$$

and therefore

$$\operatorname{Ind} T_- = 2^{-1}\operatorname{Ind} \operatorname{Op}(Y) = 2^{-1}\operatorname{Ind} \operatorname{Op}(\det Y).$$

Consequently,

$$0 = \operatorname{Ind} T_- = \frac{1}{2}\operatorname{Ind} \operatorname{Op}(\det Y) = \lim_{m\to\infty} \frac{1}{4\pi}\big\{ \arg \det Y(r,x)\big\}_{(r,x)\in\partial\Pi_m}. \tag{8.44}$$

Hence, we deduce from (8.43) and (8.44) that for all sufficiently large $m > 0$,

$$\begin{aligned} &\operatorname{Ind}_{\mathbb{R}} \det A_{\alpha^{-1}}(m,\cdot) - \operatorname{Ind}_{\mathbb{R}} \det A_{\alpha^{-1}}(m^{-1},\cdot) \\ &- \sum\nolimits_{k=1}^{N} \frac{\varepsilon_k}{2\pi}\big\{\delta\omega_k(r)/p\big\}_{r\in[m^{-1},m]} = 0. \end{aligned} \tag{8.45}$$

Combining (8.41), (8.42) and (8.45), we conclude that for all sufficiently large $m > 0$,

$$\begin{aligned} \frac{1}{2\pi}\big\{ \arg \det \mathcal{N}_\alpha(r,x)\big\}_{(r,x)\in\partial\Pi_m} &= \sum\nolimits_{k=1}^{N} \frac{\varepsilon_k}{2\pi}\big\{ \arg G_k(r)\big\}_{r\in[m^{-1},m]} \\ &+ \operatorname{Ind}_{\mathbb{R}} \det \mathcal{T}(m,\cdot) - \operatorname{Ind}_{\mathbb{R}} \det \mathcal{T}(m^{-1},\cdot). \end{aligned}$$

This implies in view of (8.35), (8.36) and (8.39) that

$$\begin{aligned} \operatorname{Ind} T = \lim_{m\to\infty} \Big(&\sum\nolimits_{k=1}^{N} \frac{\varepsilon_k}{2\pi}\big\{ \arg G_k(r)\big\}_{r\in[m^{-1},m]} \\ &+ \operatorname{Ind}_{\mathbb{R}} \det \mathcal{T}(m,\cdot) - \operatorname{Ind}_{\mathbb{R}} \det \mathcal{T}(m^{-1},\cdot)\Big), \end{aligned} \tag{8.46}$$

where functions $\det \mathcal{T}(m^{\pm 1},\cdot)$ are separated from zero for all sufficiently large $m > 1$. Then similarly to Lemma 7.2 we conclude that $\varphi_{p,\delta}(\eta(r),\lambda(r)) \notin \mathbb{Z}$ for all sufficiently large $r^{\pm 1} = m > 1$. Applying now Corollary 7.3 and Lemma 8.1 with $\eta(\xi)$ and $\lambda(\xi)$ replaced by $\eta(m^{\pm 1})$ and $\lambda(m^{\pm 1})$, respectively, we obtain (8.33) from (8.46) and (8.3). □

References

[1] A.V. Aĭzenshtat, Yu.I. Karlovich, and G.S. Litvinchuk, *The method of conformal gluing for the Haseman boundary value problem on an open contour.* Complex Variables **28** (1996), 313–346.

[2] M.A. Bastos, A. Bravo, and Yu.I. Karlovich, *Convolution type operators with symbols generated by slowly oscillating and piecewise continuous matrix functions.* Operator Theory: Advances and Applications **147** (2004), 151–174.

[3] M.A. Bastos, Yu.I. Karlovich, and B. Silbermann, *Toeplitz operators with symbols generated by slowly oscillating and semi-almost periodic matrix functions*, Proc. London Math. Soc. **89** (2004), 697–737.

[4] A. Böttcher and Yu.I. Karlovich, *Carleson Curves, Muckenhoupt Weights, and Toeplitz Operators.* Progress in Mathematics **154**, Birkhäuser, Basel, 1997.

[5] A. Böttcher, Yu.I. Karlovich, and V.S. Rabinovich, *Emergence, persistence, and disappearance of logarithmic spirals in the spectra of singular integral operators.* Integr. Equ. Oper. Theory **25** (1996), 406–444.

[6] A. Böttcher, Yu.I. Karlovich, and V.S. Rabinovich, *Mellin pseudodifferential operators with slowly varying symbols and singular integral on Carleson curves with Muckenhoupt weights.* Manuscripta Math. **95** (1998), 363–376.

[7] A. Böttcher, Yu.I. Karlovich, and V.S. Rabinovich, *The method of limit operators for one-dimensional singular integrals with slowly oscillating data.* J. Operator Theory **43** (2000), 171–198.

[8] A. Böttcher, Yu.I. Karlovich, and V.S. Rabinovich, *Singular integral operators with complex conjugation from the viewpoint of pseudodifferential operators.* Operator Theory: Advances and Applications **121** (2001), 36–59.

[9] A. Böttcher, Yu.I. Karlovich, and I.M. Spitkovsky, *Convolution Operators and Factorization of Almost Periodic Matrix Functions.* Birkhäuser, Basel, 2002.

[10] A. Böttcher and B. Silbermann, *Analysis of Toeplitz Operators.* 2nd edition, Springer, Berlin, 2006.

[11] G. David, *Opérateurs intégraux singuliers sur certaines courbes du plan complexe. Ann. Sci. École Norm. Sup.* **17** (1984), 157–189.

[12] R.V. Duduchava, *On convolution type integral operators with discontinuous coefficients.* Math. Nachr. **79** (1977), 75–98 [Russian].

[13] R.V. Duduchava, *Integral equations of convolution type with discontinuous coefficients.* Soobshch. Akad. Nauk Gruz. SSR **92** (1978), 281–284 [Russian].

[14] R. Duduchava, *Integral Equations with Fixed Singularities.* Teubner Verlagsgesellschaft, Leipzig, 1979.

[15] A.Yu. Karlovich, Yu.I. Karlovich, and A.B. Lebre, *Invertibility of functional operators with slowly oscillating non-Carleman shifts.* Operator Theory: Advances and Applications **142** (2003), 147–174.

[16] A.Yu. Karlovich, Yu.I. Karlovich, and A.B. Lebre, *Sufficient conditions for Fredholmness of singular integral operators with shifts and slowly oscillating data.* Integr. Equ. Oper. Theory **70** (2011), 451–483.

[17] A.Yu. Karlovich, Yu.I. Karlovich, and A.B. Lebre, *Necessary conditions for Fredholmness of singular integral operators with shifts and slowly oscillating data.* Integr. Equ. Oper. Theory **71** (2011), 29–53.

[18] A.Yu. Karlovich, Yu.I. Karlovich, and A.B. Lebre, *Fredholmness and index of simplest singular integral operators with two slowly oscillating shifts.* Operators and Matrices **8** (2014), 935–955.

[19] A.Yu. Karlovich, Yu.I. Karlovich, and A.B. Lebre, *On regularization of Mellin PDO's with slowly oscillating symbols of limited smoothness.* Comm. Math. Anal. **17** (2014), no. 2, 189–208.

[20] Yu.I. Karlovich, *On the Haseman problem.* Demonstratio Mathematica **26** (1993), No. 3–4, 581–595.

[21] Yu.I. Karlovich, *An algebra of pseudodifferential operators with slowly oscillating symbols.* Proc. London Math. Soc. **92** (2006), 713–761.

[22] Yu.I. Karlovich, *Pseudodifferential operators with compound slowly oscillating symbols.* Operator Theory: Advances and Applications **171** (2006), 189–224.

[23] Yu.I. Karlovich, *Algebras of pseudo-differential operators with discontinuous symbols.* Operator Theory: Advances and Applications **172** (2007), 207–233.

[24] Yu.I. Karlovich, *Nonlocal singular integral operators with slowly oscillating data.* Operator Theory: Advances and Applications **181** (2008), 229–261.

[25] Yu.I. Karlovich, *The Haseman boundary value problem with slowly oscillating data.* In: "Analytic Methods of Analysis and Differential Equations: AMADE 2006", eds. A.A. Kilbas, S.V. Rogozin, Cambridge Scientific Publishers, 2008, pp. 81–110.

[26] Yu.I. Karlovich, *An algebra of shift-invariant singular integral operators with slowly oscillating data and its application to operators with a Carleman shift.* Operator Theory: Advances and Applications **193** (2009), 81–95.

[27] Yu.I. Karlovich and V.G. Kravchenko, *On a singular integral operator with non-Carleman shifts on an open contour.* Soviet Math. Dokl. **18** (1977), 1263–1267.

[28] Yu.I. Karlovich and V.G. Kravchenko, *An algebra of singular integral operators with piecewise-continuous coefficients and a piecewise-smooth shift on a composite contour.* Math. USSR Izvestiya **23** (1984), 307–352.

[29] V.G. Kravchenko and G.S. Litvinchuk, *Introduction to the Theory of Singular Integral Operators with Shift.* Kluwer Academic Publishers, Dordrecht, 1994.

[30] G.S. Litvinchuk, *Boundary Value Problems and Singular Integral Equations with Shift.* Nauka, Moscow, 1977 [Russian].

[31] G.S. Litvinchuk, *Solvability Theory of Boundary Value Problems and Singular Integral Equations with Shift.* Mathematics and Its Applications **523**, Kluwer Academic Publishers, Dordrecht, 2000.

[32] V.S. Rabinovich, *Algebras of singular integral operators on compound contours with nodes that are logarithmic whirl points.* Izv. Math. **60** (1996), 1261–1292.

[33] V.S. Rabinovich, *Mellin pseudodifferential operators techniques in the theory of singular integral operators on some Carleson curves.* Operator Theory: Advances and Applications **102** (1998), 201–218.

[34] V. Rabinovich, S. Roch, and B. Silbermann, *Limit Operators and Their Applications in Operator Theory.* Birkhäuser, Basel, 2004.

[35] S. Roch, P.A. Santos and B. Silbermann, *Non-commutative Gelfand Theories. A Tool-kit for Operator Theorists and Numerical Analysts.* Springer, London, 2011.

[36] D. Sarason, *Toeplitz operators with piecewise quasicontinuous symbols.* Indiana Univ. Math. J. **26** (1977), 817–838.

[37] D. Sarason, *Toeplitz operators with semi-almost periodic symbols.* Duke Math. J. **44** (1977), 357–364.

Yu.I. Karlovich
Centro de Investigación en Ciencias
Instituto de Investigación en Ciencias Básicas y Aplicadas
Universidad Autónoma del Estado de Morelos
Av. Universidad 1001, Col. Chamilpa,
C.P. 62209 Cuernavaca, Morelos, México
e-mail: `karlovich@uaem.mx`

Operator Theory:
Advances and Applications, Vol. 259, 501–510

On the Norm of Linear Combinations of Projections and Some Characterizations of Hilbert Spaces

Nahum Krupnik and Alexander Markus

Dedicated to Albrecht Böttcher
on the occasion of his 60th birthday

Abstract. Let $\mathcal{B}$ be a Banach space and let P, Q $(P, Q \neq 0)$ be two complementary projections in $\mathcal{B}$ (i.e., $P + Q = I$). For $\dim \mathcal{B} > 2$ we show that formulas of the kind $\|aP + bQ\| = f(a, b, \|P\|)$ hold if and only if the norm in $\mathcal{B}$ can be induced by an inner product. The two-dimensional case needs special consideration which is done in the last two sections.

Mathematics Subject Classification (2010). 47C15, 47C99, 47A56.

Keywords. Characterizations of Hilbert spaces, projections, linear combinations of two additional projections.

1. Introduction

Let $\mathcal{B}$ denote a Banach space (real or complex) with $\dim \mathcal{B} > 1$. If the norm in the space $\mathcal{B}$ can be induced by some inner product (i.e., $\|f\|^2 = (f, f)$), it is called Hilbert space. Note, right away, that for the problems considered below the completeness of the space plays no role, and therefore, instead of a Banach (resp. a Hilbert) space one may consider below a normed space (resp. a space with the norm generated by an inner product, i.e., a pre-Hilbert space).

An operator P acting in $\mathcal{B}$ is called a projection if $P^2 = P$. If, in addition, $P \notin \{0, I\}$, we call the projection *nontrivial.*

Consider a linear combination $A = aP + bQ$, where $a, b \in \mathbb{C}$ (or $a, b \in \mathbb{R}$) and $Q = I - P$. The following statement was formulated in [7, Theorem 1] and proved in [8, Theorem 1.1] (see also [5, Example 3.12]).

The research of the first author was partially supported by Retalon Inc., Toronto, ON, Canada.

Theorem 1.1. *Let $\mathcal{H}$ be a Hilbert space and P a nontrivial projection on $\mathcal{H}$. Then*

$$\|aP + bQ\| = \frac{\gamma_+ + \gamma_-}{2}, \quad \text{where} \quad \gamma_\pm = \sqrt{(|a| \pm |b|)^2 + |a-b|^2(\|P\|^2 - 1)}. \quad (1.1)$$

It follows from (1.1) that

$$\|I - P\| = \|P\| \quad (1.2)$$

(Ljance's formula, [11]) and

$$\|S\| = \|P\| + \sqrt{\|P\|^2 - 1}, \quad \text{where} \quad S := P - Q$$

(Spitkovsky's formula [12], see also [13]). In particular,

$$\|P\| = 1 \quad \Longrightarrow \quad \|Q\| = \|S\| = 1. \quad (1.3)$$

In this paper we obtain some characterizations of Hilbert spaces in terms of projections, and study the Banach spaces for which some analogues of equality (1.1) hold.

We start with the following result.

Theorem 1.2. *If a Banach space $\mathcal{B}$ satisfies condition* (1.1)*, then $\mathcal{B}$ is a Hilbert space.*

Thus, equality (1.1) is a characterization of Hilbert space.

The proof of Theorem 1.2 will be given in Section 3. The following characterization of Hilbert spaces can be extracted from this proof.

Theorem 1.3. *If in a real reflexive Banach space $\mathcal{B}$ the implication*

$$\|P\| = 1 \quad \Longrightarrow \quad \|2P - I\| = 1$$

holds for every projection P, then $\mathcal{B}$ is a Hilbert space.

Equality (1.1) prompted us to formulate

Problem 1.4. *To describe Banach spaces $\mathcal{B}$ which satisfy the following condition:*

Condition 1.5. *There exists a function $f(a, b, x) = f_{\mathcal{B}}(a, b, x)$ such that*

$$\|aP + bQ\| = f_{\mathcal{B}}(a, b, \|P\|) \quad (Q = I - P) \quad (1.4)$$

for all $a, b \in \mathbb{C}$ $(a, b \in \mathbb{R})$ and all nontrivial projections P.

The solution of the Problem 1.4 for the case $\dim \mathcal{B} > 2$ is given by the following.

Theorem 1.6. *Let* $\dim \mathcal{B} > 2$. *There exists a function $f_{\mathcal{B}}(a, b, x)$ such that Condition* 1.5 *holds if and only if $\mathcal{B}$ is a Hilbert space.*

This theorem gives another characterization of Hilbert spaces (for $\dim \mathcal{B} > 2$). Theorem 1.6 follows directly from the following two statements.

Theorem 1.7. *Let $\mathcal{B}$ be a Banach space satisfying Condition* 1.5. *Then the equality $\|I - P\| = \|P\|$ holds for all nontrivial projections P.*

Theorem 1.8. *Let $\mathcal{B}$ be a Banach space and* $\dim \mathcal{B} > 2$. *If* $\|I - P\| = \|P\|$ *holds for all nontrivial projections* P, *then* $\mathcal{B}$ *is a Hilbert space.*

A sketch of the proof of these two theorems is given in [8, Subsection 1.3]. Below more detailed proofs of Theorems 1.7 and 1.8 are given. As for Theorem 1.8, it is more convenient for us to prove the following stronger proposition.

Theorem 1.9. *Let* $\dim \mathcal{B} > 2$. *If for every one-dimensional projection* P *in* $\mathcal{B}$ *the condition*

$$\|P\| = 1 \quad \Longrightarrow \quad \|I - P\| = 1 \tag{1.5}$$

holds, then $\mathcal{B}$ *is a Hilbert space.*

Theorem 1.9 for the real spaces is well known [1, p. 144]. We give in Section 2 another proof of this theorem which works for complex spaces as well.

Remark 1.10. Theorem 1.8 fails if $\dim \mathcal{B} = 2$. There exists a wide class of non-Hilbert norms in two-dimensional (real and complex) spaces, where equality (1.2) holds for any one-dimensional projection P. See Sections 4 and 5.

Remark 1.11. In contrast with the case $\dim \mathcal{B} > 2$, the statement converse to Theorem 1.7 fails in two-dimensional spaces. See Proposition 4.7.

We conclude the introduction with an open

Question 1.12. *Do there exist two-dimensional non-Hilbert Banach spaces for which Condition* 1.5 *holds?*

It is our pleasure to thank Ilya M. Spitkovsky for useful remarks and comments.

2. Proof of Theorems 1.7 and 1.9

2.1. We start with the proof of Theorem 1.7.

Proof. Let Condition 1.5 be fulfilled in some Banach space $\mathcal{B}$. We set in equality (1.4) $a = 0, b = 1$ and obtain $\|Q\| = f_{\mathcal{B}}(0, 1, \|P\|)$. Denote the function $f_{\mathcal{B}}(0, 1, \|P\|)$ by $h(\|P\|)$ and we obtain

$$\|I - P\| = h(\|P\|). \tag{2.1}$$

Let $y_0 \in \mathcal{B}$, $f_0 \in \mathcal{B}^*$ and $f_0(y_0) = \|y_0\| = \|f_0\| = 1$. We choose a non-zero vector z_0 such that $f_0(z_0) = 0$, and consider the one-dimensional projections $P_t = f_0(\cdot)(y_0 + tz_0)$. Then $\|P_t\| = \|y_0 + tz_0\|$ and $\|P_t\| \to +\infty$ when $t \to +\infty$. It follows that $\|P_t\|$ $(t \geq 0)$ takes all values form $[1, +\infty)$, and hence $h(x)$ is defined on $[1, +\infty)$.

Let $x \in [1, +\infty)$ and $\|P\| = x$. Then

$$h(x) = h(\|P\|) = \|I - P\| \geq \|P\| - 1 = x - 1,$$

therefore

$$\lim_{x \to +\infty} h(x) = +\infty.$$

Let us show that $h(x)$ is a monotone function. It is clear that

$$h(h(x)) = h(\|I - P\|) = \|I - (I - P)\| = \|P\| = x,$$

and hence,

$$h(a) = h(b) \quad \Longrightarrow \quad h(h(a)) = h(h(b)) \quad \Longrightarrow \quad a = b.$$

Thus, we proved that $h(x)$ is an increasing function on $[1, \infty)$ and $h(h(x)) = x$. It follows that $h(x) = x$ for all $x \in [1, \infty)$. Indeed, assume that $h(a) \geq a$ for some a. Then $a = h(h(a)) \geq h(a)$, i.e., $h(a) = a$. The same follows if we assume that $h(a) \leq a$ for some a. This proves that $h(x) \equiv x$. Using equality (2.1), we obtain that $\|I - P\| = \|P\|$. □

2.2. In this subsection we prove Theorem 1.9. We start with the following lemma.

Lemma 2.1. *If a Banach space $\mathcal{B}$ $(\dim \mathcal{B} > 3)$ satisfies the condition of Theorem 1.9 (i.e., condition (1.5)) then any three-dimensional subspace $E \subset \mathcal{B}$ also satisfies the condition of Theorem 1.9.*

Proof. Let P_0 be an arbitrary one-dimensional projection in E with $\|P_0\| = 1$. Let us write $P_0 = f_0(\cdot)z_0$. We may assume that $\|z_0\| = 1$ and $\|f_0\| = 1$. Since $z_0 \in \operatorname{Im} P_0$ we have $P_0 z_0 = z_0$, and since $P_0 z_0 = f_0(z_0)z_0$ we have $f_0(z_0) = 1$. Using the Hahn–Banach theorem we obtain a functional $f_1 \in \mathcal{B}^*$ such that

$$f_1(u) = f_0(u) \ (u \in E) \quad \text{and} \quad \|f_1\| = \|f_0\|.$$

In particular, $f_1(z_0) = f_0(z_0) = 1$.

If $P := f_1(\cdot)z_0$, then P is a one-dimensional projection in $\mathcal{B}$ and $\|P\| = \|f_1\|\|z_0\| = 1$. Obviously, for $u \in E$ the following equalities hold:

$$Pu = f_1(u)z_0 = f_0(u)z_0 = P_0 u.$$

Since condition (1.5) is satisfied in the space $\mathcal{B}$, we have $\|I - P\| = 1$. Furthermore for $u \in E$ we have $(I - P)u = (I - P_0)u$, and hence $\|I - P_0\| \leq \|I - P\| = 1$. But the norm of a non-zero projection cannot be less than 1. Hence, $\|I - P_0\| = 1$. So, the subspace E also satisfies the condition of Theorem 1.9, and the lemma is proved. □

In the proof of Theorem 1.9 we use the following statement.

Proposition 2.2. *A Banach space $\mathcal{B}$ of dimension ≥ 3 is a Hilbert space if and only if every two-dimensional subspace M of $\mathcal{B}$ is the range of a projection P of norm 1.*

For the real case this statement was proved by S. Kakutani [10] and for the complex case by F. Bohnenblust [4].

Proof of Theorem 1.9. First we prove the theorem for the case $\dim \mathcal{B} = 3$. Let M be any two-dimensional subspace of $\mathcal{B}$. There exists a functional $f \in \mathcal{B}^*$ such that $\|f\| = 1$ and $\ker f = M$. Furthermore there exists a vector $z \in \mathcal{B}$ such that $f(z) = 1$ and $\|z\| = 1$. Let $P = f(\cdot)z$. Then P is a one-dimensional projection with $\|P\| = 1$.

By (1.5) the norm of $Q = I - P$ also equals 1. Evidently, $\ker P = \ker f = M$, and hence $\operatorname{Im} Q = \ker P = M$. So, for an arbitrary two-dimensional subspace M of $\mathcal{B}$ there exists a projection P with norm 1 and with range M. By Proposition 2.2 $\mathcal{B}$ is a Hilbert space.

Consider now the case $\dim \mathcal{B} > 3$. Let E be an arbitrary three-dimensional subspace of $\mathcal{B}$. By Lemma 2.1 for each one-dimensional projection P_0 in E relation (1.5) also holds. According to the first part of the proof, E is a Hilbert space. Since every three-dimensional subspace of $\mathcal{B}$ is a Hilbert space, it follows that $\mathcal{B}$ is also a Hilbert space. □

3. Proof of Theorem 1.2

We start with the case when $\mathcal{B}$ is a complex space and we will use the following characterization of Hilbert spaces (see [3, p. 314]).

Proposition 3.1. *Let $\mathcal{B}$ be a complex Banach space. Then $\mathcal{B}$ is a Hilbert space if and only if every its one-dimensional subspace is the range of a Vidav-Hermitian projection.*

Recall that an operator $A \in L(\mathcal{B})$ is called *Vidav-Hermitian*, if

$$\|I - itA\| = 1 + o(t), \ t \in \mathbb{R}, \ t \to 0.$$

Proof of Theorem 1.2 *for complex spaces.* Let F be an arbitrary one-dimensional subspace of $\mathcal{B}$ and let P be a projection on this subspace with $\|P\| = 1$. Set in (1.1) $a = 1 + it$ and $b = 1$. Then $|a| = \sqrt{1+t^2}$, $|a - b| = t$, and (1.1) implies $\|(1+it)P + Q\| = (\gamma_+ + \gamma_-)/2$, where $\gamma_\pm = \big||a| \pm |b|\big| = \sqrt{1+t^2} \pm 1$. Hence

$$\|(1+it)P + Q\| = \sqrt{1+t^2}, \qquad \text{i.e.,} \qquad \|I + itP\| - 1 \sim \frac{t^2}{2} = o(t).$$

Thus P is a Vidav-Hermitian projection and we can use Proposition 3.1. This proves the theorem. □

We pass to the case when $\mathcal{B}$ is a real space and we will use the following characterization of Hilbert spaces (see [1, p. 34, Statement (4.8)]).

Proposition 3.2. *Let $\mathcal{B}$ be a real Banach space. Then $\mathcal{B}$ is a Hilbert space if and only if for each maximal subspace M of $\mathcal{B}$ there exists an element $z \notin M$ such that $\|z - u\| = \|z + u\|$ for all $u \in M$.*

Recall that a subspace M of $\mathcal{B}$ is called *maximal* if $\dim(\mathcal{B}/M) = 1$.

Proof of Theorem 1.2 *for real spaces.* Let $\mathcal{B}$ be a Banach space satisfying condition (1.1). It suffices to prove this theorem for any fixed two-dimensional subspace $E \subset \mathcal{B}$. Let M be an arbitrary one-dimensional (and thus maximal) subspace of E. We take $f \in E^*$ such that $\|f\| = 1$ and $\ker f = M$. By the Hahn–Banach theorem, there exists a vector z such that $\|z\| = 1$ and $f(z) = 1$. Consider the following two projections: $P = f \otimes z$ and $Q = I - P$. Note that $\|Px\| = \|f(x)z\| = |f(x)|$, therefore $\|P\| = \|f\| = 1$. Denote by S the operator $S = P - Q$. It follows from (1.3) that $\|S\| = 1$. Since, in addition, $S^2 = (P - Q)^2 = I$, it follows that S is an isometry, i.e., $\|Sx\| = \|x\|$ for all $x \in E$. Let u be an arbitrary vector from M and $x = u + z$. Then

$$Px = f(u + z)z = z, \quad Qx = x - Px = u, \quad Sx = Px - Qx = z - u.$$

Therefore

$$\|z - u\| = \|Sx\| = \|x\| = \|u + z\|.$$

It remains to use Proposition 3.2, and the theorem is proved. □

Proof of Theorem 1.3. In the proof of Theorem 1.2 for a real two-dimensional space $\mathcal{B}$ we have used only the following part of equality (1.3):

$$\|P\| = 1 \Longrightarrow \|S\| = 1 \quad (S = 2P - I).$$

The same proof can be used for any real reflexive space $\mathcal{B}$. This proves Theorem 1.3. □

We do not know if Theorem 1.3 holds for complex spaces.

4. Real two-dimensional spaces

In this section we assume that $\mathcal{B}$ is a two-dimensional space. If the space $\mathcal{B}$ satisfies Condition 1.5, then (by Theorem 1.7) equality (1.2) holds for any one-dimensional projection P acting in this space. If in a two-dimensional space $\mathcal{B}$ equality (1.2) holds, we will call this space (for short) *suitable*. In this section we will use the following.

Theorem 4.1. *Let $\mathcal{B}$ be a two-dimensional space. Then the equation $\|P\| = \|I - P\|$ holds for every one-dimensional projection P if and only if for arbitrary vectors $z \in \mathcal{B}$ and $f \in \mathcal{B}^*$ the following equality holds:*

$$\|f\|^*\|z\| = \|\tilde{z}\|^*\|\tilde{f}\|, \tag{4.1}$$

where $\tilde{f} := (-f_2, f_1)$, $\tilde{z} := (-z_2, z_1)$ and $\|\cdot\|^*$ denotes the norm in the space $\mathcal{B}^*$.

Proof. Let P be a one-dimensional projection, then it can be represented in the form $Px = f(x)z$, where $f(z) = 1$ and $\|P\| = \|f\|^*\|z\|$. Denote $Qx = \tilde{z}(x)\tilde{f}$. It is easy to see that

$$\begin{aligned} Px + Qx &= (f_1x_1 + f_2x_2)(z_1, z_2) + (-z_2x_1 + z_1x_2)(-f_2, f_1) \\ &= (f_1z_1 + f_2z_2)(x_1, x_2) = f(z)x = x, \end{aligned}$$

i.e., $Q = I - P$. Thus,

$$P = f \otimes z \Longleftrightarrow I - P = \tilde{z} \otimes \tilde{f} \quad \text{and} \quad \|I - P\| = \|\tilde{z}\|^*\|\tilde{f}\|.$$

1. Let condition (4.1) be fulfilled, then $\|I - P\| = \|\tilde{z}\|^*\|\tilde{f}\| = \|f\|^*\|z\| = \|P\|$.

2. Let $\|I-P\| = \|P\|$ and let two vectors z, f be given. First, we assume that $f(z) = d \neq 0$, denote $z_0 = z/d$. Then $f(z_0) = 1$ and $P = f \otimes z_0$ is a projection. Equality (4.1) follows from the following one:

$$\|f\|^*\|z_0\| = \|P\| = \|I-P\| = \|\tilde{z}_0\|^*\|\tilde{f}\|.$$

Now we assume that $f(z) = 0$. Denote $z_s := z + sw$, where $f(w) = 1$. If $s \neq 0$, then $f(z_s) \neq 0$ and for the pairs f, z_s we just proved equality (4.1). Passing to the limit when $s \to 0$ we obtain equality (4.1) for the pair f, z. □

Recall the following known statement (see, for example, [2, p. 10]), which will be used below in some examples.

Proposition 4.2. *Let the unit ball in $\mathcal{B}$ have only a finite number of extreme points: w_k $(k = 1, \ldots, 2n)$ and $w_{k+n} = -w_k$. Then the norm of each operator $A \in L(\mathcal{B})$ equals*

$$\|A\| = \max_{k=1,\ldots,n} \|Aw_k\|.$$

Example 4.3. Let us take for the unit ball D in $\mathcal{B}$ a regular hexagon with vertices $w_{1,2,3} = \{(1,0), (\pm 1/2, \sqrt{3}/2)\}$ and $w_{k+3} = -w_k$.

Let $f \in \mathcal{B}^*$. Using Proposition 4.2 it is not difficult to show that

$$\|f\|^* = \max\{|f_1 y_1 + f_2 y_2| : (y_1, y_2) \in \{w_{1,2,3}\}\} = \max\left\{|f_1|, \frac{|f_1| + |f_2|\sqrt{3}}{2}\right\}.$$

It follows from this that the unit ball D^* in the space $\mathcal{B}^*$ is the regular hexagon with vertices

$$\tilde{w}_{1,2,3} = \{(0, 2/\sqrt{3}), (\pm 1, 1/\sqrt{3}\}) \quad \text{and} \quad \tilde{w}_{k+3} = -\tilde{w}_k.$$

This implies, that the norm of the vector $z = (z_1, z_2) \in \mathcal{B}$ equals

$$\|z\| = \max\{|z_1 t_1 + z_2 t_2| : (t_1, t_2) \in \tilde{w}_{1,2,3}\} = \max\left\{\frac{2|z_2|}{\sqrt{3}}, \frac{|z_2| + |z_1|\sqrt{3}}{\sqrt{3}}\right\}.$$

Thus

$$\|f\|^*\|z\| = \frac{1}{2}\max\left\{2|f_1|, |f_1| + |f_2|\sqrt{3}\right\} \times \frac{1}{\sqrt{3}}\max\left\{2|z_2|, |z_2| + |z_1|\sqrt{3}\right\}. \quad (4.2)$$

Using the same arguments as above we obtain that

$$\|\tilde{z}\|^*\|\tilde{f}\| = \frac{1}{2}\max\left\{2|z_2|, |z_2| + |z_1|\sqrt{3}\right\} \times \frac{1}{\sqrt{3}}\max\left\{2|f_1|, |f_1| + |f_2|\sqrt{3}\right\}. \quad (4.3)$$

It follows from (4.2) and (4.3) that $\|f\|^*\|z\| = \|\tilde{z}\|^*\|\tilde{f}\|$ and from Theorem 4.1 that $\|P\| = \|I-P\|$, i.e., the space $\mathcal{B}$ is suitable.

Remark 4.4. There exists a general procedure, proposed by M.M. Day ([6, Section 6], see also [1, p. 77]), which turns some non-suitable spaces with the norm

$\|(x,y)\|$ into suitable spaces with the norm

$$\|(x,y)\|_1 := \begin{cases} \|(x,y)\| & \text{if} \quad xy \geq 0 \\ \|(-y,x)\|^* & \text{if} \quad xy \leq 0. \end{cases}$$

Example 4.5. Let $E = \ell_1^2$, and $P(x,y) = (x+y,0)$. Then $\|P\| = 1$ and $\|I-P\| > 1$. Thus, the space E (with the standard norm) is not suitable. But the space $\mathcal{B}$ with the norm

$$\|(x,y)\|_1 := \begin{cases} |x| + |y| & \text{if} \quad xy \geq 0 \\ \max(|x|,|y|) & \text{if} \quad xy \leq 0, \end{cases} \tag{4.4}$$

obtained by the Day procedure, is a suitable space.

Remark 4.6. One can prove directly (using the same arguments as in Example 4.3), that the space $\mathcal{B}$ with the norm (4.4) is suitable.

Proposition 4.7. *In contrast with the case* $\dim \mathcal{B} > 2$, *the statement converse to Theorem* 1.7 *fails for two-dimensional spaces.*

Proof. Let $\mathcal{B}$ denote the real two-dimensional Banach space with the norm defined by equality (4.4), and let $A \in L(B)$. The closed unit ball in $\mathcal{B}$ has only six extremal points: $\pm(1,0)$, $\pm(0,1)$, $\pm(-1,1)$. By Proposition 4.2 the norm of each operator $A \in L(\mathcal{B})$ equals

$$\|A\| = \max_{1 \leq k \leq 3} \{\|Aw_k\| : \ w_1 = (1,0); \ w_2 = (0,1); \ w_3 = (-1,1)\}.$$

Consider two collections of operators:

$$P_t := \begin{bmatrix} 1 & t \\ 0 & 0 \end{bmatrix} \quad \text{and} \quad S_t = I - 2P_t = -\begin{bmatrix} 1 & 2t \\ 0 & -1 \end{bmatrix}.$$

Direct calculations show that

$$\|P_t w_1\| = 1; \ \|P_t w_2\| = |t| \quad \text{and} \quad \|P_t w_3\| = |t-1|.$$

It is easy to check that $\max(1, |a|, |a-1|) = (1+|a|+|a-1|)/2$ for any real number a. Therefore

$$\|P_t\| = \max_{k=1,2,3} \|P_t w_k\| = (|t| + |t-1| + 1)/2.$$

Analogous calculations show that

$$\|S_t w_2\| = \begin{cases} 2|t|+1 & \text{if} \quad t < 0 \\ 1 & \text{if} \quad 0 < 2t < 1 \\ 2t & \text{if} \quad 2t > 1 \end{cases}; \quad \|S_t w_3\| = \begin{cases} 2-2t & \text{if} \quad 2t \leq 1 \\ 1 & \text{if} \quad 1 < 2t < 2 \\ 2t-1 & \text{if} \quad t > 1 \end{cases}$$

and $\|S_t w_1\| = 1$. Therefore

$$\|S_t\| = \max_{k=1,2,3} \|S_t w_k\| = |2t-1| + 1.$$

If, in particular, $t \in [0,1]$, then $\|P_t\| \equiv 1$, i.e., it does not depend on t, but $\|S_t\|$ is not constant on this segment. This proves that Condition 1.5 fails in $\mathcal{B}$. □

5. Complex two-dimensional spaces

In order to consider the complex case we need a small introduction into the complexification of real Banach spaces and operators on these spaces. Denote by $\mathcal{B}$ a real two-dimensional normed space with the norm $\|\cdot\|$. By definition its compexification $\mathcal{B}^{\mathbb{C}}$ is the direct sum of $\mathcal{B}$ and $i\mathcal{B}$. Let $\|\cdot\|^{\mathbb{C}}$ denote a norm in $\mathcal{B}^{\mathbb{C}}$.

Definition 5.1. The norm $\|\cdot\|^{\mathbb{C}}$ is a complexification of the norm $\|\cdot\|$ if

$$\|(x,y)\| = \|(x+i0, y+i0)\|^{\mathbb{C}}.$$

The complexification of a norm is not unique [9, p. 288, Problem 25].

Definition 5.2. Let $A \in L(\mathcal{B})$ and $z = u + iv$ $\left(u, v \in \mathcal{B},\ z \in \mathcal{B}^{\mathbb{C}}\right)$. The complexification $A^{\mathbb{C}}$ of operator A is defined by the equality $A^{\mathbb{C}}z := Au + iAv$.

Definition 5.3. We say that a complexification of the norm agrees with the complexification of operators if $\|A^{\mathbb{C}}\|^{\mathbb{C}} = \|A\|$ for all $A \in L(\mathcal{B})$.

Among all complexifications of a given norm there exists minimal and maximal [9, p. 289, Problems 26-28]. On the other hand, there exist complexifications which transfer any suitable real space into a suitable complex space. We will see that this property is possessed, in particular, by the complexification with minimal norm. This minimal norm is defined by equality ([9, p. 289, Problem 27])

$$\|z\|^{\mathbb{C}} = \sup_{f\in\mathcal{B}^*} \frac{|f^{\mathbb{C}}(z)|}{\|f\|}, \quad z = u + iv,\ f^{\mathbb{C}}(z) = f(u) + if(v). \tag{5.1}$$

Theorem 5.4. *If a space $\mathcal{B}$ is suitable, then the space $\mathcal{B}^{\mathbb{C}}$ with the norm* (5.1) *is also suitable. Moreover, the complexification* (5.1) *agrees with the complexification of operators.*

In the proof of this theorem we use the following lemma, proved in [14, Lemma 1 and Theorem 1]. For the convenience of the reader we give here its proof.

Lemma 5.5. *The complexification of the norm $\|\cdot\|$, defined in* (5.1), *agrees with the complexification of operators.*

Proof. Equality (5.1) can be rewritten in the following form:

$$\begin{aligned}\|z\|^{\mathbb{C}} &= \sup_{f\in\mathcal{B}^*} \frac{|f(u) + if(v)|}{\|f\|} = \sup_{f\in\mathcal{B}^*} \frac{\left(f(u)^2 + f(v))^2\right)^{1/2}}{\|f\|} \\ &= \sup_{f,\theta} \frac{|f(u)\cos\theta + f(v)\sin\theta|}{\|f\|} = \max_{\theta}(\|u\cos\theta + v\sin\theta\|).\end{aligned}$$

It follows from here that

$$\|A^{\mathbb{C}}z\|^{\mathbb{C}} = \max_{\theta}(\|Au\cos\theta + Av\sin\theta\|) \le \|A\| \max_{\theta}(\|u\cos\theta + v\sin\theta\|) = \|A\|\|z\|^{\mathbb{C}}.$$

Hence, $\|A^{\mathbb{C}}\|^{\mathbb{C}} \le \|A\|$.

The inverse inequality is evident, and we obtain $\|A^{\mathbb{C}}\|^{\mathbb{C}} = \|A\|$. □

Proof of Theorem 5.4. The second assertion of the theorem coincides with Lemma 5.5. It follows also from this lemma that $\|P^{\mathbb{C}}\|^{\mathbb{C}} = \|P\|$ for all non-trivial projections P in $\mathcal{B}$. But all non-trivial projections P in $\mathcal{B}$ are one-dimensional, and hence $I-P$ is nontrivial as well. Hence $\|I-P^{\mathbb{C}}\|^{\mathbb{C}} = \|I-P\|$, and equality $\|P\| = \|I-P\|$ implies $\|P^{\mathbb{C}}\|^{\mathbb{C}} = \|I-P^{\mathbb{C}}\|^{\mathbb{C}}$. So, if the space $\mathcal{B}$ is suitable, then $\mathcal{B}^{\mathbb{C}}$ is also suitable. The theorem is proved. □

We conclude this section with the following.

Corollary 5.6. *It follows from Remark* 4.4 *and Theorem* 5.4 *that there exist two-dimensional real and complex Banach spaces which are suitable but non-Hilbert ones.*

References

[1] D. Amir, *Characterizations of Inner Product Spaces*, Birkhäuser Verlag, Basel, 1986.

[2] G.R. Belitskii and Yu.I. Lyubich, *Matrix Norms and their Applications*, Birkhäuser Verlag, Basel, 1988.

[3] E. Berkson, *A characterization of complex Hilbert spaces*, Bull. London Math. Soc. **2** (1970), 313–315.

[4] F. Bohnenblust, *A characterization of complex Hilbert spaces*, Port. Math. **3** (1942), 103–109.

[5] A. Böttcher and I.M. Spitkovsky, *A gentle guide to the basics of two projection theory*, Linear Algebra Appl. **432** (2010), 1412–1459.

[6] M.M. Day, *Some characterizations of inner product spaces*, Trans. Amer. Math. Soc. **62** (1947), 320–337.

[7] I. Feldman, N. Krupnik, and A. Markus, *Norm of a linear combination of projectors in Hilbert space*, Funct. Anal. Appl. **23** (1989), 327–329.

[8] I. Feldman, N. Krupnik, and A. Markus, *On the norm of polynomials of two adjoint projections*, Integr. Equ. Oper. Theory **14** (1991), 69–91.

[9] I.M. Glazman and Ju.I. Ljubič, *Finite-Dimensional Linear Analysis*, The MIT Press, Cambridge, Mass., 1974.

[10] S. Kakutani, *Some characterizations of Euclidean space*, Japan. J. Math. **16** (1939), 93–97.

[11] V.E. Ljance, *Some properties of idempotent operators*, Teor. i Prikl. Mat., L'vov **1** (1959), 16–22 (Russian).

[12] I.M. Spitkovsky, *On partial indices of continuous matrix-valued functions*, Soviet Math. Dokl. **17** (1976), 1155–1159.

[13] I.M. Spitkovskii, *Some estimates for the partial indices of measurable matrix-valued functions*, Math. USSR Sbornik **39** (1981), 207–226.

[14] I. Verbitsky, *On some relations between the norms of operators and the norms of their complexifications*, In: "Non-selfadjoint operators", Matematicheskie Issledovaniya, Kishinev **42** (1976), 3–12 (Russian).

Nahum Krupnik
208 – 7460, Bathurst str.
Thornhill, L4J 7K9
Ontario, Canada
e-mail: krupnik13@rogers.com

Alexander Markus
Dept. of Mathematics
Ben-Gurion University of the Negev
Beer-Sheva, 84105, Israel
e-mail: markus@cs.bgu.ac.il

Operator Theory:
Advances and Applications, Vol. 259, 511–531

Pseudodifferential Operators in Weighted Hölder–Zygmund Spaces of Variable Smoothness

Vadim Kryakvin and Vladimir Rabinovich

Dedicated to Professor Albrecht Böttcher on the occasion of his 60th birthday anniversary

Abstract. We consider pseudodifferential operators of variable orders acting in Hölder–Zygmund spaces of variable smoothness. We prove the boundedness and compactness of the operators under consideration and study the Fredholm property of pseudodifferential operators with slowly oscillating at infinity symbols in the weighted Hölder–Zygmund spaces of variable smoothness.

Mathematics Subject Classification (2010). Primary 35S05; Secondary 47G30.

Keywords. Pseudodifferential operators, Hölder–Zygmund spaces of variable smoothness.

1. Introduction

The main aim of the paper is the study of the pseudodifferential operators of variable orders acting in Hölder–Zygmund spaces of variable smoothness. The pioneering works in the study of the Sobolev spaces of variable smoothness and pseudodifferential operators of variable orders acting in these spaces are [26] and [28]. Among the later papers in this topic we note [5, 6, 15, 16] (see also references cited in these papers). There are also many papers devoted to different classes of integral and differential operators acting in the Hölder–Zygmund spaces of variable smoothness (see [10, 11, 12, 22, 27], and references cited in these papers).

It should be noted that in the last time there is a big interest in the study of the function spaces connected with the variable exponent Lebesgue spaces

The work of the second author is partially supported by the National System of Investigators of Mexico (SNI), and the Project CB 2012-179872F.

$L^{p(\cdot)}(\mathbb{R}^n)$ and operators acting in these spaces (see for instance [1, 2, 3, 7, 20, 21], and references cited there).

The paper is organized as follows. In Section 2 we give necessary notation and define a class of pseudodifferential operators of the variable order with symbols a satisfying the estimates

$$\left|\partial_x^\beta \partial_\xi^\alpha a(x,\xi)\right| \langle\xi\rangle^{-m(x)+\rho|\alpha|-\delta|\beta|} \leq C_{\alpha,\beta}$$

for all multi-indices α, β, where $0 \leq \delta < \rho \leq 1$, and $m(x)$ is a real-valued bounded continuous function on $\mathbb{R}^n$. We denote the class of such symbols by $S^{m(\cdot)}_{\rho,\delta}(\mathbb{R}^n)$, and the appropriate class of pseudodifferential operators by $OPS^{m(\cdot)}_{\rho,\delta}(\mathbb{R}^n)$.

We prove the usual properties of operators in $OPS^{m(\cdot)}_{\rho,\delta}(\mathbb{R}^n)$ with respect to the product and the formally adjoint operator. We also introduce the class of slowly oscillating at infinity symbols $a(x,\xi) \in \tilde{S}^{m(\cdot)}_{\rho,\delta}(\mathbb{R}^n)$ of the variable order satisfying the estimates:

$$\left|\partial_x^\beta \partial_\xi^\alpha a(x,\xi)\right| \langle\xi\rangle^{-m(x)+\rho|\alpha|-\delta|\beta|} \leq C_{\alpha,\beta}(x), \tag{1}$$

where

$$\lim_{x\to\infty} C_{\alpha,\beta}(x) = 0 \tag{2}$$

for all α and $\beta \neq 0$, and we denote by $\mathring{S}^{m(\cdot)}_{\rho,\delta}(\mathbb{R}^n)$ the class of symbols satisfying (1) such that (2) holds for all α, β. We associate to the symbol a a pseudodifferential operator $Op(a)$. The corresponding classes of pseudodifferential operators are denoted by $OP\tilde{S}^{m(\cdot)}_{\rho,\delta}(\mathbb{R}^n)$, $OP\mathring{S}^{m(\cdot)}_{\rho,\delta}(\mathbb{R}^n)$. We prove that if

$$|a(x,\xi)| \geq C\langle\xi\rangle^{m(x)}, \quad \langle x,\xi\rangle \geq R > 0, \tag{3}$$

then there exists a parametrix $\mathcal{P} \in OP\tilde{S}^{-m(\cdot)}_{\rho,\delta}(\mathbb{R}^n)$ such that

$$\mathcal{P}Op(a) = I + \mathcal{R}_1, \quad Op(a)\mathcal{P} = I + \mathcal{R}_2, \tag{4}$$

where $\mathcal{R}_j \in OP\mathring{S}^{-(\rho-\delta)}_{\rho,\delta}(\mathbb{R}^n)$.

In Section 3 we introduce the Hölder–Zygmund spaces $\Lambda^{s(\cdot)}(\mathbb{R}^n)$ with variable smoothness by means of the Littlewood–Paley partition of unity, and we prove the boundedness of pseudodifferential operators in $OPS^{m(\cdot)}_{1,\delta}(\mathbb{R}^n)$ from $\Lambda^{s(\cdot)}(\mathbb{R}^n)$ into $\Lambda^{s(\cdot)-m(\cdot)}(\mathbb{R}^n)$ with a norm estimate.

In Section 4 we look at the Fredholm properties of operators in $OP\tilde{S}^{m(\cdot)}_{1,\delta}(\mathbb{R}^n)$. Applying formula (4) we prove that condition (3) provides Fredholmness of $Op(a) \in OP\tilde{S}^{m(\cdot)}_{1,\delta}(\mathbb{R}^n)$ from $\Lambda^{s(\cdot)}(\mathbb{R}^n)$ into $\Lambda^{s(\cdot)-m(\cdot)}(\mathbb{R}^n)$. Moreover in this section we consider pseudodifferential operators in $OP\tilde{S}^{m(\cdot)}_{1,\delta}(\mathbb{R}^n)$ with symbols which are analytic functions with respect to the variable ξ in some tube domains. We prove that these operators are bounded in weighted Hölder–Zygmund spaces $\Lambda^{s(\cdot)}(\mathbb{R}^n, w)$ with exponential weights w. We also prove here an analogue of the Phragmén–Lindelöf principle (see for instance [14]) for the exponentially decreasing solutions of the pseudodifferential equations in the class $OP\tilde{S}^{m(\cdot)}_{1,\delta}(\mathbb{R}^n)$. These results are an extension of results of the papers [18, 19] on the pseudodifferential operators in the class $OP\tilde{S}^{m(\cdot)}_{1,\delta}(\mathbb{R}^n)$ acting in the weighted Hölder–Zygmund spaces $\Lambda^{s(\cdot)}(\mathbb{R}^n, w)$.

Remark 1. We consider in the paper the scalar pseudodifferential operators only. But with evident changes we can reformulate results of the paper for matrix pseudodifferential operators acting in the vector-valued Hölder–Zygmund spaces.

2. Pseudodifferential operators

2.1. Notation

We will use the following notation.

- Given Banach spaces X, Y, we denote the Banach space of all bounded linear operators acting from X to Y by $\mathcal{B}(X,Y)$. In case $X = Y$, we simply write $\mathcal{B}(X)$.
- Let $x = (x_1, \ldots, x_n) \in \mathbb{R}^n$. Then we denote by $\xi = (\xi_1, \ldots, \xi_n) \in \mathbb{R}^n$ the points of the dual space with respect to the scalar product
$$x \cdot \xi = x_1\xi_1 + \cdots + x_n\xi_n,$$
and
$$\langle\xi\rangle := (1 + |\xi|^2)^{1/2}, \quad \langle x, \xi\rangle = (1 + |x|^2 + |\xi|^2)^{1/2}.$$
- For $j = 1, \ldots, n$, let $\partial_{x_j} := \frac{\partial}{\partial x_j}$. More generally, given a multi-index $\alpha = (\alpha_1, \ldots, \alpha_n)$, set $|\alpha| := \alpha_1 + \cdots + \alpha_n$, $\partial_x^\alpha := \partial_{x_1}^{\alpha_1} \ldots \partial_{x_n}^{\alpha_n}$, and
$$D_x^\alpha = (-i\partial_{x_1})^{\alpha_1} \ldots (-i\partial_{x_n})^{\alpha_n}.$$

We use the standard notation:

- $C_b(\mathbb{R}^n)$ is the space of bounded continuous functions on $\mathbb{R}^n$.
- $C^\infty(\mathbb{R}^n)$ is the set of all infinitely differentiable functions on $\mathbb{R}^n$.
- $C_0^\infty(\mathbb{R}^n)$ is the set of all functions in $C^\infty(\mathbb{R}^n)$ with compact supports.
- $C_b^\infty(\mathbb{R}^n)$ is the set of all functions $a \in C^\infty(\mathbb{R}^n)$ such that
$$\sup_{x\in\mathbb{R}^n} |(\partial_x^\alpha a)(x)| < \infty$$
for every $\alpha \in \mathbb{N}_0^n$, where $\mathbb{N}_0 = \{0\} \cup \mathbb{N}$.
- $S(\mathbb{R}^n)$ is the Schwartz space of functions $a \in C^\infty(\mathbb{R}^n)$ such that
$$\sup_{x\in\mathbb{R}^n, |\alpha|\le k} \langle x\rangle^k |\partial_x^\alpha a(x)| < \infty$$
for every $k \in \mathbb{N}_0$ and $S'(\mathbb{R}^n)$ is the class of the tempered distributions.
- $SO^\infty(\mathbb{R}^n)$ is the set of functions slowly oscillating at infinity, that is, functions $a \in C_b^\infty(\mathbb{R}^n)$ such that
$$\lim_{x\to\infty} \partial_{x_j} a(x) = 0, \quad j = 1, \ldots, n.$$

- $H^s(\mathbb{R}^n)$, $s \in \mathbb{R}$, is the Sobolev space of the distributions in $S'(\mathbb{R}^n)$ such that

$$\|u\|_{H^s(\mathbb{R}^n)} = \left(\int_{\mathbb{R}^n} |\hat{u}(\xi)|^2 \langle \xi \rangle^{2s} d\xi \right)^{1/2} < \infty,$$

where $\hat{u}$ is the Fourier transform of u.
- $\mathbb{B}_R = \{x \in \mathbb{R}^n : |x| < R\}, \mathbb{B}_R^{'} = \{x \in \mathbb{R}^n : |x| > R\}$.

2.2. Pseudodifferential operators in the class $OPS^m_{\rho,\delta}(\mathbb{R}^n)$

Standard references on the pseudodifferential operators are: [9, 13, 23, 25].

We say that a function $a \in C^\infty(\mathbb{R}^n \times \mathbb{R}^n)$ belongs to the L. Hörmander class $S^m_{\rho,\delta}(\mathbb{R}^n)$, $0 \le \delta < \rho \le 1$, if

$$|a|_{r,t} = \sup_{(x,\xi)\in\mathbb{R}^n\times\mathbb{R}^n} \sum_{|\alpha|\le r, |\beta|\le t} \left|\partial_\xi^\alpha \partial_x^\beta a(x,\xi)\right| \langle \xi \rangle^{-m+\rho|\alpha|-\delta|\beta|} < \infty$$

for $r, t \in \mathbb{N}_0$. As usual we associate with a symbol a the pseudodifferential operator (psdo) defined on the space $S(\mathbb{R}^n)$ by the formula

$$Op(a)u(x) = a(x,D)u(x) = (2\pi)^{-n} \int_{\mathbb{R}^n} d\xi \int_{\mathbb{R}^n} a(x,\xi)u(y)e^{i(x-y)\cdot\xi} dy. \quad (5)$$

We will denote by $OPS^m_{\rho,\delta}(\mathbb{R}^n)$ the class of psdo's with symbols in $S^m_{\rho,\delta}(\mathbb{R}^n)$ and we use the notation:

$$S^{-\infty}(\mathbb{R}^n) := \bigcap_{m\in\mathbb{R}} S^m_{\rho,\delta}(\mathbb{R}^n), \quad OPS^{-\infty}(\mathbb{R}^n) := \bigcap_{m\in\mathbb{R}} OPS^m_{\rho,\delta}(\mathbb{R}^n).$$

Below we need some properties of psdo's.

Proposition 2.

(i) *Let $A = Op(a) \in OPS^{m_1}_{\rho,\delta}(\mathbb{R}^n)$, $B = Op(b) \in OPS^{m_2}_{\rho,\delta}(\mathbb{R}^n)$. Then $AB \in OPS^{m_1+m_2}_{\rho,\delta}(\mathbb{R}^n)$ and $AB = Op(c)$, where*

$$c(x,\xi) = (2\pi)^{-n} \int_{\mathbb{R}^n} \int_{\mathbb{R}^n} a(x,\xi+\eta)b(x+y,\xi)e^{-iy\cdot\eta} dy\, d\eta,$$

and the double integral is understood as oscillatory. Moreover, for every $l_1, l_2 \in \mathbb{N}_0$ there exist $N_1, N_2 \in \mathbb{N}_0$ such that

$$|c|_{l_1,l_2} \le C\,|a|_{N_1,N_2}\,|b|_{N_1,N_2},$$

where $C > 0$ is independent of a, b.

(ii) *For every $N \in \mathbb{N}_0$,*

$$c(x,\xi) = \sum_{j=0}^{N-1} \sum_{|\alpha|=j} \frac{1}{\alpha!} \partial_\xi^\alpha a(x,\xi) D_x^\alpha b(x,\xi) + T_N(x,\xi),$$

where $T_N(x,\xi) \in S^{m_1+m_2-N(\rho-\delta)}_{\rho,\delta}(\mathbb{R}^n)$.

Proposition 3. *A psdo $Op(a) \in OPS^m_{\rho,\delta}(\mathbb{R}^n)$ is bounded from $H^s(\mathbb{R}^n)$ to $H^{s-m}(\mathbb{R}^n)$ and*

$$\|Op(a)\|_{\mathcal{B}(H^s(\mathbb{R}^n),H^{s-m}(\mathbb{R}^n))} \leq C\,|a|_{N_1,N_2}$$

with constants $C > 0$ and $N_1, N_2 \in \mathbb{N}_0$ independent of a.

Proposition 4 ([4]). *Let $A = Op(a) \in OPS^m_{\rho,\delta}(\mathbb{R}^n)$ be invertible from $H^s(\mathbb{R}^n)$ to $H^{s-m}(\mathbb{R}^n)$. Then $A^{-1} \in OPS^{-m}_{\rho,\delta}(\mathbb{R}^n)$.*

2.3. Pseudodifferential operators of variable order

Definition 5. Let $m \in C_b(\mathbb{R}^n)$ be a real-valued function. We say that a function $a \in C^\infty(\mathbb{R}^n \times \mathbb{R}^n)$ is a symbol of the class $S^{m(\cdot)}_{\rho,\delta}(\mathbb{R}^n)$, $0 \leq \delta < \rho \leq 1$, if

$$|a|_{l_1,l_2} = \sup_{(x,\xi)\in\mathbb{R}^n\times\mathbb{R}^n} \sum_{|\alpha|\leq l_1, |\beta|\leq l_2} \left|\partial_x^\beta \partial_\xi^\alpha a(x,\xi)\right| \langle\xi\rangle^{-m(x)+\rho|\alpha|-\delta|\beta|} < \infty \tag{6}$$

for every $l_1, l_2 \in \mathbb{N}_0$. We associate with $S^{m(\cdot)}_{\rho,\delta}(\mathbb{R}^n)$ the class $OPS^{m(\cdot)}_{\rho,\delta}(\mathbb{R}^n)$ of psdo's given by formula (5).

Let

$$m^+ = \sup_{x\in\mathbb{R}^n} m(x), \quad m^- = \inf_{x\in\mathbb{R}^n} m(x).$$

Then one can see that

$$S^{m_-}_{\rho,\delta}(\mathbb{R}^n) \subset S^{m(\cdot)}_{\rho,\delta}(\mathbb{R}^n) \subset S^{m_+}_{\rho,\delta}(\mathbb{R}^n). \tag{7}$$

Proposition 6.

(i) *An operator $Op(a) \in OPS^{m(\cdot)}_{\rho,\delta}(\mathbb{R}^n)$ is bounded in $S(\mathbb{R}^n)$.*

(ii) *Let $a_j \in S^{m_j(\cdot)}_{\rho,\delta}(\mathbb{R}^n)$, $j = 1, 2$. Then*

$$Op(a_2)Op(a_1) = Op(b) \in OPS^{m_1(\cdot)+m_2(\cdot)}_{\rho,\delta}(\mathbb{R}^n)$$

where b can be represented as a double oscillatory integral

$$b(x,\xi) = (2\pi)^{-n} \int_{\mathbb{R}^n}\int_{\mathbb{R}^n} a_2(x,\xi+\eta)a_1(x+y,\xi)e^{-iy\cdot\eta}\,dy\,d\eta.$$

Moreover, for every $M > 0$ there exists $N \in \mathbb{N}_0$ such that

$$b(x,\xi) = \sum_{j=0}^{N-1}\sum_{|\alpha|=j} \frac{1}{\alpha!}\partial_\xi^\alpha a_2(x,\xi) D_x^\alpha a_1(x,\xi) + T_N(x,\xi),$$

where $T_N \in S^{m_1(\cdot)+m_2(\cdot)-M}_{\rho,\delta}(\mathbb{R}^n)$.

(iii) *Let A^* be the formal adjoint operator to $A = Op(a) \in OPS^{m(\cdot)}_{\rho,\delta}(\mathbb{R}^n)$, that is,*

$$(A\varphi,\psi)_{L^2(\mathbb{R}^n)} = (\varphi, A^*\psi)_{L^2(\mathbb{R}^n)}$$

for every $\varphi, \psi \in S(\mathbb{R}^n)$. Then $A^* = Op(a^*) \in OPS^{m(\cdot)}_{\rho,\delta}(\mathbb{R}^n)$ where

$$a^*(x,\xi) = (2\pi)^{-n}\int_{\mathbb{R}^n}\int_{\mathbb{R}^n} \bar{a}(x+y,\xi+\eta)e^{-iy\cdot\eta}\,dy\,d\eta,$$

and for every $M > 0$ there exists $N \in \mathbb{N}_0$ such that

$$a^*(x,\xi) = \sum_{j=0}^{N-1} \sum_{|\alpha|=j} \frac{1}{\alpha!} \partial_\xi^\alpha D_x^\alpha \bar{a}(x,\xi) + T_N(x,\xi),$$

where $T_N(x,\xi) \in S_{\rho,\delta}^{m(\cdot)-M}(\mathbb{R}^n)$.

(iv) *An operator $Op(a) \in OPS_{\rho,\delta}^{m(\cdot)}(\mathbb{R}^n)$ is bounded in $S'(\mathbb{R}^n)$.*

Proof. Statements (i) and (iv) follow from the inclusion (7) and they are well known for operators in $OPS_{\rho,\delta}^m(\mathbb{R}^n)$, where m is constant.

(ii) We use the representation of the remainder T_N as an oscillatory integral (see for instance [13, 17])

$$T_N(x,\xi) = (2\pi)^{-n} \sum_{|\gamma|=N} \frac{N}{\gamma!} \int_0^1 (1-\theta)^{N-1} d\theta$$
$$\times \int_{\mathbb{R}^n} \int_{\mathbb{R}^n} \partial_\xi^\gamma a_2(x, \xi + \theta\eta) D_x^\gamma a_1(x+y,\xi) e^{-iy\cdot\eta}\, dy\, d\eta.$$

Applying the standard estimates of the oscillatory integrals ([13, 17]), we obtain the estimate

$$\left|\partial_x^\beta \partial_\xi^\alpha T_N(x,\xi)\right| \le C_{\alpha,\beta}^{T_N} \langle\xi\rangle^{m_1^+ + m_2^+ - (\rho-\delta)N - \rho|\alpha| + \delta|\beta|}.$$

Let $\Delta(m_1) = m_1^+ - m_1^-$ and $\Delta(m_2) = m_2^+ - m_2^-$. Then

$$m_1^+ + m_2^+ \le m_1(x) + m_2(x) + \Delta(m_1) + \Delta(m_2) \tag{8}$$

for every $x \in \mathbb{R}^n$. Estimate (8) implies that

$$m_1^+ + m_2^+ - (\rho-\delta)N < m_1(x) + m_2(x) - M$$

if $N > \frac{M + \Delta(m_1) + \Delta(m_2)}{\rho-\delta}$. Statement (iii) is proved as statement (ii). □

Theorem 7. *Let $a \in S_{\rho,\delta}^{m(\cdot)}(\mathbb{R}^n)$, $0 \le \delta < \rho \le 1$. Suppose there exist $C > 0$ and $R > 0$ such that for every $x \in \mathbb{R}^n$,*

$$|a(x,\xi)| \ge C\, |\xi|^{m(x)}, \quad |\xi| \ge R. \tag{9}$$

Then there exists an operator $\mathcal{P} \in OPS_{\rho,\delta}^{-m(\cdot)}(\mathbb{R}^n)$ such that

$$\mathcal{P}Op(a) = I + \mathcal{R}_1, \quad Op(a)\mathcal{P} = I + \mathcal{R}_2$$

where $\mathcal{R}_j \in OPS_{\rho,\delta}^{-(\rho-\delta)}(\mathbb{R}^n)$, $j = 1, 2$.

Proof. Estimates (6) and (9) yield that

$$\frac{\left|\partial_x^\beta \partial_\xi^\alpha a(x,\xi)\right|}{|a(x,\xi)|} \le C_{\alpha,\beta}\, |\xi|^{-\rho|\alpha| + \delta|\beta|}, \quad |\xi| \ge R.$$

Let $\varphi \in C_0^\infty(\mathbb{R}^n)$ be such that $\varphi(\xi) = 1$ for $|\xi| \le 1$ and $\varphi(\xi) = 0$ for $|\xi| \ge 2$. Put $\psi(\xi) = 1 - \varphi(\xi)$, $\varphi_R(\xi) = \varphi(\xi/R)$, and $\psi_R(\xi) = \psi(\xi/R)$. We set

$$a_R(x,\xi) = a(x,\xi)\psi_R(\xi), \quad b_R(x,\xi) = a^{-1}(x,\xi)\psi_R(\xi).$$

Estimate (9) implies that

$$\frac{\left|\partial_x^\beta \partial_\xi^\alpha a_R(x,\xi)\right|}{|a_R(x,\xi)|} \le C^1_{\alpha,\beta} |\xi|^{-\rho|\alpha|+\delta|\beta|}, \quad \frac{\left|\partial_x^\beta \partial_\xi^\alpha b_R(x,\xi)\right|}{|b_R(x,\xi)|} \le C^2_{\alpha,\beta} |\xi|^{-\rho|\alpha|+\delta|\beta|}$$

for $R > 0$ large enough. Let $C_R = Op(b_R)Op(a_R) = Op(c_R)$. Then for every $M > 0$ there exists $N \in \mathbb{N}_0$ such that

$$\begin{aligned} c_R(x,\xi) &= \sum_{j=0}^{N-1} \sum_{|\alpha|=j} \frac{1}{\alpha!} \partial_\xi^\alpha b_R(x,\xi) D_x^\alpha a_R(x,\xi) + T_{N,R}(x,\xi) \\ &= 1 + \sum_{j=1}^{N-1} \sum_{|\alpha|=j} \frac{1}{\alpha!} \frac{\partial_\xi^\alpha b_R(x,\xi)}{b_R(x,\xi)} \frac{D_x^\alpha a_R(x,\xi)}{a_R(x,\xi)} \\ &\quad + \frac{T_{N,R}(x,\xi)}{b_R(x,\xi) a_R(x,\xi)} + Q_R(x,\xi), \end{aligned} \tag{10}$$

where

$$\frac{\partial_\xi^\alpha b_R(x,\xi)}{b_R(x,\xi)} \frac{D_x^\alpha a_R(x,\xi)}{a_R(x,\xi)} \in S^{-(\rho-\delta)|\alpha|}_{\rho,\delta}(\mathbb{R}^n), \quad T_{N,R}(x,\xi) \in S^{-M}_{\rho,\delta}(\mathbb{R}^n), \quad M \ge \rho - \delta,$$

and $Q_R(x,\xi) \in S^{-\infty}(\mathbb{R}^n)$. Formula (10) yields that for $\mathcal{P} = Op(b_R)$,

$$\mathcal{P} Op(a) = I + \mathcal{R}_1, \quad Op(a)\mathcal{P} = I + \mathcal{R}_2$$

where $\mathcal{R}_j \in OPS^{-(\rho-\delta)}_{\rho,\delta}(\mathbb{R}^n)$. □

2.4. Psdo's of variable orders depending on a parameter

Definition 8. We consider a family of symbols $a_\lambda \in S^{m(\cdot)}_{\rho,\delta}(\mathbb{R}^n)$, $0 \le \delta < \rho \le 1$, depending on a parameter $\lambda \ge 0$ such that

$$|a_\lambda|_{l_1,l_2} = \sup_{(x,\xi)\in\mathbb{R}^n\times\mathbb{R}^n} \sum_{|\alpha|\le l_1, |\beta|\le l_2} \left|\partial_x^\beta \partial_\xi^\alpha a_\lambda(x,\xi)\right| \langle \xi, \lambda\rangle^{-m(x)+\rho|\alpha|-\delta|\beta|}$$

for every $l_1, l_2 \in \mathbb{N}_0$. We denote this class of symbols depending on the parameter λ by $S^{m(\cdot)}_{\rho,\delta}(\mathbb{R}^n, \lambda)$ and the associated class of psdo's by $OPS^{m(\cdot)}_{\rho,\delta}(\mathbb{R}^n, \lambda)$.

Theorem 9. *Let $a \in S^{m(\cdot)}_{\rho,\delta}(\mathbb{R}^n, \lambda)$, $0 \le \delta < \rho \le 1$. Suppose there exist $C > 0$ and $R > 0$ such that for every $x \in \mathbb{R}^n$,*

$$|a_\lambda(x,\xi)| \ge C \langle \xi, \lambda\rangle^{m(x)}, \quad \langle \xi, \lambda\rangle \ge R.$$

Then there exists an operator $\mathcal{P}(\lambda) \in OPS^{-m(\cdot)}_{\rho,\delta}(\mathbb{R}^n, \lambda)$ such that

$$\mathcal{P}(\lambda) Op(a_\lambda) = I + \mathcal{R}_1(\lambda), \quad Op(a_\lambda)\mathcal{P}(\lambda) = I + \mathcal{R}_2(\lambda)$$

where $\mathcal{R}_j(\lambda) \in OPS^{-(\rho-\delta)}_{\rho,\delta}(\mathbb{R}^n, \lambda)$, $j = 1, 2$.

The proof is similar to the proof of Theorem 7.

2.5. Psdo's of variable orders with slowly oscillating at infinity symbols

Definition 10. We say that $a \in S_{\rho,\delta}^{m(\cdot)}(\mathbb{R}^n)$ is a symbol of the class $\tilde{S}_{\rho,\delta}^{m(\cdot)}(\mathbb{R}^n)$, $0 \leq \delta < \rho \leq 1$, if

$$\left|\partial_x^\beta \partial_\xi^\alpha a(x,\xi)\right| \leq C_{\alpha,\beta}(x)\langle\xi\rangle^{m(x)-\rho|\alpha|+\delta|\beta|} \tag{11}$$

where

$$\lim_{x\to\infty} C_{\alpha,\beta}(x) = 0 \tag{12}$$

for all multi-indices α and $\beta \neq 0$. We associate with the class $\tilde{S}_{\rho,\delta}^{m(\cdot)}(\mathbb{R}^n)$ the class $OP\tilde{S}_{\rho,\delta}^{m(\cdot)}(\mathbb{R}^n)$ of psdo's given by formula (5). We say that symbols belonging $\tilde{S}_{\rho,\delta}^{m(\cdot)}(\mathbb{R}^n)$ are slowly oscillating at infinity. We say that the symbol $a \in \tilde{S}_{\rho,\delta}^{m(\cdot)}(\mathbb{R}^n)$ belongs to the class $\mathring{S}_{\rho,\delta}^{m(\cdot)}(\mathbb{R}^n)$ if (12) holds for all multi-indices α and β. We use the notation $OP\mathring{S}_{\rho,\delta}^{m(\cdot)}(\mathbb{R}^n)$ for the associated class of psdo's.

Proposition 11. *Let $a_j \in \tilde{S}_{\rho,\delta}^{m_j(\cdot)}(\mathbb{R}^n)$, $j = 1, 2$. Then $A = Op(a_2)Op(a_1) = Op(a)$ where $a \in \tilde{S}_{\rho,\delta}^{m_1(\cdot)+m_2(\cdot)}(\mathbb{R}^n)$. Moreover, for every $M > 0$ there exists $N \in \mathbb{N}_0$ such that*

$$a(x,\xi) = \sum_{j=0}^{N-1} \sum_{|\alpha|=j} \frac{1}{\alpha!} \partial_\xi^\alpha a_2(x,\xi) D_x^\alpha a_1(x,\xi) + T_N(x,\xi),$$

where $T_N \in \mathring{S}_{\rho,\delta}^{m_1(\cdot)+m_2(\cdot)-M}(\mathbb{R}^n)$.

The proof is similar to the proof of Proposition 6 (ii).

We denote by

$$\mathring{S}^{-\infty}(\mathbb{R}^n) = \bigcap_{m\in\mathbb{R}} \mathring{S}_{\rho,\delta}^{m}(\mathbb{R}^n),$$

and by $OP\mathring{S}^{-\infty}(\mathbb{R}^n)$ the associated class of psdo's.

Theorem 12. *Let $a \in \tilde{S}_{\rho,\delta}^{m(\cdot)}(\mathbb{R}^n)$, $0 \leq \delta < \rho \leq 1$. Suppose there exist $C > 0$ and $R > 0$ such that*

$$|a(x,\xi)| \geq C \langle\xi\rangle^{m(x)}, \quad |x| + |\xi| \geq R. \tag{13}$$

Then there exists an operator $\mathcal{P} \in OP\tilde{S}_{\rho,\delta}^{-m(\cdot)}(\mathbb{R}^n)$ such that

$$\mathcal{P}Op(a) = I + \mathcal{T}_1, \quad Op(a)\mathcal{P} = I + \mathcal{T}_2$$

where $\mathcal{T}_j \in OP\mathring{S}_{\rho,\delta}^{-(\rho-\delta)}(\mathbb{R}^n)$, $j = 1, 2$.

Proof. Estimates (11) and (13) yield that there exists $R > 0$ such that

$$\frac{\left|\partial_x^\beta \partial_\xi^\alpha a(x,\xi)\right|}{|a(x,\xi)|} \leq C_{\alpha,\beta}(x) \langle\xi\rangle^{-\rho|\alpha|+\delta|\beta|}, \quad |x| + |\xi| \geq R,$$

for some $C_{\alpha,\beta}(x)$ such that (12) holds for every α and $\beta \neq 0$. Let $\varphi \in C_0^\infty(\mathbb{R}^n \times \mathbb{R}^n)$ be such that $\varphi(x,\xi) = 1$ for $|x| + |\xi| \leq 1$ and $\varphi(x,\xi) = 0$ for $|x| + |\xi| \geq 2$. Put $\psi(x,\xi) = 1 - \varphi(x,\xi)$, $\varphi_R(x,\xi) = \varphi(x/R, \xi/R)$, and $\psi_R(x,\xi) = \psi(x/R, \xi/R)$. We set

$$a_R(x,\xi) = a(x,\xi)\psi_R(x,\xi), \quad b_R(x,\xi) = a^{-1}(x,\xi)\psi_R(x,\xi).$$

Let $Op(b_R)Op(a_R) = Op(c_R)$ for $R > 0$ large enough. Then

$$\begin{aligned} c_R(x,\xi) &= \sum_{j=0}^{N-1} \sum_{|\alpha|=j} \frac{1}{\alpha!} \partial_\xi^\alpha b_R(x,\xi) D_x^\alpha a_R(x,\xi) + T_{N,R}(x,\xi) \\ &= 1 + \sum_{j=1}^{N} \sum_{|\alpha|=j} \frac{1}{\alpha!} \frac{\partial_\xi^\alpha b_R(x,\xi)}{b_R(x,\xi)} \frac{D_x^\alpha a_R(x,\xi)}{a_R(x,\xi)} \\ &\quad + \frac{T_{N,R}(x,\xi)}{b_R(x,\xi) a_R(x,\xi)} + Q_R(x,\xi), \end{aligned}$$

where

$$\frac{T_{N,R}(x,\xi)}{b_R(x,\xi)a_R(x,\xi)} \in \mathring{S}_{\rho,\delta}^{-M}(\mathbb{R}^n), \quad M > \rho - \delta, \quad Q_R(x,\xi) \in \mathring{S}^{-\infty}(\mathbb{R}^n).$$

Hence

$$Op(b_R)Op(a) = I + \mathcal{R}_1 + Op(a\varphi_R) = I + \mathcal{T}_1$$

where $\mathcal{T}_1 \in OP\mathring{S}_{\rho,\delta}^{-(\rho-\delta)}(\mathbb{R}^n)$. In the same way we can show that

$$Op(a)Op(b_R) = I + \mathcal{T}_2,$$

where $\mathcal{T}_2 \in OP\mathring{S}_{\rho,\delta}^{-(\rho-\delta)}(\mathbb{R}^n)$. □

3. Hölder–Zygmund spaces of variable orders

3.1. Definition of the Hölder–Zygmund space

We introduce the Littlewood–Paley partition of unity

$$\sum_{k=0}^{\infty} \lambda_k(\xi) = 1, \quad \xi \in \mathbb{R}^n, \tag{14}$$

with $\lambda_0(\xi) = \eta_0(\xi)$, $\lambda_k(\xi) = \eta_k(\xi) - \eta_{k-1}(\xi)$, $k \in \mathbb{N}_0$ where $\eta_0 \in C_0^\infty(\mathbb{R}^n)$, so that $\eta_0(\xi) = 1$ for $|\xi| \leq 1$ and 0 for $|\xi| \geq 2$, and $\eta_k(\xi) = \eta_0(2^{-k}\xi)$.

Note some properties of the partition of unity (14) (see for instance [24]):

(i) $\operatorname{supp} \lambda_0 = \{\xi \in \mathbb{R}^n : |\xi| \leq 2\}$, $\operatorname{supp} \lambda_k = \left\{\xi \in \mathbb{R}^n : 2^{k-1} \leq |\xi| \leq 2^{k+1}\right\}$.

(ii) Let $\psi_k = \lambda_{k-1} + \lambda_k + \lambda_{k+1}$, $k \in \mathbb{N}_0$, $\lambda_{-1} = 0$. Then $\psi_k(\xi) = 1$ for every $\xi \in \operatorname{supp} \lambda_k$. Hence $\lambda_k \psi_k = \lambda_k$.

Definition 13 (see [11, 12]). Fix a real-valued function $s \in C_b(\mathbb{R}^n)$. We suppose that there is a constant $L_s > 0$ such that

$$|s(x+y) - s(x)| \leq \frac{L_s}{|\log_2 |y||}$$

for every $x \in \mathbb{R}^n$ and $y \in \mathbb{R}^n$ such that $0 < |y| < 1$. We say that a distribution $u \in S'(\mathbb{R}^n)$ belongs to the Hölder–Zygmund space $\Lambda^{s(\cdot)}(\mathbb{R}^n)$ of the variable smoothness $s(\cdot)$ if

$$\|u\|_{\Lambda^{s(\cdot)}(\mathbb{R}^n)} = \sup_{k \in \mathbb{N}_0} \left\| 2^{ks(\cdot)} \lambda_k(D) u \right\|_{L^\infty(\mathbb{R}^n)} < \infty. \tag{15}$$

Note that $\Lambda^{s(\cdot)}(\mathbb{R}^n)$ is a complete Banach space and this space coincides with the classical Hölder–Zygmund space if $s(\cdot) = s$ is a constant (see [24]).

Let

$$s_+ = \sup_{x \in \mathbb{R}^n} s(x), \quad s_- = \inf_{x \in \mathbb{R}^n} s(x), \quad \Delta s = s_+ - s_-.$$

Then the following estimates for the norm hold:

$$\|u\|_{\Lambda^{s_-}(\mathbb{R}^n)} \leq \|u\|_{\Lambda^{s(\cdot)}(\mathbb{R}^n)} \leq \|u\|_{\Lambda^{s_+}(\mathbb{R}^n)}.$$

Moreover if $s_1(x) \leq s_2(x)$ for every $x \in \mathbb{R}^n$, then

$$\|u\|_{\Lambda^{s_1(\cdot)}(\mathbb{R}^n)} \leq \|u\|_{\Lambda^{s_2(\cdot)}(\mathbb{R}^n)},$$

and hence $\Lambda^{s_2(\cdot)}(\mathbb{R}^n) \subset \Lambda^{s_1(\cdot)}(\mathbb{R}^n)$.

Note if $0 < s_- \leq s(x) \leq s_+ < 1$, then the norm (15) coincides with the norm (see [12])

$$\|u\|_{\Lambda^{s(\cdot)}(\mathbb{R}^n)} = \|u\|_{L^\infty(\mathbb{R}^n)} + \sup_{0<|h|<1} \sup_{x \in \mathbb{R}^n} \frac{|u(x+h) - u(x)|}{|h|^{s(x)}}.$$

3.2. Boundedness of psdo's of variable orders in $\Lambda^{s(\cdot)}(\mathbb{R}^n)$

Theorem 14. *Let $a \in S^{m(\cdot)}_{1,\delta}(\mathbb{R}^n)$, $0 \leq \delta < 1$. Then*

$$A = Op(a) : \Lambda^{s(\cdot)}(\mathbb{R}^n) \to \Lambda^{s(\cdot)-m(\cdot)}(\mathbb{R}^n)$$

is a bounded operator, and there exist constants $C > 0$ and $N_1, N_2 \in \mathbb{N}_0$ independent of a such that

$$\|A\|_{\mathcal{B}(\Lambda^{s(\cdot)}(\mathbb{R}^n), \Lambda^{s(\cdot)-m(\cdot)}(\mathbb{R}^n))} \leq C\, |a|_{N_1,N_2}. \tag{16}$$

The proof of Theorem 14 is based on the following propositions concerning the psdo's in the class $OPS^{m(\cdot)}_{1,\delta}(\mathbb{R}^n)$.

Proposition 15. *Let $a \in S^{m(\cdot)}_{1,\delta}(\mathbb{R}^n), 0 \leq \delta < 1, m \in C_b(\mathbb{R}^n)$. Then*

$$Op(a\lambda_j)u = \int_{\mathbb{R}^n} k_j(x, x-y) u(y) dy, \quad u \in S(\mathbb{R}^n),$$

and for all multi-indices β, γ, and $M \in \mathbb{N}_0$ there exist constants $C_{\beta,\gamma} > 0$ independent of a such that

$$\left|\partial_x^\beta \partial_z^\gamma k_j(x,z)\right| \le C_{\beta,\gamma} \left|a\right|_{M,|\beta|} |z|^{-M} 2^{j(n+m(x)-M+|\gamma|+\delta|\beta|)}. \tag{17}$$

Proof. We set $a_j(x,\xi) = a(x,\xi)\lambda_j(\xi)$. Then

$$k_j(x,z) = (2\pi)^{-n} \int_{\mathbb{R}^n} a_j(x,\xi) e^{iz\cdot\xi}\, d\xi. \tag{18}$$

By integration by parts in (18) we obtain

$$(iz)^\theta \partial_x^\beta \partial_z^\gamma k_j(x,z) = (2\pi)^{-n} \int_{\mathbb{R}^n} (-\partial_\xi)^\theta \left[(i\xi)^\gamma \partial_x^\beta a_j(x,\xi)\right] e^{iz\cdot\xi}\, d\xi.$$

Note that

$$\left|(-\partial_\xi)^\theta \left[(i\xi)^\gamma \partial_x^\beta a_j(x,\xi)\right]\right| \le C \left|a\right|_{|\theta|,|\beta|} \langle\xi\rangle^{m(x)+|\gamma|-|\theta|+\delta|\beta|}.$$

Taking into account that

$$\operatorname{supp} \lambda_0 = \{\xi \in \mathbb{R}^n : |\xi| \le 2\}, \quad \operatorname{supp} \lambda_j = \left\{\xi \in \mathbb{R}^n : 2^{j-1} \le |\xi| \le 2^{j+1}\right\},$$

we obtain the estimate

$$\left|(iz)^\theta \partial_x^\beta \partial_z^\gamma k_j(x,z)\right| \le C \left|a\right|_{|\theta|,|\beta|} 2^{j(n+m(x)+|\gamma|-|\theta|+\delta|\beta|)}, \quad j \in \mathbb{N}_0. \tag{19}$$

Applying estimate (19) for all multi-indices θ such that $|\theta| = M$ we obtain estimate (17). □

Proposition 16. *Let $a \in S_{1,\delta}^{m(\cdot)}(\mathbb{R}^n)$, $0 \le \delta < 1$. Then for every $j \in \mathbb{N}_0$,*

$$\left\|2^{j(s(\cdot)-m(\cdot))} Op(a_j) u\right\|_{L^\infty(\mathbb{R}^n)} \le C \left|a\right|_{M,0} \left\|2^{js(\cdot)} u\right\|_{L^\infty(\mathbb{R}^n)} \tag{20}$$

with constants $C > 0$ and $M \in \mathbb{N}_0$ independent of a and $j \in \mathbb{N}_0$.

Proof. We have

$$\begin{aligned}\left|2^{j(s(x)-m(x))} Op(a_j) u(x)\right| \le &\left\|2^{js(\cdot)} u\right\|_{L^\infty(\mathbb{R}^n)} \\ &\times \int_{\mathbb{R}^n} 2^{j(s(x)-m(x)-s(z))} \left|k_j(x, x-z)\right| dz.\end{aligned}$$

For fixed $x \in \mathbb{R}^n$, $j \in \mathbb{N}_0$, and $\varepsilon \in (0,1)$ we divide $\mathbb{R}^n$ into 3 sets:

$$\begin{aligned} X &= \left\{z \in \mathbb{R}^n : |x-z| \le 2^{-j}\right\}, \\ Y &= \left\{z \in \mathbb{R}^n : 2^{-j} < |x-z| \le 2^{-\varepsilon j}\right\}, \\ Z &= \left\{z \in \mathbb{R}^n : 2^{-\varepsilon j} < |x-z| < \infty\right\}. \end{aligned}$$

Note that in the domain X,

$$j\left|s(x) - s(z)\right| \le \frac{jL_s}{\left|\log_2 |x-z|\right|} \le L_s. \tag{21}$$

Hence applying estimates (21) and (17) with $M=0, \gamma=\beta=0$ we obtain

$$\int_{|x-z|\leq 2^{-j}} 2^{j(s(x)-m(x))-s(z))}\left|k_j(x,x-z)\right| dz \leq C2^{L_s}\left|a\right|_{0,0}, \tag{22}$$

with a constant C independent of a and $j \in \mathbb{N}_0$.

For the estimate of the integral over domain Z we apply estimate (17) with $M > n + \frac{\Delta s}{1-\varepsilon}$. Then

$$\begin{aligned}
&\int_{|x-z|>2^{-\varepsilon j}} 2^{j(s(x)-m(x)-s(z))}\left|k_j(x,x-z)\right| dz \\
&\quad \leq C_{0,0}2^{j\Delta s+j(n-M)}\left|a\right|_{M,0}\int_{|x-z|\geq 2^{-\varepsilon j}}\left|x-z\right|^{-M} dz \\
&\quad \leq 2^{j\Delta s+j(n-M)}C\left|a\right|_{M,0}\int_{2^{-\varepsilon j}}^{\infty} r^{-M+n-1}dr \\
&\quad = C\left|a\right|_{M,0}2^{j(\Delta s+(1-\varepsilon)(n-M))} \\
&\quad \leq C\left|a\right|_{M,0},
\end{aligned} \tag{23}$$

with a constant $C>0$ independent of a and $j \in \mathbb{N}_0$.

For $x \in \mathbb{R}^n$,

$$j\left|s(x)-s(z)\right| \leq \frac{jL_s}{\left|\log_2\left|x-z\right|\right|} \leq L_s\frac{j}{\varepsilon j} = \frac{L_s}{\varepsilon}, \quad z \in Y. \tag{24}$$

Applying the estimate (24) we obtain for $M>n$,

$$\begin{aligned}
&\int_{2^{-j}<|x-z|\leq 2^{-\varepsilon j}} 2^{j(s(x)-m(x)-s(z))}\left|k_j(x,x-z)\right| dz \\
&\quad \leq C|a|_{M,0}2^{\frac{L_s}{\varepsilon}}2^{j(n-M)}\int_{2^{-j}}^{2^{-\varepsilon j}} r^{n-1-M}dr \\
&\quad \leq C2^{\frac{L_s}{\varepsilon}}(n-M)^{-1}\left|a\right|_{M,0}2^{j(n-M)}(2^{-\varepsilon j(n-M)}-2^{-j(n-M)}) \\
&\quad = C2^{\frac{L_s}{\varepsilon}}(M-n)^{-1}\left|a\right|_{M,0}(1-2^{-j(1-\varepsilon)(M-n)}) \\
&\quad \leq C2^{\frac{L_s}{\varepsilon}}(M-n)^{-1}\left|a\right|_{M,0}.
\end{aligned} \tag{25}$$

Estimates (22), (23), (25) yield estimate (20). □

Proposition 17. *Let $a \in S_{1,\delta}^{m(\cdot)}(\mathbb{R}^n)$, $0 \leq \delta < 1$. Then for every $j \in \mathbb{N}_0$,*

$$\left\|2^{j(s(\cdot)-m(\cdot))}Op(a_j)\lambda_j(D)u\right\|_{L^\infty(\mathbb{R}^n)} \leq C\left|a\right|_{M,0}\left\|u\right\|_{\Lambda^{s(\cdot)}(\mathbb{R}^n)}$$

with constants $C>0$ and $M \in \mathbb{N}_0$ independent of a and $j \in \mathbb{N}_0$.

Proof. We change in inequality (20) u by $\psi_j(D)u$. Then since $\lambda_j(D)\psi_j(D) = \lambda_j(D)$ we obtain that

$$\left\| 2^{j(s(\cdot)-m(\cdot))} Op(a)\lambda_j(D)u \right\|_{L^\infty(\mathbb{R}^n)} \leq C\,|a|_{M,0} \left\| 2^{js(\cdot)}\psi_j(D)u \right\|_{L^\infty(\mathbb{R}^n)} \leq 3C\,|a|_{M,0}\,\|u\|_{\Lambda^{s(\cdot)}(\mathbb{R}^n)}. \qquad \square$$

Corollary 18. *Let* $a(\xi) \in S^m_{1,0}(\mathbb{R}^n)$. *Then* $a(D) : \Lambda^{s(\cdot)}(\mathbb{R}^n) \to \Lambda^{s(\cdot)-m}(\mathbb{R}^n)$ *is a bounded operator.*

Now we are able to prove Theorem 14 for arbitrary exponent $s(\cdot) \in C_b(\mathbb{R})$ and $A \in OPS^{m(\cdot)}_{1,\delta}$.

Proof of Theorem 14. Applying the partition of a unity (14) we obtain

$$Op(a)u = \sum_{l=0}^{\infty} Op(a)\lambda_l(D)u,\, u \in \Lambda^{s(\cdot)}(\mathbb{R}^n).$$

Then

$$\lambda_j(D)Op(a)u = \sum_{l=0}^{\infty} \lambda_j(D)Op(a)\lambda_l(D)u = T_{j,1}u + T_{j,2}u,$$

where

$$T_{j,1}u = \sum_{|j-l|>2} \lambda_j(D)Op(a)\lambda_l(D)u, \quad T_{j,2}u = \sum_{|j-l|\leq 2} \lambda_j(D)Op(a)\lambda_l(D)u.$$

Let $s, r \in \mathbb{R}$ be arbitrary numbers. Then

$$\begin{aligned} 2^{jr}\,\|T_{j,1}u\|_{L^\infty(\mathbb{R}^n)} &= 2^{jr} \left\| \sum_{|j-l|>2} \lambda_j(D)Op(a)\lambda_l(D)u \right\|_{L^\infty(\mathbb{R}^n)} \\ &= 2^{jr} \left\| \sum_{|j-l|>2} \lambda_j(D)\psi_j(D)Op(a)\lambda_l(D)\psi_l(D)u \right\|_{L^\infty(\mathbb{R}^n)} \\ &= 2^{jr} \left\| \sum_{|j-l|>2} \lambda_j(D)\left\langle D\right\rangle^{-L} A_{j,l}\psi_l(D)u \right\|_{L^\infty(\mathbb{R}^n)} \end{aligned}$$

where $A_{j,l} = \left\langle D\right\rangle^{L} \psi_j(D)Op(a)\lambda_l(D)$. Since supp $\psi_j \cap$ supp $\lambda_l = \varnothing$ the operator $A_{j,l} \in OPS^{-M}_{1,\delta}(\mathbb{R}^n)$ for every M. By Proposition 16 we obtain

$$2^{jr} \left\| \lambda_j(D)\left\langle D\right\rangle^{-L} A_{j,l}\psi_l(D)u \right\|_{L^\infty(\mathbb{R}^n)} \leq C2^{j(r-L)}\,\|A_{j,l}\psi_l(D)u\|_{L^\infty(\mathbb{R}^n)}.$$

Then Proposition 17 implies that

$$\begin{aligned} &2^{jr} \left\| \lambda_j(D)\left\langle D\right\rangle^{-L} A_{j,l}\psi_l(D)u \right\|_{L^\infty(\mathbb{R}^n)} \\ &\leq C'2^{j(r-L)}2^{-(M-s)l}|a|_{N_1,N_2}2^{sl}\,\|\lambda_l(D)u\|_{L^\infty(\mathbb{R}^n)}. \end{aligned}$$

We choose $L > r$ and $\gamma = M - s > 0$. Then for every $r, s \in \mathbb{R}$,

$$2^{jr} \|T_{j,1}u\|_{L^\infty(\mathbb{R}^n)} \leq C''|a|_{N_1,N_2} \left(\sum_{l=0}^{\infty} 2^{-\gamma l} \right) \sup_{l \in \mathbb{N}_0} 2^{sl} \|\lambda_l(D)u\|_{L^\infty(\mathbb{R}^n)}$$
$$\leq C'' \frac{1}{1-2^{-\gamma}} |a|_{N_1,N_2} \|u\|_{\Lambda^s(\mathbb{R}^n)}. \tag{26}$$

Estimate (26) implies that

$$\left\| 2^{j(s(x)-m(x))} T_{j,1}u \right\|_{L^\infty(\mathbb{R}^n)} \leq C |a|_{N_1,N_2} \|u\|_{\Lambda^{s(\cdot)}(\mathbb{R}^n)} \tag{27}$$

for an arbitrary exponent $s(\cdot)$ and the psdo $Op(a) \in OPS^{m(\cdot)}(\mathbb{R}^n)$. We consider now $T_{j,2}u$. In this case $j = l + k$ where $k = 0, \pm 1, \pm 2$. Then for such j, l,

$$\left\| 2^{j(s(x)-m(x))} \lambda_j(D) Op(a) \lambda_l(D) u \right\|_{L^\infty(\mathbb{R}^n)}$$
$$= \left\| 2^{l(s(x)-m(x))} 2^{k(s(x)-m(x))} \lambda_j(D) Op(a) \lambda_l(D) u \right\|_{L^\infty(\mathbb{R}^n)}$$
$$\leq H \left\| 2^{l(s(x)-m(x))} \lambda_j(D) Op(a) \lambda_l(D) u \right\|_{L^\infty(\mathbb{R}^n)},$$

where

$$H = \max \left\{ 2^{2(m_+ - s_-)}, 2^{(m_+ - s_-)}, 1, 2^{(s_+ - m_-)}, 2^{2(s_+ - m_-)} \right\}.$$

Now by Proposition 17 we obtain that

$$\sum_{|j-l| \leq 2} \left\| 2^{j(s(x)-m(x))} \lambda_j(D) Op(a) \lambda_l(D) u \right\|_{L^\infty(\mathbb{R}^n)} \leq C |a|_{N_1,N_2} \|u\|_{\Lambda^s(\mathbb{R}^n)}. \tag{28}$$

Estimates (27) and (28) yield inequality (16) with constants $C > 0$ and $N_1, N_2 \in \mathbb{N}_0$ independent of the symbol a. □

We denote by $\Lambda_0^{s(\cdot)}(\mathbb{R}^n)$ the closure in the norm $\Lambda^{s(\cdot)}(\mathbb{R}^n)$ of the linear space $\Lambda_{\text{comp}}^{s(\cdot)}(\mathbb{R}^n)$ of distributions in $\Lambda^{s(\cdot)}(\mathbb{R}^n)$ with compact supports.

Theorem 19. *An operator $A = Op(a) \in OPS_{1,\delta}^{m(\cdot)}(\mathbb{R}^n)$ is bounded from $\Lambda_0^{s(\cdot)}(\mathbb{R}^n)$ into $\Lambda_0^{s(\cdot)-m(\cdot)}(\mathbb{R}^n)$ for all $s(\cdot) \in C_b(\mathbb{R})$.*

Proof. Let $\varphi \in C_0^\infty(\mathbb{R}^n)$ be such that $\varphi(x) = 1$ if $|x| \leq 1$ and $\varphi(x) = 0$ if $|x| \geq 2$, and put $\varphi_R(x) = \varphi(x/R)$. Then if $u(\in \Lambda^s(\mathbb{R}^n))$ has a compact support, there exists R_0 such that $\varphi_R u = u$ for $R > R_0$. Hence,

$$Au = A\varphi_R u = \varphi_R A u + [A, \varphi_R]\, u$$

where $[A, \varphi_R] = A\varphi_R - \varphi_R A$.

Then by Theorem 14,

$$\|Au - \varphi_R A u\|_{\Lambda^{s(\cdot)-m(\cdot)}(\mathbb{R}^n)} \leq \|[A, \varphi_R]\|_{\mathcal{B}(\Lambda^s(\mathbb{R}^n), \Lambda^{s(\cdot)-m(\cdot)}(\mathbb{R}^n))} \|u\|_{\Lambda^{s(\cdot)}(\mathbb{R}^n)}$$
$$\leq \frac{C}{R} \|u\|_{\Lambda^{s(\cdot)}(\mathbb{R}^n)} \to 0$$

for $R \to \infty$. Thus, for every $u \in \Lambda^{s(\cdot)}(\mathbb{R}^n)$ with a compact support we deduce that $Au \in \Lambda_0^{s(\cdot)-m(\cdot)}(\mathbb{R}^n)$. Let $v \in \Lambda_0^s(\mathbb{R}^n)$. Given $\varepsilon > 0$ there exists $u \in \Lambda^{s(\cdot)}(\mathbb{R}^n)$ with a compact support such that $\|v-u\|_{\Lambda^{s(\cdot)}(\mathbb{R}^n)} < \varepsilon$. Then

$$\begin{aligned}\|Av - \varphi_R Av\|_{\Lambda^{s(\cdot)-m(\cdot)}(\mathbb{R}^n)} \leq\ & \|Av - Au\|_{\Lambda^{s(\cdot)-m(\cdot)}(\mathbb{R}^n)} \\ & + \|Au - \varphi_R Au\|_{\Lambda^{s(\cdot)-m(\cdot)}(\mathbb{R}^n)} \\ & + \|\varphi_R (Au - Av)\|_{\Lambda^{s(\cdot)-m(\cdot)}(\mathbb{R}^n)} \\ \leq\ & C\,\|u - v\|_{\Lambda^{s(\cdot)}(\mathbb{R}^n)} + \|Au - \varphi_R Au\|_{\Lambda^{s(\cdot)-m(\cdot)}(\mathbb{R}^n)} \\ \leq\ & (C+1)\,\varepsilon\end{aligned}$$

for $R > R_0$ large enough. Hence $Av \in \Lambda_0^{s(\cdot)-m(\cdot)}(\mathbb{R}^n)$. □

3.3. Compactness of psdo's of variable order in $\Lambda^{s(\cdot)}(\mathbb{R}^n)$

Theorem 20. *Let $a \in \mathring{S}_{1,\delta}^{m(\cdot)-\varepsilon}(\mathbb{R}^n)$, $\varepsilon > 0$. Then*

$$Op(a) : \Lambda^{s(\cdot)}(\mathbb{R}^n) \to \Lambda^{s(\cdot)-m(\cdot)}(\mathbb{R}^n)$$

is a compact operator.

Proof. Let $\phi_j \in C^\infty(\mathbb{R}^n)$, $j = 1, 2$, be such that $\phi_j(y) = 1$ if $|y| \leq 1$ and $\phi_j(y) = 0$ if $|y| \geq 2$. Put $\psi_j(y) = 1 - \phi_j(y)$, $\phi_j^R(y) = \phi_j(y/R)$, and $\psi_j^R(y) = \psi_j(y/R)$. Let $a \in S_{1,\delta}^{m(\cdot)}(\mathbb{R}^n)$. Then we consider the operator

$$A_R = \phi_1^R(x) Op(a) \phi_2^R(D) \in OPS^{-\infty}(\mathbb{R}^n).$$

This operator is bounded from $\Lambda^s(\mathbb{R}^n)$ into $\Lambda^r(\mathbb{R}^n)$ for every constant $s, r \in \mathbb{R}$. Moreover A_R acts from $\Lambda^s(\mathbb{R}^n)$ into $\Lambda^r(\mathbb{B}_{2R})$. In view of the compactness of the embedding $\Lambda^{r_1}(\mathbb{B}_{2R}) \subset \Lambda^{r_2}(\mathbb{B}_{2R})$ if $r_1 > r_2$ (see for instance [24]) we obtain that A_R is a compact operator from $\Lambda^s(\mathbb{R}^n)$ into $\Lambda^r(\mathbb{R}^n)$ for arbitrary $s, r \in \mathbb{R}$. Let $\{u_j\}$ be a bounded sequence in $\Lambda^{s(\cdot)}(\mathbb{R}^n)$. Then $\{u_j\}$ is a bounded sequence in $\Lambda^{s_-}(\mathbb{R}^n)$, and the sequence $\{A_R u_j\}$ is a precompact sequence in $\Lambda^{s_+ - m_-}(\mathbb{R}^n)$ and hence in $\Lambda^{s(\cdot)-m(\cdot)}(\mathbb{R}^n)$.

Now we prove that

$$\lim_{R\to\infty} \|Op(a) - A_R\|_{\mathcal{B}(\Lambda^{s(\cdot)}(\mathbb{R}^n), \Lambda^{s(\cdot)-m(\cdot)}(\mathbb{R}^n))} = 0. \tag{29}$$

Indeed,

$$\begin{aligned}Op(a) - A_R &= Op(a) - \phi_1^R(x) Op(a) + \phi_1^R(x) Op(a) - \phi_1^R(x) Op(a) \phi_2^R(D) \\ &= \psi_1^R(x) Op(a) + \phi_1^R(x) Op(a) \psi_2^R(D).\end{aligned}$$

Condition of Theorem 20 and estimate (16) yield that

$$\lim_{R\to\infty} \left\|\psi_1^R(x) Op(a)\right\|_{\mathcal{B}(\Lambda^{s(\cdot)}(\mathbb{R}^n), \Lambda^{s(\cdot)-m(\cdot)}(\mathbb{R}^n))} = 0,$$
$$\lim_{R\to\infty} \left\|\phi_1^R(x) Op(a) \psi_2^R(D)\right\|_{\mathcal{B}(\Lambda^{s(\cdot)}(\mathbb{R}^n), \Lambda^{s(\cdot)-m(\cdot)}(\mathbb{R}^n))} = 0.$$

Hence (29) holds, and $Op(a) : \Lambda^{s(\cdot)}(\mathbb{R}^n) \to \Lambda^{s(\cdot)-m(\cdot)}(\mathbb{R}^n)$ is a compact operator as the limit of the uniformly convergent sequence of compact operators. □

3.4. Reduction order operators

Note that the symbol $(\lambda^2 + |\xi|^2)^{\frac{m(x)}{2}}$ where $m \in C_b^\infty(\mathbb{R}^n)$, $\lambda \geq 1$, belongs to the algebra $S_{1,\delta}^{m(\cdot)}(\mathbb{R}^n, \lambda)$ where $\delta > 0$ is arbitrary small. We set

$$\Xi_{m(\cdot)}(\lambda) = Op((\lambda^2 + |\xi|^2)^{\frac{m(x)}{2}}).$$

It follows from Theorem 9 that there exists an operator $\mathcal{P} \in OPS_{1,\delta}^{-m(\cdot)}(\mathbb{R}^n, \lambda)$ such that

$$\mathcal{P}(\lambda)\Xi_{m(\cdot)}(\lambda) = I + \mathcal{R}_1(\lambda), \quad \Xi_{m(\cdot)}(\lambda)\mathcal{P}(\lambda) = I + \mathcal{R}_2(\lambda)$$

where $\mathcal{R}_j(\lambda) \in OPS_{1,\delta}^{-(1-\delta)}(\mathbb{R}^n, \lambda)$, $j = 1, 2$. By Proposition 3,

$$\lim_{\lambda \to \infty} \|\mathcal{R}_j(\lambda)\|_{\mathcal{B}(L^2(\mathbb{R}^n))} = 0.$$

Hence there exist λ_0 such that the operators $I + \mathcal{R}_j(\lambda_0)$ are invertible in $L^2(\mathbb{R}^n)$. By Proposition 4, $(I + \mathcal{R}_j(\lambda_0))^{-1} \in OPS_{1,\delta}^0(\mathbb{R}^n)$. This implies that

$$(I + \mathcal{R}_1(\lambda_0))^{-1} \mathcal{P}(\lambda_0) = \mathcal{P}(\lambda_0) (I + \mathcal{R}_2(\lambda_0))^{-1}$$

and they are inverse operators for $\Xi_{m(\cdot)}(\lambda_0) : \Lambda^{s(\cdot)}(\mathbb{R}^n) \to \Lambda^{s(\cdot)-m(\cdot)}(\mathbb{R}^n)$ and belong to $OPS_{1,\delta}^{-m(\cdot)}(\mathbb{R}^n)$. Hence the operator $\Xi_{m(\cdot)}(\lambda_0) \in OPS_{1,\delta}^{m(\cdot)}(\mathbb{R}^n)$ and this operator is an isomorphism: $\Lambda^{s(\cdot)}(\mathbb{R}^n) \to \Lambda^{s(\cdot)-m(\cdot)}(\mathbb{R}^n)$.

The operator $\Xi_{m(\cdot)}(\lambda_0)$ is called *the reduction order operator for the Hölder–Zygmund spaces of variable smoothness.*

4. Fredholm properties of psdo's of variable order in weighted Hölder–Zygmund spaces of variable smoothness

We consider the psdo's of variable orders with slowly oscillating at infinity symbols in the class $OP\tilde{S}_{1,\delta}^{m(\cdot)}(\mathbb{R}^n), 0 \leq \delta < 1$, acting from $\Lambda^{s(\cdot)}(\mathbb{R}^n)$ to $\Lambda^{s(\cdot)-m(\cdot)}(\mathbb{R}^n)$. The next theorem gives sufficient conditions of the Fredholmness of such operators in Hölder–Zygmund spaces of variable order. For psdo's of constant order in $OP\tilde{S}_{1,0}^m(\mathbb{R}^n)$ the similar results have been obtained in [19].

Theorem 21. *Let $a \in \tilde{S}_{1,\delta}^{m(\cdot)}(\mathbb{R}^n)$, $0 \leq \delta < 1$. Suppose there exist $R > 0$ and $C > 0$ such that*

$$|a(x, \xi)| \geq C \langle \xi \rangle^{m(x)} \tag{30}$$

for every (x, ξ) satisfying $|x| + |\xi| \geq R$. Then $Op(a) : \Lambda^{s(\cdot)}(\mathbb{R}^n) \to \Lambda^{s(\cdot)-m(\cdot)}(\mathbb{R}^n)$ is a Fredholm operator.

Proof. By Theorem 12, there exist operators $\mathcal{P} \in OP\tilde{S}_{1,\delta}^{-m(\cdot)}(\mathbb{R}^n)$ such that

$$\mathcal{P}Op(a) = I + \mathcal{T}_1, \quad Op(a)\mathcal{P} = I + \mathcal{T}_2$$

where $\mathcal{T}_j \in OP\mathring{S}_{1,\delta}^{-(\rho-\delta)}(\mathbb{R}^n)$, $j = 1, 2$. By Theorem 20, $\mathcal{T}_1 : \Lambda^{s(\cdot)}(\mathbb{R}^n) \to \Lambda^{s(\cdot)}(\mathbb{R}^n)$ and $\mathcal{T}_2 : \Lambda^{s(\cdot)-m(\cdot)}(\mathbb{R}^n) \to \Lambda^{s(\cdot)-m(\cdot)}(\mathbb{R}^n)$ are compact operators. □

Definition 22. Let B be a convex bounded domain in $\mathbb{R}^n$ containing the origin. We denote by $S_{1,\delta}^{m(\cdot)}(\mathbb{R}^n, B)$ the subclass of $S_{1,\delta}^{m(\cdot)}(\mathbb{R}^n)$ consisting of symbols $a(x,\xi)$ which have an analytic extension with respect to the variable ξ in a tube domain $\mathbb{R}^n + iB$, and such that for all $l_1, l_2 \in \mathbb{N}_0$,

$$|a|_{l_1,l_2} = \sup_{(x,\xi)\in\mathbb{R}^n\times\mathbb{R}^n, \eta\in B} \langle\xi\rangle^{-m(x)+|\alpha|-\delta|\beta|} \sum_{|\alpha|\le l_1, |\beta|\le l_2} \left|\partial_x^\beta \partial_\xi^\alpha a(x, \xi + i\eta)\right| < \infty.$$

As above we associate with a symbol $a \in S_{1,\delta}^{m(\cdot)}(\mathbb{R}^n, B)$ a psdo $Op(a)$. The class of such psdo's is denoted by $OPS_{1,\delta}^{m(\cdot)}(\mathbb{R}^n, B)$. In the same way the symbols and operators of the class $OP\tilde{S}_{1,\delta}^{m(\cdot)}(\mathbb{R}^n, B)$ are defined.

Definition 23. We denote by $\mathcal{R}(B)$ the class of positive exponential weights $w(x) = \exp v(x)$ such that

(1) $v \in C^\infty(\mathbb{R}^n)$, and $\partial_{x_j} v \in SO^\infty(\mathbb{R}^n)$, $j = 1, \ldots, n$;
(2) $\nabla v(x) \in B$ for every $x \in \mathbb{R}^n$.

Proposition 24. *Let $Op(a) \in OP\tilde{S}_{\rho,\delta}^{m(\cdot)}(\mathbb{R}^n, B)$, $0 \le \delta < \rho \le 1$, $w = \exp v \in \mathcal{R}(B)$. Then the operator $wOp(a)w^{-1} \in OP\tilde{S}_{\rho,\delta}^{m(\cdot)}(\mathbb{R}^n)$, and*

$$wOp(a)w^{-1} = Op(a_w),$$

where

$$a_w(x,\xi) = a(x, \xi + i\nabla v(x)) + r(x,\xi),$$

and $r \in \mathring{S}_{\rho,\delta}^{m(\cdot)-(\rho-\delta)}(\mathbb{R}^n)$.

Proof. For $m \in \mathbb{R}$ this proposition has been proved in [18]. For the variable $m(\cdot)$ the proof is similar. □

Definition 25. We denote by $\Lambda^{s(\cdot)}(\mathbb{R}^n, w)$ the weighted Hölder–Zygmund spaces with bounded norms

$$\|u\|_{\Lambda^{s(\cdot)}(\mathbb{R}^n,w)} = \|wu\|_{\Lambda^{s(\cdot)}(\mathbb{R}^n)}$$

where w is a positive real-valued C^∞-function on $\mathbb{R}^n$.

Proposition 26. *Let $Op(a) \in OPS_{1,\delta}^{m(\cdot)}(\mathbb{R}^n, B)$, $w(x) = \exp v(x) \in \mathcal{R}(B)$. Then*

$$Op(a) : \Lambda^{s(\cdot)}(\mathbb{R}^n, w) \to \Lambda^{s(\cdot)-m(\cdot)}(\mathbb{R}^n, w)$$

is a bounded operator.

Proof. The proof follows from Proposition 24 and Theorem 14 immediately. □

Theorem 27. *Let* $Op(a) \in OP\tilde{S}^{m(\cdot)}_{1,\delta}(\mathbb{R}^n, B)$, $w = \exp v \in \mathcal{R}(B)$. *Suppose there exist* $R > 0$ *and* $C > 0$ *such that*

$$|a(x, \xi + i\nabla v(x))| \geq C \langle \xi \rangle^{m(x)} \tag{31}$$

for every (x, ξ) *satisfying* $|x| + |\xi| \geq R$. *Then*

$$Op(a) : \Lambda^{s(\cdot)}(\mathbb{R}^n, w) \to \Lambda^{s(\cdot)-m(\cdot)}(\mathbb{R}^n, w)$$

is a Fredholm operator.

Proof. The proof follows from Proposition 24 and Theorem 14. □

The similar theorem holds for the space $\Lambda_0^{s(\cdot)}(\mathbb{R}^n, w)$, which is the closure of the linear space $\Lambda^{s(\cdot)}_{\mathrm{comp}}(\mathbb{R}^n)$ in the norm of $\Lambda^{s(\cdot)}(\mathbb{R}^n, w)$.

Remark 28. Let $B = \mathbb{B}_\varepsilon$, where $\varepsilon > 0$ is arbitrarily small, and w be a subexponential weight, that is, $w(x) = 1$ for $|x| \leq R$ and $w(x) = \exp(\alpha |x|^\beta)$ for $|x| \geq 2R$ with some $\alpha \in \mathbb{R}$ and $\beta \in (0,1)$. Then $\lim_{x\to\infty} \nabla \log w(x) = 0$, and condition (31) takes the form (30).

Theorem 29. *Let* $Op(a) \in OP\tilde{S}^{m(\cdot)}_{1,\delta}(\mathbb{R}^n, B)$, $w = \exp v \in \mathcal{R}(B)$. *Suppose there exist* $R > 0$ *and* $C > 0$ *such that*

$$|a(x, \xi + i\nabla v(x))| \geq C \langle \xi \rangle^{m(x)}$$

for every (x, ξ) *satisfying* $|x| + |\xi| \geq R$. *Then*

$$Op(a) : \Lambda_0^{s(\cdot)}(\mathbb{R}^n, w) \to \Lambda_0^{s(\cdot)-m(\cdot)}(\mathbb{R}^n, w)$$

is a Fredholm operator.

Theorem 30 (Phragmén–Lindelöf principle). *Let* $Op(a) \in OP\tilde{S}^{m(\cdot)}_{1,\delta}(\mathbb{R}^n, B)$ *and* $w = \exp v \in \mathcal{R}(B)$ *satisfy* $\lim_{x\to\infty} v(x) = +\infty$. *Suppose there exist* $R > 0$ *and* $C > 0$ *such that*

$$|a(x, \xi + i\theta\nabla \log w(x))| \geq C \langle \xi \rangle^{m(x)} \tag{32}$$

for all (x, ξ) *satisfying* $|x| + |\xi| \geq R$ *and all* $\theta \in [-1, 1]$. *Then every solution* $u \in \Lambda_0^{s(\cdot)}(\mathbb{R}^n, w^{-1})$ *of the equation*

$$Op(a)u = f \in \Lambda_0^{s(\cdot)-m(\cdot)}(\mathbb{R}^n, w)$$

actually belongs to $\Lambda_0^{s(\cdot)}(\mathbb{R}^n, w)$.

Proof. The operator $w^\theta Op(a) w^{-\theta}$, $\theta \in [-1, 1]$, can be written as

$$w^\theta Op(a) w^{-\theta} = Op(a(x, \xi + i\theta\nabla \log w)) + Op(r_\theta(x, \xi)),$$

where $r_\theta(x, \xi)$ belongs to $\mathring{S}^{m(\cdot)-(\rho-\delta)}_{\rho,\delta}(\mathbb{R}^n)$ uniformly with respect to $\theta \in [-1, 1]$. In light of Theorem 21 and condition (32) $w^\theta Op(a) w^{-\theta} : \Lambda_0^{s(\cdot)}(\mathbb{R}^n) \to \Lambda_0^{s(\cdot)-m(\cdot)}(\mathbb{R}^n)$ is a Fredholm operator for all $\theta \in [-1, 1]$. It is easy to prove that the mapping

$$[-1, 1] \to \mathcal{B}(\Lambda_0^{s(\cdot)}(\mathbb{R}^n), \Lambda_0^{s(\cdot)-m(\cdot)}(\mathbb{R}^n)) : \theta \to w^\theta Op(a) w^{-\theta}$$

is continuous. Hence, Index$(w^\theta Op(a)w^{-\theta} : \Lambda_0^{s(\cdot)}(\mathbb{R}^n) \to \Lambda_0^{s(\cdot)-m(\cdot)}(\mathbb{R}^n))$ does not depend on $\theta \in [-1,1]$. Moreover, the condition $\lim_{x\to\infty} w(x) = \infty$ implies that $\Lambda_0^{s(\cdot)}(\mathbb{R}^n, w) \subset \Lambda_0^{s(\cdot)}(\mathbb{R}^n, w^{-1})$, and the last imbedding is dense because the spaces $\Lambda_0^{s(\cdot)}(\mathbb{R}^n, w)$ and $\Lambda_0^{s(\cdot)}(\mathbb{R}^n, w^{-1}))$ have the common dense set $\Lambda_{\mathrm{comp}}^{s(\cdot)}(\mathbb{R}^n)$. Then

$$\ker(Op(a) : \Lambda_0^{s(\cdot)}(\mathbb{R}^n, w^{-1}) \to \Lambda_0^{s(\cdot)-m(\cdot)}(\mathbb{R}^n, w^{-1}))$$

coincides with

$$\ker(Op(a) : \Lambda_0^{s(\cdot)}(\mathbb{R}^n, w) \to \Lambda_0^{s(\cdot)-m(\cdot)}(\mathbb{R}^n, w))$$

(see for instance [8, Appendix, Section 1, p. 226]). Moreover, if $u \in \Lambda_0^{s(\cdot)}(\mathbb{R}^n, w^{-1})$ is a solution of the equation $Op(a)u = f \in \Lambda_0^{s(\cdot)-m(\cdot)}(\mathbb{R}^n, w))$ then $u \in \Lambda_0^{s(\cdot)}(\mathbb{R}^n, w)$. □

Remark 31. Let $B = \mathbb{B}_\varepsilon$, $\varepsilon > 0$ be arbitrarily small, w be a subexponential weight defined in Remark 28. Then the condition of Theorem 30 takes the form

$$|a(x,\xi)| \geq C \langle \xi \rangle^{m(x)}$$

for every (x,ξ) satisfying $|x| + |\xi| \geq R$.

Example 32. We consider the Helmholtz operator of variable order

$$A = Op\left((|\xi|^2 + k^2)^{m(x)} \right)$$

where $k > 0, m(x) \in SO^\infty(\mathbb{R}^n)$, $w(x) = e^{r\langle x \rangle}$, $0 < r < k$. It is evident that $A \in OP\tilde{S}_{1,\delta}^{2m(\cdot)}(\mathbb{R}^n)$ where $\delta > 0$ is an arbitrarily small number. Then Theorems 19, 29, 30 imply the following result.

(i) The operators

$$A : \Lambda^{s(\cdot)}(\mathbb{R}^n, w) \to \Lambda^{s(\cdot)-2m(\cdot)}(\mathbb{R}^n, w),$$
$$A : \Lambda_0^{s(\cdot)}(\mathbb{R}^n, w) \to \Lambda_0^{s(\cdot)-2m(\cdot)}(\mathbb{R}^n, w),$$

are Fredholm operators.

(ii) Let $u \in \Lambda_0^{s(\cdot)}(\mathbb{R}^n, w^{-1})$ be a solution of the equation

$$Au = f \in \Lambda_0^{s(\cdot)-2m(\cdot)}(\mathbb{R}^n, w).$$

Then, in fact, $u \in \Lambda_0^{s(\cdot)}(\mathbb{R}^n, w)$.

References

[1] A. Almeida and P. Hästö, *Besov spaces with variable smoothness and integrability*, J. Funct. Anal. **258** (2010), 1628–1655.

[2] A. Almeida and S. Samko, *Pointwise inequalities in variable Sobolev spaces and applications*, Z. Anal. Anwend. **26** (2007), 179–193.

[3] A. Almeida and S. Samko, *Embeddings of variable Hajlasz–Sobolev spaces into Hölder spaces of variable order*, J. Math. Anal. Appl. **353** (2009), 489–496.

[4] R. Beals, *Characterization of pseudodifferential operators and applications*, Duke Math. J. **44** (1977), 45–57.

[5] O. Besov, *On spaces of functions of variable smoothness defined by pseudodifferential operators*, Proc. Steklov Inst. Math. **227** (1999), 50–69.

[6] O. Besov, *Interpolation, embedding, and extension of spaces of functions of variable smoothness*, Proc. Steklov Inst. Math. **248** (2005), 47–58.

[7] L. Diening, P. Harjulehto, P. Hästö, and M. Růžička, *Lebesgue and Sobolev Spaces with Variable Exponents*, Springer Lecture Notes **2017**, Springer-Verlag, Berlin, 2011.

[8] I. Gohberg and I. Feldman, *Convolution Equations and Projection Methods for Their Solutions*, Transl. Math. Monographs **41**, Amer. Math. Soc., Providence, R.I., 1974.

[9] L. Hörmander, *The Analysis of Linear Partial Differential Operators, Vols. I-IV*, Classics in Mathematics, Springer-Verlag, Berlin, 2003–2007.

[10] N.K. Karapetians and A.I. Ginsburg, *Fractional integrals and singular integrals in the Hölder classes of variable order*, Integral Transforms Spec. Funct. **2** (1994), 91–106.

[11] V.D. Kryakvin, *Characterization of pseudodifferential operators in Hölder–Zygmund spaces*, Differ. Equ. **49** (2013), 306–312.

[12] V.D. Kryakvin, *Boundedness of pseudodifferential operators in Hölder–Zygmund spaces of variable order*, Sib. Math. J. **55** (2014), 1073–1083.

[13] H. Kumano-go, *Pseudo-Differential Operators*, MIT Presss, Cambridge, Mass., 1981.

[14] P.D. Lax, *Functional Analyis*, Wiley-Interscience, Chichester, 2002.

[15] H.-G. Leopold, *Embedding of function spaces of variable order of differentiation in function spaces of variable order of integration*, Czechoslovak Math. J. **49** (1999), 633-644.

[16] H.-G. Leopold and E. Schrohe, *Spectral invariance for algebras of pseudodifferential operators on Besov spaces of variable order of differentiation*, Math. Nachr. **156** (1992), 7–23.

[17] V.S. Rabinovich, *An Introductory Course on Pseudodifferential Operators*, Textos de Matemática **1**, Instituto Superior Técnico, Lisboa, 1998.

[18] V.S. Rabinovich, *Pseudodifferential operators with analytic symbols and estimates for eigenfunctions of Schrödinger operators*, Z. Anal. Anwend. **21** (2002), 351–370.

[19] V.S. Rabinovich, *Fredholm property of pseudo-differential operators on weighted Hölder–Zygmund spaces*, Operator Theory: Advances and Applications **164** (2006), 95–114.

[20] V.S. Rabinovich and S.G. Samko, *Essential spectra of pseudodifferential operators in Sobolev spaces with variable smoothness and variable Lebesgue indices*, Dokl. Math. **76** (2007), 835–838.

[21] V. Rabinovich and S. Samko, *Boundedness and Fredholmness of pseudodifferential operators in variable exponent spaces*, Integr. Equ. Oper. Theory **60** (2008), 507–537.

[22] N. Samko and B. Vakulov, *Spherical fractional and hypersingular integrals of variable order in generalized Hölder spaces with variable characteristic*, Math Nachr. **284** (2011), 355–369.

[23] M. Shubin, *Pseudodifferential Operators and Spectral Theory*, Springer-Verlag, Berlin, 2001.

[24] E.M. Stein, *Harmonic Analysis: Real-Variable Methods, Orthogonality, and Oscillatory Integrals*, Princeton Univ. Press, Princeton, New Jersey, 1993.

[25] M.E. Taylor, *Pseudodifferential Operators*, Princeton Univ. Press, Princeton, New Jersey, 1981.

[26] A. Unterberger and J. Bokobza, *Sur les opérateurs pseudo-différentiels d'ordre variable*, C. R. Acad. Sci. Paris **261** (1965), 2271–2273.

[27] B.G. Vakulov, *Spherical potentials of complex order in the variable order Hölder spaces*, Integral Transforms Spec. Funct. **16** (2005), 489–497.

[28] L.R. Volevich and V.M.Kagan, *Pseudodifferential hypoelliptic operators in the theory of function spaces*, Tr. Mosk. Mat. Obs. **20** (1969), 241–275 (in Russian).

Vadim Kryakvin
Southern Federal University
ul. Milchakova, 8A
Rostov na Donu, 344090, Russia
e-mail: vadkr@math.rsu.ru

Vladimir Rabinovich
Instituto Politécnico Nacional
ESIME Zacatenco, edif.1
Av. Luis Enrique Erro S/N
C.P. 07738, Mexico, DF, Mexico
e-mail: vladimir.rabinovich@gmail.com

Operator Theory:
Advances and Applications, Vol. 259, 533–559

Commutator Estimates Comprising the Frobenius Norm – Looking Back and Forth

Zhiqin Lu and David Wenzel

A tribute to Albrecht Böttcher on his 60th birthday

Abstract. The inequality $\|XY - YX\|_F \leq \sqrt{2}\|X\|_F\|Y\|_F$ has some history to date. The growth of the task will be highlighted, supplemented by a look at future developments. On this way, we meet different forms and give an insight into various consequences of it. The collection of results will be enriched by introductive explanations. We also cross other fields that are important for theory and applications, and even uncover less known relationships.

Mathematics Subject Classification (2010). Primary 15A45; Secondary 15-02.
Keywords. Commutator, Frobenius norm, BW and DDVV conjectures.

1. Revelations

We want to take the opportunity of Albrecht Böttcher's sixtieth birthday to look at one particular topic that started with a vague idea about a dozen years ago. First, we had a conjecture, this conjecture was proven after quite a while, and several follow-ups extended the problem thereafter.

And who, if not two of the guys responsible for developing completely different proofs, could give you some insight into this story and an overview of the achievements. There we are: Albrecht's long-time research associate David Wenzel, who went this path together with him; and Zhiqin Lu, coming from geometry and opening a (to most of us) new perspective. We thank Koenraad Audenaert for some entry points into physics that unveil links to a connatural field.

At first, let's go back to the beginning. Assume two square matrices X, Y are given. We all know that, occasionally, XY and YX do not coincide. But how different can they be? For measuring the distance between them, we can take a look onto $\|XY - YX\|_F$, where $\|\cdot\|_F$ is the easily computable Frobenius norm.

The author Z. Lu is partially supported by an NSF grant DMS-1510232.

The object inside is the famous commutator[1]. What's the deal? It can't be hard! Just using the triangle inequality and the sub-multiplicativity, one clearly has

$$\|XY - YX\|_F \leq 2\|X\|_F\|Y\|_F. \tag{1.1}$$

You're right. But, the problem is that we were unable to find *only one* example in which actually equality holds.[2] The best (meaning "biggest" in our case) we got even from extended experiments with a computer – operating systematically or at random, no matter the matrix size taken – was a factor $\sqrt{2}$ instead of the trivially obtained 2. A claim was born. The one that it's all about.

Theorem 1.1 (Böttcher–Wenzel conjecture). *Suppose $n \in \mathbb{N}$ is arbitrary. Let X and Y be two $n \times n$ matrices. Then,*

$$\|[X, Y]\|_F \leq \sqrt{2}\|X\|_F\|Y\|_F. \tag{1.2}$$

Well, you probably want to throw in the operator norm into the discussion. As many investigations are done in this one, it seems like a fixed option. However, the example $\left[\left(\begin{smallmatrix} 0 & 1 \\ 1 & 0 \end{smallmatrix}\right), \left(\begin{smallmatrix} 0 & -1 \\ 1 & 0 \end{smallmatrix}\right)\right] = \left(\begin{smallmatrix} 2 & 0 \\ 0 & -2 \end{smallmatrix}\right)$ shows that we are stuck with the trivial estimate (1.1). Here, the constant can not be improved. The Frobenius norm is more selective.

The commutator is someway strange. As a map turning two matrices into a third one, it is not injective. The kernel is non-zero. In particular, every matrix commutes with itself. For this reason, $\|[X, Y]\|_F = \|[X, Y + \alpha X]\|_F$. In consequence, we can replace the right-hand side of (1.2) by virtually any of the terms $\sqrt{2}\|X\|_F\|Y + \alpha X\|_F$. Since α is arbitrary, choosing $\alpha = -\overline{\langle X, Y\rangle}/\|X\|_F^2$ is possible. This is indeed a minimizer of $\|Y + \alpha X\|_F$ and would further reduce the (squared[3]) upper bound (if it really should be valid) by an inner product[4]:

$$\|[X, Y]\|_F^2 \leq 2\|X\|_F^2\|Y\|_F^2 - 2|\langle X, Y\rangle|^2. \tag{1.3}$$

Why should (1.2), or equivalently (1.3), be true? Here is a demonstration for $n = 2$.

Proof. Put $X = \left(\begin{smallmatrix} a & b \\ c & d \end{smallmatrix}\right)$, $Y = \left(\begin{smallmatrix} e & f \\ g & h \end{smallmatrix}\right)$. An elementary calculation delivers

$$\begin{aligned} &2\|X\|_F^2\|Y\|_F^2 - 2|\langle X, Y\rangle|^2 - \|XY - YX\|_F^2 \\ &\quad = 2|ah - de|^2 + |(a + d)f - (e + h)b|^2 + |(a + d)g - (e + h)c|^2 \end{aligned} \tag{1.4}$$

which clearly is a sum of non-negative terms. □

[1] Due to algebra, the Lie bracket notation $[X, Y]$ is a common abbreviation.

[2] ... except for the case involving zero matrices, which is pretty lame, for obvious reasons.

[3] This is always a good idea because of the definition $\|A\|_F := \left(\sum_{j,k} |a_{jk}|^2\right)^{1/2}$ ... simplifying transformations and avoiding an unnecessary flood of root symbols.

[4] An advantage of the Frobenius over the operator norm: it is induced by the scalar product $\langle A, B\rangle := \mathrm{Tr}(B^* A)$, where $\mathrm{Tr}\, C := \sum_i c_{ii}$ denotes the trace.

So easy the last lines emerged, it must be clarified that the same idea refuses to work for matrices of size 3×3 and beyond. It's not that we are unable to show anything related; an analogous statement simply is not true. Also the original attempt from [6, Theorem 4.2] cannot be transferred into higher dimensions.[5]

Of course, there was something that made us believe into the validity of (1.2). During the initial hunt for "utterly non-commuting" matrix pairs, we made overview plots for the ratio $\|[X,Y]\|_F^2/\|X\|_F^2\|Y\|_F^2$ and generated pictures like these:

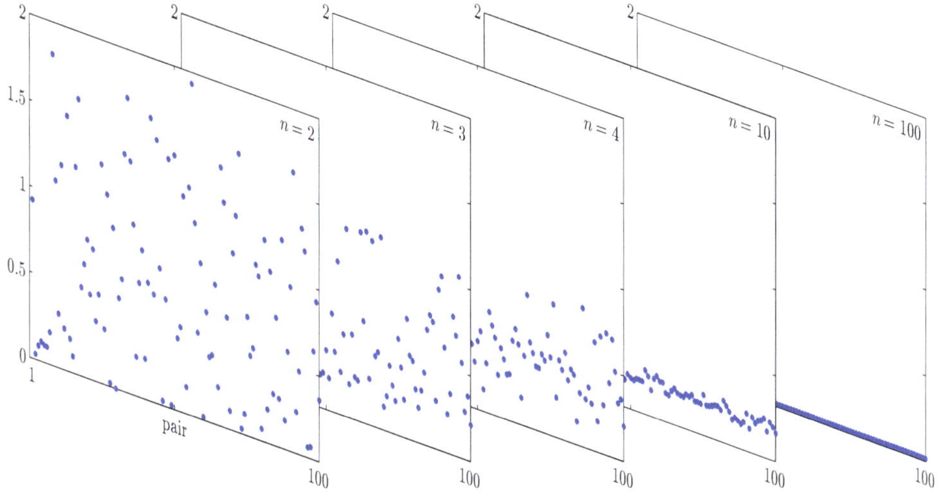

Undoubtedly, the values for $n = 2$ may reach the constant $\sqrt{2}$ (squared, of course), but won't go further. Yet, it is another point that catches the eye. Apparently, random size 3 matrices have difficulties in producing big commutators. The clustering effect at very small values becomes even stronger when n increases.

That situation left us torn between astonishment and annoyance. On the one side, the seemingly whole-range-filling 2×2 matrix case (the one where actually something happens) was tackled completely. But on the other, although large sized matrices *evidently* are far away from making any hassles, this wasn't usable for obtaining a general proof. And even though $n = 2$ later turned out to *be* the most interesting case, there should remain lots of stuff to do!

2. Widening the scope

At the start, we had a conjecture about a general norm bound, and could prove it only in some special cases (for size 2, or if one of the matrices is normal). Well, additionally, we were able to show that (1.1) must be too weak. Mastering determinants as he usually does, Albrecht could validate the estimate with 2 replaced by $\sqrt{3}$ ... that's half way on what we wished for, at least.

[5]The proof relies on the matrices' trace being spread over only two entries.

Of course, there were also the overwhelming observations that commutators typically avoid large values. So, we wanted to give this aspect a foundation. For this, we could revert to a previous work with Grudsky [5] on a bound for norms of random vectors under a linear map. Luckily, when interested in the Frobenius norm, we can look at a matrix as if it is a vector. It was shown in [6] that, taking the expected value, the ratio $\|[X,Y]\|_F^2/\|X\|_F^2\|Y\|_F^2$ tends to zero as $n \to \infty$, and the convergence is linear. In conclusion, the norm of the commutator is small compared to the norms of the involved matrices:

$$E(\|XY - YX\|_F^2) = O(\tfrac{1}{n}) \cdot E(\|X\|_F^2\|Y\|_F^2), \tag{2.1}$$

under very weak assumptions on the underlying distribution.[6]

The efforts afterwards concentrated on the minimal bound. It stayed open for quite a while. Then, simultaneously, three really distinct proofs for (1.2) saw the light of the day ([7],[26],[34]). So, the constant $\sqrt{2}$ in the inequality is now shown, and it is known to be best-possible.[7] Sure, with that, only the first step on an even longer road up to the present was done.

2.1. Illuminating the representation

In 1900, Hilbert presented the famous list with 23 problems that should heavily influence the following century of mathematics. About the half of them is completely solved by now, and still a handful of them is way too far from being understood. A particular problem, the 17th, prepares the ground for this part. It reads as follows.

> *Consider a multivariate polynomial or rational function. If it takes only nonnegative values over $\mathbb{R}$, can it be represented as a sum of squares of rational functions?*

It was confirmed "only" three decades later.[8] In the formulation, Hilbert already has taken into account that there definitely exist nonnegative polynomials that are not a SOS[9] of other *polynomials.*

So, finding a polynomial SOS is always on the agenda whenever a polynomial in n variables is nonnegative. The commutator inequality (1.2) is equivalent to $2\|X\|_F^2\|Y\|_F^2 - \|XY - YX\|_F^2 \geq 0$. On the left-hand side, there obviously is a quartic polynomial. And we claimed that it is nonnegative. The investigated form was further reduced a little more to (1.4). And in the proof for $n = 2$, we indeed obtained a sum of polynomial squares (actually of quartics, again). It is even the complex version. Naturally, this gave hope also for larger matrices. László [24] proved that, as remarked, the reduced form is not SOS. Here is his result.

---

[6] It's not even necessary that X and Y are chosen independently.

[7] A suitable example that gives equality can be found on the second page.

[8] Worked out by Artin 1927, it was just the second problem to be resolved that time! An algorithmic solution was given by Delzell in 1984. The minimal number of squares is yet not known, but limited by $2^{\#\text{variables}}$ due to Pfister 1967. In other fields, the answer may be negative. However, a generalization to symmetric matrices with polynomial entries is known. In complex analysis, the Hermitian analogue considers squared norms.

[9] In mathematics, this is not a "help me" signal like in navigation. It stands for "sum of squares."

Theorem 2.1. *Suppose $X, Y \in \mathbb{R}^{n\times n}$. The smallest value γ_n so that*

$$(2+\gamma_n)\big(\|X\|_F^2\|Y\|_F^2 - \langle X,Y\rangle^2\big) - \|XY-YX\|_F^2$$

is a sum of squares of polynomials is $\gamma_n = \frac{n-2}{2}$.

Since polynomials are of special interest, explicit sufficient conditions were developed for judging on the SOS-property.[10] Studying related optimization problems gives such a technique. The reduction to (1.4) has a benefit over regarding the form associated with the initial inequality (1.2). That only differences $x_{ij}y_{kl} - y_{ij}x_{kl}$ appear[11] is used cleverly in proofs.

In this manner, when restricted to special matrix classes, (1.4) admits a decomposition into SOS [25]. Certain prescribed structures may manage to eliminate disturbing variables. Examples of good nature are matrices having non-zero entries only in the first row and the last column. Also, if X and Y are taken from the cyclic(!) Hankel[12] or tridiagonal matrices, the original form is SOS. At last, we want to restate a case that looks like a gift made with Albrecht in mind.

Conjecture 2.2. *The form* (1.4) *generated by two arbitrary real Toeplitz matrices is SOS.*

This is currently proven for sizes up to 7×7. With $n=8$, there is a change in the properties that kills the proof's idea. Nevertheless, other indicators (esp. in the corresponding semidefinite program) sustain the assertion. The matrix structure cancels many terms that could produce troubles.[13]

2.2. Exploring new measures

There is more than one norm. Two of them were brought forward right from the start. One was interesting, as it apparently profits somehow of the commutator's structural properties. The other behaved, well, just as trivial as one would expect. Of course, a whole bunch of other norms is still waiting out there. Without any doubt, these may vary in their sensitivity in regard of the comparison of the commutator with its input matrices.

So, different metrics will yield different best constants in the inspection of

$$\|XY-YX\|_\star \le C_\star \|X\|_\star \|Y\|_\star.$$

Since any given norm may be scaled, and this factor would pop up once on the left, but twice on the right, C can be arbitrary, in principle. But, this is kind of cheating; an anchor is required for comparing the effects of different norms. For this reason, we put a natural normalization condition onto elementary matrices [35].

---

[10] Note that every real nonnegative polynomial can be approximated as closely as desired by SOS-polynomials, but we seek for finitely many summands.

[11] One easily checks $2\|X\|_F^2\|Y\|_F^2 - 2|\langle X,Y\rangle|^2 = \|X\otimes Y - Y\otimes X\|_F^2$, where $\otimes$ is the Kronecker product. This is actually the Lagrange identity.

[12] Alas, ordinary Hankel matrices will fail to produce SOS, in spite of the fact that they are symmetric.

[13] Looking at $n=2$, e.g., the first summand in the representation (1.4) will vanish.

Theorem 2.3. *Let* $E_{jk} = e_j \otimes e_k^*$. *Suppose* $\|E_{jk}\| = 1$ *and* $\|E_{kj}\| \leq 1$ *for some* $j \neq k$. *Then,* $C \geq \sqrt{2}$.

What are good norms for matrices? We could generalize the Frobenius norm to vector p norms. More important in mathematical practice are the p induced operator norms.[14] But, most important for matrices are norms that are invariant under unitary transformations. Via the singular value decomposition, such norms are in one-to-one correspondence with the norms that can be defined for the vector of singular values.[15] Prominent examples are the Ky Fan norms[16]

$$\| \cdot \|_{(m)} = \sigma_1 + \cdots + \sigma_m \qquad (\sigma_1 \geq \cdots \sigma_m \geq \cdots \geq \sigma_n \geq 0)$$

and the Schatten norms

$$\| \cdot \|_p = (\sigma_1^p + \cdots + \sigma_n^p)^{1/p}.$$

Also a mixture $\| \cdot \|_{(m),p}$ of them (counting only the m largest singular values) has been handled. The Frobenius norm fits into the scheme, too. Because of the alternative representation $\|X\|_F^2 = \|USV\|_F^2 = \sum_i \sigma_i^2$, one has $\| \cdot \|_F = \| \cdot \|_2$.

As time passed, another proof for (1.2) was discovered [2]. Going even further, the tighter inequality

$$\|[X, Y]\|_2 \leq \sqrt{2}\|X\|_2\|Y\|_{(2),2} \tag{2.2}$$

has been shown to hold true. Also the Schatten norms are fully treated. As they somehow morph from Frobenius to operator norm ($\| \cdot \|_\infty$), the resulting value C_p is also in between. For this, the Riesz–Thorin theorem[17] got a fruitful application. Knowing the commutator inequalities for 2 and 1, as well as ∞, complex interpolation yields estimates for all indices in between.[18] Luckily, these turn out to be sharp.

One has to be careful and to check many assumptions before Riesz–Thorin will release its magic, but collecting all the pieces, together with sophisticated examinations of crucial base inequalities, even an extended problem was broadly solved. Motivated by the validity of (2.2), one should think about using up to three different norms with this type of inequalities. The constants, denoted by $C_{p,q,r}$, then are determined by the indices p, q and r of the norms of $[X, Y]$, X, and Y.

Theorem 2.4. *Suppose* $n \in \mathbb{N} \setminus \{1\}$. *Let* $1 \leq p, q, r \leq \infty$ *be indices of Schatten norms in* $\mathbb{C}^{n \times n}$, *excluding the octant* $p > 2$, $q < 2$ *and* $r < 2$. *Then,*

---

[14] Simple exercise: $C = 2$ for all of them.

[15] In addition, the so-called gauge function has to be invariant under permutations of the entries and under the transformations of any single sign.

[16] Obviously, $\| \cdot \|_{(1)}$ is the classical operator norm ($p = 2$). So, it happens that $C_{(m)} = 2$, again.

[17] Originally shown as an inequality for p norms of vectors, it is often used to obtain boundedness of certain operators on L^p or Schatten classes.

[18] This covers the complete norm family. Note that direct interpolation between 1 and ∞ will result in too large constants. Two separate ranges are required.

$$C_{p,q,r} = \begin{cases} 2^{1/p} & \textit{in } ❶ \\ 2^{1-1/q} & \textit{in } ❷ \\ 2^{1-1/r} & \textit{in } ❸ \\ 2^{1+1/p-1/q-1/r} & \textit{in } ❹ \\ 2(2\lfloor\frac{n}{2}\rfloor)^{1/p-1/q-1/r} & \textit{in } ❺ \\ 2n^{1/p-1/q-1/r}\cos\frac{\pi}{2n} & \textit{in } ❻, \end{cases}$$

$$C_{\infty,1,1} = \frac{\sqrt{27}}{4}.$$

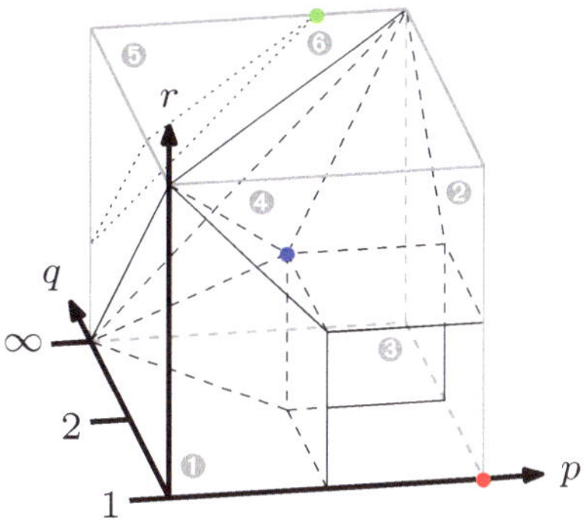

The cube $[1,\infty]^3$ of all possible norm indices (p,q,r) is divided into six connected parts. In each of these, the given value realizes the maximum over all six terms. The regions are the result from several equalities of two of them. The nature of the Riesz–Thorin theorem is convexity with respect to inverse norm indices or their duals (usually one writes $1/p' = 1-1/p$). That's the reason why the mapping $p \mapsto 1-1/p$ transforms the infinite cube to the bounded $[0,1]^3$, while orders are preserved, and the value 2 is centered. Furthermore, terms as in the exponents of the six components will yield lines and planes as borders. This pictorial ansatz helped a lot in detailing the steps of the interpolation procedure. And so, a picture also spares us to give in formulas the six regions treated by [12] and [36].

There are three congruent bodies bounding the mostly unknown octant. The numbers ❶, ❷, ❸ are pictured next to the respective outermost point. On the opposite side, they border the fourth region ❹. They all share the point $(2,2,2)$ in the cube's middle, which represents the original inequality (1.2). The last two parts are set on top. In contrast to the others, they depend on the dimension! Moreover, a flaw between even and odd sizes becomes lucid.[19] In the pyramid ❺, $2\lfloor\frac{n}{2}\rfloor$ evaluates to n or $n-1$, respectively. Additionally, if n is even, this region also occupies the last layer, so that ❻ does not exist. In the odd case, the index on the line (p,∞,∞) marking the interface is given by $p_n = \ln\frac{n-1}{n}/\ln\cos\frac{\pi}{2n} \gtrsim \frac{8}{\pi^2}n$, whence ❻ becomes thinner as $n \to \infty$.

The method also works if one norm on the right-hand side is truncated to a mix norm [18]. Thanks to (2.2), again one has, e.g., $C_{p;p;(2),p} = \max\{2^{1/p}, 2^{1-1/p}\}$. In the end, with the investigation of only a couple of inequalities (first of all, the Frobenius case in the middle), a crowd of others is delivered "for free."

Let us return to the case of utilizing only a single norms. With Theorem 2.3 in mind, the Frobenius norm is special. It realizes the minimal constant $C = \sqrt{2}$ in such kind of estimates. So far, we had no luck to meet a comparable norm. In [7], we proved that if $n = 2$, then the Frobenius norm is the only unitarily invariant norm with this property. Non-invariant norms realizing $\sqrt{2}$ exist. The invariance further imposes strong restrictions: every planar two-coordinate cut through the gauge

[19] The Riesz–Thorin approach may be used naturally within the family of vector norms (regarded on matrices). For $p < 2$, one gets the same constants C_p. However, for $p > 2$, even when all norms coincide, C_p is already size dependant (unlike the Schatten counterpart).

function's unit ball must be the usual disc.[20] That leaves $\|\cdot\|_{(2),2}$ as a suspicious candidate. Though it remains unclear if this serves the purpose for $n \geq 3$, by [17], its dual norm is an example with $C = \sqrt{2}$. The unit balls for the three-dimensional cases illustrate that the dual norm (right) is the largest possible norm[21], and that there is a huge gap to the still open $\|\cdot\|_{(2),2}$ (left) and the norms in between.

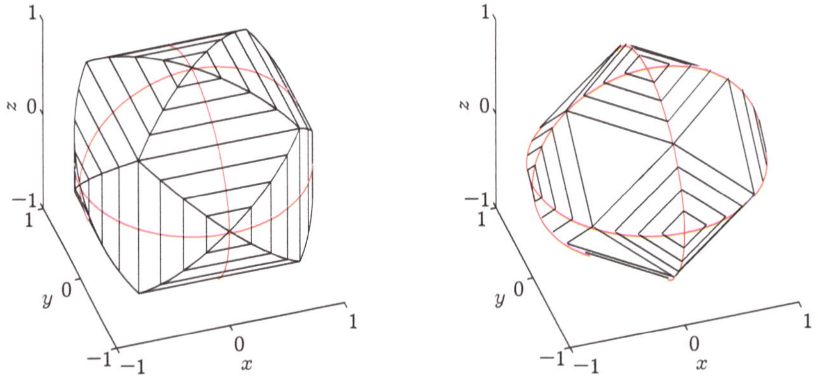

2.3. Tackling the equality cases

If one has an inequality like (1.2), a naturally appearing task is the determination of the matrices that actually result into "=". The aim is to identify matrices that are far away from commutativity.

Definition 2.5. A pair (X, Y) of matrices $\mathbb{C}^{n\times n}$ is said to be maximal if $X \neq O$, $Y \neq O$, and $\|[X, Y]\|_F = \sqrt{2}\|X\|_F\|Y\|_F$ is satisfied.

Similar definitions can be made for the inequalities subject to other norms. For the Schatten norms, we then consider (p, q, r)-maximal pairs. Early, we strengthened (1.2) to (1.3). With this observation, clearly, the additional term must be zero. So, fulfilling (1.2) as an equality will require matrices that are orthogonal to each other in the Frobenius inner product.[22] But, there is much more to come. The same can be done for any other matrix commuting with one of the two arguments. In particular, the identity matrix I yields vanishing traces. Powers X^m of one matrix are also in the respective commutant. Consequently, there are lots of restrictions for a pair to be maximal [7]. With size $n = 2$, the sufficiency is verified by a calculation as simple as (1.4).

Theorem 2.6. *Two matrices $X, Y \in \mathbb{C}^{2\times 2}$ form a maximal pair if and only if* $\operatorname{Tr} X = \operatorname{Tr} Y = 0$ *and* $\langle X, Y\rangle = 0$.

[20] In other words, if all but two singular values are zero, the norm of the vector has to equal its Frobenius norm.

[21] The mix norm directly follows the necessary cylinder restriction, whereas its dual norm possesses the tightest ball, i.e., the convex hull of the fixed circles.

[22] Recall: $\langle X, Y\rangle = \operatorname{Tr} Y^*X = 0$.

Where have all the power conditions gone? For $n = 2$, due to the theorem of Cayley–Hamilton[23], $\langle X, Y^m \rangle = 0$ with $m > 1$ is no new restriction. Moreover, taken together, the cases $m \leq 1$ already yield a sufficient condition. For $n \geq 3$, on the other side, it's not enough; they do not suffice.

In search for more restrictions, by (2.2), we have the chain of inequalities

$$\|[X,Y]\|_F \leq \sqrt{2}\|X\|_F\|Y\|_{(2),2} \leq \sqrt{2}\|X\|_F\|Y\|_F.$$

Hence, when having the equality sign in (1.2), $\|Y\|_{(2),2} = \|Y\|_F$ must hold, i.e., Y has no more than two non-trivial singular values, or equivalently, $\operatorname{Rk} Y \leq 2$. Some of the proofs of (1.2) even unveil that necessarily both, X and Y, need to have small rank at the same time. This gives a clue on why the following theorem taken from [13] is true.

Theorem 2.7. *Two matrices $X, Y \in \mathbb{C}^{n\times n}$ form a maximal pair if and only if there exists a unitary matrix U such that $X = U(X_0 \oplus O)U^*$ and $Y = U(Y_0 \oplus O)U^*$ with a maximal pair (X_0, Y_0) of 2×2 matrices.*

Surprisingly, maximality is more requiring than hinted before; the large-sized matrices need to be simultaneously unitarily similar to matrices in $\mathbb{C}^{2\times 2}$, padded with zeros.[24] Thus, they do not exceed rank two. Nevertheless, note that the inequality

$$\|[X,Y]\|_2 \leq \sqrt{2}\|X\|_{(2),2}\|Y\|_{(2),2}$$

suggested by this observation is *not* true. The best constant in the inequality is furthermore growing with the dimension.

Maximal pairs in the general Schatten case can be determined, too. The premium of interpolation (see Theorem 2.4) is that equality in an intermediate relation will hold only if the base relation also attains the upper bound (with appropriate matrices given by the method). In particular, ranks are kept. In addition, the monotonicity $\|\cdot\|_p \leq \|\cdot\|_q$ for $p \geq q$ is utilized several times. Having equality there, compels down the rank to one.[25] This applies for at least one of the matrices X, Y, or $[X,Y]$ in ➊, ➋, and ➌. Moreover, multiples of unitarities growingly appear (in ➍ all three must be of this type). Zero traces are preserved, and the orthogonality $X \perp Y$ from the central case 2 is tweaked via polar powers.[26] Note that interfaces between the regions and constellations involving the index ∞ are less restrictive, because they are not amenable to the utilized laws (cf. [36]).

Our little excursion depicting the major achievements in the direct surrounding of the Böttcher–Wenzel commutator inequality has now come to an end.

---

[23] As the characteristic polynomial (here, $\det(tI - Y)$ is of degree two) returns the zero matrix if applied to Y, indeed, Y^2 is linearly dependent of I and Y.

[24] In the real case, X_0 and Y_0 can be chosen real, and as for U, "unitary" should be replaced by "orthogonal."

[25] When looking at vector norms, the number of non-zero entries takes over the role of the rank, just as it already was in the norm definitions.

[26] The matrix entries $re^{i\varphi}$ become $r^P e^{i\varphi}$ with some fixed P.

Instead of detailing the history of the results with a long text, the development will be outlined with help of a timeline. Figure 1 at the end of the paper marks special cases along the way. An adjoining path concerned with a geometric problem will be discussed in the following section. Thereafter, we will focus our attention on some special interpretations of the original problem, as well as ongoing investigations for further generalizations. Some additional results are also summarized in the recent survey by our companions of the Macau research group [11].

3. Progressing into geometry

The inequality was connected to geometry from the very beginning. To be honest, what else could you expect when working with a norm that comes from a scalar product? Actually, the proof of (1.2) from [7] is already based on a strengthening of Cauchy–Schwarz' inequality under special restrictions. Several variants of the commutator estimate were found that have some geometric meaning.

As shown in [7], if three copies are assembled (one for every coordinate), (1.2) yields an inequality involving cross products over a family of usual vectors $v^{(jk)}$:

$$\sum_{i,j=1}^{n} \left\| \sum_{k=1}^{n} v^{(ik)} \times v^{(kj)} \right\|^2 \leq \sum_{i,k,\ell,j=1}^{n} \left\| v^{(ik)} \times v^{(\ell j)} \right\|^2 .$$

This connection wasn't too surprising, since the cross product and the commutator are close friends; they are *the* Lie products.

An interpretation in differential geometry is also given. Suppose a differentiable curve $g : (-\epsilon, \epsilon) \to \mathbb{C}^{n \times n}$ has all images invertible and is moreover normalized by $g(0) = I$. If $s(t) = g(t)Ag(t)^{-1}$ is the induced map into the so-called similarity orbit of some $A \in \mathbb{C}^{n \times n}$, then the derivatives are determined by commutators. In this context, (1.2) becomes

$$\|s'(0)\|_F \leq \sqrt{2}\, \|g'(0)\|_F .$$

We have seen already that the commutator inequality implies even better results. The estimate

$$\|XY - YX\|_F^2 \leq \Big\| \, \|Y\|_F X + \|X\|_F Y \, \Big\|_F \cdot \Big\| \, \|Y\|_F X - \|X\|_F Y \, \Big\|_F$$

fits directly between (1.3) and (1.2). The two matrices "$X + Y$" and "$X - Y$" on the right-hand side (with X and Y adjusted in their norms) are orthogonal. When they are seen just like ordinary vectors, they span a rhomb. Twice its area is written on the right-hand side.

Most notably, a geometric relative of (1.2) is crossing our ways. For quite a long time, most of us who were concerned with the commutator problem hadn't known of this link. No real wonder, as it is originated in higher geometry, where it reads like what follows:

Let a manifold be isometrically immersed into a space of constant curvature c.[27] *Then, at every point,*

$$\rho + \rho^{\perp} \leq \|H\|_F^2 + c = \|H\|^2 + c. \tag{3.1}$$

Here, ρ and $\rho^{\perp}$ denote the normalized scalar curvatures along the tangential and the normal bundle of the embedding, respectively. H is the mean curvature tensor.

Many of our fellows are no experts in this field, so we want to take the chance for pointing out how the connection is established. For those who are not that familiar with these notions, first some basic insight.

3.1. About "curvature(s)"

The curvature of a planar curve is quite well known: it is the signed reciprocal of the radius of the osculating circle. However, more dimensions provide much more freedom. One of the major definitions for arbitrary manifolds is the *Riemann curvature tensor*. Given by $R(u,v)w = \nabla_u\nabla_v w - \nabla_v\nabla_u w - \nabla_{[u,v]}w$, this fourth-order tensor measures non-commutativity of the second covariant derivatives. It quantifies the extent to which the submanifold is not locally isometric to a Euclidean space, i.e., how "non-flat" it is. When represented in a basis, the tensor is described by n^4 numbers R_{ijk}^l and can be reduced to the *scalar curvature* $R = \sum_{ik} g^{ik} \sum_j R_{ijk}^j$. The coefficients g^{ik} are derived from the *metric tensor*.[28] The number R is a quantitative characteristic for the deviation between the volumes of a geodesic ball in the manifold and the Euclidean unit ball.

Now, for an n-dimensional manifold, the (*normalized*) *scalar curvature*

$$\rho = \frac{2}{n(n-1)} \sum_{1=i<j}^{n} \langle R(e_i, e_j)e_j, e_i \rangle$$

(with $\{e_i\}$ being an orthonormal basis of the *tangent space*) is a plain multiple of R since, for an isometric immersion as under investigation, (g^{ik}) is just the identity matrix.[29] The scaling is simply done for better appearance.

Similarly, the *normal scalar curvature*[30] is declared by

$$\rho^{\perp} = \frac{2}{n(n-1)} \sqrt{\sum_{1=i<j}^{n} \sum_{1=r<s}^{m} \langle R^{\perp}(e_i, e_j)\xi_r, \xi_s \rangle^2}$$

via the curvature tensor $R^{\perp}$ of the *normal bundle* (instead of the tangential bundle), where $\{\xi_i\}$ is an orthonormal frame.[31] In contrast to ρ, the number $\rho^{\perp}$ cannot be negative by definition.

---

[27] Think of the Euclidean space $\mathbb{R}^d$ or the sphere S^{d-1} as typical examples for ambient spaces of constant (sectional) curvature.

[28] Actually, (g^{ik}) is the inverse of the Gram matrix (g_{ik}), which contains the coefficients of the first fundamental form.

[29] Clearly, we have $2 \sum_{1=i<j}^{n} \langle R(e_i, e_j)e_j, e_i \rangle = \sum_{i,j=1}^{n} R_{iji}^j$ as diagonal entries are zero.

[30] Pay attention to the absent "ized"!

[31] The number m is the embedding's codimension. So, the ambient space has dimension $n + m$.

The *second fundamental form* (or *shape tensor*) $h(v,w) = \langle \vec{n}, \nabla_v w \rangle$ consists of orthogonal projections of covariant derivatives on the normal bundle. The tensor $H = \frac{1}{n} \operatorname{Tr} h$ is obtained by contraction. It is remarkable that $\rho^\perp$ and H depend on the embedding, whereas ρ does not.[32] So, there are properties that are inherent to the object itself, and others that change when another point of view is taken.

We'll take a look onto illustrating examples in order to better understand the claim. Note that ρ is undefined for $n = 1$. Nevertheless, (3.1) makes sense even in that case if $n(n-1)$ is moved to the right-hand side. Ignoring $\rho^\perp$ for a moment, the relation then becomes $R \leq 0$. As a curve is locally flat, indeed $R = R^1_{111} = 0$. So, in the Riemann sense, a curve is not curved. Also note that, for a two-dimensional manifold, R turns out to be twice the Gaussian curvature. And analogously, if $R = 0$, the surface is developable, i.e., locally flat.

Returning to curves, obviously $\rho^\perp = 0$; so neither ρ nor $\rho^\perp$ is related to a curve's ordinary curvature. But, the idea of incorporating the normal space should be met before with a space curve's torsion – by looking at the complement, it is measuring how strong the plane is left. Also the behavior of the signs can be compared to their interplay.

The simplest non-trivial cases are surfaces ($n = 2$) in the usual three-dimensional space. And with this example, we encounter the crowd of curvature notations. As $m = 1$, still $\rho^\perp = 0$ and (3.1) shortens to

$$\rho \leq \|H\|_F^2 + c = H^2 + c. \tag{3.2}$$

Since $\rho^\perp \geq 0$ for all m, the latter is, in fact, a weakening. Back to our special case, with $c = 0$, we obtain

$$K \leq H^2, \tag{3.3}$$

relating the common *Gaussian* and *mean curvature* of the surface. Hence, (3.1) is a comprehensive extension of an old, established statement.

3.2. Connecting to formulas

How does the geometric inequality (3.1) become a statement similar to (1.2)? Well, as soon as they are represented in coordinates, tensors are manageable algebraic objects. Then long-winded calculations based on h, to which even R and $R^\perp$ can be related, result into the desired norm inequality for matrices.

For granting a brief look, think about the graph of a differentiable parameterized surface $(x, y, f(x,y))$. Then, h actually is a quadratic form and can be associated with the Hessian matrix $\left(\begin{smallmatrix} f_{xx} & f_{xy} \\ f_{xy} & f_{yy} \end{smallmatrix}\right)$.[33] Now, if one has an n-dimensional manifold of codimension m, there are m such matrices of size $n \times n$. Note that the matrices will be symmetric by the natural assumption of interchangeability of second partial derivatives, whence their traces (the components of H) are real.

[32] One says the quantities are extrinsic or intrinsic, respectively.

[33] Usually, it is written $h = L du^2 + 2M du dv + N dv^2 \sim \left(\begin{smallmatrix} L & M \\ M & N \end{smallmatrix}\right)$. In this explicit situation, L, M and N can be calculated with a Taylor expansion. A factor containing first derivatives does appear, too, but is equal for all three coefficients.

And how does the commutator come into play? When inspecting the definition of the curvature tensor, we can formally see commutators already. And it happens that they indeed correspond to commutators of matrices built from the tensor components. In the end, (3.1) admits the following algebraic reformulation. We are going to discuss later on in which ways this inequality closely interacts with the similarly looking (1.2).

Theorem 3.1 (Normal scalar curvature conjecture). *Let $A_1, \ldots, A_m$ be real symmetric $n \times n$ matrices. Then,*

$$2 \sum_{r<s} \|[A_r, A_s]\|_F^2 \leq \left(\sum_{r=1}^{m} \|A_r\|_F^2 \right)^2 . \tag{3.4}$$

The history of this claim can be traced back to (3.3), which is a classical result for immersed surfaces of $\mathbb{R}^3$. In this case, it trivially follows from the original definitions $K = \kappa_1\kappa_2$ and $H = \frac{1}{2}(\kappa_1 + \kappa_2)$.[34] In terms of geometry, the inequality can be understood as "intrinsic quantity is less than extrinsic quantity." In other words, what is depending only on the first fundamental form is not greater than that depending on both, the first and the second fundamental form. The statement naturally extends to the so-called space form $N^3(c)$, i.e., a 3D Riemannian manifold of constant sectional curvature c. Here, a surface obeys $K \leq H^2 + c$, where the right-hand side is adjusted by a new term mandated from the outer world. In 1970, Chern, do Carmo and Kobayashi [14] considered the special case of embeddings in the sphere.

There are two ways to generalize the inequality $K \leq H^2 + c$ to higher dimensions. Aiming at an answer to the questions, Chen was led to pointwise inequalities. About twenty years ago, it was proved in [8] and differently in [32] that (3.2) is true. In another direction, for an immersed surface ($n = 2$) in the space form $N^n(c)$, Wintgen [38] defined what he called normal curvature $K^\perp$, and proved $K \leq |H|^2 + c - K^\perp$. Note that, if the dimension of the manifold is 2, then ρ comes down to equal K, and $K^\perp$ is the same as $\rho^\perp$. The common generalization of the above two inequalities for an n-dimensional manifold isometrically immersed into the space form $N^{n+m}(c)$ was proposed (together with a proof for $n = 2$) by De Smet, Dillen, Verstraelen, and Vrancken [16] in 1999. That's why Theorem 3.1 is also known as *DDVV conjecture.*

There are complicated notions involved already in the formulation, so it's no wonder that verification took quite long. Also not surprising are some partial findings for specialized structures, often relying on very different solutions. The first complete proof of Theorem 3.1 was given by Ge and Tang [19] and independently by Lu [26]. The second also adapted the underlying idea to work with (the non-symmetric case of) Theorem 1.1, as well. An improved version is published with [27]. When Albrecht first saw it, he said that he can't assess the geometric parts, but the linear algebra (to which the problem was translated) is amazing.

[34] The principal curvatures κ_1, κ_2 are the eigenvalues of h's matrix representation.

A detailed list of the several steps towards the whole proof was assembled in [20]. That was not the end. The statement and its proof open applications in conjunction with the so-called comass in calibrated geometry. This is (just like norms) a point measure in manifolds defined via the supremum of an exterior form restricted to hyperplanes in the tangent space (cf. [15]).

4. Peeking into physics

4.1. Interpretations of norms and commutators

The world in which we live is not commutative. Though we got used to add and multiply numbers in any order in everyday life, physical processes usually cannot be interchanged. Hence, we can bet on meeting commutators all over, and that they are zero should be the exception. Naturally, one seeks for ways of estimating this quantity by simple means, as provided by Theorem 1.1, without the need for executing the possibly heavy calculations of XY and YX.

The form of (1.2) is of particular interest. It gives the operator norm analogue for the commutator, regarded as a bilinear map:

$$\sup\{\|[X,Y]\|_F : \|X\|_F = \|Y\|_F = 1\} = \sqrt{2}. \tag{4.1}$$

One more time, the suitability of the Frobenius norm for working with "vectors" X and Y can be recognized. For a linear operator, the unit ball is mapped to an ellipsoid with radius (largest half-axis) given by the norm. Here, $\sqrt{2}$ similarly marks the extent. The mapping $(X,Y) \mapsto [X,Y]$ is bounded – even stronger than one would expect.

As the constant is independent of the matrix size, it stays valid for infinite matrices, too. Consequently, (1.2) is moreover true in any separable Hilbert space (see [7]).[35] A concrete example with L^2-functions on $(-\pi,\pi)^2$ then reads

$$\int_{-\pi}^{\pi}\int_{-\pi}^{\pi}\left|\int_{-\pi}^{\pi}(f(x,t)g(-t,y) - g(x,t)f(-t,y))\,\mathrm{d}t\right|^2 \mathrm{d}x\,\mathrm{d}y$$
$$\leq 2\int_{-\pi}^{\pi}\int_{-\pi}^{\pi}|f(x,y)|^2\,\mathrm{d}x\,\mathrm{d}y\int_{-\pi}^{\pi}\int_{-\pi}^{\pi}|g(x,y)|^2\,\mathrm{d}x\,\mathrm{d}y.$$

The idea behind is to represent the functions in an orthonormal basis, so that Parseval's equality identifies the requested norm with the ℓ^2-norm of the obtained coefficients.[36] In the math lectures for students of physics, Albrecht always appreciates one of the major results from functional analysis (and presumably one of his favorites) in a rememberable form: $\boldsymbol{\ell^2 = L^2}$.

[35] In this context, one more commonly speaks of the Hilbert–Schmidt norm when meaning the direct analogue of $\|\cdot\|_F$.

[36] Decompose $f(t) = \sum_k \hat{f}_k \mathrm{e}^{\mathrm{i}kt}$, then $\int_{-\pi}^{\pi}|f(t)|^2\,\mathrm{d}t = 2\pi\sum_k|\hat{f}_k|^2$. Operators on the Fourier vector are given by infinite matrices, which will be approximated in norm by their finite sections.

The lighter you see on the image is a gift to Albrecht from the students of his 2008–2010 mathematics course for physicists. They gave it to him at the end of the four terms as a sign of thanks for his successful efforts to make them love and appreciate mathematics. The engraving is harmonized personally to him.

The hidden Fourier transform is of high (theoretical and practical) significance. In harmonic analysis, e.g., the inequality

$$\left\|tf(t)\right\|_{L^2(\mathbb{R})} \cdot \left\|\tau\hat{f}(\tau)\right\|_{L^2(\mathbb{R})} \geq \frac{\left\|f\right\|^2_{L^2(\mathbb{R})}}{4\pi}$$

is known. The consequence of this lower estimate is that a function and its transform cannot have small support simultaneously. In other words, a signal (like an electromagnetic or sound wave) won't be localized in both domains, time and frequency. This affects the properties of filters in signal processing.

Switching to physics, the last inequality can be re-written as

$$V^{1/2}(x) \cdot V^{1/2}(p) \geq \tfrac{1}{2}\hbar, \tag{4.2}$$

which relates the standard deviations of the position x and the momentum p to the reduced Planck constant $\hbar$. The quantities are indeed Fourier transforms of each other.[37] The inequality is a variant of Heisenberg's uncertainty principle (a.k.a. the principle of indeterminacy, which is more accurate). The interpretation: The more precisely the position of a particle is determined, the less precisely the momentum is known in this instant, and vice versa.[38] Unlike classical physics, quantum mechanics is centered around probabilities of the upshots of certain measurements instead of exact values. However, since the limiting $\hbar$ is pretty small in comparison to typical accuracies of measurements, observations often don't take note of the problems and appear as if they were classical.

[37]The position of a particle is described as a matter wave, and the momentum is its conjugate (due to the de Broglie relation $p = \hbar k$ with the wave number k).

[38]This shouldn't be confused with the observer effect, telling that measurements cannot be made without affecting the system. But, the uncertainty principle is inherent to all wave-like systems, and thus concerns a fundamental property of quantum objects. So, it's not the result of deficiencies of observation technology. Also, an observer's presence is not a determining factor.

In the mathematical formulation, observables of quantum mechanics are represented by self-adjoint operators. That's why the field is also called matrix mechanics. There exist a variety of inequalities that limit the precision for certain pairs of physical properties (known as complementary variables). The prerequisite is that they are non-commuting.[39] In the case of position and momentum, we indeed have the canonical commutation relation

$$[x, p] = \mathrm{i}\hbar$$

that issues the previously discussed (4.2). Summarizing, the bigness of the commutator indicates how "incompatible" the quantities are from a physical perspective. In view of (2.1), we could be tempted that most pairs are of (quite) good nature. Of course, the things we want are not necessarily in line with the average behavior.

4.2. From specialization to application

Similar to the mandates in geometry caused by applications, investigations in physics may be restricted to certain matrix classes. We just saw that symmetric or Hermitian matrices are of particular interest. As we know, (1.2) is an easy show in the normal case. Instead of all matrices in $\mathbb{R}^{n\times n}$ or $\mathbb{C}^{n\times n}$ (spaces of dimension n^2), smaller subspaces like the orthogonal group may be a topic of research.

In regard of the commutator, it is the prototype for the construction of so-called Lie algebras. Hence, the structure is of mathematical interest itself, but also appears naturally in applications. Bloch and Iserles [4] proposed, just like in (4.1), to look at the commutator's operator norm over a Lie algebra $\mathfrak{g}$.

Definition 4.1. The number

$$\omega(\mathfrak{g}) = \sup\left\{\frac{\|[X,Y]\|}{\|X\|\|Y\|} : X, Y \in \mathfrak{g} \setminus \{O\}\right\}$$

is called the radius of $\mathfrak{g}$ with respect to the chosen norm.

Here O is the zero element of the algebra. In the cases considered by us, this is the $n \times n$ zero matrix. For the Lie algebra $\mathfrak{so}(n)$ of real skew-symmetric matrices endowed with (a multiple of) $\|\cdot\|_F$, the problem was solved by them completely. This is not too surprising. After all, the elements are normal.[40] The constant turns out to be large only for $n \geq 4$. Of course, it's smaller than (1.2) would allow.

Theorem 4.2. *In regard of the Frobenius norm,*

$$\omega(\mathfrak{so}(n)) = \begin{cases} 0 & : \ n = 2, \\ 1/\sqrt{2} & : \ n = 3, \\ 1 & : \ \text{else}. \end{cases}$$

---

[39] For measurements, it is advantageous if the system is in an eigenstate of that observable. However, if the commutator of a pair is not zero, an eigenstate of an observable need not to be one of the other. Hence, we don't have a unique method for measuring both simultaneously.

[40] Actually, the proof in [4] is based on the same idea that was shortly thereafter used in [6]. Also note that 2×2 matrices give a commutative algebra, while larger sized won't.

Another big name in quantum mechanics is Pauli. He introduced the matrices

$$\sigma_x = \begin{pmatrix} 0 & 1 \\ 1 & 0 \end{pmatrix}, \quad \sigma_y = \begin{pmatrix} 0 & -\mathrm{i} \\ \mathrm{i} & 0 \end{pmatrix}, \quad \sigma_z = \begin{pmatrix} 1 & 0 \\ 0 & -1 \end{pmatrix}$$

into physics, whence they bear his name. The real span of the Pauli matrices and the identity is the vector space of all 2×2 Hermitian matrices.[41] So, the Pauli matrices represent the space of observables (complex, in two dimensions). These Hermitian, unitary, and involutory matrices are used for describing the angular momentum. Each of them corresponds to the spin in direction of the assigned coordinate axis of the Euclidean 3-space.

The Pauli matrices satisfy a very special relation: the commutator of two of them is (up to the multiple $\pm 2\mathrm{i}$) the respective third. Moreover, any such pair is maximal.[42] Just check them with Theorem 2.6, but we already embarked on a (scaled to real) example for getting $\sqrt{2}$. The formation of such a cycle with respect to the commutator operation is no coincidence. Whenever (X, Y) is a maximal pair, and $Z = [X, Y]$, then the commutators with Z deliver a multiple of the other [26], [35]:

$$\alpha X = [Z, Y^*] \quad \text{and} \quad \beta Y = [X^*, Z].$$

In the self-adjoint case, the star is superfluous. For this reason, the exponentially growing bound in iterated commutators is sharp [7]:

$$\big\|[X_1, [X_2, \ldots, [X_{m-1}, X_m]]]\big\|_F \leq 2^{(m-1)/2} \|X_1\|_F \cdots \|X_m\|_F \tag{4.3}$$

for all $X_i \in \mathbb{C}^{n \times n}$.

Pauli used the matrices in the statement of a differential equation concerning the spin under an external electromagnetic field.[43] Many effects in physics obey differential equations. And they play a prominent role in models of other disciplines, too. The most simple one is $f' = f$; its solution is given by the exponential function. Sadly, the identity $\mathrm{e}^{X+Y} = \mathrm{e}^X \mathrm{e}^Y$ known from scalars is no longer true for matrices (except X and Y commute). Instead, commutators appear – lots of them!

Think about the initial value problem $Y'(t) = A(t)Y(t)$ with $Y(t_0) = Y_0$. In dimension 1, we know that $Y(t) = \exp(\int_{t_0}^t A(\tau)\,\mathrm{d}\tau) Y_0$ is the (fundamental) solution. In the matrix case, this is only true if *all* the $A(t)$ commute. As pointed out, matrix exponentials can be tricky. The Magnus expansion (see [29]) expresses the solution via an infinite series based on iterated commutators:

$$Y(t) = \exp\big(\textstyle\sum_k K_k(t)\big) Y_0,$$

where the first term is like the exponent in the case of constant coefficients. The other summands are composed of terms of the form $K_2(t) \sim \iint [A(\tau_1), A(\tau_2)]\,\mathrm{d}\tau_2\,\mathrm{d}\tau_1$ and $K_3(t) \sim \iiint [A(\tau_1), [A(\tau_2), A(\tau_3)]]\,\mathrm{d}\tau_3\,\mathrm{d}\tau_2\,\mathrm{d}\tau_1$, and so on.

---

[41] This is also isomorphic to the quaternion algebra. In the Lie algebra setting, the transformations are often multiplied by i.

[42] That was the chance to meet small commutators. We are screwed in every interesting case.

[43] The Pauli equation is like a Schrödinger equation for spin $\frac{1}{2}$ particles.

The idea was adapted into numerical schemes for solving systems of linear differential equations with varying coefficients (e.g., [22]), for which the general solution is hard. The good point is that some physical structures are preserved. The bad news is that, due to (4.3), convergence may be restricted to a small time scale. But, this is also an improvement compared to the trivial bound 2^{m-1}.

Another fact is important: special matrices allow for a functional calculus. The classes usually met in applications are predestined for this. Exemplarily, we want briefly to mention a question raised in information theory. There, one is interested in estimating the entangling rate of certain systems.[44] This comes down to bound the trace norm of a commutator $\|[B, \log(A+B)]\|_1$ with positive A and B. The desired right-hand side is, driven by the application, no longer determined by the common norms. It will be governed by the primitive of the utilized function (i.e., $\int_0^x \log\xi \, d\xi$), evaluated at the traces of A, B and $A+B$ (see [30] for details). The concept is also interpreted as "incremental mixing" where (instead of the total) the immediate change of the entropy is regarded. A potential use is in approximating the running time in quantum computing.

5. Finding out about the origins

The commutator is not a newly discovered object. It is giving inspiration and motivation to many researchers for a very long time. Hence, in this section, we want to emphasize links to three other interests that have been picked out because of their direct adjacency.

We begin with bridging to the discussion presented in Section 3. What are the connections between Theorems 1.1 and 3.1? Obviously, they both give estimates for commutators of matrices in terms of the Frobenius norm. But, there is a much stronger link! The two assertions under investigation have a common ancestor. The 1970 article [14] is mentioned in Section 3.2 and also appears in Figure 1 as part of the history of (1.2). There, the authors investigate minimal immersions of submanifolds in the sphere with a certain geometric property (on the length of h). They use

$$\|[X,Y]\|_F^2 \le \frac{1}{2}\left(\|X\|_F^2 + \|Y\|_F^2\right)^2 \tag{5.1}$$

for bounding some terms of the second fundamental form. The statement and the proof are given for real symmetric matrices, fitting with the intended application.

We want to ease up notations a little:

$Q(n)$	:	(1.2) for complex $n \times n$ matrices
BW	:	all $Q(n)$ independent of the matrix size (Theorem 1.1)
$P(n, m)$	:	(3.4) for m real symmetric $n \times n$ matrices
DDVV	:	the whole normal scalar curvature conjecture (Theorem 3.1)

[44]Behind this is a dynamic version of the von Neumann entropy $S = -\operatorname{Tr}(\rho \ln \rho)$ with a density matrix ρ of a mixed state (i.e., an ensemble of several quantum states). The maximum is attained for a maximal mixing. The statistics was introduced to take care of situations in which particles interact strangely, so that the states cannot be described separately, but as a complete system.

$O(n)$: (5.1) for two real symmetric $n \times n$ matrices
CDK : the ensemble of the $O(n)$ (Chern–do Carmo–Kobayashi inequality)

It is immediately clear that $O(n) = P(n, 2)$, where we multiply by 2. Because of the estimate $4ab \leq (a + b)^2$, there is also a link to the other inequality: $Q(n)$ implies $O(n)$. So, the inequality (5.1) is weaker than (1.2). But, the strong one is meanwhile known to be true for all matrices. Stripped off of the associated symmetry restriction, we call the extended assertion of the weak inequality $\tilde{Q}(n)$, i.e., (5.1) is considered for all pairs of arbitrary complex $n \times n$ matrices.

The ties between $Q(n)$ and $\tilde{Q}(n)$ are even closer. At the first glance, the second is just a consequence of the first. But, also the reverse implication holds. Indeed, for normed matrices, the pretended weaker $\tilde{Q}(n)$ delivers a right-hand side of 2. Consequently, we have $\|[X/\|X\|_F, Y/\|Y\|_F]\|_F^2 \leq 2$. Linearity of the commutator and the homogeneity of the norm now yield $Q(n)$. So, the two are equivalent representations of the same result. The choice for the weaker statement (5.1) behind the name **CDK** may be disputed. After all, the authors of [14] apply the strong inequality (1.2) within the proof.[45] Nevertheless, our desire to make **CDK** a part of **DDVV** in the algebraic world is met by the mentioned equivalence.

CDK is linked to **DDVV** primarily from a geometrical point of view. It regards the prototypic sphere instead of general constantly curved spaces. For determining the demanded submanifolds, the equality cases of the commutator inequality were required. The results for symmetric matrices are in line with the infrequent general case.[46] It was also discovered that from a set of three such matrices, of which any pair satisfies (1.2) with equality, at least one must be zero [12], [19].

The connections between the various algebraic inequalities are summarized in the next diagram. **DDVV** concerns $m \geq 2$ matrices, but only symmetric ones, and **BW** tackles arbitrary matrices, but is restricted to pairs. They have gone separate ways, though being related in their roots, the joint special case **CDK**. The discussion on how to bring them back together (also indicated) will follow in the next section.

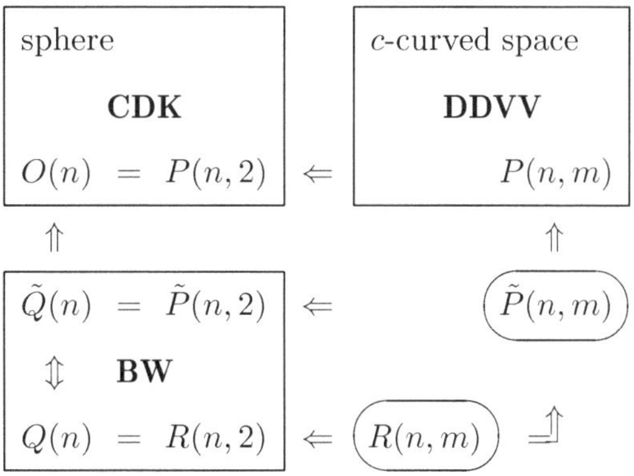

[45] The argument used is basically the same as that given in [6] to show that the claim $Q(n)$ is valid if only one matrix is normal. The diagonal eigenvalue representation is, however, fitted to the situation at hand. For the struggles in eliminating the symmetry assumption (even in the application of similar basic proof ideas) see, e.g., [26].

[46] In the symmetric situation, the non-zero pairs attaining equality are given by multiples and similarities of $X = \left(\begin{smallmatrix} 0 & 1 \\ 1 & 0 \end{smallmatrix}\right)$, $Y = \left(\begin{smallmatrix} 1 & 0 \\ 0 & -1 \end{smallmatrix}\right)$; cf. Theorem 2.6. No other choice complies with $\langle X, Y \rangle = 0$.

Parallel to the ongoing generalization process, there is an intense research on specialization. Going over to the second of the announced three fields of interest, this will be detailed on a selected example. The validity of the bound $\sqrt{2}$ for specific matrices was referred to several times throughout this paper. So, we feel urged to tell the idea of the proof, which was first found with the featured-before paper [14]: A normal X can be represented as $X = U\Lambda U^*$ with a unitary matrix and a diagonal matrix of the eigenvalues. The properties of the Frobenius norm give the quite strong estimate

$$\|XY - YX\|_F^2 = \|\Lambda(U^*YU) - (U^*YU)\Lambda\|_F^2 \leq \max_{j,k} |\lambda_j - \lambda_k|^2 \cdot \|Y\|_F^2.$$

Undoubtedly, $\max |\lambda_j - \lambda_k|^2 \leq 2\|X\|_F^2$ then proves (1.2).[47] At least with this property, it's an easy going – structure helps and made the statement available to applications very early. Moreover, if X is positive (i.e., $\lambda_i > 0$), the constant $\sqrt{2}$ can be even reduced to 1.

Like with Theorem 4.2 and the last observation, tighter estimates can be found if we restrict ourselves to certain matrix classes. This is a natural question depending on the aimed application, but the results moreover increase the understanding of the commutator. One may also modify the design of the inequalities (as noted at the end of the last section). Such an example for two positive matrices would be

$$\|XY - YX\|_F \leq \tfrac{1}{2}\|X\|_\infty \|Y \oplus Y\|_F.$$

By this, the constant in the Frobenius norm commutator inequality further diminishes to $1/\sqrt{2}$. (Comparing with the operator norm, in the same vein we get 1 if one matrix, and 1/2 if both are positive.) Amongst others, Bhatia and Kittaneh established a wealth[48] of inequalities in particular for the commutator under additional assumptions on the matrices (for different norms as well as for the singular values – before and after the publication of [6]). We here only cite their joint paper [3], from which the last inequality was taken, as an entry point into the topic.

Apart from self-adjointness or positivity an important class is given by the real matrices. Theorem 1.1 gives an inequality that is valid for all matrices. As the listed example that gave $\sqrt{2}$ consists only of real matrices, the bound cannot be improved in this case. The same is true with Theorem 2.4. This comes as a little surprise since the method of proving it is not available in the real case.[49] But, the real maximal cases (for the base inequalities) produce the same value as the complex ones. They luckily pass the complex method without fatal change, whence Theorems 2.6 and 2.7 can be restated in real terms.

[47] The similarity transformation is also the basic ingredient in most of the general proofs. The problem is that one has to work with the singular value decomposition $X = U\Sigma V$ instead, whence the second matrix Y is altered differently.

[48] That's still an understatement.

[49] Interpolation in real spaces is tricky! Riesz–Thorin works only when mapping into "high-indexed" spaces, i.e., the base indices have to be ordered by $q_i \leq p_i$. Otherwise, an additional factor 2 is necessary in the constant. The alternative Marcinkiewicz way delivers good results in the middle, but not beside the bordering indices.

Our attention now turns to an intriguing problem, the third we wanted to discuss. We know that two (large) matrices typically have a small commutator, i.e., $\|XY - YX\| \leq \delta$ in some norm. If so, they are called "almost commuting." Nevertheless, this says nothing on how close the arguments are to commutativity. Actually, one wants to find really commuting matrices in the neighborhood, i.e.,

$$\tilde{X}, \tilde{Y} : \tilde{X}\tilde{Y} = \tilde{Y}\tilde{X} \wedge \|X - \tilde{X}\| \leq \epsilon, \|Y - \tilde{Y}\| \leq \epsilon.$$

This is known in the literature as "nearly commuting."

Back in 1969, Rosenthal raised the question whether or not two almost commuting matrices also nearly commute.[50] Shortly thereafter, an affirmative answer was submitted. But, the crux is to take into account the matrix size. It was too non-uniform for passing to infinite-dimensional operators like in Section 4. Later, unitary matrices with small self-commutator were found that cannot be well approximated by normal matrices. Consequently, not all almost commuting pairs can be perturbed to an exactly commuting one. The big astonishment: for Hermitian matrices the conclusion "almost $\Rightarrow$ nearly" can be drawn (e.g., [33])! It's a big field. In general, the outcome is heavily determined by the considered norm.[51] Another result was obtained for transferring closeness to a property of commutativity, fitting well into the scheme with Theorem 2.7. If two self-adjoint matrices almost commute, they are almost jointly diagonalizable by a unitary matrix; see [21].

The observations from the paper's beginning are pretty clear: when chosen by chance, matrices appear as if they commute. But mostly, they are far away from pairs having a zero commutator. This effects numerics. It is not wise to decide about actual commutativity of X, Y by plain judgements based on the number $\|XY - YX\|$. More sophisticated stability considerations have to be done. However, whenever only a decision for XY or YX has to be made, don't mind which one you take. Practically, we may swap variables without experiencing much influence.

6. Objects in motion

Several aspects in the context exposed are not yet completely understood. This leaves back some open problems, but also enough room for extending the knowledge. We want to outline possible directions that may be taken.

The first way could lead to a reunion of the two main streams that have emerged. The statements $Q(n)$ and $P(n, m)$ are generalizations of $O(n)$, but evolved with unequal background. The former universalizes the (strengthened, but still equivalent) estimate to non-symmetric matrices, the latter respects larger families of matrices. The question for an estimate with families of arbitrary matrices[52] arises. The aim: a combined descendant that comprehends Theorems 1.1 and 3.1 as special cases. Very recently, the following problem was proposed in [28].

[50] Or equivalently, is a close-to-normal matrix (i.e., $[X, X^*]$ is small) close to a normal matrix?

[51] We have already encountered a different "uniformity-explosion switching" of C_p for Schatten and vector norms.

[52] At least, it should not be restricted to symmetric ones only.

Conjecture 6.1. *Let $A_1, \dots, A_m$ be $n \times n$ matrices subject to*

$$\operatorname{Tr}(A_\alpha A_\beta A_\gamma) = \operatorname{Tr}(A_\alpha A_\gamma A_\beta) \tag{6.1}$$

for all $1 \le \alpha, \beta, \gamma \le m$. Then,

$$2\left(\sum_{\alpha<\beta} \|[A_\alpha, A_\beta]\|_F^2\right) \le \left(\sum_{\alpha=1}^{m} \|A_\alpha\|_F^2\right)^2. \tag{6.2}$$

The condition (6.1) cannot be dropped. Simply insert the three Pauli matrices. However, this triplet is known for maximality. Then, the iterated commutator $[A_2, A_3^*]$ is basically A_1, again.[53] Hence, the two linearly dependent terms will not satisfy the orthogonality relation $A_\alpha^* \perp [A_\beta, A_\gamma]$ that (6.1) actually is. So, the example belongs to a kind of opposite, excluded case. As noted by the authors, the statement, call it $\tilde{P}(n,m)$, would be the desired unification. In the case $m = 2$ (pairs of matrices), the requirement (6.1) is automatically satisfied, and we are back in the regime of **BW**. For the case of three or more real symmetric matrices, the assumption is also not a restriction and hence **DDVV** is included, too.

The equivalence of $\tilde{Q}(n)$ and $Q(n)$ suggests to work with the strong inequality $Q(n)$, instead. This attempt yields the following claim, referred to as $R(n,m)$, which is proven in [28] to imply Conjecture 6.1, but assigns one of the matrices a special role. Whether the reverse direction holds true, or the homogeneity in the family may be reestablished, is unclear up to now.

Conjecture 6.2. *Let $A, A_2, \dots, A_m$ be $n \times n$ matrices with*

1. *$A_\alpha \perp A_\beta$ for $\alpha \ne \beta$;*
2. *$\langle A_\beta^*, [A, A_\alpha]\rangle = 0$ for $2 \le \alpha, \beta \le m$;*

Then,

$$\sum_{i=2}^{m} \|[A, A_i]\|_F^2 \le \left(\max_{i=2,\dots,m} \|A_i\|_F^2 + \sum_{i=2}^{m} \|A_i\|_F^2\right) \cdot \|A\|_F^2. \tag{6.3}$$

Similarly to the original commutator inequality, numerical tests can be done. Again, most of the visible action appears to be in the case $n = 2$, raising hopes that the new claims will turn out to be true like the initial result. Nevertheless, all the inequalities, (1.2), (6.2), and (6.3), are only worst-case estimates – often quite bad ones. In fact, giant-sized matrices very likely have a "small" commutator. The outcome of the experiments could be explained by calculating expected values and variances. For commonly assumed special cases such as X and Y drawn randomly with respect to the normal distribution one has, as shown in [6],

$$E\left(\frac{\|XY - YX\|_F^2}{\|X\|_F^2 \|Y\|_F^2}\right) = \frac{2}{n} - \frac{2}{n^3} \quad \text{and} \quad V\left(\frac{\|XY - YX\|_F^2}{\|X\|_F^2 \|Y\|_F^2}\right) \approx \frac{8}{n^4}.$$

But, why are these values so small?

In order to find a more causative reason for the behavior, investigations on the structure of the operation $XY - YX$ have been started. Though we can easily

[53] Recall the discussion in Section 4.

calculate the two numbers E and V, the more detailed distribution of the values would be more informative. Due to [31], the solution is meanwhile known for the Lie algebra of zero-trace 2×2 matrices $\mathfrak{sl}(2, \mathbb{C})$.

Theorem 6.3. *In the unit ball B of $\mathfrak{sl}(2, \mathbb{C})$, the commutator follows the law of distribution*

$$\iint_{B\times B} \phi\left(\tfrac{\|[X,Y]\|_F}{\sqrt{2}}\right) \mathrm{d}X\, \mathrm{d}Y = \int_0^1 \phi(r) f(r)\, \mathrm{d}r \quad \forall \phi \in C([0,1])$$

with the density function

$$f(r) = 9r^2 \left(\frac{\sqrt{1-r^2}}{r} + 2 \arctan \sqrt{\frac{1+r}{1-r}} - \pi \right).$$

For an illustration, take 1000 random pairs of uniformly distributed zero-trace matrices with a norm not bigger than one. Divide the possible range $[0, 1]$ into ten parts and, for each, count the pairs that have a value $\|[X, Y]\|_F/\sqrt{2}$ inside the small interval. This approximation to the density (scaled by $10/1000$ to make sure that the integral is 1) resembles the predicted distribution $f(r)$ very well.

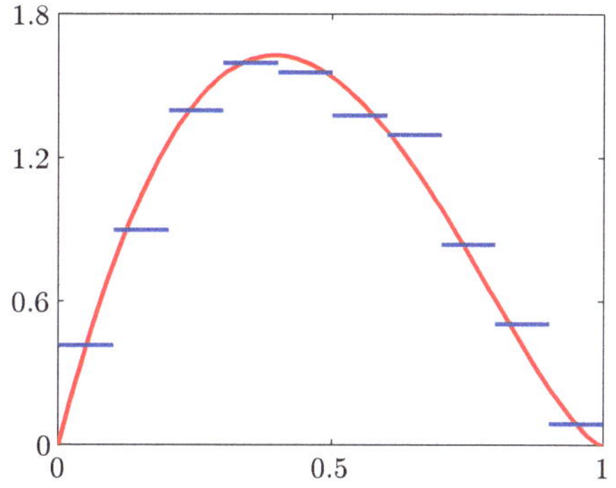

The picture demonstrates that even within the set of zero-trace matrices, maximal pairs are rare – the scalar product $\langle X, Y \rangle$ mostly is non-zero. Also note that, in contrast to the image given at the beginning, here very small values are only occasional, too. However, we looked on arbitrary matrices there.

The choice for $\mathfrak{sl}(2, \mathbb{C})$ in favor of $\mathbb{C}^{2\times 2}$ as a whole is reasonable (cf. Theorems 2.6 and 2.7). Although we know about the necessity of vanishing traces for realizing the maximal commutator value, and commutators have zero trace,[54] it is not a priori clear how $[\cdot, \cdot]$ operates on this set. One important question is whether the image of $B \times B$ is the full $\sqrt{2}B$.[55] It turns out that any matrix from the $\sqrt{2}$ sphere of $\mathfrak{sl}(2, \mathbb{C})$, indeed, can be represented as commutator of a maximal pair of normed matrices. Another point is worth mentioning. Two matrices taken from $\mathfrak{sl}(2, \mathbb{C})$ always satisfy inequality (1.3) as an equality. That's a consequence of the isometry between $\mathfrak{sl}(2, \mathbb{C})$ and $\mathbb{C}^3$ that directly links $x \times y$ and $[X, Y]/\sqrt{2}$.[56] For $n = 2$, the commutator behaves exactly like the cross-product (apart from a scaling by $\sqrt{2}$, which can be hidden in the norm as suggested in [4]).

[54] Actually, a matrix is the commutator of two matrices if and only if its trace is zero (e.g., [1]).
[55] This allows Theorem 6.3 to transfer the distributions of the arguments to one on the product space and identify it with B again. Pleasantly, the push-forward measure with respect to the mapping $(X, Y) \mapsto [X, Y]/\sqrt{2}$ is radial, whence a further reduction to the interval $[0, 1]$ is possible.
[56] In $\mathbb{R}^3$, the identity $\|x \times y\|^2 = \|x\|^2\|y\|^2 - (x, y)^2$ is well known. In $\mathfrak{sl}(2, \mathbb{C})$, a 2 appears on the right-hand side. The identification $\left(\begin{smallmatrix} a & b \\ c & -a \end{smallmatrix}\right) \sim (\sqrt{2}a, b, c)^T$ is also the key to the proof in [31].

In spaces where traces do not vanish, commutator norms are reduced even more. The average is, except for small matrices, seemingly far away from the maximum value. However, it is known that almost no pair of matrices delivers the precise value zero. For getting an impression of the image of the commutator map, the attempt in [37] is to give a systematic overview via representative snapshots. The choice was made to look at those rank r matrices with singular value vectors $(s,\dots,s,0,\dots,0)$.[57] In a first step, the commutator's possible singular value vectors for pairs of such a strict rank r matrix and a rank one matrix were determined. The only surviving, two largest singular values fill the triangle

$$\operatorname{conv}\left\{(0,0),\left(\tfrac{2}{\sqrt{r}},0\right),\left(\tfrac{1}{\sqrt{r}},\tfrac{2}{\sqrt{r}}\right)\right\},$$

except for $r=1$ where it's even less. Anyway, the results are decreasing in norm with increasing matrix size. So, to speak in the analogy (4.1) of the operator norm again: the commutator maps two unit balls to an ellipsoid, of which $\sqrt{2}$ is the maximal radius. However, it produces a very eccentric body, so that the maximal bound is substantially undercut almost every time. The majority of the half-axes is pretty short, but only a few are effectively vanishing. That explains the friction between almost and nearly commuting pairs.

Right before the finish line, we shouldn't forget to tell about the efforts that go beyond $XY-YX$ for investigating the phenomena of (non-)commutativity. The commutator itself can be generalized to higher orders. Instead of averaging pairwise products like in Conjecture 6.1, a measure for the discrepancies can be observed by multi-products. The inclusion of more factors is done similarly to the determinant by summing over all permutations:

$$\mathcal{S}_r(X_1,\dots,X_r)=\sum(\operatorname{sgn}\pi)X_{\pi(1)}\cdots X_{\pi(r)}.$$

Serre [31] has shown that, analogously to (2.1),

$$E\left(\|\mathcal{S}_r(X_1,\dots,X_r)\|_F^2\right)=O\left(\tfrac{r!}{n^{r-1}}\right)E\left(\|X_1\|_F^2\right)^r$$

for independent Gaussian variables X_i. Note that by the Amitsur–Levitski theorem, $\mathcal{S}_r$ furthermore is *identically* zero over $\mathbb{C}^{n\times n}$ for all $r\geq 2n$ (but not for smaller ones). Consequently, we have a more detailed scale than just the commutator $\mathcal{S}_2$.

In another direction, recall that $[X,Y]$ is the product in Lie algebras. One can define other products. For $XY-YX^T$, almost the same Schatten (mix) norm estimates that we got in Sections 2.2 and 2.3 can be shown (see [9]). Although the anti-commutator $XY+YX$, as another such example worth a study, is completely different (due to being non-trivial for scalars), it is also important in the theory of Lie algebras and the geometry of orbits.

---

[57] These are multiples of truncated isometries. They are extreme points for matrices with general $(s_1,\dots,s_r,0,\dots,0)$. The idea is to work on the gauge function's domain, where numerous symmetries apply.

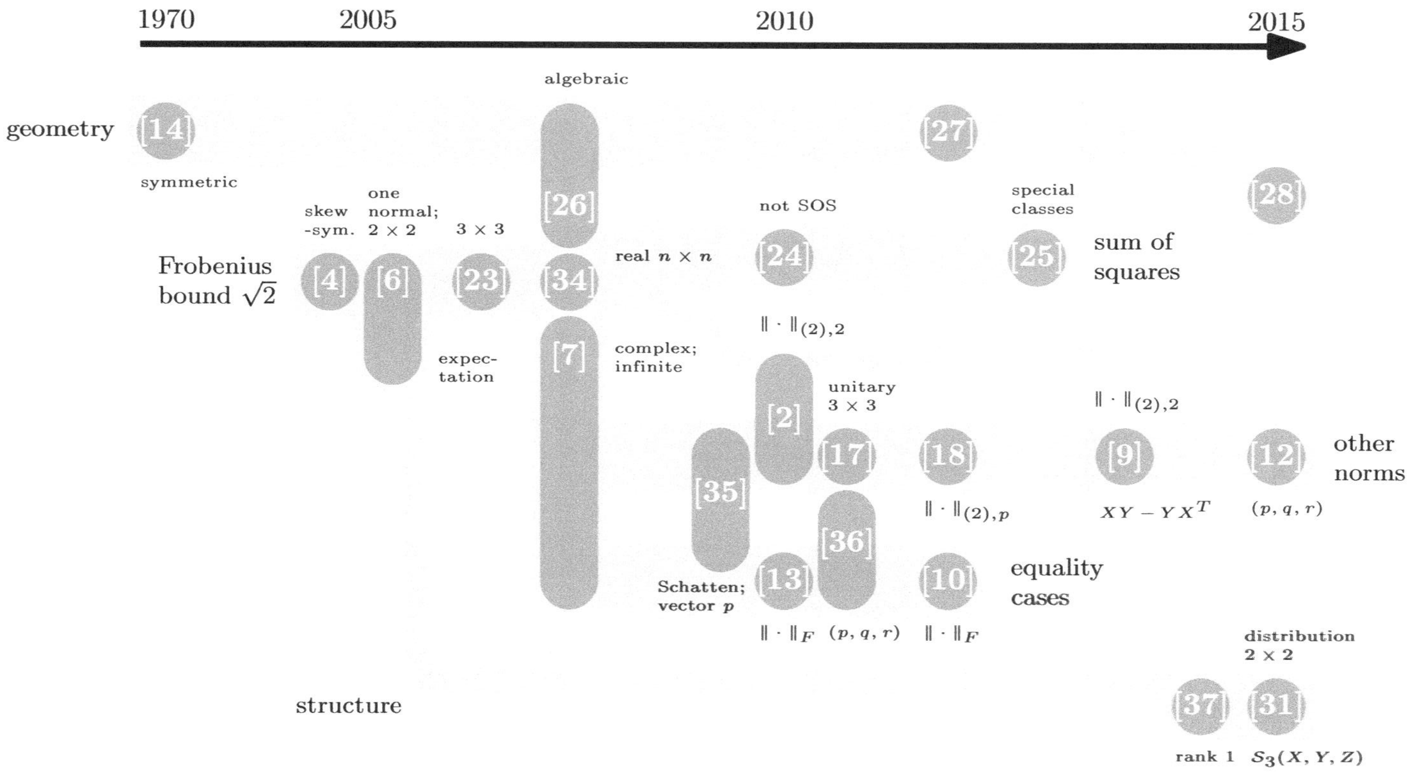

FIGURE 1. Development around the Böttcher–Wenzel commutator inequality

References

[1] A. Albert, B. Muckenhoupt. On matrices of trace zero. *Michigan Math. J.*, 4:1–3, 1957.

[2] K. Audenaert. Variance bounds, with an application to norm bounds for commutators. *Linear Algebra Appl.*, 432:1126–1143, 2010.

[3] R. Bhatia, F. Kittaneh. Commutators, pinchings and spectral variation. *Oper. Matrices*, 2:143–151, 2008.

[4] A.M. Bloch, A. Iserles. Commutators of skew-symmetric matrices. *Internat. J. Bifur. Chaos Appl. Sci. Engrg.*, 15:793–801, 2005.

[5] A. Böttcher, S. Grudsky. The norm of the product of a large matrix and a random vector. *Electr. J. Probability*, 8:1–29, 2003.

[6] A. Böttcher, D. Wenzel. How big can the commutator of two matrices be and how big is it typically? *Linear Algebra Appl.*, 403:216–228, 2005.

[7] A. Böttcher, D. Wenzel. The Frobenius norm and the commutator. *Linear Algebra Appl.*, 429:1864–1885, 2008.

[8] B. Chen. Mean curvature and shape operator of isometric immersions in real-space-forms. *Glasgow Math. J.*, 38:87–97, 1996.

[9] C. Cheng, K. Fong, W. Lei. On some inequalities involving the commutator and $XY - YX^T$. *Linear Algebra Appl.*, 438:2793–2807, 2013.

[10] C. Cheng, K. Fong, I. Lok. Another proof for commutators with maximal Frobenius norm. *Recent Adv. Sc. Comp. Matrix Anal.*, 9–14, 2011.

[11] C. Cheng, X. Jin, S. Vong. A survey on the Böttcher–Wenzel conjecture and related problems. *Oper. Matrices*, 3:659–673, 2015.

[12] C. Cheng, C. Lei. On Schatten p norms of commutators. *Linear Algebra Appl.*, 484:409–434, 2015.

[13] C. Cheng, S. Vong, D. Wenzel. Commutators with maximal Frobenius norm. *Linear Algebra Appl.*, 432:292–306, 2010.

[14] S. Chern, M. do Carmo, S. Kobayashi. Minimal submanifolds of a sphere with second fundamental form of constant length. *Functional Analysis and Related Fields*, 59–75, 1970.

[15] T. Choi, Z. Lu. On the DDVV conjecture and the comass in calibrated geometry (I). *Math. Z.*, 260:409–429, 2008.

[16] P. De Smet, F. Dillen, L. Verstraelen, L. Vrancken. A pointwise inequality in submanifold theory. *Arch. Math. (Brno)*, 35:115–128, 1999.

[17] K. Fong, C. Cheng, I. Lok. Another unitarily invariant norm attaining the minimum norm bound for commutators. *Linear Algebra Appl.*, 433:1793–1797, 2010.

[18] K. Fong, I. Lok, C. Cheng. A note on the norm of commutator and the norm of $XY - YX^T$. *Linear Algebra Appl.*, 435:1193–1201, 2011.

[19] J. Ge, Z. Tang. A proof of the DDVV conjecture and its equality case. *Pacific J. Math.*, 237:87–95, 2008.

[20] J. Ge, Z. Tang. A survey on the DDVV conjecture. arXiv:1006.5326v1, 2010.

[21] K. Glashoff, M. Bronstein. Almost-commuting matrices are almost jointly diagonalizable. arXiv:1305.2135v2, 2013.

[22] A. Iserles. Solving linear ordinary differential equations by exponentials of iterated commutators. *Numer. Math.*, 45:183–199, 1984.

[23] L. László. Proof of Böttcher and Wenzel's conjecture on commutator norms for 3-by-3 matrices. *Linear Algebra Appl.*, 422:659–663, 2007.

[24] L. László. On sum of squares decomposition for a biquadratic matrix function. *Annales Univ. Sci. Budapest, Sect. Comp.*, 33:273–284, 2010.

[25] L. László. Sum of squares representation for the Böttcher–Wenzel biquadratic form. *Acta Univ. Sapientiae, Informatica*, 4:17–32, 2012.

[26] Z. Lu. Proof of the normal scalar curvature conjecture. arXiv:0711.3510v1, 2007.

[27] Z. Lu. Normal Scalar Curvature Conjecture and its applications. *J. Funct. Analysis*, 261:1284–1308, 2011.

[28] Z. Lu, D. Wenzel. The normal Ricci curvature inequality. *Recent Advances in Submanifold Geometry, A Proceedings Volume Dedicated to the Memory of Franki Dillen* (1963–2013), *Contemp. Math.*, 674:99–110, 2016.

[29] W. Magnus. On the exponential solution of differential equations for a linear operator. *Commun. Pure Appl. Math.*, 7:649–673, 1954.

[30] M. Mariën, K. Audenaert, K. Van Acoleyen, F. Verstraete. Entanglement rates and the stability of the area law for the entanglement entropy. arXiv:1411.0680, 2014.

[31] D. Serre. Non-commutative standard polynomials applied to matrices. *Linear Algebra Appl.*, 490:202–223, 2016.

[32] B. Suceavă. Some remarks on B.Y. Chen's inequality involving classical invariants. *An. Ştiinţ. Univ. Al. I. Cuza Iaşi. Mat. (N.S.)*, 45:405–412, 1999.

[33] S. Szarek On almost commuting Hermitian operators. *Rocky Mountain J. Math.*, 20:581–589, 1990.

[34] S. Vong, X. Jin. Proof of Böttcher and Wenzel's conjecture. *Oper. Matrices*, 2:435–442, 2008.

[35] D. Wenzel. Dominating the commutator. *Oper. Theory: Adv. and Appl.*, 202:579–600, 2010.

[36] D. Wenzel, K. Audenaert. Impressions of convexity – An illustration for commutator bounds. *Linear Algebra Appl.*, 433:1726–1759, 2010.

[37] D. Wenzel. A strange phenomenon for the singular values of commutators with rank one matrices. *Electr. J. Linear Algebra*, 30:605–625, 2015.

[38] P. Wintgen. Sur l'inégalité de Chen–Willmore. *C. R. Acad. Sci. Paris Sér. A-B*, 288:A993–A995, 1979.

Zhiqin Lu
Department of Mathematics
East China Normal University
3663 N. Zhongshan Road
Shanghai 200062, China
and
Department of Mathematics
University of California, Irvine
Irvine, CA 92697, USA
e-mail: `zlu@uci.edu`

David Wenzel
Fakultät für Mathematik
Technische Universität Chemnitz
D-09107 Chemnitz, Germany
e-mail: `david.wenzel.mathematik@t-online.de`

Operator Theory:
Advances and Applications, Vol. 259, 561–591

Numerical Ranges of 4-by-4 Nilpotent Matrices: Flat Portions on the Boundary

Erin Militzer, Linda J. Patton, Ilya M. Spitkovsky and Ming-Cheng Tsai

To Albrecht Böttcher on his 60th birthday, with respect and admiration

Abstract. In their 2008 paper Gau and Wu conjectured that the numerical range of a 4-by-4 nilpotent matrix has at most two flat portions on its boundary. We prove this conjecture, establishing along the way some additional facts of independent interest. In particular, a full description of the case in which these two portions indeed materialize and are parallel to each other is included.

Mathematics Subject Classification (2010). Primary 15A60; Secondary 47A12.

Keywords. Numerical range, nilpotent matrix.

1. Introduction

We consider the space $\mathbb{C}^n$ endowed with the standard scalar product $\langle .,. \rangle$ and the norm $\|.\|$ associated with it. Elements $x \in \mathbb{C}^n$ are n-columns; however, to simplify the notation we will write them as $(x_1, \ldots, x_n)$. For an n-by-n matrix A, its *numerical range*, also known as the *field of values*, is defined as

$$F(A) = \{\langle Ax, x \rangle : x \in \mathbb{C}^n, \ \|x\| = 1\}.$$

The classical Toeplitz–Hausdorff theorem claims that the set $F(A)$ is always convex; this and other well-known properties of the numerical range are discussed in detail, e.g., in monographs [6, 7].

The first three authors (EM, LJP, and IMS) started their work on this paper during the REUF workshop at the American Institute of Mathematics (Palo Alto, California) in the Summer of 2011, and their further collaboration with MCT was enabled by IMS visiting National Sun Yat-sen University (Kaohsiung, Taiwan) in December of 2014. We gratefully acknowledge the support of the respective institutions. IMS was also supported in part by the Plumeri Award for Faculty Excellence from the College of William and Mary and by Faculty Research funding from the Division of Science and Mathematics, New York University Abu Dhabi.

As was observed by Kippenhahn ([9], see also the English translation [10]), $F(A)$ can be described in terms of the homogeneous polynomial

$$p_A(u,v,w) = \det(uH + vK + wI), \tag{1.1}$$

where

$$H = (A + A^*)/\,2 := \operatorname{Re} A, \quad K = (A - A^*)/\,2i := \operatorname{Im} A.$$

Namely, $F(A)$ is the convex hull of the curve $C(A)$ dual (in projective coordinates) to $p_A(u,v,w) = 0$. So, it is not surprising that, starting with $n = 3$, the boundary $\partial F(A)$ of $F(A)$ may contain line segments (sometimes also called *flat portions*), even when the polynomial p_A is irreducible. For a unitarily irreducible 3-by-3 matrix there is at most one such flat portion, as was first observed in the same paper [9], with constructive tests for its presence provided in [8, 13].

The phenomenon of flat portions in higher dimensions was further studied in [3]. For convenience of reference, we restate here Theorem 37 from [3].

Theorem 1.1 (Brown–Spitkovsky). *Any 4-by-4 matrix has at most 4 flat portions on the boundary of its numerical range (3, if it is unitarily irreducible).*

Of course, the upper bounds may be lower if an additional structure is imposed on A. In this paper, we will tackle the case of nilpotent matrices. For reducible 4-by-4 nilpotent matrices it is easy to see that the maximum possible number of flat portions is one; for the sake of completeness, this result is stated with a proof in Section 2 (Proposition 2.1). Unitarily irreducible 4-by-4 nilpotent matrices were considered by Gau and Wu in [4], where in particular examples of such matrices A with two flat portions on $\partial F(A)$ were given and it was also conjectured that three flat portions do not materialize. This conjecture was supported there by the following theorem, which is a special case of their result for n-by-n matrices [4, Theorem 3.4].

Theorem 1.2 (Gau–Wu). *If A is a 4-by-4 nilpotent matrix which has a 3-by-3 submatrix B with $W(B)$ a circular disk centered at the origin, then there are at most two flat portions on the boundary of $F(A)$.*

In this paper we prove the Gau–Wu conjecture. This is done in Section 5. As a natural preliminary step, necessary and sufficient conditions for a 4-by-4 nilpotent matrix to have at least one flat portion on the boundary of its numerical range are derived in Section 2. A special family of nilpotent matrices that is important for the proof of the main theorem is analyzed in Section 3. Section 4 contains necessary and sufficient conditions for a nilpotent matrix to have two parallel flat portions on the boundary of its numerical range. In addition, we show there that for a nilpotent 4-by-4 matrix A with two non-parallel flat portions on the boundary of $F(A)$ that are on lines equidistant from the origin, these are the only flat portions. The latter result is also used in Section 5, where in Theorem 5.5 it is shown that if A is a 4-by-4 nilpotent matrix, then $\partial F(A)$ contains at most 2 flat portions. The proof follows from an analysis of the locations of the singularities of the boundary generating curve (1.1). In the final Section 6 we use Theorem 5.5 to tackle the case of 5-by-5 unitarily reducible matrices.

2. Matrices with a flat portion on the boundary of their numerical range

If A is any non-zero nilpotent matrix, then the origin lies in the interior of the numerical range of A [5, Corollary 1.2]. Hence the boundary of $F(A)$ is smooth because any singularity on $\partial F(A)$ has to occur at an eigenvalue of A. So, the flat portions on the boundary, if any, are tangent to the adjacent non-flat portions of $\partial F(A)$.

We start estimating the number of flat portions with an easy case where the corresponding matrix is unitarily reducible.

Proposition 2.1. *Let A be any 4-by-4 unitarily reducible nilpotent matrix. Then its numerical range $F(A)$ has at most one flat portion on the boundary.*

Proof. Let A be unitarily similar to a direct sum $B_1 \oplus \cdots \oplus B_k$, with $k > 1$. The blocks B_j are of course also nilpotent, and the following cases are possible.

Case 1. $k = 2$. If B_1 is a 3-by-3 unitarily irreducible nilpotent matrix and $B_2 = [0]$, then $F(A) = F(B_1)$. According to [8, Theorem 4.1], $F(B_1)$ either has no flat portions on the boundary or exactly one such portion. Thus, so does $F(A)$. If there are two nilpotent 2-by-2 blocks, then the numerical range of each block is a circular disk centered at the origin, and $F(A)$ is the largest of these disks and hence has no flat portion on its boundary.

Case 2. $k = 3$, that is, B_1 is a 2-by-2 unitarily irreducible nilpotent matrix, while $B_2 = B_3 = [0]$. Then $F(A) = F(B_1)$ is again a circular disk centered at the origin, and there are no flat portions on its boundary.

Case 3. $k = 4$, implying that $A = 0$, and $F(A) = \{0\}$. Hence there are no flat portions. □

If A is not supposed to be unitarily reducible, the situation becomes more complicated. Let us establish the criterion for at least one flat portion to exist on $\partial F(A)$. To this end, some background terminology and information is useful.

First, recall the notion of an *exceptional supporting line* of $F(A)$ which for an arbitrary matrix A was introduced in [11]. Namely, let ℓ_θ be the supporting line of $F(A)$ having slope $-\cot\theta$ and such that $e^{-i\theta}F(A)$ lies to the right of the vertical line $e^{-i\theta}\ell_\theta$. Then this supporting line is exceptional (and, respectively, θ is an *exceptional angle*) if at least one $z \in \ell_\theta \cap F(A)$ is *multiply generated*, that is, there exist at least two linearly independent unit vectors x_j for which $\langle Ax_j, x_j\rangle = z$. For a given A, the angle θ is exceptional if and only if the Hermitian matrix $\mathrm{Re}(e^{-i\theta}A)$ has a multiple minimal eigenvalue [11, Theorem 2.1]; denote by $\mathcal{L}$ the respective eigenspace. The above-mentioned value z is unique if and only if the compression of $\mathrm{Im}(e^{-i\theta}A)$ (equivalently: A) onto $\mathcal{L}$ is a scalar multiple of the identity; z is then called a *multiply generated round boundary point* of $F(A)$.

On the other hand, all points in the relative interior of a flat portion on the boundary of $F(A)$ are multiply generated.We thus arrive at the following statement:

Flat portions on the boundary of $F(A)$ occur only on exceptional supporting lines, and for them to materialize it is necessary and sufficient that the compression $A|\mathcal{L}$ of A onto the eigenspace $\mathcal{L}$ (corresponding to a multiple minimal eigenvalue) is not *a scalar multiple of the identity.*

This fact, which is of course well known, will be used in several equivalent forms in what follows.

In our setting we will have to deal with 2-dimensional $\mathcal{L}$. The following test is useful in this regard.

Proposition 2.2. *Let A be such that for some $\theta \in [0, 2\pi)$ there exist two linearly independent vectors y_1, y_2 corresponding to the same eigenvalue λ of* $\mathrm{Re}(e^{-i\theta}A)$, *and let $\mathcal{L} = \mathrm{Span}\{y_1, y_2\}$. Then the compression of A onto $\mathcal{L}$ is a scalar multiple of the identity if and only if*

$$\langle Ay_1, y_1\rangle \|y_2\|^2 = \langle Ay_2, y_2\rangle \|y_1\|^2 \tag{2.1}$$

and

$$\langle Ay_2, y_1\rangle \|y_1\|^2 = \langle y_2, y_1\rangle\langle Ay_1, y_1\rangle. \tag{2.2}$$

Proof. Without loss of generality we may normalize the vectors y_1, y_2, dividing each of them by its length, and thus rewrite (2.1), (2.2) in a slightly simpler form

$$\langle Ay_1, y_1\rangle = \langle Ay_2, y_2\rangle, \tag{2.3}$$
$$\langle Ay_2, y_1\rangle = \langle y_2, y_1\rangle\langle Ay_1, y_1\rangle. \tag{2.4}$$

Observe also that (2.4) means exactly that

$$\langle A\widetilde{y}_2, y_1\rangle = 0, \tag{2.5}$$

where $\widetilde{y}_2$ is a unit vector in $\mathcal{L}$ orthogonal to y_1. So, we just need to show that $A|\mathcal{L}$ is a scalar multiple of the identity if and only if (2.3) and (2.5) hold.

The *necessity* of (2.3), (2.5) is trivial, and even holds for an arbitrary subspace $\mathcal{L}$, not consisting of eigenvectors of $\mathrm{Re}(e^{-i\theta}A)$. As for their *sufficiency*, note that $\mathrm{Re}(e^{-i\theta}A)|\mathcal{L}$, being a scalar multiple of the identity, commutes with $\mathrm{Im}(e^{-i\theta}A)|\mathcal{L}$. Thus, $A|\mathcal{L}$ is normal. As such, condition (2.5) implies that the matrix of $A|\mathcal{L}$ with respect to the orthonormal basis $\{y_1, \widetilde{y}_2\}$ is diagonal. Consequently, y_1 is an eigenvector of $A|\mathcal{L}$ corresponding to its eigenvalue $\mu = \langle Ay_1, y_1\rangle$, and the latter is an endpoint of $F(A|\mathcal{L})$.

On the other hand, (2.3) shows that this value is attained at two linearly independent unit vectors, y_1 and y_2. This is only possible if $A|\mathcal{L} = \mu I$. □

We return now to the nilpotent matrix setting. Let us first establish the criterion for an exceptional supporting line to exist, independent of whether or not it contains a proper flat portion.

Theorem 2.3. *Let A be a 4-by-4 nilpotent matrix. Then $F(A)$ has an exceptional supporting line if and only if A is unitarily similar to*

$$\alpha \begin{bmatrix} 0 & a_1 & a_2 & a_3 \\ & 0 & a_4 & a_5 \\ & & 0 & a_6 \\ & & & 0 \end{bmatrix}, \tag{2.6}$$

where

$$\alpha \in \mathbb{C}, \quad |a_j| \leq 1 \text{ for } j = 1, 2, 3, \tag{2.7}$$

$$|a_4 - \overline{a_1} a_2|^2 = (1 - |a_1|^2)(1 - |a_2|^2), \tag{2.8}$$

$$|a_5 - \overline{a_1} a_3|^2 = (1 - |a_1|^2)(1 - |a_3|^2), \tag{2.9}$$

$$|a_6 - \overline{a_2} a_3|^2 = (1 - |a_2|^2)(1 - |a_3|^2), \tag{2.10}$$

and

$$\arg(a_6 - \overline{a_2} a_3) = \arg(a_5 - \overline{a_1} a_3) - \arg(a_4 - \overline{a_1} a_2) \mod 2\pi. \tag{2.11}$$

Note that all three arguments in (2.11) are defined only if the inequalities in (2.7) are strict. If this is not the case, we agree by convention that condition (2.11) is vacuous.

Proof. The result obviously holds for $A = 0$. Indeed, then A is in the form (2.6), and every supporting line of $F(A) = \{0\}$ is exceptional. So, in what follows we will suppose that $A \neq 0$.

Necessity. Suppose $A \neq 0$ is nilpotent, and (at least) one of the supporting lines of $F(A)$ is exceptional. Multiplying A by a unimodular scalar if needed, we may without loss of generality suppose in addition that the exceptional supporting line is vertical. Let d be its distance from the origin. Then $\operatorname{Re} A + dI$ is a positive semidefinite matrix with rank at most 2.

Note that d is strictly positive because, as stated above, zero lies in the interior of $F(A)$. Multiplying A by another scalar, this time positive, we may without loss of generality suppose that $d = 1/2$, that is, $A + A^* + I$ is positive semidefinite of rank at most 2. We will show that for such matrices the statement holds with $\alpha = 1$.

To this end, use unitary similarity to put A in upper triangular form (2.6) with $\alpha = 1$, and observe that then

$$A + A^* + I = \begin{bmatrix} 1 & a_1 & a_2 & a_3 \\ \overline{a_1} & 1 & a_4 & a_5 \\ \overline{a_2} & \overline{a_4} & 1 & a_6 \\ \overline{a_3} & \overline{a_5} & \overline{a_6} & 1 \end{bmatrix}. \tag{2.12}$$

Let

$$X = \begin{bmatrix} 1 & -a_1 & -a_2 & -a_3 \\ 0 & 1 & 0 & 0 \\ 0 & 0 & 1 & 0 \\ 0 & 0 & 0 & 1 \end{bmatrix}.$$

From (2.12) we then see that $A + A^* + I$ is congruent to

$$X^* \begin{bmatrix} 1 & a_1 & a_2 & a_3 \\ \overline{a_1} & 1 & a_4 & a_5 \\ \overline{a_2} & \overline{a_4} & 1 & a_6 \\ \overline{a_3} & \overline{a_5} & \overline{a_6} & 1 \end{bmatrix} X = [1] \oplus G,$$

where

$$G = \begin{bmatrix} 1 - |a_1|^2 & a_4 - \overline{a_1}a_2 & a_5 - \overline{a_1}a_3 \\ \overline{a_4} - a_1\overline{a_2} & 1 - |a_2|^2 & a_6 - \overline{a_2}a_3 \\ \overline{a_5} - a_1\overline{a_3} & \overline{a_6} - a_2\overline{a_3} & 1 - |a_3|^2 \end{bmatrix}. \tag{2.13}$$

So, G must be positive semidefinite of rank at most 1. The former property implies the inequalities in (2.7), while due to the latter the three 2-by-2 principal minors of G are equal to zero. This is equivalent to (2.8)–(2.10).

In its turn, if $|a_1| < 1$, then choosing

$$Y = \begin{bmatrix} 1 & -\frac{a_4 - \overline{a_1}a_2}{1-|a_1|^2} & -\frac{a_5 - \overline{a_1}a_3}{1-|a_1|^2} \\ 0 & 1 & 0 \\ 0 & 0 & 1 \end{bmatrix}$$

we observe that G is congruent to

$$Y^*GY = [1 - |a_1|^2] \oplus \begin{bmatrix} u & w \\ \overline{w} & v \end{bmatrix}, \tag{2.14}$$

where

$$u = 1 - |a_2|^2 - \frac{|a_4 - \overline{a_1}a_2|^2}{1 - |a_1|^2}, \quad v = 1 - |a_3|^2 - \frac{|a_5 - \overline{a_1}a_3|^2}{1 - |a_1|^2},$$

and

$$w = a_6 - \overline{a_2}a_3 - \frac{(a_5 - \overline{a_1}a_3)(\overline{a_4} - a_1\overline{a_2})}{1 - |a_1|^2}.$$

Due to (2.8), (2.9) we have $u = v = 0$, and therefore

$$\operatorname{rank} G = \begin{cases} 1 & \text{if } w = 0, \\ 3 & \text{otherwise,} \end{cases}$$

which implies (2.11).

Sufficiency. Without loss of generality, let A be given by (2.6) with $\alpha = 1$. Then (2.12) and (2.13) hold.

If $|a_1| = 1$, then (2.8), (2.9) imply

$$G = [0] \oplus \begin{bmatrix} 1 - |a_2|^2 & a_6 - \overline{a_2}a_3 \\ \overline{a_6} - a_2\overline{a_3} & 1 - |a_3|^2 \end{bmatrix}, \tag{2.15}$$

and the second summand in (2.15) is singular due to (2.10). So, the minimal eigenvalue $-1/2$ of $\operatorname{Re} A$ has multiplicity 2. For the case $|a_1| < 1$, the same conclusion follows from the congruence of G and (2.14), since (2.8)–(2.11) imply $u = v = w = 0$. □

Note that the exceptional supporting line ℓ, the existence of which is established by Theorem 2.3, is the vertical line $x = -1/2$ scaled by α. Consequently, ℓ is at the distance $|\alpha|/2$ from the origin, and has the slope $-\cot\arg\alpha$.

Also, conditions (2.8)–(2.10) can be rewritten in an equivalent form

$$a_4 = \overline{a_1}a_2 + r_1r_2e^{i\theta_1}, \quad a_5 = \overline{a_1}a_3 + r_1r_3e^{i\theta_2}, \quad a_6 = \overline{a_2}a_3 + r_2r_3e^{i\theta_3},$$

where

$$r_j = \sqrt{1-|a_j|^2}, \quad j = 1,2,3,$$

and

$$\theta_1 = \arg(a_4 - \overline{a_1}a_2), \quad \theta_2 = \arg(a_5 - \overline{a_1}a_3), \quad \theta_3 = \arg(a_6 - \overline{a_2}a_3). \tag{2.16}$$

Consequently, the matrix (2.6) that satisfies the conditions in Theorem 2.3 can be represented more explicitly as

$$\alpha\begin{bmatrix} 0 & a_1 & a_2 & a_3 \\ & 0 & \overline{a_1}a_2 + r_1r_2e^{i\theta_1} & \overline{a_1}a_3 + r_1r_3e^{i\theta_2} \\ & & 0 & \overline{a_2}a_3 + r_2r_3e^{i\theta_3} \\ & & & 0 \end{bmatrix}, \tag{2.17}$$

where $\theta_3 = \theta_2 - \theta_1$. We will now use Proposition 2.2 to establish the additional conditions on A under which a flat portion of $F(A)$ on ℓ actually materializes.

Theorem 2.4. *Let A be unitarily similar to* (2.17). *In the notation introduced above, $F(A)\cap\ell$ is a proper line segment unless one of the following four conditions holds.*

(i) $r_1r_2r_3 \neq 0$ *and* $\tau_1 = \tau_2 = 0$, *where*

$$\tau_1 = r_3(r_1^2 + r_2^2 - r_1^2r_2^2)\operatorname{Im}\left(\overline{a}_1a_3e^{-i\theta_2}\right) + r_2(r_1^2 + r_3^2 - r_1^2r_3^2)\operatorname{Im}\left(a_1\overline{a}_2e^{i\theta_1}\right) + r_1r_2r_3|a_1|^2\operatorname{Im}\left(a_2\overline{a}_3e^{i(\theta_2-\theta_1)}\right), \tag{2.18}$$

and

$$\begin{aligned}\tau_2 = {}& \overline{a}_2a_3r_1(r_1^2 + 2r_2^2 - 2r_1^2r_2^2) - a_1\overline{a}_2^2a_3r_1^2r_2e^{i\theta_1} \\ & + \overline{a}_1a_3r_2(-r_2^2 - r_1^2 + r_1^2r_2^2)e^{-i\theta_1} + a_1\overline{a}_2r_1^2r_3|a_2|^2e^{i\theta_2} \\ & + r_1r_2r_3(1-2|a_1|^2|a_2|^2)e^{i(\theta_2-\theta_1)} + \overline{a}_1a_2|a_1|^2r_2^2r_3e^{i(\theta_2-2\theta_1)}.\end{aligned} \tag{2.19}$$

(ii) $r_1 = 0$, $r_2 = r_3 \neq 0$, *and* $\arg(a_3) = \arg(a_2) + \theta_3$.

(iii) $r_2 = 0$, $r_1 = r_3 \neq 0$, *and* $\arg(a_3) = \arg(a_1) + \theta_2 + \pi$.

(iv) $r_3 = 0$, $r_1 = r_2 \neq 0$, *and* $\arg(a_2) = \arg(a_1) + \theta_1$.

Proof. Without loss of generality we may suppose that A is in the form (2.17), not just unitarily similar to it, and that $\alpha = 1$. Then

$$\operatorname{Re}A = \frac{1}{2}\begin{bmatrix} 0 & a_1 & a_2 & a_3 \\ \overline{a}_1 & 0 & \overline{a}_1a_2 + r_1r_2e^{i\theta_1} & \overline{a}_1a_3 + r_1r_3e^{i\theta_2} \\ \overline{a}_2 & a_1\overline{a}_2 + r_1r_2e^{-i\theta_1} & 0 & \overline{a}_2a_3 + r_2r_3e^{i\theta_3} \\ \overline{a}_3 & a_1\overline{a}_3 + r_1r_3e^{-i\theta_2} & a_2\overline{a}_3 + r_2r_3e^{-i\theta_3} & 0 \end{bmatrix}.$$

Case 1. Let $r_1r_2r_3 \neq 0$. It is straightforward to check that the minimal eigenvalue $\lambda = -\frac{1}{2}$ of $\operatorname{Re} A$ has multiplicity 2, and

$$y_1 = (-a_2r_1 + a_1r_2e^{i\theta_1}, -r_2e^{i\theta_1}, r_1, 0), \quad y_2 = (-a_3r_1 + a_1r_3e^{i\theta_2}, -r_3e^{i\theta_2}, 0, r_1)$$

form a basis of the respective eigenspace. Moreover,

$$Ay_1 = (-a_1r_2e^{i\theta_1} + a_2r_1, r_1\overline{a}_1a_2 + r_1^2r_2e^{i\theta_1}, 0, 0)$$

and

$$Ay_2 = (-a_1r_3e^{i\theta_2} + a_3r_1, r_1\overline{a}_1a_3 + r_1^2r_3e^{i\theta_2}, r_1\overline{a}_2a_3 + r_1r_2r_3e^{i(\theta_2-\theta_1)}, 0).$$

Therefore

$$\begin{aligned}
\|y_1\|^2 &= 2r_1^2 + 2r_2^2 - 2r_1^2r_2^2 - a_1\overline{a}_2r_1r_2e^{i\theta_1} - \overline{a}_1a_2r_1r_2e^{-i\theta_1},\\
\|y_2\|^2 &= 2r_1^2 + 2r_3^2 - 2r_1^2r_3^2 - a_1\overline{a}_3r_1r_3e^{i\theta_2} - \overline{a}_1a_3r_1r_3e^{-i\theta_2},\\
\langle Ay_1, y_1\rangle &= -r_1^2 - r_2^2 + r_1^2r_2^2 + a_1\overline{a}_2r_1r_2e^{i\theta_1},\\
\langle Ay_2, y_2\rangle &= -r_1^2 - r_3^2 + r_1^2r_3^2 + a_1\overline{a}_3r_1r_3e^{i\theta_2}.
\end{aligned}$$

After some simplification, $\|y_2\|^2\langle Ay_1, y_1\rangle - \|y_1\|^2\langle Ay_2, y_2\rangle$ becomes $2ir_1\tau_1$ with τ_1 defined by (2.18). Hence condition (2.1) is satisfied if and only if $\tau_1 = 0$.

Next note that

$$\langle y_2, y_1\rangle = r_1^2a_3\overline{a}_2 - \overline{a}_2a_1r_1r_3e^{i\theta_2} - a_3\overline{a}_1r_1r_2e^{-i\theta_1} + (1 + |a_1|^2)r_2r_3e^{i(\theta_2-\theta_1)},$$

and

$$\langle Ay_2, y_1\rangle = a_1r_3e^{i\theta_2}(\overline{a}_2r_1 - \overline{a}_1r_2e^{-i\theta_1}).$$

Now $\|y_1\|^2\langle Ay_2, y_1\rangle - \langle y_2, y_1\rangle\langle Ay_1, y_1\rangle$ simplifies to $r_1\tau_2$, so (2.2) holds exactly when $\tau_2 = 0$.

Therefore, by Proposition 2.2, the line $x = -\frac{1}{2}$ will contain a proper line segment of $F(A)$ if and only if at least one of τ_1 or τ_2 is nonzero. This agrees with the statement of the theorem.

Case 2. At least two of r_j are equal to zero, $j = 1, 2, 3$. To be consistent with the statement of the theorem, we need to show that $F(A) \cap \ell$ is a proper line segment. But this is indeed so. For example, if $r_1 = r_2 = 0$, then it immediately follows that the vectors $y_1 = (-a_1, 1, 0, 0)$ and $y_2 = (-a_2, 0, 1, 0)$ are linearly independent eigenvectors of $\operatorname{Re} A$ corresponding to $-\frac{1}{2}$. Since $Ay_1 = (a_1, 0, 0, 0)$ and $Ay_2 = (a_2, \overline{a}_1a_2, 0, 0)$, both sides of (2.1) equal -2. However, (2.2) is not satisfied because

$$\langle Ay_2, y_1\rangle\|y_1\|^2 = 0 \neq -a_2\overline{a}_1 = \langle y_2, y_1\rangle\langle Ay_1, y_1\rangle.$$

Therefore, the flat portion will exist. All other cases where at least two r_j values are zero are treated in the same manner.

Case 3. Exactly one of r_j is equal to zero. For the sake of definiteness, let $r_1 = 0$, $r_2r_3 \neq 0$. Then $y_1 = (-a_1, 1, 0, 0)$ and $y_2 = (a_2r_3e^{i\theta_3} - r_2a_3, 0, -r_3e^{i\theta_3}, r_2)$ are linearly independent eigenvectors of $\operatorname{Re} A$ corresponding to $-\frac{1}{2}$. It still holds that $\|y_1\|^2 = 2$ and $\langle Ay_1, y_1\rangle = -1$. Now we also have

$$\langle y_2, y_1\rangle = r_2 a_3 \overline{a}_1 - a_2 \overline{a}_1 r_3 e^{i\theta_3},$$
$$\|y_2\|^2 = 2 - \overline{a}_2 a_3 r_2 r_3 e^{-i\theta_3} - a_2 \overline{a}_3 r_2 r_3 e^{i\theta_3} - 2|a_2|^2|a_3|^2,$$

and

$$Ay_2 = (-a_2 r_3 e^{i\theta_3} + r_2 a_3, -\overline{a}_1 a_2 r_3 e^{i\theta_3} + \overline{a}_1 a_3 r_2, r_2 \overline{a}_2 a_3 + r_2^2 r_3 e^{i\theta_3}, 0).$$

Therefore, we can compute the remaining quantities from Proposition 2.2:

$$\langle Ay_2, y_2\rangle = a_2 \overline{a}_3 r_2 r_3 e^{i\theta_3} - 1 + |a_2|^2|a_3|^2,$$

and

$$\langle Ay_2, y_1\rangle = 0.$$

Equation (2.1) holds if and only if $-\|y_2\|^2 = 2\langle Ay_2, y_2\rangle$. Substituting into the latter equation yields

$$-2 + \overline{a}_2 a_3 r_2 r_3 e^{-i\theta_3} + a_2 \overline{a}_3 r_2 r_3 e^{i\theta_3} + 2|a_2|^2|a_3|^2 = 2a_2 \overline{a}_3 r_2 r_3 e^{i\theta_3} - 2 + 2|a_2|^2|a_3|^2,$$

which simplifies to

$$\overline{a}_2 a_3 r_2 r_3 e^{-i\theta_3} - a_2 \overline{a}_3 r_2 r_3 e^{i\theta_3} = 0. \tag{2.20}$$

Since $a_1 \neq 0$, equation (2.2) holds if and only if

$$r_2 a_3 - a_2 r_3 e^{i\theta_3} = 0. \tag{2.21}$$

If equation (2.21) holds, then $r_3 = r_2$ and hence $\arg(a_3) = \arg(a_2) + \theta_3$. This proves the necessity of the conditions in (ii) in order for $F(A)$ to fail to have a flat portion on $x = -\frac{1}{2}$. Conversely, if the conditions in (ii) hold, then clearly (2.20) and (2.21) are true, which results in no flat portion by Proposition 2.2. This agrees with (ii) in the statement of the theorem.

The situations when $r_2 = 0$, $r_1 r_3 \neq 0$ and $r_3 = 0, r_1 r_2 \neq 0$ can be treated similarly. The only difference will be in the specific choice of the eigenvectors, namely,

$$y_1 = (-a_2, 0, 1, 0), \quad y_2 = (-a_3 r_1 + r_3 a_1 e^{i\theta_2}, -r_3 e^{i\theta_2}, 0, r_1)$$

in the former, and

$$y_1 = (-a_3, 0, 0, 1), \quad y_2 = (-a_2 r_1 + a_1 r_2 e^{i\theta_1}, -r_2 e^{i\theta_1}, r_1, 0)$$

in the latter. Direct computations show that a flat portion does not materialize if and only if, respectively, (iii) or (iv) holds. □

Conditions (i)–(iv) of Theorem 2.4 simplify somewhat if the entries a_j in (2.6) are real. Indeed, then $\theta_j = 0 \mod \pi$, and the argument conditions in (ii)–(iv) boil down to

$$a_2 a_3 (a_6 - a_2 a_3) \geq 0, \quad a_1 a_3 (a_5 - a_1 a_3) \leq 0, \text{ and } a_1 a_2 (a_4 - a_1 a_2) \geq 0,$$

respectively. In its turn, τ_1 in (i) is zero automatically, while

$$\begin{aligned}\tau_2 = {} & r_1a_2a_3(r_1^2+2r_2^2-2r_1^2r_2^2)-a_1r_2a_3(r_2^2+2r_1^2-2r_1^2r_2^2)e^{i\theta_1}\\ & +a_1a_2r_3(r_1^2+r_2^2-2r_1^2r_2^2)e^{i\theta_2}+r_1r_2r_3(1-2a_1^2a_2^2)e^{i\theta_3}.\end{aligned} \tag{2.22}$$

Example. Let A be of the form (2.6) with $\alpha = 1$ and $a_1 = \frac{\sqrt{2+\sqrt{3}}}{2}$, $a_2 = \frac{1}{2}$, $a_3 = \frac{\sqrt{2}}{2}$. Choosing a_4, a_5, a_6 in such a way that (2.8)–(2.10) hold with $\theta_j = 0$ in (2.16), $j = 1, 2, 3$ yields $a_4 = \frac{\sqrt{2}}{2}$, $a_5 = \frac{\sqrt{3}}{2}$, and $a_6 = \frac{\sqrt{2+\sqrt{3}}}{2}$. A direct substitution into (2.22) reveals that $\tau_2 = 0$. So, $F(A)$ has no flat portion on the supporting line $x = -\frac{1}{2}$ by Theorem 2.4 even though this line is exceptional by Theorem 2.3. See Figure 1.

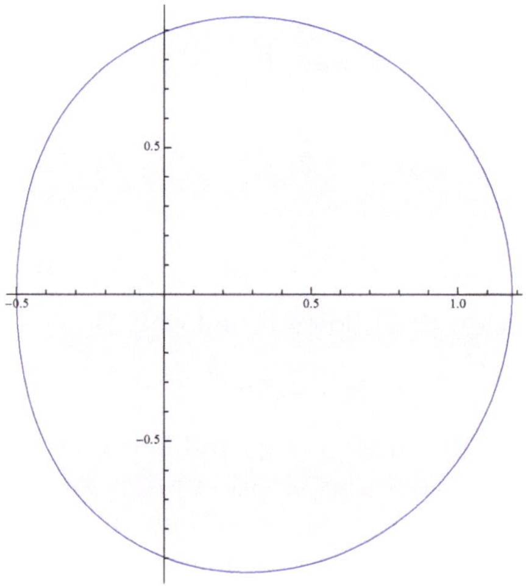

FIGURE 1. $F(A)$ with exceptional support line but no flat portion

3. Special case: An alternative approach

As can be seen from the discussion above, Theorem 2.4 provides a convenient tool for constructing specific examples of nilpotent matrices A with a prescribed exceptional supporting line ℓ, with $F(A) \cap \ell$ being just one point or a proper line segment. For another example, by setting $a_1 = 1$, $a_2 = \frac{1}{2}$, $a_3 = \frac{\sqrt{3}}{2}$, and $\theta_3 = 0$ in Theorem 2.3 we immediately obtain a matrix

$$A = \begin{bmatrix} 0 & 1 & \frac{1}{2} & \frac{\sqrt{3}}{2} \\ 0 & 0 & \frac{1}{2} & \frac{\sqrt{3}}{2} \\ 0 & 0 & 0 & \frac{\sqrt{3}}{2} \\ 0 & 0 & 0 & 0 \end{bmatrix}$$

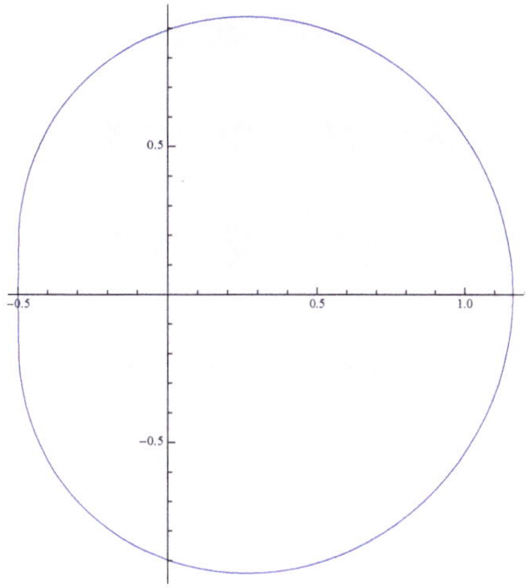

FIGURE 2. $F(A)$ with flat portion

that fails to satisfy any of the conditions (i)–(iv) in Theorem 2.4 and hence has a flat portion as shown in Figure 2.

However, the conditions in Theorem 2.4 are not as useful when considering a given (even triangular) matrix with the number and orientation of flat portions not known a priori. We present now one such case, to illustrate this point, and also since it plays an important role in Section 5.

Lemma 3.1. *Let*

$$A = \begin{bmatrix} 0 & a_1 & a_2 & a_3 \\ 0 & 0 & a_3 & a_2 \\ 0 & 0 & 0 & a_1 \\ 0 & 0 & 0 & 0 \end{bmatrix},$$

where a_1, a_2, and a_3 are real and $a_1 \neq 0$. The boundary of $F(A)$ contains a vertical flat portion if and only if $|a_1| = |a_3|$ and $|a_2| \geq |a_1|$. In this case, this is the only flat portion on $\partial F(A)$.

Proof. As is well known (see, for example, [5]), the boundary $\partial F(A)$ of the numerical range of A contains a portion of the line $\cos(\theta)x + \sin(\theta)y = d$ if and only if d is the maximum (or minimum) eigenvalue of $\operatorname{Re} e^{-i\theta}A$ and there are two eigenvectors y_1 and y_2 associated with d such that either condition (2.1) or (2.2) in Proposition 2.2 fails to hold.

A straightforward calculation shows that the characteristic polynomial of $\operatorname{Re}(e^{-i\theta}A)$ is

$$q_\theta(\lambda) = \lambda^4 - \frac{1}{2}(a_1^2 + a_2^2 + a_3^2)\lambda^2 - (a_1a_2a_3\cos\theta)\lambda + \delta_\theta,$$

where $\delta_\theta = \frac{1}{16}(a_1^4 + a_2^4 + a_3^4 - 2a_1^2a_2^2 - 2a_2^2a_3^2 - 2a_1^2a_3^2\cos 2\theta)$. This polynomial factors as

$$q_\theta(\lambda) = \left(\lambda^2 + a_2\lambda + \nu_\theta\right)\left(\lambda^2 - a_2\lambda + \rho_\theta\right),$$

where

$$\nu_\theta = \frac{a_1a_3\cos\theta}{2} + \frac{a_2^2}{4} - \frac{a_1^2}{4} - \frac{a_3^2}{4}$$

and

$$\rho_\theta = -\frac{a_1a_3\cos\theta}{2} + \frac{a_2^2}{4} - \frac{a_1^2}{4} - \frac{a_3^2}{4}.$$

Therefore the roots of q_θ, which are the eigenvalues of $\mathrm{Re}\,(e^{-i\theta}A)$, can be explicitly calculated as

$$\begin{aligned}
\lambda_1(\theta) &= \frac{1}{2}\left(-a_2 - \sqrt{a_1^2 + a_3^2 - 2a_1a_3\cos\theta}\right),\\
\lambda_2(\theta) &= \frac{1}{2}\left(a_2 - \sqrt{a_1^2 + a_3^2 + 2a_1a_3\cos\theta}\right),\\
\lambda_3(\theta) &= \frac{1}{2}\left(-a_2 + \sqrt{a_1^2 + a_3^2 - 2a_1a_3\cos\theta}\right),\\
\lambda_4(\theta) &= \frac{1}{2}\left(a_2 + \sqrt{a_1^2 + a_3^2 + 2a_1a_3\cos\theta}\right).
\end{aligned} \tag{3.1}$$

When $\theta = 0$, the eigenvalues of $\mathrm{Re}A$ simplify to

$$\begin{aligned}
\lambda_1 &= \frac{1}{2}\left(-a_2 - a_1 + a_3\right),\\
\lambda_2 &= \frac{1}{2}\left(a_2 - a_1 - a_3\right),\\
\lambda_3 &= \frac{1}{2}\left(-a_2 + a_1 - a_3\right),\\
\lambda_4 &= \frac{1}{2}\left(a_2 + a_1 + a_3\right),
\end{aligned}$$

where relabeling may have occurred based on absolute values. Furthermore, in this case it is straightforward to verify that corresponding eigenvectors of $\mathrm{Re}A$ are $y_1 = (1, -1, -1, 1)$, $y_2 = (-1, 1, -1, 1)$, $y_3 = (-1, -1, 1, 1)$, and $y_4 = (1, 1, 1, 1)$. Notice that

$$\begin{aligned}
Ay_1 &= (-a_1 - a_2 + a_3, -a_3 + a_2, a_1, 0),\\
Ay_2 &= (a_1 - a_2 + a_3, a_2 - a_3, a_1, 0),\\
Ay_3 &= (-a_1 + a_2 + a_3, a_3 + a_2, a_1, 0),\\
Ay_4 &= (a_1 + a_2 + a_3, a_2 + a_3, a_1, 0).
\end{aligned}$$

The values $\|y_j\|^2 = 4$ and $\langle Ay_j, y_j\rangle = 4\lambda_j$ for $j = 1, 2, 3, 4$. Therefore for all $1 \le j, k \le 4$, if there is a repeated eigenvalue $\lambda_j = \lambda_k$ (extremal or not), we have

$$\langle Ay_j, y_j\rangle\|y_k\|^2 = 16\lambda_j = 16\lambda_k = \langle Ay_k, y_k\rangle\|y_j\|^2.$$

Hence (2.1) will always hold, and a vertical flat portion will exist exactly when there is an extremal eigenvalue where (2.2) fails. When $j \neq k$, the eigenvectors y_j and y_k are orthogonal. Accordingly, (2.2) fails to hold for a case where $\lambda_j = \lambda_k$ if and only if

$$\langle Ay_j, y_k \rangle \neq 0. \tag{3.2}$$

All possible cases to consider can be studied with the following inner products.

$$\langle Ay_1, y_2 \rangle = 2a_2 - 2a_3, \tag{3.3}$$

$$\langle Ay_3, y_4 \rangle = 2a_3 + 2a_2, \tag{3.4}$$

$$\langle Ay_1, y_3 \rangle = \langle Ay_2, y_4 \rangle = 2a_1, \tag{3.5}$$

$$\langle Ay_1, y_4 \rangle = \langle Ay_2, y_3 \rangle = 0. \tag{3.6}$$

Now, assume the boundary of $F(A)$ has a vertical flat portion. Thus the maximal or minimal eigenvalue of ReA is repeated and the corresponding eigenvectors satisfy (3.2).

The equality $\lambda_1 = \lambda_2$ implies $a_2 = a_3$, so equation (3.3) shows condition (3.2) fails in this case. Similarly, by equation (3.4), condition (3.2) fails when $\lambda_3 = \lambda_4$ and $a_2 = -a_3$. Clearly, by (3.6), no combination with $\lambda_1 = \lambda_4$ or $\lambda_2 = \lambda_3$ is possible with a vertical flat portion.

Therefore it must hold that either $\lambda_1 = \lambda_3$ or $\lambda_2 = \lambda_4$. In the former case, we have $a_1 = a_3$; in order for this repeated λ_1 to be extremal, we must have $|a_2| \geq |a_1|$. Similarly, if $\lambda_2 = \lambda_4$, then $a_1 = -a_3$ and to be extremal $|a_2| \geq |a_1|$. Thus a vertical flat position implies the conditions $|a_1| = |a_3|$ and $|a_2| \geq |a_1|$.

Conversely, if $a_1 \neq 0$, $a_1 = a_3$ and $|a_2| \geq |a_1|$, then $\lambda_1 = \lambda_3$ is either the maximal or minimal eigenvalue of ReA and $\langle Ay_1, y_3 \rangle \neq 0$ by (3.5). Likewise, if $a_3 = -a_1 \neq 0$ and $|a_2| \geq |a_1|$, then $\lambda_2 = \lambda_4$ is an extreme eigenvalue and $\langle Ay_2, y_4 \rangle \neq 0$, again by (3.5). Hence in both cases Proposition 2.2 shows that $\partial F(A)$ has a vertical flat portion.

There are never two vertical flat portions on $\partial F(A)$ because that would require both a repeated maximal and a repeated minimal eigenvalue of ReA, which implies $a_1 = a_3$ and $a_1 = -a_3$ and we assumed $a_1 \neq 0$. To see that a vertical flat portion never coexists with any other flat portion, assume there is a vertical flat portion and there is also a flat portion on the line $x \cos\theta + y \sin\theta = d$ for some real d and $\theta \in (0, 2\pi)$ with $\theta \neq \pi$. It suffices to show that there is not a repeated maximal eigenvalue at θ, because if there was a repeated minimal eigenvalue at θ there would also be a repeated maximal eigenvalue at $\theta \pm \pi$. In the list of the eigenvalues of $\operatorname{Re}(e^{-i\theta}A)$ in (3.1), $\lambda_2(\theta) < \lambda_4(\theta)$ and $\lambda_1(\theta) < \lambda_3(\theta)$. Therefore the only possibility of a repeated maximal eigenvalue is when $\lambda_3(\theta) = \lambda_4(\theta)$. Setting $|a_1| = |a_3|$ in this equality yields

$$2|a_2| = \left| \sqrt{2a_1^2 - 2a_1^2 \cos\theta} - \sqrt{2a_1^2 + 2a_1^2 \cos\theta} \right| = 2|a_1| \left| \sin\frac{\theta}{2} - \left|\cos\frac{\theta}{2}\right| \right|.$$

The extreme values of $\sin\frac{\theta}{2} - |\cos\frac{\theta}{2}|$ on $[0, 2\pi)$ are 1 (only when $\theta = \pi$) and -1 (only when $\theta = 0$). Thus $\lambda_3(\theta) = \lambda_4(\theta)$ cannot hold at another value of θ or else $|a_2| \geq |a_1|$ is contradicted. □

A sketch of the numerical range of A when $a_1 = 1$, $a_3 = -1$, and $a_2 = 2$ is shown in Figure 3.

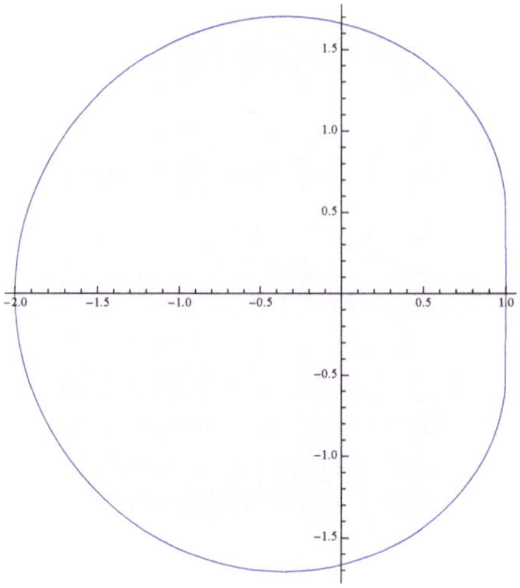

FIGURE 3. $F(A)$ with one vertical flat portion

4. Matrices with two flat portions

Proposition 4.1. *A 4-by-4 matrix A is nilpotent, with two parallel flat portions on the boundary of its numerical range, if and only if it is a product of a scalar $\alpha \in \mathbb{C} \setminus \{0\}$ and a matrix unitarily similar to*

$$\begin{bmatrix} 0 & a_1 & a_2 & a_3 \\ 0 & 0 & a_3 & -a_2 \\ 0 & 0 & 0 & a_1 \\ 0 & 0 & 0 & 0 \end{bmatrix}, \tag{4.1}$$

with $a_1, a_3 > 0$ and $a_2 \in \mathbb{R}$.

Proof. Necessity. Using Schur's lemma, we may without loss of generality suppose that the matrix A is upper triangular. Being nilpotent, it thus can be written as

$$A = \begin{bmatrix} 0 & a_{12} & a_{13} & a_{14} \\ 0 & 0 & a_{23} & a_{24} \\ 0 & 0 & 0 & a_{34} \\ 0 & 0 & 0 & 0 \end{bmatrix}. \tag{4.2}$$

Note that the general form of A has entries with double subscripts, but we will gradually simplify A so that it has the form (4.1). Multiplying matrix (4.2) by an appropriate scalar (of absolute value one, if desired), we may also suppose that the parallel flat portions on the boundary of $F(A)$ are horizontal. Yet another unitary similarity, this time via a diagonal matrix, allows us to adjust the arguments of the entries $a_{i,i+1}$ any way we wish ($i = 1, 2, 3$) without changing the absolute values. Let us agree therefore to choose these entries real and non-negative.

Having agreed on the above, we now introduce $H = \operatorname{Re} A$ and $K = \operatorname{Im} A$ as the hermitian matrices from the representation $A = H + iK$, so in particular

$$K = \frac{1}{2i}\begin{bmatrix} 0 & a_{12} & a_{13} & a_{14} \\ -a_{12} & 0 & a_{23} & a_{24} \\ -\overline{a_{13}} & -a_{23} & 0 & a_{34} \\ -\overline{a_{14}} & -\overline{a_{24}} & -a_{34} & 0 \end{bmatrix}. \tag{4.3}$$

For a matrix B of any size, two horizontal flat portions on the boundary of $F(B)$ can materialize only if the maximal and minimal eigenvalues of $\operatorname{Im} B$ both have multiplicity at least 2. For our 4-by-4 matrix A it means that K has two distinct eigenvalues, say λ_1 and λ_2, each of multiplicity two. In addition, $\operatorname{Tr} K = 0$, and so $\lambda_1 = -\lambda_2 (:= \lambda)$. Consequently,

$$K^2 = \lambda^2 I.$$

In particular, the diagonal entries of K^2 are all equal. From here and (4.3):

$$a_{12}^2 + |a_{13}|^2 + |a_{14}|^2 = a_{12}^2 + a_{23}^2 + |a_{24}|^2 = a_{34}^2 + |a_{13}|^2 + a_{23}^2 = a_{34}^2 + |a_{14}|^2 + |a_{24}|^2.$$

Equivalently (and taking into account non-negativity of a_{12}, a_{23}, a_{34}):

$$a_{12} = a_{34} := a_1, \quad |a_{14}| = a_{23} := a_3, \quad |a_{13}| = |a_{24}|. \tag{4.4}$$

Taking (4.4) into consideration, the fact that off diagonal entries of K^2 are all equal to zero boils down to

$$a_1(a_3 - a_{14}) = 0, \quad a_1(a_{13} + a_{24}) = 0, \quad a_3 a_{13} + a_{14}\overline{a_{24}} = 0, \quad a_3 a_{24} + \overline{a_{13}} a_{14} = 0. \tag{4.5}$$

On the other hand, from (4.2) and (4.4):

$$A^2 = \begin{bmatrix} 0 & 0 & a_1 a_3 & a_1(a_{13} + a_{24}) \\ 0 & 0 & 0 & a_1 a_3 \\ 0 & 0 & 0 & 0 \\ 0 & 0 & 0 & 0 \end{bmatrix},$$

which, when combined with the second equation in (4.5), yields

$$A^2 = a_1 a_3 \begin{bmatrix} 0 & 0 & 1 & 0 \\ 0 & 0 & 0 & 1 \\ 0 & 0 & 0 & 0 \\ 0 & 0 & 0 & 0 \end{bmatrix}.$$

But $A^2 \neq 0$, since otherwise $F(A)$ would be a circular disk (see, e.g., [14]) exhibiting no flat portions on the boundary. So, $a_1, a_3 > 0$. The solution to (4.5) is then given by

$$a_{14} = a_3, \quad a_{13} = -a_{24} \ (:= a_2) \in \mathbb{R}.$$

So, A is indeed in the form (4.1) up to unitary similarity and scaling.

Sufficiency. The scalar multiple α is inconsequential, so without loss of generality A is given by (4.1). The eigenvalues of $K = \operatorname{Im} A$ are then $\pm\lambda$, each of multiplicity two, where $\lambda = \sqrt{a_1^2 + a_2^2 + a_3^2}/2$. Let us apply a unitary similarity, putting K in the form $\begin{bmatrix} \lambda I & 0 \\ 0 & -\lambda I \end{bmatrix}$, and denote by $\begin{bmatrix} H_1 & Z \\ Z^* & H_2 \end{bmatrix}$ the result of applying the same unitary similarity to $\operatorname{Re} A$.

Since A is real, its numerical range $F(A)$ is symmetric with respect to the x-axis, and thus there are either two or no horizontal flat portions on its boundary. But the latter is possible only if H_1 and H_2 are scalar multiples of the identity. Applying yet another (block diagonal) unitary similarity, we can reduce A to

$$\begin{bmatrix} h_1 + i\lambda & 0 & \sigma_1 & 0 \\ 0 & h_1 + i\lambda & 0 & \sigma_2 \\ \sigma_1 & 0 & h_2 - i\lambda & 0 \\ 0 & \sigma_2 & 0 & h_2 - i\lambda \end{bmatrix}, \tag{4.6}$$

where σ_1, σ_2 are the s-numbers of Z. The matrix (4.6) is unitarily (and even permutationally) reducible, which is in contradiction with the fact that A has just one Jordan block. So, there are indeed two parallel flat portions on the boundary of $F(A)$. □

Remark. From the proof of Proposition 4.1 it is clear that the parallel flat portions of $\partial F(A)$ are on lines that are at an equal distance $|\alpha|\sqrt{a_1^2 + a_2^2 + a_3^2}/2$ from the origin, forming the angle $\arg\alpha$ with the positive direction of the x-axis. Of course, $|\alpha|$ can be changed arbitrarily via absorbing it (or part thereof) by the a_j entries in (4.1).

Corollary 4.2. *If A is a 4-by-4 nilpotent matrix with two parallel flat portions on the boundary of its numerical range, then these are the only flat portions of the boundary.*

Proof. Without loss of generality, A is in the form (4.1), and thus there are two horizontal flat portions on the boundary of $F(A)$. We also know A is irreducible from Proposition 2.1. Any other flat portion ℓ of $\partial F(A)$, if it exists, cannot be horizontal. Suppose it is not vertical either. Then, due to the symmetry of $F(A)$ with respect to the x-axis, $\partial F(A)$ would have to contain the complex conjugate of ℓ as well, bringing the number of flat portions to (at least) four. But any 4-by-4 matrix with four flat portions on the boundary of its numerical range is unitarily reducible, as stated in Theorem 1.1, while A is not.

It remains to consider the case of vertical ℓ. In order for it to exist, the matrix $H = \operatorname{Re} A$ should have a multiple eigenvalue. This eigenvalue would then have to be common for 3-by-3 principal submatrices of H. A direct computation shows, however, that the characteristic polynomials of these submatrices up to a constant multiple equal

$$-4\lambda^3+\lambda(a_1^2+a_2^2+a_3^2)+a_1a_2a_3 \quad \text{and} \quad -4\lambda^3+\lambda(a_1^2+a_2^2+a_3^2)-a_1a_2a_3.$$

So, the common eigenvalues occur only if $a_2=0$ (recall that a_1 and a_3 are strictly positive).

On the other hand, if $a_2=0$, then the matrix iA under the diagonal unitary similarity via $\operatorname{diag}[1,i,1,i]$ turns into

$$\begin{bmatrix} 0 & a_1 & 0 & a_3 \\ 0 & 0 & -a_3 & 0 \\ 0 & 0 & 0 & a_1 \\ 0 & 0 & 0 & 0 \end{bmatrix}.$$

So, $F(A)$ in this case is symmetric with respect to the y-axis as well. Thus, a vertical flat portion ℓ would have its counterpart $-\ell$ on $\partial F(A)$ which again would bring the number of flat portions up to four, in contradiction with the unitary irreducibility of A. □

Proposition 4.3. *Let A be a 4-by-4 nilpotent unitarily irreducible matrix with two non-parallel flat portions on the boundary of its numerical range $F(A)$ the supporting lines of which are equidistant from the origin. Then these are the only flat portions of $\partial F(A)$.*

Proof. Multiplying A by a non-zero scalar we may without loss of generality suppose that the given flat portions of $\partial F(A)$ lie on lines intersecting at some point on the negative real half-axis and that the distance from each line to the origin equals 1/2. Then for some unimodular $\omega=\xi+i\eta$ ($\xi\neq 0,\ \eta>0$) the imaginary part of both ωA and $-\overline{\omega}A$ will have a multiple eigenvalue $-1/2$. With this notation, these lines will be $\eta x\pm\xi y=-\frac{1}{2}$; they intersect at $(-\frac{1}{2\eta},0)$.

By an appropriate unitary similarity A can be put in the upper triangular form (4.2) and, moreover, the elements a_{12},a_{23},a_{34} of its first sup-diagonal can be all made real. Therefore the above-mentioned condition on the eigenvalues of $\operatorname{Im}(\omega A)$ and $\operatorname{Im}(-\overline{\omega}A)$ implies that the matrices

$$\begin{bmatrix} 1 & -i\omega a_{12} & -i\omega a_{13} & -i\omega a_{14} \\ i\overline{\omega}\overline{a_{12}} & 1 & -i\omega a_{23} & -i\omega a_{24} \\ i\overline{\omega}\overline{a_{13}} & i\overline{\omega}\overline{a_{23}} & 1 & -i\omega a_{34} \\ i\overline{\omega}\overline{a_{14}} & i\overline{\omega}\overline{a_{24}} & i\overline{\omega}\overline{a_{34}} & 1 \end{bmatrix} \quad \text{and} \quad \begin{bmatrix} 1 & i\overline{\omega}a_{12} & i\overline{\omega}a_{13} & i\overline{\omega}a_{14} \\ -i\omega\overline{a_{12}} & 1 & i\overline{\omega}a_{23} & i\overline{\omega}a_{24} \\ -i\omega\overline{a_{13}} & -i\omega\overline{a_{23}} & 1 & i\overline{\omega}a_{34} \\ -i\omega\overline{a_{14}} & -i\omega\overline{a_{24}} & -i\omega\overline{a_{34}} & 1 \end{bmatrix} \tag{4.7}$$

have rank at most 2. Equating the left upper 3-by-3 minors of (4.7) to zero, we see that

$$1-(|a_{13}|^2+|a_{12}|^2+|a_{23}|^2)+2\operatorname{Im}(\omega a_{12}\overline{a_{13}}a_{23})=0$$

and

$$1-(|a_{13}|^2+|a_{12}|^2+|a_{23}|^2)-2\operatorname{Im}(\overline{\omega}a_{12}\overline{a_{13}}a_{23})=0.$$

Equivalently, $a_{12}\overline{a_{13}}a_{23}$ is real, and

$$1 - (|a_{13}|^2 + |a_{12}|^2 + |a_{23}|^2) + 2\eta a_{12}\overline{a_{13}}a_{23} = 0. \tag{4.8}$$

If $a_{12} = 0$ or $a_{23} = 0$, then (4.8) implies that for the 3-by-3 matrix B located in the upper left corner of A the numerical range $F(B)$ is the circular disk $\{z\colon |z| \leq 1/2\}$ (see [12] or [8, Theorem 4.1]). Since this case is covered by Theorem 1.2, we may suppose that $a_{12}, a_{23} \neq 0$. Then a_{13} is necessarily real along with a_{12}, a_{23}, and (4.8) can be rewritten as

$$1 - (a_{13}^2 + a_{12}^2 + a_{23}^2) + 2\eta a_{12}a_{13}a_{23} = 0. \tag{4.9}$$

Repeating this reasoning for three other principal 3-by-3 minors in (4.7) we see that without loss of generality A is of the form (4.2), real, and, along with (4.9):

$$\begin{aligned} 1 - (a_{12}^2 + a_{14}^2 + a_{24}^2) + 2\eta a_{12}a_{14}a_{24} &= 0, \\ 1 - (a_{13}^2 + a_{14}^2 + a_{34}^2) + 2\eta a_{13}a_{14}a_{34} &= 0, \\ 1 - (a_{23}^2 + a_{24}^2 + a_{34}^2) + 2\eta a_{23}a_{24}a_{34} &= 0. \end{aligned} \tag{4.10}$$

Since A is real, the third flat portion of $\partial F(A)$, if it exists, must be vertical. Indeed, otherwise its reflection with respect to the real axis would be the fourth flat portion of $\partial F(A)$, implying by Theorem 1.1 the unitary reducibility of A. But this would contradict Proposition 2.1.

So, suppose now that a vertical flat portion of $\partial F(A)$ is indeed present. Then its abscissa, which it is convenient for us to denote by $-x/2$, is a multiple eigenvalue of $\operatorname{Re} A$. Equivalently, the matrix

$$\begin{bmatrix} x & a_{12} & a_{13} & a_{14} \\ a_{12} & x & a_{23} & a_{24} \\ a_{13} & a_{23} & x & a_{34} \\ a_{14} & a_{24} & a_{34} & x \end{bmatrix}$$

has rank at most two. Consequently,

$$\begin{aligned} x^3 - x({a_{13}}^2 + {a_{12}}^2 + {a_{23}}^2) + 2a_{12}a_{13}a_{23} &= 0, \\ x^3 - x({a_{12}}^2 + {a_{14}}^2 + {a_{24}}^2) + 2a_{12}a_{14}a_{24} &= 0, \\ x^3 - x({a_{13}}^2 + {a_{14}}^2 + {a_{34}}^2) + 2a_{13}a_{14}a_{34} &= 0, \\ x^3 - x({a_{23}}^2 + {a_{24}}^2 + {a_{34}}^2) + 2a_{23}a_{24}a_{34} &= 0. \end{aligned} \tag{4.11}$$

Comparing the respective lines in (4.9)–(4.10) and (4.11), we conclude that $x^3 - x + 2C(1 - x\eta) = 0$ when in place of C is plugged in any of the products $a_{12}a_{13}a_{23}, a_{12}a_{14}a_{24}, a_{13}a_{14}a_{34}, a_{23}a_{24}a_{34}$. If $1 - x\eta = 0$, then $-x/2 = -1/2\eta$, the point on the negative x-axis where the support lines containing the flat portions of $\partial F(A)$ intersect. This would be a corner point on $F(A)$, thus implying unitary reducibility of A. So, $1 - x\eta \neq 0$ and the above-mentioned permissible values of C

are all equal:

$$a_{12}a_{13}a_{23} = a_{12}a_{14}a_{24} = a_{13}a_{14}a_{34} \\ = a_{23}a_{24}a_{34}. \tag{4.12}$$

From here we conclude that the coefficients of x in all the equations (4.11) are also equal:

$$a_{13}{}^2 + a_{12}{}^2 + a_{23}{}^2 = a_{12}{}^2 + a_{14}{}^2 + a_{24}{}^2 = a_{13}{}^2 + a_{14}{}^2 + a_{34}{}^2 \\ = a_{23}{}^2 + a_{24}{}^2 + a_{34}{}^2. \tag{4.13}$$

From (4.12)–(4.13) we conclude that

$$a_{34} = \epsilon_1 a_{12},\ a_{14} = \epsilon_2 a_{23},\ a_{24} = \epsilon_3 a_{13}, \quad \epsilon_j = \pm 1,\ j = 1, 2, 3, \tag{4.14}$$

and either $a_{12}a_{13}a_{23} = 0$ or $\epsilon_1 = \epsilon_2 = \epsilon_3$. Since the former case is covered by Theorem 1.2, we may concentrate on the latter. Moreover, applying a unitary similarity via the diagonal matrix diag$[1, 1, 1, -1]$, we may change the signs of all ϵ_j simultaneously, and thus suppose that they are all equal to 1.

In other words, it remains to consider

$$A = \begin{bmatrix} 0 & a_1 & a_2 & a_3 \\ 0 & 0 & a_3 & a_2 \\ 0 & 0 & 0 & a_1 \\ 0 & 0 & 0 & 0 \end{bmatrix} \tag{4.15}$$

with $a_j \in \mathbb{R}$ and $a_1 \neq 0$. By Lemma 3.1, the numerical range of a matrix of this form can only have a vertical flat portion on its boundary if that is the only flat portion. Therefore the original matrix has only the two flat portions on lines which are equidistant from the origin. □

5. Proof of the main result

In this section, we show there are at most two flat portions on the boundary of the numerical range of a 4-by-4 unitarily irreducible nilpotent matrix. This result follows from an analysis of the singularities of the polynomial p_A defined in (1.1), so some facts about algebraic curves are reviewed next for convenience.

Define the complex projective plane $\mathbb{CP}^2$ to be the set of all equivalence classes of points in $\mathbb{C}^3 \setminus \{(0, 0, 0)\}$ determined by the equivalence relation $\sim$ where $(x, y, z) \sim (a, b, c)$ if and only if $(x, y, z) = \lambda(a, b, c)$ for some nonzero complex number λ. The complex plane can be considered a subset of $\mathbb{CP}^2$ if the point $(x, y, 1)$ for $(x, y) \in \mathbb{R}^2$ is identified with $x + iy$.

An *algebraic curve* C is defined to be the zero set of a homogeneous polynomial $f(x, y, z)$ in $\mathbb{CP}^2$. If $f(x, y, z)$ has real coefficients, the real affine part of the curve C is defined to be all $(x, y) \in \mathbb{R}^2$ such that $f(x, y, 1) = 0$.

There is a one-to-one correspondence between points $(u, v, w) \in \mathbb{CP}^2$ and lines L in $\mathbb{CP}^2$ given by the mapping that sends (u, v, w) to the line

$$\left\{(x, y, z) \in \mathbb{CP}^2 \ :\ ux + vy + wz = 0\right\}.$$

Therefore, (u, v, w) could denote either a point or a line. In the latter case, the coordinates are called *line coordinates.*

Let $f(x, y, z)$ be a homogeneous polynomial. Let C be the algebraic curve defined by f in point coordinates. That is,

$$C = \{(x, y, z) \in \mathbb{CP}^2 \,:\, f(x, y, z) = 0\}.$$

The curve C' which is *dual* to C is obtained by considering f in line coordinates. That is,

$$C' = \{(u, v, w) \in \mathbb{CP}^2 \,:\, ux + vy + wz = 0 \text{ is a tangent line to } C\}.$$

The dual curve is an algebraic curve [2] [15] except in the special case where the original curve is a line and the dual is a point; we will assume this is not the case. Therefore there exists a homogeneous polynomial $p(u, v, w)$ such that $(u, v, w) \in C'$ if and only if $p(u, v, w) = 0$. The references just noted also show that the dual curve to the dual curve is the original curve. Therefore a point on C corresponds to a tangent line to C' and vice versa.

Let $p(u, v, w)$ be a homogeneous polynomial and as usual set

$$\operatorname{grad} p(u, v, w) = (p_u(u, v, w), p_v(u, v, w), p_w(u, v, w)).$$

A *singular point* of p is a point (u_0, v_0, w_0) on the curve C' defined by $p = 0$ such that $\operatorname{grad} p(u_0, v_0, w_0) = (0, 0, 0)$. If (u_0, v_0, w_0) is not a singular point, then the curve C' has a well-defined tangent line at (u_0, v_0, w_0) with line coordinates $\operatorname{grad} p(u_0, v_0, w_0)$.

If $(u_0, v_0, 1)$ is a singular point of C', then the line $u = u_0 + \lambda t$, $v = v_0 + \mu t$ for $t \in \mathbb{C}$ intersects the curve C' at $t = 0$. If the second-order partial derivatives of p are not all zero at $(u_0, v_0, 1)$, then the Taylor expansion of $p(u_0 + \lambda t, v_0 + \mu t, 1)$ shows that

$$p(u, v, 1) = \left(p_{uu}(u_0, v_0, 1)\lambda^2 + 2p_{uv}(u_0, v_0, 1)\lambda\mu + p_{vv}(u_0, v_0, 1)\mu^2\right) t^2 + \cdots,$$

and in this case the curve C' has two tangent lines (counting multiplicity) at $(u_0, v_0, 1)$ defined by $(\lambda, \mu, 1)$ such that

$$p_{uu}(u_0, v_0, 1)\lambda^2 + 2p_{uv}(u_0, v_0, 1)\lambda\mu + p_{vv}(u_0, v_0, 1)\mu^2 = 0.$$

If the minimum order for which all the partial derivatives at $(u_0, v_0, 1)$ of order r are not identically zero is $r > 2$, we similarly obtain r tangent lines to C' at $(u_0, v_0, 1)$, counting multiplicities. Conversely, any point at which $p = 0$ has two (or more) tangent lines is a singular point of p.

Since $p_A(u, v, w)$ defined in (1.1) is a homogeneous polynomial, its zero set is an algebraic curve in $\mathbb{CP}^2$. Recall that Kippenhahn [9] showed that the convex hull of the real affine part of the curve $C(A)$ which is dual to $p_A(u, v, w) = 0$ is the numerical range of A. In terms of the description above, Kippenhahn showed that $p_A(u, v, w)$ is the polynomial p above, while the curve $C(A)$ is given by C above. He called $C(A)$ the *boundary generating curve* of $F(A)$. In the proof below, $C'(A)$ is the curve given by $p_A(u, v, w)$ in point coordinates.

Lemma 5.1. *Let A be an n-by-n matrix. If the line*

$$\{(x,y) \in \mathbb{R}^2 \, : \, u_0 x + v_0 y + 1 = 0\}$$

contains a flat portion on the boundary of $F(A)$, then the homogeneous polynomial $p_A(u,v,w)$ defined by equation (1.1) has a singularity at $(u_0, v_0, 1)$.

Proof. Any flat portion on the boundary of $F(A)$ is a line L defined by real numbers u_0, v_0 such that $p_A(u_0, v_0, 1) = 0$. Furthermore, L is tangent to two or more points on $C(A)$. Since the dual to the dual is the original curve, these points of tangency are both tangent lines to the dual curve $C'(A)$ at $(u_0, v_0, 1)$. Therefore $(u_0, v_0, 1)$ is a singular point of p_A since the tangent line there is not unique. □

Therefore the singularities of p_A help determine how many flat portions are possible on the boundary of $F(A)$. In order to study the flat portions on the boundary of a general nilpotent 4×4 matrix, we will show that the associated polynomial p_A has a special form where many of the coefficients are either zero or are equal to each other.

Note that if $z = \frac{u+iv}{2}$ and p_A is given by (1.1), then $p_A(u,v,w) = \det(zA^* + \overline{z}A - (-w)I)$. The latter expression is $q(-w)$ where $q(w)$ is the characteristic polynomial of $zA^* + \overline{z}A$. Applying Newton's identities to this matrix yields Lemma 5.2 below. To facilitate writing some lengthy expressions, we will use the following notation for certain traces of words that arise in the polynomial coefficients:

$$\beta_0 = \mathrm{Tr}(A^*AA^*A); \quad \beta_{11} = \mathrm{Tr}(AA^*); \quad \beta_{22} = \mathrm{Tr}(A^2(A^*)^2).$$

In addition, we set $\beta_{21} = \mathrm{Tr}(A^2A^*)$ and $\beta_{31} = \mathrm{Tr}(A^3A^*)$; trace properties then imply that $\overline{\beta}_{21} = \mathrm{Tr}([A^*]^2\, A)$ and $\overline{\beta}_{31} = \mathrm{Tr}([A^*]^3\, A)$.

Lemma 5.2. *Let A be a 4×4 nilpotent matrix. The boundary generating curve for A is defined by*

$$\begin{aligned} p_A(u,v,w) = c_1u^4 + c_2u^3v + c_3u^3w + (c_1+c_4)u^2v^2 + c_5u^2w^2 \\ + c_6u^2vw + c_2uv^3 + c_3uv^2w + c_4v^4 + c_6v^3w + c_5v^2w^2 + w^4, \end{aligned}$$

where the coefficients c_j are given below.

$$c_1 = -\frac{1}{16}\left(\beta_{31} + \overline{\beta}_{31} + \beta_{22} + \frac{1}{2}\beta_0 - \frac{1}{2}\beta_{11}^2\right),$$

$$c_2 = \frac{i}{8}\left(\beta_{31} - \overline{\beta}_{31}\right), \quad c_3 = \frac{1}{8}\left(\beta_{21} + \overline{\beta}_{21}\right),$$

$$c_4 = \frac{1}{16}\left(\beta_{31} + \overline{\beta}_{31} - \beta_{22} - \frac{1}{2}\beta_0 + \frac{1}{2}\beta_{11}^2\right), \quad c_5 = -\frac{1}{4}\beta_{11}, \quad c_6 = \frac{i}{8}\left(\overline{\beta}_{21} - \beta_{21}\right).$$

Proof. Let M be an $n \times n$ matrix with characteristic polynomial

$$q(w) = \sum_{j=0}^{n} (-1)^j q_j w^{n-j}.$$

By Newton's Identities (see [1]), if $q_0 = 1$, then the remaining coefficients ($m = 1, \dots, n$) satisfy

$$(-1)^m q_m = -\frac{1}{m}\sum_{j=0}^{m-1}(-1)^j \operatorname{Tr}\left(M^{m-j}\right) q_j.$$

Applying these identities to $M = zA^* + \overline{z}A$ will yield the coefficients of the polynomial

$$q(w) = q_0 w^4 - q_1 w^3 + q_2 w^2 - q_3 w + q_4$$

where each q_j will be a polynomial in u and v. The polynomial p_A will then be defined by $p_A(u,v,w) = q(-w) = q_0 w^4 + q_1 w^3 + q_2 w^2 + q_3 w + q_4$.

Note that since A is nilpotent, $\operatorname{Tr}(A^k) = \operatorname{Tr}([A^*]^k) = 0$ for $k = 1,2,3,4$. The calculations below are also simplified with the identity $\operatorname{Tr}(MN) = \operatorname{Tr}(NM)$ for all $n \times n$ matrices M and N.

Thus $q_0 = 1$ and $q_1 = \operatorname{Tr}(zA^* + \overline{z}A)q_0 = 0$. Next,

$$q_2 = -\frac{1}{2}\left(\operatorname{Tr}([zA^* + \overline{z}A]^2)q_0 - \operatorname{Tr}[zA^* + \overline{z}A]q_1\right) = -|z|^2\beta_{11} = -\frac{1}{4}(u^2+v^2)\beta_{11},$$

and

$$\begin{aligned}
q_3 &= \frac{1}{3}\left(\operatorname{Tr}([zA^* + \overline{z}A]^3)q_0 - \operatorname{Tr}([zA^* + \overline{z}A]^2)q_1 + \operatorname{Tr}(zA^* + \overline{z}A)q_2\right)\\
&= \frac{1}{3}\left(\operatorname{Tr}([zA^* + \overline{z}A]^3) - \operatorname{Tr}([zA^* + \overline{z}A]^2)0 + 0q_2\right)\\
&= \frac{1}{3}\operatorname{Tr}([zA^* + \overline{z}A]^3)\\
&= \frac{1}{3}\left(z^3\operatorname{Tr}[A^*]^3 + z^2\overline{z}\operatorname{Tr}([A^*]^2A + A[A^*]^2 + A^*AA^*)\right)\\
&\quad + \frac{1}{3}\left(\overline{z}^2 z\operatorname{Tr}(AA^*A + A^2A^* + A^*A^2) + \overline{z}^3\operatorname{Tr}(A^3)\right)\\
&= z^2\overline{z}\overline{\beta}_{21} + \overline{z}^2 z\beta_{21}\\
&= \left(\frac{u^3 + iu^2v + uv^2 + iv^3}{8}\right)\overline{\beta}_{21} + \left(\frac{u^3 - iu^2v + uv^2 - iv^3}{8}\right)\beta_{21}\\
&= \frac{u^3}{8}(\overline{\beta}_{21} + \beta_{21}) + \frac{u^2v}{8}(i\overline{\beta}_{21} - i\beta_{21}) + \frac{uv^2}{8}(\overline{\beta}_{21} + \beta_{21}) + \frac{v^3}{8}(i\overline{\beta}_{21} - i\beta_{21}).
\end{aligned}$$

Finally,

$$\begin{aligned}
q_4 &= -\frac{1}{4}\left\{\operatorname{Tr}([zA^* + \overline{z}A]^4)q_0 - \operatorname{Tr}([zA^* + \overline{z}A]^3)q_1\right.\\
&\qquad \left. + \operatorname{Tr}([zA^* + \overline{z}A]^2)q_2 - \operatorname{Tr}(zA^* + \overline{z}A)q_3\right\}\\
&= -\frac{1}{4}\left\{\operatorname{Tr}([zA^* + \overline{z}A]^4) + \operatorname{Tr}([zA^* + \overline{z}A]^2)q_2\right\}\\
&= -\frac{1}{4}\left\{\operatorname{Tr}([zA^* + \overline{z}A]^4) + 2|z|^2\operatorname{Tr}(AA^*)(-|z|^2\operatorname{Tr}(AA^*))\right\}
\end{aligned}$$

$$
\begin{aligned}
&= -\frac{1}{4}\left\{\operatorname{Tr}([zA^* + \overline{z}A]^4 - 2|z|^4\left[\operatorname{Tr}(AA^*)\right]^2\right\}\\
&= -\frac{1}{4}\left\{z^3\overline{z}\,4\overline{\beta}_{31} + |z|^4(4\beta_{22} + 2\beta_0 - 2\beta_{11}^2) + \overline{z}^3 z\,4\beta_{31}\right\}\\
&= -\frac{1}{64}\left\{(u+iv)^3(u-iv)4\overline{\beta}_{31} + (u-iv)^3(u+iv)4\beta_{31}\right.\\
&\quad \left. + (u^2+v^2)^2(4\beta_{22} + 2\beta_0 - 2\beta_{11}^2)\right\}.
\end{aligned}
$$

Expanding all powers and products involving u and v yields

$$
q_4 = c_1u^4 + c_2u^3v + (c_1 + c_4)u^2v^2 + c_2uv^3 + c_4v^4,
$$

where c_1, c_2 and c_4 are defined in the statement of the lemma.

Now $p_A(u,v,w) = w^4 + 0w^3 + q_2w^2 + q_3w + q_4$, and from this expression we can identity the coefficients of each of the degree 4 homogeneous terms in u, v, and w as stated in the lemma.

The w^4 term has coefficient 1 and all of the terms containing w^3 have coefficient 0.

The terms containing w^2 are obtained from q_2 and clearly the u^2w^2 and v^2w^2 coefficients are both $c_5 = -\operatorname{Tr}(AA^*)/4$, while there is no uvw^2 term.

The terms containing w are obtained from q_3. Note that the coefficients of u^3w and uv^2w are equal to each other with the value c_3, while the coefficients of u^2vw and v^3w are equal to each other with the value c_6.

The computation of q_4 provided the terms of p_A without w. □

Now we consider the condition where p_A has a singularity. Define the matrix

$$
Z(u,v,w) = \begin{bmatrix} 4u^3 + 2uv^2 & 3u^2v + v^3 & 3u^2w + v^2w & 2uv^2 & 2uw^2 & 2uvw \\ 2u^2v & u^3 + 3uv^2 & 2uvw & 2u^2v + 4v^3 & 2vw^2 & u^2w + 3v^2w \\ 0 & 0 & u^3 + uv^2 & 0 & 2(u^2+v^2)w & u^2v + v^3 \end{bmatrix}.
$$

The three rows of $Z(u,v,w)$ contain the terms of the partial derivatives of p_A with respect to u, v, and w, respectively.

Lemma 5.3. *The homogeneous polynomial p_A from Lemma 5.2 has a singularity at (u,v,w) if and only if*

$$
Z(u,v,w)\begin{bmatrix} c_1 \\ c_2 \\ c_3 \\ c_4 \\ c_5 \\ c_6 \end{bmatrix} = \begin{bmatrix} 0 \\ 0 \\ -4w^3 \end{bmatrix} \tag{5.1}
$$

Proof. Since the polynomial p_A has a singularity at a point (u,v,w) if and only if $\operatorname{grad} p_A(u,v,w)$ is the zero vector, the form of p_A from Lemma 5.2 yields the equations in (5.1). □

When $w = 1$, the system (5.1) becomes the non-homogeneous system

$$Z(u, v, 1)\begin{bmatrix} c_1 \\ c_2 \\ c_3 \\ c_4 \\ c_5 \\ c_6 \end{bmatrix} = \begin{bmatrix} 0 \\ 0 \\ -4 \end{bmatrix} \tag{5.2}$$

from which the following special case is immediate.

Lemma 5.4. *The polynomial p_A has a singularity at $(2, 0, 1)$ if and only if*

$$\begin{aligned} 8c_1 + 3c_3 + c_5 &= 0. \\ 2c_2 + c_6 &= 0. \\ 8c_3 + 8c_5 &= -4. \end{aligned}$$

The above condition is necessary for $F(A)$ to have a flat portion at $x = -1/2$. This system can be rewritten as

$$\begin{cases} c_3 = -4c_1 + \dfrac{1}{4}. \\ c_6 = -2c_2. \\ c_5 = 4c_1 - \dfrac{3}{4}. \end{cases} \tag{5.3}$$

We can use this necessary condition to eliminate certain possibilities involving other flat portions.

Theorem 5.5. *If A is a 4-by-4 unitarily irreducible nilpotent matrix, then $\partial F(A)$ has at most two flat portions.*

Proof. Assume A is a 4-by-4 unitarily irreducible nilpotent matrix. The associated polynomial p_A thus has the form given by Lemma 5.2. If there is at least one flat portion on the boundary of $F(A)$, we may rotate and scale A so that there is a flat portion on the line $x = -\frac{1}{2}$. This flat portion corresponds to a singularity $(2, 0, 1)$ so the system (5.3) is satisfied. Thus only the variables c_1, c_2, and c_4 are free.

Assume there is another flat portion on the line $ux + vy + 1 = 0$. By Lemma 5.1 there is a singularity at this $(u, v, 1)$ where $(u, v) \neq (2, 0)$. For any such singularity, we can eliminate c_3, c_5 and c_6 in the necessary equations (5.2) to obtain the new consistent system below.

$$\begin{bmatrix} 4u^3 + 2uv^2 + 8u - 12u^2 - 4v^2 & 3u^2v + v^3 - 4uv & 2uv^2 \\ 2u^2v - 8uv + 8v & u^3 + 3uv^2 - 2u^2 - 6v^2 & 2u^2v + 4v^3 \\ 4(2-u)(u^2+v^2) & -2v(u^2+v^2) & 0 \end{bmatrix}\begin{bmatrix} c_1 \\ c_2 \\ c_4 \end{bmatrix} = \begin{bmatrix} -\frac{1}{4}v^2 + \frac{3}{2}u - \frac{3}{4}u^2 \\ -\frac{1}{2}uv + \frac{3}{2}v \\ -4 + \left(\frac{6-u}{4}\right)(u^2+v^2) \end{bmatrix}. \tag{5.4}$$

If $v = 0$ and $u \neq 2$ for the singular point $(u, v, 1)$, then the corresponding flat portion is on a vertical line $x = -\frac{1}{u}$ and there are two parallel flat portions which must be the only flat portions by Corollary 4.2. If $u = 0$ at the singularity, then the system above is consistent if and only if $v = \pm 2$. The point $(0, 2)$ could only correspond to a flat portion on the line $y = -\frac{1}{2}$ and the point $(0, -2)$ could only correspond to a flat portion on $y = \frac{1}{2}$. Each of these support lines is at a distance of $\frac{1}{2}$ from the origin. Therefore Proposition 4.3 shows that if there are flat portions both on $x = -\frac{1}{2}$ and on either $y = \frac{1}{2}$ or $y = -\frac{1}{2}$, then there will only be these two flat portions. Therefore in the remainder of the argument, we will assume that any singular points satisfy $u \neq 0$ and $v \neq 0$.

To simplify row reductions in (5.4), put c_4 in the first column and c_1 in the third column. If the resulting matrix is row reduced using only the extra assumption that neither u nor v is zero then we get the matrix

$$\begin{bmatrix} 2uv^2 & v^3 + u(3u-4)v & 2(u-2)\left(v^2 + 2(u-1)u\right) & -\frac{v^2}{4} - \frac{3}{4}(u-2)u \\ 0 & -2v & 8-4u & -\frac{u}{4} + \frac{3}{2} - \frac{4}{u^2+v^2} \\ 0 & -2v(u^2+v^2-u) & 4(2-u)(u^2+v^2-u) & \frac{3}{4}u^2 + \frac{1}{2}v^2 - \frac{3}{2}u \end{bmatrix}.$$

If $u^2+v^2-u = 0$, the system described above is inconsistent unless $\frac{3}{4}u^2+\frac{1}{2}v^2-\frac{3}{2}u = 0$, but the combination of those equations implies $u = 0$ which has already been ruled out. Therefore we may assume $u^2 + v^2 - u \neq 0$ and thus obtain the row-equivalent matrix

$$\begin{bmatrix} 2uv^2 & v^3 + u(3u-4)v & 2(u-2)\left(v^2 + 2(u-1)u\right) & -\frac{v^2}{4} - \frac{3}{4}(u-2)u \\ 0 & v & 2(u-2) & \frac{u}{8} - \frac{3}{4} + \frac{2}{u^2+v^2} \\ 0 & 0 & 0 & t \end{bmatrix}, \tag{5.5}$$

where $t = \left(u^2+v^2-4\right)\left(u(u-2)^2 + (u-4)v^2\right)$.

The matrix (5.5) corresponds to an inconsistent system unless either $u^2+v^2 = 4$ or $v^2 = \frac{u(u-2)^2}{4-u}$. Any point (u, v) corresponding to a flat portion must satisfy at least one of these conditions so if there are two flat portions besides the one on $x = -\frac{1}{2}$, both must satisfy at least one of these conditions. When $u^2 + v^2 = 4$ the line $ux + vy + 1 = 0$ is at the same distance $\frac{1}{2}$ from the origin as the line $x = -\frac{1}{2}$ containing the original flat portion. Consequently if there is a singularity with $u^2 + v^2 = 4$, then the corresponding line contains the only other possible flat portion by Proposition 4.3.

Therefore, there could only be three flat portions if there are two different pairs (u_1, v_1) and (u_2, v_2) that form a 6×4 augmented matrix that is consistent and where each (u_j, v_j) pair satisfies $v^2 = \frac{u(u-2)^2}{4-u}$.

For a given singularity $(u, v, 1)$ with $v^2 = \frac{u(u-2)^2}{4-u}$, lengthy calculations show that the matrix (5.5) is row equivalent to

$$\begin{bmatrix} 1 & 0 & \frac{u-4}{u} & \frac{(u-6)(u-4)}{8u^2} \\ 0 & 1 & \frac{2(u-2)}{v} & \frac{(u-2)(u-8)}{8uv} \\ 0 & 0 & 0 & 0 \end{bmatrix}.$$

Therefore, if there are three flat portions on $\partial F(A)$, then there exist points (u_1, v_1) and (u_2, v_2) with neither $u_i = 0$ nor $v_i = 0$ for $i = 1, 2$ such that the matrix M below corresponds to a consistent system, and consequently satisfies $\det(M) = 0$.

$$M = \begin{bmatrix} 1 & 0 & \frac{u_1-4}{u_1} & \frac{(u_1-6)(u_1-4)}{8u_1^2} \\ 0 & 1 & \frac{2(u_1-2)}{v_1} & \frac{(u_1-2)(u_1-8)}{8u_1v_1} \\ 1 & 0 & \frac{u_2-4}{u_2} & \frac{(u_2-6)(u_2-4)}{8u_2^2} \\ 0 & 1 & \frac{2(u_2-2)}{v_2} & \frac{(u_2-2)(u_2-8)}{8u_2v_2} \end{bmatrix}.$$

Note that M has the form

$$M = \begin{bmatrix} 1 & 0 & a_1 & a_1c_1 \\ 0 & 1 & b_1 & b_1d_1 \\ 1 & 0 & a_2 & a_2c_2 \\ 0 & 1 & b_2 & b_2d_2 \end{bmatrix},$$

from which it follows that

$$\det M = b_2\left((a_2 - a_1)d_2 + a_1c_1 - a_2c_2\right) + b_1\left((a_1 - a_2)d_1 + a_2c_2 - a_1c_1\right).$$

Therefore,

$$\det M = \frac{2(u_2 - 2)}{v_2}\Delta_1 + \frac{2(u_1 - 2)}{v_1}\Delta_2,$$

where

$$\Delta_1 = \left(\frac{u_2 - 4}{u_2} - \frac{u_1 - 4}{u_1}\right)\frac{u_2 - 8}{16u_2} + \frac{(u_1 - 6)(u_1 - 4)}{8u_1^2} - \frac{(u_2 - 6)(u_2 - 4)}{8u_2^2}$$

and

$$\Delta_2 = \left(\frac{u_1 - 4}{u_1} - \frac{u_2 - 4}{u_2}\right)\frac{u_1 - 8}{16u_1} + \frac{(u_2 - 6)(u_2 - 4)}{8u_2^2} - \frac{(u_1 - 6)(u_1 - 4)}{8u_1^2}.$$

Simplifying and removing the common factor $\frac{u_1-u_2}{4u_1^2u_2^2}$ from both terms in Δ_1 and Δ_2 shows that

$$\det M = \frac{u_1 - u_2}{4u_1^2u_2^2}\left[\frac{2(u_2 - 2)}{v_2}\left(-u_1(u_2 - 8) - 12u_1 - 12u_2 + 5u_1u_2\right)\right.$$
$$\left. + \frac{2(u_1 - 2)}{v_1}\left(u_2(u_1 - 8) + 12u_1 + 12u_2 - 5u_1u_2\right)\right].$$

If $u_1 = u_2$, then $v_1^2 = \frac{u_1(u_1-2)^2}{4-u_1} = \frac{u_2(u_2-2)^2}{4-u_2} = v_2^2$ and hence the singular points (u_1, v_1) and (u_1, v_2) result in flat portions that are the same distance $1/\sqrt{u_1^2 + v_1^2}$ from the origin. Therefore these two flat portions cannot coincide with the original flat portion at $x = -1/2$ by Proposition 4.3.

So the only remaining case that could lead to three flat portions on the boundary of $F(A)$ is

$$\frac{u_2 - 2}{v_2}\left(u_1u_2 - u_1 - 3u_2\right) = \frac{u_1 - 2}{v_1}\left(u_1u_2 - u_2 - 3u_1\right), \tag{5.6}$$

because $\det M = 0$. Squaring both sides of (5.6) and replacing $\frac{(u_j-2)^2}{v_j^2}$ with $\frac{4-u_j}{u_j}$ for $j = 1, 2$ results in

$$\frac{4-u_2}{u_2}\left(u_1u_2 - u_1 - 3u_2\right)^2 = \frac{4-u_1}{u_1}\left(u_1u_2 - u_2 - 3u_1\right)^2,$$

and this implies that

$$u_1(4-u_2)(u_1u_2 - u_1 - 3u_2)^2 - u_2(4-u_1)(u_1u_2 - u_2 - 3u_1)^2 = 0.$$

However, the left side of the expression above is $4(u_1 - u_2)^3$, and as mentioned previously, $u_1 = u_2$ leads to a contradiction of Proposition 4.3. □

6. Case of unitarily reducible 5-by-5 matrices

With Theorem 5.5 at our disposal, it is not difficult to describe completely the situation with the flat portions on the boundary of $F(A)$ for nilpotent 5-by-5 matrices A, provided that they are unitarily reducible.

Theorem 6.1. *Let a 5-by-5 matrix A be nilpotent and unitarily reducible. Then there are at most two flat portions on the boundary of its numerical range. Moreover, any number from 0 to 2 is actually attained by some such matrices A.*

Proof. Suppose first that $\ker A \cap \ker A^* \neq \{0\}$. Then A is unitarily similar to $A_1 \oplus [0]$, where A_1 is also nilpotent. Consequently, $F(A) = F(A_1)$, and the statement follows from Proposition 2.1 if A_1 is in its turn unitarily reducible and Theorem 5.5 otherwise. Note that all three possibilities (no flat portions, one or two flat portions) on $\partial F(A)$ already materialize in this case.

Suppose now that $\ker A \cap \ker A^* = \{0\}$. Then A is unitarily similar to $A_1 \oplus A_2$, with one 2-by-2 and one 3-by-3 unitarily irreducible block. Multiplying A by an appropriate scalar and applying yet another unitary similarity if needed, we may without loss of generality suppose that

$$A_1 = \begin{bmatrix} 0 & r \\ 0 & 0 \end{bmatrix}, \quad A_2 = \begin{bmatrix} 0 & r_1 & r_2 \\ 0 & 0 & r_3 \\ 0 & 0 & 0 \end{bmatrix},$$

where $r, r_1, r_3 > 0$, $r_2 \geq 0$. The numerical range of A_1 is the circular disk of the radius $r/2$ centered at the origin. If $r_2 = 0$, then $F(A_2)$ also is a circular disk centered at the origin [8, Theorem 4.1], and $F(A)$, being the largest of the two disks, has no flat portions on its boundary. So, it remains to consider the case when all r_j are positive.

The distance from the origin to the supporting line of $F(A_2)$ forming angle θ with the vertical axis equals the maximal eigenvalue of $\operatorname{Re}(e^{i\theta}A_2)$, that is, half of the largest root of the polynomial

$$f_\theta(\lambda) = \lambda^3 - \lambda(r_1^2 + r_2^2 + r_3^2) - 2r_1r_2r_3\cos\theta. \tag{6.1}$$

Since f_θ is a monotonically increasing function of λ for $\lambda \geq \lambda_0 = \left(\frac{r_1^2+r_2^2+r_3^2}{3}\right)^{1/2}$, and since $f_\theta(\lambda_0) \leq 0$ due to the inequality between the arithmetic and geometric means of r_j^2, the maximal root $\lambda(\theta)$ of f_θ is bigger than λ_0. If $\cos\theta_1 < \cos\theta_2$, then $f_{\theta_1}(\lambda(\theta_2)) > 0$, and so $\lambda(\theta_1) < \lambda(\theta_2)$. In other words, the maximal root of f_θ is a strictly monotonic function of θ both on $[0,\pi]$ and $[-\pi,0]$. So, the disk $F(A_1)$ will have exactly two common supporting lines with $F(A_2)$ when $r/2$ lies strictly between the minimal and maximal distance from the supporting lines of $\partial F(A_2)$ to the origin, one common supporting line when $r/2$ equals this minimal or maximal distance, and none otherwise. Further reasoning depends on whether or not the parameters r_j are all equal.

Case 1. Among r_j at least two are distinct. According to already cited Theorem 4.1 from [8], $F(A_2)$ has the so-called "ovular shape"; in particular, there are no flat portions on its boundary. Then the flat portions on the boundary of $F(A)$ are exactly those lying on common supporting lines of $F(A_1)$ and $F(A_2)$, and so there are either two or none of them. To be more specific, the distance from the origin to the supporting line at θ discussed above is (using Viète's formula)

$$\lambda(\theta)/2 = \sqrt{s/3}\cos\left(\frac{1}{3}\arccos\left(\frac{3\sqrt{3}t\cos\theta}{s^{3/2}}\right)\right),$$

where $s = r_1^2 + r_2^2 + r_3^2$ and $t = r_1r_2r_3$. Since the distance from the origin to the tangent line of the disk $F(A_1)$ is a constant $r/2$, there will be two values of θ (opposite of each other) for which these tangent lines coincide with supporting lines of $F(A_2)$ if and only if

$$\sqrt{s/3}\cos\left(\frac{1}{3}\arccos\left(-\frac{3\sqrt{3}t}{s^{3/2}}\right)\right) < \frac{r}{2} < \sqrt{s/3}\cos\left(\frac{1}{3}\arccos\left(\frac{3\sqrt{3}t}{s^{3/2}}\right)\right),$$

and none resulting in flat portions otherwise.

Case 2. All r_j are equal. The boundary generating curve $C(A_2)$ (see Section 5 for the definition) is then a cardioid, appropriately shifted and scaled, as shown (yet again) in [8, Theorem 4.1]. Consequently, $\partial F(A_2)$ itself has a (vertical) flat portion, and we need to go into more details. To this end, suppose (without loss of generality) that $r_1 = r_2 = r_3 = 3$, and invoke the formula on p. 130 of [8], according to which $C(A_2)$ is given by the parametric equations

$$x(\theta) = 2\cos\theta + \cos 2\theta, \quad y(\theta) = 2\sin\theta + \sin 2\theta, \qquad \theta \in [-\pi,\pi]. \tag{6.2}$$

The boundary of $F(A_2)$ is the union of the arc γ of (6.2) corresponding to $\theta \in [-2\pi/3, 2\pi/3]$ with the vertical line segment ℓ connecting its endpoints. The remaining portion of the curve (6.2) lies inside $F(A_2)$.

Observe also that $|x(\theta) + iy(\theta)| = \sqrt{5 + 4\cos\theta}$ is an even function of θ monotonically decreasing on $[0,\pi]$. Putting these pieces together yields the following:

For $r \leq 3$, the disk $F(A_1)$ lies inside $F(A_2)$. Thus, $F(A) = F(A_2)$ has one flat portion on the boundary.

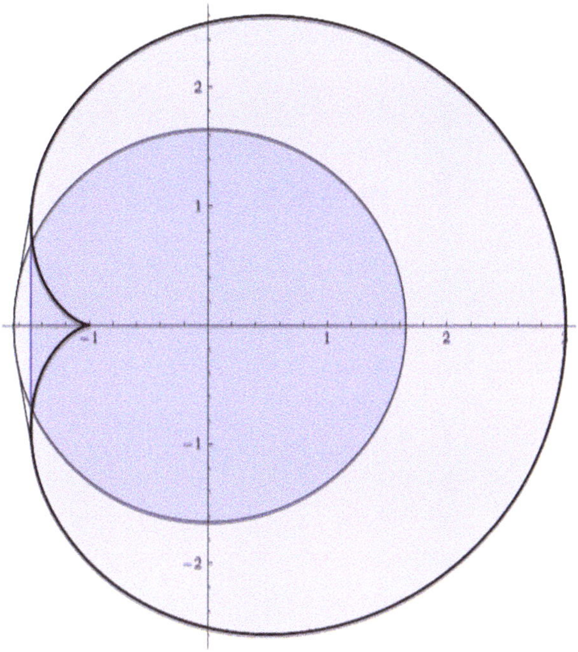

FIGURE 4. Numerical range of reducible 5-by-5 matrix with two flat portions

For $3 < r \leq 2\sqrt{3}$ the circle $\partial F(A_1)$ intersects $\partial F(A_2)$ at two points of ℓ. This results in two flat portions on $\partial F(A)$.

For $2\sqrt{3} < r < 6$ the circle $\partial F(A_1)$ intersects $\partial F(A_2)$ at two points of γ, while ℓ lies inside $F(A_1)$. This again results in two flat portions on $\partial F(A)$.

Finally, if $r \geq 6$, then $F(A_2)$ lies in $F(A_1)$, so $F(A) = F(A_1)$ is a circular disk, and there are no flat portions on its boundary. □

The case where $r_1 = r_2 = r_2 = 3$ and $r = 3.3$, which results in two flat portions caused by intersections between the circular disk and the numerical range of the 3-by-3 nilpotent matrix, is shown in Figure 4. The case where $r_1 = r_2 = r_3 = r = 3$, which results in one flat portion from the numerical range of the 3-by-3 block, with the circular disk tangent inside is shown in Figure 5. In all cases the cardioid boundary generating curve and the boundary of the numerical range of the 2-by-2 matrix is shown.

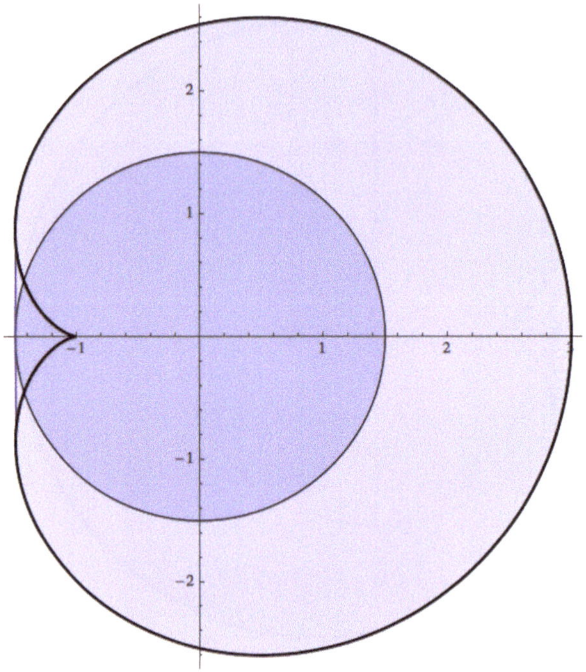

FIGURE 5. Numerical range of reducible 5-by-5 matrix with one flat portion

References

[1] G.A. Baker, *A new derivation of Newton's identities and their application to the calculation of the eigenvalues of a matrix*, J. Soc. Ind. Appl. Math **7** (1959), 143–148.

[2] E. Brieskorn and H. Knörrer, *Plane algebraic curves*, Birkhäuser Verlag Basel, Basel, 1986.

[3] E. Brown and I. Spitkovsky, *On flat portions on the boundary of the numerical range*, Linear Algebra Appl. **390** (2004), 75–109.

[4] H.-L. Gau and P.Y. Wu, *Line segments and elliptic arcs on the boundary of a numerical range*, Linear Multilinear Algebra **56** (2008), no. 1-2, 131–142.

[5] ______, *Numerical ranges of nilpotent operators*, Linear Algebra Appl. **429** (2008), no. 4, 716–726.

[6] K.E. Gustafson and D.K.M. Rao, *Numerical range. The field of values of linear operators and matrices*, Springer, New York, 1997.

[7] R.A. Horn and C.R. Johnson, *Topics in matrix analysis*, Cambridge University Press, Cambridge, 1991.

[8] D. Keeler, L. Rodman, and I. Spitkovsky, *The numerical range of* 3×3 *matrices*, Linear Algebra Appl. **252** (1997), 115–139.

[9] R. Kippenhahn, *Über den Wertevorrat einer Matrix*, Math. Nachr. **6** (1951), 193–228.

[10] ———, *On the numerical range of a matrix*, Linear Multilinear Algebra **56** (2008), no. 1-2, 185–225, Translated from the German by Paul F. Zachlin and Michiel E. Hochstenbach.

[11] T. Leake, B. Lins, and I.M. Spitkovsky, *Pre-images of boundary points of the numerical range*, Operators and Matrices **8** (2014), 699–724.

[12] M. Marcus and C. Pesce, *Computer generated numerical ranges and some resulting theorems*, Linear and Multilinear Algebra **20** (1987), 121–157.

[13] L. Rodman and I.M. Spitkovsky, 3×3 *matrices with a flat portion on the boundary of the numerical range*, Linear Algebra Appl. **397** (2005), 193–207.

[14] Shu-Hsien Tso and Pei Yuan Wu, *Matricial ranges of quadratic operators*, Rocky Mountain J. Math. **29** (1999), no. 3, 1139–1152.

[15] R.J. Walker, *Algebraic curves*, Princeton University Press, Princeton, 1950.

Erin Militzer
Mathematics Department
Ferris State University
Big Rapids, MI 49307, USA
e-mail: `militze@ferris.edu`

Linda J. Patton
Mathematics Department
California Polytechnic State University
San Luis Obispo, CA 93407, USA
e-mail: `lpatton@calpoly.edu`

Ilya M. Spitkovsky
Division of Science
New York University Abu Dhabi (NYUAD)
Saadiyat Island, P.O. Box 129188
Abu Dhabi, UAE
e-mail: `ims2@nyu.edu`
`imspitkovsky@gmail.com`

Ming-Cheng Tsai
General Education Center
National Taipei University of Technology
Taipei, 10608, Taiwan (R.O.C.)
e-mail: `mctsai2@gmail.com`

Operator Theory:
Advances and Applications, Vol. 259, 593–619

Traces on Operator Ideals and Related Linear Forms on Sequence Ideals (Part IV)

Albrecht Pietsch

Dedicated to the 60th birthday of Albrecht Böttcher

Abstract. The concept of a dyadic representation was used for the first time in 1963, when I constructed traces of operators (acting on a Banach space) whose sequence of approximation numbers is summable. Only recently, in a series of papers, those representations played the decisive role in describing *all* traces on arbitrary operator ideals over the separable infinite-dimensional Hilbert space. Now this method is extended to operator ideals on Banach spaces defined by means of generalized approximation numbers. The results are demonstrated on the example of convolution operators generated by functions belonging to certain Lipschitz/Besov spaces.

Mathematics Subject Classification (2010). Primary 47B10, 35S05. Secondary 46B45.

Keywords. Operator ideal, sequence ideal, trace, shift-invariant linear form, dyadic representation, generalized approximation number, convolution operator, Lipschitz/Besov space.

1. Notation

Given scalar-valued functions f and g on a set $\mathbb{S}$, we write $f = O(g)$ (Landau's *big-O*-condition) if $|f(s)| \le c|g(s)|$ for all $s \in \mathbb{S}$ and a constant $c > 0$.

The cardinality of any finite set $\mathbb{F}$ is denoted by $|\mathbb{F}|$. In contrast to $\subseteq$, we write $\subset$ for *strict* inclusions.

2. Operator ideals

Throughout, X and Y are (real or complex) *Banach spaces* with *duals* X^* and Y^*, respectively. We write $\langle x, x^* \rangle$ for the value of the functional $x^* \in X^*$ at the element $x \in X$.

Many thanks to Winfried Sickel and Sergey Tikhonov for helpful hints concerning approximation of periodic functions by trigonometric polynomials.

The set of all (bounded and linear) operators S from X into Y is denoted by $\mathfrak{L}(X,Y)$.

Concerning the theory of (*quasi-Banach*) *operator ideals*

$$\mathfrak{A} = \bigcup_{X,Y} \mathfrak{A}(X,Y)$$

and their history, we refer to [20, 21, 22]. The underlying quasi-norm $\|\cdot|\mathfrak{A}\|$ is supposed to satisfy the quasi-triangle inequality

$$\|S+T|\mathfrak{A}\| \le c_{\mathfrak{A}}\big[\|S|\mathfrak{A}\| + \|T|\mathfrak{A}\|\big] \quad \text{for} \quad S,T \in \mathfrak{A}(X,Y)$$

with a constant $c_{\mathfrak{A}} \ge 1$.

The ideal of *finite rank operators* is denoted by $\mathfrak{F}$. Our basic examples are the ideals $\mathfrak{P}_{q,2}$ with $q \ge 2$ formed by the *absolutely* $(q,2)$*-summing operators*, which are defined as follows:

There exists a constant $c \ge 0$ such that

$$\Big(\sum_{k=1}^{n} \|Sx_k\|^q\Big)^{1/q} \le c \sup_{\|x^*\|\le 1} \Big(\sum_{k=1}^{n} |\langle x_k, x^*\rangle|^2\Big)^{1/2}$$

for any finite family of elements $x_1,\dots,x_n \in X$. The underlying norm is given by

$$\|S|\mathfrak{P}_{q,2}\| := \inf c.$$

Instead of $\mathfrak{P}_{2,2}$ we simply write $\mathfrak{P}_2$, and $\mathfrak{P}_{\infty,2}$ is nothing but $\mathfrak{L}$, the ideal of all operators.

3. Sequence ideals

Throughout, $\mathfrak{l}_\infty(\mathbb{N})$ denotes the *algebra* of all bounded scalar sequences $s = (\sigma_m)$ indexed by $\mathbb{N} := \{1,2,\dots\}$. We refer to

$$a_n(s) := \inf_{|\mathbb{F}|<n} \sup_{m\notin\mathbb{F}} |\sigma_m|$$

as the nth *approximation number* of $s = (\sigma_m) \in \mathfrak{l}_\infty(\mathbb{N})$. If $|\sigma_1| \ge |\sigma_2| \ge \dots \ge 0$, then $a_n(s) = |\sigma_n|$.

By a *symmetric sequence ideal* we mean a linear subspace $\mathfrak{s}(\mathbb{N})$ of $\mathfrak{l}_\infty(\mathbb{N})$ such that

$$s \in \mathfrak{s}(\mathbb{N}), \quad t \in \mathfrak{l}_\infty(\mathbb{N}), \quad \text{and} \quad a_n(t) \le a_n(s) \quad \text{imply} \quad t \in \mathfrak{s}(\mathbb{N}).$$

Similarly to the preceding, $\mathfrak{l}_\infty(\mathbb{N}_0)$ denotes the *algebra* of all bounded scalar sequences $a = (\alpha_h)$ indexed by $\mathbb{N}_0 := \{0,1,2,\dots\}$. We refer to

$$o_k(a) := \sup_{h\ge k} |\alpha_h|$$

as the kth *ordering number* of $a = (\alpha_h) \in \mathfrak{l}_\infty(\mathbb{N}_0)$. If $|\alpha_0| \ge |\alpha_1| \ge |\alpha_2| \ge \dots \ge 0$, then $o_k(a) = |\alpha_k|$.

On $\mathfrak{l}_\infty(\mathbb{N}_0)$, we consider the *forward shift*

$$S_+ : (\alpha_0, \alpha_1, , \alpha_2, \dots) \mapsto (0, \alpha_0, \alpha_1, \dots)$$

and the *backward shift*

$$S_- : (\alpha_0, \alpha_1, , \alpha_2, \dots) \mapsto (\alpha_1, \alpha_2, \alpha_3, \dots).$$

By a *shift-monotone sequence ideal* we mean an S_+-invariant linear subspace $\mathfrak{z}(\mathbb{N}_0)$ of $\mathfrak{l}_\infty(\mathbb{N}_0)$ such that

$$a \in \mathfrak{z}(\mathbb{N}_0), \quad b \in \mathfrak{l}_\infty(\mathbb{N}_0), \quad \text{and} \quad o_k(b) \leq o_k(a) \quad \text{imply} \quad b \in \mathfrak{z}(\mathbb{N}_0).$$

This condition ensures that $\mathfrak{z}(\mathbb{N}_0)$ is also S_--invariant. In particular, we have $a \in \mathfrak{z}(\mathbb{N}_0)$ if and only if $(o_k(a)) \in \mathfrak{z}(\mathbb{N}_0)$. From now on, $\mathfrak{z}(\mathbb{N}_0)$ always stands for a shift-monotone sequence ideal.

Let d_k denote the characteristic sequence of

$$\mathbb{D}_k{:=}\{m \in \mathbb{N} : 2^k \leq m < 2^{k+1}\},$$

and define the stretched sequence

$$d_a := \sum_{k=0}^{\infty} \alpha_k d_k = (\alpha_0, \alpha_1, \alpha_1, \dots, \overbrace{\alpha_k, \dots, \alpha_k}^{2^k \text{ terms}}, \dots) \quad \text{for} \quad a = (\alpha_k) \in \mathfrak{l}_\infty(\mathbb{N}_0).$$

Then $a \longmapsto d_a$ is a linear map from $\mathfrak{l}_\infty(\mathbb{N}_0)$ into $\mathfrak{l}_\infty(\mathbb{N})$.

We now recall Theorem 4.6 from [25]:
The rules

$$\mathfrak{s}(\mathbb{N}) \longmapsto \mathfrak{z}(\mathbb{N}_0) := \{a \in \mathfrak{l}_\infty(\mathbb{N}_0) : d_a \in \mathfrak{s}(\mathbb{N})\}$$

and

$$\mathfrak{z}(\mathbb{N}_0) \longmapsto \mathfrak{s}(\mathbb{N}) := \{s \in \mathfrak{l}_\infty(\mathbb{N}) : (a_{2^k}(s)) \in \mathfrak{z}(\mathbb{N}_0)\}$$

define a one-to-one correspondence between all symmetric sequence ideals $\mathfrak{s}(\mathbb{N})$ *and all shift-monotone sequence ideals* $\mathfrak{z}(\mathbb{N}_0)$.

Sequence ideals $\mathfrak{s}(\mathbb{N})$ and $\mathfrak{z}(\mathbb{N}_0)$ related in this way are called *associated.*

The most important examples are the symmetric *Lorentz sequence ideals*

$$\mathfrak{l}_{r,w}(\mathbb{N}) := \{s \in \mathfrak{l}_\infty(\mathbb{N}) : (n^{1/r-1/w} a_n(s)) \in \mathfrak{l}_w(\mathbb{N})\}$$

and their *shift-monotone* counterparts

$$\mathfrak{d}_{r,w}(\mathbb{N}_0) := \{a \in \mathfrak{l}_\infty(\mathbb{N}_0) : (2^{k/r} o_k(a)) \in \mathfrak{l}_w(\mathbb{N}_0)\},$$

defined for $0 < r < \infty$ and $0 < w \leq \infty$. Using Landau's *big-O*-condition yields

$$\mathfrak{d}_{r,\infty}(\mathbb{N}_0) = \{(\alpha_h) : |\alpha_h| = O(2^{-h/r})\}.$$

4. Generalized approximation numbers

Let $\mathfrak{A}$ be any quasi-Banach operator ideal. The nth *$\mathfrak{A}$-approximation number* of $S \in \mathfrak{A}(X,Y)$ is defined by

$$a_n(S|\mathfrak{A}) := \inf\big\{\|S-F|\mathfrak{A}\| : F \in \mathfrak{F}(X,Y),\ \ \mathrm{rank}(F) < n\big\}.$$

In the case $\mathfrak{A} = \mathfrak{L}$, we obtain the usual approximation numbers, $a_n(S)$.

The following properties can easily be verified for arbitrary operators $S,T \in \mathfrak{A}(X,Y)$, $A \in \mathfrak{L}(X_0,X)$, and $B \in \mathfrak{L}(Y,Y_0)$.

(1) $\|S|\mathfrak{A}\| = a_1(S|\mathfrak{A}) \ge a_2(S|\mathfrak{A}) \ge \cdots \ge 0$.
(2) $a_{m+n-1}(S+T|\mathfrak{A}) \le c_{\mathfrak{A}}\big[a_m(S|\mathfrak{A}) + a_n(T|\mathfrak{A})\big]$.
(3) $a_n(BSA|\mathfrak{A}\| \le \|B\| a_n(S|\mathfrak{A})\|A\|$.
(4) $a_n(S|\mathfrak{A}) = 0$ if and only if $\mathrm{rank}(S) < n$.

Suppose that the sequence ideals $\mathfrak{s}(\mathbb{N})$ and $\mathfrak{z}(\mathbb{N}_0)$ are associated. Then the components

$$\mathfrak{A}^{\mathrm{app}}_{\mathfrak{s}}(X,Y) := \big\{S \in \mathfrak{A}(X,Y) : \big(a_n(S|\mathfrak{A})\big) \in \mathfrak{s}(\mathbb{N})\big\}$$

and

$$\mathfrak{A}^{\mathrm{APP}}_{\mathfrak{z}}(X,Y) := \big\{S \in \mathfrak{A}(X,Y) : \big(a_{2^k}(S|\mathfrak{A})\big) \in \mathfrak{z}(\mathbb{N}_0)\big\}$$

form one and the same operator ideal, $\mathfrak{A}^{\mathrm{app}}_{\mathfrak{s}} = \mathfrak{A}^{\mathrm{APP}}_{\mathfrak{z}}$.

The classical Lorentz operator ideal $\mathfrak{A}^{\mathrm{app}}_{\mathfrak{l}_{r,w}} = \mathfrak{A}^{\mathrm{APP}}_{\mathfrak{d}_{r,w}}$ is simply denoted by $\mathfrak{A}^{\mathrm{app}}_{r,w}$ or $\mathfrak{A}^{\mathrm{APP}}_{r,w}$. In particular, $\mathfrak{L}^{\mathrm{app}}_r := \mathfrak{L}^{\mathrm{app}}_{r,r}$.

5. Dyadic representations

The infinite series

$$S = \sum_{k=0}^{\infty} S_k \quad \text{(convergent with respect to the operator norm)}$$

is called a *dyadic representation* of the operator $S \in \mathfrak{L}(X,Y)$ if $S_k \in \mathfrak{F}(X,Y)$ such that $\mathrm{rank}(S_k) \le 2^k$.

Given a quasi-Banach operator ideal $\mathfrak{A}$ and a shift-monotone sequence ideal $\mathfrak{z}(\mathbb{N}_0)$, we refer to $S = \sum_{k=0}^{\infty} S_k$ as an *$(\mathfrak{A},\mathfrak{z})$-representation* if

$$\Big(\Big\|S - \sum_{k=0}^{h-1} S_k\Big|\mathfrak{A}\Big\|\Big) \in \mathfrak{z}(\mathbb{N}_0). \qquad (\lozenge)$$

Theorem 5.1. $\mathfrak{A}^{\mathrm{APP}}_{\mathfrak{z}}$ *consists of all operators that admit an* $(\mathfrak{A},\mathfrak{z})$*-representation.*

Proof. If $S \in \mathfrak{A}^{\mathrm{APP}}_{\mathfrak{z}}(X,Y)$, then there exist operators $F_k \in \mathfrak{F}(X,Y)$ such that

$$\|S - F_k|\mathfrak{A}\| \le 2a_{2^k}(S|\mathfrak{A}) \quad \text{and} \quad \mathrm{rank}(F_k) < 2^k.$$

Note that $F_0 = O$. Letting

$$S_0 := O, \quad S_1 := O, \quad \text{and} \quad S_{k+2} := F_{k+1} - F_k \quad \text{for} \quad k = 0,1,2,\dots,$$

we get

$$\sum_{k=0}^{h+1} S_k = F_h, \quad \left\| S - \sum_{k=0}^{h+1} S_k \middle| \mathfrak{A} \right\| = \|S - F_h|\mathfrak{A}\| \le 2a_{2^h}(S|\mathfrak{A}), \text{ and } \operatorname{rank}(S_k) \le 2^k.$$

So the S_+-invariance of $\mathfrak{z}(\mathbb{N}_0)$ yields ($\Diamond$), which means that $S = \sum_{k=0}^{\infty} S_k$ is an $(\mathfrak{A}, \mathfrak{z})$-representation.

Since

$$\operatorname{rank}\left(\sum_{k=0}^{h-1} S_k\right) < 2^h \quad \text{whenever} \quad \operatorname{rank}(S_k) \le 2^k,$$

we have

$$a_{2^h}(S|\mathfrak{A}) \le \left\| S - \sum_{k=0}^{h-1} S_k \middle| \mathfrak{A} \right\|.$$

Therefore, conversely, ($\Diamond$) implies $\big(a_{2^h}(S|\mathfrak{A})\big) \in \mathfrak{z}(\mathbb{N}_0)$ or $S \in \mathfrak{A}_{\mathfrak{z}}^{\text{APP}}(X, Y)$. □

We mention a trivial, but useful observation.

Lemma 5.2. *If*

$$S = \sum_{k=0}^{\infty} S_k \quad \textit{and} \quad T = \sum_{k=0}^{\infty} T_k$$

are $(\mathfrak{A}, \mathfrak{z})$*-representations of* $S \in \mathfrak{L}(X, Y)$ *and* $T \in \mathfrak{L}(X, Y)$*, respectively, then*

$$S + T = \sum_{k=0}^{\infty} Z_k \quad \textit{with} \quad Z_0 := O \quad \textit{and} \quad Z_{k+1} := S_k + T_k \quad \textit{for} \quad k = 0, 1, 2, \dots$$

is an $(\mathfrak{A}, \mathfrak{z})$*-representation of their sum.*

Proposition 5.3. *For every dyadic representation* $S = \sum_{k=0}^{\infty} S_k$*,*

$$\left(\left\| S - \sum_{k=0}^{h-1} S_k \middle| \mathfrak{A} \right\| \right) \in \mathfrak{z}(\mathbb{N}_0) \tag{$\Diamond$}$$

implies

$$\big(\|S_h|\mathfrak{A}\|\big) \in \mathfrak{z}(\mathbb{N}_0). \tag{$\blacklozenge$}$$

Proof. Use

$$\|S_h|\mathfrak{A}\| \le c_{\mathfrak{A}} \left[\left\| S - \sum_{k=0}^{h-1} S_k \middle| \mathfrak{A} \right\| + \left\| S - \sum_{k=0}^{h} S_k \middle| \mathfrak{A} \right\| \right]$$

and the S_--invariance of $\mathfrak{z}(\mathbb{N}_0)$. □

6. Traces

The concept of a trace makes sense only for square matrices. Similarly, in the setting of operators, we must assume that the operators act on one and the same Banach space.

A *trace* on an operator ideal $\mathfrak{A}$ is a linear form τ, defined on all components $\mathfrak{A}(X){:=}\mathfrak{A}(X,X)$, for which the following holds:

$$\tau(AS)=\tau(SA) \quad \text{whenever}$$

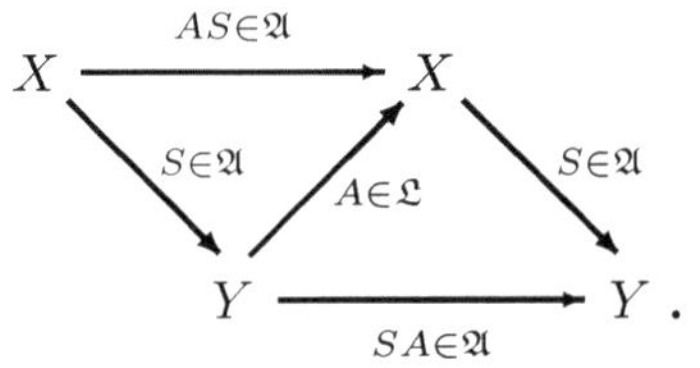

The smallest non-trivial ideal $\mathfrak{F}$ consists of all finite rank operators. Every trace on $\mathfrak{F}$ is a scalar multiple of the classical trace defined by

$$\operatorname{trace}(F):=\sum_{m=1}^{n}\langle x_m,x_m^*\rangle,$$

where $F=\sum_{m=1}^{n}x_m^*\otimes x_m$ with $x_1,\dots,x_n\in X$ and $x_1^*,\dots,x_n^*\in X^*$ is any finite representation of $F\in\mathfrak{F}(X)$.

According to [24, p. 38], a quasi-Banach operator ideal $\mathfrak{A}$ is said to be of *trace type* $\varrho\in[0,1]$ if there exists a constant $c\geq 1$ such that, for every operator $F\in\mathfrak{F}(X)$, we have

$$|\operatorname{trace}(F)|\leq cn^{\varrho}\|F|\mathfrak{A}\| \quad \text{whenever} \quad \operatorname{rank}(F)\leq n.$$

Being of trace type 1 is a universal property; see Example 6.2 (1).

An operator ideal $\mathfrak{A}$ is of *eigenvalue type* $\mathfrak{l}_{r,w}$ if, for any Banach space X, every $S\in\mathfrak{A}(X)$ is a Riesz operator and $(\lambda_m(S))\in\mathfrak{l}_{r,w}$, where $(\lambda_m(S))$ denotes the sequence of all eigenvalues counted according to their multiplicities and ordered such that $|\lambda_1(S)|\geq|\lambda_2(S)|\geq\cdots\geq 0$; see [21, 3.4.1].

In the case of a quasi-Banach ideal, there holds a principle of uniform boundedness; see [21, 3.4.6]:

$$\|(\lambda_m(S))|\mathfrak{l}_{r,w}\|\leq c\|S|\mathfrak{A}\| \quad \text{for all } S\in\mathfrak{A}(X) \text{ and some constant } c\geq 1.$$

Proposition 6.1. *Let $1<r<\infty$. If a quasi-Banach operator ideal $\mathfrak{A}$ is of eigenvalue type $\mathfrak{l}_{r,\infty}$, then it is of trace type $\frac{1}{r^*}$.*

Proof. In view of the classical trace formula, it follows from $\operatorname{rank}(F)\leq n$ that

$$\begin{aligned}|\operatorname{trace}(F)|&=\Big|\sum_{m=1}^{n}\lambda_m(F)\Big|\leq\sum_{m=1}^{n}m^{-1/r}m^{1/r}|\lambda_m(F)|\\&\leq c_r n^{1-1/r}\|(\lambda_m(F))|\mathfrak{l}_{r,\infty}\|\leq c_r c\, n^{1/r^*}\|F|\mathfrak{A}\|.\end{aligned}$$

□

Example 6.2.

(1) $\mathfrak{L}$ *is of trace type* 1,

(2) $\mathfrak{L}_{r,\infty}^{\text{app}}$ *is of trace type* $\frac{1}{r^*}$ *for* $1 < r < \infty$,

(3) $\mathfrak{P}_{q,2}$ *is of trace type* $\frac{1}{q^*}$ *for* $2 \le q < \infty$.

Proof. If $\operatorname{rank}(F) \le n$, then we infer from $|\lambda_m(F)| \le \|F\|$ that

$$|\operatorname{trace}(F)| = \left| \sum_{m=1}^{n} \lambda_m(F) \right| \le n\|F\|,$$

which proves (1).

By Proposition 6.1, the cases (2) and (3) are corollaries of [21, 3.6.3] and [21, 3.7.6], respectively. □

A linear form μ on $\mathfrak{z}(\mathbb{N}_0)$ is said to be γS_+*-invariant* with $\gamma > 0$ if

$$\mu(\gamma S_+ a) = \mu(a) \quad \text{for all} \quad a \in \mathfrak{z}(\mathbb{N}_0).$$

We now present the main result of this paper, which is a generalization of [27, Theorem 4]; see also [24, Theorem 7.12].

Theorem 6.3. *Let* $\mathfrak{A}$ *be a quasi-Banach operator ideal of trace type* ϱ, *and let* μ *be a* $2^{-\varrho}S_+$*-invariant linear form on a shift-monotone sequence ideal* $\mathfrak{z}(\mathbb{N}_0)$. *Then the expression*

$$\tau(S) := \mu\big(2^{-k\varrho}\operatorname{trace}(S_k)\big)$$

does not depend on the choice of the $(\mathfrak{A},\mathfrak{z})$*-representation* $S = \sum_{k=0}^{\infty} S_k$. *This definition yields a trace* τ *on* $\mathfrak{A}_{\mathfrak{z}}^{\text{APP}}$.

Proof. First of all, we infer from $|\operatorname{trace}(S_k)| \le c2^{k\varrho}\|S_k|\mathfrak{A}\|$ and Proposition 5.3 that $\big(2^{-k\varrho}\operatorname{trace}(S_k)\big) \in \mathfrak{z}(\mathbb{N}_0)$. Hence $\mu\big(2^{-k\varrho}\operatorname{trace}(S_k)\big)$ makes sense.

Consider different $(\mathfrak{A},\mathfrak{z})$-representations

$$S = \sum_{k=0}^{\infty} S_k^{(1)} \quad \text{and} \quad S = \sum_{k=0}^{\infty} S_k^{(2)}.$$

By Lemma 5.2, letting

$$Z_0 := O \quad \text{and} \quad Z_{k+1} := S_k^{(1)} - S_k^{(2)} \quad \text{for} \quad k = 0, 1, 2, \dots$$

gives an $(\mathfrak{A},\mathfrak{z})$-representation of the zero operator.

If the sequences $a = (\alpha_h)$ and $b = (\beta_h)$ are defined by

$$\alpha_h := 2^{-h\varrho}\operatorname{trace}(Z_h) \quad \text{and} \quad \beta_h := 2^{-h\varrho}\operatorname{trace}\Big(\sum_{k=0}^{h} Z_k\Big),$$

then it follows from

$$\alpha_0 = \beta_0 = 0 \quad \text{and} \quad \beta_h - 2^{-\varrho}\beta_{h-1} = 2^{-h\varrho}\operatorname{trace}(Z_h) = \alpha_h \quad \text{for} \quad h = 1, 2, \dots$$

that $S_+S_-a = a$ and $a = b - 2^{-\varrho}S_+b$. We have $|\alpha_h| \le c\|Z_h|\mathfrak{A}\|$ and

$$|\beta_h| = 2^{-h\varrho}\left|\operatorname{trace}\Big(\sum_{k=0}^{h} Z_k\Big)\right| \le c2^{\varrho}\left\|\sum_{k=0}^{h} Z_k\Big|\mathfrak{A}\right\| = c2^{\varrho}\left\|O - \sum_{k=0}^{h} Z_k\Big|\mathfrak{A}\right\|.$$

Therefore $a, b \in \mathfrak{z}(\mathbb{N}_0)$. We now obtain

$$\begin{aligned}\mu\big(2^{-k\varrho}\operatorname{trace}(S_k^{(1)})\big) - \mu\big(2^{-k\varrho}\operatorname{trace}(S_k^{(2)})\big) &= \mu\big(2^{-k\varrho}\operatorname{trace}(Z_{k+1})\big)\\ = 2^{\varrho}\mu(\alpha_{k+1}) = 2^{\varrho}\mu(S_-a) = \mu(S_+S_-a) &= \mu(a) = \mu(b) - \mu(2^{-\varrho}S_+b) = 0,\end{aligned}$$

which shows that $\tau(S)$ is well defined.

Using $(\mathfrak{A},\mathfrak{z})$-representations $S = \sum_{k=0}^{\infty} S_k$, $T = \sum_{k=0}^{\infty} T_k$, and

$$S + T = \sum_{k=0}^{\infty} Z_k \quad \text{with } Z_0{:=}O \text{ and } Z_{k+1}{:=}S_k + T_k \text{ for } k = 0, 1, 2, \dots,$$

we get

$$\begin{aligned}\tau(S) &= \mu\big(\operatorname{trace}(S_0), \dots, 2^{-k\varrho}\operatorname{trace}(S_k), \dots\big)\\ &= \mu\big(\quad 0 \quad, 2^{-\varrho}\operatorname{trace}(S_0), \dots, 2^{-(k+1)\varrho}\operatorname{trace}(S_k), \dots\big),\end{aligned}$$

$$\begin{aligned}\tau(T) &= \mu\big(\operatorname{trace}(T_0), \dots, 2^{-k\varrho}\operatorname{trace}(T_k), \dots\big)\\ &= \mu\big(\quad 0 \quad, 2^{-\varrho}\operatorname{trace}(T_0), \dots, 2^{-(k+1)\varrho}\operatorname{trace}(T_k), \dots\big),\end{aligned}$$

and

$$\tau(S+T) = \mu\big(\operatorname{trace}(Z_0), 2^{-\varrho}\operatorname{trace}(Z_1), \dots, 2^{-(k+1)\varrho}\operatorname{trace}(Z_{k+1}), \dots\big).$$

Thus τ is additive. The remaining trace properties are obvious. □

7. The operator ideals $(\mathfrak{P}_{p^*,2})^{\mathrm{app}}_{r,\infty}$

In this section, Theorem 6.3 will be applied to the operator ideal $(\mathfrak{P}_{p^*,2})^{\mathrm{app}}_{r,\infty}$ with $p^* \ge 2$. The use of $\mathfrak{P}_{p^*,2}$ instead of $\mathfrak{P}_{q,2}$ is motivated by the desire to stress the relationship described in Remark 9.3.

Theorem 7.1. *Let μ be a $2^{-1/p}S_+$-invariant linear form on $\mathfrak{d}_{r,\infty}(\mathbb{N}_0)$. Then the expression*

$$\tau(S) := \mu\big(2^{-k/p}\operatorname{trace}(S_k)\big)$$

does not depend on the choice of the $(\mathfrak{P}_{p^*,2}, \mathfrak{d}_{r,\infty})$-representation $S = \sum_{k=0}^{\infty} S_k$. This definition yields a trace τ on $(\mathfrak{P}_{p^*,2})^{\mathrm{app}}_{r,\infty}$.

Proof. Note that $\mathfrak{P}_{p^*,2}$ is of trace type $\frac{1}{p}$. □

For simplification, we pass from linear forms on different sequence ideals $\mathfrak{d}_{r,\infty}(\mathbb{N}_0)$ to linear forms on $\mathfrak{l}_\infty(\mathbb{N}_0)$.

Lemma 7.2. *Let $0 < p, r < \infty$. Then the map*

$$W_r : (\alpha_h) \mapsto (2^{-h/r}\alpha_h)$$

defines an isomorphism between $\mathfrak{l}_\infty(\mathbb{N}_0)$ and $\mathfrak{d}_{r,\infty}(\mathbb{N}_0)$. The formula

$$\lambda(a) = \mu(W_r a) \quad \textit{for all} \quad a \in \mathfrak{l}_\infty(\mathbb{N}_0)$$

yields a one-to-one correspondences between

the $2^{1/r-1/p}S_+$-invariant linear forms λ on $\mathfrak{l}_\infty(\mathbb{N}_0)$

and

the $2^{-1/p}S_+$-invariant linear forms μ on $\mathfrak{d}_{r,\infty}(\mathbb{N}_0)$.

Proof. With the understanding that $\alpha_{-1} = 0$, we have

$$W_r S_+ : (\alpha_h) \overset{S_+}{\longmapsto} (\alpha_{h-1}) \overset{W_r}{\longmapsto} (2^{-h/r}\alpha_{h-1})$$

and

$$S_+ W_r : (\alpha_h) \overset{W_r}{\longmapsto} (2^{-h/r}\alpha_h) \overset{S_+}{\longmapsto} (2^{-(h-1)/r}\alpha_{h-1}).$$

Hence

$$2^{1/r} W_r S_+ = S_+ W_r.$$

Assume that $\mu(2^{-1/p}S_+ b) = \mu(b)$ for $b = W_r a \in \mathfrak{d}_{r,\infty}(\mathbb{N}_0)$. Then we get

$$\lambda(2^{1/r-1/p}S_+ a) = \mu(2^{1/r-1/p}W_r S_+ a) = \mu(2^{-1/p}S_+ W_r a) = \mu(W_r a) = \lambda(a),$$

which shows that the $2^{-1/p}S_+$-invariance of μ implies the $2^{1/r-1/p}S_+$-invariance of λ. The converse direction follows similarly. □

Now Theorem 7.1 can be restated as follows.

Theorem 7.3. *Let λ be a $2^{1/r-1/p}S_+$-invariant linear form on $\mathfrak{l}_\infty(\mathbb{N}_0)$. Then the expression*

$$\tau(S) := \lambda\big(2^{k/r-k/p}\,\mathrm{trace}(S_k)\big)$$

does not depend on the choice of the $(\mathfrak{P}_{p^*,2}, \mathfrak{d}_{r,\infty})$-representation $S = \sum_{k=0}^\infty S_k$. This definition yields a trace τ on $(\mathfrak{P}_{p^*,2})^{\mathrm{app}}_{r,\infty}$.

Remark 7.4. Applying the results from [26, Sect. 7] about spectral theory of shift operators, we are able to answer the question for which parameters $\gamma > 0$ do γS_+-invariant linear forms exist on $\mathfrak{l}_\infty(\mathbb{N}_0)$.

If $0 < \gamma < 1$, then $I - \gamma S_+$ is invertible. Hence every γS_+-invariant linear form vanishes on all of $\mathfrak{l}_\infty(\mathbb{N}_0)$.

If $1 < \gamma < \infty$, then the range of $I - \gamma S_+$ has codimension 1 since it consists of all $(\alpha_h) \in \mathfrak{l}_\infty(\mathbb{N}_0)$ such that $\sum_{h=0}^\infty \gamma^{-h}\alpha_h = 0$. So every γS_+-invariant linear form is a scalar multiple of

$$\lambda(a) := \sum_{h=0}^\infty \gamma^{-h}\alpha_h.$$

Only the case $\gamma = 1$ is of interest. With regard to Theorem 7.3 this means that $p = r$. Then there exists a huge amount of continuous as well as of non-continuous S_+-invariant linear forms on $\mathfrak{l}_\infty(\mathbb{N}_0)$. Most prominent are the Banach limits.

To establish the next theorem, we need a basic inequality, which goes back to [13, p. 97]. For the convenience of the reader, we present a proof that, in contrast to [35, pp. 150–151], uses Hölder's inequality instead of complex interpolation. The main tool, however, remains unchanged.

Lemma 7.5. *Let $2 \le q^* < p^* \le \infty$. Then*

$$\|F|\mathfrak{P}_{q^*,2}\| \le \sqrt{2}\, n^{1/q^*-1/p^*} \|F|\mathfrak{P}_{p^*,2}\| \quad \textit{whenever} \quad \operatorname{rank}(F) \le n.$$

Proof. Since $\operatorname{rank}(F) \le n$, a famous theorem of Tomczak–Jaegermann [35, p. 143] says that the $\mathfrak{P}_2$-norm of F can be controlled as follows:

There exist $u_1, \dots, u_n \in X$ such that

$$\|F|\mathfrak{P}_2\| \le \sqrt{2}\Big(\sum_{k=1}^n \|Fu_k\|^2\Big)^{1/2} \quad \text{and} \quad \sup_{\|x^*\|\le 1}\Big(\sum_{k=1}^n |\langle u_k, x^*\rangle|^2\Big)^{1/2} \le 1.$$

In the case $q^* = 2$, the required estimate is easily obtained:

$$\|F|\mathfrak{P}_2\| \le \sqrt{2}\, n^{1/2-1/p^*}\Big(\sum_{k=1}^n \|Fu_k\|^{p^*}\Big)^{1/p^*} \le \sqrt{2}\, n^{1/2-1/p^*}\|F|\mathfrak{P}_{p^*,2}\|.$$

If $2 < q^* < p^* \le \infty$ and $x_1, \dots, x_m \in X$, then

$$\Big(\sum_{h=1}^m \|Fx_h\|^{q^*}\Big)^{1/q^*} \le \Big(\sum_{h=1}^m \|Fx_h\|^2\Big)^{(1-\theta)/2}\Big(\sum_{h=1}^m \|Fx_h\|^{p^*}\Big)^{\theta/p^*},$$

where

$$\tfrac{1}{q^*} = \tfrac{1-\theta}{2} + \tfrac{\theta}{p^*} \quad \Leftrightarrow \quad \tfrac{1}{q^*} - \tfrac{1}{p^*} = \big(\tfrac{1}{2} - \tfrac{1}{p^*}\big)(1-\theta) \quad \text{and} \quad 0 < \theta < 1.$$

Hence

$$\begin{aligned}\Big(\sum_{h=1}^m \|Fx_h\|^{q^*}\Big)^{1/q^*} &\le \|F|\mathfrak{P}_2\|^{1-\theta}\|F|\mathfrak{P}_{p^*,2}\|^{\theta} \sup_{\|x^*\|\le 1}\Big(\sum_{h=1}^m |\langle x_h, x^*\rangle|^2\Big)^{1/2}\\ &\le \big(\sqrt{2}\, n^{1/2-1/p^*}\big)^{1-\theta}\|F|\mathfrak{P}_{p^*,2}\| \sup_{\|x^*\|\le 1}\Big(\sum_{h=1}^m |\langle x_h, x^*\rangle|^2\Big)^{1/2},\end{aligned}$$

which yields the general case. □

Theorem 7.6. *Let $2 \le q^* < p^* \le \infty$, $0 < r < s < \infty$, and $\frac{1}{p} - \frac{1}{r} = \frac{1}{q} - \frac{1}{s}$. Then*

$$(\mathfrak{P}_{p^*,2})^{\text{app}}_{r,\infty} \subset (\mathfrak{P}_{q^*,2})^{\text{app}}_{s,\infty}.$$

Proof. In view of the inequality

$$\|S_k|\mathfrak{P}_{q^*,2}\| \le 2^{1/2+k(1/p-1/q)}\|S_k|\mathfrak{P}_{p^*,2}\|,$$

every $(\mathfrak{P}_{p^*,2},\mathfrak{d}_{r,\infty})$-representation $S=\sum_{k=0}^{\infty}S_k$ is also a $(\mathfrak{P}_{q^*,2},\mathfrak{d}_{s,\infty})$-representation, where $-\frac{1}{s}=\frac{1}{p}-\frac{1}{q}-\frac{1}{r}$. So the asserted inclusion $(\mathfrak{P}_{p^*,2})^{\text{app}}_{r,\infty}\subseteq(\mathfrak{P}_{q^*,2})^{\text{app}}_{s,\infty}$ follows from Theorem 5.1.

By [21, 2.9.10], for the diagonal operator $D_t:(\xi_n)\mapsto(\tau_n\xi_n)$, we have

$$D_t\in\mathfrak{L}^{\text{app}}_{r,\infty}(\ell_\infty,\ell_2)\quad\text{if and only if}\quad t=(\tau_n)\in\mathfrak{l}_{r_0,\infty}(\mathbb{N}),$$

where $\frac{1}{r_0}=\frac{1}{r}+\frac{1}{2}$.

Moreover, the 'little' Grothendieck theorem says that $\mathfrak{P}_2(\ell_\infty,\ell_2)=\mathfrak{L}(\ell_\infty,\ell_2)$; see [8, p. 64]. Since $\mathfrak{P}_2\subseteq\mathfrak{P}_{p^*,2}$, we get

$$\mathfrak{P}_{p^*,2}(\ell_\infty,\ell_2)=\mathfrak{L}(\ell_\infty,\ell_2).$$

Combing the preceding results yields the criterion

$$D_t\in(\mathfrak{P}_{p^*,2})^{\text{app}}_{r,\infty}(\ell_\infty,\ell_2)\quad\text{if and only if}\quad t=(\tau_n)\in\mathfrak{l}_{r_0,\infty}(\mathbb{N}).$$

Hence the component $(\mathfrak{P}_{p^*,2})^{\text{app}}_{r,\infty}(\ell_\infty,\ell_2)$, which does not depend on p^*, strictly increases with r. That is,

$$(\mathfrak{P}_{p^*,2})^{\text{app}}_{r,\infty}(\ell_\infty,\ell_2)=(\mathfrak{P}_{q^*,2})^{\text{app}}_{r,\infty}(\ell_\infty,\ell_2)\subset(\mathfrak{P}_{q^*,2})^{\text{app}}_{s,\infty}(\ell_\infty,\ell_2).\qquad\square$$

Remark 7.7. The strict monotonicity of $(\mathfrak{P}_{p^*,2})^{\text{app}}_{r,\infty}(\ell_\infty,\ell_2)$ with respect to r carries over to the components

$$(\mathfrak{P}_{p^*,2})^{\text{app}}_{r,\infty}(c_0,\ell_2),\quad(\mathfrak{P}_{p^*,2})^{\text{app}}_{r,\infty}(c_0\oplus\ell_2),\quad\text{and}\quad(\mathfrak{P}_{p^*,2})^{\text{app}}_{r,\infty}(C(\mathbb{T})).$$

The latter case is of interest for applications presented in Section 9.

Remark 7.8. The results of this section remain true if $\mathfrak{P}_{p^*,2}$ is replaced by the slightly larger operator ideal $\mathfrak{L}^{\text{weyl}}_{p^*,\infty}$ generated by the Weyl numbers; see [21, 2.4.13, 2.7.4, and 3.6.2]. Then the counterpart of Lemma 7.5 becomes trivial, since

$$\|F|\mathfrak{L}^{\text{weyl}}_{p^*,\infty}\|:=\sup_{1\le m<\infty}m^{1/p^*}x_m(F)=\max_{1\le m\le n}m^{1/p^*}x_m(F)\quad\text{if}\quad\operatorname{rank}(F)\le n.$$

8. Lipschitz/Besov spaces

This section contains a summary of subsequently needed well-known results about function spaces. Most of them are special cases of much more general theorems spread over the literature. Therefore some elementary proofs are included. I apologize for being an old-fashioned man who likes examples.

In what follows, $\mathbb{T}:=\mathbb{R}/2\pi\mathbb{Z}$ is the 1-dimensional torus $\{e^{ix}\}$ (unit circle in the complex plane). We will work with the intervals $[0,2\pi]$ or $[-\pi,+\pi]$, whose endpoints are identified.

The classical Banach spaces of 2π-periodic functions on the real line, equipped with the norms

$$\|f|L_p\|:=\left(\frac{1}{2\pi}\int_{\mathbb{T}}|f(x)|^p\,dx\right)^{1/p}\quad\text{and}\quad\|f|C\|:=\sup_{0\le x\le 2\pi}|f(x)|,$$

are denoted by $L_p(\mathbb{T})$ and $C(\mathbb{T})$, respectively.

For all $f \in L_p(\mathbb{T})$, we define the nth *trigonometric approximation number* (approximation error) by

$$E_n(f|L_p) := \inf \|f - t|L_p\|,$$

where the infimum ranges over all trigonometric polynomials

$$t(x) = \sum_{|m|\le n} \gamma_m e^{imx}$$

of degree less than or equal to $n = 0, 1, 2, \dots$. However, in asymptotic relations containing some $O(n^{-\varrho})$, the value $n = 0$ must be excluded.

The *Dirichlet projection*

$$D_n : f(x) = \sum_{m\in\mathbb{Z}} \gamma_m e^{imx} \mapsto \frac{1}{2\pi}\int_{\mathbb{T}} f(x-y)\frac{\sin\frac{(2n+1)y}{2}}{\sin\frac{y}{2}}\,dy = \sum_{|m|\le n} \gamma_m e^{imx}$$

assigns to every function $f \in L_1(\mathbb{T})$ the nth partial sum of its Fourier series.

In the limiting case $p = 1$, the norms $\|D_n : L_1 \to L_1\|$ fail to be uniformly bounded. Luckily,

$$C_p := \sup_{0\le n<\infty} \|D_n : L_p \to L_p\| \quad \text{is finite for } 1 < p < \infty;$$

see [4, pp. 350–351] and [36, Vol. I, p. 266]. The original source of this deep result is [31, p. 490]; see Remark 9.10. It follows that

$$\|f - D_n f|L_p\| \le (C_p + 1)E_n(f|L_p) \quad \text{for all} \quad f \in L_p(\mathbb{T}).$$

Assuming that $1 \le p < \infty$ and $0 < \varrho < \infty$, we let

$$B^{\varrho}_{p,\infty}(\mathbb{T}) := \{f \in L_p(\mathbb{T}) : E_n(f|L_p) = O(n^{-\varrho})\}.$$

The conditions $E_n(f|L_p) = O(n^{-\varrho})$ and $E_{2^k}(f|L_p) = O(2^{-k\varrho})$ are equivalent.

Remark 8.1. *Besov spaces $B^{\varrho}_{p,w}(\mathbb{T})$ have three parameters p, ϱ, and $0 < w \le \infty$. In this paper, however, only the limiting case $w = \infty$ is of interest. The Russian school refers to $B^{\varrho}_{p,\infty}(\mathbb{T})$ as a Nikolskiĭ space.*

The original definition of a Besov space is based on smoothness properties of its members and goes back to Hardy–Littlewood [10], who only considered the 1-dimensional case,

$$\mathrm{Lip}_1(\varrho|L_p(\mathbb{T})) := \left\{ f \in L_p(\mathbb{T}) : \left(\int_{\mathbb{T}} |f(x+t) - f(x)|^p\,dx\right)^{1/p} = O(t^{\varrho})\right\}$$

provided that $0 < \varrho \le 1$.

The extension to $0 < \varrho \le 2$ was proposed by Zygmund [37],

$$\mathrm{Lip}_2(\varrho|L_p(\mathbb{T})) := \left\{ f \in L_p(\mathbb{T}) : \left(\int_{\mathbb{T}} |f(x+t)+f(x-t)-2f(x)|^p\,dx\right)^{1/p} = O(t^{\varrho})\right\}.$$

By [4, pp. 68, 76], we have

$$\mathrm{Lip}_1\big(\varrho|L_p(\mathbb{T})\big) = \mathrm{Lip}_2\big(\varrho|L_p(\mathbb{T})\big) \quad \text{for } 0 < \varrho < 1,$$

but

$$\mathrm{Lip}_1\big(1|L_p(\mathbb{T})\big) \subset \mathrm{Lip}_2\big(1|L_p(\mathbb{T})\big).$$

Due to the described historical reasons, many people prefer the term *Lipschitz space*. Both approaches are equivalent.

Jackson–Bernstein Theorem. $B^{\varrho}_{p,\infty}(\mathbb{T}) = \mathrm{Lip}_2\big(\varrho|L_p(\mathbb{T})\big)$ *for* $0 < \varrho < 2$.

Remark 8.2. The classical results of Jackson–Bernstein concern the case $0 < \varrho < 1$ in $C(\mathbb{T})$, while Quade [30, pp. 535–538] contributed the extension to $L_p(\mathbb{T})$. For detailed historical comments, we refer to [11]. Proofs of the preceding version are given in [4, 2.2.1 and 2.3.5], [5, Sect. 4.2], [7, p. 210], and [33, pp. 43, 46]. Using differences of higher order, one can treat Besov spaces $B^{\varrho}_{p,\infty}(\mathbb{T})$ with arbitrary exponents $\varrho > 0$.

We decompose $\mathbb{Z}:=\{\,\dots,-2,-1,0,+1,+2,\dots\}$ into the subsets

$$\mathbb{Z}_0 := \{0\} \quad \text{and} \quad \mathbb{Z}_k := \{m \in \mathbb{Z} : 2^{k-1} \le |m| < 2^k\} \quad \text{for} \quad k = 1, 2, \dots .$$

The following criterion is an older brother of Theorem 5.1.

Theorem 8.3. *Let* $1 < p < \infty$. *Then a function* $f \in L_p(\mathbb{T})$ *belongs to* $B^{\varrho}_{p,\infty}(\mathbb{T})$ *if and only if there exists a representation*

$$f = \sum_{k=0}^{\infty} f_k,$$

where f_k *are trigonometric polynomials such that*

$$\deg(f_k) < 2^k \quad \textit{and} \quad \|f_k|L_p\| = O\big(2^{-k\varrho}\big).$$

A uniquely defined representation is given by

$$f_k^D(x) := \sum_{m\in\mathbb{Z}_k} \gamma_m \mathrm{e}^{imx} \quad \textit{whenever} \quad f(x) = \sum_{m\in\mathbb{Z}} \gamma_m \mathrm{e}^{imx}.$$

Proof. Obviously, $\deg(f_k^D) < 2^k$ for every $f \in L_p(\mathbb{T})$. If $f \in B^{\varrho}_{p,\infty}(\mathbb{T})$, then it follows from

$$f_k^D = D_{2^k-1}f - D_{2^{k-1}-1}f \quad \text{and} \quad \|f - D_n f|L_p\| \le (C_p+1)E_n(f|L_p)$$

that

$$\begin{aligned}\|f_k^D|L_p\| &\le \|D_{2^k-1}f - f|L_p\| + \|f - D_{2^{k-1}-1}f|L_p\| \\ &\le (C_p+1)\big[E_{2^k-1}(f|L_p) + E_{2^{k-1}-1}(f|L_p)\big] = O\big(2^{-k\varrho}\big).\end{aligned}$$

Thus the *Dirichlet representation* works.

Conversely, if $f = \sum_{h=0}^{\infty} f_h$ is any representation with the required properties, then

$$E_{2^k}(f|L_p) \le \Big\| f - \sum_{h=0}^{k} f_h \Big| L_p \Big\| \le \Big\| \sum_{h=k+1}^{\infty} f_h \Big| L_p \Big\|$$
$$\le \sum_{h=k+1}^{\infty} \|f_h|L_p\| = O(2^{-k\varrho}),$$

which proves that $f \in B^{\varrho}_{p,\infty}(\mathbb{T})$. □

Remark 8.4. The latter implication even holds for $1 \le p < \infty$.

To include the limiting case $p = 1$ in both directions, we need the *de la Vallée Poussin operator*,

$$V_k := \frac{1}{2^k} \sum_{2^k \le m < 2^{k+1}} D_m.$$

Then $\|V_k : L_p \to L_p\| \le 3$ implies that

$$\|f - V_k f|L_p\| \le 4E_{2^k}(f|L_p) \quad \text{for all} \quad f \in L_p(\mathbb{T});$$

see [4, p. 108], [15, pp. 92–93], [18, pp. 149–152], [33, pp. 27–28], and [36, Vol. I, p. 115].

The following supplement of Theorem 8.3 holds.

Theorem 8.5. *Let $1 \le p < \infty$. Then a function $f \in L_p(\mathbb{T})$ belongs to $B^{\varrho}_{p,\infty}(\mathbb{T})$ if and only if there exists representations*

$$f = \sum_{k=0}^{\infty} f_k,$$

where f_k are trigonometric polynomials such that

$$\deg(f_k) < 2^{k+1} \quad \textit{and} \quad \|f_k|L_p\| = O(2^{-k\varrho}).$$

A uniquely defined representation is given by

$$f_0^V := V_0 f \quad \textit{and} \quad f_k^V := V_k f - V_{k-1} f \quad \textit{for} \quad k = 1, 2, \ldots .$$

Proof. Replace $\deg(f_k^D) < 2^k$ by $\deg(f_k^V) < 2^{k+1}$ and adapt the rest of the preceding proof. □

The following embedding theorem is classical, as well. Its proof can be based on *Nikolskiĭ's inequality* for trigonometric polynomials t:
If $1 \le p < q < \infty$ and $\deg(t) \le n$, then

$$\|t|L_q\| \le c\, n^{1/p - 1/q} \|t|L_p\| \quad \text{with some constant} \quad c \ge 1.$$

See [7, p. 102], [19, 3.4.3], [33, p. 38], and [34, 4.9.2]. Compare with the analogous result stated in Lemma 7.5.

Theorem 8.6. *Let $1 \le p < q < \infty$, $0 < \sigma < \varrho < \infty$, and $\frac{1}{p} - \varrho = \frac{1}{q} - \sigma$. Then*

$$B^{\varrho}_{p,\infty}(\mathbb{T}) \subset B^{\sigma}_{q,\infty}(\mathbb{T}).$$

The 2π-periodic *lacunary function*

$$f_\sigma(x) := \sum_{h=0}^{\infty} 2^{-h\sigma} \sin 2^h x \quad \text{with} \quad \sigma > 0$$

shows that the preceding inclusion is actually strict.

Example 8.7. $f_\sigma \in B^{\sigma}_{q,\infty}(\mathbb{T}) \setminus B^{\varrho}_{p,\infty}(\mathbb{T})$ *for* $1 \le p, q < \infty$ *and* $0 < \sigma < \varrho < \infty$.

Proof. The upper estimate

$$E_{2^k}(f_\sigma | L_q) \le \sum_{h=k+1}^{\infty} 2^{-h\sigma} = O\big(2^{-k\sigma}\big)$$

says that $f_\sigma \in B^{\sigma}_{q,\infty}(\mathbb{T})$ for all exponents q.

The lower estimate follows from Theorem 8.5. Indeed, since

$$D_m f_\sigma(x) = \sum_{h=0}^{k} 2^{-h\sigma} \sin 2^h x \quad \text{for} \quad 2^k \le m < 2^{k+1},$$

we have

$$V_k f_\sigma(x) = \frac{1}{2^k} \sum_{2^k \le m < 2^{k+1}} D_m f = \sum_{h=0}^{k} 2^{-h\sigma} \sin 2^h x.$$

Now

$$V_k f_\sigma(x) - V_{k-1} f_\sigma(x) = 2^{-k\sigma} \sin 2^k x$$

implies that

$$\begin{aligned} 2^{-k\sigma} \frac{1}{2\pi} \int_{\mathbb{T}} |\sin x| dx &\le \|V_k f_\sigma - f_\sigma | L_1\| + \|f_\sigma - V_{k-1} f_\sigma | L_1\| \\ &\le 4\big[E_{2^k}(f_\sigma | L_1) + E_{2^{k-1}}(f_\sigma | L_1)\big] \le 8 E_{2^{k-1}}(f_\sigma | L_1). \end{aligned}$$

Therefore $f_\sigma \notin B^{\varrho}_{1,\infty}(\mathbb{T})$ whenever $\varrho > \sigma$. □

9. Convolution operators on $C(\mathbb{T})$

Convolution operators

$$C_f : g(y) \mapsto f * g(x) := \frac{1}{2\pi} \int_{\mathbb{T}} f(x-y) g(y) dy$$

generated by 2π-periodic functions f belonging to Besov spaces $B^{\varrho}_{p,\infty}(\mathbb{T})$ are convincing examples, which illustrate the preceding theory.

Remark 9.1. Using the normalized Lebesgue measure on $\mathbb{T}$ ensures that the eigenvalues of C_f are just the Fourier coefficients of f; see [21, 6.5.5].

Theorem 9.2. *Let $1 \le p \le 2$ and $0 < r < \infty$. Then*

$$f \in B^{1/r}_{p,\infty}(\mathbb{T}) \quad \textit{implies} \quad C_f \in (\mathfrak{P}_{p^*,2})^{\text{app}}_{r,\infty}(C(\mathbb{T})).$$

Proof. The reader should keep in mind an unpleasant difference of notation: For operators the approximation numbers $a_n(S)$ are indexed by $n = 1, 2, \dots$, while the $E_n(f|L_p)$'s are indexed by $n = 0, 1, 2, \dots$. Since

$$\operatorname{rank}(C_t) \le 2\deg(t) + 1 \quad \text{for all trigonometric polynomials } t,$$

we obtain

$$a_{2(n+1)}(C_f : L_{p^*} \to C) \le E_n(f|L_p);$$

compare with [21, 6.5.13], where the quantity $a_n(f|L_p) := E_{n-1}(f|L_p)$ is used.

The assertion now follows from the factorization

$$C_f : C(\mathbb{T}) \xrightarrow{\text{identity}} L_{p^*}(\mathbb{T}) \xrightarrow{C_f} C(\mathbb{T})$$

and [21, 1.2.4, 1.3.8].

We could also employ the representation

$$f = \sum_{k=0}^{\infty} f_k^D$$

provided by Theorem 8.3. Then $\operatorname{rank}\big(C_{f_k^D}\big) \le |\mathbb{Z}_k| = 2^k$. Moreover, it follows from $\|f_k^D|L_p\| = O\big(2^{-k/r}\big)$ that

$$\|C_{f_k^D} : C \to C|\mathfrak{P}_{p^*,2}\| \le \|id : C \to L_{p^*}|\mathfrak{P}_{p^*,2}\| \|f_k^D|L_p\| = O\big(2^{-k/r}\big).$$

Therefore

$$C_f = \sum_{k=0}^{\infty} C_{f_k^D}$$

is a $(\mathfrak{P}_{p^*,2}, \mathfrak{d}_{r,\infty})$-representation of C_f viewed as an operator on $C(\mathbb{T})$.

If we use Theorem 8.5, then $\operatorname{rank}\big(C_{f_k^V}\big) < 2^{k+2}$. So, to achieve that

$$C_f = \sum_{k=0}^{\infty} C_{f_k^V}$$

becomes a $(\mathfrak{P}_{p^*,2}, \mathfrak{d}_{r,\infty})$-representation, the summands must be shifted to the right by two steps, □

Remark 9.3. Letting $1 \le p < q \le 2$ and $\frac{1}{p} - \frac{1}{r} = \frac{1}{q} - \frac{1}{s}$, we illustrate the relationship between Theorems 7.6 and 8.6 by the diagram

$$\begin{array}{ccc} B^{1/r}_{p,\infty}(\mathbb{T}) & \subset & B^{1/s}_{q,\infty}(\mathbb{T}) \\ \downarrow & & \downarrow \\ (\mathfrak{P}_{p^*,2})^{\text{app}}_{r,\infty}(C(\mathbb{T})) & \subset & (\mathfrak{P}_{q^*,2})^{\text{app}}_{s,\infty}(C(\mathbb{T})) \end{array},$$

where the vertical arrows $\downarrow$ indicate the correspondence $f \mapsto C_f$ described in Theorem 9.2. Replacing $C(\mathbb{T})$ by $L_2(\mathbb{T})$ yields a different result; see Remark 11.2.

The following formula can be checked by elementary manipulations.

Proposition 9.4.

$$\operatorname{trace}(C_t) = \sum_{|m|\le n} \gamma_m = t(0) \quad \textit{if} \quad t(x) = \sum_{|m|\le n} \gamma_m \mathrm{e}^{\mathrm{i}mx}.$$

Keeping in mind Remark 7.4, we now establish the main application of our theory. The $\mathbb{Z}_k$'s are defined in front of Theorem 8.3.

Theorem 9.5. *Let $1 \le p \le 2$. If a trace τ on $(\mathfrak{P}_{p^*,2})^{\mathrm{app}}_{p,\infty}$ is associated to an S_+-invariant linear form λ on $\mathfrak{l}_\infty(\mathbb{N}_0)$, then its restriction to convolution operators on $C(\mathbb{T})$ generated from functions $f \in B^{1/p}_{p,\infty}(\mathbb{T})$ has the form*

$$\tau(C_f) = \lambda\Big(\sum_{m\in\mathbb{Z}_k} \gamma_m\Big) \quad \textit{whenever} \quad f(x) = \sum_{m\in\mathbb{Z}} \gamma_m \mathrm{e}^{\mathrm{i}mx}.$$

Proof. Apply Theorem 7.3 and Proposition 9.4 to the $(\mathfrak{P}_{p^*,2}, \mathfrak{d}_{p,\infty})$-representation

$$C_f = \sum_{k=0}^{\infty} C_{f_k^D}$$

described in the proof of Theorem 9.2. □

An S_+-invariant linear form λ on $\mathfrak{l}_\infty(\mathbb{N}_0)$ is said to be *normalized* if $\lambda(e) = 1$ for $e{:=}(1,1,1,\dots)$. We call $a = (\alpha_h) \in \mathfrak{l}_\infty(\mathbb{N}_0)$ *strictly Cesàro convergent* if there exists some $\alpha \in \mathbb{C}$ such that $\Big(\sum_{h=0}^{k}(\alpha_h - \alpha)\Big) \in \mathfrak{l}_\infty(\mathbb{N}_0)$. The terminology is justified by the following criterion:

$$\frac{1}{k+1}\sum_{h=0}^{k} \alpha_h = \alpha + O\big(\tfrac{1}{k+1}\big).$$

Remark 9.6. The related concept of an *almost convergent sequence* was introduced by Lorentz [16, pp. 169–170].

The next result is taken from [28, p. 18].

Lemma 9.7. *If*

$$\frac{1}{k+1}\sum_{h=0}^{k} \alpha_h = \alpha + O\big(\tfrac{1}{k+1}\big),$$

then $\lambda(\alpha_h) = \alpha$ for all normalized S_+-invariant linear forms λ on $\mathfrak{l}_\infty(\mathbb{N}_0)$.

Proof. Since

$$\beta_k := \sum_{h=0}^{k} (\alpha_h - \alpha) = O(1),$$

the sequence $b = (\beta_k)$ is bounded. Hence it follows from $a - \alpha e = b - S_+ b$ that

$$\lambda(a - \alpha e) = \lambda(b - S_+ b) = 0.$$ □

We now state the following consequence of Theorem 9.5.

Theorem 9.8. *Let* $1 \leq p \leq 2$. *If a trace* τ *on* $(\mathfrak{P}_{p^*,2})^{\text{app}}_{p,\infty}$ *is associated to a normalized* S_+*-invariant linear form* λ *on* $\mathfrak{l}_\infty(\mathbb{N}_0)$ *and* $f \in B^{1/p}_{p,\infty}(\mathbb{T})$, *then*

$$\tau(C_f) = \lim_{k\to\infty} \frac{1}{k+1} \sum_{|m|<2^k} \gamma_m \quad \textit{whenever} \quad f(x) = \sum_{m\in\mathbb{Z}} \gamma_m \mathrm{e}^{\mathrm{i}mx}$$

provided that the sequence $\left(\sum_{m\in\mathbb{Z}_k} \gamma_m\right)$ *is strictly Cesàro convergent.*

Remark 9.9. It would be extremely worthwhile to know whether the expression

$$\lim_{k\to\infty} \frac{1}{k+1} \sum_{|m|<2^k} \gamma_m \quad \text{(if it exists)}$$

has any analytical meaning for the function $f(x) = \sum_{m\in\mathbb{Z}} \gamma_m \mathrm{e}^{\mathrm{i}mx}$.

I am not able to rule out the existence of S_+-invariant linear forms on $\mathfrak{l}_\infty(\mathbb{N}_0)$ for which the associated traces vanish at all convolution operators C_f generated by $f \in B^1_{1,\infty}(\mathbb{T})$; see Remark 11.7 and Theorem 11.8.

To give an answer for normalized S_+-invariant linear forms, we need some preparation:

Roughly speaking, the *Hilbert transform* is defined by

$$\mathscr{H} : f(x) = \sum_{m\in\mathbb{Z}} \gamma_m \mathrm{e}^{\mathrm{i}mx} \mapsto f^\sim(x) = -\mathrm{i} \sum_{m\in\mathbb{Z}} \operatorname{sgn}(m)\, \gamma_m \mathrm{e}^{\mathrm{i}mx}.$$

In particular,

$$\mathscr{H} : \sin mx \mapsto -\cos mx, \qquad \mathscr{H} : \cos mx \mapsto \sin mx, \quad \text{and} \quad \mathscr{H} : 1 \mapsto 0.$$

A famous theorem of Riesz [31] says that the Hilbert transform provides an operator on $L_p(\mathbb{T})$ whenever $1 < p < \infty$; see [4, p. 336] and [12, p. 68].

Remark 9.10. The uniform estimate $\|D_n : L_p \to L_p\| \leq C_p$ stated in Section 8 is a corollary; see [12, pp. 47–49].

For $p = \infty$, the following result goes back to Privalov [29]. The general case is treated in [3, p. 234], [4, pp. 322, 324], and [5, p. 146].

Theorem 9.11. *If* $1 \leq p < \infty$ *and* $0 < \varrho < 2$, *then the Hilbert transform maps* $\mathrm{Lip}_2\left(\varrho|L_p(\mathbb{T})\right)$ *into itself.*

We consider the 2π-periodic functions

$$f_{\sin}(x) := \sum_{m=1}^{\infty} \tfrac{1}{m} \sin mx = \tfrac{\pi - x}{2}$$

and

$$f_{\cos}(x) := \sum_{m=1}^{\infty} \tfrac{1}{m} \cos mx = -\log|2 \sin \tfrac{x}{2}|$$

defined on the interval $(0, 2\pi)$; see [36, Vol. I, p. 5].

Example 9.12. $f_{\sin} \in \mathrm{Lip}_1\big(1|L_1(\mathbb{T})\big) \subset B^1_{1,\infty}(\mathbb{T})$.

Proof. Since

$$f(x+t)-f(x) = \begin{cases} -\frac{t}{2} & \text{if} \quad 0<x<2\pi-t, \\ \pi-\frac{t}{2} & \text{if} \quad 2\pi-t<x<2\pi, \end{cases}$$

we get

$$\int_0^{2\pi} |f(x+t)-f(x)|\,dx = \tfrac{t}{2}(2\pi-t)+(\pi-\tfrac{t}{2})t = O(t). \qquad \square$$

Example 9.13. $f_{\cos} \in \mathrm{Lip}_2\big(1|L_1(\mathbb{T})\big) = B^1_{1,\infty}(\mathbb{T})$.

Proof. Since $f_{\cos}$ is the conjugate of $-f_{\sin} \in \mathrm{Lip}_2\big(1|L_1(\mathbb{T})\big)$, the conclusion follows from Theorem 9.11.

A direct proof (compare with [33, p. 41]) can be obtained by a slight modification of the technique used in [1, Chap. 1, § 30] and [36, Vol. I, p. 183].

For $n=0,1,2,\dots$, we consider

the Dirichlet kernel $$D_n(x) := 1 + 2\sum_{m=1}^{n} \cos mx$$

and

the Fejér kernel $$F_n(x) := \frac{1}{n+1}\sum_{m=0}^{n} D_m(x).$$

In particular,

$$D_0(x) = F_0(x) = 1.$$

Obviously,

$$2\cos mx = D_m(x) - D_{m-1}(x) \quad \text{and} \quad D_m(x) = (m+1)F_m(x) - mF_{m-1}(x)$$

for $m=1,2,\dots$. If $n \ge 2$ and $|x|<\pi$, then

$$\begin{aligned} 2\sum_{m=1}^{n} \tfrac{1}{m}\cos mx &= \sum_{m=1}^{n} \tfrac{1}{m}\big[D_m(x)-D_{m-1}(x)\big] \\ &= \sum_{m=1}^{n-1} \tfrac{1}{m}D_m(x) + \tfrac{1}{n}D_n(x) - D_0(x) - \sum_{m=1}^{n-1} \tfrac{1}{m+1}D_m(x) \\ &= -1 + \sum_{m=1}^{n-1} \tfrac{1}{m(m+1)}D_m(x) + \tfrac{1}{n}D_n(x) \\ &= -1 + \sum_{m=1}^{n-1} \tfrac{1}{m}F_m(x) - \sum_{m=1}^{n-1} \tfrac{1}{m+1}F_{m-1}(x) + \tfrac{1}{n}D_n(x) \\ &= -1 + \sum_{m=1}^{n-2} \tfrac{1}{m}F_m(x) + \tfrac{1}{n-1}F_{n-1}(x) - \tfrac{1}{2}F_0(x) - \sum_{m=1}^{n-2} \tfrac{1}{m+2}F_m(x) + \tfrac{1}{n}D_n(x) \\ &= -\tfrac{3}{2} + \sum_{m=1}^{n-2} \tfrac{2}{m(m+2)}F_m(x) + \tfrac{1}{n-1}F_{n-1}(x) + \tfrac{1}{n}D_n(x). \end{aligned}$$

Recall from [4, pp. 42–43] that

$$D_n(0) = 2n+1, \quad |D_n(x)| \le \tfrac{1}{|\sin\frac{x}{2}|}, \quad \text{and} \quad \|D_n|L_1\| = \tfrac{4}{\pi^2}\log n + O(1),$$

$$F_n(0) = n+1, \quad 0 \le F_n(x) \le \tfrac{1}{n+1}\tfrac{1}{(\sin\frac{x}{2})^2}, \quad \text{and} \quad \|F_n|L_1\| = 1.$$

So

$$\tfrac{3}{2} + 2\sum_{m=1}^{\infty} \tfrac{1}{m}\cos mx = \sum_{m=1}^{\infty} \tfrac{2}{m(m+2)} F_m(x) \quad \text{pointwise for } x\neq 0.$$

We also have convergence in $L_1(\mathbb{T})$. Finally, the conclusion follows from

$$E_n(f_{\cos}|L_1) \le \|f_{\cos} + \tfrac{3}{4} - \sum_{m=1}^{n} \tfrac{1}{m(m+2)} F_m|L_1\| \le \sum_{m=n+1}^{\infty} \tfrac{1}{m(m+2)} = O(\tfrac{1}{n}). \qquad \square$$

Remark 9.14. It seems to me that a direct proof of

$$\frac{1}{2\pi}\int_{\mathbb{T}} \big|f_{\cos}(x+t) + f_{\cos}(x-t) - 2f_{\cos}(x)\big|\, dx = O(t)$$

would be rather involved. Moreover, if the asymptotic relation

$$\frac{1}{2\pi}\int_{\mathbb{T}} \big|f_{\cos}(x+t) - f_{\cos}(x)\big|\, dx = \frac{1}{2\pi}\int_{\mathbb{T}} \Big|\log\big[\big(\cot\tfrac{t}{2} + \cot\tfrac{x}{2}\big)\sin\tfrac{t}{2}\big]\Big|\, dx \asymp$$

$$\asymp \int_0^1 \Big|\log\big[\big(\tfrac{2}{t} + \tfrac{2}{x}\big)\tfrac{t}{2}\big]\Big|\, dx = \int_0^1 \big[\log(x+t) - \log x\big]\, dx \asymp t|\log t|$$

were true, then one might infer that $f_{\cos}\notin \mathrm{Lip}_1\big(1|L_1(\mathbb{T})\big)$.

Now we present the announced example, which shows that any trace on $(\mathfrak{P}_{p^*,2})^{\mathrm{app}}_{p,\infty}$ associated to a normalized S_+-invariant linear form on $\mathfrak{l}_\infty(\mathbb{N}_0)$ does not vanish identically on the convolution operators. So it is non-trivial.

Example 9.15. *Let $1 \le p \le 2$. Then $\tau(C_{f_{\cos}}) = \log 2$ for any trace τ on $(\mathfrak{P}_{p^*,2})^{\mathrm{app}}_{p,\infty}$ that is associated to a normalized S_+-invariant linear form on $\mathfrak{l}_\infty(\mathbb{N}_0)$.*

Proof. Using the asymptotic formula

$$\sum_{m=1}^{n} \tfrac{1}{m} = \log n + O(1),$$

we get

$$\sum_{0<|m|<2^k} \tfrac{1}{2|m|} = \sum_{m=1}^{2^k-1} \tfrac{1}{m} = \log 2^k + O(1).$$

It now follows from Theorem 9.8 and

$$f_{\cos}(x) := \sum_{m=1}^{\infty} \tfrac{1}{m}\cos mx = \sum_{m\in\mathbb{Z}\setminus\{0\}} \tfrac{1}{2|m|} e^{imx}$$

that

$$\tau(C_{f_{\cos}}) = \lim_{k\to\infty} \frac{1}{k+1} \sum_{0<|m|<2^k} \tfrac{1}{2|m|} = \log 2. \qquad \square$$

10. Operator ideals on the Hilbert space

Finally, we restrict our considerations to operators acting on the (real or complex) separable infinite-dimensional Hilbert space, denoted by H. In this case, an *operator ideal* $\mathfrak{A}(H)$ is just an ordinary ideal in the algebra $\mathfrak{L}(H)$.

By the Riesz representation theorem, every element $u \in H$ generates a (bounded and linear) functional $u^\star : x \mapsto (x|u)$ on H.

Every compact operator admits a *Schmidt representation*

$$S = \sum_{n=1}^{\infty} \sigma_n u_n^\star \otimes v_n,$$

where $(\sigma_n) \in c_0(\mathbb{N})$ and (u_n), (v_n) are orthonormal sequences in H. It may be arranged that $\sigma_1 \geq \sigma_2 \geq \cdots \geq 0$. In this case, $\sigma_n = a_n(S)$.

We refer to $S \in \mathfrak{L}(H)$ as a *Hilbert–Schmidt operator* if the quantity

$$\|S|\mathfrak{L}_2^{\text{app}}\| := \Big(\sum_{m=1}^{\infty} \|Se_m\|^2 \Big)^{1/2},$$

which does not depend on the choice of the *orthonormal basis* (e_m), is finite. Using any Schmidt representation, we obtain

$$\|S|\mathfrak{L}_2^{\text{app}}\| = \Big(\sum_{n=1}^{\infty} \sigma_n^2 \Big)^{1/2} = \Big(\sum_{n=1}^{\infty} a_n(S)^2 \Big)^{1/2}.$$

Hence the collection of all Hilbert–Schmidt operators is just the Hilbert space component of the operator ideal $\mathfrak{L}_2^{\text{app}}$. Moreover, the norm $\|S|\mathfrak{L}_2^{\text{app}}\|$ can be obtained form the inner product

$$(S|T) := \sum_{m=1}^{\infty} (Se_m|Te_m).$$

Remark 10.1. Note that $\mathfrak{P}_{q,2}(H) = \mathfrak{L}_q^{\text{app}}(H)$ and $\|S|\mathfrak{P}_{q,2}\| = \|S|\mathfrak{L}_q^{\text{app}}\|$ whenever $2 \leq q < \infty$; see [20, 17.5.3]. Especially, $\mathfrak{P}_2(H) = \mathfrak{L}_2^{\text{app}}(H)$ and $\|S|\mathfrak{P}_2\| = \|S|\mathfrak{L}_2^{\text{app}}\|$.

The following result goes back to Schmidt [32, pp. 467–470].

Proposition 10.2. $a_n(S|\mathfrak{L}_2^{\text{app}}) = \Big(\sum_{k=n}^{\infty} a_k(S)^2 \Big)^{1/2}$ *for all* $S \in \mathfrak{L}_2^{\text{app}}(H)$.

Remark 10.3. I am deeply grateful to a referee who informed me about a relevant paper; see [17]. Dealing with finite matrices, Mirsky showed that the proposition above holds not only for the Hilbert–Schmidt norm but even for all uniformly invariant norms.

Therefore, in the infinite-dimensional setting, one may ask for which quasi-norms $\| \cdot |\mathfrak{A}\|$ on an operator ideal $\mathfrak{A}(H)$ the generalized approximation numbers are obtained by the formula

$$a_n(S|\mathfrak{A}) = \left\| S - \sum_{k=1}^{n-1} a_k(S) u_k^\star \otimes v_k \middle| \mathfrak{A} \right\|.$$

It should be checked whether Mirsky's Theorem 2 can be used to give an affirmative answer in the case of Banach ideals in which the finite rank operators are $\|\cdot|\mathfrak{A}\|$-dense; see [20, Sect. 8.6, 13.4, and § 15.4.5]. This conjecture is certainly true for the Schatten–von Neumann ideals $\mathfrak{L}_p^{\text{app}}(H)$ with $1 \le p < \infty$.

Theorem 10.4. *Let* $1 \le p \le 2$, $0 < s_1 \le r \le s_2 < \infty$, *and* $\frac{1}{1} - \frac{1}{s_1} = \frac{1}{p} - \frac{1}{r} = \frac{1}{2} - \frac{1}{s_2}$. *Then*

$$\mathfrak{L}_{s_1,\infty}^{\text{app}}(H) = (\mathfrak{P}_{p^*,2})_{r,\infty}^{\text{app}}(H) = (\mathfrak{P}_2)_{s_2,\infty}^{\text{app}}(H).$$

In particular,

$$\mathfrak{L}_{1,\infty}^{\text{app}}(H) = (\mathfrak{P}_{p^*,2})_{p,\infty}^{\text{app}}(H) = (\mathfrak{P}_2)_{2,\infty}^{\text{app}}(H).$$

Proof. The inclusions

$$\mathfrak{L}_{s_1,\infty}^{\text{app}}(H) \subseteq (\mathfrak{P}_{p^*,2})_{r,\infty}^{\text{app}}(H) \subseteq (\mathfrak{P}_2)_{s_2,\infty}^{\text{app}}(H)$$

follow from Theorem 7.6.

On the other hand, Proposition 10.2 tells us that

$$a_n(S|\mathfrak{P}_2) = \left(\sum_{k=n}^{\infty} a_k(S)^2\right)^{1/2} \ge \sqrt{n}\, a_{2n-1}(S) \quad \text{for all} \quad S \in \mathfrak{P}_2(H).$$

Hence, in view of $\frac{1}{s_1} = \frac{1}{2} + \frac{1}{s_2}$, we have

$$(\mathfrak{P}_2)_{s_2,\infty}^{\text{app}}(H) \subseteq \mathfrak{L}_{s_1,\infty}^{\text{app}}(H).$$ □

Remark 10.5. The preceding result is a special case of the *reiteration theorem* from interpolation theory; see [21, 2.8.13] and [23, p. 123].

11. Convolution operators on $L_2(\mathbb{T})$

To begin with, we establish a counterpart of Theorem 9.2.

Theorem 11.1. *Let* $1 \le p \le 2$ *and* $0 < r < \infty$. *Then*

$$f \in B_{p,\infty}^{1/r}(\mathbb{T}) \quad \textit{implies} \quad C_f \in (\mathfrak{P}_{p^*,2})_{r,\infty}^{\text{app}}\big(L_2(\mathbb{T})\big).$$

Proof. Define $s \ge r$ by $\frac{1}{2} - \frac{1}{s} = \frac{1}{p} - \frac{1}{r}$. Then Theorem 8.6 yields

$$f \in B_{p,\infty}^{1/r}(\mathbb{T}) \subseteq B_{2,\infty}^{1/s}(\mathbb{T}).$$

As in the proof of Theorem 9.2, we infer from the (interchanged) factorization

$$C_f : L_2(\mathbb{T}) \xrightarrow{C_f} C(\mathbb{T}) \xrightarrow{\text{identity}} L_2(\mathbb{T})$$

and Theorem 10.4 that $C_f \in (\mathfrak{P}_2)_{s,\infty}^{\text{app}}\big(L_2(\mathbb{T})\big) = (\mathfrak{P}_{p^*,2})_{r,\infty}^{\text{app}}\big(L_2(\mathbb{T})\big)$. □

Remark 11.2. The formulation of the preceding theorem is somehow misleading. While the scale of the spaces $B^{1/r}_{p,\infty}(\mathbb{T})$ with $1 \le p \le 2$ and $\frac{1}{p} - \frac{1}{r} = \text{const.}$ strictly increases as $p \nearrow 2$, the components $(\mathfrak{P}_{p^*,2})^{\text{app}}_{r,\infty}\big(L_2(\mathbb{T})\big)$ do not change; all of them coincide with $(\mathfrak{P}_2)^{\text{app}}_{s,\infty}\big(L_2(\mathbb{T})\big)$, where $\frac{1}{2} - \frac{1}{s} = \frac{1}{p} - \frac{1}{r}$. In particular, if $f \in B^{1/p}_{p,\infty}(\mathbb{T})$, then we always obtain $C_f \in (\mathfrak{P}_2)^{\text{app}}_{2,\infty}\big(L_2(\mathbb{T})\big) = \mathfrak{L}^{\text{app}}_{1,\infty}\big(L_2(\mathbb{T})\big)$. So $B^{1/2}_{2,\infty}(\mathbb{T})$, the largest of these spaces, is *optimal.*

Theorem 9.5 can be modified as follows:

Theorem 11.3. *Every S_+-invariant linear form λ on $\mathfrak{l}_\infty(\mathbb{N}_0)$ yields a trace τ on $\mathfrak{L}^{\text{app}}_{1,\infty}\big(L_2(\mathbb{T})\big)$ whose restriction to convolution operators on $L_2(\mathbb{T})$ generated from functions $f \in B^{1/2}_{2,\infty}(\mathbb{T})$ has the form*

$$\tau(C_f) = \lambda\bigg(\sum_{m\in\mathbb{Z}_k} \gamma_m\bigg) \quad \textit{whenever} \quad f(x) = \sum_{m\in\mathbb{Z}} \gamma_m e^{imx}.$$

We now consider Fourier series of the form

$$f(x) = \gamma_0 + \sum_{m=1}^{\infty} \alpha_m \cos mx + \sum_{m=1}^{\infty} \beta_m \sin mx.$$

Let

$$c_m(x) := \cos mx \quad \text{and} \quad s_m(x) := \sin mx,$$
$$c^\star_m : f \mapsto (f|c_m) \quad \text{and} \quad s^\star_m : f \mapsto (f|s_m).$$

Then $(c_m|c_m) = (s_m|s_m) = \frac{1}{2}$ implies that

$$\alpha_m = 2(f|c_m) = \frac{1}{\pi}\int_{\mathbb{T}} f(x)\cos mx\, dx$$

and

$$\beta_m = 2(f|s_m) = \frac{1}{\pi}\int_{\mathbb{T}} f(x)\sin mx\, dx.$$

The following counterpart of Proposition 9.4 holds.

Proposition 11.4.

$$\text{trace}(C_t) = \sum_{m=1}^{n} \alpha_m = t(0) \quad \textit{if} \quad t(x) = \sum_{m=1}^{n} \alpha_m \cos mx + \sum_{m=1}^{n} \beta_m \sin mx.$$

Proof. For the convolution operators generated by c_m and s_m, we have

$$C_{c_m} = c^\star_m \otimes c_m + s^\star_m \otimes s_m \quad \text{and} \quad C_{s_m} = c^\star_m \otimes s_m - s^\star_m \otimes c_m.$$

Hence $\text{trace}\big(C_{c_m}\big) = 1$ and $\text{trace}\big(C_{s_m}\big) = 0$. □

The preceding result is an elementary part of Theorem 5.5 in [2], which says that every trace on an operator ideal $\mathfrak{A}(H)$ vanishes on the skew-symmetric operators. So, concerning traces of convolution operators C_f, only the case

$$f(x) = \sum_{m=1}^{\infty} \alpha_m \cos mx$$

is of interest. Those series can be decomposed by letting

$$\mathbb{D}_k := \{ m \in \mathbb{N} : 2^k \leq m < 2^{k+1} \}.$$

Theorem 11.5. *A function $f \in L_2(\mathbb{T})$ of the form $f(x) = \sum_{m=1}^{\infty} \alpha_m \cos mx$ belongs to $B^{1/2}_{2,\infty}(\mathbb{T})$ if and only if*

$$\Big(\sum_{m \in \mathbb{D}_k} |\alpha_m|^2 \Big)^{1/2} = O(2^{-k/2}).$$

Proof. Use Theorem 8.3 or

$$E_{2^k}(f|L_2) = \Big(\frac{1}{2} \sum_{m=2^k+1}^{\infty} |\alpha_m|^2 \Big)^{1/2}. \qquad \square$$

In view of the observations above, we can reformulate Theorem 11.3.

Theorem 11.6. *Every S_+-invariant linear form λ on $\mathfrak{l}_\infty(\mathbb{N}_0)$ yields a trace τ on $\mathfrak{L}^{\text{app}}_{1,\infty}(L_2(\mathbb{T}))$ whose restriction to convolution operators on $L_2(\mathbb{T})$ generated from functions $f \in B^{1/2}_{2,\infty}(\mathbb{T})$ has the form*

$$\tau(C_f) = \lambda\Big(\sum_{m \in \mathbb{D}_k} \alpha_m \Big) \quad \textit{whenever} \quad f(x) = \sum_{m=1}^{\infty} \alpha_m \cos mx.$$

According to [25, Theorem 6.2], [27, Theorem 7], and [28, Remark 5], there is a one-to-one correspondence between all S_+-invariant linear forms on $\mathfrak{l}_\infty(\mathbb{N}_0)$ and all traces on $\mathfrak{L}^{\text{app}}_{1,\infty}(L_2(\mathbb{T})) = (\mathfrak{P}_2)^{\text{app}}_{2,\infty}(L_2(\mathbb{T}))$. To complete this result we need some preparations.

For $a = (\alpha_k) \in \mathfrak{l}_\infty(\mathbb{N}_0)$, we define the function

$$f_a := \sum_{k=0}^{\infty} \frac{\alpha_k}{2^k} f^k_{\cos}, \quad \text{where} \quad f^k_{\cos}(x) := \sum_{m \in \mathbb{D}_k} \cos mx,$$

which belongs to $B^{1/2}_{2,\infty}(\mathbb{T})$, by Theorem 11.5.

Remark 11.7. As shown in [33, pp. 24–25], there holds $\|D_n|L_p\| = O(n^{1/p^*})$ whenever $1 < p < \infty$. Therefore $\|f^k_{\cos}|L_p\| = O(2^{k/p^*})$, which implies that $f_a \in B^{1/p}_{p,\infty}(\mathbb{T})$. In other words, $a \mapsto f_a$ is a map from $\mathfrak{l}_\infty(\mathbb{N}_0)$ into $B^{1/p}_{p,\infty}(\mathbb{T})$. I do not know what happens in the limiting case $p = 1$.

Theorem 11.8. *Every S_+-invariant linear form λ on $\mathfrak{l}_\infty(\mathbb{N}_0)$ can be reconstructed from the associated trace τ on $\mathfrak{L}_{1,\infty}^{\text{app}}(L_2(\mathbb{T}))$:*

$$\lambda(a) = \tau\big(C_{f_a}\big) \quad \textit{for all} \quad a \in \mathfrak{l}_\infty(\mathbb{N}_0).$$

That is, only the values of τ at certain convolution operators are needed.

Proof. Since rank $\big(C_{f^k_{\cos}}\big) = 2^{k+1}$,

$$C_{f_a} = \sum_{k=0}^{\infty} \tfrac{\alpha_k}{2^k} C_{f^k_{\cos}}$$

becomes a $(\mathfrak{P}_2, \mathfrak{d}_{2,\infty})$-representation when shifting all summands one step to the right. Moreover, trace $\big(C_{f^k_{\cos}}\big) = 2^k$, by Proposition 11.4. So it follows from Theorem 7.3 that

$$\tau\big(C_{f_a}\big) = \lambda\Big(\tfrac{\alpha_k}{2^k} \operatorname{trace}\big(C_{f^k_{\cos}}\big)\Big) = \lambda(\alpha_k). \qquad \square$$

12. Connes' trace theorem

Connes' trace theorem deals with (singular) traces of *classical* pseudo-differential operators on Riemannian manifolds; see [6, p. 307], [9, p. 293], [14, p. 372], and [28, p. 17].

In the case of the 1-dimensional torus (unit circle), the crucial symbols $p(x,n)$ are 2π-periodic functions of x and homogeneous of degree -1 with respect to n. The latter condition means that

$$p(x,n) = \tfrac{1}{|n|} p\big(x, \tfrac{n}{|n|}\big) \quad \text{whenever} \quad n \in \mathbb{Z} \setminus \{0\}.$$

For simplification purposes, we assume that $p(x,0) = 0$. Those symbols are uniquely determined by the 2π-periodic functions $p(x,+1)$ and $p(x,-1)$. The corresponding pseudo-differential operator is defined by

$$S : g(y) \mapsto \sum_{n\in\mathbb{Z}} \Big(\frac{1}{2\pi} \int_{\mathbb{T}} g(y) \mathrm{e}^{-\mathrm{i}ny}\, dy\Big)\, p(x,n) \mathrm{e}^{\mathrm{i}nx}. \qquad \text{(PDO)}$$

In particular, $S : \mathrm{e}^{\mathrm{i}ny} \mapsto p(x,n)\mathrm{e}^{\mathrm{i}nx}$ for all $n \in \mathbb{Z}$.

If S is a convolution operator, then $\mathrm{e}^{+\mathrm{i}x}$ and $\mathrm{e}^{-\mathrm{i}x}$ are eigenfunctions, which implies that $p(x,+1)$ and $p(x,-1)$ must be constant. So, in view of

$$\begin{aligned} Sg(x) = {} & \frac{p(x,+1)}{2\pi} \int_{\mathbb{T}} \Big(\sum_{n=1}^{\infty} \tfrac{1}{n} \mathrm{e}^{+\mathrm{i}n(x-y)}\Big) g(y)\, dy \\ & + \frac{p(x,-1)}{2\pi} \int_{\mathbb{T}} \Big(\sum_{n=1}^{\infty} \tfrac{1}{n} \mathrm{e}^{-\mathrm{i}n(x-y)}\Big) g(y)\, dy, \end{aligned}$$

all S obtained by (PDO) that are convolution operators form a 2-dimensional space. A basis is given by $C_{f_{\cos}}$ and $C_{f_{\sin}}$. Since every trace of $C_{f_{\sin}}$ vanishes, the only interesting convolution operators covered by Connes' trace theorem are scalar multiples of $C_{f_{\cos}}$; see Example 9.15.

References

[1] N.K. Bari, *A Treatise on Trigonometric Series*, two volumes, Pergamon Press, New York, 1964; Russian original: Fizmatgiz, Moscow, 1961.

[2] A. Böttcher and A. Pietsch, *Orthogonal and skew-symmetric operators in real Hilbert space*, Integr. Equ. Oper. Theory **74** (2012), 497–511.

[3] P.L. Butzer and H. Berens, *Semi-Groups of Operators and Approximation*, Springer, Berlin–Heidelberg–New York, 1967.

[4] P.L. Butzer and R.-J. Nessel, *Fourier Analysis and Approximation, One-Dimensional Theory*, Birkhäuser, Basel–Stuttgart, 1971.

[5] P.L. Butzer and K. Scherer, *Approximations-Prozesse und Interpolations-Methoden*, Bibliograph. Inst., Mannheim–Zürich, 1968.

[6] A. Connes, *Noncommutative Geometry*, Acad. Press, New York etc., 1994.

[7] R.A. DeVore and G.G. Lorentz, *Constructive Approximation*, Springer, Berlin, 1993.

[8] J. Diestel, H. Jarchow, and A. Tonge, *Absolutely Summing Operators*, Cambridge Univ. Press, 1995.

[9] J. Gracia-Bondía, J.C. Várilly, and H. Figueroa, *Elements of Noncommutative Geometry*, Birkhäuser, Boston, 2001.

[10] G.H. Hardy and J.E. Littlewood, *A convergence criterion for Fourier series*, Math. Z. **28** (1928), 612–634.

[11] V.I. Ivanov, *Direct and inverse theorems in approximation theory for periodic functions in S.B. Stechkin's papers and the development of these theorems*, Proc. Steklov Inst. Math. **273** (2011), Suppl. 1, 1–13.

[12] Y. Katznelson, *Introduction to Harmonic Analysis*, Wiley, New York, 1968; we refer to the Dover edition from 1976.

[13] H. König, J.R. Retherford, and N. Tomczak-Jaegermann *On the eigenvalues of $(p, 2)$-summing operators and constants associated with normed spaces*, J. Funct. Anal. **37** (1980), 88–126.

[14] S. Lord, F. Sukochev, and D. Zanin, *Singular Traces*, De Gruyter, Berlin, 2013.

[15] G.G. Lorentz, *Approximation of Functions*, Holt, Rinehart and Winston, New York etc., 1966.

[16] G.G. Lorentz, *A contribution to the theory of divergent sequences*, Acta Math. **80** (1948), 167–190.

[17] L. Mirsky, *Symmetric gauge functions and unitarily invariant norms*, Quart. J. Math. Oxford (2) **11** (1960), 50–59.

[18] I.P. Natanson, *Konstruktive Funktionentheorie*, Akademie-Verlag, Berlin, 1955; Russian original: Gostekhizdat, Moscow–Leningrad, 1949.

[19] S.M. Nikolskiĭ, *Approximation of Functions of Several Variables and Imbedding Theorems*, Springer, Berlin–Heidelberg–New York, 1975; Russian original: Nauka, Moscow, 1969.

[20] A. Pietsch, *Operator Ideals*, Deutsch. Verlag Wiss., Berlin, 1978; North-Holland, Amsterdam–London–New York–Tokyo, 1980.

[21] A. Pietsch, *Eigenvalues and s-Numbers*, Geest & Portig, Leipzig, and Cambridge Univ. Press, 1987.

[22] A. Pietsch, *History of Banach Spaces and Linear Operators*, Birkhäuser, Boston, 2007.

[23] A. Pietsch, *Approximation spaces*, J. Approx. Theory **32** (1981), 115–134.

[24] A. Pietsch, *Traces and shift invariant functionals*, Math. Nachr. **145** (1990), 7–43.

[25] A. Pietsch, *Traces on operator ideals and related linear forms on sequence ideals (part I)*, Indag. Math. (new series) **25** (2014), 341–365.

[26] A. Pietsch, *Traces on operator ideals and related linear forms on sequence ideals (part II)*, Integr. Equ. Oper. Theory **79** (2014), 255–299.

[27] A. Pietsch, *Traces on operator ideals and related linear forms on sequence ideals (part III)*, J. Math. Anal. Appl. **421** (2015), 971–981.

[28] A. Pietsch, *Traces and residues of pseudo-differential operators on the torus*, Integr. Equ. Oper. Theory **83** (2015), 1–23.

[29] I.I. Privalov, *Sur les fonctions conjugées*, Bull. Soc. Math. France **44** (1916), 100–103.

[30] E.S. Quade, *Trigonometric approximation in the mean*, Duke Math. J. **3** (1937), 529–543.

[31] M. Riesz, *Sur les maxima des formes bilinéaires et sur les fonctionnelles linéaires*, Acta Math. **49** (1927), 465–497.

[32] E. Schmidt, *Zur Theorie der linearen und nichtlinearen Integralgleichungen, Entwicklung willkürlicher Funktionen nach Systemen vorgeschriebener*, Math. Ann. **63** (1907), 433–476 (Thesis, Univ. Göttingen 1905).

[33] V.N. Temlyakov, *Approximation of Periodic Functions*, Nova Science, New York, 1993.

[34] A.F. Timan, *Theory of Approximation of Functions of a Real Variable*, Pergamon Press, Oxford, 1963; Russian original: Fizmatgiz, Moscow, 1960.

[35] N. Tomczak-Jaegermann, *Banach–Mazur Distances and Finite-Dimensional Operator Ideals*, Longman, Harlow, 1989.

[36] A. Zygmund, *Trigonometric Series*, Warszawa, 1935; since there are many revised editions in two volumes by Cambridge Univ. Press, we refer to that from 1959.

[37] A. Zygmund, *Smooth functions*, Duke Math. J. **12** (1945), 47–76.

Albrecht Pietsch
Biberweg 7
D-07749 Jena, Germany
e-mail: `a.pietsch@uni-jena.de`

Operator Theory:
Advances and Applications, Vol. 259, 621–648

Error Estimates for the ESPRIT Algorithm

Daniel Potts and Manfred Tasche

Dedicated to the 60th birthday of Albrecht Böttcher

Abstract. Let $z_j := \mathrm{e}^{f_j}$ $(j = 1, \ldots, M)$ with $f_j \in [-\varphi, 0] + \mathrm{i}\,[-\pi, \pi)$ and small $\varphi \geq 0$ be distinct nodes. With complex coefficients $c_j \neq 0$, we consider an exponential sum $h(x) := c_1\,\mathrm{e}^{f_1\,x} + \ldots + c_M\,\mathrm{e}^{f_M\,x}$ $(x \geq 0)$. Many applications in electrical engineering, signal processing, and mathematical physics lead to the following problem: Determine all parameters of h, if N noisy sampled values $\tilde{h}_k := h(k) + e_k$ $(k = 0, \ldots, N-1)$ with $N \gg 2\,M$ are given, where e_k are small error terms. This parameter identification problem is a nonlinear inverse problem which can be efficiently solved by the ESPRIT algorithm. In this paper, we present mainly corresponding error estimates for the nodes z_j $(j = 1, \ldots, M)$. We show that under appropriate conditions, the results of the ESPRIT algorithm are relatively insensitive to small perturbations on the sampled data.

Mathematics Subject Classification (2010). 65F15, 65F20, 65F35, 94A12.

Keywords. ESPRIT algorithm, exponential sum, frequency analysis problem, parameter identification problem, rectangular Hankel matrix, companion matrix, error estimates.

1. Introduction

The following *frequency analysis problem* resp. *parameter identification problem* arises in electrical engineering, signal processing, and mathematical physics (see, e.g., [15]):

Recover the positive integer M, distinct numbers $f_j \in [-\varphi, 0] + \mathrm{i}\,[-\pi, \pi)$ with small $\varphi \geq 0$, and complex coefficients $c_j \neq 0$ $(j = 1, \ldots, M)$ in the *exponential sum of order* M

$$h(x) := \sum_{j=1}^{M} c_j\,\mathrm{e}^{f_j\,x} \quad (x \geq 0)\,, \tag{1.1}$$

if noisy sampled data $\tilde{h}_k := h(k) + e_k$ $(k = 0, \ldots, N-1)$ with $N > 2M$ are given, where $e_k \in \mathbb{C}$ are small error terms with $|e_k| \le \varepsilon_1$ and $0 \le \varepsilon_1 \ll \min\{|c_j|;\, j = 1, \ldots, M\}$.

Note that $-\mathrm{Re}\, f_j \in [0,\, \varphi]$ is the *damping factor* and that $\mathrm{Im}\, f_j \in [-\pi,\, \pi)$ is the *angular frequency* of the exponential $\mathrm{e}^{f_j x}$. The *nodes* $z_j := \mathrm{e}^{f_j}$ $(j = 1, \ldots, M)$ are distinct values in the annulus $\mathbb{D} := \{z \in \mathbb{C} : \mathrm{e}^{-\varphi} \le |z| \le 1\}$.

This frequency analysis problem can be seen as a nonlinear approximation problem to recover the best M-term approximation of h in the ∞-dimensional linear space $\{\mathrm{e}^{fx};\, f \in [-\varphi,\, 0] + \mathrm{i}\,[-\pi,\, \pi),\, x \ge 0\}$. One known method to solve the frequency analysis problem is the Prony method (see, e.g., [17]). But the main drawback of the Prony method is that it may be unstable in some cases. By oversampling of the exponential sum (1.1), i.e., using N sampled data of (1.1) with $N \gg 2M$, and applying stable numerical methods, one can obtain efficient algorithms for the parameter identification [6]. A frequently used stable method is the so-called ESPRIT method (ESPRIT = Estimation of Signal Parameters via Rotational Invariance Techniques) [19]. If all nodes z_j $(j = 1, \ldots, M)$ are lying on the unit circle, then an alternative to ESPRIT is the so-called MUSIC algorithm (MUSIC = MUltiple SIgnal Classification). For a systematic numerical study of the MUSIC algorithm see [13].

The aim of this paper is to present error estimates for the ESPRIT algorithm. We show that under appropriate conditions, the results of the ESPRIT algorithm is relatively insensitive to small perturbations on the sampled data. Our study is mainly based on the research of F.S.V. Bazán [1, 3] and on perturbation theory for eigenvalues of a nonnormal $M \times M$ matrix. We extend these results and apply this method to the very popular ESPRIT algorithm. We emphasize that our numerical study is restricted to a moderate size of M.

The outline of this paper is as follows. Section 2 has preliminary character and summarizes known properties (see [1]) of the rectangular Vandermonde matrix

$$\boldsymbol{V}_{P,M}(\boldsymbol{z}) := \left(z_j^{k-1}\right)_{k,j=1}^{P,M} \tag{1.2}$$

with $\boldsymbol{z} := (z_j)_{j=1}^{M}$. For well-separated nodes $z_j \in \mathbb{D}$ $(j = 1, \ldots, M)$, the rectangular Vandermonde matrix $\boldsymbol{V}_{P,M}(\boldsymbol{z})$ is well conditioned for sufficiently large $P > M$. In Section 3, we solve the frequency analysis problem by the ESPRIT method which is mainly based on singular value decomposition (SVD) of a rectangular Hankel matrix. Instead of SVD, one can use a QR factorization of this rectangular Hankel matrix too (see [17]). In Section 4, we consider the orthogonal projection onto the signal space and the projected companion matrix. In Section 5, we present error estimates of the nodes z_j for the ESPRIT method. Finally, some numerical examples are given in Section 6.

In the following we use standard notations. By $\mathbb{C}$ we denote the set of all complex numbers. The set of all positive integers is $\mathbb{N}$. The linear space of all column vectors with n complex components is denoted by $\mathbb{C}^n$, where $\mathbf{0}$ is the corresponding zero vector. By $\boldsymbol{e}_j := (\delta_{j,k})_{k=1}^{n}$ $(j = 1, \ldots, n)$ we denote the canonical basis vectors of $\mathbb{C}^n$. The 2-norm in $\mathbb{C}^n$ is $\|\cdot\|_2$. The linear space of all complex

$m \times n$ matrices is denoted by $\mathbb{C}^{m\times n}$, where $\boldsymbol{O}_{m,n}$ is the corresponding zero matrix. For a matrix $\boldsymbol{A}_{m,n} \in \mathbb{C}^{m\times n}$, its transpose is $\boldsymbol{A}_{m,n}^{\mathrm{T}}$, its conjugate–transpose is $\boldsymbol{A}_{m,n}^{*}$, and its Moore–Penrose pseudoinverse is $\boldsymbol{A}_{m,n}^{\dagger}$. A square $m \times m$ matrix is abbreviated to $\boldsymbol{A}_m$. By $\boldsymbol{I}_m$ we denote the $m \times m$ identity matrix. The spectral norm in $\mathbb{C}^{m\times n}$ is denoted by $\|\cdot\|_2$, and the Frobenius norm is $\|\cdot\|_{\mathrm{F}}$. Further we use the known submatrix notation. Thus $\boldsymbol{A}_{m,m+1}(1:m,\,2:m+1)$ is the $m \times m$ submatrix of $\boldsymbol{A}_{m,m+1}$ obtained by extracting rows 1 through m and columns 2 through $m+1$. Note that the first row or column of a matrix can be indexed by zero. Other notations are introduced when needed.

2. Rectangular Vandermonde matrices

Let M and P be given positive integers with $P > M$. Further let $z_j \in \mathbb{D}$ $(j = 1,\ldots,M)$ be given distinct nodes and $\boldsymbol{z} := (z_j)_{j=1}^{M}$. Under mild assumptions, it was shown in [1] that the $P \times M$ Vandermonde matrix (1.2) is well conditioned for sufficiently large $P > M$. Let

$$\mathrm{cond}_2\, \boldsymbol{V}_{P,M}(\boldsymbol{z}) := \|\boldsymbol{V}_{P,M}(\boldsymbol{z})\|_2\, \|\boldsymbol{V}_{P,M}(\boldsymbol{z})^{\dagger}\|_2$$

be the *spectral norm condition number* of (1.2), where $\boldsymbol{V}_{P,M}(\boldsymbol{z})^{\dagger}$ denotes the Moore–Penrose pseudoinverse of (1.2) (see [9, p. 382]). Since $\boldsymbol{V}_{P,M}(\boldsymbol{z})$ has full rank, we have

$$\boldsymbol{V}_{P,M}(\boldsymbol{z})^{\dagger} = \big(\boldsymbol{V}_{P,M}(\boldsymbol{z})^{*}\, \boldsymbol{V}_{P,M}(\boldsymbol{z})\big)^{-1}\, \boldsymbol{V}_{P,M}(\boldsymbol{z})^{*}\,.$$

Analogously,

$$\mathrm{cond}_{\mathrm{F}}\, \boldsymbol{V}_{P,M}(\boldsymbol{z}) := \|\boldsymbol{V}_{P,M}(\boldsymbol{z})\|_{\mathrm{F}}\, \|\boldsymbol{V}_{P,M}(\boldsymbol{z})^{\dagger}\|_{\mathrm{F}}$$

is the *Frobenius norm condition number* of (1.2). Since the spectral norm and the Frobenius norm are unitarily invariant norms, we conclude that

$$\|\boldsymbol{V}_{P,M}(\boldsymbol{z})\|_2 = \|\boldsymbol{V}_{P,M}(\boldsymbol{z})^{\mathrm{T}}\|_2\,, \quad \|\boldsymbol{V}_{P,M}(\boldsymbol{z})\|_{\mathrm{F}} = \|\boldsymbol{V}_{P,M}(\boldsymbol{z})^{\mathrm{T}}\|_{\mathrm{F}}\,.$$

Further we remark that $\big(\boldsymbol{V}_{P,M}(\boldsymbol{z})^{\dagger}\big)^{\mathrm{T}} = \big(\boldsymbol{V}_{P,M}(\boldsymbol{z})^{\mathrm{T}}\big)^{\dagger}$. Then we introduce the following numbers

$$\alpha := \max\{|z_j|;\, j = 1,\ldots,M\} \in [\mathrm{e}^{-\varphi},\, 1]\,, \tag{2.1}$$

$$\beta := \min\{|z_j|;\, j = 1,\ldots,M\} \in [\mathrm{e}^{-\varphi},\, 1]\,, \tag{2.2}$$

$$\mu := \sum_{j=1}^{M} |z_j|^2 \in [M\,\mathrm{e}^{-2\varphi},\, M]\,, \tag{2.3}$$

$$\nu := \prod_{j=1}^{M} |z_j|^2 \in [\mathrm{e}^{-2M\varphi},\, 1]\,.$$

The *separation distance* of all nodes $z_j \in \mathbb{D}$ $(j = 1,\ldots,M)$ is explained by

$$\delta := \min\{|z_j - z_k|;\, j,\, k = 1,\ldots,M,\, j \neq k\} > 0\,. \tag{2.4}$$

For $P > M$, the spectral norm of $\big(\boldsymbol{V}_{P,M}(\boldsymbol{z})^{\mathrm{T}}\big)^{\dagger}$ decreases monotonously with respect to P. If $|z_j| = 1$ for all $j = 1, \ldots, M$, then by [1, Theorem 1]

$$\lim_{P\to\infty} \|\big(\boldsymbol{V}_{P,M}(\boldsymbol{z})^{\mathrm{T}}\big)^{\dagger}\|_2 = 0\,.$$

Let $\boldsymbol{q}_P \in \mathbb{C}^P$ with $P > M$ be the minimum 2-norm solution of the underdetermined linear system

$$\boldsymbol{V}_{P,M}(\boldsymbol{z})^{\mathrm{T}}\,\boldsymbol{q}_P = -\big(z_j^P\big)_{j=1}^{M} \tag{2.5}$$

such that

$$\boldsymbol{q}_P = -\big(\boldsymbol{V}_{P,M}(\boldsymbol{z})^{\mathrm{T}}\big)^{\dagger}\,\big(z_j^P\big)_{j=1}^{M}\,.$$

Then by [1, Theorem 2], the norms $\|\boldsymbol{q}_P\|_2$ are bounded with respect to P. If either $|z_j| = 1$ $(j = 1, \ldots, M)$ or $|z_j| < 1$ $(j = 1, \ldots, M)$, then

$$\lim_{P\to\infty} \|\boldsymbol{q}_P\|_2 = 0\,.$$

Let $P > M \geq 2$ and let $\boldsymbol{V}_{P,M}(\boldsymbol{z})$ be the $P \times M$ Vandermonde matrix with distinct nodes $z_j \in \mathbb{D}$ $(j = 1, \ldots, M)$. Then by [1, Theorem 6 and Lemma 7], the Frobenius norm condition number of $\boldsymbol{V}_{P,M}(\boldsymbol{z})$ can be estimated by

$$\mathrm{cond}_{\mathrm{F}}\,\boldsymbol{V}_{P,M}(\boldsymbol{z}) \leq M\,\Big(1 + \frac{\|\boldsymbol{q}_P\|_2^2 + M + \nu - \mu - 1}{(M-1)\,\delta^2}\Big)^{(M-1)/2}\,\Phi_P(\alpha,\beta) \tag{2.6}$$

with

$$\Phi_P(\alpha,\beta) := \Big(\frac{1 + \alpha^2 + \alpha^4 + \ldots + \alpha^{2(P-1)}}{1 + \beta^2 + \beta^4 + \ldots + \beta^{2(P-1)}}\Big)^{1/2}\,,$$

where α, β, μ, ν, and δ are defined by (2.1)–(2.4).

The Vandermonde matrix $\boldsymbol{V}_{P,M}(\boldsymbol{z})$ with $P > M$ satisfies the inequality

$$M - 2 + \mathrm{cond}_2\,\boldsymbol{V}_{P,M}(\boldsymbol{z}) + \big(\mathrm{cond}_2\,\boldsymbol{V}_{P,M}(\boldsymbol{z})\big)^{-1} \leq \mathrm{cond}_{\mathrm{F}}\,\boldsymbol{V}_{P,M}(\boldsymbol{z})\,. \tag{2.7}$$

This inequality (2.7) follows directly from a corresponding result in [20] for an invertible, square matrix. Since the rectangular Vandermonde matrix $\boldsymbol{V}_{P,M}(\boldsymbol{z})$ possesses full rank M, the square matrix $\boldsymbol{V}_{P,M}(\boldsymbol{z})^*\,\boldsymbol{V}_{P,M}(\boldsymbol{z})$ is positive definite. Hence its square root $\big(\boldsymbol{V}_{P,M}(\boldsymbol{z})^*\,\boldsymbol{V}_{P,M}(\boldsymbol{z})\big)^{1/2}$ is defined. The eigenvalues of $\big(\boldsymbol{V}_{P,M}(\boldsymbol{z})^*\,\boldsymbol{V}_{P,M}(\boldsymbol{z})\big)^{1/2}$ coincide with the singular values of $\boldsymbol{V}_{P,M}(\boldsymbol{z})$. Thus one obtains for the spectral resp. Frobenius norm

$$\mathrm{cond}_2\,\big(\boldsymbol{V}_{P,M}(\boldsymbol{z})^*\,\boldsymbol{V}_{P,M}(\boldsymbol{z})\big)^{1/2} = \mathrm{cond}_2\,\boldsymbol{V}_{P,M}(\boldsymbol{z})\,,$$
$$\mathrm{cond}_{\mathrm{F}}\,\big(\boldsymbol{V}_{P,M}(\boldsymbol{z})^*\,\boldsymbol{V}_{P,M}(\boldsymbol{z})\big)^{1/2} = \mathrm{cond}_{\mathrm{F}}\,\boldsymbol{V}_{P,M}(\boldsymbol{z})\,.$$

From (2.7) it follows that

$$\begin{aligned}\mathrm{cond}_2\,\boldsymbol{V}_{P,M}(\boldsymbol{z}) \leq{}& \frac{1}{2}\,\Big(\mathrm{cond}_{\mathrm{F}}\,\boldsymbol{V}_{P,M}(\boldsymbol{z}) - M + 2\Big)\\ &+ \frac{1}{2}\,\Big(\big(\mathrm{cond}_{\mathrm{F}}\,\boldsymbol{V}_{P,M}(\boldsymbol{z}) - M + 2\big)^2 - 4\Big)^{1/2}\,.\end{aligned}$$

If all nodes z_j are lying on the unit circle, i.e., $|z_j| = 1$ $(j = 1, \dots, M)$, then $\alpha = \beta = \nu = \Phi_P(1,1) = 1$, $\mu = M$, and hence by (2.6)

$$\mathrm{cond_F}\, \boldsymbol{V}_{P,M}(\boldsymbol{z}) \le M \left(1 + \frac{\|\boldsymbol{q}_P\|_2^2}{(M-1)\,\delta^2}\right)^{(M-1)/2}.$$

In the case $|z_j| = 1$ $(j = 1, \dots, M)$, better estimates for the spectral norm condition number of $\boldsymbol{V}_{P,M}(\boldsymbol{z})$ are possible (see [16, 13]).

3. ESPRIT algorithm

In practice, the order M of the exponential sum (1.1) is often unknown. Assume that $L \in \mathbb{N}$ is a convenient upper bound of M with $M \le L \le \lceil \frac{N}{2} \rceil$, where N is a sufficiently large integer with $N \gg 2M$. In applications, such an upper bound L is mostly known *a priori*. If this is not the case, then one can choose $L = \lceil \frac{N}{2} \rceil$. Suppose that N noisy sampled data $\tilde{h}_k := h(k) + e_k \in \mathbb{C}$ $(k = 0, \dots, N-1)$ of (1.1) are given, where $e_k \in \mathbb{C}$ are small error terms with $|e_k| \le \varepsilon_1$ and $0 \le \varepsilon_1 \ll 1$. Often the sequence $\{\tilde{h}_0, \tilde{h}_1, \dots, \tilde{h}_{N-1}\}$ of sampled data is called *time series of length* N. Then we form the *L-trajectory matrix* of this time series

$$\tilde{\boldsymbol{H}}_{L,N-L+1} := \left(\tilde{h}_{\ell+m}\right)_{\ell,m=0}^{L-1,N-L} \tag{3.1}$$

with the *window length* $L \in \{M, \dots, \lceil \frac{N}{2} \rceil\}$. Obviously, (3.1) is an $L \times (N-L+1)$ Hankel matrix.

The nonincreasingly ordered singular values $\tilde{\sigma}_k(\tilde{\boldsymbol{H}}_{L,N-L+1})$ of the L-trajectory matrix (3.1) possess the following property:

Lemma 3.1. *For fixed $N \gg 2\,M$, the singular values of* (3.1) *increase almost monotonously with respect to $L = M, \dots, \lceil \frac{N}{2} \rceil$, i.e., they fulfil the inequalities*

$$\tilde{\sigma}_k(\tilde{\boldsymbol{H}}_{L,N-L+1})^2 \le \tilde{\sigma}_k(\tilde{\boldsymbol{H}}_{L+1,N-L})^2 + \|\tilde{\boldsymbol{h}}_L\|_2^2 \quad (k = 1, \dots, M), \tag{3.2}$$

where $\tilde{\boldsymbol{h}}_L := (\tilde{h}_k)_{k=N-L}^{N-1}$.

Proof. For $M \le L < \lceil \frac{N}{2} \rceil$, we represent the Hankel matrices $\tilde{\boldsymbol{H}}_{L,N-L+1}$ and $\tilde{\boldsymbol{H}}_{L+1,N-L}$ as block matrices

$$\tilde{\boldsymbol{H}}_{L,N-L+1} = \left(\tilde{\boldsymbol{H}}_{L,N-L} \,|\, \tilde{\boldsymbol{h}}_L\right), \quad \tilde{\boldsymbol{H}}_{L+1,N-L} = \begin{pmatrix} \tilde{\boldsymbol{H}}_{L,N-L} \\ \tilde{\boldsymbol{h}}_{N-L}^{\mathrm{T}} \end{pmatrix}$$

with $\tilde{\boldsymbol{h}}_{N-L} := (\tilde{h}_k)_{k=L}^{N-1}$. Setting $\tilde{\boldsymbol{B}}_L := \tilde{\boldsymbol{H}}_{L,N-L}\, \tilde{\boldsymbol{H}}_{L,N-L}^*$, we obtain that

$$\tilde{\boldsymbol{A}}_L := \tilde{\boldsymbol{H}}_{L,N-L+1}\, \tilde{\boldsymbol{H}}_{L,N-L+1}^* = \tilde{\boldsymbol{B}}_L + \tilde{\boldsymbol{h}}_L\, \tilde{\boldsymbol{h}}_L^*, \tag{3.3}$$

which is a rank-one Hermitian perturbation of the Hermitian matrix $\tilde{\boldsymbol{B}}_L$, and

$$\tilde{\boldsymbol{A}}_{L+1} := \tilde{\boldsymbol{H}}_{L+1,N-L}\, \tilde{\boldsymbol{H}}_{L+1,N-L}^* = \begin{pmatrix} \tilde{\boldsymbol{B}}_L & \tilde{\boldsymbol{y}}_L \\ \tilde{\boldsymbol{y}}_L^* & \|\tilde{\boldsymbol{h}}_{N-L}\|_2^2 \end{pmatrix}$$

with $\tilde{\boldsymbol{y}}_L := \tilde{\boldsymbol{H}}_{L,N-L}\,\overline{\tilde{\boldsymbol{h}}}_{N-L}$. Using Cauchy's Interlacing Theorem (see [10, p. 242]) for the bordered Hermitian matrix $\tilde{\boldsymbol{A}}_{L+1}$, the corresponding nondecreasingly ordered eigenvalues of $\tilde{\boldsymbol{A}}_{L+1}$ and $\tilde{\boldsymbol{B}}_L$ fulfil the inequalities

$$\tilde{\lambda}_j(\tilde{\boldsymbol{B}}_L) \le \tilde{\lambda}_{j+1}(\tilde{\boldsymbol{A}}_{L+1}) \quad (j = 1, \ldots, L).$$

By (3.3) and by Weyl's Theorem (see [10, p. 239]), we obtain that

$$\tilde{\lambda}_1(-\tilde{\boldsymbol{h}}_L\,\tilde{\boldsymbol{h}}_L^*) + \tilde{\lambda}_j(\tilde{\boldsymbol{A}}_L) \le \tilde{\lambda}_j(\tilde{\boldsymbol{B}}_L) \le \tilde{\lambda}_{j+1}(\tilde{\boldsymbol{A}}_{L+1}).$$

Since the first eigenvalue of the rank-one matrix $-\tilde{\boldsymbol{h}}_L\,\tilde{\boldsymbol{h}}_L^*$ is equal to $-\|\tilde{\boldsymbol{h}}_L\|_2^2$, it follows that

$$\tilde{\lambda}_j(\tilde{\boldsymbol{A}}_L) \le \tilde{\lambda}_{j+1}(\tilde{\boldsymbol{A}}_{L+1}) + \|\tilde{\boldsymbol{h}}_L\|_2^2.$$

The non-zero eigenvalues of $\tilde{\boldsymbol{A}}_L$ resp. $\tilde{\boldsymbol{A}}_{L+1}$ are the squares of the positive singular values of $\tilde{\boldsymbol{H}}_{L,N-L+1}$ resp. $\tilde{\boldsymbol{H}}_{L+1,N-L}$. This completes the proof of (3.2). □

The convenient choice of the window length L is essential for the following ESPRIT method. By Lemma 3.1, a sufficiently large integer $L \approx \lceil \frac{N}{2} \rceil$ is a good choice. Then the L-trajectory matrix (3.1) with $L = \lceil \frac{N}{2} \rceil$ is almost square. Several numerical experiments in [8] confirm that the optimal window length L lies in the near of $\lceil \frac{N}{2} \rceil$.

The main step in the solution of the frequency analysis problem is the determination of the order M and the computation of the exponents f_j or alternatively of the nodes $z_j = \mathrm{e}^{f_j} \in \mathbb{D}$ $(j = 1, \ldots, M)$. Afterwards one can calculate the coefficient vector $\boldsymbol{c} := (c_j)_{j=1}^M \in \mathbb{C}^M$ as least squares solution of the overdetermined linear system

$$\boldsymbol{V}_{N,M}(\boldsymbol{z})\,\boldsymbol{c} = \big(\tilde{h}_k\big)_{k=0}^{N-1}$$

with the rectangular Vandermonde matrix (1.2), i.e., the coefficient vector $\boldsymbol{c}$ is the solution of the least squares problem

$$\|\boldsymbol{V}_{N,M}(\boldsymbol{z})\,\boldsymbol{c} - \big(\tilde{h}_k\big)_{k=0}^{N-1}\|_2 = \min.$$

As known, the square Vandermonde matrix $\boldsymbol{V}_M(\boldsymbol{z})$ is invertible and the matrix $\boldsymbol{V}_{N,M}(\boldsymbol{z})$ has full column rank. Additionally we introduce the rectangular Hankel matrices

$$\tilde{\boldsymbol{H}}_{L,N-L}(s) := \tilde{\boldsymbol{H}}_{L,N-L+1}(1:L,\,1+s:N-L+s) \quad (s = 0,\,1). \tag{3.4}$$

In the case of exactly sampled data $\tilde{h}_k = h(k)$ $(k = 0, \ldots, N-1)$, the Hankel matrix (3.1) is denoted by $\boldsymbol{H}_{L,N-L+1}$ and the related Hankel matrices (3.4) are denoted by $\boldsymbol{H}_{L,N-L}(s)$ $(s = 0,\,1)$.

Remark 3.2. The Hankel matrices $\boldsymbol{H}_{L,N-L+1}$ and $\boldsymbol{H}_{L,N-L}(s)$ $(s = 0,\,1)$ have the same rank M for each window length $L \in \{M, \ldots, \lceil \frac{N}{2} \rceil\}$ (see [17, Lemma 2.1]). Consequently, the order M of the exponential sum (1.1) coincides with the rank of these Hankel matrices. □

First we assume that exactly sampled data $\tilde{h}_k = h(k)$ $(k = 0, \ldots, N-1)$ of (1.1) are given. We choose $L \approx \lceil \frac{N}{2} \rceil$. Then the matrix pencil

$$z\, \boldsymbol{H}_{L,N-L}(0) - \boldsymbol{H}_{L,N-L}(1) \quad (z \in \mathbb{C}) \tag{3.5}$$

has the nodes $z_j \in \mathbb{D}$ $(j = 1, \ldots, M)$ as eigenvalues (see, e.g., [11, 17]). We start the ESPRIT method by the SVD of the exact L-trajectory matrix $\boldsymbol{H}_{L,N-L+1}$, i.e.,

$$\boldsymbol{H}_{L,N-L+1} = \boldsymbol{U}_L\, \boldsymbol{D}_{L,N-L+1}\, \boldsymbol{W}^*_{N-L+1}\,,$$

where $\boldsymbol{U}_L \in \mathbb{C}^{L \times L}$ and $\boldsymbol{W}_{N-L+1} \in \mathbb{C}^{(N-L+1)\times(N-L+1)}$ are unitary matrices and where $\boldsymbol{D}_{L,N-L+1} \in \mathbb{R}^{L \times (N-L+1)}$ is a rectangular diagonal matrix. The diagonal entries of $\boldsymbol{D}_{L,N-L+1}$ are the singular values σ_j of the L-trajectory matrix arranged in nonincreasing order $\sigma_1 \geq \cdots \geq \sigma_M > \sigma_{M+1} = \cdots = \sigma_L = 0$. Thus we can determine the order M of the exponential sum (1.1) by the number of positive singular values σ_j.

Remark 3.3. For fixed N, the size of the lowest positive singular value σ_M of $\boldsymbol{H}_{L,N-L+1}$ with $M \leq L \leq \lceil \frac{N}{2} \rceil$ depends on the choice of L by Lemma 3.1. The M positive singular values of $\boldsymbol{H}_{L,N-L+1}$ are the square roots of the M positive eigenvalues of

$$\boldsymbol{A}_L := \boldsymbol{H}_{L,N-L+1}\, \boldsymbol{H}^*_{L,N-L+1}\,.$$

All the other singular values of $\boldsymbol{H}_{L,N-L+1}$ resp. eigenvalues of $\boldsymbol{A}_L$ are zero. The trace of $\boldsymbol{A}_L$ is equal to

$$\mathrm{tr}\,(\boldsymbol{A}_L) = \sum_{\ell=0}^{L-1} \sum_{j=\ell}^{N-L+\ell} |h(j)|^2$$

and the sum of all the principal minors of size 2 amounts

$$s_2(\boldsymbol{A}_L) = \sum_{\ell=0}^{L-2} \sum_{k=1}^{L-\ell-1} \Bigg[\Big(\sum_{j=\ell}^{N-L+\ell} |h(j)|^2 \Big) \Big(\sum_{j=\ell}^{N-L+\ell} |h(j+k)|^2 \Big) \\ - \Big| \sum_{j=\ell}^{N-L+\ell} h(j)\, \overline{h(j+k)} \Big|^2 \Bigg].$$

In the case $M = 1$, i.e., $h(j) = c_1\, z_1^j$ with $c_1 \neq 0$ and $0 < |z_1| \leq 1$, the only positive eigenvalue λ_1 of $\boldsymbol{A}_L$ reads as follows

$$\lambda_1 = \mathrm{tr}\,(\boldsymbol{A}_L) = |c_1|^2 \Big(\sum_{j=0}^{N-L} |z_1|^{2j} \Big) \Big(\sum_{\ell=0}^{L-1} |z_1|^{2\ell} \Big),$$

so that the only positive singular value σ_1 of $\boldsymbol{H}_{L,N-L+1}$ fulfils the estimate

$$\sigma_1 = |c_1| \sqrt{\Big(\sum_{j=0}^{N-L} |z_1|^{2j} \Big) \Big(\sum_{\ell=0}^{L-1} |z_1|^{2\ell} \Big)} \leq |c_1|\, \sqrt{(N-L+1)\, L} \tag{3.6}$$

with equality for $|z_1| = 1$. Thus σ_1 is maximal for $L = \lceil \frac{N}{2} \rceil$.

In the case $M = 2$, i.e., $h(j) = c_1\, z_1^j + c_2\, z_2^j$ with $c_k \neq 0$ and $0 < |z_k| \le 1$ ($k = 1, 2$), there exist only two positive eigenvalues λ_1, λ_2 of $\boldsymbol{A}_L$, all the other eigenvalues of $\boldsymbol{A}_L$ vanish. Then λ_1 and λ_2 are the solutions of the quadratic equation

$$\lambda^2 - \lambda\, \mathrm{tr}\,(\boldsymbol{A}_L) + s_2(\boldsymbol{A}_L) = 0$$

(see [10, p. 54]) so that

$$\lambda_{1,2} = \frac{1}{2}\, \mathrm{tr}\,(\boldsymbol{A}_L) \pm \frac{1}{2}\, \sqrt{(\mathrm{tr}\,(\boldsymbol{A}_L))^2 - 4\, s_2(\boldsymbol{A}_L)}\,.$$

Hence the two positive singular values of $\boldsymbol{H}_{L,N-L+1}$ are $\sigma_{1,2} = \sqrt{\lambda_{1,2}}$.

In the case $M > 1$, one can estimate the positive singular values of $\boldsymbol{H}_{L,N-L+1}$ by Weyl's Theorem (see [21, p. 68]). Since

$$\boldsymbol{H}_{L,N-L+1} = \sum_{k=1}^{M} \left(c_k\, z_k^{\ell+m}\right)_{\ell,m=0}^{L-1,N-L}$$

with $c_k \neq 0$ and $0 < |z_k| \le 1$ ($k = 1, \ldots, M$), one obtains by (3.6) that

$$0 < \sigma_M \le \sigma_1 \le \sum_{k=1}^{M} |c_k|\, \sqrt{(N-L+1)\, L}\,. \tag{3.7}$$

A lower estimate of σ_M was presented in [4].

A good criterion for the choice of optimal window length L is to maximize the lowest positive singular value σ_M of $\boldsymbol{H}_{L,N-L+1}$. By Lemma 3.1, (3.6) and (3.7), one can see that $L = \lceil \frac{N}{2} \rceil$ is a good choice (see also [4, 8] and Example 6.4). For the ESPRIT Algorithm 3.4 (i.e., the determination of the numerical rank in step 1) and the corresponding error estimates (see Theorem 5.4), it is very important that σ_M is not too small. □

Introducing the matrices

$$\begin{aligned} \boldsymbol{U}_{L,M} &:= \boldsymbol{U}_M(1:L,\, 1:M) \\ \boldsymbol{W}_{N-L+1,M} &:= \boldsymbol{W}_{N-L+1}(1:N-L+1,\, 1:M) \end{aligned}$$

with orthonormal columns as well as the diagonal matrix $\boldsymbol{D}_M := \mathrm{diag}\,(\sigma_j)_{j=1}^M$, we obtain the partial SVD of the matrix (3.1) with exact entries, i.e.,

$$\boldsymbol{H}_{L,N-L+1} = \boldsymbol{U}_{L,M}\, \boldsymbol{D}_M\, \boldsymbol{W}_{N-L+1,M}^*\,.$$

Setting

$$\boldsymbol{W}_{N-L,M}(s) := \boldsymbol{W}_{N-L+1,M}(1+s : N-L+s, 1 : M) \quad (s = 0,\, 1)\,, \tag{3.8}$$

it follows by (3.8) and (3.4) that both Hankel matrices (3.4) can be simultaneously factorized in the form

$$\boldsymbol{H}_{L,N-L}(s) = \boldsymbol{U}_{L,M}\, \boldsymbol{D}_M\, \boldsymbol{W}_{N-L,M}(s)^* \quad (s = 0,\, 1)\,. \tag{3.9}$$

Since $\boldsymbol{U}_{L,M}$ has orthonormal columns and since $\boldsymbol{D}_M$ is invertible, the generalized eigenvalue problem of the matrix pencil

$$z\,\boldsymbol{W}_{N-L,M}(0)^* - \boldsymbol{W}_{N-L,M}(1)^* \quad (z \in \mathbb{C})$$

has the same non-zero eigenvalues $z_j \in \mathbb{D}$ $(j = 1, \ldots, M)$ as the matrix pencil (3.5) except for additional zero eigenvalues. Finally we determine the nodes $z_j \in \mathbb{D}$ $(j = 1, \ldots, M)$ as eigenvalues of the $M \times M$ matrix

$$\boldsymbol{F}_M^{\mathrm{SVD}} := \boldsymbol{W}_{N-L,M}(1)^* \left(\boldsymbol{W}_{N-L,M}(0)^*\right)^\dagger \tag{3.10}$$

Analogously, we can handle the general case of noisy data $\tilde{h}_k = h(k) + e_k \in \mathbb{C}$ $(k = 0, \ldots, N-1)$ with small error terms $e_k \in \mathbb{C}$, where $|e_k| \le \varepsilon_1$ and $0 < \varepsilon_1 \ll \min\{|c_j|;\, j = 1, \ldots, M\}$. For the Hankel matrix (3.1) with the singular values $\tilde{\sigma}_1 \ge \cdots \ge \tilde{\sigma}_L \ge 0$, we can calculate the numerical rank M of (3.1) by the property $\tilde{\sigma}_M \ge \varepsilon\,\tilde{\sigma}_1$ and $\tilde{\sigma}_{M+1} < \varepsilon\,\tilde{\sigma}_1$ with convenient chosen tolerance ε. Using the IEEE double precision arithmetic, one can choose $\varepsilon = 10^{-10}$ for given exact data. In the case of noisy data, one has to use a larger tolerance ε. Let

$$\boldsymbol{E}_{L,N-L+1} := \tilde{\boldsymbol{H}}_{L,N-L+1} - \boldsymbol{H}_{L,N-L+1}$$

be the *error matrix of given data*. Assuming $2\,\|\boldsymbol{E}_{L,N-L+1}\|_2 \ll \sigma_M$ and choosing $\varepsilon \approx 2\,\|\boldsymbol{E}_{L,N-L+1}\|_2/\tilde{\sigma}_1$, we find by a corollary of Weyl's Theorem (see [21, p. 70]) that

$$|\tilde{\sigma}_j - \sigma_j| \le \|\boldsymbol{E}_{L,N-L+1}\|_2 \quad (j = 1, \ldots, L)\,.$$

Thus one obtains that $\tilde{\sigma}_M \ge \sigma_M - \|\boldsymbol{E}_{L,N-L+1}\|_2 \gg \|\boldsymbol{E}_{L,N-L+1}\|_2 \approx \varepsilon\,\tilde{\sigma}_1$ and $\tilde{\sigma}_{M+1} \le \|\boldsymbol{E}_{L,N-L+1}\|_2 \approx \frac{\varepsilon}{2}\,\tilde{\sigma}_1$, i.e., $\tilde{\sigma}_M/\tilde{\sigma}_1 \ge \varepsilon$ and $\tilde{\sigma}_{M+1}/\tilde{\sigma}_1 < \varepsilon$.

For the Hankel matrix (3.1) with noisy entries, we use its SVD

$$\tilde{\boldsymbol{H}}_{L,N-L+1} = \tilde{\boldsymbol{U}}_L\,\tilde{\boldsymbol{D}}_{L,N-L+1}\,\tilde{\boldsymbol{W}}_{N-L+1}^*$$

and define as above the matrices $\tilde{\boldsymbol{U}}_{L,M}$, $\tilde{\boldsymbol{D}}_M := \mathrm{diag}\,\left(\tilde{\sigma}_j\right)_{j=1}^M$, and $\tilde{\boldsymbol{W}}_{N-L+1,M}$. Then

$$\tilde{\boldsymbol{U}}_{L,M}\,\tilde{\boldsymbol{D}}_M\,\tilde{\boldsymbol{W}}_{N-L+1,M}^*$$

is a low-rank approximation of (3.1). Analogously to (3.8) and (3.10), we introduce corresponding matrices $\tilde{\boldsymbol{W}}_{N-L,M}(s)$ $(s = 0,\, 1)$ and $\tilde{\boldsymbol{F}}_M^{\mathrm{SVD}}$. Note that

$$\tilde{\boldsymbol{K}}_{L,N-L}(s) := \tilde{\boldsymbol{U}}_{L,M}\,\tilde{\boldsymbol{D}}_M\,\tilde{\boldsymbol{W}}_{N-L,M}(s)^* \quad (s = 0,\, 1) \tag{3.11}$$

is a low-rank approximation of $\tilde{\boldsymbol{H}}_{L,N-L}(s)$. Thus the SVD-based ESPRIT algorithm reads as follows:

Algorithm 3.4. [ESPRIT via SVD]
Input: $N \in \mathbb{N}$ ($N \gg 2\,M$), M unknown order of (1.1), $L \approx \lceil \frac{N}{2} \rceil$ window length with $M \le L \le \lceil \frac{N}{2} \rceil$, $\tilde{h}_k = h(k) + e_k \in \mathbb{C}$ ($k = 0, \ldots, N-1$) noisy sampled values of (1.1), $0 < \varepsilon \ll 1$ tolerance.

1. Compute the SVD of the rectangular Hankel matrix (3.1). Determine the numerical rank M of (3.1) such that $\tilde{\sigma}_M \ge \varepsilon\, \tilde{\sigma}_1$ and $\tilde{\sigma}_{M+1} < \varepsilon \tilde{\sigma}_1$. Form the matrices $\tilde{\boldsymbol{W}}_{N-L,M}(s)$ ($s = 0,\, 1$) as in (3.8).
2. Calculate the $M \times M$ matrix $\tilde{\boldsymbol{F}}_M^{\mathrm{SVD}}$ as in (3.10) and compute all eigenvalues $\tilde{z}_j \in \mathbb{D}$ ($j = 1, \ldots, M$) of $\tilde{\boldsymbol{F}}_M^{\mathrm{SVD}}$. Set $\tilde{f}_j := \log \tilde{z}_j$ ($j = 1, \ldots, M$), where log denotes the principal value of the complex logarithm.
3. Compute the coefficient vector $\tilde{\boldsymbol{c}} := (\tilde{c}_j)_{j=1}^M \in \mathbb{C}^M$ as solution of the least squares problem
$$\|\boldsymbol{V}_{N,M}(\tilde{\boldsymbol{z}})\, \tilde{\boldsymbol{c}} - \big(\tilde{h}_k\big)_{k=0}^{N-1}\|_2 = \min\,,$$
where $\tilde{\boldsymbol{z}} := \big(\tilde{z}_j\big)_{j=1}^{M}$ denotes the vector of computed nodes.

Output: $M \in \mathbb{N}$, $\tilde{f}_j \in [-\varphi,\, 0] + \mathrm{i}\,[-\pi, \pi)$, $\tilde{c}_j \in \mathbb{C}$ ($j = 1, \ldots, M$).

Remark 3.5. One can pass on the computation of the Moore–Penrose pseudoinverse in (3.10). Then the second step of Algorithm 3.4 reads as follows (see [17, Algorithm 3.1]):

2′. Calculate the matrix products
$$\tilde{\boldsymbol{A}}_M := \tilde{\boldsymbol{W}}_{N-L,M}(0)^*\, \tilde{\boldsymbol{W}}_{N-L,M}(0)\,, \quad \tilde{\boldsymbol{B}}_M := \tilde{\boldsymbol{W}}_{N-L,M}(1)^*\, \tilde{\boldsymbol{W}}_{N-L,M}(0)$$
and compute all eigenvalues $\tilde{z}_j \in \mathbb{D}$ ($j = 1, \ldots, M$) of the square matrix pencil $z\, \tilde{\boldsymbol{A}}_M - \tilde{\boldsymbol{B}}_M$ ($z \in \mathbb{C}$) by the QZ-Algorithm (see [7, pp. 384–385]). Set $\tilde{f}_j := \log \tilde{z}_j$ ($j = 1, \ldots, M$). □

In the second step of Algorithm 3.4, the matrix (3.10) can be replaced by the matrix
$$\boldsymbol{F}_M := \boldsymbol{X}_{N-L,M}(1)^*\, \boldsymbol{X}_{N-L,M}(0)\,, \tag{3.12}$$
where
$$\boldsymbol{X}_{N-L,M}(s) := \boldsymbol{W}_{N-L,M}(s) \big(\boldsymbol{W}_{N-L,M}(0)^*\, \boldsymbol{W}_{N-L,M}(0)\big)^{-1/2} \quad (s = 0,\, 1)\,. \tag{3.13}$$
Since $\boldsymbol{W}_{N-L,M}(0)^*\, \boldsymbol{W}_{N-L,M}(0)$ is positive definite, the above matrix (3.13) is well defined. Obviously, we have
$$\boldsymbol{X}_{N-L,M}(0)^*\, \boldsymbol{X}_{N-L,M}(0) = \boldsymbol{I}_M\,, \tag{3.14}$$
i.e., the columns of $\boldsymbol{X}_{N-L,M}(0)$ are orthonormal. As later will be shown in Lemma 4.2, the new matrix (3.12) has the same eigenvalues $z_j \in \mathbb{D}$ ($j = 1, \ldots, M$) as (3.10).

4. Orthogonal projection onto the signal space

First we consider the ESPRIT method for exact sampled data $\tilde{h}_k = h(k)$ $(k = 0, \ldots, N-1)$ of the exponential sum (1.1) of order M. We choose a convenient window length $L \approx \lceil \frac{N}{2} \rceil$ such that $M \le L \le \lceil \frac{N}{2} \rceil$.

Analogously to (2.5), the vector $\boldsymbol{q}_{N-L} := (q_k)_{k=0}^{N-L-1} \in \mathbb{C}^{N-L}$ is defined as the minimum 2-norm solution of the (underdetermined) linear system

$$\boldsymbol{V}_{N-L,M}(\boldsymbol{z})^{\mathrm{T}}\, \boldsymbol{q}_{N-L} = -(z_j^{N-L})_{j=1}^{M}\,. \tag{4.1}$$

Forming the corresponding monic polynomial q_{N-L} of degree $N-L$

$$q_{N-L}(z) := \sum_{k=0}^{N-L-1} q_k\, z^k + z^{N-L} \quad (z \in \mathbb{C})\,,$$

then by (4.1) this polynomial has all nodes $z_j \in \mathbb{D}$ $(j = 1, \ldots, M)$ as roots. By (4.1) and by the factorization

$$\boldsymbol{H}_{L,N-L}(0) = \boldsymbol{V}_{L,M}(\boldsymbol{z})\, (\operatorname{diag} \boldsymbol{c})\, \boldsymbol{V}_{N-L,M}(\boldsymbol{z})^{\mathrm{T}}\,,$$

the vector $\boldsymbol{q}_{N-L}$ is also the minimum 2-norm solution of the *Yule–Walker system*

$$\boldsymbol{H}_{L,N-L}(0)\, \boldsymbol{q}_{N-L} = -\big(h(k)\big)_{k=N-L}^{N-1}\,.$$

Now we introduce the *companion matrix* of the monic polynomial q_{N-L} resp. of the vector $\boldsymbol{q}_{N-L}$

$$\boldsymbol{C}_{N-L}(\boldsymbol{q}_{N-L}) := \big(\boldsymbol{e}_2 \,|\, \boldsymbol{e}_3 \,|\, \ldots \,|\, \boldsymbol{e}_{N-L} \,|\, -\boldsymbol{q}_{N-L}\big)\,. \tag{4.2}$$

Remark 4.1. The companion matrix (4.2) has the known property

$$\det\big(z\, \boldsymbol{I}_{N-L} - \boldsymbol{C}_{N-L}(\boldsymbol{q}_{N-L})\big) = q_{N-L}(z) \quad (z \in \mathbb{C})\,.$$

All singular values of (4.2) can be explicitly determined (see [12] or [10, p. 197]). □

By [17, Lemma 2.2], the companion matrix (4.2) has the property

$$\boldsymbol{H}_{L,N-L}(0)\, \boldsymbol{C}_{N-L}(\boldsymbol{q}_{N-L}) = \boldsymbol{H}_{L,N-L}(1)\,. \tag{4.3}$$

Now we show interesting relations between the $M \times M$ matrix (3.10) resp. (3.12) and the $(N-L) \times (N-L)$ companion matrix (4.2).

Lemma 4.2. *Between the matrices* (3.10), (3.12), *and* (4.2) *there consist the following relations*

$$\boldsymbol{F}_M^{\mathrm{SVD}} = \boldsymbol{W}_{N-L,M}(0)^*\, \boldsymbol{C}_{N-L}(\boldsymbol{q}_{N-L})\, \big(\boldsymbol{W}_{N-L,M}(0)^*\big)^{\dagger}\,, \tag{4.4}$$

$$\boldsymbol{F}_M = \boldsymbol{X}_{N-L,M}(0)^*\, \boldsymbol{C}_{N-L}(\boldsymbol{q}_{N-L})\, \boldsymbol{X}_{N-L,M}(0)\,, \tag{4.5}$$

where $\boldsymbol{W}_{N-L,M}(0)^*$ *is the third factor in the factorization* (3.9) *of the Hankel matrix* $\boldsymbol{H}_{L,N-L}(0)$ *and where* $\boldsymbol{X}_{N-L,M}(0)$ *is defined by* (3.13). *Further the matrix* (3.10) *is similar to* (3.12) *by*

$$\boldsymbol{F}_M^{\mathrm{SVD}} = \big(\boldsymbol{W}_{N-L,M}(0)^*\,\boldsymbol{W}_{N-L,M}(0)\big)^{1/2}\,\boldsymbol{F}_M\,\big(\boldsymbol{W}_{N-L,M}(0)^*\,\boldsymbol{W}_{N-L,M}(0)\big)^{-1/2} \tag{4.6}$$

so that both matrices have the same eigenvalues.

Proof. 1) By (4.3) and (3.9) we obtain that

$$\boldsymbol{D}_M\,\boldsymbol{W}_{N-L,M}(0)^*\,\boldsymbol{C}_{N-L}(\boldsymbol{q}_{N-L}) = \boldsymbol{D}_M\,\boldsymbol{W}_{N-L,M}(1)^*\,,$$

since $\boldsymbol{U}_{L,M}^*\,\boldsymbol{U}_{L,M} = \boldsymbol{I}_M$. Multiplying the above equation with

$$\boldsymbol{D}_M^{-1} = \mathrm{diag}\,(\sigma_j^{-1})_{j=1}^M\,,$$

where $\sigma_1 \geq \cdots \geq \sigma_M > 0$ denote all positive singular values of $\boldsymbol{H}_{L,N-L+1}$, it follows that

$$\boldsymbol{W}_{N-L,M}(0)^*\,\boldsymbol{C}_{N-L}(\boldsymbol{q}_{N-L}) = \boldsymbol{W}_{N-L,M}(1)^*\,. \tag{4.7}$$

Thus we receive by (3.10) that

$$\begin{aligned}\boldsymbol{W}_{N-L,M}(0)^*\,\boldsymbol{C}_{N-L}(\boldsymbol{q}_{N-L})\,\big(\boldsymbol{W}_{N-L,M}(0)^*\big)^{\dagger} &= \boldsymbol{W}_{N-L,M}(1)^*\,\big(\boldsymbol{W}_{N-L,M}(0)^*\big)^{\dagger}\\ &= \boldsymbol{F}_M^{\mathrm{SVD}}\,.\end{aligned} \tag{4.8}$$

2) Formula (4.5) is an immediate consequence of (4.7) and (3.13), if we multiply (4.7) by $\boldsymbol{X}_{N-L,M}(0)$ from the right and by $\big(\boldsymbol{W}_{N-L,M}(0)^*\,\boldsymbol{W}_{N-L,M}(0)\big)^{-1/2}$ from the left.

3) Using the representations (4.4)–(4.5) as well as formula (3.13), we obtain (4.6), since

$$\big(\boldsymbol{W}_{N-L,M}(0)^*\big)^{\dagger} = \boldsymbol{W}_{N-L,M}(0)\,\big(\boldsymbol{W}_{N-L,M}(0)^*\,\boldsymbol{W}_{N-L,M}(0)\big)^{-1}$$

(cf. (4.18)). As known, similar matrices possess the same eigenvalues. This completes the proof. □

The *signal space* $\mathcal{S}_{N-L} \subset \mathbb{C}^{N-L}$ is defined as the range of the matrix $\boldsymbol{V}_{N-L,M}(\bar{\boldsymbol{z}})$, where $\bar{\boldsymbol{z}} := (\bar{z}_j)_{j=1}^M$, i.e., the signal space $\mathcal{S}_{N-L}$ is spanned by the M linearly independent vectors $(\bar{z}_j^k)_{k=0}^{N-L-1}$ $(j = 1, \ldots, M)$. Thus the M-dimensional signal space is fully characterized by the distinct nodes $z_j \in \mathbb{D}$ $(j = 1, \ldots, M)$. By the properties of the Moore–Penrose pseudoinverse $\big(\boldsymbol{V}_{N-L,M}(\boldsymbol{z})^{\mathrm{T}}\big)^{\dagger}$ it follows that

$$\boldsymbol{P}_{N-L} := \big(\boldsymbol{V}_{N-L,M}(\boldsymbol{z})^{\mathrm{T}}\big)^{\dagger}\,\boldsymbol{V}_{N-L,M}(\boldsymbol{z})^{\mathrm{T}} \tag{4.9}$$

is the *orthogonal projection onto the signal space* $\mathcal{S}_{N-L}$. Further, we remark that

$$\boldsymbol{V}_{N-L,M}(\boldsymbol{z})^{\mathrm{T}}\,\big(\boldsymbol{V}_{N-L,M}(\boldsymbol{z})^{\mathrm{T}}\big)^{\dagger} = \boldsymbol{I}_M\,, \tag{4.10}$$

since $\boldsymbol{V}_{N-L,M}(\boldsymbol{z})^{\mathrm{T}}$ has full row rank. Note that $\boldsymbol{q}_{N-L} \in \mathcal{S}_{N-L}$, since by (4.1) and (4.9)

$$\begin{aligned}\boldsymbol{P}_{N-L}\,\boldsymbol{q}_{N-L} &= -\big(\boldsymbol{V}_{N-L,M}(\boldsymbol{z})^{\mathrm{T}}\big)^{\dagger}\,\boldsymbol{V}_{N-L,M}(\boldsymbol{z})^{\mathrm{T}}\,\big(\boldsymbol{V}_{N-L,M}(\boldsymbol{z})^{\mathrm{T}}\big)^{\dagger}\,\big(z_j^{N-L}\big)_{j=1}^{M}\\ &= -\big(\boldsymbol{V}_{N-L,M}(\boldsymbol{z})^{\mathrm{T}}\big)^{\dagger}\,\big(z_j^{N-L}\big)_{j=1}^{M} = \boldsymbol{q}_{N-L}\,.\end{aligned}$$

Now we describe the properties of the *projected companion matrix*

$$\boldsymbol{P}_{N-L}\,\boldsymbol{C}_{N-L}(\boldsymbol{q}_{N-L}) \in \mathbb{C}^{(N-L)\times(N-L)}\,. \tag{4.11}$$

Note that in [3] the matrix $\boldsymbol{F}_M$ with the representation (4.5) is called "projected companion matrix".

Theorem 4.3. *The projected companion matrix* (4.11) *can be represented in the following forms*

$$\begin{aligned}\boldsymbol{P}_{N-L}\,\boldsymbol{C}_{N-L}(\boldsymbol{q}_{N-L}) &= \big(\boldsymbol{V}_{N-L,M}(\boldsymbol{z})^{\mathrm{T}}\big)^{\dagger}\,(\operatorname{diag}\boldsymbol{z})\,\boldsymbol{V}_{N-L,M}(\boldsymbol{z})^{\mathrm{T}} && (4.12)\\ &= \big(\boldsymbol{W}_{N-L,M}(0)^*\big)^{\dagger}\,\boldsymbol{W}_{N-L,M}(1)^* && (4.13)\\ &= \boldsymbol{X}_{N-L,M}(0)\,\boldsymbol{X}_{N-L,M}(1)^*\,. && (4.14)\end{aligned}$$

The signal space $\mathcal{S}_{N-L}$ coincides with the range of $\boldsymbol{H}_{L,N-L}(0)^*$ and also with the range of $\boldsymbol{W}_{N-L,M}(0)$. The columns of $\boldsymbol{X}_{N-L,M}(0)$ form an orthonormal basis of the M-dimensional signal space $\mathcal{S}_{N-L}$. Moreover, the signal space $\mathcal{S}_{N-L}$ is an invariant subspace for $\boldsymbol{X}_{N-L,M}(0)$. Further, $\boldsymbol{C}_{N-L}(\boldsymbol{q}_{N-L})^*$ maps the signal space $\mathcal{S}_{N-L}$ into itself. The orthogonal projection $\boldsymbol{P}_{N-L}$ onto $\mathcal{S}_{N-L}$ can be represented as follows

$$\boldsymbol{P}_{N-L} = \big(\boldsymbol{W}_{N-L,M}(0)^*\big)^{\dagger}\,\boldsymbol{W}_{N-L,M}(0)^* = \boldsymbol{X}_{N-L,M}(0)\,\boldsymbol{X}_{N-L,M}(0)^*\,. \tag{4.15}$$

The nonvanishing eigenvalues of the projected companion matrix (4.11) *coincide with the eigenvalues of* (3.10) *resp.* (3.12).

Proof. 1) By (4.1) and (4.2) we obtain that

$$\boldsymbol{V}_{N-L,M}(\boldsymbol{z})^{\mathrm{T}}\,\boldsymbol{C}_{N-L}(\boldsymbol{q}_{N-L}) = (\operatorname{diag}\boldsymbol{z})\,\boldsymbol{V}_{N-L,M}(\boldsymbol{z})^{\mathrm{T}}\,. \tag{4.16}$$

Note that (4.16) shows a close relationship between the Vandermonde matrix $\boldsymbol{V}_{N-L,M}(\boldsymbol{z})$ and the companion matrix $\boldsymbol{C}_{N-L}(\boldsymbol{q}_{N-L})$. From (4.16) it follows immediately that

$$\boldsymbol{C}_{N-L}(\boldsymbol{q}_{N-L})^*\,\boldsymbol{V}_{N-L,M}(\bar{\boldsymbol{z}}) = \boldsymbol{V}_{N-L,M}(\bar{\boldsymbol{z}})\,(\operatorname{diag}\bar{\boldsymbol{z}})\,,$$

i.e., $\boldsymbol{C}_{N-L}(\boldsymbol{q}_{N-L})^*$ maps the signal space $\mathcal{S}_{N-L}$ into itself. Multiplying (4.16) with $\big(\boldsymbol{V}_{N-L,M}(\boldsymbol{z})^{\mathrm{T}}\big)^{\dagger}$, we receive the factorization (4.12) of the projected companion matrix (4.11).

2) Using the factorization

$$\boldsymbol{H}_{L,N-L}(0) = \boldsymbol{V}_{L,M}(\boldsymbol{z})\,(\operatorname{diag}\boldsymbol{c})\,\boldsymbol{V}_{N-L,M}(\boldsymbol{z})^{\mathrm{T}}\,,$$

we obtain that

$$\boldsymbol{H}_{L,N-L}(0)^* = \boldsymbol{V}_{N-L,M}(\bar{\boldsymbol{z}})\,(\operatorname{diag}\bar{\boldsymbol{c}})\,\boldsymbol{V}_{L,M}(\boldsymbol{z})^*\,.$$

Consequently, $\mathcal{S}_{N-L}$ coincides with the range of $\boldsymbol{H}_{L,N-L}(0)^*$. By (3.9) for $s = 0$ it follows that

$$\boldsymbol{H}_{L,N-L}(0)^* = \boldsymbol{W}_{N-L,M}(0)\, \boldsymbol{D}_M\, \boldsymbol{U}_{L,M}^* \,.$$

Hence $\mathcal{S}_{N-L}$ coincides with the range of $\boldsymbol{W}_{N-L,M}(0)$ too. Further

$$\big(\boldsymbol{W}_{N-L,M}(0)^*\big)^\dagger\, \boldsymbol{W}_{N-L,M}(0)^*$$

is the orthogonal projection onto the range of $\big(\boldsymbol{W}_{N-L,M}(0)^*\big)^\dagger$ which coincides with the range of $\boldsymbol{W}_{N-L,M}(0)$. Since the range of $\boldsymbol{W}_{N-L,M}(0)$ is equal to the signal space $\mathcal{S}_{N-L}$, we conclude that

$$\boldsymbol{P}_{N-L} = \big(\boldsymbol{W}_{N-L,M}(0)^*\big)^\dagger\, \boldsymbol{W}_{N-L,M}(0)^* \,. \tag{4.17}$$

Multiplying (4.7) with $\big(\boldsymbol{W}_{N-L,M}(0)^*\big)^\dagger$ from the left, by (4.17) we receive the factorization (4.13) of the projected companion matrix (4.11). Formula (4.14) follows immediately from (4.13) and (3.13), since the Moore–Penrose pseudoinverse of the full rank matrix $\boldsymbol{W}_{N-L,M}(0)^*$ reads as follows

$$\big(\boldsymbol{W}_{N-L,M}(0)^*\big)^\dagger = \boldsymbol{W}_{N-L,M}(0)\, \big(\boldsymbol{W}_{N-L,M}(0)^*\, \boldsymbol{W}_{N-L,M}(0)\big)^{-1} \tag{4.18}$$

and since the inverse square root of the positive definite matrix

$$\boldsymbol{W}_{N-L,M}(0)^*\, \boldsymbol{W}_{N-L,M}(0)$$

is well defined.

By the properties of the Moore–Penrose pseudoinverse $\big(\boldsymbol{W}_{N-L,M}(0)^*\big)^\dagger$, the matrix $\big(\boldsymbol{W}_{N-L,M}(0)^*\big)^\dagger\, \boldsymbol{W}_{N-L,M}(0)^*$ is the orthogonal projection onto the range of $\boldsymbol{W}_{N-L,M}(0)$ which coincides with the signal space $\mathcal{S}_{N-L}$. Hence we conclude that

$$\boldsymbol{P}_{N-L} = \big(\boldsymbol{W}_{N-L,M}(0)^*\big)^\dagger\, \boldsymbol{W}_{N-L,M}(0)^* \,.$$

Using (4.18) and (3.13), we obtain (4.15).

3) By the property (3.14), the M columns of $\boldsymbol{X}_{N-L,M}(0)$ are orthonormal and are contained in the M-dimensional signal space $\mathcal{S}_{N-L}$, because $\mathcal{S}_{N-L}$ coincides with the range of $\boldsymbol{W}_{N-L,M}(0)$. Hence the M columns of $\boldsymbol{X}_{N-L,M}(0)$ form an orthonormal basis of $\mathcal{S}_{N-L}$. Using (3.12) and (4.14), we obtain the relation

$$\boldsymbol{P}_{N-L}\, \boldsymbol{C}_{N-L}(\boldsymbol{q}_{N-L})\, \boldsymbol{X}_{N-L,M}(0) = \boldsymbol{X}_{N-L,M}(0)\, \boldsymbol{F}_M \,.$$

From this it follows that the signal space $\mathcal{S}_{N-L}$ is an invariant subspace for $\boldsymbol{X}_{N-L,M}(0)$.

4) By simple calculations, one can see that the nonvanishing eigenvalues of the projected companion matrix (4.11) coincide with the eigenvalues of (3.10) resp. (3.12). Let $(z, \boldsymbol{y})$ with $z \neq 0$ and $\boldsymbol{y} \in \mathbb{C}^{N-L}$ ($\boldsymbol{y} \neq \boldsymbol{0}$) be a right eigenpair of (4.11), i.e.,

$$\boldsymbol{P}_{N-L}\, \boldsymbol{C}_{N-L}(\boldsymbol{q}_{N-L})\, \boldsymbol{y} = z\, \boldsymbol{y} \,.$$

Hence $z\, \boldsymbol{P}_{N-L}\boldsymbol{y} = z\boldsymbol{y}$ and thus $\boldsymbol{P}_{N-L}\boldsymbol{y} = \boldsymbol{y}$ by $z \neq 0$. For $\boldsymbol{x} := \boldsymbol{X}_{N-L,M}(0)^*\boldsymbol{y}$ we obtain $\boldsymbol{X}_{N-L,M}(0)\, \boldsymbol{x} = \boldsymbol{P}_{N-L}\, \boldsymbol{y} = \boldsymbol{y}$ by (4.15) so that $\boldsymbol{x} \neq \boldsymbol{0}$. Further by (3.14) and (4.5) it follows that

$$\begin{aligned}\boldsymbol{X}_{N-L,M}(0)^* \boldsymbol{P}_{N-L} \boldsymbol{C}_{N-L}(\boldsymbol{q}_{N-L}) \boldsymbol{y} &= \boldsymbol{X}_{N-L,M}(0)^* \boldsymbol{C}_{N-L}(\boldsymbol{q}_{N-L}) \boldsymbol{y} \\ &= \boldsymbol{X}_{N-L,M}(0)^* \boldsymbol{C}_{N-L}(\boldsymbol{q}_{N-L}) \boldsymbol{X}_{N-L,M}(0) \boldsymbol{x} \\ &= \boldsymbol{F}_M \boldsymbol{x} = z \boldsymbol{x} ,\end{aligned}$$

i.e., $(z, \boldsymbol{X}_{N-L,M}(0)^* \boldsymbol{y})$ is a right eigenpair of (3.12). Analogously, one can show that each eigenvalue of (3.12) is an eigenvalue of (4.11) too. By Lemma 4.2, the eigenvalues of (3.12) coincide with the eigenvalues of (3.10). This completes the proof. □

Remark 4.4. The singular values of the matrix $\boldsymbol{F}_M$ can be characterized by [1, Theorem 4]. Assume that $2 \le M \le L \le \lceil \frac{N}{2} \rceil$. Let $\boldsymbol{p}_{N-L} := \boldsymbol{P}_{N-L} \boldsymbol{e}_1$ be the first column of $\boldsymbol{P}_{N-L}$. Then the singular values of the matrix (3.12) are $\omega_2 = \cdots = \omega_{M-1} = 1$ and

$$\begin{aligned}\omega_1 &= \frac{1}{\sqrt{2}} \Big(2 + \|\boldsymbol{q}_{N-L}\|_2^2 - \|\boldsymbol{p}_{N-L}\|_2^2 + \sqrt{(\|\boldsymbol{q}_{N-L}\|_2^2 + \|\boldsymbol{p}_{N-L}\|_2^2)^2 - 4\,|q_0|^2}\Big)^{1/2} , \\ \omega_M &= \frac{1}{\sqrt{2}} \Big(2 + \|\boldsymbol{q}_{N-L}\|_2^2 - \|\boldsymbol{p}_{N-L}\|_2^2 - \sqrt{(\|\boldsymbol{q}_{N-L}\|_2^2 + \|\boldsymbol{p}_{N-L}\|_2^2)^2 - 4\,|q_0|^2}\Big)^{1/2} ,\end{aligned}$$

where q_0 is the first component of $\boldsymbol{q}_{N-L}$. Further, the spectral resp. Frobenius norm of the matrix (3.12) is equal to

$$\|\boldsymbol{F}_M\|_2 = \omega_1 , \quad \|\boldsymbol{F}_M\|_{\mathrm{F}} = \sqrt{M + \|\boldsymbol{q}_{N-L}\|_2^2 - \|\boldsymbol{p}_{N-L}\|_2^2} . \tag{4.19}$$

□

5. Error estimates of the nodes

From the matrix perturbation theory, the following results on the perturbation of eigenvalues are known. Let $\boldsymbol{A}_P \in \mathbb{C}^{P\times P}$ be a square nonnormal matrix with the eigenvalues $y_k \in \mathbb{C}$ $(k = 1, \ldots, P)$ and let $\tilde{\boldsymbol{A}}_P \in \mathbb{C}^{P\times P}$ be a perturbation of $\boldsymbol{A}_P$. If y_j be a simple eigenvalue of $\boldsymbol{A}_P$ with right resp. left eigenvectors $\boldsymbol{u}_j$ resp. $\boldsymbol{v}_j$, then there exists a unique eigenvalue $\tilde{y}_j$ of $\tilde{\boldsymbol{A}}_P$ (see, e.g., [22, pp. 183–184]) such that

$$\tilde{y}_j = y_j + \frac{\boldsymbol{v}_j^* (\tilde{\boldsymbol{A}}_P - \boldsymbol{A}_P) \boldsymbol{u}_j}{\boldsymbol{v}_j^* \boldsymbol{u}_j} + \mathcal{O}(\|\tilde{\boldsymbol{A}}_P - \boldsymbol{A}_P\|_2^2)$$

Note that the right and left eigenvectors of a simple eigenvalue y_j cannot be orthogonal, i.e., $\boldsymbol{v}_j^* \boldsymbol{u}_j \ne 0$. If the left and right eigenvectors of an eigenvalue of $\boldsymbol{A}_P$ are nearly orthogonal, then $\boldsymbol{A}_P$ must be near one with multiple eigenvalue (see [23]). For sufficiently small spectral norm $\|\tilde{\boldsymbol{A}}_P - \boldsymbol{A}_P\|_2$, we obtain the first-order estimate

$$|\tilde{y}_j - y_j| \le \frac{|\boldsymbol{v}_j^* (\tilde{\boldsymbol{A}}_P - \boldsymbol{A}_P) \boldsymbol{u}_j|}{|\boldsymbol{v}_j^* \boldsymbol{u}_j|} \le \frac{\|\boldsymbol{v}_j\|_2 \, \|\boldsymbol{u}_j\|_2}{|\boldsymbol{v}_j^* \boldsymbol{u}_j|} \|\tilde{\boldsymbol{A}}_P - \boldsymbol{A}_P\|_2 . \tag{5.1}$$

Thus the quantity

$$\kappa_j(\boldsymbol{A}_P) := \frac{\|\boldsymbol{v}_j\|_2 \, \|\boldsymbol{u}_j\|_2}{|\boldsymbol{v}_j^* \, \boldsymbol{u}_j|} \ge 1$$

measures the sensitivity of the eigenvalue y_j to perturbations on $\boldsymbol{A}_P$, see also [5]. Therefore this number is called the *condition number of the eigenvalue* y_j. By definition we have $\kappa_j(\boldsymbol{A}_P) = \kappa_j(\boldsymbol{A}_P^*)$. By [20], the condition number of a simple eigenvalue y_j of $\boldsymbol{A}_P$ can be estimated by

$$\kappa_j(\boldsymbol{A}_P) \le \left(1 + \frac{\|\boldsymbol{A}_P\|_{\mathrm{F}}^2 - \sum_{k=1}^{P} |y_k|^2}{(P-1)\, d_j^2}\right)^{(P-1)/2} . \tag{5.2}$$

where $d_j := \min\{|y_j - y_k|;\, k = 1, \ldots, P,\, k \ne j\}$ denotes the *separation distance for the eigenvalue* y_j. Note that

$$\Delta(\boldsymbol{A}_P) := \sqrt{\|\boldsymbol{A}_P\|_{\mathrm{F}}^2 - \sum_{k=1}^{P} |y_k|^2}$$

is the so-called *departure from normality* of $\boldsymbol{A}_P$ (see [20]). If $\Delta(\boldsymbol{A}_P) = 0$, then $\boldsymbol{A}_P$ is normal.

Now we apply these results for the perturbation of eigenvalues concerning the projected companion matrix (4.11) and the $M \times M$ matrix (3.12).

Theorem 5.1. *Let $M, L, N \in \mathbb{N}$ with $M \le L \le \lceil \frac{N}{2} \rceil$ be given. Then the projected companion matrix (4.11) has $z_j \in \mathbb{D}$ ($j = 1, \ldots, M$) as simple eigenvalues. Further, the matrix (4.11) has 0 as an eigenvalue with algebraic multiplicity $N - L - M$. Moreover, $(z_j, (\boldsymbol{V}_{N-L,M}(\boldsymbol{z})^{\mathrm{T}})^{\dagger}\, \boldsymbol{e}_j)$ ($j = 1, \ldots, M$) is a right eigenpair and $(z_j, \boldsymbol{V}_{N-L,M}(\bar{\boldsymbol{z}})\, \boldsymbol{e}_j)$ ($j = 1, \ldots, M$) is a left eigenpair of (4.11), where $\boldsymbol{e}_j \in \mathbb{C}^M$ is the jth canonical basis vector. The condition number of the eigenvalue z_j of the matrix (4.11) fulfils*

$$\kappa_j(\boldsymbol{P}_{N-L}\, \boldsymbol{C}_{N-L}(\boldsymbol{q}_{N-L})) = \|(\boldsymbol{V}_{N-L,M}(\boldsymbol{z})^{\mathrm{T}})^{\dagger}\, \boldsymbol{e}_j\|_2 \, \|\boldsymbol{V}_{N-L,M}(\bar{\boldsymbol{z}})\, \boldsymbol{e}_j\|_2 \, . \tag{5.3}$$

Corresponding to each eigenvalue $z_j \in \mathbb{D}$ ($j = 1, \ldots, M$) of $\boldsymbol{P}_{N-L}\, \boldsymbol{C}_{N-L}(\boldsymbol{q}_{N-L})$, there exists a unique eigenvalue $\tilde{z}_j$ of $\tilde{\boldsymbol{P}}_{N-L}\, \boldsymbol{C}_{N-L}(\tilde{\boldsymbol{q}}_{N-L})$ so that

$$|z_j - \tilde{z}_j| \le \kappa_j(\boldsymbol{P}_{N-L}\, \boldsymbol{C}_{N-L}(\boldsymbol{q}_{N-L}))\Big(\|\boldsymbol{P}_{N-L} - \tilde{\boldsymbol{P}}_{N-L}\|_2 + \|\boldsymbol{q}_{N-L} - \tilde{\boldsymbol{q}}_{N-L}\|_2\Big) \, . \tag{5.4}$$

Here $\tilde{\boldsymbol{q}}_{N-L} \in \mathbb{C}^{N-L}$ denotes the minimum 2-norm solution of the linear system

$$\tilde{\boldsymbol{K}}_{L,N-L}(0)\, \tilde{\boldsymbol{q}}_{N-L} = -\big(\tilde{h}_k\big)_{k=N-L}^{N-1} ,$$

where $\tilde{\boldsymbol{K}}_{L,N-L}(0)$ is the low-rank approximation (3.11) of $\tilde{\boldsymbol{H}}_{L,N-L}(0)$.

Proof. 1) By the representation (4.12) of the projected companion matrix (4.11) and the property (4.10) it follows immediately that

$$\big(\boldsymbol{V}_{N-L,M}(\boldsymbol{z})^{\mathrm{T}}\big)^{\dagger}\, \boldsymbol{e}_j \text{ resp. } \big(\boldsymbol{V}_{N-L,M}(\boldsymbol{z})^{\mathrm{T}}\big)^{*}\, \boldsymbol{e}_j = \boldsymbol{V}_{N-L,M}(\bar{\boldsymbol{z}})\, \boldsymbol{e}_j$$

is a right resp. left eigenvector of (4.11) with respect to the eigenvalue z_j. By (4.10), these eigenvectors possess the property

$$\left(\left(\boldsymbol{V}_{N-L,M}(\boldsymbol{z})^{\mathrm{T}}\right)^* \boldsymbol{e}_j\right)^* \left(\boldsymbol{V}_{N-L,M}(\boldsymbol{z})^{\mathrm{T}}\right)^{\dagger} \boldsymbol{e}_j = \boldsymbol{e}_j^{\mathrm{T}} \boldsymbol{e}_j = 1\,.$$

Then the condition number of the projected companion matrix (4.11) with respect to the eigenvalue $z_j \in \mathbb{D}$ is given by (5.3).

By assumption, it holds $N - L > M$. Since $\boldsymbol{V}_{N-L,M}(\boldsymbol{z})$ has full rank, we see that by (4.12) the null space of $\boldsymbol{P}_{N-L}\,\boldsymbol{C}_{N-L}(\boldsymbol{q}_{N-L})$ coincides with the null space of $\boldsymbol{V}_{N-L,M}(\boldsymbol{z})^{\mathrm{T}}$. Hence the null space of $\boldsymbol{V}_{N-L,M}(\boldsymbol{z})^{\mathrm{T}}$ has the dimension $N-L-M$. This means that 0 is an eigenvalue with algebraic multiplicity $N-L-M$ and that $z_j \in \mathbb{D}$ $(j = 1, \ldots, M)$ are simple eigenvalues of (4.11).

2) Let $\tilde{z}_j$ $(j = 1, \ldots, M)$ be one of the eigenvalues of (4.11) that is the closest to $z_j \in \mathbb{D}$. Our goal is to estimate the error $|\tilde{z}_j - z_j|$ $(j = 1, \ldots, M)$. Let $\boldsymbol{P}_{N-L}$ resp. $\tilde{\boldsymbol{P}}_{N-L}$ denote the orthogonal projector onto the corresponding signal space $\mathcal{S}_{N-L}$ resp. $\tilde{\mathcal{S}}_{N-L}$. Setting

$$\boldsymbol{A}_{N-L} := \boldsymbol{P}_{N-L}\,\boldsymbol{C}_{N-L}(\boldsymbol{q}_{N-L}) - \tilde{\boldsymbol{P}}_{N-L}\,\boldsymbol{C}_{N-L}(\tilde{\boldsymbol{q}}_{N-L})\,,$$

by (5.1) the following first-order estimate holds

$$|\tilde{z}_j - z_j| \le \kappa_j(\boldsymbol{P}_{N-L}\,\boldsymbol{C}_{N-L}(\boldsymbol{q}_{N-L}))\,\|\boldsymbol{A}_{N-L}\|_2$$

for $j = 1, \ldots, M$. Using the special structure of the companion matrix (4.2), one can see that

$$\begin{aligned}\boldsymbol{A}_{N-L}\,\boldsymbol{A}_{N-L}^* &= (\boldsymbol{P}_{N-L} - \tilde{\boldsymbol{P}}_{N-L})\,(\boldsymbol{P}_{N-L} - \tilde{\boldsymbol{P}}_{N-L})^* \\ &\quad + (\boldsymbol{q}_{N-L} - \tilde{\boldsymbol{q}}_{N-L})\,(\boldsymbol{q}_{N-L} - \tilde{\boldsymbol{q}}_{N-L})^* \\ &\quad - (\boldsymbol{p}_{N-L} - \tilde{\boldsymbol{p}}_{N-L})\,(\boldsymbol{p}_{N-L} - \tilde{\boldsymbol{p}}_{N-L})^*\,,\end{aligned}$$

where $\boldsymbol{p}_{N-L}$ and $\tilde{\boldsymbol{p}}_{N-L}$ are the first columns of $\boldsymbol{P}_{N-L}$ resp. $\tilde{\boldsymbol{P}}_{N-L}$. Then for each unit vector $\boldsymbol{x} \in \mathbb{C}^{N-L}$ we receive that

$$\begin{aligned}\boldsymbol{x}^*\,\boldsymbol{A}_{N-L}\,\boldsymbol{A}_{N-L}^*\,\boldsymbol{x} &= \|\boldsymbol{A}_{N-L}^*\boldsymbol{x}\|_2^2 \\ &= \|(\boldsymbol{P}_{N-L} - \tilde{\boldsymbol{P}}_{N-L})^*\boldsymbol{x}\|_2^2 + |\boldsymbol{x}^*\,(\boldsymbol{q}_{N-L} - \tilde{\boldsymbol{q}}_{N-L})|^2 \\ &\quad - |\boldsymbol{x}^*\,(\boldsymbol{p}_{N-L} - \tilde{\boldsymbol{p}}_{N-L})|^2 \\ &\le \|\boldsymbol{P}_{N-L} - \tilde{\boldsymbol{P}}_{N-L}\|_2^2 + \|\boldsymbol{q}_{N-L} - \tilde{\boldsymbol{q}}_{N-L}\|_2^2\,.\end{aligned}$$

Thus it follows that for *all* unit vectors $\boldsymbol{x} \in \mathbb{C}^{N-L}$

$$\|\boldsymbol{A}_{N-L}^*\boldsymbol{x}\|_2 \le \|\boldsymbol{P}_{N-L} - \tilde{\boldsymbol{P}}_{N-L}\|_2 + \|\boldsymbol{q}_{N-L} - \tilde{\boldsymbol{q}}_{N-L}\|_2\,,$$

and hence

$$\|\boldsymbol{A}_{N-L}\|_2 \le \|\boldsymbol{P}_{N-L} - \tilde{\boldsymbol{P}}_{N-L}\|_2 + \|\boldsymbol{q}_{N-L} - \tilde{\boldsymbol{q}}_{N-L}\|_2\,.$$

Consequently we obtain the above estimate of $|\tilde{z}_j - z_j|$. This completes the proof. □

In the next theorem, we show that $\kappa_j(\boldsymbol{P}_{N-L}\,\boldsymbol{C}_{N-L}(\boldsymbol{q}_{N-L})) = \kappa_j(\boldsymbol{F}_M)$. The matrix $\boldsymbol{F}_M$ is computed from exactly sampled data $h(k)$ $(k = 0, \ldots, N-1)$. Analogously, the matrix $\tilde{\boldsymbol{F}}_M$ is obtained from noisy sampled data $\tilde{h}_k$ $(k = 0, \ldots, N-1)$. Thus $\tilde{\boldsymbol{F}}_M$ has a similar form as $\boldsymbol{F}_M$, namely

$$\tilde{\boldsymbol{F}}_M = \tilde{\boldsymbol{X}}_{N-L,M}(1)^* \, \tilde{\boldsymbol{X}}_{N-L,M}(0)\,.$$

Theorem 5.2. *The matrix* (3.12) *has only* $z_j \in \mathbb{D}$ $(j = 1, \ldots, M)$ *as simple eigenvalues. Further,* $(z_j, \boldsymbol{X}_{N-L,M}(0)^* \left(\boldsymbol{V}_{N-L,M}(\boldsymbol{z})^{\mathrm{T}}\right)^{\dagger} \boldsymbol{e}_j)$ $(j = 1, \ldots, M)$ *is a right eigenpair and* $(z_j, \boldsymbol{X}_{N-L,M}(0)^{\dagger}\, \boldsymbol{V}_{N-L,M}(\bar{\boldsymbol{z}})\, \boldsymbol{e}_j)$ $(j = 1, \ldots, M)$ *is a left eigenpair of* (3.12). *The condition number of the eigenvalue* z_j *of the matrix* (3.12) *fulfils*

$$\begin{aligned} \kappa_j(\boldsymbol{F}_M) &= \|\boldsymbol{X}_{N-L,M}(0)^* \left(\boldsymbol{V}_{N-L,M}(\boldsymbol{z})^{\mathrm{T}}\right)^{\dagger} \boldsymbol{e}_j\|_2 \, \|\boldsymbol{X}_{N-L,M}(0)^{\dagger}\, \boldsymbol{V}_{N-L,M}(\bar{\boldsymbol{z}})\, \boldsymbol{e}_j\|_2 \\ &= \kappa_j(\boldsymbol{P}_{N-L}\,\boldsymbol{C}_{N-L}(\boldsymbol{q}_{N-L})) && (5.5) \\ &\le \Big(1 + \frac{M + \|\boldsymbol{q}_{N-L}\|_2^2 - \|\boldsymbol{p}_{N-L}\|_2^2 - \mu}{(M-1)\,\delta_j^2}\Big)^{(M-1)/2} && (5.6) \end{aligned}$$

with the first column $\boldsymbol{p}_{N-L}$ *of* $\boldsymbol{P}_{N-L}$, *with* μ *defined by* (2.3), *and with* $\delta_j := \min\{|z_j - z_k|;\, k = 1, \ldots, M,\, k \neq j\}$.

Proof. Using (4.5) and (4.12), we obtain the matrix factorizations

$$\begin{aligned} \boldsymbol{F}_M &= \boldsymbol{X}_{N-L,M}(0)^*\, \boldsymbol{C}_{N-L}(\boldsymbol{q}_{N-L})\, \boldsymbol{X}_{N-L,M}(0) \\ &= \boldsymbol{X}_{N-L,M}(0)^*\, \boldsymbol{P}_{N-L}\, \boldsymbol{C}_{N-L}(\boldsymbol{q}_{N-L})\, \boldsymbol{X}_{N-L,M}(0) \\ &= \boldsymbol{X}_{N-L,M}(0)^* \left(\boldsymbol{V}_{N-L,M}(\boldsymbol{z})^{\mathrm{T}}\right)^{\dagger} (\operatorname{diag} \boldsymbol{z})\, \boldsymbol{V}_{N-L,M}(\boldsymbol{z})^{\mathrm{T}}\, \boldsymbol{X}_{N-L,M}(0)\,. \end{aligned}$$

Consequently, $\boldsymbol{X}_{N-L,M}(0)^* \left(\boldsymbol{V}_{N-L,M}(\boldsymbol{z})^{\mathrm{T}}\right)^{\dagger} \boldsymbol{e}_j$ resp. $\boldsymbol{X}_{N-L,M}(0)^{\dagger}\, \boldsymbol{V}_{N-L,M}(\bar{\boldsymbol{z}})\, \boldsymbol{e}_j$ is a right resp. left eigenvector of (3.12) with respect to z_j. Since these eigenvectors possess the property

$$\begin{aligned} &\boldsymbol{e}_j^{\mathrm{T}}\, \boldsymbol{V}_{N-L,M}(\boldsymbol{z})^{\mathrm{T}} \left(\boldsymbol{X}_{N-L,M}(0)\, \boldsymbol{X}_{N-L,M}(0)^* \left(\boldsymbol{V}_{N-L,M}(\boldsymbol{z})^{\mathrm{T}}\right)^{\dagger} \boldsymbol{e}_j\right. \\ &= \boldsymbol{e}_j^{\mathrm{T}}\, \boldsymbol{V}_{N-L,M}(\boldsymbol{z})^{\mathrm{T}}\, \boldsymbol{P}_{N-L} \left(\boldsymbol{V}_{N-L,M}(\boldsymbol{z})^{\mathrm{T}}\right)^{\dagger} \boldsymbol{e}_j \\ &= \boldsymbol{e}_j^{\mathrm{T}}\, \boldsymbol{V}_{N-L,M}(\boldsymbol{z})^{\mathrm{T}} \left(\boldsymbol{V}_{N-L,M}(\boldsymbol{z})^{\mathrm{T}}\right)^{\dagger} \boldsymbol{e}_j \\ &= \boldsymbol{e}_j^{\mathrm{T}}\, \boldsymbol{e}_j = 1\,, \end{aligned}$$

the condition number of the simple eigenvalue z_j of the matrix (3.12) is given by

$$\kappa_j(\boldsymbol{F}_M) = \|\boldsymbol{X}_{N-L,M}(0)^* \left(\boldsymbol{V}_{N-L,M}(\boldsymbol{z})^{\mathrm{T}}\right)^{\dagger} \boldsymbol{e}_j\|_2 \, \|\boldsymbol{X}_{N-L,M}(0)^*\, \boldsymbol{V}_{N-L,M}(\bar{\boldsymbol{z}})\, \boldsymbol{e}_j\|_2\,.$$

From (5.2) it follows the estimate (5.6). Since both

$$\left(\boldsymbol{V}_{N-L,M}(\boldsymbol{z})^{\mathrm{T}}\right)^{\dagger} \boldsymbol{e}_j \quad \text{and} \quad \boldsymbol{V}_{N-L,M}(\bar{\boldsymbol{z}})\, \boldsymbol{e}_j$$

belong to the signal space $\mathcal{S}_{N-L}$ and since the columns of $\boldsymbol{X}_{N-L,M}(0)$ form an orthonormal basis of $\mathcal{S}_{N-L}$, it is clear that

$$\|\boldsymbol{X}_{N-L,M}(0)^* \left(\boldsymbol{V}_{N-L,M}(\boldsymbol{z})^{\mathrm{T}}\right)^{\dagger} \boldsymbol{e}_j\|_2 = \|\left(\boldsymbol{V}_{N-L,M}(\boldsymbol{z})^{\mathrm{T}}\right)^{\dagger} \boldsymbol{e}_j\|_2 ,$$
$$\|\boldsymbol{X}_{N-L,M}(0)^* \boldsymbol{V}_{N-L,M}(\bar{\boldsymbol{z}}) \boldsymbol{e}_j\|_2 = \|\boldsymbol{V}_{N-L,M}(\bar{\boldsymbol{z}}) \boldsymbol{e}_j\|_2 .$$

Thus we receive (5.5). From (5.2) and (4.19) it follows immediately the nice upper bound (5.6) of $\kappa_j(\boldsymbol{F}_M)$. Using (5.1), we obtain the above estimate of $|z_j - \tilde{z}_j|$. Similarly to (5.6), a corresponding estimate was also presented in [3, Proposition 3]. This completes the proof. □

Remark 5.3. The matrix (3.12) is not Hermitian in general. By balancing one can often improve the accuracy of the computed eigenvalues of (3.12). *Balancing* is a convenient diagonal scaling of (3.12), i.e., a diagonal matrix $\boldsymbol{\Delta}_M$ is computed in $\mathcal{O}(M^2)$ operations, so that the jth column and the jth row of $\boldsymbol{\Delta}_M^{-1} \boldsymbol{F}_M \boldsymbol{\Delta}_M$ for each $j = 1, \ldots, M$ have almost the same 1-norm. Since the diagonal entries of $\boldsymbol{\Delta}_M$ are chosen as powers of 2, the balanced matrix $\boldsymbol{\Delta}_M^{-1} \boldsymbol{F}_M \boldsymbol{\Delta}_M$ can be calculated without roundoff (see [14]). □

By construction, the columns of the matrices $\boldsymbol{X}_{N-L,M}(0)$ and $\tilde{\boldsymbol{X}}_{N-L,M}(0)$ form orthonormal bases for the M-dimensional signal spaces $\mathcal{S}_{N-L}$ and $\tilde{\mathcal{S}}_{N-L}$, respectively. Assume that $\boldsymbol{U}_M \boldsymbol{D}_M \boldsymbol{V}_M^*$ is the singular value decomposition of

$$\boldsymbol{X}_{N-L,M}(0)^* \tilde{\boldsymbol{X}}_{N-L,M}(0),$$

where $\boldsymbol{U}_M$ and $\boldsymbol{V}_M$ are unitary matrices and $\boldsymbol{D}_M$ is a diagonal matrix with the diagonal entries d_j $(j = 1, \ldots, M)$ arranged in nonincreasing order $1 \geq d_1 \geq \ldots \geq d_M \geq 0$. Then

$$\theta_j := \arccos d_{M-j+1} \quad (j = 1, \ldots, M)$$

are the *canonical angles* between $\mathcal{S}_{N-L}$ and $\tilde{\mathcal{S}}_{N-L}$ (see [22, p. 43 and p. 45]). We remark that

$$\frac{\pi}{2} \geq \theta_1 \geq \ldots \geq \theta_M \geq 0$$

such that θ_1 is the largest canonical angle between $\mathcal{S}_{N-L}$ and $\tilde{\mathcal{S}}_{N-L}$.

Note that $\|\boldsymbol{P}_{N-L} - \tilde{\boldsymbol{P}}_{N-L}\|_2$ is the *distance* between the M-dimensional signal spaces $\mathcal{S}_{N-L}$ and $\tilde{\mathcal{S}}_{N-L}$ (cf. [7, p. 76]). Since $\boldsymbol{P}_{N-L} (\boldsymbol{P}_{N-L} - \tilde{\boldsymbol{P}}_{N-L}) = \boldsymbol{P}_{N-L} (\boldsymbol{I}_{N-L} - \tilde{\boldsymbol{P}}_{N-L})$, we see immediately that

$$\|\boldsymbol{P}_{N-L} - \tilde{\boldsymbol{P}}_{N-L}\|_2 \leq \|\boldsymbol{P}_{N-L}\|_2 \, \|\boldsymbol{I}_{N-L} - \tilde{\boldsymbol{P}}_{N-L}\|_2 \leq 1 .$$

As known (see [22, pp. 43–44]), the largest singular value of $\boldsymbol{P}_{N-L} - \tilde{\boldsymbol{P}}_{N-L}$ is equal to $\sin \theta_1$. Hence the distance between $\mathcal{S}_{N-L}$ and $\tilde{\mathcal{S}}_{N-L}$ amounts to

$$\|\boldsymbol{P}_{N-L} - \tilde{\boldsymbol{P}}_{N-L}\|_2 = \sin \theta_1 .$$

Now we estimate $\|\boldsymbol{P}_{N-L} - \tilde{\boldsymbol{P}}_{N-L}\|_2 = \sin \theta_1$, where $\boldsymbol{P}_{N-L}$ is the orthogonal projection onto the signal space $\mathcal{S}_{N-L}$ which coincides with the range of $\boldsymbol{H}_{L,N-L}(0)^*$ by Theorem 4.3. After the construction (see step 1 of Algorithm

3.4), $\tilde{\boldsymbol{P}}_{N-L}$ is the orthogonal projection onto the signal space $\tilde{\mathcal{S}}_{N-L}$ which is the range of $\tilde{\boldsymbol{K}}_{L,N-L}(0)^*$, where $\tilde{\boldsymbol{K}}_{L,N-L}(0)$ defined by (3.11) is the rank-M approximation of the given noisy matrix $\tilde{\boldsymbol{H}}_{L,N-L}(0)$. Thus the *error matrix of low-rank approximation* can be estimated by

$$\|\tilde{\boldsymbol{H}}_{L,N-L}(0) - \tilde{\boldsymbol{K}}_{L,N-L}(0)\|_2 \le \tilde{\sigma}_{M+1} < \varepsilon\, \tilde{\sigma}_1\,, \tag{5.7}$$

where $\tilde{\sigma}_1$ is the largest singular value of (3.1) and $\varepsilon > 0$ is a convenient chosen tolerance. Let

$$\boldsymbol{E}_{L,N-L} = \tilde{\boldsymbol{H}}_{L,N-L}(0) - \boldsymbol{H}_{L,N-L}(0) = (e_{\ell+m})_{\ell,m=0}^{L-1,N-L-1}$$

be the error matrix of given data. Using the maximum column resp. row sum norm of $\boldsymbol{E}_{L,N-L}$, we obtain by $|e_k| \le \varepsilon_1$ $(k = 0, \ldots, N-1)$ that

$$\|\boldsymbol{E}_{L,N-L}\|_2 \le \sqrt{\|\boldsymbol{E}_{L,N-L}\|_1\, \|\boldsymbol{E}_{L,N-L}\|_\infty} \le \sqrt{L\,(N-L)}\,\varepsilon_1 \le \frac{N}{2}\,\varepsilon_1\,.$$

Theorem 5.4. *Let $N \in \mathbb{N}$ $(N \gg 1)$ be given. Assume that the order M of the exponential sum* (1.1) *fulfils $2\,M \ll N$ and that the coefficients c_j of* (1.1) *satisfy the condition $|c_j| \ge \rho > 0$ $(j = 1, \ldots, M)$. Let σ_M be the lowest positive singular value of $\boldsymbol{H}_{L,N-L+1}$ with $L \approx \lceil \frac{N}{2} \rceil$ $(M \le L \le \lceil \frac{N}{2} \rceil)$.*

If $2\,\|\boldsymbol{E}_{L,N-L}\|_2 \ll \sigma_M$, then the spectral norm $\|\boldsymbol{P}_{N-L} - \tilde{\boldsymbol{P}}_{N-L}\|_2 = \sin\theta_1$ can be estimated by

$$\|\boldsymbol{P}_{N-L} - \tilde{\boldsymbol{P}}_{N-L}\|_2 \le \frac{2}{\sigma_M}\, \|\boldsymbol{E}_{L,N-L}\|_2\,. \tag{5.8}$$

Further, one has

$$\|\boldsymbol{H}_{L,N-L}(0)^\dagger\|_2 = \frac{1}{\sigma_M} \le \frac{1}{\rho}\, \|\boldsymbol{V}_{L,M}(\boldsymbol{z})^\dagger\|_2^2\,,$$

where

$$\|\boldsymbol{V}_{L,M}(\boldsymbol{z})^\dagger\|_2^2 \le \begin{cases} \frac{M\,(1-\beta^2)}{1-\beta^{2L}} \left(1 + \frac{M + \|\boldsymbol{q}_L\|_2^2 - \|\boldsymbol{p}_L\|_2^2 - \mu}{(M-1)\,\delta^2}\right)^{M-1} & \text{if } \beta < 1\,, \\ \frac{M}{L} \left(1 + \frac{\|\boldsymbol{q}_L\|_2^2 - \|\boldsymbol{p}_L\|_2^2}{(M-1)\,\delta^2}\right)^{M-1} & \text{if } \beta = 1 \end{cases}$$

with the first column $\boldsymbol{p}_L$ of $\boldsymbol{P}_L$. Note that β, μ, and δ are defined by (2.2), (2.3) *resp.* (2.4).

Proof. 1) For the orthogonal projections $\boldsymbol{P}_{N-L}$ and $\tilde{\boldsymbol{P}}_{N-L}$ we obtain that

$$\begin{aligned} \|\boldsymbol{P}_{N-L} - \tilde{\boldsymbol{P}}_{N-L}\|_2 &= \|\boldsymbol{P}_{N-L}\,(\boldsymbol{P}_{N-L} - \tilde{\boldsymbol{P}}_{N-L})\|_2 = \|\boldsymbol{P}_{N-L} - \boldsymbol{P}_{N-L}\,\tilde{\boldsymbol{P}}_{N-L}\|_2 \\ &= \|\boldsymbol{P}_{N-L}\,(\boldsymbol{I}_{N-L} - \tilde{\boldsymbol{P}}_{N-L})\|_2\,. \end{aligned}$$

Since $\boldsymbol{P}_{N-L}$ is the orthogonal projection onto the range of $\boldsymbol{H}_{L,N-L}(0)^*$, this projection has the form $\boldsymbol{P}_{N-L} = \boldsymbol{H}_{L,N-L}(0)^\dagger\, \boldsymbol{H}_{L,N-L}(0)$. Analogously, the orthogonal projection $\tilde{\boldsymbol{P}}_{N-L}$ onto the range of $\tilde{\boldsymbol{K}}_{L,N-L}(0)^*$ is given by $\tilde{\boldsymbol{P}}_{N-L} = \tilde{\boldsymbol{K}}_{L,N-L}(0)^\dagger\, \tilde{\boldsymbol{K}}_{L,N-L}(0)$. Then it follows that

$$\boldsymbol{P}_{N-L}\,(\boldsymbol{I}_{N-L} - \tilde{\boldsymbol{P}}_{N-L}) = \boldsymbol{H}_{L,N-L}(0)^\dagger\, \boldsymbol{H}_{L,N-L}(0)\,(\boldsymbol{I}_{N-L} - \tilde{\boldsymbol{P}}_{N-L})\,,$$

where

$$\boldsymbol{H}_{L,N-L}(0) = \tilde{\boldsymbol{K}}_{L,N-L}(0) + (\tilde{\boldsymbol{H}}_{L,N-L}(0) - \tilde{\boldsymbol{K}}_{L,N-L}(0)) - \boldsymbol{E}_{L,N-L}\,.$$

Since $\tilde{\boldsymbol{K}}_{L,N-L}(0) = \tilde{\boldsymbol{K}}_{L,N-L}(0)\,\tilde{\boldsymbol{P}}_{N-L}$ and since $\boldsymbol{I}_{N-L} - \tilde{\boldsymbol{P}}_{N-L}$ is an orthogonal projection too, we obtain by $\|\boldsymbol{H}_{L,N-L}(0)^{\dagger}\|_2 = \frac{1}{\sigma_M}$ that

$$\|\boldsymbol{P}_{N-L} - \tilde{\boldsymbol{P}}_{N-L}\|_2 \le \frac{1}{\sigma_M}\left(\|\tilde{\boldsymbol{H}}_{L,N-L}(0) - \tilde{\boldsymbol{K}}_{L,N-L}(0)\|_2 + \|\boldsymbol{E}_{L,N-L}\|_2\right).$$

From (5.7) and $\tilde{\sigma}_{M+1} \le \|\boldsymbol{E}_{L,N-L}\|_2$ by a corollary of Weyl's Theorem (see [21, p. 70]), it follows the inequality (5.8).

2) The exact Hankel matrix $\boldsymbol{H}_{L,N-L}(0)$ has the rank M and can be factorized into the following product of full rank matrices

$$\boldsymbol{H}_{L,N-L}(0) = \boldsymbol{V}_{L,M}(\boldsymbol{z})\,(\operatorname{diag}\boldsymbol{c})\,\boldsymbol{V}_{N-L,M}(\boldsymbol{z})^{\mathrm{T}}\,.$$

Thus the Moore–Penrose pseudoinverse of $\boldsymbol{H}_{L,N-L}(0)$ has the form

$$\boldsymbol{H}_{L,N-L}(0)^{\dagger} = \left(\boldsymbol{V}_{N-L,M}(\boldsymbol{z})^{\mathrm{T}}\right)^{\dagger}(\operatorname{diag}\boldsymbol{c})^{-1}\,\boldsymbol{V}_{L,M}(\boldsymbol{z})^{\dagger}\,.$$

Hence its norm can be estimated as follows

$$\|\boldsymbol{H}_{L,N-L}(0)^{\dagger}\|_2 \le \frac{1}{\rho}\,\|\left(\boldsymbol{V}_{N-L,M}(\boldsymbol{z})^{\mathrm{T}}\right)^{\dagger}\|_2\,\|\boldsymbol{V}_{L,M}(\boldsymbol{z})^{\dagger}\|_2 \le \frac{1}{\rho}\,\|\left(\boldsymbol{V}_{L,M}(\boldsymbol{z})^{\mathrm{T}}\right)^{\dagger}\|_2^2\,,$$

since for $M \le L \le N-L$ it holds by [1, Theorem 1]

$$\|\left(\boldsymbol{V}_{N-L,M}(\boldsymbol{z})^{\mathrm{T}}\right)^{\dagger}\|_2 \le \|\left(\boldsymbol{V}_{L,M}(\boldsymbol{z})^{\mathrm{T}}\right)^{\dagger}\|_2 = \|\boldsymbol{V}_{L,M}(\boldsymbol{z})^{\dagger}\|_2\,.$$

3) Finally, we estimate $\|\left(\boldsymbol{V}_{L,M}(\boldsymbol{z})^{\mathrm{T}}\right)^{\dagger}\|_2^2$ for $L \ge M$. We start with

$$\begin{aligned}\|\left(\boldsymbol{V}_{L,M}(\boldsymbol{z})^{\mathrm{T}}\right)^{\dagger}\|_2^2 &\le \|\left(\boldsymbol{V}_{L,M}(\boldsymbol{z})^{\mathrm{T}}\right)^{\dagger}\|_{\mathrm{F}}^2 = \sum_{j=1}^{M}\|\left(\boldsymbol{V}_{L,M}(\boldsymbol{z})^{\mathrm{T}}\right)^{\dagger}\boldsymbol{e}_j\|_2^2\\ &= \sum_{j=1}^{M}\frac{\|\left(\boldsymbol{V}_{L,M}(\boldsymbol{z})^{\mathrm{T}}\right)^{\dagger}\boldsymbol{e}_j\|_2^2\,\|\boldsymbol{V}_{L,M}(\bar{\boldsymbol{z}})\,\boldsymbol{e}_j\|_2^2}{\|\boldsymbol{V}_{L,M}(\bar{\boldsymbol{z}})\,\boldsymbol{e}_j\|_2^2}\,.\end{aligned}$$

As shown in the proof of Theorem 5.2, we know that

$$\|\left(\boldsymbol{V}_{L,M}(\boldsymbol{z})^{\mathrm{T}}\right)^{\dagger}\boldsymbol{e}_j\|_2^2\,\|\boldsymbol{V}_{L,M}(\bar{\boldsymbol{z}})\,\boldsymbol{e}_j\|_2^2 \le \left(1 + \frac{M + \|\boldsymbol{q}_L\|_2^2 - \|\boldsymbol{p}_L\|_2^2 - \mu}{(M-1)\,\delta^2}\right)^{M-1}$$

with $\delta = \min\{|z_j - z_k|;\, j,k = 1,\ldots,M,\, j \ne k\}$. Finally we use the estimate

$$\|\boldsymbol{V}_{L,M}(\bar{\boldsymbol{z}})\,\boldsymbol{e}_j\|_2^2 = \sum_{k=0}^{L-1}|z_j|^{2k} \ge \sum_{k=0}^{L-1}\beta^{2k}\,.$$

This completes the proof. □

We summarize: If the error bound ε_1 in Algorithm 3.4 are very small for sufficiently large integer N ($N \gg 2\,M$) so that $2\,\|\boldsymbol{E}_{L,N-L}\|_2 \ll \sigma_M$ for a window length $L \approx \lceil \frac{N}{2} \rceil$ ($M \le L \le \lceil \frac{N}{2} \rceil$), if all nodes z_j ($j = 1, \ldots, M$) are lying near to the unit circle with $\beta < 1$ (see (2.2)) but not extremely close to each other, and if all coefficients c_j ($j = 1, \ldots, M$) fulfil $|c_j| \ge \rho > 0$, then for each node z_j ($j = 1, \ldots, M$) there exists a unique node $\tilde{z}_j$ such that

$$|z_j - \tilde{z}_j| \le \Big(1 + \frac{M + \|\boldsymbol{q}_{N-L}\|_2^2 - \|\boldsymbol{p}_{N-L}\|_2^2 - \mu}{(M-1)\,\delta_j^2}\Big)^{(M-1)/2} \times \Big(\sin\theta_1 + \|\boldsymbol{q}_{N-L} - \tilde{\boldsymbol{q}}_{N-L}\|_2\Big)\,,$$

where μ and δ_j are defined by (2.3) and in Theorem 5.2, and where

$$\sin\theta_1 \le \frac{2}{\sigma_M}\,\|\boldsymbol{E}_{L,N-L}\|_2\,.$$

If the nodes $z_j \in \mathbb{D}$ ($j = 1, \ldots, M$) are computed with low errors, then the nonvanishing coefficients $c_j \in \mathbb{C}$ ($j = 1, \ldots, M$) can be determined as solution $\boldsymbol{c} = (c_j)_{j=1}^M$ of the least squares problem

$$\|\boldsymbol{V}_{N,M}(\boldsymbol{z})\,\boldsymbol{c} - \boldsymbol{h}\|_2 = \min$$

with the vector $\boldsymbol{h} = \big(h(k)\big)_{k=0}^{N-1}$ of exact data and $N > 2\,M$ (see Algorithm 3.4). Note that $\boldsymbol{V}_{N,M}(\boldsymbol{z})$ has full rank. Let $\tilde{\boldsymbol{h}} = (\tilde{h}_k)_{k=0}^{N-1}$ be the vector of noisy data and let $\tilde{\boldsymbol{z}} = (\tilde{z}_j)_{j=1}^M$ be the vector of computed nodes $\tilde{z}_j \in \mathbb{D}$ with $\tilde{z}_j \approx z_j$. Let $\tilde{\boldsymbol{c}} = (\tilde{c}_j)_{j=1}^M$ be the solution of the least squares problem

$$\|\boldsymbol{V}_{N,M}(\tilde{\boldsymbol{z}})\,\tilde{\boldsymbol{c}} - \tilde{\boldsymbol{h}}\|_2 = \min\,.$$

For large $N \gg 2\,M$, the Vandermonde matrix $\boldsymbol{V}_{N,M}(\tilde{\boldsymbol{z}})$ has full rank and is well conditioned with respect to the spectral norm. Assume that $\varepsilon_2 > 0$ fulfils the inequalities

$$\|\boldsymbol{V}_{N,M}(\boldsymbol{z}) - \boldsymbol{V}_{N,M}(\tilde{\boldsymbol{z}})\|_2 \le \varepsilon_2\,\|\boldsymbol{V}_{N,M}(\boldsymbol{z})\|_2\,, \quad \|\boldsymbol{h} - \tilde{\boldsymbol{h}}\|_2 \le \varepsilon_2\,\|\boldsymbol{h}\|_2\,,$$
$$\varepsilon_2\,\mathrm{cond}_2\,\boldsymbol{V}_{N,M}(\boldsymbol{z}) < 1\,.$$

By the perturbation theory of the least squares problem one obtains the normwise estimate (see [9, p. 382 and pp. 400–402])

$$\frac{\|\boldsymbol{c} - \tilde{\boldsymbol{c}}\|_2}{\|\boldsymbol{c}\|_2} \le \frac{\varepsilon_2\,\mathrm{cond}_2\,\boldsymbol{V}_{N,M}(\boldsymbol{z})}{1 - \varepsilon_2\,\mathrm{cond}_2\,\boldsymbol{V}_{N,M}(\boldsymbol{z})} \times \Big[2 + (\mathrm{cond}_2\,\boldsymbol{V}_{N,M}(\boldsymbol{z}) + 1)\frac{\|\boldsymbol{V}_{N,M}(\boldsymbol{z})\,\boldsymbol{c} - \boldsymbol{h}\|_2}{\|\boldsymbol{V}_{N,M}(\boldsymbol{z})\|_2\,\|\boldsymbol{c}\|_2}\Big]\,.$$

Consequently, the sensitivity of the least squares problem can be measured by the spectral norm condition number $\mathrm{cond}_2\,\boldsymbol{V}_{N,M}(\boldsymbol{z})$ when $\|\boldsymbol{V}_{N,M}(\boldsymbol{z})\,\boldsymbol{c} - \boldsymbol{h}\|_2$ is small and by the square $\big(\mathrm{cond}_2\,\boldsymbol{V}_{N,M}(\boldsymbol{z})\big)^2$ otherwise.

6. Numerical examples

Finally we illustrate the results by some numerical experiments. All computations are performed in MATLAB with IEEE double-precision arithmetic.

First we summarize the corresponding assumptions of our study:

(A1) The number N of noisy sampled data $\tilde{h}_k = h(k) + e_k$ $(k = 0, \ldots, N-1)$ satisfies the condition $N \gg 2\,M$. In other words, we use oversampling of the exponential sum (1.1). The order M of the exponential sum (1.1) is only of moderate size.

(A2) The coefficients $c_j \in \mathbb{C}$ $(j = 1, \ldots, M)$ of the exponential sum (1.1) fulfil the condition $|c_j| \ge \rho > 0$, where ρ is not too small.

(A3) The distinct nodes $z_j = \mathrm{e}^{f_j} \in \mathbb{D}$ $(j = 1, \ldots, M)$ are lying in the near of the unit circle.

(A4) The error terms e_k $(k = 0, \ldots, N-1)$ are relatively small so that $|e_k| \le \varepsilon_1$ with $0 < \varepsilon_1 \ll \rho$ and $2\,\|\boldsymbol{E}_{L,N-L}\|_2 \ll \sigma_M$, where σ_M is the lowest positive singular value of the L-trajectory matrix $\boldsymbol{H}_{L,N-L+1}$ with the window length $L \approx \lceil \frac{N}{2} \rceil$, where $M \le L \le \lceil \frac{N}{2} \rceil$.

We start with the estimate from Theorem 5.1 and show that the estimates (5.4) are sharp for some parameters, but also useless in other cases. To this end we compute $\kappa_j(\boldsymbol{P}_{N-L}\boldsymbol{C}_{N-L}(\boldsymbol{q}_{N-L}))$ and the estimate by the RHS of (5.6) as well as the values $\|\boldsymbol{P}_{N-L} - \tilde{\boldsymbol{P}}_{N-L}\|_2$ and $\|\boldsymbol{q}_{N-L} - \tilde{\boldsymbol{q}}_{N-L}\|_2$.

Example 6.1. We choose M equispaced nodes $z_j = \exp(2\pi \mathrm{i} j/M)$ $(j = 1, \ldots, M)$ on the unit circle and set the coefficients $c_j = 1$ $(j = 1, \ldots, M)$. We form the exponential sum (1.1) so that

$$h(k) = \sum_{j=1}^{M} z_j^k \quad (k = 0, \ldots, N-1)\,. \tag{6.1}$$

We use noisy sampled data $\tilde{h}_k := h(k) + e_k \in \mathbb{C}$ $(k = 0, \ldots, N-1)$ of (1.1), where $e_k \in [-10^{-s}, 10^{-s}] + \mathrm{i}\,[-10^{-s}, 10^{-s}]$ $(s = 4, 6, 8)$ are uniformly random error terms. The corresponding results are shown in Table 6.1, where we have chosen $L = N - L + 1$. We observe that $\kappa_j(\boldsymbol{F}_M) = 1$ and furthermore that the RHS of (5.6) is also one, i.e., the estimate is sharp. The condition number $\kappa_j(\boldsymbol{F}_M^{\mathrm{SVD}})$ is only slightly larger. □

Further examples are given in [17, Examples 4.1–4.2] and [18, Example 6.1].

Example 6.2. Now we choose M nodes on an Archimedean spiral given in the form

$$z_j = \sqrt{\frac{j+M}{2M}}\, \exp \frac{8\pi \mathrm{i}\,\sqrt{j+M}}{5} \quad (j = 1, \ldots, M)$$

and the coefficients $c_j = 1$ $(j = 1, \ldots, M)$. The exact data of (1.1) are denoted by (6.1). As in Example 6.1, we use noisy sampled data $\tilde{h}_k := h(k) + e_k \in \mathbb{C}$ $(k = 0, \ldots, N-1)$ of (1.1), where $e_k \in [-10^{-s}, 10^{-s}] + \mathrm{i}\,[-10^{-s}, 10^{-s}]$ $(s = 4, 6, 8)$

M	$N-L$	s	$\max \lvert z_j - \tilde{z}_j \rvert$	$\lVert \boldsymbol{P}_{N-L} - \tilde{\boldsymbol{P}}_{N-L} \rVert_2$	$\lVert \boldsymbol{q}_{N-L} - \tilde{\boldsymbol{q}}_{N-L} \rVert_2$
10	10	4	4.733e-06	2.023e-15	3.340e-05
10	20	4	2.029e-06	1.014e-05	1.453e-05
10	30	4	1.305e-06	1.066e-05	1.039e-05
10	10	6	4.002e-08	5.793e-15	2.764e-07
10	20	6	1.587e-08	7.933e-08	1.336e-07
10	30	6	1.259e-08	1.028e-07	9.445e-08
10	100	6	1.623e-09	4.663e-08	2.177e-08
50	100	6	2.803e-09	7.009e-08	6.918e-08
50	100	8	2.562e-11	6.406e-10	7.030e-10
50	200	8	1.081e-11	6.042e-10	3.315e-10

TABLE 6.1. Maximum errors of the nodes and the related estimates for noisy sampled data in Example 6.1.

are uniformly random error terms. In Tables 6.2 and 6.3, we present maximum errors of the nodes, where we have chosen $L = N - L + 1$. The condition numbers $\kappa_j(\boldsymbol{F}_M)$ and $\kappa_j(\boldsymbol{F}_M^{SVD})$ of the eigenvalues are very similar. □

M	$N-L$	s	$\max \lvert z_j - \tilde{z}_j \rvert$	$\max \kappa_j$	RHS of (5.6)
10	10	6	9.746e-07	2.143e+00	3.330e+01
10	20	6	6.977e-07	1.749e+00	1.112e+01
10	30	6	4.991e-07	1.731e+00	9.864e+00
10	100	6	9.097e-07	1.718e+00	9.206e+00
30	100	6	2.415e-04	3.658e+01	1.036e+14
30	100	4	5.758e-04	3.658e+01	1.028e+14

TABLE 6.2. Maximum errors of the nodes and the related estimates for noisy sampled data in Example 6.2, where $\max \kappa_j$ denotes the maximal condition number of $\kappa_j(\boldsymbol{F}_M)$.

The Examples 6.1–6.2 show that our estimates for $|z_j - \tilde{z}_j|$ $(j = 1, \ldots, M)$ based on Theorem 5.1 are very precise. The estimations of the condition numbers are sharp and cannot be improved in some cases. However we observe also that the estimates of the condition numbers of the eigenvalues based on RHS of (5.6) are useless for higher order M.

In the following Example 6.3 we show that the orthogonal projection onto the signal space is essential for good error estimates for the ESPRIT Algorithm 3.4. Applying (5.1) to the matrices $\boldsymbol{F}_M$ and $\tilde{\boldsymbol{F}}_M$, we obtain the first-order error

M	$N-L$	s	$\|\boldsymbol{P}_{N-L} - \tilde{\boldsymbol{P}}_{N-L}\|_2$	$\|\boldsymbol{q}_{N-L} - \tilde{\boldsymbol{q}}_{N-L}\|_2$
10	10	6	4.145e-15	2.869e-06
10	20	6	1.879e-06	1.044e-06
10	30	6	1.486e-06	1.905e-06
10	100	6	2.703e-06	1.584e-06
30	100	6	3.445e-04	3.385e-04
30	100	4	2.476e-03	2.968e-04

TABLE 6.3. Continuation of Table 6.2, where $\max |z_j - \tilde{z}_j|$ is estimated by (5.4) for noisy sampled data in Example 6.2.

estimate

$$|z_j - \tilde{z}_j| \le \kappa_j(\boldsymbol{F}_M)\, \|\boldsymbol{F}_M - \tilde{\boldsymbol{F}}_M\|_2 \tag{6.2}$$

for $j = 1, \ldots, M$. In Example 6.3, one can see that the norm $\|\boldsymbol{F}_M - \tilde{\boldsymbol{F}}_M\|_2$ is not small also for large $N - L$. In other words, one cannot explain the good error behavior of the ESPRIT Algorithm 3.4 by the estimate (6.2). If we replace (3.12) by (3.10), then the same statement is true.

Example 6.3. As in Example 6.1, we choose the M equispaced nodes on the unit circle $z_j = \exp(2\pi \mathrm{i} j/M)$ $(j = 1, \ldots, M)$ and the coefficients $c_j = 1$ $(j = 1, \ldots, M)$. The corresponding results are shown in Table 6.4. If we use (3.10) instead of (3.12), then we obtain similar results. □

M	$N-L$	s	$\max \|z_j - \tilde{z}_j\|$	$\max \kappa_j(\boldsymbol{F}_M)$	$\|\boldsymbol{F}_M - \tilde{\boldsymbol{F}}_M\|_2$
10	20	4	2.145e-06	1.007e+00	2.014e+00
10	30	4	1.354e-06	1.004e+00	2.069e+00
10	100	4	2.317e-07	1.000e+00	1.999e+00
50	100	4	2.719e-07	1.002e+00	2.161e+00
50	100	8	2.772e-11	1.002e+00	2.101e+00

TABLE 6.4. Maximum errors of the nodes and the related estimates for noisy sampled data in Example 6.3.

Example 6.4. Finally we use the same parameters of a nuclear magnetic resonance (NMR) signal as in [2, Table 1], i.e., $M = 5$ with the nodes $z_1 = 0.6342 - 0.7463\,\mathrm{i}$, $z_2 = 0.8858 - 0.4067\,\mathrm{i}$, $z_3 = 0.9663 - 0.1661\,\mathrm{i}$, $z_4 = 0.9642 + 0.2174\,\mathrm{i}$, $z_5 = 0.8811 + 0.2729\,\mathrm{i}$ and the coefficients $c_1 = 5.8921 + 1.5788\,\mathrm{i}$, $c_2 = 9.5627 + 2.5623\,\mathrm{i}$, $c_3 = 5.7956 + 1.5529\,\mathrm{i}$, $c_4 = 2.7046 + 0.7247\,\mathrm{i}$, $c_5 = 16.4207 + 4.3999\,\mathrm{i}$. For $N = 160$, the left Figure 6.1 shows the 5 positive singular values of the exact Hankel matrix

$\boldsymbol{H}_{L,160-L+1}$ for different window lengths $L = 5, \ldots, 80$. As expected (see Lemma 3.1 and Remark 3.3), the positive singular values of $\boldsymbol{H}_{L,160-L+1}$ increase for increasing window length $L = 5, \ldots, 80$. Thus $L = N/2$ is an optimal window length, where in practical applications it may be enough to choose $L = 2M$ or $L = 4M$. Note that the computational cost of the ESPRIT Algorithm 3.4 may be better for $L = 2M$ or $L = 4M$. For exactly sampled data, the right Figure 6.1 shows the errors $|z_j - \tilde{z}_j|$ between the given nodes z_j and the reconstructed nodes $\tilde{z}_j$ for $j = 1, \ldots, 5$. Note that both axes of ordinates in Figures 6.1 and 6.2 have a logarithmic scale. In Figure 6.2 we show the corresponding results for noisy sampled data $\tilde{h}_k = h(k) + e_k$ $(k = 0, \ldots, 159)$, where $\mathrm{Re}\, e_k$ and $\mathrm{Im}\, e_k$ zero-mean Gaussian random numbers with standard deviation 1. □

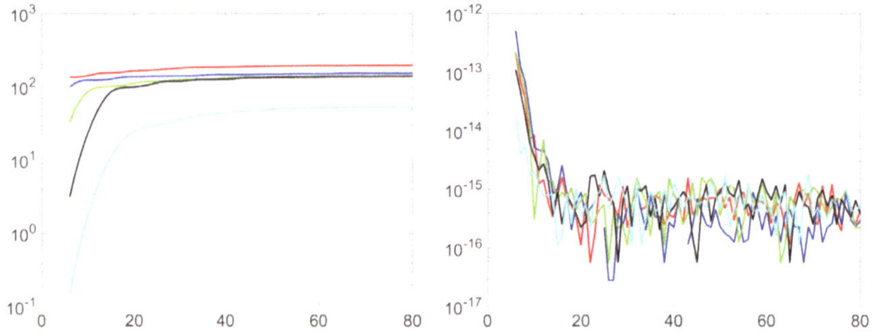

FIGURE 6.1. Singular values of $\boldsymbol{H}_{L,160-L+1}$ for different window lengths $L = 5, \ldots, 80$ (left) and the errors $|z_j - \tilde{z}_j|$ for $j = 1, \ldots, 5$ (right) in the case of exactly sampled data.

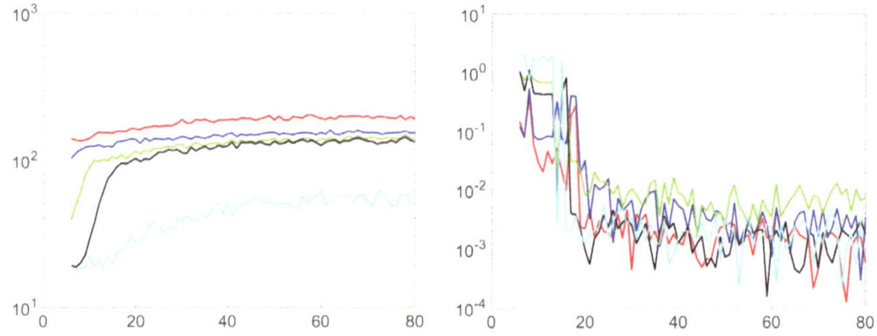

FIGURE 6.2. Singular values of $\tilde{\boldsymbol{H}}_{L,160-L+1}$ for different window lengths $L = 5, \ldots, 80$ (left) and the errors $|z_j - \tilde{z}_j|$ for $j = 1, \ldots, 5$ (right) in the case of noisy sampled data.

Acknowledgment

The first named author gratefully acknowledges the support by the German Research Foundation within the project PO 711. Special thanks are given to our friend Albrecht Böttcher for providing very valuable suggestions.

References

[1] F.S.V. Bazán. Conditioning of rectangular Vandermonde matrices with nodes in the unit disk. *SIAM J. Matrix Anal. Appl.*, 21:679–693, 2000.

[2] F.S.V. Bazán. Sensitivity eigenanalysis for single shift-invariant subspace-based methods. *Signal Process.*, 80:89–100, 2000.

[3] F.S.V. Bazán. Error analysis of signal zeros: a projected companion matrix approach. *Linear Algebra Appl.*, 369:153–167, 2003.

[4] F.S.V. Bazán and P.L. Toint. Conditioning of infinite Hankel matrices of finite rank. *Systems Control Letters*, 41:347–359, 2000.

[5] B. Beckermann, G.H. Golub, and G. Labahn. On the numerical condition of a generalized Hankel eigenvalue problem. *Numer. Math.*, 106:41–68, 2007.

[6] F. Filbir, H.N. Mhaskar, and J. Prestin. On the problem of parameter estimation in exponential sums. *Constr. Approx.*, 35:323–343, 2012.

[7] G.H. Golub and C.F. Van Loan. *Matrix Computations.* Third edn. Johns Hopkins Univ. Press, Baltimore, 1996.

[8] N. Golyandina. On the choice of parameters in singular spectrum analysis and related subspace-based methods. *Stat. Interface*, 3:259–279, 2010.

[9] N.J. Higham. *Accuracy and Stability of Numerical Algorithms.* Second edn. SIAM, Philadelphia, 2002.

[10] R.A. Horn and C.R. Johnson. *Matrix Analysis.* Second edn. Cambridge Univ. Press, Cambridge, 2013.

[11] Y. Hua and T.K. Sarkar. Matrix pencil method for estimating parameters of exponentially damped/undamped sinusoids in noise. *IEEE Trans. Acoust. Speech Signal Process.*, 38:814–824, 1990.

[12] F. Kittaneh. Singular values of companion matrices and bounds on zeros of polynomials. *SIAM J. Matrix Anal. Appl.*, 16:333–340, 1995.

[13] W. Liao and A. Fannjiang. MUSIC for single-snapshot spectral estimation: Stability and super-resolution. *Appl. Comput. Harmon. Anal.*, 40:33–67, 2016.

[14] B.N. Parlett and C. Reinsch. Balancing a matrix for calculation of eigenvalues and eigenvectors. *Numer. Math.*, 13:293–304, 1969.

[15] V. Pereyra and G. Scherer. *Exponential Data Fitting and its Applications.* Bentham Sci. Publ., Sharjah, 2010.

[16] D. Potts and M. Tasche. Parameter estimation for exponential sums by approximate Prony method. *Signal Process.*, 90:1631–1642, 2010.

[17] D. Potts and M. Tasche. Parameter estimation for nonincreasing exponential sums by Prony-like methods. *Linear Algebra Appl.*, 439:1024–1039, 2013.

[18] D. Potts and M. Tasche. Fast ESPRIT algorithms based on partial singular value decompositions. *Appl. Numer. Math.*, 88:31–45, 2015.

[19] R. Roy and T. Kailath. ESPRIT – estimation of signal parameters via rotational invariance techniques. *IEEE Trans. Acoust. Speech Signal Process.*, 37:984–994, 1989.

[20] R.A. Smith. The condition numbers of the matrix eigenvalue problem. *Numer. Math.*, 10:232–240, 1967.

[21] G.W. Stewart. *Matrix Algorithms.* Vol. I: Basic Decompositions. SIAM, Philadelphia, 1998.

[22] G.W. Stewart and J.-G. Sun. *Matrix Perturbation Theory.* Academic Press, Boston, 1990.

[23] J.H. Wilkinson. Note on matrices with a very ill-conditioned eigenproblem. *Numer. Math.*, 19:176–178, 1972.

Daniel Potts
Technische Universität Chemnitz
Faculty of Mathematics
D–09107 Chemnitz, Germany
e-mail: `potts@mathematik.tu-chemnitz.de`

Manfred Tasche
University of Rostock
Institute of Mathematics
D–18051 Rostock, Germany
e-mail: `manfred.tasche@uni-rostock.de`

Operator Theory:
Advances and Applications, Vol. 259, 649–662

The Universal Algebra Generated by a Power Partial Isometry

Steffen Roch

Dedicated to Albrecht Böttcher on the occasion of his 60th birthday

Abstract. The goal of this paper is to characterize (a slight modification of) the algebra of the finite sections method for Toeplitz operators with continuous generating function, as first described by Albrecht Böttcher and Bernd Silbermann in [2], by a universal property, namely as the universal C^*-algebra generated by a power partial isometry (PPI). A PPI is an element of a C^*-algebra with the property that every non-negative power of that element is a partial isometry.

Mathematics Subject Classification (2010). Primary 46L05; Secondary 47B35, 65R20.

Keywords. Power partial isometry, universal algebra, finite sections algebra.

1. Introduction

Let $\mathcal{A}$ be a C^*-algebra. An element $v \in \mathcal{A}$ is called a partial isometry if $vv^*v = v$. Simple examples show that a power of a partial isometry needs not to be a partial isometry. One therefore calls v a *power partial isometry* (PPI) if every non-negative power of v is a partial isometry again.

Examples

(a) In a C^*-algebra with identity element e, every unitary element u (i.e., $u^*u = uu^* = e$) is a PPI. In particular, the function $u : t \mapsto t$ is a unitary element of the algebra $C(\mathbb{T})$ of the continuous functions on the complex unit circle $\mathbb{T}$, and the operator U of multiplication by the function u is a unitary operator on the Hilbert space $L^2(\mathbb{T})$ of the squared integrable functions on $\mathbb{T}$.

(b) In a C^*-algebra with identity element e, every isometry v (i.e., $v^*v = e$) and every co-isometry v (i.e., $vv^* = e$) is a PPI. In particular, the operators $V : (x_0, x_1, \ldots) \mapsto (0, x_0, x_1, \ldots)$ and $V^* : (x_0, x_1, x_2, \ldots) \mapsto (x_1, x_2, \ldots)$

of forward and backward shift, respectively, are PPIs on the Hilbert space $l^2(\mathbb{Z}^+)$ of the squared summable sequences on the non-negative integers.

(c) The matrix $V_n := (a_{ij})$ with $a_{i+1,i} = 1$ and $a_{ij} = 0$ if $i \neq j+1$, considered as an element of the algebra $\mathbb{C}^{n\times n}$ of the complex $n \times n$ matrices, is a PPI.

(d) If v_i is a PPI in a C^*-algebra $\mathcal{A}_i$ for every i in an index set I, then $(v_i)_{i\in I}$ is a PPI in the direct product $\prod_{i\in I} \mathcal{A}_i$. In particular, the operator (V, V^*), considered as an element of $L(l^2(\mathbb{Z}^+)) \times L(l^2(\mathbb{Z}^+))$, is a PPI.

Note that the PPI V_n in (c) and (V, V^*) in (d) are neither isometric nor co-isometric. □

A C^*-algebra $\mathcal{A}$ generated by a PPI v is said to be *universal* if, for every other C^*-algebra $\mathcal{B}$ generated by a PPI w, there is a *-homomorphism from $\mathcal{A}$ to $\mathcal{B}$ which sends v to w. The universal algebra generated by a unitary resp. isometric element is defined in an analogous way. The existence of a universal algebra generated by a PPI is basically a consequence of Example (d). A first concrete description of this universal algebra was given by Conway, Duncan and Paterson in [7]. Later, Hancock and Raeburn presented an alternative proof (Theorem 1.3 in [10]), and Fowler and Raeburn gave an explicit description as a corollary of a more general result about Toeplitz algebras of Hilbert bimodules (Corollary 5.4 in [8]). The proofs in [7] and [10] rely on a structure theorem of Halmos and Wallen; for a recent proof see [1]. For a general account on C^*-algebras generated by partial isometries, with special emphasis on their relation to graph theory, see [5].[1]

It follows from the Gelfand–Naimark theorem that the universal algebra generated by a unitary element is *-isomorphic to the algebra $C(\mathbb{T})$, generated by the unitary function u. Coburn [6] identified the universal algebra generated by an isometry as the Toeplitz algebra $\mathsf{T}(C)$ which is the smallest C^*-subalgebra of $L(l^2(\mathbb{Z}^+))$ which contains the isometry V, the shift operator. This algebra bears its name since it can be described as the smallest C^*-subalgebra of $L(l^2(\mathbb{Z}^+))$ which contains all Toeplitz operators $T(a)$ with generating function $a \in C(\mathbb{T})$. Recall that the Toeplitz operator with generating function $a \in L^1(\mathbb{T})$ is given by the matrix $(a_{i-j})_{i,j=0}^\infty$ where a_k stands for the kth Fourier coefficient of a. This operator is bounded on $l^2(\mathbb{Z}^+)$ if and only if $a \in L^\infty(\mathbb{T})$ (see [3, 4]).

Since both the Toeplitz algebra $\mathsf{T}(C)$ and its symbol algebra $C(\mathbb{T})$ are universal for some relation, one might wonder whether another algebra naturally related to Toeplitz operators, the so-called finite sections algebra $\mathcal{S}(\mathsf{T}(C))$, is also distinguished by a universal property. This algebra is defined via the finite sections discretization with respect to the projections $P_n : (x_0, x_1, \ldots) \mapsto (x_0, \ldots, x_{n-1}, 0, 0, \ldots)$ on $l^2(\mathbb{Z}^+)$ as follows. Write $\mathcal{F}$ for the set of all bounded sequences $(A_n)_{n\geq 1}$ of operators $A_n \in L(\operatorname{im} P_n)$ and $\mathcal{G}$ for the set of all sequences $(A_n) \in \mathcal{F}$ with $\|A_n\| \to 0$. Provided with entry-wise defined operations and the supremum norm, $\mathcal{F}$ becomes a C^*-algebra and $\mathcal{G}$ a closed ideal of $\mathcal{F}$. Since $L(\operatorname{im} P_n)$

[1] I am grateful to Iain Raeburn for helpful comments and for bringing relevant references to my attention.

is isomorphic to $\mathbb{C}^{n\times n}$, we can identify $\mathcal{F}$ with the direct product and $\mathcal{G}$ with the direct sum of the algebras $\mathbb{C}^{n\times n}$ for $n \geq 1$. Now consider the smallest C^*-subalgebra $\mathcal{S}(\mathsf{T}(C))$ of $\mathcal{F}$ which contains all sequences $(P_nT(a)P_n)$ with $a \in C(\mathbb{T})$ and its C^*-subalgebra $\mathcal{S}_{\geq 2}(\mathsf{T}(C))$ which is generated by the sequence (P_nVP_n) (note that V is the Toeplitz operator with generating function $t \mapsto t$). With these notations, our characterization of the finite sections algebra by a universal property reads as follows.

Theorem 1. *The C^*-algebra $\mathcal{S}_{\geq 2}(\mathsf{T}(C))$ generated by the PPI (P_nVP_n) is *-isomorphic to the universal algebra generated by a PPI.*

It would be not hard to derive this statement directly from the references cited above. The goal of this paper is to give a complete proof which is inspired by the precise knowledge of the finite sections algebra $\mathcal{S}(\mathsf{T}(C))$ (Proposition 3 and Corollaries 4 and 5 below). Even if not every step in this proof is new (e.g., Lemma 12 below is also used in [7] and is attributed to Glushkin there), the hope is that it will shed some new light on this algebra and its structure.

Before going into the details, we provide some basic (and well-known) facts on the algebras $\mathcal{S}(\mathsf{T}(C))$ and $\mathcal{S}_{\geq 2}(\mathsf{T}(C))$. Since the first entry of the sequence (P_nVP_n) is zero, the first entry of every sequence in $\mathcal{S}_{\geq 2}(\mathsf{T}(C))$ is zero. So we can omit the first entry and consider the elements of $\mathcal{S}_{\geq 2}(\mathsf{T}(C))$ as sequences labeled by $n \geq 2$ (whence the notation). In fact this is the only difference between the algebras $\mathcal{S}(\mathsf{T}(C))$ and $\mathcal{S}_{\geq 2}(\mathsf{T}(C))$.

Proposition 2. *$\mathcal{S}_{\geq 2}(\mathsf{T}(C))$ consists of all sequences $(A_n)_{n\geq 2}$ where $(A_n)_{n\geq 1}$ is a sequence in $\mathcal{S}(\mathsf{T}(C))$.*

The sequences in $\mathcal{S}(\mathsf{T}(C))$ are completely described in the following theorem, where we let R_n denote the operator $(x_0, x_1, \ldots) \mapsto (x_{n-1}, \ldots, x_0, 0, 0, \ldots)$ on $l^2(\mathbb{Z}^+)$. Further we set $\tilde{a}(t) := a(t^{-1})$ for every function a on $\mathbb{T}$. This description was found by A. Böttcher and B. Silbermann and first published in their 1983 paper [2] on the convergence of the finite sections method for quarter plane Toeplitz operators (see also [9], Section 1.4.2).

Proposition 3. *The algebra $\mathcal{S}(\mathsf{T}(C))$ consists of all sequences $(A_n)_{n\geq 1}$ of the form*

$$(A_n) = (P_nT(a)P_n + P_nKP_n + R_nLR_n + G_n) \tag{1}$$

where $a \in C(\mathbb{T})$, K and L are compact operators, and $(G_n) \in \mathcal{G}$. The representation of a sequence $(A_n) \in \mathcal{S}(\mathsf{T}(C))$ in this form is unique.

Corollary 4. *$\mathcal{G}$ is a closed ideal of $\mathcal{S}(\mathsf{T}(C))$, and the quotient algebra $\mathcal{S}(\mathsf{T}(C))/\mathcal{G}$ is *-isomorphic to the C^*-algebra of all pairs*

$$(T(a) + K, T(\tilde{a}) + L) \in L(l^2(\mathbb{Z}^+)) \times L(l^2(\mathbb{Z}^+)) \tag{2}$$

with $a \in C(\mathbb{T})$ and K, L compact. In particular, the mapping which sends the sequence (1) to the pair (2) is a *-homomorphism from $\mathcal{S}(\mathsf{T}(C))$ onto $\mathcal{S}(\mathsf{T}(C))/\mathcal{G}$ with kernel $\mathcal{G}$.

It is not hard to see that the algebra of all pairs (2) is just the smallest C^*-subalgebra of $L(l^2(\mathbb{Z}^+)) \times L(l^2(\mathbb{Z}^+))$ that contains the PPI (V, V^*).

Corollary 5. *The set $\mathcal{J}$ of all pairs (K, L) of compact operators K, L is a closed ideal of $\mathcal{S}(\mathsf{T}(C))/\mathcal{G}$. The quotient algebra $(\mathcal{S}(\mathsf{T}(C))/\mathcal{G})/\mathcal{J}$ is *-isomorphic to $C(\mathbb{T})$. In particular, the mapping which sends the pair (2) to the function a is a *-homomorphism from $\mathcal{S}(\mathsf{T}(C))/\mathcal{G}$ onto $C(\mathbb{T})$ with kernel $\mathcal{J}$.*

Observe that all of the above examples (a)–(d) appear somewhere in the algebra $\mathcal{S}(\mathsf{T}(C))$ and its quotients.

2. Elementary properties of PPI

Our first goal is a condition ensuring that the product of two partial isometries is a partial isometry again.

Proposition 6. *Let u, v be partial isometries. Then uv is a partial isometry if and only if*

$$u^*uvv^* = vv^*u^*u, \tag{3}$$

i.e., if the initial projection u^*u of u and the range projection vv^* of v commute.

Proof. Condition (3) implies that

$$(uv)(uv)^*(uv) = uvv^*u^*uv = (uu^*u)(vv^*v) = uv;$$

hence, uv is a partial isometry. Conversely, if uv is a partial isometry, then a simple calculation gives

$$v^*(vv^*u^*u - u^*uvv^*)(u^*uvv^* - vv^*u^*u)v = 0.$$

With the C^*-axiom we conclude that $v^*(vv^*u^*u - u^*uvv^*) = 0$, hence $vv^*(vv^*u^*u - u^*uvv^*) = 0$, which finally gives

$$vv^*u^*u = vv^*u^*uvv^*.$$

The right-hand side of this equality is selfadjoint; so must be the left-hand side. Thus, $vv^*u^*u = (vv^*u^*u)^* = u^*uvv^*$, which is condition (3). □

In particular, if v is a partial isometry, then v^2 is a partial isometry if and only if

$$v^*vvv^* = vv^*v^*v. \tag{4}$$

Proposition 7. *Let v be a partial isometry with property* (4) *(e.g., a PPI). Then*

$$e := v^*v + vv^* - v^*vvv^* = v^*v + vv^* - vv^*v^*v$$

is the identity element of the C^*-algebra generated by v. Moreover,

$$p := vv^* - vv^*v^*v = e - v^*v \quad \textit{and} \quad \tilde{p} := v^*v - v^*vvv^* = e - vv^*$$

are mutually orthogonal projections (meaning that $p\tilde{p} = \tilde{p}p = 0$).

Proof. Condition (4) implies that e is selfadjoint. Further,

$$ve = vv^*v + vvv^* - vv^*vvv^* = v + vvv^* - vvv^* = v$$

and, similarly, $v^*e = v^*$. Taking adjoints it follows that $ev^* = v^*$ and $ev = v$, and e is the identity element. The remaining assertions are also easy to check. □

We will often use the notation v^{*n} instead of $(v^*)^n$.

Proposition 8.

(a) *If v is a PPI, then*

$$v^{*k}v^kv^nv^{*n} = v^nv^{*n}v^{*k}v^k \quad \textit{for all } k, n \geq 1. \tag{5}$$

(b) *If v is a partial isometry and if* (5) *holds for $k = 1$ and for every $n \geq 1$, then v is a PPI.*

Proof. Assertion (a) is a consequence of Proposition 6 (the partial isometry v^{n+k} is the product of the partial isometries v^k and v^n). Assertion (b) follows easily by induction. For $k = 1$, condition (5) reduces to

$$(v^*v)(v^nv^{*n}) = (v^nv^{*n})(v^*v).$$

Thus if v and v^n are partial isometries, then v^{n+1} is a partial isometry by Proposition 6. □

Lemma 9. *If v is a PPI, then $(v^nv^{*n})_{n\geq 0}$ and $(v^{*n}v^n)_{n\geq 0}$ are decreasing sequences of pairwise commuting projections.*

Proof. The PPI property implies that the v^nv^{*n} are projections and that

$$v^nv^{*n}n^{n+k}(v^*)^{n+k} = (v^nv^{*n}v^n)v^k(v^*)^{n+k} = v^nv^k(v^*)^{n+k} = n^{n+k}(v^*)^{n+k}$$

for $k, n \geq 0$. The assertions for the second sequence follow similarly. □

3. A distinguished ideal

Let $\mathcal{A}$ be a C^*-algebra generated by a PPI v. By $\operatorname{alg}(v, v^*)$ we denote the smallest (symmetric, not necessarily closed) subalgebra of $\mathcal{A}$ which contains v and v^*. Further we write N_v for the set of all non-negative integers such that $pv^n\tilde{p} \neq 0$. From Proposition 7 we know that $0 \notin \mathrm{N}_v$. Finally, we set

$$\pi_n := pv^n\tilde{p}v^{*n}p \quad \text{and} \quad \tilde{\pi}_n := \tilde{p}v^{*n}pv^n\tilde{p}.$$

Proposition 10.

(a) *The element $pv^n\tilde{p}$ is a partial isometry with initial projection $\tilde{\pi}_n$ and range projection π_n. Thus, the projections π_n and $\tilde{\pi}_n$ are Murray–von Neumann equivalent in $\mathcal{A}$, and they generate the same ideal of $\mathcal{A}$.*

(b) *$\pi_m\pi_n = 0$ and $\tilde{\pi}_m\tilde{\pi}_n = 0$ whenever $m \neq n$.*

Proof. (a) By definition,

$$\pi_n = pv^n\tilde{p}v^{*n}p = pv^n(e - vv^*)v^{*n}p = p(v^nv^{*n} - v^{n+1}(v^*)^{n+1})p.$$

Since $p = e - vv^*$ and $v^n v^{*n}$ commute by Proposition 8,

$$\pi_n = p(v^n v^{*n} - v^{n+1}(v^*)^{n+1}) = (v^n v^{*n} - v^{n+1}(v^*)^{n+1})p.$$

Being a product of commuting projections (Lemma 9), π_n is itself a projection. Analogously, $\tilde{\pi}_n$ is a projection. Thus, $pv^n\tilde{p}$ is a partial isometry, and π_n and $\tilde{\pi}_n$ are Murray–von Neumann equivalent. Finally, the equality

$$\pi_n = \pi_n^2 = (pv^n\tilde{p}v^{*n}p)^2 = pv^n\tilde{\pi}_n v^{*n}p$$

shows that π_n belongs to the ideal generated by $\tilde{\pi}_n$. The reverse inclusion follows analogously. Assertion (b) is again a simple consequence of Lemma 9. □

Let $\mathcal{C}_n$ denote the smallest closed ideal of $\mathcal{A}$ which contains the projection π_n (likewise, the projection $\tilde{\pi}_n$). We want to show that $\mathcal{C}_n$ is isomorphic to $\mathbb{C}^{(n+1)\times(n+1)}$ whenever $n \in \mathbb{N}_v$ (Proposition 17 below). For we need to establish a couple of facts on (finite) words in $\mathrm{alg}\,(v, v^*)$.

Lemma 11. *Let a, b, c be non-negative integers. Then*

$$v^{*a}v^b v^{*c} = \begin{cases} (v^*)^{a-b+c} & \text{if } \min\{a,c\} \geq b, \\ v^{b-a}v^{*c} & \text{if } a \leq b \leq c, \\ v^{*a}v^{b-c} & \text{if } a \geq b \geq c \end{cases}$$

and

$$v^a v^{*b} v^c = \begin{cases} v^{a-b+c} & \text{if } \min\{a,c\} \geq b, \\ v^a(v^*)^{b-c} & \text{if } a \geq b \geq c, \\ (v^*)^{b-a}v^c & \text{if } a \leq b \leq c. \end{cases}$$

Proof. Let $\min\{a,c\} \geq b$. Thus,

$$v^{*a}v^b v^{*c} = (v^*)^{a-b}v^{*b}v^b v^{*b}(v^*)^{c-b} = (v^*)^{a-b}v^{*b}(v^*)^{c-b} = (v^*)^{a-b+c},$$

where we used that v^{*b} is a partial isometry. If $a \leq b \leq c$, then

$$v^{*a}v^b v^{*c} = v^{*a}v^a v^{b-a}(v^*)^{b-a}(v^*)^{c-b+a} = v^{b-a}(v^*)^{b-a}v^{*a}v^a(v^*)^{c-b+a}$$

by Proposition 8 (a). Then,

$$v^{*a}v^b v^{*c} = v^{b-a}(v^*)^{b-a}v^{*a}v^a v^{*a}(v^*)^{c-b} = v^{b-a}(v^*)^{b-a}v^{*a}(v^*)^{c-b} = v^{b-a}v^{*c}.$$

Similarly, $v^{*a}v^b v^{*c} = v^{*a}v^{b-c}$ if $a \geq b \geq c$. The second assertion of the lemma follows by taking adjoints. □

Every word in $\mathrm{alg}\,(v, v^*)$ is a product of powers v^n and v^{*m}. Every product $v^a v^{*b} v^c$ and $v^{*a}v^b v^{*c}$ of three powers can be written as a product of at most two powers if one of the conditions

$$\min\{a,c\} \geq b \quad \text{or} \quad a \leq b \leq c \quad \text{or} \quad a \geq b \geq c \tag{6}$$

in Lemma 11 is satisfied. Since (6) is equivalent to $\max\{a,c\} \geq b$, such a product can not be written as a product of less than three powers by means of Lemma 11 if $\max\{a,c\} < b$. Since it is not possible in a product $v^a v^{*b} v^c v^{*d}$ or $v^{*a}v^b v^{*c}v^d$ of four powers that $\max\{a,c\} < b$ *and* $\max\{b,d\} < c$, one can shorten every product of powers v^n and v^{*m} to a product of at most three powers.

Summarizing we get the following lemma.

Lemma 12. *Every finite word in* $\mathrm{alg}\,(v, v^*)$ *is of the form* $v^a v^{*b}$ *or* $v^{*b} v^a$ *with* $a, b \geq 0$ *or of the form* $v^a v^{*b} v^c$ *or* $v^{*a} v^b v^{*c}$ *with* $0 < \min\{a, c\} \leq \max\{a, c\} < b$.

Corollary 13. *Let* w *be a word in* $\mathrm{alg}\,(v, v^*)$.

(a) *If* $pwp \neq 0$, *then* $w = v^a v^{*a}$ *for some* $a \geq 0$.
(b) *If* $\tilde{p}w\tilde{w} \neq 0$, *then* $w = v^{*a} v^a$ *with some* $a \geq 0$.

Proof. We only check assertion (a). By the preceding lemma, w is a product of at most three powers $v^a v^{*b} v^c$ or $v^{*a} v^b v^{*c}$. First let $w = v^a v^{*b} v^c$. Since $vp = pv^* = 0$, we conclude that $c = 0$ if $pwp \neq 0$. Writing

$$pwp = \begin{cases} pv^a v^{*a} (v^*)^{b-a} p = v^a v^{*a} p (v^*)^{b-a} p & \text{if } \ a \leq b, \\ pv^{a-b} v^b v^{*b} p = pv^{a-b} p v^b v^{*b} & \text{if } \ a \geq b, \end{cases}$$

we obtain by the same argument that $a = b$ if $pwp \neq 0$. Thus, $w = v^a v^{*a}$. The case when $w = v^{*a} v^b v^{*c}$ can be treated analogously. □

An element k of a C^*-algebra $\mathcal{A}$ is called an *element of algebraic rank one* if, for every $a \in \mathcal{A}$, there is a complex number α such that $kak = \alpha k$.

Proposition 14. *Let* $m, n \in \mathbb{N}_v$. *Then*

(a) π_n *is a projection of algebraic rank one in* $\mathcal{A}$.
(b) π_m *and* π_n *are Murray–von Neumann equivalent if and only if* $m = n$.

Analogous assertions hold for $\tilde{\pi}_n$ in place of π_n.

Proof. (a) Every element of $\mathcal{A}$ is a limit of linear combinations of words in v and v^*. It is thus sufficient to show that, for every word w, there is an $\alpha \in \mathbb{C}$ such that $\pi_n w \pi_n = \alpha \pi_n$. If $\pi_n w \pi_n = 0$, this holds with $\alpha = 0$. If $\pi_n w \pi_n = \pi_n pwp \pi_n \neq 0$, then $w = v^a v^{*a}$ for some $a \geq 0$ by Corollary 13. In this case,

$$\pi_n w \pi_n = \pi_n v^a v^{*a} \pi_n = p(v^n v^{*n} - v^{n+1} (v^*)^{n+1}) v^a v^{*a} (v^n v^{*n} - v^{n+1} (v^*)^{n+1}) p.$$

From Lemma 9 we infer that

$$(v^n v^{*n} - v^{n+1} (v^*)^{n+1}) v^a v^{*a} = \begin{cases} v^n v^{*n} - v^{n+1} (v^*)^{n+1} & \text{if } \ a \leq n, \\ v^a v^{*a} - v^a v^{*a} = 0 & \text{if } \ a \geq n+1. \end{cases}$$

Thus,

$$\pi_n w \pi_n = \begin{cases} p(v^n v^{*n} - v^{n+1} (v^*)^{n+1})^2 p = \pi_n & \text{if } \ a \leq n, \\ 0 & \text{if } \ a \geq n+1, \end{cases}$$

i.e., $\alpha = 1$ if $a \leq n$ and $\alpha = 0$ in all other cases.

(b) The projections π_m and π_n are Murray–von Neumann equivalent if and only if $\pi_m \mathcal{A} \pi_n \neq \{0\}$. So we have to show that $\pi_m \mathcal{A} \pi_n = \{0\}$ whenever $m \neq n$. Again it is sufficient to show that $\pi_m w \pi_n = 0$ for every word w.

Suppose there is a word w such that $\pi_m w \pi_n = \pi_m pwp \pi_n \neq 0$. Then $w = v^a v^{*a}$ for some $a \geq 0$ by Corollary 13. The terms in parentheses in

$$\pi_m w \pi_n = \pi_m v^a v^{*a} \pi_n = p(v^m v^{*m} - v^{m+1} (v^*)^{m+1}) (v^a v^{*a}) (v^n v^{*n} - v^{n+1} (v^*)^{n+1}) p$$

commute by Lemma 9. Since

$$(v^m v^{*m} - v^{m+1}(v^*)^{m+1})(v^n v^{*n} - v^{n+1}(v^*)^{n+1}) = 0$$

for $m \neq n$ we conclude that $\pi_m w \pi_n = 0$, a contradiction. □

Lemma 15.

(a) *If $a > n$ or $b > a$, then $v^b v^{*a} \pi_n = 0$.*
(b) *If $b \leq a \leq n$, then $v^b v^{*a} \pi_n = (v^*)^{a-b} \pi_n$.*

Proof. (a) One easily checks that $(v^*)^{n+1} p = 0$, which gives the first assertion. Let $b > a$. Then, since p commutes with $v^k v^{*k}$ and $vp = 0$,

$$\begin{aligned} v^b v^{*a} \pi_n &= v^{b-a} v^a v^{*a} (v^n v^{*n} - v^{n+1}(v^*)^{n+1}) p \\ &= v^{b-a} p v^a v^{*a} (v^n v^{*n} - v^{n+1}(v^*)^{n+1}) = 0. \end{aligned}$$

(b) Applying Lemma 11 to the terms in inner parentheses in

$$v^b v^{*a} \pi_n = ((v^b v^{*a} v^n) v^{*n} - (v^b v^{*a} v^{n+1})(v^*)^{n+1}) p,$$

one can simplify this expression to

$$((v^*)^{a-b} v^n v^{*n} - (v^*)^{a-b} v^{n+1})(v^*)^{n+1}) p = (v^*)^{a-b} \pi_n.$$ □

Corollary 16.

(a) *If w is a word in v and v^*, then $w\pi_n \in \{0, \pi_n, v^*\pi_n, \ldots, v^{*n}\pi_n\}$.*
(b) *For every $w \in \mathcal{A}$, $w\pi_n$ is a linear combination of elements $v^{*i}\pi_n$ with $i \in \{0, 1, \ldots, n\}$.*
(c) *Every element of the ideal $\mathcal{C}_n$ generated by π_n is a linear combination of elements $v^{*i}\pi_n v^j$ with $i, j \in \{0, 1, \ldots, n\}$.*

In particular, $\mathcal{C}_n$ is a finite-dimensional C^*-algebra. We are now in a position to describe this algebra exactly.

Proposition 17.

(a) *For $n \in \mathbb{N}_v$, the algebra $\mathcal{C}_n$ is *-isomorphic to $\mathbb{C}^{(n+1)\times(n+1)}$.*
(b) *$\mathcal{C}_m \mathcal{C}_n = \{0\}$ whenever $m \neq n$.*

Proof. (a) The elements $e_{ij}^{(n)} := v^{*i}\pi_n v^j$ with $i, j \in \{0, 1, \ldots n\}$ span the algebra $\mathcal{C}_n$ by Corollary 16 (c). Thus, the assertion will follow once we have shown that these elements form a system of $(n+1) \times (n+1)$ matrix units in the sense that $(e_{ij}^{(n)})^* = e_{ji}^{(n)}$ and

$$e_{ij}^{(n)} e_{kl}^{(n)} = \delta_{jk} e_{il}^{(n)} \qquad \text{for all } i, j, k, l \in \{0, 1, \ldots, n\}, \tag{7}$$

with δ_{jk} the standard Kronecker delta. The symmetry property is clear. To check (7), first let $j = k$. Then

$$e_{ij}^{(n)} e_{jl}^{(n)} = v^{*i}\pi_n (v^j v^{*j} \pi_n) v^l = v^{*i} \pi_n^2 v^l = e_{il}^{(n)}$$

by Lemma 15 (b). If $j > k$, then

$$e_{ij}^{(n)} e_{kl}^{(n)} = v^{*i}\pi_n (v^j v^{*k} \pi_n) v^l = 0$$

by Lemma 15 (a). Finally, if $j < k$, then

$$e_{ij}^{(n)} e_{kl}^{(n)} = v^{*i}(\pi_n v^j v^{*k})\pi_n v^l = v^{*i}(v^k v^{*j}\pi_n)^*\pi_n v^l = 0,$$

again by Lemma 15 (a). This proves (a). Assertion (b) follows from Proposition 14 (b). □

Given a PPI v, we let $\mathcal{G}_v$ stand for the smallest closed ideal which contains all projections π_n. If $\mathbb{N}_v$ is empty, then $\mathcal{G}_v$ is the zero ideal. Let $\mathbb{N}_v \neq \emptyset$. The ideal generated by a projection π_n with $n \in \mathbb{N}_v$ is isomorphic to $\mathbb{C}^{(n+1)\times(n+1)}$ by Proposition 17, and if u, w are elements of $\mathcal{A}$ which belong to ideals generated by two different projections π_m and π_n, then $uw = 0$ by Proposition 14 (b). Hence, $\mathcal{G}_v$ is then isomorphic to the direct sum of all matrix algebras $\mathbb{C}^{(n+1)\times(n+1)}$ with $n \in \mathbb{N}_v$.

If $\mathcal{A}$ is the *universal* C^*-algebra generated by a PPI v, then $\mathbb{N}_v$ is the set of *all* positive integers. Indeed, the algebra $\mathcal{S}_{\geq 2}(\mathsf{T}(C))$ introduced in the introduction is generated by the PPI $v := (P_n V P_n)$, and $\mathbb{N}_v = \mathbb{N}$ in this concrete setting.

Corollary 18. *If $\mathcal{A}$ is the universal C^*-algebra generated by a PPI v, then $\mathbb{N}_v = \mathbb{N}$, and $\mathcal{G}_v$ is isomorphic to the ideal $\mathcal{G}_{\geq 2} := \mathcal{S}_{\geq 2}(\mathsf{T}(C)) \cap \mathcal{G}$.*

4. PPI with $\mathbb{N}_v = \emptyset$

Our next goal is to describe the C^*-algebra $\mathcal{A}$ which is generated by a PPI v with $\mathbb{N}_v = \emptyset$. This condition is evidently satisfied if one of the projections $p = e - v^*v$ and $\tilde{p} = e - vv^*$ is zero, in which cases the algebra generated by the PPI v is well known:

- If $p = 0$ and $\tilde{p} = 0$, then v is unitary, and $\mathcal{A}$ is *-isomorphic to $C(X)$ where $X \subseteq \mathbb{T}$ is the spectrum of v by the Gelfand–Naimark theorem.
- If $p = 0$ and $\tilde{p} \neq 0$, then v is a non-unitary isometry, $\mathcal{A}$ is *-isomorphic to the Toeplitz algebra $\mathsf{T}(C)$ by Coburn's theorem, and the isomorphism sends v to the forward shift V.
- If $p \neq 0$ and $\tilde{p} = 0$, then v is a non-unitary co-isometry, again $\mathcal{A}$ is *-isomorphic to the Toeplitz algebra $\mathsf{T}(C)$ by Coburn's theorem, and the isomorphism sends v to the backward shift V^*.

Thus the only interesting case is when $\mathbb{N}_v = \emptyset$, but $p \neq 0$ and $\tilde{p} \neq 0$. Let $\mathcal{C}$ and $\widetilde{\mathcal{C}}$ denote the smallest closed ideals of $\mathcal{A}$ which contain the projections p and $\tilde{p}$, respectively. For $i, j \geq 0$, set

$$f_{ij} := v^{*i} p v^j \quad \text{and} \quad \tilde{f}_{ij} := v^i \tilde{p} v^{*j}.$$

Lemma 19. *If v is a PPI with $\mathbb{N}_v = \emptyset$, then $(f_{ij})_{i,j\geq 0}$ is a (countable) system of matrix units, i.e., $f_{ij}^* = f_{ji}$ and*

$$f_{ij} f_{kl} = \delta_{jk} f_{il} \qquad \textit{for all } i, j, k, l \geq 0. \tag{8}$$

If one of the f_{ij} is non-zero (e.g., if $f_{00} = p \neq 0$), then all f_{ij} are non-zero.

An analogous assertion holds for the family of the $\tilde{f}_{ij}$.

Proof. The symmetry condition is evident, and if $f_{ij} = 0$ then $f_{kl} = f_{ki} f_{ij} f_{jl} = 0$ for all k, l by (8). Property (8) on its hand will follow once we have shown that

$$pv^j v^{*k} p = \delta_{jk} p \qquad \text{for all } j, k \geq 0. \tag{9}$$

The assertion is evident if $j = k = 0$. If $j > 0$ and $k = 0$, then

$$\begin{aligned} pv^j p &= (e - v^* v) v^j (e - v^* v) \\ &= v^j - v^* v^{j+1} - v^{j-1}(vv^* v) + v^* v^j (vv^* v) \\ &= v^j - v^* v^{j+1} - v^{j-1} v + v^* v^j v = 0, \end{aligned}$$

and (9) holds. Analogously, (9) holds if $j = 0$ and $k > 0$. Finally, let $j, k > 0$. The assumption $\mathbb{N}_v = \emptyset$ ensures that

$$pv^{j-1} \tilde{p} = (e - v^* v) v^{j-1} (e - vv^*) = v^{j-1} - v^* v^j - v^j v^* + v^* v^{j+1} v^* = 0 \tag{10}$$

for all $j \geq 1$. Employing this identity we find

$$\begin{aligned} pv^j v^{*k} p &= (e - v^* v)\, v^j v^{*k} p = v^j v^{*k} p - (v^* v^{j+1} v^*)\,(v^*)^{k-1} p \\ &= v^j v^{*k} p - (v^{j-1} - v^* v^j - v^j v^*)(v^*)^{k-1} p = (e - v^* v) v^{j-1} (v^*)^{k-1} p. \end{aligned}$$

Thus, $pv^j v^{*k} p = pv^{j-1} (v^*)^{k-1} p$ for $j, k \geq 1$. Repeated application of this identity finally leads to one of the cases considered before. □

Proposition 20. *Let $\mathbb{N}_v = \emptyset$ and $p \neq 0$.*

(a) *The ideal $\mathcal{C}$ of $\mathcal{A}$ generated by p coincides with the smallest closed subalgebra of $\mathcal{A}$ which contains all f_{ij} with $i, j \geq 0$.*

(b) *$\mathcal{C}$ is *-isomorphic to the ideal of the compact operators on a separable infinite-dimensional Hilbert space.*

Analogous assertions hold for the projection $\tilde{p}$, the algebra $\widetilde{\mathcal{C}}$, and the $\tilde{f}_{ij}$.

Proof. For a moment, write $\mathcal{C}'$ for the smallest closed subalgebra of $\mathcal{A}$ which contains all f_{ij} with $i, j \geq 0$. The identities

$$f_{ij} v = v^{*i} p v^j v = f_{i,j+1}, \qquad v f_{0j} = vpv^j = v(e - v^* v) v^j = 0$$

and, for $i \geq 1$,

$$\begin{aligned} v f_{ij} &= vv^{*i} (e - v^* v) v^j = vv^{*i} v^j - (v(v^*)^{i+1} v) v^j \\ &= vv^{*i} v^j + ((v^*)^{i-1} - (v^*)^i v - vv^{*i}) v^j \\ &= (v^*)^{i-1} v^j - (v^*)^i v v^j = (v^*)^{i-1} (e - v^* v) v^j \;=\; f_{i-1,j} \end{aligned}$$

(where we used the adjoint of (10)) and their adjoints show that $\mathcal{C}'$ is a closed ideal of $\mathcal{A}$. Since $p = f_{00}$ we conclude that $\mathcal{C} \subseteq \mathcal{C}'$. Conversely, we have $f_{ij} = v^{*i} p v^j \in \mathcal{C}$ for all $i, j \geq 0$ whence the reverse inclusion $\mathcal{C}' \subseteq \mathcal{C}$. This settles assertion (a).

For assertion (b), note that every C^*-algebra generated by a (countable) system of matrix units (in particular, the algebra $\mathcal{C}'$) is naturally *-isomorphic to the algebra of the compact operators on a separable infinite-dimensional Hilbert space (see, e.g., Corollary A.9 in Appendix A2 in [11]). □

Lemma 21. *If $\mathbb{N}_v = \emptyset$, then $\mathcal{C} \cap \widetilde{\mathcal{C}} = \{0\}$.*

Proof. $\mathcal{C}$ and $\widetilde{\mathcal{C}}$ are closed ideals. Thus, $\mathcal{C} \cap \widetilde{\mathcal{C}} = \mathcal{C}\widetilde{\mathcal{C}}$, and we have to show that $f_{ij}\tilde{f}_{kl} = 0$ for all $i, j, k, l \geq 0$. Since

$$f_{ij}\tilde{f}_{kl} = (v^{*i}pv^j)\,(v^k\tilde{p}v^{*l}) = v^{*i}\,(pv^{j+k}\tilde{p})\,v^{*l},$$

this is a consequence of $\mathbb{N}_v = \emptyset$. □

Remember that $p \neq 0$ and $\tilde{p} \neq 0$. From the preceding lemma we conclude that the mapping

$$\mathcal{A} \to \mathcal{A}/\mathcal{C} \times \mathcal{A}/\widetilde{\mathcal{C}}, \quad w \mapsto (w + \mathcal{C}, w + \widetilde{\mathcal{C}})$$

is an injective *-homomorphism; thus $\mathcal{A}$ is *-isomorphic to the C^*-subalgebra of $\mathcal{A}/\mathcal{C} \times \mathcal{A}/\widetilde{\mathcal{C}}$ generated by $(v + \mathcal{C}, v + \widetilde{\mathcal{C}})$. The element $v + \mathcal{C}$ is an isometry in $\mathcal{A}/\mathcal{C}$ (since $e - v^*v \in \mathcal{C}$), but it is not unitary (otherwise $e - vv^* \in \widetilde{\mathcal{C}}$ would be a non-zero element of $\mathcal{C}$, in contradiction with Lemma 21). Analogously, $v + \widetilde{\mathcal{C}}$ is a non-unitary co-isometry. By Coburn's Theorem, there are *-isomorphisms $\mu : \mathcal{A}/\mathcal{C} \to \mathsf{T}(C)$ and $\tilde{\mu} : \mathcal{A}/\widetilde{\mathcal{C}} \to \mathsf{T}(C)$ which map $v + \mathcal{C} \mapsto V$ and $v + \widetilde{\mathcal{C}} \mapsto V^*$, respectively. But then

$$\mu \times \tilde{\mu} : \mathcal{A}/\mathcal{C} \times \mathcal{A}/\widetilde{\mathcal{C}} \to \mathsf{T}(C) \times \mathsf{T}(C), \quad (a, \tilde{a}) \mapsto (\mu(a), \tilde{\mu}(\tilde{a}))$$

is a *-isomorphism which maps the C^*-subalgebra of $\mathcal{A}/\mathcal{C} \times \mathcal{A}/\widetilde{\mathcal{C}}$ generated by $(v + \mathcal{C}, v + \widetilde{\mathcal{C}})$ to the C^*-subalgebra of $\mathsf{T}(C) \times \mathsf{T}(C)$ generated by the pair (V, V^*). The latter algebra has been identified in Corollary 4. Summarizing we get:

Proposition 22. *Let the C^*-algebra $\mathcal{A}$ be generated by a PPI v with $\mathbb{N}_v = \emptyset$ and $p \neq 0$ and $\tilde{p} \neq 0$. Then $\mathcal{A}$ is *-isomorphic to the algebra $\mathcal{S}(\mathsf{T}(C))/\mathcal{G}$ (likewise, to $\mathcal{S}_{\geq 2}(\mathsf{T}(C))/\mathcal{G}_{\geq 2}$), and the isomorphism sends v to $(P_nVP_N)_{n\geq 1} + \mathcal{G}$ (likewise, to $(P_nVP_N)_{n\geq 2} + \mathcal{G}_{\geq 2}$).*

5. The general case

We are now going to finish the proof of Theorem 1. For we think of $\mathcal{A}$ as being faithfully represented as a C^*-algebra of bounded linear operators on a separable infinite-dimensional Hilbert space H (note that $\mathcal{A}$ is finitely generated, hence separable). As follows easily from (7), $z_n := \sum_{i=0}^{n} e_{ii}^{(n)}$ is the identity element of $\mathcal{C}_n$. So we can think of the z_n as orthogonal projections on H. Moreover, these projections are pairwise orthogonal by Proposition 17 (b). Thus, the operators $P_n := \sum_{i=1}^{n} z_n$ form an increasing sequence of orthogonal projections on H. Let $P \in L(H)$ denote the least upper bound of that sequence (which then is the limit of the P_n in the strong operator topology). Clearly, P is an orthogonal projection again (but note that P does not belong to $\mathcal{A}$ in general).

Lemma 23.

(a) *Every z_n is a central projection of $\mathcal{A}$.*
(b) *P commutes with every element of $\mathcal{A}$.*

Proof. Assertion (b) is a consequence of (a). We show that

$$z_n = \sum_{i=0}^{n} v^{*i} \pi_n v^i = \sum_{i=0}^{n} v^{*i} p (v^n v^{*n} - v^{n+1}(v^*)^{n+1}) v^i$$
$$= \sum_{i=0}^{n} v^{*i} (e - v^* v)(v^n v^{*n} - v^{n+1}(v^*)^{n+1}) v^i$$

commutes with v. Indeed,

$$\begin{aligned} v z_n &= v(e - v^* v)(v^n v^{*n} - v^{n+1}(v^*)^{n+1}) \\ &\quad + \sum_{i=1}^{n} v v^{*i} (e - v^* v)(v^n v^{*n} - v^{n+1}(v^*)^{n+1}) v^i \\ &= \sum_{i=1}^{n} v v^{*i} (e - v^* v) v^n (v^{*n} - v(v^*)^{n+1}) v^i \\ &= \sum_{i=1}^{n} (v v^{*i} v^n - v(v^*)^{i+1} v^{n+1})(v^{*n} - v(v^*)^{n+1}) v^i \\ &= \sum_{i=1}^{n} ((v^*)^{i-1} v^n - (v^*)^i v^{n+1})(v^{*n} - v(v^*)^{n+1}) v^i \end{aligned}$$

by Lemma 11, and we further get

$$\begin{aligned} v z_n &= \sum_{i=1}^{n} (v^*)^{i-1} (e - v^* v)(v^n v^{*n} - v^{n+1}(v^*)^{n+1}) v^i \\ &= \sum_{i=0}^{n-1} v^{*i} (e - v^* v)(v^n v^{*n} - v^{n+1}(v^*)^{n+1}) v^i v \\ &= \sum_{i=0}^{n} v^{*i} (e - v^* v)(v^n v^{*n} - v^{n+1}(v^*)^{n+1}) v^i v \\ &\quad - (v^*)^n (e - v^* v)(v^n v^{*n} - v^{n+1}(v^*)^{n+1}) v^{n+1} \\ &= \sum_{i=0}^{n} v^{*i} (e - v^* v)(v^n v^{*n} - v^{n+1}(v^*)^{n+1}) v^i v \; = \; z_n v \end{aligned}$$

again by Lemma 11. Thus, $v z_n = z_n v$. Since $z_n = z_n^*$, this implies that z_n also commutes with v^* and, hence, with every element of $\mathcal{A}$. □

Consequently, $\mathcal{A} = P\mathcal{A}P \oplus (I - P)\mathcal{A}(I - P)$ where I stands for the identity operator on H. We consider the summands of this decomposition separately. The part $(I - P)\mathcal{A}(I - P)$ is generated by the PPI $v' := (I - P)v(I - P)$. Since

$$(I - P) p v^n \tilde{p} v^{*n} p (I - P) = (I - P)\pi_n (I - P) = (I - P) z_n e_{00}^{(n)} \pi_n (I - P) = 0,$$

we conclude that $\mathbb{N}_{v'} = \emptyset$. Thus, this part of $\mathcal{A}$ is described by Proposition 22.

The part $P\mathcal{A}P$ is generated by the PPI PvP. It follows from the definition of P that $\mathbb{N}_{PvP} = \mathbb{N}_v$ and that $\mathcal{G}_{PvP} = P\mathcal{G}_v P = \mathcal{G}_v$. We let $\prod_{n\in\mathbb{N}_v} \mathcal{C}_n$ stand for the direct product of the algebras $\mathcal{C}_n$ and consider the mapping

$$P\mathcal{A}P \to \prod_{n\in\mathbb{N}_v} \mathcal{C}_n, \quad PAP \mapsto (z_n PAP z_n)_{n\in\mathbb{N}_v} = (z_n A z_n)_{n\in\mathbb{N}_v}. \tag{11}$$

If $z_n A z_n = 0$ for every $n \in \mathbb{N}_v$, then

$$PAP = \sum_{m,n\in\mathbb{N}_v} z_m PAP z_n = \sum_{n\in\mathbb{N}_v} z_n A z_n = 0.$$

Thus, the mapping (11) is injective, and the algebra $P\mathcal{A}P$ is *-isomorphic to the C^*-subalgebra of $\prod_{n\in\mathbb{N}_v} \mathcal{C}_n$ generated by the sequence $(z_n v z_n)_{n\in\mathbb{N}_v}$.

Further we infer from Proposition 17 (a) that $\mathcal{C}_n$ is isomorphic to $\mathbb{C}^{(n+1)\times(n+1)}$ if $n \in \mathbb{N}_v$. We are going to make the latter isomorphism explicit. For we note that

$$\begin{aligned}
e_{ii}^{(n)} v e_{jj}^{(n)} &= v^{*i}\pi_n v^{i+1} v^{*j} \pi_n v^j \\
&= \begin{cases} v^{*i}\pi_n v^{i+1}(v^*)^{i+1}\pi_n v^{i+1} & \text{if} \quad i+1 = j, \\ 0 & \text{if} \quad i+1 \neq j \end{cases} \qquad \text{(by Corollary 13)} \\
&= \begin{cases} v^{*i}\pi_n v^{i+1} & \text{if} \quad i+1 = j, \\ 0 & \text{if} \quad i+1 \neq j \end{cases} \qquad \text{(by Lemma 15)} \\
&= \begin{cases} e_{i,i+1}^{(n)} & \text{if} \quad i+1 = j, \\ 0 & \text{if} \quad i+1 \neq j. \end{cases}
\end{aligned}$$

We choose a unit vector $e_i^{(n)}$ in the range of $e_{ii}^{(n)}$ (recall Proposition 14 (a)), and let $f_i^{(n)}$ stand for the $n+1$-tuple $(0, \ldots, 0, 1, 0, \ldots, 0)$ with the 1 at the ith position. Then $(e_i^{(n)})_{i=0}^n$ forms an orthonormal basis of $\operatorname{im} z_n$, $(f_i^{(n)})_{i=0}^n$ forms an orthonormal basis of $\mathbb{C}^{n+1}$, the mapping $e_i^{(n)} \mapsto f_{n-i}^{(n)}$ extends to a linear bijection from $\operatorname{im} z_n$ onto $\mathbb{C}^{n+1}$, which finally induces a *-isomorphism ξ_n from $\mathcal{C}_n \cong L(\operatorname{im} z_n)$ onto $\mathbb{C}^{(n+1)\times(n+1)} \cong L(\mathbb{C}^{n+1})$. Then

$$\xi : \prod_{n\in\mathbb{N}_v} \mathcal{C}_n \to \prod_{n\in\mathbb{N}_v} \mathbb{C}^{(n+1)\times(n+1)}, \quad (A_n) \mapsto (\xi_n(A_n))$$

is a *-isomorphism which maps the C^*-subalgebra of $\prod_{n\in\mathbb{N}_v} \mathcal{C}_n$ generated by the sequence $(z_n v z_n)_{n\in\mathbb{N}_v}$ to the C^*-subalgebra of $\prod_{n\in\mathbb{N}_v} \mathbb{C}^{(n+1)\times(n+1)}$ generated by the sequence $(V_{n+1})_{n\in\mathbb{N}_v}$, where V_n is the matrix described in Example (c). Note that V_n is just the $n \times n$th finite section $P_n V P_n$ of the forward shift operator.

If now $\mathcal{A}$ is the universal algebra generated by a PPI v, then $\mathbb{N}_v = \mathbb{N}$, as we observed in Corollary 18. Thus, in this case, the algebra $P\mathcal{A}P$ is *-isomorphic to the smallest C^*-subalgebra of $\mathcal{F} = \prod_{n\geq 1} \mathbb{C}^{n\times n}$ generated by the sequence $(P_n V P_n)$, i.e., to the C^*-algebra $\mathcal{S}_{\geq 2}(\mathsf{T}(C))$.

It remains to explain what happens with the part $(I-P)\mathcal{A}(I-P)$ of $\mathcal{A}$. The point is that the quotient $P\mathcal{A}P/P\mathcal{G}_v P$ is generated by a PPI u for which $\mathbb{N}_u$ is empty. We have seen in Proposition 22 that both this quotient and the algebra

$(I-P)\mathcal{A}(I-P)$ are canonically *-isomorphic to $\mathcal{S}_{\geq 2}(\mathsf{T}(C))/\mathcal{G}_{\geq 2}$. Thus, there is a *-homomorphism from $P\mathcal{A}P$ onto $(I-P)\mathcal{A}(I-P)$ which maps the generating PPI PvP of $P\mathcal{A}P$ to the generating PPI $(I-P)v(I-P)$ of $(I-P)\mathcal{A}(I-P)$. Hence, if $\mathcal{A}$ is the universal C^*-algebra generated by a PPI, then already $P\mathcal{A}P$ has the universal property, and $\mathcal{A} \cong P\mathcal{A}P \cong \mathcal{S}_{\geq 2}(\mathsf{T}(C))$. □

References

[1] A. an Huef, I. Raeburn, I. Tolich, *Structure theorems for star-commuting power partial isometries.* Lin. Alg. Appl. **481** (2015), 107–114.

[2] A. Böttcher, B. Silbermann, *The finite section method for Toeplitz operators on the quarter-plane with piecewise continuous symbols.* Math. Nachr. **110** (1983), 279–291.

[3] A. Böttcher, B. Silbermann, *Analysis of Toeplitz Operators.* Akademie-Verlag, Berlin 1989 and Springer-Verlag, Berlin, Heidelberg, New York 1990.

[4] A. Böttcher, B. Silbermann, *Introduction to Large Truncated Toeplitz Matrices.* Springer-Verlag, Berlin, Heidelberg 1999.

[5] I. Cho, P. Jorgensen, *C^*-Algebras generated by partial isometries.* J. Appl. Math. Comp. **26** (2008), 1-2, 1–48.

[6] L.A. Coburn, *The C^*-algebra generated by an isometry.* Bull. Am. Math. Soc. **73** (1967), 722–726.

[7] J.B. Conway, J. Duncan, A.L.T. Paterson, *Monogenic inverse semigroups and their C^*-algebras.* Proc. Royal Soc. Edinburgh **98A** (1984), 13–24.

[8] N.J. Fowler, I. Raeburn, *The Toeplitz algebra of a Hilbert bimodule.* Indiana Univ. Math. J. **48** (1999), 1, 155–181.

[9] R. Hagen, S. Roch, B. Silbermann, *C^*-Algebras and Numerical Analysis.* Marcel Dekker, Inc., New York, Basel 2001.

[10] R. Hancock, I. Raeburn, *The C^*-algebras of some inverse semigroups.* Bull. Austral. Math. Soc. **42** (1990), 335–348.

[11] I. Raeburn, *Graph algebras.* CMBS **103**, Amer. Math. Soc., Providence, R.I., 2005.

Steffen Roch
Fachbereich Mathematik
Technische Universität Darmstadt
Schlossgartenstrasse 7
D-64289 Darmstadt, Germany
e-mail: roch@mathematik.tu-darmstadt.de

Operator Theory:
Advances and Applications, Vol. 259, 663–680

Norms, Condition Numbers and Pseudospectra of Convolution Type Operators on Intervals

Markus Seidel

Dedicated to Albrecht Böttcher on the occasion of his 60th birthday

Abstract. The results in this paper describe the asymptotic behavior of convolution type operators on finite intervals as the length of these intervals tends to infinity. The family of operators under consideration here is generated (among others) by Fourier convolutions with slowly oscillating, almost periodic, bounded and uniformly continuous, and quasi-continuous multipliers, as well as operators of multiplication by slowly oscillating, almost periodic, and piecewise continuous functions. The focus is on the convergence of norms, condition numbers and pseudospectra.

Mathematics Subject Classification (2010). 65R20, 47G10, 47B35.

Keywords. Operator sequence, convolution operator, Fourier multiplier, finite section method, condition number, pseudospectrum.

1. Introduction

Over the last decades the study of convolution operators has evolved and branched in several directions and stimulated important developments: In fact, there are several types of related operators (Fourier convolutions, Wiener–Hopf operators, Hankel operators) over various classes of symbol functions (ranging from functions in the Wiener algebra over continuous, piecewise continuous, almost periodic and semi-almost periodic to slowly oscillating and piecewise quasi-continuous ones). Some of the main fields of interest are concerned with invertibility and Fredholm theory, with spectral theory, and with approximation by, e.g., finite sections of such operators. Among the tools one has factorization, algebraic (in particular C^*-algebra based) techniques, non-commutative Gelfand theories and localization. In all of these fields Albrecht Böttcher has made important and stimulating contributions and has left his footprints. Here, some of his work is revisited in order to bring a few of the ideas and tools on algebras of approximating sequences for Wiener–Hopf and related operators, on pseudo-spectral approximation as well as on norm preserving localization together, and to combine it with some recent observations.

The paper is organized as follows: we start introducing all required classes of functions and Fourier multipliers, the desired algebras of convolution type operators in Definitions 1.4 and 1.6, and in Section 2 the fundamentals of the concept of finite section sequences and their snapshots. In Section 3 we state the Main Theorem 3.1 on the asymptotics of the norms and discuss its consequences for condition numbers. Section 4 addresses pseudospectra and demonstrates how the main theorem yields convergence results on these spectral sets for operators from the algebras under consideration and their finite sections. The concept of norm preserving localization and the proof of the main theorem are subject of Section 5. Finally, in Section 6, we discuss several generalizations and particularly a modification of the previous approach which permits to include the flip, hence Hankel operators.

Our first aim is to define the classes of convolution type operators which we are interested in. Basically, these operators are build from two families of generators: operators aI of multiplication by certain essentially bounded functions a, and Fourier convolution operators $W^0(b)$ with certain symbol functions (multipliers) b.

1.1. Function classes

Within $L^\infty(\mathbb{R})$, the Banach algebra of all essentially bounded functions, we consider the following subclasses (actually closed subalgebras):

- BUC – the set of all bounded and uniformly continuous functions on $\mathbb{R}$.
- $\mathrm{C}(\overline{\mathbb{R}})$ – the set of continuous functions f with finite limits

$$f(-\infty) := \lim_{t\to-\infty} f(t), \quad f(+\infty) := \lim_{t\to+\infty} f(t).$$

- $\mathrm{C}(\dot{\mathbb{R}})$ – the subset of $\mathrm{C}(\overline{\mathbb{R}})$ of all functions f with $f(-\infty) = f(+\infty)$.
- AP – the smallest closed subalgebra of $L^\infty(\mathbb{R})$ which contains all functions $t \mapsto e^{\mathrm{i}\lambda t}$, $\lambda \in \mathbb{R}$. We say AP is generated by these $t \mapsto e^{\mathrm{i}\lambda t}$, and denote this by $\mathrm{AP} := [\{t \mapsto e^{\mathrm{i}\lambda t} : \lambda \in \mathbb{R}\}]$. The elements of AP are referred to as almost periodic functions.
- L_0^∞ – the set of all functions f in $L^\infty(\mathbb{R})$ with $f(-\infty) = f(+\infty) = 0$.
- PC – the set of all piecewise continuous functions, i.e., those functions f which have finite one-sided limits

$$f(\lambda - 0) := \lim_{t\to\lambda-0} f(t), \quad f(\lambda + 0) := \lim_{t\to\lambda+0} f(t)$$

at every point $\lambda \in \dot{\mathbb{R}}$.[1] Notice that this algebra is generated by the characteristic functions $\chi_{[\lambda,\infty)}$, $\lambda \in \mathbb{R}$, that is $\mathrm{PC} := [\{\chi_{[\lambda,\infty)} : \lambda \in \mathbb{R}\}]$.

[1] Here $\dot{\mathbb{R}} = \mathbb{R} \cup \{\infty\}$ is the one-point compactification of the real line.

- QC – the set of all quasi-continuous functions, that is

$$\mathrm{QC} := (\mathrm{C}(\dot{\mathbb{R}}) + H^\infty_+) \cap (\mathrm{C}(\dot{\mathbb{R}}) + H^\infty_-),$$

  where $H^\infty_\pm$ is the closed subalgebra of $L^\infty(\mathbb{R})$ consisting of all functions being non-tangential limits on $\mathbb{R}$ of bounded analytic functions on the upper/lower half-plane, respectively.
- SO^λ – the set of all functions f being continuous on $\dot{\mathbb{R}} \setminus \{\lambda\}$ and *slowly oscillating at* λ, i.e., for every $r \in (0,1)$

$$\lim_{t\to+0} \mathrm{osc}(f, \lambda + ([-t,-rt] \cup [rt,t])) = 0 \quad \text{in the case} \quad \lambda \in \mathbb{R}$$

$$\lim_{t\to\infty} \mathrm{osc}(f, [-t,-rt] \cup [rt,t]) = 0 \quad \text{in the case} \quad \lambda = \infty$$

  where $\mathrm{osc}(f, I) := \mathrm{esssup}\{|f(t) - f(s)| : t, s \in I\}$.
- SO – the Banach algebra of all so-called *slowly oscillating* functions, which is generated by all functions in all SO^λ, $\lambda \in \dot{\mathbb{R}}$, i.e., $\mathrm{SO} := \left[\bigcup_{\lambda\in\dot{\mathbb{R}}} \mathrm{SO}^\lambda\right]$.

In the present paper we will consider an algebra of convolution type operators which has all operators of multiplication aI by functions $a \in L^\infty_0 \cup \mathrm{PC} \cup \mathrm{SO} \cup \mathrm{AP}$ among its generators (see Definition 1.4). Actually, notice that

Remark 1.1. $[L^\infty_0 \cup \mathrm{PC}]$ already includes $\mathrm{C}(\overline{\mathbb{R}})$, $\mathrm{C}(\dot{\mathbb{R}})$ and all SO^λ, $\lambda \in \mathbb{R}$.

1.2. Fourier convolutions and multipliers

It is well known that the Fourier transform

$$F : L^2(\mathbb{R}) \to L^2(\mathbb{R}), \quad (Fu)(x) := \int_{\mathbb{R}} u(t)e^{\mathrm{i}tx}dt, \quad x \in \mathbb{R},$$

is invertible, hence for every $b \in L^\infty(\mathbb{R})$ the operator $W^0(b) = F^{-1}bF$ is a well-defined bounded linear operator on $L^2(\mathbb{R})$, and is called a Fourier convolution operator with the symbol function b.

On the spaces $L^p(\mathbb{R})$, $p \neq 2$, the situation is somewhat more involved: A function $b \in L^\infty(\mathbb{R})$ is said to be a Fourier multiplier on $L^p(\mathbb{R})$, $1 \le p < \infty$, if there is a constant c such that the operator $W^0(b)$, restricted to the dense subset $L^p(\mathbb{R}) \cap L^2(\mathbb{R})$ of $L^p(\mathbb{R})$, maps into $L^p(\mathbb{R}) \cap L^2(\mathbb{R})$ and it holds that

$$\|W^0(b)u\|_p \le c\|u\|_p \quad \text{for every} \quad u \in L^p(\mathbb{R}) \cap L^2(\mathbb{R}).$$

Then $W^0(b)$ extends to a bounded linear operator on $L^p(\mathbb{R})$, again denoted by $W^0(b)$ and again called the Fourier convolution operator with the symbol b. The set of all such multipliers on $L^p(\mathbb{R})$ is denoted by M^p and is known to be a Banach algebra when equipped with the norm $\|b\|_{\mathrm{M}^p} := \|W^0(b)\|_{\mathcal{L}(L^p(\mathbb{R}))}$.

Finally, since $L^2(\mathbb{R}) \cap L^\infty(\mathbb{R})$ is not dense in $L^\infty(\mathbb{R})$, we introduce the convolutions on $L^\infty(\mathbb{R})$ in a different but still simple way, namely as the adjoints $W^0(a) := (W^0(\bar{a}))^*$ of convolutions $W^0(\bar{a})$ on $L^1(\mathbb{R})$. Hörmander [18] already pointed out that both M^1 and M^∞ coincide with the set of all Fourier–Stieltjes transforms of bounded Borel measures, and both are proper subalgebras of M^p, $1 < p < \infty$. Actually, $\mathrm{M}^1 \subset \mathrm{BUC}$ (cf. [32, 1.3.3]). Moreover, all M^p, $p \in [1,\infty]$, are continuously embedded in $\mathrm{M}^2 = L^\infty(\mathbb{R})$.

Definition 1.2. Let $p \in (1,\infty) \setminus \{2\}$. By $\mathrm{M}^{\langle p \rangle}$ we denote the set of all multipliers $b \in \mathrm{M}^p$ for which there exists a $\delta > 0$ (depending on b) such that $b \in \mathrm{M}^r$ for all $r \in (p-\delta, p+\delta)$. If $p \in \{1,2,\infty\}$ then we set $\mathrm{M}^{\langle p \rangle} := \mathrm{M}^p$.
Now, for a subalgebra $\mathcal{B} \subset L^\infty(\mathbb{R})$ let $\mathcal{B}_p$ denote the closure of $\mathcal{B} \cap \mathrm{M}^{<p>}$ in M^p.

In Definition 1.4 we will introduce the announced algebra of convolution type operators which will cover all Fourier convolutions $W^0(b)$ with symbol functions $b \in [\mathrm{BUC} \cup \mathrm{QC}]_p$. This then covers much more, since

Lemma 1.3 (cf. [19, Theorem 4.2], [32, 1.3.3]). *Actually,* $[\mathrm{BUC} \cup \mathrm{QC}]_p$ *already includes* M^1, SO_p, $\mathrm{C}(\overline{\mathbb{R}})_p$, $\mathrm{C}(\dot{\mathbb{R}})_p$ *and* AP_p.

Also notice that convolution operators which arise from the generating functions $t \mapsto e^{ist}$ of AP actually coincide with the canonical shift operators

$$V_s : \mathcal{L}(L^p(\mathbb{R})) \to \mathcal{L}(L^p(\mathbb{R})), \quad u(x) \mapsto u(x-s),$$

for each $s \in \mathbb{R}$, respectively. Moreover, by the famous theorem of Hörmander [18], the Fourier convolution operators A on $L^p(\mathbb{R})$ are exactly the *shift invariant* operators on $L^p(\mathbb{R})$, i.e., $V_s A V_{-s} = A$ for every $s \in \mathbb{R}$.

1.3. Convolution type operators

Let $\mathcal{P} = (P_n)$ be the sequence of the canonical projections $P_n = \chi_{[-n,n]} I$ on $L^p(\mathbb{R})$.

Definition 1.4. Define the Banach operator algebra $\mathcal{A}$ as the smallest closed subalgebra of $\mathcal{L}(L^p(\mathbb{R}))$ which contains

- all operators aI of multiplication by functions $a \in [L_0^\infty \cup \mathrm{PC} \cup \mathrm{SO} \cup \mathrm{AP}]$,
- all convolution operators $W^0(b)$ with symbols $b \in [\mathrm{BUC} \cup \mathrm{QC}]_p$, and
- all operators $K \in \mathcal{L}(L^p(\mathbb{R}))$ with

$$\|K(I-P_n)\| + \|(I-P_n)K\| \to 0 \quad \text{as} \quad n \to \infty. \tag{1.1}$$

Such operators K having the property (1.1) are said to be $\mathcal{P}$-compact.

Also let $\mathcal{A}_0$ denote the closed subalgebra of $\mathcal{A}$ which is generated by the same operators K and $W^0(b)$, but only those aI with $a \in [L_0^\infty \cup \mathrm{PC}]$.

Note that for every operator $A \in \mathcal{L}(L^p(\mathbb{R}))$ and every $m \in \mathbb{N}$ the compression $P_m A P_m$ is $\mathcal{P}$-compact. Also all integral operators

$$(Hu)(x) = \int_{\mathbb{R}} h(x,y) u(y) dy, \quad x \in \mathbb{R},$$

with h a bounded measurable function with bounded support belong to this class. The key observation for the operators in $\mathcal{A}$ is the following asymptotic commutation relation w.r.t. inflated continuous functions:

Theorem 1.5. *Let* $\varphi \in \mathrm{C}(\overline{\mathbb{R}})$ *and, for every* $n \in \mathbb{N}$, *let* φ_n *denote its dilated copies* $\varphi_n(x) := \varphi(x/n)$. *Then, for every* $A \in \mathcal{A}$, *the commutator* $[A, \varphi_n I] := A\varphi_n I - \varphi_n A$ *goes to zero in the norm, i.e.,*

$$\|[A, \varphi_n I]\| \to 0 \quad \text{as} \quad n \to \infty. \tag{1.2}$$

Proof. Clearly, it suffices to consider the generators of $\mathcal{A}$. For operators aI of multiplication this is obvious. For $\mathcal{P}$-compact operators K the assertion is also almost obvious, since $\|K - P_m K P_m\| \leq \|K - KP_m\| + \|K - P_m K\| \|P_m\|$ tends to zero as m gets large and $\|(\varphi_n - \varphi(0))P_m\| = \|P_m(\varphi_n - \varphi(0))I\| \to 0$ as $n \to \infty$.

For symbol functions $b \in \mathrm{BUC} \cup \mathrm{QC}$ we conclude with [23, Corollary 8] and [22, Corollary 24] that $W^0(b)$ is a quasi-banded[2] operator on $L^2(\mathbb{R})$. Since, by [22, Theorem 11], the set $\mathcal{Q}_2$ of all quasi-banded operators over $L^2(\mathbb{R})$ is a Banach algebra, we find $W^0(b) \in \mathcal{Q}_2$ for all $b \in [\mathrm{BUC} \cup \mathrm{QC}]_2$. Finally, if $b \in [\mathrm{BUC} \cup \mathrm{QC}]_p$ then [22, Proposition 23] provides (by a Riesz–Thorin interpolation argument) that $W^0(b)$ is quasi-banded on $L^p(\mathbb{R})$.

Now the assertion (1.2) follows by [23, Theorem 9]. □

1.4. Wiener–Hopf and Hankel operators

Wiener–Hopf operators are analogues of the above convolution operators $W^0(b)$ on function spaces over the half-line, i.e., on $L^p([0,\infty))$, defined as

$$W(b) : L^p([0,\infty)) \to L^p([0,\infty)), \quad W(b)u := PW^0(b)u,$$

where P is the restriction operator $u \in L^p(\mathbb{R}) \mapsto u|_{[0,\infty)} \in L^p([0,\infty))$. Defining $\chi_+ := \chi_{[0,\infty)}$ and $\chi_- := \chi_{(-\infty,0]}$ we can identify $W(b)$ with $\chi_+ W^0(b)\chi_+ I$, hence the respective classes of Wiener–Hopf operators with $b \in [\mathrm{BUC} \cup \mathrm{QC}]_p$ can be found in $\mathcal{A}$ in the sense of this identification.

Definition 1.6. We let $\mathcal{B}$ denote the Banach algebra that is generated by all operators in $\mathcal{A}$ and the flip operator $J : L^p(\mathbb{R}) \to L^p(\mathbb{R})$, $(Ju)(t) = u(-t)$.

Setting $Q := I - P$ we get access to Hankel operators $H(b) := PW^0(b)QJ$ as well. Clearly, also these operators can be embedded into $\mathcal{B}$ in the sense of the identification of $H(b)$ with $\chi_+ W^0(b)\chi_- J$.

The standard text books [4, 6, 9, 10, 13, 14] and [19, 28, 30] offer an excellent gateway into the comprehensive and eclectic theory on invertibility, Fredholmness, approximation and factorization for various classes of such convolution and convolution type operators.

Throughout the subsequent sections the theory and the main results will be developed for the smaller algebra $\mathcal{A}$, since the required notations are simpler, the techniques are more straightforward, and the reader is probably more familiar with this setting than with the framework which is to be constructed for $\mathcal{B}$. However, once the concepts, results and proofs are available for $\mathcal{A}$, they can easily be modified and translated to the $\mathcal{B}$-case. This is shortly discussed in Section 6.

---

[2] A is quasi-banded iff $\lim_{m\to\infty} \sup_{n>0} \|(I - P_{n+m})AP_n\| = \lim_{m\to\infty} \sup_{n>0} \|P_n A(I - P_{n+m})\| = 0$.

2. The Finite Section Method and its applicability

We are now turning our attention to convolution operators over finite intervals, and we are going to study the connections between the infinite (over $\mathbb{R}$) and the finite (over intervals) case.

A common goal is to replace an operator $A \in \mathcal{L}(L^p(\mathbb{R}))$ by its compressions $A_n = P_n A P_n \in \mathcal{L}(\operatorname{im} P_n) = \mathcal{L}(L^p([-n,n]))$ and to ask what information about A can be obtained from the study of the operators of the sequence $\mathbb{A} = \{A_n\}_{n\in\mathbb{N}}$. One of the most prominent questions in this direction concerns with the (approximate) solution of an operator equation $Au = b$ on $L^p(\mathbb{R})$ by replacing it with the "simpler" truncated equations $A_n u_n = P_n b$ on the spaces $\operatorname{im} P_n = L^p([-n,n])$, hoping for the existence and the convergence of the solutions u_n of the substitutes $A_n u_n = P_n b$ to the solution u of the initial problem.

Here the convergence is to be understood in the sense of $\|P_m(u_n - u)\| \to 0$ as $n \to \infty$ for every fixed m. This coincides with the usual L^p-convergence if $p < \infty$, and offers the appropriate extension to the L^∞-case.

The above procedure which, due to its simplicity, is probably one of the most popular and most widely used approximation approaches, is called the *Finite Section Method.* It has been applied and extensively studied for many and large classes of operators on a variety of Hilbert and Banach spaces (see, e.g., the standard monographs [6, 7, 8, 13, 15, 21, 26, 30] and the references cited therein). It is well known, that the *applicability* of this method, i.e., the existence and convergence of the approximate solutions u_n of the finite section equations $A_n u_n = P_n b$ is not guaranteed, even if the full operator A is invertible. However, one typically obtains necessary and sufficient conditions for the applicability in terms of invertibility of A *and* certain additional operators.

Actually, such observations, despite being somewhat capricious and surprising at a first glance, appear to be very natural after a suitable change of perspective. For this let us start with the sequence $\{A_n\} = \{P_n A P_n\}$ of finite sections instead and then ask whether the limit[3] $A = \lim P_n A P_n$ carries (all) relevant information about the sequence. Obviously, a lot of the structure of the operators $A_n = P_n A P_n$ over the "interior of the intervals $[-n,n]$" carries over to this limit

$$A = \mathsf{W}^c\{A_n\} := \lim_{n\to\infty} A_n, \tag{2.1}$$

but the impact of the projections P_n gets completely lost. Therefore, in order to capture this impact at the "boundary points of the intervals $[-n,n]$", it is standing to reason to look at transformations/shifts of the A_n which focus on these critical points $\pm n$ and then pass to limits which display the behavior there:

$$\mathsf{W}^-\{A_n\} := \lim_{n\to\infty} V_n A_n V_{-n}, \quad \mathsf{W}^+\{A_n\} := \lim_{n\to\infty} V_{-n} A_n V_n. \tag{2.2}$$

Since each of these limit operators (if they exist) captures a certain facet of the sequence $\{A_n\}$ they are called *snapshots* of $\{A_n\}$.

[3]We specify in Definition 2.1 in what sense the limits can be understood.

After this brief introduction we are now going to clarify the details and to state the precise results about these finite section sequences $\{A_n\} = \{P_n A P_n\}$.

Definition 2.1. We say that a bounded sequence of operators $A_n \in \mathcal{L}(L^p(\mathbb{R}))$ converges $\mathcal{P}$-strongly to an operator A if, for every $m \in \mathbb{N}$,[4]

$$\|P_m(A_n - A)\| + \|(A_n - A)P_m\| \to 0 \quad \text{as} \quad n \to \infty.$$

From [22] we know that the "central snapshots" $\mathsf{W}^c\{P_n A P_n\} = A$ exist as $\mathcal{P}$-strong limits for every $A \in \mathcal{A}$. Looking at the generators of $\mathcal{A}_0$, we find that the $\mathsf{W}^{\pm}$-snapshots (2.2) of sequences $\{K\}$ with $\mathcal{P}$-compact K, of $\{W^0(b)\}$ with arbitrary multipliers b and of $\{aI\}$ with $a \in [L_0^\infty \cup \mathrm{PC}]$ obviously exist in this sense as well. For more general $\{aI\}$ this may fail. However we have that there are sufficiently many subsequences which have all snapshots. This is done as follows:

Definition 2.2. Let $\mathcal{F}$ be the Banach algebra of all bounded sequences $\mathbb{A} = \{A_n\}$ of bounded linear operators $A_n \in \mathcal{L}(L^p(\mathbb{R}))$, equipped with entry-wise defined linear operations and multiplication as well as the norm $\|\mathbb{A}\| = \|\{A_n\}\| = \sup_n \|A_n\|$.

Further, let $\mathcal{F}_\mathcal{A}$ denote the subalgebra of $\mathcal{F}$ which is generated by all finite section sequences $\{P_n A P_n + (I - P_n)\}$ of operators $A \in \mathcal{A}$. Obviously, the elements $\mathbb{A} = \{A_n\}$ of $\mathcal{F}_\mathcal{A}$ are of the form

$$\mathbb{A} = \{A_n\} = \{P_n A_n P_n\} + \mu_\mathbb{A}\{I - P_n\} \quad \text{with a certain} \quad \mu_\mathbb{A} \in \mathbb{C}.$$

Proposition 2.3 ([22, Sect. 6, 7]). *Let $\mathbb{A} = \{A_n\} \in \mathcal{F}_\mathcal{A} \cup \mathcal{A}$ and $g : \mathbb{N} \to \mathbb{N}$ be strictly increasing. Then there exists a subsequence h of g such that the snapshots*

$$\mathsf{W}^-(\mathbb{A}_h) = \mathcal{P}\text{-}\lim_{n\to\infty} V_{h_n} A_{h_n} V_{-h_n} \quad \textit{and} \quad \mathsf{W}^+(\mathbb{A}_h) = \mathcal{P}\text{-}\lim_{n\to\infty} V_{-h_n} A_{h_n} V_{h_n}$$

of the operator subsequence $\mathbb{A}_h := \{A_{h_n}\}$ exist.

Given $\mathbb{A}$ we denote by $\mathbf{H}_\mathbb{A}$ the set of all sequences $h : \mathbb{N} \to \mathbb{N}$ for which both snapshots $\mathsf{W}^{\pm}(\mathbb{A}_h)$ exist. Notice that for all sequences $\mathbb{A} \in \mathcal{F}_{\mathcal{A}_0}$ the snapshots $\mathsf{W}^{\pm}(\mathbb{A}_h)$ exist for arbitrary h and do not depend on h.

We come back to the question whether or when the Finite Section Method is applicable for the approximate solution of operator equations $Au = b$.

Theorem 2.4. *Let $\mathbb{A} = \{A_n\}$ be a sequence in the finite section algebra $\mathcal{F}_\mathcal{A}$, and set $A := \mathsf{W}^c(\mathbb{A})$. If A and all snapshots $\mathsf{W}^{\pm}(\mathbb{A}_h)$, $h \in \mathbf{H}_\mathbb{A}$, are invertible then, for sufficiently large n and every right-hand side $b \in L^p(\mathbb{R})$, the equations $A_n u_n = P_n b$ and $Au = b$ are uniquely solvable and the solutions $u_n = A_n^{-1} P_n b$ converge to the solution $u = A^{-1} b$.*

This result stands at the end of a series of developments which were made over decades and whose probably most important milestones are the deployment of the algebraic approach [13, 20, 24, 37, 29, 30] with the understanding of the shape, role and importance of the snapshots, the foundation of the $\mathcal{P}$-strong convergence [25, 26, 31, 35], and the introduction of the concept of rich sequences [21, 22, 34, 36].

[4] Profound references for the theory behind the $\mathcal{P}$-strong convergence are [21, 26, 35].

In its currently stated form it follows from [22, Theorem 17], and for $\mathcal{F}_{\mathcal{A}_0}$ it even simplifies:

Corollary 2.5. *Let* $\mathbb{A} = \{A_n\} \in \mathcal{F}_{\mathcal{A}_0}$, $A := \mathsf{W}^c(\mathbb{A})$. *If* A, $\mathsf{W}^-(\mathbb{A})$ *and* $\mathsf{W}^+(\mathbb{A})$ *are invertible then, for sufficiently large* n *and every* $b \in L^p(\mathbb{R})$, *the equations* $A_n u_n = P_n b$ *and* $Au = b$ *are uniquely solvable and the solutions* $u_n = A_n^{-1} P_n b$ *converge to the solution* $u = A^{-1} b$.

We point out that the finite section algebras $\mathcal{F}_{\mathcal{A}}$ and $\mathcal{F}_{\mathcal{A}_0}$ clearly contain the classical basic finite section sequences $\mathbb{A} = \{P_n A P_n + (I - P_n)\}$ of each single operator $A \in \mathcal{A}$ (resp. $A \in \mathcal{A}_0$) with $\mathsf{W}^c(\mathbb{A}) = A$. But working in these algebras additionally gives the freedom to, e.g., optionally choose $(P_n ABP_n)$ or $(P_n AP_n)(P_n BP_n)$ as approximations for an equation $ABu = b$.

3. Asymptotics of norms

3.1. The main theorem on convergence of norms

The primary contribution of the present paper consists in the extension of the qualitative applicability result in Theorem 2.4 to the following quantitative relation between the asymptotics of the finite sections and their respective snapshots. It generalizes the results of Böttcher, Grudsky, Silbermann [3] and Roch [27].

Theorem 3.1. *Let* $\mathbb{A} = \{A_n\} \in \mathcal{F}_{\mathcal{A}}$. *Then*[5]

$$\begin{aligned}
\limsup_{n\to\infty} \|A_n\| &= \max_{h\in\mathbf{H}_{\mathbb{A}}} \max\{\|\mathsf{W}^c(\mathbb{A}_h)\|, \|\mathsf{W}^-(\mathbb{A}_h)\|, \|\mathsf{W}^+(\mathbb{A}_h)\|\} \\
\liminf_{n\to\infty} \|A_n\| &= \min_{h\in\mathbf{H}_{\mathbb{A}}} \max\{\|\mathsf{W}^c(\mathbb{A}_h)\|, \|\mathsf{W}^-(\mathbb{A}_h)\|, \|\mathsf{W}^+(\mathbb{A}_h)\|\} \\
\limsup_{n\to\infty} \|A_n^{-1}\| &= \max_{h\in\mathbf{H}_{\mathbb{A}}} \max\{\|\mathsf{W}^c(\mathbb{A}_h)^{-1}\|, \|\mathsf{W}^-(\mathbb{A}_h)^{-1}\|, \|\mathsf{W}^+(\mathbb{A}_h)^{-1}\|\} \\
\liminf_{n\to\infty} \|A_n^{-1}\| &= \min_{h\in\mathbf{H}_{\mathbb{A}}} \max\{\|\mathsf{W}^c(\mathbb{A}_h)^{-1}\|, \|\mathsf{W}^-(\mathbb{A}_h)^{-1}\|, \|\mathsf{W}^+(\mathbb{A}_h)^{-1}\|\}.
\end{aligned}$$

If $g : \mathbb{N} \to \mathbb{N}$ *is strictly increasing, then the analogous assertions also hold for* $\mathbb{A}_g$, *where the* max *and* min *are taken over all* $h \in \mathbf{H}_{\mathbb{A}_g}$, *i.e., subsequences of* g.

The proof of this theorem is subject of Section 5. Here we just mention its obvious and beautiful simplification for the $\mathcal{A}_0$-case:

Corollary 3.2. *Let* $\mathbb{A} = \{A_n\} \in \mathcal{F}_{\mathcal{A}_0}$. *Then*

$$\begin{aligned}
\lim_{n\to\infty} \|A_n\| &= \max\{\|\mathsf{W}^c(\mathbb{A})\|, \|\mathsf{W}^-(\mathbb{A})\|, \|\mathsf{W}^+(\mathbb{A})\|\} \\
\lim_{n\to\infty} \|A_n^{-1}\| &= \max\{\|\mathsf{W}^c(\mathbb{A})^{-1}\|, \|\mathsf{W}^-(\mathbb{A})^{-1}\|, \|\mathsf{W}^+(\mathbb{A})^{-1}\|\}.
\end{aligned}$$

[5]Here and in what follows, define $\|B^{-1}\| := \infty$ if B is not invertible.

3.2. Convergence of condition numbers

Having Theorem 3.1 it becomes almost clear that the snapshots also store information about the asymptotics of the condition numbers $\operatorname{cond}(A_n) := \|A_n\|\|A_n^{-1}\|$:

Theorem 3.3. *Let $\mathbb{A} = \{A_n\} \in \mathcal{F}_{\mathcal{A}}$. Then*

$$\limsup_{n\to\infty} \operatorname{cond}(A_n) = \max_{h\in\mathbf{H}_{\mathbb{A}}} \Big(\max\{\|\mathsf{W}^c(\mathbb{A}_h)\|, \|\mathsf{W}^-(\mathbb{A}_h)\|, \|\mathsf{W}^+(\mathbb{A}_h)\|\} \cdot \max\{\|\mathsf{W}^c(\mathbb{A}_h)^{-1}\|, \|\mathsf{W}^-(\mathbb{A}_h)^{-1}\|, \|\mathsf{W}^+(\mathbb{A}_h)^{-1}\|\}\Big)$$

$$\liminf_{n\to\infty} \operatorname{cond}(A_n) = \min_{h\in\mathbf{H}_{\mathbb{A}}} \Big(\max\{\|\mathsf{W}^c(\mathbb{A}_h)\|, \|\mathsf{W}^-(\mathbb{A}_h)\|, \|\mathsf{W}^+(\mathbb{A}_h)\|\} \cdot \max\{\|\mathsf{W}^c(\mathbb{A}_h)^{-1}\|, \|\mathsf{W}^-(\mathbb{A}_h)^{-1}\|, \|\mathsf{W}^+(\mathbb{A}_h)^{-1}\|\}\Big).$$

Proof. Clearly, we can apply Theorem 3.1 to every $h \in \mathbf{H}_{\mathbb{A}}$ and obtain formulas for the limits $\lim \|A_{h_n}\|$ and $\lim \|A_{h_n}^{-1}\|$, hence also for $\lim \operatorname{cond}(A_{h_n})$.

Now we choose $g : \mathbb{N} \to \mathbb{N}$ to be a strictly increasing sequence which realizes $\limsup \operatorname{cond}(A_n) = \lim \operatorname{cond}(A_{g_n})$ (analogously for $\liminf$). By Proposition 2.3 there is a subsequence $h \in \mathbf{H}_{\mathbb{A}}$ of g which yields the asserted equality. □

Corollary 3.4. *Let $\mathbb{A} = \{A_n\} \in \mathcal{F}_{\mathcal{A}_0}$. Then*

$$\lim_{n\to\infty} \operatorname{cond}(A_n) = \max\{\|\mathsf{W}^c(\mathbb{A})\|, \|\mathsf{W}^-(\mathbb{A})\|, \|\mathsf{W}^+(\mathbb{A})\|\} \cdot \max\{\|\mathsf{W}^c(\mathbb{A})^{-1}\|, \|\mathsf{W}^-(\mathbb{A})^{-1}\|, \|\mathsf{W}^+(\mathbb{A})^{-1}\|\}.$$

4. Asymptotics of pseudospectra

Although we have the above strong relations between the operators A_n of a sequence $\mathbb{A} = \{A_n\} \in \mathcal{F}_{\mathcal{A}}$ and the family of its snapshots $\mathsf{W}^t(\mathbb{A}_h)$, $t \in \{c, -, +\}$, one cannot expect to observe convergence of the spectra $\operatorname{sp}(A_n)$ to a limiting set which is determined by the snapshots. As a basic example one may consider the shift operator $A = V_1$ or its Wiener–Hopf counterpart $B = \chi_+ A \chi_+ I$ which have the unit circle and the unit disc as spectrum, respectively, but their finite sections are nilpotent hence have spectrum $\{0\}$. For further examples one may consult [2].

However, in this pioneering paper [2], Böttcher demonstrates that the situation drastically changes and simplifies if one passes to pseudospectra instead:

Definition 4.1. For $\epsilon > 0$, the ϵ-pseudospectrum $\operatorname{sp}_\epsilon(A)$ of an operator A is defined as the set[6]

$$\operatorname{sp}_\epsilon(A) := \left\{ z \in \mathbb{C} : \|(A - zI)^{-1}\| \geq \frac{1}{\epsilon} \right\}.$$

Let $M_1, M_2, \dots$ be a sequence of non-empty subsets of $\mathbb{C}$. The uniform (and the partial) limiting set

$$u\text{-}\lim_{n\to\infty} M_n \quad \left(p\text{-}\lim_{n\to\infty} M_n \right)$$

[6]Here we again use the notation $\|B^{-1}\| := \infty$ if B is not invertible.

of this sequence is the set of all $z \in \mathbb{C}$ that are limits (resp. partial limits) of a sequence (z_n) with $z_n \in M_n$. These limiting sets are closed, as is discussed in [15, Proposition 3.2]. Moreover, [15, Proposition 3.6] states that for compact sets M_n the limits $u\text{-lim}\, M_n$ and $p\text{-lim}\, M_n$ coincide if and only if the M_n converge w.r.t. the Hausdorff distance (to the same limiting set).[7] For this, the Hausdorff distance of two compact sets $S, T \subset \mathbb{C}$ is

$$d_H(S,T) := \max\left\{\max_{s\in S} \operatorname{dist}(s,T), \max_{t\in T} \operatorname{dist}(t,S)\right\}, \quad \text{where } \operatorname{dist}(s,T) := \min_{t\in T} |s-t|,$$

and actually defines a metric on the set of all non-empty compact subsets of $\mathbb{C}$.

Theorem 4.2. *Let $\mathbb{A} = \{A_n\} \in \mathcal{F}_{\mathcal{A}}$, $h \in \mathbf{H}_{\mathbb{A}}$ and $\epsilon > 0$. Then*

$$u\text{-}\lim_{n\to\infty} \operatorname{sp}_\epsilon A_{h_n} = p\text{-}\lim_{n\to\infty} \operatorname{sp}_\epsilon A_{h_n} = \operatorname{sp}_\epsilon \mathsf{W}^c(\mathbb{A}_h) \cup \operatorname{sp}_\epsilon \mathsf{W}^-(\mathbb{A}_h) \cup \operatorname{sp}_\epsilon \mathsf{W}^+(\mathbb{A}_h).$$

Corollary 4.3. *Let $\mathbb{A} = \{A_n\} \in \mathcal{F}_{\mathcal{A}}$, and $\epsilon > 0$. Then (applying Proposition* 2.3)

$$p\text{-}\lim_{n\to\infty} \operatorname{sp}_\epsilon A_n = \bigcup_{h\in\mathbf{H}_{\mathbb{A}}} \left(\operatorname{sp}_\epsilon \mathsf{W}^c(\mathbb{A}_h) \cup \operatorname{sp}_\epsilon \mathsf{W}^-(\mathbb{A}_h) \cup \operatorname{sp}_\epsilon \mathsf{W}^+(\mathbb{A}_h)\right).$$

Corollary 4.4. *Let $\mathbb{A} = \{A_n\} \in \mathcal{F}_{\mathcal{A}_0}$, and $\epsilon > 0$. Then*

$$u\text{-}\lim_{n\to\infty} \operatorname{sp}_\epsilon A_n = p\text{-}\lim_{n\to\infty} \operatorname{sp}_\epsilon A_n = \operatorname{sp}_\epsilon \mathsf{W}^c(\mathbb{A}) \cup \operatorname{sp}_\epsilon \mathsf{W}^-(\mathbb{A}) \cup \operatorname{sp}_\epsilon \mathsf{W}^+(\mathbb{A}).$$

The proof of this theorem is based on the fact that the norm of the resolvent $z \mapsto \|(A - zI)^{-1}\|$ is nowhere locally constant on the resolvent set. This was firstly discovered by Globevnik [11, 12] in the 70s, unfortunately unnoticed by the operator theory community, and later on developed again by Böttcher and Daniluk [2, 3] in the 90s as a part of the proof of Theorem 4.2 on the asymptotics of pseudospectra. The precise result, taken in its final form from Shargorodsky [38, Theorem 2.5], reads as follows.

Theorem 4.5. *Let Ω be an open subset of $\mathbb{C}$, let $p \in [1,\infty]$, and let $A \in \mathcal{L}(L^p(\mathbb{R}))$ be such that $A - zI$ is invertible for all $z \in \Omega$. If $\|(A - zI)^{-1}\| \le M$ for all $z \in \Omega$ then $\|(A - zI)^{-1}\| < M$ for all $z \in \Omega$.*

Proof of Theorem 4.2. This is an adaption of the idea which has already been applied, e.g., in [7, Theorem 3.17], [15, Theorem 3.31], [36, Theorem 2.14].

Suppose $z \in \operatorname{sp} \mathsf{W}^t(\mathbb{A}_h)$ for some $t \in T := \{c, -, +\}$, that is $\mathsf{W}^t(\mathbb{A}_h - z\mathbb{I}_h)$ is not invertible. Then Theorem 3.1 yields that $\lim \|(A_{h_n} - zI)^{-1}\| = \infty$ which implies $z \in \operatorname{sp}_\epsilon A_{h_n}$ for all sufficiently large n. Thus $z \in u\text{-}\lim \operatorname{sp}_\epsilon A_{h_n}$.

Next, suppose that $\lim \|(A_{h_n} - zI)^{-1}\| < \infty$, but $z \in \operatorname{sp}_\epsilon \mathsf{W}^t(\mathbb{A}_h)$ for a t, which means that $\|(\mathsf{W}^t(\mathbb{A}_h) - zI)^{-1}\| \ge 1/\epsilon$. Let U be an arbitrary open ball around z such that $\mathsf{W}^t(\mathbb{A}_h) - yI$ is invertible for all $y \in U$. If $\|(\mathsf{W}^t(\mathbb{A}_h) - yI)^{-1}\|$ would be less than or equal to $1/\epsilon$ for every $y \in U$ then Theorem 4.5 would imply

[7] This observation is due to Hausdorff and can be found in his book [17] as well.

that $\|(\mathsf{W}^t(\mathbb{A}_h) - zI)^{-1}\| < 1/\epsilon$, a contradiction. Hence there is a $y \in U$ such that $\|(\mathsf{W}^t(\mathbb{A}_h) - yI)^{-1}\| > 1/\epsilon$, that is we can find a k_0 such that

$$\|(\mathsf{W}^t(\mathbb{A}_h) - yI)^{-1}\| > \left(\epsilon - \frac{1}{k}\right)^{-1} \quad \text{for all} \quad k \geq k_0.$$

Because U was arbitrary we can choose a sequence $(z_m)_{m\in\mathbb{N}}$ of complex numbers $z_m \in \mathrm{sp}_{\epsilon-1/m}\,\mathsf{W}^t(\mathbb{A}_h)$ such that $z_m \to z$. Since $\lim_n \|(A_{h_n} - z_m I)^{-1}\|$ exists and equals $\max_{t\in T} \|(\mathsf{W}^t(\mathbb{A}_h) - z_m I)^{-1}\|$, due to Theorem 3.1, it is greater than or equal to $(\epsilon - 1/m)^{-1}$. Consequently, for sufficiently large n, $\|(A_{h_n} - z_m I)^{-1}\| \geq 1/\epsilon$ and thus $z_m \in \mathrm{sp}_\epsilon A_{h_n}$. This shows that $z = \lim_m z_m$ belongs to the closed set $u\text{-}\lim_n \mathrm{sp}_\epsilon A_{h_n}$.

Finally consider the case that $\|(\mathsf{W}^t(\mathbb{A}_h) - zI)^{-1}\| < 1/\epsilon$ for all $t \in T$. Then

$$\lim_{n\to\infty} \|(A_{h_n} - zI)^{-1}\| = \max_{t\in T} \|(\mathsf{W}^t(\mathbb{A}_h) - zI)^{-1}\| < 1/\epsilon,$$

hence there are a $\delta > 0$ and an $n_0 \in \mathbb{N}$ such that $\|(A_{h_n} - zI)^{-1}\| \leq 1/\epsilon - \delta$ for all $n \geq n_0$. If $|y - z|$ is sufficiently small and $n \geq n_0$ then we have

$$\begin{aligned}
\|(A_{h_n} - yI)^{-1}\| &= \|((A_{h_n} - zI)(I + (z-y)(A_{h_n} - zI)^{-1}))^{-1}\| \\
&= \|(I + (z-y)(A_{h_n} - zI)^{-1})^{-1}(A_{h_n} - zI)^{-1}\| \\
&\leq \frac{\|(A_{h_n} - zI)^{-1}\|}{1 - |z-y|\|(A_{h_n} - zI)^{-1}\|} \\
&\leq \frac{1/\epsilon - \delta}{1 - |z-y|\|(A_{h_n} - zI)^{-1}\|} < 1/\epsilon.
\end{aligned}$$

Thus, $z \notin p\text{-}\lim \mathrm{sp}_\epsilon A_{h_n} \supset u\text{-}\lim \mathrm{sp}_\epsilon A_{h_n}$. This completes the proof. □

5. Norm preserving localization and the proof of the main result

Within this section we prove the Main Theorem 3.1 based on the concept of norm preserving localization in KMS-algebras which was developed by Böttcher, Krupnik, Silbermann [5].

We recall the algebra $\mathcal{F}$ of all bounded sequences $\mathbb{A} = \{A_n\}$ of bounded linear operators $A_n \in \mathcal{L}(L^p(\mathbb{R}))$ from Definition 2.2 and note that the subset

$$\mathcal{G} := \{\{G_n\} \in \mathcal{F} : \|G_n\| \to 0 \text{ as } n \to \infty\}$$

is a closed two-sided ideal in $\mathcal{F}$. For brevity we denote the quotient algebra as $\mathcal{F}^\mathcal{G} := \mathcal{F}/\mathcal{G}$ and the cosets as $\mathbb{A}^\mathcal{G} := \mathbb{A} + \mathcal{G}$ for $\mathbb{A} \in \mathcal{F}$, and further notice that the norm $\|\mathbb{A}^\mathcal{G}\| = \|\mathbb{A} + \mathcal{G}\| := \inf\{\|\mathbb{A} + \mathbb{G}\| : \mathbb{G} \in \mathcal{G}\}$ equals $\limsup \|A_n\|$.

Define $\mathcal{C} := \{\{\varphi_n I\} : \varphi \in \mathrm{C}(\dot{\mathbb{R}})\}$, where φ_n still denote the dilated copies $\varphi_n(x) := \varphi(x/n)$ of φ. Clearly, $\mathcal{C}^\mathcal{G} := (\mathcal{C} + \mathcal{G})/\mathcal{G} = \{\{\varphi_n I\} + \mathcal{G} : \varphi \in \mathrm{C}(\dot{\mathbb{R}})\}$ is a closed commutative subalgebra of $\mathcal{F}^\mathcal{G}$. The set

$$\mathcal{L} := \{\mathbb{A} \in \mathcal{F} : [\mathbb{A}, \mathbb{C}] := \mathbb{A}\mathbb{C} - \mathbb{C}\mathbb{A} \in \mathcal{G} \text{ for all } \mathbb{C} \in \mathcal{C}\}$$

forms a Banach subalgebra of $\mathcal{F}$. Writing $\mathcal{L}^{\mathcal{G}} := \mathcal{L}/\mathcal{G}$, we observe that all cosets $\mathbb{A}^{\mathcal{G}} \in \mathcal{L}^{\mathcal{G}}$ commute with all $\mathbb{C}^{\mathcal{G}} \in \mathcal{C}^{\mathcal{G}}$. If $\mathbb{A} \in \mathcal{L}$, $\mathbb{B} \in \mathcal{F}$ and $\mathbb{AB} - \mathbb{I}, \mathbb{BA} - \mathbb{I} \in \mathcal{G}$ then $\mathbb{B} \in \mathcal{L}$, i.e., $\mathcal{L}^{\mathcal{G}}$ is inverse closed in $\mathcal{F}^{\mathcal{G}}$. Indeed, this is clear since, for every $\mathbb{C} \in \mathcal{C}$,

$$\mathbb{B}^{\mathcal{G}}\mathbb{C}^{\mathcal{G}} = \mathbb{B}^{\mathcal{G}}\mathbb{C}^{\mathcal{G}}\mathbb{A}^{\mathcal{G}}\mathbb{B}^{\mathcal{G}} = \mathbb{B}^{\mathcal{G}}\mathbb{A}^{\mathcal{G}}\mathbb{C}^{\mathcal{G}}\mathbb{B}^{\mathcal{G}} = \mathbb{C}^{\mathcal{G}}\mathbb{B}^{\mathcal{G}}.$$

The Banach algebra $\mathcal{C}^{\mathcal{G}}$ even turns into a C^*-algebra when equipped with the involution which is induced by the mapping $\{\varphi_n I\} \mapsto \{\overline{\varphi_n} I\}$. Moreover it is isometrically $*$-isomorphic to $\mathrm{C}(\dot{\mathbb{R}})$ hence its maximal ideal space can be identified with the one-point compactification $\dot{\mathbb{R}} = \mathbb{R} \cup \{\infty\}$ of the real line. For each maximal ideal x in $\mathcal{C}^{\mathcal{G}}$ we introduce $\mathcal{J}_x$ as the smallest closed ideal in $\mathcal{L}^{\mathcal{G}}$ containing x, and let $\Phi_x : \mathcal{L}^{\mathcal{G}} \to \mathcal{L}^{\mathcal{G}}/\mathcal{J}_x$ denote the canonical homomorphism, respectively.

By this construction we can decompose each $\mathbb{A}^{\mathcal{G}} \in \mathcal{L}^{\mathcal{G}}$ into a family of local cosets $\Phi_x(\mathbb{A}^{\mathcal{G}})$, with $x \in \dot{\mathbb{R}}$ in the sense of the above-mentioned identification. Therefore the elements of $\mathcal{L}$ are said to be *localizable.* Notice that all the above definitions and observations can be made for subsequences $\mathbb{A}_g \in \mathcal{F}_g$ in an analogous way as well: $\mathbb{C}_g^{\mathcal{G}_g}$ is a C^*-algebra, its maximal ideal space is homeomorphic to $\dot{\mathbb{R}}$, and for the cosets $\mathbb{A}_g^{\mathcal{G}_g} \in \mathcal{L}_g^{\mathcal{G}_g}$ there is a local decomposition via analogously defined ideals $\mathcal{J}_x$ in $\mathcal{L}_g^{\mathcal{G}_g}$ and canonical homomorphisms $\Phi_x : \mathcal{L}_g^{\mathcal{G}_g} \to \mathcal{L}_g^{\mathcal{G}_g}/\mathcal{J}_x$. The great achievements of [5] for such a setting are summarized in the next theorem:

Theorem 5.1. *For every $\mathbb{A} = \{A_n\} \in \mathcal{L}$ and every strictly increasing $g : \mathbb{N} \to \mathbb{N}$*

$$\limsup_{n\to\infty} \|A_{g_n}\| = \|\mathbb{A}_g^{\mathcal{G}_g}\| = \max_{x\in\dot{\mathbb{R}}} \|\Phi_x(\mathbb{A}_g^{\mathcal{G}_g})\|, \tag{5.1}$$

where $$\|\Phi_x(\mathbb{A}_g^{\mathcal{G}_g})\| = \inf\Big\{ \big\|\{\varphi_n I\}_g^{\mathcal{G}_g}\mathbb{A}_g^{\mathcal{G}_g}\big\| : \varphi \in \mathrm{C}(\dot{\mathbb{R}}), 0 \le \varphi \le 1, \varphi \equiv 1 \text{ in a neighborhood of } x\Big\}. \tag{5.2}$$

Proof. Let $\varphi, \psi \in \mathrm{C}(\dot{\mathbb{R}})$ be functions with disjoint supports, i.e., $\varphi\psi \equiv 0$. Then, obviously, $\|(\varphi_n + \psi_n)A_n\| \le \max\{\|\varphi_n A_n\|, \|\psi_n A_n\|\}$ holds for every $n \in \mathbb{N}$, hence

$$\big\|\{(\varphi_n + \psi_n)I\}_g^{\mathcal{G}_g}\mathbb{A}_g^{\mathcal{G}_g}\big\| \le \max\left\{\big\|\{\varphi_n I\}_g^{\mathcal{G}_g}\mathbb{A}_g^{\mathcal{G}_g}\big\|, \big\|\{\psi_n I\}_g^{\mathcal{G}_g}\mathbb{A}_g^{\mathcal{G}_g}\big\|\right\}.$$

Thus, in terms of [5, Definition 5.2], $\mathcal{L}_g^{\mathcal{G}_g}$ is a so-called KMS-algebra with respect to $\mathcal{C}_g^{\mathcal{G}_g}$, and the main result [5, Theorem 5.3] yields the Equality (5.1) for $\mathbb{A}_g$. The Formula (5.2) for the local norm $\|\Phi_x(\mathbb{A}_g^{\mathcal{G}_g})\|$ holds by [5, Proposition 5.1]. □

Now we turn our attention to the finite section algebra $\mathcal{F}_{\mathcal{A}}$. As an immediate consequence of Theorem 1.5 and the obvious fact that $\{P_n\} \in \mathcal{L}$ we get $\mathcal{F}_{\mathcal{A}} \subset \mathcal{L}$.

The great advantages of the more particular sequences $\mathbb{A} \in \mathcal{F}_{\mathcal{A}}$ are that, by Proposition 2.3, there are "many" subsequences $\mathbb{A}_h$, $h \in \mathbf{H}_{\mathbb{A}}$, having the three snapshots $\mathsf{W}^t(\mathbb{A}_h)$, $t \in \{c, -, +\}$, and that these snapshots already determine the respective local cosets as the following lemma shows:

Lemma 5.2. *Let* $\mathbb{A} = \{A_n\} = \{P_nA_nP_n + \mu_{\mathbb{A}}(I - P_n)\} \in \mathcal{F}_{\mathcal{A}}$ *and* $h \in \mathbf{H}_{\mathbb{A}}$. *Then*

$$\Phi_x(\mathbb{A}_h^{\mathcal{G}_h}) = \begin{cases} \Phi_x(\{\mathsf{W}^c(\mathbb{A}_h)\}_h^{\mathcal{G}_h}) & : |x| < 1 \\ \Phi_x(\{V_{\pm n}\mathsf{W}^{\pm}(\mathbb{A}_h)V_{\mp n}\}_h^{\mathcal{G}_h}) & : x = \pm 1 \\ \Phi_x(\{\mu_{\mathbb{A}}I\}_h^{\mathcal{G}_h}) & : |x| > 1 \end{cases}.$$

If the three snapshots are invertible then $\mathbb{A}_h^{\mathcal{G}_h}$ *is invertible in* $\mathcal{L}_h^{\mathcal{G}_h}$ *and*

$$\Phi_x((\mathbb{A}_h^{\mathcal{G}_h})^{-1}) = \begin{cases} \Phi_x(\{\mathsf{W}^c(\mathbb{A}_h)^{-1}\}_h^{\mathcal{G}_h}) & : |x| < 1 \\ \Phi_x(\{V_{\pm n}\mathsf{W}^{\pm}(\mathbb{A}_h)^{-1}V_{\mp n}\}_h^{\mathcal{G}_h}) & : x = \pm 1 \\ \Phi_x(\{\mu_{\mathbb{A}}^{-1}I\}_h^{\mathcal{G}_h}) & : |x| > 1 \end{cases}.$$

Proof. Let $|x| < 1$ and choose $\varphi \in \mathrm{C}(\dot{\mathbb{R}})$, $0 \le \varphi \le 1$, $\operatorname{supp}\varphi \subset (-1, 1)$ and $\varphi \equiv 1$ in a neighborhood of x. Then $\{\varphi_n(A_n - \mathsf{W}^c(\mathbb{A}))\} \in \mathcal{G}$. This is clear for all sequences $\{B\}$ with B being a generator of $\mathcal{A}$ (see Definition 1.4) as well as for the sequence $\{P_n\}$, hence for all finite sums and products of these generators, and then follows for $\mathbb{A}$ by an approximation argument. Thus, for the subsequences with respect to h, with $\Phi_x(\{\varphi_nI\}_h^{\mathcal{G}_h}) = \Phi_x(\{I\}_h^{\mathcal{G}_h})$,

$$\Phi_x(\mathbb{A}_h^{\mathcal{G}_h}) = \Phi_x(\{\varphi_nA_n\}_h^{\mathcal{G}_h}) = \Phi_x(\{\varphi_n\mathsf{W}^c(\mathbb{A})\}_h^{\mathcal{G}_h}) = \Phi_x(\{\mathsf{W}^c(\mathbb{A})\}_h^{\mathcal{G}_h}).$$

Next, let $x = 1$ and $\varphi \in \mathrm{C}(\dot{\mathbb{R}})$, $0 \le \varphi \le 1$, $\operatorname{supp}\varphi \subset [1/2, 2]$ and $\varphi \equiv 1$ in a neighborhood of x. Then $\{\varphi_n(A_n - V_n\mathsf{W}^+(\mathbb{A}_h)V_{-n})\}_h \in \mathcal{G}_h$, hence the asserted representation of $\Phi_1(\mathbb{A}_h^{\mathcal{G}_h})$ holds again. To show this, just assume the contrary, i.e., that there exists a subsequence g of h and a constant $c > 0$ such that, for all n, $\|\varphi_{g_n}(A_{g_n} - V_{g_n}\mathsf{W}^+(\mathbb{A}_h)V_{-g_n})\| > c$. Now, choose a sequence $(\mathbb{A}^{(m)})_{m\in\mathbb{N}}$ of sequences $\mathbb{A}^{(m)} \in \mathcal{F}_{\mathcal{A}}$ being finite sums and products consisting of constant sequences $\{B_n\} = \{B\}$ of generators of $\mathcal{A}$ and $\{P_n\}$, and such that $\mathbb{A}^{(m)} \to \mathbb{A}$ in $\mathcal{F}_{\mathcal{A}}$. Applying Proposition 2.3 repeatedly we can pass to a subsequence g^1 of g such that the snapshots of each summand and factor in $\mathbb{A}_{g^1}^{(1)}$, hence the snapshots of $\mathbb{A}_{g^1}^{(1)}$ itself, exist. Iterating this argument, we can pass to a subsequence $g^2 \subset g^1$ of g^1 such that the snapshots of $\mathbb{A}_{g^2}^{(2)}$ exist as well, and so on. Thus, we arrive at $g^1 \supset g^2 \supset g^3 \supset \cdots$, define a "diagonal sequence" $\tilde{h} = (\tilde{h}_n)$ by $\tilde{h}_n := g_n^n$ and observe that all $\mathsf{W}^t(\mathbb{A}_{\tilde{h}}^{(m)})$ with $t \in \{c, -, +\}$ and $m \in \mathbb{N}$ exist. The relation

$$\{\varphi_{\tilde{h}_n}(B_{\tilde{h}_n} - V_{\tilde{h}_n}\mathsf{W}^+\{B_{\tilde{h}_n}\}V_{-\tilde{h}_n})\} \in \mathcal{G}_{\tilde{h}}$$

holds for all the basic sequences $\{B_{\tilde{h}_n}\}$ that we met during the above construction: it is clear for convolutions and $\mathcal{P}$-compact operators and easily checked for the multiplication operators among the generators in Definition 1.4. Hence it also holds for all $\mathbb{A}_{\tilde{h}}^{(m)}$, and it finally follows for $\mathbb{A}_{\tilde{h}}$ by an approximation argument. This contradicts our assumption.

For $x = -1$ we proceed similarly using $\mathsf{W}^-(\mathbb{A}_h)$ instead. In case $|x| > 1$ just choose φ with $\operatorname{supp}\varphi \subset \dot{\mathbb{R}} \setminus [-1, 1]$ appropriately and notice that $\varphi_nA_n = \varphi_n\mu_{\mathbb{A}}I$.

Finally, let all the snapshots $\mathsf{W}^t(\mathbb{A}_h)$, $t \in \{c, -, +\}$, be invertible. Then the $\{\mathsf{W}^t(\mathbb{A}_h)^{-1}\}^{\mathcal{G}}$ are the inverses of $\{\mathsf{W}^t(\mathbb{A}_h)\}^{\mathcal{G}}$ in the (inverse closed) Banach algebra $\mathcal{L}^{\mathcal{G}}$. Thus all local cosets $\Phi_x(\mathbb{A}_h^{\mathcal{G}_h})$ of $\mathbb{A}_h^{\mathcal{G}_h}$ which, by the above, coincide with $\Phi_x(\{\mathsf{W}^c(\mathbb{A}_h)\}_h^{\mathcal{G}_h})$, $\Phi_x(\{V_{\pm n}\mathsf{W}^{\pm}(\mathbb{A}_h)V_{\mp n}\}_h^{\mathcal{G}_h})$ or $\Phi_x(\{\mu_{\mathbb{A}}I\}_h^{\mathcal{G}_h})$, depending on $|x| < 1$, $|x| = 1$ or $|x| \geq 1$, are invertible. The local principle of Allan/Douglas [1] (see also [5, Theorem 3.2]) provides the invertibility of $\mathbb{A}_h^{\mathcal{G}_h}$ in $\mathcal{L}_h^{\mathcal{G}_h}$. By their uniqueness, the local inverses $(\Phi_x(\mathbb{A}_h^{\mathcal{G}_h}))^{-1} = \Phi_x((\mathbb{A}_h^{\mathcal{G}_h})^{-1})$ coincide with $\Phi_x(\{\mathsf{W}^c(\mathbb{A}_h)^{-1}\}_h^{\mathcal{G}_h})$, $\Phi_x(\{V_{\pm n}\mathsf{W}^{\pm}(\mathbb{A}_h)^{-1}V_{\mp n}\}_h^{\mathcal{G}_h})$ or $\Phi_x(\{\mu_{\mathbb{A}}^{-1}I\}_h^{\mathcal{G}_h})$, depending on x, respectively. □

Proposition 5.3. *Let* $\mathbb{A} = \{A_n\} = \{P_nA_nP_n + \mu_{\mathbb{A}}(I - P_n)\} \in \mathcal{F}_{\mathcal{A}}$ *and* $h \in \mathbf{H}_{\mathbb{A}}$. *Then*

$$\lim_{n\to\infty} \|A_{h_n}\| = \max\{\|\mathsf{W}^c(\mathbb{A}_h)\|, \|\mathsf{W}^-(\mathbb{A}_h)\|, \|\mathsf{W}^+(\mathbb{A}_h)\|\}$$
$$\lim_{n\to\infty} \|A_{h_n}^{-1}\| = \max\{\|\mathsf{W}^c(\mathbb{A}_h)^{-1}\|, \|\mathsf{W}^-(\mathbb{A}_h)^{-1}\|, \|\mathsf{W}^+(\mathbb{A}_h)^{-1}\|\}.$$

Proof. The norm of each snapshot $\mathsf{W}^t(\mathbb{A}_h)$ is at most $\liminf \|A_{h_n}\|$ by [26, Proposition 1.1.17]. Thus, taking Theorem 5.1 into account,

$$\max\{\|\mathsf{W}^c(\mathbb{A}_h)\|, \|\mathsf{W}^-(\mathbb{A}_h)\|, \|\mathsf{W}^+(\mathbb{A}_h)\|\} \leq \limsup_{n\to\infty} \|A_{h_n}\| = \max_{x\in\dot{\mathbb{R}}} \|\Phi_x(\mathbb{A}_h^{\mathcal{G}_h})\|.$$

Furthermore, estimating the local norms by the previous lemma and using that, e.g., $\|\Phi_x(\{\mu_{\mathbb{A}}I\}_h^{\mathcal{G}_h})\| \leq |\mu_{\mathbb{A}}| \leq \|\chi_-\mathsf{W}^+(\mathbb{A}_h)\chi_-I + \mu_{\mathbb{A}}\chi_+I\| = \|\mathsf{W}^+(\mathbb{A}_h)\|$, we get the converse estimate. Moreover, both estimates remain the same if h is replaced by any of its subsequences, hence the lim sup is even a lim, which finishes the proof of the first assertion.

Assume now that the three snapshots are invertible. Then $\mathbb{A}_h^{\mathcal{G}_h}$ is invertible in $\mathcal{L}_h^{\mathcal{G}_h}$ by the previous lemma. [35, Theorem 2.21] shows that every representative of its inverse actually possesses all snapshots which are just the inverses of the snapshots of $\mathbb{A}_h$. In particular, there is a representative $\{B_{h_n}\} \in (\mathbb{A}_h^{\mathcal{G}_h})^{-1}$ and a number n_0 such that $B_{h_n} = A_{h_n}^{-1}$ for all $n \geq n_0$. Therefore the same arguments as above also yield the limit $\lim \|A_{h_n}^{-1}\|$.

Finally assume that one snapshot is not invertible. Then neither $\mathbb{A}_h$ nor any of its subsequences can be invertible modulo $\mathcal{G}$ (again by [35, Theorem 2.21 or 2.6]), thus $\lim \|A_{h_n}^{-1}\| = \infty$. □

Having this proposition, the *proof of Theorem* 3.1 is now immediate: Clearly, the $\limsup \|A_n\|$ is not smaller than the supremum over all $\lim \|A_{h_n}\|$ with $h \in \mathbf{H}_{\mathbb{A}}$. On the other hand, one can choose a subsequence $\mathbb{A}_g$ which realizes the lim sup, and then pass to a subsequence $\mathbb{A}_h$ of $\mathbb{A}_g$ by Proposition 2.3 in order to get equality and to see that the supremum is a maximum. The remaining assertions are analogous. □

6. On some generalizations

Remark 6.1. Notice that all previous results of this paper remain true in a larger and more abstract setting: Let $\mathcal{Q}$ stand for the set of all rich quasi-banded operators, i.e., A with $\lim_{m\to\infty}\sup_{n>0}\|(I-P_{n+m})AP_n\| = \lim_{m\to\infty}\sup_{n>0}\|P_nA(I-P_{n+m})\| = 0$ and the property that every strictly increasing $g:\mathbb{N}\to\mathbb{N}$ has a subsequence h such that the three W-snapshots of $\{A\}_h$ exist. Then $\mathcal{F}_{\mathcal{A}}$ can be replaced by $\mathcal{F}_{\mathcal{Q}}$. The point is that the commutation relation (1.2) is still true by [23, Theorem 9], hence $\mathcal{F}_{\mathcal{Q}} \subset \mathcal{L}$ where the localization is done after suitable modifications in the definition of the algebra $\mathcal{C}$ similar to [36, Section 2.2].

Besides that, analogues of the observations in Section 4 can also be derived for the (N,ϵ)-pseudospectra which were introduced by Hansen [16] (see [33, 36]).

The flip. Another more particular and highly interesting generalization $\mathcal{B}$ was already announced in Definition 1.6: If one could include the flip operator $(Ju)(t) = u(-t)$, this would give access to Hankel operators $H(b) := \chi_+W^0(b)\chi_-J$ as well. Unfortunately, this cannot be done directly, since neither $\{J\}$ nor any of its subsequences possess $\mathsf{W}^{\pm}$-snapshots. However, there is an elegant solution in [22, 23] which replaces these two snapshots by one operator matrix-valued snapshot W^*, such that

$$\mathsf{W}^*\{J\} = \begin{pmatrix} 0 & J \\ J & 0 \end{pmatrix}, \quad \text{and} \quad \mathsf{W}^*(\mathbb{A}_h) = \begin{pmatrix} \mathsf{W}^+(\mathbb{A}_h) & 0 \\ 0 & \mathsf{W}^-(\mathbb{A}_h) \end{pmatrix} \tag{6.1}$$

for every $\mathbb{A} \in \mathcal{A}$ and every $h \in \mathbf{H}_{\mathbb{A}}$. This is done as follows:

1st step: Given $\mathbb{A} = \{A_n\}$, the operators

$$\tilde{A}_n := \begin{pmatrix} A_n & 0 \\ 0 & JA_nJ \end{pmatrix} \in \mathcal{L}(L^p(\mathbb{R}) \times L^p(\mathbb{R})),$$

where $L^p(\mathbb{R}) \times L^p(\mathbb{R})$ is equipped with the norm $\|(f,g)\|_p = \sqrt[p]{\|f\|_p^p + \|g\|_p^p}$ if $p<\infty$ and $\|(f,g)\|_\infty = \max\{\|f\|_\infty, \|g\|_\infty\}$, have the same invertibility properties, the same norms, the same condition numbers and the same pseudospectra as the A_n, respectively. Further set

$$T := \begin{pmatrix} \chi_+I & \chi_-I \\ \chi_-I & \chi_+I \end{pmatrix} \in \mathcal{L}(L^p(\mathbb{R}) \times L^p(\mathbb{R})) \quad \text{and} \quad \hat{P}_n := \begin{pmatrix} P_n & 0 \\ 0 & P_n \end{pmatrix}.$$

Clearly, $T = T^{-1}$ is an isometry, hence the $T\tilde{A}_nT$ still have the same properties as A_n.

2nd step: Now, define the announced snapshots of $\mathbb{A}$ or subsequences $\mathbb{A}_h$ as

$$\mathsf{W}^*(\mathbb{A}_h) := \hat{\mathcal{P}}\text{-}\lim_{n\to\infty} \begin{pmatrix} V_{-h_n} & 0 \\ 0 & V_{h_n} \end{pmatrix} T\tilde{A}_{h_n}T \begin{pmatrix} V_{h_n} & 0 \\ 0 & V_{-h_n} \end{pmatrix},$$

where the $\hat{\mathcal{P}}$-strong limit is defined as in Definition 2.1 w.r.t. $\hat{\mathcal{P}} := (\hat{P}_n)$. The asserted limits (6.1) are easily checked by straightforward computations (see also [23, Theorem 27]).

3rd step: The final task is to localize the sequence $\{T\tilde{A}_nT\}$ w.r.t. the algebra

$$\mathcal{D} := \left\{\left\{\begin{pmatrix}\varphi_n I & 0\\ 0 & \varphi_n I\end{pmatrix}\right\} : \varphi \in \mathrm{C}(\dot{\mathbb{R}}), \varphi = -\varphi\right\}$$

again in an analogously constructed appropriate KMS-algebra setting.

This leads to the following analogon of Theorem 3.1 and thus to all the other results of the above sections also for sequences in $\mathcal{F}_{\mathcal{B}}$.

Theorem 6.2. *Let* $\mathbb{A} = \{A_n\} \in \mathcal{F}_{\mathcal{B}}$. *Then*

$$\begin{aligned}\limsup_{n\to\infty} \|A_n\| &= \max_{h\in\mathbf{H}_{\mathbb{A}}} \max\{\|\mathsf{W}^c(\mathbb{A}_h)\|, \|\mathsf{W}^*(\mathbb{A}_h)\|\}\\ \liminf_{n\to\infty} \|A_n\| &= \min_{h\in\mathbf{H}_{\mathbb{A}}} \max\{\|\mathsf{W}^c(\mathbb{A}_h)\|, \|\mathsf{W}^*(\mathbb{A}_h)\|\}\\ \limsup_{n\to\infty} \|A_n^{-1}\| &= \max_{h\in\mathbf{H}_{\mathbb{A}}} \max\{\|\mathsf{W}^c(\mathbb{A}_h)^{-1}\|, \|\mathsf{W}^*(\mathbb{A}_h)^{-1}\|\}\\ \liminf_{n\to\infty} \|A_n^{-1}\| &= \min_{h\in\mathbf{H}_{\mathbb{A}}} \max\{\|\mathsf{W}^c(\mathbb{A}_h)^{-1}\|, \|\mathsf{W}^*(\mathbb{A}_h)^{-1}\|\}.\end{aligned}$$

If $g : \mathbb{N} \to \mathbb{N}$ *is strictly increasing, then the analogous assertions also hold for* $\mathbb{A}_g$, *where the* max *and* min *are taken over all* $h \in \mathbf{H}_{\mathbb{A}_g}$, *i.e., subsequences of* g.

References

[1] G.R. Allan, Ideals of vector-valued functions, *Proc. London Math. Soc.* **18** (1968), No. 3, 193–216.

[2] A. Böttcher, Pseudospectra and singular values of large convolution operators, *J. Integral Equations Appl.* **6** (1994), No. 3, 267–301.

[3] A. Böttcher, S.M. Grudsky, B. Silbermann, Norms of inverses, spectra, and pseudospectra of large truncated Wiener–Hopf operators and Toeplitz matrices, *New York J. Math.* **3** (1997), 1–31.

[4] A. Böttcher, Y.I. Karlovich, I.M. Spitkovsky, *Convolution Operators and Factorization of Almost Periodic Matrix Functions*, Birkhäuser Verlag, Basel, Boston, Berlin, 2002.

[5] A. Böttcher, N. Krupnik, B. Silbermann, A general look at local principles with special emphasis on the norm computation aspect, *Integral Equations Oper. Theory* **11** (1988), No. 4, 455–479.

[6] A. Böttcher, B. Silbermann, *Analysis of Toeplitz Operators*, Second edition, Springer-Verlag, Berlin, Heidelberg, 2006.

[7] A. Böttcher, B. Silbermann, *Introduction to Large Truncated Toeplitz Matrices*, Springer-Verlag, New York, 1999.

[8] E.B. Davies, *Linear Operators and Their Spectra*, Cambridge Studies in Advanced Mathematics **106**, Cambridge University Press, 2007.

[9] R.G. Douglas, *Banach Algebra Techniques in Operator Theory*, Academic Press, New York, London, 1972.

[10] R. Duduchava, *Integral Equations with Fixed Singularities*, Teubner Verlag, Leipzig, 1979.

[11] J. Globevnik, Norm-constant analytic functions and equivalent norms, *Illinois J. Math.* **20** (1976), 503–506.

[12] J. Globevnik, On complex strict and uniform convexity, *Proc. Amer. Math. Soc.* **47** (1975), 176–178.

[13] I.C. Gohberg, I.A. Feldman, *Convolution Equations and Projection Methods for Their Solution* Nauka, Moscow 1971; Engl. transl.: Amer. Math. Soc. Transl. of Math. Monographs 41, Providence, RI., 1974.

[14] I.C. Gohberg, N. Krupnik, *One-dimensional Linear Singular Integral Equations*, Birkhäuser Verlag, Basel, Boston, Berlin, 1992.

[15] R. Hagen, S. Roch, B. Silbermann, *C^*-Algebras and Numerical Analysis*, Marcel Dekker, Inc., New York, Basel, 2001.

[16] A.C. Hansen, On the approximation of spectra of linear operators on Hilbert spaces, *J. Funct. Anal.* **254** (2008), No. 8, 2092–2126.

[17] F. Hausdorff, *Set Theory*, Chelsea, New York, 1957.

[18] L. Hörmander, Estimates for translation invariant operators in L^p spaces, *Acta Math.* **104** (1960), 93–104.

[19] Yu.I. Karlovich, I. Loreto Hernández, On convolution type operators with piecewise slowly oscillating data, *Operator Theory: Adv. and Appl.* **228** (2013), 185–207.

[20] A.V. Kozak, A local principle in the theory of projection methods, *Dokl. Akad. Nauk SSSR* **212** (1973), 6, 1287–1289; Engl. transl.: *Sov. Math. Dokl.* **14** (1974), 5, 1580–1583.

[21] M. Lindner, *Infinite Matrices and their Finite Sections*, Birkhäuser Verlag, Basel, Boston, Berlin, 2006.

[22] H. Mascarenhas, P.A. Santos, M. Seidel, Quasi-banded operators, convolutions with almost periodic or quasi-continuous data, and their approximations, *J. Math. Anal. Appl.* **418** (2014), No. 2, 938–963.

[23] H. Mascarenhas, P.A. Santos, M. Seidel, Approximation sequences on Banach spaces: a rich approach, *in preparation.*

[24] N.I. Polski, Projection methods for solving linear equations, *Uspekhi Mat. Nauk* **18** (1963), 179–180 (Russian).

[25] V.S. Rabinovich, S. Roch, B. Silbermann, Fredholm theory and finite section method for band-dominated operators, *Integral Equations Oper. Theory* **30** (1998), No. 4, 452–495.

[26] V.S. Rabinovich, S. Roch, B. Silbermann, *Limit Operators and Their Applications in Operator Theory*, Birkhäuser Verlag, Basel, 2004.

[27] S. Roch, Spectral approximation of Wiener–Hopf operators with almost periodic generating function, Proceedings of the International Conference on Fourier Analysis and Applications (Kuwait, 1998), *Numer. Funct. Anal. Optim.* **21** (2000), No. 1-2, 241–253.

[28] S. Roch, P. Santos, Finite section approximations in an algebra of convolution, multiplication and flip operators on $L^p(\mathbb{R})$, *J. Appr. Theory* **186** (2014), 64–97.

[29] S. Roch, P. Santos, B. Silbermann, Finite section method in some algebras of multiplication and convolution operators and a flip, *Z. Anal. Anwendungen* **16** (1997), No. 3, 575–606.

[30] S. Roch, P. Santos, B. Silbermann, *Non-commutative Gelfand Theories*, Universitext, Springer-Verlag London, 2011.

[31] S. Roch, B. Silbermann, Non-strongly converging approximation methods, *Demonstr. Math.* **22** (1989), No. 3, 651–676.

[32] W. Rudin, *Fourier Analysis on Groups*, John Wiley & Sons Ltd., New York, 1990.

[33] M. Seidel, On (N, ϵ)-pseudospectra of operators on Banach spaces, *J. Funct. Anal.* **262** (2012), No. 11, 4916–4927.

[34] M. Seidel, *On some Banach Algebra Tools in Operator Theory*, Dissertationsschrift, Chemnitz, 2011.

[35] M. Seidel, B. Silbermann, Banach algebras of operator sequences, *Oper. Matrices* **6** (2012), No. 3, 385–432.

[36] M. Seidel, B. Silbermann, Finite sections of band-dominated operators – norms, condition numbers and pseudospectra, *Operator Theory: Adv. and Appl.* **228** (2013), 375–390.

[37] B. Silbermann, Lokale Theorie des Reduktionsverfahrens für Toeplitzoperatoren, *Math. Nachr.* **104** (1981), No. 1, 137–146.

[38] E. Shargorodsky, On the level sets of the resolvent norm of a linear operator, *Bull. London Math. Soc.* **40** (2008), 493–504.

Markus Seidel
University of Applied Sciences Zwickau
Dr.-Friedrichs-Ring 2a
D-08056 Zwickau, Germany
e-mail: markus.seidel@fh-zwickau.de

Operator Theory:
Advances and Applications, Vol. 259, 681–702

Paired Operators in Asymmetric Space Setting

Frank-Olme Speck

Dedicated to Albrecht Böttcher with admiration

Abstract. Relations between paired and truncated operators acting in Banach spaces are generalized to asymmetric space settings, i.e., to matrix operators acting between different spaces. This allows more direct proofs and further results in factorization theory, here in connection with the Cross Factorization Theorem and the Bart–Tsekanovsky Theorem. Concrete examples from mathematical physics are presented: the construction of resolvent operators to problems of diffraction of time-harmonic waves from plane screens which are not convex.

Mathematics Subject Classification (2010). Primary 47A05; Secondary 47A20, 47A68.

Keywords. Paired operators, matrical coupling, equivalence after extension.

Introduction

Paired operators in symmetric setting are well known from the theory of singular operators since the 1960s, particularly as to be a useful vehicle to derive properties of operators such as the Fredholm property via operator relations, see [8, 24, 27, 36]. In the standard setting a paired operator has the form

$$T_1 \;=\; PA+QB \quad \text{or} \quad T_2 \;=\; AP+BQ \tag{0.1}$$

where $A, B \in \mathcal{L}(X)$ are bounded linear operators in a Banach space and P, Q are complementary projectors, i.e., $P, Q \in \mathcal{L}(X)$, $P^2 = I$, $Q = I - P$. The operators T_j are said to be of *normal type*, if A and B are isomorphisms (linear homeomorphisms), in brief $A, B \in \mathcal{GL}(X)$. In this case they are obviously equivalent to $PAB^{-1} + Q$ and to $P + QBA^{-1}$ in the first case and to $B^{-1}AP + Q$ and $P + A^{-1}BQ$ in the second, but not to each other in general. As usual two operators S, T (in topological vector spaces) are said to be equivalent if there exist isomorphisms E, F such that $T = E\,S\,F$. In this case it often suffices to consider only one type, say $T = PA+Q$ and to refer to analogy in several other cases.

In connection with singular integral operators of Cauchy type, Riemann boundary value problems and factorization theory the following well-known formula occurs (see [24], Section 5.1 and [36], Section 2.2.1, for instance):

$$\begin{aligned} PAP + Q &= (I - PAQ)(PA + Q) \\ &= (I - PAQ)(I + QA^{-1}P)(P + QA^{-1}Q)A\,. \end{aligned} \tag{0.2}$$

In particular it shows the equivalence between $PAP+Q$, $PA+Q$ and $P+QA^{-1}Q$, and allows many further conclusions. The formula is known from various different sceneries in matrix, ring and operator theory and has received several names such as Schur complement identity, Kozak formula, Sherman–Morrison–Woodbury formula and has even roots in the works of Sylvester, see for instance [8, 10, 17, 41, 45] and the more recent book [1], Section 2.2.

In the present article we consider an extension of formula (0.2) (and related formulas) to an asymmetric space setting where $A \in \mathcal{GL}(X,Y)$ is an isomorphism acting between different Banach spaces, see Section 1. It produces heavy formulas, but a very clear and useful strategy for many purposes exposed subsequently:

- A more direct proof of the Cross Factorization Theorem (see Theorem 2.2), connected with further results for general Wiener–Hopf operators (see the Corollaries 2.4 and 2.5);
- a more direct proof of the Bart–Tsekanovsky Theorem (see Theorem 3.1) including an extension of that result with the equivalence to the existence and construction of special forms of the equivalence after extension relation (see Corollary 3.4 and Corollary 4.2);
- applications to the solution of boundary and transmission problems that appear in problems of diffraction of time-harmonic waves from plane screens in $\mathbb{R}^3$ (see Section 7).

For technical reasons we present the precise formulation of the main results only in the corresponding sections, loc. cit.

Hence the key idea of this paper is to give a proper meaning to terms such as P_2A+Q_1 where $A \in \mathcal{L}(X,Y)$, $P_1 = P_1^2 = I - Q_1 \in \mathcal{L}(X)$, $P_2 \in \mathcal{L}(Y)$ which does not make sense immediately when the two operators P_2A and Q_1 do not map into the same space. Thinking of

$$U = \begin{pmatrix} P_2A & 0 \\ 0 & Q_1 \end{pmatrix} : X \times X \to Y \times X$$

we have a proper definition but find the difficulty that U can hardly be interpreted as an invertible operator. So we switch to the truncated operator

$$T = \begin{pmatrix} P_2A|_{P_1X} & P_2A|_{Q_1X} \\ 0 & I|_{Q_1X} \end{pmatrix} : P_1X \times Q_1X \to P_2Y \times Q_1X \tag{0.3}$$

as generalization of the paired operator $PA+Q$. Now it acts in an asymmetric space setting, with a restricted image space in contrast to the previous operator U. This notion transforms many formulas directly into reasonable variants. It turns out that the resulting formulas are looking complicated but are quite powerful and allow a clear reasoning. To the authors best knowledge, the idea did not gain much interest in the literature until now.

The Cross Factorization Theorem (in asymmetric setting) is theoretically important also for other kinds of operator factorizations such as the Wiener–Hopf factorization through an intermediate space and full range factorization [9, 43]. Further it is well known that general WHOs admit plenty of realizations such as Toeplitz, classical Wiener–Hopf and Wiener–Hopf plus Hankel operators etc. which are fundamental in the solution of problems in mathematical physics, see for instance [11, 13, 16, 33, 44].

Recently the Bart–Tsekanovsky Theorem has gained much attention and resulted in an intensive discussion of the equivalence between various operator relations including also one-sided equivalence after extension and Schur coupling, see [30, 46] and further papers cited therein. Related to the notion of equivalence after one-sided extension, it was shown in [29] that this operator relation is not equivalent to equivalence after (two-sided) extension in the case of general Banach space operators, while it is equivalent to equivalence after (two-sided) extension for Hilbert space operators, as was proved in [46].

The present results concerning the Bart–Tsekanovsky Theorem are based upon a trick to derive from a given equivalence after extension relation a more special one that has the form of the basic formula (0.2) generalized to an asymmetric space setting, i.e., using paired operators in this more general sense.

The interplay of these operator relations is also of great importance in various applications. For instance certain boundary value problems in non-convex domains allow an "equivalent reduction" to associated problems in convex domains which can be solved, see [13, 42]. This idea will be exemplified in the last section.

1. A generalization of formula (0.2)

Formula (0.2) "can be written" in matrix form as

$$\begin{aligned}\begin{pmatrix} PAP & 0 \\ 0 & Q \end{pmatrix} &= \begin{pmatrix} P & -PAQ \\ 0 & Q \end{pmatrix} \begin{pmatrix} PAP & PAQ \\ 0 & Q \end{pmatrix} \\ &= \begin{pmatrix} P & -PAQ \\ 0 & Q \end{pmatrix} \begin{pmatrix} P & 0 \\ QA^{-1}P & Q \end{pmatrix} \begin{pmatrix} P & 0 \\ 0 & QA^{-1}Q \end{pmatrix} \begin{pmatrix} PAP & PAQ \\ QAP & QAQ \end{pmatrix}.\end{aligned} \tag{1.1}$$

More precisely, (0.2) and (1.1) are equivalent via the decomposition operator

$$\iota \; : \; X = PX \oplus QX \; \to \; PX \times QX \, , \quad x \mapsto \begin{pmatrix} Px \\ Qx \end{pmatrix} \tag{1.2}$$

where $PX \oplus QX$ denotes the direct sum and $PX \times QX$ denotes the topological product space, which are isomorphic (as normed spaces). Namely we have

$$PAP + Q \; = \; \iota^{-1} \begin{pmatrix} PAP & 0 \\ 0 & Q \end{pmatrix} \iota \; = \; (P \, , Q) \begin{pmatrix} PAP & 0 \\ 0 & Q \end{pmatrix} \begin{pmatrix} P \\ Q \end{pmatrix}. \tag{1.3}$$

Hence (1.1) can be understood simply as an operator acting in $PX \times QX$ whilst (1.3) is acting in $X = PX \oplus QX$, strictly speaking. Their identification is useful to avoid heavy formulas (*Convention* 1). Note that so far all these formulas hold for elements of a unital ring $\mathcal{R}$, as well, provided A is invertible and P, Q are idempotent with $P + Q = I$.

Now the first operator PAP in (1.1) (as acting in PX and not in X) is often denoted as *general Wiener–Hopf operator* (WHO) and written in the form [18, 39]

$$W \;=\; T_P(A) \;=\; PA|_{PX} \;:\; PX \to PX\,. \tag{1.4}$$

It is also referred to as abstract WHO, projection of A, truncation of A or compression of A, to mention a few notations for the same thing [8, 17, 27, 35, 36, 41]. Other terms in (1.1) are identified similarly, e.g., PAQ as $PA|_{QX}$ etc. Note that (1.4) stands for an operator restricted not only in the domain but in the image space, as well (*Convention* 2). These notational conventions are used for brevity and tradition, in order to avoid rather heavy formulas.

A *general Wiener–Hopf operator* (WHO) *in asymmetric space setting* has the form [41]

$$W \;=\; P_2A|_{P_1X} \;:\; P_1X \to P_2Y \tag{1.5}$$

provided $A \in \mathcal{L}(X, Y)$, $P_1 = P_1^2 \in \mathcal{L}(X)$ and $P_2 = P_2^2 \in \mathcal{L}(Y)$. The two conventions are applied by analogy: W is briefly written as P_2AP_1 (in certain operator matrices) and considered as a mapping into P_2Y, not as a mapping into Y.

Besides of operator equivalence, the following operator relations are very important in what follows: Two bounded linear operators S, T acting in Banach spaces are said to be *matrically coupled*, if there exist operator matrices such that

$$\begin{pmatrix} T & * \\ * & * \end{pmatrix} = \begin{pmatrix} * & * \\ * & S \end{pmatrix}^{-1} \tag{1.6}$$

where the stars stand for suitable bounded linear operators and the operator matrices represent isomorphisms (in the corresponding Banach spaces). Secondly, S and T are *equivalent after extension*, if there exist operators E and F such that

$$\begin{pmatrix} T & 0 \\ 0 & I_1 \end{pmatrix} = E \begin{pmatrix} S & 0 \\ 0 & I_2 \end{pmatrix} F \tag{1.7}$$

where I_1, I_2 are identity operators and E, F are isomorphisms acting between suitable Banach spaces.

The first relation is called a *matrical coupling relation* (MCR), the second an *equivalence after extension relation* (EAER). They were introduced in [2] and gained great attention in system theory and other applications, see [2, 5] and [14], for instance. An EAER is evidently an equivalence relation in the genuine mathematical sense (reflexive, symmetric and transitive) and therefore abbreviated by $T \overset{*}{\sim} S$. A MCR is obviously symmetric. Anyway the two relations are equivalent [6], which fact will be further discussed later (under the name Bart–Tsekanovsky Theorem).

Now we turn to consider paired operators acting between different Banach spaces. As explained in the introduction we are here only interested in a generalization of the operators (0.1) where A or B is the identity operator, i.e., in the analogues of

$$PA+Q\ ,\quad AP+Q\ ,\quad P+QA\ ,\quad P+AQ\,. \tag{1.8}$$

Definition 1.1. Consider a *basic space setting*

$$\begin{aligned} &X, Y \text{ are Banach spaces}\,,\\ &P_1 \in \mathcal{L}(X)\,,\ P_2 \in \mathcal{L}(Y) \text{ are projectors}\,,\\ &Q_1 = I - P_1\,,\ Q_2 = I - P_2\,, \end{aligned} \tag{1.9}$$

and let $A \in \mathcal{L}(X,Y)$. Then each of the following operators is referred to as a *paired operator in an asymmetric space setting* (PAO):

$$\begin{aligned} &\begin{pmatrix} P_2AP_1 & P_2AQ_1 \\ 0 & Q_1 \end{pmatrix} : P_1X \times Q_1X \to P_2Y \times Q_1X\\ &\begin{pmatrix} P_2AP_1 & 0 \\ Q_2AP_1 & Q_2 \end{pmatrix} : P_1X \times Q_2Y \to P_2Y \times Q_2Y\\ &\begin{pmatrix} P_1 & 0 \\ Q_2AP_1 & Q_2AQ_1 \end{pmatrix} : P_1X \times Q_1X \to P_1X \times Q_2Y\\ &\begin{pmatrix} P_2 & P_2AQ_1 \\ 0 & Q_2AQ_1 \end{pmatrix} : P_2Y \times Q_1X \to P_2Y \times Q_2Y \end{aligned}$$

where we apply Convention 1 and Convention 2, as well.

Remark 1.2. Any upper or lower triangular operator matrix with an identity in one of the main diagonal entries can be seen as a PAO and vice versa. Hence this formulation can serve as an alternative definition. However we preferred here to see the analogy with classical PAOs in the notation more clearly.

Remark 1.3. We could also work with a scalar writing of PAOs, for instance with "paired operators" of the form $P_2AP_1 + Q_1 \,:\, X \to Z = \widetilde{P_2}Z \oplus \widetilde{Q_1}Z$ where $Z \cong P_2Y \times Q_1X$ etc. However, for cosmetic reasons we prefer to remain with the foregoing matrix notation of paired operators in asymmetric setting.

Lemma 1.4. *Let $X = X_1 \oplus X_2\,, Y = Y_1 \oplus Y_2$ be Banach spaces, P_1, Q_1, P_2, Q_2 the corresponding projectors onto X_1, X_2, Y_1, Y_2, respectively (i.e., we have a basic setting), and let $A \in \mathcal{GL}(X,Y)$ (an isomorphism). Then the two associated asymmetric WHOs*

$$\begin{aligned} W &= P_2A|_{X_1}\ :\ X_1 \to Y_1\\ W_* &= Q_1A^{-1}|_{Y_2}\ :\ Y_2 \to X_2 \end{aligned} \tag{1.10}$$

are equivalent after extension. An EAER between the two operators is explicitly given by the following formula:

$$\begin{pmatrix} W & 0 \\ 0 & I|_{Q_1X} \end{pmatrix} = \begin{pmatrix} -P_2A|_{Q_1X} & I|_{P_2Y} - P_2AQ_1A^{-1}|_{P_2Y} \\ I|_{Q_1X} & Q_1A^{-1}|_{P_2Y} \end{pmatrix} \tag{1.11}$$
$$\begin{pmatrix} W_* & 0 \\ 0 & I|_{P_2Y} \end{pmatrix} \begin{pmatrix} Q_2A|_{P_1X} & Q_2A|_{Q_1X} \\ P_2A|_{P_1X} & P_2A|_{Q_1X} \end{pmatrix}$$
$$: \quad Q_1X \times P_2Y \ \leftarrow \ Q_1X \times P_2Y \ \leftarrow \ P_2Y \times Q_2Y \ \leftarrow \ P_1X \times Q_1X \, .$$

Proof. Generalizing formula (1.1) to the asymmetric setting we write

$$\begin{pmatrix} P_2AP_1 & 0 \\ 0 & Q_1 \end{pmatrix} = \begin{pmatrix} P_2 & -P_2AQ_1 \\ 0 & Q_1 \end{pmatrix} \begin{pmatrix} P_2AP_1 & P_2AQ_1 \\ 0 & Q_1 \end{pmatrix} \tag{1.12}$$
$$= \begin{pmatrix} P_2 & -P_2AQ_1 \\ 0 & Q_1 \end{pmatrix} \begin{pmatrix} P_2 & 0 \\ Q_1A^{-1}P_2 & Q_1 \end{pmatrix} \begin{pmatrix} P_2 & 0 \\ 0 & Q_1A^{-1}Q_2 \end{pmatrix} \begin{pmatrix} P_2AP_1 & P_2AQ_1 \\ Q_2AP_1 & Q_2AQ_1 \end{pmatrix} .$$

A straightforward verification shows that the identity is correct and the matrix factorization in the last line maps the spaces like

$$\begin{matrix} Y_1 & & Y_1 & & Y_1 & & Y_1 & & X_1 \\ \times & \leftarrow & \times & \leftarrow & \times & \leftarrow & \times & \leftarrow & \times \\ X_2 & & X_2 & & X_2 & & Y_2 & & X_2 \end{matrix} \ . \tag{1.13}$$

So we have a factorization of the extended WHO into a composition of bounded operators where the first two factors and the last one are obviously invertible and the remainder factor is an extension of the associated WHO W_*. Formula (1.12) can be regarded as an EAER between $T = W$ and $S = W_*$ which gains the form of (1.7) after composition of the first two matrices in the second line of (1.12) and a permutation of some rows and columns:

$$\begin{pmatrix} P_2AP_1 & 0 \\ 0 & Q_1 \end{pmatrix} \tag{1.14}$$
$$= \begin{pmatrix} -P_2AQ_1 & P_2 - P_2AQ_1A^{-1}P_2 \\ Q_1 & Q_1A^{-1}P_2 \end{pmatrix} \begin{pmatrix} Q_1A^{-1}Q_2 & 0 \\ 0 & P_2 \end{pmatrix} \begin{pmatrix} Q_2AP_1 & Q_2AQ_1 \\ P_2AP_1 & P_2AQ_1 \end{pmatrix} .$$

This implies (1.11) by Convention 1. □

Remarks 1.5.

1. It is also possible to work with Q_2 instead of Q_1 in the first matrix of (1.12), which will be discussed later in Section 4.

2. To the author's best knowledge Formula (1.12) can not be found in the existing literature, although its idea was already behind the remarks of [3], I guess.

3. Some common properties of the two associated WHOs W and W_* in an asymmetric setting were already discovered by the author in 1983 such as the fact that they are one-sided invertible only simultaneously, see Proposition 2 in [40]. But the EAER (1.12) implies much more, namely that the two operators have isomorphic kernels and cokernels, see Proposition 1 in [6]. This yields that they (as every two operators fulfilling an EAER) belong to the same *regularity*

class of normally solvable linear operators in Banach spaces, such as the class of invertible, Fredholm, or generalized invertible operators etc., cf. a classification in [14, 41], which is just recalled here as a convenient vehicle for clear and efficient formulations later on. Abbreviating $\alpha(T) = \dim\ker T, \beta(T) = \dim\operatorname{coker} T$, we distinguish the following regularity classes of operators:

	$\alpha(T)=0$	$\alpha(T)<\infty$	$\ker T$ complem.	$\ker T$ closed
$\beta(T)=0$	bdd. invertible	right inv. Fredholm	right invertible	surjective
$\beta(T)<\infty$	left inv. Fredholm	Fredholm	right regulariz.	semi-Fred. $\mathcal{F}_-$
$\operatorname{im} T$ complem.	left invertible	left regulariz.	generalized invertible	no name
$\operatorname{im} T$ closed	injective	semi-Fred. $\mathcal{F}_+$	no name	normally solvable

Moreover in applications it is important that the knowledge of an explicit (generalized) inverse, (Fredholm) pseudoinverse or regularizer of W implies a formula for a corresponding one of W_* and vice versa [14]. That is again evident from an EAER, but not from a MCR, see [5], Corollary 5.13, for comparison.

2. An alternative proof of the Cross Factorization Theorem

Let us recall the definition of a cross factorization from [41]. Assume a basic setting where X, Y are Banach spaces, $P_1 \in \mathcal{L}(X)$, $P_2 \in \mathcal{L}(Y)$ projectors, $Q_1 = I_X - P_1$, $Q_2 = I_Y - P_2$ and let $A \in \mathcal{GL}(X,Y)$. Then (an operator triple A_-, C, A_+ with)

$$\begin{aligned} A = \quad & A_- \quad C \quad A_+ \\ & : Y \leftarrow Y \leftarrow X \leftarrow X\,. \end{aligned} \tag{2.1}$$

is referred to as a *cross factorization* of A (with respect to X, Y, P_1, P_2), in brief CFn, if the factors $A_\pm$ and C possess the following properties:

$$\begin{aligned} A_+ &\in \mathcal{GL}(X) \quad , \quad A_- \in \mathcal{GL}(Y)\,, \\ A_+P_1X &= P_1X \quad , \quad A_-Q_2Y = Q_2Y\,, \end{aligned} \tag{2.2}$$

and $C \in \mathcal{L}(X,Y)$ splits the spaces X, Y both into four subspaces such that

$$\begin{array}{ccccccc} X & = & \overbrace{X_1 \oplus X_0}^{P_1X} & \oplus & \overbrace{X_2 \oplus X_3}^{Q_1X} \\ & & \downarrow \qquad\qquad & C \times & \qquad\qquad \downarrow \\ Y & = & \underbrace{Y_1 \oplus Y_2}_{P_2Y} & \oplus & \underbrace{Y_0 \oplus Y_3}_{Q_2Y}. \end{array} \tag{2.3}$$

This means that the factors $A_\pm$ are invertible in $\mathcal{L}(X)$ or $\mathcal{L}(Y)$, mapping P_1X onto P_1X and Q_2Y onto Q_2Y, respectively. Further C maps each X_j onto Y_j, $j = 0, 1, 2, 3$, i.e., the complemented subspaces $X_0, X_1, \ldots, Y_3$ are images of corresponding projectors $p_0, p_1, \ldots, q_3$, namely

$$\begin{aligned}
X_1 &= p_1X = C^{-1}P_2CP_1X\,, & X_0 &= p_0X = C^{-1}Q_2CP_1X\,,\\
X_2 &= p_2X = C^{-1}P_2CQ_1X\,, & X_3 &= p_3X = C^{-1}Q_2CQ_1X\,,\\
Y_1 &= q_1Y = CP_1C^{-1}P_2Y\,, & Y_2 &= q_2Y = CQ_1C^{-1}P_2Y\,,\\
Y_0 &= q_0Y = CP_1C^{-1}Q_2Y\,, & Y_3 &= q_3Y = CQ_1C^{-1}Q_2Y\,.
\end{aligned} \tag{2.4}$$

$A_\pm$ are called *strong WH factors* (or *plus/minus factors*), C is said to be a *cross factor*, since it maps a part of P_1X onto a part of Q_2Y ($X_0 \to Y_0$) and a part of Q_1X onto a part of P_2Y ($X_2 \to Y_2$), which are all complemented subspaces.

Lemma 2.1. *Let $C \in \mathcal{GL}(X, Y)$ be of the form*

$$C = \begin{pmatrix} C_{11} & C_{12} \\ C_{21} & C_{22} \end{pmatrix} = \begin{pmatrix} P_2CP_1 & P_2CQ_1 \\ Q_2CP_1 & Q_2CQ_1 \end{pmatrix} : P_1X \times Q_1X \to P_2Y \times Q_2Y \tag{2.5}$$

(applying Convention 1*) and let $W = P_2C|_{P_1X}$ be generalized invertible. Then C is a cross factor with respect to X, Y, P_1, P_2 if and only if the operators $p_0, p_1, \ldots, q_3$ mentioned in* (2.4) *are idempotent.*

Proof. Sufficiency is easily verified: The subspaces in (2.4) are complemented as shown in diagram (2.3) and the mapping properties of C and C^{-1} are satisfied.

Necessity is obvious by definition from the diagram (2.3). □

The following theorem is mainly known from [41], and here extended by another (the last) statement. The proof of the necessity part of the first statement is new and more compact in comparison with [41], pp. 27–29.

Theorem 2.2 (Cross Factorization Theorem). *Let X, Y, P_1, P_2 be a basic space setting and A be an isomorphism. Then $W = P_2A|_{P_1X}$ is generalized invertible if and only if A admits a cross factorization. In this case, a formula for a reflexive generalized inverse of W is explicitly given by the* reverse order law*:*

$$W^- = A_+^{-1}P_1C^{-1}P_2A_-^{-1}|_{P_2Y} \quad : \quad P_2Y \to P_1X\,. \tag{2.6}$$

Conversely, if V is a reflexive generalized inverse of W, then (applying our conventions) a cross factorization is explicitly given by the formulas

$$\begin{aligned}
A &= A_-\ C\ A_+ \quad : \quad P_1X \times Q_1X \to P_2Y \times Q_2Y, \qquad (2.7)\\
A_- &= \begin{pmatrix} P_2 & 0 \\ Q_2AP_1V & Q_2 \end{pmatrix},\\
C &= \begin{pmatrix} W & P_2(A - AVP_2A)Q_1 \\ Q_2(A - AVP_2A)P_1 & S \end{pmatrix},
\end{aligned}$$

$$S = Q_2(A - AVP_2A + A(P_1 - VWP_1)A^{-1}(P_2 - WVP_2)A)Q_1,$$
$$A_+ = \begin{pmatrix} P_1 & (VP_2A - (P_1 - VWP_1)A^{-1}(P_2 - WVP_2)A)Q_1 \\ 0 & Q_1 \end{pmatrix}.$$

Proof. If A admits a cross factorization $A = A_- C A_+$, the operator defined by (2.6) is shown to satisfy $WW^-W = W$ and $W^-WW^- = W^-$ by verification. For details see [41], p. 29.

The inverse conclusion is only known in the symmetric case, see [41], pp. 117–118. Here we present a similar reasoning in the asymmetric case, with the help of Lemma 1.4 and Lemma 2.1. Now concretely, if $WVW = W$ and $VWV = V$ hold, it is not hard to verify that the first line of (2.7) is an identity. Moreover $A_\pm$ are easily recognized as strong plus/minus factors (see (2.2)).

It remains to prove that C as defined in (2.7) is a cross factor. To this end we first define the projectors $p_0, p_1, \ldots, q_3$ by (instead of (2.4))

$$p_1 = VW \ , \ p_0 = P_1 - VW \ , \ p_2 = Q_1A^{-1}(P_2 - WV)A \ , \ p_3 = Q_1 - p_2 \, , \tag{2.8}$$
$$q_1 = WV \ , \ q_2 = P_2 - WV \ , \ q_0 = A(P_1 - VW)A^{-1}Q_2 \ , \ q_3 = Q_2 - q_0 \, .$$

Obviously $p_2Q_1 = p_2$, $Q_2q_0 = q_0$ are satisfied and (hence) all items are idempotent.

Second we prove (in brief) that C can be written as

$$\begin{aligned} C &= q_1(A - AVA)p_1 + q_0(A - AVA)p_0 + q_2(A - AVA)p_2 + q_3(A - AVA)p_3 \\ &= q_1\,A\,p_1 \ + \ q_0\,A\,p_0 \ + \ q_2\,A\,p_2 \ + \ q_3\,V_*\,p_3. \end{aligned} \tag{2.9}$$

where

$$V_* \ = \ Q_2(A - AVA)Q_1 \, , \tag{2.10}$$

see analogous computations for the symmetric case in [41], p. 118, which needs some calculations and where the two cited lemmas (in symmetric setting) played a crucial role. This implies the mapping properties of C in the diagram (2.3).

The foregoing two direct verifications include already the mutual calculation of a reflexive generalized inverse of W and a CFn of A. □

Remark 2.3. Note that asymmetric PAOs are substantially behind the proof, as it makes use of Lemma 1.4. Another point is the interpretation of formula (2.10), presented later in Section 6, which is not needed here but originally led us to the idea of the preceding proof. However, at the end it turns out that the formulas (2.7) can be written in scalar form, as well, by the help of (1.4).

The previous results allow further interpretations. We summarize some of them:

Corollary 2.4. *Consider a basic setting X, Y, P_1, P_2 and $A \in \mathcal{GL}(X,Y)$. Then the following statements are equivalent:*

1. *The WHO $W = P_2A|_{P_1X}$ is generalized invertible,*
2. *$W = P_2A|_{P_1X} \sim P_2C|_{P_1X}$ where C is a cross factor,*
3. *$W = P_2A|_{P_1X} = P_2C|_{P_1X}$ where C is a cross factor,*

4. *A admits a CFn,*
5. *The WHO* $W_* = Q_1 A^{-1}|_{Q_2 Y}$ *is generalized invertible,*
6. $W_* = Q_1 A^{-1}|_{Q_2 Y} = Q_1 C_*|_{Q_2 Y}$ *where* C_* *is a cross factor.*

Corollary 2.5 (Decomposition Theorem for generalized invertible WHOs). *Again consider a basic setting* X, Y, P_1, P_2 *and* $A \in \mathcal{GL}(X,Y)$. *If* V *is a reflexive generalized inverse of* $W = P_2 A|_{P_1 X}$, *then the spaces* X *and* Y *are decomposable as shown in the diagram* (2.3) *with*

$$\begin{aligned} X_0 &= A_+ \ker W &&= A_+ (I - VW) P_1 X &&= C^{-1} Q_2 C P_1 X \\ X_3 &= Q_1 A_+ \operatorname{im} W_* &&= Q_1 A_+ \; W_* V_* Q_1 X &&= C^{-1} Q_2 C Q_1 X \\ Y_1 &= P_2 A_-^{-1} \operatorname{im} W &&= P_2 A_-^{-1} \; W V P_2 Y &&= C P_1 C^{-1} P_2 Y \\ Y_0 &= A_-^{-1} \ker W_* &&= A_-^{-1} (I - V_* W_*) Q_2 Y &&= C P_1 C^{-1} Q_2 Y \end{aligned}$$

where C *is given by* (2.7) *and* $V_* = Q_2(A - AVP_2A)|_{Q_1 X}$, *which is a reflexive generalized inverse of* W_* (cf. Formula (2.3) in [41]).

Remark 2.6. As we have seen in the symmetric case (where $X = Y$ and $P_1 = P_2$), it is possible to present formulas for all existing generalized inverses of W (provided we have one of them) which is somehow equivalent to describe all complements of $\ker W$ in $P_1 X$, $\operatorname{im} W$ in $P_2 Y$, $\ker W_*$ in $Q_2 Y$, and $\operatorname{im} W_*$ in $Q_1 X$, as well as to describe all cross factors C of A, see [41], Chapter 6, for symmetric settings.

3. An alternative proof of the Bart–Tsekanovsky Theorem

H. Bart and V.E. Tsekanovsky published the following theorem in 1991 [6].

Theorem 3.1 (of Bart and Tsekanovsky). *Given two bounded linear operators* $T : X_1 \to Y_1$ *and* $S : Y_2 \to X_2$ *acting in Banach spaces, the relations* (1.6) *and* (1.7) *between* S *and* T *are equivalent, i.e., the two operators are matrically coupled if and only if they are equivalent after extension.*

The step from (1.6) to (1.7) was proved already in 1984 [2]. It can also be regarded as an interpretation of Formula (1.12) as shown below. The inverse conclusion of how to construct an MCR from an EAER between T and S is not at all evident. A prompt (maybe surprising) solution was given in [6], in the proof of Theorem 1, namely: If T, S satisfy the EAER (1.7), then one can verify the following MCR where $E_{11}, \ldots, F_{22}^-$ are the elements of E, F, E^{-1}, F^{-1}:

$$\begin{pmatrix} T & -E_{11} \\ F_{11} & F_{12} E_{21} \end{pmatrix} = \begin{pmatrix} F_{12}^- E_{21}^- & F_{11}^- \\ -E_{11}^- & S \end{pmatrix}^{-1} . \tag{3.1}$$

Clearly the solution is not unique and the variety of possible MCRs depends on the two given operators. According to the number of elements of E and F it seems that the collection of EAERs between T and S is quite large. Hence some questions appear: Is it possible to find (3.1) constructively somehow? Are there some predestined EAERs which in a sense serve for an "inversion" of the simpler conclusion MCR $\Rightarrow$ EAER? What does this have to do with the idea of paired operators as realized in (1.12)? The following result provides a partial answer to these questions.

Theorem 3.2. *Given any EAER* (1.7) *between the two bounded linear operators* $T : X_1 \to Y_1$ *and* $S : Y_2 \to X_2$*, we obtain another EAER of the form* (1.14)*, i.e., in the notation*[1] *of Bart and Tsekanovsky of the special form*

$$\begin{pmatrix} T & 0 \\ 0 & I_1 \end{pmatrix} = \begin{pmatrix} * & * \\ I_1 & * \end{pmatrix} \begin{pmatrix} S & 0 \\ 0 & I_2 \end{pmatrix} \begin{pmatrix} * & * \\ T & * \end{pmatrix} \tag{3.2}$$

by putting $I_1 = I_{X_2}$, $I_2 = I_{Y_1}$ *and*

$$\begin{pmatrix} T & 0 \\ 0 & I_1 \end{pmatrix} = \begin{pmatrix} \alpha E_{11} & I_2 - E_{11}E_{11}^- \\ I_1 & -\beta E_{11}^- \end{pmatrix} \begin{pmatrix} S & 0 \\ 0 & I_2 \end{pmatrix} \begin{pmatrix} \beta F_{11} & F_{12}E_{21} \\ T & -\alpha E_{11} \end{pmatrix} \tag{3.3}$$

where $\alpha, \beta \in \mathbb{C}$, $\alpha\beta = 1$.

Proof. Given (1.6) with the above notation, we like to solve the sudoku (3.2). Denoting the unknown matrices by

$$e = \begin{pmatrix} e_{11} & e_{12} \\ I_1 & e_{22} \end{pmatrix}, \quad f = \begin{pmatrix} f_{11} & f_{12} \\ T & f_{22} \end{pmatrix}, \quad e^{-1} = \begin{pmatrix} e_{11}^- & e_{12}^- \\ e_{21}^- & e_{22}^- \end{pmatrix}$$

we see, composing (3.2) with e^{-1} from the left, that (3.2) implies

$$\begin{aligned} e_{11}^- T &= S f_{11}\,, & e_{12}^- &= S f_{12} \\ e_{21}^- T &= f_{21} = T\,, & e_{22}^- &= f_{22}\,. \end{aligned}$$

Corresponding relations result from the original EAER:

$$\begin{aligned} E_{11}^- T &= S F_{11}\,, & E_{12}^- &= S F_{12} \\ E_{21}^- T &= F_{21}\,, & E_{22}^- &= F_{22}\,. \end{aligned}$$

We choose (tentatively, but keeping in mind the variety of operators which fulfil the same relation $T \stackrel{*}{\sim} S$)

$$e_{11}^- = E_{11}^-\,, \quad f_{11} = F_{11}\,, \quad e_{21}^- = I_1. \tag{3.4}$$

Considering (3.2) again we conclude for its 21-entry that

$$0 = S f_{11} + e_{22} T,$$

i.e., $e_{22}T = -Sf_{11} = -SF_{11} = -E_{11}^- T$, i.e., $e_{22} = -E_{11}^-$ as a possible choice. From here it is easy to fill the rest of the matrix (considering $ee^{-1} = I$, $e^{-1}e = I$ etc.) and obtain the result with $\alpha = \beta = 1$.

As we can see in (3.4), there is some freedom in the choice of e_{11}^- and f_{11}^-, namely one can also choose $e_{11}^- = \beta E_{11}^-$ and $f_{11}^- = \beta F_{11}^-$ with $\beta \neq 0$. The rest of the proof runs similarly and leads to the indicated solutions.

It is also not difficult to verify Formula (3.3) directly. □

[1] Note that the notation of the subspaces is exchanged here against [6], since it is here more related to the notation of WHOs [41].

Now we present the proof of the BT Theorem based upon formula (1.12).

Proof of the BT Theorem. As mentioned already in the introduction, the step from (1.6) to (1.7) can be regarded as an interpretation of formula (1.12). Namely, if T and S are matrically coupled, we can identify the matrix on the left of (1.6) with an invertible operator matrix A, further T with P_2AP_1 and S with $Q_1A^{-1}Q_2$. The identity (1.12) tells us that $T \overset{*}{\sim} S$.

The step from (1.7) to (1.6) runs as follows: An EAER (1.7) between T and S implies another one of the form (3.3), say with $\alpha = \beta = 1$, because of Lemma 1.4. This can be re-written by a permutation of the lines and rows as

$$\begin{pmatrix} T & 0 \\ 0 & I_1 \end{pmatrix} = \begin{pmatrix} I_2 - E_{11}E_{11}^- & E_{11} \\ -E_{11}^- & I_1 \end{pmatrix} \begin{pmatrix} I_2 & 0 \\ 0 & S \end{pmatrix} \begin{pmatrix} T & -E_{11} \\ F_{11} & F_{12}E_{21} \end{pmatrix}. \tag{3.5}$$

It has the form of (1.12), if we identify the last matrix with

$$A = \begin{pmatrix} T & -E_{11} \\ F_{11} & F_{12}E_{21} \end{pmatrix}. \tag{3.6}$$

Hence we have the MCR (1.6) with the identifications of (1.10). □

Remark 3.3. At the end we observe that the matrix (3.1) given by Bart and Tsekanovsky coincides with our matrix A in case of $\alpha = \beta = 1$. However the present approach gives an idea of how to obtain it constructively and demonstrates the usefulness of (1.12), i.e., of the idea of paired operators.

Corollary 3.4 (Extension of the BT Theorem). *Let $T : X_1 \to Y_1$ and $S : Y_2 \to X_2$ be bounded linear operators in Banach spaces. Then the following statements are equivalent:*

1. *T and S are matrically coupled;*
2. *T and S are equivalent after extension;*
3. *T and S satisfy a special EAER of the form* (3.2).

Corollary 3.5. *EAERs of the special form* (3.2) *are reflexive, symmetric and transitive, hence represent (for its own) an equivalence relation in the genuine mathematical sense.*

Remark 3.6. The three equivalence properties are obvious in EAERs.
In MCRs symmetry is obvious and reflexivity almost obvious:

$$\begin{pmatrix} T & -I_{Y_1} \\ I_{X_1} & 0 \end{pmatrix} = \begin{pmatrix} 0 & I_{X_1} \\ -I_{Y_1} & T \end{pmatrix}^{-1}, \tag{3.7}$$

see [6], page 147, where we also find a direct proof of the transitivity of MCRs which is not obvious.

Remark 3.7. In this context it should be mentioned (thanks to the referee's report) that there is another well-known operator relation referred to as Schur coupling. T and S are called Schur coupled if there exist bounded linear operators A, B, C, D, where A, D are isomorphisms, such that

$$T = A - BD^{-1}C \quad \text{and} \quad S = D - CA^{-1}B.$$

Schur coupling implies matrical coupling and equivalence after extension; actually many expect the three operator relations to be equivalent, and for Hilbert space operators this was settled in [46]. As explained on page 215 of [4], showing that Schur coupling is an equivalence relation (only transitivity is unknown) would imply that these three operator relations coincide.

4. Further alternative formulas

The first line of formula (0.2) can be seen as a relation between the general WHO $W = PA|_{PX}$ in (1.4) and a paired operator $(PA + QB)$ where $B = I$. The second line of (0.2) holds if A is invertible. Alternatively one can consider another relation with a paired operator of the form $AP + BQ$ (again $B = I$):

$$PAP + Q = (AP + Q)(I - QAP). \tag{4.1}$$

We obtain instead of (0.2), (1.1), (1.12) the following formulas. Firstly

$$PAP + Q = A(P + QA^{-1}Q)(I + PA^{-1}Q)(I - QAP) \tag{4.2}$$

which is also well known from [24, 36]. In matrix form it may be written as

$$\begin{aligned}\begin{pmatrix} PAP & 0 \\ 0 & Q \end{pmatrix} &= \begin{pmatrix} PAP & 0 \\ QAP & Q \end{pmatrix} \begin{pmatrix} P & 0 \\ -QAP & Q \end{pmatrix} \\ &= \begin{pmatrix} PAP & PAQ \\ QAP & QAQ \end{pmatrix} \begin{pmatrix} P & 0 \\ 0 & QA^{-1}Q \end{pmatrix} \begin{pmatrix} P & PA^{-1}Q \\ 0 & Q \end{pmatrix} \begin{pmatrix} P & 0 \\ -QAP & Q \end{pmatrix}.\end{aligned} \tag{4.3}$$

This identity can be seen as a special case of the factorization

$$\begin{aligned}\begin{pmatrix} P_2AP_1 & 0 \\ 0 & Q_2 \end{pmatrix} &= \begin{pmatrix} P_2AP_1 & 0 \\ Q_2AP_1 & Q_2 \end{pmatrix} \begin{pmatrix} P_1 & 0 \\ -Q_2AP_1 & Q_2 \end{pmatrix} \\ &= \begin{pmatrix} P_2AP_1 & P_2AQ_1 \\ Q_2AP_1 & Q_2AQ_1 \end{pmatrix} \begin{pmatrix} P_1 & 0 \\ 0 & Q_1A^{-1}Q_2 \end{pmatrix} \begin{pmatrix} P_1 & P_1A^{-1}Q_2 \\ 0 & Q_2 \end{pmatrix} \begin{pmatrix} P_1 & 0 \\ -Q_2AP_1 & Q_2 \end{pmatrix}\end{aligned} \tag{4.4}$$

with a mapping scheme which is now in the last line given by (cf. (1.12))

$$\begin{matrix} Y_1 & & X_1 & & X_1 & & X_1 & & X_1 \\ \times & \leftarrow & \times & \leftarrow & \times & \leftarrow & \times & \leftarrow & \times \\ Y_2 & & X_2 & & Y_2 & & Y_2 & & Y_2 \end{matrix}. \tag{4.5}$$

Formula (4.4) is the motivation for the following.

Theorem 4.1. *Given any EAER between bounded linear operators $T : X_1 \to Y_1$ and $S : Y_2 \to X_2$, we obtain another EAER of the form (4.4), i.e., (up to a permutation of lines and rows for having S in the right position) in the special form*

$$\begin{pmatrix} T & 0 \\ 0 & I_2 \end{pmatrix} = \begin{pmatrix} * & T \\ * & * \end{pmatrix} \begin{pmatrix} S & 0 \\ 0 & I_1 \end{pmatrix} \begin{pmatrix} * & I_2 \\ * & * \end{pmatrix} \tag{4.6}$$

by putting $I_1 = I_{X_1}$, $I_2 = I_{Y_2}$ *and*

$$\begin{pmatrix} T & 0 \\ 0 & I_2 \end{pmatrix} = \begin{pmatrix} \alpha E_{11} & T \\ F_{12}E_{21} & -\beta F_{11} \end{pmatrix} \begin{pmatrix} S & 0 \\ 0 & I_1 \end{pmatrix} \begin{pmatrix} \beta F_{11} & I_2 \\ I_1 - F_{11}^{-}F_{11} & -\alpha F_{11} \end{pmatrix} \tag{4.7}$$

where $\alpha, \beta \in \mathbb{C}$, $\alpha\beta = 1$.

Proof. Now we could imitate the constructive proof of Theorem 2.1 to find the analogue of (3.2). But we know already a matrix A from (3.1) that, given an EAER (1.7), satisfies $P_2AP_1 = T$, $Q_1A^{-1}Q_2 = S$ (in the sense of the above identification). Hence we have the completion problem

$$\begin{pmatrix} T & 0 \\ 0 & I_2 \end{pmatrix} = \begin{pmatrix} T & * \\ * & * \end{pmatrix} \begin{pmatrix} I_1 & 0 \\ 0 & S \end{pmatrix} \begin{pmatrix} * & * \\ * & I_2 \end{pmatrix} \tag{4.8}$$

solved by (4.4) where A is given by (3.5), i.e.,

$$\begin{pmatrix} T & 0 \\ 0 & I_2 \end{pmatrix} = \begin{pmatrix} T & -E_{11} \\ F_{11} & F_{12}E_{21} \end{pmatrix} \begin{pmatrix} I_1 & 0 \\ 0 & S \end{pmatrix} \begin{pmatrix} I_1 - F_{11}^{-}F_{11} & F_{11}^{-} \\ -F_{11} & I_2 \end{pmatrix}. \tag{4.9}$$

□

Corollary 4.2. *The list of three equivalent statements in Corollary* 2.4 *can be expanded with another one, namely with*

4. *T and S satisfy a special EAER of the form* (4.6).

Remark 4.3. The problem to determine all EAERs (or all MCRs) holding between two operators from one given EAER (or from one given MCR) remains still open. Even the problem to determine all special EAERs of the form (4.1) is not yet solved.

5. Further relations with paired operators

The combination of the BT Theorem with the paired operators reasoning leads us to the following results.

Proposition 5.1. *Let X be a Banach space, $P, Q \in \mathcal{L}(X)$ be complementary projectors, and $A \in \mathcal{L}(X)$. The following operators are matrically coupled:*

$$\begin{aligned} T_P(A) &= PA|_{PX} &&: \quad PX \to PX \\ AP+Q\,,\ PA+Q\,,\ PAP+Q & &&: \quad X \to X \end{aligned}$$

If A is boundedly invertible, each of these four operators is matrically coupled with each of the following four operators:

$$\begin{aligned} T_Q(A^{-1}) &= QA^{-1}|_{QX} &&: \quad QX \to QX \\ A^{-1}Q+P\,,\ QA^{-1}+P\,,\ QA^{-1}Q+P & &&: \quad X \to X\,. \end{aligned}$$

Proof. Formulas (0.2) and (4.2) show that the operators are equivalent after extension, hence they are matrically coupled as a consequence of the BT Theorem. Transitivity of the EAERs yields the rest. □

Corollary 5.2. *Under the assumptions of Proposition* 5.1 *all of those operators belong to the same regularity class in the sense of the diagram. In particular all* *of them are invertible only simultaneously, and this is equivalent to each of the following facts:*

(a) im AP *and* im Q *are complemented subspaces of* X*,*
(b) im $A^{-1}Q$ *and* im P *are complemented subspaces of* X*.*

Remark 5.3. Formulas (0.2) and (4.2) in fact show that several of these operators (those in the second lines of the cited formulas) are equivalent, which is much stronger than equivalence after extension, and has much further reaching implications. In the case of positive Hilbert space operators, the difference is characterized by a subtle difference between two characterizing inequalities for the spectral projections in the two spaces, see[46], Section 4.

Now the asymmetric versions of these conclusions follow by analogy.

Proposition 5.4. *Let* X, Y, P_1, P_2 *be a basic setting and* $A \in \mathcal{L}(X,Y)$*. Further let* $X_1 = P_1X$ $(= \operatorname{im} P_1 = \ker Q_1)$, $X_2 = Q_1X$, $Y_1 = P_2Y$, $Y_2 = Q_2Y$*. The following operators are matrically coupled:*

$$W = P_2A|_{P_1X} : X_1 \to Y_1$$

$$\begin{pmatrix} P_2AP_1 & 0 \\ Q_2AP_1 & Q_2 \end{pmatrix}, \begin{pmatrix} P_2AP_1 & 0 \\ 0 & Q_2 \end{pmatrix} : \begin{matrix} X_1 \\ \times \\ Y_2 \end{matrix} \to \begin{matrix} Y_1 \\ \times \\ Y_2 \end{matrix} \tag{5.1}$$

$$\begin{pmatrix} P_2AP_1 & P_2AQ_1 \\ 0 & Q_1 \end{pmatrix}, \begin{pmatrix} P_2AP_1 & 0 \\ 0 & Q_1 \end{pmatrix} : \begin{matrix} X_1 \\ \times \\ X_2 \end{matrix} \to \begin{matrix} Y_1 \\ \times \\ X_2 \end{matrix} .$$

If A *is an isomorphism, all of them are matrically coupled with*

$$W_* = Q_1A^{-1}|_{Q_2} : Y_2 \to X_2$$

$$\begin{pmatrix} Q_1A^{-1}Q_2 & 0 \\ P_1A^{-1}Q_2 & P_1 \end{pmatrix}, \begin{pmatrix} Q_1A^{-1}Q_2 & 0 \\ 0 & P_1 \end{pmatrix} : \begin{matrix} Y_2 \\ \times \\ X_1 \end{matrix} \to \begin{matrix} X_2 \\ \times \\ X_1 \end{matrix} \tag{5.2}$$

$$\begin{pmatrix} Q_1A^{-1}Q_2 & Q_1A^{-1}P_2 \\ 0 & P_2 \end{pmatrix}, \begin{pmatrix} Q_1A^{-1}Q_2 & 0 \\ 0 & P_2 \end{pmatrix} : \begin{matrix} Y_2 \\ \times \\ Y_1 \end{matrix} \to \begin{matrix} X_2 \\ \times \\ Y_1 \end{matrix} .$$

Proof. Similarly as in the previous proof, the formulas (1.12) and (4.3) give us the first block of formulas. The second follows by an exchange of the roles of A and A^{-1}. Again the BT Theorem and transitivity of the MCRs imply the rest. □

Corollary 5.5. *Under the assumptions of Proposition 5.4 all of the operators in (5.1) and (5.2) belong to the same regularity class. They are invertible only simultaneously, and this is equivalent to each of the following facts:*

(a) im AP_1 *and* im Q_2 *are complemented subspaces of* Y*,*
(b) im $A^{-1}Q_2$ *and* im P_1 *are complemented subspaces of* X*.*

Remark 5.6. The last two conditions attracted considerable attention under the name "geometric perspective of general Wiener–Hopf operators" [37, 38] and go back to [18] (see also [31]). However, the present conclusion is a little different as based upon the asymmetric version of paired operators. The very last equivalence between the two lines in the last corollary is obvious from itself.

6. Applications

The main applications of the BT Theorem consist roughly speaking in the conclusion: If S and T satisfy a MCR, then we can transfer nice properties from S to T and vice versa. Examples can be found in various fields ranging from abstract algebraic settings [41] to concrete applications in diffraction theory, for instance. Here we outline a few of them. The first two are known from [14] and [40], respectively. They coincide partly with the previous results, after an identification of the operators $T = P_2A|_{P_1X}$ etc. as done in Section 1. However the outcome is considered to deserve independent interest due to their different notation and concrete applications, cf. corresponding remarks in [3].

Theorem 6.1. *Let S and T satisfy a MCR* (1.6). *Then the two operators belong to the same regularity class in the sense of the classification in Remark 1.5.3.*

Proof. (Sketch) The BT Theorem implies that the two matrically coupled operators are equivalent after extension which yields the statement after substitution according to Corollary 5.5. □

Theorem 6.2. *Let S and T be bounded linear operators in Banach spaces which are matrically coupled, i.e., $T = W = P_2A|_{P_1X}$ and $S = W_* = Q_1A^{-1}|_{Q_2Y}$ in the above notation. Further let V be a generalized inverse of W, i.e., $WVW = W$. Then a generalized inverse of W_* is given by*

$$V_* = Q_2(A - AP_1VP_2A)|_{Q_1X}. \tag{6.1}$$

Proof. The assumptions imply that

$$L = \begin{pmatrix} P_2AP_1 & 0 \\ 0 & Q_1 \end{pmatrix} = \begin{pmatrix} P_2AP_1 & 0 \\ 0 & Q_1 \end{pmatrix} \begin{pmatrix} V & 0 \\ 0 & Q_1 \end{pmatrix} \begin{pmatrix} P_2AP_1 & 0 \\ 0 & Q_1 \end{pmatrix} = L\,L^-\,L \tag{6.2}$$

where L^- is a generalized inverse of L. Writing (1.12) briefly as

$$L = \quad E \quad K \quad F$$

$$K = \begin{pmatrix} P_2 & 0 \\ 0 & Q_1A^{-1}Q_2 \end{pmatrix}$$

we conclude that $K = E^{-1}\,L\,F^{-1}$. According to the reverse order law

$$\begin{aligned} K^- &= F \quad L^- \quad E \\ &= \begin{pmatrix} P_2AP_1 & P_2AQ_1 \\ Q_2AP_1 & Q_2AQ_1 \end{pmatrix} \begin{pmatrix} V & 0 \\ 0 & Q_1 \end{pmatrix} \begin{pmatrix} P_2 & -P_2AQ_1 \\ 0 & Q_1 \end{pmatrix} \begin{pmatrix} P_2 & 0 \\ Q_1A^{-1}P_2 & Q_1 \end{pmatrix} \end{aligned}$$

represents a generalized inverse of K, i.e., $KK^{-}K = K$. Taking the last entry (indexed 22) of that operator matrix we obtain

$$R_{22}\left\{\begin{pmatrix} P_2AP_1V & P_2AQ_1 \\ Q_2AP_1V & Q_2AQ_1 \end{pmatrix}\begin{pmatrix} P_2 - P_2AQ_1A^{-1}P_2 & -P_2AQ_1 \\ Q_1A^{-1}P_2 & Q_1 \end{pmatrix}\right\}$$
$$= Q_2AQ_1 - Q_2AP_1VP_2AQ_1$$

which coincides with (6.1) in the sense of the above-mentioned identification. □

Remarks 6.3.

1. Although the conclusion runs via an EAER, the data of E and F do not enter explicitly into (6.1). That is because the EAER is directly obtained from the MCR (by use of paired operators).
2. The two formulas (1.12) and (4.4) lead to the same result (6.1).
3. Another earlier proof of the last theorem can be found in [41], pp. 21–22. It did not make use of an EAER but of a relation with paired operators, as well.
4. Clearly the formulation of Theorem 6.2 includes the most interesting properties of one-sided invertibility, Fredholmness, and invertibility as particular cases.
5. By symmetry we obtain that
$$V = Q_2(A^{-1} - A^{-1}Q_2V_*Q_1A^{-1})|_{P_2Y} \tag{6.3}$$
is a generalized inverse of W provided $W_*V_*W_* = W_*$.
6. The operations (6.1) and (6.3) are inverse to each other in the sense that the step from V_* to V leads back to V if V_* was calculated by (6.1). This fact again results from the paired operators reasoning.

Corollary 6.4. *Let W and W_* be matrically coupled (in the notation of (1.1)) and let V and V_* be related by (6.1) or by (6.3). Then V is a reflexive generalized inverse of W if and only if V_* is a reflexive generalized inverse of W_*.*

Remark 6.5. There are plenty of realizations of general WHOs in the theory of singular operators [25, 26, 36] including, e.g., Toeplitz operators [8], classical WHOs in Sobolev spaces [19, 23, 34], and Wiener–Hopf plus Hankel operators [15, 16, 20].

Finally we like to mention also a prominent class of examples from the area of integral (and pseudo-differential) equations without going into details. It is well known that convolution operators on a finite interval can be "reduced" to matrix WHOs (convolution operators on the half-line) [7]. This step can be considered as an EAER with paired operators in various space settings, also including operators with transmission property, see [12], Section 2. Now the next step of relating, e.g., convolution operators on a finite interval with those acting on its complement (the union of two separated semi-axes) by a MCR is an evident and elegant task which admits results similar to those discussed before.

7. Examples from diffraction theory

A concrete realization of the MCR (1.6) can be found in the theory of diffraction of electromagnetic or acoustic waves from plane screens in $\mathbb{R}^3$. For precise formulation we need some preparation. Given a proper open subset $\Sigma \subset \mathbb{R}^2$ with $\operatorname{int}\operatorname{clos}\Sigma = \Sigma$, we consider the domain Ω defined by

$$\begin{aligned} \Omega &= \mathbb{R}^3 \setminus \Gamma \\ \Gamma &= \overline{\Sigma} \times 0 = \{x = (x_1, y_1, 0) \in \mathbb{R}^3 : x' = (x_1, y_1) \in \overline{\Sigma}\}. \end{aligned} \tag{7.1}$$

Problems of diffraction from a plane screen Γ are often formulated in terms of (or reduced to) the solution of the three-dimensional Helmholtz equation in Ω with Dirichlet or Neumann conditions on Γ, briefly written as

$$\begin{array}{rclll} \left(\Delta + k^2\right) u & = & 0 & \text{in} & \Omega \\ Bu & = & g & \text{on} & \Gamma = \partial\Omega. \end{array} \tag{7.2}$$

Herein k is the wave number and we assume that $\Im m\, k > 0$. B stands for the boundary operator, taking the trace or normal derivative of u on Γ (in the same direction on both banks of Γ). We think of the weak formulation looking for $u \in \mathcal{H}^1(\Omega)$, i.e., u is defined in Ω and its restriction $u|_{\Omega^*} \in H^1(\Omega^*)$ is a weak solution of the Helmholtz equation in any special Lipschitz subdomain $\Omega^* \subset \Omega$, see [13, 21, 22] for details. Boundary data $g \in H^{1/2}(\Sigma)$ (in the Dirichlet problem) or $g \in H^{-1/2}(\Sigma)$ (in the Neumann problem) are arbitrarily given (to study continuous dependence from the data in well-posed problems) and one looks for the resolvent operator as the inverse to the boundary (trace or normal derivative) operator

$$B \quad = \quad \mathcal{H}^1(\Omega) \longrightarrow H^{\pm 1/2}(\Sigma). \tag{7.3}$$

Representation formulas for a function $u \in \mathcal{H}^1(\Omega)$ in the half-spaces $\{x \in \mathbb{R}^3 : \pm x_3 > 0\}$ yield that, in case of the Dirichlet problem, B is equivalent to a boundary pseudo-differential operator which has the form of a general WHO

$$W \;=\; P_2 A_t^{-1}|_{\operatorname{im} P_1} : H_{\Sigma}^{-1/2} \to \operatorname{im} P_2 \subset H^{1/2}(\mathbb{R}^2). \tag{7.4}$$

Herein $A_t = \mathcal{F}^{-1} t \cdot \mathcal{F} : H^{1/2}(\mathbb{R}^2) \to H^{-1/2}(\mathbb{R}^2)$ is a convolution operator with Fourier symbol $t(\xi) = (\xi^2 - k^2)^{1/2}$, P_1 is a projector in $H^{-1/2}$ with $\operatorname{im} P_1 = H_{\Sigma}^{-1/2}$ denoting the subspace of distributions supported on $\overline{\Sigma}$, and P_2 a projector in $H^{1/2}$ with $\ker P_2 = H_{\Sigma'}^{1/2}$, the subspace of functions supported on the closure $\overline{\Sigma'}$ of the complementary screen $\Sigma' = \mathbb{R}^2 \setminus \overline{\Sigma}$.

By analogy we obtain an operator for the Neumann problem due to the complementary screen that has the form

$$W_* = Q_1 A_t|_{\operatorname{im} Q_2} : H_{\Sigma'}^{1/2} \to \operatorname{im} Q_1 \subset H^{-1/2}(\mathbb{R}^2) \tag{7.5}$$

with projectors $Q_1 = I - P_1$ and $Q_2 = I - P_2$ complementary to the previous. The operators (7.4) and (7.5) meet the (abstract) situation of (1.10), they are matrically coupled and consequently the inverses can be computed from each other according to Theorem 6.2 and Remark 6.3.5.

Now the point is that the resolvent operator for certain screens are much simpler to obtain than the resolvent operators for the complementary screens. This happens for instance in case of the quarter-plane [42] or other convex so-called polygonal-conical screens [13], defined as the intersection of a finite number of half-planes. By the help of Theorem 6.2, diffraction problems can be solved for screens which are the complement of a convex polygonal-conical set and, with some additional techniques, also for screens which are finite unions of such sets, see [13] for details. The basis of the last mentioned idea is the following result that has also to do with the paired operators reasoning.

Theorem 7.1. *Let $A \in \mathcal{L}(X, Y)$ be boundedly invertible and $W = P_2 A|_{P_1 X}$ defined as in* (1.10). *Then W is invertible if and only if AP_1X and $Q_2Y = (I - P_2)Y$ are complemented subspaces of Y. In this case, the inverse of W is given by*

$$W^{-1} \quad = \quad A^{-1}\Pi|_{P_2Y} \quad = \quad P_2Y \to P_1X \tag{7.6}$$

where $\Pi \in \mathcal{L}(Y)$ is the projector onto AP_1X along Q_2Y.

Proof. (Sketch) It is a consequence of a formula that appeared when deriving formula (1.12), namely the relation between a general WHO and a paired operator in the first line of (4.4) taking into account that $P_2AP_1 + Q_2AP_1 = AP_1$. Indeed the argument is more visible in the symmetric case and scalar writing, see (4.1). However it works in the general case, as well. See [13], Section 3.2, for more details. □

Remark 7.2. The idea of Theorem 7.1 originated from diffraction theory [31, 32], was formalized in an abstract setting in [37, 38] and further developed in [13, 42].

Example. Concrete applications can be found in the screen diffraction problems described before. In the simplest case, we consider the wave number $k = i$, as a model problem. Hence $t(\xi) = (\xi^2 + 1)^{1/2}$ and we take $A = A_t$. Then the above-mentioned projectors are orthogonal, since

$$H^{1/2}_{\Sigma} \perp A_t^{-1} H^{-1/2}_{\Sigma'} \quad , \quad H^{-1/2}_{\Sigma} \perp A_t H^{1/2}_{\Sigma'}$$

for any admissible Σ.

Now let Σ be a convex cone, thus intersection of two half-planes Σ_1 and Σ_2, say. In case of the Dirichlet problem, the orthogonal projector $P_{1,\Sigma}$ in $H^{-1/2}$ onto $H^{-1/2}_{\Sigma}$ can be represented as the infimum of the two orthogonal projectors P_{1,Σ_1} in $H^{-1/2}$ onto $H^{-1/2}_{\Sigma_1}$ and P_{1,Σ_2} onto $H^{-1/2}_{\Sigma_2}$, i.e.,

$$P_{1,\Sigma} \quad = \quad P_{1,\Sigma_1} \wedge P_{1,\Sigma_2} \quad = \quad \prod_{j=1}^{\infty} (P_{1,\Sigma_1} P_{1,\Sigma_2})^j \tag{7.7}$$

in the sense of strong convergence, see [28], Problem 96. For the complementary screen, this argument is not directly applicable. However, one can determine the resolvent with formula (7.6), further the resolvent for the Neumann problem and the complementary screen with formula (6.1), and even the orthogonal projector

$P_{1,\Sigma'}$ in $H^{1/2}$ onto $H^{1/2}_{\Sigma'}$ for the complementary screen. More details and further results can be found in [13, 31, 42]. Finally it has been shown that, with this technique, diffraction problems can be explicitly solved for domains Σ which belong to the set algebra generated by half-planes, i.e., for sets of the form

$$\Sigma \quad = \quad \text{int} \bigcup_{j=1,\dots,m} \text{clos}\, \Sigma_j$$

where Σ_j are intersections of finite sets of half-planes.

Acknowledgment

The present work was supported by FCT – Portuguese Science Foundation through the Center for Functional Analysis, Linear Structures, and Applications at Instituto Superior Técnico, Universidade de Lisboa and by a voluntary agreement with the Instituto Superior Técnico.

The author likes to thank Albrecht Böttcher, Luís Castro, Sanne ter Horst and André Ran for recent discussions about the topic and historical information. He also likes to express his gratitude for fruitful discussions with Harm Bart, Israel Gohberg and Rien Kaashoek about basic questions of operator relations (which started already in the 1980s).

References

[1] D.Z. Arov and H. Dym, *$\mathcal{J}$-contractive matrix-valued functions and related topics.* Encyclopedia of Mathematics and its Applications **116**. Cambridge University Press, Cambridge 2008.

[2] H. Bart, I. Gohberg, and M. Kaashoek, *The coupling method for solving integral equations.* Oper. Theory Adv. Appl. **2** (1984), 39–73.

[3] H. Bart, I. Gohberg, and M. Kaashoek, *Addendum to: The coupling method for solving integral equations.* Integral Equations Oper. Theory **8** (1985), 890–891.

[4] H. Bart, I. Gohberg, M. Kaashoek, and A.C.M. Ran, *Schur complements and state space realizations.* Linear Algebra Appl. **399** (2005), 203–224.

[5] H. Bart, I. Gohberg, M. Kaashoek, and A.C.M. Ran, *A state space approach to canonical factorization with applications.* Oper. Theory Adv. Appl. **200**, Birkhäuser, Basel 2010.

[6] H. Bart and V.E. Tsekanovskii, *Matrical coupling and equivalence after extension.* In: Operator Theory and Complex Analysis (eds.: T. Ando et al.), Oper. Theory Adv. Appl. **59**, Birkhäuser, Basel 1991, 143–160.

[7] A. Böttcher, Yu.I. Karlovich, and I.M. Spitkovsky, *Convolution operators and factorization of almost periodic matrix functions.* Birkhäuser, Basel 2002.

[8] A. Böttcher and B. Silbermann *Analysis of Toeplitz Operators.* Springer, Berlin 2006.

[9] A. Böttcher and F.-O. Speck *On the symmetrization of general Wiener–Hopf operators.* J. Operator Theory **76** (2016), 335–349.

[10] D. Carlson, *What are Schur complements, anyway?* Linear Algebra Appl. **74** (1986), 257–275.

[11] L.P. Castro, R. Duduchava, and F.-O. Speck, *Localization and minimal normalization of some basic mixed boundary value problems.* In: Factorization, Singular Operators and Related Problems (eds.: S. Samko et al.), Kluwer, Dordrecht 2003, 73–100.

[12] L.P. Castro, R. Duduchava, and F.-O. Speck, *Finite interval convolution operators with transmission property.* Integral Equations Oper. Theory **52** (2005), 165–179.

[13] L.P. Castro, R. Duduchava, and F.-O. Speck, *Diffraction from polygonal-conical screens – an operator approach.* In: Operator Theory, Operator Algebras and Applications (eds.: A. Bastos et al.), Oper. Theory Adv. Appl. **242**, Birkhäuser, Basel 2014, 113–137.

[14] L.P. Castro and F.-O. Speck, *Regularity properties and generalized inverses of delta-related operators.* Z. Anal. Anwend. **17** (1998), 577–598.

[15] L.P. Castro, F.-O. Speck, and F.S. Teixeira, *On a class of wedge diffraction problems posted by Erhard Meister.* Oper. Theory Adv. Appl. **147** (2004), 211–238.

[16] L.P. Castro, F.-O. Speck, and F.S. Teixeira, *Mixed boundary value problems for the Helmholtz equation in a quadrant.* Integr. Equ. Oper. Theory **56** (2006), 1–44.

[17] G.N. Cebotarev, *Several remarks on the factorization of operators in a Banach space and the abstract Wiener–Hopf equation.* Mat. Issled. **2** (1967), 215–218 (in Russian).

[18] A. Devinatz and M. Shinbrot, *General Wiener–Hopf operators.* Trans. AMS **145** (1969), 467–494.

[19] R. Duduchava, *Integral Equations in Convolution with Discontinuous Presymbols, Singular Integral Equations with Fixed Singularities, and Their Applications to Some Problems of Mechanics.* Teubner, Leipzig 1979.

[20] T. Ehrhardt, *Invertibility theory for Toeplitz plus Hankel operators and singular integral operators with flip.* J. Funct. Anal. **208** (2004), 64–106.

[21] T. Ehrhardt, A.P. Nolasco, and F.-O. Speck, *Boundary integral methods for wedge diffraction problems: the angle $2\pi/n$, Dirichlet and Neumann conditions.* Operators and Matrices **5** (2011), 1–39.

[22] T. Ehrhardt, A.P. Nolasco, and F.-O. Speck, *A Riemann surface approach for diffraction from rational angles.* Operators and Matrices, Operators and Matrices **8** (2014), 301–355.

[23] G.I. Èskin, *Boundary Value Problems for Elliptic Pseudodifferential Equations.* Translations of Mathematical Monographs **52**, AMS, Providence, R. I., 1981 (Russian edition 1973).

[24] I.Z. Gohberg and I.A. Feldman, *Faltungsgleichungen und Projektionsverfahren zu ihrer Lösung.* (German), Birkhäuser and Akademie-Verlag, Berlin 1974 (Russian edition 1971).

[25] I. Gohberg, S. Goldberg, and M.A. Kaashoek *Classes of linear operators. Vol. I.* Oper. Theory Adv. Appl. **49**, Birkhäuser, Basel 1990.

[26] I. Gohberg, S. Goldberg, and M.A. Kaashoek *Classes of linear operators. Vol. II.* Oper. Theory Adv. Appl. **63**, Birkhäuser, Basel 1993.

[27] I. Gohberg and N. Krupnik, *Einführung in die Theorie der eindimensionalen singulären Integraloperatoren.* Birkhäuser, Basel, 1979 (Russian edition 1973).

[28] P.R. Halmos, *A Hilbert Space Problem Book.* second edition, Springer, New York, 1982.

[29] S. ter Horst, M. Messerschmidt, and A.C.M. Ran, *Equivalence after extension for compact operators on Banach spaces.* J. Math. Anal. Appl. **431** (2015), 136–149.

[30] S. ter Horst and A.C.M. Ran, *Equivalence after extension and matricial coupling coincide with Schur coupling, on separable Hilbert spaces.* Linear Algebra Appl. **439** (2013), 793–805.

[31] E. Meister and F.-O. Speck, *Scalar diffraction problems for Lipschitz and polygonal screens.* Z. Angew. Math. Mech. **67** (1987), T 434–T 435.

[32] E. Meister and F.-O. Speck, *A contribution to the quarter-plane problem in diffraction theory.* J. Math. Anal. Appl. **130** (1988), 223–236.

[33] E. Meister and F.-O. Speck, *Modern Wiener–Hopf methods in diffraction theory.* Ordinary and Partial Differential Equations **2**, Pitman Res. Notes Math. Ser. **216** (Longman, London, 1989), 130–171.

[34] A. Moura Santos, F.-O. Speck, and F.S. Teixeira, *Minimal normalization of Wiener–Hopf operators in spaces of Bessel potentials.* J. Math. Anal. Appl. **225** (1998), 501–531.

[35] V.V. Peller, *Hankel Operators and Their Applications.* Springer, New York, 2003.

[36] S. Prössdorf, *Some Classes of Singular Equations.* North-Holland, Amsterdam, 1978 (German edition 1974).

[37] A.F. dos Santos, *Abstract Wiener–Hopf Operators – Geometric Perspective.* Thesis on the occasion of "Provas de Agregação" (in Portuguese), Departamento de Matemática, Instituto Superior Técnico, Universidade Técnica de Lisboa, Lisbon, 1988.

[38] A.F. dos Santos, *General Wiener–Hopf operators and representation of their generalized inverses.* Oper. Theory Adv. Appl. **41** (1989), 473–483.

[39] M. Shinbrot, *On singular integral operators.* J. Math. Mech. **13** (1964), 395–406.

[40] F.-O. Speck, *On the generalized invertibility of Wiener–Hopf operators in Banach spaces.* Integr. Equat. Oper. Th. **6** (1983), 458–465.

[41] F.-O. Speck, *General Wiener–Hopf Factorization Methods.* Pitman, London 1985.

[42] F.-O. Speck, *Diffraction by a three-quarter-plane using an abstract Babinet principle.* Z. Angew. Math. Mech. **93** (2012), 485–491.

[43] F.-O. Speck, *Wiener–Hopf factorization through an intermediate space.* Integr. Equ. Oper. Theory **82** (2015), 395–415.

[44] F.-O. Speck, *A class of interface problems for the Helmholtz equation in* $\mathbb{R}^n$. Math. Meth. Appl. Sciences **40** (2017), 391–403.

[45] J.J. Sylvester, *On the relation between the minor determinants of linearly equivalent quadratic functions.* Philos. Magazine **14** (1851), 295–305.

[46] D. Timotin, *Schur coupling and related equivalence relations for operators on a Hilbert space.* Linear Algebra Appl. **452** (2014), 106–119.

Frank-Olme Speck
Departamento de Matemática, Instituto Superior Técnico
Universidade de Lisboa, Avenida Rovisco Pais
P-1049-001 Lisboa, Portugal
e-mail: `fspeck@tecnico.pt`

Operator Theory:
Advances and Applications, Vol. 259, 703–718

Natural Boundary for a Sum Involving Toeplitz Determinants

Craig A. Tracy and Harold Widom

For Albrecht Böttcher on the occasion of his sixtieth birthday

Abstract. In the theory of the two-dimensional Ising model, the *diagonal susceptibility* is equal to a sum involving Toeplitz determinants. In terms of a parameter k the diagonal susceptibility is analytic for $|k| < 1$, and the authors proved the conjecture that this function has the unit circle as a natural boundary. The symbol of the Toepltiz determinants was a k-deformation of one with a single singularity on the unit circle. Here we extend the result, first, to deformations of a larger class of symbols with a single singularity on the unit circle, and then to deformations of (almost) general Fisher–Hartwig symbols.

Mathematics Subject Classification (2010). 47B35, 30E99, 82B20.

Keywords. Toeplitz determinants, natural boundary, 2D Ising susceptibility.

1. Introduction

In the theory of the two-dimensional Ising model there is a quantity, depending on a parameter k, called the *magnetic susceptibility*, which is analytic for $|k| < 1$. It is an infinite sum over M, $N \in \mathbb{Z}$ involving correlations between the spins at sites $(0, 0)$ and (M, N). It was shown in [12] to be representable as a sum over $n \geq 1$ of n-dimensional integrals. In [8] B. Nickel found a set of singularities of these integrals which became dense on the unit circle as $n \to \infty$. This led to the (as yet unproved) *natural boundary conjecture* that the unit circle is a natural boundary for the susceptibility.

Subsequently [4] a simpler model was introduced, called the *diagonal susceptibility*, in which the sum of correlations was taken over the diagonal sites (N, N). These correlations were equal to Toeplitz determinants, and the diagonal susceptibility was expressible in terms of a sum involving Toeplitz determinants.

The Toeplitz determinant $D_N(\varphi)$ is $\det(\varphi_{i-j})_{1\le i,j\le N}$, where φ_j is the jth Fourier coefficient of the symbol φ defined on the unit circle. The sum in question is

$$\sum_{N=1}^{\infty}[D_N(\varphi)-\mathcal{M}^2], \quad \text{where} \quad \varphi(\xi)=\sqrt{\frac{1-k/\xi}{1-k\,\xi}},$$

and $\mathcal{M}$, the spontaneous magnetization, is equal to $(1-k^2)^{1/8}$. This also (as we explain below) is equal to a sum of n-dimensional integrals, the sum is analytic for $|k|<1$, and the singularities of these summands also become dense on the unit circle as $n\to\infty$. This led to a natural boundary conjecture for the diagonal susceptibility, which we proved in [11].

The question arises whether the occurrence of the natural boundary is a statistical mechanics phenomenon and/or a Toeplitz determinant phenomenon. This note shows that at least the latter is true. We consider here the more general class of symbols

$$\varphi(\xi)=(1-k\,\xi)^{\alpha_+}\,(1-k\,/\xi)^{\alpha_-}\,\psi(\xi), \tag{1}$$

where ψ is a nonzero function analytic in a neighborhood of the unit circle with winding number zero and geometric mean one. We assume $\alpha_\pm\notin\mathbb{Z}$, $\operatorname{Re}\alpha_\pm<1$.[1] The parameter k satisfies $|k|<1$. We define

$$\chi(k)=\sum_{N=1}^{\infty}[D_N(\varphi)-E(\varphi)], \tag{2}$$

where

$$E(\varphi)=\lim_{N\to\infty}D_N(\varphi).\,^2$$

Each summand in (2) is analytic in the unit disc $|k|<1$, the only singularities on the boundary being at $k=\pm1$, and the series converges uniformly on compact subsets. Therefore $\chi(k)$ is analytic in the unit disc.

Theorem 1. *The unit circle $|k|=1$ is a natural boundary for $\chi(k)$.*

The result in [11] was established, and here will be established, by showing that the singularities of the nth summand of the series are not canceled by the infinitely many remaining terms of the series.[3] We shall see that a certain derivative of the nth term is unbounded as k^2 tends to an nth root of unity while the same derivative of the sum of the later terms is bounded, and if it is a primitive nth root the same derivative of each earlier term is also bounded.

---

[1]Observe that we never consider a limiting symbol with k on the unit circle, and so do not require that $\operatorname{Re}\alpha_\pm>-1$. The condition stated is one that will be needed for the integrability on the unit circle of a function that arises in the proof of the theorem.

[2]When $\psi(\xi)=1$ it equals $(1-k^2)^{-\alpha_+\alpha_-}$. In general it equals this times a function that extends analytically beyond the unit disc.

[3]A nice example [9] where such a cancelation does occur is

$$\frac{z}{1-z}=\frac{z}{1-z^2}+\frac{z^2}{1-z^4}+\frac{z^4}{1-z^8}+\cdots+\frac{z^{2^n}}{1-z^{2^{n+1}}}+\cdots.$$

To put what we have done into some perspective, we start with a symbol

$$(1-\xi)^{\alpha_+}\,(1-1/\xi)^{\alpha_-}.$$

Then we introduce its k-deformation, times a "nice" function $\psi(\xi)$, and consider their Toeplitz determinants as functions of the parameter k inside the unit circle. Is it important that we begin with a symbol with only one singularity on the boundary? It is not. We may begin instead with a general Fisher–Hartwig symbol [5]

$$\prod_{p=1}^{P}(1-u_p\,\xi)^{\alpha_p^+}\;\prod_{q=1}^{Q}(1-v_q/\xi)^{\alpha_q^-},$$

where $|u_p|$, $|v_q|=1$ and P, $Q>0$. With some conditions imposed on the α_p^+ and the α_q^-, we show that the conclusion of the theorem holds for the deformations of these symbols.

Here is an outline of the paper. In the next section we derive the expansion for $\chi(k)$ as a series of multiple integrals. In the following section the theorem is proved, and in the section after that we show how to extend the result to (almost) general Fisher–Hartwig symbols. In two appendices we give the proof of a proposition used in Section II and proved in [11], and discuss a minimum question that arises in Section IV.

2. Preliminaries

We invoke the formula of Geronimo–Case [6] and Borodin–Okounkov [2] to write the Toeplitz determinant in terms of the Fredholm determinant of a product of Hankel operators. The Hankel operator $H_N(\varphi)$ is the operator on $\ell^2(\mathbb{Z}^+)$ with kernel $(\varphi_{i+j+N+1})_{i,j\ge 0}$.

We have a factorization $\varphi(\xi)=\varphi_+(\xi)\,\varphi_-(\xi)$, where φ_+ extends analytically inside the unit circle and φ_- outside, and $\varphi_+(0)=\varphi_-(\infty)=1$. More explicitly,

$$\varphi_+(x)=(1-k\,\xi)^{\alpha_+}\,\psi_+(\xi)\quad\text{and}\quad\varphi_-(\xi)=(1-k/\xi)^{\alpha_-}\,\psi_-(\xi).$$

If $\psi(\xi)$ is analytic and nonzero for $s<|\xi|<s^{-1}$ then $\psi_+(\xi)$ resp. $\psi_-(\xi)$ is analytic and nonzero for $|\xi|<s^{-1}$ resp. $|\xi|>s$.

The formula of G-C/B-O is

$$D_N(\varphi)=E(\varphi)\,\det\Big(I-H_N\Big(\frac{\varphi_-}{\varphi_+}\Big)\,H_N\Big(\frac{\tilde\varphi_+}{\tilde\varphi_-}\Big)\Big),$$

where for a function f we define $\tilde f(\xi)=f(\xi^{-1})$. Thus, if we write

$$\Lambda(\xi)=\frac{\varphi_-(\xi)}{\varphi_+(\xi)}=(1-k\,\xi)^{-\alpha_+}\,(1-k/\xi)^{\alpha_-}\,\frac{\psi_-(\xi)}{\psi_+(\xi)},$$

$$K_N=H_N(\Lambda)\,H_N(\tilde\Lambda^{-1}),\tag{3}$$

then $\chi(k)$ equals $E(\varphi)$ times

$$\mathcal{S}(k)=\sum_{N=1}^{\infty}\left[\det(I-K_N)-1\right].$$

In [11] the following was proved. We give the proof in Appendix A.

Proposition. *Let* $H_N(du)$ *and* $H_N(dv)$ *be two Hankel matrices acting on* $\ell^2(\mathbb{Z}^+)$ *with* i,j *entries*

$$\int x^{N+i+j}\,du(x), \qquad \int y^{N+i+j}\,dv(y), \tag{4}$$

respectively, where u *and* v *are measures supported in the open unit disc. Set* $K_N = H_N(du)\,H_N(dv)$. *Then*

$$\sum_{N=1}^{\infty}[\det(I-K_N)-1]$$
$$=\sum_{n=1}^{\infty}\frac{(-1)^n}{(n!)^2}\int\cdots\int\frac{\prod_i x_i y_i}{1-\prod_i x_i y_i}\left(\det\left(\frac{1}{1-x_i y_j}\right)\right)^2\prod_i du(x_i)\,dv(y_i),$$

where indices in the integrand run from 1 *to* n.

We apply this to the operator $K_N = H_N(\Lambda)\,H_N(\tilde{\Lambda}^{-1})$ given by (3). The matrix for $H_N(\Lambda)$ has i,j entry

$$\frac{1}{2\pi i}\int\Lambda(\xi)\,\xi^{-N-i-j-2}\,d\xi,$$

where the integration is over the unit circle. The integration may be taken over a circle with radius in $(1,|k|^{-1})$ as long as $|k|>s$. (Recall that $\psi_\pm(\xi)$ are analytic and nonzero for $s<|\xi|<s^{-1}$.) We assume this henceforth.

Setting $\xi = 1/x$ we see that the entries of $H_N(\Lambda)$ are given as in (4) with

$$du(x)=\frac{1}{2\pi i}\,\Lambda(x^{-1})\,dx,$$

and integration is over a circle $\mathcal{C}$ with radius in $(|k|,1)$. Similarly, $H_N(\tilde{\Lambda}^{-1}) = H_N(dv)$ where in (4)

$$dv(y)=\frac{1}{2\pi i}\,\Lambda(y)^{-1}\,dy,$$

with integration over the same circle $\mathcal{C}$.

Hence the proposition gives

$$\mathcal{S}(k)=\sum_{n=1}^{\infty}\mathcal{S}_n(k), \tag{5}$$

where

$$\mathcal{S}_n(k)=\frac{(-1)^n}{(n!)^2}\frac{1}{(2\pi i)^{2n}}$$
$$\times\int\cdots\int\frac{\prod_i x_i y_i}{1-\prod_i x_i y_i}\left(\det\left(\frac{1}{1-x_i y_j}\right)\right)^2\prod_i\frac{\Lambda(x_i^{-1})}{\Lambda(y_i)}\prod_i dx_i\,dy_i,$$

with all integrations over $\mathcal{C}$.

We deform each $\mathcal{C}$ to the circle with radius $|k|$ (after which there are integrable singularities on the contours). Then we make the substitutions $x_i \to kx_i$, $y_i \to ky_i$, and obtain

$$\mathcal{S}_n(k) = \frac{(-1)^n}{(n!)^2} \frac{\kappa^{2n}}{(2\pi i)^{2n}} \tag{6}$$
$$\times \int \cdots \int \frac{\prod_i x_i y_i}{1 - \kappa^n \prod_i x_i y_i} \left(\det\left(\frac{1}{1-\kappa x_i y_j}\right)\right)^2 \prod_i \frac{\Lambda(k^{-1}x_i^{-1})}{\Lambda(ky_i)} \prod_i dx_i\, dy_i,$$

where integrations are on the unit circle. We record that

$$\frac{\Lambda(k^{-1}x^{-1})}{\Lambda(ky)} = \frac{(1-\kappa x)^{\alpha_-}}{(1-\kappa y)^{-\alpha_+}} \frac{(1-x^{-1})^{-\alpha_+}}{(1-y^{-1})^{\alpha_-}} \frac{\rho(k^{-1}x^{-1})}{\rho(ky)}, \tag{7}$$

where we have set

$$\kappa = k^2, \quad \rho(x) = \frac{\psi_-(x)}{\psi_+(x)}.$$

The complex planes are cut from κ^{-1} to ∞ for the first quotient in (7) and from 0 to 1 for the second quotient.

Using the fact that the determinant in the integrand is a Cauchy determinant we obtain the alternative expression

$$\mathcal{S}_n(k) = \frac{(-1)^n}{(n!)^2} \frac{\kappa^{n(n+1)}}{(2\pi i)^{2n}} \tag{8}$$
$$\times \int \cdots \int \frac{\prod_i x_i y_i}{1 - \kappa^n \prod_i x_i y_i} \frac{\Delta(x)^2\, \Delta(y)^2}{\prod_{i,j}(1-\kappa\, x_i y_j)^2} \prod_i \frac{\Lambda(k^{-1}x_i^{-1})}{\Lambda(ky_i)} \prod_i dx_i\, dy_i,$$

where $\Delta(x)$ and $\Delta(y)$ are Vandermonde determinants

3. Proof of Theorem 1

For any $\delta < 1 - s$ we can deform each contour of integration to one that goes back and forth along the segment $[1-\delta, 1]$ and then around the circle with center zero and radius $1-\delta$.[4] This is the contour we use from now on.

There will be three lemmas. In these, $\epsilon \neq 1$ will be an nth root of unity and we consider the behavior of $\mathcal{S}(k)$ as $\kappa \to \epsilon$ radially. Because the argument that follows involves only the local behavior of $\mathcal{S}(k)$, we may consider κ as the underlying variable and in (7) replace k by the appropriate $\sqrt{\kappa}$. We define

$$\mu = \kappa^{-n} - 1, \quad \beta = \alpha_+ + \alpha_-, \quad b = \operatorname{Re}\beta,$$

so that $\mu > 0$ and $\mu \to 0$ as $\kappa \to \epsilon$.

[4] To expand on this, it goes from $1-\delta$ to 1 just below the interval $[1-\delta, 1]$, then from 1 to $1-\delta$ just above the interval $[1, 1-\delta]$, then counterclockwise around the circle with radius $1-\delta$ back to $1-\delta$.

Lemma 1. *We have*[5]

$$\left(\frac{d}{d\kappa}\right)^{2n^2-[bn]} \mathcal{S}_n(k) \approx \mu^{[bn]-\beta n-1}.$$

Proof. We set

$$\ell = 2n^2 - [bn]$$

and first consider

$$\int \cdots \int \frac{\prod_i x_i y_i}{(1-\kappa^n \prod_i x_i y_i)^{\ell+1}} \, \frac{\Delta(x)^2\,\Delta(y)^2}{\prod_{i,j}(1-\kappa\, x_i y_j)^2} \prod_i \frac{\Lambda(k^{-1}x_i^{-1})}{\Lambda(k y_i)} \prod_i dx_i\, dy_i, \tag{9}$$

where all indices run from 1 to n. This will be the main contribution to $d^\ell \mathcal{S}_n(k)/d\kappa^\ell$.

For the i,j factor in the denominator in the second factor, if x_i or y_j is on the circular part of the contour then $|x_i y_j| \le 1-\delta$ and the factor is bounded away from zero; otherwise $x_i y_j$ is real and positive and this factor is bounded away from zero as $\kappa \to \epsilon$ since $\epsilon \neq 1$. So we consider the rest of the integrand.

If $\prod_i |x_i y_i| < 1-\delta$ then the rest of the integrand is bounded except for the last quotient, and the integral of that is $O(1)$ since $\mathrm{Re}\,\alpha_\pm < 1$.

When $\prod_i |x_i y_i| > 1-\delta$ then each $|x_i|$, $|y_i| > 1-\delta$, so each x_i, y_i is integrated below and above the interval $[1-\delta, 1]$. If all the integrals are taken over the interval itself we must multiply the result by the nonzero constant $(4\,\sin\pi\alpha_+\,\sin\pi\alpha_-)^n$. The factors $1-\kappa\, x_i y_j$ in the second denominator equal $1-\kappa(1+O(\delta)) = (1-\kappa)\,(1+O(\delta))$ since κ is bounded away from 1. From this we see that if we factor out $\kappa^{(\ell+1)n}$ from the first denominator, $(1-\kappa)^{n^2}$ from the second denominator, and $(1-\kappa)^{\beta n}(\rho(k^{-1})/\rho(k))^n$ from the last factor (all of these having nonzero limits as $\kappa \to \epsilon$), the integrand becomes

$$\frac{\Delta(x)^2\,\Delta(y)^2}{(\kappa^{-n}-\prod_i x_i y_i)^{\ell+1}} \prod_i (1-x_i)^{-\alpha_+}\,(1-y_i)^{-\alpha_-}\,(1+O(\delta)). \tag{10}$$

We make the substitutions $x_i = 1-\xi_i$, $y_i = 1-\eta_i$ and set $r = \sum_i(\xi_i+\eta_i)$. Then since $\prod_i (1-\xi_i)(1-\eta_i) = 1-r+O(r^2)$ this becomes

$$\frac{\Delta(\xi)^2\,\Delta(\eta)^2}{(\mu+r+O(r^2))^{\ell+1}} \prod_i \xi_i^{-\alpha_+}\,\eta_i^{-\alpha_-}\,(1+O(\delta)).$$

The integration domain becomes $r < \delta + O(\delta^2)$. Consider first the integral without the $O(\delta)$ term. By homogeneity of the Vandermondes and the product, the integral

[5] We use the usual notation $[bn]$ for the greatest integer in bn. The symbol $\approx$ here indicates that the ratio tends to a nonzero constant as $\mu \to 0$.

equals a nonzero constant[6] times

$$\int_0^{\delta+O(\delta^2)} \frac{r^{2n^2-\beta n-1}}{(\mu+r+O(r^2))^{\ell+1}}\,dr. \tag{11}$$

Making the substitution $r \to \mu r$ results in

$$\begin{aligned} &\mu^{2n^2-\beta n-\ell-1}\int_0^{(\delta+O(\delta^2))/\mu} \frac{r^{2n^2-\beta n-1}}{(1+r+O(\mu^2r^2))^{\ell+1}}\,dr \\ &\quad= \mu^{[bn]-\beta n-1}\int_0^{(\delta+O(\delta^2))/\mu} \frac{r^{2n^2-\beta n-1}}{(1+r+O(\mu^2r^2))^{2n^2-[bn]+1}}\,dr, \end{aligned} \tag{12}$$

where we have put in our value of ℓ. The integral has the $\mu \to 0$ limit the convergent integral

$$\int_0^{\infty}\left(\frac{r}{1+r}\right)^{2n^2-[bn]+1} r^{[bn]-\beta n-2}\,dr,$$

and (11) is asymptotically this times $\mu^{[bn]-\beta n-1}$.

For the integral with the $O(\delta)$ we take the absolute values inside the integrals and find that it is $O(\delta)$ times what we had before, except that the β in the exponents are replaced by b, and in footnote 6 the exponents $\alpha_\pm$ are replace by their real parts. Since δ is arbitrarily small, it follows that the integral of (10) is asymptotically a nonzero constant times $\mu^{[bn]-\beta n-1}$.

To compute the derivative of order $2n^2-[bn]$ of the integral in (8) one integral we get is what we just computed. The other integrals are similar but in each the ℓ in the first denominator is at most $2n^2-[bn]-1$, while we get extra factors obtained by differentiating the rest of the integrand for $\mathcal{S}_n(k)$. These factors are of the form $(1-\kappa x_i y_i)^{-1}$, $(1-\kappa x_i)^{-1}$, $(1-\kappa y_i)^{-1}$, or derivatives of $\rho(k^{-1}x_i^{-1})$ or of $\rho(ky_i)^{-1}$. These are all bounded. Because $\ell \le 2n^2-[bn]-1$ the integral (11) is $O(\mu^{-1+\gamma})$ for some $\gamma > 0$. The lemma follows. □

Lemma 2. *If* $\epsilon^m \neq 1$ *then*

$$\left(\frac{d}{d\kappa}\right)^{2n^2-[bn]} \mathcal{S}_m(k) = O(1).$$

Proof. If $\epsilon^m \neq 1$ all terms, aside from those coming from the last factors, obtained by differentiating the integrand in (8) with n replaced by m are bounded as $\kappa \to \epsilon$. Differentiating the last factor in the integrand any number of times results in an integrable function. □

---

[6]This is the integral of $\Delta(\xi)^2\,\Delta(\eta)^2\prod \xi_i^{-\alpha_+}\,\eta_i^{-\alpha_-}$ over $r=1$. It can be evaluated using a Selberg integral [7, (17.6.5)], with the result

$$\frac{1}{\Gamma(2n^2-\beta n)}\prod_{j=0}^{n-1}\Gamma(j+2)^2\,\Gamma(j-\alpha_++1)\,\Gamma(j-\alpha_-+1).$$

Lemma 3. *We have* $\sum_{m>n} (d/d\kappa)^{2n^2-[bn]}\, S_m(k) = O(1)$.

Proof. We shall show that for κ sufficiently close to ϵ all integrals we get by differentiating the integral for $S_m(k)$ are at most $A^m\, m^m$, where A is some constant.[7] Because of the $1/(m!)^2$ appearing in front of the integrals this will show that the sum is bounded.

As before, we first use (8) with n replaced by m, and consider the integral we get when the first factor in the integrand is differentiated $2n^2 - [bn]$ times. All indices in the integrands now run from 1 to m.

First,

$$|1-\kappa^m \prod_i x_i y_i| \geq 1 - \prod_i |x_i y_i|.$$

Next we use that either $|x_i| = 1-\delta$ or $x_i \in [0,1]$, and $\kappa \in [0,\epsilon]$, to see that $|1-\kappa x_i| \geq \min(\delta, d)$, where $d = \operatorname{dist}(1, [0,\epsilon])$. We may assume $\delta < d$. Then $|1-\kappa x_i| \geq \delta$, and similarly, $|1-\kappa y_i| \geq \delta$. It follows that the integrand in (8) after differentiating the first factor has absolute value at most A^m times

$$\frac{1}{(1-\prod_i |x_i y_i|)^{2n^2-[bn]+1}}\, \frac{\Delta(x)^2\, \Delta(y)^2}{\prod_{i,j} |1-\kappa x_i y_j|^2} \prod_i |1-x_i|^{-a_+}\, |1-y_i|^{-a_-}, \tag{13}$$

where $a_\pm = \operatorname{Re} \alpha_\pm$.

If $\prod_i |x_i y_i| < 1-\delta$ then the first factor is at most $\delta^{-2n^2+[bn]-1}$. When $\prod_i |x_i y_i| > 1-\delta$ we set, as before, $x_i = 1-\xi_i$, $y_i = 1-\eta_i$ with $\xi_i, \eta_i \in [0,\delta]$. Since we are to integrate back and forth over these intervals we must multiply the estimate below by the irrelevant factor 2^{2m}.

We have $\prod_i (1-\xi_i)(1-\eta_i) \leq (1-\xi_i)(1-\eta_i)$ for each i, and so averaging gives

$$\prod_i (1-\xi_i)(1-\eta_i) \leq \frac{1}{2m} \sum_i (1-\xi_i)(1-\eta_i),$$

and therefore

$$\begin{aligned} 1-\prod_i (1-\xi_i)(1-\eta_i) &\geq \frac{1}{2m}\sum_i (1-(1-\xi_i)(1-\eta_i)) \\ &= \frac{1}{2m}\sum_i (\xi_i+\eta_i-\xi_i\eta_i) \geq \frac{1}{2m}\sum_i (\xi_i+\eta_i)/2 \end{aligned} \tag{14}$$

if $\delta < 1/2$, since each $\xi_i, \eta_i < \delta$. From this we see that in the region where $\sum_i (\xi_i+\eta_i) > \delta$ the first factor in (13) is at most $(4m/\delta)^{2n^2-[bn]+1}$.

So in either of these two regions the first factor is at most A^m. We then use (13) with the second factor replaced by the absolute value of

$$\big(\det(1/(1-\kappa x_i y_j))\big)^2.$$

Each denominator has absolute value at least δ, so by the Hadamard inequality the square of the determinant has absolute value at most $\delta^{-2m}\, m^m$. Therefore the

[7]The value of A will change with each of its appearances. It may depend on n and δ, which are fixed, but not on m.

integral over this region has absolute value at most

$$A^m\, m^m \int \cdots \int \prod_i |1-x_i|^{-a_+}\, |1-y_i|^{-a_-} \prod_i dx_i\, dy_i.$$

The integral here is A^m, and so we have shown that the integral in the described region is at most $A^m\, m^m$.

It remains to bound the integral over the region where $x_i = 1-\xi_i$, $y_i = 1-\eta_i$ with $\xi_i, \eta_i \in [0,\delta]$, and $r = \sum_i(\xi_i+\eta_i) < \delta$. Using (14) again, we see that the integrand has absolute value at most A^m times

$$d^{-m^2}\, \frac{\Delta(\xi)^2\, \Delta(\eta)^2}{(\sum_i(\xi_i+\eta_i))^{2n^2-[bn]+1}} \prod_i \xi_i^{-a_+}\, \eta_i^{-a_-}.$$

(Recall that $d = \mathrm{dist}(1,[0,\epsilon])$, and $\kappa x_i y_j \in [0,\epsilon]$. The factor $(4m^2)^{2n^2-[\gamma n]+1}$ coming from using (14) were absorbed into A^m.) Integrating this with respect to r over $r<\delta$, using homogeneity, gives

$$\int_{r=1} \Delta(\xi)^2\, \Delta(\eta)^2 \prod_i \xi_i^{-a_+}\, \eta_i^{-a_-}\, d(\xi,\eta)$$

(where $d(\xi,\eta)$ denotes the $(2n-1)$-dimensional measure on $r=1$) times

$$d^{-m^2} \int_0^\delta r^{2m^2-2n^2+[bn]-bn-1}\, dr.$$

The first integral is given in footnote 6 with n replaced by m and $\alpha_\pm$ replaced by $a_\pm$, and is exponentially small in m. The last integral is $O(\delta^{2m^2})$ since $m>n$ and n is fixed. Since $\delta^2 < d$, the product is exponentially small in m.

So we have obtained a bound for one term we get when we differentiate $2n^2-[\gamma n]$ times the integrand for $\mathcal{S}_m(k)$. The number of factors in the integrand involving κ is $O(m^2)$ so if we differentiate $2n^2-1$ times we get a sum of $O(m^{4n^2})$ terms. In each of the other terms the denominator in the first factor has a power even less than $2n^2-[\gamma n]$ and at most $2n^2$ extra factors appear which are of the form $(1-\kappa x_i y_i)^{-1}$, $(1-\kappa x_i)^{-1}$, or $(1-\kappa y_i)^{-1}$. Also, $\rho(k^{-1}x_i^{-1})$ or $\rho(ky_i)^{-1}$ may be replaced by some of its derivatives. Each has absolute value at most δ^{-1}, so their product is $O(\delta^{-4n^2})$. It follows that we have the bound $A^m\, m^m$ for the sum of these integrals. Lemma 3 is established. □

Proof of the theorem. Let ϵ be a primitive nth root of unity. Then $\epsilon^m \neq 1$ when $m<n$ so Lemma 2 applies for these m. Combining this with Lemmas 1 and 3 we obtain

$$(d/d\kappa)^{2n^2-[bn]}\, \mathcal{S}(k) \approx \mu^{[bn]-\beta n-1}$$

as $\kappa \to \epsilon$. This is unbounded, so $\mathcal{S}(k)$ cannot be analytically continued beyond any such ϵ, and these are dense on the unit circle.

Thus the unit circle is a natural boundary for $\mathcal{S}(k)$, and this implies that the same is true of $\chi(k)$. This completes the proof of the theorem. □

4. Fisher–Hartwig symbols

In this section we show how to extend the proof of the theorem to deformations of Fisher–Hartwig symbols. We start with a Fisher–Hartwig symbol[8]

$$\prod_{p=1}^{P}(1-u_p\,\xi)^{\alpha_p^+}\prod_{q=1}^{Q}(1-v_q/\xi)^{\alpha_q^-},$$

where $|u_p|$, $|v_q|=1$ and P, $Q>0$, and then its k-deformation

$$\varphi(\xi)=\prod_{p=1}^{P}(1-ku_p\,\xi)^{\alpha_p^+}\prod_{q=1}^{Q}(1-kv_q/\xi)^{\alpha_q^-}.$$

We assume that $\operatorname{Re}\alpha_p^+$, $\operatorname{Re}\alpha_q^-<1$ and α_p^+, $\alpha_q^-\notin\mathbb{Z}$. (Plus a simplifying assumption that comes later.) Now the singularities of $D_N(\varphi)$ on the unit circle are at the $(u_pv_q)^{-1/2}$, and

$$E(\varphi)=\prod_{p,q}(1-k^2u_pv_q)^{-\alpha_p^+\alpha_q^-}.$$

Theorem 2. *With the given assumptions, the unit circle $|k|=1$ is a natural boundary for $\chi(k)$.*

Proof. Using previous notation, we have now

$$\Lambda(\xi)=\prod_{p,q}(1-ku_p\,\xi)^{-\alpha_p^+}\,(1-kv_q/\xi)^{\alpha_q^-},$$

$$\frac{\Lambda(k^{-1}x^{-1})}{\Lambda(ky)}=\prod_{p,q}\frac{(1-u_p/x)^{-\alpha_p^+}\,(1-\kappa v_q\,x)^{\alpha_q^-}}{(1-\kappa u_p\,y)^{-\alpha_p^+}\,(1-v_q/y)^{\alpha_q^-}}.$$

Again we begin by considering the integral

$$\int\cdots\int\frac{\prod_i x_iy_i}{(1-\kappa^n\prod_i x_iy_i)^{\ell+1}}\,\frac{\Delta(x)^2\,\Delta(y)^2}{\prod_{i,j}(1-\kappa\,x_iy_j)^2}\prod_i\frac{\Lambda(k^{-1}x_i^{-1})}{\Lambda(ky_i)}\prod_i dx_i\,dy_i. \quad (15)$$

For $\delta\in(0,1)$, our integrations are for the x_i around the cuts $[1-\delta,1]\,u_p$ and for the y_i around the cuts $[1-\delta,1]\,v_q$ and then both around the circle with radius $1-\delta$. If we replace integrals around the cuts by integrals on the cuts, then for a cut $[1-\delta,1]\,u_p$ we must multiply by $2\sin\pi\alpha_p^+$ and for a cut $[1-\delta,1]\,v_q$ we multiply by $2\sin\pi\alpha_q^-$. These are both nonzero. We assume that this has been done.

We now let $\kappa\to\epsilon$ radially, where ϵ is an nth root of $\prod(u_{p_i}\,v_{q_i})^{-1}$, but not equal to any $(u_pv_q)^{-1}$. We also choose it so that it is not an mth root of any product of the form $\prod(u_{p_i}\,v_{q_i})^{-1}$ with $m<n$. These ϵ become dense on the unit circle as $n\to\infty$. The last condition assures that the integrals with $m<n$ are bounded, which will give the analogue of Lemma 2. We now consider the analogue of Lemma 1.

[8]We could easily add a factor $\psi(\xi)$ to give the general Fisher–Hartwig symbol. Then the δ chosen below would depend on the region of analyticity of ψ.

The integral over $\prod |x_i y_i| < 1 - \delta$ is bounded, as before. In the region where $\prod |x_i y_i| > 1-\delta$ each x_i and y_i is integrated on the union of its associated cuts. This is the sum of integrals in each of which each x_i is integrated over one of the cuts and each y_i is integrated over one of the cuts. Suppose that x_i is integrated over $[1-\delta, 1]\, u_{p_i}$ and y_i is integrated over $[1-\delta, 1]\, v_{q_i}$. (We consider this one possibility at first. Then we will have to sum over all possibilities.)

If we factor out $\prod u_{p_i} v_{q_i}$ from the first numerator, $\prod (1 - \kappa u_{p_i} v_{q_i})^2$ from the second denominator, and $\prod (1 - \kappa\, u_{p_i} v_{q_i})^{\alpha_{p_i} + \alpha_{p_i}}$ from the last product the integrand becomes $1 + O(\delta)$ times

$$\frac{\Delta(x)^2\, \Delta(y)^2}{(1 - \kappa^n \prod_i x_i y_i)^{\ell+1}} \prod_i (1 - u_{p_i}/x_i)^{-\alpha^+_{p_i}}\, (1 - v_{q_i}/y_i)^{-\alpha^-_{q_i}}. \tag{16}$$

We make the substitutions $x_i = (1 - \xi_i)\, u_{p_i}$, $y_i = (1 - \eta_i)\, v_{q_i}$, and define

$$I_p = \{i : p_i = p\}, \quad I_q = \{i : q_i = q\}.$$

Then

$$\prod_i (1 - u_{p_i}/x_i)^{-\alpha^+_{p_i}}\, (1 - v_{q_i}/y_i)^{-\alpha^-_{q_i}} = \prod_{p,\, i \in I_p} \xi_i^{-\alpha^+_p} \cdot \prod_{q,\, i \in I_q} \eta_i^{-\alpha^-_q} \times (1 + O(\delta)).$$

As for the Vandermondes, we have

$$\Delta(x) = \pm \prod_p \Delta(x_i : i \in I_p) \cdot \prod_{p \neq p'} \prod_{j \in I_p,\, j' \in I_{p'},\, j < j'} (x_j - x_{j'}), \tag{17}$$

and similarly for $\Delta(y)$. If we define $n_p = |I_p|$, $n_q = |I_q|$, then the last double product is to within a factor $1 + O(\delta)$ equal to

$$\pm \prod_{p < p'} (u_p - u_{p'})^{n_p n_{p'}},$$

while the first product is to within a factor $1 + O(\delta)$ equal to

$$\prod_p u_p^{n_p(n_p+1)/2} \prod_p \Delta(\xi_i : i \in I_p).$$

Thus, if we factor out $(\kappa^n \prod u_{p_i}\, v_{q_i})^{\ell+1}$ from the denominator in (16), and set $\mu = \kappa^{-n} \prod (u_{p_i}\, v_{q_i})^{-1} - 1$, then (16) may get replaced by a constant times $1 + O(\delta)$ times

$$\frac{1}{(\mu + \sum_i (\xi_i + \eta_i))^{\ell+1}} \prod_{p,\, i \in I_p} \Delta(\xi_i : i \in I_p)^2\, \xi_i^{-\alpha^+_p} \cdot \prod_{q,\, i \in I_q} \Delta(\eta_i : i \in I_q)^2\, \eta_i^{-\alpha^-_q}. \tag{18}$$

If we use homogeneity the integral of (18) becomes a nonzero constant[9] times

$$\int_0^\delta \frac{1}{(\mu + r)^{\ell+1}}\, r^{-1 + \sum_p n_p(n_p - \alpha^+_p) + \sum_q n_q(n_q - \alpha^-_q)}\, dr. \tag{19}$$

[9]Also computable using the Selberg integral, it is [see next page]

This is largest when the power of r is smallest. So we minimize

$$\sum_p n_p(n_p - a_p^+), \quad (a_p^+ = \operatorname{Re} \alpha_p^+)$$

over all $\{n_p\}$ with $n_p \geq 0$, $\sum_p n_p = n$. The solution is not necessarily unique.[10] But in any case

$$M_n^+ := \min \sum_p n_p(n_p - a_p^+) = \frac{n^2}{P} + O(n), \tag{20}$$

and $M_{n+1}^+ > M_n^+$ for large enough n.[11] Similarly, with $a_q^- = \operatorname{Re} \alpha_q^-$ and

$$M_n^- := \min \sum_q n_q(n_q - a_q^-) = \frac{n^2}{Q} + O(n). \tag{21}$$

Then we choose

$$\ell = \sum_p n_p^2 + \sum_q n_q^2 - \left[\sum_p n_p \, a_p^+ + \sum_q n_q \;\, a_q^- \right] \tag{22}$$

with the minimal n_p and n_q. The integral (19) is equal to

$$\mu^{-1+[\sum_p n_p \, a_p^+ + \sum_q n_q \;\, a_q^-] - (\sum_p n_p \, \alpha_p^+ + \sum_q n_q \, \alpha_q^-)}$$

times

$$\int_0^{\delta/\mu} \frac{1}{(1+r)^{\ell+1}} \, r^{-1+\sum_p n_p(n_p - \alpha_p^+) + \sum_q n_q(n_q - \alpha_q^-)} \, dr.$$

The exponent of μ has real part in $(-2, -1]$ and the integral has a nonzero limit (a Beta function) as $\mu \to 0$.

Once we take care of the integrals with the $O(\delta)$ as in the proof of Lemma 1 we deduce that this is the asymptotic result for the integral when we choose this set of cuts.

We now assume the minimal solutions are unique.[12]

Then for the other choices of cuts the integral (19) is $O(\mu^{-1+\gamma})$ for some $\gamma > 0$, and so the integral over the chosen set of cuts dominates. We still have to allocate the x_i and y_i to the various cuts, once the numbers of each have been

---

$$\frac{1}{\Gamma(\sum_p n_p(n_p - \alpha_p^+) + \sum_q n_q(n_q - \alpha_p^-))}$$
$$\times \prod_p \prod_{j=0}^{n_p - 1} \Gamma(j+2)\,\Gamma(j - \alpha_p^+ + 1) \cdot \prod_q \prod_{j=0}^{n_q - 1} \Gamma(j+2)\,\Gamma(j - \alpha_q^- + 1).$$

[10] A similar minimum problem was encountered in [3], where an asymptotic formula was a sum over all solutions. We shall eventually assume a unique solution to avoid the possibility of cancelation of terms of the same order.

[11] See Appendix B.

[12] We shall see in Appendix B that for large n uniqueness is a condition on the a_p^+ and a_q^- that depends only on the residue classes of n modulo P and Q. It suffices for our purposes that we have uniqueness for some sequence $n \to \infty$.

chosen. The number of ways of doing this is $n!/\prod n_p!$ for the x_i and $n!/\prod n_q!$ for the y_i. (The total number of ways is at most $P^n\, Q^n$.)

This takes care of the integral (15), the main contributions to $(d/d\kappa)^\ell S_n(\kappa)$. We complete the proof of the analogue of Lemma 1 as we did at the end of the proof of that lemma. Thus, with ℓ given by (22),

$$\left(\frac{d}{d\kappa}\right)^\ell \mathcal{S}_n(k) \approx \mu^{-1+[\sum_p n_p\, a_p^+ + \sum_q n_q\;\; a_q^-]-(\sum_p n_p\, \alpha_p^+ + \sum_q n_q\, \alpha_q^-)}.$$

For the analogue of Lemma 3 we first consider the integral (15) with n replaced by $m > n$, and ℓ given by (22). As before it remains to bound the integrals over the regions where each $x_i = (1-\xi_i)u_{p_i}$ and each $y_i = (1-\eta_i)v_{q_i}$, with $\xi_i, \eta_i \in [0,\delta]$, and $r = \sum_i(\xi_i+\eta_i) < \delta$.

Replacing the first denominator in (15) by $(\sum(\xi_i+\eta_i))^{\ell+1}$ introduces a factor $(4m^2/\delta)^{\ell+1}$ as before, a factor that can be ignored. The reciprocal of the second denominator is at most d^{-m^2} where $d = \min_{p,q} \operatorname{dist}([0,\epsilon],(u_p v_q)^{-1})$. The product of the terms involving κ in the last product is $d^{-O(m)}$, and so may also be ignored. The square of the product over $p < p'$ in (17), times the square of the analogous product over $q < q'$, is at most $2^{(P^2+Q^2)m^2}$. There remains an integrand whose absolute value is bounded by

$$\frac{1}{(\sum_i(\xi_i+\eta_i))^{\ell+1}} \prod_{p,\, i\in I_p} \Delta(\xi_i : i \in I_p)^2\, \xi_i^{-a_p^+} \cdot \prod_{q,\, i\in I_q} \Delta(\eta_i : i\in I_q)^2\, \eta_i^{-a_q^-}.$$

The integral of the products over $r = 1$ (given exactly in footnote 9) is trivially at most its maximum (at most $A^m\, 4^{m^2}$) times the $(2m-1)$-dimensional measure of $r = 1$, which is $1/\Gamma(2m)$. We use the crude bound 4^{m^2}. This is to multiply

$$\int_0^\delta r^{-\ell-2+\sum_p m_p(m_p-a_p^+)+\sum_q m_q(m_q-a_q^-)}\, dr.$$

Now

$$\sum_p m_p(m_p-a_p^+) + \sum_q m_q(m_q-a_q^-)$$

is at least $M_m^+ + M_m^-$, and it follows from (20) and (21), and the strict monotonicity of the sequences $\{M_n^\pm\}$, that for large enough n and some R this greater than $\ell+1+m^2/R$ for all $m > n$. Then the integral is at most $\delta^{m^2/R}$.

This integral is one of at most $P^m Q^m$ integrals, and this factor also can be ignored. The factors we had before that could not be ignored combine to $(4\,2^{P^2+Q^2}/d)^{m^2}$. It follows that if we choose $\delta < (4\,2^{P^2+Q^2}/d)^{-R}$ the integral over $r < \delta$ of (15) with m replacing n is exponentially small.

This takes care of the integral (15) with m replacing n, the main contributions to $(d/d\kappa)^\ell S_m(\kappa)$. We complete the proof of the analogue of Lemma 3 as we did at the end of the proof of that lemma. This completes the proof of the theorem. □

Appendix A. Proof of the proposition

The Fredholm expansion is

$$\det(I-K_N)=1+\sum_{n=1}^{\infty}\frac{(-1)^n}{n!}\sum_{p_1,\ldots,p_n\geq 0}\det(K_N(p_i,p_j)).$$

Therefore its suffices to show that

$$\sum_{N=1}^{\infty}\sum_{p_1,\ldots,p_n\geq 0}\det(K_N(p_i,p_j))$$
$$=\frac{1}{n!}\int\cdots\int\frac{\prod_i x_iy_i}{1-\prod_i x_iy_i}\left(\det\left(\frac{1}{1-x_iy_j}\right)\right)^2 du(x_1)\cdots du(x_n)\,dv(y_1)\cdots dv(y_n).$$

We have

$$K_N(p_i,p_j)=\int\int\frac{x^{N+p_i}\,y^{N+p_j}}{1-xy}\,du(x)\,dv(y).$$

It follows by a general identity [1] (eqn. (1.3) in [10]) that

$$\det(K_N(p_i,p_j))$$
$$=\frac{1}{n!}\int\cdots\int\det(x_i^{N+p_j})\,\det(y_i^{N+p_j})\prod_i\frac{1}{1-x_iy_i}\prod_i du(x_i)\,dv(y_i)$$
$$=\frac{1}{n!}\int\cdots\int\left(\prod_i x_iy_i\right)^N\det(x_i^{p_j})\,\det(y_i^{p_j})\prod_i\frac{1}{1-x_iy_i}\prod_i du(x_i)\,dv(y_i).$$

Summing over N gives

$$\sum_{N=1}^{\infty}\det(K_N(p_i,p_j))$$
$$=\frac{1}{n!}\int\cdots\int\frac{\prod_i x_iy_i}{1-\prod_i x_iy_i}\det(x_i^{p_j})\,\det(y_i^{p_j})\prod_i\frac{1}{1-x_iy_i}\prod_i du(x_i)\,dv(y_i).$$

(Interchanging the sum with the integral is justified since the supports of u and v are in the open unit disc.)

Now we sum over $p_1,\ldots,p_n\geq 0$. Using the general identity again (but in the other direction) gives

$$\sum_{p_1,\ldots,p_n\geq 0}\det(x_i^{p_j})\,\det(y_i^{p_j})=n!\,\det\left(\sum_{p\geq 0}x_i^p\,y_j^p\right)=n!\,\det\left(\frac{1}{1-x_iy_j}\right).$$

We almost obtained the desired result. It remains to show that

$$\det\left(\frac{1}{1-x_iy_j}\right)\prod_i\frac{1}{1-x_iy_i},\qquad (23)$$

which we obtain in the integrand, may be replaced by

$$\frac{1}{n!}\left(\det\left(\frac{1}{1-x_iy_j}\right)\right)^2.\qquad (24)$$

This follows by symmetrization over the x_i. (The rest of the integrand is symmetric.) For a permutation π, replacing the x_i by $x_{\pi(i)}$ multiplies the determinant in (23) by $\operatorname{sgn}\pi$, so to symmetrize we replace the other factor by

$$\frac{1}{n!}\sum_{\pi}\operatorname{sgn}\pi\prod_{i}\frac{1}{1-x_{\pi(i)}y_i}=\frac{1}{n!}\det\left(\frac{1}{1-x_iy_j}\right).$$

Thus, symmetrizing (23) gives (24). □

Appendix B. The minimum problem

Changing notation, we consider

$$M_n=\min\left\{\sum_{i=1}^{k}n_i\,(n_i-a_i):n_i\in\mathbb{Z}^+,\ \sum_{i=1}^{k}n_i=n\right\},$$

and ask when this is uniquely attained. Set

$$s=k^{-1}\sum_{i=1}^{k}a_i,\quad \bar{a}_i=(a_i-s)/2,\quad \bar{n}_i=n_i-n/k,$$

and define

$$\mathbb{N}^k=\left\{(x_i)\in\mathbb{R}^k:\sum_{i=1}^{k}x_i=0\right\}.$$

Then $\bar{a}=(\bar{a}_i)\in\mathbb{N}^k$ and $\bar{n}=(\bar{n}_i)\in\mathbb{N}^k$. If $n\equiv\nu\pmod k$ the other conditions on the $\bar{n}_i$ become

$$\bar{n}_i\geq -n/k,\quad \bar{n}_i\in\mathbb{Z}-\nu/k.$$

(Think of ν as fixed and n as large and variable.) A little algebra gives

$$\sum_{i=1}^{k}n_i\,(n_i-a_i)=\sum_{i=1}^{k}(\bar{n}_i-\bar{a}_i)^2+k\,(n/k-s/2)^2-\sum_{i=1}^{k}a_i^2/4.$$

Minimizing the sum on the left is the same as minimizing the first sum on the right, with the stated conditions on the $\bar{n}_i$. Several things follow from this. First, since the minimum of the first sum on the right is clearly $O(1)$, the condition $\bar{n}_i\geq -n/k$ may be dropped when n is sufficiently large; second, $M_n=n^2/k-sn+O(1)$; third (from this), $M_{n+1}-M_n=2n/k+O(1)>0$ for sufficiently large n; and fourth, for uniqueness we may replace our minimum problem by

$$\min\left\{\sum_{i=1}^{k}(\bar{n}_i-\bar{a}_i)^2:\bar{n}\in\mathbb{N}^k,\ \bar{n}_i\in\mathbb{Z}-\nu/k\right\}.$$

This minimum is uniquely attained if and only if there is a unique point closest to $\bar{a}$ in the set of lattice points $(\mathbb{Z}-\nu/k)^k$ in $\mathbb{N}^k$. This condition depends only on the residue class of n modulo k.

When $k=2$ the subspace $\mathbb{N}^2$ is the line $x_1+x_2=0$ in $\mathbb{R}^2$. When n is even the lattice consists of the points on the line with coordinates in $\mathbb{Z}$ and $\bar{a}$ is equidistant from two adjacent ones when $a_1 - a_2 \in 4\mathbb{Z} + 2$; when n is odd the lattice consists of the points of the line with coordinates in $\mathbb{Z} + 1/2$ and $\bar{a}$ is equidistant from two adjacent ones when $a_1 - a_2 \in 4\mathbb{Z}$. Non-uniqueness occurs in these cases.

Acknowledgment

This work was supported by the National Science Foundation through grants DMS–1207995 (first author) and DMS–1400248 (second author).

References

[1] C. Andréief, *Note sur une relation des intégrales définies des produits des fonctions*, Mém. de la Soc. Sci., Bordeaux **2** (1883), 1–14.

[2] A. Borodin and A. Okounkov, *A Fredholm determinant formula for Toeplitz determinants*, Int. Eqs. Oper. Th. **37** (2000), 386–396.

[3] A. Böttcher and B. Silbermann, *The asymptotic behavior of Toeplitz determinants for generating functions with zeros of integral order*, Math. Nachr. **102** (1981), 78–105.

[4] S. Boukraa, S. Hassani, J.-M. Maillard, B.M. McCoy and N. Zenine, *The diagonal Ising susceptibility*, J. Phys. A: Math. Theor. **40** (2007), 8219–8236.

[5] M.E. Fisher and R.E. Hartwig, *Toeplitz determinants: some applications, theorems, and conjectures*, Adv. Chem. Phys. **15** (1968), 333–353.

[6] J.S. Geronimo and K.M. Case, *Scattering theory and polynomials orthogonal on the unit circle*, J. Math. Phys. **20** (1979), 299–310.

[7] M.L. Mehta, Random Matrices, 2nd ed., Academic Press (1991).

[8] B. Nickel, *On the singularity structure of the* 2D *Ising model*, J. Phys. A: Math. Gen. **32** (1999), 3889–3906.

[9] E. Stein and R. Sharkarchi, Complex Analysis, Princeton Univ. Press (2003), p. 29.

[10] C.A. Tracy and H. Widom, *Correlation functions, cluster functions, and spacing distributions for random matrices*, J. Stat. Phys. **92** (1998), 809–835.

[11] C.A. Tracy and H. Widom, *On the diagonal susceptibility of the* 2D *Ising model*, J. Math. Phys., **54** (2013) 123302.

[12] T.T. Wu, B.M. McCoy, C.A. Tracy and E. Barouch, *Spin-spin correlation functions for the two-dimensional Ising model: Exact theory in the scaling region*, Phys. Rev. **B13** (1976), 315–374.

Craig A. Tracy
Department of Mathematics
University of California
Davis, CA 95616, USA
e-mail: tracy@math.ucdavis.edu

Harold Widom
Department of Mathematics
University of California
Santa Cruz, CA 95064, USA
e-mail: widom@ucsc.edu

Operator Theory:
Advances and Applications, Vol. 259, 719–740

A Riemann–Hilbert Approach to Filter Design

Elias Wegert

An illustrated journey accompanied by and dedicated to Albrecht Böttcher

Abstract. The paper is devoted to interrelations between boundary value problems of Riemann–Hilbert type and optimization problems in spaces of bounded holomorphic functions which are motivated by optimal filter design. A numerical method of Newton type for the iterative solution of nonlinear Riemann–Hilbert problems is adapted for solving the optimization problem and its convergence is proved. The approach is illustrated by the design of a low-pass filter, which we discuss in some detail.

My long-time acquaintance with Albrecht Böttcher, and in particular our journey to a conference on H^∞ control at lake Como in 1990, form the background of the paper.

Mathematics Subject Classification (2010). Primary 93B36; Secondary 30E25, 49K30, 49M15, 93B51, 93E11.

Keywords. Riemann–Hilbert problem, filter design, low-pass filter, optimization, control theory, H^∞ control, Newton's method.

1. From Mathematical Olympiads to Mathematics

My acquaintance with Albrecht dates back to 1969, when we were at the age of 14. The year before we both successfully participated in local Mathematical Olympiads some 60 km apart, and subsequently we were invited to a central training group. After doing some (in fact a lot of) home work, we met regularly in Karl-Marx-Start, which then was the name of our district's capital – now and before it was named Chemnitz.

Albrecht was a shooting star. Though he had no academic family background and just recently became seriously interested in mathematics, he impressed us deeply by his knowledge and intuition. If Albrecht could not solve a problem, nobody else could do it. And most importantly: next time he was not only able to cope with the problem of last month, but developed techniques which allowed him to answer any similar question.

In 1971 we became classmates, after passing the entrance exams for the "Spezialklasse", an elite class for mathematically gifted school students. It was affiliated with the Technische Hochschule Karl-Marx-Stadt, and several of our teachers (in mathematics, physics and chemistry) were university staff members. These two years gave us an enormous push and were decisive for our mathematical development. Our mathematics teacher Gerhard Eichler took not only care for mathematical education, but also shaped our personalities. This wasn't always easy for him, because we often were in high spirits and played jokes. Once, when we had a calculus test, Albrecht solved all extremal problems by elementary techniques, thus *completely avoiding* calculus.

FIGURE 1. Albrecht's first participation in a local Mathematical Olympiad (left), Albrecht and I at the (East-)German Mathematical Olympiad in Berlin 1973 (right)

After these formative years and successful participation in the International Mathematical Olympiad in 1973 (where we both won Silver medals) it became clear that mathematics will play a prominent role in our lives, and in 1975 (after a break by the obligatory military service) we enrolled at the TH Karl-Marx-Stadt to study mathematics.

We were lucky to have excellent teachers from the very beginning: Bernd Silbermann read the analysis course, with tutorials supervised by Georg Heinig. Thanks to the good preparation in the Spezialklasse, we freshmen even managed to participate in an advanced course on differential topology by Wolfgang Sprößig.

Each professor was interested in integrating the most talented students into his or her research group as early as possible, and hence Albrecht received several offers. Because he was taken with Bernd Silberman's lectures and research profile since his first year as a student, Albrecht chose him for jointly tackling problems beyond the standard university program. At that time, Bernd Silbermann started working on Toeplitz determinants, and it did not take long until Albrecht had fallen in love with this topic, too. It was probably in our third year at the university – we were sitting in a pub accompanied by a third fellow and some glasses of beer – when Albrecht announced that he became fascinated by the Fisher–Hartwig conjecture and that Bernd Silbermann and he decided to make a serious attempt to prove it. This gave me a chill, and although we always had a friendly competition as students, at this point I felt it would be better to choose a different topic.

And they indeed accomplished the adventure successfully seven years later: delving deeper and deeper into the matter and persistently developing the necessary requisites and techniques, Albrecht and Bernd Silbermann confirmed the conjecture for the most important symbol classes in [1], [2]. It turned out that the conjecture was not true in its full generality as expected, and some more delicate situations still withstood their attempts, but the breakthrough was made. For the continuation of the story after Albrecht and Bernd Silbermann see Torsten Ehrhardt's paper [5] and the recent survey [3].

2. From Chemnitz to Como

In the late 1980s we finished our Habilitation (a postdoctoral qualification which was essential for an academic career), Albrecht in 1987 in Karl-Marx-Stadt, I one year later in Freiberg. At that time we had no idea about the dramatic changes soon to come. In autumn 1989 the German wall collapsed, the border was opened, and the GDR (which means German Democratic Republic) begun to fade away like the DDR-sign at the back of the car in Figure 2. So we were entering a dream-like period with unimaginable opportunities to travel and establish personal contacts with colleagues from "The West". In spring 1990, after some years in which I met Albrecht only sporadically, we somewhat surprisingly found ourselves at Stuttgart University, where we spent several weeks working as guests at the institute of Wolfgang Wendland.

This visit helped us greatly to manage a smooth transition from the East German system to unified Germany and gave us the unique chance to earn some Deutsche Mark before the monetary union. Saying that "we've never been so far west, why not go a bit further", we used part of this money for excursions to places in Germany and France we never before could hope to visit. The stickers at the rear window of my car are witnesses of these travels.

The final goal of our journey was Italy, where we had an invitation for a workshop on H^∞ control and optimization (the C.I.M.E. Session "Recent Developments in H^∞ Control Theory"). Traveling across the Alps through Liechtenstein and Switzerland with a 20-year-old car and crossing the Italian border with an East German passport was an unforgettable experience. I remember that Albrecht joyfully shouted "It's unbelievable, now we are in Italy, the home country of Giuseppe

Verdi!" Somehow we survived the heavy traffic in Milan and finally arrived at Lake Como, where the conference was held in the noble Villa Olmo.

FIGURE 2. Albrecht (left) and our car (right) on the travel from Stuttgart to Como in 1990

Participation in this event, and meeting renowned mathematicians whom we knew by name, but had never seen before, was clearly the culmination of the journey. Needless to say that we used a two days break of the conference to travel further south: accompanied by Georg Heinig, who was also invited to the workshop, we managed to visit Genua, Elba, Pisa, and Florence. Going back by train, we met William Helton and Bruce Francis who were both lecturing in Como.

The main topic of Helton's talks was H^∞ control, and in particular a general optimization problem which will be the subject of the rest of this paper. I became interested in this problem shortly before, working on the geometric theory of nonlinear Riemann–Hilbert problems. Not knowing from each other that Helton and Marshall [13] and myself [26] just proved (almost) the same result we now met in Como and could discuss the different approaches. I still consider this personal experience of parallel developments in two parts of the world which, at this time, were lightyears away as a convincing example of the unity of mathematics. What this problem is about will be described in the next section.

3. Helton's OPT-problem

For the convenience of the reader we present some background material. A *linear, time invariant, causal,* and *stable* (DIDO-stable, to be more precise) system is a shift-invariant bounded linear operator S in $L^2(\mathbb{R})$ which maps the subspace $L^2(\mathbb{R}_+)$ of functions supported on $\mathbb{R}_+$ (the "signals") into itself. For the sake of brevity we shall often omit the four attributes, and just speak of systems.

It is one of the astonishing results that systems are completely characterized by a single function T, called the *transfer function* of S. This can be best described in the *frequency domain*, consisting of the Laplace transforms Lx of signals x in $L^2(\mathbb{R}_+)$. According to Payley–Wiener theory, these transforms form the Hardy space $H^2(\mathbb{C}_+)$ of holomorphic functions in the right half-plane. In this setting, the (linear, time invariant, causal, stable) systems are precisely the operators acting by $Sx = L^{-1}TLx$, where T is (the multiplication operator by) a function in the Hardy space $H^\infty(\mathbb{C}_+)$ of bounded holomorphic functions in the right half-plane $\mathbb{C}_+$. In other words, the transplanted system LSL^{-1} acts in $H^2(\mathbb{C}_+)$ as multiplication with a bounded holomorphic function T. If one additionally requires that S preserves real signals, T must satisfy the symmetry condition $T(\overline{z}) = \overline{T(z)}$.

As a member of $H^\infty(\mathbb{C}_+)$, a transfer function has a boundary function on the imaginary axis (defined almost everywhere as non-tangential limit). The function $f : \omega \mapsto T(\mathrm{i}\omega)$ is called the *frequency response* of S and has a physical meaning: operating at a harmonic input $\exp(\mathrm{i}\omega t)$ (here $t \in \mathbb{R}$ denotes time) with frequency ω, the system produces the output $f(\omega) \cdot \exp(\mathrm{i}\omega t)$, which has the same frequency as the input, but may have different amplitude and phase. The functions $|f|$ and $\arg f$ and are said to be the *magnitude response* (or gain) and the *phase response* of the system. Since T can be reconstructed from its boundary function f, systems are completely determined by their frequency response.

Looking for systems with specific properties, such as filters which suppress certain frequencies, the designer usually knows which values of $f(\omega)$ are considered as "good performance" of the system at frequency ω. However, since the frequency response is the boundary function of a holomorphic function, it cannot be prescribed arbitrarily: *design of systems is an optimization problem.*

In the eighties of the last century, William Helton and his co-authors proposed the following general framework for the design of systems (see [10], [11]): Assume that the performance of a (virtual) system with frequency response f is evaluated in terms of a *penalty function*

$$F\colon\ \mathbb{R} \times \mathbb{C} \to \mathbb{R}_+,\ (\omega, w) \mapsto F(\omega, w),$$

where $F(\omega, w)$ attains small values if $w = f(\omega)$ is considered a "good response" at frequency ω, while these values are large if $w = f(\omega)$ is a "bad response". A reasonable choice of F could be $F(\omega, w) := |w - f_0(\omega)|$, which measures the Euclidean distance from an "ideal" frequency response $f_0(\omega)$, but it is clear that there may be much better options for special purposes.

To give an example (which may or may not be of practical relevance), we consider the design of a low-pass filter. The ideal frequency response of this filter would be $f_0(\omega) = 1$ if $|\omega| < 1$ and $f_0(\omega) = 0$ if $|\omega| > 1$. When measuring the performance of a (realizable) filter, magnitude response is often considered to be more important than phase. This aspect cannot be modeled by a penalty function based on the Euclidean distance to f_0, since then restrictions on the amplitude automatically induce (strong) conditions for the phase (see Figure 3, left). A more general penalty function F allows one to relax these requirements, as illustrated in the right picture of Figure 3, showing the contour lines and the sub-level sets for a suitable choice of F in the pass-band. Compared with the penalty $|w - f_0(\omega)|$, the conditions for the modulus are unchanged, but phase restrictions are much weaker.

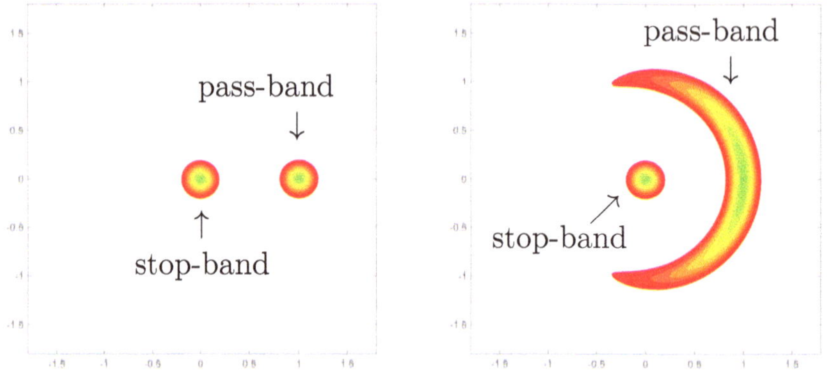

FIGURE 3. Color coded penalty functions in pass band and stop band

Once the penalty function F has been chosen, the over-all rating of a system is determined by *worst case analysis* with respect to all frequencies, i.e.,

$$P(f) := \sup \{F(\omega, f(\omega)) : \ \omega \in \mathbb{R}\}.$$

Optimal design of a system then requires to find the *optimal performance*

$$P^* := \inf \{P(f) : f \text{ is a frequency response function}\} \qquad \text{(OPT)}$$

and all functions f (if such exist) for which the infimum is attained. This is Helton's OPT-problem.

For our purposes it is convenient to transplant the transfer functions T from the right half-plane to the unit disk, pre-composing them with a conformal mapping of $\mathbb{D}$ onto $\mathbb{C}_+$, $T \mapsto w := T \circ C$, where $C(z) := (1 - z)/(1 + z)$ is (a variant of) the Cayley transform. This transformation establishes a bijection between the space $H^\infty(\mathbb{C}_+)$ of transfer functions and the Hardy space $H^\infty := H^\infty(\mathbb{D})$ of bounded holomorphic functions in the unit disk. So the OPT-problem is equivalent to finding the infimum p^* of all values p for which the set

$$\mathbb{A}(p) := \{w \in H^\infty(\mathbb{D}) : \ F(t, w(t)) \le p \ \text{ a.e. on } \mathbb{T} := \partial\mathbb{D}\}$$

of bounded analytic functions is non-void, as well as all functions w^* in $\mathbb{A}(p^*)$. In order to rephrase the problem in a geometric setting, we define the level sets

$$\Gamma(p) := \big\{(t, w) \in \mathbb{T} \times \mathbb{C} : F(t, w) = p\big\}, \qquad p \in \mathbb{R},$$

of F and the corresponding contour lines

$$\Gamma_t(p) := \big\{w \in \mathbb{C} : F(t, w) = p\big\}, \qquad t \in \mathbb{T}, \ p \in \mathbb{R}.$$

Definition 1. An OPT-problem is called *admissible* if the following conditions are satisfied:

(i) All $\Gamma_t(p)$ are simple closed curves depending continuously on t and p.
(ii) For all $p \in \mathbb{R}$, the set $G(p)$ is a smooth submanifold of $\mathbb{T} \times \mathbb{C}$ of class C^1 which is transversal to the planes $\{t\} \times \mathbb{C}$ for all $t \in \mathbb{T}$.
(iii) The family $\Gamma(p)$ is nested: if $p < q$ then $\Gamma(p) \in \operatorname{clos} \operatorname{int} \Gamma(q)$.

Here and throughout the following $\operatorname{int} \Gamma(p)$ denotes the bounded component of $(\mathbb{T} \times \mathbb{C}) \setminus \Gamma(p)$ and $\operatorname{clos} \operatorname{int} \Gamma(p)$ is its topological closure. We want to point out that $\Gamma(p) = \Gamma(q)$ for some $p < q$ is *not excluded.* The continuity condition in (i) means that $(t, p) \mapsto \Gamma_t(p)$ is a continuous mapping of $\mathbb{T} \times \mathbb{R}$ into the metric space of compact subsets of $\mathbb{C}$ endowed with the Hausdorff distance.

To visualize a (frequency response) function $f : \mathbb{T} \to \mathbb{C}$ in this geometric setting, we define its trace by

$$\operatorname{tr} f := \{(t, f(t)) \in \mathbb{T} \times \mathbb{C} : t \in \mathbb{T}\}.$$

The trace of a bounded holomorphic (transfer) function in $\mathbb{D}$ is defined as the trace of its boundary function (existing almost everywhere on $\mathbb{T}$).

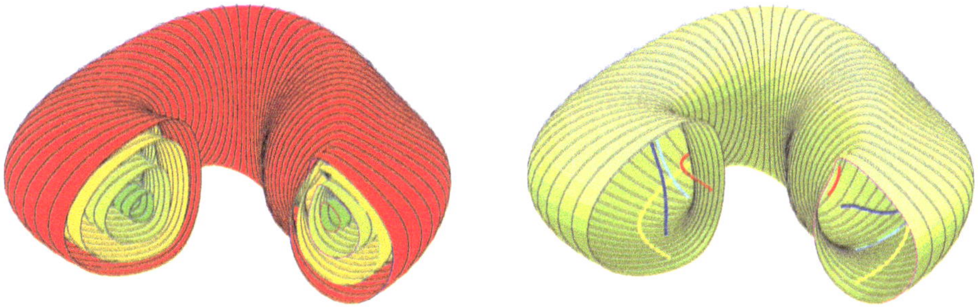

FIGURE 4. Nested contour sets $\Gamma(p)$ and traces of functions in $\mathbb{A}(p)$

Figure 4 depicts (about three quarters of) a nested family of manifolds $\Gamma(p)$ for four values of p (left) and the traces of some functions in $\mathbb{A}(p)$ for fixed p (right). The manifolds $\Gamma(p) \subset \mathbb{T} \times \mathbb{C}$ are shown in a perspective view, with $\mathbb{T}$ lying in a horizontal plane and fibers $\{t\} \times \mathbb{C}$ orthogonal to it. The planes where $\Gamma(p)$ is cut off are two such fibers. The contour lines $\Gamma_t(p)$ are the fibers of $\Gamma(p)$ lying over $t \in \mathbb{T}$ and appear as thin black lines. The values of the penalty function are represented by colors (in the online version).

Let $|\mathbb{A}(p)|$ denote the number of elements in $\mathbb{A}(p)$. We say that an OPT-problem is *nontrivial,* if $|\mathbb{A}(p)| = 0$ for sufficiently small p, and $|\mathbb{A}(p)| \geq 2$ for sufficiently large p (just meaning that $\mathbb{A}(p)$ contains at least two elements).

A characterization of solutions to the OPT-problem needs the topological concept of winding numbers. Let $f : \mathbb{T} \to \mathbb{C}$ be a continuous function satisfying $\operatorname{tr} f \subset \Gamma(p)$ for some fixed p. Then its *winding number* $\operatorname{wind}_{\Gamma(p)} f$ about $\Gamma(p)$ is defined as the (usual) winding number of $f - g$ about zero, where $g : \mathbb{T} \to \mathbb{C}$ is any continuous function with $\operatorname{tr} g \subset \operatorname{int} G(p)$.

After these preparations we can state the main result about the solvability of the OPT-problem.

Theorem 1. *Any nontrivial admissible* OPT*-problem has a unique solution*

$$p^* = \min\{p \in \mathbb{R} : \mathbb{A}(p) \neq \emptyset\}, \quad w^* \in \mathbb{A}(p^*). \tag{1}$$

More precisely:

(i) *There exists a uniquely determined real number p^* such that $|\mathbb{A}(p)| = 0$ if $p < p^*$, $|\mathbb{A}(p^*)| = 1$, and $|\mathbb{A}(p)| = \infty$ if $p > p^*$.*

(ii) *A function w in H^∞ is a solution of* (1) *if and only if there exists a number p in $\mathbb{R}$ such that*

$$w \in H^\infty \cap C, \qquad \operatorname{tr} w \subset \Gamma(p), \qquad \operatorname{wind}_{\Gamma(p)} w < 0. \tag{2}$$

(iii) *For generic problems the winding number* $\operatorname{wind}_{\Gamma(p^*)} w^*$ *of the solution about $\Gamma(p^*)$ is equal to -1.*

Assuming the existence of *continuous solutions*, their characterization (2) was given by Helton and Howe [12], and Helton, Schwartz, and Warschawski [16]. Assertions (i) and (ii) were proved by Helton and Marshall [13] under somewhat stronger assumptions. The first (independent) proof of the theorem in its current form is in Wegert [26]. For more details on the early history we refer to [15] and Section 6.1 of [28].

It follows from (i) and (ii) that condition (2) for some p and w implies $w = w^*$, but we point out that then not necessarily $\Gamma(p) = \Gamma(p^*)$ or even $p = p^*$. However, both implications are correct if the $\Gamma(p)$ form a *strongly nested* family, which additionally requires that $\Gamma_t(p) \in \operatorname{int} \Gamma_t(q)$ for at least one $t \in \mathbb{T}$ and all $p, q \in \mathbb{R}$ with $p < q$.

Translated back to the design problem, the condition $\operatorname{tr} w^* \subset \Gamma(p^*)$ tells us that optimal systems have one and the same performance $P(\omega) = p^*$ for all frequencies; their frequency response "flattens the penalty", as some authors say. In the context of *boundary value problems*, this means that the optimal transfer function is the solution of a *nonlinear Riemann–Hilbert problem.* This observation is the key to the proof given in [26] (which will be sketched in the next section), as well as to the construction of numerical methods in Sections 5 and 6.

4. From OPT to RHP

The second condition of (2), $\operatorname{tr} w \subset \Gamma(p)$, is a nonlinear boundary condition for the holomorphic function w in the disk algebra $H^\infty \cap C$. Let us forget about the OPT-problem for a moment, and just study this nonlinear Riemann–Hilbert problem (RHP for short) for some fixed p. Omitting the parameter p, we write the boundary condition as $\operatorname{tr} w \subset \Gamma$, which means that $w(t) \in \Gamma_t$ for all $t \in \mathbb{T}$. The *target manifold* Γ and the *target curves* Γ_t (i.e., the *fibers* of Γ) are supposed to satisfy assumptions (i) and (ii) of Definition 1, in particular all Γ_t are Jordan curves and Γ is a (totally real) submanifold of $\mathbb{T} \times \mathbb{C}$ of smoothness class C^1. If these conditions are satisfied we say that Γ is an *admissible target manifold.*

The basic result about the solvability of nonlinear Riemann–Hilbert problems in the considered class (but with $\Gamma \in C^2$) was proved by Alexander Shnirel'man [20] in 1972. Alternative proofs were given later by Franc Forstnerič [6] (still under the assumption that $\Gamma \in C^2$), and the author [22] (for $\Gamma \in C^1$).

Shnirel'man made the additional assumption that the origin lies in the interior of all target curves. When this condition is dropped, the solvability of the Riemann–Hilbert problem may change dramatically. It turns out that the set $\mathcal{M}$ of admissible target manifolds Γ splits into three classes with different properties.

Definition 2. An admissible target manifold Γ is said to be *regularly traceable* if there exists a holomorphic function $w_0 \in H^\infty \cap C$ with $\operatorname{tr} w_0 \subset \operatorname{int} \Gamma$; it is called *singularly traceable* if it is not regularly traceable but there still exists w_0 in $H^\infty \cap C$ with $\operatorname{tr} w_0 \subset \operatorname{clos} \operatorname{int} \Gamma$; otherwise Γ is called *non-traceable.* The classes of regularly, singularly, or non-traceable target manifolds are denoted by $\mathcal{R}$, $\mathcal{S}$, and $\mathcal{N}$, respectively.

A complete description of the solvability of Riemann–Hilbert problems in all three cases and a characterization of their solutions by an extremal principle is given in [24]. We summarize the relevant results in the next theorem. For details see also [28], Chapters 2 and 3, in particular Theorem 2.5.1, Theorem 2.5.3, Theorem 3.3.2, and Theorem 3.4.2.

Theorem 2. *Let Γ be an admissible target manifold. Then the Riemann–Hilbert problem* $\operatorname{tr} w \subset \Gamma$ *has the following properties.*

(i) *If $\Gamma \in \mathcal{R}$ the traces of solutions with winding number zero about Γ cover the target manifold in a schlicht manner. If w_1 and w_2 are two such solutions, then $w_1(0) = w_2(0)$ implies $w_1 = w_2$. Moreover, the problem has solutions w with arbitrarily given positive winding number* $\operatorname{wind}_\Gamma w$, *while no solutions with* $\operatorname{wind}_\Gamma w < 0$ *exist.*

(ii) *If $\Gamma \in \mathcal{S}$ the problem has exactly one solution, and the winding number of this solution about Γ is negative,* $\operatorname{wind}_\Gamma w < 0$.

(iii) *If $\Gamma \in \mathcal{N}$ the Riemann–Hilbert problem has no solution.*

Figure 5 shows a regular target manifold carrying traces of solutions with winding number zero (right) and a singular target manifold with its unique solution (left).

Unfortunately there is no simple (practical) criterion which allows one to decide to which of the three classes a target manifold belongs, but there is at least a nice theoretical result (see Wegert [23] or Theorem 3.3.2 in [28]).

Theorem 3. *An admissible target manifold Γ belongs to $\mathcal{R}, \mathcal{S}$, or $\mathcal{N}$ if and only if the corresponding set* $\mathbb{A} := \{w \in H^\infty : \operatorname{tr} w \subset \operatorname{clos} \operatorname{int} \Gamma\}$ *of bounded holomorphic functions contains none, exactly one, or at least two elements, respectively,*

$$\Gamma \in \mathcal{N} \Leftrightarrow |\mathbb{A}| = 0, \quad \Gamma \in \mathcal{S} \Leftrightarrow |\mathbb{A}| = 1, \quad \Gamma \in \mathcal{R} \Leftrightarrow |\mathbb{A}| \geq 2.$$

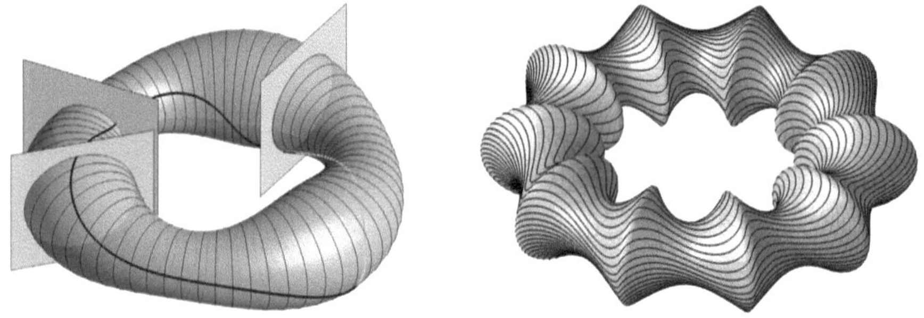

FIGURE 5. Target manifolds $\Gamma \in \mathcal{S}$ (left) and $\Gamma \in \mathcal{R}$ (right) with solutions

To convince the reader that the result is deeper than it looks like, we state some corollaries.

Corollary 1. *Let* Γ *be an admissible target manifold and define* $\mathbb{A}$ *as in Theorem* 3.

(i) *If* $|\mathbb{A}| = 1$ *and* $w \in \mathbb{A}$, *then* w *is in the disk algebra* $H^\infty \cap C$.

(ii) *If* $|\mathbb{A}| \geq 2$, *then* $\mathbb{A}$ *contains functions* $w \in H^\infty \cap C^\infty$ *with* $\operatorname{tr} w \subset \operatorname{int} \Gamma$.

For studying parameter-depending problems, we endow the set $\mathcal{M}$ of all admissible target manifolds Γ with the topology induced by the Hausdorff distance of compact subsets in $\mathbb{T} \times \mathbb{C}$. Properties of $\mathcal{R}, \mathcal{S}$, and $\mathcal{N}$ in this topological space are described in the next theorem. For a proof we refer to [26] (see also [28], Section 3.4, especially Theorem 3.4.2).

Theorem 4. *The sets* $\mathcal{R}$ *and* $\mathcal{N}$ *are open subsets of* $\mathcal{M}$, $\mathcal{S}$ *is closed,* $\partial\mathcal{N} = \partial\mathcal{R} = \mathcal{S}$, *and* $\mathcal{R} \cup \mathcal{N}$ *is dense in* $\mathcal{M}$.

Thus singular problems are "degenerate", and "generic" Riemann–Hilbert problems either belong to $\mathcal{R}$ or $\mathcal{N}$.

In order to finish the proof of assertions (i) and (ii) of Theorem 1 we first observe that the existence (as an infimum!) and uniqueness of the optimal performance p^* is clear from its definition and the fact that the problem is nontrivial. If the target manifold $\Gamma(p^*)$ would belong to $\mathcal{N}$ or $\mathcal{R}$, continuity of $p \mapsto \Gamma(p)$ would imply that $\Gamma(p)$ is in the same class for every p sufficiently close to p^*. Since this is in conflict with the definition of p^*, we must have $\Gamma(p^*) \in \mathcal{S}$. Moreover, the single (continuous) function in $\mathbb{A}(p^*)$ is the (unique!) solution of the OPT-problem and satisfies (2) for $p = p^*$.

Conversely, any function $w \in H^\infty$ which satisfies the conditions (2) is a solution of the Riemann–Hilbert problem $\operatorname{tr} w \subset \Gamma(p)$ with negative winding number. This is only possible if $\Gamma(p) \in \mathcal{S}$. Since $p < p^*$ can be excluded, we have $p \geq p^*$ and hence w and w^* are members of $\mathbb{A}(p) \in \mathcal{S}$, which is a singleton, so that $w = w^*$.

Before proving assertion (iii) of Theorem 1, we introduce some notation. For $a, b \in \mathbb{C}$ and $A \subset \mathbb{C}$ we set $a\,A + b := \{a\,z + b \in \mathbb{C} : z \in A\}$. If $\varphi, \psi : \mathbb{T} \to \mathbb{C}$ are functions and Γ is a target manifold with fibers Γ_t, we let

$$\varphi\Gamma + \psi := \{(t,z) \in \mathbb{T} \times \mathbb{C} : t \in \mathbb{T},\ z \in \varphi(t)\,\Gamma_t + \psi(t)\}.$$

In a somewhat sloppy way, we shall also write this set as $\varphi(t)\,\Gamma + \psi(t)$. Moreover, the family $\{\Gamma(p)\}_{p\in\mathbb{R}}$ is briefly denoted by $\{\Gamma\}$, and we adopt the notation $\{\varphi\Gamma + \psi\}$ as a shorthand for the family $\{\varphi\,\Gamma(p) + \psi\}_{p\in\mathbb{R}}$. Finally we let $\kappa\{\Gamma\}$ denote the (negative) winding number of the solution w^* to the OPT-problem about $\Gamma(p^*)$, $\kappa := -\mathrm{wind}_{\Gamma(p^*)}\, w^*$.

Theorem 5. *If p^*, w^* is a solution to the OPT-problem with $\kappa := \kappa\{\Gamma\}$, we have:*

(i) *The number κ is the smallest positive integer k for which $t^k\,\Gamma(p^*) \in \mathcal{R}$, i.e.,*

$$\Gamma(p^*) \in \mathcal{S}, \quad t\,\Gamma(p^*) \in \mathcal{S}, \quad \ldots, \quad t^{\kappa-1}\,\Gamma(p^*) \in \mathcal{S}, \quad t^{\kappa}\,\Gamma(p^*) \in \mathcal{R}.$$

(ii) *The Riemann–Hilbert problem $\mathrm{tr}\, w \subset t^\kappa \Gamma(p^*)$ has a unique solution w in $H^\infty \cap C$ with $w(0) = 0$, this solution has (automatically) a zero of order κ at the origin,*

$$w(0) = w'(0) = \cdots = w^{(k-1)}(0) = 0,$$

and $w^*(z) = z^{-\kappa} w(z)$ is the solution of the OPT-problem.

The last theorem and the following corollary imply the third assertion of Theorem 1. The corollary also indicates (a strict proof is somewhat more technical) that the real codimension of OPT-problems with $\kappa \geq 2$ within the class of non-symmetric problems is 2, while it equals 1 when we only admit *symmetric* problems satisfying the symmetry condition $\Gamma_{\bar t} = \overline{\Gamma}_t$.

Corollary 2. *If the OPT-problem for $\{\Gamma\}$ has a solution with $\kappa\{\Gamma\} = 1$, the same holds for every sufficiently small perturbation of $\{\Gamma\}$. If $\kappa\{\Gamma\} \geq 2$ then, for all $c \in \mathbb{C} \setminus \{0\}$ and $d \in \mathbb{R} \setminus \{0\}$,*

$$\kappa\{\Gamma + c\,\bar t\} = 1, \qquad \kappa\{\Gamma + d\,(t + \bar t)\} = 1.$$

Proof. Let $\kappa = 1$. Then $\Gamma(p^*) \in \mathcal{S}$, and the RHP $\mathrm{tr}\, w \subset t\,\Gamma(p^*) \in \mathcal{R}$ has a solution w with winding number zero and $w(0) = 0$. By Theorem 2, this implies that $t\,\Gamma_\varepsilon(p) \in \mathcal{R}$ for all Γ_ε sufficiently close to Γ and $|p - p^*|$ sufficiently small. Moreover, $\mathrm{tr}\, w \subset t\,\Gamma_\varepsilon(p)$ has a unique solution $p = p_\varepsilon$ and $w = w_\varepsilon$ with winding number zero about $t\,\Gamma_\varepsilon(p_\varepsilon)$ and $w_\varepsilon(0) = 0$. Then $z^{-1} w_\varepsilon(z)$ is a solution of $\mathrm{tr}\, w \subset \Gamma_\varepsilon(p_\varepsilon)$ with winding number -1, and consequently $\Gamma_\varepsilon(p_\varepsilon) \in \mathcal{S}$. Hence $\kappa(\Gamma_\varepsilon) = 1$.

Let $\kappa \geq 2$ and assume that (p^*, w^*) is a solution for Γ. For fixed $c \in \mathbb{C} \setminus \{0\}$ we define $\{\Gamma_c\} := \{\Gamma + c\,\bar t\}$. Then w_c given by $w_c(z) := z\, f^*(z) + c$ is a solution of the Riemann–Hilbert problem $\mathrm{tr}\, w \subset t\,\Gamma_c(p^*)$ with winding number $-\kappa + 1 < 0$, and hence $t\,\Gamma_c(p^*) \in \mathcal{S}$.

Let now p_c^* be the optimal performance for $\{\Gamma_c\}$. Then the Riemann–Hilbert problem $\mathrm{tr}\, w \subset t\,\Gamma_c(p_c^*)$ admits a solution w_c^* with $w_c^*(0) = 0$. Since the family $\{t\,\Gamma_c\}$ of nested manifolds cannot have a second member in $\mathcal{S}$ which admits a solution different from w_c (then we would have two solutions of a singular problem), either $t\,\Gamma_c(p_c^*) \in \mathcal{S}$ and $w_c^* = w_c$, or $t\,\Gamma_c(p_c^*) \in \mathcal{R}$. The first case can be excluded since $w_c(0) = c \neq 0 = w_c^*(0)$; in the second case we get $\kappa\{\Gamma_c\} = 1$ as desired.

The proof for the symmetrically perturbed manifolds $\{\Gamma + d\,(t+\bar{t})\}$ with $d \in \mathbb{R} \setminus \{0\}$ is similar; it suffices to replace the function w_c by w_d defined by $w_d(z) := z\,f^*(z) + d(z^2+1)$. □

In summary we have shown that for *each value* of $\kappa\{\Gamma\}$ the OPT-problem for $\{\Gamma\}$ is equivalent to the parameter-depending regular Riemann–Hilbert problem

$$\operatorname{tr} w \subset t^{\kappa}\Gamma(p), \qquad w(0) = 0. \tag{3}$$

This suggests the following strategy for solving the OPT-problem by reducing it to a regular Riemann–Hilbert problem: First try to solve the RHP (3) in the generic case $\kappa = 1$. If this fails (since the problem is singular), increase κ step-by-step until $t^{\kappa}\Gamma(p)$ becomes regular and has a solution w with $w(0) = 0$. When this happens for the first time, the solution w of (3) has a zero of order κ at the origin and $w^*(z) := z^{-\kappa}\,w(z)$ is a solution of the OPT-problem.

5. From RHP to OPT

Transforming the optimization problem to a (parameter depending) boundary value problem actually allows one to solve it numerically. In this section we describe the adaptation of an iterative method which has originally been developed for numerical conformal mapping (Wegmann [29]) and was extended to nonlinear Riemann–Hilbert problems by the author [25] (see also [28], Chapter 7). Alternative numerical methods for solving (OPT) are introduced and studied in [14], [15].

Though we could work directly with the penalty function F, we prefer to modify the setting, in particular to make it more convenient for practical applications. In fact the explicit construction of an "appropriate" penalty function F may cause serious technical problems, but at least some of these difficulties can be avoided replacing F by a parametric representation γ of the level curves,

$$\Gamma_t(p) := \big\{w \in \mathbb{C} : F(t,w) = p\big\} = \big\{\gamma(t,s,p) : s \in \mathbb{R}\big\}, \qquad t \in \mathbb{T},\ p \in \mathbb{R}.$$

Definition 3. A function $\gamma : \mathbb{T}\times\mathbb{R}^2 \to \mathbb{C}$ is said to be an *admissible parametrization* for the OPT-problem if it satisfies the following conditions (i)–(iv).

(i) The function γ is continuously differentiable on $\mathbb{T}\times\mathbb{R}^2$, $\gamma \in C^1(\mathbb{T}\times\mathbb{R}^2)$.

(ii) For all $t \in \mathbb{T}$ and $p \in \mathbb{R}$, the function $s \mapsto \gamma(t,s,p)$ is twice continuously differentiable on $\mathbb{R}$, 2π-periodic, and injective on $(-\pi,\pi]$. In addition the derivative $\partial_s\gamma$ does not vanish, and the mapping $\mathbb{T} \to \mathbb{C}$, $\mathrm{e}^{\mathrm{i}s} \mapsto \partial_s\gamma(t,s,p)$ has winding number 1 about the origin.

(iii) For all $t \in \mathbb{T}$ and $s \in \mathbb{R}$, the function $p \mapsto \gamma(t,s,p)$ belongs to $C^2(\mathbb{R})$ and

$$\operatorname{Im}\big(\partial_p\gamma(t,s,p)/\partial_s\gamma(t,s,p)\big) \geq 0, \tag{4}$$

with strict inequality for at least one $t \in \mathbb{T}$ and all $s \in \mathbb{R}$.

(iv) The mapping $\mathbb{T} \to \mathbb{C}, t \mapsto \partial_s\gamma(t,0,0)$ has winding number zero about 0.

Condition (ii) tells us that all $\Gamma_t(p)$ are C^2-smooth Jordan curves which are traced by $\gamma(t,s,p)$ in counter-clockwise direction for increasing s.

Condition (iii) implies that the family $\{\Gamma(p)\}$ is strongly nested, i.e., whenever $p<q$ we have $\Gamma_t(p) \subset \operatorname{clos}\operatorname{int}\Gamma_t(q)$ for all $t \in \mathbb{T}$, and $\Gamma_t(p) \subset \operatorname{int}\Gamma_t(q)$ for at least one $t \in \mathbb{T}$.

The winding condition (iv) can always be satisfied by an appropriate choice of the "initial point" $\gamma(t,0,0)$ on $\Gamma_t(0)$. It implies that for every $p \in \mathbb{R}$ and every continuous mapping $\varphi : \mathbb{T} \to \mathbb{R}$ the function

$$\psi : \mathbb{T} \to \mathbb{T} \times \mathbb{C},\ t \mapsto \gamma\big(t, \varphi(t), p\big)$$

has winding number zero about the (target) manifold

$$\Gamma(p) := \big\{\big(t, \gamma(t,s,p)\big) \in \mathbb{T} \times \mathbb{C} : t \in \mathbb{T}, s \in \mathbb{R}\big\}.$$

More generally, if φ is continuous on $[-\pi,\pi]$ with $\varphi(\pi) - \varphi(-\pi) = 2\pi\kappa$, and ψ is defined by

$$\psi : \mathbb{T} \to \mathbb{T} \times \mathbb{C},\ \mathrm{e}^{\mathrm{i}\tau} \mapsto \gamma\big(\mathrm{e}^{\mathrm{i}\tau}, \varphi(\tau), p\big)$$

then $\operatorname{wind}_{\Gamma(p)} \psi = \kappa$. All these assertions follow from assumption (ii) by a simple homotopy argument.

Returning to the OPT-problem, we are faced with the determination of a solution to the singular Riemann–Hilbert problem $\operatorname{tr} w \subset \Gamma(p^*)$. This seems to be a delicate task, since any perturbation changes the solvability of singular Riemann–Hilbert problems dramatically. Fortunately, as we have seen in the preceding section, it suffices to solve the *regular* Riemann–Hilbert problem (3). The "twisted" target manifolds $t^\kappa\Gamma(p)$ can be described by the transformed (admissible) parameter function μ defined by

$$\mu\big(\mathrm{e}^{\mathrm{i}\tau}, s, p\big) = \mathrm{e}^{\kappa \mathrm{i}\tau}\, \gamma\big(\mathrm{e}^{\mathrm{i}\tau}, s - \kappa\tau, p\big), \tag{5}$$

which also satisfies the conditions (i)–(iv). So we arrive at the following variant of a *parameter-depending nonlinear Riemann–Hilbert problem*:

Find a bounded holomorphic function $w \in H^\infty \cap C$, a bounded 2π-periodic real-valued function s, and a real parameter p which satisfy

$$w(t) = \mu\big(t, s(t), p\big) \quad \text{for all } t \in \mathbb{T}, \quad \text{and} \quad w(0) = 0. \tag{RHP}$$

Recall that the boundary functions of H^∞-solutions automatically belong to the Sobolev space $W_C := W_2^1(\mathbb{T};\mathbb{C})$, i.e., $w \in W_H := H^\infty \cap W_C$. Thanks to the assumptions on γ and μ, this implies that the corresponding function s is in the Sobolev space $W_R := W_2^1(\mathbb{T};\mathbb{R})$ (see [28], Lemma 1.8.2).

In order to rewrite (RHP) as operator equation, we introduce the Riesz projection P, which acts as a bounded linear operator from W_C on its subspace W_H of functions admitting a holomorphic extension into $\mathbb{D}$. The complementary projection to P is $Q := I - P$, and we let P_0 denote the operator which maps a function in W_C to its mean value on $\mathbb{T}$.

With these definitions, (RHP) can be rewritten as $A(s,p) = (0,0)$, where A is the mapping

$$A : W_R \times \mathbb{R} \to W_C \times \mathbb{C}, \quad (s,p) \mapsto \big(Q\,\mu(.,s,p), P_0\,\mu(.,s,p)\big). \tag{6}$$

Since $\mu \in C^2$, the operator A is continuously Fréchet differentiable (see [28], Lemma 1.3.5) and its derivative $DA(s,p)$ at a point $(s,p) \in W_R \times \mathbb{R}$ acts on $(\Delta s, \Delta p) \in W_R \times \mathbb{R}$ by

$$DA(\ldots)(\Delta s, \Delta p) = \big(Q\,(\Delta s \cdot \partial_s\mu + \Delta p \cdot \partial_p\mu),\ P_0\,(\Delta s \cdot \partial_s\mu + \Delta p \cdot \partial_p\mu)\,\big), \tag{7}$$

where we omitted the variables s and p in the notation. We now study the operator equation

$$DA(\ldots)(\Delta s, \Delta p) = (\Delta f, \Delta c), \tag{8}$$

with $\Delta f \in W_C$ and $\Delta c \in \mathbb{C}$. Equation (8) is equivalent to the system

$$\Delta s \cdot \partial_s\mu + \Delta p \cdot \partial_p\mu = \Delta f + w, \qquad P_0\,\big(\Delta s \cdot \partial_s\mu + \Delta p \cdot \partial_p\mu\big) = \Delta c, \tag{9}$$

where w is some (holomorphic) function in the kernel W_H of Q. The first equation can be transformed to a (parameter depending) linear Riemann–Hilbert problem: Dividing it by $\partial_s\mu(.)$, which is a strictly positive function, we get

$$\Delta s = (\partial_s\mu)^{-1} \cdot (\Delta f + w - \Delta p \cdot \partial_p\mu)\,.$$

Setting $g := 1/\partial_s\mu$ and using that Δs and Δp are real-valued, we can split this into two equations

$$\operatorname{Im}\,(g \cdot w) = \Delta p \cdot \operatorname{Im}\,(g \cdot \partial_p\mu) - \operatorname{Im}\,(g \cdot \Delta f), \tag{10}$$

$$\Delta s \qquad = \Delta p \cdot \operatorname{Re}\,(g \cdot \partial_p\mu) - \operatorname{Re}\,(g \cdot \Delta f). \tag{11}$$

A little thought shows that here only (10) is relevant, which is a linear Riemann–Hilbert problem for w. Note that $\partial_s\mu \in C^1$ and $\partial_s\mu \neq 0$ guarantee that $g \in W_C$. Once this problem is solved, the second equation (11) determines the corresponding function Δs. Finally, the second condition in (9) can be transformed to

$$w(0) = \Delta c - P_0(\Delta f). \tag{12}$$

After these preparations we can prove the following result.

Lemma 1. *Let γ be an admissible parametrization. Then for every $s \in W_R$ and $p \in \mathbb{R}$ the operator $DA(s,p) : W_R \times \mathbb{R} \to W_C \times \mathbb{C}$ is invertible.*

Proof. By Banach's theorem, it suffices to show that (10), (12) has a unique solution. Since the function $s \in W_R$ is continuous on $\mathbb{T}$, the winding number of $t \mapsto \big(t, \mu(t, s(t), p)\big)$ about $\Gamma(p)$ is zero (see the comment to condition (iv) in Definition 3), which implies that $\operatorname{wind} g = 0$, and hence the linear Riemann–Hilbert problem (10) has index $\operatorname{wind} \overline{g} = 0$. Consequently, its general solution has the form

$$w = w_f + d \cdot w_h + \Delta p \cdot w_p, \quad \text{with} \quad d,\ \Delta p \in \mathbb{R}, \tag{13}$$

where w_h denotes a non-trivial solution of the homogeneous problem, while w_f and w_p are (arbitrary) special solutions of the inhomogeneous problems

$$\operatorname{Im}\,(g \cdot w_f) = -\operatorname{Im}\,(g \cdot \Delta f), \qquad \operatorname{Im}\,(g \cdot w_p) = \operatorname{Im}\,(g \cdot \partial_p\mu), \tag{14}$$

respectively. The following lemma shows that there are unique parameters $d, \Delta p$ in $\mathbb{R}$ such that w satisfies (12), which then completes the proof of Lemma 1. $\square$

Lemma 2. *The values $w_h(0)$ and $w_p(0)$ are linearly independent over $\mathbb{R}$.*

Proof. Since $g \in W_C$ has winding number zero about the origin, it can be represented as $g = |g|\,\mathrm{e}^{-\mathrm{i}\varphi}$, with a (Hölder) continuous real-valued function $\varphi \in W_R$. Denoting by H the Hilbert transform, it is easily seen (and well known) that

$$w_h := \exp\left(H\varphi + \mathrm{i}\varphi\right)$$

is the boundary function of a nontrivial solution to the homogeneous problem $\mathrm{Im}\,(g\,w_h) = 0$. Since w_h and $\psi := H\varphi + \mathrm{i}\varphi$ have holomorphic extensions into $\mathbb{D}$ (denoted by the same letters), and $P_0 H\varphi = 0$, we have

$$w_h(0) = \exp\big(\psi(0)\big) = \exp\left(\mathrm{i}P_0\varphi\right) \neq 0.$$

Knowing a solution w_h of the homogeneous problem, the inhomogeneous problem (14) can be solved by the ansatz $w_p = w_h\,w$ (note that w_h does not vanish on $\overline{\mathbb{D}}$). Using that $w_h = \mathrm{e}^{\mathrm{i}\varphi}\,|w_h|$ and $g = \mathrm{e}^{-\mathrm{i}\varphi}|g|$, the boundary condition (14) is transformed to

$$\mathrm{Im}\,w = |g|^{-1}\,|w_h|^{-1}\,\mathrm{Im}\left(g\,\partial_p\mu\right) = |g|^{-1}\,|w_h|^{-1}\,\mathrm{Im}\left(\partial_p\mu/\partial_s\mu\right).$$

It follows from condition (4) that

$$\mathrm{Im}\left(\partial_p\mu(t,s,p)/\partial_s\mu(t,s,p)\right) \geq 0, \tag{15}$$

with strict inequality for at least one t (and hence on an interval), and using the mean value property of harmonic functions we get $\mathrm{Im}\,w(0) > 0$. Hence $w_h(0)$ and $w_p(0) = w_h(0)\cdot w(0)$ are linearly independent over $\mathbb{R}$. $\square$

We propose to solve the nonlinear Riemann–Hilbert problem (RHP) by a standard Newton method, starting with an initial solution $p_0 \in \mathbb{R}$, $s_0 \in W_R$ and iterating

$$\left(s_{k+1}, p_{k+1}\right) = \left(s_k, p_k\right) - DA(s_k, p_k)^{-1}\,A(s_k, p_k).$$

for $k = 0, 1, 2, \ldots$. The implementation of the kth step with given s_k and p_k requires the following computations:

(1) Define $f_k := \mu(., s_k, p_k)$, $g_k := 1/\partial_s\mu(., s_k, p_k)$, and determine w_k and Δp_k as the (unique) solution of the linear RHP

$$\mathrm{Im}\,(g\,w_k) = \Delta p_k \cdot \mathrm{Im}\left(g\,\partial_p\mu(., s_k, p_k)\right) - \mathrm{Im}\left(g\,f_k\right), \qquad w_k(0) = 0. \tag{16}$$

(2) Let $p_{k+1} := p_k + \Delta p_k$ and compute

$$\Delta s_k := g_k \cdot \left(w_k - f_k - \Delta p_k \cdot \partial_p\mu(., s_k, p_k)\right), \qquad s_{k+1} := s_k + \Delta s_k. \tag{17}$$

Note that Δs_k is automatically real-valued.

The procedure yields four sequences p_k, s_k, f_k, and w_k. The functions f_k satisfy the boundary condition $\mathrm{tr}\,f_k \subset \Gamma(p_k)$, but they are not necessarily holomorphic. In contrast, the w_k are holomorphic functions with $w_k(0) = 0$, but they need not fulfil the boundary condition. The next result shows that eventually both sequences f_k and w_k converge to the solution w^*, provided the initial value (s_0, p_0) is sufficiently close to (s^*, p^*) in $W_R \times \mathbb{R}$.

Theorem 6. *Let γ be an admissible parametrization of the OPT-problem and assume that its solution p^*, w^* has winding number $-\kappa := \operatorname{wind}_{\Gamma(p^*)} w^*$. Then the boundary function of w^* admits a representation $w^*(e^{it}) = \gamma\big(e^{i\tau}, s^*(e^{i\tau}) - \kappa\tau, p^*\big)$ with $s^* \in W_R$. For all initial pairs (s_0, p_0) in $W_R \times \mathbb{R}$ which are sufficiently close to (s^*, p^*), the Newton iterates (s_k, p_k) defined by (16), (17) converge in $W_R \times \mathbb{R}$ to (s^*, p^*). Convergence is quadratic if γ is three times continuously differentiable.*

The (super-linear) convergence of Newton's method follows from Lemma 1 and the fact that A is continuously Fréchet differentiable (see [19], Theorem 10.1.6 and the comment on p. 307, or Paragraph 1.2.7 of [28]). If γ is three times continuously differentiable, the Fréchet derivative DA is Lipschitz continuous, and then quadratic convergence of Newton's method is a standard result.

As always with Newton's method, it is crucial to find a good starting value, in our case (s_0, p_0). This can be achieved by an *embedding method*: introduce an additional parameter, say $\lambda \in [0, 1]$, defining a family of OPT-problems depending continuously on λ, such that $\lambda = 0$ corresponds to a problem with a known solution, while $\lambda = 1$ represents the actual problem to be solved. Appropriate assumptions (see Chapter 4 in [28]) guarantee that the solutions of Riemann–Hilbert problems depend continuously on λ, so that one can find a sequence $\lambda_0 = 0, \lambda_1, \dots, \lambda_n = 1$ such that the solution for the problem with parameter λ_k is a good initial guess for solving the problem with λ_{k+1}. We demonstrate this in the next section.

6. Numerical experiments

Each step of the iterative procedure described above requires the solution of a linear Riemann–Hilbert problem with two right-hand sides. The solutions are given in explicit form, involving the Hilbert transform H. A simple and efficient way to solve these problems numerically consists in discretizing the unit circle $\mathbb{T}$ by a mesh $\mathbb{T}_N$ of N (uniformly distributed) points and replacing the Hilbert transform by a discrete version H_N. The latter is defined via the discrete Fourier transform F_N by $H_N := F_N^{-1} J_N F_N$, where J_N is a diagonal matrix with entries $\pm i$ (and two zeros) on the main diagonal (see [7], for instance). If N is chosen as a power of 2, the method is extremely fast and runs within a second on current computers for N up to 2^{20}.

We have implemented this method in MATLAB and made some tests for designing a low-pass filter with penalty functions as shown on the right-hand side of Figure 3. The homotopy parameter λ controls shape and position of the moon-like sub-level domains, for small λ they are (almost) circular and contain the origin, as $\lambda \to 1$ they degenerate to circular arcs. The formulas for the parametrization $\mu(t, s, p, \lambda)$ are a bit too involved to be presented here. Instead of computing the derivatives $\partial_s \mu$ and $\partial_p \mu$ explicitly (by hand or computer algebra), they were approximated by finite differences.

Figure 6 shows the optimal target manifold $\Gamma(p^*)$ for $\lambda = 0.6$ with the trace of the solution w^* of the singular Riemann–Hilbert problem $\operatorname{tr} w \subset \Gamma(p^*)$.

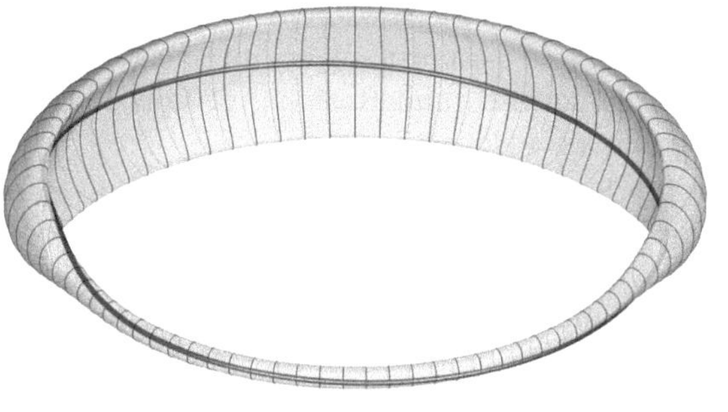

FIGURE 6. Manifold $\Gamma(p^*)$ with trace of solution w^* for $\lambda = 0.6$

The twisted target manifold $t\,\Gamma(p^*)$ for $\lambda = 0.96$ and the corresponding solution of the regular Riemann–Hilbert problem are depicted in Figure 7. Here the moon-shaped sub-level sets in the pass-band are almost self-intersecting and form sort of a hollow sausage.

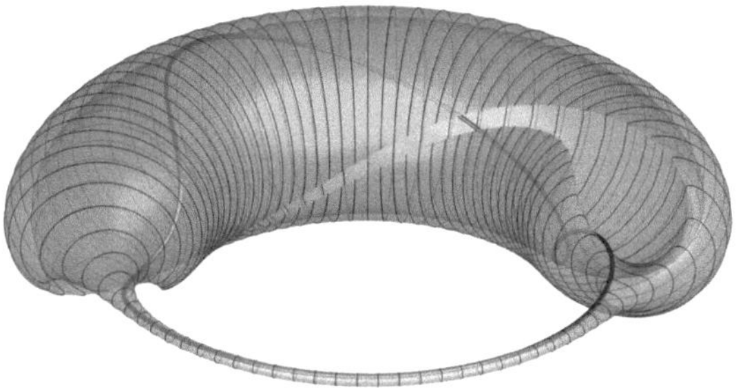

FIGURE 7. Manifold $t\,\Gamma(p^*)$ with trace of solution zw^* for $\lambda = 0.96$

The accuracy of the approximate solutions can easily be verified: interpolate s_k on a finer grid, form $f_k = \gamma(.,s_k,p_k^*)$ and apply a (fast) Fourier transform to check how much this function differs from a polynomial.

One possible complication inherent in this approach should not been hidden: As has been observed by several authors in similar situations, discretization may destroy convergence of the iteration. Initially the (discrete) iterates approximate the non-discrete functions s_k, but after some steps oscillations may occur which are enforced in subsequent iterations, and finally become dominating.

There are several options which allow one to overcome this phenomenon. The most elegant one works with *discrete Riemann–Hilbert problems* instead of a *discretized solution formula* for the continuous RHP. It was proposed by Wegmann [30], [31] in the context of numerical conformal mapping and extended to general nonlinear Riemann–Hilbert problems in [27]. This approach has higher numerical complexity: each step of the Newton method requires the solution of several Toeplitz systems. Even if one uses fast solvers for such systems (for instance those developed by Heinig, Jankowski, and Rost [9], or Gutknecht and Hochbruck [8]) the procedure is significantly slower.

Much simpler is a brute force method, which relies on smoothing the involved functions during the iteration (by truncating their Fourier coefficients) and damping the Newton step (by updating $s_{k+1} = s_k + d \cdot \Delta s_k$ and $p_{k+1} = p_k + d \cdot \Delta p_k$ with $0 < d < 1$). The damping is especially important during the first steps, when it helps to stabilize the approximate solutions, which otherwise often rotate wildly about $\Gamma(p)$. Of course convergence is not longer quadratic, but since iterations are cheep this is not a real issue.

Note that all involved parameters (homotopy, smoothing, damping) can be controlled automatically by observing the behavior of the residue (error of the right-hand side) during the iteration.

For some problems, solutions may behave quite tame on most part of the unit circle, while their values change rapidly in some small regions. In those cases it may be reasonable not to use the FFT-based discrete Hilbert transform, but other discretizations of H, admitting local refinement of the underlying mesh, for instance the (adaptive) wavelet-based methods developed in [17], [18].

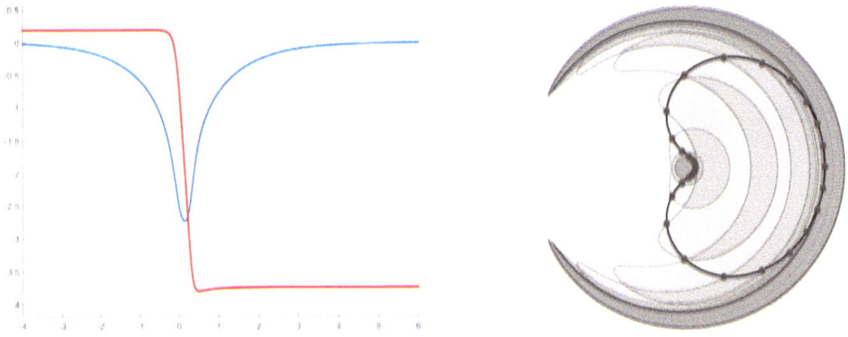

FIGURE 8. A Bode plot and a Nyquist plot with sub-level sets

Once the solution w^* to the OPT-problem on the unit disk is determined, the frequency response f^* of the optimal filter is given by the (inverse of the) Cayley transform. The Bode plot of the optimal low-pass filter for $\lambda = 0.98$ is depicted in Figure 8. It shows the gain $\log |f^*(\omega)|$ (in dB) and the phase $\arg f^*(\omega)$ (in rad) as functions of $\log \omega$. The image on the right-hand side of Figure 8 depicts a Nyquist plot for $\lambda = 0.95$. The bold black line is the image curve of the frequency response function. The small dots indicate the values $f^*(\omega)$ for some distinguished frequencies ω. The sub-level sets of the penalty function corresponding to these values are depicted as shaded regions, containing the respective dot on their boundary.

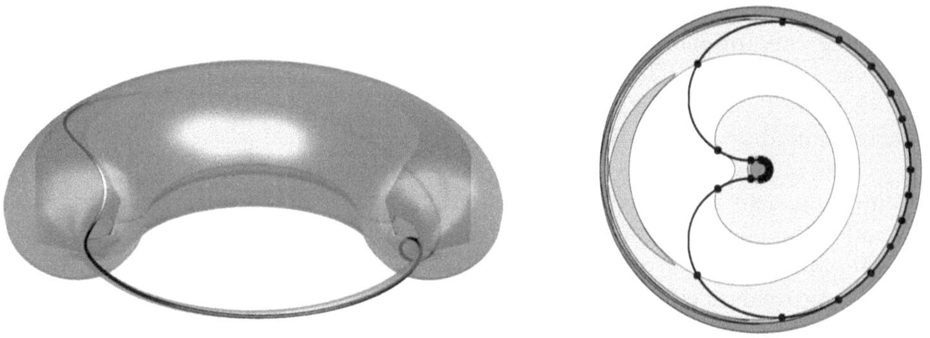

FIGURE 9. Solution on $\Gamma(p^*)$ and Nyquist plot for $\lambda = 0.98$

Defining the performance of a system by a parametric representation γ of the level sets has great advantage against determining it by a penalty function F: it also works *without* the assumption that $s \mapsto \gamma(t, s, p)$ is *injective*, thus not excluding *self-overlapping* sub-level sets. In fact this happens in the example for $\lambda = 0.98$, as is illustrated in Figure 9.

7. Discussion

Readers with some background in systems design might be disappointed by the unspectacular frequency response of the optimal filter depicted in Figure 8. The reason for that is the special meaning of "optimal". So let us interpret and evaluate the results to put them into the right context.

First of all, the penalty function was chosen *ad hoc* and with no *special purpose*, and the computed filter with frequency response f^* is optimal with respect to this choice. There may be (and there are) many filters with only slightly worse performance which have better properties in practice. Among these candidates are all small perturbations of f^*, and it needs further criteria to select a good one – which then is "optimal" for a slightly perturbed penalty function.

What is perhaps more important than the construction of an optimal filter is another feature of the approach: it allows one to *decide* whether or not for *given* p (respectively for prescribed sub-level sets $F(\omega, .) \le p$) a filter with performance *not worse than* p *exists.*

It is also useful to look at the result the other way around: once the optimal performance p^* is determined, we *know for sure* that there is *no filter* with frequency response $f(\omega)$ in the sub-level sets $F(\omega, .) < p^*$ for all ω.

And one can even say much more. Assume that the penalty function of an OPT-problem is defined by a parametrization γ^* and that f^* is the corresponding solution with performance p^*. Suppose further that γ^* is modified, resulting in an admissible parameter function γ, with the only restriction that (for all $t \in \mathbb{T}$) the modified level curve $\Gamma_t(p^*)$ intersects the original curve $\Gamma_t^*(p^*)$ at $f^*(t)$, and that their (non-oriented) angle of intersection is less than $\pi/2$ (these assumptions can even be relaxed). Then it follows from Theorem 1 (ii) that the solution to the modified OPT-problem is again f^*. Consequently, even with respect to *all penalty functions* modified in this way, no filter with performance better than p^* can exist.

This observation may help searching for "improved" penalty functions. For instance one can try to relax the restrictions in frequency regions which are considered to be "uninteresting". The location of the value $f^*(t)$ on the curve $\Gamma_t(p^*)$ delivers information on which modifications of the penalty function may be effective. In particular, increasing the sub-level domains $\Gamma_t(p)$ has no effect unless at least some of the points $f^*(t)$ are involved.

For instance, applying this observation to the example in the preceding section, we see that it makes no sense to blow up the "horns" (the ends of the moon-like sub-level sets for low frequencies), since this will not affect the optimal solution at all.

The approach described above is restricted to scalar systems. In the meantime there is also a mathematical foundation of the OPT-problem for vector-valued functions (Vitaev [21], Whittlesey [32], [33]). Related numerical methods have been developed by Dym, Helton, and Merino [4].

There are many practical issues which are beyond the scope of this paper. In real applications, solutions to the OPT-problem should be considered as raw material which needs some postprocessing (rational approximation, model reduction, application of other criteria, ...). In particular, the efficient *interactive modification of penalty functions* and the development of more elaborate numerical methods are challenging problems.

8. Epilogue

Above I repeatedly referred to my book [28], and this makes me coming back to Albrecht again. He is a very accurate person, a prominent example of the rule that good mathematicians often have great gift for languages, and a reliable and extremely strong proof-reader with eagle eyes. (It is not without reason that for many years he is on the editorial boards of several renowned journals.) Thus, in the early 1990s, when I was writing the book, we had many fruitful discussions, and I was pleased that Albrecht agreed to proof-read the entire manuscript. As a kind of return, I produced the LaTeX masters of an article by him. In those times, most paper drafts were *manuscripts* in the proper sense of the word, then given to a secretary for writing the text with a typewriter, and after that formulas were inserted with Indian ink by hand. However, LaTeX was already in its coming, and as at this time my LaTeX skills were a little more advanced than Albrecht's, we made that trade. Clearly, composing a paper in LaTeX is considerably less effort than proof-reading an entire book, and so the trade was very much to my advantage.

After so many years of acquaintance it is a little strange that Albrecht and myself never wrote a joint paper, but life continues and this is not yet the end of the story.

References

[1] A. Böttcher and B. Silbermann, *Toeplitz matrices and determinants with Fisher–Hartwig symbols.* J. Funct. Analysis **63** (1985), 178–214.

[2] A. Böttcher and B. Silbermann, *Toeplitz operators and determinants generated by symbols with one Fisher–Hartwig singularity.* Math. Nachr. **127** (1986), 95–124.

[3] P. Deift, A. Its, and I. Krasovsky, *Toeplitz matrices and Toeplitz determinants under the impetus of the Ising model: some history and some recent results.* Comm. Pure Appl. Math. **66** (2013), 1360–1438.

[4] H. Dym, J.W. Helton, and O. Merino, *Multidisk problems in H^∞ optimization: A method for analyzing numerical algorithms.* Indiana Univ. Math. J. **51** (2002), 1111–1159 (2002).

[5] T. Ehrhardt, *A status report on the asymptotic behavior of Toeplitz determinants with Fisher–Hartwig singularities.* Oper. Theory Adv. Appl. **124** (2001), 217–241.

[6] F. Forstnerič, *Analytic disks with boundaries in a maximal real submanifold of $\mathbb{C}^2$.* Ann. Inst. Fourier **37** (1987), 1–44.

[7] M.H. Gutknecht, *Fast algorithms for the conjugate periodic function.* Computing **22** (1979), 79–91.

[8] M.H. Gutknecht and M. Hochbruck, *Look-ahead Levinson and Schur algorithms for non-Hermitian Toeplitz systems.* Numer. Math. **70** (1995), 181–227.

[9] G. Heinig, P. Jankowski, and K. Rost, *Fast inversion algorithms of Toeplitz-plus-Hankel matrices.* Numer. Math. **52** (1988), 665–682.

[10] J.W. Helton, *Worst case analysis in the frequency domain: The H^∞-approach to control.* IEEE Trans. Automat. Contr. **39** (1985), 1154–1170.

[11] J.W. Helton, *Operator Theory, Analytic Functions, Matrices, and Electrical Engineering.* Reg. Conf. Ser. Math. **68** (1987), 134 p.

[12] J.W. Helton and R.E. Howe, *A bang-bang theorem for optimization over spaces of analytic functions.* J. Approx. Theory **47** (1986), 101–121.

[13] J.W. Helton and D.E. Marshall, *Frequency domain design and analytic selections.* Indiana Univ. Math. J. **39** (1990), 157–184.

[14] J.W. Helton and O. Merino, *Novel approach to accelerating Newton's method for sup-norm optimization arising in H^∞-control.* J. Optim. Theory Appl. **78** (1993) 553–578.

[15] J.W. Helton and O. Merino, *Classical Control Using H^∞ Methods. Theory, Optimization, and Design.* Philadelphia, SIAM (1998), 292 p.

[16] J.W. Helton, D.F. Schwartz, and S.E. Warschawski, *Local optima in H^∞ produce a constant objective function.* Complex Variables **8** (1987), 65–81.

[17] F. Martin and E. Wegert, *Computing the Hilbert transform using biorthogonal spline wavelets.* J. Math. Sci. (N.Y.) **189** (2013), 150–163.

[18] F. Martin and E. Wegert, *Computing the Hilbert transform in wavelet bases on adaptive grids.* Oper. Theory Adv. Appl. **236** (2014), 337–356.

[19] J.M. Ortega, and W.C. Rheinboldt, *Iterative Solution of Nonlinear Equations in Several Variables.* San Diego, Academic Press (1970).

[20] A.I. Shnirel'man, *The degree of a quasi-linearlike mapping and the nonlinear Hilbert problem (Russian).* Mat. Sb. **89** (1972), 366–389; English transl.: Math. USSR Sbornik **18** (1974), 373–396.

[21] A.E. Vityaev, *Uniqueness of solutions of a H^∞ optimization problem and complex geometric convexity.* J. Geom. Anal. **9** (1999), 161–173.

[22] E. Wegert, *Topological methods for strongly nonlinear Riemann–Hilbert problems for holomorphic functions.* Math. Nachr. **134** (1987), 201–230.

[23] E. Wegert, *Nonlinear Riemann–Hilbert problems and their relationship to extremal problems for holomorphic functions.* Math. Nachr. **137** (1988), 144–157.

[24] E. Wegert, *Boundary value problems and extremal problems for holomorphic functions.* Complex Variables **11** (1989), 223–256.

[25] E. Wegert, *An iterative method for solving nonlinear Riemann–Hilbert problems.* J. Comput. Appl. Math. **29** (1990), 311–327.

[26] E. Wegert, *Boundary value problems and best approximation by holomorphic functions.* J. Approx. Theory **16** (1990), 322–334.

[27] E. Wegert, *Iterative methods for discrete nonlinear Riemann–Hilbert problems.* J. Comput. Appl. Math. **46** (1993), 143–163.

[28] E. Wegert, *Nonlinear Boundary Value Problems for Holomorphic Functions and Singular Integral Equations.* Berlin, Akademie-Verlag (1992), 240 p.

[29] R. Wegmann, *Convergence proofs and error estimates for an iterative method for conformal mapping.* Numer. Math. **44** (1984), 435–461.

[30] R. Wegmann, *Discrete Riemann–Hilbert problems, interpolation of simply closed curves, and numerical conformal mapping.* J. Comput. Appl. Math. **23** (1988), 323–352.

[31] R. Wegmann, *Discretized versions of Newton type iterative methods for conformal mapping.* J. Comput. Appl. Math. **29** (1990), 207–224.

[32] M.A. Whittlesey, *Polynomial hulls and H^∞ control for a hypoconvex constraint.* Math. Ann. **317** (2000), 677–701.

[33] M.A. Whittlesey, *Polynomial hulls and an optimization problem.* J. Geom. Anal. **14** (2004), 545–556.

Elias Wegert
Institute of Applied Analysis
TU Bergakademie Freiberg
D-09596 Freiberg, Germany
e-mail: `wegert@math.tu-freiberg.de`

MIX
Papier aus verantwortungsvollen Quellen
Paper from responsible sources
FSC® C105338

If you have any concerns about our products,
you can contact us on
ProductSafety@springernature.com

In case Publisher is established outside the EU,
the EU authorized representative is:
Springer Nature Customer Service Center GmbH
Europaplatz 3, 69115 Heidelberg, Germany

Printed by Libri Plureos GmbH
in Hamburg, Germany